AA001017

Proceedings of the 2007 Asia Optical Fiber Communication and Optoelectronics Conference

Shanghai, China
17-19 October 2007

Pages 1-319

IEEE Catalog Number: CFP0739B-PRT
ISBN 10: 0-9789217-3-9
ISBN 13: 978-0-9789217-3-6

Copyright © 2007 by WEN GLOBAL SOLUTIONS
All Rights Reserved

IEEE Catalog Number: CFP0739B-PRT

ISBN 10:   0-9789217-3-9
ISBN 13:   978-0-9789217-3-6

Additional Copies of This Publication Are Available From:

IEEE Service Center
445 Hoes Lane
Piscataway, NJ 08854
Phone: (800) 701-4333
                (732) 981-1393
Fax: (732) 981-9667
E-mail: customer-service@ieee.org

AOE 2007 COMMITTEES

General Co-Chairs

Thomas L. Koch
Lehigh University
Bethlehem, PA, USA

Yi-Xin Chen
Shanghai Jiao Tong University
Shanghai, China

Program Chair
Sailing He
Zhejiang University
Hangzhou, China

SC1: Optical Fibers, Fiber Components and Subsystems
Lead Co-Chair: John Zyskind, *JDSU, , NJ, USA*

Co-Chair: Ping-kong Alexander Wai, *Hong Kong Polytechnic University, Kowloon, Hong Kong*

Committee Members: John Feng, *Avanex Corporation, Fremont, CA, USA*

Philippe Grelu, *Université de Bourgogne, Dijon, France*

Ruxiang Jin, *Lightelli Corporation, Shanghai, China*

Jie Luo, *Yangtz Optical Fiber and Cable Co., Ltd., Wuhan, China*

Sanjai Parthasarathi, *Avanex Corporation, Fremont, CA, USA*

Chester Shu, *Chinese University of Hong Kong, Shatin, Hong Kong*

Ping Shum, *National Technological University, , Singapore*

Yikai Su, *Shanghai Jiao Tong University, Shanghai, China*

Xu Wang, *Heriot-Watt University, Edinburgh, UK*

SC2: Optoelectronic Devices and Materials
Lead Co-Chair: Katsunari Okamoto, *University of California - Davis, Davis, CA, USA*

Co-Chairs: Jian-Jun He, *Zhejiang University, Hangzhou, China*

El-Hang Lee, *Inha University, Nam-ku, Korea*

Anders Olsson, *Finisar Corporation, Sunnyvale, CA, USA*

Committee Members: John E. Bowers, *UCSB, Santa Barbara, CA, USA*

Wei-Ping Huang, *McMaster University, Hamilton, ON, Canada*

Emil Koteles, *Lightip Technologies, Inc., Ottawa, ON, Canada*

Min Qiu, *KTH, Stockholm, Sweden*

Rang-Chen Yu, *Fiberxon, Inc., Santa Clara, CA, USA*

SC3: Optical Sensors and Biophotonics
Lead Co-Chair: Chinlon Lin, *Chinese University of Hong Kong, Shatin, Hong Kong*

Co-Chairs: Hui Ma, *Tsinghua University, Shenzhen, China*

Yun-Jiang Rao, *UESTC, Chengdu, China*

Hwa-Yaw Tam, *Hong Kong Polytechnic University, Kowloon, Hong Kong*

Committee Members: Chien Chou, *National Yang-Ming University, Taipei, Taiwan, R.O.C.*
Ho-pui Aaron Ho, *Chinese University of Hong Kong, Shatin, Hong Kong*
Hongdu Liu, *Peking University, Beijing, China*
Qingming Luo, *Huazhong University of Science and Technology, Wuhan, China*
Ammasi Periasamy, *University of Virginia, Charlottesville, VA, USA*
Steffen B. Petersen, *Aalborg University, Aalborg, Denmark*
Jianan Qu, *Hong Kong University of Science and Technology, Kowloon, Hong Kong*
Weihong Tan, *University of Florida, Gainesville, FL, USA*
Yin Yeh, *UC-Davis, Davis, CA, USA*
Libo Yuan, *Harbin Engineering University, Harbin, China*

SC4: Displays, Solid-State Lighting & Optoelectronics in Energy

Lead Co-Chair: Shin-Tson Wu, *University of Central Florida, Orlando, FL, USA*

Co-Chair: Wenzhong Shen, *Shanghai Jiao Tong University, Shangai, China*

Committee Members: Kyung Cheol Choi, *KAIST, Daejon, Korea*
Wallace C. H. Choy, *University of Hong Kong, , Hong Kong*
Rongqiang Cui, *Shanghai Jiao Tong University, Shangai, China*
Sin-Doo Lee, *Seoul National University, Seoul, Korea*
Yung S. Liu, *National Tsing Hua University, Hsinchu, Taiwan, R.O.C.*
Yong Qiu, *Tsinghua University, Beijing, China*
Franky So, *University of Florida, Gainesville, FL, USA*
Deng-Ke Yang, *Kent State University, Kent, OH, USA*

Industrial Forum on FTTH Technologies

Co-Chairs: Shoichi Hanatani, *Hitachi Communication Technologies, Ltd., Shinagawa, Japan*
Wei-Ping Huang, *McMaster University, Hamilton, ON, Canada*
Anders Olsson, *Finisar Corporation, Sunnyvale, CA, USA*

Joint Symposium on Enabling Technologies for Next Generation

Co-Chairs: Jian-Jun He, *Zhejiang University, Hangzhou, China*
John Zyskind, *JDSU, NJ, USA*

Slow-Light Workshop

Chair: Yikai Su, *Shanghai Jiao Tong University, Shanghai, China*

Table of Contents

Silicon Photonics And Lasers 1
John Bowers

Recent Research Activities On Photonic Network Technologies..................................4
Yuichi Matsushima

Prospects For Nanophotonics Circuits 7
Lars Thylén

Advanced Liquid Crystal Displays 10
Shin-Tson Wu

Advanced Technologies For High Quality LC Display 12
Yi-Pai Huang, Fang-Cheng Lin, Cheng-Yumr Liao, Ya -Ting Hsu, Wei-Kai Huang, Cheng-Han Tsao, Lin-Yao Liao, Chun-Ho Chen, Han-Ping D Shieh

Response Times In Pi-Cell Liquid Crystal Displays...15
Hongmei Ma, Li Jiang and Yubao Sun

Specialty Fibers As Key Components For Dispersion Management 18
Hans Damsgaard

Micro/Nano-Scale Optical Circuits And Networks For Information And Telecommunication Applications...21
El-Hang Lee

The Transition From Discrete Optics To Optical Integration.......................................24
A. P. Janssen

Recent Progress On PLC Technologies For Large-Scale Integration...........................27
Shinji Mino

Reflective Cholesteric Display: Principle And Progress 30
Deng-Ke Yang

Polarizer-Free Liquid Crystal Displays..31
Yi-Hsin Lin, Jhih-Ming Yang, Shin-Tson Wu, Chi-Chang Liao

The Color Temperature Adjusting Method For Multi-Primary Display Using Nonlinear Programming Problems..34
Yan Cheng, Xu Liu, Haifeng Li

Amplitude-Sensitive Interferometric Ellipsometer On TN-LCD Optical Parameters Measurement 37
H. C. Wei, C. C. Tsai, C. Chou

Biophotonics - A Tutorial Overview 40
Arthur Chiou

Reduced Dispersion Fiber Extends Reach For Dispersion Tolerant Systems...............43
Chris Towery and Tidd Zhang

The Breakthrough Of Specialty Fiber Fabricated By PCVD Based Process 47
Han Qingrong, Tu Feng, Luo Jie and R.Matai

Online RIC Process For G.652.D Fiber Production 50
Ralph Sattmann and Jan Vydra

Optical Ultra-Wideband Pulse Generation Using Air-Guiding Photonic Bandgap Fiber And A Semiconductor Optical Amplifier 53
Shangyuan Li, Xiaoping Zheng and Bingkun Zhou

Polarization Changes Of Partially Coherent Pulses Propagating In Optical Fibers 56
Weihong Huang, Sergey A. Ponomarenko and Michael Cada

Mid-Infrared Optoelectronic Devices And Applications 58
Zhang Yong-gang and Li Ai-zhen

Microwave And Millimeter-Wave Photonic Devices For Communications And Measurement Applications 61
Tadao Nagatsuma and Yuichi Kado

Optically Controllable Millimeter-Wave Oscillator Using Inp-Based Hemts 64
Hiroshi Murata, Noriyo Kobayashi, Toshihiko Kosugi, Takatomo Enoki and Yasuyuki Okamura

Wavelength-Tunable Slow Light Of Fs Laser Pulse By Quadratic Nonlinear Cascading Process 67
Wenjie Lu, Yuping Chen, Lihong Miu, Weirui Dang, Feng Lu, Xianfeng Chen and Yuxing Xia

Characteristics Of All-Optical Ultra-Fast Retiming Switches Using Cascaded Second-Order Nonlinear Effect In Periodically Poled Lithium Niobate Waveguides 70
Yutaka Fukuchi and Joji Maeda

Overview Of Research Activities At The NSF Center For Biophotonics Science And Technology (CBST) 73
Yin Yeh

Least-Invasive Harmonic Generation Microscopy For Intravital Imaging 76
Chi-Kuang Sun

The Purcell Effect Of Silver Nanoshell On The Fluorescence Of Nanoparticles 79
Wallace C.H. Choy, X.W. Chen, S.L. He and P.C. Chui

Nanoparticle-Assisted DNA Nanosensor 82
Xin Li1, Jun Qian, Lili Chen, Ying Zhu, Qun Fang and S. He

DNA Hybridisation Biosensor Based On Dual-Peak Long-Period Grating 85
Xianfeng Chen, Kaiming Zhou, Marcus Hughes, Edward Davies, Lin Zhang, Anna Hine, Kate Sugden and Ian Bennion

Wettability Patterning Technology For Organic Displays 88
Yu-Jin Na, Sung-Jin Kim and Sin-Doo Lee

Synthesis Of High Birefringence Liquid Crystals For Display Application 91
Chain-Shu Hsu and Yung-Ming Liao

Stereo Viewing Zone In Autostereoscopic Display Based On Parallax Barrier.................94
Qiong-Hua Wang, Ren-Liang Zhao, Wu-Xiang Zhao, Da-Hai Li, Yan-Xia Xin and Ai-Hong Wang

Ultrafast Laser Direct-Writing Of Bragg -Glass Photonic Devices 97
G. Marshall, N. Jovanovic, D. Kan, A. Fuerbach, A. Asatryan, L. Botten and M. Withford

Advanced Modulation Techniques In OCDMA System 98
Xu Wang, N. Wada, T. Miyazaki, G. Cincotti and K. Kitayama

2.5Gbps 60km OCDMA Transmission Experiment Using EPS-SSFBG En/Decoder 101
Lin Lu, Weilei Wu, Hui Peng, Tao Pu and Yuquan Li

Experimental Study On The Spectral Behavior Of An Asymmetric Long Period Fiber Grating Via Erosion 104
Minwei Yang, Jianping Chen, Yiping Wang and Xinwan Li

A Review Of The Effects Of High Refractive Index Overlays On Tunable Long Period Fiber Gratings....................107
J. Lee, Q. Chen, Q. Zhang, and S. Yin

High-Speed Versatile Modulator For Huge-Capacity Transmission....................110
Tetsuya Kawanishi

Recent Advances In Commercial Electro-Optic Polymer Modulator 113
Bing Li, Raluca Dinu, Dan Jin, Diyun Huang, Baoquan Chen, Anna Barklund, Eric Miller, Merly Moolayil, Guomin Yu, Yun Fang, Lixin Zheng, Hui Chen and Jeevan Vemagiri

All-Optical Inverted Triode Based On Cross-Gain Modulation Using Inas Quantum Dot Semiconductor Optical Amplifiers....................116
Yoshinobu Maeda, Sayaka Maki, Yasuhiko Kuroki, Hideki Nakayama and Jae-Hoon Huh

Modulation Properties Of Erbium Doped Silicon Laser Diode....................119
Md. Zahid Hossain, Samia Subrina and Md. Quamrul Huda

In-Line Fiber-Optic Etalon Formed By Hollow-Core Photonic Crystal Fiber 122
Y. J. Rao

Two-Core Fiber Based In-Fiber Integrated Interferometers And Its Sensing Applications 125
Libo Yuan, Jun Yang and Zhihai Liu

A Nonimaging Optics Approach For Photoelectric Sensor Applications....................128
Jun Jiang

Fiber-Optic Interferometric Temperature Sensor Using A Hollow Fiber 131
Jeung-Hwan Bae, Jaehee Park and Chomsik Lee

Transverse-Load Sensor Based On A Distributed Bragg Reflector Fiber Laser....................134
Li-Yang Shao, Xinyong Dong and Hwa-Yaw Tam

Bistable Reflective Displays For Paper-Like Displays 137
Liang-Chy Chien

Fabrications Of Mechanically Stable Plastic Liquid Crystal Displays 140
Kwang-Soo Bae, Yoonseuk Choi, Se-Jin Jang, Ji-Hong Bae, Jong-Wook Jung and Jae-Hoon Kim

The Electrolytic Polishing Of Flexible Display Steel Substrate 143
Li Yuqiong, Yu Zhinong, Xue Wei and Leng Jian

The Bending Properties Of Flexible ITO Films 146
Yu Zhinong, Xiang Longfeng, Xue Wei and Wang Huaqing

Characterization Of Polymer Microtip Array Coated Gan Thin Film Using Femtosecond Pulsed Laser Deposition 149
X.L. Tong, H.Q. Wen, D.S. Jiang and L. Liu

Dissipative Solitons For Real World Optical Solitons 152
Philippe Grelu, Ludovic Rapp, Jose M. Soto-Crespo and Nail Akhmediev

Generation Of Energetic Wavelength Tunable Femtosecond Pulses In Higher-Order-Mode Fiber 154
Chris Xu

Phase Noise Tolerant & Real Time Multilevel Homodyne 157
Tetsuya Miyazaki, Moriya Nakamura and Yukiyoshi Kamio

Photonic Crystal And Plasmonic Devices For Photonic Integration 160
Min Qiu

Light Confinement At Interfaces And Talbot Effect Using Optical Surface Modes 162
F.J. García de Abajo, R. Sainidou, T. V. Teperik, M. Dennis and N. I. Zheludev

Optical Polarization Beam Splitting Through Anisotropic Metamaterial Slab Realized By Layered Metal-Dielectric System* 163
Junming Zhao, Yan Chen and Yijun Feng

Asymmetric Hybrid Three-Arm Coupler With Long Range Surface Plasmon Polariton And Dielectric Waveguides 166
Fang Liu, Ruiyuan Wan, Yi Rao, Yuxin Zheng, Yidong Huang, Wei Zhang and Jiangde Peng

Single-Beam Self-Referenced Phase-Sensitive Surface Plasmon Resonance Sensor With High Detection Resolution 169
Shu-Yuen Wu and Ho-Pui Ho

All Fiber Optic Coal Mine Safety Monitoring System 172
Tongyu Liu

Research On Optical Sensor For Pulsed Magnetic Field Measurement 175
Su Yang and Li Yu-quan

Distributed Bragg-Reflector Fiber-Laser Sensor For Lateral Force Measurement 178
Yang Zhang, Bai-Ou Guan and HwaYaw Tam

Application Of Microplasma Modes To A Highly Efficient Light Source For Displays 181
Kyung Cheol Choi, Seung Hun Kim and Kwan Hyun Cho

Dielectric Superlattice And Its Potential Applications In Display Technology 184
Yan-qing Lu, Shi-ning Zhu, Yong-yuan Zhu, Yan-feng Chen, Hui-tian Wang and Nai-ben Ming

Organic Light Emitting Devices From OLED To Organic Laser Diode 187
Chihaya Adachi, Toshinori Matsushima, Hajime Nakanotani, Daisuke Yokoyama and Masayuki Yahiro

Integrated Photonics: Enabling Optical Component Technologies For Next Generation Access Networks 190
Valery I. Tolstikhin

PLC Based Bi-Directional Optical Module For Access Fiber Networks 193
N. Kitamura

The Low Cost Single Mode Laser Technology For Mass Deployment 196
Bo Cai

Tunable Optical Delay Schemes Using All-Optical Processing In A Highly Nonlinear Bismuth Oxide Fiber 198
Chester Shu and Mable P. Fok

Recent Advances In The Practical Fiber Optical Parametric Amplifiers 201
K. K. Y. Wong, B. P. P. Kuo, M. E. Marhic, G. Kalogerakis and L. G. Kazovsky

Single Polarisation Fibre Ring Laser By Utilising Intracavity 45° Tilted Fibre Bragg Grating 204
Kaiming Zhou(1), Chengbo Mou, Xianfeng Chen, Lin Zhang, Ian Bennion, Shenggui Fu and Xiaoyi Dong

Brillouin/Erbium Fiber Laser With Pre-Amplified Brillouin Pump Using Ring-Cavity Configuration 207
1N. Md Samsuri, A. K. Zamzuri and A. Mahdi

Impairment In Amplification Of Optical Packets Regarding The Gain Transient And Nonlinear Effect Depending On Peak Power Of NRZ Payload 210
Y. Awaji, H. Furukawa and N. Wada

Inp-Based Photonic Integrated Devices 213
Shinji Matsuo, Hiroyuki Ishii, Toru Segawa, Takaaki Kakitsuka and Hiromi Oohashi

Analysis Of Deep Etched Trench In Planar Optical Waveguide By FDTD Method 216
Jun Wang, Mingyu Li and Jian-Jun He

Fabrication Of Polymer Integrated Optical Microring Resonator With Photobleaching Method 219
Jun Zhou, Anan Pyayt, Jingdong Luo, Antao Chen and Nam Quoc Ngo

Image Resolution Analysis Of Different Super Lenses 222
P. Andalib and N. Granpayeh

Photoacoustic And Thermoacoustic Imaging For Biomedical Applications 225
Da Xing and Liangzhong Xiang

Optical Coherence Tomography For Oral Cancer Diagnosis 228
Meng-Tsan Tsai, Hsiang-Chieh Lee, Chih-Wei Lu, Yih-Ming Wang, Cheng- Kuang Lee, Chun-Ping Chiang and C. C. Yang

Raman Signal Enhancement In A Liquid-Core Optical Fiber Based On Hollow-Core Photonic Crystal Fiber 231
Li Huo, Chinlon Lin, Yick Keung Suen and Siu Kai Kong

Time-Of-Flight Laser Spectroscopy In Biomedical Diagnostics........................234
Stefan Andersson-Engels, Johan Axelsson, Ann Johansson, Jonas Johansson, Sune Svanberg and Tomas Svensson

Extraction Efficiency Enhancement Of An OLED Using Surface Plasmon Resonance 237
Shou-Yu Nien, Nan-Fu Chiu, Yao-Chou Tsai, Chii-Wann Lin, Kou-Chen Liu and Jiun-Haw Lee

Real-Time Voltage Controlled Color Tunable Oleds 240
Wallace C.H. Choy, C.J. Liang and H.M. Zhang

High-Performance Passive-Matrix OLED Display By Colour Conversion Method 243
Y. Terao, M. Kobayashi, N. Kanai, R. Makino, C. Li, Y. Kawamura, K. Kawaguchi, T. Saito, H. Hashida and H. Kimura

Improvement Of Electrical Characteristics Of Fluorinated Perylene Diimide Thin-Film Transistors By Gate Dielectric Surface Treatment 246
Li-Gong Yang, Jia-Chi Huang, Rong-jin Li, Min-Min Shi, Yan Gao, Mang Wang, Wen- Ping Hu and Hong-Zheng Chen1

Electrochemical Polyaniline/Polypyrrole Composite Film With Novel Nanostructure And High Biosensitivity...249
Yunan Chenga, Gang Wua, Gustaaf Borghsb, Mang Wanga and Hong-Zheng Chena

Overview Of Japanese FTTH Market And NTT's Strategies For Entering Full-Scale FTTH Era..252
Hiromichi Shinohara

On-Line Optical System Performance Monitoring Using Coherent Detection.................253
Rongqing Hui

Waveguide Structure Evaluation Based On A Photon-Counting OTDR 256
Takamichi Aiba, Nori Shibata and Masaharu Ohashi

Measurements Of Multimode Fiber PON Bandwidth 259
L. Maksymiuk, G. Stepniak and J. Siuzdak

Novel Technique For Measuring Raman Gain Efficiency Distribution By Conventional OTDR...262
Masaharu Ohashi, Yabu Tetsuro and Ikuo Yamashita*

EPON Deployment Challenges - Now And In The Future 265
Bill McDonald

Fault Location For Fiber Links In PON By Means Of FSF Fiber Laser And Fbgs............268
NianyZuo and Yoshinori Namihira

VCSEL Photonics - Athermalization And Slowing Down - 271
Fumio Koyama

Threshold Analysis Of A Novel Dispersive Grating Distributed Feedback Laser Diode....274
Xun Li, Yanping Xi and Wei-Ping Huang

40 Ghz Self-Pulsation In Two-Section DFB Lasers With Varied Ridge Width 277
Dingbo Chen, Hongliang Zhu, Song Liang, Huan Wang and Yali Zhang

Emission Characteristics Of A Surface-Emitting Organic Photonic Crystal Laser 280
Sidney S. Yang, Li-Wen Chang and Chong-Jie Huang

Uncooled Submarine Pump Laser Module At 980 Nm 283
Wenjuan Shen, Stefan Mohrdiek, Bing Guo, Tomas Pliska, Mark Ives, Shaun Quinlan, Warren Grace, Andrew Miller, Thomas Goodall, Jeffrey Greatrex, Robert Cann and Wen Ma

Investigating The Cortical Hemodynamics With High Spatiotemporal Resolution By Optical Imaging Techniques 286
Pengcheng Li and Qingming Luo

Photonics And Immobilisation Of Biomolecules 289
M. Durouxab, E. Skovsenab, M. T. Neves-Petersenab, L. Durouxab and S. B. Petersenab

Methane Concentration Monitoring System Based On A Pair Of Fbgs 292
Bin Zhou and Zuguang Guan

Emerging Fiber-Optic Microendoscopy Technologies For High-Resolution Biomedical Imaging 295
Xingde Li

Advances Of Lighting Technologies - From Light Bulbs To Solid State Light Sources 298
Yung S. Liu

Organic Light-Emitting Devices For Solid State Lighting 301
Jie Liu and Anil R Duggal

In-Situ Fabrication Of Highly-Fluorescent Nanohybrids Based On Carbon Nanotubes And Gold Nanoparticles 302
Renjia Zhou, Mang Wang and Hong-Zheng Chen

Planar Lightwave Circuits For FTTH And Photonic Networks 303
Katsunari Okamoto

Convergence Of Rof And Access Systems Employing Dualparallel Modulator In The Central Station 306
Yikai Su and Qingjiang Chang

160 Gbit/S/Port Colored Optical Packet Switching System 309
Naoya Wada

Modified Duobinary Signals With Tunable Duty Cycle And Its Application In A Label Switching Optical Network 312
Yufeng Shao, Shuangchun Wen, Lin Chen, Huiwen Xu and Jin He

Dual Band Optical Receiver For Video Broadcasting Services Over Fiber-To-The-Home Network 314
Young Cheol Kim, Young Ho Jang and Hyun Deok Kim

Novel Distributed All-Optical Multicast WDM Fiber Network: Design And Implementation 317
Dan Lu, Xi Qi, Feng Zhang, Bo Lv, Ming Chen and Shui-sheng Jian

Burst Mode Receiver Based On SOA 320
  Xiaobin Hong, Weiping Huang and Jian Wu

Microring Resonator Devices 323
  Yasuo Kokubun

Athermal AWG Multiplexer/Demultiplexer For E/C-Band WDM-PON Application 326
  Tae Hoon Kim, Byung Gwon You, Hyung Jae Lee and Tae Hyung Rhee

Experimental Demonstration Of Cross-Order Arrayed Waveguide Grating Triplexer 329
  Tingting Lang, Liu Yang, Jing Hu, Zhe-Chao Wang, Zhen Sheng, Jian-Jun He and Sailing He

Lateral Leakage In Symmetric SOI Rib-Type Slot Waveguides 332
  Rainer Hainberger, Paul Müllner and Norman Finge

Analysis On Curved Waveguide Grating (CWG) With Rowland Circle Construction 335
  Yinlei Hao, Jianyi Yang, Xiaoqing Jiang, Wei Zheng, Jianying Zhou, Haifeng Zhou and Minghua Wang

Paired Surface Plasmon Waves Biosensor 338
  Chien Chou, Hsieh-Ting Wu and Ying-Chang Li

Temperature-Insensitive Pressure Sensor Using A Polarization-Maintaining Photonic Crystal Fiber Based Sagnac Interferometer 341
  H Y Tam, Sunil K. Khijwania and X.Y. Dong

Fiber Bragg Grating Interrogating System Employing An Arrayed Waveguide Grating 344
  Zhou Qinfeng and Xu Tiefeng

Phosphor-Free White-Light Light-Emitting Diodes Based On Ingan/Gan Quantum Wells 347
  Chi-Feng Huang, Chih-Feng Lu, Dong-Ming Yeh, Yung-Sheng Chen, Wen-Yu Shiao and C. C. Yang

Comprehensive Investigation Of Light Emission Of Oleds: From Absolute Optical Properties To The Purcell Effect 350
  Wallace C.H. Choy, X.W. Chen, H.H. Fong and S.L. He

Mutual Thermal Effects Of Light-Emitting Diode With Wafer-Level Packages 353
  Jae-Wan Choi, Jeung-Mo Kang, Jae-Wook Kim, Jeong-Hyeon Choi, Du-Hyun Kim, Geun-Ho Kim and Jeong-Soo Lee

Enhance The Extraction Efficiency Of Zns:Mn TFEL By Photonic Crystals Structure 356
  Yurong Jiang, Jinwei Li, Xia Li and Wei Xue

A Facile Route To Synthesize Three-Dimensional Cds Nanocrystals 359
  Fei Chen, Renjia Zhou, Mang Wang and Hongzheng Chen

High Power Fiber Sources: More Than Kilowatts 360
  Johan Nilsson

Dispersion Controlled In A Birefringent Modified Octagon Photonic Crystal Fiber For Optical Communication Applications 363
  S. F. Kaijage, Y. Namihira, N. H. Hai, F. Begum, S. M. A. Razzak, T. Kinjo and N. Zou

Full-Vector Effective Index Method For Modeling Endlessly Single-Mode And Large Mode Area Of Photonic Crystal Fiber 366
Wang Liwen, Lou Shuqin,Chen Weiguo and Fang Hong

High Negative Dispersion And Low Confinement Loss Photonic Crystal Fiber 369
Lei Yao, Shuqin Lou, Hong Fang, Tieying Guo, Honglei Li and Shuisheng Jian

Adiabatic Compression Of Quadratic Solitons And Frequency Shift By Using Cascading Nonlinearities 372
Zeng Xianglong, Satoshi Ashihara, Chen Xianfeng, Tsutomu Shimura and Kazuo Kuroda

New Microwave Up-Conversion Solution Using An Optical Phase Modulator In Radio-Over-Fiber Networks 375
Haiyan Ou, Hongyan Fu and Biao Chen

Photonic Frequency Down-Conversion For Millimeter-Wave-Band Radio-Over-Fiber Systems By Directly Modulating A Dual-Wavelength Fiber Laser 378
Shiming Gao, Ying Gao, Hongyan Fu, Daru Chen and S. He

Analysis Of Dispersion Properties In Highly Nonlinear Photonic Crystal Fibers 381
Honglei Li, Shuqin Lou, Hong Fang, Lei Yao and Shuisheng Jian

Micro-Structured Photonic Crystal Fibers With Large Mode Area And High Negative Dispersion 384
Nguyen Hoang Hai, Y. Namihira and Shubi Kaijage

C+L-Band Erbium-Doped Fiber ASE Source Using Dual-Forward Pumping Configuration 387
Wencai Huang, Chaohong Huang, Xiulin Wang, Benrui Zheng, Huiying Xu and Zhiping Cai

Enhancement Of Multi-Wavelength Brillouin-Erbium Fibre Laser Utilizing Fibre Bragg Grating Filter 390
M. N. Mohd Nasir, M. H. Al-Mansoori, H. A. Abdul Rashid, P. K. Choudhury and Z. Yusoff

A Novel Millimeter-Wave Generation Technique For Mm-ROF System Based On Harmonic Generation Principle 393
Meiwei Zhu, Jiajun Ye and Rujian Lin

Fiber Ring Based Microwave Photonic Filters Implemented In A Radio-Over-Fiber Link 396
Kun Zhu, Hongyan Fu and Yun Xiao

Study On Optical Digital Phase Modulation Applied To Millimeter-Wave Radio-Over-Fibre System 399
Meiwei Zhu, Rujian Lin and Jiajun Ye

A Simplified Model Of Multi-Wavelength Fibre Lasers Based On Hybrid Fibre Raman And Erbium Fibre Amplifications 402
Shan Qin and Daru Chen

Lagrange Multiplier Optimization Synthesis Of Long-Period Fiber Gratings 405
Cheng-Ling Lee, Ray-Kuang Lee and Yee-Mou Kao

Feasibility Study Of A Simple 100Gb/S Transmitter With Lowspeed Electronics And 0.8bit/S/Hz Spectral Efficiency 408
Junming Gao, Xinyu Xu and Yikai Su

Controlling Chaos In An Erbium -Doped Fiber Dual-Ring Laser Via Modulating Its Loss And Phase **411**
    Yan Sen Lin

A Shared Sub-Path Protection Strategy In Multi-Domain Optical Networks **414**
    Xuejuan Xie, Weiqiang Sun, Weisheng Hu and Jun Wang

Novel Multi-Channel Temporal Phase En/Decoder Used In OCDMA Over WDM PON **417**
    Ying-xun Zhu, Rong Wang and Tao Pu

Untraditional All-Optical Chromatic Dispersion Compensating Elements - Experimental Verification **420**
    J. Vojtech, M. Karasek and J. Radil

Simultaneously Realizing Optical Millimeter-Wave Generation And Photonic Frequency Down-Conversion Employing Optical Phase Modulator And Sidebands Separation Technique **423**
    Hong Wen, Lin Chen, Jing He and Shuangchun Wen

Analysis Of Photonic Band-Gaps Of A Novel PBGF Structure **426**
    Xiao Yueyu

Application Of Lambert W Function To Raman Fiber Laser **429**
    Chaohong Huang, Wencai Huang, Xiulin Wang, Huiying Xu and Zhiping Cai

A New Technique For Side Pumping Of Double-Clad Fiber Lasers **432**
    Wang Da-zheng, Feng Xiao-ming, Wang Yong-gang, Wang Cui-luan, Lan Yong-sheng, Liu Su-ping and Ma Xiao-yu

Design Of A Doubly Grooved Binary Metallic Diffraction Grating For Efficient Side-Pumping Of High-Power Fiber Lasers **435**
    Fan Zhang, Chuncan Wang, Zhi Tong, Geng Rui and Ning Tigang

Combined FEC/ SOP Scrambling With Delay Line PMD Mitigation Scheme **438**
    J. Ferreira and J. P. von der Weid

Tailoring Confinement Losses Of Photonic Crystal Fibers **441**
    Lei Yao, Shuqin Lou, Hong Fang, Tieying Guo, Honglei Li and Shuisheng Jian

Two-Stage Hermite-Gaussian Function Method With Perfectly Matched Layers For Analyzing Microstructured Optical Fibers **444**
    Xue-Wen Chen, Sailing He, Zhi Wang and P.K.A Wai

Liquid Crystal Optical Modulator Based On In-Plane Switching **447**
    Yubao Sun and Li Jiang

Analysis Of The Scalability Of The Video-Overlay System **450**
    Ick Chang Choi, Byoung-Ju Yun and Hyun Deok Kim

An All-Optical Frequency Down-Converter Based On Four-Wave-Mixing In A Highly Nonlinear Fiber For Radio-Over-Fiber Systems **453**
    Ying Gao, Shiming Gao, Hongyan Fu, Daru Chen and Xiangrui Miao

Shape Influence On The Two-Dimensional Photonic Crystal Devices **456**
    Ya-Zhao Liu, Shuai Feng, Zhi-Yuan Li, Bing-Ying Cheng and Dao-Zhong Zhang

Tunable Artificial Birefringence In Woodpile Photonic Crystals 459
Ming Che, Zhi-Yuan Li and Rong-Juan Liu

Wavelength Assignment Algorithm For Hybrid WDM/TDM Passive Optical Network 462
Shaofeng Qiu, Weitao zhu, Jun Huang and Zhizhong zhang

Effect Of Source Parameters On Beam Self-Collimation In 2D Photonic Crystal 465
Xia Li, Wei Xue and Yurong Jiang

Improvement Of Automatic Alignment Algorithm For Butterflylaser Module Packaging 468
Hao Shen, Xin Wang, Bernard Leung and Hongdu Liu

**Low Loss Performances Of Long Range Surface Plasmon Polariton Waveguides With
Buffer Layer Structures** 471
Yi Rao, Fang Liu, Yidong Huang, Wei Zhang and Jiangde Peng

**Measurement Of Small Aspheric Surface Using Interferometric System For Spherical
Surface Test** 474
Nam-Young Jang, Pyung-Suk Choi and Jae-Jeong Eun

**Design And Realization Of Strip-Loaded Waveguide Electro- Optic Modulators In
Barium Titanate** 477
Jiansheng Tang, Shujun Yang and Apichai Bhatranand

Evolution Of Partially Coherent Solitons In Optical Lattices 480
Hui Zhuo, Shuangchun Wen and Yonghua Hu

**A Proposal For Passive Optical Network Architecture (WDM-PON) Based On Array Of
Ring Resonators** 483
G. Rostami, R. Faraji-Dana and M. Shahabadi

**An Improved Selective Area Growth Method In Fabrication Of Electroabsorption
Modulated Laser** 486
*Huan Wang, Hongliang Zhu, Yuanbing Cheng, Dingbo Chen, Wei Zhang, Liesong Wang,
Yunxiao Zhang, Yu Sun and Wei Wang*

**Experiments And Simulations Of Infrared Transmission By Transverse Electric Mode
Through Au Gratings On Silicon With Various Au Widths** 489
Yan-Ru Chen and C. H. Kuan

**All-Optical Switch In Alkoxysilane Dye Doped Waveguides Based On M-Line
Spectroscopy Technique** 492
*Weirui Dang, Yuping Chen, Rui Wu,Dandan Pan, Xianfeng Chen, Yuxing Xia and
Qinghua Meng*

**Experimental Demonstration Of All Optical Wavelength Full Conversion Based On
Quadratic Cascading Effect In Periodically Poled Mgo-Doped Lithium Niobate** 495
Junfeng Zhang, Feng Lu, Yuping Chen, Wenjie Lu, Xianfeng Chen and Yuxing Xia

The Application Of The Wavelet Transform To The Continuous Wave Terahertz Imaging 498
Yu Fei, Hui Mei, Song Qian and Zhao Yue-jin

**A Novel And Simple Power Splitter Utilizing Two-Branches Of Equal-Frequency Contours
Of A Dielectric Periodic Structure** 501
Yuan Zhang, Yurong Jiang, Wei Xue and S. He

Modelling And Numerical Analysis Of Carrier Transport Effects On The Wavelength Chirp Of SCH-QW Lasers..504
Farzan Gity, M. Naser Moghaddasi and Lida Ansari

The Iterative Ranked Phased-Array Method..507
Pojamarn Pojanasomboon and Okan Ersoy

A Rigorous Vectorial Gaussian Beam Modeling Of Virtually-Imaged- Phased-Array..510
A. Mokhtari and A. A. Shishegar

Fabrication And Characterization Of Deeply-Etched Sio2 Waveguides..513
Zhen Sheng, Liu Yang, Daoxin Dai, Tingting Lang and Zhechao Wang

Transmissive Properties And Faraday Rotation Of Tunable Photonic-Band-Gap System Containing Liquid Crystal..516
Ping Xu, Lei Gao and Z. Y. Li

Resonance-Induced Transmissions Through Waveguides Below Cut-Off Frequencies: An Effective-Medium Model For Waveguide..519
Hao Xu, Jiaming Hao, Jiajie Dai and Lei Zhou

Modeling And Optimization For Segmented Transmission-Line Electroabsorption Modulators With Asymmetrical Electrodes..521
Yongbo Tang, Yichuan Yu and Yuqian Ye

Study Of Optical Phased-Array Technology Based On PLZT Electro-Optic Ceramic..524
Qing Ye, Zuoren Dong, Ronghui Qu and Zujie Fang

Controlling Chaos In An Injection Multi-Quantum Well Laser Via Modulating The Injection Light..527
Yan Sen Lin

Analysis And Simulation Of A Channel Add-Drop Filter Composed Of Two Dimensional Photonic Crystal..530
N. Nozhat and N. Granpayeh

Numerical Research On Quality Factor Q Of 2D Photonic Crystal Microcavity With Modulation Of Localized States..533
Ziqiang Wang, Lieming Li, Dan Wang and Wenbin Cao

Ultra-Fast All-Optical Switch And Its Nonlinear Dynamical Process..536
Ye Liu, Zhi-Yuan Li and Dao-Zhong Zhang

New PON Add/Drop Multiplexer To Support Next-Generation PON..539
Sahrul Hilmi Ibrahim and Abu Bakar Mohammad

A Novel Method To Measure Brillouin Frequency Shift For Brillouin-Based Sensing Application Incorporating A Dual- Wavelength Single-Longitudinal-Mode Fibre Laser..542
Yizhen Wei, Yongbo Tang, and Daru Chen

Application Of Half-Cycle Phase-Stepping Algorithm In Eliminating Or Diminishing Errors Of Phase Measurement..545
Hui Mei, Yu Fei and Zhao Yue-jin

The Annealing Process Of R.F. Magnetron Sputtered Zno:Al Films..548
Yu Zhinong, Xu Jin, Xue Wei and Li Jinwei

Uniform Color Space For Color Storage 551
Yan Cheng, Xu Liu and Haifeng Li

Chalcogenide Glass Photonic Chips For All-Optical Signal Processing 554
V.G. Ta'eed, M.R.E. Lamont, M.D. Pelusi, M.A.F. Roelens, D.J. Moss, S. Madden,
D-Y. Choi, B. Luther-Davies and B.J. Eggleton

Low Cost Integrated Optical Mux/Demux For LX4 Transceiver 557
Hongtao Han, Jim Morris and Keith Main

Tuneable Photonic Millimetre Wave Generation Using An Optical Phase Modulator And
DWDM Thin Film Filters 560
P. Shen, J. James, N. J. Gomes and P. G. Huggard

An Ultrasmall Polarization Rotator Based On Si Nanowire 563
Zhechao Wang and Daoxin Dai

Bistable Device Based On The Kerr Effect In A Microfiber Resonator 566
G. Vienne, Ph. Grelu, Y. Li, X. Chen-Perdereau and L. Tong

Recent Progress In The Integration Of MGY-Based Tunable Lasers And
Mach-Zehnder Modulators 569
P-J Rigole

Wavelength And Space Switchable Semiconductor Laser 572
Jian-Jun He

Widely Tunable Slow-Light Delay Line Using Parametricamplification Assisted Silicon
Microring Resonator 575
Fangfei Liu, Chun Jiang and Yikai Su

Proposal Of A Thermally-Tunable Silicon-On-Insulator Microring Resonator Filter 578
Daoxin Dai and Liu Yang

Rotating Linear Differential Polarization Imaging For Quantitative Characterization Of
Superficial Tissues 580
Xiaoyu Jiang, Wei Li, Tianliang Yun, Nan Zeng, Yonghong He and Hui Ma

Applications Of Total Internal Reflection Fluorescence (TIRF) Microscopy In
Cellular Bio-Imaging 583
L. Jin, R.K. Lee, S. K. Kong, W. Yuan, H. P. Ho and Chinlon Lin

Fluid Sensor Based On Transmission Dip Caused By Mini Stop-Bands In 2D Photonic
Crystal Waveguides 586
Lei Cao, Yidong Huang, Xiaoyu Mao, Kaiyu Cui, Wei Zhang and Jiangde Peng

Fabrication And Photochromic Properties Of Ag/Ag+- Codoped Germano-Silicate
Glass Fiber 589
Aoxiang Lin, Sung-Ho Kim, Youngjoo Chung and Won-Taek Han

Organic Photovoltaics 592
Jiangeng Xue

Development Of Solar Photovoltaic In China 595
Cui Rongqiang, Wang Jianqiang, Ye Qinghao, Yan Shiquan, Shi Yang, Du Jiabing,
Yang Le, Meng Fanying, Xu Lin and Shen Wenzhong

Control Of Slow Light In Coupled Resonator Optical Waveguide Structures With Highly Dispersive Media **598**
  Min Qiu

Slow Light And Its Potential Applications **599**
  Li Zhengbin, Peng Chao and Xu Anshi

Delay-Bandwidth Product Of A Novel Slow Light Waveguide **600**
  Chun Jiang

All-Fiber Acousto-Optic Tunable Filters **601**
  Byoung Yoon Kim

A Simple Implementation Of Tunable All-Optical Microwave Notch Filter With A Negative Tap Based On A Semiconductor Optical Amplifier **604**
  Hongyan Fu, Haiyan Ou, Kun Zhu and Sailing He

Demonstration Of Optical Line Terminal For Full Colorless Bidirectional WDM-Passive Optical Networks Using Injection-Locked Fabry Perot Laser And Optical Carrier Suppression **607**
  Yong-Yuk Won, Dong-Hyeon Kim, Sang-Kil Roh, Yin-Xing Piao and Sang-Kook Han

Patterned Photonic Crystals For Novel Applications **610**
  Y. Ohtera, T. Sato and S. Kawakami

Novel Glasses And Glass-Ceramics For Broadband Optical Amplification **613**
  Jianrong Qiu, Shifeng Zhou, Jinjun Ren and Botao Wu

Raman Enhancement Of TO-520cm-1 Mode Of Si By Off-Plane One-Dimensional Grating Etched On Si Substrate **616**
  Ling-Chung Choua and C.-H. Kuan

Modeling Of Spontaneous Emission From Erbium Incorporated Silicon Nanocrystal **619**
  Samia Subrina, Md. Zahid Hossain and Md. Quamrul Huda

Surface Plasmonic Microscopy For Live Cell Membrane Imaging **622**
  Ruei-Yu He, Yuan-Deng Su, Kuo-Chih Chiu, Hua-Lin Wu, Chi-Hung Lin, Guan-Liang Chang and Shean-Jen Chen

Laser Ultrasound Detecting Experiment With Fiber Michelson Interferometer **625**
  Zhang Jian-liang, Sheng Xin-zhi, Wu Chong-qing and Zhang Li-jun

Spectroscopic Applications To Environmental Monitoring And Nanobiophotonics **628**
  Gabriel Somesfalean

Multiplexing Of Fiber Bragg Grating Pairs For Sensing Based On Optical Low Coherence Technology **631**
  Weisheng Liu

Bio-Inspired Nanodevices For Artificial Solar Energy Conversion **634**
  Mamoru Nango

Challenges In Luminescent Materials For Lighting And Medical Applications **637**
  Cees Ronda

Spectrally Broadened Optical Pump Source Via Phase Modulation For Wideband SBS Slow Light 638

    Chester Shu, Alan Cheng and Mable P. Fok

Slow Light In Silicon Nano-Waveguide 639

    Fangfei Liu

Storage Capacity Of Slow-Light Based On Fiber Brillouin Amplifiers 640

    Li Zhan

This page intentionally left blank.

Silicon Photonics and Lasers

John Bowers
Dept of ECE
University of California, Santa Barbara
Santa Barbara, CA 93106
bowers@ece.ucsb.edu
Plenary Paper at AOE 2007

Abstract

Tremendous advances are being made in silicon photonics. Complicated photonic integrated circuits combining active and passive devices are becoming available for a large variety of applications from telecommunications to imaging.

Introduction

Silicon is the principal material used in semiconductor manufacturing today because it is plentiful, inexpensive, well suited for nanometer scale processing, and well understood by the semiconductor industry. Silicon photonics aims to provide inexpensive silicon building blocks that can be integrated to produce optical products that solve real communication problems for consumers. Silicon is an especially useful material for photonics components because it is transparent at the infrared wavelengths at which optical communication systems operate.

Photonic devices have historically been relatively expensive, largely due to the low volume of devices produced. Fig. 1 illustrates the volume scaling that has been achieved with a variety of photonic technologies. One of the goals of silicon photonics is to use CMOS technology to bring a number of transmitter, receiver and other photonic technologies to a high volume, low cost capability, thereby transforming what has been a relatively low volume, high cost photonics industry.

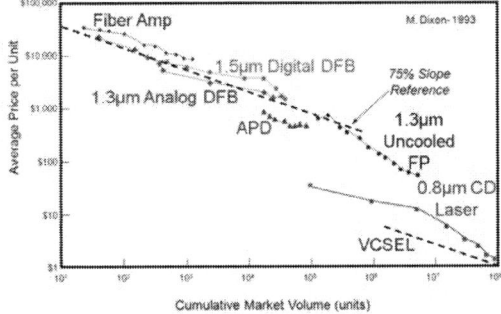

Fig. 1. Volume price reduction for several photonic devices (after R. Leheny and M. Dixon[1]).

Prior to 2000, silicon had been considered useful in photonics primarily for passive waveguides and devices. Fig. 2 summarizes some of the many tremendous innovations in silicon photonics and the many breakthroughs in passive and active devices, from modulators and photodetectors to light sources and amplifiers.

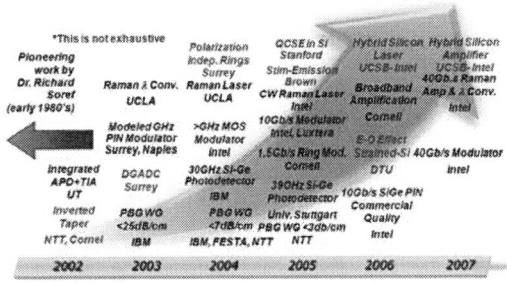

Fig. 2. Partial list of significant advances in silicon photonics[2].

Silicon Modulators

One key problem has been making high speed modulators, since silicon is not electro-optic due to its inversion symmetry. This leaves carrier-optic and thermo-optic effects as the primary mechanisms for modulation in silicon and due to the relatively long carrier lifetime in silicon are slow under carrier injection. The first demonstration of a gigabit modulator was in 2004[3], utilizing an MOS based structure under depletion. Progress in this area has been very rapid by many groups around the world. A recent, exciting advance[4] is the 40 Gbit/s silicon modulator demonstrated by Intel. Further progress in this area is anticipated.

978-0-9789217-3-6/07/$25.00
©2007 WEN GLOBAL SOLUTIONS

Silicon Light Emitters

A key hurdle for realizing practical silicon based photonic integrated circuit is achieving electrically pumped silicon lasers and amplifiers due to the indirect bandgap of silicon. Silicon is capable of routing, modulating, and detecting light, but an external light source has been needed to provide the initial light. These external light sources are generally discrete lasers and require careful alignment to the silicon waveguides. The problem is that accurate alignment is difficult and expensive to achieve. The realization of lasers on silicon has been pursued through two major approaches.

The first area of research is to make silicon itself emit light. Nanopatterning of silicon has shown band to band emission in silicon at cryogenic temperatures[5]. Another approach is to use the Raman effect. The first silicon Raman laser was demonstrated by Boyraz et al.[6] and was limited to pulsed operation due to the free carrier losses caused by two photon absorption (TPA) of the pump laser. A continuous wave silicon Raman laser was demonstrated by H. Rong et al. by incorporating a P-I-N structure to sweep away the TPA generated carriers[7]. Raman gain relies on an external optical pump and does not lead to electrically pumped operation.

Pavesi et al. first showed optically pumped optical amplification in the 750 nm regime through the quantum confinement of excitons in silicon nanocrystals embedded in silicon dioxide[8]. Electrical injection schemes have been demonstrated by R. Walters et al. by independently injecting holes and electrons into the silicon nanocrystal[9]. Although silicon nanocrystal light emission is typically in the silicon absorption regime, rare-earth doping has been utilized to achieve emission at other wavelengths such as 1.5 μm with erbium dopants.

The second area of research to realize lasers on silicon is to transfer compound semiconductor structures to silicon wafers to either 1) make compound semiconductor lasers and couple them to silicon waveguides, or 2) to make hybrid lasers on silicon that utilize silicon waveguides to form cavities while obtaining gain from the compound semiconductor layers. Campenhout et al.[10] demonstrated electrically pumped InP microdisk lasers with thresholds as low as 0.5 mA with a max output power coupled to a silicon waveguide of 10 microwatts. Optically pumped DFB lasers were demonstrated by Okumura et al.[11].

Electrically pumped lasers where the optical supermode lies primarily in the silicon region with a small portion evanescently coupled to the compound semiconductor region, have been demonstrated with threshold currents of 65 mA[12] and maximum operating temperatures of 60 °C[13]. This latter approach is the basis of the hybrid silicon evanescent integrated circuit technology described in the rest of this paper.

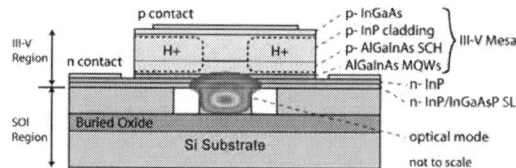

Fig.3. Cross sectional structure of silicon evanescent devices.

Silicon Evanescent Lasers

The silicon evanescent device is a hybrid structure that consists of an offset multiple quantum well region bonded to a silicon waveguide which is fabricated on a silicon-on-insulator (SOI) wafer. With this architecture, the optical mode can obtain electrically pumped gain from the III-V region while being guided by the underlying silicon waveguide region. Fig. 3 shows the structure used for hybrid silicon evanescent lasers, photodetectors, and amplifiers. Lasers made with the structure have lased up to 60 °C and emit up to 30 mW at 15 °C. (Fig. 4). Ring and linear cavities have been demonstrated.

Silicon Evanescent Amplifiers

Optical amplifiers are also important components in realizing high levels of photonic integration as they compensate for optical losses from individual photonic elements. Using this platform, a maximum chip gain of 13 dB at 1575 nm with a spectral full-width at half-maximum of 62 nm was demonstrated[14]. The 3 dB output saturation power was measured to be 11 dBm. The internal noise figure of the device varied between 8 dB and 5 dB. These amplifiers were characterized at 2.5, 10 and 40 Gbit/s, and a low power penalty of 0.5 dB for all three data rates was achieved, limited by ASE noise.

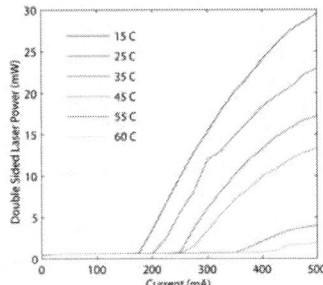

Fig. 4. The LI curves for a racetrack laser.

Silicon Evanescent Photodetectors

Silicon evanescent photodetectors have been fabricated separately and integrated with lasers and amplifiers. The leakage current is typically below 100 nA, with a quantum efficiency above 90% [15].

Silicon Evanescent Photonic Integrated Circuits

One example of the potential applications of this technology is shown in Fig. 5. It should be possible to make an integrated 1 Tbit/s transmitter by integrating an array of DFB lasers with an array of silicon modulators and an arrayed waveguide multiplexer. With the power of CMOS processing capability, such sources could be quite low cost, ushering in a new era in data and telecommunications.

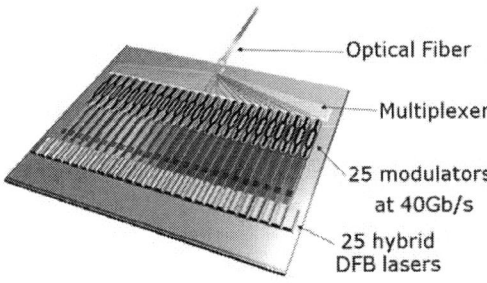

Fig. 5. Schematic diagram of a 1 Tbit/s transmitter made using silicon evanescent technology[2].

Conclusions

Silicon photonics has been an area of tremendous innovation and advances during the past decade. The performance of many devices, from modulators to amplifiers, far exceeds what most people thought possible just a few years ago. Once the remaining technical challenges are resolved, the ability of high volume, low cost CMOS fabrication promises to revolutionize photonics and transform it into a ubiquitous technology. Important new applications such as chip to chip interconnections and even on-chip communications then become possible, solving critical communication and power bottlenecks in next generation computing.

Acknowledgements

This work was supported by DARPA through contracts W911NF-05-1-0175 and W911NF-04-9-0001, and by Intel, IBM and the UC Discovery Program. The author thanks Mario J. Paniccia, Mike Haney, Jag Shah, Alex Fang, Hyundai Park, Ying-hao Kuo, Richard Jones, Jeff Kash, Oded Cohen, Omri Raday, and Wayne Chang for useful discussions.

References

1. R. Leheny. STAR meeting, Monterey, CA July 25, 2008.
2. M. Paniccia, personal communication (2007).
3. A. Liu, et al., Nature, 427, 615-618 (2004).
4. A. Liu et al., IPNRA, 2007.
5. Cloutier et al., Nat. Mater. 4, 887 (2005)
6. O. Boyraz, et al., Opt. Express 12, 5269 (2004).
7. H. Rong, et al., Nature 433, 725 (2005).
8. Pavesi et al., Nature 408, 440-444 (2000).
9. Waters et al., Nature Materials 4, 143-146 (2005).
10. Campenout et al., Optics Express, Vol. 15, Issue 11, pp. 6744-6749 (May 2).
11. Okumura et al., IPRM 2007
12. A. W. Fang, et al., Opt. Express 14, 9203, (2006).
13. A. W. Fang, et al., Opt. Express 15, 2315, (2007).
14. H. Park, et al., IEEE Photon. Technol. Lett. 19, 230 (2007).
15. H. Park, et al., Opt. Express 15, 6044, (2007).

Recent Research Activities on Photonic Network Technologies

Yuichi Matsushima

National Institute of Information and Communication Technology (NICT)
4-2-1, Nukui-Kitamachi, Koganei, Tokyo 184-8795, Japan
Phone: +81-42-327-7485, E-mail: matsushima@nict.go.jp
http://www.nict.go.jp

Abstract *Recent activities on optical network R&D promoted by NICT will be reviewed. Topics include an optical packet network system, an ultra-fast optical communication system as well as advanced optical devices. Open laboratory, Japan giga-bit optical network (JGN-II), where industry, government and education are collaborating will also be presented.*

Introduction: What is NICT?

To maintain and reinforce precedence in technology which realized one of the foremost global broadband environments so called ubiquitous network society, NICT promotes technical R&D that will be the foundation for next generation networks in Japan. NICT is Japan's sole national research institute involved in information and communications field and is conducting R&D and offering comprehensive project support in the field of information and communications as shown in Fig.1.

Photonic R&D promoted by NICT

In optical networking research field, NICT has been promoting R&D toward new generation broadband network based on new technologies from basic research to applied research. Major research field is from high quality universal access technologies including fixed-mobile-convergence to optical core network technologies with high performance control methods and service application layer including ultra wideband network systems like grid computing as shown in Fig.2.
For NICT own research, we are mainly promoting three research fields as shown in Fig.3,
1. Optical packet switching :OPS networks
2. Ultra-fast optical transmission systems
3. Application to optical GRID networks.
We have already proposed and experimentally demonstrated OPS system based on optical label processing and also investigated electronic scheduling for optical buffering based on the road map in Fig.4.

The first OPS prototype has 40Gbit/s interface and ultra-fast functions such as optical label processing and transparent optical buffering. Recently, a 160Gbit/s/port OPS prototype with new techniques of optical code label processing, high-extinction-ratio optical buffering, optical MUX/DEMUX, optical packet 3R receiver is proposed and experimentally demonstrated. We also demonstrated highly-efficient optical transmission systems at more than 160 Gbit/s per wavelength and exploited field trial using JGN-II (Japan Gig-bit Network II) test bed. We also investigate high performance wavelength (lambda) path networking for distributed processing system (GRID) based on GPMLS protocol.

NICT is offering national project support in the field of key technologies for next generation optical networks such as
1. Ultra high capacity photonic node technology
2. Lambda utility technology
3. Lambda access technology
4. Optical RAM system technology.
Those are carried out by the collaboration among industry, university and telecommunication carriers.
JGN-II is operated by NICT which is 20/10 Gbit/s backbones performed by the OXC/WDM network. The test bed system has over 60 access point nationwide including two optical field trial test line for current inspection research and is connected with overseas networks.

Current R&D topics will also be presented.

978-0-9789217-3-6/07/$25.00
©2007 WEN GLOBAL SOLUTIONS

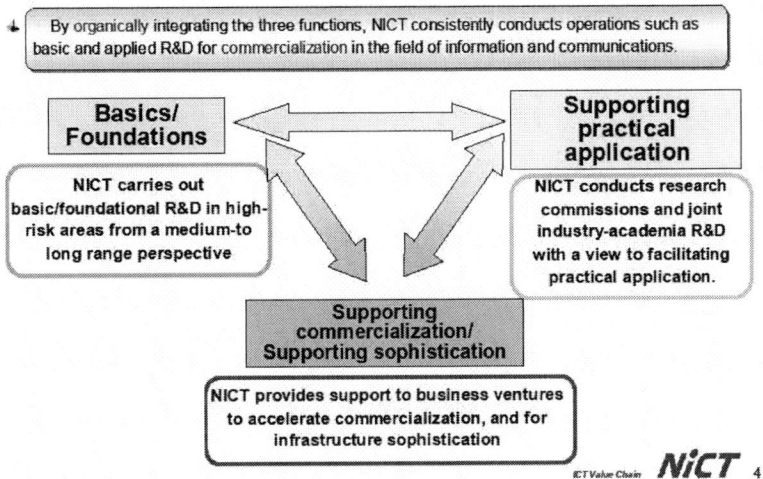

Fig.1 Outline of mission of NICT

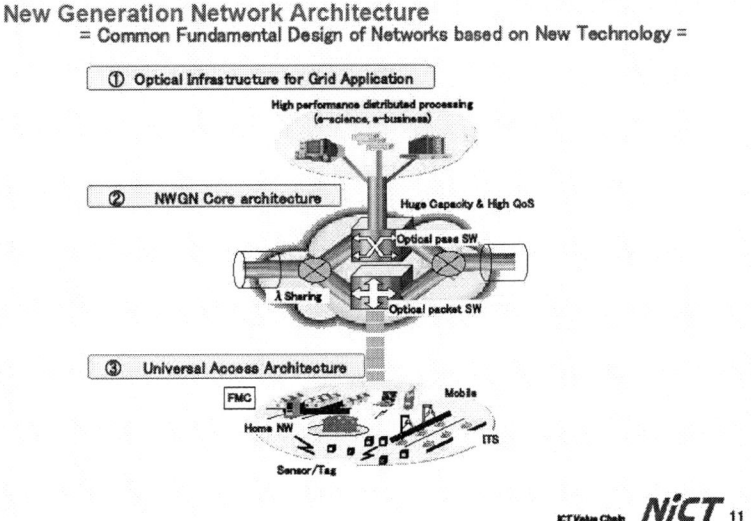

Fig.2 R&D for fundamental technologies on new generation network

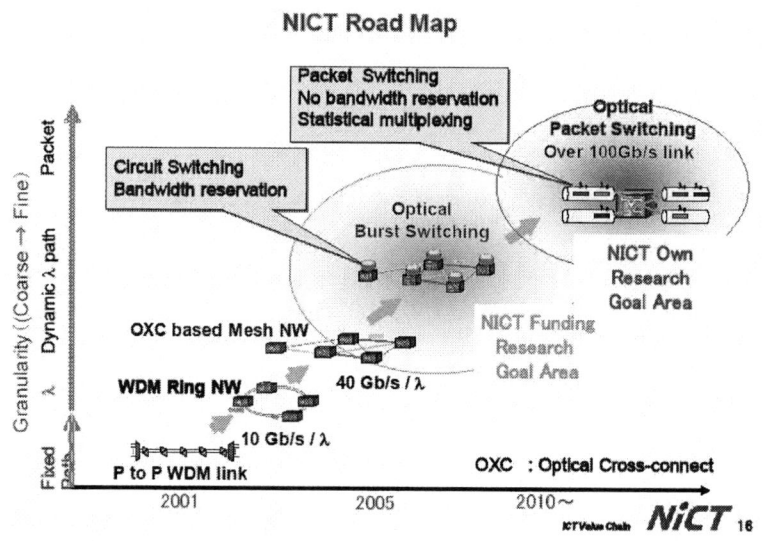

Fig.3   Core research field for photonic networks

Fig.4   Road map of optical network architecture

Prospects for nanophotonics circuits

Lars Thylén

Royal Institute of Technology (KTH) and Kista Photonics Research Center (KPRC) ,
Electrum 229, SE-164 40 Kista, Sweden
Joint Research Center of Photonics of the Royal Institute of Technology and Zhejiang
University, Zhejiang University, Yuquan, Hangzhou 310027, P. R. China
Email: lthylen @imit.kth.se

Abstract *We discuss from a theoretical viewpoint the required advances in material technology and device structures needed to continue the rapid development of photonic circuit spatial integration density. Metamaterials and various waveguide device structures are discussed.*

Nanophotonics, although seemingly a *contradictio in adjecto*, has received much attention in recent years, fuelled by a general interest in nanotechnology but also by the rapid advances in photonics. The reason for the apparent contradiction is that nanotechnology is generally defined as a technology of feature sizes in the nanometer range, whereas conventional photonics exhibits characteristic sizes at least on the order of the wavelength, around 1 µm and in many cases orders of magnitude more. However, there has been a rapid progress in photonics integration density in recent years, primarily brought about by using silicon/ air or quartz interfaces, giving a larger index step [1]. One can show that the minimum field lateral spatial width for a planar silicon waveguide in air is ~300 nm, with a wavelength in the medium of ~500 nm, at a vacuum wavelength of 1.55 µm. Thus, any attempts in nanophotonics integration should be below these values, as measured in proportion to the relevant vacuum wavelength. In fact, to increase spatial integration further, it is most likely

necessary to find a successor to the current silicon nanowire technology, which, as noted, has to a large extent enabled the recent progress in integration density [2]. Such successors seem to have to rely in some way on materials with negative ε [3] since they offer a possibility for increasing the integration density in photonic lightwave circuits in two ways: (i) By using principles other than total internal reflection (limited as noted above by the refractive index step) and (ii) employing e g metamaterials to generate artificially very large index steps.

In this paper we discuss several ways of increasing such modal confinement (for lateral packing density) as well as high effective indices (e g for ultra short resonators), essential for creating large lateral and longitudinal confinement

For planar waveguides, the first method suffices, one could e g use metamaterials in the shape of a diluted metal, e g silver nanospheres in an embedding dielectric, or use pure metals such as silver to generate a negative ε [4]. This structure is then interfaced to a dielectric with positive ε

of a slightly smaller magnitude to give a near surface plasmon resonance and a tightly confined plasmon mode with the desired high effective index [5]. Now, starting from this, a channel waveguide is created by confinement in the orthogonal dimension, employing the high effective index to achieve a highly confined field in this orthogonal dimension as well as offering the possibility of making subwavelength resonators due to the large resulting effective index. But in principle, also the second method could be used to generate waveguides in the conventional way, planar as well as channel ones, by using the high (positive) ε of metamaterial waveguides. Some ideas are shown in fig 1: to the left a surface plasmon waveguide in the vertical direction, accomplished by Ag nanospheres in different dielectric hosts, and "conventional" guiding in the horizontal direction. Light is propagating in the direction of the longest side. To the right, quantum dots are added to (partly) offset losses.

However, in all cases, material issues pertaining to loss (determined by ε'') remain the critical issue impeding the generation of ubiquitously useful photonics light wave circuits. Even though the losses per wavelength in the medium will be lowered by the high effective index, this is not enough to bring about useful (reasonably high Q values) ultra short resonators [5].

The talk will address the properties of some representative silver nanosphere metamaterials and the corresponding near resonant planar plasmon waveguides, as well the prospects for using the very high positive effective index to generate "conventional" waveguides. Hybrids of these phenomena to create channel waveguides will be discussed.

Conclusions

In this paper we have discussed several ways of increasing modal confinement (for increased lateral packing density) as well as effective index (for e g ultra short resonators), essential for creating large lateral and longitudinal confinement and further advance the density of integration in integrated photonic circuits.

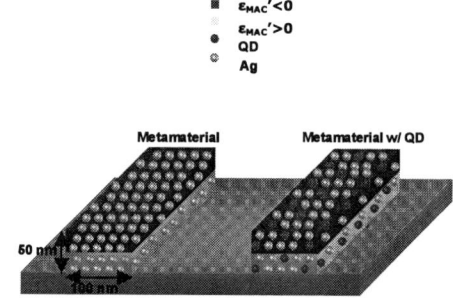

Fig. 1 Metallodielectric metamaterials with and without quantum dots shown in waveguide configuration. The waveguides have dimensions of e g 50 x 100 nm^2

References:

1. L. Thylén et.al, "The Moore's Law for photonic integrated circuits", J. Zhejiang Univ. SCIENCE 2006 7(12) p.1961-1964

2. D. Dai, L. Liu, L. Wosinski, and S. He , " Design and Fabrication of an Ultra-small Overlapped AWG Demultiplexers Based on - Si Nanowire Waveguides",

Electron. Lett., vol.42/7, pp.400-4002, 2006.

3. Stefan A. Maier, Mark L. Brongersma,Pieter G. Kik, Sheffer Meltzer, Ari A. G. Requicha,and Harry A. Atwater: "Plasmonics-A Route to Nanoscale Optical Devices", Adv. Mater. 2001, 13, No. 19

4. see e g A N Oraevskii et al, "Optical properties of heterogeneous media", Quantum Electronics Vol 31 , pp 252-256 (2001)

5. L. Thylen, E. Berglind, "Plasmonics, coherent light matter interactions and photonic crystals: Shaping the future of photonics?" Paper Mo.A.1, (Invited), Proc Internat. Conf on Transparent Optical Networks, ICTON 2006, Nottingham, June 2006

Advanced Liquid Crystal Displays

Shin-Tson Wu

College of Optics and Photonics, University of Central Florida, Orlando, FL 32816

Tel: (407) 823-4763, Fax: (407)823-6880, Email: swu@mail.ucf.edu

Abstract:

This tutorial will describe the basic operating physics of transmissive and transflective TFT-LCDs. Methods for improving the viewing angle, response time, color gamut, color shift, contrast ratio, and optical efficiency will be discussed.

Technical Summary:

Direct-view liquid crystal displays (LCDs) have been widely used in notebook computers, desktop monitors, high definition televisions, and mobile displays. For LCD TVs, wide view, high contrast ratio, fast motion picture response time, wide color gamut and weak color shift, and low cost are required. To meet these technical challenges, transmissive LCDs are favorable choices. On the other hand, mobile displays require good indoor and outdoor readabilities. To achieve such a wide dynamic range, a hybrid transflective LCD which combines the advantages of transmissive LCD for indoor applications and reflective LCD for sunlight readability has been developed.

In this tutorial, we will describe the basic operation principles and review the recent advances of transmissive and transflective TFT-LCDs. For transmissive LCDs, we will emphasize the following performance aspects:

1. Wide viewing angle: We will compare the performance of two major approaches: film-compensated multi-domain vertical alignment (MVA) and film-compensated in-plane switching (IPS) or fringing field switching (FFS).
2. High contrast ratio: We will discuss the light leakage mechanisms within an LCD system and describe approaches for reducing light leakage. The concept of dynamic local dimming using LED backlight will be introduced.
3. Fast motion picture response time: A TFT LCD is a holding type of display which results in image blurring. For notebook and desktop computers which mainly display static images in which image blur is not a big concern. However, for TVs which often display motion pictures, image blur becomes a serious issue. To reduce image blurring, impulse type LCDs and fast LC response time are needed.
4. Color: The color saturation, color gamut, and color shift of MVA and IPS using cold cathode fluorescent lamp and RGB LED backlight units will be compared. Methods for reducing color shift will be reviewed.

For transflective LCDs, we will emphasize the following performance aspects:

1. Dual cell gap approach: We will introduce the double cell gap approach to compensate the optical path length differences between the transmissive backlight and reflective ambient light. The pros and cons of this approach will be discussed.

978-0-9789217-3-6/07/$25.00

©2007 WEN GLOBAL SOLUTIONS

2. Single cell gap approach: In this approach, the ambient light would traverse the LC layer twice in the reflective region, but the transmissive light will pass only once. Therefore, we need to develop methods for reducing the phase retardation of the reflective sub-pixels in order to balance the phase retardation with that of the transmissive sub-pixels.

3. Wide viewing angle: The transmissive part can achieve wide view through compensation films, but for the reflective part the viewing angle is limited by the surface reflection and bumpy reflectors. Multi-domain structure is still needed for expanding the viewing angle for transflective LCDs.

4. High contrast ratio: To obtain high contrast ratio for the reflective part, a circular polarizer or in-cell phase retarder is needed. Moreover, surface reflection plays a key role affecting the contrast ratio of the reflective display. A broadband and robust anti-reflection coating is critically needed. High contrast ratio will also improve the color saturation.

References:
1. S. T. Wu and D. K. Yang, *Reflective Liquid Crystal Displays* (Wiley, New York, 2001).
2. D. K. Yang and S. T. Wu, *Fundamentals of Liquid Crystal Devices* (Wiley, New York, 2006).

Advanced Technologies for High Quality LC Display

Yi-Pai Huang[1], Fang-Cheng Lin[1], Cheng-Yumr Liao[1], Ya-Ting Hsu[1], Wei-Kai Huang, Cheng-Han Tsao,
Lin-Yao Liao[1], Chun-Ho Chen[1], Han-Ping D. Shieh[1]

1 : Dep't of Photonics and Display Inst., National Chiao Tung University,
Hsin-Chu, Taiwan, R.O.C., boundshuang@mail.nctu.edu.tw

2 : AU Optronics Technology Center, Science Park, Hsin-Chu, Taiwan, R. O. C.

Abstract *For achieving high quality LC displays, Advanced-MVA mode was proposed to wide the viewing angle, 4-color filed sequential method was utilized to increase the optical efficiency, and the local controlled LED backlight was applied to much improve the contrast and lower the power consumption.*

Introduction

The market of liquid crystal display (LCD) has dramatically increased in recently. However, the viewing angle, contrast, optical efficiency, and power consumption are the issues need to be improved. In this paper, we reported three developing technologies, (1) Advanced-MVA mode, (2) 4-color filed sequential method, and (3) local controlled LED backlight to yield a high quality LCD.

Advanced-MVA(A-MVA) Mode for Wide Viewing Angle

Multi-domain vertical alignment (MVA) technology[1], which separates a sub-pixel into 4-domains, has been the most applicable mode applying to LCD-TVs. However, color washout at large viewing angle is the issue that will degrade the image quality of MVA mode.

The Advanced-MVA(A-MVA) technology[2-3] generates an 8-domains sub-pixel to reduce the color washout. A-MVA is a novel pixel design technology which divided a sub-pixel into main- and sub-regions to yield 8-domains (4 azimuthal x 2 polar), as the pixel circuit shown in Fig. 1.

A-MVA utilized an additional TFT(Sub-TFT), which has different W/L and charging ratio to the main-TFT, to refresh(recharge) the voltage of sub-region(Vsub) in each frame. Therefore, two different voltage are applied on a sub-pixel automatically to generate 8-domains.

Fig. 2(b) illustrates the A-MVA mode reduced the color difference dramatically compare to that of conventional MVA (Fig. 2 (a)). With A-MVA, all the special color difference can be lower than $\Delta u'v' < 0.02$, which should be almost indistinguishable by the human eyes.

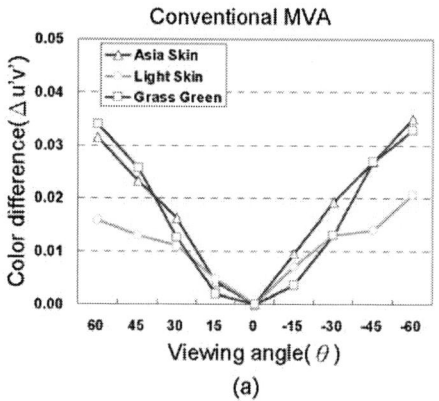

(a)

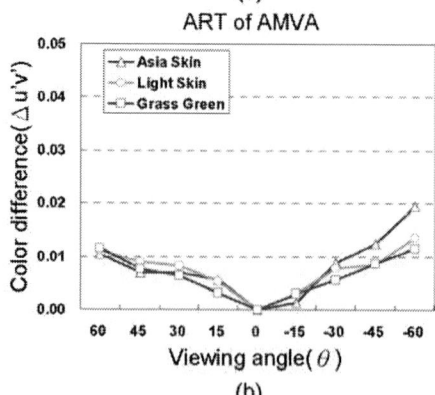

(b)

Fig. 2. The color difference of conventional MVA and Advanced-MVA.

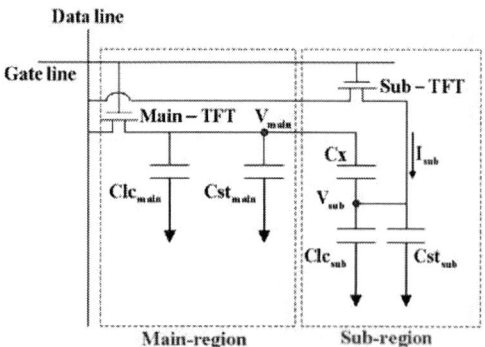

Fig. 1. The pixel circuit of A-MVA mode.

A-MVA technology had already been successfully implemented into the AUO's LCD-TV products. A 32" WXGA pixel image, as shown in Fig. 3, proved the practicability of the proposed "A-MVA mode" for wide viewing angle with less color-washout.

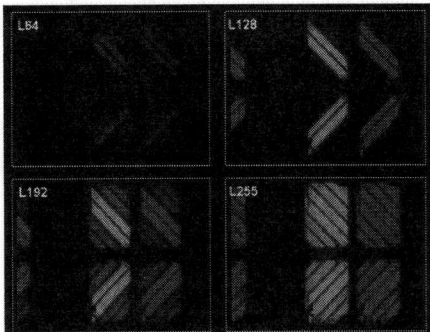

Fig. 3. The 32" WXGA pixel demonstration image.

4-Color Field Sequential LCD for High Optical Efficiency

Field Sequential Color(FSC) [4] LCD, which displays multi-primary color fields in temporal sequence to form a full-color frame, is an effective way to generate full color images. FSC-LCD utilized color LED backlight, thus can remove the color filter to much increase the optical efficiency. Color BreakUp (CBU) and flicker; however, has appeared intrinsically in FSC-type displays to degrade visual quality.

We proposed a 4-color field arrangement (4-CFA) method[5] to suppress the flicker phenomenon and to eliminate CBU of dynamic image. The color fields with order of RGBR, GBRG, and BRGB in three consecutive frames is shown in Fig. 4(b). By the re-arrangement of the 4-color filed, the The flicker phenomenon is eliminated due to the 80 Hz sub-frame frequency.

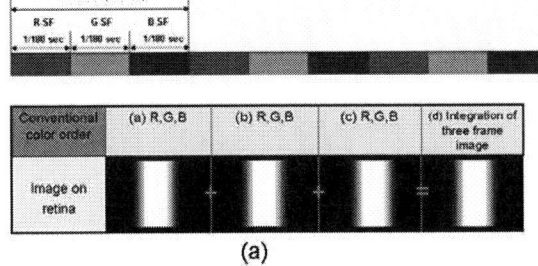

(a)

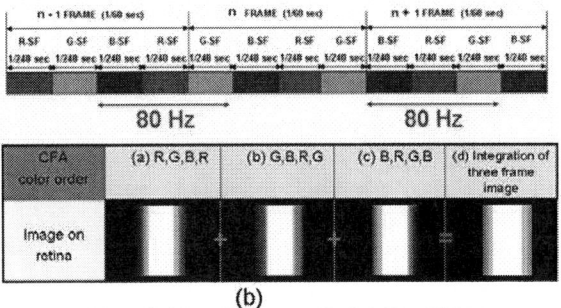

(b)

Fig. 4. Color field with order of RGBR, GBRG, and BRGB in three frames to eliminate flicker and color-breakup.

A demonstrated photograph of a 5.6-inch FSC-LCD panel with 4-CFA driving method is shown in Fig. 5. From the demonstrated results, the CBU in 4-CFA driving method was successfully eliminated which was independent to the image color and moving velocity. The optical power efficiency of demonstrated 5.6" LCD is 7.8Lm/W, which is almost a factor of 3 than that of conventional LCD (2.86Lm/W).

Fig. 5. Demonstrated Photo of the 4-CFA 5.6" FSC-LCD.

Local Controlled Backlight for High Contrast and Low Power Consumption

A drawback of conventional LCDs is a poor image contrast ratio (about 1,000:1) due to light leakage from liquid crystals. After dynamic-backlight-related technologies proposed, the backlight signal could be modulated to extend the image dynamic range (contrast) of LCDs [6] named high dynamic range LCD(HDR-LCD).

In order to further enhancing the image quality on a LCD, we utilized LED and LCD as a dual-panel display (Fig. 6). One is LED backlight panel and the other one is LC panel. LED Backlight panel is a low resolution panel for controlling the contrast ratio of images. According to each input image, a mapping curve for backlight was decided by frame to control the output

13

backlight signal. For the LC panel, it is a high resolution panel for maintaining the detail of the image

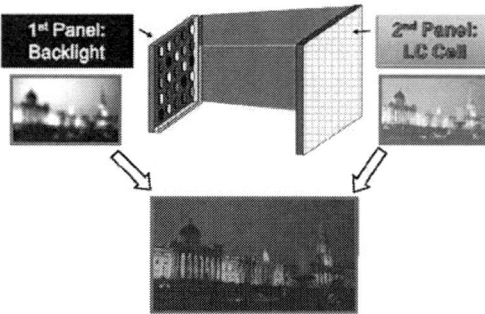

Fig 6. The cumulative function and Inverse of Mapping Function (IMF)

We proposed Inverse of Mapping Function (IMF) method[7] to optimize the backlight "image" according to the information of each frame, thus can not only keep high contrast ratio but also reduce the power consumption. The basic procedure of the IMF method is to obtain the global histogram for obtaining the mapping function (cumulative distribution function or CDF) from the histogram equalization (HE). According to the image mapping function of image, the backlight mapping function was inverted to the line y=x, and results in a new curve for backlight modulation named "Inverse of Mapping Function" (IMF) as shown in Fig. 7.

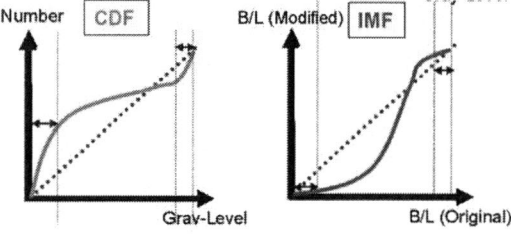

Fig 7. The cumulative function and Inverse of Mapping Function (IMF) response of backlight curve for high contrast ratio.

By IMF method, backlight signal can be controlled by frame to perform high quality images. For example, in a high contrast ratio image, the inverse of mapping function will be similar as Fig. 7(b) shown; the curve can modulate a higher contrast image due to the enhancement of bright/dark ratio. Fig. 8 shows the images of conventional LCD and the LCD with IMF method, which demonstrated a much obvious improvement of the image quality. The measured contrast

with IMF method also can yield ~20,000:1, as shown in the table. 1. Additionally, the LEDs at the dark area will be dimmed to much reduce the power consumption become 50% only.

| (a) | (b) |

Fig 8. The cumulative function and Inverse of Mapping Function (IMF)

Table. 1. The measured contrast and power consumption of conventional LCD and HDR-LCD with IMF method.

	Contrast	Power Consumption(Watt)
Conventional LCD	1230	175
IMF Method HDR-LCD	19600	88

Conclusions

Three methods were proposed to improve the quality of conventional LCDs. 8-domain A-MVA mode can reduce the color-shift at 60° viewing angle to $\Delta u'v' < 0.02$, which should be indistinguished by the human. 4-Color Field Sequential LCD improves the optical efficiency of a factor of 3 (7.8lm/W) without color-breakup and flicker. Moreover, HDR-LCD with inverse of mapping function can increase the contrast to 20.000:1 and lower the power consumption to 50% only. The results successfully demonstrated the enabling advnced technologies for high quality LCDs.

References

1. S. S. Kim, et al., SID'06 Symposium Digest, pp. 1938-1941, 2006.
2. Y. P. Huang, et al., SID'07 Symposium Digest, pp. 1010-1013, 2007.
3. Y. P. Huang, et al., IDW05 Symposium Digest, pp. 787-788, 2005.
4. T. Kurita, et al., IDW00 Symposium Digest, pp. 69-72, 2000.
5. Y. T. Hsu, et al., IDW07 Symposium Digest submitting.
6. H. Seetzen, et al., SID'03 Symposium Digest, pp. 1450-1453, 2003.
7. F. C. Lin, et al., SID'07 Symposium Digest, pp. 1343-1346, 2007.

Response Times in Pi-cell Liquid Crystal Displays

Hongmei Ma, Li Jiang, Yubao Sun

Department of Applied Physics, Hebei University of Technology, Tianjin, 300401, P. R. China,
Email: hmtj450@vip.sina.com

Abstract The rise times of Pi-cell with arbitrary pretilt angle are simulated under normal driven methods. The off times of the cell driven by under-shoot driven method is faster than that under normal driven method.

Introduction

Many applications of liquid crystal displays require fast response speed. For example, color sequential LCD requires the LC response speed to be fast. A ferroelectric LC has very fast response speed but it lack gray and requires difficult processing [1]. The Pi-cell [2] (also known as optically compensational bend (OCB) [3] cell) shows the fast response characteristic. The Pi-cell operates between the bend deformation at the critical voltage (Uc) and the near homeotropic state at high voltage. Many researchers [4-10] only studied the response times of the cell with various pretilt angles, which is switched between Uc and on-state voltage. However, the response time is sensitive with the different driven voltage and pretilt angle, but it has not analyzed in detail before.

In this letter, we calculate the response time of the Pi-cell with arbitrary pretilt angle driven by different voltage, and using normal and under-shoot method in the off process. Our results show that the response times strongly depend on the voltage and the pretilt angle. The response times (both rise and off times) of the cell driven by large voltage are faster than that of the cell driven by low voltage using normal driven method. However, for the off time, it is very fast for the cell driven by low voltage than that of the cell driven by high voltage using under-shoot method.

Simulation

In the following numerical calculation, the LC E7 material parameters are chosen as n_e=1.65, n_o=1.5, $\Delta\varepsilon$=13.745 (=19.24-5.495), K_{11}=10.6pN, K_{33}=15.5pN, γ=0.15Pa·s, η_1=0.1732 Pa·s, η_2=0.0232 Pa·s, and η_{12}=0.0Pa·s [7,8]. The same cell gap (3μm) is used in our simulation to compare with the various pretilt angle and voltage's effect on the response times.

For the terse discussion, the relationship between the electric field in LC layer and the driven voltage can be written as $E \approx U/d$. Taking into account the flow effect, the dynamic equation for the cell switching in the bend state can be described by[7-10]

$$\gamma * \frac{d\theta}{dt} = \left(K_{11}\cos^2\theta + K_{33}\sin^2\theta\right)\frac{\partial^2\theta}{\partial z^2}$$
$$+ \left[\left(K_{33} - K_{11}\right)\left(\frac{d\theta}{dz}\right)^2 + \varepsilon_0\Delta\varepsilon E^2\right]\sin\theta\cos\theta \quad (1)$$

where the relation between the effective rotational viscosity ($\gamma*$), the rotational viscosity (γ), and Miesowicz viscosity constants (η_i) is determined by

$$\gamma* = \gamma\left(1 - \frac{\left[\frac{1}{2}\left(1-\frac{\eta_1-\eta_2}{\gamma}\right)\cos^2\theta + \frac{1}{2}\left(1+\frac{\eta_1-\eta_2}{\gamma}\right)\sin^2\theta\right]^2}{\frac{\eta_{12}}{\gamma}\sin^2\theta\cos^2\theta + \frac{\eta_1-\eta_2}{\gamma}\sin^2\theta + \frac{\eta_2}{\gamma}}\right) \quad (2)$$

The response time is evaluated as the variety of retardation normalizing the difference of the optical design condition which is caused by the different pretilt angles. The response time, τ_{on} is defined as the transition time of normalized retardation from 1.0 to exp(-1.0), and τ_{off} as the transition time of that from 0 to (1-exp(-1.0)). Uc for the cell with arbitrary pretilt can be calculated using our primary work [12,13]. Using the LC parameters, the critical pretilt angle is 47.7°, the bend state is the stable state as the pretilt is larger than the critical pretilt. That is to say, the cell has zero critical voltage if the pretilt is larger than 47.7°.

978-0-9789217-3-6/07/$25.00

©2007 WEN GLOBAL SOLUTIONS

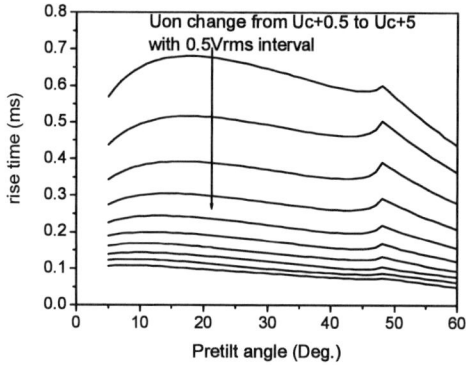

Figure1 The rise times of the cell switched between Uc and Von of the cell with different pretilt angle.

We set the on-state voltage (Von) from Uc to Uc+5V with 0.5V interval, the voltage change from Uc to Von (on case) and from Von to Uc (off case). The rise times, simulated by Eq. (1), are shown in figure 1 for the Pi-cell with different pretilt angle, respectively. From figure 1, we see that the rise time decreases as the driven voltage increases, but has no more change for low driven voltage and decreases fast as the pretilt angle increases, which is also confirmed partially by the previous experiment data [5, 10]. For achieving faster rise time, it is driven by a high voltage, as the LC profile arrive the appropriate state of a grey-level, the voltage for this grey-level is added to hold the state, this method is also called as over-drive method.

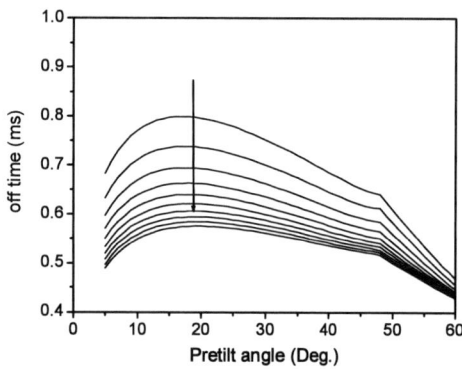

Figure 2 The off times of the cell switched between Uc and Von of the cell with different pretilt angle using normal method.

The off times are shown in Fig. 2 for the cell driven by normal method. The off time

increases firstly and decreases as the pretilt increases, and decreases with the increasing voltage. The off time cannot remain a constant. The fluid directions are same at both sides of the bend cell, and the fluid velocity increases with approaching the center of the cell, so the torque induced by this flow accelerates the relaxation of liquid crystal molecules, and gets stronger as the pretilt angle and driven voltage increases [10]. For this reason, the off time become very faster for high pretilt angle and high driven voltage.

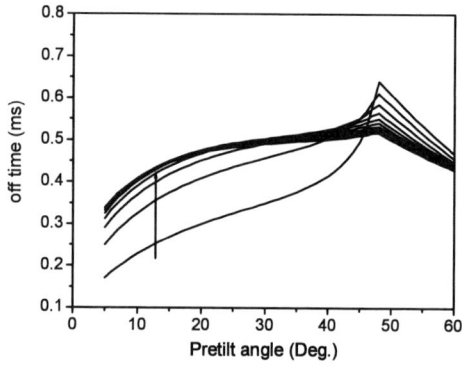

Figure 3 The off times of the cell switched between Uc and Von of the cell with different pretilt angle using under-shoot method.

To achieve faster off time, the under-shoot method need to be used [4]. The off times of the cell switched between Von and Uc by using the under-shoot method are shown in figure 3. From figure 3, we can see that the fastest response appears at the low pretilt angle which has been confirmed by the experimental data when the cell is driven by the under-shoot method [4], and the off time of the cell which has nonzero critical voltage is faster than that of the cell driven by normal method.

Conclusions
In summary, the response times are calculated in detail. As the cell driven by normal method, the response times (both rise and off times) increase as the voltage increases. As the cell driven by under-shoot method in the decay process, the off time is improved and increases as the voltage increases for the cell with nonzero critical

voltage. These results have more potential application in designing Pi-cell LCDs with faster response speed in the future.

This research was supported by Key Subject Construction Project of Hebei Province University and Natural Science Foundation of Hebei Province (No. A2006000675 and No. 103002), P. R. China.

References

1 J. Someya, SID'05 Dig., 2005, p.1018.
2 P. J. Bos et al Mol. Cryst. Liq. Cryst. 133 (1984), 329.
3 T. Miyashita et al Proc. 13th Int. Display Res. Conf. (Eurodisplay'93, Strabourg, France, 1993) p.149.
4 X. D. Mi et al SID'99 Dig., 1999, p.24.

5 M. Xu et al SID'98 Dig., 1999 p.139.
6 F. S. Yeung et al Appl. Phys. Lett., 88 (2006), 063505.
7 E. J. Acosta et al Liq. Cryst. 27 (2000), 977.
8 H. G. Walton et al Liq. Cryst. 27 (2000), 1329.
9 E. J. Acosta et al J. Appl. Phys. 97 (2005), 093106.
10 K. Wako et al SID'05 Digest, 2005, p.666.
11 Y. Sun et al Appl. Phys. Lett., 89 (2006), 041110.
12 Y. Sun et al Appl. Phys. Lett., 90 (2007), 091103.
13 Y. Sun et al Jpn. J. Appl. Phys., 45 (2006), 5810.

Specialty Fibers as Key Components for Dispersion Management

Hans Damsgaard
OFS Fitel Denmark, 680 Priorparken,
DK-2605 Brondby, Denmark

Enabling technologies based on highly advanced optical fibers for high-speed fiber communication systems are discussed with special attention to dispersion management and amplification. History, recent progress, and possible future directions will be discussed.

Introduction

25 years ago, when the single-mode fiber was developed and introduced to the market, researchers considered this type of fiber as capable of transporting almost infinity of bandwidth. At that time the SM fibers were operated at 1310 nm. In 1985 the Er-doped fiber amplifier (EDFA) was introduced for amplification around 1550 nm (1) and this created a technology revolution, which basically resulted in repeatered all optical DWDM long distance systems at 2.5 Gb/s, and later at higher bit rates. The systems operate primarily in the C-band from 1535nm to1565 nm, with channel spacing down to 25 GHz.
During this period a lot of engineering was accomplished in order to achieve all optical Tb/s transmission over several 1000 km all optical transmission.

There are two types of dispersion of the optical fibers, which heavily impact the performance of such systems: Polarization Mode Dispersion (PMD) and Chromatic Dispersion (CD). Basically, PMD needs to be controlled very carefully and maintained at a consistent low level when manufacturing fibers and cables (2,3), whereas **Chromatic Dispersion is needed** in the transmission fiber to avoid non-linear effects like e.g. four wave mixing (FWM) (4). Consequently, chromatic dispersion management has become an integral part of modern systems, and an ensemble of advanced fibers have been developed and introduced to facilitate management of Chromatic Dispersion. These highly advanced fibers are used both as transmission fibers and as optical components in the terminal equipment. Simultaneous optimization of cost, performance, simplicity, amplification schemes, and CD management creates new opportunities for such highly advanced optical fibers to be used in cables and in components. CD management is crucial all across the band in which the system is to operate – in future systems more than 100 nm!

Advanced fibers to be Used in Cables

Installation of optical fiber cables is expensive, time consuming, and complicated to organize practically because of "right of ways" and interference with other types of infrastructure. As a result, the lifetime of the cables must be long (20 years). Mechanically, it is well known how to obtain this. However, to make sure that the fibers are well suited for futuristic optical transmission gear is not trivial. Merits of different types of transmission fiber will be reviewed with respect to performance in future 40 Gb/s + systems, with special attention to PMD, amplification scheme, and CD management scheme.

First, PMD needs to be controlled and maintained at a low consistent level. This statement is validated by millions of km of fiber in the ground today, which cannot be upgraded to 10Gb/s – and some fibers have so bad PMD that it cannot operate even at 2.5Gb/s. To compensate for PMD is possible to some extent. It is, however, expensive, and it requires a device for each channel and adds complexity to the transmission gear.

Next, the CD map needs to be optimized for performance. Standard transmission fibers have a Chromatic Dispersion of about 17 ps/(nm·km), which is more than sufficient to avoid typical non-linear effects like FWM etc. However, the accumulated CD needs to be compensated together with the loss associated with the CD compensation. Thus, to have more CD in the transmission fiber than needed to combat non-linearities is not always desirable. That is why a number of non-zero dispersion fibers (NZDF) have been developed, with dispersion in the range 4 to 8 ps/(nm·km).

Third, when stepping up to higher bit rates, the Optical Signal to Noise Ratio (OSNR) is decreased, and more amplification is needed. As has been pointed out in many papers distributed, amplification is an efficient way of enhancing the OSNR in a long amplifier spacing configuration in long haul systems. Hence, a modern transmission fiber also needs to be optimized for Raman pumping with a multitude of e.g. 14xx nm pumps. For example, this means that the transmission fiber must be optimized for consistent Raman gain efficiency to reduce pump power (5), it must be optimized to avoid energy transfer between the pumps, and it must be optimized to minimize non-linearities like Self Phase Modulation (SPM).

Finally, the fiber must be capable of operating in a 100 + nm wide band. In the talk, a practical example of a product where all these requirements were simultaneously optimized will be reviewed together with possible future designs with distributed amplification and full in line Dispersion Compensation (11).

Advanced Fibers to be Used as Components in the Terminal Equipment

Amplification and dispersion management is crucial for metro and long haul optical networks. CD compensation is for practical purposes needed for 10 Gb/s systems extending more than 100 km (4), and the first commercial dispersion compensators were deployed 10 years ago. Since then, more than 200,000 broad band CD compensators based on highly advanced optical fiber with negative dispersion and negative dispersion slope have been installed (6,7). Many different technologies for CD compensation have been researched including fiber based technologies, FBG technology, free space optics etalon, planar waveguides, microstructured fiber, ring resonators, mid span spectral inversion, and more recently channel by channel electronics based dispersion compensation. Recent developments and the merits and commercial status of these different technologies will be reviewed with primary focus on technologies that have been deployed in the field. Bandwidth, dispersion, dispersion slope, dispersion curvature, loss, size, PMD, PDL, tuneability, non-linearity, cost, reliability, and ability to upgrade to higher bit rates are among the merits for such devices. In addition an example of significant cost reduction by using cheap directly modulated laser sources together with modern small size low cost dispersion compensators will be shown(7,8).

Also possible future developments will be discussed including how Raman pumping can transform a fiber-based dispersion compensator with loss into a 100 nm broadband amplifier conducting simultaneously amplification and dispersion compensation (9). Thereby complex banding of C- and L-band windows can be avoided, and one device can simultaneously amplify and dispersion compensate hundreds of channels showing the ultimate strength and value of all optical transmission technology. Also examples of recent installations of negative dispersion fiber in ocean systems will be shown (10), and it will be discussed how this technology may migrate into terrestrial systems soon (11).

Finally, recent development of Er-doped fibers used in EDFAs will be reviewed. Focus will be on efficiency, consistency, splicing, PMD, co-design with CD modules, and reliability (12,13,14,15). Examples of high power broad band EDFAs based on double clad microstructured fibers will also be shown.

List of references

1. S.B. Poole, D.N. Payne, F.E. Fermann, "Fabrication of low-loss optical fibres containing rare.earth ions", Electro. Lett. 21, pp 737-738, 1985.

2. T.Geisler; Low PMD Transmission Fibers; Mo3.3.1 ECOC, 2006

3. T. Geisler, O. Lumholt; A. Sørby, R. Ortega; Large Volume Fiber and Cable Results for low Slope NZDF for 40 Gb/s; NWE4, OFC/NFOEC 2005

4. *L. Nelson and B. Zhu, Raman Amplifiers for Telecommunications. New York: Springer Verlag; 2004*

5. B. Pálsdóttir, C.C. Larsen:
"Raman gain efficiency measured on 16 Mm of Raman optimized NZFD fiber"
OFC 2005, paperOThF7

6. L. Grüner-Nielsen, M. Wandel, P. Kristensen, C. Jørgensen, L. V. Jørgensen, B. Edvold, B. Pálsdóttir and D. Jakobsen; Dispersion Compensating Fibers; Journal of Lightwave Technology, Vol. 23, No. 11, pp. 3566-3579; 2005

7. P. Kristensen, M. N. Andersen, B. Edvold, T. Veng, and L. Grüner-Nielsen; Dispersion and Slope Compensating Module for G.652 Fiber with x4 Reduced Physical Dimensions; Proceedings of ECOC'03; paper We4P.15; 2003

8. M. Du, L. Grüner-Nielsen, C. Jørgensen, and D. DiGiovanni; Dispersion Compensated 10 Gb/s Directly Modulated Lasers for 6x80km, DWDM Metro Network Applications; Proceedings of ECOC'06; paper We3.P.133; 2006

9. L. Grüner-Nielsen and Y. Quian; Dispersion-Compensating Fibers for Raman Applications, Chapter 6 of "Raman Amplifiers for Telecommunications 1", Mohammed N. Islam, Editor, Springer Verlag; 2004

10. Lars Grüner-Nielsen, David Peckham, Robert Lingle and Ole A. Levring; Dispersion managed fibre spans optimised for submarine links; Proceedings of SubOptic 2004, paper Th B1.2.; 2004

11. E. Pincemin, A. Tan, A. Tonello, S. Wabnitz, J.D. Ania-Castañón, V. Mezentsev, S.K. Turitsyn , Y. Jaouën and L. Grüner-Nielsen; Performance Comparison of SSMF and UltraWave Fibers for Ultra-Long Haul 40-Gb/s WDM Transmission; Photonics Technology Letters, vol. 19, no. , pp.; 2007

12. T. Veng, B. Pálsdóttir:
"Investigation and optimization of fusion splicing abilities between erbium-doped optical fibres and standard singlemode fibres".
El. Lett., vol 41, no1, p10, 2005

13. B. Pálsdóttir, S. Primdahl, P. Gaarde, G. Puc, B. Wang:
"Consistency in Gain Spectra of Erbium Doped Fibers".
Technical Digest of OFC 2003, paper ThZ1.

14. P.B. Gaarde, T. Geisler, P. Kristensen, and B. Pálsdóttir:
"Bending Induced PMD in Spun Erbium Doped Fiber".
OFS Fitel Denmark , proceedings of ECOC 2005

15. Bera Pálsdóttir, Peter B Gaarde and Torben Veng:
Requirements for erbium doped fibers and dispersion
compensating fibers in high performance amplifiers
in *Optical Amplifiers and Their Applications and
Coherent Optical Technologies and Applications on CD-ROM* (The Optical Society of America, Washington, DC, 2006), OTuB1.

Micro/Nano-Scale Optical Circuits and Networks for Information and Telecommunication Applications

Prof. Dr. El-Hang Lee, Director

OPERA (Optics and Photonics Elite Research Academy)
National Research Center for Photonic Integration Technology
Micro/Nano-Photonics Advanced Research Center (μ-PARC)
Graduate School of Information and Communication Engineering,
INHA University, 253 YongHyun-Dong, Nam-Ku, Incheon, South Korea
www.opera.re.kr   ehlee@inha.ac.kr

Abstract *We discuss on the theory, design, and fabrication of high-density micro/nano-photonic circuits and networks on a platform called optical printed circuit board for applications in telecommunications and information technology.*

Introduction

In the historical and global perspective, much of the macro-scale electrical networks based on copper wires and copper cables which dominated the world over the many decades from the early 20th century through the 1970s, has now been replaced with the optical fibers and optical cables, which cover long distance communications through the terrains and the oceans around the world, carrying Gigabits and Terabits per second information capacities. While much of the optical networks exist only in the macro-scale and mega-scale dimensions covering tens of thousands kilometers down to tens of meters, there is still much room to explore in terms of optical networks down to the centimeter, millimeter, micron and nano-scale domain. [1-9] This new frontier is to be explored in order to make the telecommunications possible down to the ever-increasing micro/nano-scale information storages.

We perform a systematic study on the micro/nano-scale optical circuits and networks. It is an extension of the macro-scale optical fiber networks that have revolutionized the world of telecommunication into the domain of micro/nano-scale network. [1-9] Micro/nano-scale photonic wires and devices are interconnected on a planar board or on a chip to form a network of various functions that are compact, high-speed, intelligent, light-weight, low-energy and environmentally friendly, low-cost, and high-volume applications. Photonic devices include micro/nano-scale lasers, switches, couplers, detectors, sensors, actuators, modulators, and related devices. Applications can vary from telecom applications to datacom applications, biosensor applications, medical applications and environmental applications, and aero-space applications.

Design Considerations for Micro/Nano-Photonic Integration

As the sizes of devices become small, the proximity effect, the energy confinement effect, the microcavity effect in microlasers, single photon effect, and the optical interference effect between devices become important issues. Nonlinear effects can become pronounced. Interface physics and environmental issues also become important. The high field intensity of the light within the devices can cause nonlinear interactions. Understanding and controlling the dynamics of noise in small photonic devices becomes important. Quantum chaotic phenomena and quantum interference effect become important. Quantum optics is expected to become an important part of micro/nano-photonics study.

In the design of micro/nano-photonic devices, parameters to be put into consideration include device dimension, size, electrical current, light output power, input power, wavelength, refractive index, mode, polarization, and device proximity, in addition to all the effects discussed above. We worked out some design rules for the minimization of the waveguide sizes and the maximization of the integration density of the optical wires by way of theoretical and

978-0-9789217-3-6/07/$25.00

©2007 WEN GLOBAL SOLUTIONS

simulation studies. For a crosstalk of -30dB for a family of different waveguide with refractive index difference of 0.02, for example, the minimum waveguide width and the optimum separation between waveguides comes out to be 7μm and 14μm, respectively, allowing for 46 waveguides within a 1mm x 100 mm substrate. We also calculate the bandwidths for various sizes of waveguide width and for cross sections varying from 1 micron to 150 microns with index difference as 0.2, for the refractive indices of the core and the cladding fixed for 1.47 and 1.45. For an O-PCB with 100 micron width and 100 micron height, for example, the bandwidth is 62Gbps for 10cm length. As the cross-sectional area diminishes, the bandwidth increases exponentially, even up to several Tera Gbps range.

Micro/Nano-scale Photonic Components and Circuit Integration

We design and integrate arrayed waveguide devices, directional couplers, multi-mode interference devices, micro-ring resonant devices with the polymer optical wires on the O-PCB board. We use mode adapters for the integration of these devices. And we also examine the compatibility of integration between many different devices. The scientific and technical issues and challenges for the integrations include the size mismatch, the polarization mismatch, the thermal mismatch, the mechanical mismatch, and the shape mismatch

Silicon-based Micro/Nano-Photonic Circuits and Networks

We design and integrate silicon based micro-ring resonant devices and photonic circuits for sensor applications. We measure the changes in the refractive indices of the molecules and/or bio-materials coming in to touch with the micro-ring surfaces. We also design and fabricate silicon based nano-wires and photonic crystal devices and circuits for nano-scale optical interconnection and integration. We also design silicon microstructures in which we can confine the light between two waveguides. We then extend this to a multiple array of such waveguide pairs for confinement of multiple beams of light. Silicon light amplifiers are also designed and have been demonstrated using rare earth dopants.

Photonic Crystal Micro/Nano-Circuits and Networks

Photonic crystals can be utilized for ultra-small optical devices such as waveguides, add/drop filters, switches, and modulators. To realize these nano-scale optical networks, we have designed power-splitting devices using 2-dimensional photonic crystals. In order to increase the density of integration of the waveguides, we used parallel-shaped directional couplers in photonic crystals, and have found this structure superior to other structures, such as Y-branch (or junction) 2-dimensional photonic crystal power splitters, which seem to take up too much space for high-density integration. We also designed a two-dimensional passive optical network triplexer by using the photonic crystal. By varying the sizes and the refractive indices of the localized defects in the photonic crystal, we have been able to generate curves that can determine the transmission wavelengths. Here, we used a two-dimensional photonic crystal with a square pattern of rods and have been able to use three kinds of lights with free space wavelengths, λ_1=1550nm, λ_2=1490nm, and λ_3=1310nm for the multiplexing. We set the lattice constant, a=490nm, and obtained the corresponding frequencies and radii for each wavelength by using FDTD.

Plasmonic Micro/Nano-Photonic Circuits and Networks

We also use surface plasmon-polaritons (SPPs) formed on a flat metal surface like silver or gold for micro/nano-photonic circuits and networks. In our study, the stripe waveguides were formed by using 20nm thin and 5um wide gold strips sandwiched between 12um thick spin coated layer of polymer supported by silicon wafer. Fundamental SPP mode, excited by end-fire coupling at telecommunication wavelengths (1.52~1.58um) with polarization controlled (PC) single mode fiber, for example, showed the guiding propagation loss of ~19dB/cm and the coupling loss with the fiber of about ~1.8dB per facet. Vertical directional couplers (VDCs) of 20nm thin and 5um wide gold stripes, consisting of metal stripe waveguide embedded in polymer, revealed the extinction ratio of about 28dB and the separation distance of about 4.2um, indicating a possibility of high density integration. For a strip separation of 4.2um, the power transfer was as much as ~28dB with the interaction length of 260um and the total insertion loss was about ~ 24dB.

Assembly of Micro/Nano-Networks

The networks were assembled with an array of 1 x 12 polymer waveguides fabricated by UV embossing. The refractive index of the waveguide is 1.47 at 850nm. The dimensions of the waveguides are 50 micron in width and height with a pitch of 250 micron. Arrays of vertical cavity surface emitting lasers (VCSELs) and photodiodes (PDs) are wire-bonded or flip-chip bonded to the silicon devices for optical input–output interconnection. When the light emitted from the VCSEL was set to +3dBm, the fiber's light receiving power through the waveguide was found to be -4.9 dB and the total loss, including the propagation loss and coupling loss between optical transmitter and waveguide, was -7.9dB. We have been able to measure the transmission speed of up to 10 Gps for each channel using the eye diagram. We also have been able to fabricate the circuits on hard and flexible substrates and have been able to measure the speed up to 10Gbps. Micro/nano-circuits and networks are then interconnected and integrated using these arrays of waveguides. We tested the assembled micro-scale by watching a moving video stored in a memory chip by interconnecting the memory chips and the CPU chips using optical wires. What we have done is to cut the copper wires connecting the memory chip and the CPU chip and replace all the copper wires with the optical wires. We used the electrical signal from the memory device to provide the optical signals from a semiconductor laser (VCSEL, for example) and guided these optical signals from the VCSEL laser to the detectors by way of polymer optical wires. We have been able to watch the moving images through the optical wires.

Summary and Conclusion

We presented an overview on the theory, design, fabrication, and integration of micro/nano-scale optical circuits and networks for information and telecommunication applications. Theoretical, experimental and simulation studies suggest that the micro/nano-scale network is a new challenge.

Acknowledgements

This work has been supported by the Korea Science and Engineering Foundation (KOSEF) through the Grant for the Integrated Photonics Technology Research Center (R11-2003-022) at the Optics and Photonics Elite Research Academy (OPERA), Inha University, Incheon, South Korea.

References

1. E. H. Lee, Invited Talk, SPIE Photonics West, San Jose, CA, USA. Jan. 22-26, 2007
2. E. H. Lee, Keynote Lecture, "Optical Micro/Nano-Networks" IASTED2006, July 2-4, 2006, Banff, CANADA
3. E. H. Lee, Plenary Lecture, "VLSI Photonics," ICSEP2006, April 1-4, 2006, Taipei, TAIWAN
4. E. H. Lee, Invited Lecture, SPIE-Europe, April 3-7, 2006, Strasbourg, FRANCE
5. E. H. Lee, Invited Talk, SPIE Photonics West, Proceedings, Vol. 5729, pp. 118, 2005
6. E. H. Lee, Invited Talk, SPIE Photonics West, Proceedings, Vol. 4652, 2002, p.1.
7. E. H. Lee, Invited Talk, SPIE Photonics West, Proceedings, Vol. 5356, 2004
8. E. H. Lee, Invited Talk, IEEE/LEOS, Summer Topical Meeting, June 29-30, 2004, San Diego, CA, USA.
9. Inha University, (2005) Integrated Photonics Technology Research Center, Optics and Photonics Elite Research Academy (OPERA), Home Page Address: http://www.opera.re.kr

The Transition from Discrete Optics to Optical Integration

A. P. Janssen

Bookham Inc. Brixham Road, Paignton, Devon TQ4 7BE

Tel: +44 1803 662607, email: apjanssen@bookham.co.uk

Abstract: Integration has provided evolutionary routes to achieving improved opto-electronic component cost-performance ratios at many levels. We outline some of the technology choices and strategies with examples relevant to today's systems and future systems.

Index terms: integrated optics, micro-optics, tuneable laser, MZ modulator, opto-electronics.

1. Introduction

Over the last two decades the evolution of optical components for optical communications has shown growth from simple directly modulated semiconductor laser sources to multi-function sources which are capable of wavelength tuning, multi-level coding at greater than 10Gb/s and a variety of additional optical system enablers. In comparison to the silicon IC industry, the materials technologies required to combine both optical and electronic functions are significantly more diverse. This diversity includes a wide variety of III-V quantum well materials, special glasses for fibres, polymer bonding materials, magneto-optic materials, multi-layer optical coatings, electro-optic materials such as lithium niobate and thermo-electric materials. The quest for increased functionality in conjunction with reduced costs of manufacture has prompted rationalisation of many of the technologies that have been developed over this period. Integration methods at both the chip level and the package level have allowed much simplification of material and design and thus reduction in cost and increased density. Following the exponential density growth observed for silicon ICs over several decades (Moore's Law), comparable growth for opto-electronic devices is apparent despite the diversity of material technologies and smaller global markets.

2. Technology choices.

The requirements to control optical as well as electronic functionality have led to the development of many kinds of solution to specific performance objectives.

Examples of this are the use of silica on silicon waveguide technology for passive optical routing and design of optical filters. These functions may also be obtained using free-space micro-optics, multi-layer filters and fibre grating technologies. The choice is governed by more specific factors such as functional performance, size, cost, ease of manufacture and the ability to interface to other technologies. As manufacturers of O-E components, Bookham has experienced transitions through a number of these technologies and it is clear that some have shorter lifetimes and market applications than others. The ability to pre-empt these transitions and identify leading technologies is clearly of strategic advantage.

3. Laser-modulator Integration

Mach-Zehnder interferometers are one of the most appropriate methods of providing optical modulation with high extinction at 10Gb/s and beyond. Lithium niobate MZ modulators have fulfilled these requirements for many years. However the different material technology from that used for semiconductor laser sources means that the optical interface is managed via fibre. The implications of this are the additional costs of coupling optics (lenses), isolators and precision manufacturing processes as well as package costs. MZ fabrication using the same compatible III-V technology provides a means of efficient co-packaging both laser source and modulator using common optical elements. The Bookham 'Compact' design allows an etalon wavelength locker as well as power monitor to be built in.

Figure 1. Package integrated laser – InP MZ modulator.

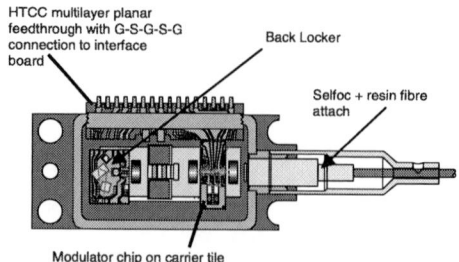

The 'Compact design shown in figure 1 is also the basis of the fully tuneable DSDBR laser and modulator, (figure2).

Figure 2. Co-packaged tuneable laser MZ (ITLA) .

Fabrication is facilitated by automated chip bonding, piecepart attachment and optical

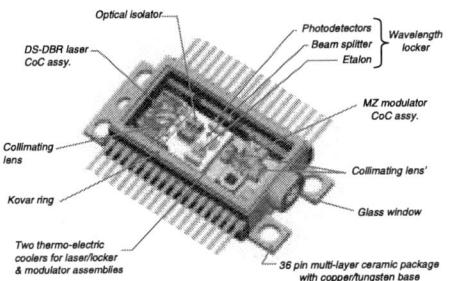

alignment giving a consistently high degree of assembly control and high yields. Manufacturing stability and efficiency is demonstrated over 100,000 field devices with high reliability.

The fully tuneable C and L band laser chip requires integration of a number of optical sections (double grating, phase adjust, shutter, SOA, and increased number of process sequences.(figure 2).

Figure 2. DSDBR C,L band tuneable laser.

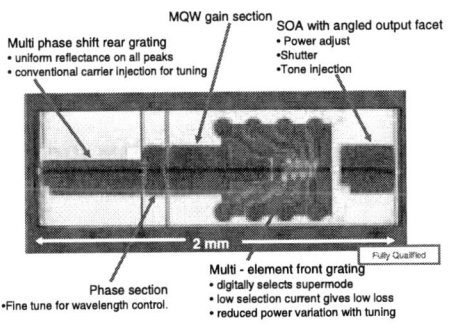

The laser has >40nm tuning range, >15dBm power, SMSR >40dB.

The InP MZ, (figure 3), involves 3-stage growth and deep etched ridge and comprises several optical elements; the main modulator arms fed by MMI couplers are supplemented by phase and control electrodes. Operating point and power are monitored and controlled by PIN detectors and a VOA. Coupling efficiency and alignment are optimised with the inclusion of a spot-size converter.

Figure 3. Wide band InP MZ chip on carrier.

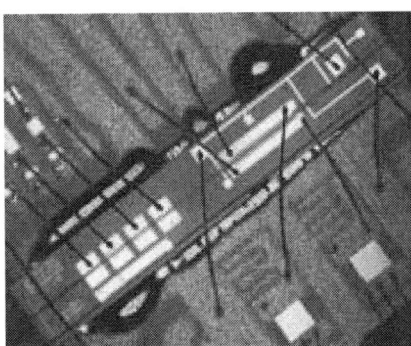

The transition from single wavelength source to tuneable laser over complete C or L band is enhanced by a second stage integration of the laser with a MZ into a single chip. Development of the processes to achieve this has taken several years. The advantages of this are; dramatically reduced piece-part count and manufacturing assembly stages, improved performance via reduced dissipation, higher optical powers, reduced footprint allowing access to TOSA and

Figure 4. Fully integrated tuneable laser + MZ modulator.

small form-factor modules, improved temperature performance at lower combined cost.

Figure 4 shows a fully integrated single chip C-band tuneable laser with MZ 10Gb modulator. The increased growth and processing steps required to achieve this are balanced by a reduction in piece-parts; 2 lenses, isolator, TEC, as well as the significantly less complex optical assembly.

In addition, the integrated design allows the use of a back–locker and the ability to shutter the laser during wavelength tuning and channel locking. The integrated 5mm chip incorporates back SOA, DSDBR laser, multi-contact tuning section, front SOA, waveguide laser-MZ interface, MMI section, imbalance electrodes, RF electrodes, MMI and power tap guide. Further generations include 20Gb – 40Gb, integrated photodiodes and parallel MZM's for multi-level modulation. The additional possibility of pm wavelength control via an integrated locker is a future possibility but dictated by material stability considerations.

An alternative technology for modulator integration is the electro-absorption (EA-laser) suited for small footprint applications; semi-cooled, pluggable for intermediate reach (40km). Low capacitance design absorption section is interfaced with an etched MQW DFB laser waveguide with low loss transition to ion implanted EA section, figure 5. This has a process platform based on the DSDBR laser.

Figure 5. EA-laser modulator

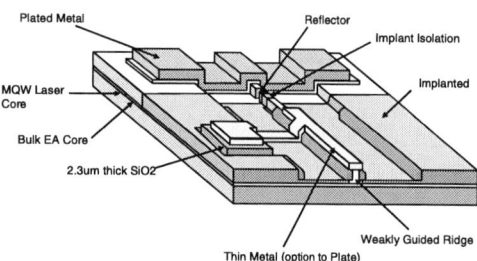

4. Silicon based integration.

The stability, fabrication precision and thermal properties of silicon also make it an attractive candidate for opto-electronic integration. Passive alignment to sub-micron precision between fibre and chip can be achieved by using silicon micro-machined parts and high precision control of chip bonding as shown in figure 6. Optical coupling and routing between elements could also be achieved using Si/SiO$_2$ waveguides and on-board devices.

Figure 6. Integration on silicon precision carrier.

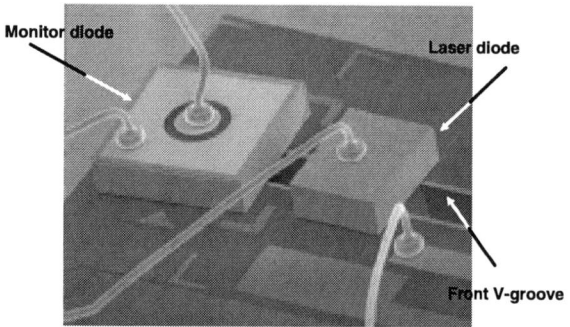

The technology lends itself to automatic chip placement, assembly for high volume and high density manufacture.

Silicon based MEMS technology can also provide additional functionality. Figure 7 shows a high speed PIN receiver with an electrically controlled MEMs shutter providing overload protection from +10dBm with a sensitivity down to -18.5dBm and 22dB of attenuation..

This example shows the specific advantage of a technology combination to give low optical losses and high performance capability and reliability.

Figure 7. MEMS high speed integrated detector.

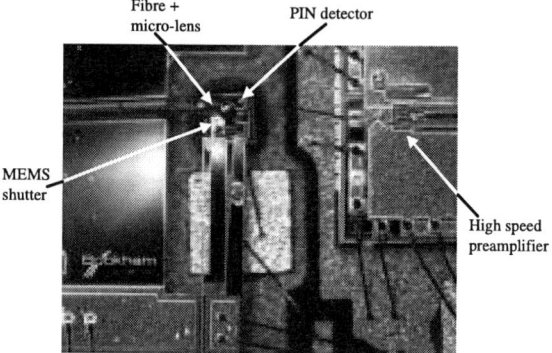

5. Summary.

Integrated technologies provide a path to lower component costs, increased functional complexity and densities. Material and process developments coupled with a need for increasing functional diversity in optical systems will provide the drivers to maintain technology on the growth curve already established in electronic integrated circuits, (figure 8).

Figure 8. Comparison of the density evolution for electronic IC's and Opto-electronic integration.

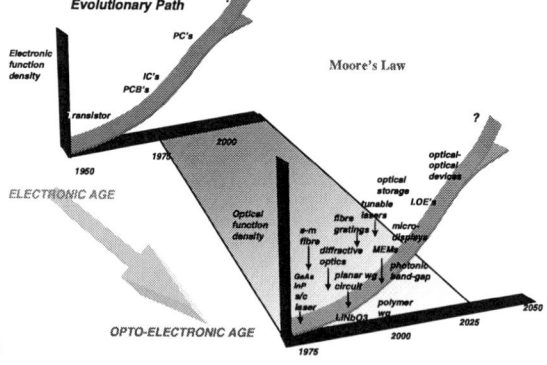

Acknowledgements.

The author would like to thank Andy Carter and the Caswell and Paignton development teams for many helpful discussions and much of the material contributing to this paper and Bookham Shenzhen manufacturing team for their help and contributions.

Recent Progress on PLC Technologies for Large-scale Integration

Shinji Mino

NTT Photonics Laboratories, NTT Corporation, 3-1, Morinosato Wakamiya, Atsugi-shi,
Kanagawa, 243-0198, Japan

minos@aecl.ntt.co.jp

Abstract *We review recent progress on PLC technologies for large-scale integration. We focus on three topics: the reduction of the switching power of thermo-optic switches, the hybrid-integration of chip-scale-package (CSP)-PDs for optical channel monitoring, and the hybrid integration of LiNbO3.*

1. Introduction

The recent widespread use of photonic networks has accelerated the development of various kinds of large-scale and highly functional optical modules. In addition, various integration technologies are needed to make these large-scale modules compact and low cost.

We have been developing a variety of PLC technologies for large-scale integration to satisfy these demands [1], and have demonstrated various PLC modules. Figure 1 shows a conceptual view of these technologies.

In this review, we focus on three of these technologies: (1) the reduction of the switching power of thermo-optic switches (TOSWs) or variable optical attenuators (VOAs) to achieve the compact integration of large-scale switches, (2) hybrid-integration technology for chip-scale-package (CSP)-PDs for optical channel monitoring, and (3) multi-chip integration technology for a PLC-LiNbO3 (LN) module.

2. Reduction in TOSW switching power

We have worked on large-scale PLC TOSWs, and demonstrated a 32 x 32 matrix TOSW [2] and arrayed SW for ROADM

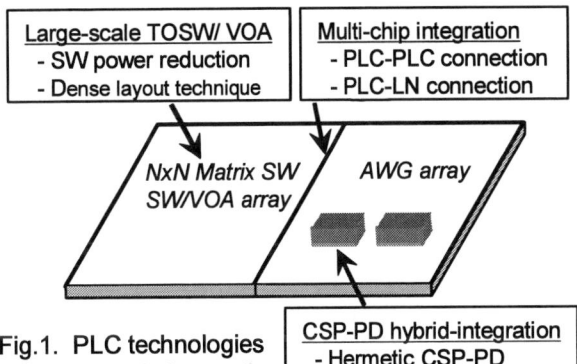

Fig.1. PLC technologies for large-scale integration

applications. The switching power of these TOSWs or VOAs is a few hundred mW per Mach-Zehnder interferometer. This value is undesirable for large-scale TOSWs. We must reduce the switching power to increase both the switch element density and the number of integrated switch elements. To solve this problem, we proposed a VOA with a suspended narrow ridge structure [3].

Figure 2(a) shows the top view of a PLC-VOA using the proposed structure and (b) shows the schematic structure of a TO phase-shifter. The PLC-VOA consists of a

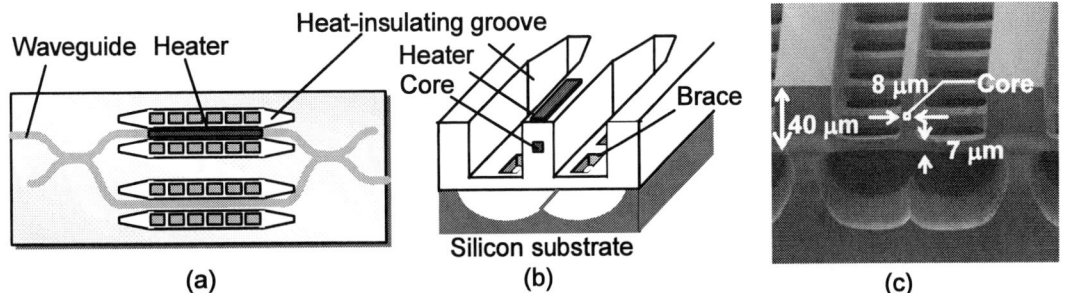

Fig. 2 Schematic configuration of PLC-VOA using suspended narrow ridge structure.
(a) top view, and (b) schematic structure of TO phase shifter, and (c) SEM image

978-0-9789217-3-6/07/$25.00

©2007 WEN GLOBAL SOLUTIONS

Mach-Zehnder interferometer (MZI) and a thin film heater. Heat-insulating grooves designed to suppress horizontal heat diffusion are formed beside the arm waveguides. The cladding between the heat-insulating grooves is suspended above the silicon substrate in order to suppress the flow of heat into the silicon substrate.

The switching power and response time decrease as the ridge becomes narrower. However, too narrow a ridge may induce optical radiation loss. We can achieve a switching power of 20 mW/channel when we employ a ridge width of 8 μm with a small excess loss [3]. This value is one tenth the conventional switching power.

3. Hybrid integration of CSP-PD on PLC

A compact assembly for optical channel monitoring is important for the PLC-ROADM. To achieve this, we developed two key components, namely a hermetically sealed chip-scale-package (CSP) PD and an integrated micro-mirror with high reflection

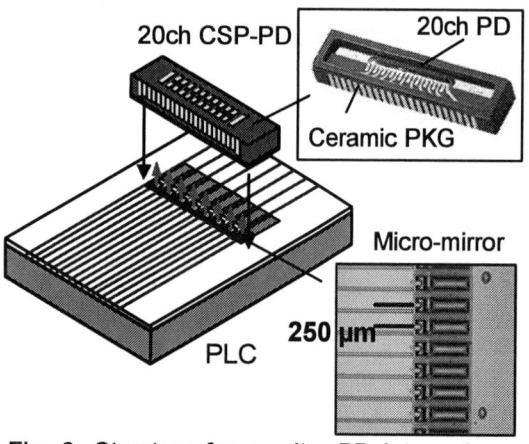

Fig. 3. Structure for monitor PD integration, and a photograph of developed 20-ch CSP-PD (above) and micro-mirror array (below)

angle accuracy [4]. This structure is superior to conventional in-plane bare chip integration, because it provides high reliability and sufficient flexibility for the PLC circuit layout.

The top photograph in Fig. 3 shows the developed 20-ch CSP-PD. The CSP-PD consists of a PD array chip with a detection area diameter of 250 μm and a pitch of 80 μm. We mounted the chip in an alumina case and hermetically sealed it with a sapphire window using AuSn solder. The package size is only 13 x 1.5 x 3 mm. This corresponds to less than one tenth that of a conventional CAN-PD with the same channel number. To assemble the CSP-PD on a PLC surface, we employ active alignment and UV curable adhesive.

To achieve high responsivity despite the large distance between the waveguide and the PD, we developed a micro-mirror with a resin slope [4]. The fabricated mirror array with a 250-μm pitch is shown in the lower photograph in Fig. 3.

Figure 4 (a) shows the measured reflection angle distribution, which was less than ± 3.5 degrees. Figure 4 (b) shows the responsivity characteristics of a CSP-PD mounted on a PLC. We obtained high responsivities of 0.7 to 1.1 A/W, and low adjacent crosstalk values ranging from –25 to –30 dB.

4. Hybrid integration of LiNbO₃ with PLC

We have worked on the hybrid integration of a silica-PLC and an LN waveguide to utilize the various passive functions of the PLC and the excellent EO effect of LN [5]. As regards the connection between the PLC

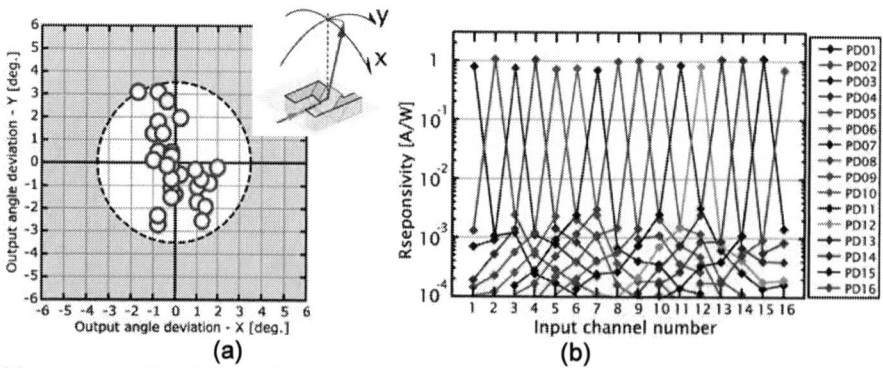

Fig. 4. (a) Measured reflection angle distribution of fabricated micro-mirrors, (b) Responsivity characteristics of CSP-PD mounted on PLC

and the LN waveguide, this is possible to achieve with a small excess loss because the mode field is almost the same and both materials have very good stiffness. So we can butt-couple a PLC and an LN waveguide in the same manner as PLC-fiber attachment.

Recently we reported optical modulators that employ a PLC-LN connection technique for differential quadrature phase-shift-keying (DQPSK) format transmissions. Figure 5 shows the configuration and a photograph of the PLC-LN hybrid DQPSK modulator. The module consists of two super-high 1.5%-Δ PLCs and an LN waveguide with four arrayed optical phase modulators. The optical coupling loss between the PLC and an LN waveguide is 1 dB. The 3-dB optical bandwidths of the electro-optic response of both the sub MZ modulators exceeded 26 GHz for all four driving conditions.

We undertook an 86-Gbit/s CSRZ-DQPSK transmission experiment using this modulator. As seen in Fig. 6, we obtained a clearly open eye diagram, and its bandwidth is around 129 GHz [5].

Furthermore, using this PLC-LN modulator, we reported a 14-Tb/s (111 Gb/s x 2 x 70ch) CSRZ-DQPSK transmission experiment over 160 km in a single optical fiber [6]. This huge total capacity of 14 Tb/s was the largest reported for a single optical fiber transmission before 2006.

Conclusions
We described the three integration technologies required for highly functional and large-scale integration, namely the reduction of the switching power of thermo-optic switches, the hybrid-integration of CSP-PDs for optical channel monitoring, and the hybrid integration of LiNbO3 for a modulator. These technologies are promising in terms of realizing higher levels of performance in next-generation photonic networks.

References
1. I. Ogawa, H. Yamazaki, and A. Kaneko, "Highly integrated PLC-type devices with surface-mounted monitor PDs for ROADM," Proc. of OFC2007, OWO2 (2007).
2. S. Sohma, T. Watanabe, N. Ooba, et al., "Silica-based PLC type 32 x 32 optical matrix switch," Proc. of ECOC2006, Tu4.4.3.
3. Y. Hashizume, K. Watanabe, et al., "Silica PLC-VOA using suspended narrow ridge structures and its application to V-AWG," Proc. of OFC2007, OWO4.
4. I. Ogawa, H. Yamazaki, et al., "32ch reconfigurable optical add multiplexer using technique for stacked integration of chip-scale-package PDs on silica-based PLC," Proc. ECOC2006, Tu4.4.2. (2006)
5. T. Yamada, Y. Sakamaki, et al., "86-Gbit/s differential quadrature phase-shift-keying modulator using hybrid assembly technique with planar lightwave circuit and LiNbO3 devices," *Proc. of LEOS2006,* ThDD4 (2006).
6. A. Sano, Y. Miyamoto, T. Yamada, et al., "14-Tb/s (140 x 111-Gb/s PDM/WDM) CSRZ-DQPSK transmission over 160 km using 7-THz bandwidth extended L-band EDFAs,", *Proc. ECOC 2006,* PD, Th4.1.1.

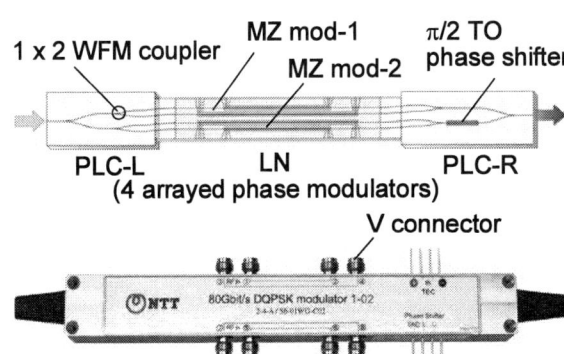

Fig.5. Configuration and photograph of PLC-LN integration module (DQPSK modulator)

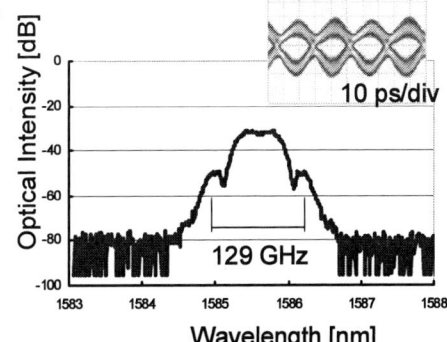

Fig. 6. Measured 86Gbit/s CSRZ-DQPSK spectra and waveform

Reflective Cholesteric Display: Principle and Progress
Deng-Ke Yang
Chemical Physics Interdisciplinary Program
Liquid Crystal Institute
Kent State University, Kent, OH 44242, USA

Cholesteric liquid crystals exhibit two stable states at zero field: the reflecting planar texture and non-reflecting focal conic texture. Because of the bistability, Ch liquid crystals can be used to make multiplexed displays on passive matrices. They do not need power-hungry backlights and reflect ambient light. Also because of the bistability, they do not need to be addressed constantly to display static images. Ch liquid crystals can also be encapsulated to make flexible displays with flexible plastic substrates, making the displays rugged and light-weight. Furthermore they can be fabricated in roll-to-roll process to make large area displays. Cholesteric reflective displays have the merits of energy-saving and low manufacturing cost low

Polarizer-free Liquid Crystal Displays

Yi-Hsin Lin (1), Jhih-Ming Yang (1), Shin-Tson Wu (2), Chi-Chang Liao (3)

1 : Department of Photonics and Institute of Electro-Optical Engineering, National Chiao
Tung University, 1001 Ta Hsueh Rd., Hsinchu 30050, Taiwan
e-mail: yilin@mail.nctu.edu.tw

2 : College of Optics and Photonics/ CREOL, University of Central Florida, USA
e-mail: swu@mail.nctu.edu.tw

3: Electronics & Optoelectronics Research Laboratories,
Industrial Technology Research Institute, Hsinchu 310, Taiwan
e-mail: ccliao@itri.org.tw

Abstract *A reflective polarizer-free liquid crystal display (LCD) using dye-doped LC gels is demonstrated. The main mechanism is to combine dye absorption and polydomain scattering. The potential applications are electronic papers and flexible displays.*

Introduction

Typically LCDs require at least one polarizer. The optical efficiency is low. Polarizer-free LCDs can increase the optical efficiency and also lower the fabrication cost of LCDs. Several approaches can achieve polarizer-free LCDs. [1-4] In this paper, we demonstrated a polarizer-free LCD using dye-doped LC gels. This polarizer-free LCD is mainly based on the dye-absorption and polydomain scattering.

Mechanism

The main operation principles are shown in Fig. 1 (a) and (b). The LC cell consists of two ITO glass substrates over coated with PI alignment layers only in order to provide vertical alignment for LC directors. The alignment layers do not require any rubbing process. Dye-doped LC gels are sandwich between two substrates. The whole fabrication process of dye-doped LC gels is based on thermal-induced and photo-induced phase separations [3]. The materials we employed are negative liquid crystals, a diacrylate monomer and a dichroic dye. The cell gap is 4 μm. The polymer networks are perpendicular to the glass substrates along z direction. The LC directors and dye molecules are perpendicular to the glass substrates as well at voltage-off state, as shown in Fig. 1(a). The cell does not scatter light and the absorption is rather weak. Therefore, the display has the highest reflectance. When we apply a high voltage at f= 1 kHz in the dye-doped LC gel, the liquid crystals and dye molecules are reoriented in the x-y

plane, as Fig. 1(b) depicts. The polymer network scatters light strongly. Since the alignment layer has no rubbing treatment, the absorption has no preferred direction. Therefore, the light scattering and dye absorption efficiency reaches their maxima. As a result, the display appears black.

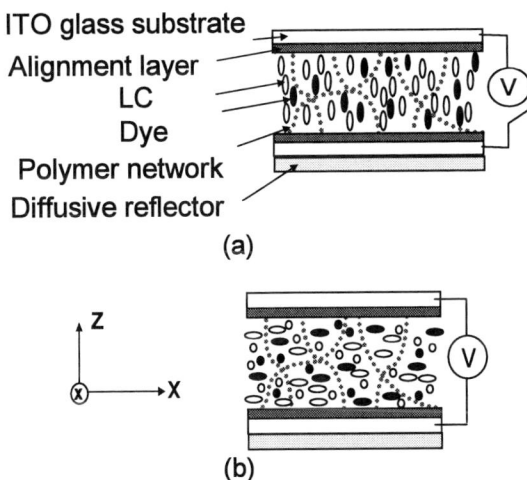

ITO glass substrate
Alignment layer
LC
Dye
Polymer network
Diffusive reflector

(a)

(b)

Fig. 1 Reflective polarizer-free LCD at(a) voltage-on state and (b) voltage-off state.

Experimental Setup and Results

We observed the image of our LC cell under a microscope, as shown in Fig. 2. At 0 V_{rms}, the cell appears good bright state. It appears red and we can see the polymer network structure at 30 V_{rms}. We also observed the image of red ink on a white paper as shown in Fig. 2(b). The image of our dye-doped LC gels mimics the color ink on a paper. The mechanism of our dye-doped LC gels is also similar to the color ink

978-0-9789217-3-6/07/$25.00

©2007 WEN GLOBAL SOLUTIONS

on a paper.

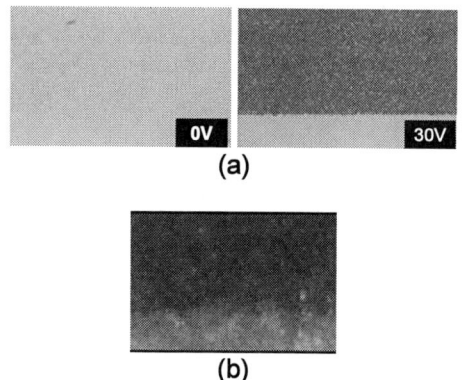

(a)

(b)

Fig. 2 Image of (a) dye-doped LC gels and (b) red ink on a white paper observed under a microscope.

We used a unpolarized green He-Ne laser as a light source for characterizing the device performances because the dye-doped LC gels we employed appears dark red rather than black. A dielectric mirror was placed behind the cell so that the laser beam passed through the cell twice. A large area photodiode detector was placed at ~25 cm (the normal distance for viewing a mobile display) behind the sample which corresponds to ~2° collection angle. A computer controlled LabVIEW data acquisition system was used for driving the sample and recording the light reflectance.

The contrast ratio as function of curing temperature is shown in Fig. 3. The contrast is around 450: 1 at 10 °C. The contrast ratio decreases when T< 30°C and then increases when T>30°C. That is because the decrease of a curing temperature results in larger polydomain; therefore, the contrast decreases. When the temperature is higher than 30°C, we found some polydomains have dynamic scatterings to help reboot the contrast ratio. The contrast ratio at 30 V_{rms} is higher than that at 20 V_{rms} because the LC directors and dye molecules are further along x-y plane well in Fig. 1 (b) to increase the scattering and absorption. The average reflectance at 0 volt is around 50% when the curing temperature between 10 °C and 40 °C. The threshold voltage is around 6 V_{rms} at 10 °C. The threshold voltage decreases with curing temperature. Besides the curing

temperature, the UV curing intensity, monomer concerntation and dye concerntration also affect the performance of the cell.

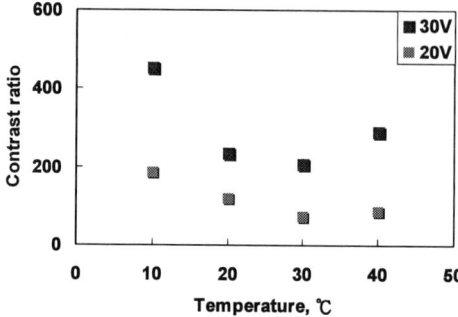

Fig.3 Contrast ratio as function of curing temperature at 20 V_{rms} and 30 V_{rms}.

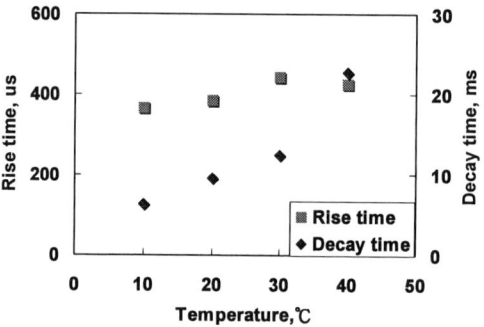

Fig.4 Response time as function of curing temperature.

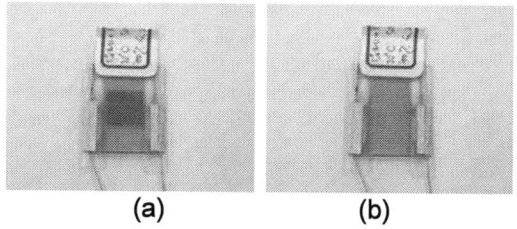

(a)   (b)

Fig.5 Single pixel reflective polarizer-free LCD using dye-doped LC gels at (a) 30 V_{rms} and (b) 0 V_{rms}.

Response time is also an important issue for guest-host displays. A typical response time of a guest-host display is around 50 ms. The response time of our dye-doped LC gel is fast because polymer network helps LC director to relax back. The rise time is ~0.4 ms and decay time is ~6 ms at 10 °C when the applied voltage is from 0 to 30 V_{rms} at f=1 kHz. When the curing temperature

32

increases, the rise time is similar, but the decay time increases with curing temperatures, as shown in Fig. 4. That is because the rise time depends on the applied electric field. The decay time increases due to the polymer domain size increases.

We also fabricated a single pixel reflective polarizer-free LCD using the dye-doped LC gel. To avoid specular reflection, we laminated a diffusive reflector on the backside of the bottom glass substrate in order to widen the viewing angle. The ambient white light was used to illuminate the sample. Fig. 5 shows the display using a 4-μm dye-doped LC gel. The bright part represents the state of V=0. The dark area represent the ITO electrodes with V=30 V_{rms} at f=1 kHz in dye-doped LC gel.

Conclusions
We have demonstrated a polarizer-free LCD using dye-doped LC gels. The contrast ratio is 450:1 at 30 V_{rms} and the response time is ~6 ms. The threshold voltage is ~6 V_{rms}. The performance depends on the fabrication parameters, such as curing temperature, UV curing intensity, material concentrations. The driving voltage is still high, but the low temperature process and non-rubbing process are suitable for fabrication of flexible displays.

References
1 S. T. Wu and D. K. Yang, *Reflective liquid crystal displays*, Wiley: New York, (2001).
2 Y. H. Lin et al, J. Display Technology **1**, (2005) 230.
3 Y. H. Lin et al, SID Tech. Digest **37**, (2006) 780
4 Y. H. Lin et al, Mol. Cryst. Liq. Cryst. **453** (2006) 371.

The Color Temperature Adjusting Method for Multi-primary Display Using Nonlinear Programming Problems

Yan CHENG, Xu LIU, Haifeng LI

1 : Yan CHENG , Doctor candidateAddress: #406, Building 3, Yuquan Campus, Zhejiang University, Zhejiang Province, PRC. chinacheng404@163.com
2 : Xu LIU, (correspond author)Professor, liuxu@zju.edu.cn
3 : Haifeng LI, Professor, lihaifeng@zju.edu.cn

Abstract *Multi-primary display are now wildly researched and used. There are few researches concerning with color temperature design. It is easy in three-primary display. But it is difficult when the number of primaries gets more than three, because the coefficient matrix is not in a square form, which always yields an infinite solution set. This paper presents a method to design the relative primary luminances of multi-primary color display.*

Introduction

Recently, color display market emphasized on two points: enlarging the display panel size and enlarging the display color gamut. A color display having a larger color gamut can show a more vivid image. But traditional color reproduction technologies, like CRT, PDP, LCD, use three low saturation colors to reproduce colors, which are limited in a very small triangle color gamut.

The methods for enlarging color gamut can be divided into two classes. One is using high saturation colors to widening the triangle composed by the three primary colors. The other is using four or more multiple primaries in color display. This method changes the triangle to a polygon by adding the additional primaries. When the number of the primaries is more than 3, the display system is called multi-primary color display (MPD) [4].

There are many studies on MPD, such as color space conversion [1,5], the relationship between color gamut and brightness [3], the relative luminances [2]. And there are many application studies on implementation the MPD [4,6].

If we look at the color management system (CMD), which is now wildly used in the color reproduction, we can find that one of the key points of CMD is color temperature adjusting, which is determining the color gamut coordinates of the white point. The white point is determined by the relationship of the primaries' luminance. So color temperature design is solving a set of linear equations. It is easy to design the relationship in three-primary display, because the coefficient matrix is in a square form and full of rank. It is easy to calculate by simply doing a matrix inversion, and then, one can get the solution. But it is difficult when the number of primaries gets more than three, because the coefficient matrix is no more in a square form, which yields an infinite solution set.

There are few researches on color temperature design of MPD [2]. This paper presents a new method, using nonlinear program, to solve this problem, and the designing method of the relative primary luminances of multi-primary color display will be presented in the paper. This method is also easy to implementation.A template is a set of styles and page layout settings that determine the appearance of a document. This template matches the printer settings that will be used to the proceeding and publication on Internet. Use of the template is mandatory.

White Point Design of Three-primary Color Display

We all know that the color coordinates of the white point of three-primary color display determines the relative luminance controlling signals uniquely. It can be written in a matrix form, shown as Eq.1.

$$\begin{bmatrix} X_w \\ Y_w \\ Z_w \end{bmatrix} = \begin{bmatrix} X_r & X_g & X_b \\ Y_r & Y_g & Y_b \\ Z_r & Z_g & Z_b \end{bmatrix} \cdot \begin{bmatrix} C_r \\ C_g \\ C_b \end{bmatrix} \quad (1)$$

Appling $X=(x/y)Y$, $Z=(z/y)Y$ to Eq.1, and make three notations

$$RC_r = C_r \frac{Y_r}{Y_w}$$

$$RC_g = C_g \frac{Y_g}{Y_w}$$

$$RC_b = C_b \frac{Y_b}{Y_w}$$

Eq.1 can be calculated into

$$\begin{bmatrix} x_w \\ y_w \\ 1 \\ \dfrac{z_w}{y_w} \end{bmatrix} = \begin{bmatrix} \dfrac{x_r}{y_r} & \dfrac{x_g}{y_g} & \dfrac{x_b}{y_b} \\ 1 & 1 & 1 \\ \dfrac{z_r}{y_r} & \dfrac{z_g}{y_g} & \dfrac{z_b}{y_b} \end{bmatrix} \cdot \begin{bmatrix} RC_r \\ RC_g \\ RC_b \end{bmatrix} \quad (2)$$

The coefficient matrix is

$$\begin{bmatrix} \dfrac{x_r}{y_r} & \dfrac{x_g}{y_g} & \dfrac{x_b}{y_b} \\ 1 & 1 & 1 \\ \dfrac{z_r}{y_r} & \dfrac{z_g}{y_g} & \dfrac{z_b}{y_b} \end{bmatrix}$$

This is a 3x3 matrix and is full of rank normally. There must exist a multiplicative inverse of this matrix. This means when the color coordinates of the three primaries and the white point, four points total, are given, the relative luminance controlling signals can be calculated by Eq.2 uniquely.

White Point Design of Multi-primary Color Display

The process for a multi-primary display is similar to the one for three-primary display. For N-primary colors display, the color controlling signals can be solved by Eq.3

$$\begin{bmatrix} X_w \\ Y_w \\ Z_w \end{bmatrix} = \begin{bmatrix} X_1 & X_2 & \dots & X_n \\ Y_1 & Y_2 & \dots & Y_n \\ Z_1 & Z_2 & \dots & Z_n \end{bmatrix} \begin{bmatrix} C_1 \\ C_2 \\ \vdots \\ C_n \end{bmatrix} \quad (3)$$

Doing the same process, we get

$$\begin{bmatrix} \dfrac{x_w}{y_w} \\ 1 \\ \dfrac{z_w}{y_w} \end{bmatrix} = \begin{bmatrix} \dfrac{x_1}{y_1} & \dfrac{x_2}{y_2} & \dots & \dfrac{x_n}{y_n} \\ 1 & 1 & \dots & 1 \\ \dfrac{z_1}{y_1} & \dfrac{z_2}{y_2} & \dots & \dfrac{z_n}{y_n} \end{bmatrix} \cdot \begin{bmatrix} RC_1 \\ RC_2 \\ \vdots \\ RC_n \end{bmatrix} \quad (4)$$

The coefficient matrix is

$$\begin{bmatrix} \dfrac{x_1}{y_1} & \dfrac{x_2}{y_2} & \dots & \dfrac{x_n}{y_n} \\ 1 & 1 & \dots & 1 \\ \dfrac{z_1}{y_1} & \dfrac{z_2}{y_2} & \dots & \dfrac{z_n}{y_n} \end{bmatrix}$$

This is a 3xn matrix and has no multiplicative inverse. The number of equations is larger than the number of variables. This means the solution set of Eq.4 is infinite. We should take other considerations into account to get a unique solution.

Designing White Point of MPD Using Nonlinear Programming

The nonlinear programming problem is an optimization problem. In our problems, the solution should obey Eq.3, so it is constraint. The objective function is the summation the luminance of each light source.

A numerical simulation is performed in order to confirm the color temperature adjusting using nonlinear programming. The tri-stimulus values of DLP projector are shown in Table 1.

Table 1 tri-stimulus values of CIE XYZ

Color	X	Y	Z
Red	933.1	569	66.7
Green	1402.4	2472	442
Blue	511.3	334	2531.9
Yellow	5059.9	7135	283.3

The ITU-R BT.709 designates the white point coordinates as $(x_w, y_w)=(0.3127, 0.3291)$. From Eq.5, the objective function can be specialized to

maximize

$$f = 15.9C_r + 60.47C_g + 23.05C_b + 18.57C_o + 37.42C_c$$

constraints

$$0.3127 = \frac{36.6C_r + 16.13C_g + 47.16C_b + 25.97C_o + 8.49C_c}{159.9C_r + 86.1C_g + 359.46C_b + 44.579C_o + 103.27C_c}$$

$$0.3291 = \frac{15.9C_r + 60.47C_g + 23.05C_b + 18.57C_o + 37.42C_c}{159.9C_r + 86.1C_g + 359.46C_b + 44.579C_o + 103.27C_c}$$

$$0 \le C_r, C_g, C_b, C_o, C_c \le 1$$

The result is

$$C_r = 1, C_g = 0.5, \quad C_b = 1, \quad C_y = 0.066$$

If the primaries are modulated as the C_r, C_g, C_b, C_y, the white point composed of these four primaries is on (0.3128, 0.3292) in color gamut, Fig.2 (b). While the white point composed of these four primaries without modulating is on (0.3637, 0.4834) in color gamut, Fig. 2 (a). It is clear that the white point reproduced by the display is corrected to match the ITU-R BT.709 requirement.

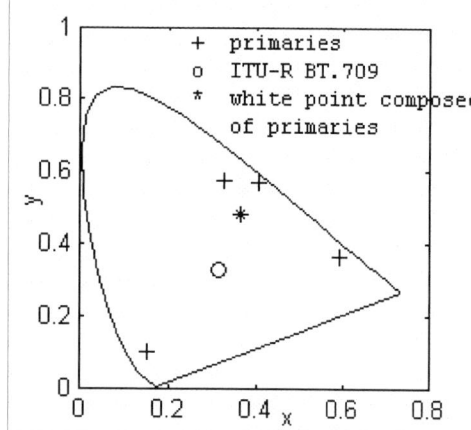

(a) Primaries without modulating

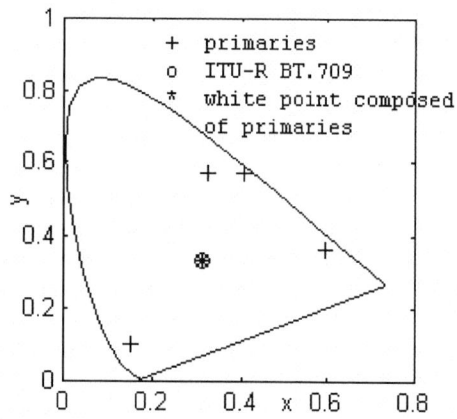

(b) Primaries with modulating
Fig. 1 The Coordinates of the two White Points

Conclusions

A color temperature adjusting method for MPD is proposed in this paper. To solve the linear equations, a nonlinear programming problem is adopted. The nonlinear programming problem is an optimization problem, which finds a solution by minimizing or maximizing an objective function under some specified conditions. Maximizing the summation the luminance of each light source is regarded as the optimal consideration. By adopting the nonlinear programming problem, the best solution of Eq.3 comes out from the set of infinite solutions.

The coordinates of the white point composed of modulated primaries match the ITU-R BT.709 requirement, which means the color temperature need the demand. By given the different white point, it is easy to get the appropriate luminance controlling signals by this method. And a different required color temperature adjusting is thus achieved.

References

1. M. Takaya, K. Ito, G. Ohashi, Y. Shimodaira, "Color-conversion method for a multi-primary display to reduce power consumption and conversion time", Journal of the SID 13(8) pp.658-690 (2005)
2. S. Wen, "Design of relative primary luminances for four-primary display", Displays 26(4-5) pp. 171-176 (2005)
3. M. Ou-Yang, SW. Huang, "Design Considerations Between Color Gamut and Brightness for Multi-primary Color Displays", Journal of Display Technology, 3(1) pp.71-82 (2007)
4. MC. Kim, TC. Shin, JS. Um, "Wide Color Gamut Five Channel Multi-primary Display for HDTV Application", Journal of Imaging Science and Technology, 49(6) pp.594-604 (2005)
5. T. Ajito, K. Ohsawa, T. Obi, "Color Conversion Method for Multiprimary Display Using Matrix Switching", Optical Review, 8(3) pp.191-197, (2001)
6. David R., Mitchell R., "Color Management of Four-Primary Digital Light Processing Projectors", Journal of Imaging Science and Technology, 50(1) pp17-24 (2006)

Amplitude-Sensitive Interferometric Ellipsometer on TN-LCD Optical Parameters Measurement

H. C. Wei, C. C. Tsai, C. Chou

1 : Institute of Biophotonics, National Yang Ming University, Taipei, Taiwan 112
e-mail: g39426008@ym.edu.tw
2 : b8731114@student.nsysu.edu.tw , 3 : cchou@ym.edu.tw

Abstract This research proposes an amplitude–sensitive heterodyne interferometric ellipsometer to determine TN-LC cell parameters precisely based on single wavelength at normal incidence. The advantage is the capability of two dimensional distribution measurement using CCD camera.

Introduction

Different methods to evaluate the liquid crystal (LC) cell parameters including cell gap d and pre-tilt angle θ_{pre} are successfully proposed and demonstrated. [1-3] All of these methods are based on intensity measurement via conventional photometric ellipsometer, or polarimeter. Kawamura et al. [3] used a broad band light source that is able to build 2-D distributions of pre-tilt angle and cell gap of a twisted nematic LC (TNLC). Tsai et al. [4] proposed a phase-sensitive heterodyne interferometric ellipsometry successfully able to determine LC parameters at single point precisely. However, a limitation on 2-D distribution detection of cell parameters is resulted by conventional phase lock-in detection technique. In this research, we propose an amplitude-sensitive heterodyne interferometric ellipsometry in contrast to phase-sensitive interferometric ellipsometer in order to extend the ability on 2-D measurement. A quarter wave plate (QWP) place prior to TN-LC and then rotate QWP to modulate the polarization state of incident laser beam onto TN-LC. Thus, an amplitude ratio curve is measured. As results, four parameters of TN-LC including twisted angle Φ, untwisted phase retardation Γ, direction angle α and cell gap d by using least square curve fitting method properly. Extensively, 2-D distribution of cell parameters are available too by scanning the tested TN-LC linearly and precisely. In the experiment, the amplitude ratio stability is 0.25% within 6 minutes.

Principle

In Fig. 1, a polarized common-path optical heterodyne ellipsometer is setup in which two acousto-optic modulators (AOMs) are driven at frequencies of $\omega_1 = 80.000 MHz$ and $\omega_2 = 80.0329 MHz$ in the signal and reference channels of the interferometer respectively. A linearly polarized He-Ne laser beam is incident into the Mach-Zehnder interferometer such that the P_1 and S_1 waves in the reference beam present ω_1 frequency shifted whereas the P_2 and S_2 waves are ω_2 frequency shifted in the signal channel. Both P waves (P_1+P_2) are mixed together at BS_2 to produce a P-polarized heterodyne signal with a beat frequency $\Delta\omega = 33.2 kHz$ at photo detector D_p. Simultaneously, a S-polarized heterodyne signal is generated by S_1 and S_2 waves at photo detector D_s accordingly. Then, the heterodyne signals with respect to P and S polarizations are expressed by

$$I_p = I_{p1} + I_{p2}^{(o)} + 2\sqrt{I_{p1}I_{p2}^{(o)}} \cos\left(\Delta\omega t + \delta_{p2}^{(o)}\right) \qquad (1)$$

$$I_s = I_{s1} + I_{s2}^{(o)} + 2\sqrt{I_{s1}I_{s2}^{(o)}} \cos\left(\Delta\omega t + \delta_{s2}^{(o)}\right) \qquad (2)$$

where I_{p1} and I_{s1} are the intensities of P_1-wave and S_1-wave in the reference channel and $I_{p2}^{(o)}$ and $I_{s2}^{(o)}$ are intensities of P_2-wave and S_2-wave in the signal channel of emerging beam of the tested sample. $\delta_{p2}^{(o)}$ and $\delta_{s2}^{(o)}$ are the phases of P_2-wave and S_2-wave in the signal and reference channels of emerging beam of the tested sample. The heterodyne signals of interest are the ac part in Eqs. (1) and (2) which are

978-0-9789217-3-6/07/$25.00

©2007 WEN GLOBAL SOLUTIONS

the detected amplitude-sensitive data we process in this experiment.

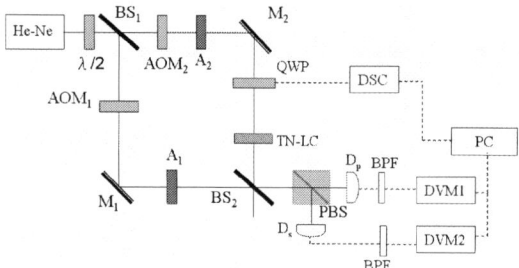

Fig. 1. Experimental setup

In order to analyze the state of polarization quantitatively, a parameter X is defined by

$$X = \frac{E_s}{E_p}\exp[i(\delta_s - \delta_p)] = |X|\exp[i\delta] \qquad (3)$$

where E_p and E_s are the amplitudes of the P and S waves respectively and δ_p and δ_s are their phases accordingly. Therefore, the output elliptical polarization $X^{(o)}$ can be determined by the elliptical polarization at input $X^{(i)}$ of the optical setup such that the bilinear transformation can be described as

$$X^{(o)} = |X^{(o)}|e^{-i\delta^{(o)}} = \frac{t_{21} + t_{22}|X^{(i)}|e^{-i\delta^{(i)}}}{t_{11} + t_{12}|X^{(i)}|e^{-i\delta^{(i)}}} \qquad (4)$$

where T= [t_{11} t_{12} ; t_{21} t_{22}] is the transmission transfer matrix of tested sample. Finally, the amplitude ratio of heterodyne signals of Eqs. (1) and (2) becomes

$$\frac{I_{sac}}{I_{pac}} = \frac{E_{s1}E_{s2}^{(o)}}{E_{p1}E_{p2}^{(o)}} = |X_1||X_2^{(o)}| \qquad (5)$$

where I_{sac} and I_{pac} are the amplitudes of ac part heterodyne signals with respect to S and P polarizations.

In order to keep the relative orientation unchanged between the sample and photo-detector in Fig.1, we rotate a QWP which locates in front of the tested sample is able to modulate the input polarization state of laser beam incident onto the sample. According to Scierski et al [5], the transmission transfer matrix in the amplitude-sensitive detection can be modified to be

$$M_{QWP} = \begin{bmatrix} \cos^2\beta + \sin^2\beta e^{i\gamma} & \sin\beta\cos\beta(1-e^{-i\gamma})e^{-i\delta_f} \\ \sin\beta\cos\beta(1-e^{-i\gamma})e^{i\delta_f} & \sin^2\beta + \cos^2\beta e^{-i\gamma} \end{bmatrix} \qquad (6)$$

where β is the azimuth angle of QWP, γ is the phase retardation between two elliptical eigenvector and δ_f is the phase retardation in one of the eigen elliptical polarizations. In Fig. 2, it is shown the a twisted nematic liquid crystal (TN-LC) cell in which the optical parameters, (Φ,Γ,α,d) are presented.

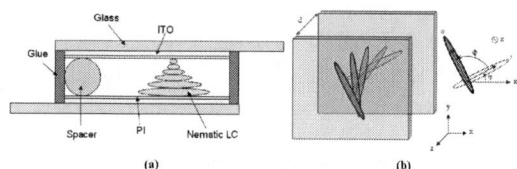

Fig.2. (a) the structure of a TNLC cell (b) the structure of a TN-LC with twisted angle Φ, untwisted phase retardation Γ, direction angle α.

The transmission transfer matrix related to these parameters is then expressed by [6]

$$M_{TNLC} = \begin{bmatrix} p\cos\Phi + qr\sin\Phi - iqs\cos(2\alpha+\Phi) \\ p\sin\Phi - qr\cos\Phi - iqs\sin(2\alpha+\Phi) \end{bmatrix}$$
$$\begin{bmatrix} -p\sin\Phi + qr\cos\Phi - iqs\sin(2\alpha+\Phi) \\ p\cos\Phi + qr\sin\Phi + iqs\cos(2\alpha+\Phi) \end{bmatrix} \qquad (7)$$

where $p = \cos\chi, q = \sin\chi, r = \Phi/\chi, s = \Gamma/2\chi$.

Finally, the total transmission transform matrix including QWP and TN-LC becomes

$$M_{QTNLC} = T_{TNLC}T_{QWP} = \begin{bmatrix} m_{11} & m_{12} \\ m_{21} & m_{22} \end{bmatrix} \qquad (8)$$

Where

$$m_{11} = \left[p\cos\Phi + qr\sin\Phi - iqs\cos(2\alpha+\Phi)\right]$$
$$\times\left(\cos^2\beta + \sin^2\beta e^{-i\gamma}\right)$$
$$+\left[-p\sin\Phi + qr\cos\Phi - iqs\sin(2\alpha+\Phi)\right]$$
$$\times\sin\beta\cos\beta\left(1-e^{-i\gamma}\right)e^{i\delta_f}$$

$$m_{12} = \left[p\cos\Phi + qr\sin\Phi - iqs\cos(2\alpha+\Phi)\right]$$
$$\times\sin\beta\cos\beta\left(1-e^{-i\gamma}\right)e^{-i\delta_f}$$
$$+\left[-p\sin\Phi + qr\cos\Phi - iqs\sin(2\alpha+\Phi)\right]$$
$$\times\left(\sin^2\beta + \cos^2\beta e^{-i\gamma}\right)$$

$$m_{21} = \left[p\sin\Phi - qr\cos\Phi - iqs\sin(2\alpha+\Phi)\right]$$
$$\times\left(\cos^2\beta + \sin^2\beta e^{-i\gamma}\right)$$
$$+\left[p\cos\Phi + qr\sin\Phi + iqs\cos(2\alpha+\Phi)\right]$$
$$\times\sin\beta\cos\beta\left(1-e^{-i\gamma}\right)e^{i\delta_f}$$

$$m_{22} = \left[p \sin \Phi - qr \cos \Phi - iqs \sin \left(2\alpha + \Phi \right) \right]$$
$$\times \sin \beta \cos \beta \left(1 - e^{-i\gamma} \right) e^{-i\delta_f}$$
$$+ \left[p \cos \Phi + qr \sin \Phi + iqs \cos \left(2\alpha + \Phi \right) \right]$$
$$\times \left(\sin^2 \beta + \cos^2 \beta e^{-i\gamma} \right)$$

Then, we could use the transform matrix to characterize optical parameters of TN-LC accordingly.

Experimental setup and results

Fig. 1 is the experimental setup for measurement. There are two digital voltmeters (DVM 34401) are used in the interferometer in order to read the ac voltage of heterodyne signals in real time. At first, we rotate the QWP from $\beta = \beta_{ini}$ to $\beta_{ini} + 360°$ under the condition of no TN-LC in the interferometer. The measured result is shown in Fig. 3(a) in which the theoretical calculation I_{sac} / I_{pac} at different β in least square fit the experimental data by using transfer matrix M_{QWP}. Thus, the parameters of (γ, δ_f) = (84.86°, -4.13°) of QWP is obtained. Then, we rotate QWP again whereas a tested TN-LC is located behind QWP in Fig.1. Then, I_{sac} / I_{pac} at different β is fit experimental data by using transfer matrix M_{QTNLC}. The result is shown in Fig. 3(b). Then, the parameters of tested TNLC (Φ,Γ,α) = (-88.81°, 224.25°, -163.18°) are obtained.

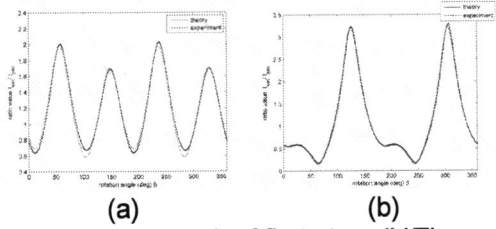

(a) (b)

Fig. 3. (a) The result of first step. (b)The result of second step. Blue dot line is the experiment data. Red line is the theory data.

In order to prove that the capability of 2-D distribution of all parameters by this method, we use a precision linear moving stage to move TNLC by 4x4 testing points. Fig. 6

shows the 2D distributions of Φ, Γ and α experimentally.

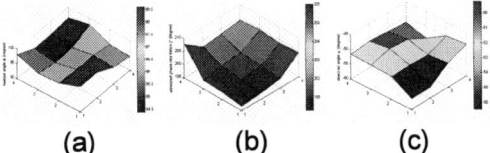

(a) (b) (c)

Fig. 6. The 2D distributions of (a) twisted angle (b) untwisted phase retardation (c)direction angle

Discussions and conclusions

This proposed method based on polarized common-path amplitude-sensitive detection and synchronized detection technique enables to reduce the common phase noise including environmental disturbance and laser frequency noise simultaneously. In the meantime, an amplitude ratio detection is measured further improves the stability of amplitude sensitive detection in this setup. These result is 0.25% stability on amplitude ratio measurement within 6 minutes in this experiment. The scattering and the absorbance of TN-LC is overcome by this method because optical heterodyne detection able to reduce the scattering effect effectively as well. Additionally, 2-D capability on cell parameters is available by using CCD camera which is amplitude dependent rather than phase dependent. Thus, this proposed method of amplitude-sensitive detection shown the potential not only on cell gap and phase retardation of TN-LC but also extended into pre-tilt angle of TN-LC simultaneously.

References

1. Yablonskiĭ et al J. Appl. Phys., 85 (1999), 2556-2561.
2. Baba et al Jpn. J. Appl. Phys., 37 (1998), 2581-2586.
3. Kawamura et al Jpn. J. Appl. Phys. 43 (2004), 709-714.
4. Tsai et al Appl. Opt. 44 (2005), 7509-7514.
5. Scierski et al Optik 68 (1984), 121-125.
6. P. Yeh and C. Gu, *Optics of Liquid Crystal Display* (Wiley,1999), pp. 129–130

Biophotonics – A Tutorial Overview

Arthur Chiou

National Yang-Ming University
No. 155, Sec. 2, Li-Nong Street, Taipei, Taiwan
aechiou@ym.edu.tw

Abstract: A tutorial overview of biophotonics is given followed by the highlights of some recent progresses in potential biomedical applications of optical microscopy, spectroscopy, and manipulation.

Optics (and photonics) has been well-recognized as an enabling technology that plays an indispensable role in the advancement of a wide spectrum of modern sciences and technologies. The critical roles photonics have played in modern information technology (IT), such as those in the telecommunication and the display industries, represent no more than a small sampling of the ubiquity of modern photonics technology. Following the success of photonics in IT, the roles of photonics in biomedical science and technology have received more and more attention in recent years. Many technical conferences and symposiums focusing on "biophotonics" have been held in almost every continent in recent years, and the number of researchers attending these meetings has been increasing at a tremendous pace. Biomedical applications of photonics science and technology have become the focus of numerous research institutes worldwide owing to the enormous amount of valuable information that can be extracted via the interactions of photons with biological materials including nucleic acids, proteins, cells, and tissues.

Biophotonics, which can be defined (in a broad sense) as the science and technology dealing with the interaction of photons with biological materials, covers a wide spectrum of principles and applications [1-7]. Although some researchers prefer to draw a line between "Biomedical Optics" and "Biophotonics", to my opinion, the boundary line is rather fuzzy [1-11], and I have been (and will be in this lecture) using "Biophotonics" as a generic term that encompasses biomedical optics. As is displayed in the block diagram in Fig. 1, in basic research, biophotonics includes the use of optical techniques to better understand the fundamental physics and chemistry of the functions and the interaction of all biological samples at the molecular, genomic, proteomic, sub-cellular and cellular, and tissue and organ levels. In clinical applications it can probably be loosely classified into two broad categories, namely, diagnostics and therapeutics.

Historically, the critical role of microscopy and spectroscopy in biomedical science and technology has long been well-recognized. With the continuous advancement of innovative IT (including digital signal and image processing, storage, and transmission) in conjunction with the availability of a wide range of laser sources, the functionality of conventional microscopy and spectroscopy has been integrated to a higher dimensionality as the fusion of spatial, temporal, spectral, and polarization information becomes practical [Fig. 2]. In addition, fiber optics and nanotechnology have further facilitated in-vivo diagnostic and therapeutic applications with reduced invasiveness. In the last decade, many novel approaches to overcome the classical optical diffraction limit [12] with promising biomedical applications have been demonstrated. Although the roles of microscopy (including optical coherence tomography) and spectroscopy and their integration are expected to increase tremendously in medical diagnosis, noninvasiveness and early detection (of diseases) remain the major challenges.

978-0-9789217-3-6/07/$25.00

©2007 WEN GLOBAL SOLUTIONS

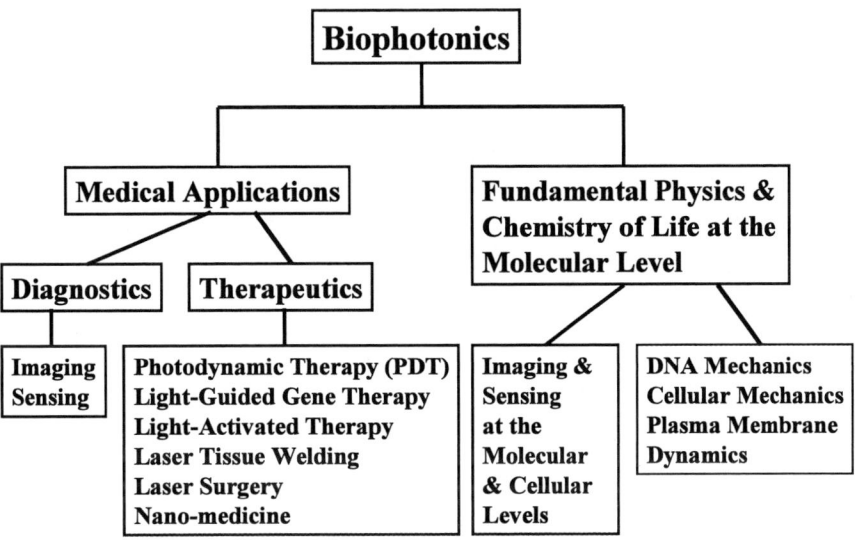

Fig. 1. A block diagram showing a generic overview of Biophotonics

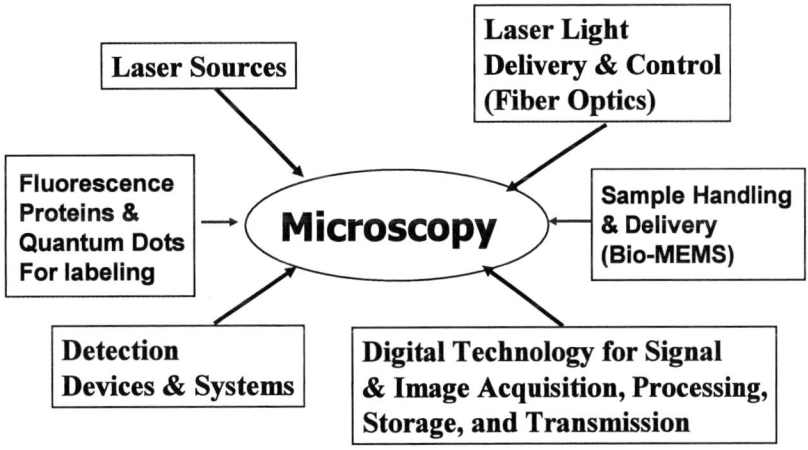

Fig. 2. A block diagram showing recent technological advances that have revolutionized the functional capability of microscopy

In medical diagnosis, the applications of optical coherence tomography (OCT) in ophthalmology have advanced to the points that several commercial systems are now available [13]. In parallel, photodynamic therapy (PDT) represents an excellent example of translational research for cancer therapy [14, 15].

In the study of the fundamental physics and chemistry of life at the molecular level, optical trapping and manipulation [16, 17] has emerged as one of the most important tools for the measurements of the dynamics of molecular interactions including protein-protein interaction and protein-DNA interactions. In the August 2005 supplementary issue, "The Scientist" listed "Optical Tweezers" as one of the seven most important technologies that have revolutionized research and development in life sciences [18].

Another recent advance in optical manipulation is optical stretching of biological cells by either fiber-optical dual-beam trap [19, 20] or by oscillatory optical tweezers [20]. Specifically, we have recently demonstrated that optical trapping and stretching via oscillatory tweezers with the aid of an acousto-optic modulator (AOM) provide a versatile platform for the study of either the steady-state or the dynamic viscoelastic property of micron-size soft particles including biological cells and other living biological samples [20]. Measurements of the elastic-coefficients of normal, cancerous, and metastatic breast epithelial cells by optical stretcher indicated that the elastic-coefficient of the cell may serve as an inherent cell maker that offers a sensitive cellomic alternative to current proteomic techniques [21].

In this talk, I will give a brief overview of biophotonics followed by a few selected illustrative examples in recent advances in microscopy, spectroscopy, and optical manipulation. It is obvious that it can hardly cover even a small portion of the massive and continually growing field of research and development in biophotonics. It is, at best, a biased sampling of some of the recent activities that I am familiar with.

References

1. Introduction to Biophotonics, P. N. Prasad, John Wiley & Sons, Inc., Hoboken, New Jersey (2003).
2. Advances in Biophotonics, B. C. Wilson, V. V. Tuchin, and S. Tanev, Ed., IOS Press, Amsterdam (2005).
3. www.photonics.com/bioPhotonicsHome.aspx
4. Introduction to Nanobiophotonics, P. N. Prasad, ieeexplore.ieee.org/xpls/abs_all.jsp.
5. http://cbst.ucdavis.edu
6. www.biophotonics-journal.org
7. biophotonicsWorld.org
8. Handbook of Optical Biomedical Diagnostics, V. V. Tuchin, Ed., SPIE Press, Bellingham (2002).
9. Biomedical Photonics Handbook, T. Vo-Dinh, Ed., CRC Press, Boca Raton (2003).
10. An Introduction to Biomedical Optics, R. Splinter and B. A. Hooper, Taylor and Francis, New York (2007).
11. Biomedical Optics: Principles and Imaging, L. V. Wang and H.-I. Wu, John Wiley & Sons, Inc., Hoboken, New Jersey (2007).
12. S. Bretschneider, C. Eggeling, S. W. Hell (2007), "Breaking the Diffraction Barrier in Fluorescence Microscopy by Optical Shelving." Phys. Rev. Lett. **98**(21): 218103.
13. http://www.rle.mit.edu/rleonline/ProgressReports/1995_12.pdf
14. www.elsevier.com/locate/pdpdt.
15. www.bmb.leeds.ac.uk/pdt/PDToverview.htm
16. A. Ashkin, "History of Optical Trapping and Manipulation of Small-Neutral Particle, Atoms, and Molecules," IEEE J. Selected Topics in Quantun Electron. 6(6), 841-856 (2000).
17. K. C. Neuman and S. M. Block, "Optical Trapping," Review of Scientific Instruments, 78(9), 2787-2809 (2004).
18. http://www.the-scientist.com
19. J. Guck, R. Ananthakrishnan, H. Mahmood, T. J. Moon, C.C. Cunningham, and J. Käs, "The Optical Stretcher: A Novel Laser Tool to Micromanipulate Cells," Biophys. J. **81**, 767-784 (2001).
20. G. B. Liao, Y.-Q. Chen, et al., "Recent Progresses in Optical Trap-and-Stretch of Red Blood Cells," SPIE's Proc. Vol. 6633 (2007).
21. J. Guck, S. Schinkinger, B. Lincoln, et al., "Optical Deformability as an Inherent Cell Marker for Testing Malignant Transformation and Metastatic Competence," Biophys. J. **88**, 3689–3698 (2005).

Reduced Dispersion Fiber Extends Reach for Dispersion Tolerant Systems

Chris Towery, Product Line Manager, Corning Optical Fiber, towerycr@corning.com
Tidd Zhang, Marketing Manager, Corning Optical Fiber, China, ZhangH@corning.com.cn

Introduction

As the demand for additional bandwidth continues to grow across all segments of the telecommunication network and across all major geographic markets, the challenge that has faced the industry for the past several decades remains. That being how to cost effectively deploy and manage the network infrastructure to carry that additional traffic demand with systems that are both robust in terms of capacity yet are less complex to deploy and plan for. As traffic demand in the long haul and regional networks has grown world wide by 50 to 100 percent year for the past several years. Telecommunications carriers must be continuously upgrading sections of their network to keep pace with demand and grow revenues. Recently, dispersion tolerant telecommunications technologies have been very much in the spotlight. Specifically, the capability of duobinary modulation and receiver based-electronic dispersion compensation to enable networks to operate without in-line dispersion compensation has been well publicized and each technology is considered to be vying to become the preferred alternative for dispersion compensation [Ref.1, 2] reducing both cost and complexity in the network.

Advanced Modulation Schemes

The modulation format used to encode data streams for transmission over optical networks can have a significant impact on the data stream's tolerance to distortion due to non-linear transmission effects and chromatic dispersion. Traditional data encoding systems in optical networks like non-return to zero (NRZ) and return to zero (RZ) are characterized by a signal with two signal amplitude levels representing the 1's and the 0's. Advanced modulation formats such as duobinary display a high degree of tolerance to pulse distortion through dispersion than the traditional modulation schemes like NRZ. By introducing alternative phase states as a third level of signal encoding, duobinary reduces dispersion induced inter-symbol interference and yield a much increased dispersion tolerance. In fact, with commercially available duobinary transponders the typical uncompensated reach of G.652 fibers of 80km at 1550nm, can be extended to almost 200 km for 10 Gb/s operation [Ref. 1].

Electronic Dispersion Compensation

Electronic dispersion compensation (EDC) is another recent technology that works to extend the uncompensated transmission reach of optical networks and is now attaining commercial maturity. EDC systems have evolved into two forms, transmitter-based EDC (Tx-based EDC) and receiver-based EDC (Rx-based EDC). Rx-based EDC uses advanced signal processing techniques at the receiver to successfully decode signals that have been distorted due to dispersion. There are three common Rx-based EDC techniques. They are Decision Feedback Equalisation (DFE), Feed Forward Equalisation (FFE) and Maximum Likelihood Sequence Estimation (MLSE). Corning research has shown that DFE and FFE do not have as strong a dispersion correction capability as MLSE. It has been demonstrated that Rx-based EDC with MLSE can provide up to 60% reach extension when using traditional NRZ modulation and a 10 to 15% additional reach extension when using duobinary modulation [Ref.3].

Non-Zero Dispersion Shifted Fibers

In the late 1990s, as 10Gb/s transmission rates and wavelength division multiplexing (WDM) were adopted and the limitations of dispersion with multi-channel operation at those data rates were realised, Non-Zero Dispersion Shifted Fibers with lower dispersion in the C-band were introduced. These products are now broadly categorised by the G.655 standard. Both G.655 and G.652 fibers need dispersion compensation to span conventional distances in regional and long haul networks. However, a lower dispersion G.655 fiber like Corning's LEAF fiber has about 25% of the dispersion of a G.652 fiber. As a result, LEAF fiber alone can enable DCM free networks up to 320 km (in the linear regime reducing the fiber dispersion to 25% of G.652 fiber dispersion enables 4 times the uncompensated reach) with NRZ modulation. If we apply duobinary modulation to LEAF fiber we can further extend the uncompensated reach to 600 km or more [Ref. 4]. If we apply Rx-EDC to NRZ signals transmitted over LEAF fiber, Corning research has concluded that an uncompensated reach of approximately 600 km is also possible. This distance will enable many of today's optical networks to operate without in-line dispersion compensation, but still will not deliver that goal for extended reach networks.

System Integration: Rx based EDC, duobinary modulation and NZDS fiber
Corning has studied the performance capabilities of using a lower dispersion fiber with the new dispersion tolerant technologies. Research at Corning's Sullivan Park research laboratories has demonstrated 10.7 Gb/s transmission of 38 channels with 50 GHz spacing in the long-wavelength half of the C-band (worst case condition) using duobinary modulation and MLSE Rx-based EDC over 900 km of LEAF fiber without using any optical dispersion compensation [Ref. 4,5]. Further studies have shown that with the use of one optical pre-compensation module at the transmitter with fixed pre-compensation of -3360 ps/nm at 1550 nm, this distance can be extended to 1500 km without any in-line or post-compensation as shown in Figures 2a and 2b [Ref. 5]. At 1500 km, the system is dispersion-limited by the longest wavelength channels, which have the greatest accumulated dispersion. However, it is noteworthy that the system performance is relatively flat with distance out to approximately 1200 km across all channels, such that any channel can be dropped at any distance up to 1500 km without receiver modifications. This feature is highly advantageous for transparent and reconfigurable optical networks. The integration of MLSE Rx-based EDC with duobinary modulation and LEAF fiber enables a major new concept in telecommunications networks: high data rate 10 Gb/s optical transmission free of in-line compensation for all networks up to 1500 km.

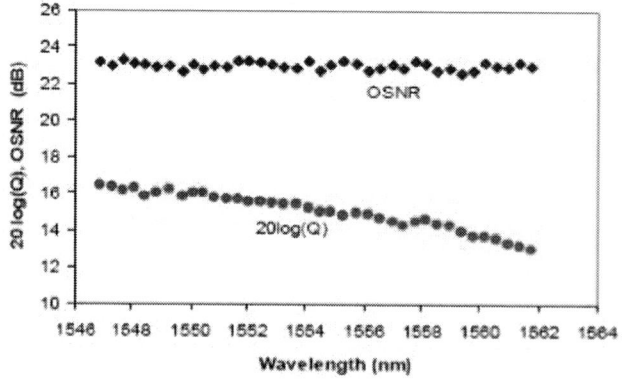

Figure 2a: OSNR and Q values of 38 channels after 10.7 Gb/s transmission over 1500 km of NZDS fiber using duobinary modulation and Rx-based EDC.

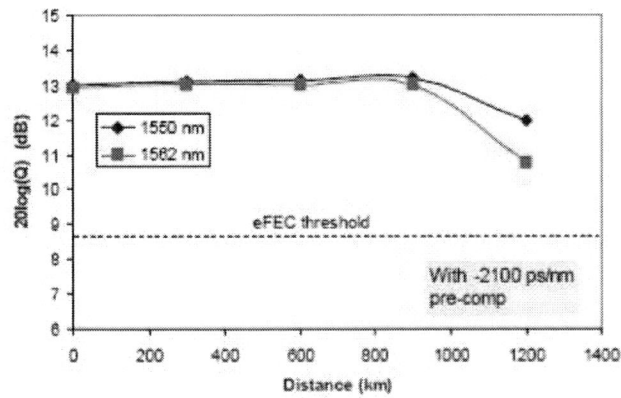

Figure 2b: Edge and central channel Q values as a function of distance over NZDS fiber at 10.7 Gb/s with duobinary modulation and Rx-based EDC.

Based on this data, it is expected that using Rx-based EDC and duobinary modulation with one fixed pre-compensation module, the uncompensated reach of G.652 fibers can be extended to 450 km and for medium dispersion fibers (like ITU-T G.656 compliant NZDS fibers) an uncompensated reach of 900 km is possible as shown in Table 1.

Table 1

Reach (km) Comparison Table					
Fiber Type	NRZ (km)	Duobinary (km)	Rx-EDC MLSE + NRZ (km)	Duobinary + Rx-EDC (km)	Duobinary + Rx-EDC + pre-comp (km)
G.652	80 – 100[*]	200[*]	135-170[*]	250[*]	450[†]
G.656	160-200[†]	400[†]	270-340[†]	500[†]	800-900[†]
LEAF	350[*]	700-750[†]	550-600[†]	900[*]	1500[*]

[*] Experimental data.

[†] Derived from experimental data, assumes 1550 nm dispersion of 8 ps/nm.km for G.656 fiber and 17 ps/nm.km for G.652 fiber.

Conclusion
Dispersion-tolerant technologies like duobinary modulation and Rx-based EDC have gained significant media attention recently. These technologies have important implications for the future development of uncompensated short reach networks and are now maturing with duobinary products available from multiple companies.
With Corning LEAF fiber, the integrated application of duobinary modulation with Rx-based EDC can deliver 1500 km of transmission reach without any in-line dispersion compensation and with attractive cost savings. So finally the goal of cost-effective extended reach transmission without any in-line dispersion compensation can be achieved, and this itself translates into a significant advantage for transparent and reconfigurable networks.

[ref. 1.] "Optical duobinary promises improved reach, greater flexibility", Stephen Hardy, Lightwave May 2006.

[ref. 2.] "Dispersion Tolerant Technologies Battle for Attention", Hardy, Lightwave, June 2006.

[ref.3]: " Experimental measurements of uncompensated reach increase from MLSE-EDC with regard to measurement BER and modulation format," J. D. Downie, M. Sauer, and J. Hurley, Opt. Express **14**, 11520-11527 (2006)
http://www.opticsinfobase.org/abstract.cfm?URI=oe-14-24-11520

[ref. 4]: "Flexible 10.7Gb/s DWDM Transmission over up to 1200km without Optical In-Line or Post-Compensation of Dispersion using MLSE-EDC", Downie, Sauer,Hurley

[ref. 5]: "1500km Transmission over NZ-DSF without in-line or post-compensation of dispersion for 38 10.7Gbit/s channels", Downie, Sauer, Hurley, Electronics Letters, 25th May 2006, Vol.42, No.11.

The Breakthrough of Specialty Fiber Fabricated by PCVD Based Process

Han Qingrong   Tu Feng   Luo Jie   R.Matai

(R&D Centre of Yangtze Optical Fiber and Cable Co.Ltd., Wuhan, China, 430073)

Abstract: In this paper, the principle of PCVD process was introduced briefly. The material composition and structure design combining with waveguide design of PCVD preform were analyzed. Besides, the application, key features and advantages of some specialty fibers fabricated by PCVD process were introduced.

Key Words: PCVD   material composition and structure   specialty fiber

1. Introduction

There are three fundamental aspects for specialty fiber design and fabrication: glass composition, waveguide design and coatings. Glass composition is one of the most basic fiber parameters and variables used for the design of specialty fibers [1]. PCVD process has following advantages: hyperfine refractive index profile control, high levels of incorporation of Ge, F, B and etc., high material homogeneity，high deposition efficiency and controllable OH content for deposition doped silica, which is flexible for specialty fiber design and fabrication. In this paper, the principle of PCVD process was introduced briefly. Meanwhile, the material composition and structure design combining with waveguide design of PCVD preform were analyzed. Besides, the key features and advantages of some specialty fibers fabricated by PCVD process, such as dispersion compensation fiber (DCF), high bandwidth hard polymer cladding optical fiber, polarization maintaining fiber (PMF), bending insensitive single mode fiber, photosensitive single mode fiber and radiation resistance fiber, were introduced.

2. PCVD Process

For PCVD process, the reaction energy comes from high frequency microwave energy from the magnetron microwave generator system (Figure 1). An important feature of PCVD is that the microwave energy is transferred almost without any energy loss to the substrate tube itself. The total deposition thickness has little effect on deposition process and the substrate tube will not deform easily, which can ensure that the uniformity of preform. Besides, the chemical reaction and vitrification happen in a split second and the speed of the resonator along the tube can be as high as above 20,000mm/min, so the thickness of each layer can be very thin. Besides, during each layer the gas composition can be changed. All these can ensure the precision of waveguide and material structure as well as composition control.

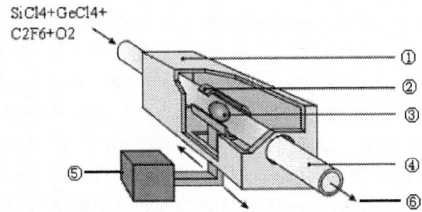

Fig.1 Plasma Activated Chemical Vapour Deposition (PCVD) Scheme: □Furnace; □Resonator; □Plasma; □Substrate tube; □Magnetron; □ Pumping system.

3. Material Composition and Structure Design in PCVD Fiber

The composition of PCVD preform or fiber is synthetic silica with needed GeO_2, B_2O_3 and F (Cl). The co-depositions of GeO_2 and F as well as the functional graded material design ensure that the viscosity match in each layer of optical fiber performs [2]. Meanwhile, the thermal stress can be released because there is no evident interface inside the preform. For PCVD synthetic silica, the viscosity-matching design can be realized by following for simple stepped index fiber [3]:

$$\Delta_{F2} = \Delta_{F1} - \frac{3}{17} * \Delta_{Total} \tag{1}$$

Where Δ_{F1} and Δ_{F2} represent fluorine contribution in core and cladding respectively. Δ_{Total} is the relative index difference of the core relative to cladding.

4. Typical PCVD Specialty Fiber Introduction

4.1 Dispersion Compensation Fiber (DCF)

The main applications of DCFs lie in WDM networks based on single-mode fiber and channel-by-channel dispersion compensation in submarine networks. With PCVD process, it has following key features and advantages: a. high figure of merit (FOM) with low loss and high negative dispersion, b. full slope compensation, low PMD, no MPI due to high order modes and high reliability. It has been proved that PCVD is the best process for DCF fabrication with large scale production

4.2 High Bandwidth Hard Polymer Cladding Fiber (HPCF)

For multimode fibers, the large core makes it much easier to capture light from a transceiver, allowing source costs to be kept down. For short distance data transmission, larger core and higher NA is expected for easy and low cost connectors. With PCVD process, untra-fine graded index RI profile can ensure the fiber have bandwidth above 50MHz.km with even 200µcore and the low refracting index fluoroacrylate coating will ensure the fiber with very high NA above 0.40 (Figure 2).

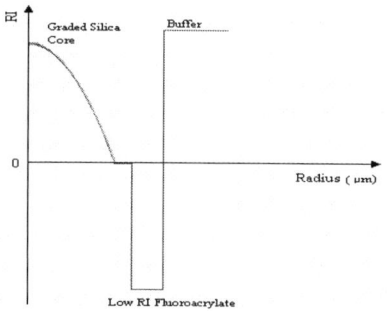
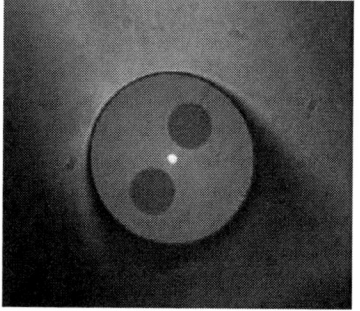

Fig.2 The RI Profile of high bandwidth HPCF   ***Fig.3*** The cross section of PCVD PANDA PMF

4.3 Polarization Maintaining Fiber (PMF)

High quality preform and stress rod can be fabricated with PCVD process, which can ensure that the PCVD PMFs have following key features and advantages: a. excellent polarization maintaining properties with very short beat length, b. low attenuation, c. low bending-induced attenuation, d. high environmental stability and reliability (Figure 3).

4.4 Bending Insensitive Single Mode Fiber

With PCVD process, the waveguide design and the material composition and structure design can be combined through controlling the supplied gaseous reactants. FGM design combining with optimized waveguide designs have been adopted for mechanical reliability. Especially the trench-assisted cladding design can be improve the bending resistance properties much. The main properties are shown in Table 1.

Table 1: The main parameters of PCVD bending insensitive fiber

Fiber Category	A	B	C
MFD@1310nm	8.8±0.4 μm	6.5±0.4 μm	5.6±0.4 μm
Attenuation@1310nm (dB/km)	≤0.35	≤0.39	≤0.52
Attenuation@1550nm (dB/km)	≤0.21	≤0.23	≤0.33
Attenuation@1625nm (dB/km)	≤0.23	≤0.25	≤0.33
Cable cut-off wavelength	≤1260nm	≤1260nm	≤1260nm
1turn×15mm diameter @1550nm	≤0.5dB	≤0.05dB	≤0.01dB
1turn×15mm diameter @1625nm	≤1.0dB	≤0.10dB	≤0.02dB
1turn×20mm diameter @1550nm	≤0.1dB	≤0.02dB	≤0.005dB
1turn×20mm diameter @1625nm	≤0.2dB	≤0.05dB	≤0.005dB
10turn×30mm diameter @1550nm	≤0.03dB	≤0.01dB	≤0.002dB
10turn×30mm diameter @1625nm	≤0.10dB	≤0.02dB	≤0.005dB

4.5 Photosensitive Single Mode Fiber

With PCVD process, the silica structure can be adjusted by modifying the process and the supplied gaseous reactants especially the defects can be tailored for photosensitivity. PCVD photosensitive fiber has many advantages, such as enhanced photosensitivity, high reliability and thermal stability and flexible NA and etc..

4.6 Radiation Resistance Fiber

Many researches have proven that fluorine doping in optical fiber can increase the radiation resistance properties of optical fiber through scaring over defects [4]:

$$\equiv Si - Cl + F \rightarrow \equiv Si - F + Cl \qquad (2)$$
$$\equiv Si - Si \equiv +2F \rightarrow 2 \equiv Si - F \qquad (3)$$

PCVD is the best process for fluorine doped silica deposition with least defects, so PCVD fiber has good radiation resistance through optimizing process. PCVD radiation resistance fibers have lowest attenuation changes under radiation exposure. It can be used for transmission applications in irradiation threatened environments.

5. Conclusion

Making full use of the advantages of PCVD process and combining the material composition and structure design together with the wave-guide design of optical fiber are the basics which can ensure that PCVD process can be used for specialty fiber fabrication with good characteristics. PCVD based process is also good for highly nonlinear fiber, high index fiber, UV-NIR transmission fiber and high power transmission fiber design and fabrication besides above mentioned specialty fibers.

References

[1] ALEXIS MENDEZ, T.F.MORSE, "Specialty Optical Fibers Handbook", Elsevier Academic Press Publication, 2007, Preface

[2] Han Qingrong, Zhao Xiu-jian, Tu Feng et al. Material Composition and Structure Design in PCVD Silica-Based Single-Mode Fiber. Proc. of SPIE, 2006（10）, Vol.6352,635236

[3] Qingrong Han, Xiujian Zhao, Shuqiang Zhang et al. Functional graded material design in PCVD single-mode fiber. Proc. of SPIE, 2005. 6019: 996-1000

[4] M O Zabezhailov, A L Tomashuk, I V Nikolin, et al. The role of fluorine-doped cladding in radiation-induced absorption of silica optical fibers[J]. IEEE Transaction on Nuclear Science, 2002,49(3): 1410-1413

Online RIC process for G.652.D fiber production

Ralph Sattmann, Jan Vydra

Heraeus Quarzglas GmbH & Co. KG, Quarzstrasse 8, D-63450 Hanau, Germany,

ralph.sattmann@heraeus.com

Abstract The Online-RIC process for the production of G652.D single mode fiber is presented. This process utilizes large synthetic fused silica hollow cylinders for multi-thousand kilometres preform size, combining the highest technical and economic benefits.

Introduction

Since the beginning of optical fiber production, there has been a steady growth in the size of preforms. As single mode preform diameters have increased so has the amount of fiber being drawn from these preforms. Using jacket tubes up to an outer diameter of 90 mm, the Rod-in-Tube (RIT) technology facilitated preforms up to 800 km in size. The new Rod in Cylinder (RIC) technology can result in preforms equivalent to fiber lengths of up to 6,000 km. These large preforms improve manufacturing cost and fiber quality.

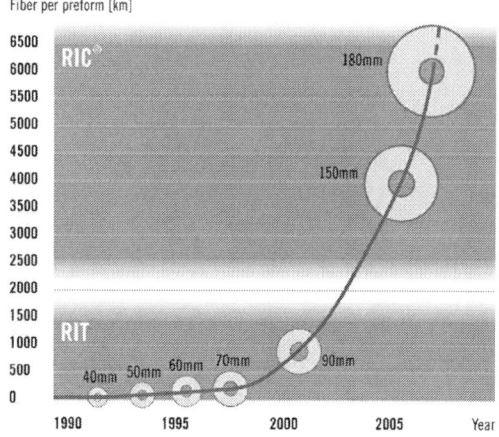

Figure 1: Preform size trend for RIT and RIC technologies.

Online-RIC process

Generally, the RIC process is the evolution of the well established RIT technology. The major part of the cladding material of the fiber is provided in form of a fused silica tube which is overclad onto the core rod. In practice, Online-RIC is a quantum leap not only in preform size, which is shown in figure 1, but also in process efficiency and fiber manufacturing cost.

In the Online-RIC process, core rods are loosely stacked inside a large machined hollow fused silica cylinder. The overcladding process takes place during the fiber drawing process, as shown schematically in figure 2.

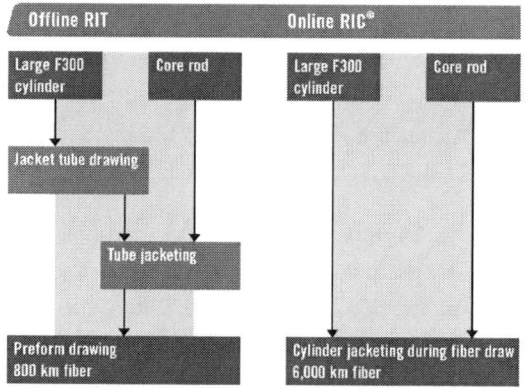

Figure 2: Process flow Offline RIC and Online-RIC .

This lean process has manifold advantages:
- No offline jacketing process is needed.
- No equipment for soot overcladding is needed.
- OH free jacket material and interface enables small b/a ratio core rods.
- Minimum handling and processing of the core rods needed.
- Large preform sizes ensure high up-time of the draw towers.

978-0-9789217-3-6/07/$25.00

©2007 WEN GLOBAL SOLUTIONS

To illustrate the advancement from RIT to RIC, in Figure 3 a jacket tube and a RIC cylinder are shown.

Figure 3: Comparison of large jacket tube and RIC cylinder.

Although the large overclad cylinder size clearly sets some requirements on handling equipment and fiber draw, these are far outweighed by the great benefits of a lean, efficient and high yield process with large preform sizes.

RIC cylinders

RIC cylinders are produced by a highly efficient multi-burner OVD soot process. After subsequent dehydration, chlorination, and sintering, they are precisely shaped using CNC controlled grinding lathes and honing machines. The ground outer and inner surface of the cylinders is suitable for direct fiber draw.

The F300 high purity RIC cylinder material ensures that preforms will result in production of world class optical fiber. Since the F300 cladding material is essentially free of OH, low and zero water peak fibers can be readily made. In production, break rates below 10 breaks per 1,000 km of fiber can be achieved due to the high material purity.
The CNC machining yields outstanding geometrical performance, as summarized in table 1. The tight tolerances enable the production of fibers easily meeting the international standards for fiber geometry.

OD	150 ... 200 mm
OD max – OD min	≤0.6 mm
ID tol	+/- 0.5 mm
Length	1,500 & 3,000 mm
Ovality	≤0.1 mm
Siding	≤1 mm
Bow	≤0.5 mm/m

Table 1: Geometric parameters of RIC cylinders

Since the cylinder geometry can be adjusted in a wide range, the Online-RIC process can be utilized with any established core rod production technique (MCVD, VAD, OVD). Figure 4 shows the geometry parameter range of core rods and cylinders. Since the overcladding process does not introduce any moisture on the surface of the core rod, low b/a values can be utilized, thus dramatically increasing core rod production efficiency. The gap between core rod and cylinder is in the range of 1 to 3 mm. The most recommended working range is shaded.

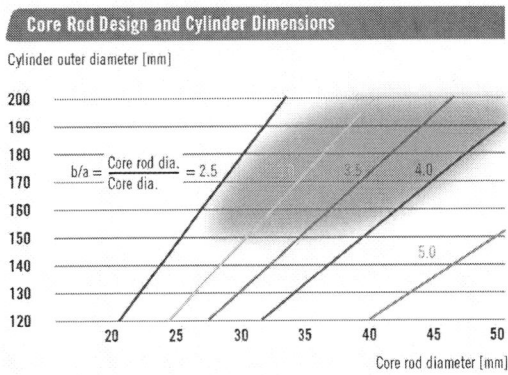

Figure 4: Diagram for matching core rods and RIC cylinders.

Fiber results

Characteristics of fibers produced with the RIC technology have been reported before [1]. The international fiber standards are readily achieved. The excellent performance in terms of geometry, mainly core eccentricity and clad non-circularity is due to the precise geometry of the RIC cylinders, whose relative tolerances in siding, ID, OD and cross sectional area are much tighter than those for drawn RIT overclad tubes.

Optical parameters depend primarily on the core rod quality, however in low b/a concepts purity of the cladding glass is also essential. Due to the high purity and low OH of the bulk cladding material and the OH-free interface between core rod and cylinder, very low attenuation can be achieved [1, 2]. With the intrinsic purity of the cylinders and keeping strict cleaning procedures, break rates below 10 breaks per 1000 km can be readily achieved.

Conclusions

The Online-RIC technology for the manufacturing of single mode optical fibers is very cost competitive. The cost benefits compared to Offline RIT are significant: economy of scale, no off-line jacketing process, reduction of b/a ratio, and improved efficiencies of core rod production equipment.

References

1 Chang et al, OFC (2005), JWA5
2 Matthijsse et al, OFC(2004), TuB5

Optical Ultra-Wideband Pulse Generation Using Air-Guiding Photonic Bandgap Fiber and a Semiconductor Optical Amplifier

Shangyuan Li (1), Xiaoping Zheng (2), Bingkun Zhou(3)

1: Department of Electronic Engineering, Tsinghua University, 100084, Beijing, P. R. China,
lisy@mails.tsinghua.edu.

2: Department of Electronic Engineering, Tsinghua University, xpzheng@tsinghua.edu.cn

3: Department of Electronic Engineering, Tsinghua University, zbk-dee@tsinghua.edu.cn

Abstract *A method to generate ultra-wideband monocycle and doublet pulses using air-guiding photonic bandgap fiber and cross-gain modulation in a semiconductor optical amplifier is proposed. The fractional bandwidth for monocycle and doublet are 157% and 163%.*

Introduction

Ultra-wideband (UWB) is a fast emerging technology for short-range broadband wireless communication at weak spectrum intensity and high data-rate. UWB has many applications such as high-throughput wireless communication, personal area network (PAN), sensor networks, etc. Recently, two international standards, ISO/IEC 26907: 2007 and ISO/IEC 26908: 2007 are released and aimed at high rate ultra wideband PHY and MAC design [1]. Part 15 of the Federal Communications Commission (FCC) regulations defined a UWB signal as a signal that has a fractional bandwidth larger than 20% of the centre frequency, or occupies a 10-dB bandwidth of 500MHz or more [2].

Generating UWB monocycle or doublet pulses in electrical domain attracts many researchers for years. The implementations based on integrated circuit are the best candidates for hand-held UWB terminals [3]. However, considering the cross-network design between UWB access network and fixed wired networks, UWB-over-fiber technology is an excellent candidate for future applications, not only to distribute UWB signal, but to combine UWB with fiber-to-the-home topology [4]. Several methods to generate UWB pulses are proposed, for example, using phase modulator [5-8] or cross-phase modulation in dispersive shifted fiber [9]. Fiber Bragg gratings (FBG), a long piece fiber are commonly used in these schemes to covert the phase modulated signal to intensity modulated. Wang adopted cross-gain modulation (XGM)

in semiconductor optical amplifier (SOA) to generate a pair of polarity-reversed Gaussian pulses, which were time-delayed by two specially designed cascade FBGs to generate UWB pulses [10].

In this paper, we take advantages of XGM in SOA and incorporate a new time-delay method based on air-guiding photonic bandgap fiber (PBGF) in the system to produce UWB monocycle and doublet pulses. Compared to [10], PBGF takes place of the cascade FBGs to realize the time delay. The large dispersion coefficient of PBGF leads to shorter length requirement of the fiber, which gains more stability and less weight compared to schemes with longer single mode fiber or polarization maintaining fiber (PMF). It also provided more feasibility for the system to generate two kinds of monocycle pulses and doublets by changing the wavelength of the laser source.

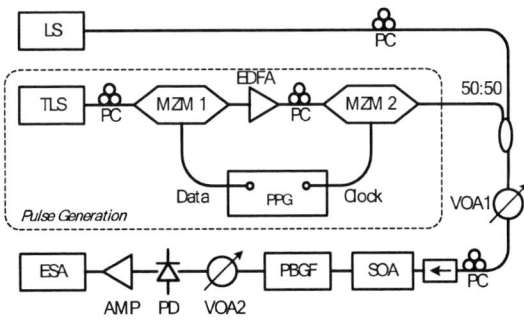

Fig. 1 System block diagram for UWB pulses generation system.

Principle

The system block diagram is shown in Fig.

1. A tunable laser source (TLS) serves as the pump light, while another DFB laser source (LS) as the probe light. The output of TLS is firstly modulated in Mach-Zehnder modulator (MZM) 1 by electrical data sequence. It is then carved through the clock signal in MZM2. A time delay between the data sequence and the clock should be carefully set. The CW probe light is coupled with the narrow optical pulse through a 50:50 coupler. They are attenuated by a variable optical attenuator (VOA) and then injected to a SOA. Because of the XGM effect in SOA, the pump pulse modulates the probe light and generates a dark pulse at the probe wavelength.

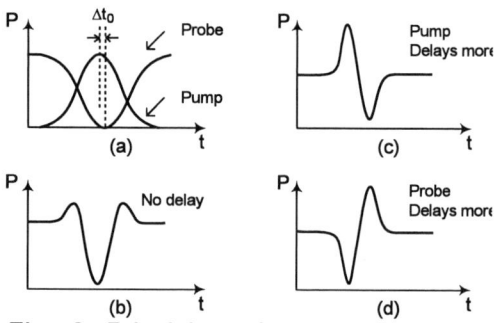

Fig. 2 Principles of the proposed UWB monocycle and doublet pulses generation.

A peak-valley delay Δt0 between these two pulses at the SOA output is observed and illustrated in Fig. 2(a). Also because of the XGM, the probe pulse is wider than the pump pulse. If the TLS is adjusted to compensate the existed Δt0, the wider pulse plus the narrower pulse results a UWB doublet at PD, shown in Fig. 2(b). If the wavelength of pump light is shorter than the probe light, the pump pulse delays more. Then the direct detection of these two pulses would shape a positive portion lead monocycle, shown in Fig. 2(c). If the wavelength of pump light is longer, a negative portion lead monocycle is obtained and shown in Fig. 2(d). A VOA is placed between the PBGF and PD. The generated UWB pulses is amplified by a wideband microwave amplifier and sent to an electrical spectrum analyzer (ESA, HP 8593E). Also, the waveforms of the pulses are measured by a sampling oscilloscope (HP 54120B).

The PBGF used in the system is introduced

in reference [11]. The cross-section image of this piece PBGF by scanning electron micrograph (SEM) is shown in Fig. 3(a). The dispersion coefficient D is measured as approximately 820 ps/nm·km at 1550 nm and 1090 ps/nm·km at 1560 nm. The length of the fiber is 20.13m, so a wavelength dependent time delay between different wavelengths is existed, as illustrated in Fig. 3(b).

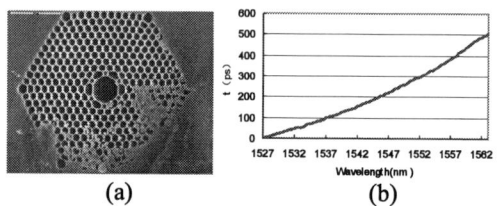

(a)   (b)

Fig. 3 (a) Cross-section image of PBGF. (b) Wavelength dependent time delay between different wavelengths of the PBGF

Experiments

In the experiment, the electrical data applied to MZM1 is generated by pulse pattern generator (PPG) with a fixed 12-bit pattern "1000 0000 0000" at 7.5 Gbps. This signal has a repetition rate of 625MHz. The TLS is set to 1550.0 nm. After the modulation via MZM2 by the clock, the pulse has a full width at half maximum (FWHM) of 54.0 ps, with a shape close to Gaussian. The PD used in the system has an inverse output polarity with the light, which means an optical pulse generates an electrical dark pulse and vice versa. The probe light has a wavelength of 1546.05 nm and an output power of 2.8 dBm. After the 50:50 coupler and VOA1, the coupled light is injected into the SOA. After the SOA, the probe light pulse has a FWHM of 104.6 ps and the pump pulse has 51.6 ps. The output power of SOA is 8.04 dBm at a driving current of 248 mA. The waveforms of the two pulses at the SOA output are shown in Fig. 4. It is observed that a time delay between the two pulses is 22.0 ps. The delay is invariant while TLS is changed.

When the TLS is tuned to 1547.70 nm, the delay between the two polarity-reversed pulses is fully compensated. The dispersion coefficient around 1547nm is about 740 ps/nm·km. Thus the dispersion induced delay for a 1.65nm wavelength difference is

54

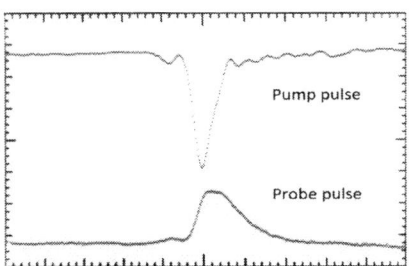

Fig. 4 Waveforms of the pump pulse and probe pulse at the output of SOA

24.6 ps, $\Delta t0$ is almost compensated. The waveform and electrical spectrum this pulse is shown in Fig. 5. The measured data are listed in Table 1, row 2. The ESA used in the experiment quite an old one which has a high noise level of -65 dBm at a resolution bandwidth (RBW) of 1MHz at 5GHz. Thus a wideband amplifier with an approximate 20dB gain (at 0.5GHz to 20GHz) is applied to raise the signal level.

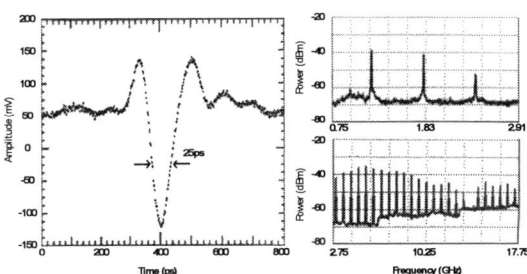

Fig. 5 Waveform and electrical spectrum (ES) of doublet pulse.

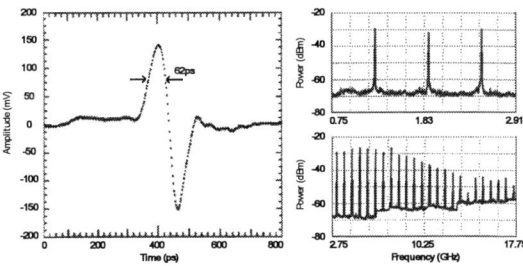

Fig. 6 Waveform and ES of positive portion leading monocycle pulse

When the TLS is tuned to 1551.60 nm, the delay between TLS and LS caused by PBGF is calculated as 86 ps based on the dispersion data of PBGF. Because of the wavelength relationship, the probe pulse is 66 ps prior to the pump pulse. Count in the 22 ps delay caused by SOA, a total delay change is 88.0 ps, which is close to the calculation. The positive portion leading monocycle pulse is generated at PD. The

waveform and ES is shown in Fig. 6. Measured data is shown in Table 1. When the TLS is tuned to 1542.3 nm, a negative portion leading pulse is generated. The data of it is listed in table 1. The frequency spacing between the two neighboring lines in the frequency spectrum is 0.625 GHz, which equals the repetition rate of the UWB pulses.

Table 1

Pulse Type	FWHM (ps)	-10 dB BW(GHz)	CF (GHz)	Fractional BW
Doublet	24.6	10.1	6.2	163%
Monocycle(+)	62	9.4	6.0	157%
Monocycle(-)	46	8.8	5.6	157%

Conclusion

In conclusion, UWB monocycle and doublet pulses are generated in one system based on large dispersion of air-guiding PBGF and XGM in a SOA. The monocycles have FWHMs of 62ps and 46ps, both with a fractional BW of 157%. The doublet has a FWHM of 24.6ps, with a fractional BW of 163%.

This work is supported by Natural Scientific Foundation of China grants 60432020 and 60520130298 and National 973 Project No. 2006CB302805.

References

1 ISO/IEC 26907:2007, 26908:2007,.
2 G. R. Aiello et al, IEEE Microw. Mag., v4, (2003), pp 36–47.
3 Y. Jeong et al, in Proc. 2004 IEEE Int. Symp. Circuit and Systems (2004), pp IV-129–IV-132.
4 S. Kim et al, in Proc. Int. Workshop Ultra Wide-Band Systems (2004), pp 187–191.
5 T. Kawanishi et al, in Proc. IEEE Int. Topical Meeting Microw. Photon. (2004) pp 48–51.
6 F. Zeng et al, IEEE Photon. Technol. Lett. v18 (2006), pp 823-825.
7 F. Zeng et al, IEEE Photon. Technol. Lett. V18(2006), pp 2062-2064.
8 H. Chen et al, Elec. Lett. v43(2007), pp 542–543.
9 F. Zeng et al, Elec. Lett. v43(2007), pp 119–121.
10 Q. Wang et al, Optics Lett., v31 (2006), pp 3083-3085.
11 Z. Liu, et al, Optics Lett., v31 (2006), pp 2789-2791.

Polarization changes of partially coherent pulses propagating in optical fibers

Weihong Huang, Sergey A. Ponomarenko, and Michael Cada

Dalhousie University, Electrical and Computer Engineering
Halifax, NS, B3J 2X4, Canada

w.huang@dal.ca

Govind P. Agrawal

The Institute of Optics and Department of Physics and Astronomy
University of Rochester, Rochester, NY 14627

gpa@optics.rochester.edu

Compiled July 4, 2007

We consider polarization changes of statistical pulses in single-mode fibers. We show that the evolution of the degree of polarization is determined by the interplay between the coherence properties of the pulse and fiber birefringence. © 2007 Optical Society of America

OCIS codes: (030.0030) Coherence and statistic optics; (060.2430) Single mode fibers

1. Introduction

The degree of polarization reflects vectorial nature of statistical electromagnetic fields, and it is an important measurable quantity that characterizes the evolution of partially coherent light fields.[1] The polarization of light may change on propagation due to various factors, including the nonlinearity and random birefringence of the medium.[2] We here focus on polarization changes of light induced by the source fluctuations. While coherence-induced polarization changes have been extensively studied in spatial domain for stationary fields, a few studies on statistical optical pulses are available and have been so far limited to the scalar case.

Propagation of fully coherent pulses along optical fibers has been studied extensively in the presence of dispersion as well as nonlinear effects.[2] Random birefringence of optical fibers has also attracted much attention in the last decade because the resulting polarization mode dispersion (PMD) has become a limiting effect in high bit rate and long haul communication systems. Such effects can be suppressed in the so-called polarization-maintaining fibers by introducing a relatively large constant birefringence. In this paper we consider propagation of partially coherent pulses along such fibers in the linear regime, *i.e.*, we neglect the nonlinear effects but include both the group-velocity dispersion and birefringence. The evolution of the degree polarization of light in such fibers is determined by the interplay between the pulse spreading, which depends on the coherence properties of the two polarization components, and the spatial walk-off resulting from a birefringence-induced group-velocity mismatch. This circumstance makes the polarization dynamics quite different from the previously studied correlation-induced polarization changes of sta-

tistically stationary fields propagating in free space.[3]

2. Formulism of the Problem

To describe partially coherent pulses, we choose a realization from the statistical ensemble of optical pulses and employ the Jones-vector notation to write its fluctuating electrical field at a time t and position z in the form

$$\mathbf{E}(t, \mathbf{r}) = \begin{pmatrix} A_x(t, z) \\ A_y(t, z) \end{pmatrix} F(x, y) e^{i(\beta_0 z - \omega_0 t)}, \quad (1)$$

where ω_0 is the carrier frequency and β_0 is an effective propagation constant at this frequency of the single mode supported by the fiber with the spatial profile $F(x, y)$. Physically, $A_x(t, z)$ and $A_y(t, z)$ are the slowly varying amplitudes of the two mutually orthogonal polarization components of the field $\mathbf{E}(t, \mathbf{r})$. The correlation properties of a statistical pulse can then be characterized by a second-order correlation tensor, defined as[1]

$$\Gamma_{jl}(t_1, t_2, z) = \langle A_j^*(t_1, z) A_l(t_2, z) \rangle, \quad (2)$$

where j and l take on values x or y and the angle brackets denote ensemble averaging; $A_j(t, z)$ satisfies the following wave equation[2]

$$\frac{\partial A_j}{\partial z} + \beta_{1j} \frac{\partial A_j}{\partial t} + \frac{i}{2} \beta_{2j} \frac{\partial^2 A_j}{\partial t^2} = i(\beta_{0j} - \beta_0) A_j \quad (3)$$

where $\beta_{mj} = (d^m \beta_j / d\omega^m)_{\omega = \omega_0}$ and β_j is the corresponding propagation constant. Physically, $\beta_{1x} \neq \beta_{1y}$ due to fiber birefringence and $\beta_{2x} \simeq \beta_{2y} \equiv \beta_2$ accounts for the group-velocity dispersion (GVD).

We now focus on a particular GSM source which generates the field with the correlation tensor of the form

$$\Gamma_{jl}(t_1, t_2, 0) = A^2 \delta_{jl} \, e^{-\frac{(t_1^2 + t_2^2)}{4\sigma_t^2}} \, e^{-\frac{(t_1 - t_2)^2}{2\sigma_{cj}^2}}. \quad (4)$$

978-0-9789217-3-6/07/$25.00

©2007 WEN GLOBAL SOLUTIONS 56

Solving Eqs. (2), (3) and (4) yields

$$\Gamma_{jl}(t_1, t_2, z) = \frac{A^2 \delta_{jl}}{\Delta_j(\xi)} e^{-\frac{\gamma_j \tau_{j1}^2 + \gamma_j^* \tau_{l2}^2 - 2\beta_j \tau_{j1} \tau_{l2}}{\sigma_t^2 \varrho_j^2 \Delta_j^2(\xi)}}, \quad (5)$$

where we have introduced the following notations:

$$\tau_{j1} = t_1 - z\beta_{1j}, \qquad \tau_{l2} = t_2 - z\beta_{1l}, \quad (6)$$

$$\frac{1}{\varrho_j^2} = \frac{1}{4\sigma_t^2} + \frac{1}{\sigma_{cj}^2}, \qquad \gamma_j = \alpha_j - \frac{i\xi}{2\omega_0}, \quad (7)$$

$$\xi = \omega_0 \beta_2 z, \qquad \Delta_j(\xi) = \sqrt{1 + \xi^2/(\omega_0 \sigma_t \varrho_j)^2}, \quad (8)$$

$$\alpha_j = \frac{a_j}{4(a_j^2 - b_j^2)}, \qquad \beta_j = \frac{b_j}{4(a_j^2 - b_j^2)}, \quad (9)$$

$$a_j = \frac{1}{4\sigma_t^2} + \frac{1}{2\sigma_{cj}^2}, \qquad b_j = \frac{1}{2\sigma_{cj}^2}. \quad (10)$$

Equations (5)-(10) show how the coherence properties of the pulse change with propagation. The ensemble-averaged 2×2 matrix $\mathbf{J}(t, z)$, characterizing the polarization properties of the pulse at some point (t, z), is referred to as the coherency tensor with its matrix elements defined as $J_{jl}(t, z) = \langle A_j^*(t, z) A_l(t, z) \rangle$. The degree of polarization can be expressed in terms of the coherency tensor as follows[1]

$$P(t, z) = \left(1 - \frac{4\det[\mathbf{J}(t, z)]}{\{\mathrm{tr}[\mathbf{J}(t, z)]\}^2}\right)^{1/2}, \quad (11)$$

where det and tr denote the determinant and the trace of a matrix, respectively. The coherency tensor is related to the second order correlation tensor by the expression

$$J_{jl}(t, z) = \Gamma_{jl}(t, t, z). \quad (12)$$

3. Results

Transforming the field $\mathbf{E}(t, \mathbf{r})$ from the fixed coordinate system to a coordinate system moving with the average group velocity $v_g = 2/(\beta_{1x} + \beta_{1y})$, Eqs. (5)-(12) lead to the simplified expression of the degree of polarization as a function of the retarded time $\tau = t - \bar{\beta}z$,

$$P(\tau, z) = \frac{|I_x(\tau, z) - I_y(\tau, z)|}{I_x(\tau, z) + I_y(\tau, z)}, \quad (13)$$

where

$$I_j(\tau, z) = \frac{A^2}{\Lambda_j(z)} e^{-\frac{(\tau \pm \delta z)^2}{2\sigma_t^2 \Lambda_j^2(z)}}, \quad (14)$$

and

$$\Lambda_j(z) = \sqrt{1 + \left(\frac{\beta_2 z}{2\sigma_t^2}\right)^2 + \left(\frac{\beta_2 z}{\sigma_t \sigma_{cj}}\right)^2}. \quad (15)$$

Here $+$ and $-$ signs correspond to the x- and y-polarized modes, respectively, and $\delta = (\beta_{1y} - \beta_{1x})/2$ represents the extent of group-velocity mismatch. It follows from Eqs. (13) - (15) that the different dynamics of the intensity profiles of the two polarization modes is caused by the competition between the walk-off effect and the

polarization-dependent pulse spreading. The walk-off effect results from the fiber birefringence, while the widths of the intensity profiles increase at different rates due to different correlation times of the two polarization components at the source.

Variations in the degree of polarization across the pulse are calculated for partially coherent light with $\sigma_{cx} = 12$ ps and $\sigma_{cy} = 16$ ps. The results are exhibited in Fig. 1(a) for $\delta = 0$, and (b) for $\delta = 0.26$ ps/km. The propagation of the degree of polarization is also calculated with and without birefringence, and the results are shown in parts (c) and (d) of Fig. 1.

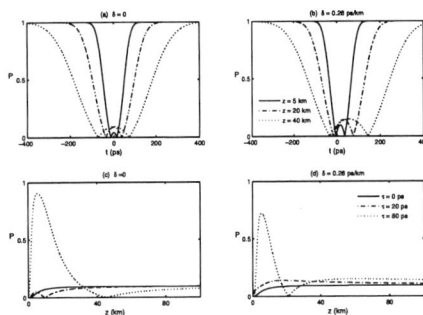

Fig. 1. Variations of the degree of polarization across the pulse for (a) $\delta = 0$ and (b) $\delta = 0.26$ ps/km. The degree of polarization as a function of z for (c) $\delta = 0$ and (d) $\delta = 0.26$ ps/km.

References

1. L. Mandel and E. Wolf, *Optical Coherence and Quantum Optics,* (Cambridge University Press, Cambridge, 1995).
2. G. P. Agrawal, *Nonlinear Fiber Optics,* 4th ed. (Academic Press, New York, NY, 2007).
3. D. James, "Change of polarization of light beams on propagation in free space," J. Opt. Soc. Am. A **11**, 1641-1643 (1993).

Mid-infrared optoelectronic devices and applications

Zhang Yong-gang, Li Ai-zhen

State Key Laboratory of Functional Materials for Informatics, Shanghai Institute of
Microsystem and Information Technology, Chinese Academy of Sciences.
865 Chang Ning Road, Shanghai 200050, China.
Emai: ygzhang@mail.sim.ac.cn

Abstract *The semiconductor sources and detectors in mid-infrared band, including quantum cascade lasers, antimonide quantum well lasers and photovoltaic detectors grown using MBE in our laboratory are introduced, gas detection using those devices are demonstrated.*

Introduction

The mid-infrared (MIR) band of 2-25 μm has attracted much attention for its particular spectroscopic features including atmospheric window, characteristic absorption of gases, spectral or thermal imaging, night sight, infrared countermeasure, etc. The applications related to those features makes the light source and detectors in this band very important. Compare to other type of thermal devices, the semiconductor optoelectronic devices have the superiorities of high speed, high efficiency and favorable spectral features, the demands of semiconductor sources and detectors in this band are increasing.

Among various applications, the tunable diode laser absorption spectroscopy (TDLAS) for gas sensing in MIR band have attracted much attention with the appearance of different kind of MIR semiconductor lasers, including antimonide multi-quantum-well (MQW) lasers, quantum cascade lasers (QCL) and interband cascade lasers (ICL), as well as suitable photodetectors. The progress of such devices have accelerated the TDLAS in MIR dramatically, the main motivation should be creating compact and robust gas sensors with appropriate performance, especially the specificity. Keeping this target in mind, we have made continuous efforts on developing the semiconductor lasers and detectors in MIR band, and towards applications as shown below.

InP based quantum cascade lasers

InP based InAlAs/InGaAs QCLs are most suitable for the 4~10 μm band, our study on

those devices has lasted one decade, including the MBE of the microstructure with hundreds of layers down to ~1nm, processing and characterization of the F-P and DFB lasers as well as analysis and optimize of their performance[1-8]. Pulse operation of the QCL above room temperature with moderate performance have been reached, Fig. 1 shows the typical characteristics of our gas source molecular beam epitaxy (GSMBE) grown DFB-QCL with lasing wavelength around 7.7 μm, this laser is suitable for TDLAS application of green house gas N_2O detection.

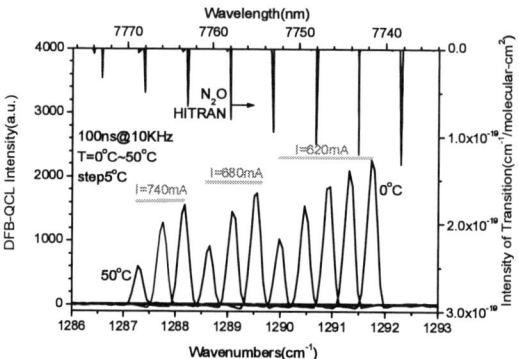

Fig.1 Measured lasing srcttra of GSMBE grown InAlAs/InGaAs DFB-QCL at different heat sink temperatures, tte HITRAN data of N_2O are also shown for reference.

Antimonide quantum well lasers

GaSb based AlGaAsAs/InGaAsSb MQW lasers are most suitable for the 2~3 μm band, our study on those devices has lasted more than one decade[9-14]. For those laser the CW lasing have been reached at temperature exceeding 80°C, Fig. 2 shows the typical performance of our solid source molecular beam epitaxy (SSMBE) grown MDW-LD with lasing wavelength around 2.1 µm using a home made MBE system, the demonstration of N_2O gas detection have been done using this laser.

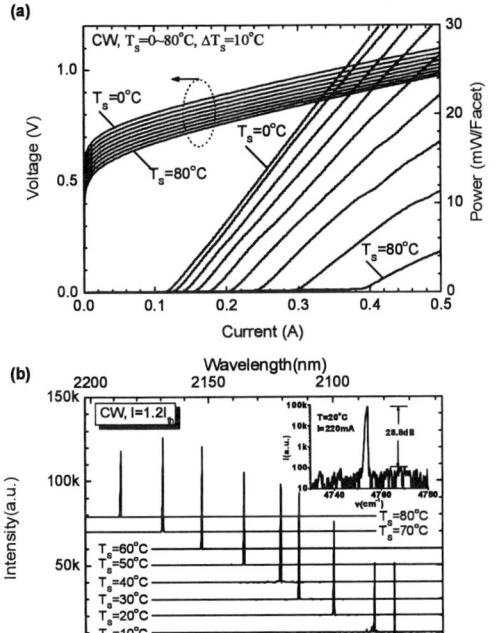

Fig.2 Measured characteristics of SSMBE grown AlGaAsSb/InGaAsSb MQW laser. (a): I-P and I-V performance at CW operation; (b): Lasing spectra at different heat sink temperatures.

Mid-infrared III-V photodetectors

Our research on III-V MIR detectors could be traced to 1980's, at the time liquid phase epitaxy (LPE) are used for grown InAsPSb/InAs photovoltaic diode in 2-3 µm band[15-16]. Then SSMBE are used to fabricate InGaAsSb/GaSb photovoltaic diode with good performance in this band [17]. At longer wavelength of 3-5 µm and 8-14 µm atmospheric window, the photo-conductive quantum well infrared photodetectors (QWIP) on InP based InAlAs/InGaAs[18-19] and GaAs based AlGaAs/GaS system also have been fabricated using GSMBE. For the photovoltaic detectors cut-off at 2~3 µm, the wavelength extended InGaAs photodiode with higher indium content on InP substrate also have been fabricated using GSMBE

with different kinds of buffer and window layers[20-21], Fig.3 shows the response spectra of the devices.

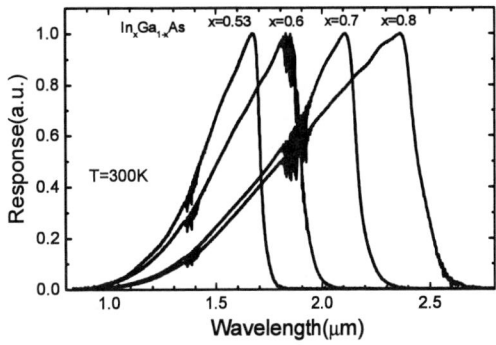

Fig.3 Measured response spectra of GSMBE grown $In_xGa_{1-x}As$ photovoltaic detectors at room temperature with dfifferent indium compositions.

Applications

Using our home made MIR optoelectronic devices, the TDLAS gas detection have been demonstrated.

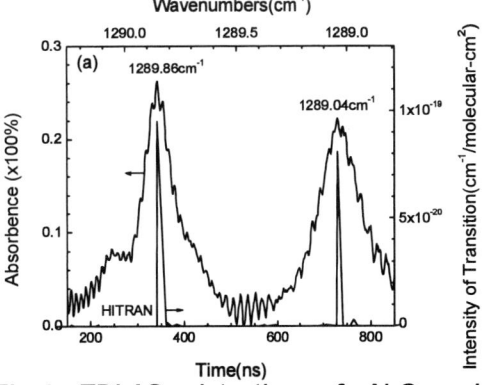

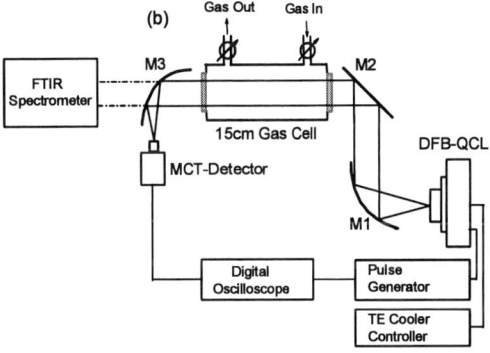

Fig.4 TDLAS detection of N_2O using InAlAs/InGaAs DFB-QCL around 7.7 µm. (a): Measured absorbance using pulse wavelength scan scheme; (b): Schematic diagram of the measurement system.

In our first experiments, the green house gas N_2O is used as target gas, a home made 7.7 µm DFB-QCL is used for TDLAS purpose[22], Fig.4 show the results and the system. After that, home made 2.1 µm AlGaAsSb/InGaAsSb MQW-LD and InGaAs PD were also have been used for TDLAS purpose[23], Fig.5 show the results and the system. In this system the detection limit below 1 ppmv have been reached using a 10 m gas cell.

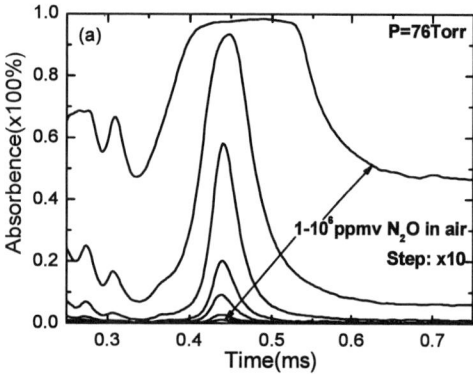

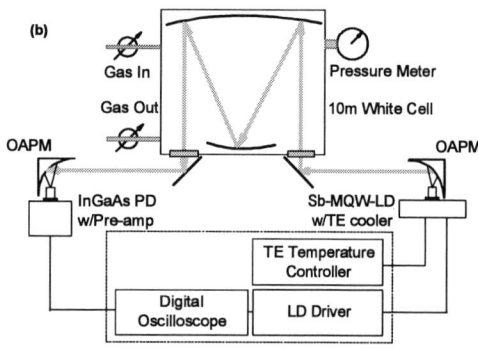

Fig.5 Demonstration of TDLAS N_2O detection using AlGaAsSb/InGaAsSb MQW-LD around 2.1 µm. (a): Measured absorbance using a simple pulse wavelength scan scheme at different N_2O concentrastions from 1 ppmv to 100% at preasure of 76 Torr; (b): Schematic diagram of the measurement system.

Conclusions

In conclusion, through continuous efforts of more than a decade in the development of various MIR optoelectronic device, their performance have boosted drimatically, some of which have been suitable for real applications, especially the TDLAS in MIR band. The TDLAS in MIR band have the superiorities of excellent sensitivity and specificity, so further demonds of the devices in this band could be expected.

References

1 A. Z. Li et al J. Crystal Growth **201/202** (1999), 901

2 A.Z. Li et al J. Crystal Growth **227/228** (2001), 313

3 Y. G. Zhang et al Spectrochimica Acta Part A, **58** (2002), 2323

4 ZHANG Yong-Gang et al Chin. Phys. Lett., **20** (2003), 678

5 A.Z. Li et al J. Crystal Growth **278** (2005), 770

6 Cheng Zhu et al J. Appl. Phys., **100** (2006), 053105

7 Gangyi Xu et al Appl. Phys. Lett., **89** (2006), 161102

8 A. Z. Li et al J. Crystal Growth, **301/302** (2007), 129

9 A. Z. Li et al J. Crystal Growth, **127** (1993), 566

10 A. Z. Li et al J. Crystal Growth, **175/176** (1997), 873

11 Y. G. Zhang et al J. Crystal Growth, **227/228** (2001), 582

12 C Zhu et al Semicond. Sci. & Technol. **20** (2005), 563,

13 C. Zhu et al J. Crystal Growth **278** (2005), 173,

14 ZHANG Yong-Gang et al Chin. Phys. Lett., **23** (2006) , 2262

15 Zhang Yonggang et al Chin J. Rare Metals, **9** (1990), 46

16 Zhang Yonggang et,al. Chin. J. Semiconductors, **13** (1992), 624

17 A. Z. Li et al J. Crystal Growth, **150** (1995), 1375

18 ZHANG Yong-Gang et al Chin. Phys. Lett., **14** (1997), 443

19 ZHANG Yong-Gang et al Chin. Phys. Lett., **16** (1999), 747

20 ZHANG Yong-Gang et al Chin. Phys. Lett., **22** (2005), 250

21 Yonggang Zhang et al Infrared Physics & Technology, **47** (2006), 257

22 ZHANG Yong-Gang et al Chin. Phys. Lett., **23** (2006), 1780

23 ZHANG Yong-Gang et al Chin. Phys. Lett.,**24** (2007), to be published.

Microwave and Millimeter-wave Photonic Devices for Communications and Measurement Applications

Tadao Nagatsuma [1,2] and Yuichi Kado [1]

1: NTT Microsystem Integration Laboratories, NTT Corporation
3-1 Morinosato Wakamiya, Atsugi, Kanagawa 243-0198, Japan
2: Graduate School of Engineering Science, Osaka University
ngtm@aecl.ntt.co.jp, nagatuma@ee.es.osaka-u.ac.jp

Abstract Microwave and millimeter-wave photonics, which merges radio-wave and photonics technologies, has recently attracted increasing interest. This paper provides an overview on the status of microwave and millimeter-wave photonic devices and some of their latest applications.

Introduction

The recent explosive growth in communications has been brought about by wired (fiber-optic) and wireless (radio-wave) communications technologies. These two technologies have started to merge to create a new interdisciplinary area called **M**icrowave and **M**illimeter-**W**ave **P**hotonics (MWP) [1]. In addition, viewing the electro-magnetic spectrum with wavelengths progressively decreasing to the millimeter and submillimeter-wave bands on the radio-wave side and wavelengths progressively increasing to the infrared region on the light-wave side, we see that there is a large gap in utilization on the boundary between radio waves and light waves, i.e., the frequency band between 100 GHz and 10 THz. This untapped region represents a major resource for humankind in the 21st century.

MWP technology aims at achieving advancement and improved functions in telecommunications systems that cannot be achieved by extension of individual technologies, mainly through the combination of radio-wave technology and photonic technology. At the same time, MWP technology is also expected to open up unused frequency bands through the fusion of different fields. The opening of new application fields other than communications is also expected.

This paper provides an overview of the status of MWP technology, focusing on a system concept and enabling devices, and describes latest applications.

MWP system concept

Typical radio-wave application system is illustrated in Fig. 1. Wireless communication link consists of a transmitter (Tx) and a receiver (Rx) as shown in Fig. 1(a), and some kind of object is placed between the Tx and Rx in applications to measurement, testing, and sensing as shown in Fig.1(b).

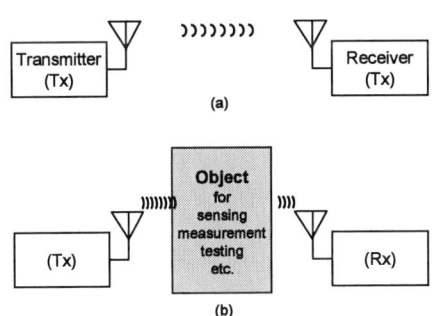

Fig. 1. Radio-wave system for (a) communication, and (b) measurement.

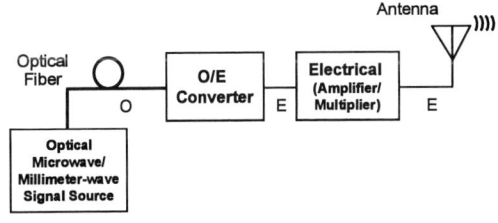

Fig. 2. Block diagram of the photonically-assisted microwave/millimetre-wave transmitter.

Figure 2 shows a block diagram of MWP-based transmitter, that is, a photonically assisted radio-wave transmitter. First, the optical (O) signal, whose intensity is modulated at microwave and/or millimetre-wave frequencies, is generated by the optical microwave/millimetre-wave signal source, and is delivered through optical fiber cables, and converted to the electrical (E) signal by a high-

978-0-9789217-3-6/07/$25.00
©2007 WEN GLOBAL SOLUTIONS

frequency O-E converter such as a photodiode. The converted signal is followed by a power amplifier and/or a frequency multiplier, and is finally radiated into free space by an antenna. The antenna unit can be separated and remotely controlled by optical fiber cables.

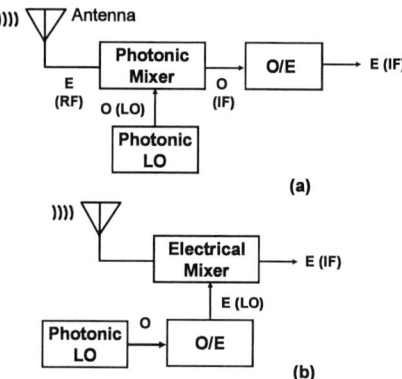

Fig. 3. Block diagram of the photonically-assisted microwave/millimetre-wave receiver using (a) photonic mixer and (b) electrical mixer.

Figure 3 shows two types of photonically-assisted radio-wave receivers; one employs a photonic mixer pumped by photonic local oscillator (LO) signals. Typical photonic mixer is a bulk electro-optic (EO) crystal, and optical modulator devices such as a LiNbO$_3$ waveguide modulator and a semiconductor electro-absorption modulator. Here, the optical intermediate frequency (IF) signal is converted to the electrical IF signal by a slow photodiode. The other type is based on a nonlinear electrical mixer such as a Schottky-diode mixer, and a superconducting (SIS) mixer. The LO signal is generated by a high-frequency photodiode followed by the optical microwave/millimetre-wave signal source, as is used in the transmitter (Fig. 2).

Enabling devices technology

As for the optical microwave/millimetre-wave source in Fig. 2, there are lots of options such as optical heterodyning using two frequency-tunable laser diodes, optical heterodyning using two modes filtered from a multi-frequency (wavelength) optical source or optical frequency comb generator (OFCG), the combination of a continuous-wave (CW) laser with an external modulator, and semiconductor mode-locked lasers. Low-phase-noise and frequency-tunable optical millimeter-wave generators based on the optical heterodyning

technique is shown in Fig. 4 [2].

An O-E converter is a key device in the system. Since optical amplifiers with a high gain of over 30 dB and a large bandwidth of over 1 THz are now readily available, we need a high-power O-E converter to boost the signal generator performance. We used an ultrafast photodiode called a uni-traveling-carrier photodiode (UTC-PD) [3]. Figure 5 depicts an example of photonic millimetre-wave emitter, where the UTC-PD and the antenna are integrated [4].

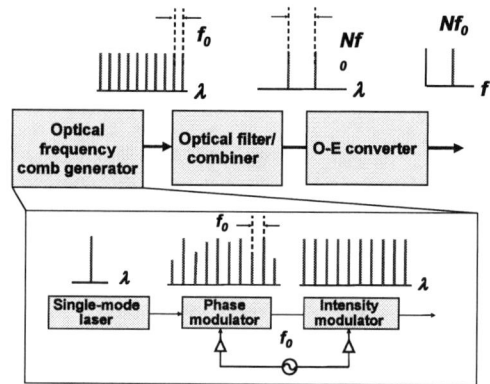

Fig. 4. Block diagram of the millimeter-wave generator based on optical heterodyning technique.

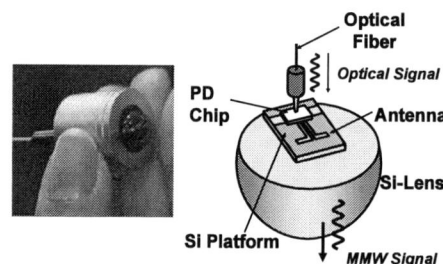

Fig. 5. Example of the millimeter-wave emitter using photonic technique.

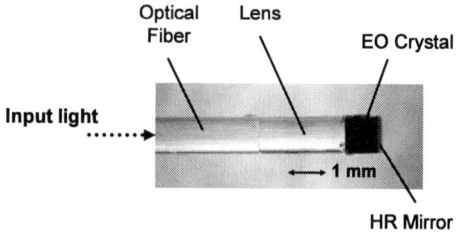

Fig. 6. Example of the microwave/millimeter-wave sensor using the electro-optic crystal.

As an example of the optical millimeter-wave receiver or detector, the EO sensor made of a bulk EO crystal offers the largest bandwidth extending to the terahertz frequency region. Figure 6 shows the EO sensor attached to the optical fiber [5]. Highly sensitive EO materials used at an optical wavelength of 1.55 µm are CdTe and DAST. This EO sensor is also applicable to microwave regions, and is proven to be useful in the specific absorption rate (SAR) measurement at cellular phone frequency (1.5 GHz ~ 2 GHz) [5].

System applications

We have applied the photonic millimeter-wave transmitter to the 120-GHz-band wireless link system to realise a 10-Gbit/s transmission capacity [6]. A photograph of the photonics-based MMW transmitter for a long-distance (>1 km) link using a high-gain Cassegrain antenna is shown in Fig. 7. This wireless link can support the optical network standards of both 10 GbE (10.3 Gbit/s) and OC-192 (9.95 Gbit/s) with a bit error rate of 10^{-12}. We have also been successful in the wireless transmission of 6-channel uncompressed high-definition television (HDTV) signals using the link.

The ultralow-noise characteristics of the photonically generated MMW/THz-wave signal have been verified through their application to the LO for superconducting mixers in receivers used for radio astronomy. Radioastronomical signals from the universe have been successfully observed using a 97.98-GHz photonic LO [7].

A great advantage of photonic LOs in spectroscopic measurement systems is their wide tunability. For this purpose, a wideband receiver has been tested with the same combination of superconducting mixers and a photonic LO at frequencies from 260 to 340 GHz [8].

MMWs/THz waves generated by the optical heterodyning using the OFCG and UTC-PD are successfully applied to the spectroscopy measurement [9, 10].

Conclusions

We described a brief overview of microwave and millimeter-wave photonics systems, and key devices incorporated in the system. The fusion of "wireless" and optical-fiber-based "wired"

telecommunications technologies will continue to steadily advance in a form that will support the need for high speed and ubiquity in communications. Technology for the optical generation and detection of radio waves will become essential for various fields of measurement, as it facilitates the handling of ultra-high-frequency radio waves, which has been difficult with previous technologies.

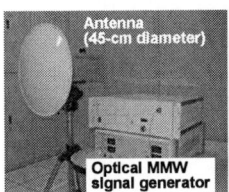

Fig. 7. Photographs of the transmitter front end, and its setup for the field trial.

Acknowledgement

The authors wish to thank Drs. A. Hirata, R. Yamaguchi, H. Takahashi, N. Kukutsu, H. Togo, N. Shimizu, H.-J. Song, T. Kimura, H. Ito, T. Furuta, T. Kosugi, K. Murata, K. Iwatsuki, H. Suzuki and M. Fujiwara for their contribution and support., and Professor K. Okamoto for his valuable discussion and comments.

References

[1] A. Seeds, *IEEE Trans. Microwave Theory and Tech.*, 50, pp. 877-887, 2002.

[2] A. Hirata *et al.*, *IEICE Trans. Electron.*, Vol. E88-C, pp. 1458-1464, 2005.

[3] H. Ito *et al.*, *IEEE J. Lightwave Technology*, Vol. 23, pp. 4016-4021, 2005.

[4] A. Hirata *et al.*, *IEEE Trans. Microwave Theory Tech.*, 49, pp.2157-2162, 2001.

[5] H. Togo *et al.*, *IEICE Trans. Electron.*, Vol. E90-C(2), pp. 436-442, 2007.

[6] A. Hirata *et al.*, *IEEE Trans. Microwave Theory Tech.*, 54, pp.1937-1944, 2006.

[7] S. Takano *et al.*, *Publ. Astron. Soc. Japan*, Vol. 55, pp. L53–L56, 2003.

[8] S. Kohjiro *et al.*, *Tech. Digest of Intern. Workshop on Terahertz Technology*, 18B-6, Osaka, pp. 119-120, 2005.

[9] Ho-Jin Song *et al.*, *to be presented at IEEE/LEOS Summer Topicals 2007*.

[10] N. Shimizu *et al.*, *to be presented at IRMMW/THz 2007*.

Optically Controllable Millimeter-Wave Oscillator Using InP-based HEMTs

Hiroshi Murata[1], Noriyo Kobayashi[1], Toshihiko Kosugi[2], Takatomo Enoki[2], and Yasuyuki Okamura[1]

[1]*Graduate School of Engineering Science, Osaka University*
1-3 Machikaneyama, Toyonaka, Osaka 560-8531 Japan.
Tel: +81-6-6850-6306, Fax:+81-6-6850-6341, E-mail:murata@ee.es.osaka-u.ac.jp

[2]*NTT Photonics Laboratories*
3-1 Morinosatowakamiya, Atsugi, Kanagawa 234-0198 Japan

Abstract— Optical response in InP-based HEMT by focusing a laser beam onto a surface was studied in detail. The changes in Drain current, gain, and capacitance were observed clearly. We propose an optically controllable millimeter-wave oscillator.

Index Terms— Millimeter-wave, HEMT, Opto-Electronic Mixing, Optical Control, Photo detection

I. INTRODUCTION

The High-Electron Mobility Transistor (HEMT) is a promising device for lightwave to microwave transducers in microwave photonic systems because of its excellent microwave characteristics and light absorption characteristics. Many theoretical and experimental studies about opto-electronic mixing in HEMT have been reported [1]-[5]. We have also studied the optical response of the HEMT by irradiating a tightly focused laser beam onto its surface, and have reported on the positive and negative responses according to the irradiating position of the focused laser beam [6], and on the frequency conversion of the photo-detected signals to ~40GHz by opto-electronic mixing [7]. In this report, the changes in Drain current, small signal gain, and Gate capacitance by the focused laser beam are presented. By utilizing these features, a 60GHz oscillator, whose oscillation frequency is controllable by a laser beam, is proposed.

II. EXPERIMENTAL SET-UP

An InP-based InAlAs/InGaAs ultra-fast HEMT of f_t=170GHz and f_{max}=350GHz, developed at NTT photonics laboratories [8], was used in the experiments. Figure 1 shows the experimental set-up for the measurement of the optical response in the HEMT. The HEMT was fixed on a precisely-controllable X-Y mechanical stage and this was installed into a microscope. DC power supplies for Gate and Drain bias voltages and a microwave network analyser/spectrum analyser were connected to the HEMT by use of two precise microwave probes and bias-Tees. An optical beam from a DFB laser of 1550nm wavelength was fed to the microscope by use

of a half mirror. The laser beam incident to the microscope was tightly focused with a spot size of ~10μm by use of an objective lens. It was fed onto the surface of the HEMT. The photograph of the surface of the HEMT with a focused laser beam is shown in Fig. 2. The position of the focused laser beam spot was observed by use of an IR/CCD camera and precisely controlled by use of the X-Y stage.

The position of the focused laser beam spot was scanned near these electrodes and the optical response was measured in detail. As a result, when the beam spot was set to the position between the Drain and Source electrodes, a large positive optical response was observed, and when the beam spot was set to the side-gating position, a negative optical response was observed [6], [7].

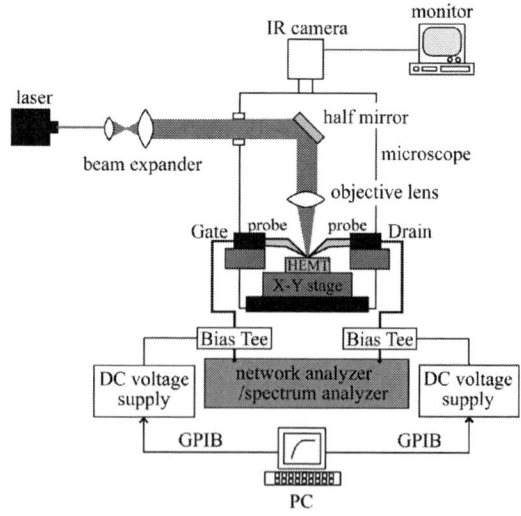

Fig. 1. Experimental set-up for the measurement of the optical response in the HEMT.

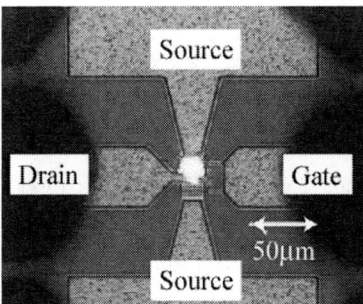

Fig. 2. A photograph of the surface of the HEMT with a focused laser beam.

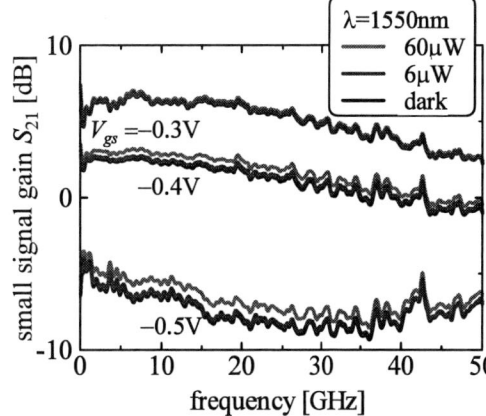

Fig. 4. Change of the small signal gain by focusing the laser beam onto the surface of the HEMT.

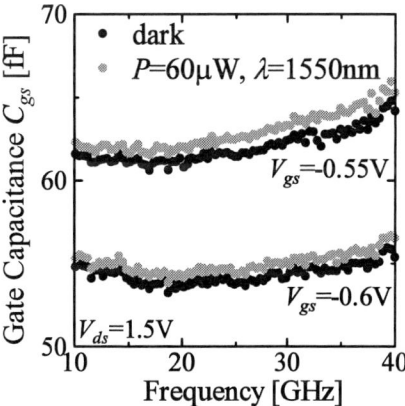

Fig. 5. Change of the Gate capacitance by focusing the laser beam onto the surface of the HEMT.

III. OPTICAL RESPONSE OF THE HEMT

Figure 3 shows the change in the I_{ds}-V_{ds} characteristics caused by focusing the CW laser beam of 1550nm onto the position between the Drain and Source electrodes. By focusing the laser beam, not only the rejection of kinks in I_{ds}-V_{ds} characteristics, but also the increase in the Drain current was obtained, as shown in Fig. 3.

Figure 4 shows the change in the small signal gain of the HEMT by focusing the CW laser beam. An increase in the small signal gain of 2dB was obtained in the frequency range from DC to 50GHz by focusing the 60μW laser beam of 1550nm on the position between the Drain and Source electrodes.

The increase in the Gate capacitance C_{gs} was also observed by focusing the CW laser beam to the HEMT near the Gate electrode. Figure 5 shows the measured C_{gs} from 10GHz to 40GHz using a network analyser. The C_{gs} increased by ~1fF with the illumination of the 60μW laser beam near the Gate electrode. It is believed that it was caused by the decrease of the depletion layer in the HEMT by the injection of photo-carriers generated by the laser beam irradiation.

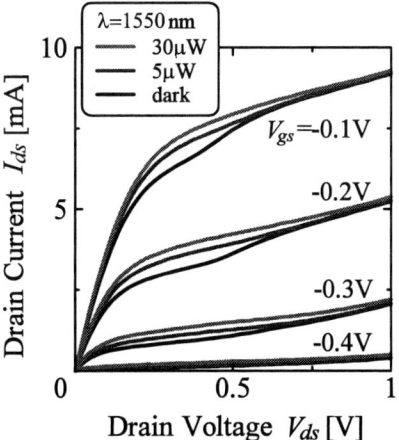

Fig. 3. Change of the Drain current by focusing the laser beam onto the surface of the HEMT.

IV. OPTICALLY CONTROLLED MMW OSCILLATOR

By using the change in C_{gs} caused by the laser beam irradiation, we proposed an optically controlled millimeter-wave (MMW) oscillator. Figure 6 (a) shows an equivalent circuit of the proposed MMW oscillator composed of two HEMTs; HEMT-1 is used as an active transistor with an appropriate Drain bias voltage for generating appropriate gain at a designed frequency range for MMW oscillation. HEMT-2 is used as an optically controlled variable capacitor with a large negative Gate bias voltage. The bias and feedback circuits for HEMT-1 are composed of HEMT-2, Z_1, and Z_2. The oscillation frequency can be controlled by the laser beam illumination to the Gate electrode of HEMT-2, since the increase in the Gate capacitance by laser beam illumination causes a change in the MMW oscillation frequency determined by $f_{osc} = 1/2\pi\sqrt{LC}$.

We designed an MMIC for an optically controlled oscillator operating at 60GHz using the HEMT and coplanar waveguides (CPWs). The detailed MMIC design parameters were determined using MMW circuit design software. Figure 6 (b) shows a photograph of the fabricated MMIC for the optically controlled 60GHz oscillator.

The fabricated MMIC was fixed on the precise X-Y stage and installed into the microscope of the experimental setup in Fig. 1. Figure 7 shows an example of the measured MMW spectrum from the fabricated MMIC. The MMW oscillation with an output power of ~0dBm at 60GHz was clearly observed. By the irradiation of a focused 1550nm laser beam onto HEMT-2 of the MMIC, the MMW oscillation frequency was decreased. Figure 8 shows the change in the MMW oscillation frequency caused by the laser beam irradiation power with the several different bias voltages. An optically induced oscillation frequency shift of ~0.2MHz was observed.

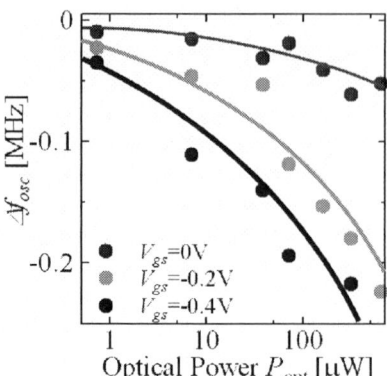

Fig. 8. Shift of the oscillation frequency by the optical power.

V. CONCLUSIONS

By tuning the operation condition of the HEMT used in the 60GHz oscillator, a larger frequency tuning range (~100MHz) is expected. It is also possible to obtain other functions of optically locking and MMW output power control by focusing another laser beam onto HEMT-1. These are attractive for applications in Radio-On-Fiber systems. The proposed oscillator is also applicable for converting an optical ASK signal to a MMW FSK signal.

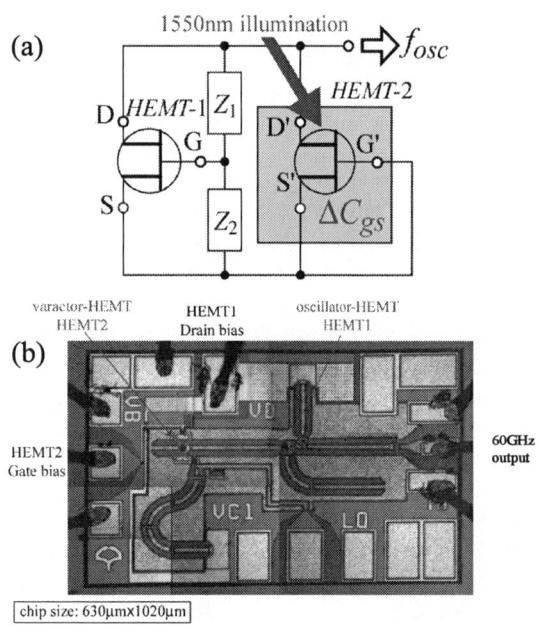

Fig. 6. An equivalent circuit of designed optically controlled 60GHz oscillator (a), and a photograph of the fabricated MMIC (b).

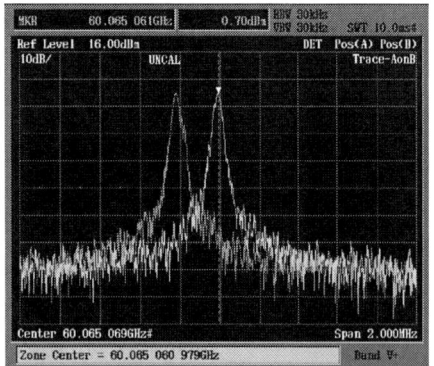

Fig. 7. Measured millimetre-wave spectrum from the MMIC with (blue) and without (yellow) a laser beam irradiation.

REFERENCES

[1] A. J. Seeds, A. A. A. Salles, "Optical control of microwave semiconductor devices," *IEEE Trans. Microwave Theory & Tech.*, **vol.38**, pp.577-585, 1990.

[2] M. A. Romero, M. A. G Martinez, and P. H. Herczfeld, "An analytical model for the photo-detection mechanism in High-Electron Mobility Transistors," *IEEE Trans. Microwave Theory & Tech.*, **vol.44**, pp.2279-2287, December 1996.

[3] Y. Takanashi, K. Takahata, and Y. Muramoto, "Characteristics of InAlAs/InGaAs high-electron mobility transistors under illumination with modulated light," *IEEE Trans. Electron Devices*, **vol.46**, pp.2271-2277, 1999.

[4] M. E. Ali, K. S. Ramesh, H. R. Fetterman, M. Matloubian, G Boll, "Optical mixing with difference frequencies to 552GHz in ultrafast high electron mobility transistors," *IEEE Photon. Technol. Lett.*, **vol.12**, pp.879-881, 2000.

[5] Y. Miyake, and K. Hoshino, "A light-controlled oscillator using InAlAs/InGaAs high-electron mobility transistor," *IEICE Trans. Electron.*, **vol.E84-C**, pp.1356-1360, 2001.

[6] A. Ishikawa, H. Murata, T. Tanaka, H. Shiomi, Y. Okamura, and S. Yamamoto, "Positive and negative optical responses in high-electron mobility transistors and their applications to optically controlled microwave oscillators," *J. J. Appl. Phys.*, **vol.43**, pp.997-1001, 2004.

[7] H. Murata, N. Kobayashi, Y. Okamura, T. Kosugi, and T. Enoki, "Optical mixing in InP-based high-electron mobility transistors by use of a focused laser beam," *Tech. Digest of CLEO2006*, JThC62, 2006.

[8] T. Enoki, H. Ito, K. Ikuta, and Y. Ishii, "0.1-μm InAlAs/InGaAs HEMT's with an InP-recess-etch stopper grown by MOCVD," *Tech. Digest of IPRM'95*, pp.81-84, 1995.

Wavelength-tunable Slow Light of fs Laser Pulse by Quadratic Nonlinear Cascading Process

Wenjie Lu, Yuping Chen, Lihong Miu, Weirui Dang, Feng Lu, Xianfeng Chen and Yuxing Xia

State key lab of advanced optical communication systems and networks, Shanghai Jiaotong University, 200240, Shanghai, China

lu_wenjie@sjtu.edu.cn

Abstract *Wavelength-tunable all optical delay of femtosecond laser pulse demonstrated theoretically through SHG and DFG quadratic nonlinear cascading interactions, in which group velocity of signal pulse can be controlled by pump beam.*

Introduction

In recent years, great interest has been focused on how to control the propagation velocity of light pulses through optical materials. Slow and fast light in communication band has been demonstrated in EDFA [1]. Other ways such as SBS [2] and SRS [3] have also been reported. But all of these methods have some inevitable restraints. Group velocity of ultrashort pulse can also be controlled in the SHG process under phase mismatch [4]. In the cycle of up- and down-conversion between fundamental frequency wave (FF) and second harmonic wave (SH) in PPLT crystal, FF pulse can be dragged by the SH pulse. Thus, FF pulse will decelerate (or accelerate) corresponding to the situation that the SH pulse is slower (or faster) than the FF pulse. In this paper, we simulated the SHG-DFG cascading process of the signal and pump pulse propagation through PPMGLN. Time delay more than one single pulse width is attained.

Theoretical model

Figure 1 shows the experimental scheme. The pump pulse (central wavelength at ω_p) firstly incident into the PPMGLN to generate the second harmonic wave and then the SH pulse (central wavelength at ω_{SH}) difference frequency with the signal pulse (central wavelength at ω_s) along with the idle wave (central wavelength at ω_c). The coupling equation to describe the interaction among the pump, signal, SH and idle wave inside the crystal are given [5,6] below:

$$\frac{\partial A_p}{\partial z} = -\beta_1^p \frac{\partial A_p}{\partial t} - \frac{j}{2}\beta_2^p \frac{\partial^2 A_p}{\partial t^2} - j\omega_p \kappa_{pp} A_p^* A_{SH}$$

$$\times \exp\left(-j\Delta k_p z\right) - \frac{\alpha_p}{2} A_p \qquad (1)$$

$$\frac{\partial A_{SH}}{\partial z} = -\beta_1^{SH} \frac{\partial A_{SH}}{\partial t} - \frac{j}{2}\beta_2^{SH} \frac{\partial^2 A_{SH}}{\partial t^2}$$

$$- j\omega_p \kappa_{pp} A_p A_p \exp\left(j\Delta k_p z\right)$$

$$- 2j\omega_p \kappa_{sc} A_s A_c \exp\left(j\Delta k_c z\right) - \frac{\alpha_{SH}}{2} A_{SH} \qquad (2)$$

$$\frac{\partial A_s}{\partial z} = -\beta_1^s \frac{\partial A_s}{\partial t} - \frac{j}{2}\beta_2^s \frac{\partial^2 A_s}{\partial t^2} - j\omega_s \kappa_{sc} A_c^* A_{SH}$$

$$\times \exp\left(-j\Delta k_c z\right) - \frac{\alpha_s}{2} A_s \qquad (3)$$

$$\frac{\partial A_c}{\partial z} = -\beta_1^c \frac{\partial A_c}{\partial t} - \frac{j}{2}\beta_2^c \frac{\partial^2 A_c}{\partial t^2} - j\omega_c \kappa_{sc} A_s^* A_{SH}$$

$$\times \exp\left(-j\Delta k_c z\right) - \frac{\alpha_c}{2} A_c \qquad (4)$$

$$\Delta k_p = \beta\left(\omega_{SH}\right) - 2\beta\left(\omega_p\right) - \frac{2\pi}{\Lambda}$$

$$= \frac{4\pi}{\lambda_p}\left[n\left(\lambda_{SH}\right) - n\left(\lambda_p\right)\right] - \frac{2\pi}{\Lambda}$$

$$\Delta k_c = \beta\left(\omega_{SH}\right) - \beta\left(\omega_s\right) - \beta\left(\omega_c\right) - \frac{2\pi}{\Lambda}$$

$$= \frac{2\pi}{\lambda_{SH}}n\left(\lambda_{SH}\right) - \frac{2\pi}{\lambda_s}n\left(\lambda_s\right) - \frac{2\pi}{\lambda_c}n\left(\lambda_c\right) - \frac{2\pi}{\Lambda}$$

$$\kappa_{pp} = \frac{\sqrt{2\mu_0/c}\, d_{eff}}{\sqrt{n\left(\lambda_{SH}\right)n\left(\lambda_p\right)^2 A_{eff}}},$$

$$\kappa_{sc} = \frac{\sqrt{2\mu_0/c}\, d_{eff}}{\sqrt{n\left(\lambda_{SH}\right)n\left(\lambda_s\right)n\left(\lambda_s\right)}\, A_{eff}},$$

$$d_{eff} = \frac{2}{\pi}d_{33}$$

978-0-9789217-3-6/07/\$25.00

©2007 WEN GLOBAL SOLUTIONS

Where A_p, A_{SH}, A_s, and A_c, as functions of the time t and position z, represent the complex electric fields of the pump, the SH, signal, and idle waves under the slowly varying envelope approximation. λ_p, λ_{SH}, λ_s, and λ_c are their central wavelengths. β_1^P and β_2^P are the first and second derivatives of the propagation constants with respect to the angular frequency ω, calculated at ω_p for the pump wave; while β_1^{SH} and β_2^{SH} (β_1^s, β_2^s and β_1^c, β_2^c) are defined similarly for the SH (signal and idle) wave and calculated at ω_{SH} (ω_s and ω_c). κ_{pp} (κ_{sc}) is the SH (idle wave) coupling coefficient in the waveguide, deff is the effective nonlinear coefficient, and Aeff is the effective interaction area. α_p, α_s, α_c, α_{SH} are the attenuation coefficients of waveguide for the pump, the SH, signal, and idle waves, respectively. Δk_p (Δk_c) refers to the phase mismatching in the SHG–DFG process, and n is the effective refractive index of waveguide, whose dependence on wavelength can be approximated by the Sellmeier equation[7].

QPM in 5mol% MgO:PPLN

Fig.1 Experiment scheme

Simulation results

The above coupled-mode differential equations were solved using the split-step Fourier method [8]. In this work, we introduced a new time variable $T=t-z\beta_1^s$ measured in the reference frame moving with the input pulse. In practical opinion, we set the temperature at 25℃ in our simulation. In addition, we took $\alpha_p=\alpha_s=\alpha_c=\alpha_{SH}=0$ dB/cm for simplify and L=30 mm with Λ=19.6μm, A_{eff}=45μm^2, d_{eff}=16.5pm/V in the simulations. The central wavelength of the input pump (signal) pulse is 1550nm (1600nm) with the intensity about 50 GW/cm^2 (1GW/cm^2). Both widths of two pulses are 80fs. Our simulation result is presented below:

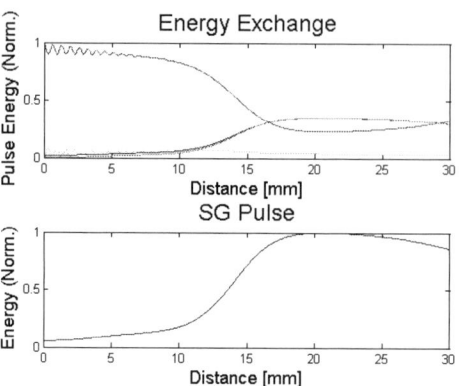

Fig. 2 Normalized energy exchange of the waves propagate in the crystal

Fig 2 shows the energy exchange of the waves interacting inside the crystal. After a short transition period of less than 10mm, the oscillating energy exchange stops, solitons are formed. During the whole process, the SH difference frequency with the signal and exchange the energy with the signal and the idle wave. Thus, after propagate for a long length; the energy of the pump was transferred to the signal and the idle. (The energy of the SH didn't increase a lot because of its role as an interim.)

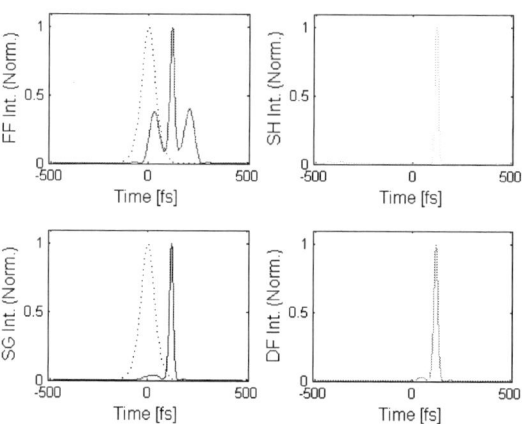

Fig. 3 Normalized pulse intensity of the input and output pulse

In Fig. 3, output signal pulse has been delayed for 120fs (fractional delay of 1.5 is obtained, i.e., over 1 bit) compared with the input. In addition, the energy of the signal increased because of the supplement from the pump. The compression of the pulse width is similar as the soliton compression mentioned by Zeng et al [9].

Conclusions

In summary, over 1 bit of fractional time delay of fs laser pulse has been obained through quadratic nonlinear cascading interactions in our theoretical simulation. Group-delay up to 120 fs is obtained with 80 fs input pulse. More time delay or advancement of signal pulse can be expected by changing the intensity of pump beam, as well as phase matching temperature and signal wavelength.

Acknowledge

This research is supported by the National Natural Science Foundation of China (60407006 and 60477016).

References

1 A. Schweinsberg et al Europhys. Lett. 73(2006), 218-224
2 Yoshitomo Okawachi et al Phys. Rev. Lett. 94 (2005), 153902
3 Jay E. Sharping et al Optics Express, 13(2005), 6092-6098
4 Marco Marangoni et al OPTICS LETTERS, 31 (2006), 534-536
5 F. Baronio et al Optics Express, 14 (2006), 4774-4779
6 C. Balslev Clausen et al Phys. Rev. Lett, 78 (1997), 4749-4752.
7 Y. R. Shen, The Principles of Nonlinear Optics (Wiley, 1984).
8 G. P. Agrawal, Nonlinear Fiber Optics, 2nd ed. (Academic, 1995).
9 X Zeng et al Optics Express, 14(2006), 9358-9370

Characteristics of All-Optical Ultra-Fast Retiming Switches Using Cascaded Second-Order Nonlinear Effect in Periodically Poled Lithium Niobate Waveguides

Yutaka Fukuchi, Joji Maeda

Department of Electrical Engineering, Faculty of Engineering, Tokyo University of Science
1-3 Kagurazaka, Shinjuku-Ku, Tokyo 162-8601, Japan
E-mail: fukuchi@ee.kagu.tus.ac.jp

Abstract *We numerically analyze characteristics of all-optical retiming switches employing the cascaded second-order nonlinear effect in periodically poled lithium niobate waveguides. A time offset between the signal and clock pulses can improve the timing-jitter transfer characteristics.*

Introduction

The cascaded second-order nonlinear effect in periodically poled lithium niobate (PPLN) waveguides has recently realized efficient wavelength conversion and optical phase conjugation in the optical communication band at 1550 nm [1]. In these devices, the quasi-phase matching (QPM) wavelength can be arbitrarily controlled by the modulation period of the second-order nonlinear susceptibility. When the wavelength of a continuous-wave (CW) pump is set around the QPM wavelength, its second harmonic (SH) is first generated, and difference frequency mixing (DFM) between the CW SH and the input signal creates the wavelength-converted or phase-conjugated signal.

Since the cascade of second harmonic generation (SHG) and DFM in the PPLN devices has the same effect as four-wave mixing (FWM) in the third-order nonlinear devices, optical fibers or semiconductor optical amplifiers (SOAs) for example, the cascade can also be applied to all-optical ultra-fast gate switches [2]-[6]. In these devices, the gating pulse, the center wavelength of which is set to the QPM wavelength, switches the gated pulse through the FWM-like process, and the wavelength-converted output pulse is generated. For example, in all-optical demultiplexers for the optical time-division multiplexed (OTDM) systems, received OTDM signal pulses and restored base-clock pulses are used as the gated pulses and the gating pulses, respectively. The PPLN switches feature ultra-fast response, high efficiency, low noise, compactness, integration compatibility, and high stability, only a part of which have been achieved either in fiber-based switches or in SOA-based ones.

In our previous papers, we have both numerically and experimentally investigated the performance of all-optical ultra-fast demultiplexing switches using the PPLN waveguides, and have shown that the available bit rate is limited by crosstalk between neighboring symbols [3],[4]. Because of the large group-velocity mismatch (GVM) between the fundamental and the SH, there exists walk-off between the fundamental pulse and the generated SH pulse. Due to this walk-off, the crosstalk is induced in the bit duration succeeding the demultiplexed channel. For a given device length, the crosstalk increases as the bit rate increases. In other words, the crosstalk limits the bit rate.

In this paper, we numerically study characteristics of the all-optical ultra-fast retiming switch using the PPLN waveguide with consideration for the crosstalk effect. It is found that the switching efficiency and the timing-jitter suppression can be significantly improved when we apply an appropriate time offset between the signal and clock pulses at the input port of the switch.

Principle of all-optical retiming and effect of walk-off compensation

We consider an all-optical retiming switch using a PPLN waveguide, where we use transmitted signal pulses in return-to-zero (RZ) format as gating pulses. The bit rate of the signal is given by $1/T_{bit}$, where T_{bit} is the bit interval. The signal pulses with timing jitter are launched on the PPLN device together with a clean clock pulse train with a fixed repetition rate of $1/T_{bit}$. The clean clock pulses are restored from the signal pulses, and used as gated pulses.

Figure 1 illustrates how retiming is achieved in the PPLN device. In Fig. 1 (a), we consider a train of three incoming symbols, space, mark, and space. When the center wavelength of the input signal pulse is set to the QPM wavelength, the SH of the input signal is first generated (Fig. 1 (b)). Hereafter, we refer to this frequency-doubled signal pulse as the SH signal pulse. Then, DFM between the SH signal pulse and the center clock pulse to be switched generates the wavelength-converted signal pulse (Fig. 1 (c)), which is the output from the switch. The center wavelength of the wavelength-converted signal pulse is different from those of the fundamental signal, SH signal, and clock pulses. Therefore, the wavelength-converted signal pulse can be filtered out by an optical bandpass filter with an appropriate bandwidth. The timing jitter of the wavelength-converted output signal can be suppressed because the input signal switches the clean clock pulse train, which have a fixed repetition rate [5],[6].

Since the PPLN device is operated below saturation regime, the switching efficiency (in this paper, this term will be used as the ratio of the output signal power to the input clock power) is improved as the device length becomes longer. However, in this case, crosstalk induced in the bit duration succeeding the switched bit also increases (Fig. 1 (c)) [3],[4]. The origin of the crosstalk can be explained as in the

978-0-9789217-3-6/07/$25.00

©2007 WEN GLOBAL SOLUTIONS

following. In the case of LN devices using the maximum nonlinear optical tensor element d_{33}, the GVM between the fundamental pulse in the 1550nm-band and its SH is as large as 350 ps/m. Due to the GVM, the SH signal pulse is delayed with respect to the fundamental signal pulse (Fig. 1 (b)). The delayed SH signal pulse then overlaps with the clock pulse that succeeds the switched clock pulse, and produces a certain amount of signal in the succeeding bit duration (Fig. 1 (c)). Hereafter, we will refer to this undesirable signal as a crosstalk pulse. As the device becomes longer, the crosstalk increases and causes the power penalty in the output. Therefore, given an operated bit rate $1/T_{bit}$, the device length is limited by the crosstalk. We define the maximum device length L_{max}, at which the crosstalk reaches a prescribed critical level. The switching efficiency then becomes a function of L_{max} and the input signal power.

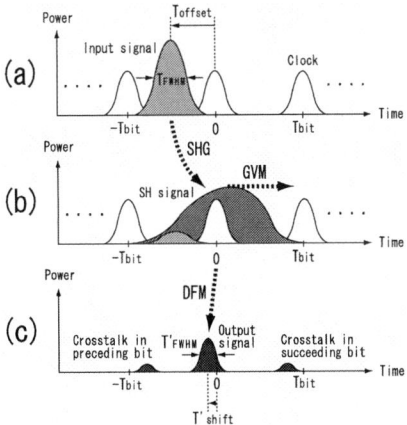

Figure 1. Principle of all-optical retiming in PPLN devices and walk-off compensation by applying initial time-offset.

The switching efficiency can be improved when we compensate for the walk-off delay by applying an appropriate time offset between the signal and clock pulses [3],[4]. At the input port of the switch, the fundamental signal pulse is set to precede the target clock pulse (center) by T_{offset} (Fig. 1 (a)). Therefore, nearby the launching end, the generated SH signal pulse also precedes the target clock pulse. During propagation, the SH signal pulse is delayed due to the GVM, overlaps the target clock pulse, and finally moves past the target (Fig. 1 (b)). As T_{offset} becomes larger, the interaction time between the SH signal pulse and the target clock pulse also becomes longer, resulting in an improvement in the switching efficiency. However, the switching efficiency decreases when T_{offset} becomes too large. Moreover, the crosstalk increases in the bit duration preceding the switched bit, causing the power penalty in the output (Fig. 1 (c)). Therefore, T_{offset} must be optimally adjusted to maximize the switching efficiency while keeping the total crosstalk below a critical level.

Numerical results and discussions

We consider a PPLN waveguide device with a domain inversion period of 16.2 μm. This period is required for SHG using d_{33} when the center

wavelength of the input fundamental signal pulse is 1550 nm. We assume the value of d_{33} to be 25.9 pm/V [4]. The GVM between the fundamental and SH pulses is assumed to be 350 ps/m. The effective cross-section of the waveguide is 8 μm^2. The center wavelength of the input clock pulse is set to 1520 nm. We consider three succeeding pulses as the input clock and a single pulse as the input signal, where, as shown in Fig. 1 (a), the incident time of the signal pulse is shifted from that of the target clock pulse (center) by T_{offset}. All input pulses are assumed to be chirp-free Gaussian pulses having the same width T_{FWHM} in the full width at half-maximum. The bit interval T_{bit} of the clock pulses is assumed to be $3T_{FWHM}$. The bit rate of the signal is given by $1/T_{bit}$. In the following analyses, we calculate evolution of pulse envelopes of the fundamental signal, SH signal, clock, and wavelength-converted signal by using the nonlinear coupled-mode equations under a plane-wave approximation. The analyses consider the sufficient number of frequency components contained in each optical pulse. Details are given in Ref. 4.

We measure the extinction ratio by comparing the peak power of the crosstalk and that of the output signal pulse converted from the target clock pulse. We denote R_p and R_s as the extinction ratio for the crosstalk induced in the preceding bit duration, and that induced in the succeeding bit duration, respectively. Our previous studies on all-optical retiming systems using the PPLN devices have revealed that the total crosstalk power must be at least 30 dB smaller than the output signal power to suppress the power penalty in the output below 0.1 dB [5]. We use the same criteria, and allow the total crosstalk $R_p + R_s$ up to −30 dB. Our calculations reveal that $R_p + R_s$ is a function of T_{offset}, and that the input time offset should be in the range of $0.1T_{FWHM} \leq T_{offset} \leq 1.2T_{FWHM}$. This range of offset is almost independent either of the device parameters or of the powers of the input signal and clock pulses. As qualitatively discussed above, L_{max} is revealed to be a function of $1/T_{bit}$ and an approximate relation

$$\frac{1}{T_{bit}} \times L_{max} = 2.0 \text{Gbps} \cdot \text{m} \tag{1}$$

holds for the given GVM value, regardless of other device parameters or the input pulse powers; for $1/T_{bit}$ = 200 Gbps, L_{max} = 10 mm; for $1/T_{bit}$ = 400 Gbps, L_{max} = 5 mm.

Figure 2 (a) shows P_{out}, the peak power of the output signal pulse as a function of T_{offset}. We vary $1/T_{bit}$, P_c, the peak power of the input clock pulse, and P_{in}, the peak power of the input signal pulse. The maximum device length L_{max} is given by equation (1). We find that the metric $P_{out}/(P_c \times P_{in}^2 \times L_{max}^4)$ is almost independent of $1/T_{bit}$, P_c, or P_{in}. The optimum T_{offset} maximizing $P_{out}/(P_c \times P_{in}^2 \times L_{max}^4)$ is about $0.5T_{FWHM}$. In addition, Fig. 2 (a) also indicates that the fluctuation of the input timing can be converted into the fluctuation of P_{out}. The fluctuation of P_{out} can also be suppressed effectively when T_{offset} is optimally set to $0.5T_{FWHM}$. Figure 2 (b) shows the output time shift T'_{shift} as a function of T_{offset}, where T'_{shift} is defined as the shift of the pulse peak position from that of the original pulse (images drawn in Fig. 1 (c)), and T'_{FWHM}

71

is the full width at half-maximum of the output signal pulse. We find that the timing of the output signal pulses can be pulled near to the clock pulses to prevent the fluctuation of the input timing. The suppression ratio of the timing jitter is about 50 % when $0.1T_{FWHM} \leq T_{offset} \leq 2.4T_{FWHM}$. For example, an all-optical ultra-fast retiming switch operated at $1/T_{bit} = 200$ Gbps can be realized using a 10-mm-long device; optimum T_{offset} is about 0.8 ps, and the switching efficiency P_{out}/P_c is about 1.3 % for $P_{in} = 100$ mW.

Finally, we investigate the impact of the all-optical retiming function of this device on the transmission system. In the analyses, the RZ pulses having normally distributed timing jitter of ΔT_{in} are launched on the PPLN device as the input signal, where the pulses are patterned by the 2^7-1 pseudo-random bit sequence. We set T_{offset} to the optimal value of $0.5T_{FWHM}$. The input powers P_c and P_{in} are 10 mW and 100 mW, respectively. We vary ΔT_{in} and $1/T_{bit}$, with which L_{max} is calculated by equation (1). An eye pattern of the input signal pulses and that of the output signal pulses are shown in Fig. 3 (a) and (b), respectively, where $1/T_{bit} = 200$ Gbps. The timing jitter ΔT_{in} in Fig. 3 (a) is 0.25 ps, whereas in Fig. 3 (b), ΔP_{out}, the standard deviation of P_{out} and the timing jitter ΔT_{out} are calculated to be 4.1 μW and 0.12 ps, respectively. Figures 4 (a) and (b) show ΔP_{out} and ΔT_{out}, respectively as a function of ΔT_{in} for various $1/T_{bit}$ values. These results show effectiveness of the retiming function of the switch, showing good agreement with those estimated from Fig. 2 (a) and (b).

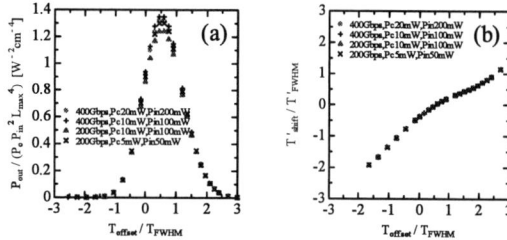

Figure 2. Retiming characteristics for various bit rates $1/(3T_{FWHM})$, where T_{FWHM} is the input pulse width (full width at half-maximum: FWHM). The bit rate - maximum device length L_{max} product is 2.0 Gbps·m. (a): Output signal peak power P_{out} as a function of the input time offset T_{offset}, where P_c and P_{in} are the input clock and signal peak powers, respectively. (b): Output time shift T'_{shift} as a function of T_{offset}, where T'_{FWHM} is the width of the output signal.

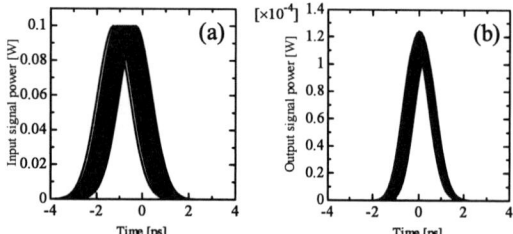

Figure 3. Eye pattern of the input signal at 200 Gbps (a) and that of the output signal (b). The device length and the input clock peak power are 10 mm and 10 mW, respectively.

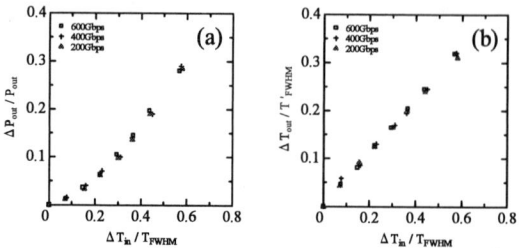

Figure 4. Transfer characteristics of the timing jitter for various bit rates $1/(3T_{FWHM})$, where T_{FWHM} is the input pulse width (FWHM). The bit rate - maximum device length product is 2.0 Gbps·m. The input time offset is $0.5T_{FWHM}$. The input clock and signal peak powers are 10 mW and 100 mW, respectively. (a): Fluctuation of the peak power of the output signal ΔP_{out} as a function of the timing jitter of the input signal ΔT_{in}. (b): Timing jitter of the output signal ΔT_{out} as a function of ΔT_{in}, where T'_{FWHM} is the width of the output signal.

Conclusions

We have numerically calculated the performance of the all-optical ultra-fast retiming switch using the cascade of SHG and DFM in the PPLN waveguide with consideration for the crosstalk between the neighboring symbols. We have shown that both the improvement of the switching efficiency and the suppression of the timing jitter will be available when the time offset between the signal and clock pulses is properly given at the input port of the switch. An all-optical ultra-fast retiming switch operated at the bit rate of 200 Gbps can be realized using a 10-mm-long device; by optimizing the time offset, the switching efficiency of 1.3 % will be available with the peak power of the input signal pulses as small as 100 mW.

References

1 M. H. Chou, I. Brener, M. M. Fejer, E. E. Chaban, and S. B. Christman, "1.5-μm-band wavelength conversion based on cascaded second-order nonlinearity in LiNbO3 waveguides," IEEE Photon. Technol. Lett., vol. 11, pp. 653–655, June 1999.

2 H. Ishizuki, T. Suhara, M. Fujimura, and H. Nishihara, "Wavelength-conversion type picosecond optical switching using a waveguide QPM-SHG/DFG device," Opt. Quantum Electron., vol. 33, pp. 953–961, July 2001.

3 Y. Fukuchi, T. Sakamoto, K. Taira, and K. Kikuchi, "All-optical time-division demultiplexing of 160 Gbit/s signal using cascaded second-order nonlinear effect in quasi-phase matched LiNbO3 waveguide device," Electron. Lett., vol. 39, pp. 789–790, May 2003.

4 Y. Fukuchi, M. Akaike, and J. Maeda, "Characteristics of all-optical ultrafast gate switches using cascade of second-harmonic generation and difference frequency mixing in quasi-phase-matched lithium niobate waveguides," IEEE J. Quantum Electron., vol. 41, pp. 729–734, May 2005.

5 Y. Fukuchi, T. Kawashima, M. Akaike, and J. Maeda, "Characteristics of all-optical ultra-fast retiming switches using cascade of second harmonic generation and difference frequency mixing in periodically poled lithium niobate waveguides," in Proc. Tech. Dig. Optical Amplifiers and Their Applications (OAA'2006), Whistler, British Columbia, Canada, June 25–28, 2006, Paper JWB39.

6 T. Hasegawa, X. Wang, and A. Suzuki, "Retiming of picosecond pulses by a cascaded second-order nonlinear process in quasi-phase-matched LiNbO3 waveguides," Opt. Lett., vol. 29, pp. 2776–2778, December 2004.

Overview of Research Activities at the NSF Center for Biophotonics Science and Technology (CBST)

Yin Yeh, Ph.D., Assoc. Dir., Science and Technology, CBST, Department of Applied Science, University of California, Davis, CA, 95616, USA, *yyeh@ucdavis.edu*

Abstract *We describe research highlights in three CBST theme areas: Advanced Microscopy, Molecular and Cellular Biophotonics, and Medical Biophotonics. Highlights include coherent x-ray diffraction and image reconstruction, engineering of switchable phytochromes, and time-gated CARS imaging modalities.*

Introduction

The Center for Biophotonics Science and Technology is a major research center funded by the US NSF. The participants of this center, CBST, is composed of 8 US universities with UC Davis the lead institution, LLNL, and 14 affiliated US and international university/laboratory partners. The science and technology mission of CBST is to bring photonics tools to the study of life sciences and medicine. Advanced imaging, molecular and cellular biophotonics and medical biophotonics are the three focal areas. Three topics of current research are highlighted in this presentation: Ultrahigh resolution molecular structure determination, development of molecular switchable fluorophores, and time-gated, ratio-metric CARS for cellular imaging.

Ultrahigh Resolution Imaging Using Diffraction Reconstruction

Virtually all the known information about the molecular structure of biomolecules has been obtained by x-ray crystallography. Unfortunately, only 2% of the human proteome structures have been determined because of the extreme difficulty in producing high-quality crystals. An idea is that an x-ray source of sufficient brightness and with a sufficiently short pulse could be used to produce a diffraction pattern using scattered x-rays from a *single molecule* before it is destroyed by the Coulomb explosion following rapid photoionization. The required brightness of ~10^{33} photons / (s · mm^{-2} · mrad^2 · (0.1% bandwidth)), x-ray energy of 5 keV, and pulse width of 10 femtoseconds are expected to be achieved by the Linac Coherent Light Source (LCLS) 4th generation x-ray source funded by DOE and scheduled to be in operation by 2009. Successful implementation of this approach would represent a quantum leap in structural biology and proteomics research. This approach will be used to solve structures of proteins, complexes, and viruses by coherent X-ray diffraction of molecules. We are addressing the key challenges of recording ultrafast single shot coherent diffraction patterns of injected particles with low noise, and developing robust image reconstruction algorithms.

Using the free-electron laser, FLASH, at Deutsches Elektronen-Synchrotron (DESY) in Hamburg, CBST member Henry Chapman and Janos Hajdu of Uppsala University, were able to record a single diffraction pattern of an isolated nanostructured object before the 25 fs laser pulse destroyed the sample. Upon solving the inverse scattering problem, the flash images could resolve features 50 nanometers in size, which is about 5 times smaller than what is achievable with a conventional optical microscope[1], in this case strictly limited by the wavelength of FLASH.

The Linac Coherent Light Source (LCLS) is currently under construction at SLAC, and will start operations in 2009. This hard-X-ray FEL will produce pulses with 0.15 nm wavelength. We will apply the inverse scattering methods developed here to perform spatial imaging at near atomic resolution.

978-0-9789217-3-6/07/$25.00
©2007 WEN GLOBAL SOLUTIONS

Most recently, this team has developed a novel apparatus to record coherent X-ray diffraction patterns from both fixed samples and injected objects "on the fly." We have measured the first ultrafast diffraction from test objects, single injected particles, and biological cells (Fig. 1). Reconstructed images show no evidence of damage, even though the object is ultimately completely vaporized by the pulse. Polarized-laser orientation of particles will be required for the first LCLS experiments. This coming year we plan to demonstrate CBST-developed laser alignment with single-particle diffraction imaging at FLASH.

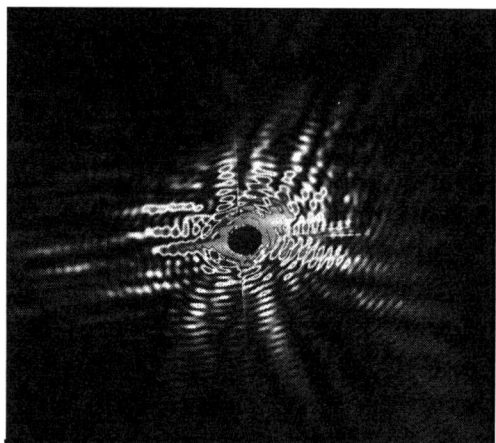

Fig. 1. We recorded single-pulse diffraction patterns of injected particles, including whole cells (as shown), latex spheres, micelles, DNA balls, and sucrose particles. We have recorded over 1000 separate diffraction patterns from injected particles. The mass spectrometer determines which pulse hits a particle.

Phytochrome Engineering

CBST researchers led by J. Clark Lagarias in the Dept of Biochemistry at UC Davis, has been pursuing the fundamental mechanism for the photochemistry and signaling mechanism of phytochromes – which represent a widespread family of protein molecules that regulate plant growth in response to red/far-red light. This research team has discovered that by modifying a single amino acid in the phytochrome protein, they are able to "engineer" the optical characteristics of these fluorescent molecules. They are able to produce proteins that emit light of different wavelengths (colors) including infrared wavelengths and can make the proteins "switchable" (illumination by one color light makes the protein fluorescent and another color makes the protein non-fluorescent). These are important properties that can make these versatile proteins extremely useful in cellular microscopy where one of the most important tools for visualizing cellular structure and function are genetically encoded fluorescent proteins. Coincidentally, very important finding from this work is a much deeper understanding of light dependent altering of plant behavior such as shade avoidance.

Lagarias and his postdoctoral research fellow, Nathan Rockwell, recently published two articles explaining how their structural breakthroughs (made through detailed saturation mutagenesis studies) and biochemical results have defined the P_r (red) state of the molecule and provided new insight into the structure of the P_{fr} (far red) state[2]. In the course of developing methods and protocols to transfect the gene for fluorescent phytochrome peptides in eukaryotic cells, the Lagarias group made a very intriguing discovery by examining the phenotypic expression of fluorescent phytochrome in *Arabidopsis*. While it was confirmed that YH mutants were indeed expressed and could be localized in plants based on their fluorescence, it was also found that YH alleles could rescue plants with a null mutation. Furthermore, these seedlings when grown in the dark, exhibited the morphology of normal plants grown in light. This finding could have a major impact on agriculture, in terms of reducing shade-avoidance in crops and increasing the yield in germinating seeds, thus have important consequences for the world food supply.

Time-gated CARS Imaging of Lipoprotein-Vascular Cell Structure

Atherosclerotic cardiovascular disease is the single largest cause of morbidity and mortality in the U.S. Triglyceride-rich lipoproteins, in particular very low density lipoproteins (VLDL) that facilitate interactions with cells and the uptake of fats by vascular cells are believed to be central to the causing initial stages of atherosclerosis. The mechanisms by which VLDL cause atherosclerosis, however, remain unclear. Our lack of understanding of the mechanisms of triglyceride-induced artery disease severely limits our ability to attack this very important health care problem and develop therapeutic interventions to attenuate and prevent arterial disease. This project aims at providing this critical information through a careful study of VLDL interaction with vascular cells, i.e. endothelial cells and monocytes.

In order to obtain new information about these particles without introducing extrinsic probes, CBST has focused on implementing the CARS (coherent anti-Stokes Raman scattering) microscope. CARS is a molecule-specific optical imaging instrument, based on probing intrinsic molecular vibrations by examining the coherent Raman signatures, and is approximately 5-6 orders-of-magnitude more efficient than spontaneous Raman scattering. Using efficient detectors, the reduced incident power needed permits long-term observations (optionally at video rates) of living cells and their interactions without adverse heating effects. We have completed the CBST CARS system which uses two Stokes pump beams (ω_0) and one Stokes beam (ω_s), thus enabling ratiometric, simultaneous imaging of two The system is capable of slow-scan imaging using the APD of up to 4 channels/signal types, multi-photon excited fluorescence lifetime imaging, point spectroscopy, simultaneous optical trapping and CARS spectroscopy, and CARS correlation spectroscopy. Initial imaging experiments have examined lipoprotein-endothelial cell interactions.

Lipoproteins also lead to lipid-rich deposits in arteries that are densely packed with fatty acids. Recently, we imaged the aliphatic stretch Raman mode of fatty acids (~2845 cm^{-1}) in a healthy rat artery section by time-correlated CARS

microscopy. The high laser peak intensities and tight focusing conditions lead to a strong multiphoton-excited autofluorescence contributions from tissue. Fig. 2a shows photon arrival ime plots for lipid-rich areas as well as autofluorescent tissue. CARS signals are instantaneous, resulting in a short lifetime, while fluorescence leads to longer-lived contributions. By imaging signals that arrived between 0-0.6 ns after the laser pulses, we could isolate most of the CARS signal (Fig. 2b), while setting the time gate to 2-10 ns results in background fluorescence (Fig. 2c)[3].

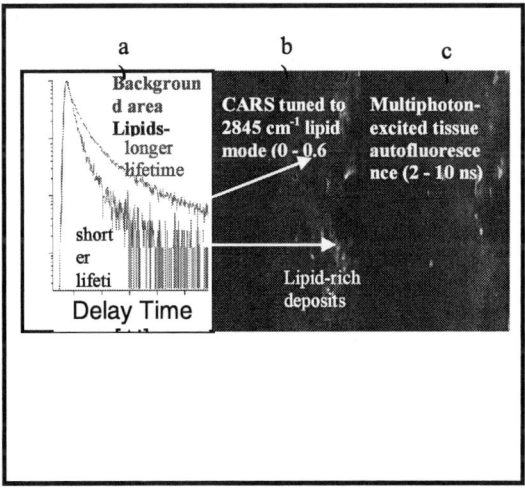

Fig. 2. Photon arrival time histograms for lipid-rich areas (CARS contributions), as well as endogenous fluorescence from a rat artery cross-section. The images indicate time-gated photons arriving within b) 0-0.6 ns and c) 2-10 ns after the laser pulse.

Conclusions

These studies are providing biologists with more tools to examine the functioning sub-cellular structure and dynamics, leading to an overall better understanding of nature.

---

[1] Chapman, H.N. et al. (2006). *Nature Physics* 2, 839-843

[2] Rockwell, N.C. & Lagarias, J.C. (2006). *The Plant Cell* 18, 4-14; Rockwell, N.C., Su, Y.S. & Lagarias, J.C. (2006). *Annual Review of Plant Biology* 57, 837-858

[3] Ly, Sonny, (2007) *A Combined Multiphoton Fluorescence Lifetime and Coherent Anti-Stokes Raman Microscope For Life Sciences Applications At The Single Cell Level*, MS thesis, Univ. of Cal. Davis.

Least-Invasive Harmonic Generation Microscopy for Intravital Imaging

Chi-Kuang Sun

Graduate Institute of Electro-Optical Engineering and Department of Electrical Engineering, National Taiwan University, Taipei,10617, Taiwan, and Research Center for Applied Sciences, Academia Sinica, Taipei 115, Taiwan,

sun@cc.ee.ntu.edu.tw

Abstract *With a virtual-transition characteristic, harmonic generation microscopy provides high-penetration non-invasive intravital optical images with a submicron 3D resolution, ideal for in vivo disease diagnoses and longterm live animal studies.*

Introduction

Optical higher harmonic generations, including second-harmonic-generation (SHG) and third-harmonic-generation (THG) processes are known to leave no energy deposition to the interacted matters due to the virtual-transition characteristic [1]. In contrast to the absorption-induced-fluorescence processes that require energy deposition and electron transitions, the higher harmonic generation processes provide the optical noninvasive nature desired for microscopy applications, especially for long-term observation of the dynamic changes of live samples, including small animals and clinical patients [2,3]. Different from single-photon and multi-photon fluorescence, no cell damage and photobleaching effect is expected from the optical harmonic-generation process due to the fact that there is no real electron-transition involved and the total generated harmonic photon energy has to be equal to the total annihilated photon energy. With a nonlinear nature similar to the multi-photon excited fluorescence, the generated SHG intensity depends on the square of the incident light intensity, while the generated THG intensity depends on the third power of the incident light intensity. These nonlinear dependencies allow localized excitations to enable intrinsic optical sectioning and a high three-dimensional resolution similar or better than that of the two-photon fluorescence microscopy. Combining with a Cr:forsterite laser operating in the biological penetration window [2], harmonic generation microscopy is ideal fir longterm in vivo imaging with high cell viability, high 3D spatial resolution, and high penetration capability [2,3]. In this presentation, we review our recent development of harmonic generation microscopy, which provide high-penetration non-invasive intravital optical microscopic images. Contrast mechanisms of higher harmonic generation microscopy will be discussed [1], with a focus on intravital imaging. Examples of longterm small animal imaging, including noninvasive embryonic studies and virtual biopsy, will be presented. Molecular imaging capability based on nanoparticles and acetic acid will be demonstrated. Fiber-optical endoscopic system development will also be discussed.

System setup

Figure 1(a) shows the setup of our harmonic generation microscope. For the purpose to assist future clinical examinations, we collect the backward propagating signals. The system is based on a home-built femtosecond Cr:forsterite laser centered at 1230 nm with a 110 MHz repetition rate and a 100-fs pulsewidth. The laser output was initially shaped and collimated by a telescope and then coupled into a modified beam scanning system (Olympus Fluoview300) connecting to an upright or an inverted microscope. An IR water-immersion objective was used to focus the laser beam into the test animals with or without fixation. For a 512×512 resolution, the maximum scanning rate of Fluoview300 is 1000 lines per second, corresponding to about two frames per second. The employed objective also acts as the collection lens to collect the backward propagating THG, SHG, and multi-photon excited fluorescence including endogenous or exogenous signals, which are separated from the laser beam with a dichroic mirror. A color filter in the system further ensures filtering out the fundamental laser wavelength. Different fluorescence wavelengths and optical harmonics are further separated with another dichroic mirror and directed into several different PMTs with appropriate interference filters. The average laser power was 100-150 mW corresponding to 0.8-1.3nJ pulse energy illuminated on the test tissue or animal. In theory, the lateral resolution of THG microscopy is 410nm with an 0.9 NA objective.

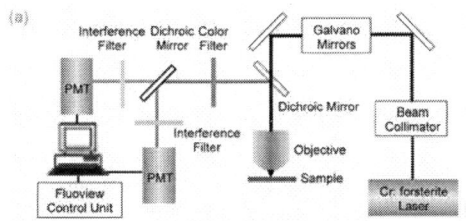

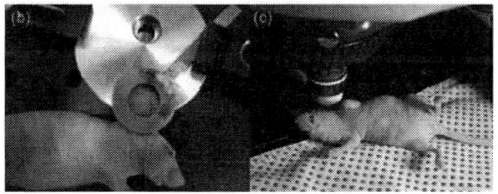

Fig. 1. System setup. (a) Schematic diagram of the harmonic generaiton microscope. (b) Example of fixation of the nude mouse ear. (c) Acquiring intravital images of the test animal under the harmonic generation microscope with a thermal blanket.

Cell viability

Table 1 shows a comparison between different light sources for vertebrate embryo viability. In previous studies, with a 80 MHz Ti:sapphire laser and a NA~1 objective, less than 2 mW average power and 50mJ accumulated energy can be applied to live samples to prevent optical damages [4]. As the excitation laser wavelength is increased to 1047 nm, the maximum average illumination power can only be increased to 13 mW to provide the cell viability [5]. However, to enable the long-term in vivo observation deep inside a thick embryo, stronger illumination power and deeper penetration depth is desired. Recently, we show that by using a laser at 1230 nm, deep penetration (> 1.3 mm) and high cell viability [2] can be provided with a strong

illumination power of 100 mW. It is thus highly desired to apply the 1230 nm light for long-term observation in live animals, including embryonic development *in vivo*.

Table 1

Light Source	CW visible light [5]	NIR fs light [4]	IR fs light [5]	IR fs light [2]
Wavelength (nm)	514nm 532nm 568nm	730nm 760nm 800nm	104 7nm	1230 nm
Allowed accumulated energy	280µJ	50mJ	2J	>2KJ
Highest average power	10-30µW	2mW	13mW	>150 mW

Example Images

Figure 2 shows example in vivo THG microscopic images taken inside live hamster oral cavities based on a femtosecond Cr:forsterite laser with a central wavelength of 1230nm corresponding to the THG resonance of hemoglobin. In both normal (Fig. 2(a)) and cancerous (Fig. 2(b)) oral cavities, not only cell membranes reflecting the basal cell distributions, strong resonance-enhanced THG signals of red blood cells can be observed. Through in vivo observation of the moving erythrocytes, the distribution of capillary can be clearly identified, through continuous imaging. Significantly increased density of capillaries is evident in cancerous oral tissue, while *angiogenesis* is one of early signs of cancer. Our study indicates that through molecular-resonant THG of hemoglobin, in vivo molecular THG microscopy of erythrocytes can be realized without using fluorescence and exogenous contrast agents. Studies in live hamster oral cavity indicate its superiority to image angiogenesis.

Figure 3 shows another example. To demonstrate the utility of HGM microscopy can be used for future clinical cancer diagnosis, we perform HGM on the freshly biopsied human lung cancerous tissue. Fresh lung cancerous tissues from different patients were observed under HGM right after the resection surgery or biopsy and then the tissues were followed with the pathological sections after the observation. Fig. 3 shows an example HGM image of the fresh lung tissue rinsed by 6% acetic acid in isotonic PBS and its corresponding pathological section. Our study shows that THG microscopy provides the detailed cell nuclei information resembling the pathological section while extracellualr matrix can be reflected by the SHG signals. It is important to notice that the use of acetic acid does not interfere with the followed pathological sections. Detailed studies will be reported in the conference with comparison images of healthy and cancerous human specimens.

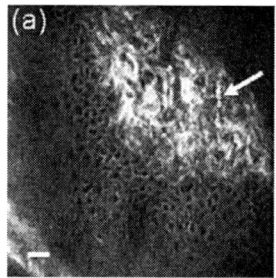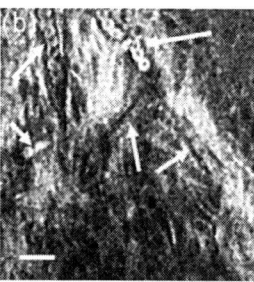

Fig. 2 In vivo THG images inside a live hamster oral cavity. (a) Healthy mucosa. (b) Cancerous mucosa. Image size: 240X240μm.

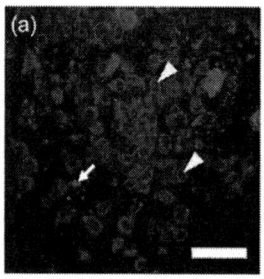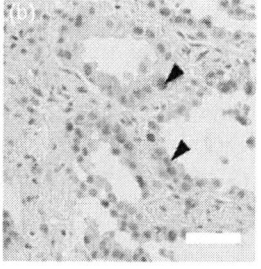

Fig. 3 (a) THG (denoted by red) image of acetic enhanced human lung cancer tissue rinsed in 6% solution of acetic acid, in comparison with (b) the followed pathological section. Lipid bodies (arrow) and cell nuclei (arrowheads) are identified. The image size is 240 μm x 240 μm with a 50 μm scale bar.

Conclusions

Higher harmonic-generation, including second harmonic generation and third harmonic generation, leaves no energy deposition to the interacted matters due to its virtual-level transition characteristic, providing a truly noninvasive modality and is ideal for in vivo imaging of live specimens without any preparation. Second harmonic generation microscopy provides images on stacked membranes and arranged proteins with organized nano-structures due to the bio-photonic crystalline effect or membrane potential imaging. Third harmonic generation microscopy provides general cellular or subcellular interface imaging due to optical inhomogeneity. With proper experimental design, molecular imaging capability can be achieved also with THG. Due to its virtual nature, no saturation or bleaching in the generated signal is expected. With no energy release, contineous viewing without comprosing sample viability can thus be achieved. Combined with its nonlinearity, higher harmonic generation microscopy provides sub-micron three-dimensional sectioning capability and millimeter penetration in live samples without using fluorescence, offering morphological, structural, molecular, and cellular information of biomedical specimens without modifying their natural biological and optical environments.

References

1. C.-K. Sun, "Higher Harmonic Generation Microscopy" in Series in Advances in Biochemical Engineering/Biotechnology, Special Volume 95: Microscopic Techniques, J. Rietdorf Ed., (SPRINGER-VERLAG, Berlin, 2005).
2. C.-K. Sun, et al Journal of Structural Biology, 147 (2004) 19-30
3. S.-P. Tai, et al Optics Express 14, (2006) 6178-6187
4. K. Konig, et al Optics Letters 22 (1997) 135-136
5 J. M. Squirell, et al Nature Biotechnology 17 (1999) 763-767

The Purcell Effect of Silver Nanoshell on the Fluorescence of Nanoparticles

Wallace C.H. Choy (1*), X.W. Chen (1,2), S.L. He (2), P.C. Chui

1 : Department of Electrical and Electronic Engineering, University of Hong Kong, Pokfulam Road, Hong Kong, China. *Corresponding author: chchoy@eee.hku.hk
2 : Centre for Optical and Electromagnetic Research, Zhejiang University; Joint Research Centre of Photonics of the Royal Institute of Technology (Sweden) and Zhejiang University, Zhijingang campus, Hangzhou 310058, China.

Abstract: *The Purcell effect on the spontaneously emission rate and fluorescence efficiency of nanoparticles with and without a silver nanoshell will be investigated which are important for nanoparticle applications in biomedical diagnostics, information storage and optoelectronics*

Introduction

Fluorescent nanomaterials, including organic and metallorganic dye molecules, fluorescent proteins, II-VI and III-V compound semiconductor nanoparticles, polymer/dye-based nanoparticles and silica/dye hybrid particles, have been the subject of intensive research in recent years for their vast applications ranging from biomedical therapeutics and diagnostics to information storage and optoelectronics. A wide range of physical and chemical methods has been developed for the synthesis of nanoparticles and nanoscale core-shell structures with controllable core radius and shell thicknesses [10,11]. For fluorescence based applications, fluorescence efficiency, i.e. the external quantum efficiency (η_{ext}) of the emitter, is an important issue. Due to the existence of pronounced nonradiative decay of excitons in the nanoscale structure, low η_{ext} is an often-observed feature. Most of the strategies so far employed aim to reduce the nonradiative decay rate [12-14] for improving the fluorescence efficiency. The direct and effective approach to improve η_{ext} is to increase radiative decay rate[15] and the out-coupling efficiency since η_{ext} is the product of internal quantum efficiency and the outcoupling efficiency. The enhancement of radiative decay rate results in the Purcell enhancement of the internal quantum yield [16,17]. However, there are few studies on increasing the radiative decay rate and out-coupling efficiency simultaneously in core-shell nanoparticles to enhance the fluorescence efficiency. Here, we will address this issue through a rigorous theoretical study on the spontaneous emission (SE) rate and the out-coupling efficiency of emitters in nanoscale structure. The study can provide a better physical understanding and optimal design of the nano-structure to achieve high-efficiency fluorescence.

Theoretical model

The structure and parameters of a multilayered sphere are displayed on Fig. 1. An emitting shell is sandwiched between two stacks of shells, i.e., P outer shells and Q inner shells. The relative permittivity and the boundary of the ith shell are denoted as ε_i ($i = -Q,...,P$) and r_{i+1} ($i = -Q,..., P-1$), respectively. The emitting medium and the outermost layer is assumed to be non-absorbing at the emitting wavelength. The quantum emitter can be modeled as incoherent classical electric dipoles. The SE of the emitter can be characterized by using the averaged total radiation power F and the power radiated to farfield U (normalized by the total radiation power of an electric dipole in infinite medium) of the dipoles in the spherically multilayered structure. As a consequence of Fermi's golden rule, the radiative decay rate is optical-environment-dependent as the radiative decay rate is optical-environment-dependent as [17,28] $\Gamma_r^s = F \cdot \Gamma_r^0$ where Γ_r^0 and Γ_r^s are the exciton radiative decay rate in free space and spherically multilayered media, respectively. The averaged total radiation power, F, also called as the Purcell factor, can be considered as a normalized SE rate. The change of radiative decay rate will result in the change of the internal quantum efficiency since the radiative and non-radiative recombinations are competing processes. Assuming that the non-radiative decay rate in the spherically multilayered structure is Γ_{nr}, the internal quantum efficiency in the spherically multilayered media η_q^s is modified as

$$\eta_q^s \equiv \frac{\Gamma_r^s}{\Gamma_r^s + \Gamma_{nr}} = \frac{F}{F\eta_q^0 + \left(1 - \eta_q^0\right)} \cdot \eta_q^0 \qquad (1)$$

where η_q^0 is the initial internal quantum efficiency when F=1. η_{ext} is given by

$$\eta_{ext} \equiv \eta_q^s \cdot \frac{U}{F} = \frac{U}{F\eta_q^0 + \left(1 - \eta_q^0\right)} \cdot \eta_q^0 \qquad (2).$$

In the presence of ultra-thin metallic shells, nonlocal effect due to excitations of longitudinal plasmon modes may come into play [22-24]. Then an additional boundary condition is required to characterize this effect [22-24] in determining F and U. We find that the nonlocal effect is prominent for silver shells thinner than 20 nm. The scheme for calculating F here is general regardless of the permittivity of the other shells, while the scheme used in ref. [27] depends on the material of the other shells and becomes increasingly complicated when the number of lossy shells increases.

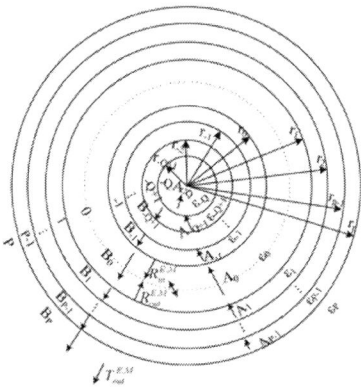

Fig.1 Geometry and parameters of a spherically multilayered structure. $R_{out,l}^{E/M}$, $R_{in,l}^{E/M}$ and $T_{out,l}^{E/M}$ are the total reflection coefficient from the outer shells, the total reflection coefficient from the inner shells and the total transmission coefficient of TE/TM polarization, respectively.

Result and Discussion

The dependences of the fluorescence efficiencies on the structural parameters are investigated for two initial internal quantum yields, i.e. $\eta_q^0 = 0.25$ and $\eta_q^0 = 0.75$. In the calculation, the refractive index of the nanoparticle (NP) is set to be 2.0 over the wavelength of interest. The dielectric function of bulk silver [32] is used in the calculation.

A. SE in a bare NP

SE properties of the emitters in a bare spherical NP with a radius r is studied and discussed in this subsection. Fig. 2(a) plots the wavelength dependence of the normalized SE rate of emitters located at the center of the NP with a radius of 20nm. One sees that the SE rate is nearly one tenth of the SE rate in free space and decreases gradually as the wavelength increases. The size dependence of the normalized SE rate at the wavelength of 500nm is shown in Fig. 2(b) as the solid line with filled squares. The SE rate decreases as the radius of the NP decreases. As shown in Fig. 2(b), due to the low SE rate (F), η_{ext} are much smaller than their initial quantum yields according to Eq. (2) where U equals F in the case of a bare NP. The degradation of η_{ext} for $\eta_q^0 = 0.25$ is more pronounced than that for $\eta_q^0 = 0.75$. This is because the ratio of radiative decay rate to nonradiative decay rate for $\eta_q^0 = 0.25$ drops faster as F decreases. Consequently, a bare NP usually shows inefficient fluorescence.

B. SE in the nanoparticle with a silver nanoshell

The SE rate can be greatly enhanced when the NP is encapsulated with a silver nanoshell due to the resonant excitation of surface plasmons. Unlike the previous case of a bare NP where U equals F, U is smaller than F in the present case since a considerable part of power is absorbed in the silver shell. Thus both F and U should be calculated for characterizing the SE properties. Here the thickness effect of the silver shell on the SE properties is studied for the silver encapsulated NP with various shell thickness s and a fixed core radius r of 30nm as shown in Fig. 3(a). Firstly, the spectrum of the SE rate shows a resonant structure and the SE rate at the resonant wavelength is several orders of magnitude larger than the SE rate in a bare NP due to the resonant excitation of surface plasmons.

The large increment of the SE rate will result in a great enhancement of the internal quantum efficiency due to the Purcell enhancement according to Eq. (1). Secondly, the resonant wavelength of the spectrum of the SE rate shows a blue shift as s increases. The results show that the peak value of the normalized SE rate (hereafter named "the peak SE rate") is reduced as s increases. To understand the reduction of the peak SE rate, we define an effective radius as $R = 2\pi r / \lambda_r$, where λ_r is the resonant wavelength, to characterize the strength of surface plasmon effects. The surface plasmon effect decreases exponentially as R increases. The inset of Fig. 3(a) shows the dependence of the effective radius on the shell thickness. It can be observed that the effective radius firstly increases as s increases. Thus in Fig. 3(a) the peak SE rate is reduced and the resonant spectrum broadens as s increases from 5nm to 30nm. Fig. 3(b) shows the dependence of the fluorescence efficiency at the resonant wavelength on the shell thickness. As compared with Fig. 2(b) for a bare NP, the fluorescence efficiency of the NP encapsulated with a suitable thickness of silver shell is much larger. Moreover, the fluorescence efficiency for the core-shell structure with $r = 30$nm and $s = 25$nm is much larger than the initial internal quantum efficiency. This is due to a combination of the Purcell enhancement of the internal quantum yield and a high out-coupling efficiency. From Fig. 3 (a) and 3(b), one sees that for the NP with ultra-thin silver shell although the SE rate is very large the fluorescence efficiency is still low since the out-coupling efficiency U/F is small, i.e. most of the emitted power is absorbed in the silver shell. As shown in Fig. 3(b), η_{ext} generally increases as s increases from 5nm to 25nm. As s further increases fluorescence decreases gradually since the absorption loss in the Ag shell increases again. Consequently, from Fig. 3(a) and 3(b), one sees that the fluorescence efficiency can be maximized through

80

simultaneously achieving a large Purcell enhancement and high out-coupling efficiency by carefully designing the Ag thickness.

Conclusions

In this paper, we have formulated a general approach for calculating the SE rate and η_{ext} of the emitters in a spherically multilayered structure with arbitrary permittivity. In the presence of silver nano-shell, nonlocal effect has been taken into account. By using the method, the SE rates and fluorescence efficiencies of the emitters in a bare NP and the NP with a silver nanoshell are investigated. The SE rate in a bare NP is only one tenth of the SE rate in free space and consequently the fluorescence is usually inefficient. The SE rate can be enhanced by orders of magnitude through encapsulating the NP with a silver nanoshell due to the resonant excitation of the surface plasmon. The spontaneous emission properties of the emitters in the NP with various thickness of silver nanoshell are studied. Through a proper design of the core-shell layer structure, highly efficient fluorescence can be achieved as a result of the combination of the large Purcell enhancement of quantum yield and a high outcoupling efficiency.

References

1. S.J. Oldenburg, R.D. Averitt, S.L. Westcott and N.J. Halas, "Nanoengineering of optical resonances," Chem. Phys. Lett. 288,

243 (1998).
2. C. Graf, A. van Blaaderen, "Metallodielectric colloidal core-shell particles for photonic applications," Langmuir 18, 524 (2002).
3. M.A. Hines and P. Guyot-Sionest, "Synthesis and characterization of strong luminescing ZnS-capped CdSe nanocrystals," J. Phys. Chem. 100, 468 (1996).
4. S. Haubold, M. Haase, A. Kornowski and H. Weller, "Strongly luminescent InP/ZnS core-shell nanoparticles," Chemphyschem, 2, 331 (2001).
5. N. Gaponik, D.V. Talapin, A.L. Rogach, K. Hoppe, E.V. Shevchenko, A. Kornowski, A. Eychmuller and H. Weller, "Thiol-Capping of CdTe nanocrystals:an alternative to organometallic synthetic routes," J. Phys. Chem. B 106, 7177 (2002).
6. J.R. Lakowicz, "Radiative Decay Engineering: Biophysical and Biomedical Applications," Anal.Biochem. 298, 1 (2001).
7. E. M. Purcell, "Spontaneous emission probabilities at radio frequencies," Phys. Rev., 69, 681 (1946).
8. W. Lukosz, "Theory of optical-environment-dependent spontaneous emission rates for emitters in thin layers,"'Phys. Rev. B 22, 3030 (1980).
9. K. Neyts, "Simulation of light emission from thin-film microcavities," J. Opt. Soc. Am. A 15, 962 (1998).
10. A.R. Melnyk, M.J. Harrison,"Theory of optical excitation of plasmons in metals," Phys. Rev. B 2, 835 (1970).
11. P.T. Leung, "Decay of molecules at spherical surfaces: nonlocal effects," Phys. Rev. B 42 7622 (1990).
12. A. Pack, M. Hietschold and R. Wannemacher, "Failure of local Mie theory: optical spectra of colloidal aggregates," Optics Comm. 194, 277 (2001).
13. A. Moroz, "A recursive transfer-matrix solution for a dipole radiating inside and outside a stratified sphere," Anna. Phys. 315, 352 (2005).
14. E. D, Palik, Handbook of Optical Constants of Solids (Academic press, Boston, 1985)

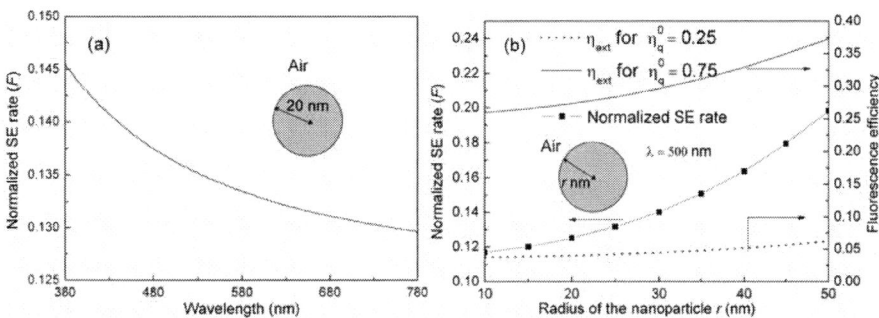

Fig. 2(a) Wavelength dependence of the normalized SE rate for a fixed nanoparticle $r = 20$nm (b) Variance of the normalized SE rate and the fluorescence efficiency for $\eta_a^0 = 0.25$ and $\eta_a^0 = 0.75$ with the increase of the size of the nanoparticle at the wavelength of 500nm.

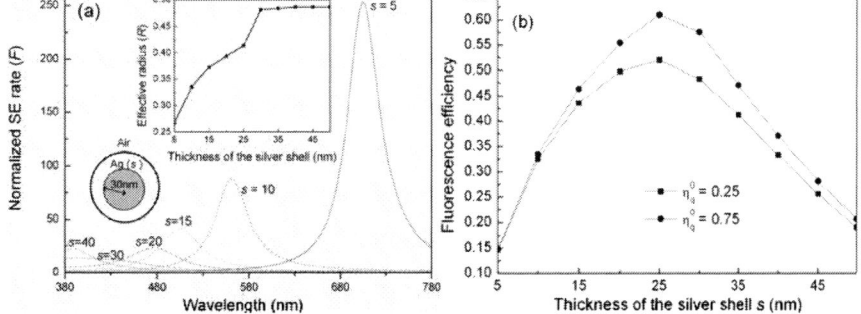

Fig. 3 (a) Wavelength dependence of the normalized SE rate in silver encapsulated nanoparticles of various Ag shell thicknesses s (nm) and fixed core radius; inset shows the variance of the effective radius with the increase of the shell thickness. (b) Variance of the fluorescence efficiency with the increase of Ag shell thickness and fixed core radius.

Nanoparticle-assisted DNA nanosensor

Xin Li[1]*, Jun Qian[1], Lili Chen[1], Ying Zhu[2], Qun Fang[2], and S. He[1]

[1]Centre for Optical and Electromagnetic Research, Zhejiang University, Joint Research Center of Photonics of the Royal Institute of Technology (Sweden) and Zhejiang University, Zijingang Campus, Hangzhou, China

[2]Institute of Microanalytical Systems, Zhejiang University, Hangzhou, China

*Email: lixin@coer.zju.edu.cn

Abstract- **We report a sensitive nanosensor based on a micro-fluidic chip and nanoparticles to detect low concentrations of DNA. The emission of CdSe/ZnS-QDs linked with single strand DNAs was quenched by gold nanoparticles linked with the complementary sequences after hybridization. Sensitively detected signal of DNA was obtained from a 100μm capillary.**

Introduction

Detection of DNA is motivated by their potential applications in diagnosis and genome mutation.[1] Rapid and sensitive DNA biosensor is critical in diagnosing genetic diseases. DNA hybridization and optical technique have become very useful for high specific and sensitive DNA biosensor.[2,3] Recently the fluorescence resonance energy transfer (FRET) based molecular probes such as molecular beacons[4], Scorpion primers[5] and TaqMan probes[6] have been developed to improve the efficiency of DNA biosensor.

FRET provides a powerful ability for probing very small changes in the separation distance between donor and acceptor fluorophores[7] which can sensitively detect DNA hybridization as a molecular binding manner.[8] Zhang *et al.* reported an ultrasensitive DNA nanosensor based on quantum dots (QDs)-Cy5-FRET in a 50μm fluidic micro-capillary.[9,10] They used CdSe/ZnS QDs as a donor and Cy5 as an acceptor to take advantage of the great photophysical properties of QDs. QDs, which have size-tunable photo-luminescence spectra, board absorption spectra, narrow emission wavelength ranges, high photostability and great quantum yields[11,12] have been proved to be a good tool for cellular and *in vivo* bioimaging.[13,14,15] Making QDs as FRET donor can minish the spectral crosstalk and direct acceptor excitation. QDs and nanometer-sized gold (Au) cluster as two novel nano-materials which forms another energy transferring system can build a sensitive quenching nano-sensor. Utilizing DNA molecules as donor-acceptor binding materials can study the distance-dependence

quenching efficiency between QDs and Au particles[16], while utilizing QDs-Au quenching nano-sensor can sensitively detect DNA sequence.[17] Our goal is to exploit QDs-Au quenching DNA nano-sensor in micro-fluidic capillary to simplify and sensitize the detection system. We will also try different QDs/Au-to-DNA ratios to get the optimized quenching ratio and discussed the opportunities of DNA-based nano-composites and building blocks materials.[18,19]

Experimental Protocol

The QDs are CdSe/ZnS core-shell structures (Wuhan Jiayuan QDs Co. Ltd.) which have an emission peak wavelength around 600nm and have been surface-functionalized by –COOH. The average diameter of the QD600-COOH is about 20-30nm. Single stranded DNA (5'-TGA CAT AGA TCA TGC-3') was functionalized with aminolinker C6 at 5' end of the oligonucleotide (Invitrogen Biotechnology Co. Ltd.). To achieve QD-DNA linkage, EDC was added to the carboxyl modified QD600 and the mixture was stirred for several hours. AMN-5'-TGA CAT AGA TCA TGC- 3' DNA were then added into the solution to react with the EDC activated QDs. Finally, the conjugated QD-ssDNA was dissolved at the concentration of 50nM in the water and stored at 4 °C for further use.

Gold nanoparticles were made by a typical method. 200μl 25mM HAuCl$_4$ and 10ml 2mM citrate sodium were mixed in 10ml water. After heated for 10 min, reddish solution was formed, that is the 10nm dia. gold nano-

978-0-9789217-3-6/07/$25.00
©2007 WEN GLOBAL SOLUTIONS

particles. The complementary DNA sequence (5'-GCA TGA TCT ATG TCA-3') was functioned with thiol C6 at 3' end of the oligonucleotide (Invitrogen Biotechnology Co. Ltd.). Citrate-stabilized gold particles can easily be covalently linked to –SH by substitution reactions[18], so Au-DNA (5'-GCA TGA TCT ATG TCA- 3'-THL) was achieved after mixing Au nanoparticles and 3'-THL-DNA for several hours. The conjugated Au-ssDNA was dissolved at a concentration of 50nM in the water and stored at 4 °C for further use.

Hybridization reaction experiments were performed in a buffer solution containing 0.1M Tris, 0.15M NaCl, 0.5% Tween 20, pH 7.5. The reaction was carried out by mixing QD-ssDNA and Au-ssDNA (the ratio of QDs to Au nanoparticles is 1:1), and hybridization buffer at 25 °C for 60 minutes.

Result and Discuss

Before adding hybridization buffer, the emission of QDs in the mixed solution of QD-DNA and Au-DNA was not quenched apparently (data is not showed) just as mixing QDs and Au particles solutions together without DNA conjugation (Fig 1, red plot). While the hybridization reaction was efficiently completed, significant quenching phenomenon appeared sequently (Fig 1, green plot).

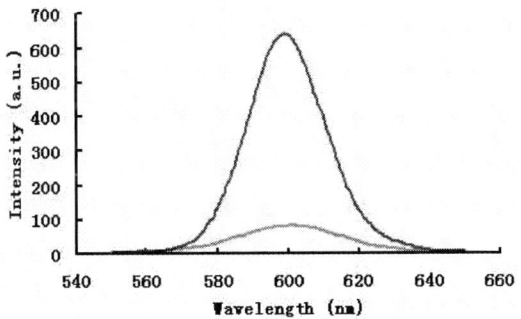

Fig. 1 Emission spectra of QDs and Au nanoparticles solution (in red) and of QD-DNA-Au solution after complete hybridization (in green).

The quenching of emission in Fig 1 is about 85% of the emission of QDs and Au nanoparticles mixed solution. This hybridization experiment was carried out at the ratio 1:1 of QDs/Au to DNA which is studied before[16,17,20]. Commonly one QD molecule can capture dozens of linking molecules[9], to achieve one emitter (or quencher) per ssDNA. Libchaber *et al* utilized 10-fold excess QDs and gold

nanoparticles to ssDNA.[16] Although there were excess QDs/Au left which were not linked to ssDNA, it was hard to say whether one nanoparticle only conjugates with one ssDNA sequence. Obviously the number of linking molecules per nanoparticle makes some influences on the efficiency of QD-Au quenching. More linkers per QD/Au nanoparticle, more opportunities to get QD-DNA-Au conjugations, and the quenching efficiency will be improved. Zhang *et al* showed that the QD-Cy5 FRET efficiency was significantly improved by increasing the ratio (R) of Cy5/QD.[9,10] However, in the QD-DNA-Au case, excess ssDNA molecule results in recovery of the emission by substitution Au-ssDNA from Qd-ssDNA.[17] Therefore, more ssDNA linked on QD/Au nanoparticles' surfaces and no excess ssDNA in the solution can improve the quenching efficiency, otherwise if the excess ssDNA in the hybridization solution is not removed, the quenching efficiency is weakened theoretically. We exploited 4 ratios of QD/Au to ssDNA: 1:1, 1:10, 1:100 and 1:500, also the excess ssDNA molecules were not removed from the hybridization solution. In the case of 1:10 QD/Au to DNA ratio, the solution turned to be transparent after the hybridization, while the solution's initial color was thin red which is the same color of 10nm Au nanoparticles. For other three ratios, the colors turned to be little heliotrope which means Au nanoparticles' aggregation happened. 1:10 is the ratio which is closed to the actual ratio of linked ssDNA per QD/Au nanoparticle, so there would be little ssDNA left in the solution to weaken the QD-Au conjugation. Therefore, heavy aggregation (in Fig 2) of QDs and Au nanoparticles formed at the ratio 1:10 to make

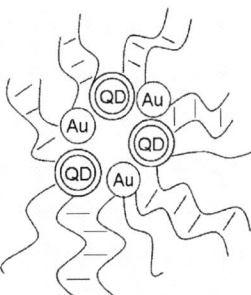

Fig. 2 Schematic diagram showing the heavy aggregation of QDs and Au nanoparticles after DNA hybridization reaction.

the solution transparent. For 1:1, 1:100 and 1:500 ratios, blue-shift colors proved that little aggregation of QDs-Au

83

happened at these ratios and there were still several DNA linkers on one QD/Au even at 1:1, 1:100, 1:500 ratios. This aggregation provides large opportunities of DNA-based nano-composite and nano-networks for new structures of hybrid materials research.[18]

To improve the detection sensitivity and simplify the detection devices, a silica micro-fluidic capillary was utilized as this DNA nanosensor channel. Micro-fluidic chip is convenient to design as a sensitive biosensor which can efficiently prevent photobleaching and has the ability of single molecule dectection.[9] Dynamic process of the hybridization still can be probed in the micro-fluidic chip by utilizing small quantity of samples. A 100μm- internal diameter capillary was designed for this experiment. The sample passed a laser-focused detection volume through hydraulic pressure-driven flow. Sensitive confocal microscopy images of QDs-DNA and Au-DNA solution before and after hybridization were taken in the capillary (Fig 3). The excitation light source is a 473nm semi-conductor laser.

Fig. 3 Confocal microscopy image of (left) QDs-DNA and Au-DNA solution before the hybridization and (right) after complete hybridization. Green spots just means the lighteness but not the wavelength which QD600 emits. The signals were collected through a ×40/0.65 NA objective lens.

Conclusion

QDs-Au quenching DNA nano-sensor has been exploited in micro-fluidic capillary by utilizing confocal scanned microscopy, and the detection sensitivity has been improved in the micro-fluidic chip. Also we found the hybridization of QDs-ssDNA and Au-ssDNA can cause nanoparticles' aggregation which improves the efficiency of the quenching biosensor. Utilizing different ratios of nanoparticles (QDs or gold nanoparticles) to DNA we can build up DNA-associated nano-composites and nano-networks for new structures of hybrid materials.

Acknowledgements

This work was supported by a multidisciplinary project of Zhejiang University and a quantum-dot collaboration project of Swedish Strategic Research (SSF).

References

1. A. H. Uddin, P. A. E. Pinunno, R. H. E. Hudson, M. J. Damha, and U. J. Krull, *Nucleic Acids Res.* 25, 4139 (1997)

2. Y. Okahata, M. Kawase, K. Niikura, F. Ohtake, H. Furusawa, and Y. Ebara, *Anal. Chem.* 70, 1288 (1998)

3. Y. F. Li, W. Q. Shu, P. Feng, C. Z. Huang, and M. Li, *Anal. Sci.* 17,583 (2001)

4. S. Tyagi & F. R. Kramer, *Nature Biotechenol.* 14, 303-308 (1996)

5. S. A. Lange et al., *Anal. Chem.* 76 1641 (2004)

6. S. W. Howell et al., *Langmuir* 19 436 (2003)

7. Aaron R. Clapp, Igor L. Medintz, Brent R. Fiosher, George P. Anderson, and Hedi Mattoussi, *J. Am. Chem. Soc.* 127, 1242-1250 (2005)

8. K. E. Sapsford, L. Berti, and I. L. Medintz, *Angew. Chem, Int. Ed.* 45, 4562-4588 (2006)

9. C. Y. Zhang, H. C. Yeh, M. T. Kuroki and T. H Wang, *Nat. Mater.* 4 826-831 (2005)

10. C. Y. Zhang, L. W. Johnson, *Anal. Chem.* 78 5532 (2006)

11. M. Bruchez, M. Moronne, P. Gin, S. Weiss, A. P. Alivisatos, *Science* 281, 2013-2016 (1998)

12. W. C. W. Chan, S. M. Nie, *Science* 281 2016-2018 (1998)

13. X. Y. Wu, H. J. Liu, J. Q. Liu, K. N. Haley, J. A. Treadway, J. P. Larson, N. Ge, F. Peale, and M. P. Bruchez, *Nat. Biotechnol.* 21 41-46 (2003)

14. J. K. Jaiswal, H. Mattoussi, J. M. Mauro, and S. M. Simon, *Nat. Biotechnol.* 21 47-51 (2003)

15. Jun Qian, K.-T Yong, I. Roy, T. Y. Ohulchanskyy, E. J. Bergey, H. H. Lee, K. M. Tramposch, S. He, A. Maitra, P. N. Prasad, *J. Phys. Chem. B.* 111 6969-6972 (2007)

16. Zoher Gueroui and Albert Libchaber, *Phy. Rev. Lett.* 15 93 (2004)

17. L. Dyadyusha, H. Yin, S. Jaiswal, T. Brown, J. J. Baumberg, F. P. Booy and T. Melvin, *Chem. Commun.* 25 3201 (2005)

18. G. P. Mitchell, C. A. Mirkin, and R. L. Letsinger, *J. Am. Chem. Soc.* 121 8122-8123 (1999)

19. C. C. You, A. Verma and V. M. Rotello, *Soft Matter* 2 190-204 (2006)

20. R. Wargnier, A. V. Baranov, V. G. Maslov, V. Stsiapura, M. Artemyev, M. Pluot, A. Sukhanova, and I. Nabiev, *Nano Lett.* 4 451-457 (2004)

DNA hybridisation biosensor based on dual-peak long-period grating

Xianfeng Chen[1], Kaiming Zhou[1], Marcus Hughes[2],
Edward Davies[1], Lin Zhang[1], Anna Hine[2], Kate Sugden[1], Ian Bennion[1]
[1] School of Engineering and Applied Science, Aston University, Birmingham, B4 7ET, UK
[2] School of Life and Health Sciences, Aston University, Birmingham, B4 7ET, UK
chenx2@aston.ac.uk

Abstract Using an optical biosensor based on dual-peak long-period fibre grating, we demonstrate the detection of interactions between DNA biomolecules in real-time, showing a high sensitivity and reusability function.

Introduction

During the last decade, the development of optical biosensor has become increasingly important for applications in biochemical, biomedical, and environmental areas. Various types of biosensors have been presented by using microarray, microchips, optical ring-resonator, plane waveguide and optical fibre [1-3]. Specially, biosensors based on in-fibre grating have attracted considerable attention [4,5]. However, some of these demonstrated optical biosensors have limitations for real-time monitoring. Here, we report an implementation of an optical biosensor using dual-peak long-period fibre grating (LPFG) for detecting the hybridisation of DNA with the advantages of high sensitivity, real-time monitoring and reusability.

Characteristics of dual-peak LPFG

In order to achieve high sensitivity detection of the designed DNA interaction, the most sensitive LPFG structures should be used. It has been reported that the coupling condition of LPFGs with relatively short periods is close to the dispersion turning points. Phase match condition close to these points results in coupling to conjugate dual-peak cladding modes which are extremely sensitivity to the external perturbations [6].
Several LPFGs with relatively small periods

were UV-inscribed in hydrogenated SMF-28 fibre employing the point-by-point method.
Fig.1 depicts the spectral evolution of a dual-peak LPFG with a period of 161μm under UV-inscription. With increasing UV exposure, the dual-peak resonances were increasing in strength and moving close to each other as the coupling approaching the dispersion turning point.

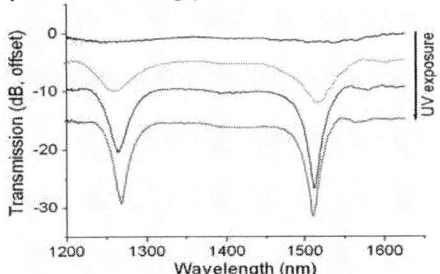

Fig.1 Spectral evolution of a dual-peak LPFG with a period of 161μm under increasing UV exposure (arrow direction).

Scheme for biosensor formation

The ability of LPFGs to couple light from fibre core mode to cladding modes allows optically detecting the change in refractive index at the grating surface, which provides an optical detection method to monitor biomolecular interactions. Fig. 2 displays the procedure of biosensor based on the silanisation of LPFG, covalent activation, DNA immobilisation and hybridisation. The hybridisation process will modify refractive

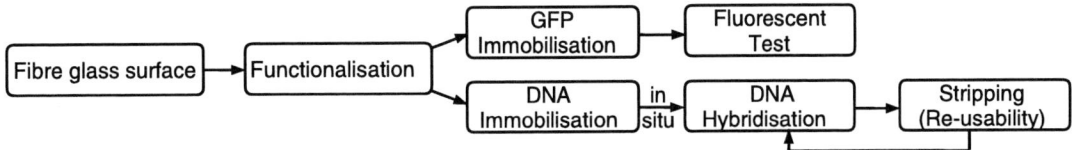

Fig. 2   Basic scheme of the functionalisation of glass surface for the generation of biosensors.

978-0-9789217-3-6/07/$25.00

©2007 WEN GLOBAL SOLUTIONS

index of the LPFG surface, resulting in its spectral shift. By demodulating the shift, the DNA hybridisation can be monitored in-situ with high sensitivity. All the biochemical experiments were performed in a fume cupboard. To minimise the bend cross-sensitivity, the LPFG sensors were placed straight in a V-groove container on a Teflon plate and all the chemicals and solvents were added and withdrawn from the container by carefully pipetting.

Surface silanisation and activation

Prior to silanisation, LPFGs were cleaned by immersion in 5M hydrochloric acid (HCl) for 30min at room temperature followed by rinsing in deionized (DI) water three times and drying in the air. Silanisation of glass surface was implemented by immersion in fresh 10% 3-Aminopropyl-triethoxysilane (APTS) for 30min at room temperature.

Fibre — linker — NH — C(=NH) — (CH$_2$)$_6$ — C(=NH) — OCH$_3$

Fig.3 Activation of the silanised surface using DMS.

To immobilise biomolecules covalently to the glass surface, a chemical bond has to be formed between a functional group of biomolecule and the amino-group of the linker [2]. As it well known in bioconjugate chemistry, Dimethyl suberimidate (DMS, the molecular structure shown in Fig.3) is water soluble, membrane permeable and is one of the best crosslinking agents to convert the amino-groups into reactive imidoester cross-linkers [7,8]. In addition, DMS does not alter the overall charge of the protein, potentially retaining the native conformation and activity of the protein. For activation of glass surface, the silanised LPFGs were immersed in 25mM DMS in phosphate buffered saline solution (PBS) for 35min at room temperature. Then the activated LPFGs were rinsed by DI water three times and dried in the air.

GFP immobilisation & fluorescent test

In order to provide a simple method to determine whether biomolecules are able to be successfully immobilised on the fibre glass surface, Green Fluorescent Protein (GFP), which is an intrinsically fluorescent protein that has been used extensively as a tool in biology to enable imaging, was employed to detect the attachment of protein onto the fibre surface. A DMS activated fibre, as described above, was incubated in 1mg/ml GFP in PBS for 16hrs at room temperature. The GFP-deposited fibre surface was observed under optical microscope with UV light source using appropriate filters for GFP fluorescence detection and the image was captured and shown in Fig.4a, exhibiting successful protein immobilisation. For comparison, an untreated fibre was also observed under microscope, no fluorescence was observed (data not shown).

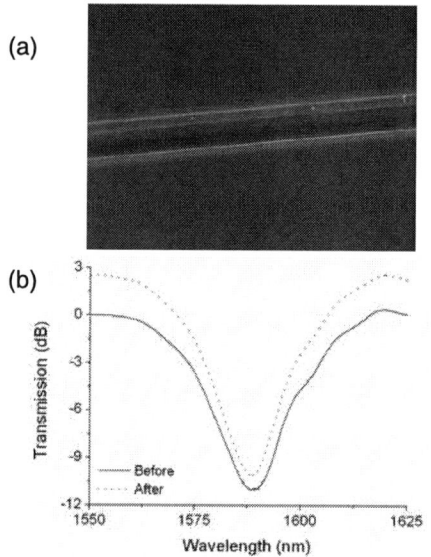

Fig.4 (a) The image of GFP fluorescence on the fibre surface; (b) Spectra of LPFG sensor before & after probe DNA immobilisation.

Immobilisation of probe DNA:

The immobilisation process was carried out by incubation of an activated LPFG in 1µM probe DNA (as shown in Table 1) in PBS for

Table 1. Sequences and modifications of the Probe and Target Oligonucleotides

Oligonucleotide	5′ end modification	Sequence	3′ end modification
Probe	none	GCA CAG TCA GTC GCC	NH$_2$
Target	none	GGC GAC TGA CTG TGC	none

86

16hrs at room temperature. The spectra of LPFG as shown in Fig. 4b were measured at the beginning & end of the immobilisation process, respectively, by an optical spectrum analyser. The grating wavelength was defined by the centroid calculation method. After 16hrs deposition, a blue-shift in wavelength of 254pm was observed, showing the fibre surface has been modified successfully. Also, the intensity of attenuation resonance increased 1.3dB, which could be due to the fibre surface roughness caused by immobilised probe DNA resulting in more evanescent wave loss.

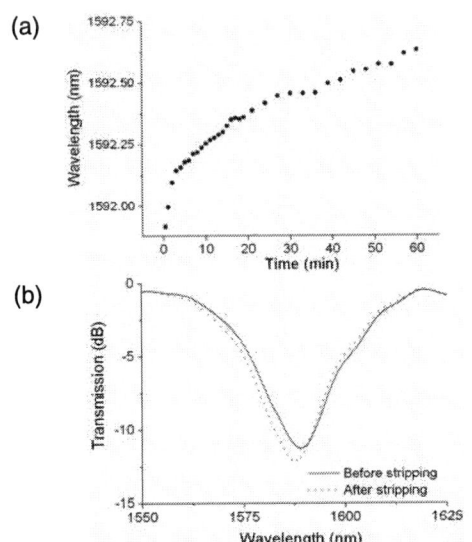

Fig.5 (a) Wavelength evolution of grating sensor against time during hybridisation of target DNA; (b) Resonance shifting by the stripping procedure.

Hybridisation of target DNA

Hybridisation was executed with target DNA. After cleaning with DI water, the grating sensor was rinsed in 6xSSPE (0.9M NaCl, 0.06M NaH_2PO_4, and 0.006M EDTA) then immersed in fresh 1µM target DNA in 6xSSPE buffer for 60min at room temperature. The grating wavelength shift, as shown in Fig.5a, was monitored in situ through whole hybridisation process. An increase of 715pm was observed in wavelength from the start of hybridisation process until the end and most of the change takes place in the first 20min showing that hybridisation takes place very quickly. Hybridisation of target DNA has been monitored successfully in real-time by this grating sensor.

Stripping procedure and reusability

For re-use, grating sensor was incubated in a freshly prepared stripping buffer of 5mM Na_2HPO_4 and 0.1%(w/v) Sodium dodecyl sulfate at 95°C for 30s [2], three times, then was washed with DI water and dried for the re-hybridisation. The spectra in Fig.5b were measured in DI water before and after the stripping procedure. A blue-shift of 1257pm has been observed, which is caused by the stripping procedure. After stripping, the sensor was re-hybridised by immersion in 2µM target DNA in 6xSSPE buffer for 60min. As shown in Fig.6, an 1165pm increase has been observed, demonstrating the re-usability of the biosensor and a significantly higher sensitivity than the reported sensor based on core-etched FBG [5].

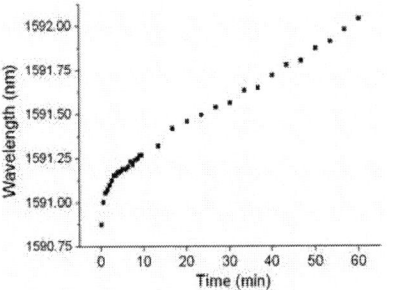

Fig.6 Wavelength shift during re-hybridisation.

Conclusion

A novel optical biosensor based on LPFG has been demonstrated and used for detection of DNA hybridisation succefully. The hybridisation reactions were monitored in situ by monitoring the grating wavelength shift. An additional advantage of this biosensor is that it offers an opportunity to reuse.

1. H. R. Luckarift et al, Nature Biotechnol., 22, 211-213 (2004).
2. M. Beier et al, Nucleic Acids Res., 27, 1970-1977 (1999).
3. A. Ksendzov et al, Opt. Lett., 30, 3344-3346 (2005)
4. M. P. DeLisa et al, Anal. Chem., 72, 2895-2900 (2000).
5. A. N. Chryssis et al, IEEE J. Sel. Top Quantum Electron. 11, 864-872 (2005).
6. X. Shu et al, J. Lightwave Technol., 20, 255-266 (2002).
7. E. S. Hand et al, J. Am. Chem. Soc., 84, 3505-3514 (1962).
8. G. Mattson et al, Mol. Biol. Rep., 17, 167-183 (1993).

Wettability Patterning Technology for Organic Displays

Yu-Jin Na, Sung-Jin Kim, Sin-Doo Lee

School of Electrical Engineering, Seoul National University

Kwanak P. O. Box 34, Seoul 151-600, Korea

e-mail: sidlee@plaza.snu.ac.kr

Abstract *We developed a simple and versatile patterning method for advanced organic displays with high resolution and uniformity in large area. Our patterning method involves the optical formation of selective wetting regions, where a regular array of functional polymer patterns is obtained by a standard spin-casting process. It is demonstrated that polymeric semiconductors and light-emitting polymers can be patterned to produce periodic arrays with feature resolution down to sub-micrometers using the selective wettability. This wettability-based patterning technique of soluble functional polymers would be directly applicable for organic displays and plastic electronics.*

1. Introduction

The potential of organic functional materials is a low-cost alternative to inorganic materials like silicon for electronic and optoelectronic applications such as flexible displays, electronic papers, and smart cards. In particular, organic light emitting diodes (OLEDs) have attracted great attention due to bright, highly-efficient, and thin characteristics [1,2]. Moreover, substantial progress has been currently made in the realization of an array of organic thin-film transistors (OTFTs), which provides an essential backplane for flexible displays in the roll-to-roll process [3].

Although several material parameters of organic semiconductors such as the carrier mobility and the operation voltage meet the requirements for electronic and display applications over large area [4], some key technologies including a simple and reliable patterning method applicable for the roll-to-roll process have not been well established yet. For polymeric materials, a solution-casting approach is more promising for the reduction in cost and process time than the vacuum deposition method which is usually employed for of low-molecular materials. It is also noted that recent methods of ink-jet printing [5] and micro-contacting [6] suffer from the limited patterning capability, such as limited resolution and poor uniformity, so that they need to be more refined for the use in a high resolution format over large area. A direct patterning of the OTFTs was demonstrated by spin-casting a conductive water-soluble polyaniline onto a substrate with hydrophilic and hydrophobic regions [7]. The hydrophobic patterns were generated through a micro-contacting process which involves inevitably multi-level errors over large area due to the deformation of an elastomeric stamp.

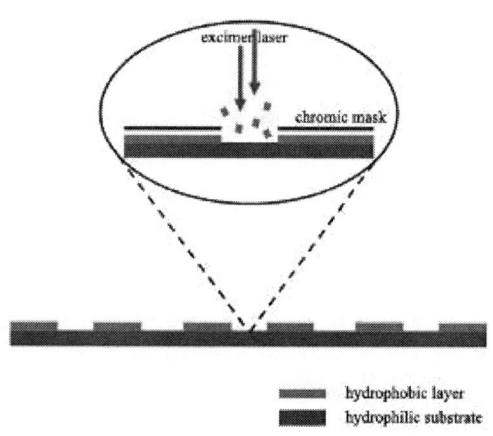

Fig. 1: Schematic diagram of forming selective wetting regions by laser ablation.

Therefore, it is extremely important to develop a reliable and versatile patterning method of soluble functional polymers for organic displays and plastic electronics.

2. Wettability Control by Laser Ablation

We developed an optical method of forming 2-dimensional selective wetting regions on

978-0-9789217-3-6/07/$25.00

©2007 WEN GLOBAL SOLUTIONS

a substrate to produce an array of functional polymer patterns from a solution by a standard spin-casting as shown in Fig. 1.

A laser ablation [8] process through a photomask was employed on a hydrophobic layer to produce an array of selective wetting regions. Note that the laser ablation process is suitable for defining high resolution patterns over large area.

For the hydrophobic layer, we examine two kinds of hydrophobic fluoropolymers, the EGC-1700 (3M) and the CYTOP (Asahi Glass), each of which can be spin-coated or dip-coated from a solution onto a glass substrate. The EGC-1700 directly absorbs ultraviolet light of an excimer laser (Kr-F 248 nm). Under the UV irradiation, a direct ablation of the EGC-1700 occurs. Whereas, the CYTOP needs a buffer layer (TDUR-P015, Tokyo Ohka Kogyo), being able to absorb the UV light, under it for laser ablation.

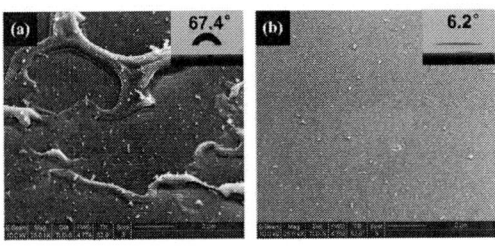

Fig. 2: SEM images of two ablated surfaces of the CYTOP on the buffer layer under the UV intensity of (a) 150 mW/cm^2 and (b) 450 mW/cm^2. The inserts show the contact angles on the two surfaces.

Figure 2 shows the images of two ablated surfaces of the CYTOP on the buffer layer observed by a scanning electron microscope (SEM). The inserts present contact angles on the two surfaces. It is clear that the CYTOP surface remains hydrophobic (the contact angle of about 98° to 67.4°) at a low UV intensity (150 mW/cm^2) due to the remaining CYTOP and the TDUR-P015, and becomes hydrophilic (the contact angle of 6.2°) as the UV intensity increases. In the high intensity range, the surface is hydrophilic since the CYTOP and the photo-thermally generated TDUR-P015 are completely removed. This means that the direct ablation of the

hydrophobic layer through a predefined photomask is a simple and powerful way of forming selective wetting regions. This process provides the feature resolution down to 2 μm, limited only by the photomask used.

3. Patterns of Functional Polymers

First, we show an array of perodic patterns of a soluble semiconducting polymer, poly(9-9-dioctyfluorene-co-bithiophene (F8T2), produced on a wettability patterned substrate having the CYTOP on a buffer layer of the TDUR-P015 by spin-casting. The F8T2 was dissolved in xylene which has an anisotropic wetting behavior and spin-coated on the substrate. As clearly seen in Fig. 3, the polymer solution of the F8T2 (1 wt %) was completely wet only in the hydrophilic regions of the substrate and formed well-defined patterns in a 2-dimensional array. Such patterns of the F8T2 can be used for fabricating the OTFTs in flexible displays and plastic electronics.

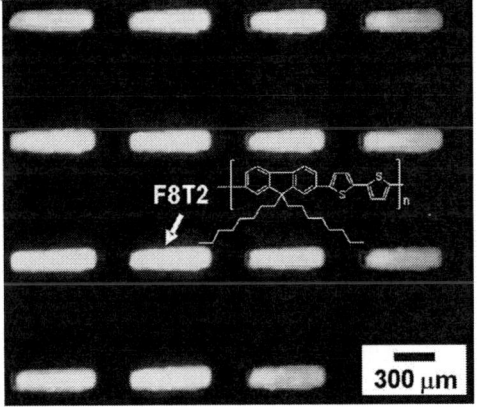

Fig. 3: Fluorescence micrograph of the well-defined patterns of the F8T2 produced on the wettability patterned substrate by spin-coating.

In a similar way, a light-emitting polymer (LEP), commercially available from Dow Corning, was used for producing an array of periodic patterns on a wettability patterned substrate by dip-coating. The patterned substrate through a direct laser ablation process was dipped into the LEP solution (2 wt %) and pulled out at the velocity of 300 µm/sec. From the photoluminescence micrograph shown in Fig. 4, it is clear that red patterns of the LEP were well-defined on the wettability patterned substrate. Each red pattern of the LEP, serving as a sub-pixel of the OLED, was about 60 nm. It is interesting to note that the local volume of the solution covered in the wetting regions, which depends critically on the coating velocity and the solution concentration, governs the height profile of each pattern.

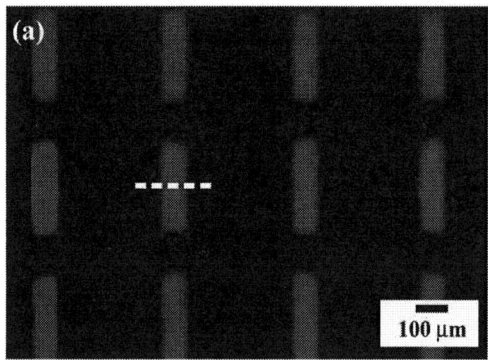

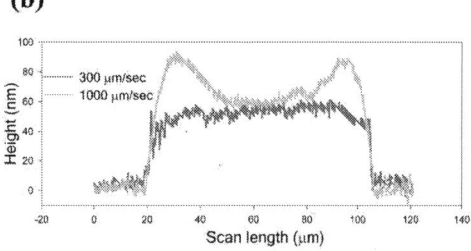

Fig. 4: (a) Photoluminescence micrograph of the LEP patterns produced on the wettability patterned substrate by dip-coating and (b) the height profiles of each pattern at two dipping velocities, measured along the dotted line in (a).

4. Concluding Remarks

We demonstrated a wettability patterning technology for fabricating an array of soluble functional polymers for the use in organic displays and plastic electronics. Two typical examples of semiconducting polymer patterns for the OTFTs and the OLED were shown over large area. Beyond the electronic and optoelectronic applications, the patterning technique presented here will be directly applicable for fabricating bio-assay chips.

Acknowledgements

This work was supported in part by Samsung SDI-Seoul National University Display Innovation Program and Inter-University Semiconductor Research Center at Seoul Nation University.

References

1. T.-W. Lee, et al., PNAS, **101**(2004), 429.
2. K. M. Vaeth, OLED-display technology. Inform. Display,**19** (2003), 12.
3. R. D. McCullough, Adv. Mater., **10** (1998), 93.
4. S. R. Forrest, Nature, **428** (2004), 911.
5. J. Bharathan, et al., Appl. Phys. Lett., **72** (1998), 2660.
6. J.-H. Choi, et al., Nanotechnology, **17** (2006), 2246.
7. K. S. Lee, et al., Appl. Phys. Lett., **86** (2005), 074102.
8. S.-J. Kim, et al., Jpn. J. Appl. Phys., **35** (2005), L1109.

Synthesis of High Birefringence Liquid Crystals for Display Application

Chain-Shu Hsu, Yung-Ming Liao

Department of Applied Chemistry, National Chiao Tung University

1001, Ta-Hsueh Rd., Hsinchu 30010, Taiwan

e-mail: cshsu@mail.nctu.edu.tw

Abstract: *Four series of high birefringence bistolane liquid crystals were synthesized and characterized. Biphenyl and naphthyl moieties were introduced to enhance the birefringence. These bistolane compounds exhibit reasonably low melting points and high birefringence of 0.5-0.8.*

Introduction

High birefringence ($\triangle$n) liquid crystals (LCs) are useful not only in conventional display devices such as STN-LCDs, but also in scattering-type PDLCDs as a reflective LCD, and in spatial light modulators. They are also of interest as componets of LCDs, for example, compensation films for improving the viewing angle, reflectors and polarizers. A number of LCs have been studied because of these applications.

It is well known that a high n values can be achieved by increasing the molecular conjugation length. A considerable number of π-conjugated compounds have been developed as high $\triangle$n LCs. Molecules that contain highly polarizable groups with high electron densisity, such as benzene rings or acetylene linking groups, will therefore have large optical anisotropies.

In this paper, we report the synthesis procedures and physical propertied of highly birefringent bistolane LCs having terminal isothiocyanato group. Phenyl, biphenyl and naphthyl moieties linked by ethynyl unit were applied as the core structure. Lateral methyl substitution in the middle phenyl ring significantly reduces the melting temperatures of bistolane LCs. By introducing different lateral substituted short alkyl chains and fluorine at various positions, the synthesized compounds were characterized to have low melting point, relatively low viscosity, and high optical anisotropy for immediate practical applications.

Experimental section

In this study the synthesis of the final materials is greatly facilitated by the use of palladium-catalysed cross-coupling reaction. Indeed, in the synthesis of some materials this synthesis procedure is virtually essential.

Despite the development of palladium-catalysed cross-coupling reaction to high levels, their exceptional versatility and their extremely wide tolerance to a whole range of functional groups, those coupling involving isothiocyanato substituents fail. The use of thiophosgene and chloroform in aqueous calcium carbonate on an aromatic amine is a very useful and efficient method of introducing the isothiocyanato group and this method is used in this research.

1a-1d: n=2~5, X=H, Y=F; 1e: n=5 X=Y=H

2a: X=Y=H; 2b: X=H, Y=F; 2c: X=F, Y=H

3a: X=Y=H; 3b: X=H, Y=F; 3c: X=H, Y=C$_2$H$_5$; 3d: X=CH$_3$, Y=H

4a-4d: n=2-5, X=Y=H; 4e-4h: n=2-5, X=H, Y=F

Results and Discussion

For the first series of bistolane LCs, methyl lateral substituent was introduced at the cental phenyl ring to get lower melting

978-0-9789217-3-6/07/$25.00

©2007 WEN GLOBAL SOLUTIONS

point LCs. All these compounds exhibit an enantiotropic nematic phase. The first series compounds contain a same terminal iscthiocyanato group on the right side and different alkyl chain length on the left . Both melting and clearing temperatures are decreased gradually with increasing carbon number. Compounds **1e** contains the similar structures to those of compound **1d**, without an additionally lateral fluoro group at C-3 position of the right side phenyl ring. For comparison purpose, we synthesized compound **1d** which contains a lateral fluoro group at C-3 position of the phenyl ring. Its clearing points are lower than those of compound **1e**. This is reasonable lateral group can hinder molecular packing effectively and thus decrease transition temperatures.

In the second series, all these bistolanes fluoro lateral substituent was introduced at the cental phenyl ring and contain a same terminal iscthiocyanato group. Compound **2a** reveals enantiotropic nematic and smectic phases. Introduction of the lateral fluoro group at C-3 position of the phenyl ring in compound **2b** did not suppress the smectic phase by destroying the symmetry infact it enhanced the melting point to 150 °C. The position of the fluorine atom was changed from C-3 to C-2 to provide pure nematic phase in compound **2c**, and it worked out as excepted exhibiting a wide nematic range from 127 °C to 259 °C.

In the third series pentyl group was placed on biphenyl ring and, in the other side, different lateral groups were introduced on right side containing a isothiocyanato phenyl ring. All the listed compounds exhibit a pure nematic phase within the whole mesomprphic region. Moreover, the smectic phases which normally appear in the NCS-based biphenyl tolanes or tolanes are completely suppressed. Such an unusual behavior is believed due to the laterally substituted ethyl chain(s) and/or the single fluorine atom. The lateral substitutions increase the molecular separation and break the molecular symmetry so that the smectic phase is difficult to form. Compound **3a** shows the highest transition temperature because of its right side phenyl ring without lateral substitutents. Compound **3b** fluoro lateral substituent was introduced at C-3 position of the right side phenyl ring, the

melting point was reduced from 115.9 °C to 108 °C. Compound **3c** ethyl group was introduced at C-3 position of the right side phenyl ring, both melting and clearing points were reduced clearly. Compound **3d** methyl lateral substituent was introduced at C-2 position of the right side phenyl ring, the transition temperature was similar to that of compound **3a**.

In the fourth series pentyl group was placed on biphenyl ring and, in the other side, different lateral groups were introduced on right side containing a isothiocyanato phenyl ring. Compounds **4a-4d** contain a same terminal isothiocyanato group on the right side phenyl ring and different alkoxy chain length on the naphthyl ring. It can be seen that melting points show an odd-even effect with increasing carbon number, while clearing temperatures decrease gradually. Compounds **4e-4h** contain the similar structures to those of compounds **4a-4d**, with an additionally lateral fluoro group at C-3 position of the right side phenyl ring. Both melting and clearing points decrease gradually with increasing carbon number. To further reduce the transition temperatures lateral fluoro group was introduced at C-3 position of the right side phenyl ring, almost compounds transition temperatures were not reduced as expected. Only compound **4f** melting point was decreased to that of compound **4b** from 157.0 to 135.3 °C.

The Δn value defined as the difference between the two principal refractive indices of an uniaxial material was estimated by a guest-host method. The Δn value of a guest-host system can be approximated from the following equation:

$$(\Delta n)_{gh} = x(\Delta n)_g + (1-x) (\Delta n)_h \qquad (1)$$

In equation (1), the subscripts g, h, and gh denote guest, host, and guest-host cells, respectively, and x is the concentration (in wt%) of the guest compound. By comparing the measured results for the guest-host mixtures to those of the host mixture, the Δn values of the guest compounds can be extrapolated.

The Δn values of some synthesized compounds are listed in Table 1. Commercial LCs ZLI- 1132 or E-63 was used as host mixture. It can be seen that Δn

values of these compounds are in the range from 0.5 to 0.8. For those compounds with biphenyl and naphthyl moieties, their Δn value is higher than that of the corresponding phenyl moiety. This is because biphenyl and naphthyl moieties have an elongated electron conjugation than the phenyl moiety. The lateral substitutions affect the LCs Δn values significantly. By comparing compounds **1e**, **3b** and **3c**, the Δn values were similar approaching to 0.5. Compound **3c** has the lowest birefringence among the compounds **2a-2d** we studied due to its two laterally substituted ethyl chains would lead to a higher Δn value due to the molecular packing density effect. The Δn values of the compounds **3b**, **3d** and **4d** is 0.67, 0.62 and 0.61, respectively. Compound **3a** has the highest Δn of 0.73 and is expected to be useful for PDLC, cholesteric display, and laser beam steering applications.

Table 1. Δn value for single compound.

Cpd. No.	Structure of the compounds	Δn
1e		0.47*
2b		0.49*
2c		0.49*
3a		0.73
3b		0.67
3c		0.55
3d		0.62
4d		0.61

Data calculated from the guest-host systems. Host mixture ZLI 1132.* Host mixture E-63 was used

Conclusions

Four series of novel super high Δn bistolane laterally substituted liquid crystals with terminal isothiocyanato groups were synthesized. Lateral alkyl or fluoro group was introduced to modify their LC properties. The extrapolated Δn of some of these compounds is greater than 0.7. Some of these compounds exhibit an odd-even effect in phase transition temperatures. A eutectic mixture consisting of these compounds and some NCS tolanes was developed. The Δn value was determined by the guest-host method using a commercial LC as host matrix. Compounds containing biphenyl and naphthyl moieties have a high Δn value. Compound **3a** has the highest Δn of 0.73, which can be a good candidate for many display applications.

References

1 Chain-Shu Hsu et al., Liq. Cryst. 27 (2000) 283.
2 S. T. Wu et al., Appl. Phys. Lett. 39 (2000) L38.
3 Yung-Ming Liao et al., Liq. Cryst. 33 (2006) 1199.
4 Yung-Ming Liao et al., Liq. Cryst., in press (2007).

Stereo Viewing Zone in Autostereoscopic Display Based on Parallax Barrier

Qiong-Hua Wang, Ren-Liang Zhao, Wu-Xiang Zhao,
Da-Hai Li, Yan-Xia Xin, Ai-Hong Wang

*School of Electronics and Information Engineering, Sichuan University,
Chengdu 610065, China, qhwang@cdnet.edu.cn*

Abstract *A well-designed autostereoscopic display based on parallax barrier can produce high-quality stereo images without crosstalk in the stereo viewing zones. The concept of stereo image quality factor was defined to indicate the stereo viewing zone of the display. The stereo viewing zone was analyzed using geometry principle. Simulated program was developed to calculate the stereo viewing zone. The calculated results of an autostereoscopic display monitor were given and the stereo viewing zone was obtained by defining threshold value of stereo image quality factor.*

Introduction

Nowadays, many kinds of autostereoscopic display devices have been developed. But the two main optical technologies for autostereoscopic display are parallax barriers and lenticular lens arrays. The two autostereoscopic displays have superior characteristics and provide the low cost necessary for mass-market adoption and they have the ability to switch between 2D display mode and 3D display mode. Both approaches can also produce high quality 3D displays utilizing either two views or multiple views for extended viewing freedom. Especially autostereoscopic display [1-3].

The key factor in consumer acceptance of autostereoscopic display based on parallax barrier is that the display without stress for users with normal stereo vision. Although this is a complicated psychophysical challenge, we should try to produce significant improvements in user acceptance by minimizing image crosstalk, essentially the leakage of the left-eye images to the right-eye, and maximizing viewing freedom. So research on the viewing zone is an important issue for the autostereoscopic display based on parallax barrier.

This paper analyzes and calculates the stereo viewing zone of the autostereoscopic display based on parallax barrier.

Principle of stereo viewing zone

The structure and principle of an autostereoscopic display based on parallax barrier are shown in Fig.1(a) and (b), respectively. The autostereoscopic display consists of a flat panel display and a parallax barrier. The parallax barrier with slit width W_s and black strip width W_B places at D from the flat panel display with pixel pitch W_P. The distance between the right and left viewing position must be equal to the interpupillary distance (usually 65mm), which is denoted by E. L is the optimal viewing distance and K is the view number. Then the following relations are obtained.

$$W_S = \frac{E \, W_P}{E + W_P} \qquad (1)$$

$$D = \frac{W_P \, L}{E + W_P} \qquad (2)$$

$$W_B = W_S(K - 1) \qquad (3)$$

Well-designed parallax barrier display system can produce high-quality stereo images with little crosstalk or even without crosstalk in the stereo viewing zones, i.e. a viewer's left eye only see left-eye images and his or her right eye only see right-eye images. Therefore we definite stereo image quality factor Q as follows.

$$Q = \frac{S_s}{S_W} \qquad (4)$$

where S_s is area of the left-eye or right-eye images that viewer's one eye (left or right)

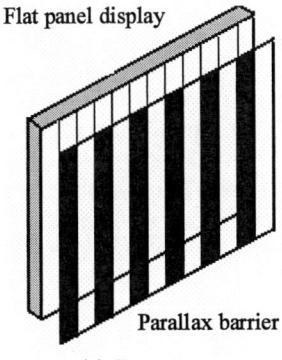

(a) Structure

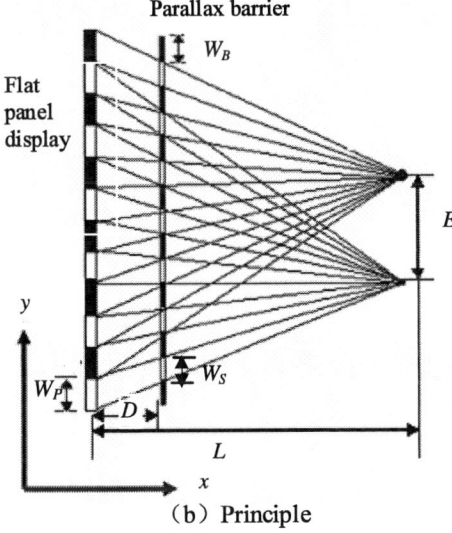

（b）Principle

Fig. 1 Structure and principle of autostereoscopic display based on parallax barrier

can see at a fixed position and S_w is area of the left-eye or right-eye images that the display screen presents. The value of Q at a position denotes if the position is good or bad for viewing stereo images. When $Q=1$, the position is best for viewing stereo images and when $Q=0$, the position is worst for viewing stereo images.

For an autostereoscopic display with view number $K=4$ in Fig.1 (b), the flat panel display's coordinate is given as follows.

$$x=0 \qquad (5a)$$
$$(n-1)W_P \le y \le nW_P$$
$$n=1,2,3,4 \qquad (5b)$$

Remainder 1, 2, 3, 0 of $n/4$ denote the 1st, 2nd, 3rd and 4th view images.

The parallax barrier's coordinate is given as follows.

$$x=D \qquad (6a)$$
$$(m-1)(W_S+W_B) \le y \le (m-1)W_S+mW_B$$
$$m=1, 2, 3, 4\ldots\ldots \text{(black strip)} \quad （6b）$$
$$(m-1)W_S+mW_B \le y \le m(W_S+W_B)$$
$$m=1, 2, 3, 4\ldots\ldots \text{(slit)} \qquad (6c)$$

For a view point $O(p,q)$ in front of the display screen, two lines are obtained when point $O(p,q)$ and the edges of the parallax barrier's slit are connected. Their equations are

$$y_{up}=\frac{x+p}{D+p}[m(W_S+W_B)+q]-q$$
$$m=1, 2, 3, 4\ldots \qquad （7a）$$
$$y_{dw}=\frac{x+p}{D+p}\{[(m-1)W_S+mW_B]+q\}-q$$
$$m=1, 2, 3, 4\ldots \qquad （7b）$$

when $x=0$, $y_{up}-y_{dw}$ is width of one-view (left-eye or right-eye) images seen through the mth slit of the parallax barrier when the viewer is located at $O(p,q)$. There are four kinds of y_{up} and y_{dw} on slit of the parallax barrier and image pane of the flat panel display as shown in Fig.2 (a）, (b）, (c）, and (d). Width W of one-view (left-eye or right-eye) images is indicated as follows, respectively.

$$W=y_{up}-y_{dw} \qquad （8a）$$
$$W=2kW_P-y_{dw} \qquad （8b）$$
$$W=W_p \qquad （8c）$$
$$W=y_{up}-(2k-1)W_P \qquad （8d）$$

Total width of all one-view images seen through all slits of the parallax barriers when the viewer is located at $O(p,q)$ can be calculated. Then total width multiplexes height of the flat panel display, and total area of the one-view images S_S at $O(p,q)$ is obtained. For a given autostereoscopic display, total area of one-view images S_W is given. From Eq.(4), stereo image quality factor Q is calculated.

A simulated program was developed to calculate stereo image quality factor Q in the viewing zones according to the preceding principle and analysis.

Simulation of stereo viewing zone

An autostereoscopic display monitor is used for simulation. Its parameters are: $W_{sc}=304.128mm$, $H_{sc}=228.096mm$, $D=0.7mm$, $W_p=0.099mm$,

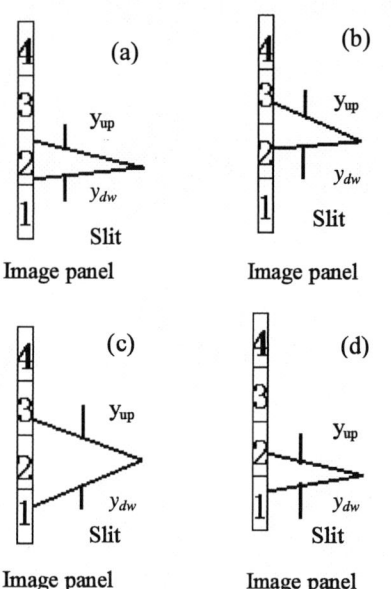

Fig. 2 Four kinds of y_{up} and y_{dw} on the parallax barrier and the image panel

$W_s=0.09885mm$, $W_B=0.29655mm$, $L=460mm$, where W_{sc} is width of the flat panel display, H_{sc} is height of the flat panel display. Fig. 3 shows the simulation results of stereo image quality factor Q in the stereo viewing zones. We can see that $Q=1$ at only few positions where viewers can see very good stereo images without crosstalk. In fact viewers can see stereo images in many positions because people can feel good stereo images although the images have some crosstalk. We definite a

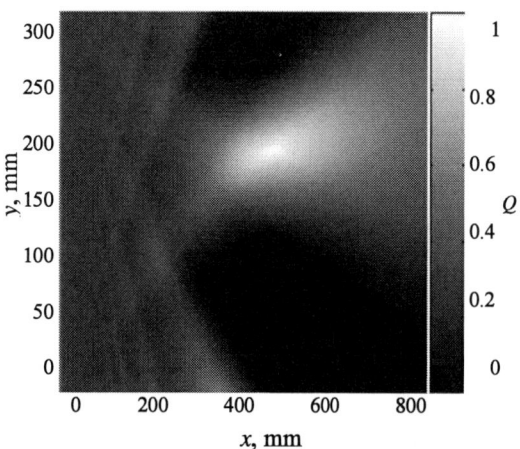

Fig.3 Simulated results of stero image quality factor Q in the stereo viewing zones

threshold value of Q as $Q_{th.}$, and the stereo viewing zones are those areas where $Q > Q_{th}$.

Fig. 4 is the distribution of stereo image quality factor Q along the horizontal direction of image panel of flat panel display when $x=300mm$, $460mm$, $600mm$, $800mm$. It is obvious that Q is the best at

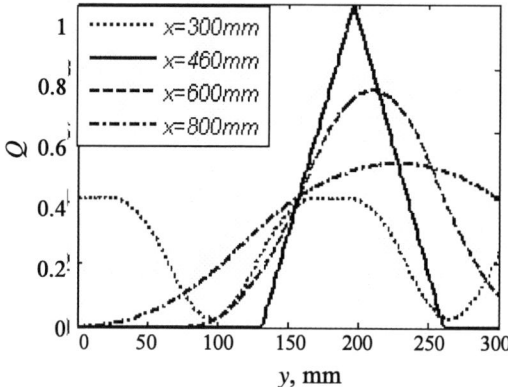

Fig. 4 Distribution of stero image quality factor Q along the horizontal direction of image panel

the optimal viewing distance $x=460mm$.

Conclusions

A well-designed parallax barrier autostereoscopic display system can produce high-quality stereo images with little crosstalk or even without crosstalk in the stereo viewing zones. The concept of stereo image quality factor was defined to indicate the stereo viewing zone of an autostereoscopic display. According to the structure and principle of the display, the stereo viewing zone was analyzed using geometry principle. Simulated program was developed to calculate stereo image quality factor in viewing zone. The calculated results of an autostereoscopic display monitor were given and the stereo viewing zone was obtained by defining threshold value of stereo image quality factor. The research of theoretical simulation helps the design of such autostereoscopic display devices.

References

1 Paul May, Information Display, No.3, 2003, 26-30.
2 Qionghua Wang et al, Proc. of Asia Display, Shanghai, 2007, 453-455.
3 K. Sakamoto, Proc. of the 2005 International Conference on Active Media Technology, 2005, 469-474.
4 J.Y, Son, Applied Optics, 43 (2004), 4985-4992.

Ultrafast laser direct-writing of Bragg –glass photonic devices.

G. Marshall[1], N. Jovanovic[1], D. Kan[2], A. Fuerbach[1], A. Asatryan[2], L. Botten[2], M. Withford[1]

[1]*Centre for Ultrahigh bandwidth Devices for Optical Systems, Centre for Lasers and Applications, Macquarie University, Sydney, NSW 2109, Australia*

[2]*Centre for Ultrahigh bandwidth Devices for Optical Systems, Department of Mathematical Sciences, University of Technology Sydney, Sydney, NSW 2007, Australia*

Email: withford@ics.mq.edu.au

Abstract: Ultrafast lasers can induce localised refractive index changes in a wide range of glass types and geometries. The challenges and developments in direct-writing of Bragg components in bulk and fibre glass forms will be presented.

Summary:

In 1996 it was first shown by Davis et al [1] that tightly focused output beams from a femtosecond pulsed laser could modify the optical properties of silica glass. In particular, femtosecond laser irradiation was reported to induce refractive index changes between 0.01 and 0.035 in pure and Ge-doped silica. Even in this seminal work the authors were quick to recognize the possibilities for using laser direct write methods for inscribing 3D lightwave circuits in bulk glass that would challenge conventional planar waveguide based optical processors. This technology has also been applied to a wide range of glass types and geometries (both fibre and planar). In the last 5 years the number of research groups working in this field has risen significantly, coinciding with the ready, commercial availability of femtosecond lasers. These groups currently span the US, UK Germany France, Italy, Spain, Canada, Lithuania, Korea, Japan, Singapore, China and Australia. The body of effort in this field is reminiscent of the significant international efforts developing silicon processing methodologies during the 70's, reflecting the high stakes involved.

A driving factor behind research in this field is the pursuit of a single fabrication platform that can produce the key components for integrated photonic devices (namely low loss waveguides, gratings for wavelength discrimination and signal division via splitters and couplers) in both passive and active glass hosts. In contrast to successful demonstrations of low loss waveguides and splitters, Bragg gratings written with ultrafast laser direct write techniques have only recently been realised in fibres [2] and bulk samples [3,4]. In this presentation we will review research regarding ultrafast laser direct-writing of Bragg –glass photonic devices. New results including demonstrations of high fidelity waveguide Bragg gratings in bulk material and the development of high power fibre lasers with intra-active core gratings will be presented. In addition, the challenges and limitations to inscribing gratings in micro-structured optical fibre will be discussed.

1. K. M. Davis, K. Miura, N. Sugimoto and K. Hirao, Opt. Lett., **21**, 1729-1731 (1996).
2. A. Martinez, M. Dubov, I. Khrushchev and I. Bennion, Electron. Lett. **40**, 1170 (2004).
3. G. D.Marshall,M. Ams andM. J.Withford, Opt. Lett. **31**, 2690–2691 (2006).
4. H. B. Zhang, S. M. Eaton, J. Z. Li and P. R. Herman, Opt. Lett., **31**, 3495-3497 (2006).

978-0-9789217-3-6/07/\$25.00

©2007 WEN GLOBAL SOLUTIONS

Advanced Modulation Techniques in OCDMA System

Xu Wang*(1,2), N. Wada(1), T. Miyazaki(1), G. Cincotti (3), and K. Kitayama(4)

1 : Photonic Network Group, New Generation Network Research Center, NICT, 4-2-1 Nukui-Kitamachi, Koganei, Tokyo 184-8795 Japan

2: School of Engineering and Physical Sciences, Heriot Watt University, Riccarton, Edinburgh, EH14 4AS, UK

* Email: x.wang@hw.ac.uk

3: Department of Applied Electronics, University Rome Tre, via della Vasca Navale 84, I-00146, Rome, Italy,

4: Department of Electronics and Information Systems, Osaka University, 2-1 Yamadaoka, Suita, Osaka 565-0871, Japan

Abstract *Advance modulation techniques such as differential phase-shift keying (DPSK), differential quaternary phase shift keying (DQPSK) and code-shift-keying (CSK) modulation with balanced detection has been proposed and demonstrated in high capacity asynchronous optical code division access (OCDMA) system using different encoding/decoding schemes to beat the MAI noise and enhance the security.*

Introduction

Optical code division multiple access (OCDMA) technique, which allows multiple users share the same transmission media by assigning different optical codes (OCs) to different users, is an attractive candidate for next generation broadband access networks [1]. Figure 1(a) illustrates the basic architecture and the working principle of an OCDMA passive optical network (PON) network.

Partricularly, coherent OCDMA technique is receiving much attention for the overall superior performance over incoherent OCDMA and the development of compact and reliable en/decoders (E/D) like spatial light phase modulator (SLPM), superstructured fiber Bragg grating (SSFBG) and multi-port array waveguide grating (AWG)-type E/D .

Previously, on-off-key (OOK) is mostly used as modulation format for payload data in OCDMA system, which is refered as OOK-OCDMA. In a multi-user asynchronous coherent OOK-OCDMA system, the major noise sources are the coherent signal-interference (SI) beat noise (coherent noise) and the incoherent MAI (incoherent noise), which limit the number of active users in a network [1]. Therefore, it is essential to combat the SI beat and MAI noises in such a network. Time gating and optical thresholding techniques can be used to

suppress the MAI enabling data-rate detection. Meanwhile, the SI noise can be mitigated by proper timing coordination in either chip- or slot-level to carefully avoid the overlaps between signal and interferences in synchronous OCDMA [2, 3]. Up to 32-user, 10 Gbps/user synchronous OCDMA has been demonstrated by combining these techniques together with polarization and time division multiplexing [3]. However, for practical access network application, the capability of multi-user asynchronous access is of key attribute, while, the beat noise is still a big issue in asynchronous OCDMA [1].

Using ultra-long OCs [1, 4] and AWG-type E/D with very high power contrast ratio (PCR) between auto-/cross-correlation [5] is one effective way to suppress the interference level in asynchronous environment. Another approach is to use forward-error-correction (FEC) technique to

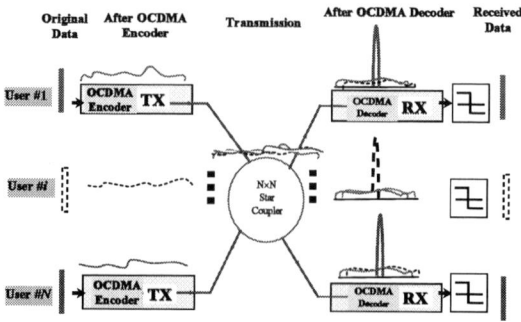

Fig. 1 Working principle of OCDMA PON

enhance the noise tolerance of the system. However, these are still not a cost-effective solution.

Meanwhile, in a multi-user asynchronous OOK-OCDMA system, with the changing of the active users' number, dynamic threshold level setting is required to maintain a wider power margin in the decoder/receiver setup [1]. However, both the real-time active users number estimating and dynamical threshold setting are still practical issues in OCDMA receivers and will result in additional cost.

In addition, the security issue has been brought forward recently that the OOK-OCDMA is vulnerable in terms of the security that could be easily broken by simple power detection without any knowledge of the code [6]. Therefore, much advanced modulation techniques are required in OCDMA in order to ensure the confidentiality of the network.

In this paper, we will introduce multi-user OCDMA systems with differential phase-shift keying (DPSK), *differential quaternary phase shift keying (DQPSK)* and code-shift-keying (CSK) modulation and balanced detection.

Multi-user DPSK- and DQPSK-OCDMA

Coherent OCDMA with DPSK modulation format and balanced detection (DPSK-OCDMA) is superior over the OOK-OCDMA with advantages of improved receiver sensitivity, better tolerance to beat noise and MAI noise without OT, and no need for

dynamic Th level setting [5].

Fig. 2 shows an experimental setup of a field trial for high capacity WDM/DPSK-OCDMA. The field trial was done on an optical testbed of JGNII (Japan Gigabit Network II). Three mode-lock laser diodes (MLLD) generated 3 WDM pulse signals with about 3.2 nm (400 GHz) channel spacing. The ~1.8 ps optical pulses were generated at a repetition rate of 10.709 GHz with central wavelengths of 1550.2 nm, 1553.4 nm and 1556.6 nm, respectively. Inset (β) in Fig. 2 shows the waveform of the mixed signals of 3 WDM, 12 OCDMA users. This signal was then launched into 100 km installed SMF. Insets (θ) and (ζ) in Fig. 2 show the decoded signal and the electrical signal after the balanced detector respectively. The data was finally tested by the BER tester.

The measured BER performances are shown in Fig. 3. Error free transmission has been successfully achieved for all the 4 decoders with 3-WDM, up to 12 OCDMA users in the B-to-B experiment and 10 OCDMA users in the field trail. The spectral efficiency (ξ) is about 0.32 and 0.27 bit/s/Hz , respectively.

For further enhancing the spectral efficiency, DQPSK-OCDMA has been demonstrated in synchronous condition together with FEC and polarization multiplexing [7]. Figure 4(a) and (b) show the experimental setup and BER performance, respectively. Total spectral efficiency is 0.87 bit/s/Hz.

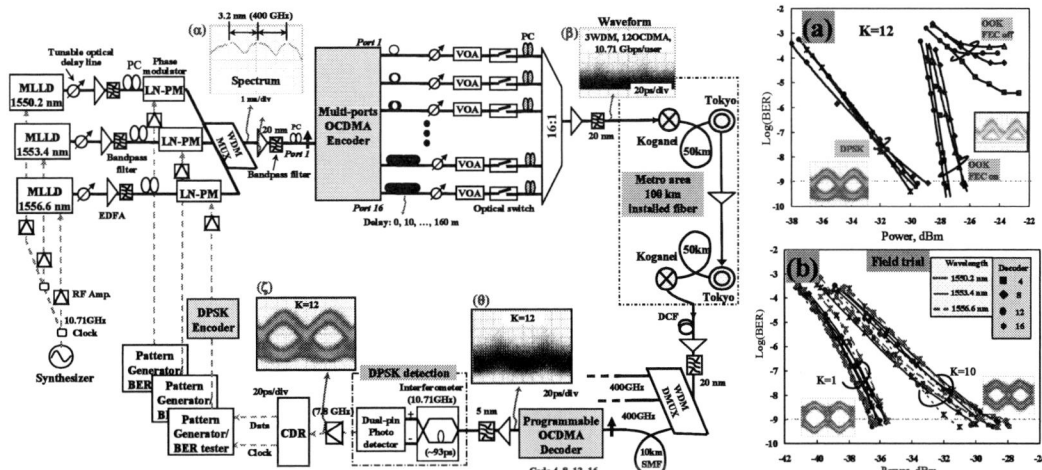

Fig. 2 Field trial of WDM/DPSK-OCDMA experiment   *Fig. 3 BER performance*

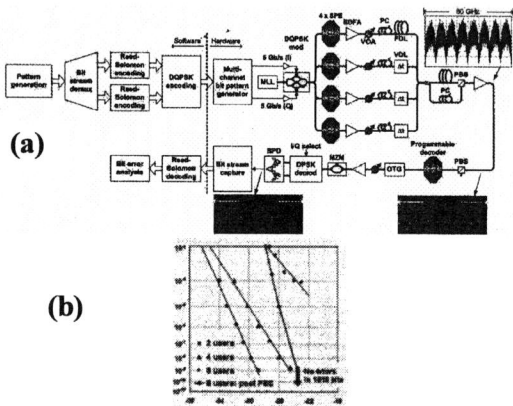

(a)

(b)

Fig. 4(a) DQPSK experiment (b) BER

Asynchronous CSK-OCDMA system

CSK-OCDMA with balanced detection (BD) can also improve multi-user capability with respect to OOK-OCDMA. The requirement for real-time K estimation and dynamic threshold level setting are relaxed in CSK-OCDMA scheme, in the same way as in the DPSK-OCDMA with balanced detection. And more importantly, the confidentiality can be significantly improved in the CSK-OCDMA system because an eavesdropper could not decipher the signal without knowing the OCs [8].

Fig. 5 shows the experimental setup of a multiuser CSK-OCDMA system. Inset η shows the eye diagram of the multiplexed signals of 8 CSK-OCDMA users. Fig. 6(a) shows the measured BER performance vs. the received optical power for different K.

BER<10^{-9} has been achieved with up to 8 active users without the using of forward-error-correction (FEC) and optical thresholding. Fig. 6(b) shows the power penalty vs. K at BER=10^{-9}. This result is largely better than in OOK-OCDMA, where asynchronous multi-user OCDMA transmission was assisted by FEC or optical thresholder.

Conclusion

Comparing to conventional OOK-OCDMA, using other advance modulation techniques, such as DPSK, DQPSK and CSK in OCDMA system has advantages of (1) Improved receiver sensitivity; (2) Better tolerance to beat noise and MAI noise; (3) No need for optical thresholding; (4) No need for dynamic threshold level setting; and (5) Enhanced security. High capacity, spectral efficient OCDMA systems have been demonstrated using these techniques.

References

1 X. Wang et al, JLT, 22(2004), 2226-2235.
2 Z. Jiang et al, IEEE PTL, 17(2005), 929.
3 V. J. Hernandez et al, OFC06 PDP45 (2006).
4 X. Wang et al, OFC05, PDP 33, (2005).
5 X. Wang et al, JLT, 25(2007), 207-215.
6 T. H. Shake JLT, 23(2005), 1652-1663.
7. J. Jackel et al, OFC07 PDP7, (2007).
8. X. Wang et al, IEEE JSTQE, to appear.

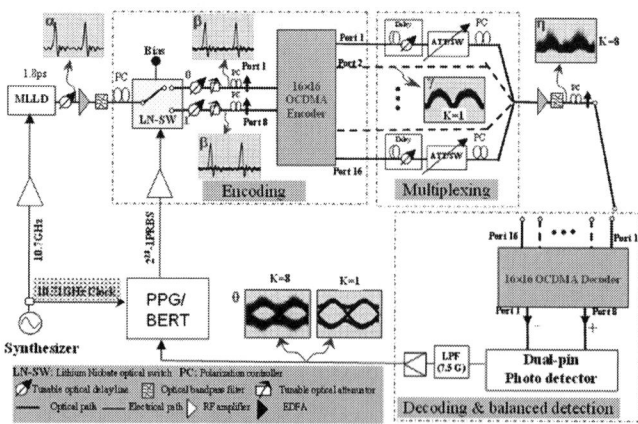

Fig. 5 Experimental setup

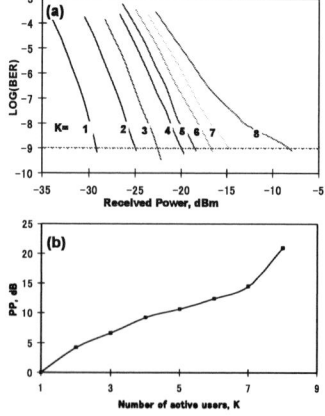

Fig. 6 BER performances

2.5Gbps 60km OCDMA Transmission Experiment Using EPS-SSFBG En/decoder

Lin Lu, Weilei Wu, Hui Peng, Tao Pu, Yuquan Li

Institute of Communication Engineering, PLA University of Science and Technology,
No.2 Biaoying, Yudao Street, Nanjing, 210007, P.R.China.
Email: goodlulin_163@163.com

Abstract *EPS-SSFBG OCDMA en/decoders were experimentally demonstrated. 2.5Gbps OCDMA experiment is shown to achieve 60km error free transmission using 63-chip ,300Gbps chiprate EPS-SSFBG en/decoder and threshold adjustable receiver.*

Introduction

Optical code division multiple access (OCDMA) is considered as one of the most promising technologies for the next generation broad-band access network. Coherent OCDMA using ultra-short optical pulse is receiving increasing attention with the progress of reliable and compact en/decoder devices, such as planar light-wave circuits (PLCs) [1], arrayed waveguide gratings (AWGs) [2], fiber Bragg gratings (FBGs) [3-7], and so on.

The super-structured FBG (SSFBG) based phase en/decoders are particularly attractive for their the longest code length, the highest chip rate and better correlation property [4,5]. Recently, the equivalent phase shift (EPS) has been proposed to realize the OCDMA en/decoders with high phase precision only based on sub-micro precision [6], and then a 1023-chip en/decoder has been demonstrated [8]. EPS-SSFBG en/decoder has the same properties as SSFBG in theory, but it has not been demonstrated in transmission experiment.

In this paper, we demonstrate 2.5Gbps error free OCDMA transmission experiment over 60km using 63-chip EPS-SSFBG en/decoder and narrowband threshold adjustable receiver.

System Design Discussion

There are three key modules in coherent OCDMA system: source of ultra-short optical pulse, en/decoder and receiver.

Pulse train generated by optical source is modulated by transmission data to converting NRZ signal to RZ signal. It not only provides spectrums for en/decoder but also affects system stability and reliability.

En/decoder executes signal mapping and correlation operation in optical domain. It can be consider as LTI (Linear and Time Invariant) system in most case. Response of en/decoder is decided by its physical parameters such as micro structure, material distribution etc. Fig1 (a) is the reflection amplitude spectrums for 63chip EPS-SSFBG used in experiment, The +1th and the -1th order spectrum are usually utilized to perform encoding/decoding. The encoding band is ~2.2nm. Since the decoder grating is the spatially reversed form of encoder grating, both encoder and decoder have similar amplitude spectrums but reversing phase spectrums.

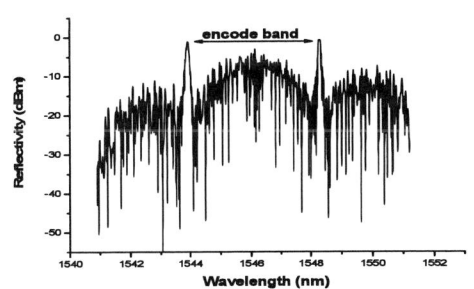

Fig. 1(a).Amplitude spectrums of 63 EPS-SSFBG en/decoder

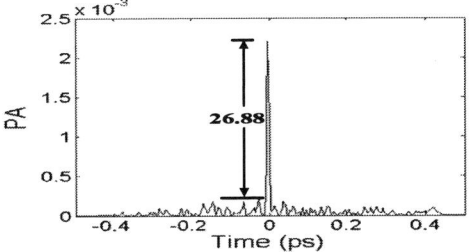

Fig. 1(b).Decode signal in matched condition

978-0-9789217-3-6/07/$25.00

©2007 WEN GLOBAL SOLUTIONS

En/decoder is a frequency and phase sensitive device, therefore two issues must be concerned in system design: first, the match of wavelength between optical source and en/decoder, that is: source spectrum should cover en/decoding spectrums suitably (shown in Fig1) to avoid sideband interference and achieve maxim frequency efficiency; second, frequency and phase chirp from optical source should be suppressed, while dispersion and PMD in transmission fiber should be compensated. Fig1 (b) is simulated back to back decoded signal in best condition which means maxim spectrum efficiency, no frequency and phase chirp. The P/W (peak to wing ratio) is 26.88.

OCDMA receiver is different from traditional NRZ receiver, since ultra-short optical pulse should be received and recovered into NRZ signal; meanwhile, it is also different from an OTDM receiver, since large sideband and noise generated by MAI, ISI and beat noise should be suppressed. Furthermore, a practical OCDMA receiver's bandwidth is preferred to be in the order of the data-rate (~GHz) instead of chip rate (~hundreds GHz). This bandwidth limitation will result in severe performance degradation in multi-user coherent OCDMA systems which use ultra-short optical pulse [9]. Optical threshold and optical time gating techniques, which are based on optical nonlinear effect, are used in coherent OCDMA system to suppress sideband, noise and system jitter [9,10]. In addition to optical domain techniques, electronic processing, such as equalizer, threshold adjusting, soft decision and error correction coding, can suppress sideband and noise to some extent. Since electronic processing is limited by "electronic speed bottleneck" and hardware complexity, both optical and electronic processing should be used to improve system performance in multi-user OCDMA systems.

Experiment and Result
Figure 2 shows the experimental setup: Driven by clock from PPG (MP1763C), gain switch laser diodes (GSLD) generated ~20ps optical pulses at a repetition rate of 2.488 GHz with central wavelengths of 1545.52 nm. The optical signal was

modulated by a LiNiO$_3$ intensity modulator to a RZ (return to zero) signal with $2^{31}-1$ pseudo random bit sequence (PRBS). Then the modulated signal was encoded by a 63chip EPS-SSFBG which was shown in Fig 1. After amplified by EDFA, the encoded signal was launched into 60km G.652 SMF. DCF fiber was placed at the end of transmission fiber to compensate dispersion completely. 10G PIN+TIA was used as a narrowband detector to receive ~20ps decoded signal. Threshold adjustable CDR (Clock and Data Recovery) module converted RZ signal to NRZ and recovered timing clock. BER curves were measured by BERT (MP1764C). Waveforms at the five measurement points (A-E) were observed by sampling oscilloscope Lecroy NR9000 which has ~30GHz optical detector and ~20GHz electrical receiver.

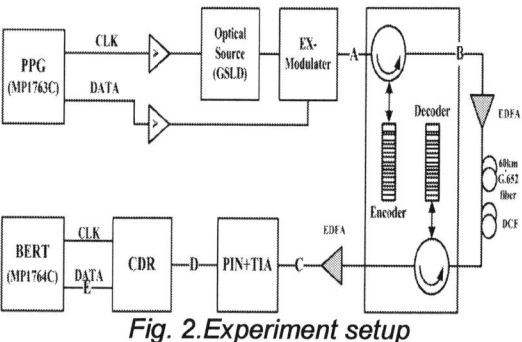

Fig. 2.Experiment setup

The external modulated RZ signals are shown in Fig.3(a),the FWHM of these pulses is ~20ps, corresponding to ~0.65nm 3dB bandwith. The waveforms of encoded signals and its eye diagram are shown in Fig. 3(b), the encode duration is about 200ps, corresponding to ~2cm grating length and ~300Gbps chip rate. The decoded signal and its eye diagram are shown in Fig. 3(c), the practical P/W≈6 and the FWHM of decoded signal is ~ 20ps. Comparing with calculated result shown in Fig. 1(b), The P/W degradation are derived from low spectrums efficiency (0.65nm vs 2.2nm) and frequency/phase chirp from GSLD. The outputs of 10G narrowband PIN+TIA are shown in Fig. 3(d), which demonstrates signal degradation due to narrowband detector: the P/W decreases to 3. Finally, the recovered NRZ data are shown in Fig. 3(e).

102

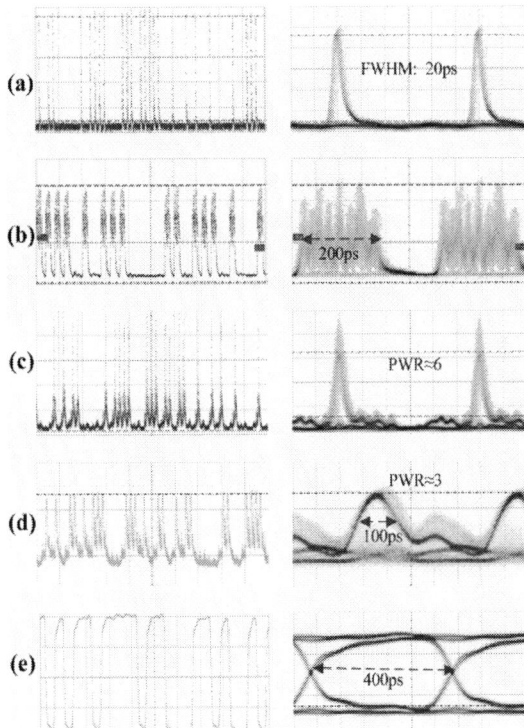

Fig. 3.Waveforms at measurement points

Figure 4 shows the measured BERs of the experiment. Error free ($<10^{-12}$) transmissions have been achieved. Comparing to that in back-to-back (B-to-B) cases, ~1.1 dB power penalty was measured for 60km transmission ,which is mainly due to residual high order despersion. Comparing to the no en/decoder case, ~1.7 dB en/decoding power penalty was measured. The receiver sensetivity is ~22.5dBm at 10^{-9} BER.

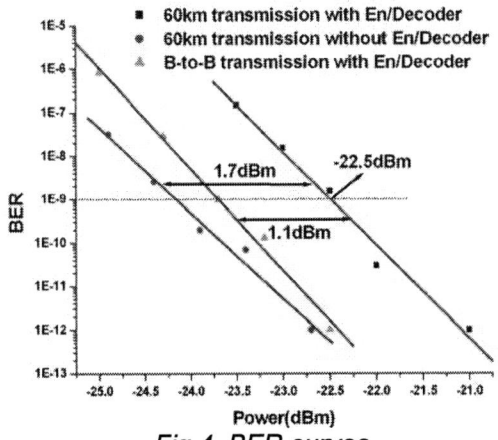

Fig.4. BER curves

Conclusions

We have experimentally demonstrated EPS-SSFBG OCDMA en/decoder. 2.5Gbps 60 km error free transmission was achieved using 63-chip en/decoder and narrowband threshold adjustable receiver. Further improvement of the better optical source ,long code and optical thresholding will enable system to accommodate more users and support longer distance.

Acknowledgments

This work was supported by the National Natural Science Foundation of China under GRANT NSFC-60502003 and 60132020, and NNSF of Jiangsu Province under GRANT BK2007501. We would like to thank Prof. Jian Wu of BUPT for providing GSLD in experiment and useful discussion.

References

1 N.Wada JLT 17(1999), 1758-1765.
2 Hiroyaki Tsuda, IEEE Electronics Letters, Vol.35(1999), No.14, Pp.1186-1188.
3 Peh Chiong Teh, Periklis Petropoulos, etc, IEEE/OSA JLT Vol.19(2001), Pp.1352-1365.
4 X. Wang, Opt. Letters, vol. 30(2005), No. 4, pp. 355–357.
5 X. Wang, Opt. Express, vol. 12(2004), No. 22, pp. 5457–5468.
6 Yitang Dai, IEEE PTL, Vol.16(2004), No.10, Pp.2284-2286.
7 Yitang Dai, Optical Letters, Vol.31(2006), No.11.
8 Yitang Dai, OFC2007, JWA28#
9 Xu Wang, OFC 2005, PDP33, 2005.
10 Hideyuki Sotobayashi, IEEE JQE, VOL. 10(2004),No.2,Pp.250-257.

Experimental study on the spectral behavior of an asymmetric long period fiber grating via erosion

Minwei Yang, Jianping Chen*, Yiping Wang, Xinwan Li

State Key Laboratory of Advanced Optical Communication Systems and Networks, Shanghai Jiao Tong University, Shanghai 200240, People's Republic of China

*Corresponding Author Email jpchen62@sjtu.edu.cn

Abstract *An asymmetric long period grating with period grooves is chemically etched. Experimental results show that though the grooves are ablated, the period structure still remains. Analysis is given to explain the related mechanism.*

Introduction

Long period fiber gratings (LPFG) have been widely used in optical communication and sensing since its ability to couple energy from the core mode to different cladding modes with the same propagation direction. Various kinds of LPFG based optical devices are reported [1]. Recently, a new kind of asymmetric LPFG is fabricated by using a focused CO_2 laser beam to carve period grooves on the fiber cladding, whose longitudinal photo is shown in Fig. 1 [2].

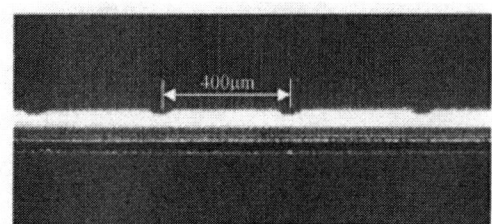

Fig. 1 The longitudinal photo of an asymmetric LPFG

It was formerly believed that the photo-elastic effect induced by these grooves is the main reason for gratings [2]. According to our experiment as will be stated below, there might be other explanation of the grating formation. That is, besides the physical deformation to the fiber, the peak CO_2 beam energy may penetrate the cladding and induce the index modulation in the fiber core. Since the grooves are about 15 μm in depth [2], we design an experiment to observe the transmission spectrum of the LPFG after the grooves are totally ablated by etching.

Experiment of etching

Before etching, we first measured the transmission spectrum of the LPFG and its center wavelength shift in different media. Fig. 2 and Fig. 3 show the results when it is put in the air and in water, respectively. We noticed that as the external refractive index changes from n=1.0 (the air) to n=1.33 (water), the resonant wavelength is shifted from 1526nm to 1522nm and the resonant depth is changed from -24.29dB to -22.43dB (the resonant depth is define as the minimum transmission minus the maximum transmission).

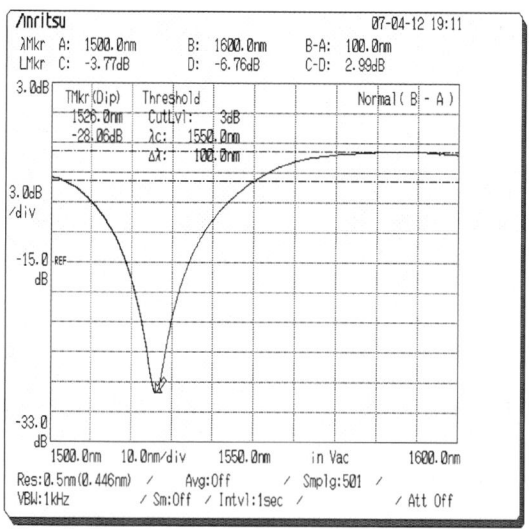

Fig. 2 The transmission spectrum of LPFG in the air before etching

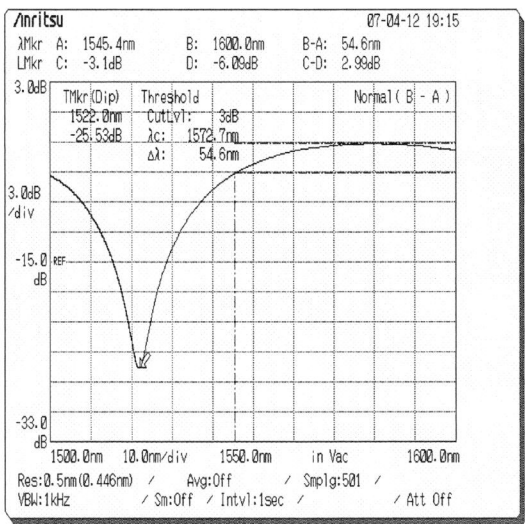

Fig. 3 The transmission spectrum of LPFG in water before etching

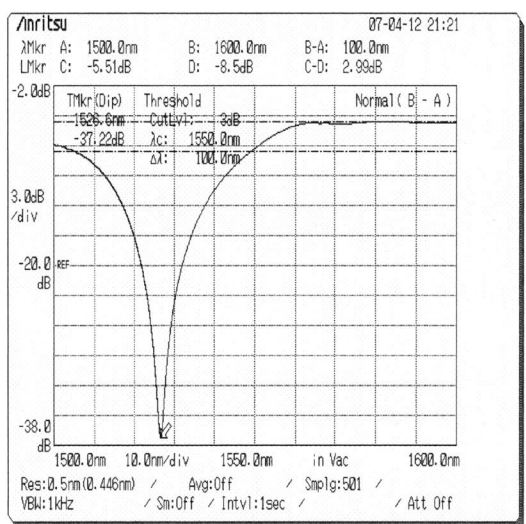

Fig. 5 The transmission spectrum of LPFG in water after etching

Then, we immersed the LPFG into a 40% hydrofluoric acid solution and used a white light source to monitor the transmission spectrum. We etch the LPFG for about 30 minutes. Measurement shows that the radial etching speed was about 15.5 μm/10min. The above etching time ensures that the grooves are totally removed.

After etching, we cleaned the fiber surface and measured the transmission spectrum again. The results are shown in Fig. 4 and Fig. 5, respectively. The wavelength is shifted from 1532.4nm to 1526.6nm and the resonant depth is changed from -32.92dB to -31.71dB.

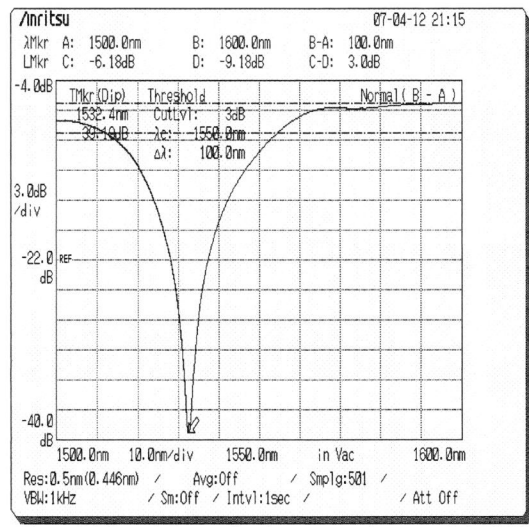

Fig. 4 The transmission spectrum of LPFG in the air after etching

Analysis of the experiment result

Generally, the resonance wavelength is sensitive to the external refractive index since it will change the effective refractive index of the cladding modes. Through a coreless approximation, the wavelength shift according to the external refractive index, the resonance change can be explained by the following expression [3]:

$$\Delta\lambda = \Delta n_{cl,m}\Lambda = \frac{u_\infty^2 \lambda_{m0}^3 \Lambda}{8\pi^3 n_{cl} r_{cl}^3}(\frac{1}{\sqrt{n_{cl}^2 - n_{ex0}^2}} - \frac{1}{\sqrt{n_{cl}^2 - n_{ex}^2}}) \quad (1)$$

Here, Λ is the pitch of the grating, λ_{m0} is the initial resonance wavelength of LP0m coupling, $\Delta n_{ex} = n_{ex} - n_{ex0}$ is the change of the external refractive index, r_{cl} is the radius of the cladding and u_∞ is the mth root of the Bessel function J_0 which correspond to the coupling between LP0m and LP01. Before etching the observed resonant peak is caused by the coupling between LP01 and LP06 and after etching, it is caused by the coupling between LP01 and LP04. The experiment results agree with the theoretical calculation well.

After etching, the grooves are ablated and the deformation of the fiber cladding is not enough to provide the necessary index modulation in the core. Therefore, there must be other mechanism that still meets the phase matching condition. We believe that the laser beam has caused two kinds of index change in the fiber: the period index modulation δn_{eff} and the average index change $\overline{\delta n_{eff}}$. The period index modulation is caused by the peak of the laser beam energy during repeated scanning within the fiber core while the average index change is caused by the residual beam energy in the fiber cladding. Normally, we have $\delta n_{eff} > \overline{\delta n_{eff}}$ and $\overline{\delta n_{eff}}$ decreases along the radius from the fiber core to the cladding surface because the energy of the laser beam decreases. Once the two index changes are formed, they will not be removed during the chemical etching progress and thus the periodical structure still remains in the fiber. Moreover, during the erosion progress, the average index change $\overline{\delta n_{eff}}$ becomes larger since the fiber cladding is reduced. This can also explain the increase in the resonant depth since a larger $\overline{\delta n_{eff}}$ will result in a deeper resonant depth according to the grating theory [4].

Conclusions

In conclusion, a new explanation of the mechanism of the asymmetric LPFG is brought forward and experiment of etching is conducted to validate it. After etching, the resonant depth is deeper and the resonant wavelength is more sensitive to the external refractive index. This means erosion might be an effective way for the improvement of the LPFG spectrum.

References

1 X. W. Shu, et al, J. Lightwave Technology, "Sensitivity characteristics of long-period fiber gratings", Vol.20 (2002): page 255-266

2 Y. P. Wang, et al, APPL PHYS LETT, "Asymmetric long period fiber gratings fabricated by use of CO2 laser to carve periodic grooves on the optical fiber", Vol.89 (2006): page 151105-1-3

3 K. S. Chiang, et al, ELECTRON LETT, "Analysis of etched long-period fibre grating and its response to external refractive index", Vol.36 (2000): page 966-967

4 T. Erdogan, J. Lightwave Technology, "Fiber grating spectra", Vol.15 (1997): page 1277-1294

A Review of the Effects of High Refractive Index Overlays on Tunable Long Period Fiber Gratings

J. Lee (1), Q. Chen (2), Q. Zhang (3), and S. Yin (4)

1: Department of Electrical Engineering, The Pennsylvania State University
University Park, PA 16802, Email: jel236@psu.edu
2: Department of Electrical Engineering, The Pennsylvania State University
Email: quc101@psu.edu
3: Department of Electrical Engineering, The Pennsylvania State University
Email: qxz1@psu.edu
4: Department of Electrical Engineering, The Pennsylvania State University
Email: sxy105@psu.edu

Abstract *An optimized high refractive index overlay can be coated on the surface of a thin cladding long period fiber grating to achieve an enhanced tuning range and a stable resonant peak depth.*

Introduction

Long period fiber gratings (LPGs) fabricated in single mode fibers couple energy between the core mode and cladding modes traveling in the same direction. The wavelengths at which an LPG couples energy from the core are roughly proportional to the grating period and the differences between the effective refractive indices of the core and the cladding modes to which the core is coupling [1-4]. By increasing the ambient refractive index that surrounds the LPG, the effective indices of the cladding modes increase, and the resonant peaks can be tuned to shorter wavelengths [2-8].

Tunable LPGs are good candidates for a variety of applications, but there are a number of hurdles in the way of realizing their full potential. LPGs fabricated in 125 μm diameter single mode fibers have multiple resonant peaks in their output spectra and limited tuning ranges. They also experience severe peak degradation during tuning. Many publications on tunable LPGs have addressed and presented potential solutions to these issues. The authors of this paper created an LPG with single resonant band and an expanded tuning range by reducing the thickness of the fiber cladding to 35-40 μm [9-11]. A thin, high index overlay (i.e. an overlay with a refractive index higher than that of the cladding) can be coated onto an LPG to further enhance the tuning performance [12-

14]. By properly coordinating the thickness and refractive index of the overlay, a stable peak depth can also be maintained across the enhanced tuning range.

A single resonant band LPG

Since multiple resonant peaks can potentially limit the maximum tuning range of an LPG, a single resonant band would be preferred in many tunable LPG applications. A single resonant band can be achieved by reducing the thickness of the LPG cladding from 125 μm to 35-40 μm. The thinner cladding reduces the number of propagating cladding modes, and the differences between the effective indices of cladding modes are significantly increased [9-11]. The thin cladding also increases the LPG's sensitivity to ambient index changes. Unfortunately, the thin cladding does not eliminate the peak depth reduction that occurs as the ambient index is increased.

High index overlays and mode transitions

The tuning behavior of LPGs coated with overlays has been reported and explained by a number of authors [12-14]. In these publications, high index overlays are used to induce mode transitions in the cladding [12-13]. Mode transitions cause a redistribution of the cladding modes and occur when a fiber coated with a high index overlay is exposed to a specific ambient refractive index range. During a transition, the lowest order cladding mode transitions

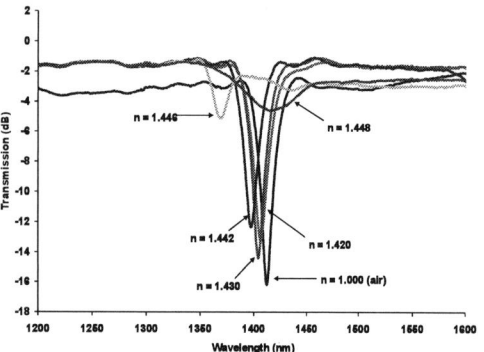

Fig. 1. Measured spectra of an LPG being tuned with refractive index oils ranging from 1.420 to 1.448. (cladding diameter = 40 µm; grating period = 400 µm; total length = 8 mm (20 periods))

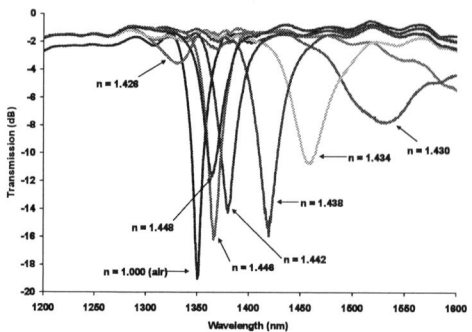

Fig. 2. Measured spectra of an ITO coated LPG being tuned with index matching oils ranging from 1.426 to 1.448. (cladding diameter = 40 µm; grating period = 400 µm; total length = 8 mm (20 periods))

to an overlay mode, and a higher order cladding mode moves in to replace it [12-13]. The effective index of the cladding mode shifting to the overlay increases sharply, and its coupling coefficient almost falls to zero [12]. As the lower order mode shifts out, a higher order cladding mode takes the effective index and coupling coefficient values held by the lower order mode prior to the transition.

When an LPG is coupling energy into a cladding mode that transitions to the overlay, the resonant peak associated with the transitioning cladding mode shifts to shorter wavelengths and becomes shallower as the ambient index is increased. As this occurs, the core mode begins to couple energy into the higher order mode that is moving in to replace the lower order mode, and a new resonant peak begins to form at a longer wavelength. Since the mode transitioning to the overlay vacates its effective index and coupling coefficient values rapidly as the ambient refractive index is increased, the higher order mode must do the same in order to fill the vacancy. The rapid change in these values relative to the increasing ambient index causes the new resonant peak to shift over a large range of wavelengths and grow in depth. The measured tuning spectra of a thin cladding LPG before and after it had been coated with a thin (< 100 nm) high index ITO overlay are shown in Figures 1 and 2, respectively.

A four-layer model and overlay optimization

The ambient index range over which the mode transition occurs is a function of the thickness and refractive index of the overlay [12-13]. An increase in either of these parameters causes the transition to move to lower ambient refractive index. The overlay can be optimized by properly coordinating the thickness and refractive index of the overlay with the parameters of the LPG (i.e. fiber diameter, grating period, etc.). The optimized overlay will induce a mode transition over the ambient refractive indices where the LPG is intrinsically most sensitive (i.e. indices that are slightly less that of the fiber cladding). It will not only maximize the LPG tuning range, but it will also provide for a stable peak depth, which is critical in tunable LPG filtering applications.

Using the measured refractive index of the ITO deposited by our sputtering system (~1.88 @ 633 nm) and a four-layer model, we determined the optimal ITO overlay thickness for an LPG with a 40 µm diameter cladding and a 400 µm grating period. When the overlay was 50 nm thick, the mode transition occurred in the 1.430-1.440 ambient index range, and by operating the ITO coated LPG over these indices, small changes in the ambient refractive index would cause large shifts in the resonant wavelength without significantly reducing the resonant peak depth. Figure 3 shows the measured tuning spectra of a 40 μm diameter LPG with a 400 μm grating period that has been coated with approximately 50 nm of ITO.

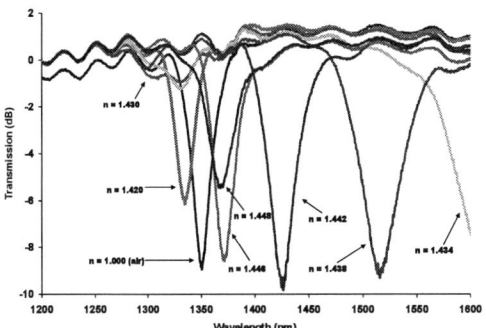

Fig. 17. Spectrum of an LPG coated with 50 nm of ITO being tuned with refractive index matching oils ranging from 1.420 to 1.448. (cladding diameter = 40 μm; grating period = 400 μm; LPG length = 8 mm (20 periods), ITO refractive index (@633nm) ~ 1.88)

After the mode transition occurs, the resonant peak tunes over 200 nm as the ambient index increase from 1.434 to 1.446. Nearly 100 nm of tuning is achieved for as small an index change as .004 (from 1.438 to 1.442) while keeping the peak depth stable within 1 dB.

Conclusions

In this paper, the effects of a high index overlay on an LPG's tuning behavior are reviewed. LPGs fabricated in single mode fibers with 125 μm diameter cladding have multiple resonant peaks in their output spectrum, and a small tuning range. They also experience severe peak degradation during tuning. As a result, they have limited use in practical applications that require tuning. However, a number of these limitations can be overcome by reducing the thickness of the fiber cladding and coating it with a high index overlay. By reducing the cladding thickness to 35-40 μm, the separation between the cladding modes increases, and a single resonant band in the output spectrum can be achieved. The high index overlay induces mode transitions in the cladding that lead to an enhanced tuning range. The mode transitions occur over a range

of ambient refractive indices that are determined by the thickness and refractive index of the overlay. If the properties of the overlay are optimized to the LPG parameters, the resonant peak can be tuned over a large range of wavelengths with a small change in the ambient index, and a uniform peak depth can be maintained across the majority of the tuning range. Experimental data is presented to support these conclusions. The tuning performance enhancements that result from the use of an optimized high index overlay make tunable LPGs a more viable technology in real world applications and device designs.

References

1. C. Tsao, Optical Fiber Waveguide Analysis, Oxford, New York (1992).
2. T. Erdogan, J. Opt. Soc. Am. A, 14 (1997) 1760-1773.
3. A. M. Vengsarkar, et al, J. Lightwave Tech., 14, (1996) 58-64.
4. T. Erdogan, J. Lightwave Tech., 15 (1997) 1277-1294.
5. V. Bhatia, Ph.D. dissertation, Virginia Polytechnical Institute and State University, (1996).
6. B. Lee, et al, Opt. Lett., 22 (1997) 1769-1771.
7. H. J. Patrick, et al., J. Lightwave Tech., 16 (1998) 1601-1612.
8. A. Abramov, et al., Elect. Lett., 35 (1999) 81-82.
9. S. Yin, et al., "Wavelength tuning range enhanced single resonant band fiber filter using a long period grating (LPG) with ultra thin cladding layer," OFC 2000, Baltimore, MD.
10. K. Chung and S. Yin, Microwave and Optical Tech. Lett., 12 (2001) 178-181.
11. K. Chung and S. Yin, Opt. Lett., 29 (2004) 812-814.
12. A. Cusano, et al., Opt. Exp., 14 (2006) 19-34.
13. I. Del Villar, et al., Opt. Exp., 1 (2005) 56-69.
14. I. Ishaq, et al., Sensors and Actuators B: Chemical, 107 (2005) 738-741.

High-speed versatile modulator for huge-capacity transmission

Tetsuya Kawanishi

National Institute of Information and Communications Technology (NICT)

4-2-1 Nukui-kita, Koganei, Tokyo 184-8795, Japan, e-mail: kawanish@nict.go.jp

Abstract *We describe high-speed versatile modulators, which can provide various types of lightwave modulation, such as differential quadrature phase shift keying, quadrature amplitude modulation, for huge-capacity transmission systems.*

Introduction

High-speed optical modulators play an important role in high-speed optical transmission technologies [1-3]. Optical modulators with traveling wave electrodes are commonly to achieve effective modulation in high-frequency region, where the velocity mismatch between lightwave and electric signals can be very small. However, frequency response of the modulators is limited by intrinsic loss in the electrodes. Typical 3-dB bandwidth of the response is about 30 GHz, which is not high enough for over 100 Gb/s serial lightwave signals in conventional on-off-keying (OOK) modulation format. Duobinary modulation is one of the solutions to overcome this problem, where the modulation signal bandwidth is a quarter of the bit rate [1]. Differential quadrature phase-shift-keying (DQPSK) is also useful to achieve over 100 Gb/s transmission [2, 3]. The symbol rate is half of the bit rate. 50 Gb/s electronics can be used both in transmitter and receiver sides, while 100 Gb/s electronic components are needed for duobinary. In addition, DQPSK modulation format is robust against chromatic dispersion and polarization mode dispersion in optical fibers. Thus, transmission systems using DQPSK have been attracted increasingly. Over 20 Tb/s huge-capacity DQPSK transmission was demonstrated by using wavelength-domain-multiplexing (WDM) and polarization multiplexing [4, 5]. Quadrature amplitude modulation (QAM) is also promising technique for high bit rate transmission, where required frequency bandwidth for modulators and electronics can be reduced [6]. In this paper, we describe optical modulation techniques for high-capacity transmission systems. Recently, we reported an integrated modulator based on a high-speed FSK modulator, where the modulator has six traveling wave electrodes and a pair of sub Mach-Zehnder interferometers (MZIs) embedded in a main MZI [7, 8]. This modulator is called a versatile lightwave modulator, because it can generate various types of lightwave signals, such as, return-to-zero (RZ) OOK, DQPSK, QAM, wideband FSK, continuous-phase FSK [9], etc.

Versatile modulator

The versatile modulator consists of two sub MZIs (MZ$_A$ and MZ$_B$) as shown in Fig. 1. The device structure is almost the same in the FSK modulator [3,

5], but the versatile modulator has six electrodes for chirp control and low half-wave voltage. The versatile modulator is composed of six optical phase modulators, where induced phases under the electrodes are denoted by ϕ_{A1}, ϕ_{A2}, ϕ_{B1}, ϕ_{B2}, ϕ_{C1} and ϕ_{C2}.

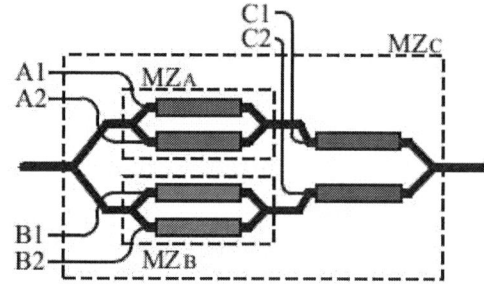

Fig. 1 Schematic structure of versatile modulator.

For simplicity, we assume that the two MZIs are in balanced push-pull operation, where the amplitude of the signal on the electrode A1 (B1) is equal to that of A2 (B2), but there is 180° phase difference ($\phi_{A1}=-\phi_{A2}$, $\phi_{B1}=-\phi_{B2}$). When the electrodes C1 and C2 are also in balanced push-pull operation ($\phi_{C1}=-\phi_{C2}$), the function would be similar to that of the FSK modulator [3]. When we apply single-tone rf-signals of the same frequency (f_m) on MZ$_A$ and MZ$_B$ with 90° phase difference, a frequency shifted lightwave can be generated at the output port of the modulator. The sub MZIs should be in null-bias point, where the dc-bias can be controlled by the electrodes A1, A2, B1 and B2. To eliminate upper sideband (USB) or lower sideband (LSB), 90° optical phase difference should be induced between the optical paths under the electrodes C1 and C2. The amplitudes of USB and LSB are, respectively, described by $[1+i\exp(-i\phi_C)]/2$ and $[1+i\exp(i\phi_C)]/2$, where $\phi_C=\phi_{C1}-\phi_{C2}$, and $\phi_C = +90°$ corresponds to an optimal condition for USB generation. Thus, by feeding a non-return-to-zero (NRZ) signal (source signal, henceforth), whose zero and mark levels respectively correspond to $\phi_C = +90°$ and -90°, to RFC, we can generate an optical FSK signal, without parasitic intensity modulation. On the other hand, if the single-tone rf-signals are in phase, we can generate a return-to-zero (RZ) OOK signal, where the zero and mark levels of the source signal should be $\phi_C = +180°$ and 0°, respectively. When MZ$_A$

978-0-9789217-3-6/07/$25.00

©2007 WEN GLOBAL SOLUTIONS   110

and MZ$_B$ are set to be in null-bias points, carrier-suppressed RZ (CSRZ) signals would be generated. The duty cycle of the RZ signals can be controlled by the bias of MZ$_A$ and MZ$_B$. When there are rf-signal amplitude differences between A1 (B1) and A2 (B2), the output would be a chirped RZ signal. For BPSK signals, the amplitude induced phase difference of the source signal should be 360°. We can also generate QPSK signals by feeding the in-phase and quadrature signal components to MZ$_A$ and MZ$_B$, respectively, where ϕ_C should be +90° or -90°. In addition, there are some other useful setups for advanced modulation, as shown in table 1. By changing electric feeding circuit configurations, the versatile modulator can generate various types of modulated signals.

Table 1 Modulation configurations for various formats.

	MZ$_A$ ϕ_{A1}, ϕ_{A2}	MZ$_B$ ϕ_{B1}, ϕ_{B2}	MZ$_C$ ϕ_{C1}, ϕ_{C2}
OOK	SS: ϕ_{A1}=-ϕ_{A2} =**0°, 180°**	SS: ϕ_{B1}=-ϕ_{B2}= **0°, 180°**	ϕ_{C1}-ϕ_{C2} = 0° for NRZ = PS for RZ
BPSK	SS: ϕ_{A1}=-ϕ_{A2} = **0°, 180°**	SS: ϕ_{B1}=-ϕ_{B2} = **0°, 180°**	ϕ_{C1}=-ϕ_{C2} = 0° for NRZ = PS for RZ
OOK	ϕ_{A1}-ϕ_{A2} =0° for NRZ = PS for RZ	ϕ_{B1}-ϕ_{B2} = 0° for NRZ = PS for RZ	SS: ϕ_{C1}=-ϕ_{C2} = **0°, 180°**
BPSK	ϕ_{A1}-ϕ_{A2} =0° for NRZ = PS for RZ	ϕ_{B1}-ϕ_{B2} = 0° for NRZ = PS for RZ	SS: ϕ_{C1}=-ϕ_{C2} = **0°, 180°**
QPSK	SS (Inphase) : ϕ_{A1} =-ϕ_{A2} =**0°, 180°**	SS (Quadrature) : ϕ_{B1} =-ϕ_{B2} = **0°, 180°**	ϕ_{C1}-ϕ_{C2} = 90°
QPSK	SS (MSB) : ϕ_{A1} =-ϕ_{A2} =**0°, 180°**	SS (MSB) : ϕ_{B1} =-ϕ_{B2} = **0°, 180°**	SS(LSB): ϕ_{C1}=-ϕ_{C2} = **0°, 90°** C1 and C2 are inphase
FSK	cos 2πf_mt	sin 2πf_mt	SS: ϕ_{C1}-ϕ_{C2} = **-90°, 90°**

When $|\phi_{A1}| \neq |\phi_{A2}|$, $|\phi_{B1}| \neq |\phi_{B2}|$, $|\phi_{C1}| \neq |\phi_{C2}|$, the output would be chirped.
SS: source signal, PS: pulse shape signal for carving,
MSB: most significant bit, LSB: least significant bit

High-speed DQPSK modulation using versatile modulator

For DQPSK modulation, two data signals are applied to the two sub MZIs, to achieve control of in-phase and quadrature components [7], as shown in Fig. 2. Each of the sub MZIs was biased for minimum dc transmission, where optical phase difference between the two sub MZIs was adjusted to π/2 by using the electrode C1 or C2. The electrode lengths of the main and sub MZIs were respectively 16 mm and 32 mm. Half-wave voltages (Vπ) of the main and sub MZIs were, respectively, 4.9 V and 2.5 V in push-pull operation at low frequency. Optical 3 dB bandwidth of each electrode was larger than 27 GHz, as shown in Fig. 3. The insertion loss of the modulator was 5.1 dB. To demonstrate 80 Gbit/s optical DQPSK modulation, a pair of non-return-zero (NRZ) electric signals were fed to the modulator for I and Q modulation, where the voltage was 6.5 V (peak-to-peak) corresponding to 2Vπ at 40 Gbit/s. We used a single one-bit delay receiver to decode each 40 Gb/s tributary by adjusting the differential optical phase in the one-bit delay interferometer ($\Delta\phi$) at π/4 or -π/4. Fig. 4 shows eye diagrams measured at the electric output of the balanced photodetector, and a bit-error-rate (BER) curve of a sub channel extracted from each tributary by a 1:4-demultiplexer. In back-to-back transmission,

clear eye openings and error-free operation were observed for the two tributaries whose symbol rate was 40 Gbaud.

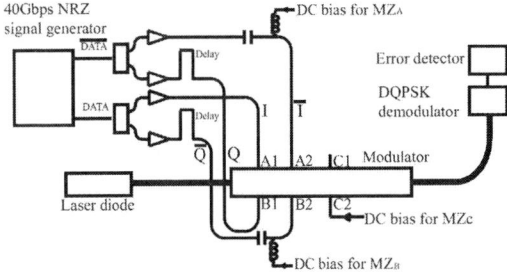

Fig. 2 Setup for DQPSK modulation.

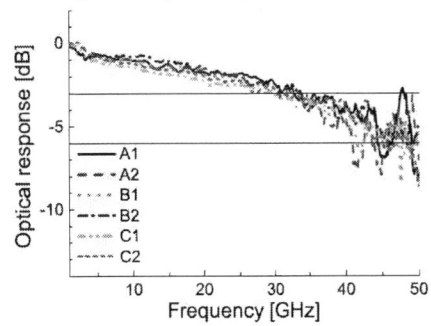

Fig. 3 Frequency response of versatile modulator.

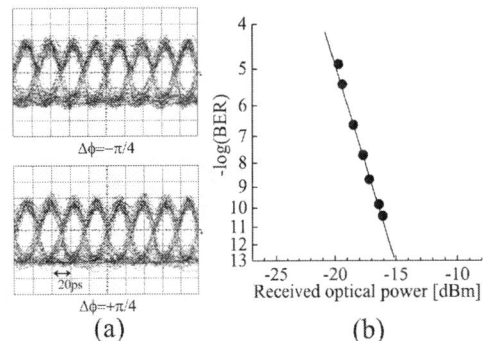

Fig. 4 80 Gb/s optical DQPSK signal: (a) eye diagrams of demodulated signals, and (b) BER curves.

Bias monitor technique

In the versatile modulator, the MZIs have Y-junctions, as shown in Fig. 1. For DQPSK modulation, the sub MZIs should be at a minimum transmission bias point. Lightwaves in two arms have π/2 phase difference at the Y-junctions of the output port side, and would be converted into radiative waves. To monitor bias condition of the sub MZIs, we propose an integrated modulator device as shown in Fig. 5, where two directional couplers were used as junctions of the sub MZIs at the output port side. Each directional coupler has two outputs. One was connected to the Y-junction of the main MZI, and the other was used for monitor purpose. The optical intensity profiles in the two

111

output ports of each directional coupler would be inverted each other. Thus, by maximizing the intensity of the monitor ports, the sub MZI can be at a minimum transmission bias point which is an optimal condition for DQPSK. We measured an optical response on voltage applied to MZ_B, by feeding 20 Vpp 1 kHz triangle waveform signal, where the optical intensity of the monitor port for MZ_A was maximized, so that MZ_A was in a minimum transmission bias point. As shown in Fig. 6, the optical response of the monitor port was almost ideally intensity inverted from that of the optical output. The extinction ratio of the main MZI was 23 dB, while that of the sub MZI (MZ_B) was 28 dB. The optical intensity at the optical output port was smaller than that of the monitor port. This is due to 3 dB intrinsic optical loss at the Y-junction of the main MZI.

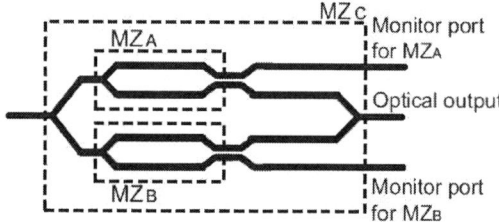

Fig. 5 Bias condition monitor.

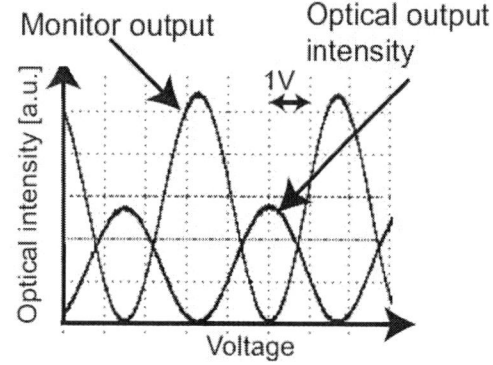

Fig. 6 Response of MZ_B monitor port

Conclusions

We described high-speed DQPSK modulation using an integrated modulator. 80 Gb/s DQPSK back-to-back transmission with -20 dBm receiver sensitivity was demonstrated by feeding 40 Gsymbol/s I and Q signals to two sub MZIs. We also investigated a bias condition monitor technique using directional couplers. In a fabricated modulator with two monitor ports, a sub MZI was successfully adjusted at a minimum transmission bias point, where the optical response of the monitor port was almost ideally intensity inverted from that of the main optical output.

Acknowledements

This study was supported by Industrial Technology Research Grant Program in 2004 from New Energy and Industrial Technology Development Organiza-tion of Japan. The authors wish to thank Dr. M. Tsuchiya of NICT for his encouragement and fruitful discussion.

References

1 P. J. Winzer et al, ECOC 2005, Th4.1.1
2 M. Daikoku et al, OFC 2006, PDP36
3 P. J. Winzer et al, OFC 2007, PDP24
4 A. H. Gnauck et al, OFC 2007, PDP19
5 H. Masuda et al, OFC 2007, PDP20
6 M. Nakazawa et al. OFC 2007, PDP26
7 T. Kawanishi et al, ECOC 2005, Th2.2.6
8 T. Kawanishi et al, JSTQE, 13,79 (2007)
9 T. Sakamoto et al, OFC 2005, OFG2
10 T. Kawanishi et al, ECOC 2005, Th1.6.6

Recent Advances in Commercial Electro-Optic Polymer Modulator

Bing Li, Raluca Dinu, Dan Jin, Diyun Huang, Baoquan Chen, Anna Barklund, Eric Miller,
Merly Moolayil, Guomin Yu, Yun Fang, Lixin Zheng, Hui Chen, Jeevan Vemagiri
Lumera Corporation, 19910 N. Creek Pkwy, Bothell, WA, USA
bli@lumera.com; raluca_dinu@lumera.com;

Abstract *We present the state-of-art polymer-based electro-optic modulator, which, for the first time, can be used in real commercial systems. The devices work up to 100Gbps and possess the Vπ as low as 1.6V.*

Introduction

For years, electro-optical (EO) polymers have been researched and developed. However, although the electro-optical co-efficient, r_{33}, has been continuously increased [1-2], poor stability, generally high optical loss, and process difficulties have hindered commercial success. The high r_{33} value from the single layer poling usually can not be realized in the device, due to the relatively high conductivity of the EO polymer compared to the polymer clads needed in typical 3-layer optical waveguides. Additionally, the low velocity mismatch between the lightwave and RF signals, which is one of the most attractive properties of EO polymers, often failed to deliver the promised bandwidth in practical devices due to high propagation loss of the traveling wave microstrip electrode.

In this paper, we review the recent efforts and achievements towards commercial electro-optic modulators that we have conducted in Lumera Corporation. We have developed the most advanced EO polymers that possess high thermal and photo-stability. A thin clad poling process has been developed to realize the high r_{33} potential of the material while not increasing optical loss in the device. At last, an innovative electrode and its launch structure design combined with a specially developed molded plating process improved the traveling wave electrode performance to fully realize the bandwidth of the EO polymer modulator.

Photo and thermal stability

We have established detailed material structure-stability relationship as guidance and continue to develop device level chromophores with high thermal stability. Figure 1 shows that chromophore B10 in APC settles at 90% of the r_{33} retention after thermal treatment at 85° C for 120 hours.

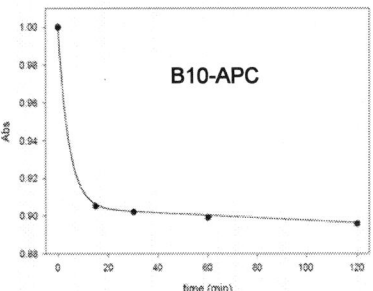

Figure 1, thermal stability of the B10-APC host guest EO polymer: r_{33} vs. time in 85°C

The long term photo stability of the device is also achieved by not only more stable molecule structure but also sealing the device in a non-oxygen environment. Figure 2 shows the device photo stability test under 100mW optical input for a variety of EO polymers developed in Lumera. It is the highest stability ever achieved in the field.

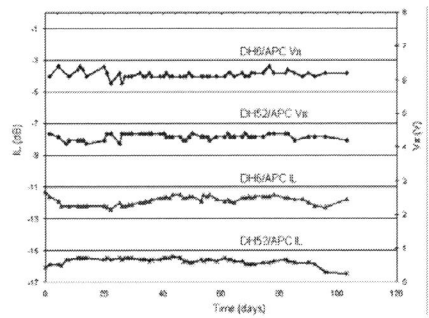

Figure 2, Device photo stability of Lumera's EO polymers at 100mW continuous optical input

978-0-9789217-3-6/07/$25.00
©2007 WEN GLOBAL SOLUTIONS

Thin clad poling for ultra-low Vpi

The difficulty of the 3-layer poling rises from the conductivity of the EO polymer. Highly conductive cladding materials have been researched and developed to improve the poling efficiency [3]. However, highly conductive clad polymers penalize the performance of the device by increasing RF loss and optical loss.

We noticed that even if the "resistivity" of the cladding material is high, if the cladding layer is thin enough during poling, a high E-field in the core layer can still be obtained. In this thin clad poling process, a thin but highly resistive cladding material is used and actually preferred. The thin cladding increases the capacitance of the layer and therefore the charging time of that layer during poling. When the layer is thin enough, the charging time will be longer than the time needed to fully rotate the molecules. Thus, this long charging time allows the voltage to drop mainly across the EO core layer in a time frame that allows the chromophores to become oriented. Additionally, the highly resistive clad polymer significantly reduces the leak current during poling, which helps prevent breakdown of the EO polymer and concomitant high optical loss, even as high voltage is applied.

In a real device, how thin the cladding can be is also restricted by the vertical index contrast, which determines the vertical confinement of the optical waveguide mode. Therefore, a high index contrast is preferred so that the electrode can be close to the core without causing high optical loss through plasmon absorption. For that reason, we chose a bottom clad with a refractive index as low as possible. For the top clad, a dual-clad process flow that integrates the thin clad poling process (Figure 3) was used to provide a dual-layer top clad with proper total thickness for the separation between the waveguide core and the metal, while the poling efficiency can still be ensured. Thus, we decoupled the waveguide loss with poling ($V\pi$) and achieved the best result for both parameters

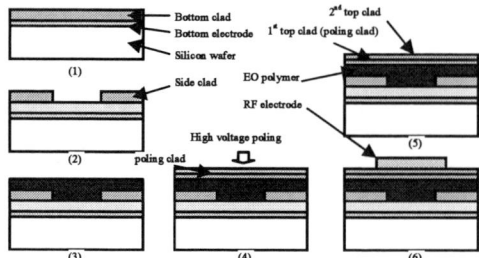

Figure 3, dual-cladding process the top cladding of the device is constructed in 2 steps. In the 1st step, the layer is only coated to the thickness optimized for poling. For different core and cladding, this poling clad thickness can vary from 0.5µm to 1.8µm in our device. After poling, the poling electrode will be removed, and the 2nd top clad layer will be coated, so that the total thickness is enough to isolate the light from the later deposited electrode.

Utilizing this thin-clad poling technique, a 1.6V $V\pi$ with single arm driving has been realized for a commercially usable EO polymer modulator, which has a bandwidth of 17GHz. The core material is DH80-APC with 50% loading. The DH80 is a Lumera developed chromophore that is both photostable and thermally stable. It also has a refractive index of 1.67, allowing us to use a 1.4µm bottom cladding and 1.8µm top cladding without considerable metal-induced optical loss. The mode profile of the device and the $V\pi$ measurement are shown in Fig.4.

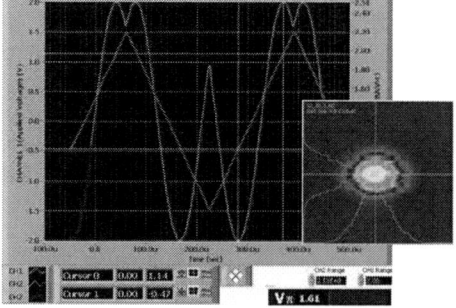

Figure 4, Vpi measurement of the commercial EO polymer modultor. The r_{33} is 80% obtained.

Realizing ultrabroad bandwidth

Because of the excellent velocity match between the RF signal and the lightwave in polymers, in principle the device can operate with ultra-broad bandwidth, and the electrode can be long to decrease $V\pi$ without degrading the bandwidth. However, due to the difficulty of plating thick metal on polymers, the traveling wave electrode of the device suffers significant loss at high frequency. The loss results from the skin effect, which can be depressed by plating thick Au with a shape that has high surface-to-volume ratio.

To achieve this, a molded plating process has been developed in Lumera. With that process, we can plate up to $20\mu m$ thick Au with a very high aspect ratio. Using the in-house developed UV curable polymer LP156, which has excellent mechanical strength, as the plating mold, both the shape and dimension of the resulted electrode can be precisely controlled (Fig.5).

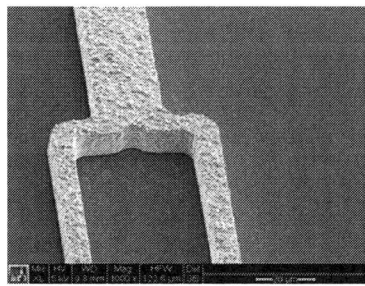

Figure 5, SEM image of Au electrode with molded plating process

Figure 6 shows the packaged device with the bird's view of the modulator chip. The embedded picture shows the newest design of the signal launch structure from the coaxial connector of the package to a coplanar waveguide (CPW) and then to the microstrip electrode of the polymer modulator. The measured S-parameter and EO bandwidth of the polymer modulator is shown in Fig.7 along with the 10Gbps eye-diagram. 60GHz bandwidth was obtained with 1cm long active length and push-pull $V\pi$ of 1.6V.

The eye diagram test was only performed at 10Gbps due to the limitation of the instruments. The device can potentially work up to 100Gbps data rate.

Figure 6, 40G commercial EO polymer modulator in package and the signal launch structure for the electrode

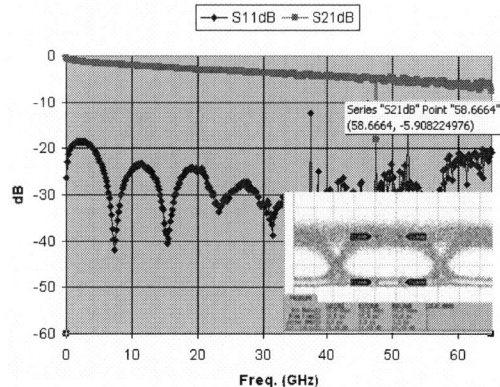

Figure 7, measured bandwidth performance of the 40Gbps EO polymer modulator; 10Gbps eye-diagram measured only due to the restriction of the instruments.

Conclusions

We present the state-of-art commercial EO polymer modulator. The device is stable over 100 days with 100mW of continuous laser input. Modulation of 40Gbps and a 60GHz bandwidth have been achieved with 1.6V $V\pi$ at push pull condition.

References

1 Yongqiang Shi, et al, Science, Vol. 288 (2000), page 119-122.

2 Min-Cheol Oh, et al, IEEE J. of Selected Topic of Quantum Electronics, Vol. 7 (2001), page 826-835.

3 Enami Y., et al, J. of Lightwave Technology, Vol. 21 (2003), page 2053-2060.

All-Optical Inverted Triode Based on Cross-Gain Modulation using InAs Quantum Dot Semiconductor Optical Amplifiers

Yoshinobu Maeda, Sayaka Maki, Yasuhiko Kuroki, Hideki Nakayama, Jae-Hoon Huh
Advanced Science and Technology Department, Toyota Technological Institute
2-12-1 Hisakata, Tempaku, Nagoya 468-8511, Japan. [†]E-mail: ymaeda@toyota-ti.ac.jp

Abstract *We designed active layer of 15 stacks of InAs quantum dots, AlGaAs/GaAs double heterostructure and fabricated QD-SOAs for optical inverted triode. Our results demonstrate input, control and output waveforms, and input-output characteristics of the optical triode. It was obtained high speed response time at higher bit rate of 40 Gbps.*

Introduction

We have been developed various semiconductor optical amplifiers (SOAs) for amplifying a high-speed optical signal processing and realizing an all-optical triode [1-5]. Commercially available SOAs are mainly used in the 1.5μm and 1.3μm bands optical fiber communication today. The optical loss of the 1.5μm band is the least, and the dispersion of the 1.3μm band is least. In the 1.5μm band, the signal light has been amplified by using an erbium-doped fiber amplifier (EDFA) as an optical amplifier. However, the dispersion of the 1.5μm band is larger than that of the 1.3μm band, so the load of plastic operation on the shape of waves of the 1.5μm band is larger than that of the 1.3μm band. Because the amplification level of the 1.3μm band by using a neodymium doped optical fiber amplifier is insufficient, the 1.5μm band is chiefly used in the present optical fiber communication. The achievement of a steady temperature characteristic, low power consumption, and a high-speed response to the input signal are expected by using quantum dot (QD) structure for SOA in a 1.3μm band. In this research, the aim is to propose and produce an optical triode with QD-SOAs that can be used with a 1.3μm band.

InAs quantum dot SOA

The feature obtained by making SOA a quantum dot structure is that the density of state becomes the delta function and many of various advantages, for example, low power consumption, in QD-SOA originate from this. In the quantum dot structure, the electron takes discrete energy so there is no extension of thermal energy. Therefore, a steady output within the wide range of temperature is obtained by using the QD structure. In this research, the aim is to propose and produce an optical triode with QD-SOAs that can be used with a 1.3μm band.

Fig.1 shows an internal structure of InAs QD-SOA and the band structure of the conduction band [6]. Fig.1 (a) shows the layer structure of QD-SOA. The difference between past SOA and QD-SOA is whether the active layer consists of a lot of quantum dot layers. QD-SOA of Fig.1 (a) consists of two layers. QD is made on Wetting layer (WL), and the career is injected from WL into QD. Fig.1 (b) shows the band structure of the conduction band of single QD. In this case, QD contains two discrete electronic states. One is ES (exited state), and the other is GS (ground state). GS relates to the stimulated emission when light is amplified. When the light is injected to the QD-SOA, careers that exist in GS are consumed by radiative recombination. The consumed career is supplied by ES by the intraband transition, and recovers at the time of 150fs order. QD-SOA can obtain higher speed response compared with past SOA because the supply time of the career from ES to GS is too short. The career is supplied to ES by WL at the time of the ps order.

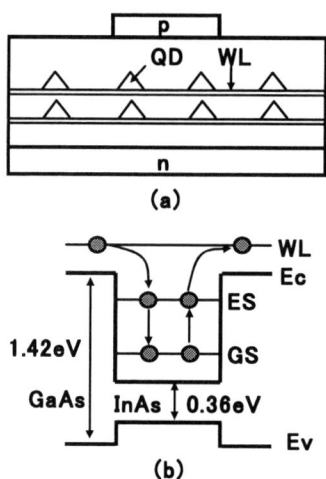

Fig.1 (a) Internal structure of InAs QD-SOA (b) Band structure of the conduction band

QD-SOA optical triode

Fig.2 (a) is a composition photograph of the InAs QD-SOA optical triode used in this experiment. Moreover, the block diagram of the optical triode is shown in Fig.2 (b). The wavelength of the band pass filters (BPF-1 BPF-2) are 1297.2±1.3nm because the BPFs are inclined. The active layer of QD-SOA consists of AlGaAs/GaAs double hetero structure that piles InAs QD by 15 layers. The center wavelength and gain are 1285nm and about 15dB, respectively, when an injection current of QD-SOA is 380mA.

978-0-9789217-3-6/07/$25.00

©2007 WEN GLOBAL SOLUTIONS

(a)

(b)

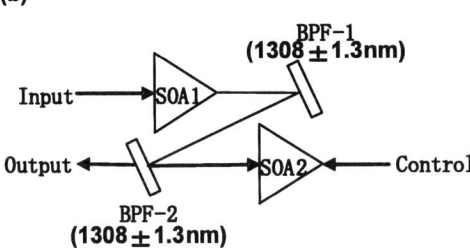

Fig. 2 (a) Structure of QD-SOA optical triode (b)Block diagram of the optical triode

Characteristics of QD-SOA optical triode

Fig.3 (a) and (b) show the input and control signal waveforms, respectively. The modulation frequencies of the input and control signals are 0.1GHz and 0.01GHz, respectively. The wavelengths of input and control lights are 1291nm and 1298nm. The input average power (P_{In}) is adjusted to a constant 0.5mW. Fig.3 (c) shows the output waveforms for six values of the control average power (P_C) from 0.03 to 0.23mW. The information of the input signal is transmitted to the output only when the power of the control signal is at the high logical level, whereas it is blocked when the control power is low. In addition, the output waveform reverses to the input wave form. This is because the light from SOA1 and control light causes cross gain modulation (XGM) in SOA2.

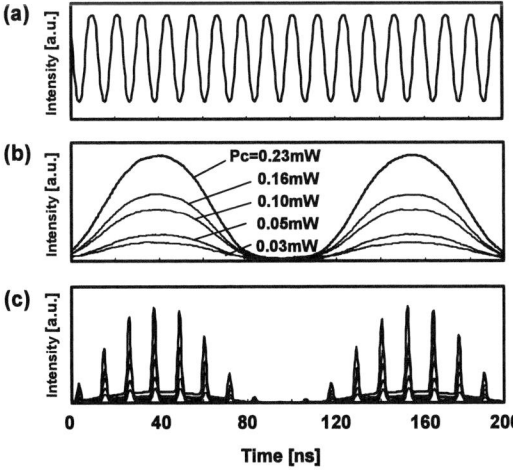

Fig. 3 (a) Input (b) Control and (c) Output signal waveforms of QD-SOA optical triode

Fig.4 shows the relationship between the input and output power in relation to control average power (P_C) from 0.03 to 0.23mW. Output power decreases gradually as input power increases. This is also because the light from SOA1 and control light causes cross gain modulation in SOA2. From this result, we can control output intensity by changing average control power.

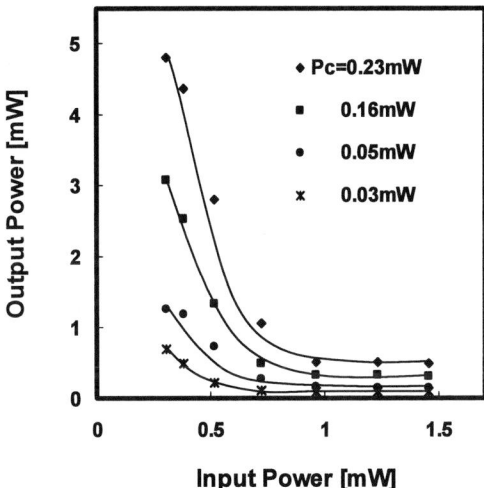

Fig.4 Relationship between Input and output power in relation to average control power.

Response time of optical triode

Fig.5 shows relationship between the modulation ratio (M_{out}/M_{in}) and the bit rate of the input signal using QD SOA structures. InAs QD-SOA is appeared to have high modulation ratio up to at least 40Gbps. The clearly revealed result shown in this experiment indicates that structure of the material play a most important role for high-bit-rate optical processing. We conclude that QD structure is effective method for high-speed optical signal response. This experimental demonstration of amplification over 40 Gbps signal in QD-SOA can predict that QD-SOA will have high impact on the telecommunication field in view of next generation application.

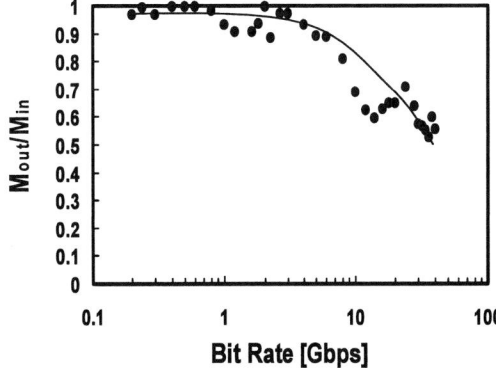

Fig.5 Dependence of modulation ratio on input signal bit rate

Conclusions

We demonstrated an optical triode using two InAs QD-SOAs. It was confirmed that the optical triode, which the output waveform reverses to the input one and we can control output power by changing control power, can respond to a higher bit rate. This experimental demonstration of amplification over 40 Gbps signal in QD-SOA can predict that QD-SOA will have high impact on the telecommunication field. Furthermore, all-optical triode developed for this QD-SOA over 40Gbps will be attractive candidate for the realization of future ultra high-speed photonic networks in view of next generation application.

Acknowledgements

This work was supported in part by the Ministry of Education, Culture, Science and Technology of Japan, Grant-in-Aid for Scientific Research (C) and Academic Frontier Project, and Japan Science and Technology Agency, Creation and Support program for Start-ups from University.

References

1 Y. Maeda et al IEEE Photon. Technol. Lett., **15** (2003) 257-259.

2 Y. Maeda Jpn. J. Appl. Phys., **43** (8B) (2004) 5907-5909.

3 Y. Maeda Appl. Phys. Lett., **88**, (2006) 101108.

4 Y. Maeda et al Rev. Laser Engineering, **34**, (2006) 71.

5 Y. Maeda Proceedings of 2006-Photonics in Switching Conference, (Capsis beach hotel, Heraklion, Crete, Greece, 2006), pp. 255-257.

6 J. Mork et al Optics and photonics News, July, 2003) 43-48.

Modulation Properties of Erbium Doped Silicon Laser Diode

Md. Zahid Hossain (1), Samia Subrina (2), Md. Quamrul Huda (3)

Department of Electrical and Electronic Engineering, Bangladesh University of Engineering and Technology, Dhaka 1000, Bangladesh

Email: (1) zahidhossain@eee.buet.ac.bd, (2) samiasubrina@eee.buet.ac.bd, (3) mqhuda@eee.buet.ac.bd

Abstract *Frequency response is of interest in determining the bandwidth of laser for application in communication networks. In this paper we investigate both amplitude and frequency modulations of erbium doped silicon laser using rate equations formalism.*

Introduction

Generation of light from silicon is a long quest in order to meet the increasing demand of high-speed optoelectronics and optical communications within the silicon chip. But the indirect band gap nature of silicon makes it unsuitable for radiative electron-hole recombination. However, various approaches such as band-gap engineering and the use of porous silicon have been adopted for efficient generation of light from Si [1]. Among them, incorporation of erbium in crystalline silicon draws enormous attention in recent years as it emits light at 1.54µm, which belongs to the minimum loss window of fibre optic communications [2-3].

Different aspects of Si:Er system have studied and light emiting diode have been reported several years ago. But very small power level of Si:Er LED makes it unsuitable for practical applications [4]. However sustained stimulated emission make Si:Er system feasible for laser operation and optical modulator [4-5]. In this paper we show analytically the performance of silicon erbium laser for direct amplitude modulation and frequency modulation.

Device Structure for laser operation

Excitation of erbium from silicon host is an electron-hole mediated process that depends on the background doping. This excitation process can be described by Shockley-Read-Hall recombination kinetics [6-7]. In our model, we have assumed erbium is doped in the p-region of a junction diode as excitation efficiency of erbium is higher in p-type material than n-type

material [5]. Upon the forward biasing, recombination energy of electron-hole pair near the erbium level excites erbium atom from ground state to the first excited state. The radiative decay of erbium atom in presence of population inversion produces stimulated light.

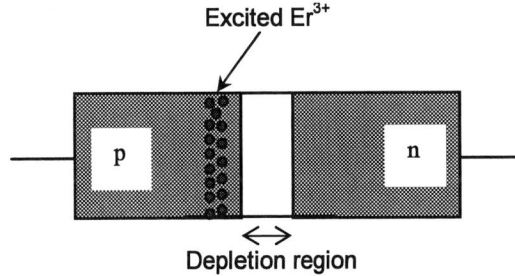

Figure 1: *Incorporation of Erbium in the p region of the junction diode.*

Theory and Analysis

We have assumed that erbium in silicon host forms a quasi two level state where the $^4I_{13/2}$ and $^4I_{15/2}$ states are involved. The single mode rate equations for excited erbium atom and photon density between these two states can be written as

$$\frac{dN^*_{Er}}{dt} = f_t c_p p \left(N_{Er} - N^*_{Er} \right) - \frac{N^*_{Er}}{\tau_{Er}} - v_g g S \quad (1)$$

$$\frac{dS}{dt} = \Gamma v_g g S + \beta \frac{N^*_{Er}}{\tau_{rad}} - \frac{S}{\tau_p} \quad (2)$$

where N^*_{Er} is the excited erbium atom, S is the photon density in the cavity, f_t is the occupation probability of erbium trap, c_p is the hole capture coefficient, N_{Er} is the concentration of erbium atom in the p-region, τ_{Er} is the life time of erbium atom

978-0-9789217-3-6/07/$25.00

©2007 WEN GLOBAL SOLUTIONS

both radiative and nonradiative, v_g is the group velocity of light, g is the gain of the cavity medium, β is the spontaneous emission factor coupled to the lasing mode, Γ is the optical confinement factor given by the ratio of active region volume to the modal volume, τ_{rad} is the radiative life time of erbium, τ_p is the photon life time and p is the injected carrier concentration. The carrier concentration is converted to current density by substituting $n = p = J\tau_{Si}/L_{eq}q$ where, τ_{si} is the silicon life time and L_{eq} is the thickness of erbium doped region. The gain of the lasing medium is a function of excited erbium atoms and is given by $g = g_0\left(2N_{Er}^* - N_{Er}\right)$ where g_0 is the gain cross section [4]. At threshold, the gain reaches to its threshold value g_{th} and the erbium atoms at the excied state pinned down at its threshold value ΔN_{Erth}^*. For simplicity in our analysis we have assumed that the value of optical confinement factor is unity and occupation probability of erbium trap by an electron is constant. Steady state solutions of the rate equations yield the laser output power as a function of drive current density.

For small signal intensity modulation, we consider the driving current density as $J(t) = J_0 + \Delta J(t)$ where J_0 is the dc current density above the threshold value and $\Delta J(t)$ is the ac modulating current. The time variation of current density leads to the time variation of $N_{Er}^*(t)$ and $S(t)$ accordingly. The small signal response of Si:Er laser is obtained by linearizing the rate equations. The transfer function for intensity modulation is expressed as

$$M(\omega) = \frac{\Delta S(\omega)}{\Delta J(\omega)} = \frac{A_{14}}{\left(A_{13} - \omega^2\right) + j\omega A_{12}} \quad (3)$$

where, $\Delta S(\omega)$ and $\Delta J(\omega)$ are the sinusoidal variation of photon density and modulaitng current density respectively. The normalized transfer function is given by

$$H_{AM}(\omega) = \frac{M(\omega)}{M(0)} = \frac{\omega_r^2}{\left(\omega_r^2 - \omega^2\right) + j\omega\gamma} \quad (4)$$

where, ω_r is the frequency of relaxation oscillation and γ is the damping coefficient of the resonance. Both of these are functions of photon density and system parameters.

Fig:2 shows the normalized modulation response of Si:Er laser against frequency. If we take the frequency of the peak amplitude of modulation response as the modulation bandwidth for simplicity, the figure shows that the bandwidth increases with the increase of power level. Another observation can be made that the peak amplitude of the response is high at low power level but decreases and becomes more flatter as the power level increases. This is because as the power level increases, the damping coefficient increases and hence suppresses the height of the resonance peaks.

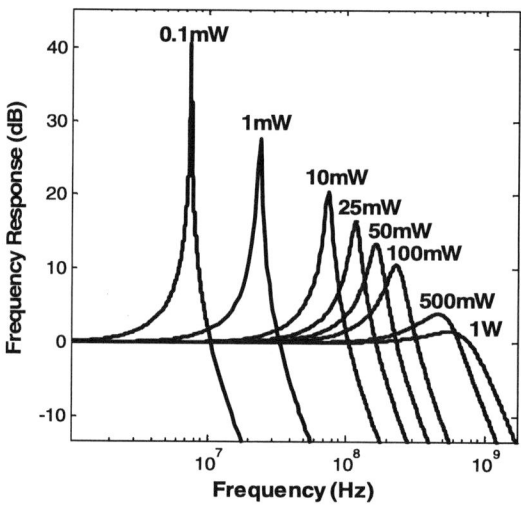

Figure 2: *Normalized modulation response of Si:Er laser for different operating power level. Doping density of Er is $10^{19}/cm^3$.*

Frequency modulation or chirp is caused due to the modulation-induced variation in the excited erbium atom. This variation leads to the variation in the refractive index of the active media which has been assumed constant for intensity modulation analysis. The variation of refractive index ultimately leads to the variation in the optical frequency. This frequency modulation of laser is desirable if we wish to dynamically tune the laser. The optical frequency shift due to change in ΔN_{Er}^* in the active layer is given by [8]

$$\Delta \vartheta = \frac{\alpha \Gamma g_0 \Delta N_{Er}^*}{4\pi} \qquad (5)$$

where, α is the linewidth enhancement factor and is given by the ratio of change in the real component to the imaginary component of refractive index of the active media. The small signal FM transfer function is given by

$$F(\omega) = \frac{\Delta \vartheta(\alpha)}{\Delta J(\alpha)} \qquad (6)$$

Using equations (1-2, 5-6), the normalized transfer function is given by

$$H_{FM}(\omega) = \frac{F(\omega)}{F(0)} = \frac{A_{03}(A_{04} - j\omega)}{A_{01} - \omega^2 + jA_{02}\omega}$$

where A_{01}, A_{02}, A_{03} and A_{04} are system parameters.

Figure 3 illustrates the normalized FM characteristics as a function of modulating frequency. Frequency chirping increases linearly with the optical modulation depth. At

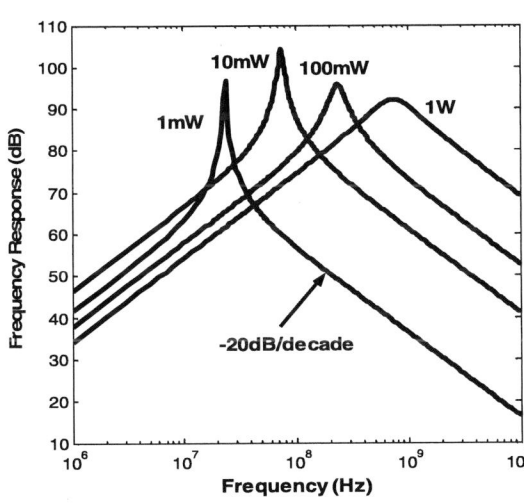

Figure 3: *Frequency modulation response with variations of the laser output power.*

high frequencies, the response drops at a slope of -20dB/decade above the peak as opposed to -40dB/decade for the IM transfer function. The peak of the resonance is much higher than the corresponding peak in intensity modulation. The figure also shows that frequency modulation broadens the modulated spectrum of the laser.

Conclusions

In this paper we have studied analytically, by using small signal analysis, the modulation response of Si:Er laser diode. The analysis shows that direct modulation in GHz level is possible to achieve for laser outputs in the range of watt. At high power level frequency chirp is also high.

References

1 Canham et al Mater. Res. Soc. Bull, Volume 18 (1993), page #22.
2 Ennen et al Applied Physics Letter, Volume 43 (1983), page # 943.
3 Coffa et al Physical Review B, Volume 48 (1993), page # 11782.
4 Xie et al Journal of Applied Physics,, Volume 70, (1991), page # 3223.
5 Huda et al Materials Science and Engineering B, Volume 105 (2003), page # 146.
6 Huda et al Solid State Commun.,Volume 118 (2001), page # 2001.
7 Kik et al Applied Physics Letter, Volume 70 (1997), page # 1721.
8 Tucker et al Journal of Lightwave Technology, volume LT-3 (1985), page # 1180.
9 Md. Zahid Hossain M.Sc. Thesis, Dept. of EEE, BUET, 2006.

In-Line Fiber-Optic Etalon Formed By Hollow-Core Photonic Crystal Fiber

Y. J. Rao

Key Lab of Broadband Optical Fiber Transmission & Communication Networks Technologies (Education Ministry of China), University of Electronic Science & Technology of China, Chengdu, Sichuan 610054, China, and Key Laboratory of Optoelectronic Technology and Systems (Education Ministry of China), Chongqing University, Chongqing 400044, China
yjrao@uestc.edu.cn

Abstract: A novel fiber-optic in-line etalon is proposed and demonstrated, formed by splicing a section of hollow-core photonic crystal fiber (HCPCF) in between two single-mode fibers, for the first time to our knowledge.

Fiber-optic interferometric sensors have found numerous industrial, military and civil applications in recent years as they have a number of outstanding advantages over conventional electrical sensors, such as immunity to electromagnetic interference, capability of responding to a wide variety of measurands, very high resolution, high accuracy, small size, etc.[1] The fiber-optic extrinsic Fabry-Perot interferometric (EFPI) and in-line etalon sensors have been successfully commercialized and widely used for measurement of various parameters, such as strain, pressure, vibration, acceleration, refractive index, etc, in many fields.[2-6] However, it is hard to realize a dense fiber-optic sensor network, which can multiplex a large number of sensors, based on these sensor approaches due to their poor multiplexing capability. A modified EFPI sensor structure called the *Fizeau* cavity allows the cavity length to be enlarged up to several millimeters, this means that more than 10 EFPI or etalon sensors can be multiplexed simultaneously with spatial-frequency division-multiplexing (SFDM).[7-8] But, in the *Fizeau* configuration, the increase in the cavity length will make the fringe visibility, i.e. signal-to-noise ration (SNR) of the interferometric signal worse. In addition, further increase in the cavity length becomes

impossible with rapid degradation of the SNR of the interferometric signal. In this letter, a novel fiber-optic in-line etalon is proposed and demonstrated, which is constructed by splicing a section of hollow-core photonic crystal fiber (HCPCF) in between two standard single-mode fibers (Corning SMF-28) to form a Fabry-Perot interferometer. Such an in-line HCPCF etalon can greatly enhance the SFDM capability due to the substantial increase in the cavity length.

The schematic diagram of the in-line etalon HCPCF sensor is showed in Fig. 1, which is fabricated by splicing the ends of two SMF-28 fibers to the cleaved end of a HCPCF fiber (Blanze Photonics: HC-1550).

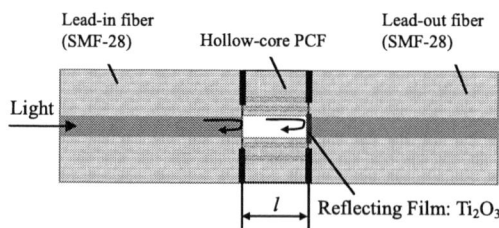

Fig.1. Configuration of an in-line HCPCF etalon

The core diameter and the distance between the centers of the cladding holes of the HCPCF are ~10.9 m and ~3.8 m, respectively. The fabrication of the etalon is simple and straightforward, i.e. splice the cleaved ends of the HCPCF to the cleaved ends of two SMF-28 fibers with an electric-arc fusion splicer (Fitel: S176). The etalon length can be cleaved down to the order of micrometers with inspection under a microscope. In addition, the etalon length could be extended up to several centimeters because the transmission loss (<0.1dB/m) of the HCPCF is much lower than the hollow core fiber configuration reported previously.[2]

Figure 2(a) is the reflective spectrum of the HCPCF etalon, which is obtained by using a high accuracy optical spectrum analyzer (OSA) (Micron Optics: Si720) with a wavelength resolution of 0.25pm and a wavelength precision of 1pm over a spectral range of 1520-1570nm. It can be seen from Fig. 2(a) that the fringe visibility is relative low due to the splicing loss between the two single-mode fibers and the HCPCF, which was measured to be ~1dB for each joint in our experiment. In order to compensate such a joint loss, a reflective film (Ti$_2$O$_3$) was coated on the end surface of the lead-out fiber to form a *Fizeau*-type etalon. Figure 2(b) shows the reflective spectrum of the *Fizeau* HCPCF etalon. It can be seen from Fig. 2(b) that a fringe visibility of ~0.3 was obtained and it was improved by ~5 times.

The fringe visibility V of this kind of *Fizeau* etalons can be given by:[9]

$$V = \frac{(\beta R_1)^{1/2}}{(R_1 + \beta)} \qquad (1)$$

where β is the intensity coupling efficiency of the incident beam reflected back from the coated end of the reflective fiber into the incident fiber and R_1 is the reflection coefficient of the incident fiber. In this work, β is mainly determined by the splicing loss between the single-mode fiber and the HCPCF as the very low transmission loss of the HCPCF can be ignored, rather than the cavity length in a conventional *Fizeau* cavity with an air gap. Hence, V varies little with the cavity length, this is verified by our experiment even when the HCPCH etalon length was extended to 2cm, as showed in Fig. 2(b). Assuming that the splicing loss is equal to 1dB, i.e. β=2dB, due to the round trip of the incident light, and R_1 is 0.04, V≈0.3 according to Eq.(1), which is in a good agreement with the experimental results.

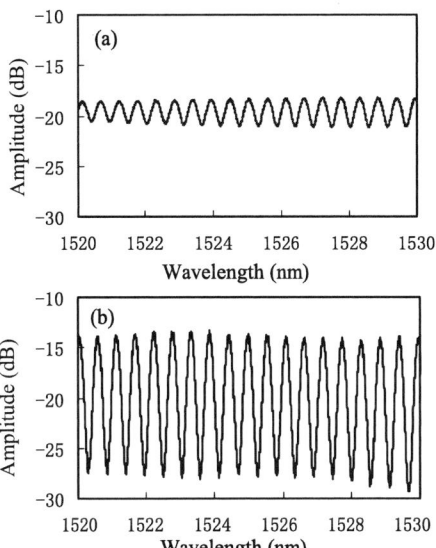

Fig.2. Reflective signals from a HCPCH etalon: (a) without reflective film, (b) with reflective film.

The novel *Fizeau* HCPCF etalon proposed in this work is demonstrated for strain measurement. The experimental setup is shown in Fig. 3.

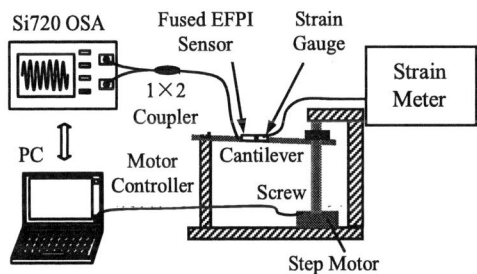

Fig.3 Schematic diagram of the experimental setup

A HCPCF etalon sensor and an electrical strain gauge were mounted on a steel cantilever, respectively. The initial length of the HCPCF cavity was 2.1mm. When a load was applied to the end of the cantilever by means of a precise step-motor controlled by a PC, the strain due to the deflection of the cantilever was measured by the two strain sensors, respectively. High-accuracy strain measurement was achieved by directly measuring the optical phase shift induced by the strain to determine the real cavity length change. The experimental results are shown in Fig.4 and it can be seen that the strain accuracy is better than ±5με for the etalon sensor and the strain results obtained from the etalon and strain gauge agreed very well. The temperature response for the etalon sensor is shown in Fig.5. It can be seen that the change in the cavity length of the etalon was ~0.25 m, corresponding to a thermal drift of ~2nm/℃ within a temperature range from -20℃ to 100℃.

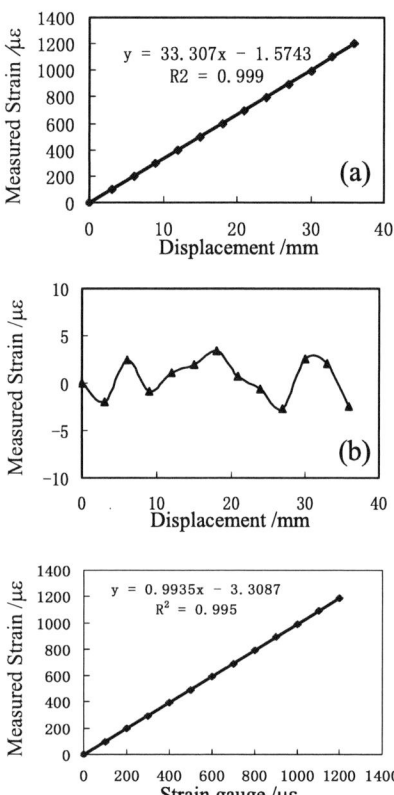

Fig.4. Comparison between measured strain and electrical strain gauge.

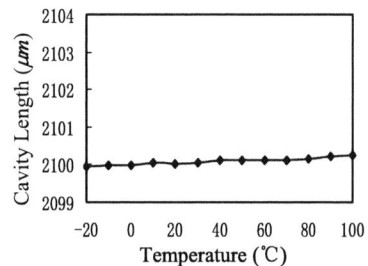

Fig.5. Temperature response of the HCPCF etalon

In this paper, a novel in-line fiber-optic etalon sensor is constructed by using a HCPCF as a low-loss optical waveguide to form a Fabry-Perot interferometer. This long-cavity etalon sensor allows the cavity length to be as long as several centimeters with good fringe visibility, offering the feasibility to realizing a practical dense fiber-optic strain sensor network if spatial-frequency division-multiplexing is adopted. The experimental results show that a strain accuracy of better than ±5με is achieved. It is anticipated that such a HCPCF-based etalon sensor system could find important applications for health monitoring of large structures.

This work was supported by the National Natural Science Foundation of China under grant 60477030, and the Key Project of the Natural Science Foundation of Chongqing under grant 8415.

References

1 D. A. Jackson, J. Phys. E: Sci. Instrum, 18, 981 (1985).

2 J. S. Sirkis, D. D. Brennan, M. A. Putman, T. A. Berkoff, A. D. Kersey and E. J. Friebele, Opt. Lett., **18**, 1973 (1993).

3 V. Bhatia, K. A. Murphy, R. O. Claus, M. E. Jones, J. L. Grace, T. A. Tran and J. A. Greene, Meas. Sci. Technol., **7**, 581 (1996).

4 Taylor H. F, in Fiber Optic Sensors, F. T. Y. Yu, eds. (Marcel Dekker, New York, 2002), pp. 41.

5 Y. J. Rao, Measur. Sci. & Technol., **7**, 981 (1996).

6 B. Yu, D. Kim, J. Deng, H. Xiao, and A. Wang, Appl. Opt. **42**, 3241 (2003).

7 Y. J. Rao, C. X. Zhou and T. Zhu, IEEE Photon. Technol. Lett., **17**, 1259(2005)

8 Y. J. Rao, Opt. Fiber Technol., **12**, 227 (2006).

9 Y. J. Rao, D. A.Jackson, R. Jones, and C. Shannon, IEEE J. Lightwave Technol., **12**, 1685(1994).

Invited Paper:

Two-core fiber based in-fiber integrated interferometers and its sensing applications

Libo Yuan, Jun Yang and Zhihai Liu
Photonics Research Center, College of Science, Harbin Engineering University,
Harbin 150001, P. R. China, Email: lbyuan@vip.sina.com

Abstract *Based on two-core optical fiber, compact in-fiber integrated Michelson and Mach-Zehnder interferometers are proposed and demonstrated, which have the potential to be exploited in a variety of wide sensing applications.*

Introduction

Fiber optic interferometric structures have been largely explored in sensing due to the high sensitivities that they exhibit on the measurement of a broad range of parameters [1,2]. Independently of their alternative topological configurations: Michelson, Mach-Zehnder, Fabry-Perot, Sagnac, etc., different implementations have been considered aiming to optimize the device performance for a particular set of measurands. In most sensor applications it is desirable that the fiber optic interferometric sensor is more compact and reliable. For this purpose, several techniques have been developed to build interferometer inside a string of fiber [3-5]. That is so-called in-fiber integrated interferometers. The group of in-fiber integrated interferometers where the interferometric phase difference is built up considering the difference in the effective refractive index of different fiber modes [6-8], the difference in the optical path [9,10] as well as the difference in multi-core of one fiber [11,12], have been widely researched in the context of environmental sensing, curvature sensing and others. These structures are attractive for several reasons, including small size and deployment flexibility, as well as the presence of a reduced thermal sensitivity in view of the usually small difference of the thermo-optic coefficients of the fiber core and cladding under concern. In this paper, two-core fiber based in-fiber integrated Michelson and Mach-Zehnder interferometers and its sensing applications have been presented.

Two-core fiber based in-fiber integrated interferometers

The proposed in-fiber integrated interferometers are based on a two-beam element formed from a two-core fiber. A compact in-fiber integrated Michelson and Mach-Zehnder interferometer was easily implemented from the fused-tapering single-mode fiber to two-core fiber coupling technique [5]. The experimental setup is shown in Fig.1. In the case of Michelson configuration (Fig.1 (a)), a segment two-core fiber was fused with the standard single mode fiber and tapered at the splicing point, and at the end face of the two-core fiber a dielectric mirror directly was deposited on it. The light from a LD light source is launched into the fiber and passing though a 3 dB coupler directly going to the in-fiber integrated Michelson interferometer. And the reflected signals from the mirror end of the two-core fiber are combined in the tapered zone and received by the photo detector. In this manner a Michelson interferometer was simply implemented from a single fiber. Similarly, in the case of Mach-Zehnder configuration, between the two segments of standard single-mode fiber, a segment of two-core fiber has been used. The three segments of fiber are fused together and tapered at the two splice pinots, then the so-called in-fiber integrated Mach-Zehnder interferometer has been built, as shown in Fig.1 (b).

Sensing applications

In order to establish the basic equations relating the optical path change induced in the two-core fiber to the fiber curvature, it is assumed that the fiber is a homogeneous and isotropic fused silica rod of radius r. For the case of weak curvature, i.e. the curvature radius R is much larger than r. A simple analysis model can be established as shown in Fig.2. And the optical phase differential of the two cores can be expressed as

$$\delta\phi = \frac{k_0 nLd}{R}\left[1 - c_2\frac{n^2}{2}\right]\cos[\theta_0 + \frac{\pi}{4} + \theta(z)] \quad (1)$$

978-0-9789217-3-6/07/$25.00

©2007 WEN GLOBAL SOLUTIONS

where, $k_0 = 2\pi / \lambda$ is the wave number and λ is the wavelength; n is the refractive index of the fiber core; L represents the length of the two-core fiber; d is the distance between the two cores.

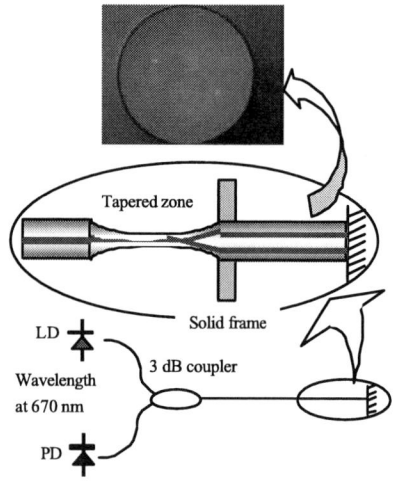

(a) In-fiber integrated Michelson interferometer

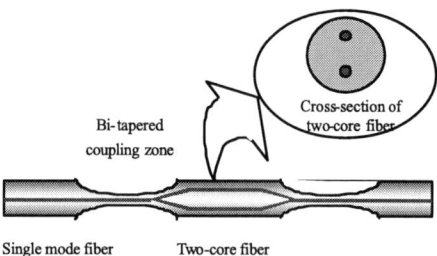

(b) n-fiber integrated Mach-Zehnder interferometer

Fig.1 In-fiber integrated compact Michelson and Mach-Zehnder interferometers built with twin-core fiber.

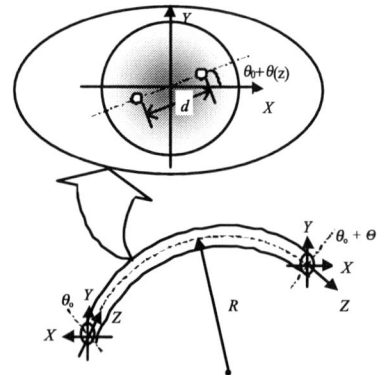

Fig.2 Bend induced phase change in the two-core fiber.

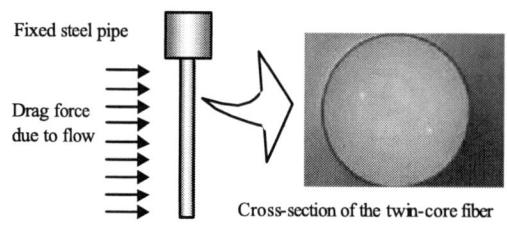

(a) Setup of compact two-core fiber flow velocity sensor

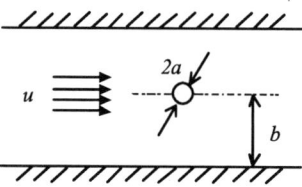

(b) Configuration of a fiber dragged by the flow

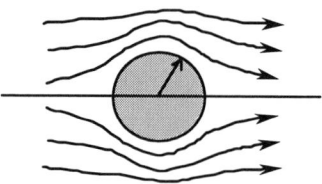

(c) Enlarged view of drag force on the optical fiber

Fig.3 Experimental setup of the flow velocity sensor.

The in-fiber integrated interferometers could be wildly used for sensing, such as a displacement sensor, an accelerometer and a flow-velocity sensor. Unlike many other sensors, theirs would be independent of environmental temperature and pressure changes because both arms of the interferometers would be affected equally by such changes. Here, for example, a flow-velocity sensor based on the in-fiber integrated Michelson interferometer has been demonstrated. It consists of a segment of twin-core fiber with steel pipe fixed on the solid frame, forming a micro-bending cantilever beam. The back reflected light from the two-core fiber is implemented by coating a reflecting silver surface onto the end of the twin-core fiber as shown in Fig.3 (a). For better understanding the bending sensing system of the twin-core fiber beam, the sensor head has been enlarged as shown in Fig.3 (b) and (c). When the cantilever beam of the

126

twin-core fiber is putted into a flow field, the phase difference between the twin cores of the fiber will be corresponding to the drag force of the flow exerted on the fiber cantilever beam, which the flow velocity can be measured by the signal-processing unit.

Conclusions

In conclusion, based on the two-core fiber, compact and stable in-fiber integrated Michelson and Mach-Zehnder interferometer as basic sensing device have been realized. The sensing characteristic of bending or curvature based is discussed. The sensitivity of the interferometer can be easily adjusted by changing the two-core fiber length. A lots of fiber optic sensors may be built up based on the in-fiber integrated interferometers, such as displacement sensor, curvature sensor, compact accelerator and flow velocity sensor.

Acknowledgements

This work was supported by the National Natural Science Foundation of China, under grant number 60577005, and partially supported by the Specialized Research Fund for the Doctoral Program of Higher Education Institute of MOE, China, to Harbin Engineering University.

References

1. D. A. Jackson, Monomode optical fiber interferometers for precision measurement, *J. Phys. E: Sci. Instrum*. **18**, (1985), 981-1001
2. D. A. Jackson, Recent progress in monomode fiber-optic sensors, *Meas. Sci. Technol.*, **5**, (1994), 621-638
3. X. Daxhelet, J. Bures, R. Maciejko, Temperature independent all-fiber modal interferometer, *Optical Fiber Technology*, **1**, (1995), 373-376
4. O. Duhem, J. F. Henninot, M. Douay, Study of in fiber Mach-Zehnder interferometer based on two spaced 3-dB long period gratings surrounded by a refractive index higher than that of silica, *Optics Communications*, **180**, (2000), 255-262
5. L. B. Yuan, Z. H. Liu, J. Yang, Coupling characteristics between single core fiber and multi-core fiber, *Optics Letters*, **31**(22), (2006), 3237-3239
6. T. A. Eftimov, W. J. Bock, Sensing with a LP_{01}-LP_{02} intermodal interferometer, *Journal of Lightwave Technology,* **11**(12), (1993), 2150-2156
7. A. Kumar, N. K. Goel, R. K. Varshney, Studies on a few-mode fiber-optic strain sensor nased on LP_{01}-LP_{02} Mode interference, *Journal of Lightwave Technology*, **19**(3), (2001), 358-362
8. D. Kacik, I. Turek, I. Martincek, J. Canning, N. A. Issa, K. Lyytikainen, Intermodal interference in a photonic crystal fiber, *Optics Express*, **12**(15), (2004), 3465-3470
9. P. L. Swart, Long-period grating Michelson refractometric sensor, *Measurement Science and Technology,* **15**, (2004), 1576-1580
10. J. Villatoro, V. P. Minkovich, D. Monzon-Hernandez, Compact modal interferometer built with tapered microstructured optical fiber, *IEEE Photonics Technology Letters*, **18**(11), (2006), 1258-1260
11. L. B. Yuan, J. Yang, Z H. Liu, J. X. Sun, In-fiber integrated Michelson interferometer, *Optics Letters*, **31**(18), (2006), 2692-2694
12. L. B. Yuan, X. Wang, Four-beam single fiber optic interferometer and its sensing characteristics, *Sensors and Actuator,* **138**, (2007), 9-15

A Nonimaging Optics Approach For Photoelectric Sensor Applications

Jun Jiang

Motorola Labs, 1301 E. Algonquin Road, Schaumburg, IL 60196 U.S.A

email:junjiang@motorola.com

Abstract *This paper discusses the application of a hybrid TIR lens, a nonimaging optics, in photoelectric sensors. The performances and design are evaluated using a commercial non-sequential ray tracing software. The results of the TIR lens in a conceptual thru-beam photoelectric sensor system are compared with the results from a CPC and an imaging optics.*

Introduction

Photoelectric sensing as a means of industrial control has been available for many decades. These sensors almost always have lenses built into the sensor housing. The large lenses common on photoelectric sensors direct light energy toward a target and collect any returned energy.

Most of the photoelectric sensors today use imaging lenses due to the fact that geometrical optics development has been closely linked to imaging optics since its very beginning. However, in photoelectric sensing applications, the most important problem to be addressed is "how much energy falls on the detector?" not the quality of image. It was not until the 60´s when it was recognized that the image formation constrain is not needed to solve some optical design problems, and in particular the problem of maximum radiation transfer between a source and a target. The elimination of this constrain led to an additional degree of freedom which in turn led to more effective designs suitable for photoelectric sensors application. This new types of optics were called Non-imaging Optics. A classical book in this field is written by Welford and Winston [1].

The Compound Parabolic Concentrator (CPC) probably is the simplest among all of the non-imaging optical components. It can achieve high concentration ratio and efficiency. However the total length of CPC typically is very large and thus may limit the application of CPC to photoelectric sensors where overall sensor size is very demanding.

Recent development in LED illumination industry has led to some novel structures of non-imaging optics. Analysis of different concentrator designs shows, that Etendue conserving behaviour can also be fulfilled by concentrators with folded Total Internal Reflections (RX and RXI like structures)[1]. Compared to the CPC, these types of optics can achieve very high compact ratio (total optics length/diameter ratio of approximately 1/3). The only drawback of these structures is that the photoelement has to be either embedded in the optics or a reflective coating has to be applied to the central portion of the leading surface, which may results in complicated manufacturing process. We used a TIR assisted concentrator with central lens. This structure would be much simpler to manufacture. In this paper, we will report the structure of a TIR lens and its advantage over a CPC or an imaging lens when used in photoelectric sensors application.

Characteristics of the TIR lens

We have modelled a TIR lens based photoelectric sensor with LightTools, a commercial simulation program. Conventional imaging lens parameters such as aberrations are no longer important for optimization in this TIR lens design since we only interest in the maximum energy transfer efficiency, not the image formation. In our study, a pair of TIR lenses are used: one as collimator for the VCSEL and the other as a concentrator for the photodiode.

978-0-9789217-3-6/07/$25.00

©2007 **WEN GLOBAL SOLUTIONS**

The TIR lens is composed of 3 segments: a central lens with a cylindrical cut out surrounded by a TIR surface. A secondary hemispheric optics is implemented. As shown in Figure1, when used for receiver, rays incident on the central lens portion is directed into the hemispheric lens and focused on photodiode; rays incident on the non-central portion is first refracted by the leading surface and then reflected by TIR from the back surface and directed to the cylindrical cut out; the taper angle of the cylinder wall is designed in such a way that TIR rays will be refracted into the hemispheric lens and eventually focused on photodiode. Likewise, the TIR lens can also be used as a collimator for emitter as shown in Figure 1 (a) and (b).

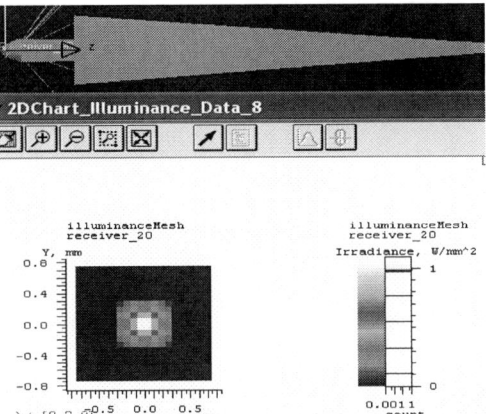

(b) lens as concentrator for receiver (PD)

Figure 1.TIR lens structure and the use as emitter collimator and receiver concentrator

Figure 2 shows some characteristics of the TIR lens. The TIR lens has a very sharp cut off acceptance angle.

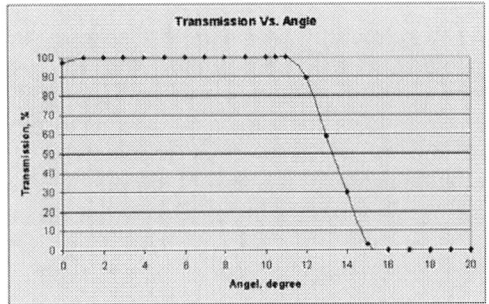

Figure 2. Angular transmission of TIR lens

A Conceptual photoelectric sensor system

The technology incorporated into photoelectric controls and sensors since the 1970s has revolutionized the market with significant advances in the state of the art in optoelectronics, optics, and packaging. Today's light sources are very compact, a typical SMD LED for illumination is around mm size (with die of about 100um), more efficient SMD VCSEL around mm size is also available (with die size of 10s of um) and cost is comparable to LED, a typical photodiode with mm size can be highly sensitive and widely available.

Figure 3 shows a conceptual thru-beam photoelectric sensor configuration. Light is

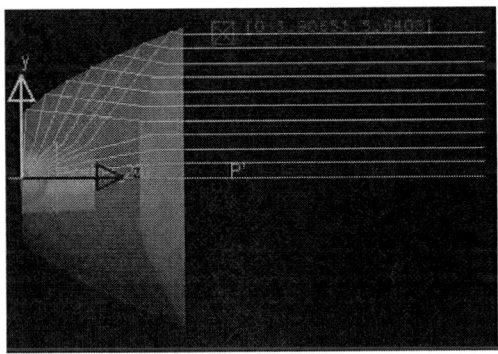

(a) Raytracing of TIR lens structure

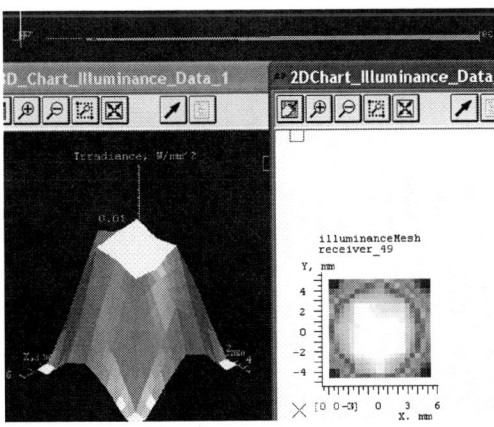

(b) Lens as collimator for emitter (VCSEL, LED etc.)

emitted by a source, e.g. a VCSEL, towards an objective lens that bends its rays and transfers the light energy with maximum efficiency to a given target some distance away (lens as collimator); a focusing lens receives the radiation and concentrates the light on a photodiode while making allowance for an angular aiming error of ±θo (lens as concentrator). Both imaging and non-imaging lenses can be used. In our study, we also included CPCs, comparison of these lenses are summarized in Table1.

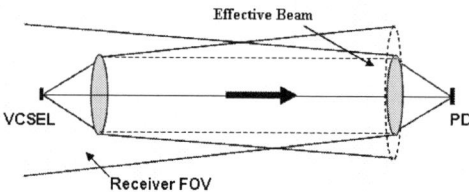

Figure3. Thru-beam configuration of photoelectric sensor

As we know, increasing the size of the photodiode will improve the sensitivity, but in the mean time this also increases the internal noise and slows the response time of the photodiode, thus there exist an optimal value for the size of the photodiode, for applications with bandwidth of about several tens of MHz, a typical Si Pin photodiode has diameter of 1mm~1.5mm.

The lenses can magnify the apparent size of the photoelements and will improve the ease of alignment or simply direct more light energy toward the photodiode to improve overall sensor sensitivity. Figure 4 illustrates how the lens magnifies the apparent size of the photoelements. The lens is completely blackened by the magnified image of the photodiode chip.

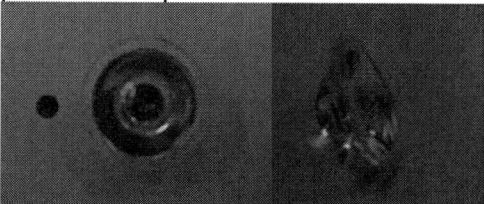

Figure4. The apparent size of photoelement

magnified by a non-imaging optics

Given the parameters of optics and photoelements, the light intensity at the receiving aperture can be calculated and thus the SNR. For simplicity we assume the rotational symmetry and no reflection, absorption or dispersion losses. From the conservation of Etendue[1] we can obtain the following parameters for a 10 degree accepting receiver sub-systems with 1.5mm diameter photodiode as summarized in Table 1.

Table 1

	TIR lens	CPC	Imaging Lens (0.5N.A.)
Aperture Diameter (mm)	8.6	8.6	4.0
Total Length (mm)	4	28.7	4
Compact Ratio	0.46	3.33	1.00
Concentration Ratio	33	33	7

One can quickly see that the TIR lens achieves the same concentration ratio within the theoretical limits as CPC while maintaining a high compact ratio of overall receiver sub-system. A high N.A. conventional imaging lens can only achieve 7X optical gain, which is much less than those of non-imaging optics.

Conclusions
A standard ray tracing computer program is used for analysis and characterization of the behaviour of a TIR lens, non-imaging optics, in a conceptual photoelectric sensor application. Performance of the TIR lenses shows advantages of achieving high concentration ratio while maintaining the compactness of the system, when compared to results of CPC and conventional imaging lenses.

More detailed results of the study will be presented at the time of the conference.

References
1 W. T. Welford, R. Winston, High Collection Nonimaging Optics, Academic Press Inc., (1989)

Fiber-Optic Interferometric Temperature Sensor Using a Hollow Fiber

Jeung-Hwan Bae, Jaehee Park, and Chomsik Lee*

Keimyung University, Dept. of Electronic Engineering, 1000, Sindang-Dong, Daegu, S. Korea

*3I 335-1, Jamsil, Songpa-Gu, Seoul, S. Korea

Abstract

A fiber-optic interferometric sensor for measuring temperature in the range from 28 °C to 100 °C is developed using a hollow optical fiber with an air- hole around the center axis. This sensor is formed with a 13 mm long hollow optical fiber with one end joined to the single mode fiber by a fusion splicing technique and the other cleaved end. The fabrication procedure provides easier and cheaper technologies for fiber-optic interferometric temperature sensors. This temperature sensor has a linear response and high resolution.

Key words: fiber-optic sensor, temperature, hollow optical fiber

Fiber-optic temperature sensors possess inherent advantages over conventional temperature sensors, such as immunity to electromagnetic interference, rapid response and suitability for remote sensing; i.e., locating the sensor head far from the signal processing electronics[1]. Most fiber-optic temperature sensors with high sensitivity reported to data are classed as interferometric. The temperature sensors using the interrferometric configurations expect the Fabry-Perot configuration experience undesirable external perturbations in a sensing cable and can not provide an accurate measurement. Several methods are used to produce the Fiber-optic Fabry-Perot interferometers for the measurement of temperature. The Fabry-Perot temperature sensor is fabricated using two internal mirrors made by joining a single mode fiber to a TiO_2 coated single mode fiber[2]-[3]. This sensor has high sensitivity but the fabrication procedure is not easy technology. The temperature sensor using the Fabry-Perot interferometer configuration is produced by arc fusing a section of multimode fiber between two single mode fibers[4]. The sensor is much easier and cheaper to fabricate but the reflectance of two partial reflectors of this temperature sensor generated at each splicing point is low because of small refractive index difference between a single mode fiber and a multimode fiber. The intrinsic Fabry-Perot temperature sensor using splicing method to fuse different core diameter fibers and to make a reflective mirror is proposed. This sensor performs very well in detecting temperature in the range 30 °C ~ 250 °C but the reflective mirrors has low reflectance due to the small refractive index difference between different core diameter fibers[5]. In this paper, a novel interferometric temperature sensor using a hollow optical fiber is proposed. This sensor is much easier

978-0-9789217-3-6/07/$25.00

©2007 WEN GLOBAL SOLUTIONS

and cheaper to fabricate, and has moderate reflectance mirrors.

The fiber-optic temperature sensor using a hollow optical fiber with an air- hole around the center axis, as shown in Fig. 1, is made by splicing the hollow optical fiber to the single-mode fiber and cleaving the end of the hollow fiber. The single-mode fiber is used as the guiding fiber and the hollow fiber is used as the sensing fiber. The Fresnel reflections are generated at the splicing point and the end of the hollow fiber. The interference fringe of the two reflections is obtained by the sensor integrator. The interference signal and the phase shift are

$$I_r = I_i(R_1 + R_2 + 2\sqrt{R_1 R_2} \cos \Phi) \tag{1}$$

$$\Phi = \frac{4\pi nL}{\lambda} \tag{2}$$

where I_i is the incident light power, R_1 is the reflectance at the splicing point, R_2 is the reflectance of the end of a hollow fiber, n is the refractive index, L is the length of the sensing fiber, and λ is the wavelength of the laser source. The output interference signal is a function of n and L depending on the temperature change.

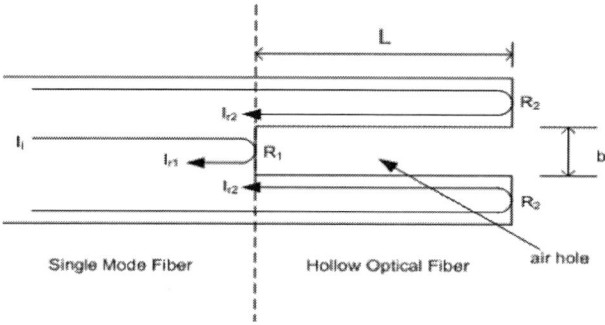

Fig. 1 Schematic diagram of the fiber-optic temperature sensor using a hollow fiber

In order to evaluate the fiber-optic sensor measurement performance, the temperature sensor is fabricated using a 13 mm long hollow fiber with an 8 um air hole. The reflectance of the reflector at the splicing point is 1% and the reflectance of the reflector at the hollow fiber end is 0.7%. The fabricated sensor is copositioned with a thermocouple and placed in a furnace. The fiber-optic temperature sensor is operated over temperature range from 28 °C to 100 °C. The response of the fiber-optic temperature sensor is illustrated in Fig. 2. The results show that the relationship between temperature change and the phase shift of the output interference signal is approximately linear. The sensitivity of this sensor is 2.7 radinas/°C and is about 5 times higher than the previous reported fiber temperature sensors[2].

The novel interferometric temperature sensor using a hollow optical fiber with an air-hole around the center axis is developed. This sensor is much easier and cheaper to fabricate, and has moderate reflectance mirrors. The phase change of the sensor output interference signal shifts linearly according to the change of temperature in the range 28 °C ~ 100 °C. The resolution of this sensor is 2.7 radinas/°C and is about 5 times higher than the previous reported fiber temperature sensors.

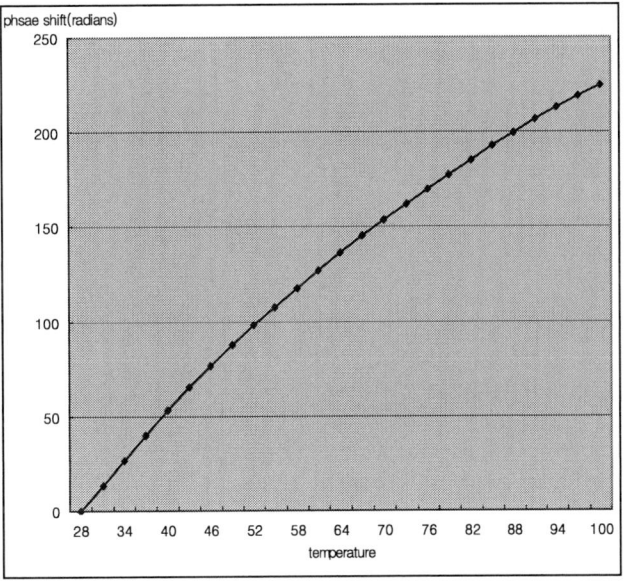

Fig. 2 Phase changes of the output signal according to the change of temperature

Acknowledgement

The present research has been conducted by the Bisa Research Grant of Keimyung University in 2006.

References

[1] X. Wan and H. Taylor, "Intrinsic fiber Fabry-Perot temperature sensor with fiber Bragg grating mirrors," *Opt. Lett.*, vol. 27, no. 16, pp. 1388-1390, 2002.

[2] C. Lee and H. Taylor, " Interferometric optical fibre sensors using internal mirrors," *Electron. Lett.*, vol. 24, no. 4, pp. 193-194, 1988.

[3] Y. Yeh, C. Lee, R. Atkins, W. Gibler, and H. Taylor, " Fiber optic sensor for substrate temperature monitoring," *J. Vac. Sci. Technol. A*, vol 8, no 4, pp. 3247-3250, 1990.

[4] Z. Huang, Y. Zhu, X. Chen, and A. Wang, " Intrinsic Fabry-Perot fiber sensor for temperature and strain measurements," *IEEE Photon. Technol. Lett.*, vol 17, no 11, pp. 2403-2405, 2005.

[5] W. Tsai and C. Lin, " A Novel structure for the intrinsic Fabry-Perot Fiber-optic Temperature sensor," *J. Lightwave Technol.*, vol. 19, no. 5, pp. 682-686, 2001.

Transverse-Load Sensor Based on a Distributed Bragg Reflector Fiber Laser

Li-Yang Shao (1, 2), Xinyong Dong (1) and Hwa-Yaw Tam (1)

1: Department of Electrical Engineering, Hong Kong Polytechnic University, Kowloon, Hong Kong, P. R. China,

Email: eelyshao@polyu.edu.hk, eexydong@polyu.edu.hk, eehytam@polyu.edu.hk,

2: Department of Optical Engineering, Zhejiang University, Hangzhou 310058, Zhejiang, P. R. China.

Abstract *We demonstrate a dual-polarization distributed Bragg reflector (DBR) fiber laser sensor for the measurement of transverse-load. Experimental results show that the beat frequency of the laser output has an orientation-dependent load sensitivity of up to 4.07 MHz/Nm^{-1}.*

Introduction

Fiber Bragg gratings (FBGs) have been demonstrated as useful fiber-optic sensors because of their compactness, high sensitivity and multiplexing capability. Many parameters including temperature, strain, pressure, force, refractive index etc, can be measured effectively with properly designed FBG sensors [1-2]. Recently, there is increased interest in measuring transverse load using fiber optic sensors [3-7]. Transverse load can induce significant birefringence in optical fibers, thus, causing the reflection spectrum of an FBG to split [3] or changing the spectral separation of two polarization modes of a FBG written in polarization-maintaining fiber (PM-FBG) [4]. A high sensitivity load measurement has also been demonstrated by the use of a π-phase-shifted FBG with narrow transmission peak [5]. Sensors based on fiber grating lasers offer the advantages of high signal-to-noise ratio and high resolution. Polarimetric Fabry-Perot (FP) fiber laser [6] and distributed-feedback (DFB) laser for measuring transverse-load induced birefringence have been reported [7]. These sensors allow less expensive detection technique to be used. Beat frequency measurement can be achieved using a relatively low-cost photodetector together with some RF electronics circuitry. The multimode operation of the FP fiber laser in [6], due to the long-cavity length, limited its performance as a sensor.

In this work, we proposed and demonstrated a short-cavity dual-

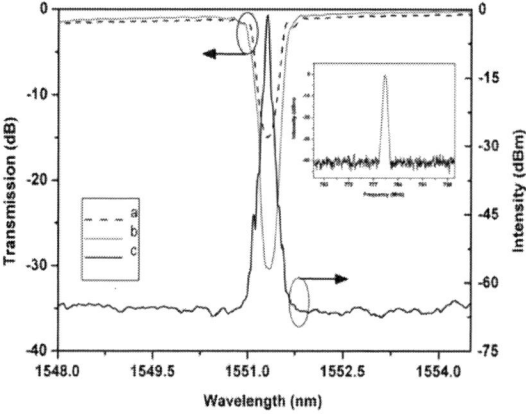

Fig. 1 Measured spectrum of 3-mm FBG (a), two FBGs (b), and the DBR laser (c). Inset is RF spectrum of the laser output.

polarization distributed Bragg reflector (DBR) fiber laser sensor to measure transverse-load. The laser cavity is only 16.5-mm long, which ensures single-frequency operation and only one beat frequency is generated [8]. This beat frequency changes with transverse-load applied to the fiber laser sensor. Sensor operation is studied at applied load up to 108 Nm^{-1}. In addition, we also investigated the effects of load orientation and temperature of the sensor.

Fabrication

A pair of wavelength-matched FBGs were written in a fiber with Er-Yb-doped core and photosensitive B-Ge-doped ring to construct the laser. To enhance the photosensitivity of the fiber, it was soaked in hydrogen gas at 1500 psi and 70 °C for 7 days. We first

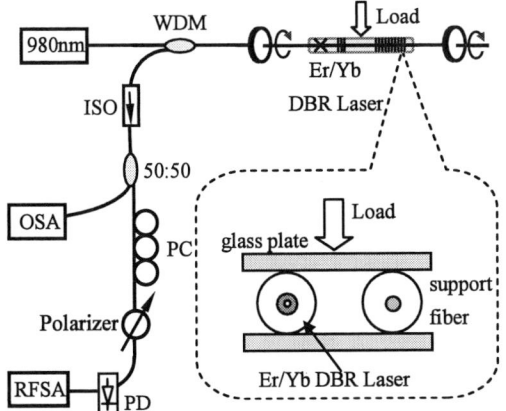

FIG.2 Schematic diagram of the experimental setup. Insert shows the loading configuration.

made the 3-mm FBG with a reflectivity of ~95%. Fig.1.a shows the transmission spectrum of the 3-mm FBG. Then, a 10-mm FBG was fabricated with a reflectivity over 99%. The distance between the two FBGs was 10mm. Fig.1.b is the transmission spectrum of light passing through both FBGs.

Measurements and Discussions
Fig.2 illustrates the schematic diagram of experimental setup and loading configuration. The fiber laser was pumped by a 980-nm laser diode via a wavelength division multiplexer (WDM) coupler from the 3-mm FBG side. The output of the laser was split into the two arms by a 3-dB coupler. An optical spectrum analyzer (OSA) was employed to observe the optical spectrum of the laser output at one arm. Fig.1.c shows the measured spectrum with lasing wavelength of 1551.33 nm. At the other arm, the laser output passed through a polarization controller (PC) and a linear polarizer. The polarization beat frequency is generated at the photo-detector (PD) and measured by a radio-frequency (RF) spectrum analyzer. The inset of Fig.1 shows a typical RF spectrum with a center frequency of 780 MHz.

After grating inscription, the fiber laser part, with its original protective coating removed before grating inscription, was recoated with a layer of UV-curable acrylate so that a diameter of 250 μm was achieved. The laser, together with a parallel support fiber,

was laid between two flat glass plates, on which a transverse-load was applied. To study the orientation-dependence of the sensor response to transverse load, both ends of the fiber laser were fixed to the centres of two rotatable disks. By rotating the disks, the direction of the applied transverse-load relative to the fiber sensor can be adjusted.

Fig.3 shows the measured beat frequency as a function of applied load (per unit fiber length) for five different orientations of the load applying. The definition of orientation θ is shown in the inset of Fig.3. The beat frequency increases with the applied load along the orientations in the range of 0°~45° and decreases in the range of 45°~90°. The load sensitivities of the beat frequency for different load orientations are different. For instance, the beat frequency is insensitive to the transverse load at the load orientation of 45°. For the orientations of 0° and 25°, the beat frequency has positive load sensitivities of 4.07 and 2.67 MHz/Nm^{-1}, respectively, whereas, for the orientations of 75° and 90°, the load sensitivities are negative of -2.28 and -3.64 MHz/Nm^{-1}, respectively.

In a separate experiment, we measured simultaneously the temperature responses of the beat frequency and average lasing wavelength (without an applied load) of the fiber laser by immersing the laser sensor in a temperature-tunable water bath. The change in wavelength was measured to be 9.12 pm/°C, and the change in beat frequency was -1.34 MHz/°C in the range of 10~50 °C (see Fig.4).

Conclusion
We have demonstrated a sensitive transverse-load sensor based on DBR fiber laser. The load-induced birefringence in the fiber laser is converted into shift of the polarization beat frequency, such that the applied load on the fiber laser can be determined from the RF signal measurement instead of the more expensive wavelength measurement. The load sensitivity exhibits orientation-dependence due to differential axial strain on the fiber cross section. The laser

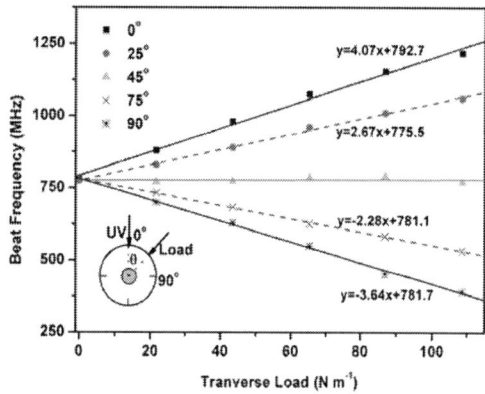

Fig.3 Measured beat frequency as a function of applied load for five different orientations of the laser sensor. Inset is the definition of load orientation θ.

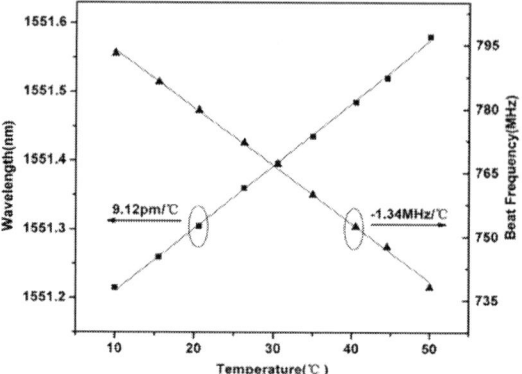

Fig. 4 Wavelength and polarization beat frequency as a function of temperature (without an applied load).

operated robustly at two orthogonal modes with some slightly intensity fluctuations. There was no obvious wavelength shift with the increase of applied transverse load. The fiber laser sensor was tuned in temperature range of 10~50 °C without any mode hopping.

In summary, we have demonstrated that the proposed DBR fiber laser sensor is a good candidate for transverse-load sensing, which also permits independent measurements of both transverse-load and temperature.

Acknowledgement
This work is partially supported by Natural Science Foundation of China (Grant No: 60607011) and the Hong Kong Polytechnic University - Zhejiang University Joint Research Project G-U224.

References
1 H. Y. Tam et al, Proc. SPIE, 5634 (2005), 85

2 L.-Y. Shao et al IEEE Photon. Technol. Lett., 19 (2007), 30

3 B.K.A. Ngoi et al Optics Communications, 242 (2004), 425

4. C.-C. Ye et al Meas. Sci. Technol., 13 (2002), 1446

5 M. Leblanc et al Opt. Lett., 24 (1999), 1091

6 H. K. Kim et al Opt. Lett., 18 (1993), 317

7 J. L. Kringlebotn et al Opt. Lett., 21 (1996), 1869

8 B. O. Guan et al IEEE Photon. Technol. Lett., 16 (2005), 16

BISTABLE REFLECTIVE DISPLAYS FOR PAPER-LIKE DISPLAYS

Liang-Chy Chien

Liquid Crystal Institute and Chemical Physics Interdisciplinary Program

Kent State University, Kent, Ohio 44242 USA

Email: lcchien@lci.kent.edu

ABSTRACT

Many types of bistable reflective displays have been demonstrated so far to target the mobile application market, electronic books and paper-like displays. To ensure as a vital display technology in this market, displays with the characteristics of full color, light weight, low power consumption, high reflectivity and motion picture capability are necessary. In addition, a display technology has to have a cost-effective production method such as using roll-to-roll continuous manufacturing process. This paper will give a comparison on the emergence and evolution of bistable reflective display technologies with bistable cholesteric liquid crystal displays.

Introduction

Paper-like electronic displays or electronic paper displays represent a new class of bistable displays in which the image can be unlimited recorded and rewritten and the image remains after the removal of power. Being a bistable device, it requires no power at all to read an image The energy saving by increasing the battery lifetime is a decisive factor for selecting an appropriate display mode. Bistable displays have at least two stable states; namely, the bright and dark states at zero voltage. Once displayed, the information is memorized extensively. This intrinsic memory is the major advantage of the bistable reflective displays especially for mobile display applications.

The other key feature of electronic paper (*e*-paper) displays is that the *e*-papers can be easily manufactured like paper into different size by using flexible substrates. To ensure as a vital display technology in display market, a display technology has to have polarizer-less and backlight-less as well as a cost-effective roll-to-roll continuous manufacturing process. Display materials such as electronic liquid powder and electrophoretic ink have been applied in the fabrication of paper-like displays. Finally, the motion picture and full color capability are essential elements to guarantee to succeed in the strong competition. Reflective cholesteric displays are the other strong contender because of the intrinsic low power consumption, high reflective brightness and large viewing angle. Bistable cholesteric displays were first introduced in 1991. Since that time, there have been numerous development to advance the performance of the displays including full color, black and white and on plastic substrates. A high brightness color CLCD can be prepared by using stacking technique. The

978-0-9789217-3-6/07/$25.00

©2007 WEN GLOBAL SOLUTIONS

stacked-panels technique provides a feasible solution of reflective cholesteric LCDs to display full color. In this paper, display technologies such as liquid powder displays, electrophoretic displays and cholesteric liquid crystal displays are closely scrutinized for their performance as potential candidate for electronic papers.

Powder Displays

Bridestone group developed a powder-based display using powder materials with liquid behavior that realize a bistable memory effect [1-3]. The display contains black and white particles with opposite charges in a cell filled with air. The display can have either a black-on-white appearance or vice verse with high contrast and quick response (less than 0.2 msec) and it can be driven by a passive-addressing method. For bistable displays, the passive driving method is adequate because of its hysteretic characteristics. To have motion-picture quality images, the device relies on an active-addressing method. The short coming of such a device is that the drive voltage required to move the particles across the cell gap is slightly under 100V. Furthermore, the color display has poor brightness because it uses a color filter.

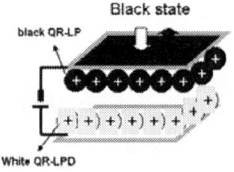 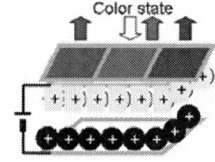

Fig. 1a Schematic representation of a QPD panel and black state image.

Fig.1b White state on polarity reversal of applied voltage.

Fig. 1c Color QPD with color filter.

Electrophoretic Displays

The operation principle and configuration of EPDs illustrated in Fig.2 is based on the microencapsulated electrophoretic displays developed by E Ink and coalition companies [4-6]. The EPD display containing micro-encapsulated spheres with black and white dye particles suspended in oil. The polarity of the voltage chooses which particles are localized to the viewer, and thus the selected color array is presented to the viewer. The particle containing solution is laminated between two substrates with patterned electrodes. The device may be driven using conventional passive matrix addressing scheme, which is very slow and require high voltage. In additional to black-and-white EPDs, monochrome and segmented color EPDs are also demonstrated with colored particles.

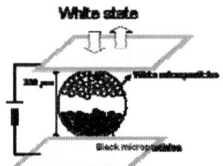 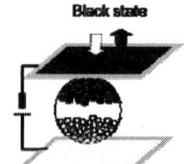 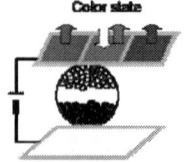

Fig.2a. The EPD configuration and operation principal of E-Ink with white appearance.

Fig.2b The black state of an EPD display by reversing the polarity of applied voltage.

Fig. 2c Color EPD image with color filter.

Cholesteric Liquid Crystal-Based Displays

Bistable reflective liquid crystal displays have received remarkable attention because of their superior performance in power consumption and memory effect to conventional polarizer based displays for electronic papers [7-10]. The intrinsic reflective nature of the cholesteric materials does not require the use of filters, polarizers, and back lighting. The cholesteric liquid crystal in the cell is switched in between two optical states: the focal conic state is a weak scattering state or transparent, depending on the helical pitch and cell treatment, while the planar state is the reflective state and reflects a pre-selected wavelength of light. Monochrome is the major product application for current electronic books and papers. Multiple colors can be achieved either by using the stacked color panels or pixilated color method. The device provides bistable switching with gray scale capability though partial domain switching of individual pixels. Fig.3 shows the display structure and flexible displays.

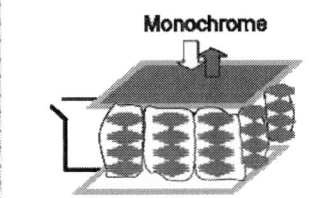
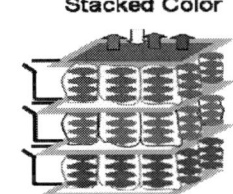

| Fig. 3a The structure of micro-encapsulated bistable cholesteric displays. | Fig. 3b The stacked color panels. | Fig. 3c The microencapsulated cholesteric displays on plastic substrates. |

References

1. R. Hattori, S. Yamada, Y. Masuda, N. Nihei, *SID 03 Digest* 846-849 (2003).

2. R. Sakurai, S. Ohno, S.-i. Kita, Y. Masuda, R. Hattori, *SID 06 Digest* 1922-1925 (2006).

3. S. H. Kwon, S. G. Lee, W. K. Cho, B. G. Ryu, M.-B. Song, *SID 06 Digest* 1838-1841 (2006).

4. P. F. Evans, H.D. Lees, M.S. Maltz, J.L. Dailey, U.S. Patent 3,612,758, Oct. 12, 1971.

5. I. Ota, U.S. Patent 3,668,106, Jun. 6, 1972.

6. J.D. Albert, B. Comiskey, P. Drzaic, J.M. Jacobson, Patent WO9910768-A1, April 3, 1999.

7. D.-K. Yang, L.-C. Chien, and J.W. Doane, *Proc. Int'l. Disp. Res. Conf.*, 49-52 (1991).

8. T. Schneider, F. Nicholson, A. Khan, and J. W. Doane, L.-C. Chien, *SID 05 Digest* 1568-1571 (2005).

9. L.-C. Chien, et al. US Pat. 5,668,614, 1997.

10. J. W. Doane, et al. US Pat. 7,170,481, 2007.

Fabrications of Mechanically Stable Plastic Liquid Crystal Displays

Kwang-Soo Bae[1], Yoonseuk Choi[2], Se-Jin Jang[3], Ji-Hong Bae[3], Jong-Wook Jung[4], and Jae-Hoon Kim[1][2][3]*

1 : Department of Information Display Engineering, Hanyang University,
2 : Research Institute of Information Display, Hanyang University,
3 : Department of Electronics and Computer Engineering, Hanyang University
Haengdang-dong, Seongdong-gu, Seoul 133-791, Korea
email : jhoon@hanyang.ac.kr

4 : LG Phillips LCD, Gumi-si, Gyungsangbuk-do, 730-731, Korea

Abstract *We present fabrications of mechanically stable plastic liquid crystal displays. The micro-structures support the stable molecular alignment of liquid crystals. Various tight bonding techniques are applied for enhancing the durability of the device.*

Introduction

In recent years, liquid crystal (LC) devices using plastic substrates have drawn much attention for their versatile applications such as smart cards, PDA, and head mount displays because of their flexibility, lighter weight, thinner packaging, and lower manufacturing cost through continuous roll-to-roll processing[1]. Different LC modes have been proposed in plastic substrates including twisted nematic (TN), cholesteric, PDLC, and bistable FLC modes. But, the mechanical stability of these devices were not satisfactory except for PDLC, since a solid mechanical support for preserving the molecular alignment of LCs is insufficient due to the lack of sustaining structure. Also, the device using PDLC has the limited applications since it only can use the scattering effect of the light. Therefore, the key technology to realize a practical device for flexible applications is to keep the uniform gap between flexible substrates against external deformations.

In this presentation, we proposed various fabrications to produce plastic liquid crystal display (LCDs) with enhanced mechanical stability by using the polymer micro-structures and new bonding techniques. The electro-optic (EO) characteristics of the samples by these methods are comparable to that of conventional LCDs and are not varied significantly when we applied high external deformations.

Pixel-Isolated Liquid Crystal Structure for Plastic LCD

The polymer walls and/or networks as supporting structures have been proposed and successfully demonstrated[2]. But, these methods have the weaknesses such as requiring high electric field to initiate the anisotropic phase separation or reduced optical properties and increased operating voltage due to the remaining residual polymers. We proposed the stability enhanced LC mode using anisotropic phase separation[3-6]. In these modes, LC molecules are isolated in pixels surrounded by interpixel vertical walls and horizontal polymer layers on the upper substrate, namely, the pixel-isolated LC (PILC) mode. The mechanical support provided by the rigidity of surrounding structures and the adhesive property of polymer maintain the uniform gap under bended circumstances.

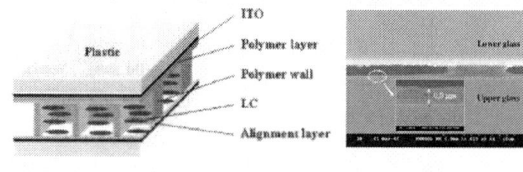

(a) (b)

Fig. 1. (a)Schematic diagram of PILC structure. (b)Cross-sectional images of PILC sample by SEM.

Device configuration of PILC structure is shown in Fig. 1(a). The polymer walls can be made by various methods such as 3-D phase separation[3,4], photolithography using photoresist[5], or stamping technique[6]. The 3-D anisotropic phase separation supports the ease of fabrication, while the other methods can create more fine structures. After fabricating walls on the bottom substrate, we dropped LC/polymer mixture and induced the phase separation by UV irradiation. The thin polymer layer formed on the upper substrate support the tight binding of two substrates. The cross-sectional images are shown in Fig. 1(b).

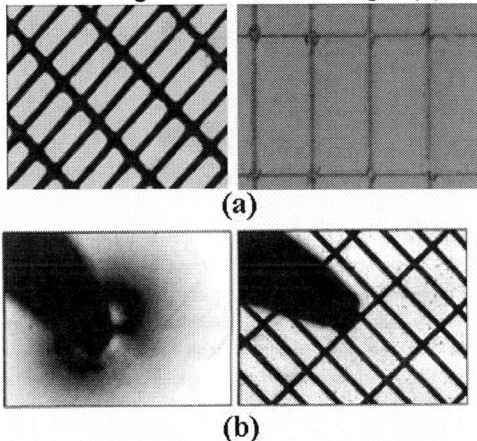

(a)

(b)

Fig. 2. Alignment textures (a)of 3-D phase separation method with nematic (left) and ferroelectric (right) LCs, (b)of a normal (left) and PILC sample using photoresist wall (right) in the presence of external point pressure with a sharp tip.

It is notable that this PILC mode can be applicable to realize any LC modes using nematic and ferroelectric LCs as we demonstrated the resultant textures in Fig. 2(a). With the pressure test in Fig. 2(b), it is clear that the PILC structure can support the mechanical stability of the device.

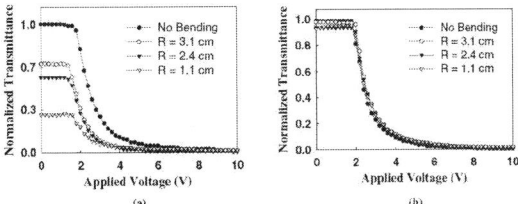

Fig. 3. Transmittance vs applied voltage (a)for normal and (b)for PILC cell at various bending states.

The EO characteristics of normal and PILC samples are shown in Fig. 3. Decreasing the curvature (R) means increasing the degree of bending. For the normal plastic sample, the transmittance and contrast ratio are reduced about 70% at the maximum degree of bending. However, the PILC cell shows almost the same behavior except for a minor decrease in the low voltage regime. It is clear that this mode shows not only good mechanical stability but also equivalent optical behavior with respect to the normal mode without a polymer.

Plastic LCD by Patterned Rigid Spacers and Micro-contact Bonding Technique

In the PILC structure, the applicable LC modes are limited due to the lack of alignment layer at upper substrate. To overcome this problem, we developed plastic LCDs supported by rigid spacers[7]. In this device, the UV curable adhesive are placed on top of the rigid spacers by the micro-contacting technique and irradiated to bind two plastic substrates tightly (Fig. 4(a)). Since the alignment layer can be used on the top substrate, the different LC modes such as TN mode can be applied in this plastic LCD (Fig. 4(b)).

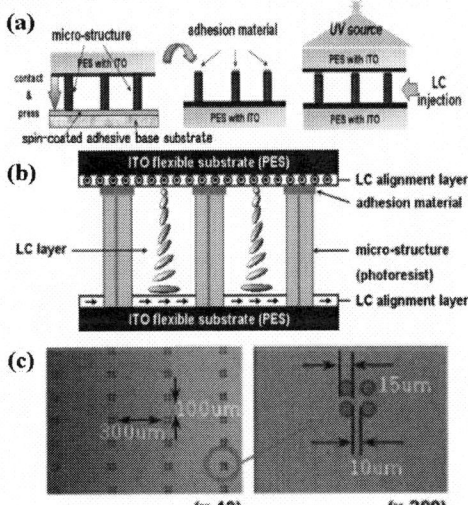

Fig. 4. (a)Fabrication process using micro-contact bonding technique. (b)Device configuration with TN mode. (c)Design of patterned rigid spacers.

To prevent the overflow of adhesive during the bonding process, we designed the rigid spacers as the assembly of four micro-pillars (Fig. 4(c)). The excessive adhesive is confined into the rigid spacers and results in the fine optical properties of the device. The measured EO characteristic of our plastic

141

TN sample was comparable to the conventional case (threshold voltage: 2V, driving voltage: 6V).

Table 1. Mechanical stability test by increasing weight of the loads before breaking the sample.

(Dimension: N / cm²)

Test	1	2	3	4
Max Loads	4.35	5.00	5.00	3.01

In our mechanical stability test, the maximum capable loads without breaking sample were measured as 4.56N/cm².

Single Substrate Plastic LCD by Laminating Technique

One of the advantages of plastic LCD is the use of cost-effective roll-to-roll process. The single substrate LCD is regarded to be very suitable one to apply this process. Recently, we developed the simple technique for fabricating mechanically stable plastic LCD with a single substrate[8].

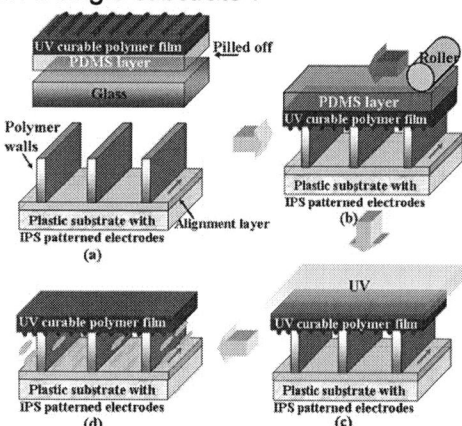

Fig. 5. Fabrication of single substrate plastic LCD (a)Preparation of cover film and bottom substrate with polymer walls. (b)Laminating process. (c)UV irradiation for solidification. (d)LC injection.

In our structure, a cover film of UV epoxy was tightly attached to the bottom wall structure of photoresist by laminating technique, and LCs were uniformly aligned by Berreman effect of micro-grooves formed on the cover film (see, Fig. 5).

Uniform LC alignment was verified by the microscopic textures. In the field-off state, the initial texture showed the dark state because LCs were aligned homogeneously along the rubbing directions. As increasing the applied voltages, the textures became brighter due to in-plane reorientation of LCs along the field direction. All textures under the applied voltages were also highly uniform.

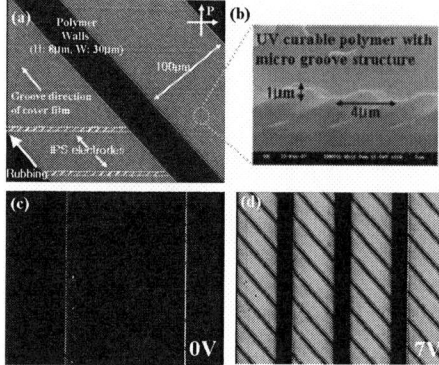

Fig. 6. (a)Microscopic textures of the single substrate LCD. (b)Cross-sectional SEM image of the micro-grooves. Polarizing microscopic images at applied voltages of (c)0V, and (d)7V.

Conclusions

We reported various fabrications of mechanically stable plastic LCDs. The proposed devices are expected to play a critical role in the next-generation flexible displays.

Acknowledgement

This research was supported by a grant from the Information Display R&D Center, one of the 21st Century Frontier R&D Program funded by the Korean Government (MOCIE).

References

1 J. L. West et al, Asia Display, (1995), 55
2 V. Vorflusev et al, Science, 283 (1999), 1903
3 J.-W. Jung et al, Jpn. J. Appl. Phys., 43 (2004), 4269
4 J.-W. Jung et al, Jpn. J. Appl. Phys., 44 (2005), 8547
5 J.-W. Jung et al, J. Inform. Display, 6 (2005), 1
6 S.-J. Jang et al, Jpn. J. Appl. Phys., 44 (2005), 6670
7 S.-J. Jang et al, IMID'06 Dig., (2006), 1189
8 K.-S. Bae et al, SID'07 Dig., (2007), 661

The electrolytic polishing of flexible display steel substrate

LI Yuqiong(1), YU Zhinong (2), XUE Wei(3), LENG Jian(4)

Department of Optical Engineering, School of Information Science and Technology,
Beijing Institute of Technology, Beijing 100081, China

(1) Beijing Institute of Technology, liyuqiong@bit.edu.cn, 010-68913259-13
(2) Beijing Institute of Technology, znyu@bit.edu.cn
(3) Beijing Institute of Technology, xuewei@bit.edu.cn
(4) Beijing Institute of Technology, lengjian@bit.edu.cn

Abstract *The electrolytic polishing process of stainless steel sheet which is used as the substrate for flexible electroluminescence display is investigated. The surface roughness of steel sheet decreases from 2µm to 0.13µm successfully.*

Introduction

Higher demand of display technology has been brought forward. And Flexible Electroluminescence Display (FELD) devices are flexible, light, easy to carry and so on, and these features greatly expanded the scope and area of the Electroluminescence (EL) devices' application.

Metal foils, besides the excellent flexibility, its temperature resistance is better than that of polymer materials and ultra-thin glass, at least 1,000□[1], and its coefficient of thermal expansion is low, and also metal foils don't have the problems of obstructing the water vapour and oxygen. But making the metal foil as a substrate material for flexible displays, the biggest problem is overcoming of the material surface roughness (Ra), in order to reduce it, currently the traditional mechanical polishing technology is adopted, but the traditional mechanical polishing will scratch the surface of the substrate, so many dongas are produced on the surface[2]. But the electrochemical polishing technique can avoid the above shortcomings. And in this paper, electrolytic polishing will be adapted.

Experimental

When stainless steel used as the anode, first, the oxide layer of the stainless steel surface will be removed because of the stainless steel chemical dissolution and the physical washing of the oxygen-bubbles which are formed on the stainless steel surface, then the surface of the stainless steel surface will become slippery and shining[4].

The electrolytic polishing process of the stainless steel surface can be explained by the sticky film theory[3]: When the stainless steel used as the anode in electrolytic polishing, if the anode dissolution rate is higher than that of the anodic dissolution products diffusing from anode surface toward the centre of the electrolyte, dissolving products will accumulate near the anode surface, forming a viscous liquid film with large resistance. And its distribution on the anode surface is uneven: in the raised area, the film is thinner, but in the concave area, the film is thicker. So the raised area has the following traits: thinner sticky film, smaller resistance, bigger current density, more oxygen, liquid updating easier, so the raised area's dissolution ratio is higher; however the concave area's traits are opposite, compared with the raised area: thicker sticky film, bigger resistance, smaller current density, so the dissolution ratio is lower. With the increasing of the electrolytic polishing time, the anode surface is sharpened gradually, and that makes the steel surface very smooth[5].

Electrolytic polishing process: normal temperature chemical degreasing →cold water washing →acid corrupting→ distilled water washing →electrolytic polishing →cold water washing → neutralisation →cold water washing →passivation →cold water cleaning →drying →testing

Results and discussing

Electrolytic polishing time、distance between cathode and anode and current density three have great impacts on polishing effect

978-0-9789217-3-6/07/$25.00

©2007 WEN GLOBAL SOLUTIONS

on the surface of steel substrate. Therefore, in this paper the relation between the above three factors and the polishing effect will be considered, and related experiments and discussions will be carried out.

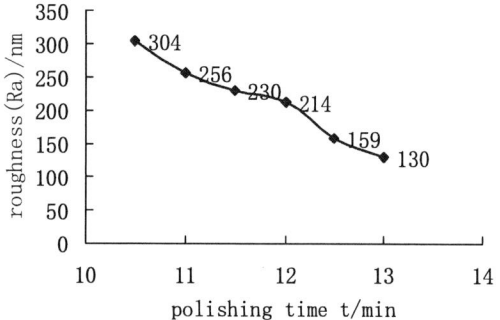

Fig.1　Relation between the surface roughness and polishing time
（I=0.56A/cm², d=4cm）

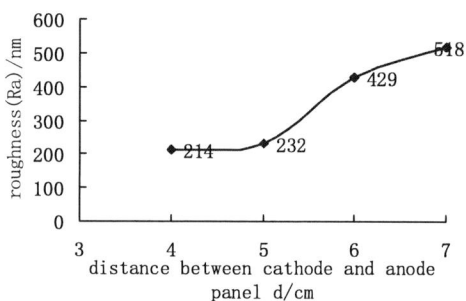

Fig.2　Relation between the surface roughness and distance of cathode and anode panel
（I=0.56A/cm², t=12min）

Under the condition of I = 0.56A/cm², d = 4cm, it is shown that different polishing time which changes from 10.5min to 13min affects the polishing effect differently, and the experimental results are shown in figure 1. We can see with the increasing of time, the polishing effect becomes better, that is to say that the surface roughness becomes lower.

Under the condition of I=0.56A/cm², t=12 minutes, the experimental results are shown in Figure 2. From Figure 2 we can conclude that when we control the distance between cathode and anode panel at 4 cm, the polishing effect is the best, and when we make the distance change between 4cm and 5cm, the effect change little.

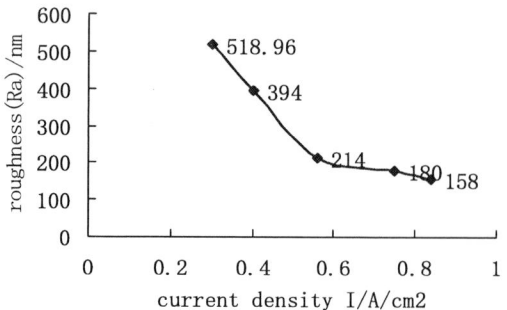

Fig.3 Relation between the surface roughness and current density（d=4cm, t=12min）

Fig.4 The surface shape of polished steel sample

Under the condition of d=4cm, t=12min, the experimental results are shown in Figure 3. We can see from the Figure 3, with the increasing of current density, the polishing effect become better, and when the current density changes from 0.56A/cm² to 0.8A /cm² the polishing effect is the best. So when I ≥ 0.56A/cm², the impact of increasing current density to the polishing effect is very little.

Taking the above three factors into consideration, when I = 0.56A/cm², t = 12min, d = 4cm, the polishing effect is the best. The surface topography is shown in Figure 4. From the Figure 4, we can see that rugged ravines are formed on the surface. There are two factors that contribute to it: First, it is the structure itself of the SUS304 stainless steel which has its own orientation marks on the surface; Secondly, uneven current directional movement、oxygen precipitation and other factors have contributed to the phenomenon.

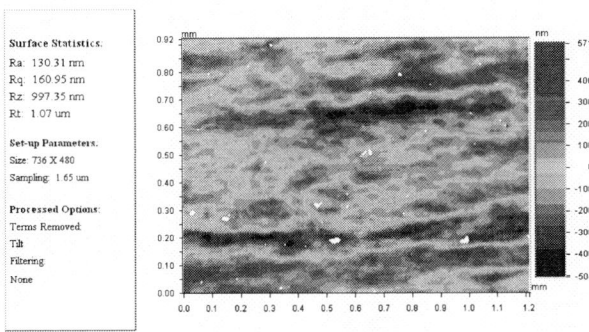

Surface Statistics:
Ra: 130.31 nm
Rq: 160.95 nm
Rz: 997.35 nm
Rt: 1.07 um

Set-up Parameters:
Size: 736 X 480
Sampling: 1.65 um

Processed Options:
Terms Removed:
Tilt
Filtering
None

Fig.5 The surface roughness of polished steel sample

The substrate surface roughness is shown in Figure 5, from it we know that surface roughness Ra is about 0.13 μm, but the Rt is 1.07 μm (the highest and lowest points of the height difference). This is largely due to the rugged ravines on the steel surface and non-metallic inclusions in it. It is further to say that it is very necessary to improve the steel substrate material, and the polished surface of SUS304 steel is bright as a mirror, as shown in figure 6.

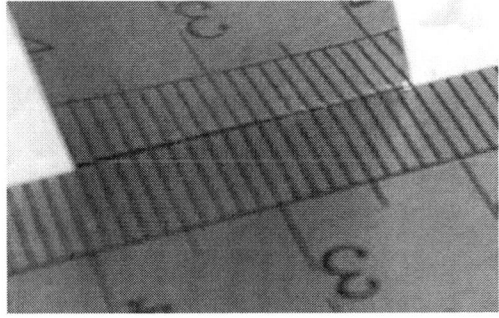

Fig.6 The reflection of polished steel sample

Conclusions

1. In this paper, the polishing effect of this electrolyte is good and the cost is low. Besides, there are other advantages of this solution: the composition is simple, besides, there is no chromate pollution and it is not volatile.

2. The SUS304 stainless steel sample surface after being polished is smooth and lustrous, and the corrosion resistance is strong. The surface roughness of the sample which is not polished is about 2μm, but after polishing it reduced to 0.13 μm, at the same time the size of the polished substrate is homogeneous, and the surface is bright as a mirror with a high metallic lustre.

3. From the experiment results we can obtain the optimum process parameters: I= 0.56A/cm^2, t=12min, d=4cm.

4. To avoid the rugged ravines to formed on the steel surface, we should do as follow: firstly, non-directional roll steel substrate should be adopted; Secondly, surface current should be as uniform as possible to avoid uneven surface stripe to be caused by the movement of oxygen gas.

References

1. Nigel Shepherd, David Morton, Eric Forsythe, and Dave Chiu. Flexible Infrared Emitting ZnS:ErF3 Alternating Current Thin Film Electroluminescent Devices [J].SID, 2005,8.3:116-119.
2. Li Weiming,Zhou You,Chen Jiongshu,Xu Qiang. A impulse electrolytic polishing solution that without cauterization and technics[p].China:1249367A,2004.
3. Zhang Shulin,Wang Xiaobo,Chen Shibo. A stainless steel electrolytic polishing technics[J].Electroplating & Pollution Control,2006,Vol.26 No.3:34-35.
4. Shuo-Jen Lee, Yu-Ming Lee, Ming-Feng Du. The polishing mechanism of electrochemical mechanical polishing technology. Journal of Meterials Processing Technology 140(2003) 280-286
5. Shuo-Jen Lee, Jian-Jang Lai. The effects of electropolishing (EP) process parameters on corrosion resistance of 316L stainless steel. Journal of Materials Processing Technology 140 (2003) 206-21

The bending properties of flexible ITO films

YU Zhinong[1*], XIANG Longfeng[2], XUE Wei[3], WANG Huaqing[4]

1. 1. Department of optical Engineering, School of Information Engineering, Beijing Institute of Technology, Beijing 100081, P.R.China，*Corresponding author, Tel.: 010-68913259-11; fax: 010-68912550. E-mail address: znyu@bit.edu.cn
2. Beijing Institute of Technology, xianglongfeng@bit.edu.cn
3. Beijing Institute of Technology, xuewei@bit.edu.cn
4. Beijing Institute of Technology, wanghuaqing@bit.edu.cn

ABSTRACT The electrical and optical properties of flexible ITO films as a function of the radiuses of the films to be bent has been investigated. The threshold radius, keeping the resistivity almost constant, is 0.75cm and 1cm for the film without buffer layer and one with buffer layer, respectively. The transmittance almost keeps same regardless of the radius.

1. Introduction

Indium tin oxide (ITO) film is a wide-band, n-type doped semiconductor material. Due to high transmission between 380nm to 800nm and low resistivity, ITO film, as a kind of transparent conductive oxide (TCO) film, was widely used in all kinds of photoelectric devices ,such as, LCD、TFEL、solar battery and so on [1]. With the rapid development of the flexible optoelectronic devices, the flexible TCO film is widely investigated. It is required that the flexible film should keep a constant high transmission and low resistivity while bent to some extent [2-6].

In the paper, ion beam-assisted reactive evaporation technology was used to prepare the ITO films using 90In-10Sn (wt%) alloy at room temperature. The properties of flexible ITO films as a function of the radius of the films to be bent has been investigated.

2. Experimental

In the experiment, the automatic coating machine whose model is ZZSX—800ZA was used. The type of the ion source is "Kaufman". An alloy of Indium and Tin, the mass proportion of Tin is about 10%, was used to deposit ITO film with bombardment of oxygen and argon ion, the detailed process is indicated in other paper [7].

PET was used as substrate and SiO_2 as buffer layer to improve the sheet resistivity. The thickness of ITO film and SiO_2 is 150nm and 200nm, respectively. According to Fig.1, the resistivity (or resistance) was measured by four–point probe method (ohmmeter) when the film was bent. The thickness, the surface features and the transmittance were examined by ellipsometer, optical profiller and spectrophotometer, respectively.

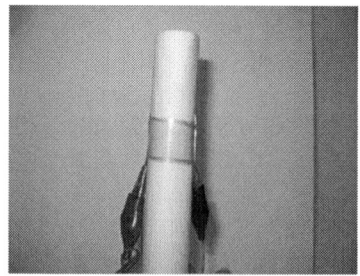

Fig.1 the measure method of the resistivity of ITO film when bent into a column

3. Results and discussion

Table 1 shows the resistance of the deposited films on PET without and with SiO_2 buffer layer with the films to be bent to some extent. With the radius of flexible film decreasing, a threshold radius occurs, and the threshold radius for the films deposited on PET without buffer layer and with buffer layer is 0.75cm and 1cm, respectively. That is, for the films deposited on PET without buffer layer, the resistance increases slowly from 500Ω to 780Ω with the radius decreasing at a radius range from ∞ to 0.75cm, and increases rapidly at a radius range from 0.75cm to 0. For the films deposited on PET with SiO_2 buffer layer, the resistance increases slowly from 359Ω to 450Ω with the radius decreasing at a radius range from ∞ to 1cm and increases rapidly at a radius range from 1cm to 0. The threshold radius corresponds to a fatigue limit of ITO film, that is, a rapid increase of resistance at a radius less than threshold radius indicates the ITO film has been destroyed or broken due to additional intensive stress. The difference between the threshold radiuses of the films deposited on PET without and with SiO_2 buffer layer is attributed to the function of SiO_2. Although the SiO2 buffer layer can improve the electrical properties of ITO film, the fatigue limit of the film may be affected unfavorably.

Table1 the comparisons of resistance of the films deposited on PET and SiO_2 buffer layer

PET substrare		SiO_2 buffer layer + PET substrate	
radius（cm）	resistance（Ω）	radius（cm）	resistance（Ω）
∞	500	∞	359
2.5	530	2.5	368
2	580	2	389
1.5	620	1.5	416
1	720	1	450
0.75	780	0.75	3700
0.5	4500	0.5	12000
0.25	10000	0.25	∞

The optical properties of ITO films deposited on PET without and with SiO_2 buffer layer were measured, and the transmittance of all the films keeps almost constant, regardless of the radius of ITO films to be bent. Fig.1 shows the transmittance of the deposited ITO films on PET without and with SiO_2 buffer layer.

Fig.2 shows the surface features of the deposited films on PET without and with SiO_2 buffer layer. The flatting function of SIO_2 to the rough surface of PET makes the surface of the deposited films on PET with SiO_2 buffer layer smoother than one without buffer layer (as shown in Fig.2), and leads to a decrease of the resistance of the deposited films on PET with SiO_2 buffer layer, compared with one without buffer layer.

4. Conclusions

The electrical and optical properties of flexible ITO films as a function of the radiuses of the films to be bent has been investigated. The threshold radius, keeping the resistivity almost constant, is 0.75cm and 1cm for the flexible ITO film without buffer layer and one with buffer layer, respectively. The transmittance almost keeps same regardless of the radius.

References

[1] Tian Minbo.Thin film technology and thin film material. Beijing：Tsinghua University Press,2006.

[2] John C C Fan. Preparation of Sn doped In2O3(ITO) films at low deposition temperature by ion beam sputtering[J]. Appl. Phys. Lett., 1979,34(8):515-517.

[3] Abhai Mansingh, Vasant CVR, Kumar R F. Sputtered indium tin oxide films on water cooled substrates [J]. Thin Solid Films, 1988，167:11-13.

[4] Karasawa T, Miyata Y. Electrical and optical properties of indium tin oxide thin films deposited on unheated substrates by DC reactive sputtering [J].Thin Solid Films, 1993, 223:135-139.

[5] Wu Wen-Fa, Chiou Bi-Shiou. Properties of radio-frequency Magnieton sputtered ITO films without tin-situ substrate heating and post-deposition annealing[J].Thin Solid Films, 1994, 247:201-207.

[6] Danson N, Safi I, Hall G W, Howson R P. Techniques for the sputtering of optimum indium tin oxide films onto room temperature substrates[J].Surface and coatings technology, 1998, 99:147-160.

[7] YU Zhi-nong, Xiang Longfeng, XUE Wei, Wang Huaqing, Lu Weiqiang. Preparation of ITO films deposited at room temperature by ion beam-assisted reactive evaporation[J]. Journal of Beijing Institute of Technology (accepted)

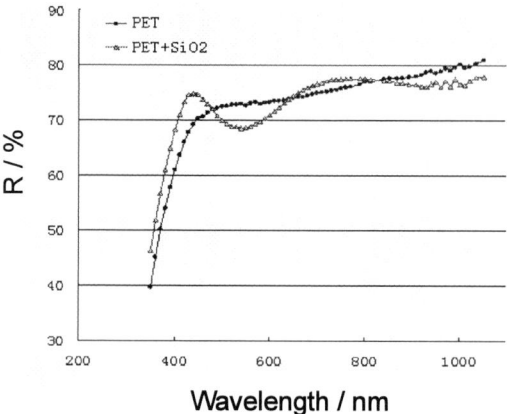

Fig.1 the transmittance of the deposited ITO films on PET without and with SiO_2 buffer layer.

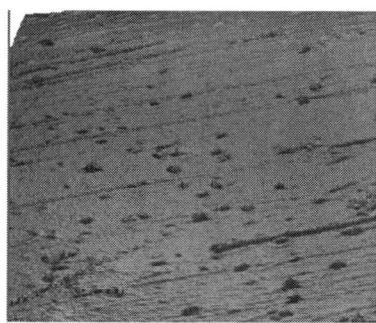

PET without buffer layer

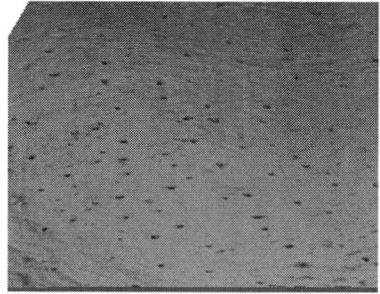

PET with buffer layer

Fig.2 the surface features of the deposited films on PET without and with SiO_2 buffer layer

Characterization of polymer microtip array coated GaN thin film using femtosecond pulsed laser deposition

X.L.Tong; H.Q. Wen(1); D.S.Jiang; L.Liu

1. Key Laboratory of Fiber Optic Sensing Technology and Information Processing, Ministry of Education. whq@whut.edu.cn

Abstract The characterizations of polymer microtip array coated GaN thin film using femtosecond pulsed laser deposition have been studied. The results indicate that the GaN thin film deposited on polymer array is hexagonal polycrystalline, and the GaN microtip FEA has uniform size and well-defined profile, which shows a field emission characteristics.

Keywords: GaN; microtip field-emitter array; folding flat-panel display

Introduction

Electron-field emitters with low operating field and high stable emission current are desirable for FED applications. Therefore potentially chemical inert emitter–tip overcoatings with wide-band-gap materials are preferred for a field emission system [1]. GaN is proved to be good choice in low vacuum applications [2]. The advantages of gallium nitride for these tips are its relatively low electron affinity and capability for n-type doping energy necessary to remove the electron from the material surface into vacuum. The low electron affinity of the semiconductor material, coupled with the high electric field present at the tips of a GaN emitter when formed in the shape of a cone, leads to efficient extraction of electrons [3]. The above advantages indicate that the GaN field emitters are promising. However, due to the substantial limitations in the fabrication of the emitters, the FEA fabrication is still a critical problem hampering the commercial application of FED. Large flat-panel displays are key devices for hang-on-wall, large area, high definition applications. In order to realize lower cost, larger area, portability, and folding field emission flat-panel display, a fabrication process of GaN microtip field-emitter array is presented in this paper. The fabrication process has low cost and good reproducibility. The Obtained results show that the FEA fabrication process could be a promising candidate for the development of FEDs aiming at folding flat-panel display because the flexible substrate.

Experimental

The details of the fabrication of polymer microtip array coated GaN thin film using femtosecond pulsed laser deposition have been described previously[4].In order to check the quality of fabrication, scanning electron microscopy (SEM, Sirion 200 FEI) was used to characterize the surface morphology of sample. The crystal structure of the GaN coating layer was characterized by a high-resolution X-ray diffraction (XRD), which was performed on a Philips X'Pert X-ray diffractometer using a thin film attachment. In order to ensure that the diffraction pattern originated mainly from the thin films and not its substrate, the angle of incidence was kept constant at $\omega=1°$. And the peaks' location of XRD pattern is approximately in accord with the reference pattern in JCPDS file no. 02-1078. It justifies that the penetration depth of X-rays in this configuration was less than the thickness of the GaN coating layer. Raman spectra was investigated at the room temperature by unpolarized Raman measurements in backscattering geometry Z(-,-)Z. Excitation wavelength was 514.5nm(Ar^+ ion laser). Field emission properties were measured for the polymer FEA coated with GaN thin film in a chamber evacuated to 5×10^{-7} Torr. The distance between the sample and the anode was 200 μm which was maintained using glass fiber rods.

Results and discussion

Fig.1. shows SEM micrographs of the polymer FEA coated with GaN thin film. The uniform GaN microtip FEA was achieved. Comparing our experiment result with the experiment result in literature [2], the GaN thin film grown on polymer FEA forms a stack of grains which were spread over the whole polymer surface in our experiment result. The polymer microtips coated with GaN thin film became sharper than the polymer microtips. Table 2 shows the compare of the aspect ratio of the polymer microtip with the polymer microtip coated with GaN thin film. Obviously, the aspect ratio of the polymer microtip coated with GaN thin film became larger than that the

978-0-9789217-3-6/07/$25.00

©2007 WEN GLOBAL SOLUTIONS

polymer microtip coated with GaN thin film, which means that less potential is required for electron emission. It is most likely that the tips were formed by deposition of GaN thin film. These types can be advantageous because emission is spread out over a large area and higher average currents [3]. Since the ture-on voltage largely depends on the surface work function as well as on the field enhancement factor. Therefore, the femtosecond pulse lasers can be used to deposit the GaN thin film on the polymer microtip array.

Fig.1. SEM picture of polymer microtip arrays coated with GaN thin film

Fig.2. shows the typical XRD pattern of the GaN coating layer. The crystal orientation and purity of the GaN coating layer are related to the field emission performance of the microtip field-emitter array. The XRD pattern exhibits nine diffraction peaks located at $2\theta=32.4°$, $34.5°$, $36.8°$, $48.1°$, $57.7°$, $63.4°$, $67.7°$, $69.1°$ and $70.6°$, which are attributed to GaN (100), GaN (002), GaN (101), GaN (102), GaN (110), GaN (103), GaN (200), GaN (113)and GaN (201), respectively. The XRD results show that GaN thin films are dominantly hexagonal structure, and the GaN thin film deposited by the femtosecond PLD is polycrystalline.

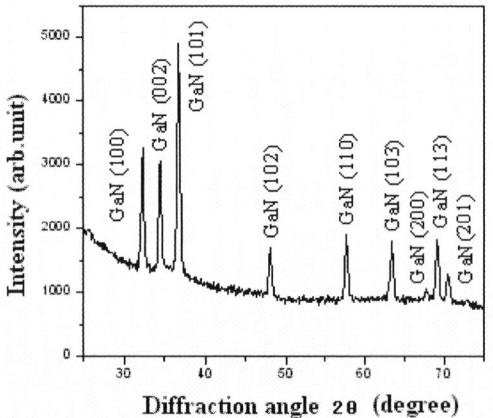

Fig.2. XRD patterns of GaN coating layer

Fig.3. shows Raman spectra acquired in unplorization configuration from the GaN thin film (all signals from the polymer substrates had been removed). Two peaks can be discerned, which are GaN E_2 at 566.1cm^{-1}and GaN A_1(LO) at 703.9cm^{-1}. It has been reported that the wurtzite GaN E_2 mode is attributed to the deformation potential scattering [5], and the LO phonons frequency is related to Phonon-plasmon interaction [6]. Raman measurement of the E_2 Raman mode is known to be shifted by stress, and the E_2 Raman mode of unstressed film appears at a wave number of 566.2cm^{-1}[7]. The E_2 Raman mode in our experiment result is close to 566.2cm^{-1}. It shows that have small stress in the GaN coating layer. For the GaN thin film, the stress is strongly linked to the interaction between the GaN thin film and polymer substrate. Stress determines the stability of a thin film /substrate composite and the life time of a component. The stress may likewise arise from a mismatch in the coefficient of thermal expansion between the substrate and the film if deposited at high temperature and measured at a lower temperature or from a mismatch in the elastic modulus if deposited on a substrate under stress. Another contribution to residual stress in the film may be caused by a variety of effects such as contamination by impurities, the presence of defects after deposition and occurrence of solid state transformations. From the tiny stress in the GaN coating layer, it can be inferred that the GaN thin film is well and steadily adhered on the surface of the polymer microtip array.

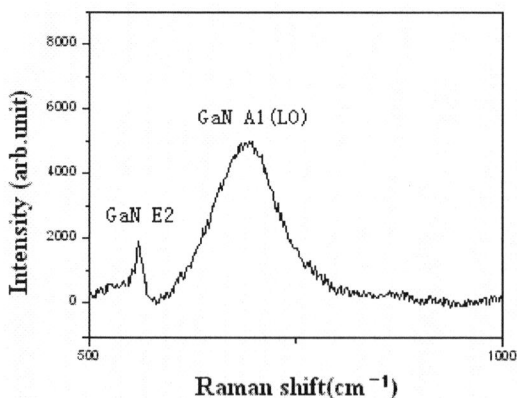

Fig.3. Raman spectra of GaN coating layer.

Fig.4. shows the characteristics of emission current versus electric field for the polymer FEA coated with GaN thin film. The Fowler–Nordheim（FN）plot is also shown in the inset of Fig.4. The FN polt is approximately linear. This indicates that the emitted current is

150

apparently due to field emission. However, the applied voltage is relatively high. The overall emission behavior of the present polymer FEA coated with GaN thin film should be attributed to the microtip structure and the electron affinity of GaN coating layer. The emission current due to (FN) tunneling is indicated by

$$I = \alpha \, (\beta V)^2 \exp(-b\,\phi^{3/2}/\beta V),$$

Where V is applied voltage, β is the field enhancement factor which is associated with the taper and density of the emitter, ϕ is the barrier height for the emission surface which is associated with the band structure of the material of the emitter, α and b are constants [8]. The field enhancement factor β is decided by the geometry of material. In order to improve the field emission performances of the GaN thin film to meet the forecasted applications, the operating voltage must be decreased. Effort will be paid out on the improvement of the sharpness and aspect ratio of the microtips, the enhancement of the microtip density, and the decrease of the barrier height for the GaN emission surface by n-type doping.

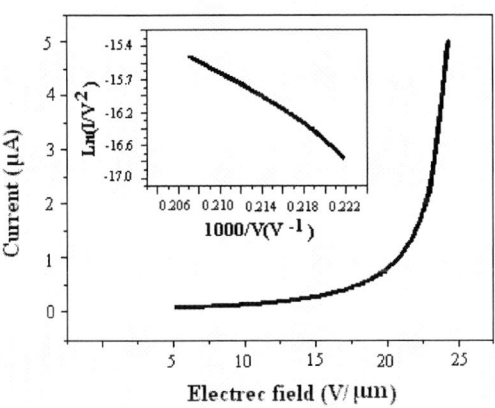

Fig.4. Emission current of the polymer microtips array coated with GaN thin film as a function of electric fields. Insert corresponding Fowler-Nordheim plot of the polymer microtips array coated with GaN thin film.

Conclusions

In summary, microtip FEA was achieved by growing GaN thin film onto like-conical-type polymer array by using femtosecond PLD. The polymer microtip could be further sharpened by GaN coating layer. The FEA fabrication process is a promising candidate for the development of FEDs aiming at folding flat-panel display because the flexible substrate. However, further work is needed to sharpen microtips, n-type dope, improve the emissive site density and the quality of the GaN coating layer.

References

[1] M. Hajra, N. N. Chubun, A. G. Chakhovskoi, C. E. Hunt, K. Liu A. Murali, S. H. Risbud, T. Tyler , V. Zhirnov, J. Vac. Sci. Technol. B 21, 458 (2003).

[2] Czarczyński, P. Kieszkowski, St. Łasisz, R. Paszkiewicz, M. Tłaczała, . Znamirowski, E. Zołnierz. Vac. Sci. Technol. B 19, 47 (2001).

[3] P. B. Shah, B. M. Nichols, M. D. Derenge, K. A. Jones, J. Vac. Sci. Technol. A 22, 1847 (2004).

[4] X.L.Tong; D.S.Jiang; Y. Li; Z.M.Liu; M.Z.Luo. Appl. Phys. Lett. 2006, 89 (6): 61108.

[5] D.Behr, R.Niebuhr, J.Wanger, K.-H.Bachem, U.Kaufmann, Appl.Phys.Lett.70, 363 (1997).

[6] G.Popovici, G.Y.Xu, A.Botchkarev, W.Kim, H.Tang, A.Salvador, H.Morkoc, R.Strange and J.O.White, J. Appl. Phys.82, 4020 (1997).

[7] C. Kisielowski, J. Krüger, S. Ruvimov, T. Suski, J.W. AgerIII, E. Jones, Z. Liliental-Weber, M. Rubin, E.R. Weber, M.D.Bremser, R.F. Davis, Physical Review B 54, 17745 (1996).

[8] C. Kimur, T. Yamamoto, T. Sugino, J. Vac. Sci. Technol. B 21, 445 (2003).

Dissipative Solitons for real world optical solitons

Philippe GRELU (1), Ludovic RAPP (1), Jose M. SOTO-CRESPO (2), Nail AKHMEDIEV (3)

1 : Institut Carnot de Bourgogne, UMR 5209 CNRS,
Université de Bourgogne, BP 47870, 21078 Dijon, France
philippe.grelu@u-bourgogne.fr

2 : Instituto de Optica, CSIC, Serrano 121, 28006 Madrid, Spain
iodsc09@io.cfmac.csic.es

3 : The Australian National University, Canberra ACT 0200, Australia.
nna124@rsphy1.anu.edu.au

Abstract *Efficient use of nonlinear dissipation enables the formation of robust optical pulses, with unique soliton properties. This way, exceptional stability of mode-locked lasers, formation of "soliton molecules", and optical regeneration schemes can be better understood.*

Introduction

If one uses well designed dissipative processes in nonlinear optics, innovative applications in all-optical information processing can be imagined. Indeed, the tailoring of nonlinear dissipation is able to support the formation of localized structures that have been called "dissipative solitons", and possess remarkable robust features as reported in recent works [1]. When several dissipative solitons are formed, the control of nonlinear dissipation allows a wide range of dynamical behaviors, from regular ones such as the assembly of "soliton molecules", to chaotic ones, along with numerous possible interaction schemes [2-4].

"Soliton molecules" in lasers

A passively mode-locked fiber ring laser can serve as a prototype experiment for the manipulation of picosecond or sub-picosecond optical pulses. Indeed, it is possible to form various stable aggregates of these pulses, or "soliton molecules", and modify some of their properties with the tuning of cavity parameters. Stable, self-phase-locked soliton molecules [2], colliding [3] and vibrating soliton molecules [4] have been reported. Figure 1 is an illustration of some of these states that circulate round the fiber cavity, as observed via a 30 GHz sampling oscilloscope. In these examples, pulse duration is typically 0.5 ps, while pulse spacing is close to 20 ps, close to the limit of the temporal resolution of the oscilloscope. Even more compact "molecules" can be analyzed via purely optical techniques such as optical

autocorrelation and spectral recording. The most stable molecules can remain unchanged for hours in a 10-meter long fiber laser cavity, without the need of active stabilization, which is an illustration of the remarkable robustness achievable with dissipative solitons.

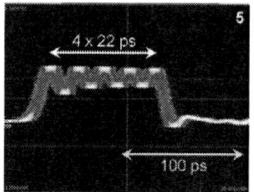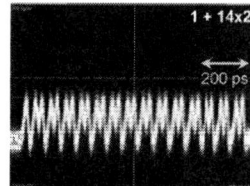

Fig. 1 Examples of stable "soliton molecules" circulating in a fiber ring laser, monitored with a 30-GHz oscilloscope. (a) Compact "pentasoliton" molecule, and (b) large periodic bit sequence $10(110)_{14}$.

Link with optical regeneration

Optical regeneration in ultrahigh bitrate, long-haul, optical communication link is a subject of intense research. When 2R-regeneration schemes based on nonlinear dissipation are considered, they can share common physical mechanisms with passively mode-locked fiber lasers via the use of dissipative soliton attractors to reshape optical pulses [5].

Cavity solitons and "light bullets" for future parallel processing?

The realization of spatio-temporal solitons that are confined in all spatial dimensions and localized temporally is among the mythical problems in nonlinear optics. Both transverse spreading due to diffraction and temporal spreading coming from chromatic

978-0-9789217-3-6/07/$25.00

©2007 WEN GLOBAL SOLUTIONS

dispersion need to be balanced. These solitons, called "light bullets", have not yet been obtained experimentally. Their realization would allow the development of parallel optical processing functions, for application such as image processing.

A novel approach to this problem consists in adding nonlinear dissipation to control pulse diffusion in space and time [6]. The extended complex Ginzburg-Landau equation provides a useful distributed model to analyze the fundamental features of "dissipative light bullets". It serves as a simplified propagation model for an active optical medium. This equation includes the effects of chromatic dispersion, spatial diffraction, Kerr nonlinearity and its (weak) saturation, linear losses, saturation of a nonlinear gain and spectral filtering. Large domains of equation parameters have been found, in which stable light bullets exist. It is also important to notice that adequate nonlinear dissipation can overcome the issue of temporal confinement for both anomalous and normal chromatic dispersion regimes [7].

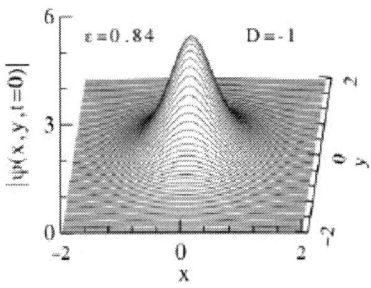

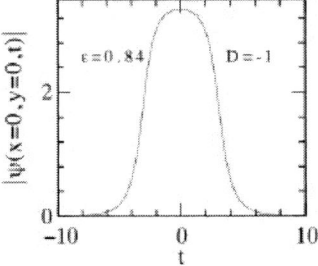

Fig. 2. Example of simulated light bullet which propagates stably in a normally dispersive medium (adapted from [7]). The transverse and temporal field intensities correspond to smooth bell-shaped profiles, as appear in upper and lower plots, respectively. Dimensionless units are used.

Latest theoretical advances include the study of interaction cases between light bullets. The large number of degrees of freedom supports many possible dynamics, as it also raises the issue of the delicate control of such multisoliton dynamics.

The experimental implementation of light bullets could be on the way via the operation of large area vertical cavity semiconductor devices. Spatial dissipative solitons obtained in these devices, or "cavity solitons" are subject to intense research [8]. The control of the temporal confinement is however a difficult task which is still to be demonstrated.

Conclusions

Since nonlinear dissipative systems can provide dynamical attractors, optical data bits evolving in these systems can be well defined, and restore their shapes after being moderately perturbed. In that sense, dissipative solitons could prove more useful for applications than Hamiltonian solitons of the well-known nonlinear Schrödinger equation that rules propagation in a passive fiber.

Dissipative solitons have been involved for a long time in the dynamics of mode-locked lasers, and make us understand additional complex dynamics, such as the formation of soliton molecules. More applications in optical data processing should be developed in future. They concern for instance the stabilization of optical data transfer at ultra-high rates and the realization of optical buffer memories, and in a longer-term perspective, ultrafast parallel optical processing.

References

1 Akhmediev and Ankiewicz (Eds.), "Dissipative Solitons", Springer (2005).
2 Grelu et al., J.O.S.A. B 20 (2003) 863.
3 Akhmediev et al. Opt. Fib. Technol. 11 (2005) 209
4 Grapinet et al. Opt. Lett. 31 (2006) 2115.
5 Peschel et al., Chapter in "Dissipative Solitons", Springer (2005).
6 Grelu et al., Opt. Express 13 (2005) 9352.
7 Soto-Crespo et al., Opt. Express 14 (2006) 4013.
8 Paulau et al., Phys. Rev. E 75 (2007) 056208.

Generation of energetic wavelength tunable femtosecond pulses in higher-order-mode fiber

Chris Xu

School of Applied and Engineering Physics, Cornell University, Ithaca, NY 14853

Invited Paper

Abstract: We demonstrate soliton-self frequency shift and Cerenkov radiation in higher-order-mode solid, silica-based fibers below 1300 nm. A wavelength shift of 240 nm and pulse energy of 1 nJ were achieved for sub 100-fs pulses.

The phenomenon of soliton self-frequency shift (SSFS) in optical fiber in which Raman self-pumping continuously transfers energy from higher to lower frequencies[1-3] has been exploited over the last decade in order to fabricate widely frequency-tunable, femtosecond pulse sources with fiber delivery.[4-9] Because anomalous (positive) dispersion (β_2 <0 or D>0) is required for the generation and maintenance of solitons, early sources which made use of SSFS for wavelength tuning were restricted to wavelength regimes > 1300 nm where conventional silica fibers exhibited positive dispersion.[4, 5] The recent development of index-guided photonic crystal fibers (PCF) and air-core photonic band-gap fibers (PBGF) relaxed this requirement with the ability to design large positive waveguide dispersion and therefore large positive net dispersion in optical fibers at nearly any desired wavelength.[10] This allowed for a number of demonstrations of tunable SSFS sources supporting input wavelengths as low as 800 nm in the anomalous dispersion regime.[6-9]

Unfortunately, the pulse energy required to support stable Raman-shifted solitons below 1300 nm in index-guided PCFs and air-core PBGFs is either on the very low side, a fraction of a nJ for silica-core PCFs,[7, 8] or on the very high side, greater than 100 nJ (requiring an input from an amplified optical system) for air-core PBGFs.[9] The low-energy limit is due to high nonlinearity in the PCF. In order to generate large positive waveguide dispersion to overcome the negative dispersion of the material, the effective area of the fiber core must be reduced. For positive total dispersion at wavelengths < 1300 nm this corresponds to an effective area, A_{eff}, of

2-5 μm^2, approximately an order of magnitude less than conventional single mode fiber (SMF). The high-energy limit is due to low nonlinearity in the air-core PBGF where the nonlinear index, n_2, of air is roughly 1000 times less than that of silica. These extreme ends of nonlinearity dictate the required pulse energy *(U)* for soliton propagation, which scales as $U \propto D \cdot A_{eff}/n_2$. In fact, most microstructure fibers and tapered fibers with positive dispersion are intentionally designed to demonstrate nonlinear optical effects at the lowest possible pulse energy, while air-core PBGFs are often used for applications that require linear propagation, such as pulse delivery. For these reasons, previous work using SSFS below 1300 nm were performed at soliton energies either too low or too high (by at least an order of magnitude) for many practical applications.

In this paper we show soliton self-frequency shift from 1064 nm to 1300 nm [11, 12] with excellent power efficiency in a higher-order-mode (HOM) fiber.[13] This new class of fiber shows great promise for generating Raman solitons in intermediate energy regimes of 1 to 10 nJ pulses that cannot be reached through the use of PCFs and PBGFs. Due to the dispersion characteristics of the HOM fiber, we also observe red-shifted Cherenkov radiation in the normal dispersion regime for appropriately energetic input pulses [14]. HOM fibers, with its higher tolerance to nonlinearities and engineerable dispersion characteristics, will allow for energetic femtosecond sources at wavelengths where sources are not currently available.

References

1. Dianov, E.M., A.Y. Karasik, P.V. Mamyshev, A.M. Prokhorov, V.N. Serkin, M.F. Stelmakh, and A.A. Fomichev, *Stimulated-Raman conversion of multisoliton pulses in quartz optical fibers.* JETP. Lett., 1985. **41**(6): p. 294-297.
2. Mollenauer, L.F. and F.M. Mitschke, *Discovery of soliton self-frequency shift.* Opt. Lett., 1986. **11**: p. 659-671.
3. Gordon, J., *Theory of the soliton self-frequency shift.* Opt. Lett., 1986. **11**: p. 662-664.
4. Nishizawa, N. and T. Goto, *Compact system of wavelength-tunable femtosecond soliton pulse generation using optical fibers.* IEEE Photon. Technol. Lett., 1999. **11**(3): p. 325-327.
5. Fermann, M.E., A. Galvanauskas, M.L. Stock, K.K. Wong, D. Harter, and L. Goldberg, *Utrawide tunable Er soliton fiber laser amplified in Yb-doped fiber.* Opt. Lett., 1999. **24**(20): p. 1428-1430.
6. Liu, X., C. Xu, W.H. Knox, J.K. Chandalia, B.J. Eggleton, S.G. Kosinski, and R.S. Windeler, *Soliton self-frequency shift in a short tapered air-silica microstructured fiber.* Opt. Lett., 2001. **26**(6): p. 358-360.
7. Washburn, B.R., S.E. Ralph, P.A. Lacourt, J.M. Dudley, W.T. Rhodes, R.S. Windeler, and S. Coen, *Tunable near-infrared femtosecond soliton generation in photonic crystal fibres.* Electron. Lett., 2001. **37**(25): p. 1510-1512.

8. Lim, H., J. Buckley, A. Chong, and F.W. Wise, *Fibre-based source of femtosecond pulses tunable from 1.0 to 1.3 μm.* Electron. Lett., 2004. **40**(24): p. 1523-1525.
9. Luan, F., J.C. Knight, P.S. Russell, S. Campbell, D. Xiao, D.T. Reid, B.J. Mangan, D.P. Williams, and P.J. Roberts, *Femtosecond soliton pulse delivery at 800nm wavelength in hollow-core photonic bandgap fibers.* Opt. Express, 2004. **12**(5): p. 835-840.
10. Knight, J.C., J. Arriaga, T.A. Birks, A. Ortigosa-Blanch, W.J. Wadsworth, and P.S. Russell, *Anomalous dispersion in photonic crystal fiber.* IEEE Photon. Technol. Lett., 2000. **12**(7): p. 807-809.
11. van Howe, J., J.H. Lee, S. Zhou, F. Wise, C. Xu, S. Ramachandran, S. Ghalmi, and M.F. Yan, *Demonstration of soliton self-frequency shift below 1300 nm in higher-order mode, solid silica-based fiber.* Optics Letters, 2007. **32**(4): p. 340-342.
12. Lee, J.H., J. van Howe, C. Xu, S. Ramachandran, and S. Ghalmi. *Energetic soliton self-frequency shift below 1300 nm over a 240 nm range in a solid silica-based fiber.* in *OFC2007.* 2007. Annaheim, CA.
13. Ramachandran, S., S. Ghalmi, J.W. Nicholson, M.F. Yan, E. Monberg, and F.V. Dimarcello, *Anomalous dispersion in a solid, silica-based fiber.* Opt. Lett., 2006. **31**(17): p. 2532-2534.
14. Lee, J.H., J. van Howe, C. Xu, S. Ramachandran, S. Ghalmi, and M.F. Yan, *Generation of femtosecond pulses at 1350 nm by Cerenkov radiation in higher-order-mode fiber.* Opt. Lett., 2007. **32**(9): p. 1053-1055.

Phase Noise Tolerant & Real Time Multilevel Homodyne

Tetsuya Miyazaki, Moriya Nakamura and Yukiyoshi Kamio

National Institute of Information and Communications Technology

4-2-1 Nukui-kita, Koganei, Tokyo 184-8795, Japan, e-mail : tmiyazaki@nict.go.jp

Abstract *Ultimately phase-noise tolerant multilevel homodyne technique without using offline signal processing has been demonstrated. The technique will allow us to use 30-MHz linewidth LD even for 80-Gb/s 256-QAM format so long as DGD is compensated.*

Introduction

Spectral efficient multi-level modulation/demodulation formats have been opening a new era of optical fiber communication systems to accommodate endlessly growing internet traffic. As a prominent example, a differential quadrature phase-shift-keying (DQPSK) format has been already implemented into commercial system in this fiscal year 2007[1]. To enhance efficiency further, M-ary PSK or quadrature amplitude modulation (QAM) has been regarded as one of the promising candidates[2-4]. However, these previously reported multilevel schemes require a narrow laser linewidth of less than about 100 kHz by costly external-cavity laser diodes (EC-LDs) or special fiber lasers[2-4], and so far only off-line signal processing technique have been utilized for carrier recovery in demodulation. We have proposed and demonstrated a self-homodyne multi-level format using a pilot carrier with polarization multiplexing as a phase-noise tolerant scheme[5] without using any spectrally narrowed light source and offline signal processing.

In this paper, at first, we experimentally investigate the tolerance of our homodyne scheme to polarization mode dispersion (PMD)[6]. Then, we demonstrated ultimately phase-noise tolerant QPSK-homodyne using a spectrum-sliced ASE light source[7]. Finally, we investigated and discuss the feasibility for M-ary QAM using a distributed-feedback laser diode (DFB-LD) with a linewidth of 30 MHz based on proposed self-homodyne scheme[8].

Principle and Experimental Setup

As shown in Fig.1(a)[5], in the proposed self-homodyne scheme, a pilot carrier, which provides an absolute optical phase reference, is polarization-multiplexed with an optical signal modulated in a multi-bit-per-symbol format. At the receiver side, self-homodyne using the pilot carrier achieves phase noise

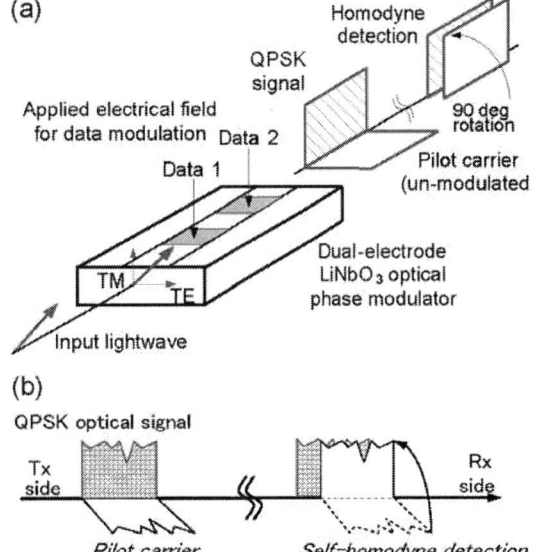

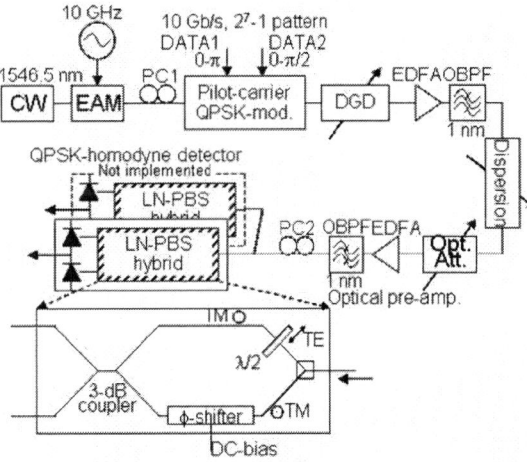

Fig.1 (a):Principle of phase noise tolerant homodyne, (b) : Inmpact of DGD on it.

Fig.2 Setup for investigating various tolerance DGD, dispersion, etc.

cancellation because the pilot carrier includes identical phase noise to the optical signal. However, if the system components and/or transmission lines have differential group delay (DGD) as shown in Fig.1 (b), this phase noise cancellation can not perform well. Figure 2 shows the experimental setup used for

978-0-9789217-3-6/07/$25.00

©2007 WEN GLOBAL SOLUTIONS

investigating DGD and dispersion-tolerance characteristics of our proposed QPSK homodyne technique[6]. Light from a continuous-wave (CW) light source at 1546.5 nm was introduced into the dual-electrode optical phase modulator, which acted as a pilot-carrier-generating QPSK modulator. QPSK modulation with a 2^7-1 pseudo-random binary sequence was applied to the modulator (DATA1 for 0-π, DATA2 for 0-π/2). A 20-bit delay between DATA1 and DATA2 was employed to de-correlate the two modulation signals, which originated from the same data generator. A polarization controller (PC1) was used to adjust the polarization state of the CW light for pilot-carrier generation. At the receiver side, an optical pre-amplifier (EDFA) followed by an optical band-pass filter (OBPF; 1 nm) were employed, and a $LiNbO_3$-based polarization beam splitter (PBS) hybrid module (LN-PBS hybrid) was employed for homodyne detection by 90-degree polarization rotation of the pilot carrier. In the LN-PBS hybrid module, a PBS, a phase shifter (ϕ-shifter), a half-wave plate (λ/2) for pilot carrier rotation, and a 3-dB coupler for balanced detection were integrated. Here, a set of the LN-PBS module and a balanced detector was implemented, although the other half of the set (shown in a dotted box as "Not implemented") was required for simultaneous demodulation of the two components of QPSK signals. Prior to the experiment, we measured the DGD of all the optical components. As a result, the pilot-carrier QPSK modulator had a DGD of 17 ps, whereas the others had negligibly small DGDs of less than about 1 ps. The fast axis of the variable DGD generator was adjusted to be coincident with the slow axis of the modulator so that we could observe the influence of the DGD-induced time shift between the modulated signal and the pilot carrier. The performance of the QPSK homodyne detection was optimized by manually adjusting a second polarization controller (PC2) before the LN-PBS hybrid. Fist, we varied the introduced DGD and investigated the influence on the receiver sensitivity. Here, the dispersion was set to 0 ps/nm. Figure 3 shows the receiver sensitivity penalty (defined at BER of 10^{-9}) of the detected QPSK signal versus the introduced DGD. In case of RZ (open circles), the power penalty increased as the DGD was

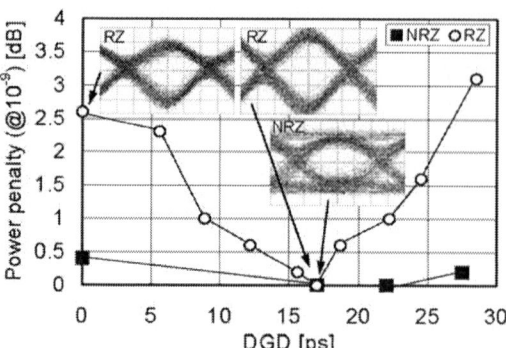

Fig. 3   DGD tolerance for the prposed scheme with RZ and NRZ cases.

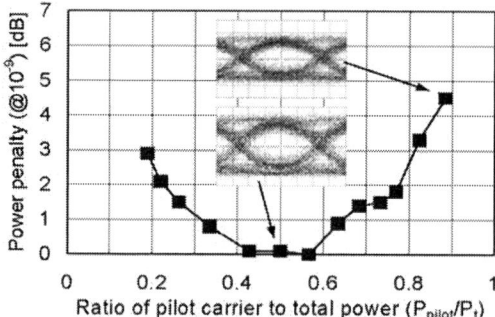

Fig. 4   Power ratio tolerance of pilot carrier

varied from an optimum point of 17 ps where the modulator-DGD was completely compensated (i.e., the gross-DGD was 0 ps). The eye diagrams in the inset show that the eye opening was improved by the compensation in the RZ case. The DGD-tolerance to attain a power penalty of less than 1 dB was 13.3 ps. In the case of NRZ (closed squares), however, the receiver sensitivity was almost insensitive to the DGD within the range of $\pm$15 ps. We also investigated the receiver sensitivity penalty versus ratio of pilot-carrier optical power (P_{plot}) to total optical power (P_t), to determine the optimum splitting ratio (P_{plot}/P_t) and tolerance against polarization detuning from 45° before the modulator. The ratio was measured for the NRZ format with both dispersion and gross-DGD set to zero. The measurement results are shown in Fig. 4. We found that a 50:50 splitting ratio between the pilot-carrier and the QPSK signal achieved the minimum power penalty. To attain a power penalty of less than 1 dB, the ratio should be controlled within about $\pm$15% at around the optimum point of 50%. This corresponds to a polarization control of $\pm$10° around 45° in the linear polarization state. This large tolerance against polarization misalignment is available so long as dispersion and DGD are compensated.

To demonstrate ultimate phase noise tolerance thanks to the phase noise cancellation performance, an erbium-doped fiber amplifier (EDFA) serving as a broadband ASE light source and a continuous-wave (CW) light source with a linewidth of 200 kHz (instead of CW and EAM in Fig. 2) were prepared to generate a lightwave with an arbitrary optical signal-to-noise ratio (OSNR) [7]. Figure 5 shows the receiver sensitivity (at a BER of 10[-6]) for 20-Gbps QPSK (QPSK-H : closed circles) and 10-Gbps BPSK (BPSK-H : closed triangles) versus the OSNR of the light source in a BTB condition. Here, the OSNR of the light source was measured with an optical spectrum analyzer (resolution bandwidth of 0.1 nm) at the output of the polarizer. The receiver sensitivities for both QPSK-H and BPSK-H degraded gradually by decreasing the OSNR below 30 dB. However, a clear eye-opening was observed in the region where the OSNR was negative and even when the CW light source was turned off because of the phase-noise cancellation capability. For comparison, we also attempted DPSK detection using the same experimental setup but with the LN-PBS hybrid module replaced by a one-bit delay line interferometer; here, the pilot-carrier power was eliminated by adjusting PC1. The results are indicated by the open squares in Fig. 5. In this case, DPSK detection was not possible in a region where the OSNR was less than about 30 dB. If DQPSK was employed, the receiver sensitivity must have been degraded at a higher OSNR than 30-dB. Finally we theoretically investigated the phase noise-tolerance of our coherent detection for high-efficiency optical fiber transmission system using QAM formats by computer simulation[8]. Figure 6 shows the I-Q constellation of 256-QAM at 80-Gb/s (10G Symbol/bit) with a received power level of -14 dBm and a spectral line-width of 200-kHz (left) and 30-MHz (right). It should be noted that this proposed detection scheme provides no degradation due to increasing of spectral line-width from 200 kHz to 30 MHz.

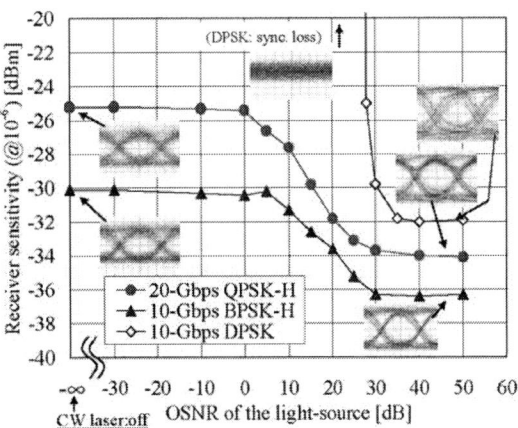

Fig. 5 Receiver sensitivity vs. OSNR using ASE sliced light source.

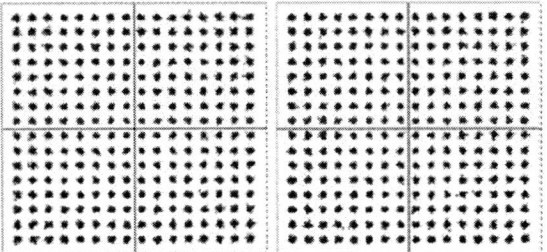

Fig. 6 The I-Q constellation of 256 QAM 200-kHz (left) and 30-MHz (right).

Conclusion

we demonstrated phase-noise tolerant multi-level homodyne using a polarization-multiplexed pilot carrier generated in a transmitter and a LiNbO$_3$-based hybrid module in a receiver. We believe that an employment of this kind of new modulation format including the choice of forward error correction (FEC) technique is expected to support future ultrafast and highly efficient photonic network without traffic jams.

References

1 T.Kataoka et al., OFC 2007, JthA45.
2 S. Tsukamoto, et al., OFC'06, OThR5.
3 M. Nakazawa, et al., Electron. Lett., vol.42 (2006), P. 710.
4 N. Kikuchi, et al., OFC'07, PDP21.
5 T.Miyazaki, Photon.Tech.Lett.,, vol.18 (2006), P. 388.
6 M. Nakamura, et al., ECOC'06, Mo4.2.5.
7 M. Nakamura, et al., OFC'07, OthD4.
8 Y.Kamio, et al., LEOS Summer Topicals '07, TuE3.2

Photonic crystal and plasmonic devices for photonic integration

Min Qiu

Department of Microelectronics and Applied Physics, Royal Institute of Technology (KTH)
Electrum 229, 164 40 Kista, Sweden
Email: min@kth.se

Abstract *Our recent research results on photonic crystal and plasmonic devices for photonic integration are reviewed here. Silicon-based photonic crystal waveguides, microcavities, and filters, as well as plasmonic waveguides will be presented.*

Introduction

Photonic crystal devices have their potentials in large scale photonic integrated circuits. In particular, photonic crystals may provide ultra-compact small-modal-volume microcavities, optical filters, which give many important applications for photonic devices.

Plasmonic waveguides have recently also attracted lots of attention due to their capability to guide the light in a sub-wavelength region.

In the present paper, we will review our recent work of photonic crystal and plasmonic devices, in particular, silicon-based photonic crystal waveguides, microcavities, and filters, as well as plasmonic waveguides.

Photonic crystal devices

Surface waves are propagating electromagnetic waves, which are bound to the interface between materials and free space. They usually do not exist on dielectric materials. However, due to the existence of photonic band gaps, it has been shown that dielectric photonic crystals may support surface waves for some cases, e.g., some truncated or deformed photonic crystal at the interface. We have recently proposed [1] novel photonic crystal microcavities with high Q, which are based on zero-group-velocity surface modes in photonic crystals. These cavities are open cavities in the sense that one of the in-plane boundaries is exposed to exterior.

We have then designed, fabricated, and characterized an optical filter in a SOI structure based on the surface mode microcavities. The PhC surface mode cavity is side coupled to a silicon wire waveguide situated in the same plane, as shown in Fig. 1. The experimental transmission spectrum clearly demonstrates the filtering characteristics [2].

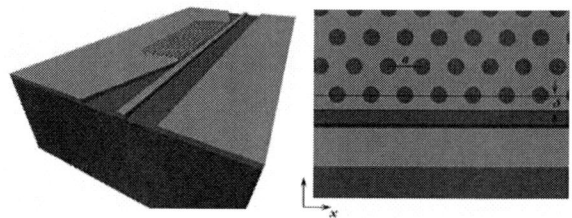

Fig. 1 Schematic of photonic crystal surface mode cavity side coupled to silicon wire waveguide in SOI structure.

Other designs of photonic crystal microcavities and filters [3] will also be presented.

Plasmonic waveguides

We have used both the FDTD method and the finite element method (FEM) to model plasmonic media [4]. In particular, our current FEM code deals with all longitudinally-invariant optical waveguides, with either lossless or lossy materials. We have then studied dispersion and loss of the guided fundamental mode in V-channel plasmon polariton waveguides at the optical and near-infrared regime. It is noticed that better confinement is always at the expense of higher propagation loss for such waveguides. Other types of plasmonic waveguides, including the wedge type

978-0-9789217-3-6/07/$25.00
©2007 WEN GLOBAL SOLUTIONS

structures, are also studied (see Fig. 2). Knowledge of the guided modes in such structures is useful for designing realistic plasmonic channel waveguides.

References

1 S. Xiao et al, Appl. Phys. Lett, 87 (2005), 111102

2 Z. Zhang et al, Appl. Phys. Lett., 90 (2007), 041108

3 Z. Zhang et al, Opt. Express, 12 (2004), 3988

4 M. Yan et al, J. Opt. Soc. Am. A, in press.

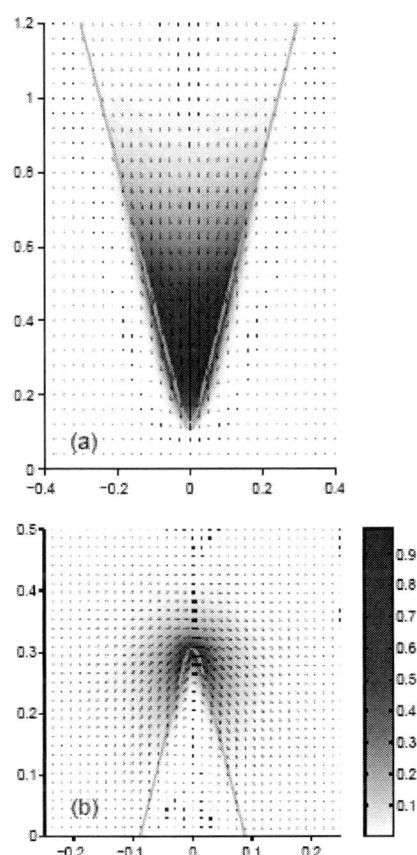

Fig. 2 Fundamental mode guided by two types of corner waveguides. (a) V-channel waveguide; (b) Λ-wedge waveguide.

Light confinement at interfaces and Talbot effect using optical surface modes

F.J. García de Abajo,[1,2] R. Sainidou,[1] T. V. Teperik,[2] M. Dennis,[3] and N. I. Zheludev[4]

[1]Instituto de Óptica – CSIC, Serrano 121, 28006 Madrid, Spain
[2]DIPC, Aptdo. 1072, 20080 San Sebastián, Spain
[3]School of Mathematics, University of Southampton, SO17 1BJ, UK
[4]Optoelectronics Reseearch Center, University of Southampton, SO17 1BJ, UK

Fax: +34-915645557; email: jga@io.cfmac.csic.es

Abstract

Light confinement to surfaces is at the heart of many recent advances in nanophotonics, for instance in the field of plasmonics, which relies upon plasmon polaritons at metallic surfaces. Various ways of confining light to an interface will be reviewed and new ones demonstrated in this talk. A common description of such modes will be offered, which leads to a global understanding of their interference at interface features. In particular, the interference of surface plasmons emanating from a row of holes in a metal film will be shown to lead to complex plasmonic structures that present self-reconstruction of the hole array at distances up to tens of wavelengths away from the holes. This is the plasmonic version of the optical Talbot effect. Interestingly, subwavelength hot plasmonic spots are observed at those distances, suggesting the possibility of using them for far-field patterning and for long-distance plasmon-based interconnects in plasmonic circuits. Some examples of such applications will be also presented.

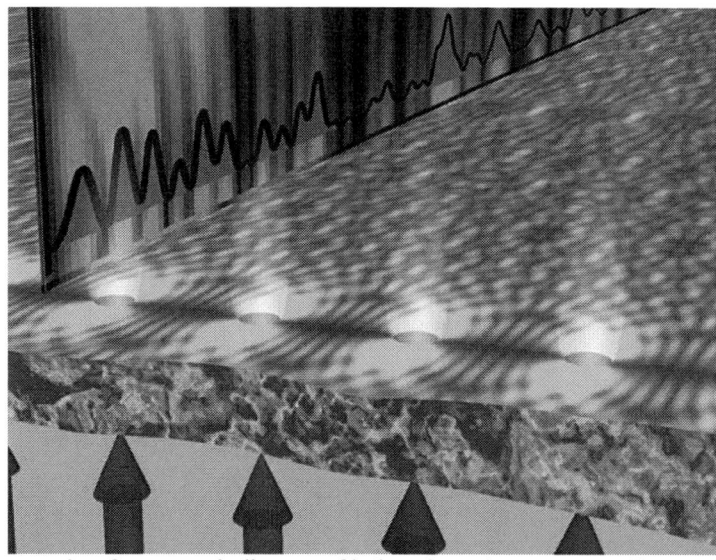

Figure 1: Light emanating from a periodic row of holes gives rise to a complicate pattern of plasmons on the exit side of the metallic film on which they are perforated. In particular, self-reconstruction of the array is observed many wavelengths away from the holes, in the spirit of the optical Talbot effect.

Optical polarization beam splitting through anisotropic metamaterial slab realized by layered metal-dielectric system[*]

Junming Zhao, Yan Chen, Yijun Feng[†]

Department of Electronic Science and Engineering, Nanjing University,
Nanjing, 210093, China.
[†]email: yjfeng@nju.edu.cn

Abstract *We report a novel polarization beam splitter (PBS) utilizing the anomalous reflection and transmission of an anisotropic metamaterial slab. By properly design of the constitutive tensors, PBS is achieved with little dependence on the incident angle and slab thickness, and realized in the optical range by layered metal-dielectric nano-structured system.*

Introduction

Polarization beam splitters (PBSs) are important optical devices that separate the two orthogonal polarizations of light beam into different propagation directions, which have broad applications in liquid crystal displays, optical communication and optical recording. Recently, PBSs have been achieved through photonic crystals, which utilize negative/positive refraction for two different polarizations [1-3]. These PBSs show good beam splitting characteristics, but they still depend on the incident angle or the slab thickness.

Anisotropic metamaterials (AMMs) having permittivity and permeability tensors with parts of the elements being negative have drawn a lot of attentions recently due to their extraordinary wave propagation properties [4-5]. The AMMs have been identified into four types based on their wave propagation properties, which are called cutoff, always-cutoff, never-cutoff (NCM) and anti-cutoff media (ACM). Anomalous electromagnetic wave refraction and reflection have been demonstrated at the interface between normal medium and the NCM or ACM medium [6]. We propose a PBS based on the anomalous wave reflection and transmission of an AMM slab. By properly designing the constitutive tensors of the slab, PBS is achieved by total reflection of one polarization beam while full transmission of the other polarization beam. The most interesting property of the PBS is

its less dependence on the incident angle and slab thickness. To realize such PBS at optical wavelength, we have designed a layered metal-dielectric nano-structured system, which could be modelled as an effective anisotropic metamaterial with the desired constitutive tensors. Full-wave electromagnetic simulation based on the finite-differential time-domain method (FDTD) has clearly confirmed the characteristics of the PBS.

PBS based on an anisotropic metamaterial slab

The basic idea of the proposed novel PBS is to utilize the anomalous wave propagating properties of AMM. Consider an AMM slab characterized by permittivity and permeability tensors with only non-zero diagonal elements as $\varepsilon = diag[\varepsilon_x, \varepsilon_y, \varepsilon_z,]$ and $\mu = diag[\mu_x, \mu_y, \mu_z,]$. By properly designing the permittivity and permeability tensors, TM (or TE) waves could be totally reflected at the interface associated with the slab whereas TE (or TM) waves could be fully transmitted through it. We take the total reflection of TM wave as example. Considering the dispersion relations of different polarization waves in an anisotropic medium,

$$k_z^2 + k_y^2 \mu_y / \mu_z = k_0^2 \varepsilon_x \mu_y \text{, (for TE wave)}$$

$$k_z^2 + k_y^2 \varepsilon_y / \varepsilon_z = k_0^2 \mu_x \varepsilon_y \text{. (for TM mode)}$$

---

[*] Supported by the National Basic Research Program of China (2004CB719800) and the National Nature Science Foundation (60671002).

978-0-9789217-3-6/07/$25.00

©2007 WEN GLOBAL SOLUTIONS

Obviously, the propagation of TE or TM wave is dominated by different permittivity and permeability tensors, which allow us to manipulate the two orthogonally polarized waves separately. If we choose, for example, $\varepsilon_x = \varepsilon_0$ and $\mu_y = \mu_z = \mu_0$, the slab behaves like vacuum to the TE waves with full transmission. To TM waves, if we choose $\mu_x \varepsilon_y < 0$, and $\varepsilon_y / \varepsilon_z > 0$, the slab behaves like an anti-cutoff medium [4]. Theoretical analysis shows that the TM waves will be totally reflected at the slab interface so far as the thickness of the slab satisfies $L > 3\lambda_0 / \pi\sqrt{|\varepsilon_y \mu_x|}$, where λ_0 is the wavelength in vacuum. Therefore it is possible to have the TE wave totally transmitted while the TM wave totally reflected by the slab, which results in an ideal PBS.

Fig. 1(a) and 1(b) shows the polarization beam splitting through an AMM slab at different incident angles. The most interesting property is that unlike other PBS the splitting does not severely depend on the incident angle and slab thickness.

In the presentation, we will show the detailed analysis of the proposed PBS, including how to design the AMM slab, the conditions that the material tensors of the AMM should satisfy. We will also describe in details the less incident angle and slab thickness dependence of this kind of PBS, as well as the loss effect on the PBS properties.

Realization the PBS by layered metal-dielectric system

AMMs with desired permittivity properties can be produced by a layered structure of thin, alternating dielectric layers (or metal and dielectric layers). Such planar systems have been proposed to demonstrate the subwavelength imaging [7], and to build photonic funnels for sub-diffraction light compression and propagation [8]. The proposed PBS which requires some of the permittivity tensors to be negative could be constructed by layered metal-dielectric system at the optical wavelength. Consider such a layered system with alternating metal layer 1 and dielectric layer 2, as shown in Fig. 2. When the layers are

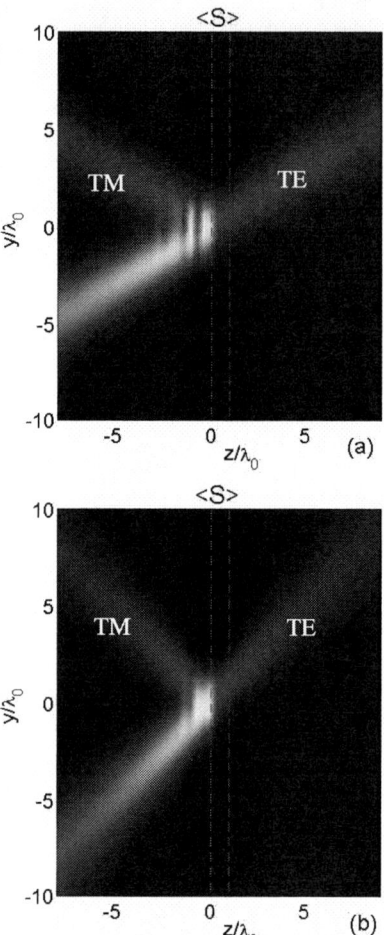

Fig. 1. Gaussian beam polarization splitting through an AMM slab (with $\varepsilon_x = \varepsilon_0$, $\mu_y = \mu_z = \mu_x = \mu_0$, $\varepsilon_y = -2\varepsilon_0$, $\varepsilon_z = 2\varepsilon_0$, $L = \lambda_0$) for different incident angles, (a) 30°, (b) 45°.

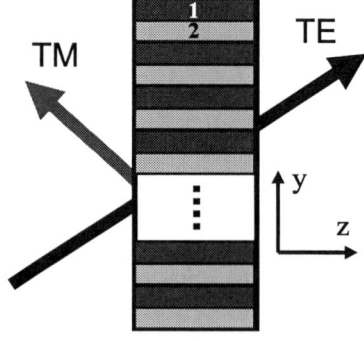

Fig. 2 Schematic of the PBS realized by layered silver/silica system. Each layer is about 10 nm thick, and the slab is about 1 μm thick along the z direction.

164

sufficiently thin compared with the wavelength, the whole layered structure can be treated as a single anisotropic medium with the dielectric permittivity as

$$\varepsilon_x, \varepsilon_z = \frac{\varepsilon_1 + \varepsilon_2}{2}, \quad \varepsilon_y = \frac{2\varepsilon_1\varepsilon_2}{\varepsilon_1 + \varepsilon_2},$$

where ε_1, ε_2 are the permittivities of the layer 1 and layer 2, respectively. We have also assumed the same thickness for the metal and dielectric layers. If we use silver film as the metal layer and the silica film as the dielectric layer, at optical waves the permittivity of the silver film is modelled by the plasmalike Drude model and is negative below the plasma frequency [9]. For example, at wavelength of 460 nm, we have

$\varepsilon_1 = -0.5\varepsilon_0$, $\varepsilon_2 = 2.5\varepsilon_0$, which results in an effective permittivity tensor of $\varepsilon_x = 1$, $\varepsilon_y = -1.25$, $\varepsilon_z = 1$. The performance of a slab of this layered system is verified by commercial FDTD based full-wave electromagnetic simulation shown in Fig. 3. TE wave has been totally transmitted with a power transmittance of 95.5% (Fig. 3a), while the TM wave is reflected with little transmission (Fig. 3b) at an incident angle of 45°.

In the presentation we will describe in details the simulation results which demonstrated a less incident angle dependence of this kind of PBS.

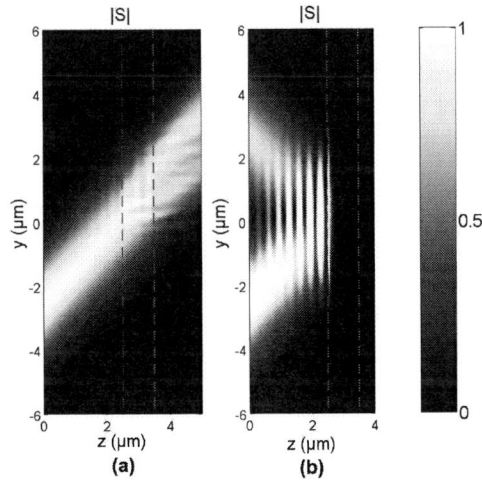

Fig. 3. (a) Total transmission of the TE wave, and (b) total reflection of the TM wave through a silver/silica layered PBS. The dashed line indicated the boundary of the slab.

References

1 V. Mocella et al Opt. Express, 13 (2005), 7699-7707.

2 Xianyu Ao et al Appl. Phys. Lett, 89 (2006), 171115.

3 J. She et al J. Opt. A: Pure Appl. Opt, 8 (2006), 345.

4 D. R. Smith et al Phys. Rev. Lett, 90 (2003), 077405.

5 T.M. Grzegorczyk et al IEEE Tran. Microwave Theory and Techniques, 53 (2005), 1443-1450.

6 Yijun Feng et al J. Appl. Phys, 100 (2006), 114901.

7 B. Wood et al Phys. Rev. B, 74 (2006), 115116.

8 A. Govyadinov et al Phys. Rev. B, 73 (2006), 155108.

9 E. D. Palik Handbook of Optical Constants of Solids, (Academic Press, London, 1985).

Asymmetric Hybrid Three-Arm Coupler with Long Range Surface Plasmon Polariton and Dielectric Waveguides

Fang Liu, Ruiyuan Wan, Yi Rao, Yuxin Zheng, Yidong Huang, Wei Zhang, and Jiangde Peng

State Key Lab. of Integrated Optoelectronics, Department of Electronic Engineering, Tsinghua University, Beijing 100084, China
yidonghuang@tsinghua.edu.cn

Abstract *The characteristics of the asymmetric hybrid three-arm coupler, which consists of one middle long range surface plasmon polariton waveguide and two outside conventional dielectric waveguides, are analyzed numerically with finite element method.*

Introduction

Surface Plasmon Polariton (SPP) is transverse-magnetic surface electromagnettic excitation that propagates in a wave like fashion along the metal and dielectric interface.[1] The metal strip guided long range surface plasmon polariton (LRSPP) mode attracts much attention for its low loss[2,3] and capability of carrying optical signals and electrical signals simultaneously[4].

We have proposed the hybrid coupler, which consists of dielectric waveguide(s) and metal strip.[5,6] Simulation results show that the hybrid coupler has a high efficient coupling between LRSPP and dielectric waveguide mode with significantly reduced loss. Different from the conventional three-arm dielectric coupler[7], it was demonstrated that, for the three-arm hybrid coupler, the asymmetric structure can improve the extinction ratio remarkably by introducing an offset of the middle arm. In this paper, the structure parameter dependences of the extinction ratio are analyzed for the three-arm hybrid coupler. It is found that the Phase Match Factor (PMF) is the determinant factor. Small PMF, which can be obtained by adjusting the structure parameters, leads to a high extinction ratio.

Structure and Simulation Method

Figure 1(a) shows the simulation model for the proposed three-arm hybrid coupler. The propagation direction is along z axis. Different from the conventional three-arm dielectric coupler,[7] the middle arm is replaced by a metal strip (LRSPP mode waveguide) with thickness T_m. The two T_d thick outside arms are the same dielectric waveguides with distance D. These three arms have the same width W. We first consider the symmetric structure with the metal strip offset $d=0$. Coupled eigenmodes supported by the hybrid coupler were calculated directly to analyze the coupling characteristics among the three arms instead of solving the coupled-mode equations,[8] just as that reported in Ref. 5 and 6. The software FEMLAB,[9] which implements the Finite Element Method (FEM) to calculate the SPP mode, was used in the simulation.

Here, we assume that the two outside dielectric waveguides (n_d=1.54) and the Au (ε_m=-132+i×12.65) strip are surrounded by SiO$_2$ (n_s=1.444) at λ_0=1.55μm,[10] with width W=2μm. The thickness of the dielectric arms T_d is fixed at 393nm.

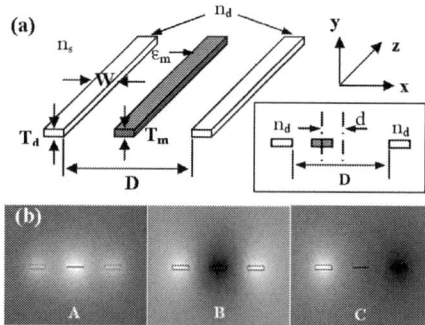

Fig. 1 (a) Three-arm hybrid coupler, dark arm (middle) stands for the metal strip and white arms (left and right) stand for the dielectric waveguide. In the cross section, d is the offset of the middle arm. (b) Magnetic field of three eigenmodes in the x-y plane.

Simulation Results

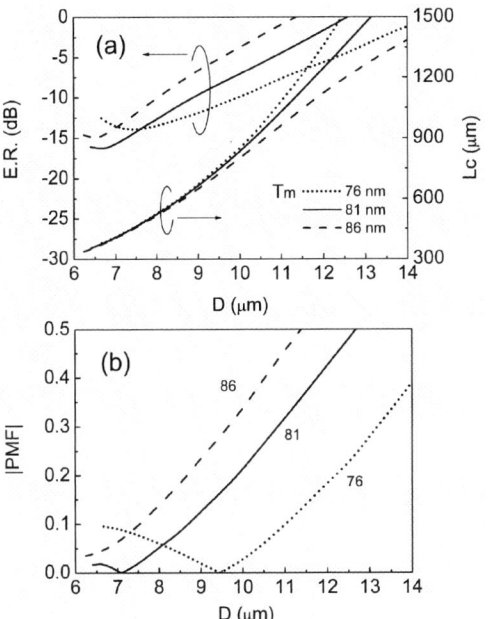

Fig. 2 (a) Extinction ratio and coupling length with different T_m. (b) The Phase Match Factor (PMF).

Figure 2(a) shows the extinction ratio[6] and coupling length (L_c) of the symmetric three-arm hybrid coupler with different metal strip thickness T_m. For small D, the L_c is rather short, while the different losses of the eigenmodes result in the bad extinction ratio.[6]

Another important factor, which decides the extinction ratio (shown in Fig. 2(b)), is the Phase Match Factor. It is defined as $|(2\beta_C - \beta_A - \beta_B)/(\beta_A - \beta_B)|$ and describes the relation between the propagation constants of three eigenmodes.

Different from conventional dielectric coupler,[7] the hybrid three-arm coupler should introduce a middle arm offset to achieve a high extinction ratio.[6] The following analysis indicates that PMF decides the maximum extinction ratio when adjusting the offset d.

Figure 3 shows the extinction ratio of the asymmetric structure with different middle arm thickness T_m. The dielectric arm distance D and the corresponding PMF value at offset d=0 in the bracket are marked near the curve. In Fig. 3(a), T_m is fixed at 86 nm with different D. We find smaller D corresponds to higher extinction ratio. This is because that the PMF is cutoff before it decreases to zero, as shown by the dashed line in Fig. 2(b), and smaller D corresponds to smaller PMF. In this case, because PMF can not be zero, extinction ratio can only reach -25 dB when D=6.4 µm, d=170 nm.

When T_m=81 nm, the PMF crosses the zero point near D = 7 µm, as shown by solid line in Fig. 2(b). Thus the solid line in Fig. 3(b) illustrates that by adjusting the position of middle LRSPP arm to d=170 nm, extinction ratio as large as -40 dB can be obtained. Of course, an unsuitable D with large PMF will worsen the maximum extinction ratio. For example, the dash dotted line rather near the cutoff point with D=6.6µm and PMF=1.8% results in maximum -30 dB extinction ratio. While for D=8 µm (the dotted line with PMF=5.2%), the maximum extinction ratio can only be -20 dB with a

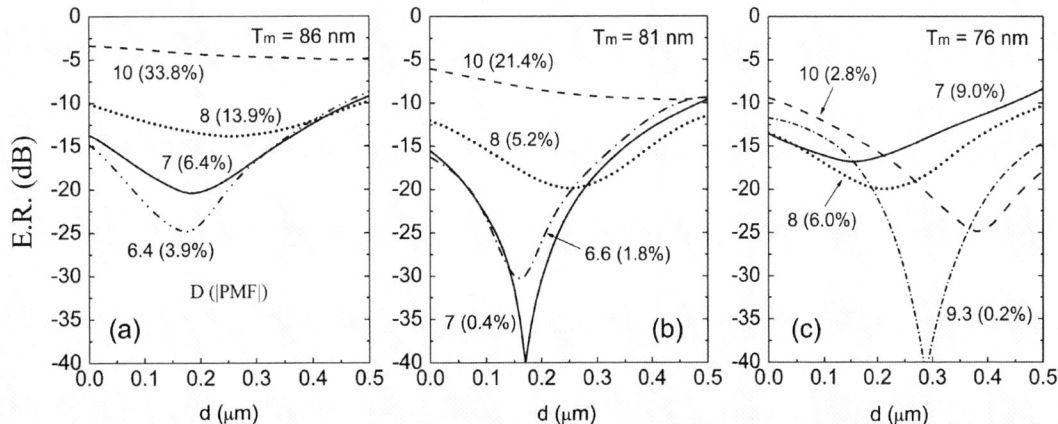

Fig. 3. Extinction Ratio of asymmetric structure with (a) T_m = 86 nm, (b) 81 nm and (c) 76 nm. The dielectric arm distance D and the corresponding PMF value in the bracket are labeled near each curve.

167

larger offset d=250 nm.

In the case of T_m=76 nm, the ideal D is about 9.3 μm with PMF=0.2% according to the dotted line in Fig. 2 (b). With the ideal D, the simulation results in Fig. 3(c) shows extinction ratio can be larger than -40 dB when d is near 290 nm. Here, other D can not reach such high extinction ratio no matter how to adjust the offset d because of their larger PMF value. Since PMF is insensitive to d, from the results in Fig. 3, it is noticed that small PMF value of symmetric structure (d=0) corresponds to high maximum extinction ratio in asymmetric structure.

Therefore, by introducing the asymmetric of the eigenmodes and utilizing their different transmission loss, high extinction ratio can be achieved for asymmetric structure. Although thin LRSPP arm (T_m=76 nm shown in Fig. 3(c)) can also reach high extinction ratio, larger D and d means larger coupling length and coupling loss. On the other hand, thick LRSPP arm (T_m=86 nm shown in Fig. 3(a)) can not reach high extinction ratio before cutoff. Therefore, T_m should be well selected (i.e. T_m=81 nm in Fig. 3(b)) to match the phase of three eigenmodes (PMF=0) and let the corresponding D close to the cutoff point.

Conclusions

We demonstrate numerically that the extinction ratio depends on the structure parameters of the three-arm hybrid coupler. Small PMF leads to larger extinction ratio. Considering the coupling length and coupling loss, optimization for structure parameters, including T_m, D, and d, makes it possible to reach extinction ratio higher than -40dB.

Acknowledgement

This work was supported by the National Basic Research Program of China (973 Program) under Contract No. 2007CB307004. The authors would like to thank D. Ohnishi, H. Takasu, and A. Kamisawa of ROHM Corporation for their valuable discussions and helpful comments.

References

1 A. V. Zayats, et al, Phys. Rep., 408 (2005), 131

2 P. Berini, et al, Phys. Rev. B, 61 (2000), 10484

3 T. Nikolajsen, et al, Appl. Phys. Lett., 82 (2003), 668

4 T. Nikolajsen, et al, Appl. Phys. Lett., 85 (2004), 5833

5 F. Liu, et al, Appl. Phys. Lett., 90 (2007), 141101

6 F. Liu, et al, Appl. Phys. Lett., 90 (2007), 241120

7 J. P. Donnelly, et al, IEEE J. Quantum Electron., QE-23 (1987), 401

8 W. P. Huang, J. Opt. Soc. Am. A, 11 (1994), 963

9 COMSOL AB, Sweden. FEMLAB Electromagnetics Module Manual, 3.0 edition, 2003

10 E. D. Palik, Handbook of Optical Constants of Solids (Academic, Orlando, 1985), 286-297

Single-beam self-referenced phase-sensitive surface plasmon resonance sensor with high detection resolution

Shu-Yuen Wu and Ho-Pui Ho*

Department of Electronic Engineering,

The Chinese University of Hong Kong, Hong Kong SAR, China

*email: hpho@ee.cuhk.edu.hk

Abstract We present a new surface plasmon resonance (SPR) sensing system exhibiting a root-mean-square resolution of $\pm 5.2 \times 10^{-9}$ refractive index units. This brings non-labeling SPR biosensing closer to that achievable by florescence-based techniques.

Introduction

Over the past two decades, optical sensors based on the surface-plasmon resonance (SPR) phenomenon have gradually evolved to become a standard tool for quantitative measurement of chemical and biological binding reactions[1]. In order to fulfill the ever-increasing demand of application requirements, SPR sensing instrumentation has been improving continuously in terms of resolution, system stability, measurement throughput and cost of ownership. One approach for achieving ultra-high resolution is to measure the SPR phase response. The SPR phase undergoes a steep jump across the resonance dip, which means that a small refractive index change near the sensing surface may lead to a large signal response. Amongst the several different approaches for detecting the SPR phase, we report herein the use of a liquid crystal modulator (LCM) performing low frequeucy retardation modulation, thereby enabling phase measurement with very respectable resolution,

In spite of its simplicity as a retardation modulator, the LCM has its own shortcomings, namely high temperature dependence, high non-linearity, low operation frequency and limited retardation modulation range. Small variations in temperature will result in sizeable measurement drift. Indeed the situation will be much more problematic for the phase-shifting approach reported by Su et al.[2], as the accuracy of the final phase measurement relies entirely on the repeatability of the five data points obtained upon shifting of the retardation within a 2π circle. With this as the objective, we have built a single beam self-referenced phase-sensitive SPR sensing system that have extremely high stability through the incorporation of a temperature control to stabilized the retardation drift in the LCM, and also the use of differential phase measurement for further elimination of unwanted phase errors from the ambient. We also incorporated a novel beam-folding device to increase the retardation modulation depth to multiples of 2π to facilitate better accuracy in the phase extraction process.

Experimental Set-up

Our experimental setup is depicted in Figure 1. Large retardation modulation is obtained by placing the LCM element between two mirrors so that the optical beam travels through the modulator multiple times before leaving the cavity. The fast axis of the LCM is aligned to the p-polarization of the laser beam in order to obtain maximum retardation modulation depth. The light source is a 12mW linearly polarized He-Ne laser at the wavelength of 632.8nm. The optical polarization axis of the laser beam is set at 45° from both p- and s-polarization orientation so that the optical power is initially divided into two equal halves along the p- or s-polarization axes. A 9:1 beam splitter is placed after the LCM

978-0-9789217-3-6/07/$25.00

©2007 WEN GLOBAL SOLUTIONS

for producing a reference signal. While all subsequently captured signals are compared against this reference, any fluctuation in the characteristics of the LCM will be eliminated through a differential measurement approach. More importantly, we have implemented a temperature-controlled chamber for stabilizing the temperature of the LCM to within ±0.01°C. We found that this is very important in achieving the expected performance from the sensor system. As the LCM is placed between two mirrors so that the beam gets folded a number of times and experiences many times larger retardation modulation depth, the total retardation depth is 14π (i.e. going through the LCM 7 times). The significance of this technique is that we need to capture as many cycles of phase modulation as possible in every signal trace in order to increase the phase extraction accuracy as the subsequent data processing algorithm relies on comparing the phase locations of truncated sine waves. If we take p-polarized light as the probe beam, while s-polarized light as the self-reference beam, then they system will naturally eliminate common-mode noise not due to the phase difference between the two polarizations (i.e. retardation). The SPR sensing head consists of two elements: a prism coupler and a sensor chip covered by a flow-cell. The dove prism coupler is made from BK7 glass and the sensor chip is a microscope glass slide (25mm × 25mm × 1mm) coated with a nominally 46nm thick gold film. Further details of the sensor chip and the flow-cell may be found from our previous paper[3]. The sensor chip is optically coupled to the back side of the dove prism through a layer of refractive index matching oil. During the experiment, sample fluid was injected into the flow-cell using a syringe. A digital storage oscilloscope (LeCroy LT264 DSO) is used for recording the signal waveforms of the probe and reference channel and the raw data will be processed on a computer. A software program has been constructed for real-time phase exaction and monitoring. The signal processing procedures including data averaging, signal trace conversion and normalization to truncated sine wave, band-pass filtering and point-wise differential phase extraction. Further details of our

phase extractive algorithm may be found in our earlier paper[4]. In order to verify the performance of this system, we performed a set of refractive index measurement experiments on glycerin/water mixtures in various weight ratios.

Results

The samples were glycerin-water mixtures with 0% to 16% of glycerin in weight percentage, which correspond to a refractive index rang from 1.3330 to 1.3521 refractive index unit (RIU)[5]. Figure 3 shows the measured SPR phase versus concentration plot. As seen from this plot, we could only observe approximately 50% of the maximum SPR phase change across the SPR resonance dip. This is because of the fact that our experiments started at the minimum point of the SPR resonance dip, which was halfway through the phase jump across resonance. The system continuously shifted away from resonance as the refractive index of the sample increased from 1.3330 to 1.3521. In order to calculate the best resolution limit of our system, we use the first two data points from the plot, which were obtained in the most sensitive region (near the resonance dip). They correspond to two sample mixtures, i.e. pure water and 0.05% glycerin in water, with a change of refractive index equivalent to 6×10^{-5} RIU and a SPR phase change of 32.12°. Moreover, as shown in Figure 3, the system stability was also monitored and we obtained a phase measurement fluctuation of ±0.0028° (root-mean-square) or 0.013° (peak-to-peak) within 45 minutes. It should be mentioned that the stability of the present system is significantly better than that reported from a similar LCM-based setup[2], which is around 0.1° (peak-to-peak). Using ±0.0028° as the baseline of our system stability, the calculated resolution limit is $\pm5.2\times10^{-9}$ RIU.

Conclusions

A single-beam self-referenced phase-sensitive SPR sensor based on differential measurement between the probe and reference channel has been demonstrated. In order to address the limitations of LCM as a retardation modulator, we proposed two simple techniques, to improve its modulation range and stability, i.e. the use of multi-pass and active temperature control. Errors due to temperature dependence and non-linearity of the LCM can also be removed by taking the differential change between the probe and reference signal obtained immediately before and after the sensor head. The experimental resolution limit of our system is estimated to be $\pm 5.2 \times 10^{-9}$ RIU.

Acknowledgment

The authors wish to acknowledge funding support of this project from the Research Grants Council (under CERG project number 411906) and the Shun Hing Institute of Advanced Engineering, The Chinese University of Hong Kong. We are also grateful to The Chinese University of Hong Kong for providing research student support to S.Y. Wu.

References

1. R.L. Rich, D. G. Myszka, J. Mol. Recognit. 13, 388 (2000).
2. Y.D. Su, S.J. Chen and T.L. Yeh, Optics Letter 30, 1488 (2005)
3. S.Y. Wu, H.P. Ho, Proc. SPIE 6099, 60990R (2006).
4. H.P. Ho, W.C. Law, S.Y. Wu, Chinlo Lin, S.K. Kong, Biosensors and Bioelectronics 20, 2177(2005).
5. Weast R C (Ed.) 1987 CRC Handbook of Chemistry and Physics 68th ed (CRC Press, Boca Raton, FL) D-232.

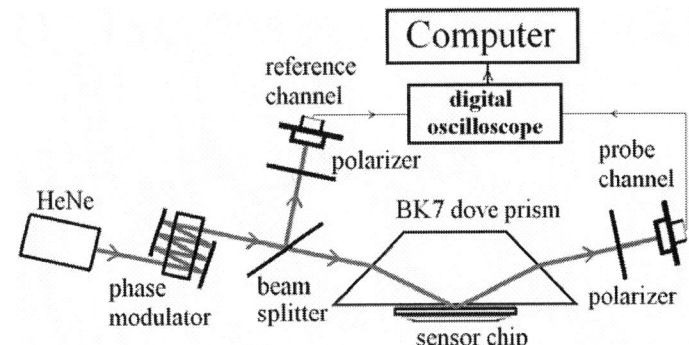

Figure 1. Experimental setup of LCM-based SPR sensing system.

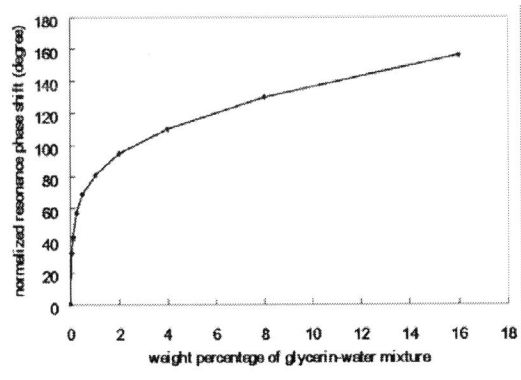

Figure 2. Variation of relative differential phase (reference to pure water) versus glycerin concentration

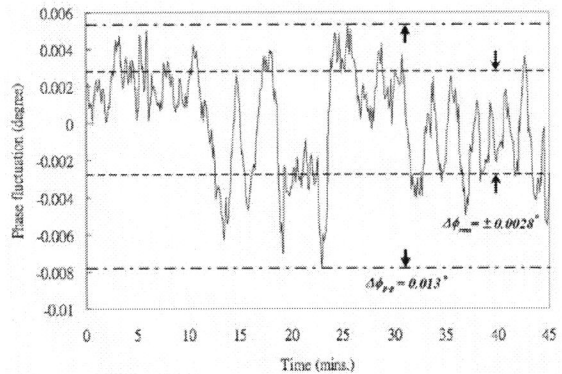

Figure 3. System stability measurement over 45 minutes

All Fiber Optic Coal Mine Safety Monitoring System

Tongyu Liu

Laser Institute, Shandong Academy of Science, 19 Keyuan Road, Jinan, Shandong province, China 250014， Email: t.liu@micro-sensor.com.cn

Abstract *This paper reports on the concept and field test results of the first all fiber optic sensor-based comprehensive coal mine safety monitoring system for remote monitoring of methane concentration, mining pressure, water pressure and temperature.*

Introduction

Coal mine safety has been a major social and economic issue in china in recent years and thousands of deaths and injuries have been caused each year due to coal mine accidents. Advanced technologies which can enhance the capability of coal mine safety condition monitoring, accidents prediction and warning are most desirable. The major causes of serious accidents include methane gas outburst caused explosion, roof collapsing, water burst and flooding, rock and coal burst and mine fire. Real time monitoring of coal mine underground ambient and structural conditions relating to the above hazards are critical to mine safety. Various systems have been developed to address some of the issues, however their effectiveness has been severely hampered by the limitation of conventional sensor technologies. For example, conventional electro-chemical methane gas sensors which are used in remote methane monitoring systems, suffer from drift with time and limited measurement range. Fiber optic sensors offer unique advantages including intrinsic safety in explosive environment, small drift, high sensitivity, and remote and multi-point multi-parameter sensing, hence they are ideal for coal mine safety monitoring. Although the fiber optic sensors are increasingly used in oil well, electrical power industry, rather few reports have been made on their application in coal mines[1]. The motivation of this work was to explore the full potential of the fiber optic sensors by demonstrating a multi-parameter comprehensive coal mine safety monitoring system.

2. The Fiber Optic Coal Mine Safety Monitoring System

The all fiber optic coal mine safety monitoring system consists of the following sub-systems: (1) fiber optic remote methane gas monitoring system, (2) mining pressure and temperature monitoring, (3) water pressure monitoring, and (4) seismic event detection.

2.1.1. Fiber optic methane gas sensor

Methane gas has harmonic and combinational absorption bands at around 1.65 and 1.3 μm respectively, which can be conveniently exploited by fiber optic sensor designs. The methane sensor consists of fiber optic gas cells made of a pair of collimated grade index (GRIN) rod lenses separated with 5 cm air gap. To protect the gas cell from dust and moisture the gas cell is covered with metallic micro-ball air filter housing and molecular filter balls. The signal interrogation was carried out by using a 1650 nm distributed feedback laser diode which was driven by a saw-tooth wave to directly modulate the laser wavelength. A reference gas cell was used to identify the peak absorption wavelength and signal normalization was used to achieve self-compensation of intensity fluctuations due to fiber lead loss and light source.

2.1.2. Fiber optic mining pressure

Mining pressure is the internal stress in rocks in the coal mine. It is an important indication for mine structural health monitoring. Conventional mining pressure monitoring sensors were typically based on strain gauges and displacement sensors, which suffer from drift, low accuracy and not convenient to monitor over long distances. Fiber Bragg grating (FBG) strain and

978-0-9789217-3-6/07/$25.00

©2007 WEN GLOBAL SOLUTIONS

temperature sensors can be conveniently adapted for mining pressure sensing. The fiber optic mining pressure prototype design consists of embedded FBG element inside slot in standard stainless steel mining reinforcement rods. Signal processing was achieved using broadband amplified spontaneous emission (ASE) light source and a high accuracy wavelength interrogator.

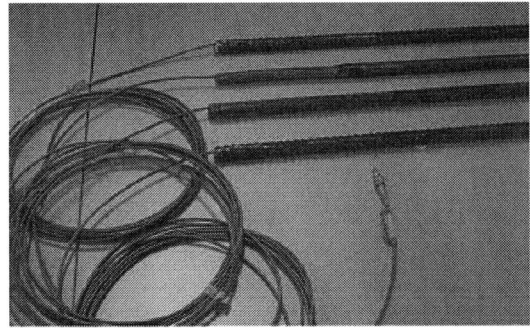

Fig. 1. Photograph of FBG-based mining pressure sensors

2.1.3. Fiber optic water pressure sensors

Fiber optic pressure sensor was made of an FBG strain sensor surface mounted to a bellows type elastic structure. Temperature effect was compensated by a separate FBG housed inside the sensor housing but in a strain-free condition. Fiber optic water pressure sensor offers the advantage of multiplexing, high accuracy and long term stability.

2.1.4. Fiber optic seismic detection sensors

Seismic detection is a powerful method for monitoring mine roof collapsing, rock and coal burst as well prediction of gas outburst. Conventional seismic sensors are made of electric conductive coils, which suffer from interference of EM noise, high transmission loss and small bandwidth. Fiber optic seismic sensors are made of FBG mounted on cantilever structure, which offer a broader bandwidth. Signal detection was achieved by matched-FBG method[2] and the advanced design was based on a distributed feedback fiber laser strain sensor, which offers higher sensitivity.

3. Underground coal mine field trials

The all fiber optic sensor-based coal mine comprehensive safety monitoring system was deployed in Linzi coal mine, Shandong province, China. Four fiber optic methane gas sensors were installed in coal production site where dust contamination and humidity level were very high. The sensors have been in reliable operation without recalibration, which was a major improvement over conventional methane gas sensors. Ambient gas concentration changes due to mining blasts were successfully recorded.

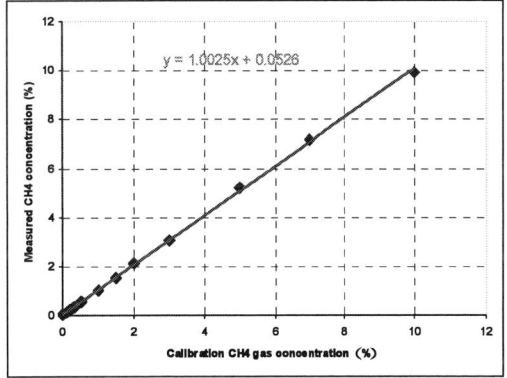

Fig. 2. Fiber optic methane sensor response to standard gas.

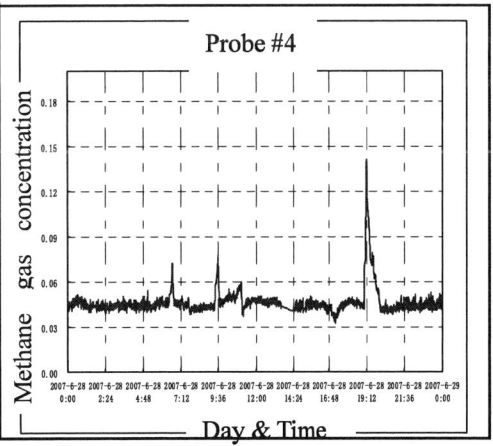

Fig. 3. Field recorded methane gas concentration changes due to mining blasts.

More than 12 FBG temperature sensors, two seismic sensors and two mining pressure sensors were installed in various locations in the underground coal mine 7 km away from the monitor centre on the ground. The ambient temperature, rock pressure, water pressure and seismic vibration signals were successfully recorded.

4. Conclusions

The first all fiber optic multi-parameter coal mine safety condition monitoring system has been demonstrated in Shandong Province China. The feasibility of applying fiber optic sensors in hazardous mining environment has been successfully proved. The fiber methane sensors offered much improved long-term stability and accuracy, intrinsic safety and ease of multiplexing. Fiber optic seismic, water pressure, mining pressure sensors provide invaluable information for real time remote mine safety condition monitoring and potentially can be used to build database and expert system which is capable of prediction and early warning of mine accidents such as gas outburst, coal and rock burst as well as water flooding. Further work are on the way to evaluate the sensor system in varied coal mine conditions and improve the ruggedness to suit for the heavy engineering mining conditions

References

1. H. Naruse1, et al., Underground Mine Monitoring Using Distributed Fiber Optic Strain Sensing System, Proceedings OFS-19, paper THD5, Cancun, 2006.

2. M. A. Davis, A. D. Kersey, Matched-filter interrogation technique for fiber Bragg grating array, Electronics Letters, Vol. 31, pp. 822-823, 1995.

Research on Optical Sensor for Pulsed Magnetic Field Measurement

SU Yang (1), LI Yu-quan (2)

1 : Institute of Communications Engineering , PLA Univ. of Sci.& Tech,
P.O. Box 32 No. 2 Biaoying Rd Yudao St, Nanjing Jiang Su, qieziyangyang@163.com

Abstract *In this paper the key technologies in optical sensor for pulsed magnetic field measurement including the effect of linear birefringence on the precision, directionality of the sensor and the reconstruction of ultra wideband analog signal are studied.*

Introduction

High-power electromagnetic (HPEM) signals are a serious threat for modern electronic systems. Intentionally generated electromagnetic interferences might cause upsets or permanent defects in electronics, particularly in the electronics' interface. There are numerous military and especially civil targets, which might be interesting for criminals or terrorists to interfere. Therefore, there is a need to characterize the overall problem by means of simulations and measurements[1].

Optical sensor is based on optical technology extracting the sensitive parameter in the form of optic signal. It has many advantages such as immunity to EMI, all dielectric construction, electrical 'fail-safe', small bulk, light, potential for DC to GHz measurement bandwidth and be easy to be connected with fiber optical communication system [2-4].

The most key technologies are the effect of linear birefringence, the sensor directionality and the reconstruction of UWB signal. This paper focuses on the above three problems.

The structure of the optical sensor

The principle of fiber optical sensor in this paper is Faraday magneto-optical effect, under which a linearly polarized light ray propagating through a magneto-optic medium in the presence of an external magnetic field undergoes a rotation of the plane of polarization proportional to the strength of the magnetic field. The rotation angle related to the Verdet constant of the material V, the magnetic field intensity B and the length of the crystal L. That is

$$\theta = VBL \tag{1}$$

Fig 1 shows the basic structure of the fibre optical sensor. The beam from a light source is guided by an optical fibre to the polarizer, and then passes through faraday material and analyzer (Wollaston prism) finally gets to the signal process module.

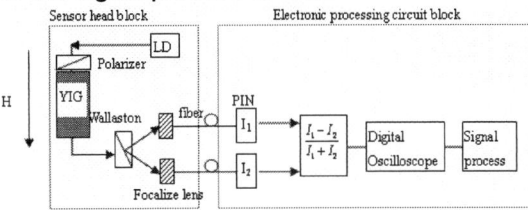

Fig. 1: The basic structure of the fibre optical sensor

The out beam from the sensor head is split into its orthogonal components by a Wollaston prism. To extract the Faraday rotation from I_1 and I_2, the ratio of the difference over sum is usually taken,

$$U_O = \frac{I_1 - I_2}{I_1 + I_2} = \sin 2\theta \approx 2\theta = 2VHL \tag{2}$$

The effect of linear birefringence

In the discussion above the linear birefringence effect is not considered. But in the practice the performance of the optical sensors is limited by the presence of the linear birefringence caused by the temperature, the stress and the intrinsic characteristic of the magneto-optical crystal. It causes phase difference between two components of the light within the magneto-optic material and so causes errors. At the same time the effect of the incident polarizing angle ϕ upon the output should be considered to discuss if or not it can exert the influence on the linear birefringence.

Using Jones Matrix[5] we get the output which including linear birefringence δ and polarizing angle ϕ,

$$U_O = 2\theta \frac{\sin \Delta}{\Delta} - \frac{\delta}{\Delta} \sin^2 \frac{\Delta}{2} \sin 4\phi \tag{3}$$

Where $(\frac{\Delta}{2})^2 = (\frac{\delta}{2})^2 + \theta^2$. The simulation is shown in figure 2. The less the slope of the

978-0-9789217-3-6/07/$25.00

©2007 WEN GLOBAL SOLUTIONS

curve the less the sensitivity of system to linear birefringence is. From figure 2 the slope of the curve marked A,B is minimal so the corresponding sensitivity is minimal. The slope of the curve marked C is maximal. The corresponding polarizing angle of A, B, C, D are $45.368°$, $89.632°$, $22.5°$ and $67.5°$ respectively.

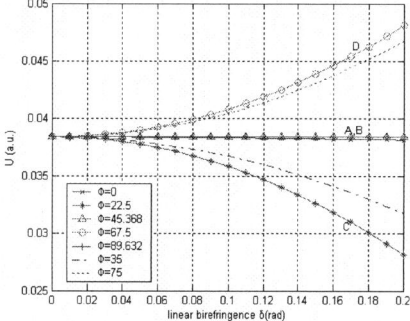

Fig.2: the effect of linear birefringence upon the output.

The further analysis show that the less the applied magnetic is the closer the optimal polarizing angle to $45°$ (or $0°$, $90°$). And the worst case is always $22.5°$ as shown in Fig 3.

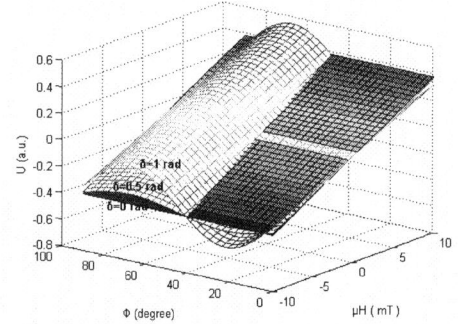

Fig.3: The optimal incident polarizing angle under different magnetic field

Sensor directionality

In practice the sensor directionality is an important issue when the direction of the applied field couldn't coincide with the propagation direction of the optical beam.

The magnetic linear birefringence (MLB) effect （also called Cotton-Mouton effect） is the factor which affect the sensor directionality. When the optical path is vertical to the magnetic field direction, the phase speed of the two linear polarized components are different. So MLB is produced. The phase difference δ_{MLB} is in proportion to refractive index difference Δn, which is in proportion to the square of

transversal magnetic field B_t [6], So,

$$\delta_{MLB} = \frac{2\pi L}{\lambda} \Delta n = 2\pi C L B_t^2 \qquad (4)$$

C is called Cotton-Mouton constant.

Many experiments have done to measure the MLB. Δn_s of (TbY)IG, Ce:YIG and YIG under saturated magnetic field are -4.2×10^{-5} [7], -1.7×10^{-4} and -0.5×10^{-4} [8] respectively.

Here the three-dimension measurement scheme is proposed. It uses three sensor heads which are vertical to each other to measure the x, y, z components of the magnetic field. Now the problem is the interfere of transverse components (such as x and y) to the one which will be measured (such as z).

Using Equation (3) and (4) the errors caused by MLB are calculated for the above mentioned crystals. The range of the magnetic field is up to 10mT. The maximal errors induced by MLB at 10mT are 0.004%, 0.0012% and0.0008% respectively. The errors can be neglected. So the feasibility of three-dimension measurement is confirmed in theory.

The experiment is performed by MR-4 crystal. If MLB can be neglect the relationship between applied-field angle β and Faraday rotation θ should be

$$\theta = VLB \times \cos\beta \qquad (5)$$

The experimental data are shown in figure 4. the measured response closely follows the ideal cosine response curve.

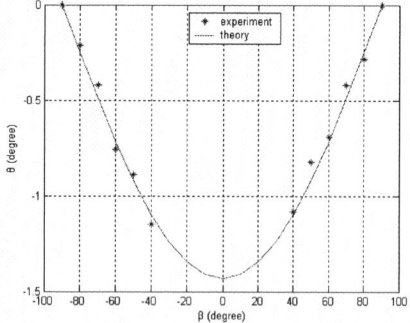

Fig. 4: Experiment of sensor directionality.

Reconstruction of UWB analogue signal

The frequency spectrum of the pulsed electromagnetic field covered from DC to GHz. That brings a challenge to the data acquisition and signal reconstruction. So we use the intersection and shift of frequency spectrum and sub-bands to decrease the requirement of ADC.

The key technology is the design of M-bands filter banks. The amplitude error and aliasing error should be low enough. The cosine-modulated filter banks are used and the prototype filter is constructed using Parks-McClellan arithmetic.

The 4-bands filter banks and the errors are shown as Figure 5(a), (b) and (c). The stopband attenuation is up to 200 (dB). The maximal aliasing error is 7.8×10^{-22} and maximal amplitude error is 1.4×10^{-3}. They are satisfied the near -prefect - reconstruction characteristic with high stopband attenuation.

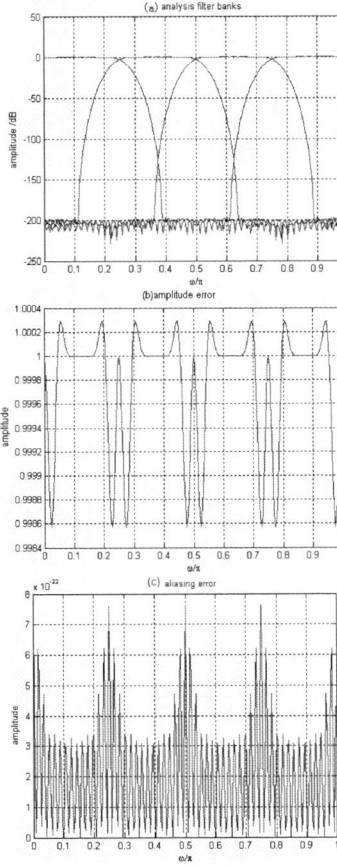

Fig.5: Four-channel cosine-modulated filter banks.

Conclusions

In this paper we study on the key technologies in optical sensor for pulsed magnetic field measurement including the effect of linear birefringence on the precision, directionality of the sensor and the reconstruction of ultra wideband analog signal.

The results show that when incident polarizing angle has some special values the sensitivity of the system to the linear birefringence will be minimized and the stability of the system will be enhanced. Three-dimension measurement scheme is proposed which is theoretical and experimentally verified. And the design of 4-bands filter banks show that amplitude error and aliasing error meet the requirements of near-perfect-reconstruction which solve the reconstruction of ultra wideband analog signal.

References

1 Thomas Weber, et al，Measurement Techniques for Conducted HPEM Signals，IEEE Transactions on Electromagnetic compatibility 2004，46(3): 431~438

2 Culshaw B. Optical Fiber Sensor Technologies: Opportunities and Perhaps Pitfalls[J]. IEEE Journal of lightwave technology, 2004, 22(1):39~50.

3 Cruden A, et al. Optical Crystal Based Devices For Current and Voltage Measurement[J], IEEE Transactions on Power Delivery.1995,10(3):1217~1223.

4 Pedja M, et al. Development of a Portable Fiber-Optic Current Sensor for Power Systems Monitoring[J]. IEEE Transactions on instrumentation and measurement, 2004, 53(1):24~30.

5 Tabor W.J. et al, Electromagnetic Propagation through Materials Possessing Both Faraday Rotation and Birefringence: Experiments with Ytterbium Orthoferrite, J.of Appl. Phys, 1969, 40(7):2760-2765.

6 Stephen R. et al, A Simple Apparatus for the Measurement of the Cotton-Mouton Effect in Particulate Suspensions[J], IEEE Transactions on magnetics, 1997, 33(5): 4349~4358.

7 Kinya Okubo et al, Magnetic Field Optical Sensors Using (TbY)IG Crystals With Stripe Magnetic Domain Structure[J], IEEE Transactions on magnetics, 2005, 41(10): 3640~3642.

8 Osamu Kamada et al, Magnetic Field Sensors Using Ce:YIG Single Crystals as a Faraday Element[J], IEEE Transactions on magnetics, 2001, 37(4): 2013~2015

Distributed Bragg-Reflector Fiber-Laser Sensor for Lateral Force Measurement

Yang Zhang (1), Bai-Ou Guan(1), HwaYaw Tam(2)

1 : School of Physics & Optoelectronic Technology, Dalian University of Technology, Dalian 116024, China. guanboo@yahoo.com

2 : Photonics Research Centre and Department of Electrical Engineering, The Hong Kong Polytechnic University, Hong Kong, China

Abstract: *We demonstrate a lateral force sensor based on dual-polarization DBR fiber laser by measuring the polarization beat frequency in ratio-frequency domain. It shows high sensitivity ~10GHz/ (N/mm) and potential for multiplexing.*

Introduction

Fiber Bragg grating (FBG) sensors have been applied in an increasing range of sensing fields because of their unique advantages like remote sensing, compact size, and multiplexing capabilities. Lateral pressure or force measurement is of great important and interesting because measurement of many other parameters, such as weight, liquid or gas pressure, acoustic pressure, etc, can be transformed to lateral force detection. Many lateral pressure sensors based on fiber Bragg gratings have been demonstrated [1-4]. Recently, fiber grating lasers, including distributed Bragg-reflector (DBR) fiber lasers [5-6] and distributed-feedback (DFB) fiber lasers [7-8], have been great attractable for their high signal-to-noise ratio and high resolution in sensing application. A polarimetric Er3+-doped DFB fiber laser was demonstrated for lateral pressure measurement by Kringlebotn [9]. However, little research has been reported on lateral pressure sensor used DBR fiber lasers.

In this paper, we report the characteristics of the lateral force sensor based on a dual-polarization DBR fiber laser by measuring the polarization beat frequency in ratio-frequency (RF) domain. It shows high sensitivity to transverse force up to ~10GHz/ (N/mm) and the potential for sensor multiplexing. The sensitivity depends on the direction of applied lateral force as cosine function because of the original birefringence in the fiber. The influence of the position of the applied force on the sensor sensitivity was also investigated.

Principle

The DBR fiber laser consists of a pair of wavelength-matched Bragg gratings written in an active fiber with appropriate separation. It operates in single longitude mode and two orthogonal polarization modes due to the fiber birefringence introduced during fiber fabrication and grating inscription. The resonance wavelengths of the two polarization modes are given by

$$\lambda_x = 2n_x \Lambda = \frac{2n_x L_{eff}}{M} \qquad (1)$$

$$\lambda_y = 2n_y \Lambda = \frac{2n_y L_{eff}}{M} \qquad (2)$$

where $n_{x, y}$ are the effective refraction index of the two polarization modes, Λ is the grating period, and M is the order number of the resonance mode, L_{eff} is effective length of the laser cavity which is given by [10]

$$L_{eff} = L_0 + l_{eff1} + l_{eff2} \qquad (3)$$

where l_{eff1} and l_{eff2} are the effective lengths of the two gratings, given by

$$l_{eff} = l_g \frac{\sqrt{R}}{2\mathrm{atanh}(\sqrt{R})} \qquad (4)$$

where l_g is the length of the grating, and R is the grating reflectivity.

When the laser output is detected with a high speed photodetector and RF spectrum analyzer, fiber laser produces a polarization mode beat signal. The beat frequency is given by

$$\Delta \nu = \nu_x - \nu_y = \frac{c}{\lambda_x} - \frac{c}{\lambda_y} = \frac{c}{n_0 \lambda_0} B \qquad (5)$$

where n_0 is average index of the fiber, λ_0 is Bragg wavelength of the fiber grating, and B is the fiber birefringence.

978-0-9789217-3-6/07/$25.00

©2007 WEN GLOBAL SOLUTIONS

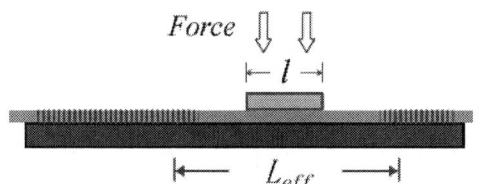

Fig.1 Part of laser cavity was subjected to lateral force.

When external lateral force applies to the laser cavity, the birefringence will change because of the photo-elastic effect therefore the beat frequency will shift accordingly. Given l as the action length of the applied transverse force, as shown in Fig.1, the resonance wavelengths of the two polarization modes changes to

$$\lambda'_x = \frac{2n_x L_{eff} + 2\delta n_x l}{M} = \lambda_x + \frac{2\delta n_x l}{M} \quad (6)$$

$$\lambda'_y = \frac{2n_y L_{eff} + 2\delta n_y l}{M} = \lambda_y + \frac{2\delta n_y l}{M} \quad (7)$$

Accordingly the beat frequency changes to

$$\Delta v' = v'_x - v'_y = \Delta v + \frac{l}{L_{eff}}\frac{c}{\lambda_0 n_0}\delta B \quad (8)$$

where $\delta B = 2n_0^{3}(p_{11} - p_{12})(1+v_p)f\cos(2\theta)/\pi r E$ [11], f is linear force (force per unit length), E is the Young's modulus of the fiber material, v_p is Poisson's ratio, r is the fiber radius, p_{11} and p_{12} are the components of strain-optical tensor of the fiber material, and θ is the angle between the direction of the applied force and the fast eigen axis, respectively.

Then the beat frequency shift can be expressed as

$$\delta(\Delta v) = \frac{l}{L_{eff}}\frac{2cn_0^{2}(p_{11} - p_{12})(1+v_p)\cos(2\theta)}{\lambda_0 \pi r E}f \quad (9)$$

Experiment and Results

The experiment arrangement is showed in Fig.2. The active fiber used in our experiment is the Er/Yb co-doped fiber [12]. The two matched Bragg gratings were photowritten in the Er/Yb fiber using an 193nm ArF excimer laser and phase mask method. The pump light from an 980nm laser diode was launched from the low reflectivity grating side through a WDM system. The DBR laser was fixed on a stage, and a silica plate was used to impose pressure on the DBR laser by loading

different weights. In the experiments, we made sure that the laser was free and untwisted. The fiber laser signal is launched to the high speed photodetector (PD) through a polarization controller (PC). The beat signal generated by the two polarization modes was monitored with a RF spectrum analyzer.

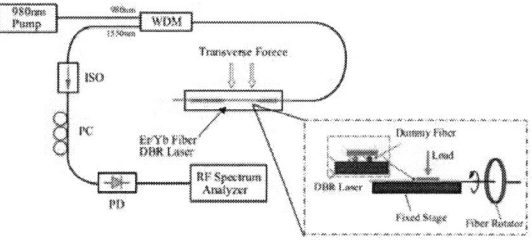

Fig.2 Lateral force measurement system based on beat frequency shift measurement.

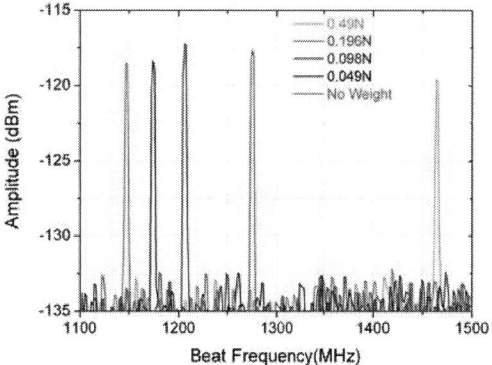

Fig.3 Beat frequency single shift on the RF spectrum analyzer

A DBR fiber laser with 9-mm-long high reflectivity grating (36.5dB), 5.6-mm-long low reflectivity grating (20dB), and 6-mm-long cavity length, was fabricated. The laser operates at 1545nm with threshold of 27.5mW. An 4-mm-long silica plate was used to apply lateral force. The beat frequency showed high sensitivity to lateral force as shown in Fig.3. We rotated the fiber from 0° to 360° by rotator with 30° interval. The response sensitivity as a function of the force direction is plotted in Fig. 4. The experimental results are in well agreement with the theoretical curve from equation (11). The maximum sensitivity happened as the force direction consistent with two fiber's inherent polarization axes. When the force orientated at 45° between the two axes, the minimums sensitivity (almost 0 at 240°) were found, since the variation of effective refractions on the two axes are almost the same ($\delta n_x - \delta n_y = 0$).

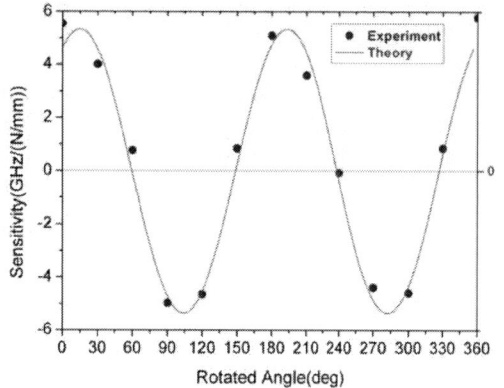

Fig.4 Response sensitivity changes with direction of the applied force

For the action length of the applied force is shorter than the laser cavity, the influence of the force action position on the sensor sensitivity was also investigated. We fixed the angle θ at 300° and moved the 4-mm-long silica plate from the low-reflectivity-grating side to the high-reflectivity-grating side with 2mm interval. Fig.5 shows the sensitivity changes with the position of the applied force. From Fig.5, the effective length of the DBR laser is estimated to be 8mm, which is very close to the calculated result about 7.9mm obtained from equations (3) and (4).

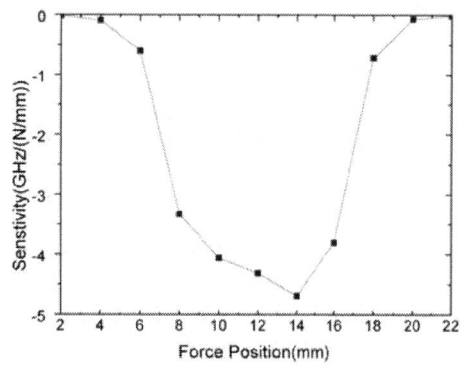

Fig.5 Response sensitivity changes with position of the applied force.

Conclusion

We have investigated the characteristics of the lateral force sensor based on a dual-polarization DBR fiber laser. It shows high sensitivity to transverse force up to ~10GHz/ (N/mm) and the potential for sensor multiplexing. The sensitivity depends on the direction of applied lateral force as cosine function because of the original birefringence in the fiber. The influence of the position of the applied force on the sensor sensitivity was also investigated.

Acknowledgements

This work was supported in part by the Program for New Century Excellent Talents in University (NCET-06-0271) and the Dalian Science and Technology Foundation (2006J23JH017).

References

1. C. M. Lawrence, et al., Experimental Mechanics, 39 (1999), 202-209.
2. A. P. Zhang, et al., Opt. Commun., 206(2002),81-87.
3. S. P. Yam, et al., OFS-18 (2006), TuE47.
4. R. B. Wagreich, et al., Electron. Lett., 32 (1996), 1223-1224.
5. G. A. Ball, et al., Opt. Lett., 18 (1993), 1976-1978.
6. K. Hsu, et al., J. Lightwave Technol., 15 (1997), 1438-1441.
7. E. Ronnekleiv, et al., J. Quantum Electron., 36(2000), 656-664.
8. O. Hadeler, et al., Appl. Opt., 38 (1999), 1953-1958.
9. J. L. Kringlebotn, et al., Opt. Lett., 21 (1996), 1869-1871.
10. Y. O. Barmenkov, et al., Opt. Lett., 14 (2006), 6394-6399.
11. S.C Rashleigh, J. Lightwave Technol., It-1 (1983), 312-331.
12. L. Dong, et al., Opt. Lett., 22 (1997), 649-696.

Application of Microplasma Modes to a Highly Efficient Light Source for Displays

Kyung Cheol Choi, Seung Hun Kim, Kwan Hyun Cho

School of Electrical Engineering and Computer Science, KAIST

Daejeon 305-701, Republic of Korea

e-mail:kyungcc@ee.kaist.ac.kr

Abstract *Microplasma modes generated in a three electrode structure, including an auxiliary electrode, were analyzed and application of this structure to PDPs and the BLU of TFT-LCDs was investigated. Highly efficient microplasma was generated in modes 1 and 2. Luminous efficacy was improved by 350% for AC PDPs and 50% for the BLU, respectively.*

1. Introduction

Currently, large-area display devices such as Thin Film Transistor–Liquid Crystal Displays (TFT-LCDs) and Plasma Display Panels (PDPs) are being widely used for digital TV screens and information displays. Because the TFT-LCD is not a self-emissive display, it requires a BLU (Back-Light Unit). Mercury-free flat fluorescent lamps are a promising BLU candidate because of qualities including good colour gamut, long-life time, and good uniformity [1]. However, because of their low luminous efficacy mercury-free flat fluorescent lamps are not leading BLU technology. In the case of PDPs, microplasma is directly used for display cells and its energy conversion rate from input electrical energy to output optical energy is relatively low [2]. With the spread of large-area display devices, power consumption has become important issue, because it is proportional to display screen size. In order to reduce power consumption, the luminous efficacy of the PDP itself and the BLU of TFT-LCDs should be improved. In particular, the luminous efficacy of mercury flat fluorescent lamps is considerably lower than to that of any other BLU. Several efforts to improve luminous efficacy in the field of flat fluorescent lamps have been reported[1,4]. Researchers of PDP technology have also made notable attempts to improve luminous efficacy [5,6]. Both PDPs and flat fluorescent lamps for BLUs utilize plasma generated inside the panel and employ the same mechanism to emit visible light. In order to improve the luminous efficacy of PDPs and flat fluorescent lamps, it is necessary to develop a highly efficient plasma. In this work, highly efficient micro-sized plasma is

investigated as a light source and its potential for application to PDPs and flat fluorescent lamps is valuated.

2. Basic structure for producing micro-plasma

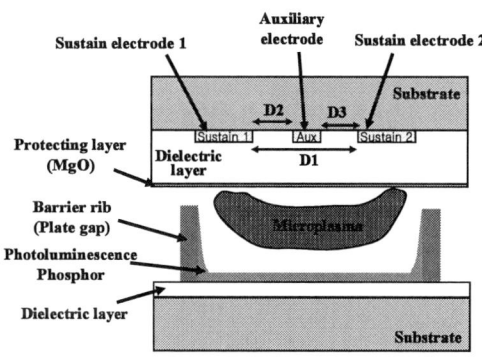

Fig.1 Schematic drawing of the basic three electrode structure

Microplasma belongs to the general category of plasma. However, its size and volume is in the range of several hundreds nanometers to several hundreds micometers and its characteristics are not well known. In the present work, a structure comprised of three electrodes is suggested for generation of highly efficient microplasma. Fig. 1 shows a schematic drawing of the basic three electrode structure. This structure has two sustain electrodes for main discharge and an auxiliary electrode for enhancing the excitation rate of the main discharge. This kind of three electrode structure is not wholly new. However, the microplasma modes generated in this structure are based on dielectric barrier discharges and a novel approach that is assessed herein. Previous works have established that an auxiliary

978-0-9789217-3-6/07/$25.00

©2007 WEN GLOBAL SOLUTIONS

electrode can influence excited particles indirectly and enhances the excitation rate of microplasma [7]. The coplanar-gap, D1, between the two sustain electrodes is in a range of 100 to 2000 μm and the height of the barrier rib is in a range of 100 to 2000 μm. The distances between the auxiliary and sustain electrode, D1 and D2, can be varied.

3. Microplasma modes

In order to generate microplasma in the three electrode structure, square pulses are applied to each electrode, as shown in Fig. 2. Main discharges occur between sustain electrodes and the auxiliary electrode control the distribution of wall charges and excited particles including meta-stable species during glow and afterglow. For example, when a square pulse with a width of T_{on} is applied to sustain electrode1, negative wall charges are accumulated on sustain electode1 and positive wall charges are accumulated on both sustain electrode2 and the auxiliary electrode. A positive square pulse is subsequently applied to the auxiliary electrode and erases some of the wall charges accumulated on this electrode and sustain electrode1 due to reciprocal interaction. This influences the following sustain discharge occurring as a result of a square pulse applied to sustain electrode2. By repeating this process, main discharges are determined by the square pulse applied to the auxiliary electrode. There are several microplasma modes due to the auxiliary pulse.

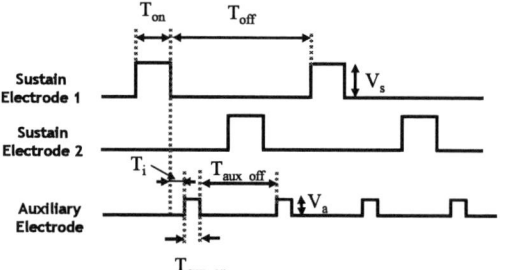

Fig.2 Pulse waveforms applied to the sustain and auxiliary electrodes

Through experiments, three types of microplasma modes are defined in accordance with the auxiliary pulse voltage (V_a), the discharge current, and the discharge delay. In the first mode, both the discharge current and delay decrease as V_a increases. In the second mode, the discharge current decreases and the discharge delay increases as V_a increases. In the third mode, the discharge current increases and the discharge delay decreases as V_a increases. Fig.3 (a) shows an example of the discharge current of modes 1 and 2 in accordance with the auxiliary pulse voltage. Fig.3(b) shows an example of the discharge current of mode 3 in accordance with the auxiliary pulse voltage. Here, Neon +Xenon gas-mixtures are used as the discharge gas.

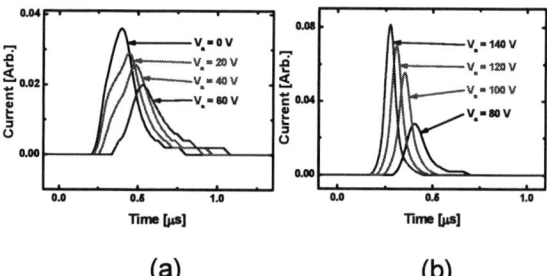

(a)   (b)

Fig. 3 (a) Discharge current of modes 1 and 2, and (b) mode 3 in accordance with the auxiliary pulse voltage (D1=200μm, D2=D3= 50μm, plate-gap= 220μm, V_s=250 V, T_{on}=4μsec, T_{off}=16μsec, T_i=0μsec, T_{aux_on}= 2 μsec, T_{aux_off}=8 μsec)

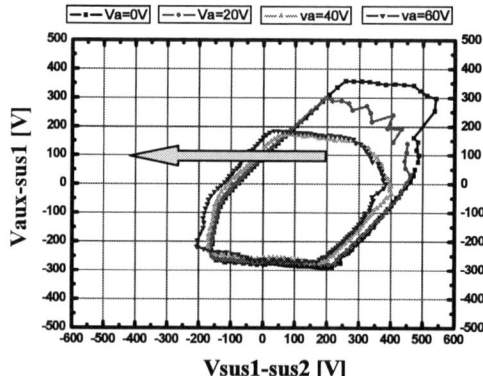

Fig. 4 Voltage transfer closed curve in mode 2 (D1= 200μm, D2=D3=50μm, plate-gap= 150μm, V_s=260 V, T_{on}=4μsec, T_i=0μsec, T_{off}=16μsec, T_{aux_on}= 2 μsec, T_{aux_off}=8 μsec)

Luminous efficacy is dependent on the microplasma modes. In the first and second modes, the luminous efficacy increases with an increase in the voltage of the auxiliary pulse. The improved efficacy resulted from a reduction of the discharge current and the

182

priming effect by the auxiliary pulse in the first and second modes. Fig. 4 shows the voltage transfer closed curve (VTC) of mode 2. The VTC shifts to the left-hand side with an increase in the voltage of the auxiliary pulse, which means the wall charges are erased by the auxiliary pulse. The reduction of wall charges results in reduced discharge current. The priming effect was also investigated. It is found that the priming effect increases with an increase in the voltage of the auxiliary pulse in the second mode. Therefore, not only the second but also the first mode is suitable for highly efficient light sources for displays. In the third mode, there is a region within which the luminous efficacy decreases as the voltage of the auxiliary pulse increases. This third mode is not recommended for developing a highly efficient light source for displays.

4. Application of Microplasma modes

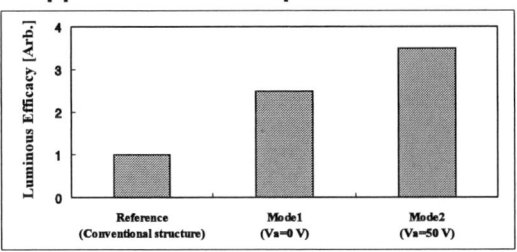

Fig. 5 Luminous efficacy of AC PDP adopting highly efficient microplasma modes (D1= 200μm, D2=D3=50μm, plate-gap= 150μm, V_s=255 V, T_{on}=4μsec, T_i=0μsec, T_{off}=16μsec, T_{aux_on}= 2 μsec, T_{aux_off}=8 μsec)

Highly efficient microplasma modes, in particular modes 1 and 2, generated in the suggested structure are adopted for a light source of PDPs and BLUs for TFT-LCDs, respectively. Fig. 5 shows the luminous efficacy of display cells of a PDP with adopting a electrode structure. Compared to conventional PDPs, the luminous efficacy of the AC PDP adopting the suggested structure is improved by 350 %. Fig. 6 shows the luminous efficacy of a flat fluorescent lamp adopting the suggested structure. The luminous efficacy of the lamp is improved by 50% in comparison with that obtained from microplasma generated without implementation of an auxiliary.

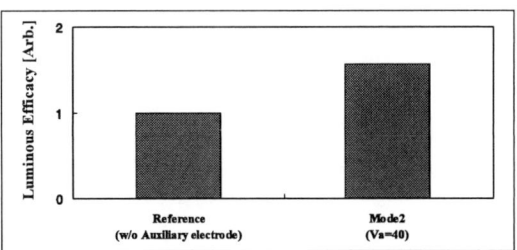

Fig. 6 Luminous efficacy of a flat fluorescent lamp adopting highly efficient microplasma modes (D1= 500μm, D2=D3=200μm, plate-gap= 1,100μm, V_s=280V, T_{on}=4μsec, T_i=0μsec, T_{off}=16μsec, T_{aux_on}= 2 μsec, T_{aux_off}=8 μsec)

5. Conclusions

A structure generating highly efficient microplasma was suggested and its application to display devices such as PDPs and BLU of TFT-LCDs was investigated. The proposed structure features an auxiliary electrode located between two sustain electrodes and shows improved luminous efficacy of microplasma in modes 1 and 2. In the case of application for PDPs, the luminous efficacy was improved by 3.5 times compared to that of the conventional structure. In the case of application to a flat fluorescent lamp, the luminous efficacy was improved by 50 % compared to that obtained from microplasma generated without an auxiliary electrode.

6. Acknowledgements

This work was partially supported by grant No.R01-2006-000-10075-0 from the Basic Research Program of the Korea Science & Engineering Foundation.

References

1 S.E.Lee et al, *J. App. Physics*, **98** (2005), 093306.
2 J.S.Seo et al, *Pro. 6th Int. Display Workshop*, **6**(1999), 667.
3 S.Mikoshiba, *Int. Symp. SID Digest Tech. papers*, **32**(2001), 287.
4 Y.Wu et al, proc. 13th *Int. Display Workshop*, **13**(1006), 1735.
5 M.hur et al, *J. App. Physics*, **99** (2006), 13301.
6 K.W.Whang et al, *Int. Symp. SID Digest Tech. Papers*, **38**(2007), 1526.
7 K.C.Choi et al, *IEEE Tran. Plasma Science*, **23**(1995), 399.

Dielectric Superlattice and Its Potential Applications in Display Technology

Yan-qing Lu, Shi-ning Zhu, Yong-yuan Zhu, Yan-feng Chen, Hui-tian Wang and Nai-ben Ming
National Laboratory of Solid State Microstructures, Nanjing University,
Nanjing 210093, P. R. China Email: yqlu@nju.edu.cn

Abstract *Dielectric superlattice has artificial microstructures in micrometer span. Cascade quasi-phase-matched frequency conversion could be realized in certain specially-designed superlattices. Efficient red-green-blue lights thus were obtained simultaneously, which may be used in laser display. On the other hand, light filtering and modulation also could be achieved though acousto-optic effect by using a dielectric superlattice.*

Introduction

Although flat panel displays increase their sizes very fast, projection displays still have merits in manufacturing cost and simplicity of structure especially when the screen size is getting larger. Laser scanning display is being developed as one of the future projection displays because of its excellent brightness, high contrast ratio and natural color expression. [1] However, excellent chromatic characteristics of laser display rely on laser-related technologies. Suitable high power red, blue and green (RGB) laser sources are the prerequisitions. Spectral filtering and spatial light modulation are also important technologies that should be developed.

On the other hand, dielectric superlattice materials are attracting more and more attentions in the fields on nonlinear optics, electro-optics and piezo-electrics. [2-8] In this report, we will discuss the possible applications of dielectric superlattice materials in display technology.

Dielectric superlattice

The concept of superlattices was initiated by Esaki and Tsu (1969) in semiconductor. In dielectric materials, including nonlinear optic, piezoelectric, and ferroelectric crystals, the most important physical processes are the propagation classical waves (light and ultrasonic waves). If some microstructures are introduced into a dielectric crystal, forming a superlattice, and if the reciprocal vectors of the superlattice are comparable with the wave vectors, the

situation will be quite different from that in homogeneous media. The interactions between wave vectors of classical waves and reciprocal vectors of a superlattice may generate some new physical effects. These interactions have led to new methods of laser frequency generation with quasi-phase-matching (QPM) schemes, [2] optical bistability with new mechanisms and ultrasonic generations with frequency up to several GHz. [8]

Dielectric superlattices can be fabricated by the modulation of microstructures, such as domain structures, phase structures, compositions, crystallographic orientations and heteroepitaxy structures. Among them, the domain modulation is the most popular way to induce superlattice in a ferroelectric since its nonlinear optical coefficient and piezoelectric coefficient could be extremely modulated. In the past years, we developed two effective ways to generate ferroelectric domain modulation, the room-temperature electric-poling [2] and the modified Czochralski crystal growth techniques. [6]

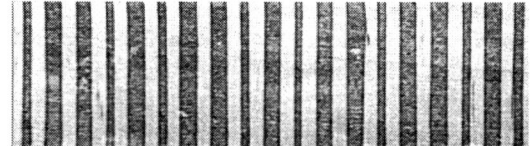

Fig. 1. Domain structure of a quasi-periodic LiTaO$_3$ superlattice.

Fig. 1 shows a typical domain structure of a quasi-periodic LiTaO$_3$ (LT) superlattice made with electric poling technique. Up to date, various LiNbO$_3$ (LN), LT and BNN superlattices with the modulation span from several to tens micrometers have been fabricated in our group. Some interesting optical and acoustic properties were demonstrated, which have wide promising applications in various fields including display.

RGB light generation

There have been a lot of reports to generate new laser wavelengths though nonlinear optical processes. A traditional way for efficient frequency conversion is using birefringence phase-matching (BPM). However, tight BPM condition limits the applications of many nonlinear crystals. For example, LT has the nonlinear coefficient d_{33} of ~21pm/V and transparency range across whole visible band. However, BPM condition may not be satisfied due to its small birefringence.

Quasi-phase-matching (QPM) in a domain modulated superlattice, e.g., periodically poled LT, is a viable way to realize the frequency conversion in any nonlinear crystal within its whole transparency range. As a pioneering group, we have studied QPM materials and related devices for 20 years. New QPM radiations from ultra-violet to mid-infrared were obtained. Even dual-wavelength and multiple wavelength frequency conversions were demonstrated. [2] Since there should be 3 primary colors for laser display, we may take this advantage to generate RGB lasers though cascade-QPM process in a single well-designed superlattice.

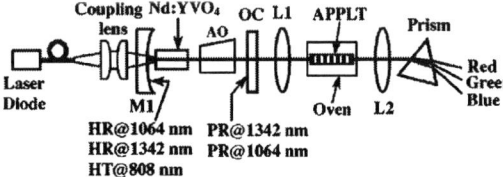

Fig. 2. Schematic of experimental setup for simultaneous RGB generation.

Fig. 2 shows the schematic diagram of an aperiodically poled LT (APPLT) based RGB laser. A two-mirror resonant cavity was used for the Q-switched dual-wavelength operation of Nd:YVO4 at 1342nm and 1064nm. Both these two waves were frequency doubled in an APPLT superlattice to generate red and green lights at 671nm and 532nm, respectively. A 447nm blue light was further obtained though the sum-frequency-generation (SFG) of 1342nm and 671nm still in this APPLT.

In our experiment, the first difficulty was the dual-wave laser oscillation. An 808nm fiber coupled laser diode was used as the pumping source. The laser crystal was a 0.5-at.%-Nd doped 7.7 mm-long YVO$_4$. The input mirror was coated to be antireflective at pump wavelength; the second surface was coated to be highly reflective at both 1342nm and 1064nm while highly transmitted at pump wavelength. An acousto-optic Q-switch was inserted into the cavity, with a repetition rate of 10 kHz. In the experiment, dual-wavelength operation under a certain pump power was successfully achieved. Output powers of two wavelengths increased monotonically as the pump power increased.

Another key part of the RGB simultaneous generation is the APPLT because two second-frequency generation (SHG) and one SFG processes should be realized together in it. This function is impossible for normal commercial nonlinear crystals. We designed and fabricated an APPLT that could realize all these three QPM processes. It is neither a period nor a quasi-periodic superlattice, while having complicated structures that supplies three reciprocal vectors for QPM SHGs and SFG.

After the domain structures were designed, a 1-cm-long APPLT was fabricated accordingly and installed in the laser cavity shown in Fig. 2. 18mW simultaneous RGB lights were generated. To display the lights and measure their power ratio, a dispersive prism is used to separate the 3 primary colors. (Fig. 3) The relative output power among RGB colors was about 9:2:1, which still needs improvement.

Fig. 3. RGB lights generation from a LT superlattice based solid-state laser

In addition to the configuration above, we also demonstrated several other schemes for RGB light generation, e.g., cascade optical parametric oscillation and SFG *etc.*.

Acousto-optic applications

In addition to RGB sources, beam filtering and modulation are also important to a projection display. Acousto-optic tunable filter (AOTF) has been proposed for this application. [9, 10] High-frequency and powerful bulk wave thus is needed for acousto-optic beam deflection. However, the resonance frequency of a normal ultrasonic-transducer is inversely proportional to the piezoelectric material's thickness. For example, a 35MHz LN transducer should use a 0.1 mm-thick wafer. It is thus difficulty to obtain high frequency bulk waves from hundreds Mega Hertz to several Giga Hertz. However, in a domain modulated superlattice, its acoustic properties exhibit some unique features. High-frequency and high power ultrasonic generation is easier achievable. This would be helpful for the laser filtering and modulation of an AOTF.

According to the acoustic superlattice theory proposed Zhu and Ming, [7] the domain boundary could be viewed as a δ-function-like sound source under an alternating external electric field. As a consequence, a domain modulated superlattice is just like a δ sound source array. If all these sources are in-phase at a certain driving frequency, the generated sound waves could interference constructively.

Strict analytical results could be obtained by using the Green's function method to solve the elastic wave equations, the electric impedance of a resonator can be derived, and then the resonance frequency can be obtained as $f_n = nv/\Lambda$, n=1, 2, 3, ..., where v is the sound velocity, Λ is the superlattice period. It is clear that the resonance frequency is determined only by the period of the superlattice, not by the total thickness of the wafer like normal piezoelectric materials. It is easy to prepare a superlattice with several micrometers period using the modified Czochralski or poling technique. Therefore, high frequency bulk wave acoustic devices could be obtained. In our group, resonance frequencies from 100MHz to 10GHz have been demonstrated in various acoustic superlattics. [8] Several prototype acousto-optic deflectors were also developed.

In addition to high frequency, another unique feature of acoustic superlattice is its adjustable impedance, which is a function of the number of laminar domains N and the electrode area. 0dB insertion loss is achievable for a 50Ω driving circuit at the resonance frequency, which means all driving electric power could be transformed to generate acoustic wave. Other advantages of acoustic superlattice include wide and adjustable frequency bandwidth, cross-field excitation *etc.*. These merits are very helpful for the acousto-optic filtering and modulation. If we use a GHz level wide-band acoustic superlattice, even acousto-optic beam scanning can be realized in a laser display with no moving mechanical parts.

Conclusions

We introduced the nonlinear optic and acoustic properties of dielectric superlattice materials. Potential applications in display technology were discussed.

References

1 J. Kranert, *et. al.*, Laser display technology, in: Proceedings of IEEE MEMS Workshop'98, pp. 99–104

2 S. N. Zhu, *et. al.*, Science, 278 (1997) 843

3 Z. D. Gao, *et. al.*, Appl. Phys. Lett., 89 (2006) 181101

4 J. Liao, *et. al.*, Appl. Phys. Lett., 82 (2003) 3159

5 Y. Q. Lu, *et. al.*, Appl. Phys. Lett., 77 (2000) 3719.

6 Y. Q. Lu, *et. al.*, Science, 284 (1999) 1822

7 Y. Y. Zhu and N. B. Ming, J. Appl. Phys. 72 (1991) 904

8 Y. F. Chen, *et. al.*, Appl. Phys. Lett., 70 (1997) 592

9 D. R. Suhre and J. G. Theodore, Appl. Opt., 35 (1996) 4494

10 J. L. Lambert, US Patent 5410371

Organic light emitting devices from OLED to organic laser diode

Chihaya Adachi, Toshinori Matsushima, Hajime Nakanotani, Daisuke Yokoyama and
Masayuki Yahiro
Center for Future Chemistry, Kyushu University and JST-CREST 819-0395 Fukuoka, Japan
e-mail: adachi@cstf.kyushu-u.ac.jp

Abstract *We mention some significant progresses on organic light emitting diode (OLED), organic laser diode (OLD) and organic field effect transistor (OFET). We discuss prospect of OLD based on recent developments of organic light emitting devices.*

Organic Light Emitting Diode

Organic light-emitting diodes (OLEDs) (**FIG1**) are being developed as alternative light sources to inorganic LEDs. Although OLEDs have advantages of making low-cost, lightweight, flexible, large-area display applications possible, driving voltages for OLEDs are much higher than those for inorganic LEDs. Therefore, lowering driving voltages is crucially important to improve power conversion efficiencies and lifetimes of OLEDs. To solve the driving voltage problem, OLEDs with *p*-doped hole transport and *n*-doped electron transport layers, called *p-i-n* OLEDs, have been emerging as low-voltage efficient light sources. Lowered driving voltages in *p-i-n* OLEDs are derived from factors including: (1) Charge transfer from matrix to dopant molecules induces generation of free charge carriers, leading to high electrical conductivities in these layers; (2) Generated carriers fill deep carrier traps and deep states in the density-of-states distribution in a disordered carrier hopping system, enhancing effective carrier mobilities; (3) Doping of transport layers induces Fermi level shift, leading to efficient carrier tunnel injection across metal/organic interfaces, i.e., ohmic contacts.

In addition to improvement in OLED characteristics, *p-i-n* OLED structures will open the way to achieving electrically pumped organic laser diodes (OLDs). *p-i-n* OLEDs with low driving voltages would make attaining high threshold current densities on the order of 4 kA/cm^2 under electrical pumping possible, which was calculated from a threshold energy under optical pumping.

For development of *p-i-n* OLEDs, we recently proposed OLED having a 40-nm-thick alpha-sexithiophene (α-6T) hole transport layer doped with a strong electron acceptor of 2 mol% 2,3,5,6- tetrafluoro-7,7,8,8-tetracyanoquinodimethane (F$_4$-TCNQ) and a 20-nm-thick phenyl-dipyrenylphosphine oxide (POPy$_2$) electron transport layer doped with a strong electron donor of 30 mol% Cs[1]. Our *p-i-n* OLED using the DCM:Alg$_3$ emitting layer sandwiched with the

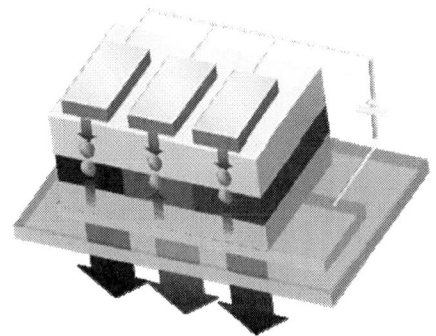

FIG1: Organic Light Emitting Diode

F$_4$-TCNQ:α-6T and Cs:POPy$_2$ transport layers exhibited an extremely low driving voltage of 2.9 V at 100 mA/cm^2, which is the lowest value ever reported (**FIG2**). Moreover, we observed bright EL from the DCM:Alg$_3$ layer at a low driving voltage: e.g., 1000cd/m^2 at 2.4 V.

We further demonstrate that mixing of organic/organic heterojunction interfaces leads to a reduction in the driving voltage required for organic light-emitting diodes (OLEDs). We manufactured multilayer OLEDs with mixed heterojunction interfaces composed of α-6T and POPy$_2$ carrier-transporting layers. In the interface mixed OLED, we observed a low driving voltage of 3.6 V at a current flow of 100 mA/cm^2 and improved power conversion efficiency. We investigated how much current can flow through this OLED with the aim of

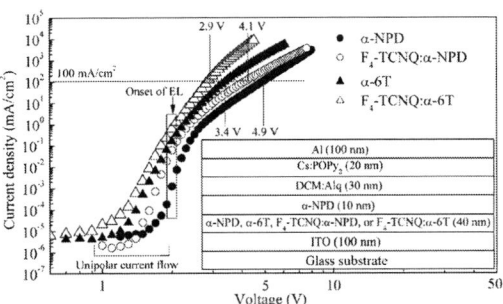

FIG2: J-V-L characteristics of p-i-n OLEDs.

978-0-9789217-3-6/07/$25.00

©2007 WEN GLOBAL SOLUTIONS

fabricating electrically pumped organic laser diodes. We found that an OLED of this type with a small active area of 625 μm^2 on a high thermal conductivity sapphire substrate can sustain high current densities of 1.1 kA/cm^2 and emits bright electroluminescence of 7.9 Mcd/m^2 under direct current.

Organic Laser Diode

Over recent years, a wide variety of challenges aiming for electrical pumping of organic laser diodes have been addressed. However, organic laser diodes (**FIG3**) have still not been realized owing to their high lasing threshold under electrical pumping. To reduce lasing threshold, a variety of organic semiconductors, including not only small molecular materials but also conjugated polymers, have been developed. Potential materials for low amplified-spontaneous-emission (ASE) thresholds are spiro derivatives, which show high photoluminescence (PL) quantum efficiency (Φ_{PL}) and low ASE thresholds in solid-state thin films. For example, 2,2',7,7'-tetrakis(9,9'-spirobifluoren-2-yl)-9,9'-spirobifluorene) (4-spiro^2) and 2',7'-Bis-(biphenyl-4-yl)-2-(5-(4-*tert*-butylphenyl)-1,3,4-oxadiazol-2-yl)-9,9'-spirobifluorene (spiro- SPO) show low ASE thresholds of E_{th}=3.2 and 1 $\mu J/cm^2$, respectively.

We recently demonstrated that bis-styrylbenzene derivatives (BSBs) have excellent PL and ASE characteristics in solid-state thin films under optical excitation. In particular, 4,4'-bis[(*N*- carbazole)styryl] biphenyl (BSB-Cz) showed a high Φ_{PL}=90±2%, short fluorescence lifetime of τ_{Flu}=of 1.0 ns, and a low ASE threshold of E_{th}=0.32 $\mu J/cm^2$ when doped into a wide energy gap 4,4'-bis(9-carbazole)-2,2'-biphenyl (CBP) host at a concentration of 6 wt%. Further, we designed and synthesized novel spiro concept derivatives, namely, 2,7-bis(*N*-carbazolyl)-9,9'- spirobifluorene (spiro-Cz) and 2,7-bis[4-(*N*-carbazole)phenylvinyl]-9,9'-spirobi-fluorene (spiro-SBCz), and report on their PL, ASE and electroluminescence (EL) characteristics. In particular, we observed an extremely low ASE threshold of E_{th}=0.11±0.05 $\mu J/cm^2$ (220 W/cm^2) in the spiro-SBCz-doped CBP film[2] (**FIG4**). We also fabricated an organic light-emitting diode (OLED) and the light-emitting organic field-effect transistor (OFET) using the 9,9'-spirobifluorene derivatives. We confirmed that the spiro-SBCz thin film functions as an active light emitting layer in OLED and FET. The OLED showed a high external EL quantum efficiency of η_{ext}=3.2±0.2%. Further, the OFET showed significant linear EL under p-type operation which will lead to future organic laser diodes.

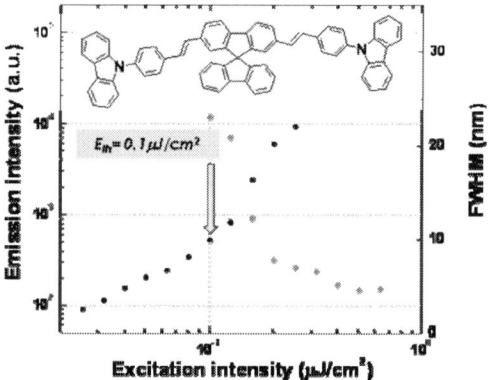

FIG4: ASE characteristics of Spiro-BSB

Light Emitting Field Effect Transistor

The demand for fusing OLEDs and OFETs to produce organic light-emitting FETs (OLEFETs) (**FIG5**) has been increasing aimed at achieving simplified organic active matrix displays. OLEFETs not only contribute to increased apertures in the pixels of light-emitting elements but also the inexpensive fabrication of active matrix displays due to the reduced number of switching thin-film transistors (TFTs). Further, controlling the carrier accumulation and successive carrier injection from both source and drain electrodes by applying a gate bias is a unique method of providing charge carriers in

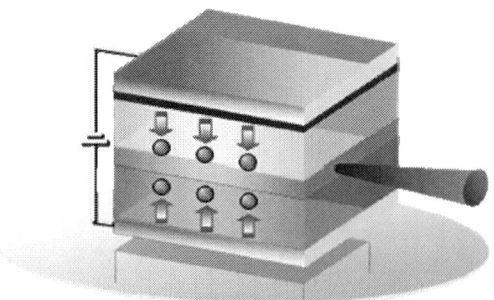

FIG3: Organic Laser Diode

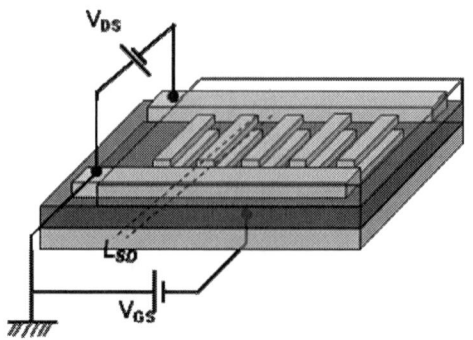

FIG5: Light Emitting FET

188

organic layers. Elucidation of the detailed mechanism has attracted a great deal of attention in this research field.

Recently, there have been several reports on OLEFETs using tetracene, oligo-thiophene, fluorine-based polymers, carbon nanotubes, and phenylene-vinylene-based polymers. We also reported on OLEFETs using 2,4-bis(4-(2'-thiophene-yl)phenyl)thiophene (TPTPT) as an active layer, demonstrating a maximum external EL quantum efficiency of $\eta_{ext}= 6.3 \times 10^{-3}$ % with a short channel length ($L_{SD}= 0.8$ µm) between the source and drain (S-D) electrodes. Although these reports demonstrated appreciable electroluminescence (EL), η_{ext} is still very low, which is probably due to inefficient electron accumulation and injection from the drain electrode. To achieve efficient OLEFETs, it has been necessary to prepare adequate organic materials with both electroluminescence and transistor characteristics. However, it is rather difficult to find candidates from well established OLED materials, since most OLED active materials have no high-performance FET characteristics probably due to their amorphous morphologies. The recent progress in OFETs has revealed that highly packed molecular thin films with tight inter-molecular π-stacking have demonstrated pronouncedly high carrier mobilities exceeding ~1 cm²/V·s. In particular, condensed aromatic compounds such as rubrene have demonstrated a high FET mobility of μ_{FET} ~10 cm²/V·s in the form of a single crystal. Instead of having a high performance TFT function, however, most TFT active materials demonstrate rather weak photoluminescence (PL) due to their strong molecular packing, i.e., concentration quenching, meaning they are useless in light-emitting applications. Therefore, we need to satisfy the demand and find organic materials that provide both light-emitting and transistor characteristics. Since condensed aromatic polyacene derivatives generally demonstrate rather good FET characteristics and some acene derivatives, e. g., anthracene, have demonstrated high PL efficiency over ~50 % even in single crystal form, we explored various polyacene derivatives and discovered that a tetraphenylpyrene (TPPy) vacuum-deposited film has typical p-type FET characteristics with a high PL quantum efficiency of 68±3 % even in a neat vacuum deposited film[3]. To achieve higher PL efficiency, we doped rubrene into a TPPy host. We observed that doping various highly fluorescent molecules into a TPPy host resulted in significant improvements to PL efficiency,

although some dopants completely eliminated TFT characteristics, and adding only specific dopants effectively maintained both TFT and PL characteristics. It is likely that no significant changes to TPPy aggregated morphology, i.e., molecular packing or grain size, by rubrene doping will contribute to retaining transistor characteristics.

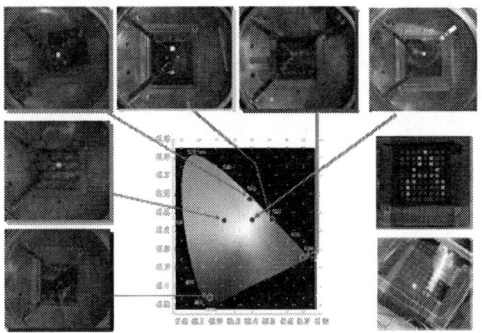

FIG6: Demonstration of full color light emitting FET with active driving.

We observed that the formation of a short channel length between the source and drain electrodes significantly increases carrier injection and EL efficiencies. We investigated the OLEFET characteristics of a 1wt%-rubrene:TPPy co-deposited layer using various channel lengths ($L_{SD}= 0.4$ µm-10 µm), and found that the OLEFET characteristics were strongly dependent on the channel length. We also employed double-layered MgAu/Au and Al/Au S-D electrodes, aimed at achieving both efficient hole and electron injection. By optimizing active layer's materials and device parameters, we obtained various emission colors with rather high EL efficiency around 0.1~1%[4] (**FIG6**).

Based on recent progresses in organic light emitting devices, we discuss the prospect of organic laser diodes.

References

1) T. Matsushima and C. Adachi, Appl. Phys. Lett. **89**, 253506 (2006).
2) S. Akiyama, M. Yahiro, T. Yoshihara, S. Tobita and C. Adachi, Advanced Functional Materials (in press).
3) T. Oyamada, H. Uchiuzou, S. Akiyama, Y. Oku, N. Shimoji, K. Matsushige, H. Sasabe, and C. Adachi, J. Appl. Phys., **98**, 074506 (2005).
4) T. Oyamada, H. Uchiuzou, H. Sasabe and C. Adachi, J. SID, **13**, 869 (2005).

Integrated Photonics: Enabling Optical Component Technologies for Next Generation Access Networks

Valery I. Tolstikhin

OneChip Photonics Inc., 46 Antares Dr., Suite 200, Ottawa, Ontario, Canada K2E 7Z1,
valery.tolstikhin@onechipphotonics.com

Abstract. *Integrated photonics technologies are reviewed in application to access network transceivers. Low manufacturing cost and high volume scalability, achievable through monolithic integration by means of semiconductor wafer fabrication techniques, identified as enablers for mass deployment.*

Introduction

Integrated photonics is an approach to optical circuit design and manufacturing, which is based on optical waveguides as a device building platform and planar technologies as a means to implement it. Optical alignment by means of lithography, reduced optical subassemblies (OSA), small footprint, high reliability and unparalleled volume scalability associated with wafer fabrication as a method of manufacturing, all featured by the integrated photonics, enable significant cost savings and ramp flexibility in a mass production. However, circuit design and fabrication complexity rapidly grows with its functional diversification and scale of integration, which reduces fabrication yields and compromises manufacturing cost/scalability. It may be argued, therefore, that the integrated photonics' best fit is in the applications focused on mass production of the low cost components with relatively simple functionality, modest performance requirements and low scale of integration.

Perfect examples of such applications are found in modern optical broadband access networks, which are becoming the core of new triple-play telecommunication services with data, video and voice delivered on the same optical fiber right to the user's end. The key optical component at the interface between optical and electrical domains, a transceiver, needs to be installed at every optical network terminal (ONT). In accordance with international standards, the ONT transceiver has to transmit and receive data signals at a bit rate up to 2.5Gb/s in 1310nm±50nm and 1490±10nm ranges, respectively, with an optional 1.7GHz bandwidth analog (video) signal overlay for

the downstream traffic in 1555nm±5nm range. Neither diplexer (without analog overlay) or triplexer (with analog overlay) ONT transceiver's functionality and performance requirements are particularly challenging compare to other applications, e.g. in long-haul wavelength division multiplexing (WDM) transceivers, but their cost/volume scalability requirements are and, as it is broadly accepted across the industry, an integrated photonics approach, possibly, is the only way to meet them.

This paper attempts to review the current status and future persepctives of the integrated photonics technologies, in application to ONT transceivers. General aspects of photonic integration and ONT transceivers are not addressed since these are well discussed elsewhere, e.g. [1,2]. Referring to different technologies, hereafter (a) photonic lightwave circuit (PLC) is a term used for the intergrated photonics components limited to only passive waveguide circuitry, (b) photonic integrated circuit (PIC) is a term reserved for the integrated components having both active and passive waveguide devices monolithically integrated in one chip, and (c) optoelectronic integrated circuit (OEIC) stands for a circuit combining active and passive optics with elements of electronics, all monolithically integrated in one chip.

PLC based ONT transceivers

PLC technology is the first step towards an integrated photonics ONT transceiver, in which a single photonic chip provides some or all passive optics functions. In the absence of the active optics, most commonly PLCs are based on the silica on silicon (SiO_2/Si) material system, which allows for very low loss optical waveguides with a large mode size – an advantage in terms of insertion loss and coupling to the optical fiber. Such PLC components can be used for the functions of optical WDM and interconnect with the active components, which still have to be hybrid integrated to form the optical di-/triplexer OSA.

There are many solutions for a WDM function in the PLC environment, which range from Mach-Zehnder interferometer [3] to lensed diffraction grating [4] to arrayed waveguide grating (AWG) [5] to reflective echelle grating (REG) [6] to photonic crystal [7] wavelength filter designs. It is worth noting, however, that even though actually fabricated PLC filters [3, 6] may deliver a competitive cost per performance in the case of a triplexer with three distinct passbands around 1310nm, 1490nm and 1555nm wavelengths, the advantage of their use in the case of a diplexer with only the first two passbands is less evident. The reasons are that while the footprint size and manufacturing cost of a 2-wavelength PLC filter are comparable to that of 3-wavelength counterpart, there are more cost efficient alternatives in the case of 2-wavelength filtering, e.g. thin film filter (TFF) technology.

Optical interconnect functions provided by a PLC are perhaps their major advantage, especially if combined with the use of surface mount on a common silicon optical bench (SiOB) for hybrid integration with active components. Indeed, interconnect PLC based optical di-/triplexer designs featuring embedded TFF solutions for WDM demonstrated by many [8-10]. Further possibilities open up when silicon on insulator (SOI) rather than SiO_2/Si chosen as a material system for PLC [11], since this allows for use of the CMOS technology and the combination of electronic components along with the passive optical ones. However, SOI has higher index contrast, which means higher insertion loss, even though in a smaller footprint size PLC.

Regardless of the material system, functionality and design choices, all PLC based OSA's still need the active optical components, such as a laser and a photodetector, to be purchased, tested, optically aligned and assembled in a di-/triplexer individually. This added complexity certainly restricts their applicability for the reduction of labour-intensive processes and improvement of cost and volume scalability factors in ONT transceiver manufacturing.

PIC based ONT transceivers
PIC's provide an alternative to PLC's, and specifically address the issue of combining the active and passive optical components onto the same substrate, therefore further advancing the level of integration. Given the spectral ranges of the receive and transmit functions in the ONT transceiver, from a practical standpoint there is only one material system that allows for the monolithic integration of all the optical components in one chip: Indium Phosphide (InP) and related III-V semiconductors.

There are three elements of an optical di-/triplexer to be monolithically integrated in an InP-based PIC: a WDM, a laser and a photodetector. With regards to WDM, note that some of the designs recently proposed for SiO_2/Si PLC's, have actually been implemented earlier in InP PIC's. For example, $1.3\mu m$/$1.55\mu m$ filtering in a PIC diplexer transceiver by using Mach-Zehnder interferometer and AWG has been reported in 1990 [12] and 1996 [13], respectively, whilst a 1310nm/1490nm/1555nm triplexer filter based on a planar REG and featuring two diffraction order design has been demonstrated in a commercial PIC product in 2005 [14]. In addition, a number of in-line [15-17], Y-junction [17], lateral [18] and vertical [19] directional coupler designs of $1.3\mu m$/$1.55\mu m$ diplexer have been realized. The major problem with WDM in InP of almost every type is its high insertion loss, as compared to their SiO_2/Si counterparts.

Monolithic integration of the active optics components into the WDM waveguide circuit partially compensates for the loss of optical power therein. For a photodetector operating at a speed of 2.5Gb/s or below, a simple and easy to fabricate active-passive waveguide integration design is based on vertical evanescent field coupling, e.g. in a single-mode vertical integration (SMVI) configuration [20]. This integration method enables polarization and wavelength independent quantum efficiency (relative to a passive waveguide coupled power) close to 100%, in conditions of nearly perfect wave impedance match between the active and passive waveguides [21]. In application to the ONT transceiver, it has been utilized in a PIC triplexer reported in [14]. The device has two broadband PIN waveguide photodetectors for data/analog receiving in 1490nm/1555nm wavelengths, respectively, and the third PIN waveguide photodetector for transmitter power monitoring in 1310nm range, all monolithically integrated into the output channels of a cross-order REG demultiplexer by using the SMVI technique. The PIC also features a spot-size converter at its input/output optical port and, perhaps, remains the most complete monolithically integrated optical triplexer to date.

Monolithic integration of a 1310nm laser with an otherwise perfectly passive WDM circuit, on one hand, and heavily absorbing 1490nm/1555nm photodetecors, on the other hand, is less trivial. The problem is that WDM, laser and photodetector waveguides have to be made from different yet compatible materials. The simplest design and perhaps the best performance solutions are associated with multiple epitaxial growth step techniques, in which unneeded epitaxial material from the previous growth step(s) is selectively etched out and replaced with a required material in the following growth step(s). Such techniques may work well for the high-end WDM applications, such as large-scale integration PICs for digital optical nodes reported in [22], but they are not particularly efficient from the fabrication yield/cost point of view and hence unsuitable for the ONT transceivers. However, fabrication yield/cost figures required by the ONT transceiver economics can be delivered by using an alternative PIC design/fabrication approach, based on a vertical integration of the active and passive components into a common multi-guide structure grown in one epitaxial step (e.g. [14]). There are several techniques enabling on-chip laser within such an approach, e.g. those featuring the etched facet FP [23], evanescent-field vertically coupled DBR [24] and laterally coupled DFB [25] laser designs, but there is little [26] reported on their utilization for the

cost-efficient PIC di-/triplexer to date.

OEIC based ONT transceivers
Fully integrated ONT transceivers, which include all the optical and electronic components monolithically integrated onto the same substrate, remain a remote target. Not so much due to the technological complexity of the OEIC, but rather because of the cost inefficiency associated with such complexity. Still, it is conceivable that certain electronic components, e.g. HBT based pre-amplifier, are added to the PIC chip. Alternatively, an InP-based PIC with all the optical functions on it may be hybrid integrated within the CMOS environment featuring all the required electronics.

Summary
In conclusion, whereas photonic integration at large is a technology well suited to serve the access network transceiver markets, this is an InP-based PIC technology, which may be the most attractive option, provided a proper way for a vertical integration of the low-loss WDM, efficient DFB laser and broadband photodetector is found within the one-step epitaxial growth approach.

References
1 Communications techn. roadmap, MIT (2005)
2 Huang W.-P. et al, JLT, **25** (2007), 11
3 Chen W. et al, OFC'06 (2006), PDP12
4 Li X. et al, IEEE PTL, **17** (2005), 1214
5 Lang T. et al, IEEE PTL, **18** (2006), 232
6 Pearson M. et al, LEOS, (2005), WX1
7 Shi Y. et al, IEEE PTL, **18** (2006), 2293
8 Hashimoto T. et al, ECTC, **53**, (2003), 279
9 Yanagisawa M. et al, OFC'04 (2004), TuI4
10 Blauvelt H. et al, OFC'05 (2005), OThU7
11 Bidnyk S. et al, PW'07 (2007), 64770F
12 Walker R. et al, IEE Proc. **J-137** (1990), 51
13 Mestric R. et al, IEEE STQE, **2** (1996), 251
14 Tolstikhin V. et al, ECOC'05, **3** (2005), 525
15 Nakajima H. et al, IEEE PTL, **8** (1996),1561
16 Mallecot F. Et al, OFC'98 (1998), ThS5
17 Hamacher M. et al, OFC'99 (1999), ThN3
18 Matz R. et al, IEEE PTL, **6** (1994), 1327
19 Magnin V. et al, IEEE PTL, **17** (2005), 459
20 Tolstikhin V., IPR'02 (2002), IFC4
21 Tolstikhin V. et al, IEEE PTL, **15** (2003), 843
22 Welch D. et al, IEEE JSTQE **13** (2007), 22
23 Whitaker H., Compound Semi., **10** (2004), 14
24 Studenkov P. et al, IEEE PTL, **12** (2000), 468
25 Reid B. et al, ECOC'03 (2003), Th1.6.6
26 Behfar A. et al, OFC'05 (2005), OTuM5

PLC based bi-directional optical module for access fiber networks

N.Kitamura (1)

1 : Fiber Optic Devices Division, NEC Corporation,1131 Hinode, Abiko,Chiba 270-1198, Japan

n-kitamura@bk.jp.nec.com

Abstract *Optical modules based on PLC and passive optical alignment technologies has been developed for access network. These technologies can realize the cost and size reduction of the optical modules, with high productivity.*

Introduction

Optical access network system such as PON system and Point-to-Point system has been introduced and commonly used, using bi-directional optical transceivers. In order to develop these systems successfully, reducing the cost of optical transceivers is one of the most important factors. Besides, mass productivity is another important factor to accommodate the rapid growth of production volume requirement. By using PLC (Planar Lightwave Circuit) as an optical bench, optical parts such as Laser Diode, Photo Diode and optical fiber can be mounted on PLC by passive alignment technology. Passive alignment enables us to apply the automated assembling machine, which can realize cost reduction and mass productivity. Another advantage of PLC platform is the reduction of module size, which enables to reduce the equipment size and to realize the effective systems.

In this paper the structure of PLC based optical module, assembly technique and products with these technologies are introduced.

PLC chip structure

A structure of PLC chip for bi-directional optical module is shown in Fig.1. This PLC chip is composed of a waveguide pattern, LD mounting parts, PD mounting parts, a slit for WDM filter assembling, and a V-groove which is used for self alignment of fiber. The waveguide core and clad layer is composed of Ge and P doped SiO_2, and B and P doped SiO_2 respectively. These films are deposited on Si substrate using atmospheric pressure chemical vapor deposition (APCVD) method. This method enables to deposit at low temperature, thick film with low stress, which is suitable for the device for fiber communication. [1]

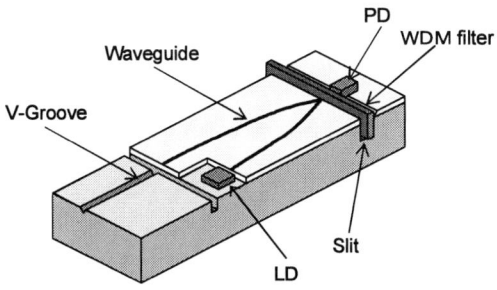

Fig.1 Structure of PLC chip

Figure 2 shows the process flow for forming the PLC chip. After a mask for the fiber guiding V-groove is fabricated on the Si substrate, an under clad layer is deposited by APCVD. A mask for LD mounting pedestals and alignment marks are fabricated on the under clad layer (Fig.2 (a)). Since the masks for fiber guiding V-groove and the LD mounting pedestals are formed on the flat layers, it is possible to realize precise alignment with waveguide mask pattern. In the next process, waveguide and upper clad layers are deposited by APCVD (Fig. 2 (b)). Low temperature process by APCVD can prevent the degradation of mask. After core and upper clad layer are deposited, the silica layers on the V-groove part are removed, and V-groove is fabricated by Si anisotropic chemical etching process (Fig. 2 (c)). Finally, silica layers on the LD mounting part are etched by reactive ion etching (RIE) method (Fig. 2 (d)). Precise pedestals are formed because the mask for LD mounting pedestals can stand for the RIE.

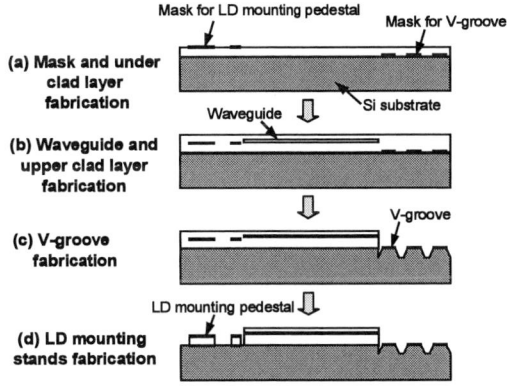

Fig.2 Process flow for PLC chip fabrication

Figure 3 shows the LD mounting structure. The precise alignment (<1um) is required for optical coupling between LD and waveguide. In vertical direction, the LD is automatically aligned to the suitable position by mounting on the pedestals which is fabricated so as to fit the height of the LD active layer and the waveguide core layer. In the horizontal direction, the LD is positioned by detecting each alignment marks on the LD and PLC chip using infrared light. After finishing mechanical alignment, LD is fixed by heating Au-Sn solder mounted on PLC electrode. Through these methods, the LD mounting accuracy of within 0.5um was achieved successfully.

978-0-9789217-3-6/07/$25.00

©2007 WEN GLOBAL SOLUTIONS

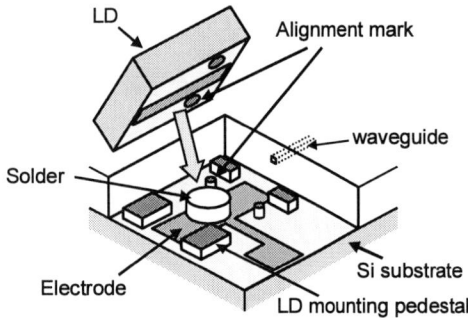

Fig.3 Structure of LD mounting part

Figure 4 shows the PD and WDM filter mounting structure. PD and WDM filter mounting parts are composed of a slit, an electrode and a mirror formed by metallizing a slope. The slit and slope are formed by dicing saw during the PLC wafer process. The mirror contributes to reflecting the input signal light to the upward direction. The WDM filter is inserted in the slit and affixed at the waveguide facet. After the PD is aligned by fitting the PD detecting area to the position of the alignment mark on the PLC chip, the PD chip is fixed by gold bumps mounted on the PLC electrode. This structure is effective for the conventional top illuminated PD whose active area is large (~80um) because there is large gap between waveguide facet and PD surface. In case of higher bit rate application such as 2.5Gbps, PD with small active area is used, because low capacitance is required. In this case, the structure shown in Fig.5 is effective. In Fig.5 bottom illuminated PD and PD carrier is used. After PD is mounted on the PD carrier, PD carrier is affixed to PLC facet by passive alignment.

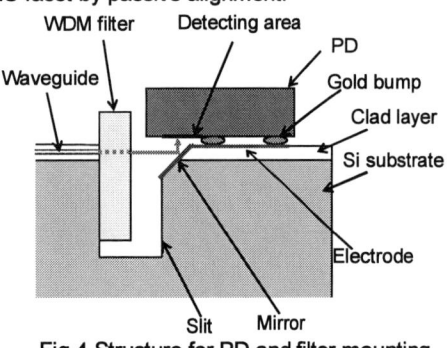

Fig.4 Structure for PD and filter mounting

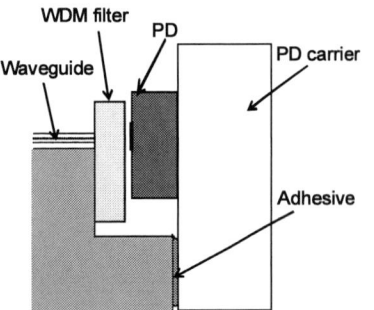

Fig.5 Another structure for PD and filter mounting

The optical fiber is self-aligned by using V-groove which is fabricated on the PLC chip. [2] In this method, the excess coupling loss between the fiber and the waveguide caused by displacement is as low as 0.2dB.

Optical module structure and performances

LD, PD and WDM filter are assembled by fully automated assembling machines. Using such machines, mass productivity and cost reduction can be obtained. [3]

Figure 6 shows the structure of the optical module which is composed of a package, Pre-Amp IC, ferrule and PLC chip on which optical components was mounted. In module assembling process, index matching gel is used between optical coupling parts for the reduction of optical coupling loss and reflection.

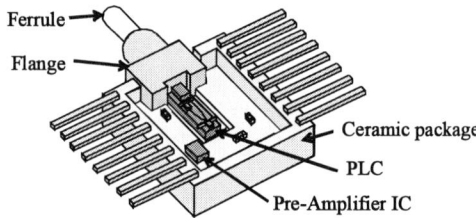

Fig.6 Structure of optical module

By selecting the type of LD, PD and WDM filters, various types of optical modules are realized.[4] DFB-LD and APD is commonly used for longer reach application such as over 20km and for higher bit rate application such as 2.5Gbps. Table1 shows the optical characteristics of PLC modules. And Fig.7 shows the temperature dependency of I-L characteristics of PLC modules.

Table1 Performance of PLC modules

Transmitter

LD	1.3um FP-LD	1.3um DFB-LD	1.5um DFB-LD
Output Power @Ith+20mA, CW	2.0mW	1.5mW	1.25mW
Threshold current	5mA	8mA	10mA
SMSR	-	50dB	50dB
Spectral width	2.0nm	-	-
Tracking error	+/-1.0dB		

Receiver

PD	PIN-PD	APD
Sensitivity	-27dBm 1.25Gb/s,PRBS2^7-1 BER=10^{-12}	-31dBm 2.4Gb/s,PRBS2^{23}-1 BER=10^{-10}

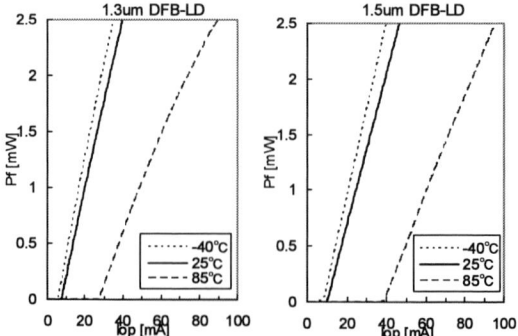

Fig.7 I-L characteristics of PLC module

PLC applications

By applying these techniques, various types of products can be obtained. We have already developed transceivers with SFF and SFP structures for Point-to-Point system and B/GE/G-PON systems. [5][6] Fig.8 shows the structure of triplexer module for G-PON applications. The optical signals with different three wavelengths are divided by two WDM filters which are affixed at the waveguide facets. By modifying PLC waveguide pattern, new functions can be easily obtained. [7]

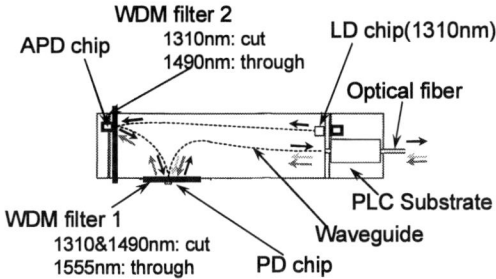

Fig.8 Structure of PLC for triplexer

Another advantage of PLC based optical module is the reduction of the module size. PLC itself is composed of silica glass deposited on silicon substrate whose thickness is 800um. Therefore the very thin optical module can be achieved. By using this advantage, we have newly developed compact size SFP and SFF transceivers shown in Fig.9 (a) and (b). The compact SFP has the function of 2 channel of bi-directional transceiver in the standard SFP size. The compact SFF was designed to offer bi-directional transmission in a half size in comparison to the standard SFF package. By connecting these compact SFFs, 2ch or 4ch SFF can be easily realized. These products can contribute to the reduction of the space for systems.

Conclusions

We have developed PLC based optical module for bi-directional transceiver which is used in access network systems. Applying PLC to optical module enables to mount optical element by passive alignment, which is suitable for mass production and cost reduction. Furthermore by utilizing the merit of PLC thickness, we have developed thin optical modules, which lead to reduction of transceiver size. These advantages based on PLC technologies will contribute to the progress of optical access networks.

References

1. M.Kitamura, et al., CLEO'98, (1998), pp.480-481
2. N.Kitamura, et al., Integrated Photonics Research B2, (1996), pp.608-611
3. K.Yamauchi, et al., Electronic Components & Technology Conference 50, (2000), pp.15-20
4. M.Oguro, et al., Electronic Components & Technology Conference 52, (2002), pp.305-310
5. N.Kimura, et al., Electronic Components & Technology Conference 53, (2003), pp.290-295
6. H.Yanagisawa, et al., ECOC'00, (2000), pp.85-86
7. K.Yamauchi, et al., ECOC'06, Vol.3 (2006) pp.495-498

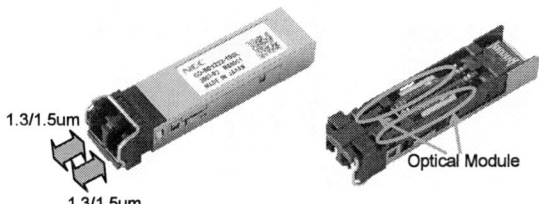

Fig.9 (a) Structure of Compact SFP

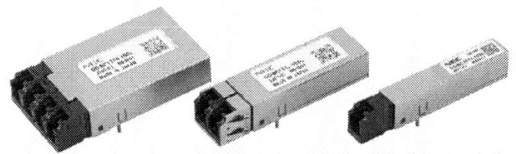

Fig.9 (b) Structure of Compact SFF
(4,2,1channel version)

The low cost single mode laser technology for mass deployment

Bo Cai, PhD, Eblana Photonics Ltd

Abstract

DM laser is a breakthrough for low cost single mode laser. Designed around standard semiconductor process, the DM laser shares the simplicity and high yield of an FP laser, yet provide single spectrum emission with noise much lower than current DFB laser.

Summary

With increasing demand for longer transmission distance and higher bit rate in applications such as GPON, the single mode laser becomes a key component. The current DFB solution is struggling to provide either sufficiently low cost or performance consistency for low cost mass deployment.

In DFB design, specialized and highly complicated process such as re-growth is used. The process and facility will be inevitably less mature, low throughput, complicated and high cost in semiconductor IC standard. Coupled with some design intrinsic weakness such as random cleaving grating phase, DFB laser suffered from low yield and wide distribution of parameters. To integrate such a device with a highly consistent electronic circuit in a system, a layer of electronics with costly fine tuning is required to mask the wide distribution. The DFB is also likely to be sensitive to feedback and with wide linewidth due to its distributed low order resonator structure.

To overcome the intrinsic problems faced by DFB, DM laser is designed around a few simple standard semiconductor process steps with a cost structure parallel to FP laser. Similar to FP laser, DM laser is designed to osculate with a single high order resonator but with spectrum selective gain. A single high order resonator and spectrum selective gain structure. In such a resonator, spontaneous and feedback noise will be averaged down by hundreds of modes maintained by the resonator. As a result, the laser has a linewidth and feedback insensitive performance 2 order of magnitude better than standard DFB.

In this presentation, the performance of DM lasers and their impact to application such as GPON is discussed in details. Device performance, production statistics and extensive reliability data are presented to demonstrate its low noise/feedback insentive nature, robustness, consistency and manufacturability. The potential for integrating more functionality on chip level based on this technology is also discussed briefly.

Tunable Optical Delay Schemes Using All-Optical Processing in a Highly Nonlinear Bismuth Oxide Fiber

Chester Shu and Mable P. Fok

Department of Electronic Engineering and Center for Advanced Research in Photonics,
The Chinese University of Hong Kong, Shatin, N.T., Hong Kong.
E-mail: ctshu@ee.cuhk.edu.hk

Abstract Widely tunable delay schemes are demonstrated using nonlinear phase modulation or four-wave mixing in a 32-cm bismuth oxide fiber followed by group velocity dispersion in a chirped fiber Bragg grating or a dispersion-compensated fiber.

Introduction

There has been an increasing interest in demonstrating a tunable optical delay line for applications in optical synchronization, optical buffer, and optical control of microwave/RF signals. With the use of a mechanically adjustable fiber delay line, optical delay can be easily achieved; however, the tuning control is very limited. Recently, many approaches have been used to demonstrate an electrically or optically controlled tunable delay. Examples are the recirculation of light in a waveguide/ring resonator [1], slow light by changing the gain and group index in a nonlinear medium [2], and change of wavelength followed by propagation in a dispersive fiber [3]. The above approaches exhibit different characteristics in terms of the maximum delay, the delay continuity, the pulse shape degradation, and the compactness.

In this paper, we demonstrate different fiber-optic approaches to achieve widely tunable optical delay using all-optical processing in a highly nonlinear bismuth oxide fiber (Bi-NLF). Our work is based on self-phase modulation (SPM), cross-phase modulation (XPM), and four-wave mixing (FWM) in a 32-cm Bi-NLF. In the SPM and XPM approaches, offset filtering from the nonlinearly broadened spectrum is used together with group velocity dispersion provided by a chirped fiber Bragg grating (CFBG). In the FWM approach, we study the cases of both single pump and dual-pump FWM. The dispersion compensated fiber (DCF) is used to introduce the group velocity dispersion and to provide a large bandwidth of operation. A cascadable scheme that uses chirped FBGs has also been demonstrated to increase the total delay achievable with a limited amount of wavelength shift. Owing to the extremely large nonlinear coefficient of Bi-NLF ($\sim$1100 $W^{-1}km^{-1}$ at 1550 nm), a very short fiber segment is sufficient to introduce significant nonlinear phase modulation and FWM. The reduced fiber length also leads to a substantial increase in the stimulated Brillouin scattering (SBS) threshold [4]. Hence, an intense signal or CW light can be used to introduce the required optical nonlinearities.

Tunable Delay Using Self-Phase Modulation and Cross-Phase Modulation

Figure 1 shows our experimental setup. A 10-GHz pulsed source at 1551 nm is generated from a fiber laser. The output is amplified to 25 dBm and is launched to a 32-cm Bi-NLF to introduce self-phase modulation (SPM). The Bi-NLF is fusion-spliced to a standard single mode fiber at each end using an intermediate fiber with an ultrahigh numerical aperture. The spectrally broadened output is sliced with a 0.4 nm optical bandpass filter and the extracted component is launched to a chirped fiber Bragg grating (CFBG). The CFBG has a dispersion of 21 ps/nm and a bandwidth of 30 nm. The delayed pulse is obtained at the output of the CFBG through an optical circulator. By selecting frequency components from different portions of the broadened spectrum, a variable delay can be obtained using group velocity dispersion in the CFBG.

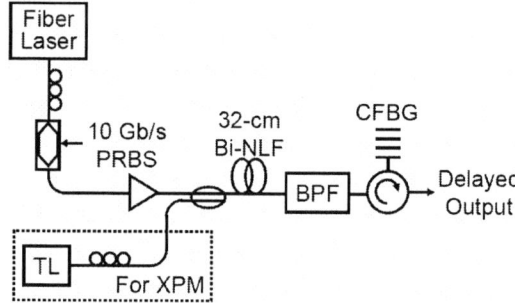

Figure 1: Experimental setup of the SPM/XPM based tunable optical delay scheme.

Figure 2(a) displays the optical spectrum of the 10-GHz mode-locked fiber laser. The 3-dB linewidth is about 2 nm. During the propagation of the mode-locked pulse in a 32-cm Bi-NLF, self-phase modulation takes place and subsequently a broadened spectrum is obtained as shown in Figure 2(b). By filtering out the desired wavelength component and directing it to the CFBG, a tunable optical delay is obtained at the output. The 0.6-nm optical bandpass filter is tuned over a range of 30 nm from 1537 to 1567 nm. The optical spectra obtained

978-0-9789217-3-6/07/$25.00

©2007 WEN GLOBAL SOLUTIONS

during the tuning are shown in Figure 2(c). By selecting the output from ~1542 to 1562 nm, we obtain a tunable delay of 415 ps after the CFBG.

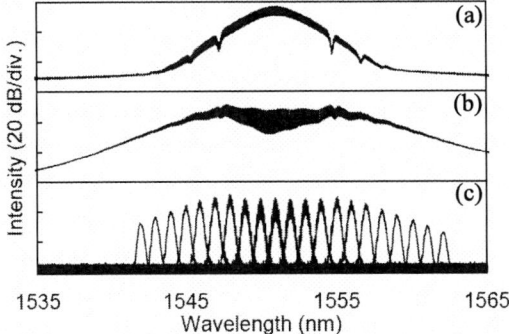

Figure 2: Optical spectra of (a) fiber laser output (b) SPM output (c) filtered output at different wavelengths.

The spectral broadening in the case of SPM depends on the signal power and occurs around the signal wavelength. A more flexible approach to control the tuning is by XPM. The input pulse now introduces a spectrum broadening of a CW probe light at another desired wavelength. To further increase the range of the total delay, one can also tune the probe wavelength in order to increase the amount of wavelength shift.

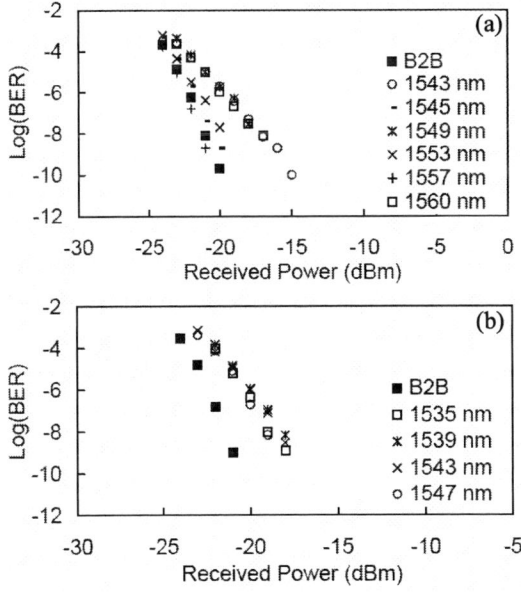

Figure 3: BER measurement results on different tunable optical delay schemes (a) SPM based (b) XPM based.

A drawback of using XPM instead of SPM is that the walk-off between the signal and the probe light has to be minimized in order to reduce pulse distortion. With the use of the 32-cm Bi-NLF in our setup to provide the nonlinear phase modulation, the walkoff can be significantly reduced. In our work, we have demonstrated tunable optical delay using both a fixed CW probe wavelength at 1540 nm and a tunable probe that scans over the range of 1535 to 1547 nm. After optical filtering and directing the output to the CFBG, we obtain a variable delay range of 104 ps and 297 ps for the two respective tuning schemes

The 10-Gb/s BER performance has been measured for tunable delay obtained with pseudorandom ASK data using SPM and XPM. The result is shown in Figure 3 (a) and (b). At a BER of 10^{-9}, we obtain a power penalty of less than 4 dB for the SPM scheme. The XPM scheme that uses a variable probe wavelength also shows a similar system performance.

Tunable Delay Using Four-Wave Mixing

FWM provides a different means of changing the wavelength of the signal for the purpose of tuning its delay. Instead of relying on the nonlinear spectral broadening effect, FWM generates a wavelength-converted signal through the interaction between the signal and a CW pump. An advantage that is offered by FWM is that the delay scheme is applicable not only to pulses and ASK data, but also to differential phase-shift keying data and other advanced modulation formats [5]. Such a transparency to data format is an intrinsic property of FWM. However, owing to the requirement of phase matching in FWM, the 3-dB wavelength conversion range is usually limited to about 12 nm in Bi-NLF that will restrict the achievable tunable delay range [6]. Here, we describe our approach that exploits dual-pump FWM to broaden the conversion range, thus resulting in a substantially wider conversion range.

In the FWM scheme, we first study the delay of a 10-GHz pulsed source with an output pulse width of 37 ps. Two optical pumps are used in the experiment. One of the pumps (pump 1) is fixed at 1551.2 nm while the other pump (pump 2) is wavelength-tunable. The pumps together with the 10-GHz pulses are amplified to 27 dBm with an erbium-doped fiber amplifier (EDFA). Subsequently, the amplified pumps and signal are fed to a 32-cm Bi-NLF where four-wave mixing takes effect. The SBS threshold is found to be at least over 30 dBm in our experiment. The four-wave mixing output is extracted by a 0.6 nm optical bandpass filter and is launched to a DCF. The delayed pulse is obtained at the output of the DCF. By adjusting the wavelength of pump 2, a widely tunable delay can be obtained using GVD in the DCF.

Figure 4 shows the wavelength conversion efficiency obtained for the cases of single and dual-pump FWM. While the former shows a 3-dB bandwidth of 12 nm, the latter approach greatly enhances the tuning range to over 30 nm. The conversion range is limited by the gain bandwidth of the EDFA in our experiment. The improvement in the conversion bandwidth results in a wider tunable optical delay range.

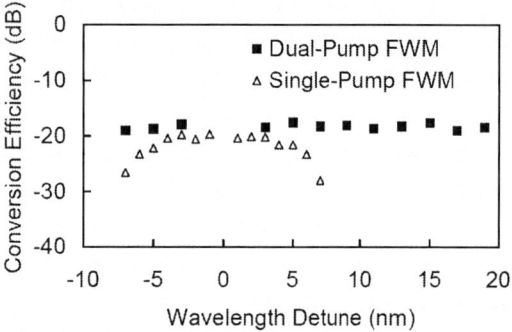

Figure 4: Conversion efficiency of conventional single pump FWM (hollow triangle) and dual-pump FWM (dark square).

The wavelength converted pulse is filtered out and is fed to a DCF of length that varies from 100 to 300 m. We study the linearity of the tuning approach by plotting the achieved delay against the wavelength of pump2. The result is shown in Figure 5. It is observed that the linearity maintains for different DCF lengths. Maximum optical delays of 840, 570, and 280 ps are obtained with 300, 200, and 100 m of DCF, respectively. The tuning slopes agree well with the values predicted from the fiber dispersion coefficient of D ~ -70 ps/(nm.km)$^{-1}$.

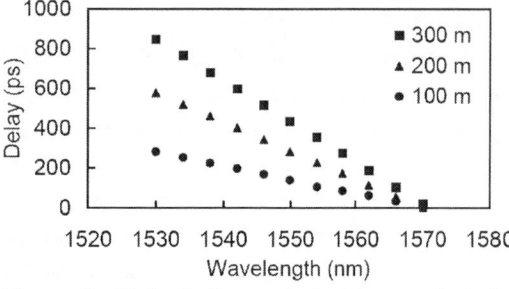

Figure 5: Plot of the optical delay against the wavelength of pump 2 using different lengths of DCF for GVD.

While the DCF provides a large GVD over a broad wavelength range, the setup is relatively bulky and may result in instability caused by environmental disturbances on the long coil of fiber. To reduce the size of the setup, we can use a CFBG to provide the required GVD. However, since the amount of dispersion in a CFBG is usually limited over a given range of wavelength, it is desirable to develop a

cascadable scheme to increase the achievable tuning range at a given amount of wavelength shift. We demonstrate such a cascadable scheme by serial connection of identical FBGs using a set of optical circulators. By extracting the wavelength converted output from 1542 to 1562 nm and directing it to 1 and 2 pieces of CFBG, each providing a GVD of ~ 20 ps/nm, we obtain a maximum delay of 400 and 800 ps, respectively. The additional power penalty associated with the second stage of delay is less than 0.5 dB.

Conclusions

Through the use of self-phase modulation, cross-phase modulation, and four-wave mixing in a 32-cm Bi-NLF, different optically controlled variable delay schemes have been demonstrated. In SPM/XPM, the desired frequency component is extracted from the broadened spectrum and is fed to a CFBG that provides the group velocity dispersion. In FWM, we have demonstrated the use of dual-pump approach together with different lengths of DCF to broaden the range of achievable delay. A compact and cascadable approach has also been discussed to maximize the delay for a given amount of wavelength shift using serial connection of CFBGs

Acknowledgment

The authors would like to thank Dr. Sugimoto and Dr. Ohara of Asahi Glass Co., Ltd. in providing the highly nonlinear bismuth oxide fiber. This work is supported by the RGC of Hong Kong (CUHK 4184/04E and 415705).

References

1 G. Lenz et al, IEEE J. Quantum Electron, vol. 37 pp. 525 – 532 (2001).
2 C. J. Chang-Hasnain et al, Proc. IEEE, vol. 91 pp. 1884 – 1897 (2003).
3 J. E. Sharping et al, Opt. Express, vol. 13 pp. 7872 – 7877 (2005).
4 J. H. Lee et al, Opt. Lett., vol. 30 pp. 1698 – 1700 (2005).
5. M. P. Fok and C. Shu, OFC/NFOEC Proc., JWA60, Anaheim, USA, 2007.
6. M. P. Fok and C. Shu, ECOC Proc., We1.3.3, Cannes, France, (2006).

Recent Advances in the Practical Fiber Optical Parametric Amplifiers

K. K. Y. Wong (1), B. P. P. Kuo (1), M. E. Marhic (2), G. Kalogerakis (3), and L. G. Kazovsky (3)

1 : Department of Electrical & Electronic Engineering, The University of Hong Kong,
Pokfulam Road, Hong Kong, Email: kywong@eee.hku.hk

2 : Institute of Advanced Telecommunications University of Wales, UK

3 : Department of Electrical Engineering, Stanford University, Stanford, CA 94305, USA

Abstract *In recent years, impressive performance of fiber optical parametric amplifiers (OPAs) have been demonstrated in different respects. We describe these recent advances and discuss some of the challenges should be addressed before OPAs can be practical.*

Introduction

Fiber optical parametric amplifiers (OPAs) are based on the third-order susceptibility $\chi^{(3)}$ of the glasses making up the fiber core. While this nonlinearity is relatively weak in silica-based glasses, the small cross-sections, low loss, and large lengths available with glass fibers can lead to sizeable effects. Fiber OPAs have the following important characteristics, which make them potentially interesting for a variety of applications:

- Bandwidth increasing with pump power
- Arbitrary center wavelength
- Large gain
- Idler generation
- High-speed optical signal processing
- High-power capability

Recent advances

Due to the convergence of several factors, research on fiber OPAs has increased over the past few years. This had led to the accomplishment of important milestones, which demonstrated that fiber OPAs can match and sometimes exceed, the performance of other types of optical amplifiers in several respects. We begin with pulsed devices, for which it is easy to obtain high pump powers, suitable for obtaining large bandwidths and high gains. We then cover CW devices, whose performance is important for applications such as optical communications.

Pulsed OPAs

OPA with 400 nm gain bandwidth. A fiber OPA with a gain bandwidth of 400 nm was reported in [1]. With such a large bandwidth, it was actually difficult to find enough signal lasers at different wavelengths to perform gain measurements. For that reason, the OPA amplified spontaneous emission (ASE) spectrum was used as a substitute for the gain spectrum (as the two are closely related). We were able to eliminate the influence of EDFA ASE from the measurements by effective filtering. In this manner we obtained a 20-dB gain bandwidth slightly over 400 nm. This value is in good agreement with the theoretical value that can be calculated from the experimental parameters.

TDM Demultiplexing. To make a time-division-multiplexing (TDM) demultiplexer from a pulsed-pump OPA, one needs to have only one pump pulse per frame, but its duration should be that of a signal bit. The pump pulses must be synchronized with the bits to be extracted. Then, at the OPA output the signal pulses coinciding with the pump pulses will be amplified. If the gain is large, say 20 to 30 dB, most of the energy in a frame could correspond to the desired channel. On the other hand, if the gain is too low, it is preferable to use the idler since only the desired pulses are present, which leads to a better OSNR. TDM demultiplexing using the signal gain has been demonstrated for 10 Gb/s signals [2].

2R or 3R regeneration. Fiber OPAs can be used for making regenerators. The idea is to use a pump which is pulsed at the same rate as the data, and synchronized with it. Since the OPA gain is a strong function of instantaneous pump power, if the pump pulses have short rise and fall times, the amplified signals pulses will be strong only during the pump pulses. This will provide reshaping and retiming of the '1' pulses. This type of regeneration is desirable because the optical frequency of the pulses is not altered, which implies that such regenerators can be cascaded. This type of regeneration has been experimentally demonstrated [3].

Optical sampling. Lasers can generate short pulses at high repetition rates. These can be used for sampling in the optical domain. This requires a very fast optical device, which responds much faster than the sampling pulses. Fiber OPAs can respond on a femtosecond time scale, and therefore can be used for optical sampling. The idea is similar to that of an optical DMUX, except that now the pump consists of pulses much shorter than the bit period. There is one pump pulse for every N_p periods, occurring at a slightly different time. Then at the OPA output the idler consists of pulses coinciding with the pump pulses, with about the same duration. These pulses, with a relatively low repetition rate, can now be converted to low-frequency electrical signals, which can be displayed on a conventional oscilloscope. The result is a display of the detailed structure of the periodic optical signal of interest, over a one-period interval. This idea has been experimentally demonstrated [4]: by using OPA pump pulses with picosecond duration, the authors were able to implement optical sampling, and to display a time-resolved periodic 300 GHz waveform.

CW OPAs

CW OPA with 60 dB gain. Advanced stimulated Brillouin scattering (SBS) suppression techniques must be used for obtaining high CW gains. In Ref. [5], an isolator was used in combination with pump phase modulation for SBS suppression. With the phase modulation only, it was not possible to exceed the performance of the previous OPAs. After adding the isolator, a maximum gain of 60 dB was measured. The gain was actually no longer limited by SBS, but by saturation of the gain by ASE. Recent effort has recorded 70 dB gain [6].

CW OPAs with low optical noise figure. While OPAs in principle have the potential for approaching the 3 dB quantum limit, a number of practical difficulties have to be overcome to do so. A major difficulty is EDFA ASE. If it is not reduced by optical filtering prior to combining, it is then amplified to a very high level. To avoid this, the pump EDFA ASE must be greatly reduced by filtering before combining with the signal. This idea was first implemented in Ref. [7]. By using an FBG, the EDFA ASE was reduced essentially to a level equal to that of vacuum fluctuations near 1560 nm, the peak of the EDFA gain. NF measurements over the OPA gain bandwidth gave a fairly constant NF value, with an average of 4.2 dB. This is larger than the 3 dB standard quantum limit. The difference is attributed to the effect of Raman gain [8].

Polarization-insensitive OPAs. A severe difficulty for fiber OPAs is that in their simplest forms they exhibit gain which is a function of the SOP of the incident signal, which is undesirable in optical communication systems. A variety of means have been used to overcome polarization sensitivity in various devices, and variations on these have been investigated for fiber OPAs. Good results have been obtained [9][10].

Amplitude-noise reduction. This application is based on the fact that the gain of fiber OPAs saturates as the signal input power increases beyond a level P_{sat}. As a result, fluctuations around the high input level result in relatively small fluctuations around the high output level. Thus, if one considers a binary input signal, with a low level near 0, and a high level above P_{sat}, the resulting binary output waveform has noise for the high level which is compressed. In other words, we have amplitude-noise reduction for the high level. This mechanism has been studied theoretically and experimentally [11], and it has been verified that it does lead to a reduction in amplitude noise. This feature has also been used in combination with timing jitter reduction [12].

Amplifier nonlinearities. For use in WDM systems, fiber OPAs will have to be designed to keep unwanted nonlinear interactions within the OPA itself at tolerable levels, such as the FWM, XGM [13]. We have proposed few schemes to reduce nonlinear crosstalk significantly in both one-pump and two-pump OPAs [14][15][16].

Challenges

SBS suppression

Several techniques have been demonstrated for reducing or suppressing SBS. Currently in most CW OPA work, SBS is suppressed by making the pump spectrum much broader than the Brillouin gain bandwidth (about 50 MHz). Broadening of up to several GHz is obtained by means of phase or frequency modulation [5]. SBS can also be suppressed by using a pulsed pump, with a low duty cycle. For example, 1-ns pulses will have a spectral width of the order of 1 GHz. This can be an incentive for performing experiments with pulses. It would be desirable to seek alternative effective SBS suppression techniques, preferably passive.

Longitudinal fluctuations of zero-dispersion wavelength

The impact of $\lambda_0(z)$ on a particular OPA depends a great deal on some other OPA parameters. The pump power, P_0, is particularly important, in several ways. If it is large, one only needs a short fiber to achieve a given gain; then the probability of encountering large λ_0 variations is minimized. In addition, the nonlinear phase shifts which enter into the phase matching are large, and dominate over the linear phase shifts which are the only ones affected by dispersion. As a result, one should be able to obtain near-theoretical performance with high-power pulses in short fibers; this has indeed been verified [17]. On the other hand, the impact of $\lambda_0(z)$ on long fiber OPAs can be substantial [18].

High-γ fibers

If everything else is the same, a large nonlinear coefficient, γ, leads to a larger gain bandwidth, and a reduced value for the product P_0L (N.B. L is the fiber length). The advent of HNLFs [19], with $\gamma = 20$ W^{-1} km^{-1}, has facilitated a number of experiments, particularly demonstrations of wideband and high-gain amplifiers. When available, they are currently the preferred medium for OPA research in the C-band.

In recent years much work has been done to develop new classes of fibers, using an array of holes around the core for confining the light, instead of a uniform low-index cladding as in conventional fibers. By changing the size, shape and spatial distribution of the holes, in principle one can tailor most fiber properties. For fiber OPAs, the potential for achieving a small mode-field diameter (and hence to increase γ) and to tailor dispersion are particularly attractive. It is anticipated that this approach will provide holey fibers suitable for advanced fiber OPAs.

Conclusions

Substantial progress has been made in recent years with the development of fiber OPAs. Important features have been demonstrated, including: high gain; gain bandwidth of several hundred nanometers; polarization-independent configurations; low noise figure, etc. We anticipate that further progress with high-power pumps, highly-nonlinear fibers with tailored dispersion properties, and SBS suppression techniques, fiber OPAs and related devices will find practical applications in areas such as high-power wavelength conversion and optical communication.

References

[1] M. E. Marhic et al, *IEEE J. of Selected Topics in Quantum Electron.*, **10** (2004), pp. 1133-1141.

[2] J. Hansryd et al, *IEEE Photon. Technol. Lett.*, **13** (2001), pp. 732-734.

[3] Y. Su et al, *OFC'01*, paper MG4.

[4] J. Li et al, *IEEE Photon. Technol. Lett.*, **13** (2001), pp. 987-989.

[5] K. K. Y. Wong et al, *IEEE Photon. Technol. Lett.*, **15** (2003), pp. 1707-1709.

[6] T. Torounidis et al, *IEEE Photon. Technol. Lett.*, **18** (2006), pp. 1194-1196.

[7] J. L. Blows et al, *Opt. Lett.*, **27** (2002), pp. 491-493.

[8] P. L. Voss et al, *Opt. Lett.*, **29** (2004), pp. 445-446.

[9] K. K. Y. Wong et al, *IEEE Photon. Technol. Lett.*, **14** (2002), pp. 1506-1508.

[10] K. K. Y. Wong et al, *IEEE Photon. Technol. Lett.*, **14** (2002), pp. 911-913.

[11] K. Inoue et al, *J. Lightwave Technol.*, **20** (2002), pp. 969-974.

[12] K. Shimizu et al, *OFC'03*, paper TuH5, pp. 197-198.

[13] T. Torounidis et al, *IEEE Photon. Technol. Lett.*, **15** (2003), pp. 1061-1063.

[14] K. K. Y. Wong et al, *Optics Comm.*, **270** (2007), pp. 429-432.

[15] K. K. Y. Wong et al, *Opt. Express*, **15** (2007), pp. 56-61.

[16] B. P. P. Kuo et al, *OFC'07*, paper OWB3.

[17] M.-C. Ho et al, *J. Lightwave Technol.*, **19** (2001), pp. 977-981.

[18] F. Yaman et al, *IEEE Photon. Technol. Lett.*, **16** (2004), pp. 1292-1294.

[19] M. J. Holmes et al, *IEEE Photon. Technol. Lett.*, **7** (1995), pp. 1045-1047.

Single Polarisation Fibre Ring Laser by Utilising Intracavity 45° Tilted Fibre Bragg Grating

Kaiming Zhou(1), Chengbo Mou(1), Xianfeng Chen(1), Lin Zhang(1), Ian Bennion(1)
Shenggui Fu(2), Xiaoyi Dong(2)

1: Photonic Research Group, School of Engineering and Applied Sciences, Aston University, Aston Triangle, Birmingham, U.K. , B4 7ET, email:k.zhou@aston.ac.uk
2: Institute of Modern Optics, Nankai University, Tianjin, P.R.China, 30007

Abstract Single polarisation operation of fibre ring laser has been realised by employing an intracavity 45°-tilted fibre Bragg grating (45°-TFBG). The degree of polarisation upto 99.94% of the laser was demonstrated with good stability.

Introduction

Fibre lasers play important roles in optical communication and sensing applications, for which single mode and single polarisation operation of the laser are desirable. However, due to low-birefringence of standard and active fibres, the output of a fibre laser is unpolarised generally. To realise single polarisation operation for fibre lasers, several methods have been developed. One scheme involves integrating an integral polariser [1] and the other approach incorporates high-birefringence rare-earth doped fibre with two anisotropic fibre Bragg gratings [2]. Though these two methods can provide very high degree of single polarisation oscillation, the integral polarisers are difficult to fabricate and may induce high insertion loss to the laser cavity decreasing its efficiency. In previous work, we reported the unique polarisation discrimination property of the 45°-tilted Bragg grating structure, i.e. it transmits only p-polarisation light in the fibre and completely outcouples s-polarisation from the side of the fibre [3]. In this paper, we report the utilisation of a 45° tilted fibre Bragg grating (45°-TFBG) structure in cavity to realise single polarisation operation of a fibre ring laser.

Polarisation Characteristics of 45°-TFBGs

The 45°-TFBGs used in our experiment were UV-inscribed in hydrogen loaded B/Ge fibres using scanning phase mask technique with a 244nm UV source from a cw frequency doubled Ar+ ion laser. A phase mask with a period of 1800nm was rotated 33.7° with respect to the fibre axis to produce titled fringes of 45° in the fibre core and induce radiation response around 1550nm range. Limited by the size of the phase mask, the maximal length of the 45°-TFBG is only about 4mm. In order to obtain high polarisation extinction ratio, multiple short gratings were produced in tandem. Since most outcoupled s-polarised light does not return to the fibre, there should be no interference for light from different sects and the over-all effect will be simple accumulation of the attenuation to the s-polarised light. With such a concatenating technique, TFBG as long as 5cm was achieved.

For an initial polarisation dependent loss (PDL) investigation, we used a system as illustrated in figure 1a, which consists of a broadband source, a fibre polariser and polarisation controller, and an optical spectrum analyser. Figure 2a shows the PDLs for a 4.6mm long and a 4cm long grating measured with this system. The maximal PDL of 4.6mm grating is only 6dB at ~1.55µm, whereas the 4cm TFBG achieved a maximum value of 28dB. The reason why the entire PDL profile over 80nm is near-Gaussian-like is because the system can only be optimised at one wavelength at one time. In order to measure the true maximum PDL profile, the gratings were further investigated employing a commercial EXFO PDL characterisation kit with measuring principle shown in figure 1b which incorporates a tuneable laser allowing optimisation of the maximum PDL over 100nm tuning range, Figure 2b plots comparatively the maximum PDL profiles of a 4.6mm and a 5cm 45°-TFBG. It can be seen clearly that the PDL extinction ratio of

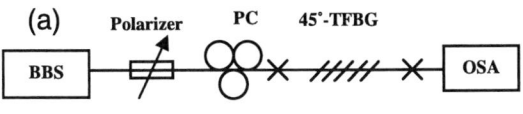

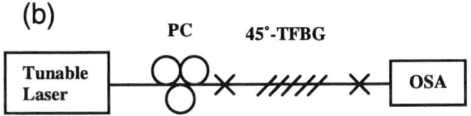

Figure.1. Schematic of PDL measurement system employing (a) a broadband source and (b) a tunable laser

the 5cm grating achieved ~33dB across entire 100nm range. This performance exceeds typical commercial devices which often be quoted with 30dB polarisation extinction ratio. By optimising the UV-inscription and increasing the grating length further, even higher (~40dB) PDL extinction ratio should be achievable. The remaining reflection at the fibre's cladding boundary can still causes some resonance to the PDL of the TFBGs, as demonstrated by both measurements to the 4.6mm grating. Consequently, it's no wonder that some unevenness exists for the PDL spectra of the concatenated gratings. It should be minimised by using large size phase masks or precisely controlling the phase stitches in the fabrication.

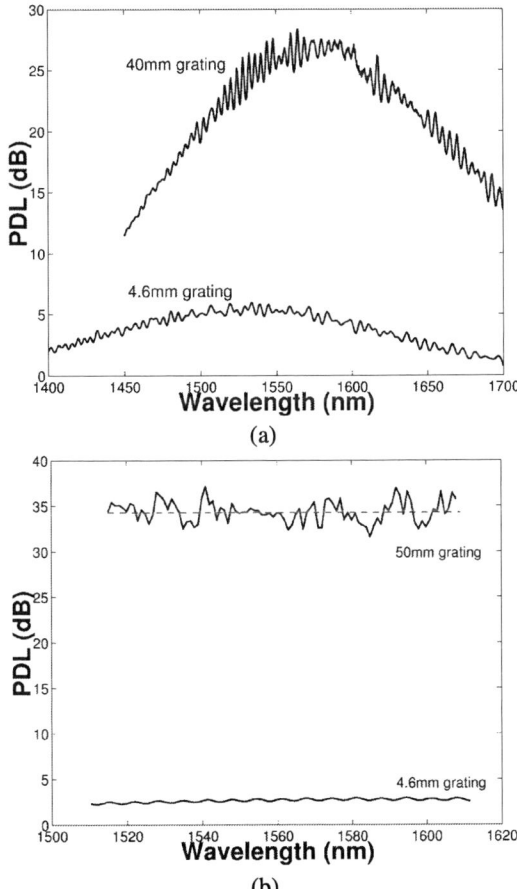

Figure.2.(a) PDL profiles of a 4.6mm and 4cm TFBGs measured using system shown in figure 1a. (b) PDL profiles of 4.6mm and 5cm TFBGs measured using EXFO PDL characterisation kit.

Degree of polarisation improvement to fibre ring laser

In order to evaluate the polarising function of 45°-TFBGs, we constructed a fibre ring laser (as shown in fig. 3) employing 10m Er fibre, two optical isolators, one 980/1550nm WDM to pump the ring laser with a 980nm diode and a 10:90 coupler for laser output. A circulator connected to a 1550nm FBG was also incorporated in the cavity to provide single frequency oscillation at 1550nm.

The degree of polarisation (DOP) measurement of the output of the fibre laser is conducted by the setup shown in the dotted line box in figure 3 and DOP calculation is based on the following expression:

$$\text{Degree of Polarisation (DOP)} = \frac{P_{POL}}{P_{POL} + P_{UNPOL}} \times 100 \ \%$$

where P_{POL} and P_{UNPOL} stand for the power of polarised and un-polarised parts of the light. Therefore, 100% DOP represents a single polarisation output.

Similar to the PDL measurement, adjusting polarisation controller can give the maximum ($P_{POL}+\frac{1}{2}P_{UNPOL}$) and minimum ($\frac{1}{2}P_{UNPOL}$) of the fibre ring laser. Without 45°-TFBG in the laser cavity,

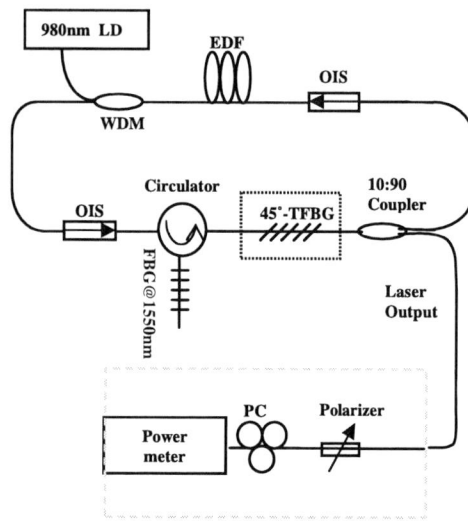

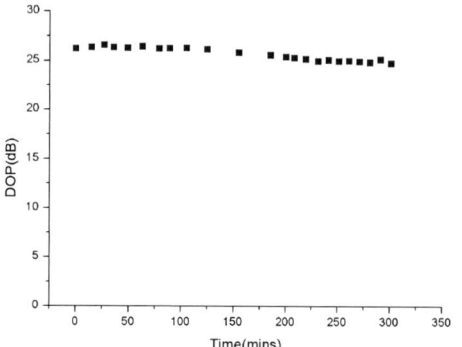

Figure.4. DOP long time stability experiment

Conclusions

We have characterised 45° in-fiber TFBGs which show strong PDL over 100nm spectral range. A 5cm 45°-TFBG was incorporated in a fibre ring laser and single polarisation operation with high DOP was achieved. Without the grating, the laser was randomly polarised indicated by its measured DOP of 22.48%. By inserting a 45°-TFBG, the laser output became highly polarised achieving a DOP larger than 99%. It was also demonstrated that the single polarisation state of the laser was very stable over 5 hours monitoring period.

Figure.3. Schematic diagram of the fibre ring laser structure. The polarisation degree of the laser output is measured using the setup shown in dotted box.

the fibre ring laser produces output with DOP only about 2dB, i.e. 22.48% in percentage term. This indicates that the laser output is almost unpolarised. When incorporating the 5 cm 45°-TFBG into the cavity, the laser output showed DOP of 35dB. They correspond to 99.94% in percentage term, which clearly show that the output of the laser is highly polarised and almost at single polarisation state. In comparison with 22.48% of the laser without intracavity 45°-TFBG, this is a remarkable improvement.

We also examined the stability of polarisation state of the fibre ring laser with the intracavity TFBG. The laser was monitored over 5 hours and the DOP variation against the monitoring period is plotted in figure 4. It can be seen that over 5 hours, the DOP value dropped by only ~2dB, which in percentage term corresponds to the change of 0.13%: an insignificant reduction.

References

[1] J.T.Lin and W.A.Gambling, "Polarization effects in fibre lasers: Phenomena, Theory and Applications," in Fiber Laser Sources and Amplifiers II Vol 1373 (SPIE 1990), pp.42-53.
[2] D.Pureur, M.Douay, P.Bernage, P,Niay, and J.F.Bayon, "Single-Polarization Fiber Using Bragg Gratings in Hi-Bi Fibers," IEEE Jour.Ligh.Techno. Vol.13, No.3 350-355(1995)
[3] K. Zhou, G. Simpson, X. F. Chen, L. Zhang, I. Bennion, Opt. Lett. 30, 1285(2005)

BRILLOUIN/ERBIUM FIBER LASER WITH PRE-AMPLIFIED BRILLOUIN PUMP USING RING-CAVITY CONFIGURATION

[1]N. Md Samsuri, [2]A. K. Zamzuri Member, IEEE and [3]M. A. Mahdi Senior Member, IEEE,

[1,2]Photonics Transmission Cluster, Telekom Research and Development Sdn. Bhd., Idea Tower1, UPM-MTDC, UPM 43400 Serdang, Selangor, Malaysia. E-mail:
[1]norhakimah@tmrnd.com.my
[2]zamzuri@tmrnd.com.my

[3]Photonics and Fiber Optic System Lab., Department of Computer and Communications Systems Engineering, Universiti Putra Malaysia. E-mail: mdadzir@upm.edu.my

Abstract *A Brillouin/Erbium fiber laser using a ring laser configuration with a pre-amplified Brillouin pump is demonstrated. This configuration generated 7 Brillouin Stokes with 0.086 nm spacing and the first-Stokes peak power is at 5.638 dBm.*

Introduction

Multiple wavelength fiber lasers have generated interest of many applications in the areas of optical sensors and WDM communication systems [1]. Apart of erbium-doped fiber laser (EDFL) and stimulated brillouin scattering (SBS) configuration techniques developed to generate multiple wavelength application, brillouin/erbium fiber laser (BEFL) has recently considered in multi wavelength laser source design. In BEFL, two types of media gain, non-linear Brillouin gain in SMOF and linear gain in erbium-doped fiber (EDF) are combined to generate a laser comb with ~10Ghz line spacing in various types configuration [2]. In this paper, we present the experimental results of a BEFL design using a ring cavity with a pre-amplified Brillouin pump. The effects of 980 nm pump power and Brillouin pump power parameters towards wavelength tuning range characteristics of this BEFL is investigated .

Experimental setup

The experimental setup for the ring-cavity Brillouin-Erbium fiber laser is illustrated in Fig. 1. The setup consists of 9.5m of Erbium-doped fiber (EDF), 10km of SMF-28, two optical circulators (Cir1 and Cir2), two 3 dB coupler (C1 and C2) and a 980/1550nm wavelength selective coupler (WSC). The linear gain medium is provided by the 9.5m Erbium-doped fiber and pumped by 980nm laser pump using a forward pumping scheme. The pump and the signal lights are multiplexed by the 980/1550nm wavelength selective coupler (WSC). Meanwhile, the Brillouin gain medium is provided by 10km of single-mode fiber, SMF-28.

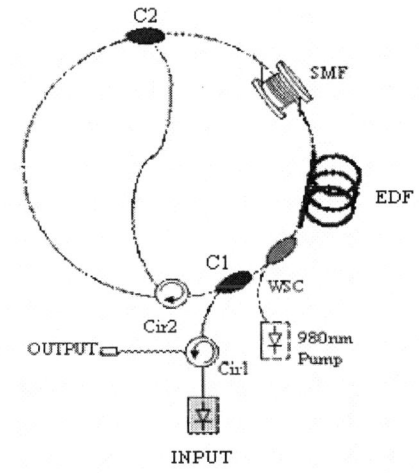

Fig. 1: Experimental setup for BEFL with pre-amplified BP in ring-cavity design.

A narrow linewidth signal from an external cavity tunable laser source (TLS) is used to provide a Brillouin pump (BP) from 1530 nm to 1569 nm. The BP is injected through port1 to port2 of Cir1 into the 3dB coupler (C1) in a counter clockwise (CCW) direction and being amplified by a forward pumping EDFA. The amplified BP is injected into the SMF-28 to create a narrow bandwidth Brillouin gain. This first-order Brillouin-Stokes (BS) signals generated at a frequency shifted by 0.086 nm from the BP frequency propagates in clockwise direction into a backward pumping scheme EDFA.

Through a 3dB coupler (C1), about 50% of the first-order BS signal is redirected into a ring resonator. If the total gain produced by stimulated Brillouin scattering (SBS) and EDF equivalent to the configuration cavity loss, a laser is formed and oscillates in the ring cavity between Cir2 and C2. This lower-order Brillouin-stokes signal can be utilized as a BP for a cascading generation of additional higher-order Brillouin-Stokes signal. The output of the system is measured and characterized by an optical spectrum analyzer through port 3 of Cir1.

Result and discussion

The output spectra of the proposed BEFL configuration, measured as a function of 980 nm pump power at fixed BP power of 3.25 mW are shown in Fig. 2. The first-order Stokes line emerges at 980 nm pump power of 20.04 mW and the wavelength of the Stokes signal generated shifted by +0.086 nm from the Brillouin pump with about - 0.744 dBm Stokes power. At a pump power of 101.31 mW, a total number of 7 Stokes lines are generated with the first-Stokes peak power at 5.638 dBm. Meanwhile about 4 anti Stokes signal are produced. This anti Stokes lines are due to the bidirectional operation and four-wave mixing in the SMF. The optical output spectra from this BEFL configuration is measured by an optical spectrum analyzer with a resolution of 0.01nm.

The wavelength tuning range characteristic of the proposed BEFL system as a function of 980 nm pump power is illustrated in Fig. 3. The BP was tuned at 1560nm with 3.25 mW signal power. At 20.04 mW pump power, the Stokes signals range was tuned to about 14.82 nm. However, the tuning range of the given BP decreases as the 980 nm pump power increases. This decrease is due to strong mode competition between the EDFL in cavity modes and the Stokes signals as the 980 nm pump power increases. In order to obtain a wider tuning range, the 980 nm pump power should be reduced to a lower value so the mode competition can be

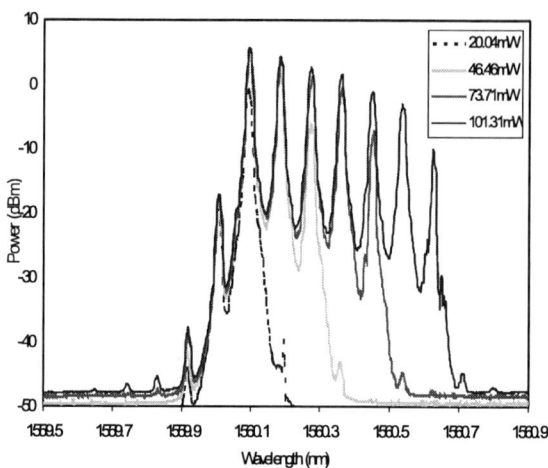

Fig. 2: Output spectra of BEFL at different EDF pump power.

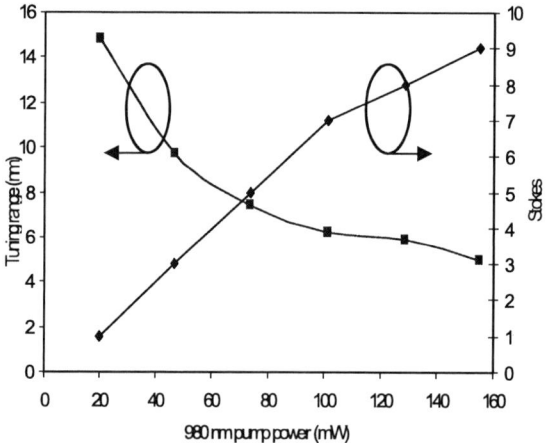

Fig. 3: Tuning range and number of Stokes signals as a function of 980nm pump power.

minimized [3]. Although higher 980nm pump power decreases the wavelength tuning number of Stokes signals generated. At 980 nm pump power of 155.06 mW, the number of Stokes generated was 9 Stokes lines. Increases in the pump power initiate cascaded Stimulated Brillouin Scattering (SBS) process therefore consecutively generating the high order Stokes signal [2].

The effects of different BP power on the wavelength tuning range and number of Stokes generated is depicted in Fig.4. This BP signal was set at 1560 nm and vary from 1.06 mW up to 3.25 mW. At a fixed 980 nm pump power of 46.46 mW, the lines of Stokes generated by a BP power of 1.06 mW are 8 lines of Stokes. Meanwhile at BP power of 3.25 mW about 3 lines of Stokes

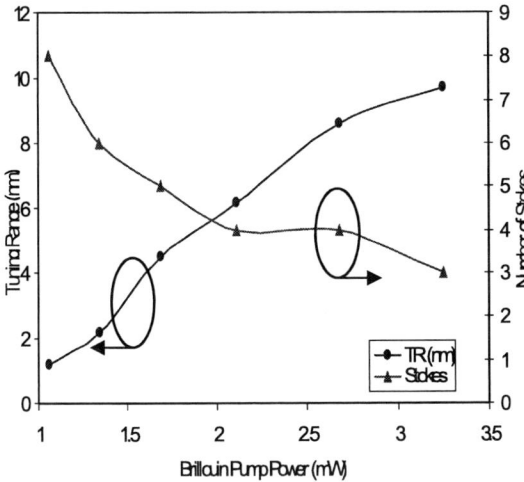

Fig. 4: The tuning range and number of Stokes signal measured for Brillouin pump from 1.06 mW to 3.25 mW.

created. A higher BP power resolves in a higher threshold level for the higher-order Stokes to become the next BP and to generate the next Stoke. This is due to stronger BP leads to gain saturation in the cavity. Hence, a higher 980 nm pump power is needed to compensate the gain saturation and fulfill the requirement to generate the next consequent Stoke [1]. As illustrated in Fig.4, the wavelength tuning range increases as the BP power increases. At BP of 3.25 mW, the wavelength can be tuned to a range of 8.60 nm while at BP of 1.06 mW the tuning range is about 1.20 nm. This explains by a higher BP results in a higher Brillouin gain. This stronger Brillouin gain will be suppress the EDFL in the cavity and hence allowed the Stimulated Brillouin Scattering to take place at a wider spectral range.

Conclusion

A laser comb consists of 7 lines multiple wavelengths with a 0.086 nm shift and a peak power of 5.638 dBm is demonstrated. The BEFL tuning range and Stokes signal characteristics of this ring cavity configuration is investigated. A tuning range of 14.82 nm was achieved at Brillouin pump of 3.25 mW and EDF pump of 20.04 mW. The tuning range is much wider at a smaller EDF pump power and a higher value of Brillouin pump at the cost of the number of Stokes generated. This is due to smaller EDF pump power generate a smaller oscillating EDFL in the ring cavity thus reduces the mode of competition between EDFL and Stokes signals. Meanwhile, higher BP power results in a lower number of Brillouin Stokes signal due to higher threshold level for the next Stokes signal.

References

1. G. J. Cowle et al J. Lightwave Technol., vol. 15, no. 7 (1997), pp. 1198-1204.
2. D. Y. Stepanov et al IEEE J. of Quantum Electron., vol.3, no. 4 (1997), pp.1049-1057.
3. D.S. Lim et al Electron. Lett., 34 (1998), pp. 2406-2407.
4. M. K. Abd-Rahman et al, TENCON 2000, vol. 3, pp. 71-74.
5. K.-D. Park et al OFC 2000, vol. 3, pp. 11-13.
6. M. H. Al-Mansoori et al Opt. Express, 13 (2005), pp. 3471-3476.
7. M. H. Al-Mansoori et al ICON 2005, pp. 74-76.

Impairment in Amplification of Optical Packets Regarding the Gain Transient and Nonlinear Effect Depending on Peak Power of NRZ payload

Y. Awaji, H. Furukawa, N. Wada

National Institute of Information and Communications Technology (NICT), 4-2-1, Nukuikita, Koganei, 184-8795 Tokyo, Japan, e-mail: {yossy, furukawa, wada}@nict.go.jp

Abstract *We classified impairments in amplification of optical packet by using EDFA which are derived from the gain transient and third order nonlinearity. Significant nonlinearity was observed even for NRZ formatted packet with moderate average power.*

Introduction

The fact that optical granular signal suffer dynamic gain transient from amplification by EDFA has been well known [1]. The most practical candidate solution to the transient was automatic gain controlling (AGC) of EDFA for long time [2], while bit rate was not so high and granularity was not so small. However, as a result of recent progress of optical packet switching, which treats smaller granularity with higher bit rate, it was revealed that conventional AGC was no longer sufficient for amplification of optical packets because it was necessary to take rather longer response time than average duration of optical packet [3].

We have proposed effective mitigation of the transient by static approach which could suppress the transient on optical packet shorter than 400ns, successfully [3]. We also found that the residual transient was affected by the variation of traffic density.

In this paper, we observed a dependency of the transient on average power and extinction ratio (ER) of optical-packet-modulation, which are important specifications, same as the traffic density, to describe the condition of traffic on physical layer of optical packet switching (OPS) network. These three aspects of traffic condition can be considered as concentration of optical energy on packet from the point of view of the transient controlling. Thus, it is important to take care of these aspects in the case of designing optical transport stratum.

We also observed significant third order nonlinearity probably caused in output

interface (I/F) of EDFA, which is ordinary standard single mode fiber (SSMF), by using extreme case of traffic condition. The nonlinearity occurred because of extensive concentration of optical energy even on NRZ formatted pulses with 100ps length. It means that there is concrete restriction for traffic condition especially in OPS network.

Principle of the gain transient and issue of amplification of optical packet

The principle of the gain transient of EDFA is approximately described by following formula [4],

$$G'(0) = \frac{[G(\infty) - G(0)]}{\tau_0}[1 + \sum_j \frac{P^{out}(\lambda_j)}{P^{IS}(\lambda_j)}] \quad (1)$$

$$P^{IS}(\lambda_j) = \frac{h\nu S}{[\sigma_a(\lambda_j) + \sigma_e(\lambda_j)]\Gamma_j \tau_0} \quad (2)$$

, where $G(0)$: gain before the transient, $G(\infty)$: steady gain after the transient, τ_0: intrinsic lifetime of the upper level of Er^{3+}, P^{IS}: intrinsic saturation power, S: active erbium area of EDF, σ_a and σ_e: absorption and emission cross section, Γ: confinement factor.

According to these formulas, the transient is the result of saturation of amplification. It is also obvious that increase of output power (P^{out}) tend to enhance the transient.

We have already reported the effect of the change of traffic density on the transient [3] which can be considered as follows. If traffic density becomes lower, the optical energy supplied from EDFA concentrate on optical packet while only the average power is monitored and kept constant. This is equivalent to increase of P^{out}. Of course, difference between $G(\infty)$ and $G(0)$ tends to increase in the case of lower traffic density which also nurture the transient.

Dependency on average power and ER

We show the case of increasing P^{out} depending on the increase of average power and increasing of ER in following section.

At first, we show comparison of two commercially available EDFAs and our prototype [3] with two cases of input average power injected into EDFA on single wavelength (1550.12nm) as shown in Fig. 1.

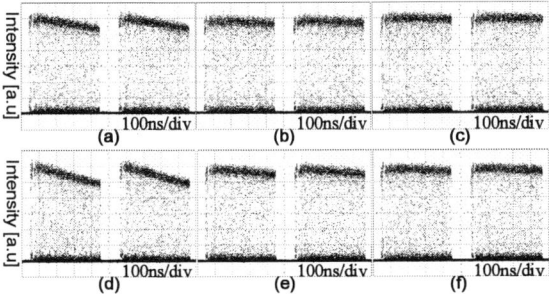

Fig. 1 Gain transient on single λ
(a,d) Maker A, (b,e) Maker B, (c,f) Our Prototype
(a)~(c) : input power -13dBm, traffic density 0.4%
(d)~(f) : input power -3dBm, traffic density 0.4%

The payload of packet was 10Gbps NRZ format and the length of packet was 4000bit (400ns). The traffic density was kept as 0.4%.

Our prototype could successfully mitigate the transient, hence the difference between two cases of average input power was almost negligible. On the other hands, the change of average input power affected significantly on two conventional EDFAs, in spite that -3dBm input is not so high value concerning that the maximum output power of these sample EDFA were almost +16dBm.

It was because the traffic density was quite low. In practical network, it is difficult to distinguish two traffic states, lower peak power with dense traffic and higher peak power with scarce traffic, by means of conventional optical performance monitoring which can capture only the average optical power. It is desirable to monitor average power and traffic density simultaneously to maintain the peak power of packet to appropriate value by adjusting of average power. However, it has to be noticed that these measures have limited

effect to mix of packets with various powers arriving from different paths.

Secondly, we observed about the impact of extinction ratio (ER) of packet-modulation to the transient. There is logically no signal on the period between packet and packet including guard time, but that blank time was physically occupied by residual DC component and ASE from EDFA. In such case, these garbage lights act as gain clamping light. Higher ER tends to weaken the garbage light.

If ER of injected packet stream into EDFA is extremely high, the optical energy from excited erbium ions tend not to be mostly consumed by the garbage light but to concentrate on packet in the case of lower traffic density. The transient will increase as the same manner that the traffic density affected to the transient.

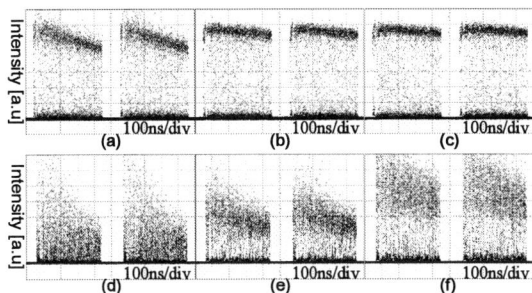

Fig. 2 Gain transient on WDM
(a,d) Maker A, (b,e) Maker B, (c,f) Our Prototype
(a)~(c) : EA modulator, 1550.12nm of 8λ x 100GHz
(d)~(f) : LN modulator, 1554.13nm of 9λ x 100GHz

Fig. 2 shows the comparison of two cases of packet modulators which generate optical packet stream on transmitter. One was electro-absorption (EA) modulator with ER of 10dBm and another was lithium-niobate (LN) modulator with ER of 23dBm. The total injection power was -3dBm and the traffic density was 0.4% for each case. In this experiment, we used 8-λ and 9-λ WDM with 100GHz spacing. The number of wavelength channel and location were slightly different for each case of modulator, however, the transient condition can be fairly compared concerning emission and absorption cross sections and average power per channel. In these cases, the ratio of total energy of amplified packet and amplified garbage light could be estimated as 1:2.5 for LN and 1:50 for EA, respectively.

The effect of high ER was serious in the case of lower traffic density, either. It was not only significant enhancement of the transient on each EDFA but also the nonlinear effect. Even our prototype could not mitigate these impairments. These discussions about ER are also applicable to signal-to-noise ratio (SNR) of the traffic as same manner.

As a matter of fact, it was rather unfair to compare nonlinear impairment by using Fig. 2 because EDFAs of Maker A and B have had output interface of dispersion-shifted fiber (DSF). Hence, such nonlinearity was not so strange if extreme high power concentrate on pulses. And what is more, it is not related with elementary process of the transient occurring inside EDF as shown later. Until this section, the important point is the degree of transient. About nonlinearity, we present further investigation in the next section.

Nonlinearity beside amplification of optical packet

We observed nonlinearity with our prototype which had output I/F of about 10m of SSMF as shown in Fig. 3.

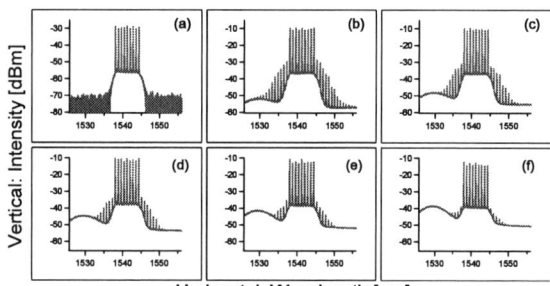

Fig. 3 Nonlinear effect in output I/F
(a) Input spectrum to EDFA, (b)~(f) Output spectra
Average input powers were, (b) -3dBm, (c) -6dBm,
(d) -9dBm, (e) -12dBm, (f) -15dBm, respectively

The spectra shows that four wave mixing occurred in EDFA. We also observed that the change of spectra according to input power behaved like spectral broadening caused by self-phase modulation (SPM). We confirmed that these nonlinear phenomena occurred mainly in SSMF-I/F by using cut-back method.

We were rather astonished by the result and considered the reason as follows. The payload was modulated by 10Gbps (~0.08nm between both sidebands) NRZ format and the one symbol has 100ps duration while residual dispersion derived from SSMF I/F was about 0.17ps/nm. Therefore, walk-off was ignorable and phase-match was maintained well.

The important point is that such nonlinearity does not related with EDFA directly but closely related with the modulation format and density of traffic and ER (and sometimes SNR). In other words, network designer have to take more care of this issue than manufacture of EDFA.

Conclusions

We observed a dependency of the transient on average power and extinction ratio (ER) of optical packet, which are also important specifications to describe the condition of traffic on physical layer of optical packet switching (OPS) network as same as traffic density.

We also observed significant third order nonlinearity caused in output I/F of EDFA, which is ordinary SSMF, by using extreme case of traffic condition. The NRZ modulation format is more effective to avoid such concentration of energy than RZ format because of longer duration of bit symbol, however, it resulted in tolerable phase-matching of FWM crosstalk ironically. It means that there is concrete restriction for traffic condition especially in OPS network.

The amplification of optical packet is sensitive issue than ordinarily expected. It is strongly recommended to design the optical transport stratum as to monitor the density of traffic simultaneously with the average power.

References
1 A. A. M. Saleh et al IEEE Photon. Technol. Lett., vol.2, no. 10 (1990), pp714-717.
2 H. Nakaji et al Optical fiber technology, vol. 9 (2003), pp25-35.
3 Y. Awaji et al Proc. CLEO2007, (2007), JtuA133.
4 Y. Sun et al Applied optics, vol. 38, no. 9 (1999), pp1682-1685.

InP-based Photonic Integrated Devices

Shinji Matsuo, Hiroyuki Ishii, Toru Segawa, Takaaki Kakitsuka, and Hiromi Oohashi
NTT Photonics Laboratories, NTT Corporation,
3-1 Morinosato Wakamiya, Atsugi, Kanagawa, Japan,
mash@aecl.ntt.co.jp

Abstract *Monolithically integrated tunable lasers have been developed for use in R-OADM, burst switching, and packet switching systems. By using monolithic integration technologies, the optimal device configuration can be constructed for each application.*

Introduction

Widely tunable lasers have been developed because they are important devices for wavelength division multiplexing (WDM) networks. For example, tunable lasers are expected to provide back-up sources and replacements for fixed wavelength sources in current dense-WDM systems to reduce inventory costs. Furthermore, they are the key to realizing the next generation of intelligent all-optical networks employing wavelength routing, such as reconfigurable optical add/drop multiplexer (R-OADM) systems, optical burst switching (OBS) systems and optical packet switching (OPS) systems. The required device characteristics, such as the wavelength switching time and wavelength accuracy during the wavelength switching operation, differ depending on the application.

Recent progress on monolithic integration techniques for InP-based photonic integrated devices, such as selective regrowth and dry etching techniques, enable us to construct a variety of optical components on the same wafer. A 100-ch WDM channel selector [1] and an 8-ch transmitter array [2] consisting of AWG filters and SOAs have been developed. By using these technologies, we can fabricate a device that has the most suitable tunable laser configuration for each application.

In this presentation, we describe three types of monolithically integrated tunable lasers; a tunable DFB laser array (TLA) [3] for R-OADM systems, a digitally tunable laser using a ladder-type filter and a ring resonator [4, 5] for OBS systems, and a double-ring-resonator-connected laser [6, 7] for OPS systems.

Tunable DFB Laser Array (TLA)

The TLA is suitable for use as a back-up light source for dense-WDM systems and as a wavelength routing light source for R-OADM systems. With these applications, the required

switching time is in the millisecond or second range. Thus, the TLA is the most promising tunable source since it is based on commercially available DFB laser technology. Wide tunability is achieved by means of appropriate LD selection and temperature control. The simple tuning scheme and mode-hop free tuning ensure its stable and reliable wavelength operation.

Figure 1 is a photograph of the device. Twelve 450 µm-long λ/4-shifted DFB lasers, a 12 x 1 funnel combiner, and a 1200 µm-long SOA with an angled output port are monolithically integrated on an InP substrate. Figure 1 also shows the structure of the funnel combiner, which consists of twelve funnel-shaped input waveguides, a slab waveguide, and an output waveguide. The width of the outer input waveguides is designed to be wider than that of the inner input waveguides to realize a constant coupling efficiency in the ports. The total size of the TLA chip is 620 x 2666 µm².

The TLA chip was packaged in a 26-pin butterfly-type module together with a wavelength locker. A fiber output power of more than 40 mW (16 dBm) was obtained even at a high LD temperature of 50 °C. The wavelength tuning characteristics are shown in Fig. 2 (a). The lasing frequency spacing of the two adjacent lasers is set at 400 GHz, where the grating pitch is precisely controlled by EB lithography. By selecting an LD and changing the LD temperature from 15 to 50 °C, we

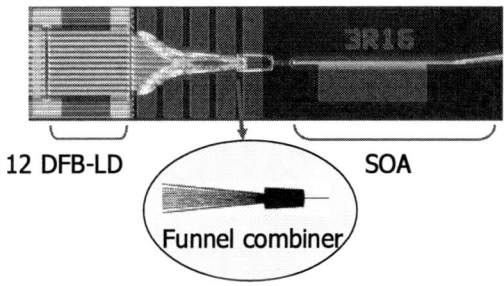

Fig. 1 Photograph of tunable DFB laser array.

978-0-9789217-3-6/07/$25.00

©2007 WEN GLOBAL SOLUTIONS

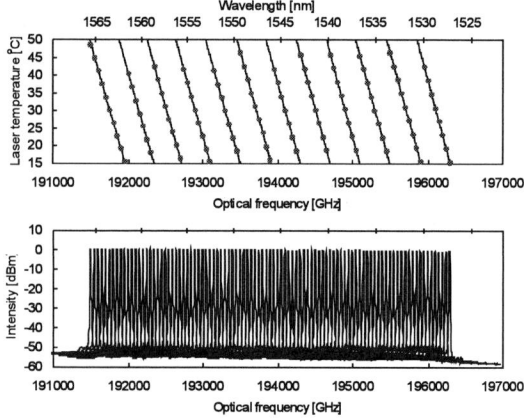

Fig. 2 Wavelength tuning characteristics of the device.

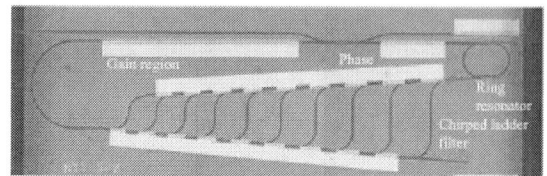

Fig. 3 Photograph of a digitally tunable laser with a ladder filter.

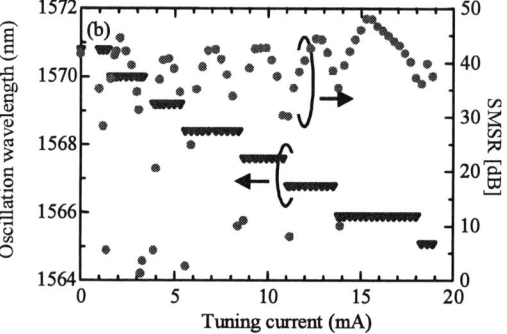

Fig. 4 Lasing wavelength and SMSR dependence of the injection current of the ladder filter.

achieve a laser module that covers a ~38 nm range. This enables us to set 97 ITU-T grid channels with a spacing of 50 GHz in the C-band. The solid circles in Fig. 2 (a) indicate the conditions for the 97 channels. The lasing spectra of the 97 channels are shown in Fig. 2 (b). Single-mode operation with SMSR >50 dB is confirmed under all conditions.

Digitally tunable laser using ladder-type filter and ring resonator

A tunable laser for an OBS system requires a fast switching time in the microsecond range and there must be little wavelength change during the wavelength switching operation because the optical burst is transmitted in backbone and metro-networks. Current injection into the wavelength control region is the most promising way to obtain fast tuning with a wide tuning range. However, this presents such problems as complicated current control for multi-electrodes and frequency drift caused by thermal transients. Digitally tunable lasers have great advantages in terms of simple operation and stable wavelength control. Thus, we have developed such a laser in which a widely tunable ladder filter is combined with a ring resonator.

Figure 3 is a photograph of our digitally tunable laser. It consists of a chirped ladder filter, a gain region, a phase control region, and a ring resonator. The widely tunable ladder filter selects one channel from the periodic channels of the ring resonator. Thus, it is easy to control the lasing wavelength. The device was fabricated monolithically by butt-coupling the active and passive waveguides. The gain region has a ridge waveguide structure whereas the ladder filter, the ring resonator, and the phase control region have a deep-ridge waveguide structure. The deep-ridge waveguide structure was formed by Cl_2 ICP-RIE, and the waveguide was 1.6 μm wide and 4.0 μm high.

Figure 4 shows the lasing wavelength and the SMSR as a function of the injection current of the ladder filter. In this experiment, the currents of the gain region and the phase control region were kept constant at 99.0 and 0 mA, respectively. The lasing wavelength abruptly shifts by 100 GHz and the SMSR is more than 40 dB except in the wavelength transition region. This means that this device provides a robust wavelength tuning mechanism. By changing the injection current to the upper or lower waveguide, we successfully achieved 31-channel digitally tunable laser operation with a 100-GHz-spacing.

Figure 5 shows experimental results for the oscillation frequency change caused by the thermal transient. We changed the injection current of the ladder filter from 0 to 14.5 or 40.3 mA at intervals of 300 μsec. In this experiment, we used a Mach-Zehnder interferometer with an FSR of 1 GHz to monitor the oscillation frequency change. This change was 4.9 nm for a current swing of 14.5 mA and 12.6 nm for a current swing of 40.3 mA. The oscillation frequency change was increased with the

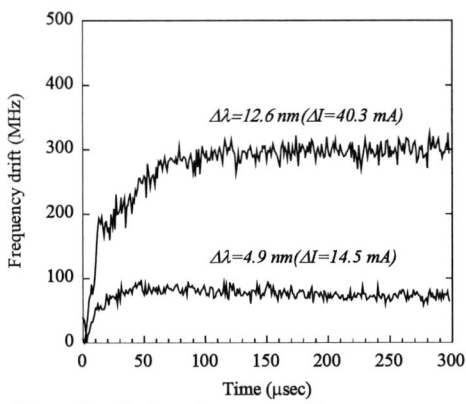

Fig. 5 Oscillation frequency change caused by the thermal transient.

injection current. However, the freqency change was about 300 MHz when the injection current was 40.3 mA.

Double-Ring-Resonator-Connected Laser

Tunable lasers for OPS systems are used by combining wavelength converters (WCs) and an arrayed waveguide grating (AWG) filter. In this context, the monolithic integration of tunable lasers, WCs and AWG filters is desired. Therefore, it is important to develop a small tunable laser with a simple structure. Furthermore, it is very important in terms of the monolithic integration of these components that the same fabrication process be used for both the filter section of the tunable laser and the AWG filter.

Figure 6 shows a photograph of the fabricated tunable laser. The laser consists of a gain region, a phase control region, and two microring resonators with different FSRs to achieve a wide tuning range. Each ring resonator is coupled with a straight waveguide via an optical coupler. In this device, only light at the resonant frequency of the two ring resonators is reflected by the back facet and then input again into the gain region through the ring resonators. Thus, the lasing frequency can be selected by controlling the resonant frequency of the ring resonators. The FSRs of the two ring resonators are 483 and 520 GHz, which correspond to ring radii of 13.6 and 11.9 μm, respectively.

The tuning characteristics of the fabricated laser are shown in Fig. 7. When the injection current of the gain region was 94 mA and the current for both ring resonators was 0 mA, lasing was observed at 1532.7 nm. The injection current of Ring2 was increased from

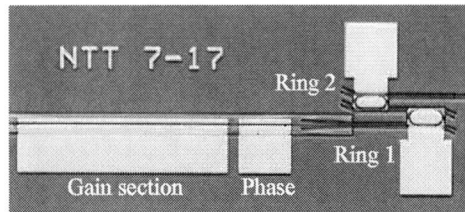

Fig. 6 Photograph of double-ring-resonator-connected laser.

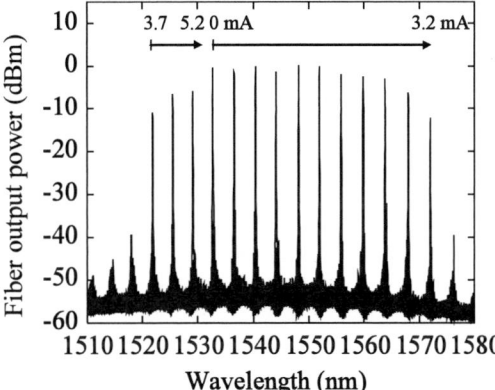

Fig. 7 Tuning characteristics of the device.

0 to 5.2 mA, while keeping that of Ring1 at 0 mA. As shown in this figure, the total tuning range was 50.0 nm and every lasing channel exhibited an SMSR larger than 30 dB. This small tuning current is important when fabricating a large scale integration device in terms of low power consumption and temperature management for maintaining the lasing wavelength.

Conclusion

We have proposed and developed three types of tunable lasers. The TLA is the most suitable tunable laser for use in dense-WDM and R-OADM systems. For future systems, monolithic photonic integration is an important technology for constructing functional devices.

References

1. N. Kikuchi et al., *IEEE PTL*, vol. 16, no. 11, pp. 2481-2483
2. Y. Suzaki et al., *JSTQE*, vol. 11, no. 1, 2005, pp. 43-49.
3. H. Ishii et al., *ISLC2006*, pp. 13-14
4. S. Matsuo et al., *JSTQE*, vol. 11, no. 5, 2005, pp. 924-930.
5. S. Matsuo et al., *OFC2006*, OWL3.
6. S. Matsuo et al., *ISLC2006*, pp. 21-22.
7. T. Segawa et al., *IPRM2007*, FrB2-3, pp. 598-601.

Analysis of Deep Etched Trench in Planar Optical Waveguide by FDTD Method

Jun Wang, Mingyu Li, and Jian-Jun He

State Key Laboratory of Modern Optical Instrumentation, Center for Integrated Optoelectronics,
Zhejiang University, Hangzhou, PR China 310027
dyingsun_1985@163.com

Abstract *A finite-difference time-domain (FDTD) method is presented for simulating deep etched trench structures in planar optical waveguides. Numerical results obtained from FDTD simulations and from the Transfer Matrix Method (TMM) are compared. The effect of sidewall verticality is analyzed.*

Introduction

With the advancement of dry etching technology, deep etched trecnhes have received increasing interest in the design of integrated photonic devices such as Q-modulate semiconductor lasers [1], photonic crystals, and optical filters. Therefore, it is necessary to study rigourously the optical performances of the trenches. In previous researches Transfer Matrix Method (TMM) has been commonly used to simulate these structures [2]. However, as demonstrated in this paper, it only produces approximate results. Besides, it cannot treat fabrication defects such as sidewall non-verticality and roughness. In this paper, we present a more accurate model and simulation results by using the finite-difference time-domain (FDTD) method [3].

The deep etched trench considered in this paper is schematically shown in Fig.1. The layer structure is a typical InP-based laser waveguide consisting of a buffer (InP), an InGaAsP multiple quantum well waveguide core and an upper cladding layer (InP). The operating wavelength is 1550nm.

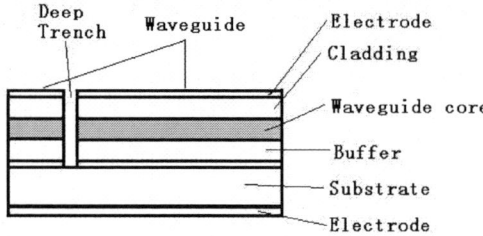

Fig.1. Schematic of the deep etched trench and the waveguide structure.

Simulations by TMM

A popular method widely used in the simulation of semiconductor lasers is the transfer matrix method [3], in which the waveguide strcuture is divided into segments along the propagation direction. Each of the segments is characterized by a transfer matrix. The optical performance of an air trench with vertical sidewalls can be easily obtained if the loss due to the beam divergence in the air trenched is neglected or is linearly proportional to the propagation distance.

To obtain the relationship between the reflectivity/transmission characteristics of the trench and the trench width, the multilayer structure is first translated into a waveguide with an effective index of 3.215. The real part of the effective index of the air trench is n=1. To account for the beam divergence loss of the air trench, a linear loss parameter α is introduced which is derived from the single pass loss using the beam propagation method. It varies from α = 280cm^{-1} for 0.4μm trench to α = 655 cm^{-1} for 5 μm trench. Fig. 2 shows the reflectivity, transmission and loss as a function of the trench width. As expected, the reflectivity reaches a maximum value and the transmission reaches a minimum value when the trench width is an odd number of quarter wavelength, and the opposite is true when the trench width is an even number of quarter wavelength.

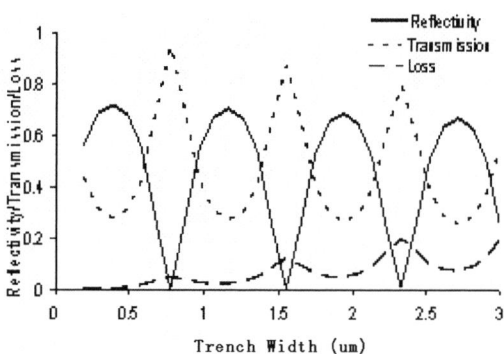

Fig.2. Reflectivity, transmission and loss as a function of the trench width simulated by TMM.

Simulations by FDTD

The FDTD method is also a well-known method in the study of electromagnetic problems. It is based on the Maxwell's equations with the computational space and time discretized into rectangular cells and intervals, respectively. It can simulate almost arbitrary structures with a high precision.

The time-dependent Maxwell's equations in a two-dimensional 2D case can be written in the following differential forms:

$$\nabla \times \boldsymbol{E} = -\mu_0 \frac{\partial \boldsymbol{H}}{\partial t} \tag{1}$$

$$\nabla \times \boldsymbol{H} = \varepsilon(\boldsymbol{r})\frac{\partial \boldsymbol{E}}{\partial t} \qquad (2)$$

where μ_0 is the permeability of the material and t is the time. Each discretization cell has lengths $\Delta x = \Delta y = 5nm$ along the x and y directions respectively. In the discretization space each grid point is denoted by i and j, where both i and j are intergers. The field components are denoted by $E_z\big|_{i,j}^n, H_x\big|_{i,j+1/2}^{n+1/2}$ and $H_y\big|_{i+1/2,j}^{n+1/2}$ for the TE polarization case, and by $H_z\big|_{i,j}^n, E_x\big|_{i,j+1/2}^{n+1/2}$ and $E_y\big|_{i+1/2,j}^{n+1/2}$ for the TM polarization case. Finally we obtain the following FDTD formulas for time stepping for TE polarization:

$$H_x\big|_{i,j+1/2}^{n+1/2} = H_x\big|_{i,j+1/2}^{n-1/2} - \frac{\Delta t}{\mu_0}\left(\frac{E_z\big|_{i,j+1}^n - E_z\big|_{i,j}^n}{\Delta y}\right)$$

$$(3)$$

$$H_y\big|_{i+1/2,j}^{n+1/2} = H_y\big|_{i+1/2,j}^{n-1/2} + \frac{\Delta t}{\mu_0}\left(\frac{E_z\big|_{i+1,j}^n - E_z\big|_{i,j}^n}{\Delta x}\right)$$

$$(4)$$

$$E_z\big|_{i,j}^{n+1} = E_z\big|_{i,j}^n + \frac{\Delta t}{\varepsilon_{i,j}}\left(\frac{H_y\big|_{i+1/2,j}^{n+1/2} - H_y\big|_{i-1/2,j}^{n+1/2}}{\Delta x}\right.$$

$$-\left.\frac{H_x\big|_{i,j+1/2}^{n+1/2} - H_x\big|_{i,j-1/2}^{n+1/2}}{\Delta y}\right) \qquad (5)$$

In a similar manner, we can obtain the FDTD formulas for time stepping for the TM polarization case.

FDTD results for reflectivity and transmission are obtained by launching a sinusoidal wave of a certain frequency into the input waveguide with a fundamental mode distribution that is calculated using FDM [4]. The computation domain contains 400*400 grid points with perfectly matched layer (PML) absorbing boundary condition imposed on the edges of computational domain. The total power flux through the output waveguide cross-sectional plane is monitored. An overlap integral with the fundamental mode of the waveguide is used to obtain the coupling efficiency.

$$\eta = \frac{\left|\int E_b(y,z)E_a^*(y)dy\right|^2}{\int E_b(y,z)E_b^*(y)dy \int E_a(y,z)E_a^*(y)dy} \qquad (6)$$

where E_a is the mode distribution of the waveguide, E_b is the electric fields along the output plane. This integral represents the proportion of energy coupled into the waveguide after a long distance propagation. This allows us to reduce the computational area and consequently save time. The amplitudes and phases of fields are obtained by using phase-delay method. The propagation of power is given by the Poynting vector $S = E \times H$. Because most detectors

measure power averaged over some time and a finite space, we use a time average of power flow $\langle S \rangle$ and then integrate it over a surface to obtain the total power $\langle S' \rangle$ flowing through that surface as the power of the wave in that surface:

$$\langle S \rangle = \frac{1}{T}\int_0^T S(t')dt' = \frac{1}{T}(E \times H^\bullet) \qquad (7)$$

$$= \frac{n}{2Z_0}E^2$$

$$\langle S' \rangle = \oint_A \langle S \rangle dA = \int \langle S'(y,z)\rangle dy \qquad (8)$$

where T is the time period of the wave and Z_0 is the impedance of free space. The reflectivity and transmission are caculated as:

$$R = \frac{\langle S' \rangle_r}{\langle S' \rangle_a}\eta_r \qquad (9)$$

$$T = \frac{\langle S' \rangle_t}{\langle S' \rangle_a}\eta_t \qquad (10)$$

Fig. 3 shows the reflectivity, transmission and loss as a function of the trench width from the FDTD simulations. One can see that when the trench is narrow (e.g. less than a quater-wavelength), the results obtained from FDTD and TMM are almost the same. For wider trench, the dissimilarity becomes notable. Particalularly, the transmission peaks from the FDTD methd are much lower compared to the TMM results, while the loss peaks are higher. The differences in the reflectivity peaks are less significant.

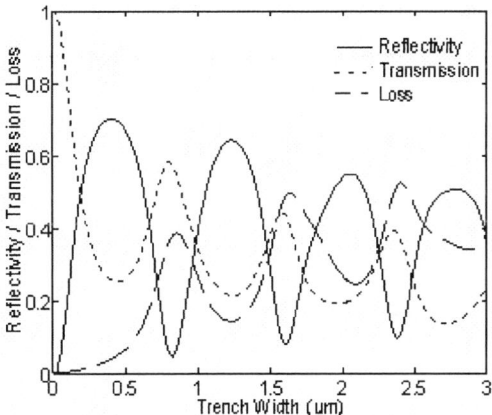

Fig.3. Reflectivity, transmission and loss as a function of the trench width simulated by FDTD.

In the TMM simulations, the incident wave is simplified as a plane wave. The multilayer waveguide is treated as equivalent to a homogeneous material characterized by an effective index, and the deep etched trench is simplified as an air slab. The loss due to the beam divergence in the air trench can only be characterized by the α parameter which treats the loss as linearly proportional to the progation distance. It is derived from the propagation loss in a sigle pass in the trench. These underestimate the loss because the beam diverges in both x and y directions. In the case of FDTD simulation, the actual structure is considered and the wave propagation back and forth in the trench is taken into account accurately, thus leading to a larger loss compared to the TMM results. The transmission peaks corresponds to the resonance condition of the trench for which the beam undergoes a larger number of reflections in the trench as compared to the anti-resonant condition corresponding to the reflectivity peaks. Therefore, the effect of beam divergence loss as reflected in the difference between the TMM and FDTD results is more significant for the transmission peaks compared to the reflectivity peaks.

From Fig.3, one can also see that the transmission peaks are not located exactly at the same position as the reflectivity minima, contrarily to the TMM result of Fig. 2. This is because the waveguide fundamental mode is used in the FDTD simulation instead of a plane wave in the case of TMM. When the fundamental mode propagates into the trench, the wave will diverge along different directions, and therefore have different optical path lengths in the trench. For a given trench, the averaged path length is slightly larger for the transmission waves as compared to the reflected waves. Therefore, the transmission peaks are located at slightly narrower trench widths in Fig. 3 as compared to the reflectivity minima, and this difference increases with the trench width.

Simulations of trenches with sloped sidewalls

Due to the limitation of fabrication technology, perfect trenches are hardly realized. The sidewalls of trenches are usually sloped, with wider opening at the top and narrower bottom. Sloped sidewalls may reflect or refract a part of the energy towards the substrate and the air, thus causing extra losses. TMM cannot be used to simulate such complicated structures. To simulate the influence of the sloped sidewalls, the FDTD method is necessary. Fig. 4 illustrates the variations of the reflectivity, transmission and loss as a function of the sidewall angle (deviation from the vertical) for a trench width equal to $5/4\lambda$. It shows that if the angle is smaller than $2°$, the influence is negligible; otherwise the extra loss becomes notable and should be considered. It is also found that this sidewall angle tolerance decreases with the trench width.

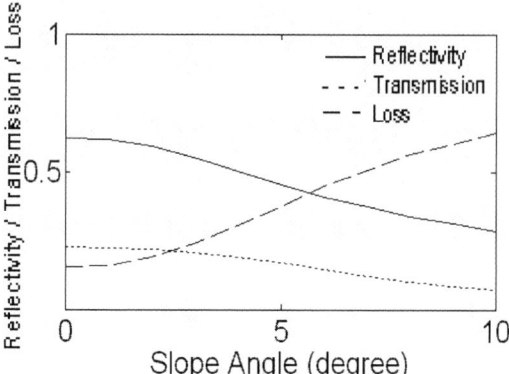

Fig.4. Reflectivity, transmission and loss versus sidewall angle simulated by FDTD.

Conclusion

FDTD simulations for deep etched trenches in optical waveguides are presented. The results are compared with those obtained from TMM method for the case where the trench has vertical sidewalls. The analysis shows that the TMM only gives approximate results. The difference between the results obtained using TMM and FDTD increases with the trench width due to the fact that the TMM method cannot treat accurately the beam divergence in the trench. The effect of the sidewall angle is also analyzed using the FDTD method, which cannot be modeled by TMM.

Acknowledgement

This work is supported by open project foundation of state key laboratory of modern optical instrumentation (No. LMOI-0601)

References

[1] Jian-Jun He, "Proposal for Q-Modulated Semiconductor Laser," IEEE Photon. Technol. Lett., vol. 19, no. 5, pp. 285-287, March, 2007.

[2] M. G. Davis and R. F. O'Dowd, "A Transfer Matrix-Based Analysis of Multielectrode DFB Lasers," IEEE Photon. Technol. Lett., vol. 3, no. 7, pp. 603-605, July 1991.

[3] Allen Taflove and Susan C. Hagness,, "Computational Electrodynamics: The Finite-Difference Time-Domain Method," 2nd ed., Artech House, Inc., 2000.

[4] Chen Fangrong, Liu Shuzhe and He Sailing, "A Full-vectorial Finite-difference Method Based on Interpolation for Modeling Step-index Optical Waveguides," Acta Photonica Sinica, vol. 32, no. 7, pp. 777-781, July, 2003.

Fabrication of polymer integrated optical microring resonator with photobleaching method

Jun Zhou[1], Anan Pyayt[2], Jingdong Luo[2] and Antao Chen[2], Nam Quoc Ngo[3]

1. Department of Physics, Ningbo University, Ningbo 315211, Zhejiang, P. R. China
2. Applied Physics Laboratory, University of Washington, Seattle WA, 98105, USA
3. Photonics Research Centre, School of Electrical & Electronic Engineering, Nanyang Technological University, Nanyang Avenue, Singapore 639798

Abstract An optical-quality polymer film based on the AJL8/APC material is used for the fabrication of a single-mode microring resonator with photobleaching method. The measured transmission responses of the microring resonator present a high Q-factor of device.

Introduction

In the past two decades, polymer optical integrated circuits (POICs) have recieved more attention and investigation towards the goal of available commercially applications. Recently polymeric micro-ring resonators have attracted a great deal of intersting and demonstrated their applications fiber optical networks, for example, wavelength filter, multiplexing, demultiplexing, routing and high-speed modulation with low switching voltage [1]. To obtain high figures of merits for polymeric micro-ring resonators, several fabrication techniques such as reactive ion etching (RIE) [2], photolocking [3], photobleaching [4] and electron beam etching [5] have been used in the process. Among them, photobleaching is one of the most favorable techniques that can be used to change the physical and optical properties of nonlinear polymeric films such as, color, refractive index, electro-optic (OE) coefficient and absorption coefficients by ultraviolet (UV) irradiation. Compared with the RIE technique, one unique advantage of the photobleaching method is its smaller fabrication steps that allow the fabrication of high-quality optical waveguides with smaller scattering losses, which can be large using the photolithography and dry etching processes. The mechanism of the photobleaching method relates to a trans-cis conformational transformation, which reduces the refractive index of the polymer film that is doped with the nonlinear chromophore [6]. In this paper, we report the polymer materials, namely, AJL8/amorphous polycarbonate (APC) and photobleaching technique have been used to fabricate an all-polymer optical microring resonator.

Design and fabrication

In our design, the microring resonator consists of a racetrack ring waveguide and a straight waveguide that acts as both input and output ports and is closely coupled with the ring waveguide, as shown in Fig. 1. Based on the configuration, some optimal parameters of the resonator have been determined by the design process and numerical simulation with the software BeamPro 5.0. Using the relative refractive index difference of $\Delta = 3.1\%$ (which is a typical value of the polymer film of the AJL8/APC material) between the core and cladding of the straight waveguide and the ring, the widths of the waveguide and the ring must be selected as $4 \ \mu m$ to achieve single-mode operation. To obtain a critical coupling condition, the following design parameters of the microring resonator have been chosen, namely, the ring radius is $R = 200 \ \mu m$, the lengths of the side waveguides are $80 \ \mu m$ and the gap between the ring and the straight waveguide is $d = 1.2 \ \mu m$.

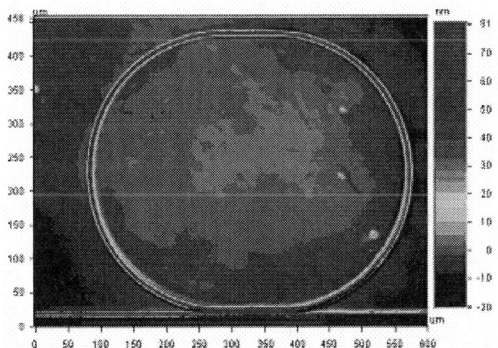

Fig. 1 Configuration of the microring resonator

The microring resonator was fabricated based on the polymer film of the AJL8/APC material using the photobleaching technique [7]. The polymer film was prepared by the spin-coating method. The polymer solution consists of 14wt.% concentration of the AJL8/APC material in cyclopentanone. The ratio of AJL8 to APC is 25wt.% chromophore loading in a polymer matrix. The thickness of the film was measured to be $2.3 \ \mu m$ after the solution was coated on a

978-0-9789217-3-6/07/$25.00
©2007 WEN GLOBAL SOLUTIONS

silica-on-silicon substrate with spinning rate 900 rpm and then baked for two days at 85^{0}C. Figure 2 shows the dependence of the refractive index of the polymer film on the UV light energy was measured using a prism coupling technique (Metricon 2010). From Fig. 2, a particular value of the refractive index of the film (depending on the resonance requirements such as its spectral response) can thus be achieved by using an appropriate amount of the UV light energy. Using the designed parameter values as described above, a chromium-quartz mask was fabricated and used to transfer the pattern of the designed microring resonator to the prepared polymer film during the photobleaching process [8,9]. The fabricated microring resonator and its surface profile are shown in Fig. 3.

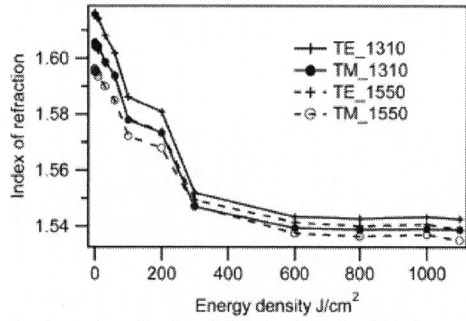

Fig. 2. Change of the refractive index of the AJL8/APC film as a function of the UV light photobleaching energy

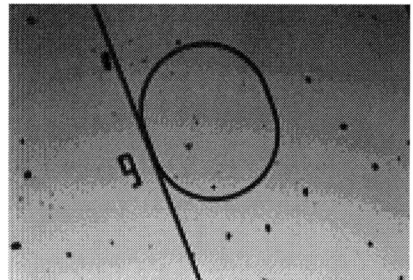

Fig. 3 Fabricated sample of the polymer microring resonator

Measurement results and discussion

Figure 4 shows the experimental setup used for measuring the characterization of the fabricated microring resonator. A single-polarization light from a tunable laser diode was coupled through a single-mode fiber into the input straight waveguide of the resonator and the output light from the resonator was coupled through a single-mode fiber out to an optical spectrum analyzer (OSA) (which is connected to a power

meter) so that the spectral response of the resonator can be measured. The two segments of the single-mode fibers used have a mode-field diameter of about 9 μm. The polarization controller was used to adjust the polarization state of the light from the laser diode to ensure that the light going into the resonator is of single polarization.

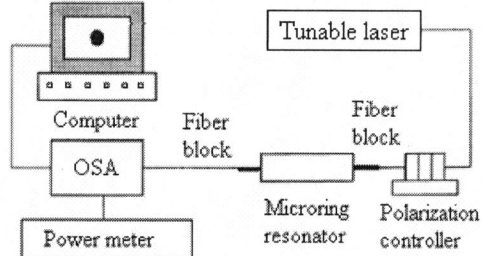

Fig. 4 Schematic diagram of the measuring experimental setup

A straight waveguide was tested and confirmed to be single moded operation in accordance with the design. And the propagation loss of the straight waveguide was measured using the cut-back method [10]. A polarizer was added to distinguish the TE and TM modes launched into the waveguide and a Newport optical power meter was used to record the transmitted-light intensity. The propagation loss of the waveguide is given by

$$T_{Loss} = \left|10\log(P_1 / P_2)/(L_1 - L_2)\right|$$

where L_1, L_2 and P_1, P_2 are the lengths and transmitted power levels before and after the waveguides have been cut, respectively. Fig. 5 shows the dependence of the measured propagation loss on the length of the waveguide (with a cross-sectional area of about 4 μm) for both TE and TM polarizations, from which the propagation losses are found to be about 2.3 dB/cm and 2.0 dB/cm for the TE and TM polarizations at the wavelength of 1.31 μm, respectively.

The transmission spectrum of the microring resonator was measured over an input wavelength range of about 10 nm from 1540 nm to 1550 nm and the results are shown in Fig. 6. As expected from the design described in section 2, the transmission spectrum should be a periodic function of wavelength which repeats itself at every FSR. The dips in the transmission response occur periodically with spacing equal to the FSR of 1.08 nm (137 GHz), which is in good agreement with the theoretical FSR of $FSR \approx \lambda_n^2 / nL$ where L is the total length of the racetrack ring. The extinction ratios or depths of the dips are measured to be -14 dB which can be further improved by fine tuning of both the fabrication process and the measurement accuracy. However, the experimental results of the microring resonator demonstrated here do meet the practical requirements of commercial components. Hence, the polymer microring resonator reported here can be used as a good and low-cost building block of many practical devices for a wide variety of applications.

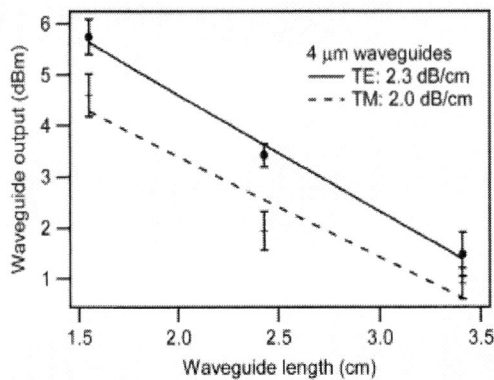

Fig. 5 Propagation losses of the waveguide for both TE and TM modes.

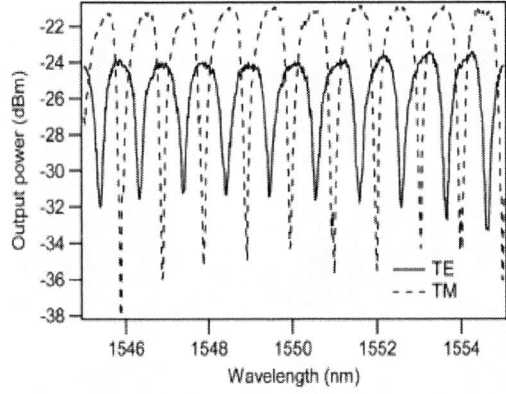

Fig. 6. Transmission spectrum of the microring resonator.

Conclusions

An optical-quality polymer film using the AJL8/APC material was prepared and used to fabricate a single-mode microring resonator. The experimental results are consistent with the theoretical design requirement. The microring resonator has a periodic transmission response with a large free spectral range (FSR) of 1.08

nm (or 137 GHz) and dips with very good extinction ratios of -14 dB under resonance condition. The microring resonator is a low-cost device because it is based on the polymer material and fabricated with a simple single-step fabrication process using the photobleaching method. Furthermore, the microring resonator is compact and can be potentially integrated with other passive and active integrated optical devices on the same chip to provide greater functionality at a reduced cost for a wide variety of applications.

Acknowledgments

The authors would like to acknowledge the support of the National Science Foundation (NSF Grant Number ECS-0437920). One of the authors Jun Zhou would also like to acknowledge the sponsored by K.C.Wong Magna Fund in Ningbo University, China.

References

1. P. Rabiei, W. H. Steier, Z. Cheng, and L.R. Dalton, *J. of Lightwave Technol.*, Vol. 20, 1968-1975, 2002.
2. D. Rezzonico, A. Guarino, C. Herzog, M. Jazbinsek, and P. Gunter, *IEEE Photonics Technology Letters*, Vol. 18, 865-867, 2006.
3. H. Tazawa and W. H. Steier, *Electronics Letters*, Vol. 41, 55-56, 2005.
4. C.-Y. Chao, W. Fung, and L.J. Guo, *IEEE Journal of Selected Topics in Quantum Electronics*, Vol. 12, 134-142, 2006.
5. Y. Huang, G.T. Paloczi, A. Yariv, C. Zhang, and L. R. Dalton, *Journal of Physical Chemistry B*, Vol. 108, 8606-8613, 2004.
6. J. Ma, S. Lin, W. Feng, R. J. Feuerstein, B. Hooker, and A. R. Mickelson, *Appl. Opt.*, Vol. 34, 5352–5360, 1995.
7. A. Chen, V. Chuyanov, F. I. Marti-Carrera, S. Garner, W. H. Steier, S. S. H. Mao, Y. Ra, L. R. Dalton, and Y. Shi, *IEEE Photonics Technology Letters*, Vol. 9, 1499-1501, 1997.
8. W.-Y. Hwang, J.-J. Kim, T. Zyung, M.-C. Oh, and S.-Y. Shin, *Appl. Phys. Lett.*, Vol. 67, 763–765, 1995.
9. J.K.S. Poon, Y. Huang, G.T. Paloczi, A. Yariv, C. Zhang, and L.R. Dalton, *Optics Letters*, Vol. 29, 2584-2586, 2004.
10. Nishihara H, Haruna M, Suhara T. Optical integrated circuits. New York: McGraw-Hill; 1989.

Image Resolution Analysis of Different Super lenses

P. Andalib and N. Granpayeh

Faculty of Electrical Engineering, K. N. Toosi University of Technology, Tehran, Iran.
Emails: p.andalib@ee.kntu.ac.ir; granpayeh@kntu.ac.ir

ABSTRACT *In this paper, we have analyzed the dependence of Quality of the image in triangular lattice PC slab with circular air holes, elliptical rods and cylindrical PC lens with circular air holes. The results have been compared.*

INTRODUCTION

Left handed material (LHM) planar slab with ability of negative refraction and therefore focusing the wave of a point source suggested by veselago in 1968 [1]. Amplification of evanescent waves by the lossless non-dispersive LHM and hence its behavior as a perfect lens proposed by Pendry in 2000 [2]. Negative refraction in optical regime and its application in the subwavelength imaging by photonic crystals (PCs) were demonstrated by Notomi [3] and Cubukcu *et al.* [4].

Image of a point source is an intensity peak in the perpendicular direction to the slab, formed by all angle negative refraction (AANR) focusing [5]. The resolution of the image can be derived by the ratio of the full width at half maximum (FWHM) of the image intensity peak to the central wavelength of the input pulse [6]. The resolution of the conventional imaging systems are limited by the wavelength of the incident light. Some information, giving the resolutions comparable to the wavelength is carried by the source evanescent waves, which are decayed in the conventional imaging systems before arriving the image plane. But amplification of the evanescent waves in LHM slabs yields subwavelength imaging.

Negative refraction could be happened by two mechanisms. First, when the frequency of the lightwave inside the PC is in the frequency range which equifrequency surface (EFS) contours in K-space become convex and their radii decrease with increasing the frequency, therefore the group velocity and the wave vector are antiparallel, so the permittivity and permeability and therefore, effective refraction index become negative which means left handed behavior ($\mathbf{S} \cdot \mathbf{K} < 0$ where **S** is the Poynting vector and **K** is the wave vector). Second, when the medium is anisotropic, therefore although the effective index of the PC is

positive ($\mathbf{S} \cdot \mathbf{K} > 0$), the lightwaves see negative refraction. Here the light refraction at the air-PC interface does not satisfy the conventional Snell's law and the lens can only focus the near field lightwaves. Therefore source should be placed near the slab [7]-[8].

NUMERICAL ANALYSIS

We have used numerical two dimensional (2D) finite difference time domain (FDTD) method to analyze the light propagation and point source imaging in two different structures of the PC slabs. Perfectly matched layers (PMLs) are used for modeling open space regions. The number of PML cells has been assumed to be 10 in all around the simulation area and Yee-cell technique used for discretizing fields in the space and time. we have simulated the propagation of TM modes; The result can easily be modified and extended for TE modes.

RESULTS AND DISCUSSIONS

The distance between each edge of the slab and the center of the nearest column rod axis (or holes in crystals with air holes in dielectric substrate), is called the surface termination [9].

So far, researches have been shown that excitation of surface waves in the photonic crystal as the surface plasmon polaritons (SPP) depends on the surface termination. This causes that transmission or coupling, resolution and position of the image, to be strongly sensitive to the surface termination [13]. It has been shown that appropriate surface termination of the PC slab, can cause efficient excitation of the surface waves and hence increases the quality of the images [9]-[10]. Surface waves are localized waves parallel to the surface and decay exponentially away from the surface in either perpendicular direction[13].

We have simulated negative refraction with the first mechanism given in section 1, hence the central frequency of the source should be

978-0-9789217-3-6/07/$25.00

©2007 WEN GLOBAL SOLUTIONS

in the frequency range which all angle negative refraction occurs in the second photonic band near the Brillouin zone center (Γ) of the PC [7]. To Study the wave propagation in PC slab, the group velocity should be considered. Since the gradient of the equifrequency contours (EFCs) in **K** space gives the group velocity of photonic modes, the EFC of the structure must be obtained by plane wave expansion method (PWEM). In PWEM a short pulse (wide spectrum) is launched to the slab as summation of plane waves and output pulse is processed to get photonic bands of the structure and EFCs.

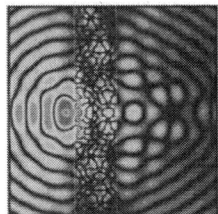

Fig. 1 Field distribution and formation of the image in a triangular lattice air hole PC slab.

The quality of the image also depends on the lattice constant of the crystal. The optimum resolution of each structure increases by reducing lattice constant. The reason is that by reducing lattice constant, the coarseness of the medium decreases [6].

Figure 1 demonstrates the TM mode field distribution in a triangular lattice PC slab with air holes, r=0.4a and ε=12.96 for a modulated Gaussian source with central normalized frequency of 0.305. The field intensity pattern is plotted in Fig. 2.

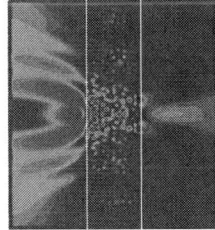

Fig. 2 Field intensity pattern of triangular lattice air hole PC slab.

The image resolution versus termination of the triangular lattice with dielectric rods is depicted in Fig. 3. There is an optimum surface termination, that the best resolution can be obtained. The appropriate termination is Δ=0.4a, which resolution reaches its optimum value of 0.3363.

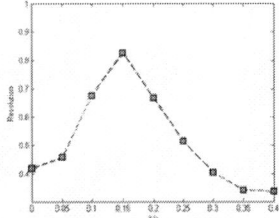

Fig. 3 The image resolution vs surface termination for triangular lattice air hole PC slab. Resolution is defined as the ratio of full width at half maximum of the image intensity and central wavelength.

The Second structure is a cylindrical lens made from exactly the same photonic crystal used for above triangular lattice PC slab with air holes. Its field distribution, intensity pattern and image resolution versus surface termination demonstrated in figure 4, 5 and 6 respectively. The appropriate termination of this structure is Δ=0.25a, which resolution reaches its optimum value of 0.2366.

Fig. 4 Field distribution and formation of the image in a triangular lattice air hole PC cylindrical lens.

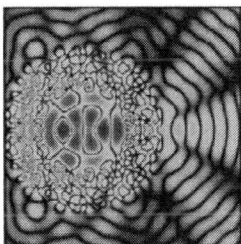

Fig. 5 Field intensity pattern of triangular lattice air hole PC cylindrical lens.

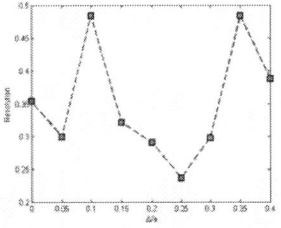

Fig. 6 The image resolution vs surface termination for triangular lattice air hole PC cylindrical lens.

Another structure which was investigated is a triangular lattice PC slab with elliptical dielectric rods with r_x=0.5a, r_y=a and ε=14. For a modulated Gaussian source with the central normalized frequency of 0.313, the effective index becomes $n_{eff} = -1$ [12].

Propagation of the TM mode in the structure and the light focusing has been demonstrated in Fig. 7. The field intensity pattern is plotted in Fig. 8.

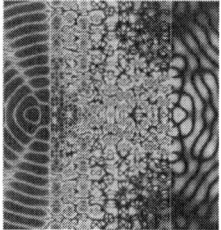

Fig. 7 Field distribution of triangular lattice with elliptical dielectric rod PC slab.

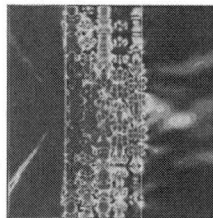

Fig. 8 The resolution as function of surface termination for triangular lattice with elliptical dielectric rod PC slab.

Variation of the image resolution versus surface termination is depicted in Fig. 9. The resolution varies linearly with surface termination and the optimum value occurs at Δ=0.5, which resolution is 0.627.

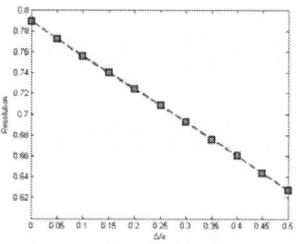

Fig. 9 Field intensity pattern of triangular lattice PC slab with elliptical dielectric rods.

As shown in Figs. 3 and 6 the sensitivity of the surface waves to the surface termination in cylindrical PC lens is more than that of PC slab, because the parallel momentum $k_{\parallel}$ in curved surface changes more by variation of the surface termination.

By comparison of Figs. 6 and 9, we conclude that if the rods become elliptical, the sensitivity of the surface waves to the surface termination decreases.

CONCLUSION

In this paper, we have analyzed the light propagation in three different structures of two dimensional PC slabs. Superlensing in second photonic band where all angle negative refraction occurs, was investigated by finite difference time domain (FDTD) method. The equifrequency contours (EFCs) of the structure were derived by plane wave expansion (PWE) method. The dependence of resolution to the surface have been compared for these structures.

REFRENCES

1. V. G. Veselago, Sov. Phys. Usp. Vol. 10 (1968), page. 509-514.

2. J. B. Pendry, Phys. Rev. Lett. Vol. 85 (2000), page 3966-3969.

3. M. Notomi, Phys. Rev. B, Vol. 62 (2000), page. 10696-10705.

4. E. Cubukcu, K. Aydin, E.ozabay, S. Foteinpolou, and C. M. Soukoulis, Phys. Rev. Lett., Vol. 91 (2003), page. 207401 (1-4).

5. J. B. Pendry, C. Luo, Steven G. Johnson, and J. D. Joannopoulos. Phys. Rev. B, Vol. 68 (2003),.page. 045115 (1-4).

6. X. Wang, Z. F. Ren, and K. Kempa "Improved superlensimg in two-dimensional photonic crystals with a basis," Appl. Phys. Lett, Vol. 86, pp. 061105 (1-3), 2005.

7. K. Ren, S. Feng, Z. Feng, Y. Sheng, Z. Li, B. Cheng, and D. Zhang. Phys. Lett. A, Vol. 348 (2006), page. 405-409.

8. Z. Tang, H. Ahang, R. peng, Y. Ye, C. Zhao, S. Wen, and D. Fan, J. Opt. A: Pure Appl. Opt., Vol. 8 (2006), page. 831-834.

9. F. AbdelMalek, W. Belhadj, and H. Bouchriha. Photon. and Nanostruc.–Fund. and Appl., Vol. 3 (2005), page. 19-24.

10. X. Wang, Z. F. Ren, and K. Kempa, Opt. Exp. Vol. 12 (2004), page 2919-2924.

11. Z. Tang, H. Ahang, R. peng, Y. Ye, C. Zhao, S. Wen, and D. Fan, Solid state commun. Vol. 141 (2007), page. 183-187.

12. Z. Tang, R. Peng, and D. Fan, Opt. Exp., Vol. 13 (2005), page 9796-9803.

13. R. Moussa, Th. Koschny and C. M. soukoulis, Phys. Rev. B, Vol. 74 (2206), pp. 115111(1-5).

"Invited Talk"

Photoacoustic and Thermoacoustic Imaging for Biomedical Applications

Da Xing, Liangzhong Xiang

MOE Key Laboratory of Laser Life Science and Institute of Laser Life Science, South China Normal University, Guangzhou 510631, China. Email: xingda@scnu.edu.cn

Abstract Laser-*based photoacoustic imaging and microwave-based thermoacoustic imaging could be the next successful generation imaging techniques in biomedical application. It can provide an effective approach of tissue structure and functional images to study the architectures, physiological and pathological properties and metabolisms of biological tissues.*

Introduction

Laser-based photoacoustic imaging and microwave-based thermoacoustic imaging, combining the advantages of both the high image contrast that results from electromagnetic absorption and the high resolution of ultrasound imaging, could be the next successful generation imaging techniques in biomedical application. It can provide an effective approach of tissue structure and functional images to study the architectures, physiological and pathological properties and metabolisms of biological tissues.

The absorption of electromagnetic energy causes thermal expansion and induces acoustic waves. This effect was first discovered in 1880, referred to as photoacoustic (PA) effect which was induced by light stimulation. In biomedical engineering, PA signal can be utilized to image biological tissues. And thermoacoustic imaging was also developed by replacing laser pulse with microwave pulse in our research. Both microwave and visible or near infrared (NIR) laser light are non-ioning radiation and safe for biomedical applications. In this paper, we have designed and assembled the integrative prototype B-scan photoacoustic tomography system, which can fast obtain the PA tomography sequence by conveniently moving the hybridized scanning head. It has the potential to provide a novel and effective approach for high contrast vascular imaging. We also have developed a fast MITT system with the

MLTAS, and gained two-dimensional thermoacoustic tomography imaging of biological tissues, respectively.

Photoacoustic imaging system

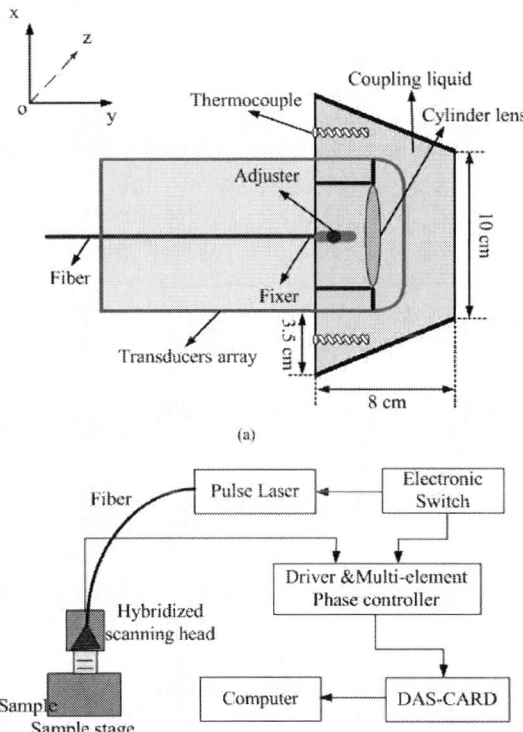

Fig.1 (a) Planform of the novel hybridized scanning head. (b). Experimental setup of the integrative prototype B-scan photoacoustic tomography system.

The integrative prototype B-scan photoacoustic tomography system includes a fiber (core diameter 600, NA=0.22), a transducer array (PL-21; SIUI, China) with

978-0-9789217-3-6/07/$25.00

©2007 WEN GLOBAL SOLUTIONS

320 vertical transducers and an ultrasonic coupling medium (glycerin and water). The dimensions of the hybridized scanning head are 3.5 cm, 8 cm and 10 cm, as shown in Figure. 1(a); and the height of the hybridized scanning head along the z-axis is 4 cm. The schematic of the experimental setup is shown in Figure.1 (b). A ND: YAG laser (Brilliant B, Bigsky) with wavelength of 1064 nm, output of 8 ns pulse width at 40 mJ/pulse, and a repetition rate of 20 Hz is used to irradiate the sample. The resonance frequency of the transducer array is 7.5 MHz and the scanning width of the array is 49mm. A built-in cylinder acoustic lens, made of silicon rubber with a focal length of 3.5 cm, was used to select the two-dimensional image plane and suppress out-of-plane signals. The fiber and the transducer array were fixed in a box and the angle between them was adjustable. The laser symmetrically irradiated the target tissues through a cylinder optical lens that focused the laser beam along the z-axis to effectively reduce potential damage to tissues. The ultrasound velocity in the ultrasonic coupling medium was adjusted to be as the same as that in the fatty tissue (1547 m/s). Glycerin (Glycerine, glycerol) is a colorless, odorless, noncorrosiveness and nontoxic liquid and can dissolve in water in any proportion, so the glycerin was chosen as the acoustic coupling medium. The glycerin concentration used in our experiments was 11.2%. The coupling medium is transparent relative to the wavelength of the laser. The coupling medium at the front of the hybridized scanning head was sealed with a waterproof polyethylene film; the gap between the film and the sample surface was filled with ultrasonic gel (DOVE, China) to ensure a good acoustic coupling; the distance between the film and the transducer array was 3 cm. The laser and ultrasonic waves can pass through the film with little attenuation. There were two thermocouples placed at the side of the transducer array to maintain a steady temperature during the experiment.

Thermoacoustic imaging system

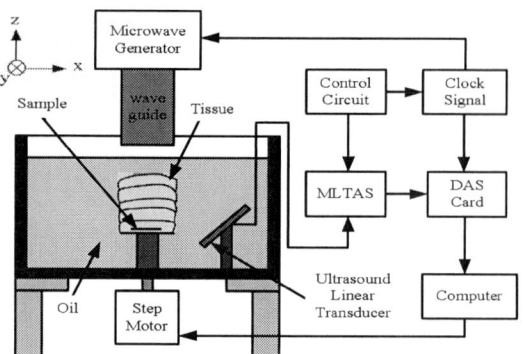

Fig.2. Diagram of the thermoacoustic imaging system

The schematic of the thermoacoustic imaging system is shown in Fig. 2. A microwave generator (BW-1200HPT, China) provides microwave pulses for thermoacoustic auditory generation. The pulses have the following properties: pulsewidth of 0.5 us, pulse-frequency of 1.2 GHz and incident energy density of 0.45 mJ/cm^2. The microwave pulses are coupled into a rectangular waveguide with a cross section of 12.7 mm x 6.3 mm and irradiated to a sample uniformly. The sample is placed on a plastic stage and immersed in a plexiglass tank, which is filled with transformer oil for better coupling of acoustic waves.

Results

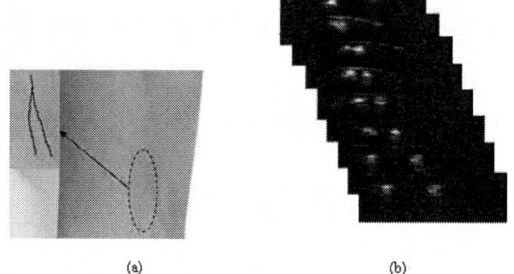

(a)   (b)

Fig.3 The photoacoustic image of two joining blood vessels on the forearm of a human volunteer

Fig.3 (a) is the photograph of the scanned area, and Fig.3 (b) shows the reconstructed PA image of the blood vessels, which consists of eight slices. The image sequence was acquired by moving the hybridized scanning head 2.1 cm with a step of 3 mm along the forearm at a position 8.7 cm away from the wrist. It can be seen that the cross section of the blood vessels can be imaged with high contrast and spatial resolution, even the joined location of the vessels. Besides the blood vessels, the skin has also been imaged, due to the difference in optical absorption between the skin and the coupling medium.

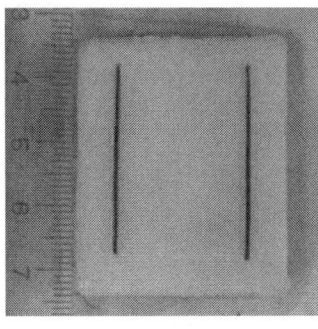

(a)

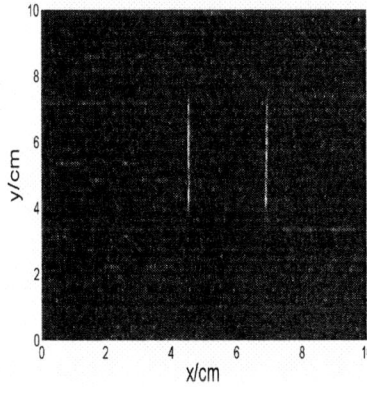

(b)

Fig. 4. (a) Cross section of the tissue sample; and (b) Thermoacoustic tomography image of the simulating blood vessels.

We made a piece of quadrate homogeneous pork fat tissue with a thickness of 4.5 cm. Next, we placed parallel two simulating blood vessels on the fat tissue as shown in Fig. 5(a). The simulating blood vessels were made of silicon rubber tubes with 0.8 mm diameter, filled with chicken blood. The length of the blood vessels and the distance between them was 3.1 cm and 2.3 cm, respectively. Then we covered the simulating blood vessels with a same thickness fat tissue. In order to finalize the design, the sample was cooled at room temperature for several minutes. Fig. 5(b) is the thermoacoustic tomography image of the simulating blood vessels with phase-controlled reconstructed algorithm. From the reconstructed image, the image of simulating blood vessels was corresponded well with the picture of sample. But the background noise of the reconstructed image is not as well as that of the images above, because the thick fat tissue results in the attenuation of thermoacoustic signal intensity and microwave energy.

Conclusions

This paper is focused on photoacoustic and thermoacoustic imaging application in biomedical research. We have designed and assembled the integrative prototype B-scan photoacoustic tomography system, which can fast obtain the PA tomography sequence by conveniently moving the hybridized scanning head. It has the potential to provide a novel and effective approach for high contrast vascular imaging.

We also have developed a fast MITT system with the MLTAS, and gained two-dimensional thermoacoustic tomography imaging of biological tissues. Comparing to other existing technologies and algorithm, our system was characterized by rapidness and convenience. The experimental results demonstrates that both photoacoustic and thermoacoustic images with multiple contrasts can reflect the absorption of electromagnetic energy. It can provide an effective approach of tissue structure and functional images to study the architectures, physiological and pathological properties and metabolisms of biological tissues which are stimulated using laser and microwave pulses, respectively.

References
1 Yang D W, et.al, Appl. Phys. Lett, 88 (2006), 174101.
2 L.zh Xiang, et.al.,J. Biomed.Opt,12(1) (2007),014001
3 Zeng L M, et.al. Chin.Phys.Lett.23 (5) (2006), 1215.

Optical Coherence Tomography for Oral Cancer Diagnosis

Meng-Tsan Tsai (1), Hsiang-Chieh Lee (1), Chih-Wei Lu (1), Yih-Ming Wang (1), Cheng-Kuang Lee (1), Chun-Ping Chiang (2), and C. C. Yang (1)

1: Graduate Institute of Electro-Optical Engineering, National Taiwan University,
No. 1, Roosevelt Road, Section 4, Taipei, Taiwan, R.O.C.
e-mail: ccy@cc.ee.ntu.edu.tw
2 : Department of Dentistry, National Taiwan University, Taipei, Taiwan, R.O.C.

Abstract Incorporating with a flexible-holder probe, a time-domain optical coherence tomography system is built to achieve 7 microns in axial resolution, 80-90 dB in sensitivity, and sub-sec imaging rate for clinical oral cancer diagnosis.

1. Introduction

Oral squamous cell carcinoma (SCC) is the most common type of oral carcinoma. SCC is a tumor arising from the oral surface epithelium, it usually form a mass of cancerous tissue connected to the surface epithelium when it invades into the underlying connective tissue. A majority of oral cancers are found to develop from oral premalignant lesions such as leukoplakia, erythroplakia, erythroleukoplakia, dysplasia, and carcinoma in situ. The malignant transformation rates of oral premalignant lesions are reported to be 1-7% for homogenous, thick leukoplakia, 4-15% for granular or verruciform leukoplakia, 18-47% for erythroleukoplakia, 4-11% for moderate dysplasia, and 20-35% for severe dysplasia [1]. The high malignant transformation rates for oral premalignant lesions also indicate the importance of early diagnosis and early treatment. In addition, these oral premalignant lesions are not homogeneous lesions, i.e., some part of the lesion may show only hyperkeratosis and acanthosis, while others may show epithelial dysplasia, carcinoma in situ, or invasive carcinoma. Therefore, one lesion may need multiple biopsies to avoid misdiagnosis of the most severe part of the lesion. To reduce the patients' suffering from multiple biopsies, one of the best non-invasive ways to help us to choose the most appropriate site for biopsy is to use the optic coherence tomography (OCT) to detect oral precancers and cancers.

By using interferometric cross-correlation techniques to detect the coherent backscattered components of short coherence length light, OCT has become a powerful tool for the diagnoses of various diseases [2-4]. Pioneered by Fujimoto and co-workers to perform *in vivo* clinical imaging of the human eye [5], OCT has been an emerging bio-imaging technology that promises to have broad and significant impact on clinical diagnostic imaging. In this paper, we report the design and fabrication of a probe of a flexible holder for the clinical diagnosis of oral cavity precancer and cancer with OCT. In a conventional probe for oral cavity scanning, a stepping motor for implementing lateral OCT scanning is installed in the holder of the probe. In this situation, the holder is quite bulky leading to the limited reachable portions in an oral cavity. Also, because the shaft of the motor forms the major body of the probe, the probe must be straight in shape such that the reachable portion in an oral cavity of complex geometry is further limited. In our newly designed probe, we used a motor of small volume to form the scanning box of around 35 x 15 x15 mm^3 in size. This scanning box is attached to a flexible holder such that it can be bent for reaching corners in an oral cavity. The OCT system together with the probe provides the axial scanning resolution of 8 microns.

2. OCT system and the scanning probe

Fig. 1 shows the layout of the OCT system. It is a standard time-domain, fiber-based OCT configuration. We used a super-luminescence diode of 1300 nm in central wavelength and 100 nm in spectral full-width at half maximum (FWHM). The maximum output power is 5 mW. An optical phase delay line is used in the reference arm for fast axial scanning. The theoretical

978-0-9789217-3-6/07/$25.00

©2007 WEN GLOBAL SOLUTIONS

axial resolution of the OCT system is 7 microns. However, because of the residual dispersion effect of the system, the implemented axial resolution is 8 microns. The scanning speed of the OCT system is 0.5 sec per frame of 1 x 1 mm^2. The system sensitivity (including the probe to be described below) can reach about 80 dB.

Fig. 2 shows the layout of the probe. On the right of the scanning box (35 x 15 x 15 mm^3), a prism is glued to the shaft end of the motor for retro-reflecting the optical beam coming in from the fiber collimator on the left. As the motor shaft swings, the light beam will scan in the direction perpendicular to the paper surface in Fig. 2, achieving the OCT lateral scanning. The scanned light beam is focused with a focusing lens of 35 mm in focal length onto the tissue samples through a quartz plate. The quartz plate is used to seal the scanning box such that the optical components inside will not be damaged during the cleaning process after the use of the probe for a patient. The collimator is connected to the fiber, which is attached to the flexible holder for reaching the OCT interferometer.

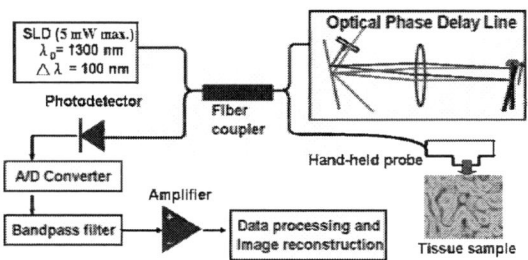

Fig. 1 Layout of the OCT system for oral cavity scanning.

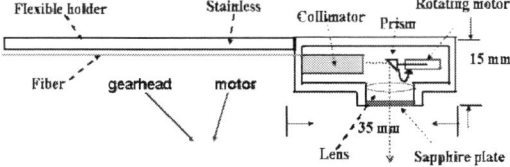

Fig. 2 Layout of the probe.

By using an algorithm, the interfaces between different layers of an OCT image can be marked. We can thus determine the thicknesses of various tissue layers. The first step in this algorithm is to determine the existence of a surface. For this purpose, the first local intensity maximum point in each

A-scan is evaluated. Then, the local intensity minimum point between the first and second local intensity maximum points in each A-scan is determined. This local minimum corresponds to the interface between the first and second layer.

3. OCT scanning results
Figs. 3 and 4 show two OCT images obtained with the OCT and the probe system. For Fig. 3, we scan an oral cancer sample from a patient *in vitro* including normal and abnormal tissues. Fig. 3(a) and Fig. 3(b) show the OCT scanning results of the normal tissue and the corresponding histology image, respectively. Here, the layer of epithelium and lamina propria can be clearly identified. To demonstrate the boundaries between the different layers, we use the aforementioned algorithm to evaluate the thickness of each layer at different locations. Fig. 4 was obtained by scanning an abnormal tissue sample from the same patient. Here, only one interface can be identified through the software process. This result is due to the down growth of the hyperplastic epithelium into the underlying lamina propria layer, mimicking the early invasion of an oral cancer into the underlying connective tissue.

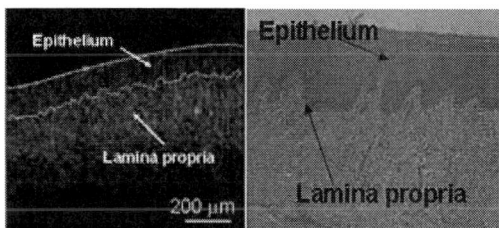

Fig. 3 OCT and histology images of normal mucosa.

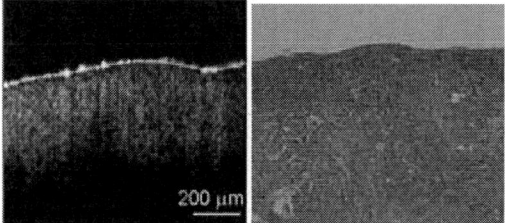

Fig. 4 OCT and histology images of abnormal mucosa.

4. Conclusions
We have designed and fabricated a probe of a flexible holder for oral cavity scanning.

With a time-domain, fiber-based OCT system, this probe is expected to increase the reachable portions in an oral cavity. This probe consisted of a scanning box, in which a small motor is installed to swing the incoming light beam for OCT lateral scanning. The light beam opening of the probe was sealed with a quartz plate for protecting the optical components inside during cleaning. With the probe and the OCT system, we could achieve 8 microns in axial resolution, 80 dB in sensitivity, and 0.5 sec in scanning speed for a 1 x 1 mm^2 image. Through a software process, the interfaces between layers can be identified to assist the diagnosis.

References

1. B. W. Neville et al., *Oral Maxillofacial Pathology*, pp. 259-321, W. B. Sauders, Philadelphia, PA, U.S.A. 1995.
2. D. Huang et al., *Science*, **254**, 1178, 1991.
3. M. R. Lee et al., *IEEE Eng. Med. Biol. Mag.* **14,** 67, 1995.
4. J. G. Fujimoto et al., *Nature Medicine*, **1**, 970-972, 1995.
5. C. A. Puliafito et al., *SLACK, Thorofare, NJ*, 1996.

Acknowledgement

This research was supported by National Health Research Institute, The Republic of China, under the grant of NHRI-EX94-9220EI.

Raman Signal Enhancement in a Liquid-Core Optical Fiber Based on Hollow-Core Photonic Crystal Fiber

Li Huo (1), Chinlon Lin (1), Yick Keung Suen(2), Siu Kai Kong (2)

1 : Center for Advanced Researches in Photonics, the Chinese University of Hong Kong, Hong Kong SAR, China. lihuo@ie.cuhk.edu.hk

2 : Department of Biochemistry, the Chinese University of Hong Kong

Abstract *Raman scattering of glycerol in a liquid-core optical fiber based on hollow-core photonic crystal fiber was demonstrated. Compared with conventional method, the Raman signal was enhanced over 20 times in a 2.5-cm liquid-core fiber.*

Introduction

Clinical test of human bio-fluids such as blood, urine, and saliva can provide rich information on the health condition. Liquid-core optical fibers based on capillary tubes such as Teflon capillaries haven been shown to provide enhanced fluorescence and Raman signal intensity over the conventional method because of the large overlap of optical field with liquid sample and the long interaction length in the fiber [1].

The recent invention of silica hollow-core Photonic-Crystal Fibers (PCFs), consisting of a large hollow core surrounded by small air holes running down the length of the fiber, presents new opportunities in the realization of liquid-core fibers.

Previously, we have shown fluorescence signal enhancement of fetal calf blood serum in liquid-core fibers based on hollow-core PCFs [2][3]. In this paper, we present the result of Raman signal enhancement in one type of such fibers. The result shows its great potential to analysis the contents of human bio-fluids both qualitatively and even quantitatively by Raman spectroscopy.

Working Principle

The working principle of liquid-core optical fiber is based on total internal reflection (TIR). When the refractive index of the cladding layer of the fiber is smaller than that of the liquid core, light can be guided stably in the liquid-core region. However, the light collection efficiency of such fibers is limited by the refractive contrast between the cladding and the liquid core. For Teflon tubes, the refractive indexes are n~1.31 for Teflon AF 1600 and n~1.29 for Teflon AF 2400. The numerical apertures (N.A.) are 0.23 and 0.32 respectively, assuming the core liquid is water (n~1.33).

The Photonic crystal fiber we used in the experiment was commercially available from Crystal Fiber A/S, Denmark with the model type of HC19-532-01. The cross section of the fiber is shown in Fig.1. The cladding has an air fill ratio of about 90% and a pitch distance of 1.5 μm. The diameter of the central hole is 9.5 μm. According to the theory of Knight [4], the equivalent refractive index was calculated to be 1.20 using plane wave expansion method. If the hollow-core is filled with water, the N.A. of the fiber is 0.57 round the wavelength of 500 nm. The higher N. A. of the liquid-core fiber based on hollow-core photonic compared with Teflon based one provides a better scattering light collection efficiency.

Fig.1. the microscopic picture of the cross section of the hollow-core photonic crystal fiber

978-0-9789217-3-6/07/$25.00

©2007 WEN GLOBAL SOLUTIONS

Experiment setup

The liquid sample tested was glycerol and water mixture with volume ratio of 1:1. The refractive index is estimated to be 1.40. The resultant N.A. of the fiber with the mixture as the core liquid is 0.72. The length of the fiber was 2.5 cm and the liquid sample consumption was estimated to be 3.3 nl.

One end of a hollow-core PCF was well cleaved for the purpose of observation. A drop of the liquid sample was loaded on a syringe tip and moved using a 3D translation stage to contact with the uncleaved end of the PCF. The liquid was draw into the HC-PCF by capillary force. The droplet was removed as the meniscus in the central hollow core reached the other end of the fiber. Because the filling speed in the larger central hole was much faster than that in the smaller cladding holes, there was a portion in the fiber where the central hollow core was filled but the cladding holes were empty. This part of the fiber was utilized in our experiment as a waveguide to obtain an enhanced measurement.

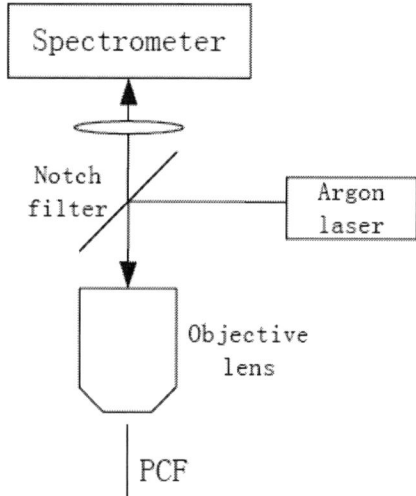

Fig.2. Experimental setup of Raman signal collection based on the liquid-core PCF. Excitation laser was launched from the cleaved end facet of the PCF and signals are collected in the backward direction.

The experimental setup is shown in Fig. 2. The filled PCF was adhered to a translation stage and examined by a Renishaw Raman microscopic spectrometer equipped with a 10X/0.65NA objective lens and a 1200-line/mm grating. An Argon ion laser with a wavelength of 514 nm was used as the excitation source and the beam was focused to a 1-μm diameter spot size by the objective lens. The laser was fed into the PCF from the well cleaved facet of the liquid-core PCF with 1-mw power. The integration time of the spectrometer was fixed as 10 s. The excited fluorescence and Raman signal in the liquid core could be collected by the TIR and transported in both the forward and the backward directions. In our micro-Raman configuration, we observed the backward signals.

Experimental result

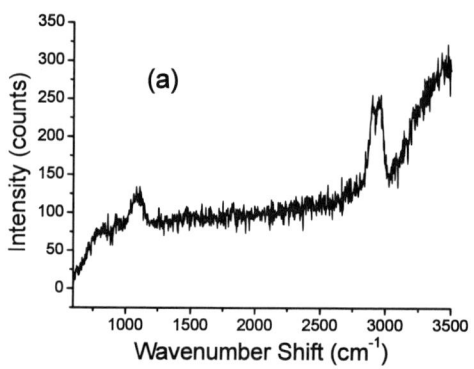

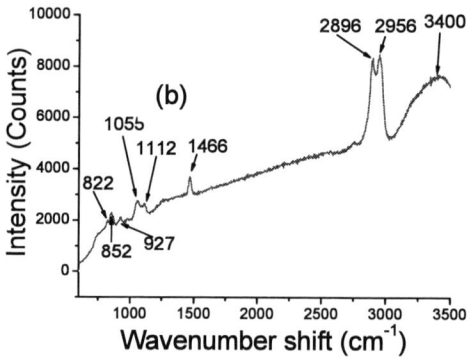

Fig.3 Observed Raman spectrum in the droplet (a) and in the 2.5-cm liquid-core PCF (b)

The Raman spectrum of the water glycerol mixture in the PCF was recorded between the wavelength shift of 600 cm^{-1} to 3500 cm^{-1}. As a reference, we also drop a droplet of the same liquid sample onto a silica glass slide and observed its Raman signal using the same objective lens and the same integration time.

Raman signals obtained are shown in Fig.3. In the reference case, the collected Raman signal intensity was weak. Some of the Raman peaks were too weak to be observed and the two peaks located at 1055 and 1112 cm^{-1} could not be resolved. In the PCF, a much stronger Raman signal was observed because of the collective light guiding effect in the optical fiber. The major characteristic peaks whose positions were consistent with the published literatures were clearly resolved. The positions, physical origins of the observed peaks and their signal enhancement by using the PCF were listed in Table.1.

Table 1 major characteristic peaks observed on the liquid sample

Characteristic peak wavenumber shift (cm^{-1})	Origin	Enhance-ment by PCF
822	CC stretch	------
852	CC stretch	------
927	CH_2 rock	------
1055	CO stretch form C-1	22.3
1112	CO stretch from C-2	22.1
1466	CH_2 deformation	------
2896	Symmetric CH stretch from CH_2	33.7
2956	Antisymmetric CH stretch from CH_2	33.6
3400	Antisymmetric OH stretch	26.4

We noticed that the enhancement factor was not uniform for different peaks. The difference might be caused by the different collection efficiencies of different wavelengths in the liquid-core optical fiber or by the re-absorption characteristics of glycerol. Despite of this difference, it could be concluded that at least 22 times of

Raman signal enhancement was obtained in a 2.5-cm PCF.

We also found if the focus of the laser fell near the edge of the liquid core, there would be a Raman peaks at 460cm^{-1}. This was induced by the silica PCF itself. When the excitation position was moved to the center of the PCF, the silica peak could be minimized. It is also observed that the polymer coating of the PCF at the excitation end, if not properly removed, introduced a fluorescence background thousands of counts high which buried the interested Raman signal. However, the background level could be reduced to below 100 counts when the coating was totally removed, enabling the detection of molecules inside the liquid-core.

Conclusion

In conclusion, we have demonstrated Raman scattering of glycerol in a liquid-core fiber based on hollow-core PCF with a backward micro-Raman configuration. Over 20 times signal intensity enhancement was obtained in a 2.5-cm long PCF. We believe a much stronger enhancement could be obtained if we use longer fiber and improved laser-fiber coupling scheme. The result revealed the great potential of such hollow-core PCF as a highly sensitive biosensor based on Raman spectroscopy.

Acknowledgement

This work is supported by RGC/CU project 414405.

References

1 Dahu Qi et al, Applied Optics, Vol.46 (2007), page 1726
2 Li Huo et al, in Proc. The 9th International Conference on Optics Within Life Sciences (OWLS'09), Taipei, 2006, page 186
3 Li Huo et al, CLEO2007, Baltimore, MA, 2007, paper JTuA46
4 J. C. Knight et al, Optics Letters, Vol. 22 (1997), page 961

Time-of-flight laser spectroscopy in biomedical diagnostics

Stefan Andersson-Engels[1], Johan Axelsson[1], Ann Johansson[1], Jonas Johansson[2], Sune Svanberg[1] and Tomas Svensson[1]

[1]Department of Physics, Lund University, PO Box 118, SE-221 00 Lund, Sweden,
stefan.andersson-engels@fysik.lth.se
[2]AstraZeneca Research and Development, Mölndal, Sweden

Optical spectroscopy is becoming a very valuable diagnostic tool in biomedical research. Time-of-flight spectroscopy is a tool providing information regarding scattering and absorption properties of the tissue, valuable for several applications as seen below. In Lund we have in the past developed the technique of supercontinuum generation for spectroscopic investigations of highly scattering media such as tissue[1-3], plant material[4], and pharmaceutical samples[5].

More recently the group in Lund has initiated the development of a unique broadband time-resolved spectroscopy system for turbid media, based on a mode-locked Ti:Sapphire laser pumping a photonic crystal fibre (PCF) and a streak-camera in syncroscan mode.[5,6] The low dispersion of the ultra short laser pulses inside the photonic crystal fibre combined with the small core diameter results in a high peak power of the light in the entire length of the fibre, yielding a high non-linear efficacy resulting in widely spectrally broadened light emission. As a result of this, light pulses with a spectral width spanning from 500 nm to at least 1200 nm were accessible. The streak camera allows recording of time-resolved data with a time resolution of approximately 30 ps. The system is stationary but still relatively flexible and permits both free space as well as fibre coupling of the light onto the sample. This allows different samples and sample geometries to be used. So far direct transmittance (pharmaceutical tablets), fibre coupled transmittance and fibre coupled reflectance (fruits) have been tested. The system has convincingly been used to demonstrate the capability of an analytical instrumentation that could separate absorption and scattering spectra in the evaluation of active substance in pharmaceutical preparations.[7-9] The technique has also been successfully demonstrated for determining the condition of fruits.[10,11]

In contrast to the bulky white light system we have also in parallel been developing a portable diode laser based system for time-resolved measurements.[12] This system is based on the technique called time-correlated single photon counting (TCSPC). This equipment is designed to allow *in vivo* measurement of tissue optical properties of small tissue volumes (less than 1 cm^3) at four wavelengths within the tissue optical window (660, 785, 915 and 970 nm). A lot of effort has gone into making the system portable and suitable for a clinical environment. It is now contained within a box of the approximate size of 50*50*30 cm and the light is coupled via optical fibres to and from the patient/sample. This allows the use of sterile fibres to be unpacked and inserted into the

978-0-9789217-3-6/07/$25.00

©2007 WEN GLOBAL SOLUTIONS

system inside an operation theatre. The system has been used at clinics for unique local characterisation of breast tissue and the prostate gland.[13,14]

Since both systems, produce similar data (although the diode laser based system is confined to only 4 wavelengths) a general evaluation toolbox based on diffusion theory has been developed within the Lund group. The diffusion models allow the calculation of the scattering and absorption coefficients for each wavelength independently. To get the most from these calculated values several approaches have been developed and tested. For example, in the case of the diode based system, the four different absorption coefficients can be used to calculate the concentration of four tissue constituents (oxy-haemoglobin, deoxy-haemoglobin, fat and water).

The multispectral information provided in fluorescence measurements of lesions located at a certain depth in tissue would, in addition to the diagnostic information that there exist a lesion, also provide depth information useful for reconstruction of fluorescent inclusions in tissue[15,16]. This is one possible concept to improving the robustness and accuracy of the fluorescence tomography reconstructions.

One can now foresee a rapidly increasing interest in time-resolved spectroscopic techniques for medical diagnostics, with the development of molecular markers. Most of the genetically specific markers for optical detection will most probably be based on fluorescence. The computer models for tissue fluorescence are not as developed as for remitted light. An accelerated Monte Carlo simulation routine was developed that is approximately 100 times faster than the conventional model used.[17] This model can be used to partly overcome the lack of other appropriate computer models.

Using very similar ideas we are also developing a new technique to perform on-line dosimetric measurements during interstitial photodynamic therapy[18-20]. This development is conducted in close collaboration with SpectraCure AB, a spin-off company from the group. The company has now integrated a system for interstitial photodynamic therapy of prostate cancer. The system is approved for first clinical studies. Our group has developed the technology concept as well as the dosimetry aspects of the treatments to allow treatment feed-back from the measurements.[21,22]. The main challenge is now to further develop the dosimetry aspects of IPDT and to evaluate the procedure by studying treatment outcome.

References

1. S. Andersson-Engels, R. Berg, A. Persson, and S. Svanberg, "Multispectral tissue characterization with time-resolved detection of diffusely scattered white light," *Opt. Lett.* **18**, 1697-1699 (1993).
2. O. Jarlman, R. Berg, S. Andersson-Engels, S. Svanberg, and H. Pettersson, "Time-resolved white light transillumination for optical imaging," *Acta Radiol.* **38**, 185-189 (1997).
3. S. Svanberg, S. Andersson-Engels, R. Cubeddu, E. Förster, M. Grätz, K. Herrlin, G. Hölzer, L. Kiernan, C. af Klinteberg, A. Persson, A. Pifferi, A. Sjögren, and C.-G. Wahlström, "Generation, characterization and medical utilization of laser-produced emission continua," *Laser and Particle Beams* **18**, 563-570 (2000).
4. J. Johansson, R. Berg, A. Pifferi, S. Svanberg, and L. O. Björn, "Time-resolved studies of light propagation in *Crassula* and *Phaseolus* leaves," *Photochem. Photobiol.* **69**, 242-247 (1999).

5. C. Abrahamsson, T. Svensson, S. Svanberg, S. Andersson-Engels, J. Johansson, and S. Folestad, "Time and wavelength resolved spectroscopy of turbid media using light continuum generated in a crystal fiber," *Opt. Express* **12**, 4103-4112 (2004).

6. C. Abrahamsson, A. Lowgren, B. Stromdahl, T. Svensson, S. Andersson-Engels, J. Johansson, and S. Folestad, "Scatter correction of transmission near-infrared spectra by photon migration data: Quantitative analysis of solids," *Appl. Spectrosc.* **59**, 1381-1387 (2005).

7. C. Abrahamsson, J. Johansson, A. Sparén, and F. Lindgren, "Comparison of Different Variable Selection Methods Conducted on NIR Transmission Measurements on Intact Tablets.," *Chemom. Intell. Lab. Syst.* **69**, 3-12 (2003).

8. J. Johansson, S. Folestad, M. Josefson, A. Sparen, C. Abrahamsson, S. Andersson-Engels, and S. Svanberg, "Time-resolved NIR/Vis spectroscopy for analysis of solids: Pharmaceutical tablets," *Appl. Spectrosc.* **56**, 725-731 (2002).

9. J. Johansson, S. Folestad, C. Abrahamsson, T. Svensson, and S. Andersson-Engels, "Time-resolved NIR spectroscopy for analysis of solid pharmaceuticals," *NIR News* **17**, 4-6 (2006).

10. F. Chauchard, S. Roussel, J. M. Roger, V. Bellon-Maurel, C. Abrahamsson, T. Svensson, S. Andersson-Engels, and S. Svanberg, "Least-squares support vector machines modelization for time-resolved spectroscopy," *Appl. Opt.* **44**, 7091-7097 (2005).

11. F. Chauchard, J. M. Roger, V. Bellon-Maurel, C. Abrahamsson, S. Andersson-Engels, and S. Svanberg, "MADSTRESS: A linear approach for evaluating scattering and absorption coefficients of samples measured using time-resolved spectroscopy in reflection," *Appl. Spectrosc.* **59**, 1229-1235 (2005).

12. A. Pifferi, A. Torricelli, A. Bassi, P. Taroni, R. Cubeddu, H. Wabnitz, D. Grosenick, M. Möller, R. Macdonald, J. Swartling, T. Svensson, S. Andersson-Engels, R. van Veen, H. J. C. M. Sterenborg, J.-M. Tualle, H. L. Nghiem, S. Avrillier, M. Whelan, and H. Stamm, "Performance assessment of photon migration instruments: the MedPhot protocol," *Appl. Opt.* **44**, 2104-2114 (2004).

13. T. Svensson, J. Swartling, P. Taroni, A. Torricelli, P. Lindblom, C. Ingvar, and S. Andersson-Engels, "Characterization of normal breast tissue heterogeneity using time-resolved near-infrared spectroscopy," *Phys. Med. Biol.* **50**, 2559-2571 (2005).

14. T. Svensson, M. Einarsdóttír, K. Svanberg, and S. Andersson-Engels, "In vivo optical characterization of human prostate tissue using near-infrared time-resolved spectroscopy," *J. Biomedical Optics* **12**, 014022-1-014022-10 (2006).

15. J. Svensson and S. Andersson-Engels, "Modeling of spectral changes for depth localization of fluorescent inclusion," *Opt. Express* **13**, 4263-4274 (2005).

16. J. Swartling, J. Svensson, D. Bengtsson, K. Terike, and S. Andersson-Engels, "Fluorescence spectra provide information on the depth of fluorescent lesions in tissue," *Appl. Opt.* **44**, 1934-1941 (2005).

17. J. Swartling, A. Pifferi, A. M. K. Enejder, and S. Andersson-Engels, "Accelerated Monte Carlo model to simulate fluorescence spectra from layered tissues," *J. Opt. Soc. Am. A* **20**, 714-727 (2003).

18. A. Johansson, T. Johansson, M. S. Thompson, N. Bendsoe, K. Svanberg, S. Svanberg, and S. Andersson-Engels, "In vivo measurement of parameters of dosimetric importance during interstitial photodynamic therapy of thick skin tumors," *Journal of Biomedical Optics* **11**, (2006).

19. M. S. Thompson, A. Johansson, T. Johansson, S. Andersson-Engels, S. Svanberg, N. Bendsoe, and K. Svanberg, "Clinical system for interstitial photodynamic therapy with combined on-line dosimetry measurements," *Applied Optics* **44**, 4023-4031 (2005).

20. M. S. Thompson, S. Andersson-Engels, S. Svanberg, T. Johansson, S. Palsson, N. Bendsoe, A. Derjabo, J. Kapostins, U. Stenram, J. Spigulis, and K. Svanberg, "Photodynamic therapy of nodular basal cell carcinoma with multifiber contact light delivery," *Journal of Environmental Pathology Toxicology and Oncology* **25**, 411-424 (2006).

21. A. Johansson, N. Bendsoe, K. Svanberg, S. Svanberg, and S. Andersson-Engels, "Influence of treatment-induced changes in tissue absorption on treatment volume during interstitial photodynamic therapy," *Medical Laser Application* **21**, 261-270 (2006).

22. A. Johansson, T. Johansson, M. Soto Thompson, N. Bendsoe, K. Svanberg, S. Svanberg, and S. Andersson-Engels, "In vivo measurement of parameters of dosimetric importance during photodynamic therapy of thick skin tumors," *J Biomed Opt.* **11**, (2006).

Extraction Efficiency Enhancement of an OLED using Surface Plasmon Resonance

Shou-Yu Nien[1], Nan-Fu Chiu[2], Yao-Chou Tsai[3], Chii-Wann Lin[2], Kou-Chen Liu[3], and Jiun-Haw Lee[1]

[1]Graduate Institute of Electro-Optical Engineering and Department of Electrical Engineering, National Taiwan University, 1, Sec. 4, Roosevelt Rd., Taipei, Taiwan
jhlee@cc.ee.ntu.edu.tw
[2]Institute of Electrical Engineering, and Institute of Biomedical Engineering, National Taiwan University,
[3]Graduate Institute of Electro-Optical Engineering, University of Chang Gung, Tao-Yuan 333, Taiwan, Republic of China

Abstract *In this paper, organic light-emitting devices with non-planar and periodic structures, fabricated by e-beam lithography, were demonstrated which exhibits improved extraction efficiency by coupling out the plasmonic mode. Their optical characteristics were presented and analyzed.*

Introduction

Organic light-emitting devices (OLEDs) are one of the most promising technologies for display and lighting applications due to their advantages of low power consumption, large viewing angle, high contrast ratio, fast response and flexible-substrate capability [1,2]. However, the outcoupling efficiency of a conventional OLED is typically less than 30%, which means that most of the light is trapped in the glass substrate [3]. To improve the OLED's outcoupling efficiency, many techniques based on eliminating the waveguiding phenomena between the glass substrate and the air have been studied [4–10]. Here, we demonstrate the effect of coupled active surface plasmon polaritons on the plasmonics response of a lamellar grating nanostructure with organic material on the surface for improving the extraction efficiency of an OLED.

Surface Plasmon Resonance (SPR) is a charge-density oscillation that can exist on the interface of two media with dielectric constants of opposite signs [11], for instance, a metal and a dielectric. This phenomenon was first observed by R. W. Wood [12] in metal grating in the early 1900s. Since SPR is very sensitive to the refractive index, it is suitable to use this phenomenon for the biosensor applications [13-15].

In this paper, we present and discuss the physics of enhancing efficiency of light emission from organic dielectric films on a lamellar grating nanostructure by using the coupled long-range SPRs in different dielectric structures.

Experiments

In our experiments, we use a thermal evaporator under high vacuum ($<1 \times 10^{-6}$ torr) for organic and metal deposition. The deposition rate is controlled at 0.1 nm/sec. All the organic materials were sublimed before deposition. 1-D pattern of nanostructure was prepared by an Electron-Beam Lithography system (ELS-7500EX, ELIONIX Co.). Fig. 1 (a) and (b) show the AFM and SEM pictures of the samples with the structure Au (40 nm)/ tris-(8-hydroxyquinoline)aluminum (Alq3) (80 nm) on the silicon substrate. The photoresist thickness is 100nm on silicon substrate with an exposure area of 1.2×1.2mm. The grating structure has 400 nm line width and 800 nm pitch width. In OLED fabrication, we used ITO glass substrates with low sheet resistivity (10 Ohm/sqr) and flat surface roughness (Ra < 1 nm). After organic and metal deposition, we encapsulated the devices in the glove box with O_2 and H_2O concentrations below 1 ppm.

978-0-9789217-3-6/07/$25.00
©2007 WEN GLOBAL SOLUTIONS

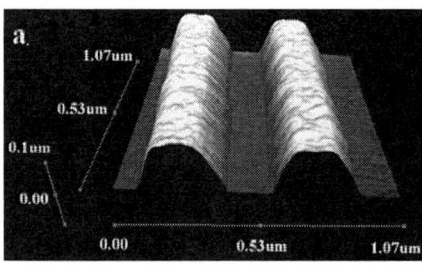

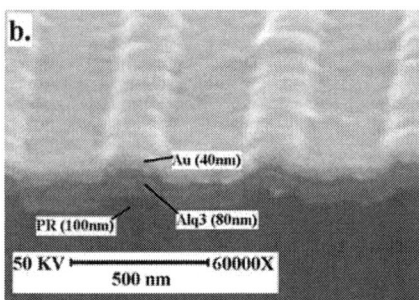

Fig. 1 (a) AFM and (b) SEM pictures of the sample: Au (40nm)/Alq3(80nm)/Si substrate.

Electrical and optical characteristics of OLED was performed by using a Keithley 2400 source meter for current-voltage measurement and a Minolta CS-1000 photometer for the brightness measurement. We also set up a reflectivity measurement system to measure the angular dependent photoluminescence (PL) of the surface plasmon coupled emission spectra produced by designated grating. All the experiments were performed at room temperature. Incident light with the wavelength of 405 nm from a white light source (Newport Oriel Spectral Luminator 69050) illuminates the sample through a polarization filter at fixed incident angle (θ_i) of 45°. The sample was kept at a fixed azimuthal angle (φ) of 0° incidence to pumping the organic molecules to excite metal grating and cross coupler produce SP-coupled emission. The emission beam is collected through a 2-inch lens with focal distance of 5cm on a spectrometer (USB2000-VIS-NIR, Ocean Optics Co.) or a CCD camera (HAMAMATSU C2400). The spectrometer moved to measure the SP-coupled emission intensity at different emission angles (θ_e) in response to the measured optical signal change in SPPs as well as the surface refractive index of the SPPs-grating. This emission is grating-scattered re-radiation from Alq3-excited SPPs modes rather than direct radiation from Alq3 molecules.

Results and discussions

Fig. 2 shows the photoluminescence (PL) of the planar and corrugated sample with the thin films Au (40nm)/Alq3(80nm). We can see that for the planar sample, the PL intensity has a Lambertian-like distribution. On the other hand, the PL intensity from the corrugated OLED is much improved by four times at zero degree and can be observed only within 10°. One has to note that there is Au layer on the top of Alq3 which impedes parts of light emission from the organic layers. However, in the corrugated sample, the strong PL signal reveals the SPR coupling exists at the interfaces which improves the extraction efficiency of the sample.

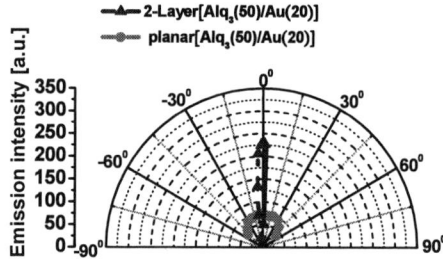

Fig. 2 PL intensities at different viewing angles with planar and corrugated samples.

Besides, we also measured the spectra at different viewing angles. For the planar sample, there is no obvious spectral shift with different viewing angles. On the other hand, the corrugated sample changes color even with a small viewing angle difference. Fig. 3 shows the pictures at different viewing conditions of the corrugated sample. We can see that the color shifts from the red to blue within the viewing angles of only 20°. Such a highly sensitive characteristic may have a potential application for bio-sensor.

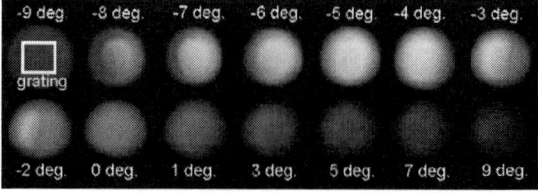

Fig. 3 PL emission pictures of the corrugated sample at different viewing angles.

Fig. 4 shows the OLED samples under electrical pumping. The corrugated area is marked with dashed red line with the size of 1.2mmx1.2mm. One can see a clear brighter image of this area which shows improved extraction efficiency by the SPR.

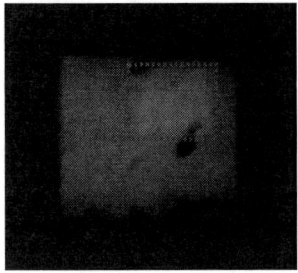

Fig. 4 Picture of the OLED under electrical pumping. The right up portion, marked with red line, is the corrugated structure.

Summary
We have shown experimentally that strong coupling photonic resonances in SP-coupled emission for Au/Alq$_3$ and Alq$_3$/air symmetric mode leads to the formation of active plasmon devices. The highly directional emission of the device is due to the dispersion relation of the SPPs and multilayer to have narrow spectral bandwidth at emission angles. Further investigations will be performed on SPPs with the integration of optimized organic electroluminescent plasmonic for active biosensor devices.

Acknowledgement
This project is supported in part by National Science and Technology Program in Pharmaceuticals and Biotechnology, National Science Council, Taiwan, R.O.C., under Grant NSC 95-2323-B002-001, NSC 95-2323-B002-004. MOEA 95-EC-17-A-05-S1-0017, and NSC 95-2221-E-002-305.

References
1. C. W. Tang et al, Appl. Phys. Lett. 51 (1987) 913
2. C. W. Tang et al, J. Appl. Phys. 65 (1989) 3610
3. M.-H. Lu et al, J. Appl. Phys. 91 (2002) 595
4. T. Yamasaki et al, Appl. Phys. Lett. 76 (2000) 1243
5. J. J. Shiang et al, J. Appl. Phys. 95 (2004) 2889
6. L. Lin et al, J. Micromech. Microeng. 10 (2000) 395
7. S. Möller et al, J. Appl. Phys. 91 (2002) 3324
8. M.-K. Wei et al, Opt. Exp. 12 (2004) 5777
9. H. Peng et al, J. Display Technol. 1 (2005) 278
10. M.-K. Wei et al, J. Micromech. Microeng. 16 (2006) 368
11. H. Raether, *Surface Plasmons on Smooth and Rough Surface and on Gratings* (Springer-Verlag, Berlin, 1988).
12. R. W. Wood, Phil. Mag., **4**, 396 (1902).
13. C.-W. Lin et al, Sensors and Actuators B: Chemical, **113**: 169 (2006).
14. C.-W. Lin et al, Sens. Actuators B: Chem., **117**, 219 (2006).
15. C.-W. Lin et al, OQE **37**, 1423 (2005).

Real-time Voltage Controlled Color Tunable OLEDs

Wallace C.H. Choy (1*), C.J. Liang (1), H.M. Zhang (1)

1 : Department of Electrical and Electronic Engineering, University of Hong Kong, Pokfulam Road, Hong Kong, China. *chchoy@eee.hku.hk

Abstract *Using the rare-earth and phosphorescent materials, voltage-controlled color tunable OLEDs has been achieved. The mechanisms of color-tuning have been investigated. The color purity and efficiency of the tunable OLEDs will be discussed.*

Introduction

Organic light-emitting device (OLED) is one of the promising technologies for lighting and display applications. In order to achieve full-color emission, complicated techniques have to be used to accurately fabricate three OLED structures for red, green, and blue emission in small areas. It will be highly desirable if a simple light-emitting device can emit light with a wide range of color, continuously tunable by changing the voltage. It may greatly simplify the fabrication techniques of multi- and full- color devices.

Voltage-controlled color tuning can be realized in a single unit OLED and has been reported in the devices using small molecules [1,2] and polymers [3,4]. In this paper, we will report a voltage-controlled color-tunable OLED based on a europium complex, in which the emitting color is tunable from red to blue or other designable color for two color tunable OLEDs. By using the vertically stacked two OLED units, full color tunable OLEDs are obtained. The colour purity and efficiency of the tunable OLEDs will also be addressed.

Experiment and Theory

In fabricating OLEDs, the ITO substrates are cleansed prior to loading into the evaporation chamber through scrubbing by detergent and soaking into de-ionized water for 10min in each step. The organic materials are purified by gradient sublimation prior to thin-film coating. The deposition rate is typically 1-2 Å/s and the substrate is Si. The evaporation chamber is operated at ~10^{-6} Torr. Film thickness is monitored *in situ* using the quartz crystal monitor and *ex situ* by a stylus profilometer (Tencor α-step 500). The OLEDs were biased a programmable Keithley source meter 2400.

In order to understand the color tuning spectrum and design the cavity structures, the light emission of multilayered OLED structures is rigorously modeled through classical approach with an emitting layer sandwiched between two stacks of films. The randomly oriented electric dipole is located in the recombination zone. The two stacks of films above and below the emitting layer can be considered as two effective interfaces characterized by the total reflection and transmission coefficients. The total radiation power F, normalized by the radiation power

of the dipole in an infinite medium ε_e, can be obtained [5-7]. Similarly, the normalized power U transmitted to outermost region (air) can also be obtained. With the consideration of the Purcell effect, the emission spectrum can be determined as

$$I(\alpha,\lambda) = s(\alpha,\lambda)L(\lambda) \quad (1)$$

where $s(\alpha,\lambda)$ is the angular power density in the outermost layer (air) and $L(\lambda)$ is the intrinsic spectrum of the emissive materials.

Results and Discussion
(a) Two Color Tunable OLEDs

The basic device structure (Device A) is ITO/ NPB(40nm)/ Eu(DBM)3bath(40nm)/ Alq3 (20nm)/ LiF/ Al. NPB is the hole-transport materials and also used as the host of emission layer in this study. Eu(DBM)3bath has been shown to have good electron-transport and emitting property in the earlier studies [8,9]. The multilayer of Alq3/LiF/Al acts as a composite electron-injection electrode. The EL spectra of Device A at various driving voltages are shown in Fig. 1. At the voltage of 6 V, the EL spectrum shows solely sharp emission from the Eu complex. At higher voltages an emission in the shorter-wavelength region appears and the intensity continuously strengthens. Comparing with the photoluminescent (PL) spectrum of NPB, the board emission band can be divided into two parts: one is the emission at ~450nm from NPB; the other at ~500nm can be due to either Alq3 or exciplex which formed at the interface between two organic layers. As a result, continuously reversible color tuning from pure red to blue is obtained.

The first reason is the efficiency decrease of Eu(DBM)$_3$bath at high voltages; the other is the shift of recombination zone in the device. Transient EL has been used to understand the biexcitonic quench and efficiency decrease of the Eu^{3+} emission in the process of color tuning. Considering that interaction between two excited Eu^{3+} ions causes a quenching channel in the decay process of the excited Eu^{3+} ions, the transient intensity $L(t)$ violate the typical exponential decay rule and follows [9]:

$$L(t) = \frac{L(0)}{(1+K\tau)e^{t/\tau} - K\tau} \quad (2),$$

where τ is the lifetime of the excited ions due to emissive decay, $L(0)$ is the initial emission intensity and K is the biexcitonic quench rate. Thus the quantum efficiency η of the Eu^{3+} ions is:

$$\eta = 1/(1 + K\tau) \qquad (3),$$

where τ can be characterized by transient PL or EL measurement under weak excitation intensity. K can be determined by fitting the transient EL decay data with Eq (2).

The decay of Eu^{3+} ions after excitation with pulse voltages are shown in Fig. 2. We set the measured wavelength at 614nm such that the detected signals only describe the decay of Eu^{3+} ions. The inset of Fig. 2 shows the PL decay of $Eu(DBM)_3$bath and the exponential decay fit gives a lifetime of ~300 μs. Using this value for τ in Eq (2) the decay process was fitted as shown in Fig. 2. The values of K are 0.7×10^3, 3.5×10^3 and 1.8×10^4 sec^{-1} for 6, 8 and 14 V, respectively. According Eq (2), η of Eu^{3+} emission are 0.82, 0.49, and 0.16 for 6, 8 and 14 V, respectively. These clarify that the biexcitonic quench do happen in the device. The quenching make the efficiency of Eu^{3+} emission decrease significantly at higher voltages and therefore is useful for achieving color tuning.

The second reason is the shift or extension of recombination zone at the interface with the driving voltages is greatly possible. As shown in Fig. 1, since the emission of $Eu(DBM)_3$bath still presents at high driving voltage, it is ready to consider that extension of recombination zone occurs in the device, i.e. the carrier recombination zone mainly locate in the $Eu(DBM)_3$bath layer at low bias and it extends across the interface into the NPB layer at higher voltage. This is the second major reason for the color tuning of the device.

(b) Color Purity

In order to improve the colour purity from NPB or othe dye, a microcavity structure is applied to the OLED structure. The optimized structure of ITO/Ag(x nm)/ CuPc(10nm)/ NPD(50 nm)/ $Eu(DBM)_3$bath (20nm)/ Alq$_3$ (40nm)/ LiF (0.5nm)/ Al (150nm) wiht x =0 (no cavity) and 40nm (with cavity) shown an improved color purity red-to-green OLEDs (see Fig. 3). Similarly, red-to-blue OLEDs can be obtained by reducing the device thickness to become ITO/Ag(40 nm)/ CuPc(10nm)/ NPD(40 nm)/ $Eu(DBM)_3$bath (15nm)/ Alq$_3$ (25nm)/ LiF (0.5nm)/ Al (150nm). The result is shown in Fig. 4.

(c) Full Color Tunable OLEDs

An unit of OLED is stacked to the two color tunable OLED discussed in (a) to achieved the full color tuning. The electrical and optical properties of Al/Au thin film as the independently-addressable intermediate electrode in the color-tunable stacked organic light-emitting device (SOLED) are examined. High brightness color tunable SOLED can be achieved by properly optimizing electron and hole injection and light transmission of the Al/Au electrode. The performance of the full color tunable device of ITO/NPB (40 nm) /Eu(DBM)$_3$Bath (20 nm) /Alq$_3$ (30 nm) /LiF (0.5 nm) /Al (16 nm) /Au (10 nm) /m-MTDATA:F4-TCNQ (7 nm) /NPB (40 nm) /Alq$_3$ (55 nm) /LiF (0.5 nm) /Al (70 nm) is shown in Fig.5.

(d) Efficiency of tunable OLEDs

To improve the efficiency of the OLEDs, we have changed Eu(DBM)3bath to phosphorescent material of Ir(piq)$_3$ in the stack OLEDs of (c). The results of ITO/NPB(40 nm) /NPB:Ir(piq)$_3$(30 nm, Weight ratio 10:1) /Bphen(20 nm) /Alq$_3$(40 nm) /LiF(0.5 nm) /Al(16 nm) /Au(10 nm) /m-MTDATA:F4-TCNQ(7 nm, Weight ratio 50:1) /NPB(50 nm) /Alq$_3$(60 nm) /LiF(0.5 nm) /Al(70 nm) are shown in Fig. 6. In addition, we further improve the efficiency by adding an oxide layer between Al/Au intermediate electrode. The result of a green stack OLED of repeated NPB/Alq3:C545T/Alq3 with Al/WO$_3$/Au as the intermediate electrode as shown in Fig. 7.

Conclusions

Full color tunable OLEDs have been demonstrated. The color purity and efficiency of the tunable OLEDs have been improved.

References

[1] M. Yoshida, A. Fujii, Y. Ohmori, and K. Yoshino, "Three-layered multicolor organic electroluminescent device", Appl. Phys. Lett., vol. 69, pp.734-746, 1996.

[2] R. Reyes, M. Cremona, E.E.S. Teotonio, H.F. Brito, and O.L. Malta, Chem. Phys. Lett. 396, 54 (2004)

[3] Y. Z. Wang, R. G. Sun, F. Meghdadi, G. Leising, and A. J. Epstein, "Multicolor multilayer light-emitting devices based on pyridine-containing conjugated polymers and para-sexiphenyl oligomer", Appl. Phys. Lett., vol. 74, pp.3613-3615, 1999.

[4] C. C. Huang, H. F. Meng, G. K. Ho, C. H. Chen, C. S. Hsu, J. H. Huang, S. F. Horng, B. X. Chen and L. C. Chen, "Color-tunable multilayer light-emitting diodes based on conjugated polymers", Appl. Phys. Lett., vol. 84, pp.1195-1197, 2004.

[5] X. W. Chen, W. C. H. Choy and S. He, "Efficient and Rigorous Modeling of Light Emission in Planar Multilayer Organic Light-Emitting Diodes", IEEE J. Display Technol., vol.3 pp.110-117, 2007

[6] W. Lukosz, ''Theory of optical-environment-dependent spontaneous emission rates for emitters in thin layers,'' Phys. Rev. B vol. 22, pp. 3030–3038, 1980.

[7] K. Neyts, "Simulation of light emission from thin-film microcavities," J. Opt. Soc. Amer. A, vol. 15, pp. 962–970, 1998.

[8] C. J. Liang, T. C. Wong, L. S. Hung, S. T. Lee, Z R Hong and W.L. Li, "Self-quenching of excited europium ions in Eu(DBM)3bath-based

organic electroluminescent devices", J. Phy. D, vol. 34, pp.L61-L64, 2001.

[9] C.J. Liang, W. C. H. Choy, "Color-tunable organic light-emitting diodes by using europium

organometallic complex", Appl. Phys. Lett., vol. 89, pp.251108-1-3, 2006.

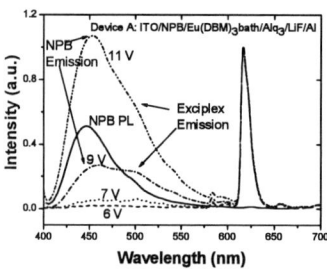

Fig.1 The EL spectra of Eu(DBM)₃bath tunable OLED. All the EL spectra are normalized to the Eu peak at 615 nm for comparison. The solid line is the PL spectrum of NPB film.

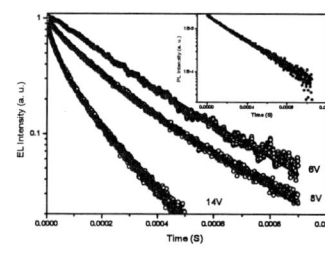

Fig. 2 Transient EL of the Eu^{3+} ions by the pulse excitations: experimental data (open circles) and fitting curves (full curves) obtained by using Eq (2) with $K = 0.7 \times 10^3$, 3.5×10^3 and 1.8×10^4 s^{-1} for 6, 8 and 14 V, respectively. The fitting curves are normalized at $t = 0$ for comparison. The inset shows the photon lifetime decay of Eu(DBM)₃bath.

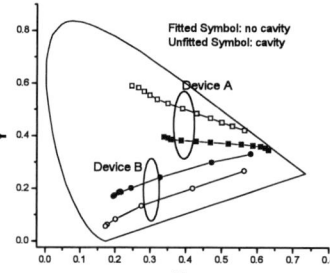
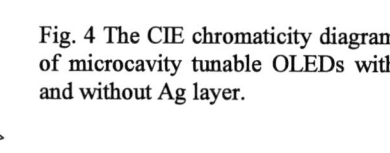

Fig. 4 The CIE chromaticity diagram of microcavity tunable OLEDs with and without Ag layer.

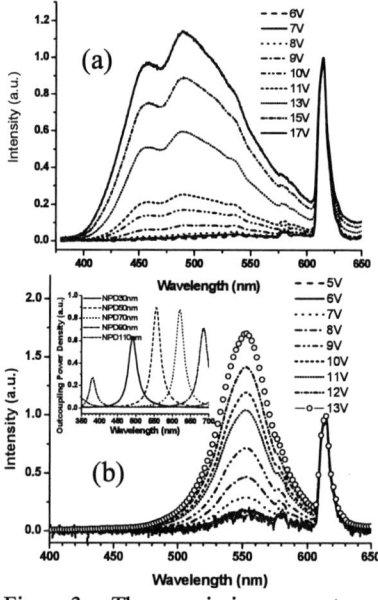

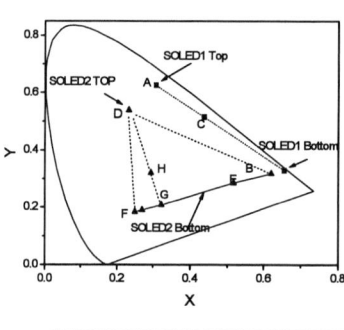

Fig. 5 The color tuning of full color tunable OLEDs on CIE coordinates.

Fig. 3 The emission spectrum microcavity tunable OLED (a) without Ag layer and (b) with Ag layer of 40nm. The inset of (b) is the zero degree outcoupling power density of the cavity structure with 40nm of Ag layer and various NPD thicknesses.

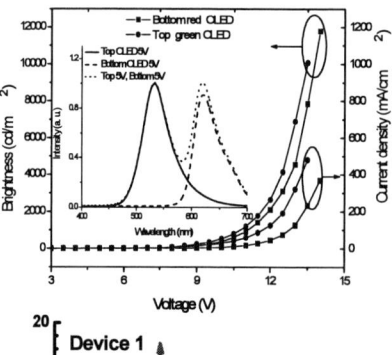
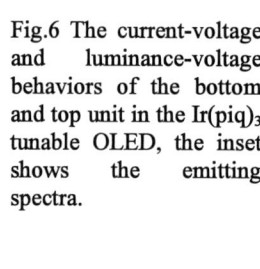

Fig.6 The current-voltage and luminance-voltage behaviors of the bottom and top unit in the Ir(piq)₃ tunable OLED, the inset shows the emitting spectra.

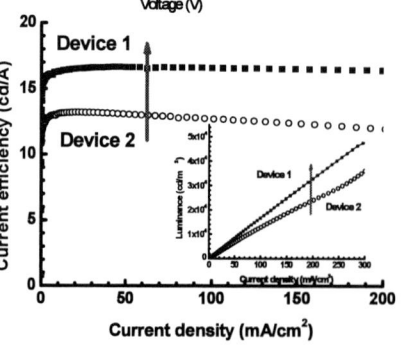

Fig.7 The improvement of efficiency by using Al/WO₃/Au intermediate electrode demonstrated in green stack OLED.

242

High-Performance Passive-Matrix OLED Display by Colour Conversion Method

Y. Terao (1), M. Kobayashi (1), N. Kanai (1), R. Makino (1), C. Li (1), Y. Kawamura (1),
K. Kawaguchi (1), T. Saito (1), H. Hashida (2) and H. Kimura (1)

1 : Fuji Electric Advanced Technology Co., Ltd., 4-18-1 Tsukama, Matsumoto, Nagano
390-0821 JAPAN, email: terao-yutaka@fujielectric.co.jp

2 : Fuji Electric Device Technology Co., Ltd., 4-18-1 Tsukama, Matsumoto, Nagano
390-0821 JAPAN

Abstract *Highly power efficient white OLEDs for PM-OLED displays were developed. The operating voltage and the power-efficiency of the optimised device are 6.4V and 5.9lm/W at 1A/cm². The prototype PM-OLED display with Advanced-CCM was made.*

Introduction

The organic light emitting diode (OLED) displays are potentially more energy efficient than LCDs, therefore the products meet with the global demand for low power consumption. Much effort has been done for the high efficiency, low-voltage and thus power efficient OLEDs to be applied to displays as well as lightings.

Colour conversion method, which typically uses a single blue OLED to pump the fluorescent wavelength down-converters, also known as colour-changing media (CCM) [1], is a possible solution for the efficient, low-cost full-colour OLED displays. In this technology the deposition process with precision metal masks to pattern the individual colours is not necessary, which permits using large substrates and therefore enables cost-effective production. But there have still been a few issues in conventional CCM: the low conversion efficiency, the instability of the materials especially for red CCM, the costly process to incorporate thick CCM layers into the display panel [2].

Lately Advanced-CCM, which is composed only of organic dyes, has been proposed [3]. It might overcome those issues because of its extremely small thickness and superior conversion efficiency. The prototype of an active-matrix OLED (AM-OLED) display with this technology was made using top-emitting architecture and showed the possibility [4].

In this paper the layer structures of OLEDs were investigated from the viewpoint of passive-matrix OLED (PM-OLED) display and a 2.8inch diagonal PM-OLED panel with QVGA resolution was made with Advanced-CCM.

OLEDs performance optimization

Figure 1 shows the schematic diagram of the device. Indium zinc oxide (IZO) is used for a transparent anode and 1nm thick LiF between an electron transport layer (ETL) and Al cathode is used for an electron injection layer. The organic layers are composed of hole injection layer (HIL), hole transport layer (HTL), emitting layer (EML) and ETL. To reduce the influence of relatively thick HIL on the drive-voltage 2,3,5,6-tetrafluoro-7,7,8,8-tetracyano-quinodimethane (F_4-TCNQ) is doped into a commercial HIL material. The EML contains guest fluorescent dyes, which are selected to produce optimum electroluminescence (EL) spectrum for CCM system. All the OLED materials studied in this paper are commercially available through each vendor.

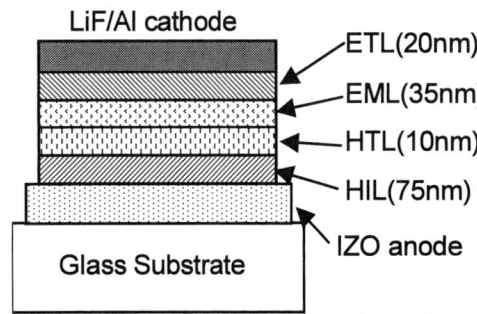

Fig. 1. *Schematic structure of OLEDs*

First the influence of EML host materials was studied. Three EML hosts were used with the same light-blue guest dye of which concentration was 5vol%. Figure 2 shows the luminance efficiency (L/J) vs. current density (J) characteristics. The efficiency roll-off of the device with EML2 is less than those of others, which indicates that EML2 is suitable for PM-OLED application.

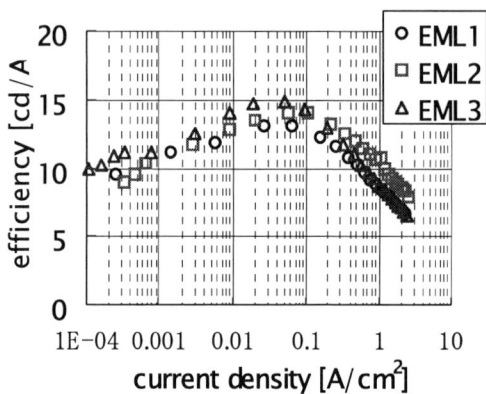

Fig. 2. *L/J-J characteristics of devices with different EMLs*

Next three kinds of ETLs were tested. In this experiment two guest dyes were co-evaporated with EML2; 5vol% light-blue dye and 0.08vol% orange dye were used for white radiation. The current density (J) vs. voltage (V) characteristics and L/J-J characteristics are shown in Figure 3 and Figure 4 respectively.

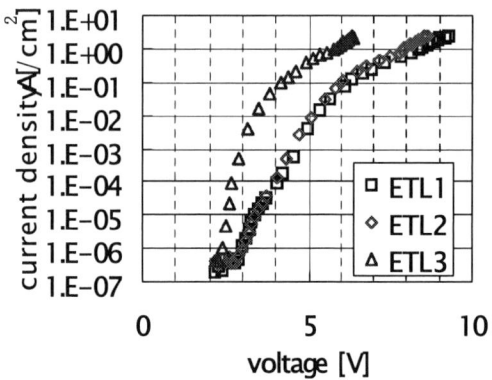

Fig. 3. *J-V characteristics of devices with different ETLs*

ETL3 significantly reduces the drive voltage of the OLED, while it also decreases luminance efficiency. These results are attributed to the excess electrons that may be easily injected into EML with ETL3. This was confirmed by the OLEDs with partially dye-doped EML with which the

recombination area could be specified. The recombination area was just at the HTL/EML interface, which suggests that the more holes could be injected into EML the more efficient the devices can be.

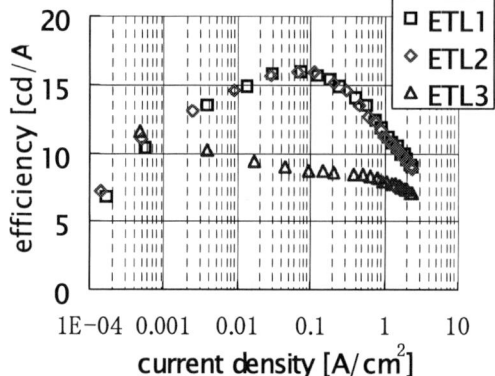

Fig. 4. *L/J-J characteristics of devices with different ETLs*

Finally HIL/HTL layers were studied to match the EML2/ETL3 system. When designing the OLED layers to achieve more holes in EML, the highest occupied molecular orbital (HOMO) level offset between HTL and EML was rather important. The HOMO level of EML2 and a candidate HTL are 5.8eV and 5.7eV below vacuum level respectively. These values are measured by AC-2 (Riken Keiki).

Figure 5 and Figure 6 show J-V characteristics and L/J-J characteristics of the optimised device. The device showed the operating voltage of 6.4V, the luminance efficiency of 12.1cd/A and the power efficiency of 5.9lm/W at $1A/cm^2$.

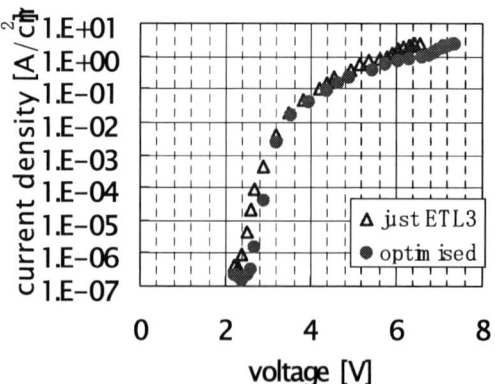

Fig. 5. *J-V characteristics of the optimised device*

244

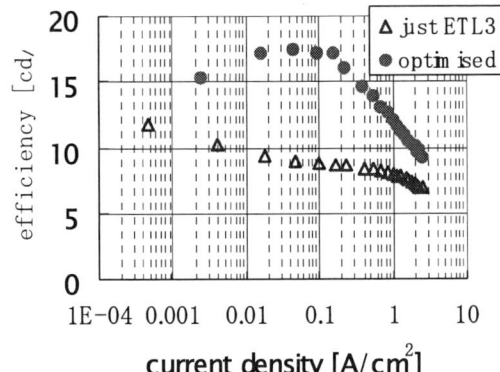

Fig. 6. *L/J-J characteristics of the optimised device*

PM-OLED display with Advanced-CCM

The schematic structure of PM-OLED with Advanced-CCM is shown in Figure 7. This time only red CCM was utilized. Filtering EL light produces blue and green light.

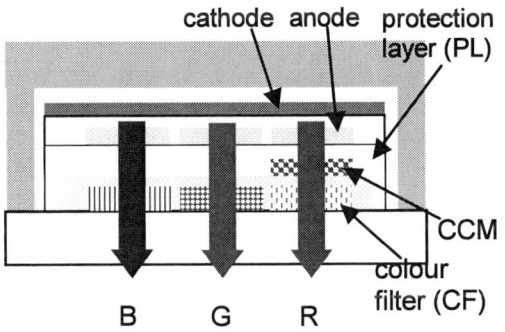

Fig. 7. The schematic structure of PM-OLED display with Advanced-CCM

Since Advanced-CCM is made of organic dyes, it should not be exposed to the air or water during the process. Therefore the protection layer (PL) is very essential for defect free displays. Even if the usual wet processes have done on the PL, Advanced-CCM has not been damaged.

Figure 8 shows the pictures of the prototype panel and the display specification is summarized in Table 1.

Conclusions

Low-voltage, highly efficient OLED was developed for PM-OLED displays. Electron-hole balance was important for high efficiency. The operating voltage and the

power-efficiency of the device are 6.4V and 5.9lm/W at 1A/cm^2. Combined with Advanced-CCM a 2.8inch diagonal PM-OLED panel with QVGA resolution was made. Advanced-CCM can be compatible with the bottom-emitting architecture.

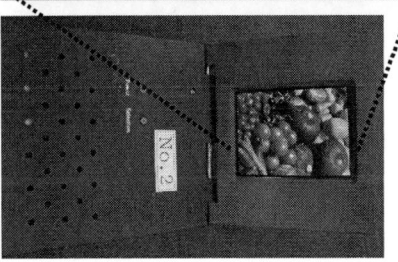

Fig. 8. The picture of PM-OLED panel

Table 1. PM-OLED display specification

Display size		2.8in
Resolution		QVGA
CIE coordinate	Red	0.64, 0.35
	Green	0.25, 0.59
	Blue	0.11, 0.21
Luminance efficiency	Red	1.8cd/A
	Green	4.5cd/A
	Blue	2.6cd/A
Peak Luminance		150cd/m^2
White point		0.31, 0.33

References

1 C. Hosokawa et al SID Symposium Digest of technical papers, 28 (1997), 1073
2 K. Sakurai et al Proceedings of the 11th International Display Workshops (2004), 1269
3 C. Li et al SID Symposium Digest of technical papers, 37 (2006), 1372
4 Y. Terao et al Proceedings of the 13th International Display Workshops (2006), 457

Improvement of electrical characteristics of fluorinated perylene diimide thin-film transistors by gate dielectric surface treatment

Li-Gong Yang[1,2], Jia-Chi Huang[1], Rong-jin Li[3], Min-Min Shi[1], Yan Gao[1,2], Mang Wang[1], Wen-Ping Hu[3], Hong-Zheng Chen[1,2]*

1 : Department of Polymer Science and Engineering, Zhejiang University, Hangzhou, 310027, P. R. China, hzchen@zju.edu.cn

2 : Zhejiang-California International Nanosystem Institute, Hangzhou, P.R. China

3 : Organic Solids Laboratory, Institute of Chemistry, Chinese Academy of Sciences, Beijing, 100086, P. R. China

Abstract *The structural and electrical properties of n-channel OTFTs based on N,N'-(4-monofluorophenyl)-3,4,9,10-perylene tetracarboxylic diimide (D4MFPP) were investigated. The influence of surface treatment on the mobilityand the interface traps were quantitatively evaluated.*

Introduction

Organic thin-film transistors (OTFTs) have been intensively studied for decades since their potential applications such as in flat panel display and electronic circuits on flexible substrates, with the advantages of low cost and low temperature fabrication [1-3]. By the request of complementary components in logic circuits, more efforts have been transferred to develop n-channel organic thin film transistors (OTFTs) by synthesizing novel n-type materials [4-5] and improving performance of devices [6]. Surface treatment of gate dielectric layer is an important step for higher performance of OTFTs [6-8], and mostly the surface modification materials form a self-assemble monolayer (SAM) through solution processes or gas phase deposition [9]. In this report, the influence of surface-treated gate dielectric on the electrical characteristics of OTFTs based on mono-fluorinated perylene diimide, an air-stable n-type material [6], was studied. We will show that the molecular stacking order on treated dielectric layer is significantly improved as well as the channel conductance.

Experimental

Fig.1 shows the molecular structure of N,N'-(4-monofluorophenyl)- 3,4,9,10-perylene tetracarboxylic diimide (D4MFPP) and the device structure used in this work. Its synthesis and purification processes have been described previously [10]. Thin-film transistors were fabricated on highly n-doped silicon wafers covered with a 500-nm-thick layer of thermally grown silicon oxide. This highly doped silicon served as gate electrode. The surface of silicon oxide layer was ultrasonically cleaned by acetone and alcohol respectively before further treatments. Then drops of octadecyltriethoxysilane (OTS, from Aldrich Chem. Co.) on a glass substrate were heated, the Si substrates were exposed to OTS vapour for around 3 hours. After that fluoroperylene diimides were deposited by vacuum thermal evaporation at a steady deposition rate of ca. 0.2Å/s under a base pressure smaller than 4×10^{-4} Pa. Then gold electrodes of interdigited geometry were evaporated through a shadow mask onto the organic film to serve as source and drain electrodes. The channel length was 110 μm and the width is 5300 μm . The thickness of channel material is 35 nm.

Influence of surface treatment on film structure and morphology

Fig.2(a) shows the XRD pattern of D4MFPP film deposited on OTS-treated SiO_2 layer. Two peaks exhibit with multiple d spacing, indexed as (*00l*) peaks with 16.98Å of d_{001}. The absence of any other (*hkl*) peaks indicates highly textured film in which all the microscopic crystallites are oriented parallel to the substrate surface. Based on this pattern it implies that D4MFPP molecules prefer "end-on" to "lie-on" deposition orientation and their π-stacking is parallel to the dielectric surface.

978-0-9789217-3-6/07/$25.00

©2007 WEN GLOBAL SOLUTIONS

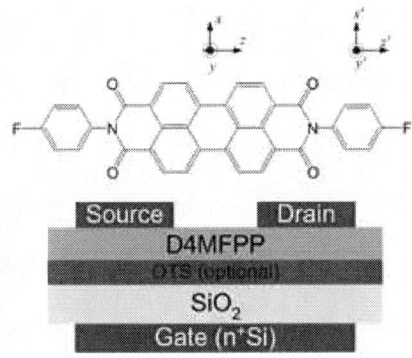

Fig.1 Molecular structure of D4MFPP and schematics of the top-contact TFT structure used in this study.

Fig.2 (b) presents the XRD pattern of D4MFPP film deposited on an untreated SiO₂ substrate. It is found that there are six reflection peaks corresponding to different crystal phases. The *d* spacings of the first two peaks are much smaller than those in Fig.2 (a) and the crystal structure is more compact, which are attributed to the crystal bulk phase of the material and are denoted with (00*l*). The middle three peaks correspond to different *d* spacings, and no apparent relationship between them and the first *d* spacings of the thin film phase or the crystal bulk phase can be found. This means they are new microcrystalline orientations. Evidently, the microscopic structure of D4MFPP film deposited on SiO₂ surface without OTS treatment is complicated. The regular stacking form is unexpected in these films.

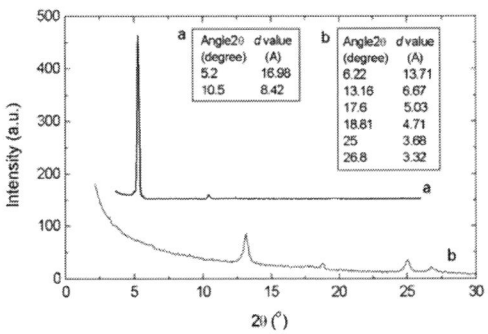

Fig.2 X-ray diffraction patterns of D4MFPP films deposited on a SiO₂ substrate (a) with and (b) without surface treatment. The corresponding d spacings have been calculated.

Further morphology investigation shows that, in the films deposited on bare SiO₂ surface, the grain size is rather small and disordered packing form can be identified. Without OTS treatment, the high reactivity of fluorine groups at the ends of D4MFPP molecules with OH groups leads to the high preference of the "lie-on" orientation of D4MFPP molecules. Therefore π-π stacking is interrupted and disordered morphology can be observed. After surface treatment, OTS is expected to form a well-ordered (2-dimensional crystalline) monolayer on the exposed SiO₂ surface [11]. The high reactivity of the chlorine groups with OH groups leads to an extended network through interchain crosslinking, which effectively reduces the surface roughness and the surface energy [12]. These modify the nucleation condition of D4MFPP molecules, which leads to a better microscopic orientation of the initial several layers of molecules. Therefore the local packing order is prominently improved and the grain size is enlarged. It's more suitable for charge transporting between source and drain electrodes.

Effect of surface treatment on channel conductance

Fig.3 shows the transfer characteristics of D4MFPP TFTs with and without OTS treatment. It is found that the devices with OTS treatment work well (see Fig.3(a)) while no conductivity can be observed for the devices without OTS treatment in Fig.3(b). By increasing the gate voltage, the devices without OTS treatment are still in "off" state. According to the analysis in the previous section, this poor performance should be attributed to the disordered packing form of D4MFPP molecules at the semiconductor-dielectric interface and a large number of traps. These traps could stem from the local defects or Si-OH electron trapping centers at the surface of SiO₂ layer [11].

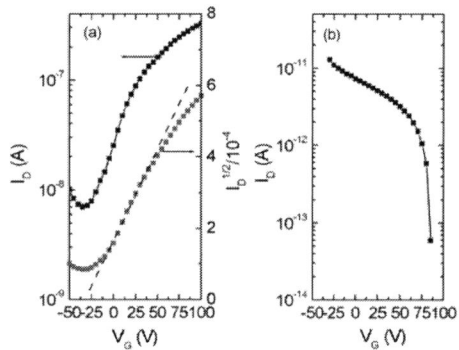

Fig.3 Drain-source current (I_D) vs. gate-source voltage (V_G) transfer characteristics (V_D=100V) of D4MFPP OTFTs (a) with and (b) without OTS treatment.

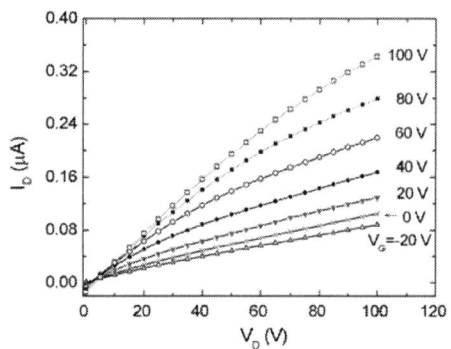

Fig.4 Drain-source (I_D) vs. drain-source voltage (V_D) output characteristics of D4MFPP TFTs.

Fig.4 is fhe output characteristic of the devices with OTS treatment. The field-effect mobility μ_{FET} and threshold voltage V_{th} of D4MFPP film can be derived from conventional field-effect transistor equations [13]. A saturation mobility of $4.3 \times 10^{-5}\,\mathrm{cm}^2/\mathrm{Vs}$ can be obtained from the plots in Fig.3(a), whereas in line regime (Fig.4) a higher mobility value of $1.03 \times 10^{-4}\,\mathrm{cm}^2/\mathrm{Vs}$ can be calculated.

The interface traps in OTS-treated devices are also estimated by utilizing the subthreshold swing ($S = \partial V_G / \partial(\log I_D)$)[14].

Thus a density of interface traps of $4.8 \times 10^{12}\,\mathrm{cm}^{-2}$ is approximately evaluated for OTS-treated D4MFPP TFTs.

Conclusions
The influence of dielectric surface treatment with OTS on the film structure of monofluorinated perylene diimide and on the electrical performance of OTFTs based on this material was investigated. The packing order of films and grain size were prominently improved after the OTS-treatment.

Financial supports from NSFC and MOST are appreciated.

References
1 A.Tsumura, et al, Appl. Phys. Lett. 49 (1986) 1210.
2 Z. T. Zhu, et al, Appl. Phys. Lett. 81 (2002) 4643.
3 C. D. Dimitrakopoulos et al, Adv. Mater. 14 (2002) 99.
4 A. Facchetti, et al, Adv. Mater. 15 (2003) 33.
5 P. R. L. Malenfant, et al, Appl. Phys. Lett. 80 (2002) 2517.
6 I. H. Campbell, et al, Appl. Phys. Lett. 71 (1997) 3528.
7 H. Klauk, et al, Solid State Technol. 43 (2000) 63.
8 H. Sirringhaus, et al, Science 290 (2000) 2123.
9 M. M. Lin, et al, Chem. Mater. 16 (2004) 4824.
10 M. M. Shi, et al, Chem. Comm. (2003) 1710.
11 D. J. Gundlach, Ph.D thesis, The Pennsylvania State University, 2001.
12 J. D. Le Grange, et al, Langmuir 9 (1993) 1749.
13 M. Shur, Physics of Semiconductor Devices, Prentice Hall, New Jersey, 1990.
14 O. Marinov, et al, J. Va, Sci. Technol. B 24 (2006) 1728.

Electrochemical polyaniline/polypyrrole composite film with novel nanostructure and high biosensitivity

Yunan Cheng[a], Gang Wu[a], Gustaaf Borghs[b], Mang Wang[a]*, Hong-Zheng Chen[a]*

[a]Department of Polymer Science and Engineering, State Key Lab of Silicon Materials,

Qiushi Academy for Advanced Studies, Zhejiang University, Hangzhou 310027, China,

Email: hzchen@zju.edu.cn

[b]IMEC, Kapeldreef 75, B-3001, Leuven, Belgium

Abstract

Conductive polymer, such as polypyrrole (PPY) and polyaniline (PANI), have shown a special ability to rapidly and reversibly switch between different oxidation states [1,2], which plays a critical role in biocommunication. In addition, various biomolecules can be incorporated into conducting polymer by electrochemical polymerization. These two abilities make conducting polymers potential candidates for application in the area of biosensors [3,4]. The synthesis of conducting polymer films thus has recently been an increasingly important subject of intensive research. [5,6]

PPY is the most thoroughly investigated one among various conducting polymers for biological applications because of its high electrical conductivity in wide range of pH values, flexible method of preparation, ease of surface modification, excellent environmental stability, ion exchange capacity and biocompatibility [3,7-9,10-11] . But one of the properties that restrict the development of electrochemically synthesized PPY film is its flat surface with limited surface area. This has been overcome to some extent by template methods. Method of polymerizing pyrrole within porous materials was used to enhance the surface area of PPY[10-11]. PPY films with sub-100-nm features were synthesized on atomically flat surfaces using adsorbed surfactant molecules as templates [12]. Arrays of PPY dots (80-180 nm diameter) were prepared by using diblock copolymer surface micelle arrays as the reaction template[13]. PPY nanotubes with average pore diameter of 6 nm were synthesized by using $FeCl_3$ oxidant and V_2O_5 nanofibers as the sacrificial template [14]. In addition, the method of preparing multilayer polymer composite film on the base of PANI[15], polythiophene [PMET-PPY], polysiloxane[16], etc., was also employed to enhance the surface area of PPY and the continuous films with microscaled structure were obtained. While, few researchers reported the synthesis of continuous PPY films with nanostructrural surface morphology with a templateless method.

In this study, we synthesized PANI/PPY bilayer composite films on ITO substrate by two-step constant potential electrochemical method. Our ultimate goal is to produce continuous PPY films with large surface area to ease the immobilization of enzyme and to enhance the concentration of enzyme redox centers. The PPY and PANI films were also separately prepared on ITO for a comparison. Scanning Electron Microscope and Atomic Force Microscope were used to observe the surface morphology. The PANI/PPY bilayer composite film on ITO was found to be made up of micro/nano domains, which may contribute in a remarkable increase of the surface area of the film. The H_2O_2 amperometric biosensor was constructed by coating horseradish peroxidase (HRP) on the PANI/PPY bilayer composite film. Compared with the PPY or PANI film on an ITO substrate, the HRP-containing micro/nano structural PANI/PPY bilayer composite film could act as highly sensitive sensor for the enzymatically generated H_2O_2. The proposed biosensor permitted reliable determinations of H_2O_2 by chronoamperometry analysis. The detection limit and sensitivity were found to be on the range of 0.3 ~1.0 mmol/l and 38.81 mA cm^{-1} mol^{-1}, respectively.

This work was financially supported by the National Natural Science Foundation of China (No.50433020, 50520150165 and 50673083). The authors also would like to thank the financial support from the Research Fund for the Doctoral Program of Higher Education in China (No. 20060335078).

References:

[1] G.G.Wallace, L.A.P.Kane-Maguire, *Adv.Mater*. 2002,14,(13-14):953-960

[2] G.G.Wallace, G.M.Spinks, P.R.Teasdale, in *Conductive Electroactive Polymers: Intelligent Materials Systems,* Technomic, Lancaster, UK 1997

[3] E.Smela *Adv.Mater*.2003,15(6):481-494

[4] Y.Li, K.G.Neoh, L.Cen, E.T.Kang, *Langmuir*,2005,21:10702-10709

[5] J.X.Huang, S.Virji, B.H.Weiller, R.B.Kaner, *Chem.Eur.J*.2004,10,1314-1319

[6] J.Joo, B.H.Kim, D.H.Park, H.S.Kim, D.S.Seo, J.H.Shim, S.J.Lee, K.S.Ryu, K.Kim, J.I.Jin, T.J.Lee, C.J.Lee, *Syn. Met*.153(2005):313-316

[7] A.Azioune, A.B.Slimane, L.A.Hamou, A.Pleuvy, M.M.Chehimi, C.Perruchot, S.P.Armes, *Langmuir*,2004, 20:3350-3356

[8] G.J.Ashwell, *Molecular Electronics*; Research Studies Press:New York,1992:pp 65.

[9] K.Ramanathan, M.A.Bangar, M.Yun, W.Chen, N.V.Myung, A..Mulchandani, J.Am.Chem.Soc.2005,127: 496-497

[10] S.D.Champagne, E.Ferain, C.Jerome, R.Jerome, R.Legras, *Eur.Polym.J*.1998,34(12):1767-1774

[11] N.Hermsdorf, M.Stamm, S.Forster, S.Cunis, S.S.Funari, R.Gehrke, P.Muller-Buschbaum, *Langmuir*,2005,21: 11987-11993

[12] A.D.W.Carswell, E.A..O'Rear, B.P. Grady, *J. Am. Chem. Soc.*2003,125:14793-14800

[13] M.Goren, R.B.Lennox, *Nano Letters,*2001,1(12):735-738

[14] X.Y.Zhang, S.K.Manohar, *J.Am.Chem.Soc*.2005,127:14156-14157

[15] J.Morales, M.G.Olayo, G.J.Cruz, R.Olayo, *J.Poly.Sci. Part B:Polymer Physics*,2002,40:1850-1856

[16] G.Cakmak, Z.Kucukyavuz, S.Kucukyavuz, *Synthetic Metals,*2005,151(1):10-18

Overview of Japanese FTTH market and NTT's strategies for entering full-scale FTTH era

Hiromichi Shinohara
NTT Information Sharing Laboratory Group

In Japan, FTTH market is growing very rapidly. The number of FTTH users surpassed 8.8 million in March 2007. The transition from ADSL to FTTH service is remarkable in NTT. The number of FTTH user in NTT has already outstripped the one of ADSL. This presentation gives an overview of the Japanese FTTH market. It will touch on FTTH services that NTT currently offers and typical applications on FTTH. It also describes driving force behind FTTH growth. NTT has been introducing EPON on a large scale to offer IP service such as high-speed Internet and high-quality IP telephone. In April 2006, NTT started to provide broadcast re-transmission service by RF format in conjunction with satellite television provider by using Video-PON overlay. This presentation gives configuration of commercial FTTH systems. It also gives lessons learned from actual FTTH deployment and latest technologies used in massive FTTH deployment, highlighting easy construction technologies. NTT set the target of moving 30 million customers to FTTH by 2010. Technologies required for entering full-scale FTTH era is also discussed.

On-line Optical System Performance Monitoring Using Coherent Detection

Rongqing Hui
Department of Electrical Engineering and Computer Science
The University of Kansas, Lawrence, KS 66045
(rhui@ku.edu, (785) 864-7740)

Abstract Techniques of *optical system performance monitoring using coherent detection will be discussed. Especially, recent field test of fiber PMD performances in traffic-carrying optical systems will be presente.*

Introduction

Optical network management is gaining importance as the number of channels is increasing in WDM optical systems and the optical networks are becoming more complex. One of the key components in optical network management is the quality monitor [1], which determines whether the network is within the specified performance limitations [2]. Performance monitoring in the optical domain is expected to be flexible and cost effective in future optical networks. It will also significantly simplify the architecture of optical networks.

Coherent detection is an attractive technique for optical system performance monitoring. In addition to performing high-resolution optical spectrum analyzing, coherent heterodyne detection enables the monitoring of chromatic dispersion (CD) and PMD at each wavelength channel. Utilizing data traffic carried on the system, this technique can also be used to extract fiber characteristics without the need to access to the transmitter.

Ultra-high resolution optical spectrum analyzing:

Optical coherent detection was developed a few decades ago in order to achieve quantum noise-limited detection sensitivity [3]. In recent years, tunable laser technology has been significantly improved and high quality small footprint, highly reliable tunable lasers are commercially available, which makes coherent detection a practical approach. In particular, coherent detection is an excellent choice for optical signal processing and optical measurement thanks to the wide wavelength coverage,

continuous wavelength tunability and narrow spectral linewidth of tunable lasers. By linearly sweeping the wavelength of the local oscillator, a coherent detection optical receiver can be used to analyze signal optical spectrum [4]. Compared to conventional grating-based optical spectrum analyzer (OSA), coherent OSA has much higher spectral resolution which is only limited by the spectral linewidth of the local oscillator.

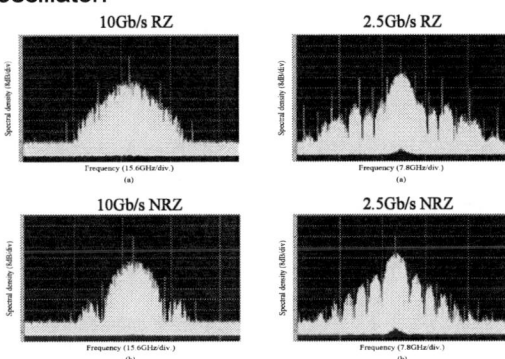

Fig.1 Measured Optical spectrum of 10Gb/s(left) and 2.5Gb/s (right) signals with RZ (top) and NRZ (bottom) modulation. Horizontal: 7.8 GHz/div. Vertical: 8 dB/div

As an example, Fig.1 shows the optical spectrum of 10Gb/s and 2.5Gb/s optical systems with RZ and NRZ modulation formats, respectively measured by a coherent OSA, which clearly revealed the detailed spectral features of the optical signals. This enables precise identification of datarate and modulation format of the signal in the optical domain, which cannot be achieved by conventional grating-based OSAs.

Monitoring chromatic dispersion

978-0-9789217-3-6/07/$25.00
©2007 WEN GLOBAL SOLUTIONS

The basic idea behind CD monitoring in digital fiber-optic system is based on the fact that, the spectrum of an optical signal typically has two redundant clock frequency components and an optical carrier. Due to CD, these two clock components propagate in different speeds creating a differential delay at the receiver. In the performance monitor, coherent detection down-converts the spectrum of the optical signal into RF domain and the relative phase-delay information of the optical signal is preserved. In RF domain, the carrier and the two clock sidebands can be selected separately by three bandpass filters. As shown in Fig.2, the carrier component is further split into two and used to mix with the upper and the lower sidebands independently to generate two independent clocks. After the RF mixer, a narrowband RF filter is used in each branch to purify the recovered clocks. The chromatic dispersion can be evaluated from the relative time delay Δt between these two recovered clocks by:

$$\Delta t = DL\lambda^2 R_b / c \qquad (1)$$

where D is the fiber chromatic dispersion parameter, L is the fiber length, λ is the signal wavelength, c is the speed of light and R_b is the data rate.

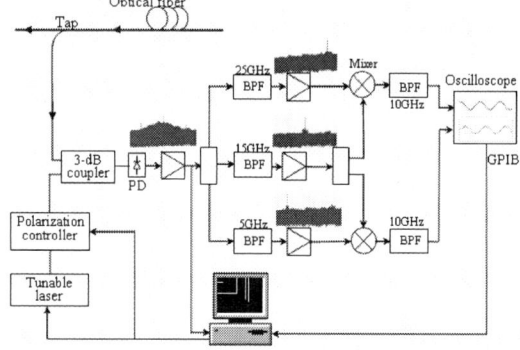

Fig.2 Block diagram of CD monitoring of a 10Gb/s system using coherent detection.

The measured results showed reasonably good accuracy using this technique [5].

Monitoring of PMD
The PMD monitoring technique was based on the measurement of differential polarization walk-off between any two different frequency components within the optical signal [5,6]. In fact, Due to first-order

DGD ($\Delta\tau$) of the fiber system, the differential polarization walk-off between two frequency components separated by Δf can be represented as their relative angular walk-off on the Poincare sphere:

$$\Delta\varphi = \pi\Delta f \cdot \Delta\tau \qquad (2)$$

Since coherent heterodyne detection shifts the optical spectrum linearly into the RF domain, this measurement can be performed by simultaneously measuring RF powers selected at two frequencies. Thanks to the fact that coherent detection is inherently polarization-selective, the efficiency of coherent detection depends on the SOP mismatch between the signal and the local oscillator. The generated RF powers of the two selected frequency components f_1 and f_2 are, respectively:

$$P_1 = \eta_1 P_{lo} P_{sig}(f_1)\cos^2(\xi)\cos^2(\varphi)$$
$$P_2 = \eta_2 P_{lo} P_{sig}(f_2)\cos^2(\xi)\cos^2(\varphi + \Delta\varphi)$$

Where, ξ is angle between the local oscillator and the $E(f_1)/E(f_2)$ plane, φ is the angle between the projection of E_{LO} on the $E(f_1)/E(f_2)$ plane and $E(f_1)$, and $\Delta\varphi$ is the SOP angle between the two selected frequency components. η_1 and η_2 represent the combined effects of the responsivities of the two detectors and the relative amplitudes of the selected two frequency components, P_{lo} and P_{sig} are the optical powers of the local oscillator and the signal. Although direct measurement of P_1 and P_2 may indicate the angular polarization walk-off $\Delta\varphi$, uncertainties may be introduced because power spectral densities of the optical signal at the two selected frequencies may be different. In addition to the uncertainties in the responsivity of the photodiod and the RF amplifiers followed, this difference may also change with the signal modulation formats and modulator characteristics in the transmitter. This problem can be solved by randomly scrambling the SOP of the local oscillator, which made both ξ and φ random variables while $\Delta\varphi$ is relatively stable assuming that the DGD variation in the fiber system is much slower than the polarization scrambling of the local oscillator. After a few cycles of P_1 and P_2 passing through their maxima and minima, and the RF power measured at the two branches can easily be normalized. The phase difference $\Delta\varphi$ between the two power traces can be obtained and the 1st order DGD of the fiber system can be derived.

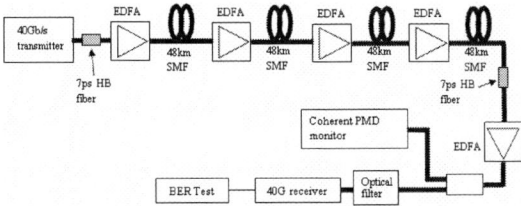

Fig.3 40Gb/s optical system for PMD monitoring

A PMD field test was setup first in a 48-km fiber link between Lawrence and Topeka Kansas and the fiber was looped back twice providing an approximately 192-km installed fiber link. The system has 40Gb/s datarate with duo-binary modulation. In order to produce a sufficient amount of PMD to observe system BER outages, two pieces of high birefringence (HB) fibers each having 7 ps of DGD were added in the system, one placed immediately after the transmitter and the other one placed before the receiver. In the test, the intermediate frequency of homodyne detection in the PMD monitor was set at 20GHz so that there is sufficient signal energy at 15GHz and 25 GHz frequencies which can be picked up by the two bandpass filters. Fig. 4 shows an example of comparison between measured DGD and system margin in dBQ for a 30-hour period. The results clearly show that the system margin is inversely proportional to the instantaneous DGD measured by the monitor. The transmission failed whenever the DGD level was higher than 15 ps.

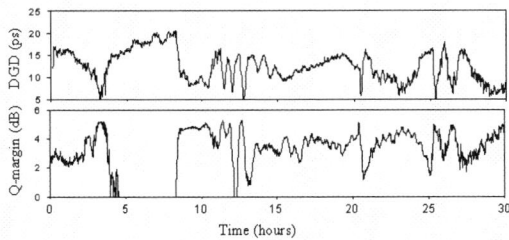

Fig.4, Comparison between measured DGD values and measured system dBQ margin

We have also performed a number of field tests on long-distance WDM fiber-optic networks which were in-service and carrying 10Gb/s commercial live traffics. Fig.5 shows two examples of the measured DGD each for a period of approximately 24

hours on a 700km and a 900km fiber system, respectively, in different locations. As expected, in these long distance systems, DGD varies much faster than in a short distance system. Fig.5 also shows the statistic distributions of the measured DGD on these two fiber links, which fit reasonably well to Maxwellian stastics. The absolute DGD values in these fiber links are not shown due to the confidentiality agreement with the user.

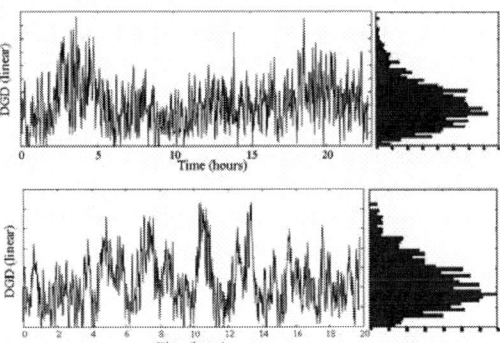

Fig.5, Examples of measured DGD in a 700km (top) and 900km (bottom) fiber links and the histogram of the data.

Conclusions

Coherent detection is an effective technique for performance monitoring of live optical systems. Techniques of ultrahigh resolution optical spectrum analyzing, chromatic dispersion monitoring and PMD monitoring have been demonstrated. The results of PMD monitoring field test on long-distance traffic-carrying fiber systems are presented.

The author would like to thank B. Fu, S. Sundhararajan P. Adany, R. Saunders, B. Heffner, D. Richards, and M.O'Sullivan for their contributions to this work, and thank Sprint-Nextel and Nortel-networks for their support.

References
1. Y. C. Chung, LEOS 2005 annual meeting. pp. 699 - 699
2. C. Pinart and G. J. Giralt, J. Lightwave Technol., 2005, p.2868 – 2876
3. *Coherent Optical Communications Systems*, S. Betti, G. De Marchis, E. Iannone, Wiley publish, 1995
4. D. M. Baney, et al, Photonics Technol. Lett., 2002, pp. 355
5. B. Fu and R. Hui, Photonics Technol. Lett., 2005, p, 1561
6. R. Hui, et al, Electron. Lett., 2007, p. 53

Waveguide Structure Evaluation Based on a Photon-Counting OTDR

Takamichi Aiba, Nori Shibata, and Masaharu Ohashi*

YAZAKI Research and Technology Center, 3-1 Hikarino-oka, Yokosuka-shi, 239-0847, JAPAN,

t-aiba@ytc.yzk.co.jp

*Osaka Prefecture University, 1-1 Gakuen-cho, Sakai, Osaka, 599-8531, JAPAN,

ohashi@uopmu.ees.osakafu-u.ac.jp

Abstract *Utility of a photon-counting OTDR for evaluating waveguide parameters along a sequentially-spliced single-mode fiber is confirmed quantitatively. Mode-field diameter and relative-index difference evaluated by the OTDR at 835 nm are in good agreement with those obtained by the reference test methods.*

Introduction

Automotive applications such as *telematics* and X-by-wire (e.g. X=Steer, Brake, and Suspension) for car weight reduction are very interesting, since these application protocols are expected to provide a powerful countermeasure to the increasing number of electronic control units connected to automobile LAN systems which provide a simple user interface. Therefore, there is strong demand for increasing the transmission capacity of such LANs. A graded-index polymer-clad silica-core (GI-PCS) fiber operating in the 850 nm-wavelength region has been studied for the deployment of ~Gb/s level automotive LANs[1]. Taking the ultimate goal of the automotive LAN system into consideration, single-mode fiber (SMF) operating in the vicinity of the 850 nm is ideal for supporting automotive fiber-optic links with the transmission capacity required.

Chromatic dispersion evaluations along the SMF link are of interest from the viewpoint of automotive fibering applications. In weakly guiding fibers[2], it is known that the chromatic dispersion is given by the sum of the material and waveguide dispersions relating to the relative-index difference Δ and the wavelength-dependent mode field diameter $2w$, respectively[3]. An optical time-domain reflectometer (OTDR) has been applied for evaluating Δ and $2w$ of SMF links operating at the 1550 nm[3]. Considering that fiber-lengths inside an automobile will be less than 100 m, an OTDR method with high spatial resolution such as a photon-counting OTDR[4] is a powerful tool for the evaluation of the waveguide parameters. In this paper, we use a single-photon-counting OTDR with 6 cm spatial resolution to evaluate Δ and $2w$ along the SMF at 835 nm. The values of Δ and $2w$ evaluated by the OTDR method are compared with those obtained by the conventional test methods.

Theoretical Background

Di Vita and Rossi[5] were the first to find a simple method in OTDR technique for obtaining the terms of the structural imperfection contribution and power decay along a multimode fiber from the backscattered waveforms. Ohashi and Tateda[6] applied the method to evaluate the chromatic dispersion distribution along an SMF. The method is based on a suitable comparison of two backscattered waveforms obtained by launching optical pulses into an SMF from both ends. For OTDR signals $S_1(\lambda,z)$ and $S_2(\lambda,L-z)$ obtained from opposite ends of a fiber of length L, the imperfection contribution $I(\lambda,z)$ is expressed as[3]

$$I(\lambda,z) = \frac{S_1(\lambda,z) + S_2(\lambda,L-z)}{2}$$

$$= C + 10 \log[\alpha(z)B(\lambda,z)] - (10 \log e) \int_0^L \xi(z)dz \ \text{[dB]} \quad (1)$$

where λ is a wavelength, C a constant independent of distance z, $\alpha(z)$ the local scattering coefficient, $\xi(z)$ the local attenuation coefficient, and $B(\lambda,z)$ the backscattered capture ratio given by

$$B(\lambda,z) = \frac{3}{2} \left\{ \frac{\lambda}{2\pi n(z)w(\lambda,z)} \right\}^2 . \quad (2)$$

Here $n(z)$ and $2w(\lambda,z)$ represent the

978-0-9789217-3-6/07/$25.00

©2007 WEN GLOBAL SOLUTIONS

refractive-index of the core and mode field diameter (MFD), respectively. The imperfection contribution $I(\lambda,z)$ can be normalized by the value at a reference point $z=z_0$, and using Eqs.(1) and (2) yields the normalized imperfection contribution $I_n(\lambda,z)$ as functions of the relative-index difference Δ and MFD as follows:

$$I_n(\lambda,z) \equiv I(\lambda,z) - I(\lambda,z_0)$$

$$\cong 10\log\left[\frac{1+k\Delta(z)}{1+k\Delta(z_0)}\right] + 20\log\left[\frac{2w(\lambda,z_0)}{2w(\lambda,z)}\right].$$

(3)

In the derivation of Eq.(3), we assume that the variation of $n(z)$ along the SMF is negligible, that is, $n(z)^2/n(z_0)^2 \approx 1$, and that the Rayleigh scattering coefficient R for GeO$_2$-doped core is expressed as $R=R_0(1+k\Delta)$[4]. When the MFD distribution and $\Delta(z_0)$ are known, $\Delta(z)$ along the fiber can be derived from Eq.(3) as follows:

$$\Delta(z)= \frac{1}{k}\left[\{1+k\Delta(z_0)\}10^{\frac{I_n(\lambda,z) - 20\log\left[\frac{2w(\lambda,z_0)}{2w(\lambda,z)}\right]}{10}} -1\right].$$

(4)

The additional reference point of $z=z_1$ is required for the evaluation of MFD along the fiber[7], and the MFD distribution is given by

$$2w(\lambda,z)= 2w(\lambda,z_0)\left[\frac{2w(\lambda,z_1)}{2w(\lambda,z_0)}\right]^{\frac{I(\lambda,z) - I(\lambda,z_0)}{I(\lambda,z_1) - I(\lambda,z_0)}}.$$

(5)

As found from Eqs.(4) and (5), $\Delta(z)$ and $2w(\lambda,z)$ can be obtained from the imperfection contribution $I(\lambda,z)$ and the relevant quantities given at $z=z_0$ and z_1.

Experimental Verification

To verify the effectiveness of the photon-counting OTDR method, Δ and $2w$ were measured along a three section SMF link (SMF-1, -2, and -3 with lengths of 10 m, 10 m, and 90 m, respectively). The waveguide structural parameters of the three SMFs are listed in Table I. SMFs-1 and -3 have almost the same waveguide parameters; only their lengths differ. A laser diode that emitted 112-ps pulses with 100 kHz-repetition rate and peak power of 40 mW at 835 nm was used as the light source for a single-photon-counting OTDR. The optical pulses were launched into the arc-fusion spliced SMF through a 2×1 fiber-coupler. The backscattered power from the test SMF-link was detected by a Si-APD through the coupler, and we set the averaging time

of 30-min in the OTDR waveform processing. Two Fresnel reflections 6 cm apart were well resolved by the OTDR scheme.

Table I: Parameters of test fibers

Parameters	SMF-1	SMF-2	SMF-3
Mode Field Diameter* $2w(\mu m)$ at 835 nm	5.0	5.2	5.0
Relative-Index Difference** Δ (%)	0.46	0.45	0.46
Cut-off Wavelength λ_c (nm)	535	759	535
Fiber Length l (m)	10	10	90

* Measured by near field pattern (NFP) method
** Measured by refractive near field (RNF) method

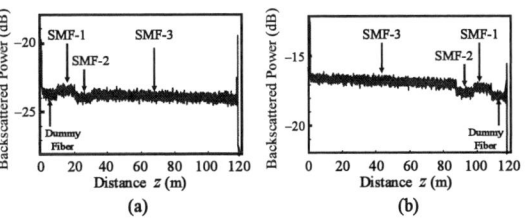

Fig.1:OTDR waveforms obtained by launching pulses (a) from SMF-1 to -3 and (b) from SMF-3 to -1.

OTDR signals $S_1(\lambda,z)$ and $S_2(\lambda,L-z)$ obtained from opposite ends of the SMF-link are shown in Fig.1(a) and (b), respectively, and the relevant imperfection contribution $I(\lambda,z)$ is displayed in Fig.2. Fig.3 plots the MFD distribution $2w(\lambda,z)$ of the SMF-link. When the MFDs of SMFs-1 and -2 are given as shown in Table I, the MFD of SMF-3 can be evaluated from Fig.2. The MFD distribution $2w(\lambda,z)$ evaluated by the OTDR method is found to be in good agreement with that measured with the near field pattern (NFP) method. From the MFD distribution shown in Fig.3, we can evaluate the relative-index difference distribution $\Delta(z)$ in the SMF-link. The result is shown in Fig.4; we used the k-value of 0.62[3] for the GeO$_2$-doped core. As seen from Fig.4 and Table I, the Δ-values of 0.46±0.003 and 0.45±0.003 % for the respective SMFs-2 and -3 with the OTDR method are in good agreement with 0.46 and 0.45 % with refractive near field (RNF) method. The experimental results shown in Figs.3 and 4 reveal that the photon-counting OTDR method is a powerful tool for the evaluation of $\Delta(z)$ and $2w(\lambda,z)$ with high spatial resolution.

257

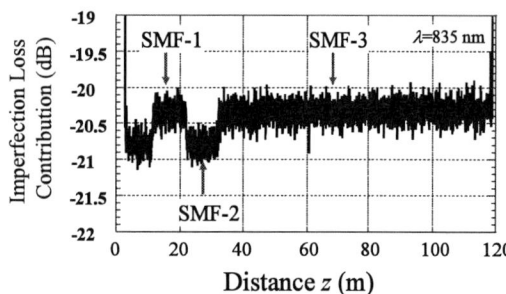

Fig.2: Structural imperfection contribution obtained from the waveforms shown in Fig.1.

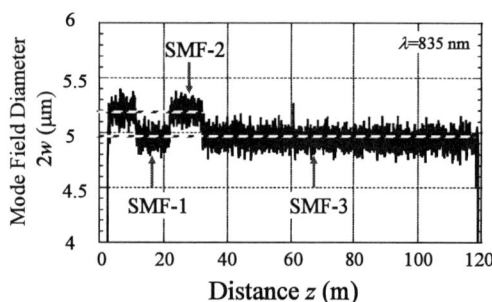

Fig.3: MFD distribution of the SMF-link at 835 nm.

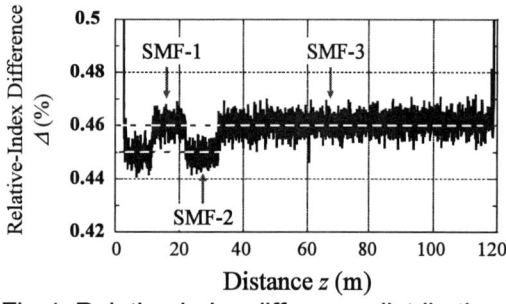

Fig.4: Relative-index difference distribution of the SMF-link.

Conclusions

A photon-counting OTDR is applied for the evaluation of MFD and relative-index difference Δ of the SMF-link at 835 nm. MFD and Δ evaluated by the OTDR method are in good agreement with those obtained by NFP and RNF methods, respectively. Experimentally obtained results reveal that the OTDR method is a powerful tool for the evaluation of structural parameters along SMF-links with high spatial resolution.

References

1 R. Bräuerle, et al., Proc. LEOS Ann. Meeting, Rio
  Grande, PR, (2000), p.496.
2 D. Gloge, *Appl. Opt.*, vol.10, (1971), p.2252.
3 M. Ohashi, *IEEE Photon. Technol. Lett.*, vol.18, (2006), p.2584.
4 A. L. Lacaita, et al., *Opt. Lett.*, vol.18, (1993),p.1110.
5 P. Di Vita and U. Rossi, *Electron. Lett.*, vol.15,
  (1979), p.467.
6 M. Ohashi and M. Tateda, *Electron. Lett.*, vol.29,
  (1993), p.426.
7 A. Rossaro, et al., *IEEE J. Sel. Topics Quantum*
  Electron., vol.7, (2001), p.475.

Measurements of multimode fiber PON bandwidth

L. Maksymiuk (1), G. Stepniak (2), J. Siuzdak (3)

1 : Warsaw University of Technology, Institute of Telecommunications, Nowowiejska 15/19
Warsaw Poland, maksymiuk@tele.pw.edu.pl
2 : Warsaw University of Technology, Institute of Telecommunications,
stepniak@tele.pw.edu.pl, 3 : Warsaw University of Technology, Institute of
Telecommunications, siuzdak@tele.pw.edu.pl,

Abstract *Employment of MM PON appears to be natural for short-range application such as intra-office or home networks. We have presented the results of the measurements of MM fiber PONs frequency responses in various configurations.*

Introduction

Passive optical network (PON) is a well-established standard of access network. It is based on single mode (SM) fiber and passive optical devices such as splitters. A few types of PON have been introduced (BPON, EPON, GPON etc.) [1], [2], [3], with the differences in transmission protocol (ATM, Ethernet, GEM) rather than in physical layer. Employment of similar PON but based on multimode (MM) fiber – MM PON appears to be natural for short-range application such as intra-office or home networks. However, little is known about transmission parameters of such networks especially about transmission bandwidth between MM fiber PON nodes. To the best of authors' knowledge no such study has been performed yet. Very extensive work has been done both theoretically [4], [5], [6] and experimentally [7], [8], [9] only with regard to point to point MM fiber links. This paper is meant to cover this gap at least partially. Here we shall present the results of bandwidth measurements of MM fiber PONs employing typical, comercially available MM fiber (62.5 µm core) and couplers/splitters. The bandwidth is the most critical parameter as it usually limits transmission speed.

Measurements

We measured frequency responses with the use of a measurement setup designed according to FOTP-204 [10]. The testing procedure standardizes two types of excitation: RML (Restricted Mode Launch), and OFL (Overfilled Launch). The first one is to emulate laser-like excitation, the second LED-like, respectively. Both RML and OFL require launch patch-cords made

of specially designed fibers [10]. In the testing procedure described here, the frequency response of the measurement setup was compensated. Here we present results for two configurations: 1 km fiber interconnected with a coupler, and passive symmetrical tree network. All presented measurements are optical frequency responses with the response of the measurement setup corrected.

In the first part of this section we present a measurements at wavelength 850 nm for a 1km long fiber connected to a coupler/splitter. We performed this measurements for two standardized launching conditions RML, and OFL. We tested fiber with 62.5 µm core diameter, and numerical aperture NA=0.275 interconnected with a 2x2 3dB coupler. We tested the case when the coupler is at the front end and at the far end of the fiber, respectively. To obtain a greater sample of different frequency responses we additionally repeated all the measurements for different couplers supplied by the same manufacturer. The measurement results are depicted in Figs. 1 and 2. First of all, as it its predicted by the theory for a regular fiber profile, RML gives better bandwidth than OFL. This is caused by the fact that RML excite fiber modes in a favorable way (only some subset of lower order modes is excited), whereas OFL excites all the modes within a fiber. Secondly, RML launch condition produces greater variations of frequency responses than OFL (standard deviation of the averaged 3dB bandwidth for RML is about 10%, whereas for OFL only about 4%). This conclusion is also valid for other wavelengths (i.e. 660 nm and 980 nm)

978-0-9789217-3-6/07/$25.00

©2007 WEN GLOBAL SOLUTIONS

that we tested in our laboratory. This may be explained by the fact that RML excites fewer modes and therefore it is more sensitive to the mode filtering in couplers/splitters. Additionally for comparison, in Figs. 1 and 2 dotted curves represent a measurement of a fiber without any coupler attached. It is very significant that these frequency responses of a fiber are neither the limit from top or bottom, they are placed more or less in the middle.

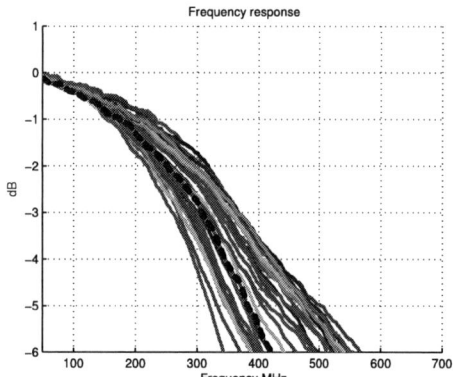

Fig.1. Frequency responses of a 1km fiber spool interconnected with a 2x2 coupler, RML

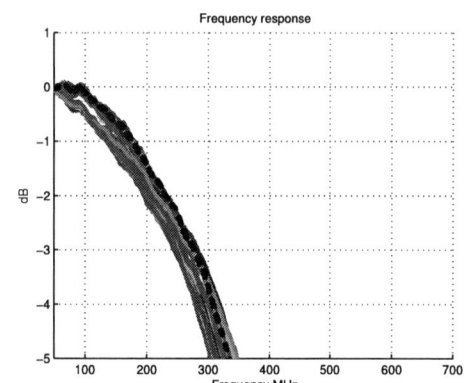

Fig.2. Frequency responses of a 1km fiber spool interconnected with a 2x2 coupler, OFL

This is an important conclusion, because when one wants to design a passive network based on multimode fibers one has to be aware of the fact that the bandwidth may be higher or lower than those predicted for the fiber alone.

To show that frequency response variations occur also in a more complex network, we investigated symmetrical passive network in tree configuration, in which all the paths have the same length. Network configuration is depicted in Fig. 3.

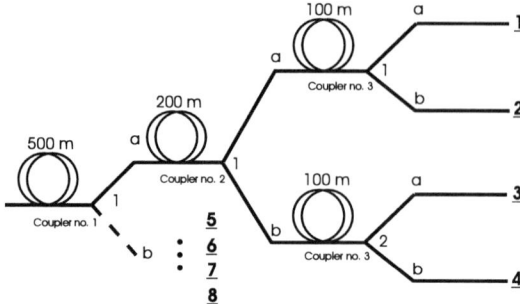

Fig. 3. Symmetrical MM PON topology

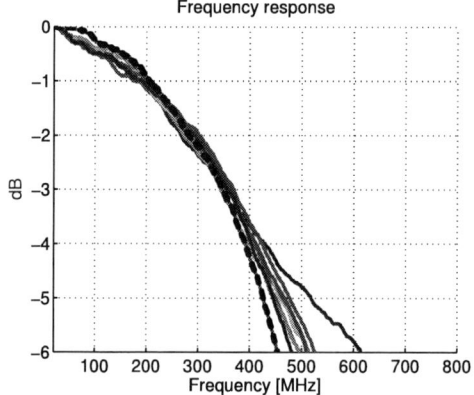

Fig.4. Frequency responses in different ending nodes of the MM PON depicted in Fig. 3, OFL

Couplers provided by the same manufacturer exhibit slightly different physical properties that influence the mode coupling and consequently the frequency response. Therefore, different paths and different network nodes have different frequency responses and bandwidths. Furthermore, the same coupler connected in a different way may change the frequency response of the network. In Fig. 3 the coupler in each stage of the network is signed with a number and denoted with port names (1, and 2 are the input ports and a, and b output ports). Similarly to the previous subsection, RML (see Fig. 5) excitation gives both the better bandwidth and higher variations in frequency response characteristics. The presented results prove that although the network is completely symmetrical the frequency responses at different nodes maybe different. This is undoubtedly caused by mode filtering properties of couplers/splitter. Therefore, we have to apply a safety margin on bandwidth requirement to avoid a problems caused by a frequency response drop at a particular node.

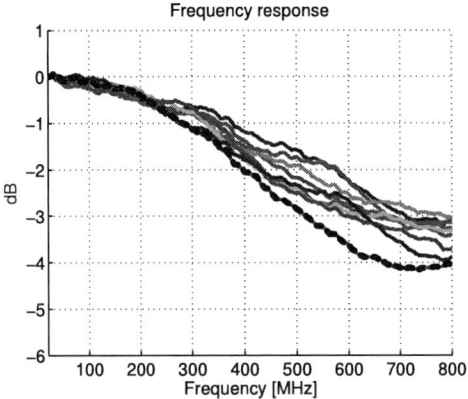

Fig.5. Frequency responses in different ending nodes of the MM PON depicted in Fig. 3, RML

Conclusions

We have presented the results of the measurements of MM fiber PONs in various configurations. As expected, the frequency responses of MM fiber PON bandwidth depends on the light launching conditions with RML bandwidth superior to (for a regular index profile) OFL bandwidth due to the fewer modes excited in the former case. The insertion of the couplers/splitters into the network causes differences between frequency responses of nodes of otherwise symmetrical network. This effect is obviously due to the mode filtering properties of couplers/splitters and is the most pronounced for a single device connected to either end of a long fiber. If more couplers/splitters are added this effect is less visible most probably due to the averaging properties of such connection. Although 3dB bandwidth variations may be fairly high, especially for RML where they fluctuate of about 10%, they do not exclude the potential use of a multimode fiber in passive optical networks. A rough estimation of the MM PON path bandwidth is the bandwidth of one strand MM fiber, which has the same length as the path. However, appropriate margins up to 10% are indispensable, especially for RML. Moreover, in the more complex topologies (with a cascade of couplers) passive network path usually exhibits higher bandwidth than a single strand of fiber, so the latter may be used as a conservative estimate of the bandwidth.

Acknowledgement

This work was done under grant no 49131005 from France Telecom. The authors gratefully acknowledge France Telecom for the permission of this publication.

References

1 G. Kramer Ethernet Passive Optical Networks, McGraw-Hill, New York (2005),
2 TU-T Broadband optical access systems based on Passive Optical Networks (PON) (2005),
3 ITU-T Gigabit-capable Passive Optical Networks (GPON): General characteristics (2003),
4 P. Pepeljugoski et al IEEE Journal of Lightwave Technology, Volume 21, No. 5, (2003), page 1242-1255,
5 G. Yabre IEEE Journal of Lightwave Technology, Volume 18, No. 2 (2003), page 166-177,
6 K. Kitayama et al IEEE Journal of Quantum Electronics, Volume 16, No. 3 (1980), page 356-362,
7 C. W. Oates et al IEEE Journal of Lightwave Technology, Volume LT-7, No. 3 (1989), page 530-532,
8 P.M. Rodhe IEEE Journal of Lightwave Technology, Volume LT-3, No. 1 (1985), page 154-158,
9 L. G. Cohen Bell System Technical Journal, vol. 55 (1976),
10 ANSI/TIA/EIA-455-204-2000.

Novel technique for measuring Raman gain efficiency distribution by conventional OTDR

Masaharu Ohashi, Yabu Tetsuro and Ikuo Yamashita*
Osaka Prefecture University, 1-1 Gakuen-cho, Sakai, Osaka, 599-8531 JAPAN
E-mail: ohashi@eis.osakafu-u.ac.jp
*Kansai Electric Power, 3-11-20 Nakoji, Amagasaki, Hyogo, 661-0974 JAPAN
E-mail: yamashita.ikuo@e3.kepco.co.jp

Abstract *A novel technique is proposed for measuring Raman gain efficiency distribution in optical fibers by using a conventional OTDR. Raman gain efficiency distributions for a single-mode fiber and a fiber link have been successfully estimated.*

Introduction

Raman amplification is an attractive technology whereby the amplification wavelength region can be adjusted by changing the pump light wavelength. When designing the ultra wide band transmission systems, it is important to estimate the Raman gain characteristics of the transmission lines. Techniques have been reported for measuring the Raman gain efficiency distribution.[1]-[3] However, there have been few reports[2],[3] on the measurement technique for the Raman gain efficiency distribution by using a conventional OTDR as far as we know.

In this paper, we propose a novel technique for measuring Raman gain efficiency distribution in optical fibers by using a conventional OTDR. The Raman gain efficiency distributions of a test fiber, and a fibre link composed of 22 single-mode fibers are estimated by using our technique.

Measurement principle

Figure 1 shows the schematic diagram for measuring the Raman gain efficiency distribution in the optical fibers by using an OTDR.

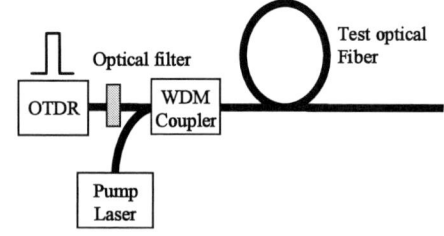

Figure 1 Schematic diagram of the proposed Raman gain efficiency measurement method

The pulsed signal and the continuous wave pump lights are launched into the test fiber through the WDM coupler. These lights co-propagate through the test optical fiber. An optical filter is inserted between the OTDR and the WDM coupler to eliminate the Rayleigh backscattering of the pump light.

Here, we derive the signal power at the distance of z. The signal power P_s can be obtained by solving the coupled power equation with regard to the signal P_s and the pump powers P_p. If the pump is un-depleted, the pump power can be expressed by the following equation.

$$\frac{dP_s}{dz} = \frac{g_R}{A_{eff}} P_p P_s - \alpha_s P_s \qquad (1)$$

$$\frac{dP_p}{dz} = -\alpha_p P_p , \qquad (2)$$

where g_R is the Raman gain coefficient. α_s and α_p are the attenuation coefficients of signal and pump wavelengths, respectively. A_{eff} denotes the effective area, which corresponds to the overlapping area between pump and signal lights.

From (2), the pump power P_p at the position of z can be obtained as

$$P_p(z) = P_p(0) \exp(-\alpha_p z) . \qquad (3)$$

Substituting (3) into (1), the signal power $P_s(z)$ at the position of z can be obtained as

$$P_s(z) = P_s(0) \exp\left[\int_0^z \left(\frac{g_R P_p(0)}{A_{eff}} \exp(-\alpha_p z) - \alpha_s \right) dz \right] . \qquad (4)$$

The signal power $P_s(z)$ at the position of z is reflected and it travels toward the input direction. The backscattered light can be expressed as the product of the signal power $P_s(z)$ and $B(z)$ α, where α is the scattering coefficient and $B(z)$ is the capture ratio. Then, the backscattered signal light $P_s(z)B(z)\alpha$ is amplified by the counter propagating pump light. Therefore, the

978-0-9789217-3-6/07/$25.00

©2007 WEN GLOBAL SOLUTIONS

backscattered power $P(z,P_p)$ from the position of z can be expressed as

$$P(z,P_p)=P_s(0)\alpha B(z)$$

$$\times \exp\left[2\int_0^z\left(\frac{g_R P_p(0)}{A_{eff}}\exp(-\alpha_p z)-\alpha_s\right)dz\right] \quad (5)$$

$$=P_s(0)\alpha B(z)\exp[2P_p(0)G(z)]\times\exp[-2\alpha_s z]$$

where G(z) is defined as

$$G(z)=\int_0^z\frac{g_R(z)}{A_{eff}(z)}\exp(-\alpha_p z)dz \cdot \quad (6)$$

On the contrary, when the pump light is off, the backscattered power P(z,0) can be obtained by substituting $P_p(0)$ =0 into (5) as

$$P_s(z,P_p)=P_s(0)\alpha B(z)\exp[-2\alpha_s z] \cdot \quad (7)$$

The backscattered power of OTDR, S (z,P$_p$) [=10log {P (z,P$_p$)}] can be expressed as

$$S(z,P_p)=10\log[P_s(z,P_p)]$$
$$=10\log[P_s(0)]+10\log[\alpha B(z)] \quad (8)$$
$$+2P_p(0)G(z)\cdot 10\log(e)-2\alpha_s z 10\log(e)$$

The backscattered power at z=z+Δz can be also expressed as

$$S(z+\Delta z,P_p)=10\log[P_s(z+\Delta z,P_p)]$$
$$=10\log[P_s(0)]+10\log[\alpha B(z+\Delta z)] \quad (9)$$
$$+2P_p(0)G(z+\Delta z)\cdot 10\log(e)$$
$$-2\alpha_s(z+\Delta z)10\log(e)$$

The following equation can be derived from (8) and (9).

$$\frac{dS(z,P_p)}{dz}=\frac{d\{10\log[\alpha B(z)]\}}{dz} \quad (10)$$
$$+2P_p(0)\cdot 10\log(e)\frac{dG(z)}{dz}-2\alpha_s 10\log(e)$$

On the contrary, when P$_p$=0, the following equation can be also derived in the same manner.

$$\frac{dS(z,0)}{dz}=\frac{d\{10\log[\alpha B(z)]\}}{dz}-2\alpha_s 10\log(e) \quad (11)$$

From the definition of G, dG(z)/dz can be expressed as

$$\frac{dG(z)}{dz}=\frac{g_R(z)}{A_{eff}(z)}\exp(-\alpha_p z) \cdot \quad (12)$$

Therefore, the Raman gain efficiency $g_R(z)/A_{eff}(z)$ at the position of z can be derived from (10) to (12) as

$$\frac{g_R(z)}{A_{eff}(z)}=\frac{dS_d(z)}{dz}\cdot\frac{1}{2P_p(0)\cdot 10\log(e)\exp(-\alpha_p z)} \quad (13)$$

Here, S$_d$(z) corresponding to the backscattered power difference between with and without pumping is defined as

$$S_d(z)=S(z,P_p)-S(z,0) \cdot \quad (14)$$

Therefore, the Raman gain efficiency distribution can be estimated from the length dependence of the pump power and the derivative of S$_d$(z) with regard to the fiber length z. It is also found from (13) that the Raman gain efficiency distribution $g_R(z)/A_{eff}(z)$ can be estimated by using the conventional OTDR.

Experimental results

The Raman gain efficiencies for a conventional single-mode fiber with a length of 25 km and a fibre link installed in the field with a length of 11km composed of 22 conventional single-mode fibers with a piece length of 500 m were measured to confirm the effectiveness of our technique.

Fig. 2 shows the backscattered powers with

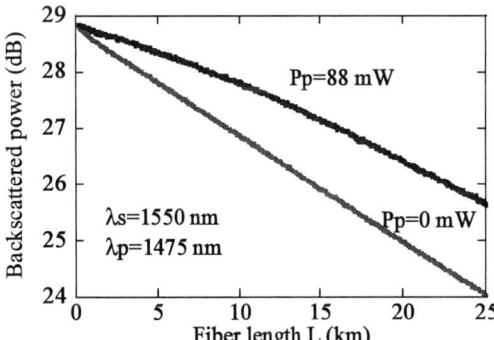

Figure 2 Backscattered powers with and without pumping

and without pumping for the conventional single-mode fiber. OTDR (Anritsu) with a wavelength of 1550 nm was used to measure the backscattered power. The OTDR pulse width was 1μs and the averaging time was 5 min. The pump laser with a wavelength of 1475 nm was used. It is found that the signal power is amplified by the pump power.

Figure 3 shows the backscattered power difference S$_d$ defined in (14) plotted as a function of fiber length. S$_d$(z) was best fitted to the polynomial function as to calculate its derivative.

$$S_d(z)=0.085+0.22\cdot z$$
$$-0.0052\cdot z^2+5.0\times10^{-5}\cdot z^3 \quad (15)$$

The attenuation coefficient of the test fiber at λ=1475 nm was 0.217 dB/km. The pump power P$_p$(0) was 88 mW. The Raman gain efficiency distribution along the fiber length

263

can be estimated by using (13) and (15).

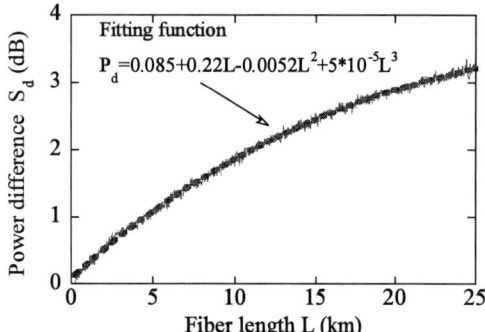

Figure 3 Relationship between power difference S_d and the fiber length

Figure 4 shows the Raman gain efficiency distribution estimated by our technique.

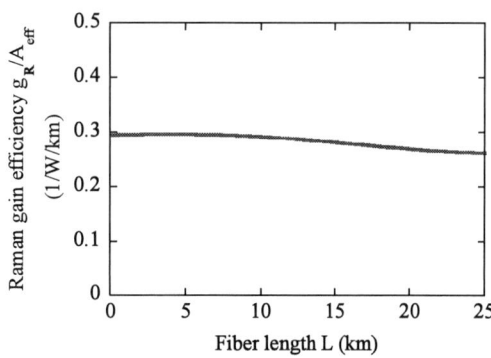

Figure 4 Measured Raman gain efficiency

It is seen that Raman gain efficiency along the fiber length is almost the same as 0.3 $W^{-1}km^{-1}$.

Next, the Raman gain efficiency for the fiber link composed of 22 conventional single-mode fibers was measured.
Figure 5 shows the backscattered powers with and without pumping and the

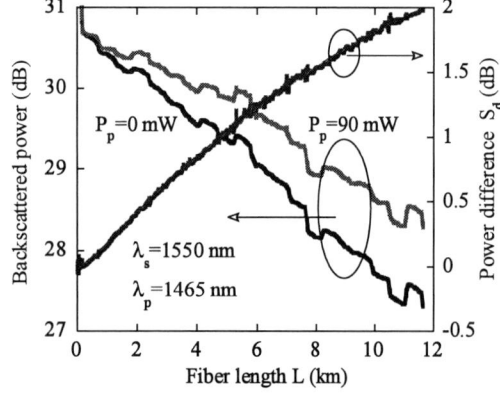

Figure 5 OTDR traces and backscattered power difference S_d along the fiber length

the backscattered power difference S_d

plotted as a function of fiber length. The attenuation coefficient α_p was measured by the OTDR with a wavelength of 1450 nm. The length dependence of the pump power $P_p(z)(= P_p(0)\exp(-\alpha_p z))$ was estimated from the OTDR trace at the pump wavelength λ_p. By using the length dependence of the pump power, and the derivative of S_d with regard to the fiber length z, the Raman gain efficiency distribution of the fiber link was estimated. The Raman gain efficiency of the fiber link is shown in Fig. 6. It is found that the Raman gain efficiency distribution in the fiber link varies from 0.22 to 0.32 $W^{-1}km^{-1}$.

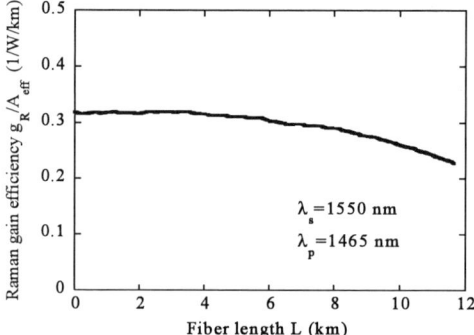

Figure 6 Raman gain efficiency of the fiber link

The Raman gain efficiency distribution of the fiber link installed in the field shows the appropriate value. As a result, it is confirmed that our technique can be applied to the fiber link. The accuracy will be further studied.

Conclusions

We proposed a new technique for measuring Raman gain efficiency distribution using a conventional OTDR. The Raman gain efficiency in the 25km long single-mode fiber and the fiber link installed in the field with a length of 11 km composed of 22 conventional single-mode fibers were successfully estimated experimentally. We express our sincere thanks to Mr. Oro, Mr. Inoguchi and Mr. Hatada for fruitful discussions.

References

1 M. Wuilpart et al., Electron. Lett., 39 (2003), 88.

2 K. Toge et al., IEEE Photon. Technol. Lett., 14(2002), 974.

3 H. Hatada et al., OECC2006, (2006) 3P-11-1.

EPON Deployment Challenges – Now and in the Future

Bill McDonald

Director of Optical Product Marketing, Centillium Communications, Inc., 215 Fourier Avenue, Fremont, CA 94539, U.S.A.; billm@centillium.com

Abstract *As several Asian telcos role out EPON en masse, they are grappling with a number of challenges that affect present and future deployments. This paper identifies and examines these challenges.*

Introduction

Ethernet passive optical networks (EPON) are being massively deployed in the Asia Pacific region. The primary objective of these EPON deployments has been to provide broadband data services at higher data rates than ADSL or CATV. Japanese service providers have been deploying PONs since 2002, initially with BPON, then ramping up on a truly massive scale with EPON beginning in late 2004. Service providers in other Asia Pacific countries, including Korea and China, have been conducting EPON trials and are gearing up for mass deployments of EPON. This paper will explore the current and future challenges of these large-scale deployments.

Current Deployment Challenges

Telecommunications services providers (telcos) recently shifted their focus from delivering higher bandwidth data services to the delivery of Triple Play services (TPS) – bundled voice, data and digital TV services over a common IP network. Their strong desire to offer TPS is rooted in their need to compete with CATV companies that are already offering TPS, and to generate new revenue streams. Triple Play services have not yet been deployed by many of the major telcos due to a number of challenges, including:

- Delivery and network neutrality
- High cost of TPS CPE
- Digital rights management and security
- Regulatory and business barriers
- Installation hurdles
- Interoperability issues

Delivery and Network Neutrality

Three approaches have emerged as leading candidates for the delivery of IPTV services to the user: IGMP snooping, VLAN tagging and a combination of the two.

IGMP snooping involves multicast MAC filtering, where the the PON optical line termination (OLT) looks for IP multicast group associations and then configures the multicast MAC address filter table at the optical network unit. The ONU allows the multicast MAC addresses that match entries in the table to be forwarded to the user. The user can access any content provider's IPTV services and the user's home network is not overwhelmed with unwanted multicast streams; however, the telco has little control over the content providers that deliver IPTV services to the user.

The second method employs VLAN tagging. In this case, the ONU filters each multicast frame based on its VLAN identifier, allowing only those multicast frames with VLAN identifiers that match entries in its VLAN filtering table to be forwarded to the user. The telco exerts control over the multicast streams that are delivered to the user by configuring the user's ONU with the appropriate VLAN identifiers. This approach limits the user's choice of content providers and IPTV services to those of the telco's selection. Also, this approach might overwhelm the user's home network with unwanted multicast streams if multiple multicast streams are supported per VLAN.

With a combined approach IGMP snooping is used to identify the multicast MAC addresses of the multicast streams that are requested by the user via his IP STB. All multicast streams are then filtered by a VLAN identifier and multicast MAC address at the ONU as described in the first and second approaches. Scalability is assured because multiple multicast streams can be supported per VLAN, and the user's home network is not overwhelmed with unwanted multicast streams since the network delivers only those multicast streams that the user selects. However, this approach also limits the user's choice of content providers and IPTV services to those of the telco's selection, similar to the VLAN tagging approach.

The IGMP snooping approach is consistent with the principles of network neutrality that promote equality with regard to the processing of multicast streams from all content providers, whether they are from the telco itself or an independent content provider. In contrast, the VLAN tagging and combined approaches give absolute control of content to the telco, enabling it to favor or disfavor content providers at will. It is unclear as to whether the advocates of network neutrality or the telcos will win this battle, but it is apparent that this issue will have a dramatic impact on the way that streamed content is delivered to the user.

CPE Cost

The cost of PON ONUs is the single largest contributor to the cost of PON equipment on a per user basis because one ONU is required per user. The main contributors to the cost of an ONU are the optical transceiver, the Gigabit PHY at the user network interface (UNI), the ONU protocol system-on-chip (SoC), and peripheral memory devices. Additionally, the configuration of an OSG contributes greatly to its cost. Today, a user requires up to three CPE "boxes" to support OSG functionality:

- An ONU to provide PON termination and bridging functionality
- A gateway router to provide routing and security functionality such as firewall, NAT, VPN and encryption
- An analog terminal adaptor (ATA) to provide telephony services end-point functionality

This OSG configuration is expensive and cumbersome, resulting in an unsightly and unmanageable conglomeration of boxes.

Digital Rights Management and Security

Content providers demand that telcos guarantee a minimum of the following before they will permit telcos to stream their content to users:

- Quality of service
- Level of security from theft and piracy
- Level of control over who can watch their content and the type of devices that their content can be viewed from

The content provider's quality of service demand may become a major obstacle to the delivery of IPTV services. Telcos base their business case for the deployment of PONs on their ability to deliver a vast number of video streams to the user at a low cost per stream. However, it is unclear whether their compression techniques and data rates are sufficient to support the quality of service that content providers demand.

Content providers also require that telcos encrypt their content so that it cannot be stolen and pirated by people or groups that have not paid to view the content. Additionally, they mandate that telcos exert a high degree of control with regard to the storage of content by the user, the length of time that a user can view a video segment, and other aspects of the user's experience that may impact the content provider's bottom line.

Business and Regulatory Barriers

There are a number of business and regulatory issues that are being addressed by telcos. The integration of traditional Layer 1 transport and Layer 2 switching functionality with higher layer routing and services requires cooperation and integration between traditionally separate "silos" within telcos' organizations. Additionally, in some markets, telcos have been prohibited from offering entertainment and information services/applications by governmental regulatory bodies.

Interoperability

The IEEE 802.3ah Ethernet in the First Mile (EFM) standard provides a solid foundation for multi-vendor EPON equipment interoperability, but does not address all aspects of interoperability required for telco deployments. Specifically, it does not address encryption and a category of operations, administration and maintenance (OAM) functions referred to as enhanced OAM (EOAM) that include the remote configuration of ONU parameters and remote firmware download. As a result, telcos must proactively fill these gaps in the standard by preparing detailed specifications for encryption and EOAM.

Installation
PON installation presents a broad spectrum of challenges to telcos. These challenges include the method of deploying fiber to the home or to the building, network demarcation, in-home cabling and lifeline telephony-related issues, to name a few. These issues have been addressed by some telcos, like NTT and Verizon, but they still represent major hurdles to many other telcos.

Future Deployment Challenges
New high-bandwidth services – such as large-screen digital imagery (LSDI) at 40-160 Mbps per channel, 3-D online interactive games and ultra high-speed Internet access – will drive the need for even greater bandwidth to the home in the near future. The IEEE 802.3 working group recently formed a new task force, IEEE 802.3av, to develop specifications for a 10Gbps PHY for EPON that will address this need.

Since 1G EPON will have been deployed to tens of millions of homes by the time that 10G EPON is commercially available for deployment, telcos will demand that 10G EPON be compatible with 1G EPON so that 10G and 1G EPON OLTs and ONUs can coexist in the same PON without interfering with each other. This capability will enable telcos to introduce 10G EPON gradually into existing networks without forcing them to abandon existing 1G EPON equipment or "flash cut" their existing 1G network to a 10G network.

It is anticipated that this constraint will compel the taskforce to specify wavelengths for downstream and upstream transmission for 10G EPON that don't interfere with 1G EPON.

Similarly, since G.652 SMF fiber will have been deployed to tens of millions of homes by the time that 10G EPON is commercially available for deployment, the taskforce will also be compelled to specify that 10G EPON will support transmission over this fiber type.

The cost of 10G laser transmitters and receivers is significantly higher than that of 1G. It is anticipated that these costs will fall over time as deployments of 10G SONET, SDH and Ethernet ramp up. However, the IEEE 802.3av taskforce is proactively addressing this issue by investigating different strategies that will decrease costs.

Conclusions
Current EPON deployment challenges include:
- The debate on network neutrality, the outcome of which will determine the preferred method for IPTV services delivery to the user
- The high cost of TPS CPE (OSG)
- Digital rights management and security
- Regulatory and business barriers
- Interopability
- Installation

These challenges must be addressed before telcos can progress to the next phase of EPON deployment – the deployment of IPTV services.

While telcos are addressing these issues, they must not lose sight of the ball by continuing to address generic deployment challenges such as installation and interoperability, as well as preparing for future challenges including the migration to 10G EPON.

References
1   First author et al Journal, Volume (year),   page

Fault Location for Fiber Links in PON by Means of FSF Fiber Laser and FBGs

[1]Nianyu Zou and [2]Yoshinori Namihira

[1]Research Institute of Photonics and Optics, DaLian Politechnic University
No.1 Qinggongyuan, Ganjingzi Distr., Ddlian 116034,P.R.China
n_y_zou@dlpu.edu.cn
[2]Department of Electrical & Electronics, University of the Ryukyus
1 Senbaru, Nishihara, Okinawa 903-0213, Japan
namihira@eee.u-ryukyu.ac.jp

Abstract *A novel fault location method for fiber links in PON is proposed based on heterodyne detection. Experiments were demonstrated by using a self- devoloped wavelength tunable FSF fiber laser and FBGs with different reflective wavelengthes.*

Introduction

Fiber To The premises (FTTP) networks have been growing rapidly in recent years. Among the several options of optical network architecture, passive optical fiber networks (PON) is low-cost, easy installation and being able to meet the needs of various services, therefore is one of the favorite types for service providers wishing to push fiber closer to the end users [1]. Practically, in PON it is highly required to locate the fault of fiber links from optical line terminal (OLT) to optical network unit (ONU) without interfering the transmission services. Therefore in network maintaining and troubleshooting potential problems, testing PON systems present great challenges to technicians. Optical time domain reflectometry (OTDR) technique is popularly utilized to monitor such PON. There exist two problems, however. One is the high power of the short pulse easily leading to non-linearity of fibers, and another one is the dynamic range requirement of the OTOR equipment. In this work, we propose a novel fault location method for PON based on optical frequency domain reflectometry (OFDR) technique. In the demonstration experiments, a frequency-shifted feedback (FSF) [2] fiber laser has been developed as light source and fiber Bragg gratings (FBGs) with different reflective wavelengths were utilized to reflect the monitor signals in fiber branches, the test are carried by heterodyne detection.

FSF laser

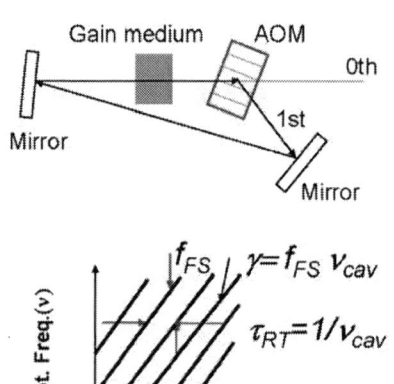

Fig.1 Schematic and output frequency of FSF laser

FSF laser is a novel light source suitable for OFDR technique, and the diagram is shown in Fig.1.The gain medium can be Nd:YVO4 or semiconductor optical amplifier or erbium-doped fibers. The FSF laser cavity can be a ring or Fabry-Perot configuration, and was closed via the first order diffracted light of an inter-cavity acousto-optic modulator (AOM). The laser output consisted of a chirped frequency comb can be expressed as

$$v_i(t) = \gamma t - q/\tau_{RT}, \quad \gamma = f_{FS}/\tau_{RT} \quad (1)$$

978-0-9789217-3-6/07/$25.00

©2007 WEN GLOBAL SOLUTIONS

Where τ_{RT} is the cavity round trip time, f_{FS} is the intracavity frequency shift (AOM driving frequency). and q is an integer. With high chirp rate, good linearity and wide chirp range, FSF laser is a proper source for heterodyne detection. In fact, it has been reported of applications in characterization of chromatic dispersion and polarization mode dispersion in optical fibers [3-4] as well as 3D measurements [5].

Proposed method

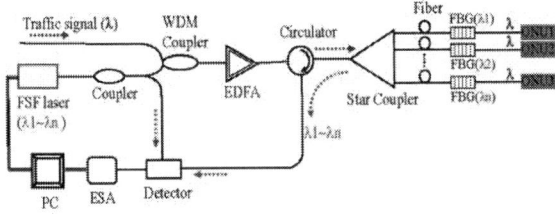

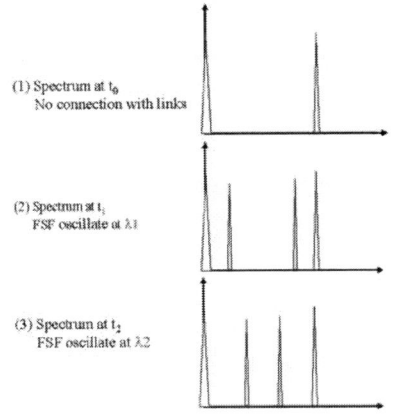

Fig.2.Measurement diagram and principle
ESA: Electronic Spectra Analyzer
EDFA: Erbium-doped Fiber Amplifier
FBG: Fiber Bragg Grating

The measurement diagram of the proposed method is shown in Fig.2. Here a laser diode (LD) source is utilized to present the traffic signals. The output of LD and a wavelength tunable FSF laser are combined by a WDM coupler, and then injected into the PON by passing through the erbium-doped fiber amplifier (EDFA), circulator and star coupler, with the wavelength of FSF laser ($\lambda 1 \sim \lambda n$) selected to be shifted from the service wavelength (λ). FBG at each

fiber branch has high reflectivity at wavelength of $\lambda_1, \lambda_2, \ldots \lambda_n$ respectively, therefore the service light (λ) pass to each ONU and $\lambda_1, \lambda_2, \ldots \lambda_n$ are reflected by each corresponding FBG. Balanced photo detector combines the light from both coupler and circulator, and generates beat signal with the beat frequency proportional to the optical distance differences, and so on the other branches. In a result, the fault in fiber branches can be located according to heterodyne detection.

Experimental results

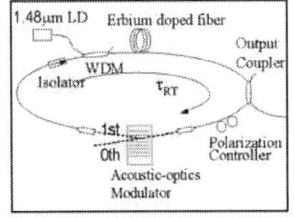 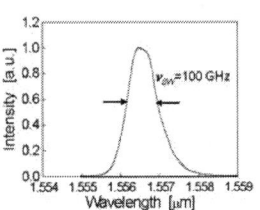

Fig.3(a)Diagram of FSF fiber laser
(b)Optical spectrum of FSF fiber laser

The proposed method was demonstrated previously by using a FSF laser with semiconductor optical amplifier as gain medium[6]. To make the cavity more stable and compact, a FSF fiber laser was developped and utilized in the experiments. The schematic cavity and optical oscillation spectrum of the FSF fiber laser are shown in Fig.3 (a) and (b), respectively. The acousto-optic frequency modulator worked at around 120MHz with a 1st diffraction efficiency of 36%, and the free spectra range was found to be 1.1MHz, therefore the chirp rate of the output was about 132 PHz. The frequency chirp range of each comb component was found to be 100 GHz in full width half maximum (FWHM). The output wavelength was able to be tuned by changing the driving frequency of the acousto-optic frequency modulator and the wavelength tunable range was found to be between 1545~1576nm.
With such a FSF fiber laser as light source, experiments were demonstrated on the system shown in Fig.2. To simulate the fiber

links in practical PON, a 5km single mode fiber (SMF) was inserted between the star coupler and $FBG_1(\lambda_1)$, and a 10km SMF was inserted between the star coupler and $FBG_2(\lambda_2)$ respectively. Here FBG_1 and FBG_2 had reflectivity of 99.5% at optcal wavelength of 1542nm and 1546nm respectively. The experimental results were shown in Fig.4, which gave the observed beat signal reflected by two SMFs when the FSF fiber laser was tuned to oscillate at 1542nm and 1546nm, respectively. It also was confirmed that such beat signals disappeared when FBG1 and FBG2 are removed, while FSF laser oscillating at 1542nm and 1546nm. Therefore the appearance or disappearance of beat signals can be utilized to monitor the status of fiber links.

Since the traffic wavelength is selected at $1.31\mu m$ at prsent PON, therefore there is no problem to totally seperate the traffic light from the test signal. From practical point of view, this is very important.

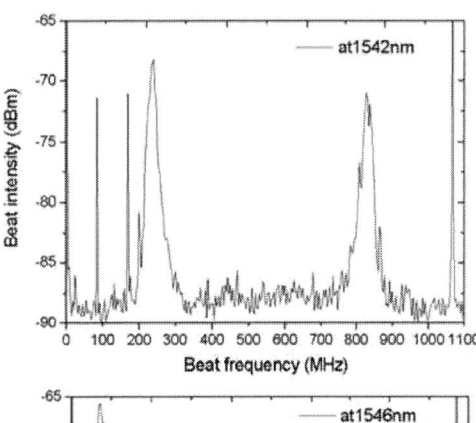

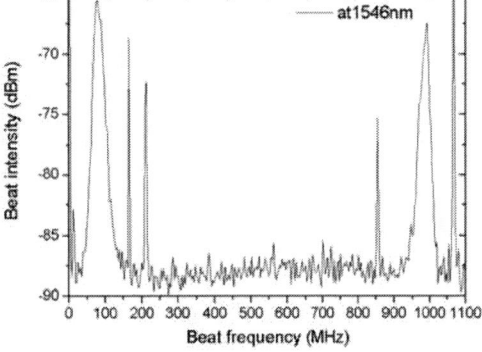

Fig.4 Observed beat signal by the reflected light
(a) from 5km SMF and FBG1 when the FSF
   was tuned to oscillate at 1542nm
(b) from 10km SMF and FBG2 when the FSF
   was tuned to oscillate at 1546nm

Conclusions

A new fault location method for fiber links in PON is proposed and experimentally demonstrated with a FSF fiber laser as light source. The fault in fiber branches can be monitored according to beat signals by heterodyne detection. The results verified the method as a potential selection for PON test.

Acknowledgements

The authors acknowledge Dr.Hromasa Ito and Dr. Chcikh Ndiaye of Tohoku University in Japan for the valuable cooperation.

References

[1] Frank Effengerger et al, "An introduction to PON technology", IEEE Optical Communications, Vol.45, s17-s25 (2007).

[2]Koichiro Nakamura et al, "Optical frequency domain ranging by a frequency-shifted feedback laser", IEEE Journal of Quantum Electronics, Vol.36, 305-316 (2000).

[3] Masato Yoshida et al, "A new method for measurement of group velocity dispersion of optical fibers by using a frequency-shifted feedback fiber laser", IEEE Photonics Technol. Lett, Vol.13, No.3, 227-229, (2001)

[4] Nianyu Zou et al, "PMD measurement based on delayed self-heterodyne OFDR and experimental Comparison with ITU-T round robin measurements", Electron. Lett., Vol.38,115~116(2002).

[5] Cheikh Ndiaye et al, "A novel 3D measurement technique using a frequency chirped laser", Proceeding of IQEC/CLEO-PR 2005, CFK2-4

[6] Nianyu Zou et al, "Fault location for branched optical fiber networks based on OFDR technique using FSF Laser as Light Source",Tech. Digest of OFC/NFOEC2007, paper NWC2, Anaheim, U.S.A.

VCSEL Photonics - Athermalization and Slowing Down -

Fumio Koyama
Microsystem Research Center, Tokyo Institute of Technology
4259-R2-22 Nagatsuta, Midori-Ku, Yokohama 226-8503, Japan
Email: koyama@pi.titech.ac.jp

Abstract

Our recent research activities on VCSEL photonics will be reviewed. This talk explores the potential and challenges for new functions of VCSELs, including the athermalization with MEMS technology and slowing light for ultra-compact photonic devices.

1. Introduction

A vertical cavity surface emitting laser (VCSEL) was invented 30 years ago [1]. A lot of unique features have been proven, such as low power consumption, wafer-level testing, small packaging capability and so on. The market of VCSELs has been growing up rapidly in recent years and they are now key devices in local area networks using multi-mode optical fibers. Also, long wavelength VCSELs are currently attracting much interest for use in single-mode fiber metropolitan area and wide area network applications. In addition, a VCSEL-based disruptive technology enables various consumer applications such as a laser mouse and laser printers.

In this talk, the potential and challenges for new functions of VCSELs will be described, which include the wavelength athermalization. The MEMS-based VCSEL technology enables "athermal operations" of semiconductor lasers with avoiding temperature controllers for uncooled WDM applications. The temperature dependence of long-wavelength VCSELs could be reduced by a factor of 50 with a novel thermally-actuated membrane mirror. Also, highly reflective periodic mirrors commonly used in VCSELs enables us to manipulate light with a slow light effect. This new scheme provides us ultra-compact intensity modulators, optical switches and so on for VCSEL-based photonic integration. We present our latest results on the modeling and experiments of slow light devices and circuits.

2. Athermal Operations of VCSELs

The temperature dependence of semiconductor lasers, which is typically 0.1nm/K even for single-mode semiconductor lasers, is a remaining problem to be solved. The elimination of costly thermoelectric controllers is desirable for use in low cost WDM networks. If it is realized, we expect low power consumption as well as small packaging. We proposed an athermal VCSEL with a fixed wavelength even under temperature changes using the self-compensation based on a thermally actuated cantilever structure [2]. We have demonstrated small temperature dependence in micromachined vertical cavity optical filters and light emitters of GaAs/GaAlAs materials [2]. It is a challenge to realize an athermal VCSEL based on the proposed concept.

We fabricated a micromachined VCSEL with an athermal operation. The base structure of the devices was grown in Corning Incorporated, which is similar to that of InP-based VCSELs with tunnel junction [3]. Because GaInAsP has a larger thermal expansion coefficient than InP, we are able to obtain the thermal actuation of the cantilever for compensating the temperature dependence of wavelength. The SEM view of a micromachined InP-based VCSEL is shown in Fig. 1 [4]. The cantilever length is varied from 65 μm to 95 μm, which gives us different temperature dependences. The threshold is 1.3 mA and the maximum power is 0.3 mW under room temperature cw operation. The device was operated at a constant current of 4mA to avoid the effect of self-heating. The measured lasing spectra of VCSELs are shown in Fig. 2 for cantilever length of 95 μm. The lowest temperature dependence we achieved is as low as 0.0016nm/K, which is 50 times smaller than that of single-mode semiconductor lasers [4].

3. Slow Light Devices

The manipulation of the speed of light has been attracting much interest in recent years. In particular, slow light appearing in photonic crystals, semiconductor amplifiers and micro-resonators has been studied for optical buffer memories, optical delay lines and so on [5]-[7]. Also, the slow group velocity of

978-0-9789217-3-6/07/$25.00

©2007 WEN GLOBAL SOLUTIONS

light dramatically reduces the size of various optical devices such as optical amplifiers, optical switches, nonlinear optical devices and so on [8], [9]. We have also observed large waveguide dispersion and slow light [10] in Bragg waveguides where light is confined with highly reflective Bragg reflectors [11]. We recently proposed a slow light modulator with a Bragg waveguide [12], which shows a possibility of low modulation voltage (< 1V) even for ultra-compact waveguide modulators (< 20 μm). We expect high speed modulation of such an ultra-compact waveguide modulator, which enables us to avoid velocity-matching traveling-wave schemes. An important issue is the coupling between free-space propagation light and slow light. .

We demonstrated an electroabsorption modulator consisting of Bragg waveguides with slow light enhancement as shown in Fig. 3 [13]. The base structure is similar to that of a conventional InP-based VCSEL without tunnel junction [3]. The bottom mirror is AlGaInAs quarter-wavelength stack mirror. At first, 1.5-pairs Si/SiO$_2$ dielectric mirror was deposited over the entire surface except top electrodes and then a 5-pair Si/SiO$_2$ dielectric mirror was partly deposited to form a 20 μm long Bragg waveguides. The role of 1.5-pair Si/SiO$_2$ dielectric mirror is the efficient excitation of slow light in a Bragg waveguide. The absorption layer consists of AlGaInAs MQWs in a 1.5 μm wavelength band. With applying reverse bias voltages in its p-n junction, an electro-absorption takes place.

The group velocity decreases with increasing the waveguide dispersion when the wavelength approaches to the cut-off wavelength. The slow-down factor, which is defined as the ratio of the group velocity of slow light versus that in conventional semiconductor waveguides, is over 10 in the wavelength range of 1550-1560 nm. Thus, the electro-absorption effect is enhanced by a factor of more than 10 in this wavelength range and we are able to reduce the size. Even for an ultra-compact modulator, we expect an extinction ratio of 7 dB over 1550 nm, which will be large enough for short-reach optical links. We also expect low polarization dependence, which is very difficult for that of photonic crystal slab waveguides.

The important issue is how to couple with slow light in a Bragg waveguide. We proposed a simple and practical method of a tilt-coupling scheme as shown in Fig. 4 [13]. The input beam is off from the vertical axis and the tilt angle is typically 30 degrees. We carried out the full-vectorial numerical simulation using the film-mode-matching method. Figure 5 shows the model and the calculated field distribution [13]. The coupling loss is less than 1.5 dB for TE and TM modes with a 4 μ m-spot-size Gaussian beam input. This coupling scheme enables us to excite slow light propagating in a Bragg waveguide where light is confined by Bragg mirrors.

We measured the zero-biased insertion loss from the measured near-field intensity. We could achieve a minimum coupling loss of 1 dB, indicating the low coupling loss of the proposed coupling scheme. We achieved an extinction ratio of 7 dB and an insertion loss of 2 dB for a 20 μm long compact waveguide modulator. The proposed structure can be monolithically integrated with VCSELs. The proposed modulator would be useful for ultrahigh speed short reach optical links. We also proposed slow light switches and detectors. Our simple coupling scheme with slow light in Bragg waveguides would be useful for slow light photonic circuits involving optical switches, amplifiers, lasers and so on.

4. Conclusion

Our recent advances on VCSEL photonics were reviewed, including the wavelength engineering and new functions of VCSELs. The small footprint of VCSELs allowed us to form a densely packed VCSEL array both in space and in wavelength. The wavelength engineering of VCSELs may open up ultra-high capacity networking. In addition, a new function of VCSEL structures was addressed, exhibiting a potential of ultra-compact slow light optical devices.

References

[1] K. Iga, IEEE J. Select. Top. Quantum Electronvol. 6, no. 6, pp.1201-1215, Dec. 2000.

[2] F. Koyama and K. Iga, Quantum Optoelectronics of 1997 OSA Spring Topical Meeting, OSA Spring Topical'97, vol. 9, QTh-14, pp. 90-92, 1997.

[3] N. Nishiyama, C. Caneau, B. Hall, G. Guryanov, M. Hu, X. Liu, M. J. Li, R. Bhat, and C.-E. Zah, IEEE J. Sel. Topics Quantum Electron., vol.11, no.5, pp.990-998, 2005.

[4] W. Janto, K. Hasebe, N. Nishiyama, C. Caneau, T. Sakaguchi, A. Matsutani, P. Babu Dayal, F.

Koyama and C.E. Zah, IEEE International Semiconductor Laser Conference, PD1.1, Hawaii, 2006.

[5] M. Notomi, K. Yamada, A. Shinya, J. Takahashi, C. Takahashi, and I. Yokohama, Phys. Rev. Lett., 87, 235902 (2001).

[6]A. Yariv, Y. Xu, R. K. Lee, and A. Scherer, Opt. Lett., 24, 711713 (1999).

[7]X. Zhao, P. Palinginis, B. Pesala, C.J. Chang-Hasnain, P. Hemmer, ECOC 2005, postdeadline paper, Th4.3.6 (2005).

[8] M. Soljacic, S.G. Johnson, S. Fan, M. Ibanescu, E. Ippen and J.D. Joannopoulas, J. Opt. Soc. Am. B, 19, 2052 (2002).

[9] E. Mizuta, H. Watanabe and T. Baba, Japanese Journal of Applied Physics, Vol. 45, No. 8A, 2006, pp. 6116–6120 (2006).

[10] Y. Sakurai and F. Koyama, Jpn. J. Appl. Phys., vol. 43, no. 8B, pp. 5828-5831 (2004).

[11] P. Yeh and A. Yariv, J. Opt. Soc. Am, vol.68, no. 9, 1196-1201 (1978).

[12] K. Kuroki and F. Koyama: 12th Microoptics Conference, Seoul, J-1, pp.222- 223, (2006).

[13] G. Hirano, F. Koyama, K. Hasebe, T. Sakaguchi, N. Nishiyama, C. Caneau and Chung-En Zah, OFC 2007, PDP34, Anaheim, (2007).

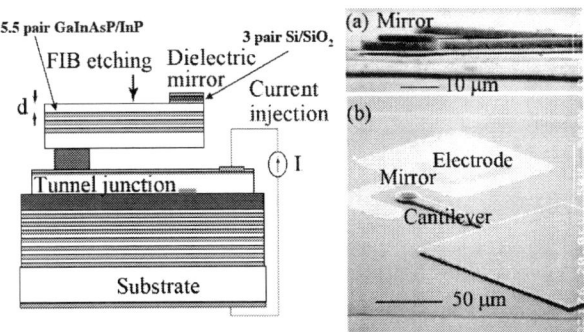

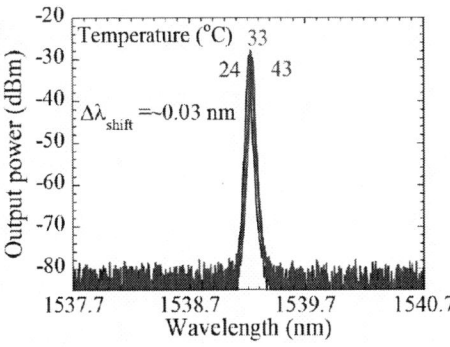

Fig. 1 Schematic structure and SEM image of athermal InP-based VCSEL with a thermally-actuated cantilever structure [4].

Fig. 2 Temperature dependences of lasing spectra for L=95 μm with a fixed bias current of 4 mA [4].

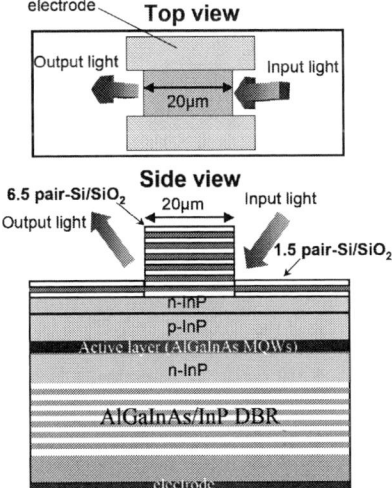

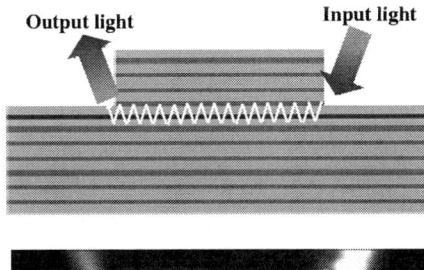

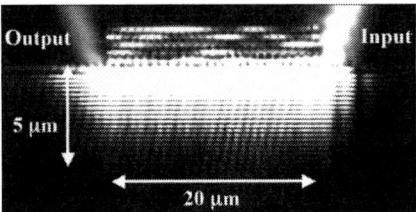

Fig. 3 Schematic structure of a slow light AlGaInAs MQW electro-absorption modulator [13].

Fig. 4 Calculation model and calculated intensity distribution with tilt light input for efficient excitation of slow light [13].

Threshold Analysis of a Novel Dispersive Grating Distributed Feedback Laser Diode

Xun Li, Yanping Xi, Wei-Ping Huang

Department of Electrical and Computer Engineering
McMaster University, 1280 Main St. W., Hamilton, Ontario L8S 4K1 Canada

Abstract

This work proposed a novel distributed feedback (DFB) laser design with a dispersive grating having its coupling strength dependence on the operating wavelength detuning from the Bragg wavelength. This structure guarantees single mode operation as there is an inherent threshold gain discrimination on the two otherwise degenerated modes.

I Background

The lasing condition for conventional DFB laser with uniform grating can be concluded as [1]:

$$\alpha - j\delta = \gamma ch(\gamma L) / sh(\gamma L) \ \text{(1a) and} \ \gamma^2 = \kappa^2 + (\alpha - j\delta)^2 \ \text{(1b)}$$

where κ denotes the coupling strength and is a real constant number for index coupled grating without phase shift; α the threshold gain; δ the propagation constant detuning; L the grating length; and γ an intermediate complex variable. δ can be linked to the lasing wavelength detuning from the Bragg wavelength through: $\delta = 2\pi n_{eff} / \lambda - \pi / \Lambda$ (2), where λ denotes the lasing wavelength; n_{eff} and Λ the laser waveguide effective index and grating period, respectively.

For any given κ, by eliminating $\alpha - j\delta$ in (1a) through (1b), we solve for the intermediate variable γ from: $\sqrt{\gamma^2 - \kappa^2} = \gamma ch(\gamma L) / sh(\gamma L)$ (3) to obtain the threshold gain and propagation constant detuning from: $\alpha - j\delta = \sqrt{\gamma^2 - \kappa^2}$ (4). Once δ is obtained, we can readily find the lasing wavelength through (2). Actually, if we notice $(\gamma^2)^* = (\gamma^*)^2$, $[ch(\gamma L)]^* = ch(\gamma^* L)$, and $[sh(\gamma L)]^* = sh(\gamma^* L)$, we find that if γ is a solution of (3), $-\gamma$ and $\pm \gamma^*$ are all solutions of (3). Therefore, if (α, δ) is a solution set, by taking complex conjugate on both sides of (1a), we know that $(\alpha, -\delta)$ is also a solution set, which means that for any lasing wavelength obtained with detuning δ and threshold gain α, there is always an accompanying lasing wavelength at detuning $-\delta$ with identical threshold gain. If we further notice that $\delta = 0$ cannot be a solution of the above lasing condition, since a real γ (due to $\delta = 0$) makes the left hand side of (3) smaller, whereas the right hand side bigger than γ, we find that the lasing wavelength corresponding to δ will never be the same as the lasing wavelength corresponding to $-\delta$. Hence we draw the conclusion that conventional DFB laser with uniform grating can never achieve single mode operation due to the threshold gain degeneracy at two different lasing wavelengths.

978-0-9789217-3-6/07/$25.00

©2007 WEN GLOBAL SOLUTIONS

In this work, through introducing a dispersive grating [2], we intend to have the grating coupling strength dependent on detuning, i.e., $\kappa = \kappa_0 + \eta\delta$ (5). Although κ remains real, the change on the sign of δ changes κ due to the explicit dependence shown in (5), which leads to a different solution γ according to (3) hence different threshold gain α according to (4). Therefore, we expect that the threshold gain degeneracy on a pair of modes $\pm\delta$ will break.

Actually, if we assume that, along the laser cavity, the effective index of the waveguide changes periodically between: $n_1 = n_{10} + (dn_1/d\lambda)(\lambda - \lambda_0)$ (6a) and $n_2 = n_{20} + (dn_2/d\lambda)(\lambda - \lambda_0)$ (6b) as illustrated in Figure 1 with not only $n_{10} \neq n_{20}$, but also $dn_1/d\lambda \neq dn_2/d\lambda$, a dispersive grating will be formed with the coupling strength given by:

$$
\begin{aligned}
\kappa &= (\pi/\lambda)(n_1 - n_2) = (\pi/\lambda)\{n_{10} - n_{20} + [d(n_1 - n_2)/d\lambda](\lambda - \lambda_0)\} \\
&= \kappa_0 + \pi[d(n_1 - n_2)/d\lambda](1 - \lambda_0/\lambda)
\end{aligned} \quad (7)
$$

where $\kappa_0 \equiv (\pi/\lambda)(n_{10} - n_{20})$ (8) denoting the conventional coupling strength when either there is no waveguide dispersion at all or the waveguide dispersions in area 1 and 2 are identical. From (2) and $\lambda_0 = 2n_{eff}\Lambda$ we also have: $\delta = 2\pi n_{eff}/\lambda - 2\pi n_{eff}/\lambda_0 = (2\pi n_{eff}/\lambda_0)(\lambda_0/\lambda - 1)$ (9). Replacing $1 - \lambda_0/\lambda$ in (7) with the one obtained from (9) yields (5), where η is defined as:

$$
\eta \equiv -(1/2)[d(n_1 - n_2)/d\lambda](\lambda_0/n_{eff}) = -[d(n_1 - n_2)/d\lambda]\Lambda \quad (10)
$$

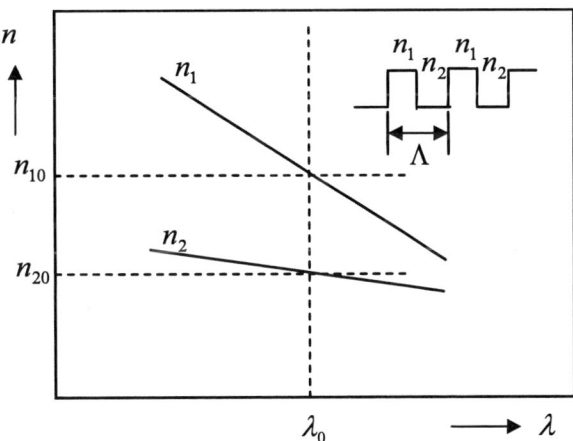

Figure 1 A sketch of a dispersive grating

Therefore, we find that in (5), κ_0 is the "background" coupling strength whereas the detuning coefficient η in quantity is the effective index dispersion difference within one grating period, given as (8) and (10), respectively.

II Design example

Through a root searching routine that solves for (1a), (1b) and (5) numerically, we find the threshold condition (α, δ) of two dispersive grating designs as shown in Figure 2.

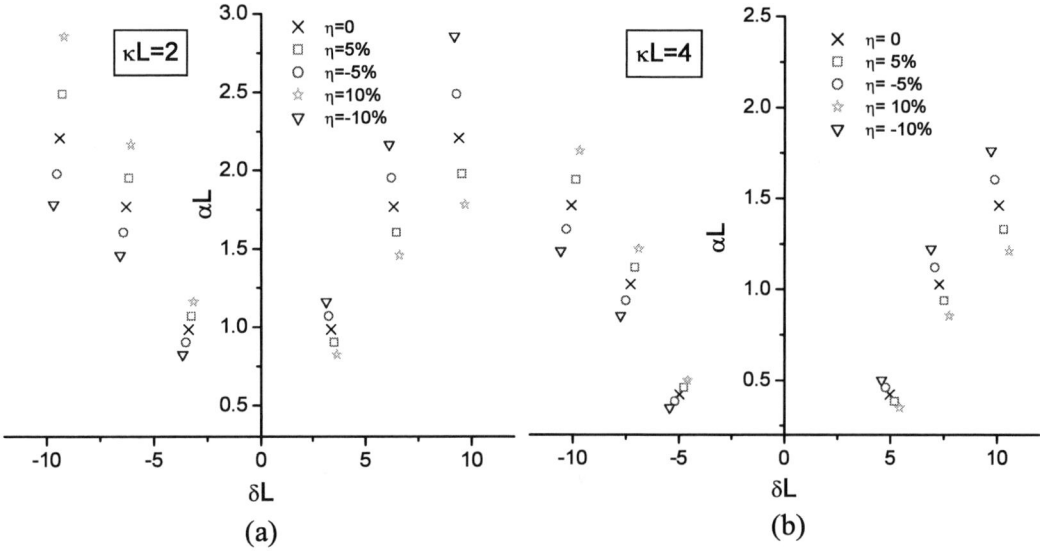

Figure 2 Threshold condition of dispersive grating DFB lasers with positive detuning coefficients $\eta = 5\%, 10\%$ (square, star) and negative detuning coefficients $\eta = -5\%, -10\%$ (circle, triangle) under normalized background coupling strength (a) $\kappa L = 2.0$ and (b) $\kappa L = 4.0$. For comparison purpose, threshold condition of conventional DFB laser ($\eta = 0$) with the same normalized coupling strengths are also plotted (cross).

It is well known that in conventional DFB laser with uniform grating, both modes at the two stopband edges have the lowest, but identical threshold gain, so in principle it works at a dual-mode operation scheme. From Figure 2, we find, however, that there is only one mode at one side of the stopband edge takes the lowest threshold gain in either case (with the positive or negative detuning coefficient η). Hence the degeneracy on the threshold gain is broken up and single mode operation is achieved. This is obviously attributed to the threshold gain discrimination brought by the grating coupling strength dependence on the detuning, as we previously analyzed. Therefore, we may draw the conclusion that, as an alternative approach, the dual-mode operation problem in uniform grating DFB laser can be cured by incorporating a grating dispersion, rather than by implementing the existing methods such as quarter-wavelength shifted or complex-coupled gratings. A straightforward way to generate the grating dispersion is to introduce non-equal effective index dispersions (i.e., non-equal waveguide dispersions) in the two different sections of every grating period.

References

[1] H. Kogelnik and C. V. Shank, "Coupled-wave theory of distributed feedback lasers," *J. Appl. Phys., vol. 43, no. 5, pp. 2327-2335, May 1972.*
[2] P. Yeh, "Christiansen-Bragg filters," *Opt. Comm., vol. 35, no. 1, 9 (1980).*

40 GHz Self-Pulsation in Two-Section DFB Lasers with Varied Ridge Width

Dingbo Chen (1), Hongliang Zhu (2), Song Liang (3), Huan Wang (4), Yali Zhang (5)

Key Laboratory of Semiconductors Materials, CAS & State Key Laboratory on Integrated Optoelectronics, Institute of Semiconductors, CAS, P. O. Box 912, Beijing 100083, P. R. China, boydchen@semi.ac.cn

Abstract *1.55-μm InGaAsP-InP two-section DFB lasers with varied ridge width have been fabricated. Self-pulsations with frequencies around 40GHz are observed. The related mechanism and the tunability of generated self-pulsations is studied.*

Introduction

Self-pulsation (SP) DFB lasers are an attractive source for the generation of millimeter-wave carriers over fiber and can be widely used in a WDM configuration for wireless networks operating at 40 GHz[1]. In addition, self-pulsating DFB lasers have many promising applications in future optical network, especially for all-optical clock recovery[2, 3], which is a key function in all-optical 3R regeneration. Self-pulsating DFB laser diodes with multiple DFB regions were fabricated for this purpose. Generally, such multisection DFB lasers can be divided as two-section type [4, 5] and three section type[6, 7], and three different SP mechanisms have been proposed, including spatial hole burning (SHB) [8, 9], dispersive self-Q-switching (DQS)[6], and beating type (BT)oscillations [5, 8].

As is shown in early reports, the detuning of lasing wavelength in each laser section of a multi section SP DFB laser plays an outstanding role in the generation of self pulsation[10]. Recently, a theoretical investigation[11] showed that a variation in the ridge width of a two-section DFB laser would lead to a slight wavelength difference of the lasing mode in each DFB section, and would in turn give rise to the beating between the two lasing modes. This method is relatively simple compared with other methods for realizing the wavelength difference, such as varying the grating period[10] and introducing phase-shift into one of the DFB sections[12].

In this paper, we fabricate two-section DFB lasers with varied ridge width and self-pulsations with frequencies around 40GHz

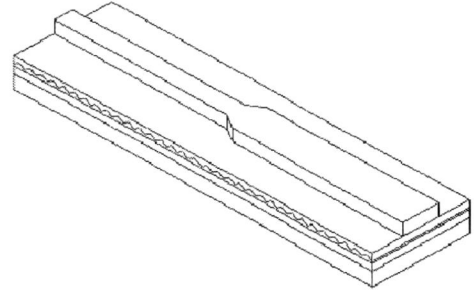

Fig. 1. structure of the DFB LD

are successfully detected. The possible pulsation mechanism is discussed. To our knowledge, this is the first demonstration generation of self-pulsation in multi-section DFB lasers with different ridge width.

Device structure

As is shown in Fig.1, the device consists of two individually injected gain-coupled DFB sections with all the same parameters but different ridge widths, which are designed to be 3μm and 1.5μm and are marked as section 1 and section 2, respectively. The device material was grown on an n-InP substrate by the metal organic chemical vapor deposition, and the active region is strain-compensated InGaAsP multiple quantum wells. The electrical isolation between the two DFB sections is realized by the implantation of He$^+$. Both sections are 325 μm long and the front and back facets are left as cleaved.

978-0-9789217-3-6/07/$25.00
©2007 WEN GLOBAL SOLUTIONS

Device characterization

In the laser spectrum measurements, an optical spectrum analyzer with a resolution of 0.01 nm is used. The measurements show that, when operated individually, the emitting wavelengths of section 1 and section 2 are 1524nm and 1521nm, respectively. The threshold current of section1 and section 2 are 29.5 mA and 34.5 mA, respectively. The analysis of the wavelength difference between the two DFB sections which share the same material and grating parameters is as following: the variation of ridge width gives rise to a difference of the effective refractive index in these two DFB sections, which in turn results in a lasing wavelength difference. In our device, the ridge widths of the two sections are designed as 1.5μ m and 3μ m, respectively, and this variation, by calculation, would result in a effective refractive index difference of about 0.006, and thus a wavelength difference $\Delta\lambda$ of 2.84nm is estimated by the equation $\Delta\lambda/\lambda=\Delta n_{eff}/n_{eff}$. This is in a good agreement of the experimental results. Here λ is assumed as the average value of the two lasing modes' wavelength, that is 1522.5 nm, and n_{eff} is assumed as 3.21, which is also a average value of the two sections' effective refractive index.

When both sections are injected with certain currents, self pulsations are generated, which are investigated with both an electrical spectrum analyzer and an optical spectrum analyzer. In the RF measurements with an electrical spectrum analyzer, the laser's emission is coupled into a single mode fiber. After passing an isolator, the signal is detected by a high speed photodiode and then analyzed with a spectrum analyzer. This system can operate in a spectral range up to 45 GHz. We operate the devices at a wide range of injection currents. The results are encouraging and we find that the devices show SP whose frequency is around 40 GHz.

Fig.2 shows the optical and RF spectra of RF signal when both sections are injected with the currents above the threshold. In

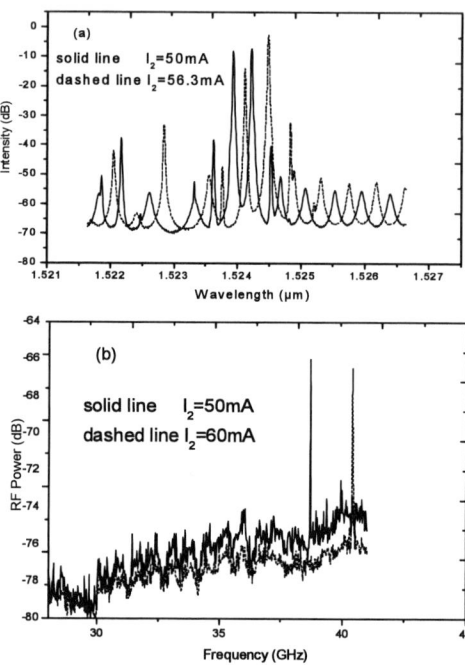

Fig.2 Optical (a) and RF (b) spectra of self-pulsation around 40GHz, when I_1 fix at 55.3 mA, I_2 fix at 50 mA (solid line) and 56.3 mA(dashed line), respectively.

Fig.2b, when I_1 is fixed at 55.3mA , RF signals with frequency of 38GHz and 40GHz are observed when I_2 is 50mA and 56.3mA, respectively. As can be seen in Fig. 2a, dual modes around 1524nm were observed in the optical spectrum, both the two modes have narrow optical line width. We attribute the observed self-pulsation to BT oscillation mechanism, in which self-pulsation is generated as a result of the beating of two modes which are slightly detuned if two DFB sections are of comparable threshold current and are both injected above the threshold current, and the pulsation frequency corresponds to the mode spacing. The following facts of our devices support the underlying BT mechanism. First, the two dominant peaks have similar intensity in the optical spectra Fig. 2a, which is a typical feature for beating type oscillation. Second, the mode spacing is in agreement with the RF frequency. Increasing the current of I_2 from 50mA to 56.3mA there is a 0.27nm red shift of the long wavelength mode, while only 0.19 nm red shift of the short wavelength mode at the same time. This in turn gives rise to an increase of the modes spacing, and thus an increase of the corresponding RF frequency. In this sense, the two-section devices can not be seen as a combination of two independent DFB lasers, instead, each section gets feedback from the other section, both modes coexist and interact with each other.

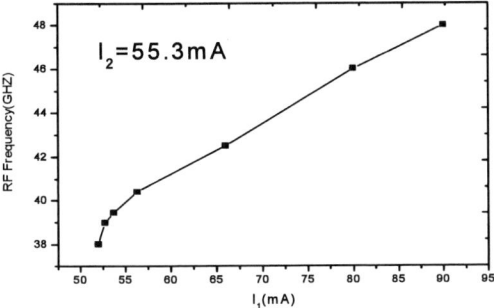

Fig.3 the tunability of RF frequency as the function of I_1, while I_2 is fixed at 55.3mA

Finally, when I_1 = 55.3mA and I_2 = 56.3mA, (the dashed line in Fig. 2), the mode spacing is about 0.32nm, and is in good agreement with the corresponding RF frequency of 40GHz according to the formula $f = c^* \Delta \lambda / \lambda^2$, here c is the velocity of light.

The pulsation frequencies of the devices can be tuned by varying the injection currents, as is shown in Fig. 3. For the 40 GHz range, the RF frequency increases from 38 GHz to 48 GHz when I_2 is increased of from 50 mA to 90 mA, and I_1 is fixed at 55.3 mA. It should be noticed that Frequencies beyond 45 GHz, which is the limit of the RF measure system, is estimated by the modes spacing in the optical spectra using the formula $f = c^* \Delta \lambda / \lambda^2$.

Conclusion

In summary, two-section DFB lasers with a variation in the ridge width have been fabricated and measured. Self-pulsation with frequency of 40GHz range are detected and discussed. The results of our measurement verify that this simple structure is liable to the generation of self-pulsation. The electrical frequency tunability around 40GHz makes DFB lasers with such structure very promising devices for high speed wireless network and all-optical signal processing.

References

1   M. Al-Mumin, L. Yuhua, and L. Guifang, 11(2001), 476.

2   C. Bornholdt et al., Electronics Letters 36 (2000), 327.

3   B. Sartorius et al., Electronics Letters 34 (1998), 1664.

4   U. Bandelow, H. J. Wunsche, and H. Wenzel, Photonics Technology Letters, IEEE 5(1993), 1176.

5   B. Sartorius, M. Mohrle, and U. Feiste, Selected Topics in Quantum Electronics, IEEE Journal of 1(1995), 535.

6   U. Bandelow et al., Selected Topics in Quantum Electronics, IEEE Journal of 3(1997), 270.

7   M. Radziunas et al., Quantum Electronics, IEEE Journal of 36(2000), 1026.

8   H. Wenzel et al., Quantum Electronics, IEEE Journal of 32 (1996), 69.

9   D. D. Marcenac, and J. E. Carroll, Electronics Letters 30(1994), 1137.

10   M. Mohrle et al., Selected Topics in Quantum Electronics, IEEE Journal of 7(2001), 217.

11   C.-Z. Sun et al., Guangdianzi Jiguang/Journal of Optoelectronics Laser 18(2007), 396.

12   S. Nishikawa et al., Applied Physics Letters 85(2004), 4840.

Emission Characteristics of a Surface-emitting Organic Photonic Crystal Laser

Sidney S. Yang, Li-Wen Chang, Chong-Jie Huang

Institute of Photonics Technologies, National Tsing Hua University,
101, Section 2 Kuang Fu Road, Hsinchu, Taiwan 30013, Republic of China
ssyang@ee.nthu.edu.tw

Abstract *Lasing action of a composite organic thin-film laser with a 2nd-order two-dimensional photonic crystal structure of triangular lattice is investigated. The analysis of band theory is also adopted to confirm the lasing mode.*

Introduction

For the last two decades, organic luminescent materials have been attracting attention in applications like organic light-emitting devices and organic lasers because of their flexibility, low processing temperature, and broad emitting spectrum in the visible range[1]. Characteristics of photonic crystals are another intriguing research topic due to the specific band gap structure[2]. By using the structures as laser cavities, light emission and propagation can be engineered. Although most researches are done with semiconductor lasers, Meier *et al.* in 1999 reported the optical pumped lasing action in two-dimensional (2D) photonic crystal structure with organic gain media[3] and showed the importance of a saddle point in the *k* space concerning photonic crystal laser[4]. The detail of feedback in the photonic crystal laser has been demonstrated by Notomi *et al.* later[5].

In this paper, we report a study of lasing characteristics of a composite organic thin-film laser with gain medium of 4-(Dicyanomethylene)-2-tertbutyl-6-(1, 1, 7, 7-tetramethyljulolidin-4-yl-vinyl)-4H-pyran (DCJTB) doped in Poly(N-vinylcarbazole) (PVK) on a 2D photonic crystal structure of triangular lattice.

Fabrication and measurement

Fig. 1(a) shows the schematic structure of an organic thin-film laser with 2D photonic crystal structure. We doped DCJTB in PVK (3:100 by weight) and spin-casting process was employed to form the organic thin-film with thickness of 470 nm on the patterned

SiO$_2$ substrate. The structure of 2D photonic crystal shown in Fig. 1(b) is the triangular lattice of circular holes which has lattice constant of 405±5 nm and radius of 150±2.5 nm fabricated by electron beam lithography and dry etching. The etching depth is 107 nm and surface profile measured by AFM is shown in Fig. 1(c). Optical Excitation is provided by a pulsed Q-switched second-harmonic generated Nd:YAG laser with wavelength of 532 nm and repetition rate of 50 Hz. The pumping laser beam is focused on the organic thin-film laser at a slating angle of 45° and forms an elliptic spot with long axis length of 180 µm and short axis length of 123 µm.

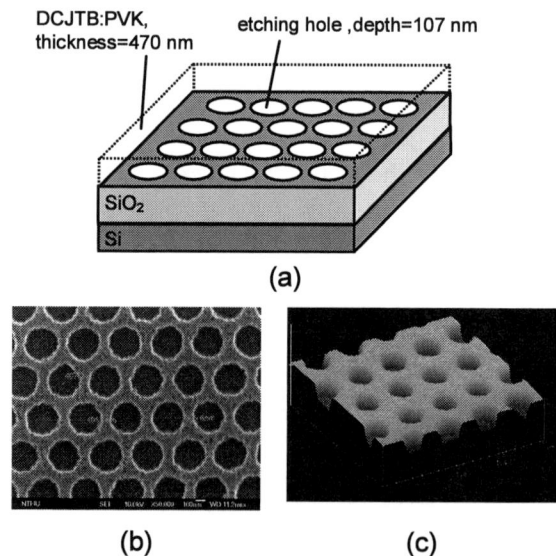

(a)

(b)   (c)

Fig. 1(a) Schematic drawing of an organic thin-film laser with 2D photonic crystal structure of (b) triangular lattice and its lattice constant of 405 ± 5 nm, radius of 150 ± 2.5 nm. (c) Surface profile of 2D photonic crystal structure measured by AFM.

978-0-9789217-3-6/07/$25.00

©2007 WEN GLOBAL SOLUTIONS

Gaining from the feedback within the planar 2nd-order 2D photonic crystal structure, part of the lasing radiation is coupled into radiation modes perpendicular to the surface via Bragg diffraction. The surface emitted light is collected by a fiber-coupled CCD spectrometer (CDI Spec 32) to produce the emission spectrum.

Lasing characteristics and theoretical analysis of band diagram

Fig. 2(a) shows the output intensity and the FWHM of the emission peak versus the pumping energy. It is apparent that there are three distinct regimes labelled I, II, and III as the pumping energy increases. At low pumping energy (region I), the output intensity rises gradually. As the pumping energy increases beyond the threshold energy of 4.67 µJ (region II), the lasing phenomenon is observed. Furthermore as the pumping energy reaches 6.89 µJ (region III), the gain saturation occurs. This phenomenon could be attributed to the increased temperature of the substrate and the organic medium. The measured emission linewidth of 4.67 nm, as shown in Fig. 2(a), is limited by the dynamic range of the CCD spectrometer. Fig. 2(b) shows the emission spectra of this device pumped below and above the lasing threshold of 4.67 µJ. The lasing wavelength is located around 621 nm.

To attain a deeper understanding of 2D-DFB laser action, we plot the band diagram of figure 1(a) for TE mode by simulation with program RSoft (parameters: free space wavelength = 630 nm, background index = 1.457, index difference = 0.18, waveguide width = 305 nm, period = 405 nm). The result is shown in Fig. 3(a). Fig. 3(b) shows the reciprocal lattice and the Brillouin zone with high symmetry points of Γ, M, and K. It is expected that lasing occurs at saddle points of the band diagram. At these saddle points, waves propagate in different direction leading to distinct coupling mechanism which can be explained by Bragg conditions. Our design is based on the saddle points of **s** indicated in Fig. 3(a). The lasing action is expected to be observed at a/λ = 0.651 as shown in Fig 3(c). With a=405 nm, the lasing wavelength

is calculated to be 622 nm, in good agreement with measured wavelength of 621 nm. The other saddle point of **s** which corresponds to longer emission wavelength has lower materimal gain based on the emission spectrum of DCJTB. The lasing action is not observed.

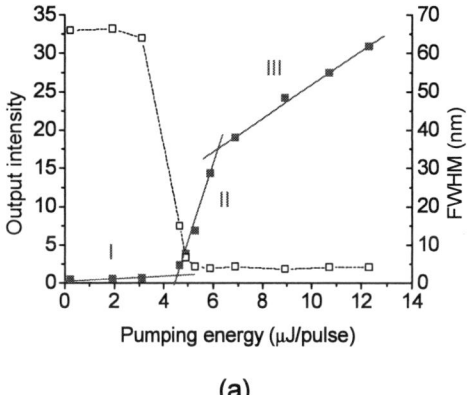

(a)

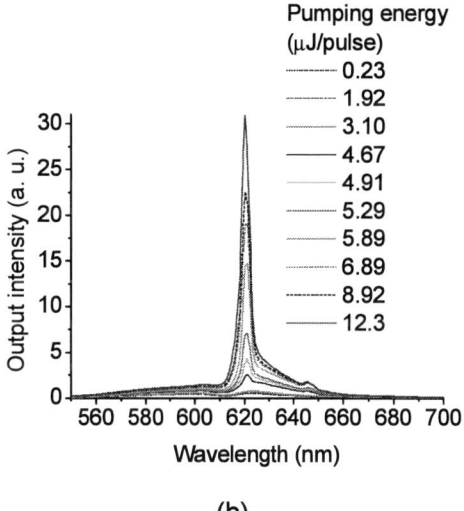

(b)

Fig. 2(a) Output intensity and the FWHM of the emission peak plot as a function of pumping energy. The labelled I and II indicate the regions below and above lasing threshold of 4.67 µJ. The III is the saturation region of lasing action. (b) Emission spectra at pumping energy below and above lasing threshold of 4.67 µJ.

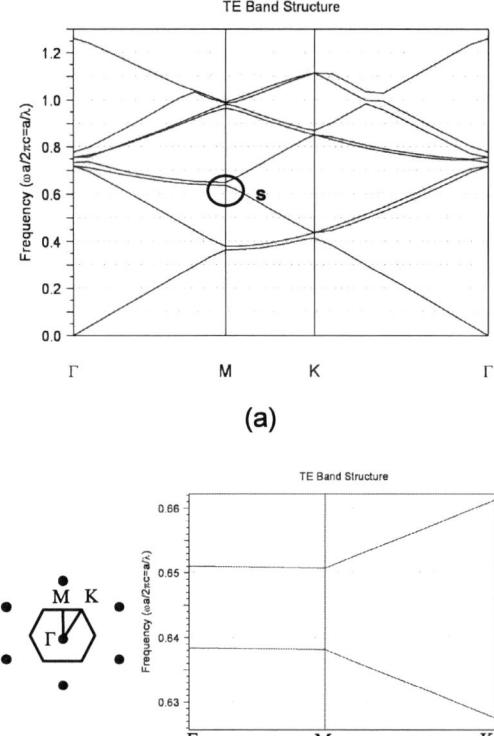

(a)

(b)   (c)

*Fig. 3(a) Band diagram of the device shown in Fig. 1(a) for TE mode operation. (b) The first Brillouin zone of triangular lattice and its reciprocal lattice with high symmetry points of Γ, M, and K. (c) Detailed band diagram of point **s** in (a).*

Conclusions

We use the organic dye DCJTB doped in the matrix of PVK as the gain medium. The organic thin-film laser with 2D photonic crystal is made by spin casting an organic thin film onto a 2D photonic crystal structure with triangular lattice fabricated by E-beam lithography and dry etching. The lattice constant is 405 ± 5 nm with air holes of radius 150 ± 5 nm. Lasing phenomenon is observed at threshold pumping energy of 6.7 μJ at the peak wavelength of 621 nm. The FWHM is indicated as 4.22 nm due to the limitation on the dynamic range of the CCD spectrometer. The calculated band diagram of lasing action at the peak wavelength of 622 nm shows good agreement with measured wavelength of 621 nm.

The investigation of 2D photonic crystal laser has not yet completed, like determine the influence of hole size (or shape) and the polarization of lasing beam. The fabrication of 2D photonic crystal structure by e-beam lithography requests a lot of time and cost. If we can develop the nano-imprint lithography[6], the procedure of 2D photonic crystal structure can be much simplified. Another fascinating issue is the possibility of three dimensional distributed feedback lasers[7]. It might have more effective output beam, lower threshold energy, and good collimation of lasing beam.

Acknowledgments

The authors are grateful to National Science Council of Taiwan (NSC 95-2221-E-007-199-) and Ministry of Economic Affairs, R.O.C (95-EC-12-A-08S1-042) for financial Support and Dr. Chao for providing access to the RSoft simulation program.

References

1 K. Kobayashi et al., IEEE J. Quantum Electronics, 39 (2003), 664.
2 E. Yablonovitch, Phys. Rev. Lett. 58 (1987), 2059.
3 M. Meier et al., Appl. Phys. Lett. 74 (1999), 7.
4 A. Mekis et al., Appl. Phys. A: Mater. Sci. Process. 69 (1999), 111.
5 M. Notomi et al., Appl. Phys. Lett. 78 (2001), 1325.
6 S .Y. Chou et al., Appl. Phys. Lett. 67 (1995), 3114.
7 K. Sakoda, Optics Express 4 (1999), 167.

Uncooled Submarine Pump Laser Module at 980 nm

Wenjuan (Janet) Shen (1), Stefan Mohrdiek (2), Bing (Bruce) Guo(3), Tomas Pliska, Mark Ives, Shaun Quinlan, Warren Grace, Andrew Miller, Thomas Goodall, Jeffrey Greatrex, Robert Cann Wen (Wendy) Ma

1 Was born in 1978, received the Ph.D degree in engineering of material physics and chemistry from Semiconductor institute of Chinese Academy of Science, Beijing, China (2006.2). She is responsible for uncooled pump laser module in Bookham as product engineer. Bookham Shenzhen Futian free trade zone, P.R.C.518038, Janet.shen@Bookham.com 2 Stefan.Mohrdiek@bookham.com, 3 Bruce.guo@bookham.com

Abstract *We present a new generation 980 nm submarine pump module that consists of a hermitically sealed 8-pin ceramic MiniDIL package without thermo-electric cooler.*

Introduction

The increasing demand for broadband internet services requires the installation of new submarine DWDM backbones over the next years. 980 nm pump laser modules with very low failure rates at high output powers and reduced overall electrical power consumption are necessary for these new submarine cable systems.

The 980nm pump module for subsea applications presented in this work is packaged in a miniature dual-inline (miniDIL) package and operated without a thermoelectric cooler (TEC). This offers the advantage of small form factor, minimal heat generation, and reduced consumption of electrical power [1]-[5]. The module incorporates our generation-08 (G08) laser chip, which represents a continuation of previously introduced and submarine qualified pump laser generations. By combining high reliability standards with materials and processes used in high volume manufacturing of terrestrial pump modules, our technology represents an attractive solution for submarine pump module,manufacturing at reduced cost. This approach is made possible as a result of Bookham experiencing with 400,000 devices shipped, accumulating 16 billion device hours, with less than 20 FIT in the field.

Uncooled Submarine Module Design

A robust, yet simple packaging technology is a key element for realising reliable high power pump laser modules. 8-pin miniDIL package preferred for coolerless devices.

The OceanBright™ submarine pump laser module comprises an industry standard eight-pin ceramic Mini-DIL package, a laser chip-on-carrier (CoC) assembly, a fiber assembly with a wedge polished, antireflection coated (AR) fiber lens, and a fiber Bragg grating (FBG) for wavelength and power stabilization incorporated in the fiber pigtail. Fig.1 shows a photograph of the submarine pump laser module. The laser chip is attached onto aluminium nitride (AlN) carrier with a thermistor for temperature monitoring and a photodiode mounted on the back side of the laser for alarm purposes. This chip-on-carrier assembly is then mounted into the package. The fiber is attached to carrier in front of the laser by using a solder technology; it is fixed in the fiber feed-through port with soldering technique that provides a hermetic seal to the package. The package lid is sealed by resistance welding.

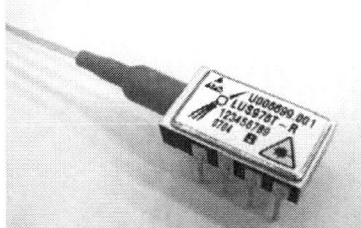

Fig.1.Mini-DIL package for submarine pump

Semiconductor laser technology

The design of our G08 980 nm laser diode is the result of continual improvement to (previously undersea qualified) successive generations of laser chips [6-8], using the same MBE growth InGaAlAs material system, proprietary E2 facet passivation technology providing COMD-free laser chips,

978-0-9789217-3-6/07/$25.00

©2007 WEN GLOBAL SOLUTIONS

and ridge waveguide manufacturing technology. The epitaxial structure of our G08 laser has been designed for reduced waveguide losses, decreased series resistance, and improved wave guiding properties along the laser cavity. These measures increase the kink free light output power. These design improvements enable the device to meet the stringent reliability requirements for undersea applications.

The capability of this technology is demonstrated by the data shown in Fig. 2, where we plot the operating current required for maintaining a constant optical output power over 16 years of continuous operation for first generation 980 nm lasers with E2-facet passivation. The noise on the data is not a result of laser instabilities, but rather a result of test set instability caused by power outages, temperature variations and moves into new locations during the last 16 years. More than one million device hours are accumulated at various stress conditions. Based on the standard reliability model [9] and data of these 9 devices the "only" fail gives and estimate of less than 32 FIT failure rate at operation conditions. Consistent with these first generation 980nm pump lasers, today's 980nm pump lasers, such as our G08-type chip, have inherited the advantages of all previous design optimizations and show lower failure rates at much higher power levels.

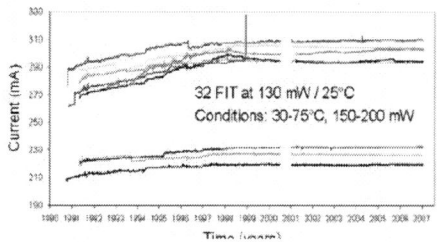

Fig. 2 Nine first generation of E2-facet lasers under test for 16 years at various stress conditions

Power-current characteristics of the G08 laser are shown in Fig. 3. A high roll-over and kink free power level are prerequisites for realizing a higher power pump laser module. The epitaxial structure and waveguide as well as the thermal and electrical properties of the G08 laser chip have been designed for high power and thermal stability. At a heat sink temperature

of 5 ℃, the laser chip achieves a continuous-wave (CW) roll-over power of more than 2 W. The roll-over power at 25 and 75 °C is 1.7 and 1.3 W, respectively, proving the thermal stability of the overall device design. The characteristic temperature T0 of the threshold current increase is 155 K for this particular design, with a threshold current of about 40 mA at 25 °C. The far field pattern is optimized for direct coupling into a lensed single-mode fiber. It has a full-width-half-maximum (FWHM) of 6-7° and 20-22° for the lateral and vertical direction, respectively. The maximum coupling efficiency into a single-mode fiber exceeds 80%. Moreover, the laser waveguide is optimized for low internal losses of below 1 cm-1 and a low series resistance of 0.4   . The slope efficiency of the laser is 0.95 W/A, and the maximum wall-plug efficiency is 60%.

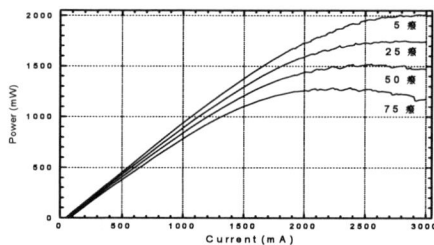

Fig. 3: Optical output power from the laser facet as a function of injection current for a 980 nm G08-type laser at four different heat sink temperatures. The laser diode is mounted junction side up and operated in cw mode.

Submarine performance
Temperature variations over 45 K in case of undersea applications causes a wavelength shift of 13.5 nm due to the natural bandgap narrowing at a shift rate of 0.3 nm/K of the pump laser diode. Since the EDFA absorption spectrum is sharply peaked at 980 nm wavelength, it is vital to efficient amplifier operation to prevent wavelength drift. Fiber Bragg gratings (FBG) are used in the Bookham uncooled pump modules to lock the pump laser wavelength over a wide range of operating conditions.

In Fig. 4, we plot the power-current curves of the FBG stabilized miniDIL module measured in a temperature range of 0 through 70 °C in steps of 10 °C. Power well in excess of 500 mW at 1 A are obtained.

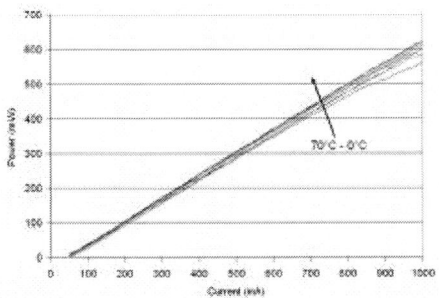

Fig.4 Power versus current characteristics for various temperatures demonstrates up to 600 mW ex-fiber at 1 A

Fig. 5 shows the effect of wavelength stabilization by an FBG over an extended temperature range of 70 K. A side lobe suppression ratio greater than 30 dB over all operating conditions to more than 98% of power within the useful pump band is obtained for all operating currents and temperatures with a total wavelength shift of less than 0.4 nm.

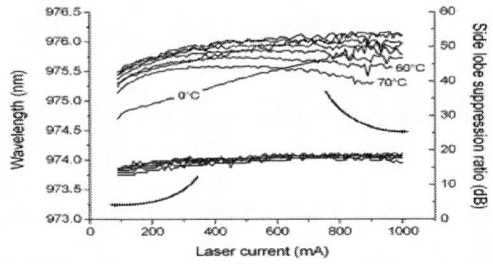

Fig.5 Peak wavelength (left scale) and side lobe suppression ratio (right scale) as a function of laser drive current measured in steps of 10 K over a total temperature range 0 to 70 °C

Device Reliability

The described package platform is used in both in Butterfly-type and MiniDIL packages for three laser generations (G06-G08) and has undergone a number of qualifications and re-qualifications according to the Telcordia GR-468-CORE standards for use in Central Office environments and beyond. More than 165 devices have been exposed to mechanical integrity tests against MIL-STD-883 standards for mechanical shock, vibration and thermal shock. Endurance tests, such as accelerated ageing for 2000 and 5000 hours, high temperature storage, temperature cycling and damp heat were performed with more than 280 devices. A number of other tests were dedicated to prove the effectiveness of the incorporated

getter. As a result, the package wear out related failure rate is estimated to be less than 10 FIT. The majority of failures are seen to be random failures due to sudden chip failure.

Since 1996 a known number of 980 nm pump modules have been shipped into the field by Bookham (formerly Nortel Networks Optical Components) with the same basic laser diode technology. This can be used to estimate cumulative field hours. Then together with the cumulative number of field failures it allows for the calculation of a field failure rate. There is reducing trend towards failure rates below 20 FIT at 95% confidence level with increasing statistics from 1996 till 2006.

Conclusions

We have introduced a cost effective Mini-DIL 980 nm pump module technology in undersea applications requiring high reliability by taking advantage of proven volume production techniques for terrestrial applications. The submarine 980 nm pump module in a MiniDIL package incorporates our G08 laser and is well suited for 500 mW wavelength stabilized power in the fiber at operating temperatures from 0°C - 45°C and a failure rate of less than 100 FIT.

References

1 R. Baettig et al, Proc. IEEE Lasers Electrooptics Society, vol.2, pp. 542-543, 2002

2 J. Yang et al ,Proc. IEEE Electronic Components and Technology Conf. pp.811-814, 2002

3 S. Mohrdiek et al., Optical Fiber Communication Conf. vol.3, WDD77, 2001

4 T. Pliska et al., Lasers and Electro-Optics Society, vol.1, pp.139-140, 2001

5. S. Mohrdiek et al., Electron, Lett., vol. 39, pp. 1105-1107, 2003

6. A. Moser et al., Applied Physics Letters, vol. 55, p. 1152, 1989

7. A. Oosenbrug, Proc. of SPIE, pp.20-27, 1998

8. H.U. Pfeiffer et al., Optical Fiber Communication Conf. pp. 483- 484, 2002

9. B. Schmidt et al., "Chapter:Pump Laser Diodes"in Optical Fiber Telecommunications IVA, Editors: Kaminov & Li, Academic Press, ISBN 012-395172-0, 2002

Investigating the cortical hemodynamics with high spatio-temporal resolution by optical imaging techniques

Pengcheng Li, Qingming Luo*

Britton Chance Center for Biomedical Photonics, Wuhan National Laboratory for Optoelectronics, Huazhong University of Science and Technology, Wuhan 430074, China

*Corresponding author: qluo@mail.hust.edu.cn

Abstract *Optical imaging techniques have been shown to be powerful tools for investigating cortical functional architecture and dynamics with high spatio-temporal resolution in normal and disease brain. In this paper, the technical background and present applications of two optical imaging techniques, intrinsic optical signal imaging and laser speckle imaging are reviewed.*

Introduction

Mapping the changes in cortical hemodynamics induced by neuronal activity with high spatial and temporal resolution is essential for understanding the complex mechanism of neurovascular coupling and the relationship between the underlying metabolic activity and the hemodynamic response observed by neuroimage techniques, as well as for investigating the disease mechanism in brain which may result in new treatments. Optical imaging of intrinsic signals has proven to be a powerful tool for the visualization of functional architecture and hemodynamics in exposed mammalian cortex. Imaging with multiple wavelengths can further provide the quantitative information on the relative changes in concentration of oxygenated hemoglobin and deoxygenated hemoglobin. Laser speckle imaging is recently used to reveal the full field two-dimensional cerebral blood flow with high temporal resolution. In this review, we summarized the technical background and present applications of how these two optical imaging methods are being applied to help giving new insight into understanding the complex mechanisms underlying the function and pathology of brain.

Laser speckle flowmetry for imaging the cerebral blood flow with high spatial and temporal resolution

Laser speckle is the random diffraction pattern produced when the lasers radiation is scattered by a rough surface or a diffuse medium such as biological tissue. If the scattering object moves, the speckle pattern also changes in time randomly. Such a case is referenced as speckle dynamics, and the speckle dynamics contains information about the velocity distribution of the scattering object. In laser speckle flowmetry, the time-integrated dynamic speckle pattern that arises from biological tissue is acquired by a CCD camera with an exposure time of around 10 milliseconds, and the blurring of the time-integrated speckle pattern is then used to be related with the velocity of the blood flow in the tissue. Comparing with the conventional laser doppler method used for the investigation of blood flow, laser speckle imaging can obtained the full field two-dimensional changes in blood flow without scanning so that it can provide much higher imaging speed than the scanning laser doppler method does.

The laser speckle flowmetry has attracted extensive interests recently in the studies of brain activities because of its high temporal resolution. However, this method also suffer from several limits, such as the compromise of spatial resolution due to the operation of spatial average, the influence of the static speckle on quantitative determination of the blood flow velocity for in vivo experiments, and so on. To prompt the solution of these problems, we proposed a laser speckle temporal contrast analysis method to improve the spatial resolution[1], and to reduce the influence of static speckle on blood flow estimation when imaging the CBF through the intact rat skull (as shown in Fig.1)[2]. The optical clearing agent was also attempted to improve the depth of penetration through the rabbit dura when imaging the CBF[3]. Preliminary applications of the laser speckle imaging technique were validated on investigating the CBF response induced by sensory stimulation and drug stimulation[4,5].

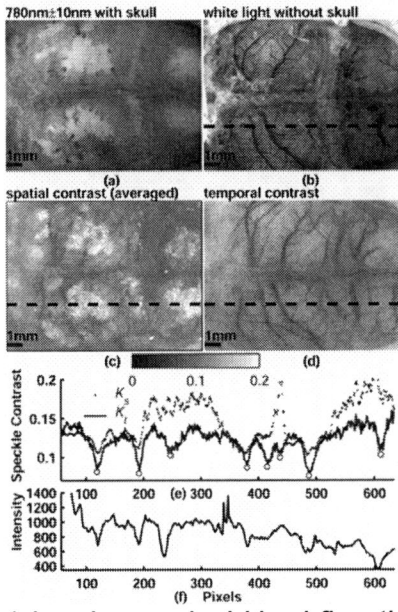

Fig. 1 Imaging cerebral blood flow through intact rat skull with laser speckle imaging temporal contrast analysis [2]

Combining multi-wavelength intrinsic optical signal imaging with laser speckle flowmetry

For the purpose of imaging cerebral hemodynamic on animal cortex, PET, fMRI and Optical imaging (such as laser doppler flowmetry, single wavelength intrinsic optical signal imaging) techniques are available as the conventional methods. But they can only get the information about changes in cerebral blood flow, blood volume and oxygenation separately. Furthermore, the spatial and temporal resolution of fMRI and PET are not high enough to study the fine mechanisms of neurovascular coupling on the level of small blood vessels.

To realize the multi-parameter imaging, we developed an optical imaging instrument that combines the multi-wavelength intrinsic signal imaging technique and laser speckle imaging technique[6]. The instrument uses a multi-wavelength light-emitting diode (LED) and laser diode (LD) as light sources. The different light sources are controlled by a microprocessor, and are time division multiplexed to share one CCD camera for signal detection. The graphic user interface software used for the imaging data analysis was also developed. This instrument provides the capability of imaging the changes in cerebral blood flow, blood volume, blood oxygenation, metabolic rate of oxygen and morphology of blood vessel following the cortical activation simultaneously in mammalian cortex with high spatial and temporal resolution. The optical imaging system is also sychronized with the instruments used for recording the neural electrophysiological signals so that the relationship between the hemodynamic response and neural activity can be investigated.

Spatial and temporal analysis of cortical activation from optical imaging data

When imaging the cortical activation by using the intrinsic optical signal, the changes in light intensity due to neuronal activity are often very small, but the noise, which arise from either the biological noise associated with the respiration, circulation and irrelevant physiological activity or the instrumentation noise such as digitization noise, illumination noise, movement artifacts, etc. are usually large. To improve the determination of the spatial pattern and temporal dynamics of the cortical activation from the optical imaging data, a couple of statistical signal processing methods such as independent component analysis (ICA), principal component analysis (PCA) and temporal cluster analysis (TCA), are attempted to suppress these large background noises and extract the small signal of interest from the noisy raw data. Both the advantages and limitations of these methods for detecting the cortical activation are evaluated using the simulated data and experimental data from rat somatosensory cortex during the electrical stimulation at contralateral sciatic nerve[4,7,8].

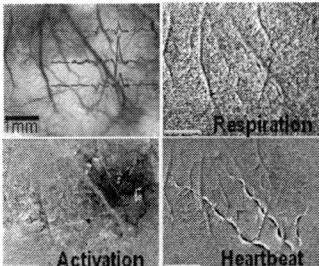

Fig.2 Extracting cortical activation pattern from intrinsic optical signal data by ICA

Optical imaging of cortical spreading depression

Cortical spreading depression (CSD) is believed to be involved in some important neurological diseases such as migraine, ischemia and brain injury. However the exact mechanisms underlying SD are still not fully understood. By using the technique of intrinsic optical signal imaging we investigated the spatiotemporal pattern and wavelength dependence of optical signals during CSD[9]. The spectroscopic analysis was applied to investigate the physiological source of optical intrinsic signal during CSD. Time-varying spatial pattern of CSD was observed, and the time interval between CSD waves was found to affect the spatial pattern of its propagation[10]. ATP sensitive potassium ion channel was found to be involved in the regulation of the constriction and dilation of the pial arteries during CSD[11].

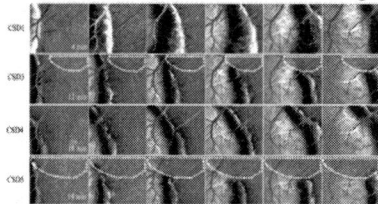

Fig.3 Time-varing spatial pattern of CSD[10]

Monitoring cerebral ischemic penumbra by optical imaging

In the focal cerebral ischemia rats, a series of spontaneous CSD waves were observed by optical intrinsic signal imaging. Optical reflectance signals during each CSD episode presented three types of regional variation significantly: flat, increase and decrease. The changes were corresponding to the inhomogeneous ischemic cortex: infarct area, penumbra and normal area[12]. With the reoccurrence of spontaneous CSD waves, the origins of CSD migrated in the ipsilateral hemisphere with a general trend towards the medial cortex[13]. The un-invaded area of CSD in the lateral was enlarging, the penumbra was attenuating but the infarct was expanding. Intrinsic optical signal imaging data suggested that the spatio-temporal characteristics of spontaneous CSD waves can be sued to reveal the dynamic evolution of focal cerebral ischemia.

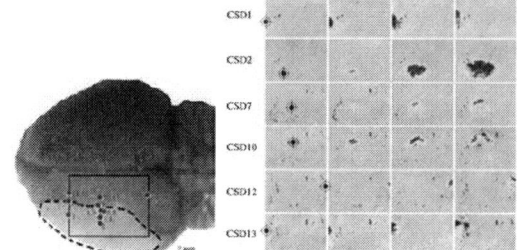

Fig. 4 Migration of origin sites of spontaneous CSD during focal cerebral ischemia [13]

References

1. Haiying Cheng et al Journal of Biomedical Optics, 8(2003), 559-564
2. Pengcheng Li et al Optics Letters, 31(2006), 1824-1826
3. Haiying Cheng et al Applied Optics, 43(2004), 5772-5777
4. Qian Liu et al Journal of Biomedical Optics, 10(2005), 024019
5. Haiying Cheng et al Applied Optics, 42(2003), 5759-5764
6. Songlin Ni et al Proceeding of SPIE, 6026 (2006), 602607
7. Pengcheng Li et al Proceeding of SPIE, 5254 (2003), 542-551
8. Weihua Luo et al Brain Research, 1131 (2007), 67-77
9. Pengcheng Li et al Progress in Biochemistry and Biophysics, 30 (2003), 605-611
10. Shangbin Chen et al Neuroscience Letters, 396 (2006), 132-136
11. Yuanyuan Yang et al Progress in Biochemistry and Biophysics, 33 (2006), 902-907
12. Shangbin Chen et al Neuroscience Letters, 403 (2006), 266-270
13. Shangbin Chen et al Journal of Biomedical Optics, 11 (2006), 034002

Photonics and Immobilisation of Biomolecules

M. Duroux[ab], E. Skovsen[ab], M. T. Neves-Petersen[ab], L. Duroux[ab], and S. B. Petersen[ab*]

[a]NanoBioTechnology Group, Dep. of Physics and Nanotechnology, University of Aalborg,
Skjernvej 4A, DK-9220 Aalborg Øst, Denmark;
[b]BioNanoPhotonics A/S, Niels Jernes vej 10, DK-[9220]Aalborg Øst, Denmark.
*Telephone: 00 45 96358469; Fax: 00 45 96341599; Email: sp@nanobio.aau.dk

Abstract

A new photonic technology is demonstrated that allows for precise immobilisation of proteins in any wanted pattern onto activated surfaces. Molecular immobilisation is limited to the μm sized focal point of the focused UV beam.

Introduction

The present paper describes a protein based system that is actively harvesting photons in the mid-UV range (260-300 nm), as well as the applications of such a reaction for immobilizing proteins potentially with micrometer precision. A new and rather interesting effect of UV-light interaction with biomolecules is that UV-radiation absorbed by aromatic residues induces disruption of nearby disulphide bridges and attaches the molecule covalently to a chemically activated, quartz or glass surface [1, 2, and 3].

This technology "light assisted immobilisation" (LAPI) can be used to immobilise molecules with micrometer resolution on a surface according to a bitmap provided to a computer. The expected resolution, taken into account the size of the focused laser beam, the precision of our stage and the scanner resolution of our laser scanner approximates to a few microns. In this way arbitrary patterns of immobilized biomolecules can be created. Posible applications of this new technology include creation of biosensors [4] and security labelling through immobilisation with unique diffraction- or interference patterns.

Materials and Methods

Setups used for light-induced protein immobilisation:
Illumination with focused laser beam of light
A schematic of the setup used for light-induced immobilisation can be seen in figure 1.

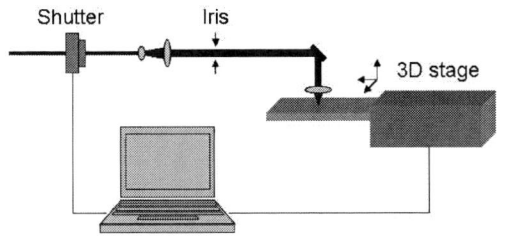

Figure 1. Immobilisation Setup:
A beam of UV pulses was sent through a computer

controlled electronic shutter, a beam expander, an iris diaphragm, and a focusing lens, which focused the beam onto the sample with a spot size of 20 μm in diameter in the focal plane. A thiolized optically flat glass slide was mounted on a computer controlled translation stage with the derivatised surface of the slide in the focal plane of the UV-light.

The Tsunami is a Mode-locked Ti:Sapphire femtosecond laser (Tsunami 3960, Spectra-Physics) pumped by a high-power diode laser (Millennia V, Spectra-Physics). The output wavelength was centred at 840 nm (FWHM = 12 nm), the average power of the ~80 fs laser pulses was 910 mW at a repetition rate of 80 MHz. The infrared femtosecond laser pulses from the Tsunami were sent through a pulse picker (Model 3980, Spectra-Physics) that reduced the repetition rate to 8 MHz. After the pulse picker, the pulses were passed through a frequency doubler/tripler (GWU, Spectra-Physics) to convert the pulses into 280 nm UV pulses (i.e. the third harmonic of 840 nm). The average power of the pulse train of ~200 fs long, 280nm pulses was varied between 0.1 mW and 1.0 mW (using a neutral density filter) at a repetition rate of 8 MHz. The beam of UV pulses was passed through a computer controlled shutter, a telescope to expand the beam to about 4mm diameter, and an iris diaphragm before it was turned 90 degrees by a mirror and focused onto the sample with a 25mm focal length quarts lens (the spot size in the focus has been estimated to ~20 μm). The light beam reached its focal point on the surface of an optically flat quartz slide (average flatness of 2.0 nm, purchased from ArrayIt Microarray Technology, Telechem International Inc), where upon sensor molecules were immobilized. The quartz slide was mounted on a three-axis parallel-flexure translation stage (Model 17MAX303, Melles Griot) equipped with three computer controlled high-resolution stepper-motor actuators (Model 17DRV001, Melles Griot). Together, this combination yields a resolution of 50 nm and a bidirectional repeatability of 2 μm. The shutter and translation stage were controlled by home-made software programmed in LabView 8.0. For immobilizing a given pattern, a black and white bitmap image of the pattern can be loaded and when given the distance between pixels in µm in each direction and an exposure time multiplier, the stage will move through the bitmap pixel-by-pixel and irradiate the sample for a number of ms given by the pixel value (i.e. 0 or 1 for a black and white bitmap) times the exposure time multiplier. The default value used for the exposure time multiplier was 100 ms.

Immobilisation of arbitrary patterns of proteins according to a bitmap

Immobilisation was carried out on a dry film, using 1 µl of 1 µM fluorescently labeled cutinase (with Alexa fluor 488). The stage moved through the bitmap pixel-by-pixel and irradiated the sample for a given number of ms given by the pixel value (i.e. 0 or 1 for a black and white bitmap) times the exposure multiplier of 100ms, using a 60 µW focused laser beam. After illumination, the slides were washed with, 1 x PBS plus 1% Tween 20 detergent and water overnight. Slides were scanned with a Tecan LS 200 scanner green laser (excitation 532nm, cy3 filter), 6µm resolution.

Protein labelling.

Labelling of cutinase with Alexa Fluor 488 (Molecular Probes, InVitrogen) was done according to the manufacturer's instructions. Un-reacted dye was removed from the labeled protein using a spin column filled with Sepharose G25 fine grade (Amersham Biosciences) equilibrated in 1x PBS buffer and spun for 2 minutes at 3000 x g.

Surface Preparation of optically flat slides.

Chemical modification of the quartz slides was done as described previously [5].

Protein array and protein pattern image processing and analysis

MATLAB v7.1 SP3 was used to develop a versatile software package, BNIP-Pro. [www.bionanophotonics.dk] that allows for advanced analysis of the protein array images and patterns. The BNIP-Pro software package allows for a complete analysis of the acquired fluorescence images of immobilised patterns of biomolecules.

Results and Discussion

Light induced immobilisation allows precision, using a focused beam of light to illuminate a thin film of biomolecules on a derivatised surface. The ultimate size of the immobilized spots are in principle only limited by the focal area of the light, which for a diffraction limited beam of UV-light, can be less than one micrometer in diameter. Light-induced immobilisation has the added benefit that the immobilised molecules will be spatially oriented on the surface. For biosensor applications it has been shown that the activity of the immobilised molecules are retained, and protein arrays fabricated using

light–induced immobilisation demonstrate successful protein and antigen binding. [4, 6] The patterns of immobilized molecules on the surface are not restricted to conventional array formats; any pattern that can be created by the UV illumination source can be immobilised onto the derivatised surface. This is illustrated in figure 2 where the 'BioNanoPhotonics' logo has been written with the enzyme cutinase onto a derivatised quartz surface. The image covers an area of 1 mm². What is observed is the fluorescence from the biomolecules immobilised with light.

Figure 2.
The Logo 'BioNanoPhotonics', written with the enzyme cutinase onto a quartz surface. The image covers an area of 1 mm². What is observed is the fluorescence from biomolecues immobilised with light on a glass surface.

Another application of LAPI allows for the creation of micron sized, complex patterns of biomolecules that for example could be used for security labeling. Using complex diffraction or interference patterns of light for immobilisation, unique patterns can be created. These patterns will be very hard to counterfeit having well defined features and aspect ratios determined by the diffraction pattern used for immobilisation. (Figure 3).

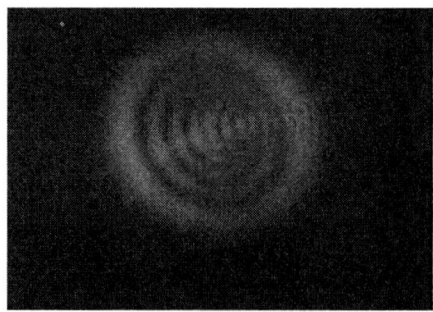

Figure 3 Molecules immobilized using a UV diffraction pattern.

The laser light source used to immobilise the sensor molecules (cutinase labelled with Alexa Fluor 488)

290

had a Gaussian profile along x and y. The sensor molecules were immobilised according to the light profile with so called "airy rings". The image of the spot measures 20μm across with the inner rings measuring 1.2μm.

Conclusions

The work presented in this paper shows how one can use light-induced immobilisation of bio-molecules onto thiol reactive surfaces to make patterns without the need for traditional micro-dispensing technologies. Using light-induced immobilisation, patterns of bio-molecules can be created with a high degree of control. The size of the smallest spatial features that can be created in the printed patterns are defined by UV pattern used to immobilize the molecules.

References

[1]. M. T. Neves-Petersen, Z. Gryczynski, J Lakowicz, P. Fojan, S. Pedersen, E. Petersen, and S. B. Petersen, "High probability of distrupting a disulphide bridge mediated by an endogenous excited

tryptophan residue," Protein Science, 11, 588-600 (2002).

[2]. M. T. Neves-Petersen, T. Snabe, S. Klitgaard, M. Duroux, and S. B. Petersen, "Photonic activation of disulphide bridges achieves oriented protein immobilization on biosensor surfaces," Protein Science, 15, 343-351 (2006).

[3]. M. T. Neves-Petersen, and S. B. Petersen, "Light induced immobilization," PCT application WO 2004/065928 (2004).
 (2005).

[4]. M. Duroux, L. Duroux, M.T. Neves-Petersen E. Skovsen and S.B. Petersen, Appl. Surf. Sci., doi:10.1016/j.apsusc.2007.02.131 (2007)

[5]. Snabe, T., Røder, G. A., Neves-Petersen, M. T., Buus, S., and Petersen, S. B. 2006. Oriented coupling of major histocompatibility complex (MHC) to sensor surfaces using light assisted immobilization technology. Biosensors and Bioelectronics 21:1553-1559.

[6]. Duroux, M., Skovsen, E., Neves-Petersen, M. T., Duroux, L., Gurevich, L., and Petersen, S. B. 2007. Light-induced immobilization of bio-molecules as an attractive alternative to micro-droplet dispensing based arraying technologies. Proteomics. In press

Methane concentration monitoring system based on a pair of FBGs

Bin Zhou (1), Zuguang Guan

1: Centre for Optical and Electromagnetic Research, Zhejiang University, East Buliding No.5, Zijingang campus, Zhejiang University, Hangzhou 310058, China. zhoubin@coer.zju.edu.cn

Abstract *In this paper, we propose a simple fiber-optic sensing system to measure the concentration of the methane based on a special catalyst and fiber Bragg gratings (FBGs). The temperature around the catalyst rises rapidly as the concentration of methane increases, and hence can be detected by an FBG-type sensor through monitoring the wavelength shift of the reflection spectrum. By employing a well-matched FBG as a reference, the wavelength shift of the sensing FBG can be transformed effectively into the variation in the measured optical intensity. An experiment is demonstrated to show the good performances of our system, such as the high resolution and good stability to the environmental temperature.*

Introduction

Methane is a greenhouse gas and it is also explosive when the concentration is over 5%. It is important to monitoring the methane-concentration under the coal mine or around the nature-gas pipe in real time.

Based on the Beer-Lambert law, the absorption sensing technique with an open path cell is widely applied to measure the concentration of a specific gas[1][2]. In order to improve the sensitivity, an open cell with a long length (tens of centimeters) is often required. In this kind of structure, the light path is difficult to adjust and its stability is influenced by the variation of the environmental temperature.

In this work, we propose a simple fiber-optic sensor to measure the concentration of the methane in the air. Since the reflection spectrum of a FBG shifts towards a longer wavelength as the temperature around increases, the concentration of the methane in the air can be detected by monitoring the spectrum of the FBG which is close to a special catalyst[3]. In order to reduce the system cost and the crosstalk induced by the variation of the environmental temperature, we employ a reference FBG to interrogate the sensing signal from the sensing FBG with an edge-detection technology. As the methane concentration in the environment increases, the temperature around the special catalyst

rises rapidly, and therefore the spectrum of a fiber Bragg grating (FBG) attached to the catalyst shifts towards a longer wavelength. By employing another FBG (far away from the catalyst) as a reference (filter), the wavelength shift of the sensing FBG can be transformed effectively into the variation in the measured optical intensity. The experimental results show that the resolution of our system is pretty high (a minimal value of 64ppm can be detected) and crosstalk induced by the environmental temperature is quite low (The variation of the output voltage is less than 0.52 mV as the temperature in gas cell changes from 30 ℃ to 60℃.). Compared with the absorption method, our system has higher stability, space-resolution and compact size.

Principles and analysis

FBG is a periodic perturbation of the refractive index in the fiber core. This wavelength position is proportional to the temperature around the FBG[4][5] and the relationship can be expressed as following,

$$\Delta\lambda = 2(n_{eff}\frac{\partial\Lambda}{\partial T} + \Lambda\frac{\partial n_{eff}}{\partial T})\Delta T \quad (1)$$

where n_{eff} is the effective refractive index of the fiber core, and Λ is the period of the FBG. Based on this principle, FBG can be used as a temperature sensor.

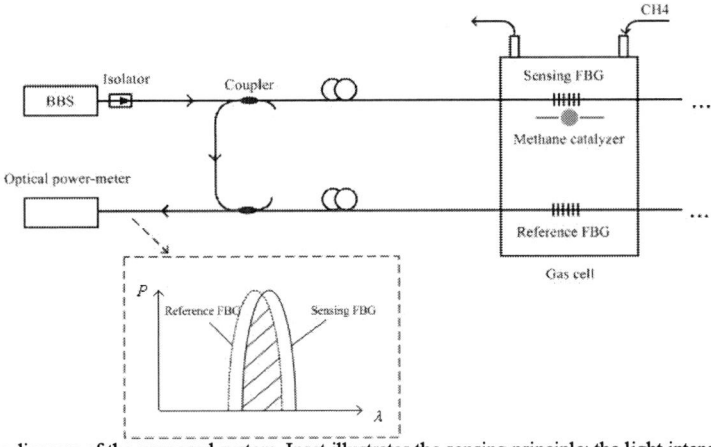

Fig.1: Schematic diagram of the proposed system. Inset illustrates the sensing principle: the light intensity received by the optical power-meter decreases as the mismatch of the reflection spectra between sensing and reference FBGs increases.

In this work, a simple fiber-optic methane concentration sensor is proposed. As shown in Fig. 1, a broadband light passes through an isolator (avoid the reflective light to damage the light source) and a fiber coupler, and then is reflected by an (sensing) FBG fixed on a special methane catalyst. The reflected light (with a narrow-band spectrum) will pass through another fiber coupler and meet a reference FBG (with a spectrum well-matched to that of the sensing FBG). Finally, we employ a photodiode to detect the intensity of the

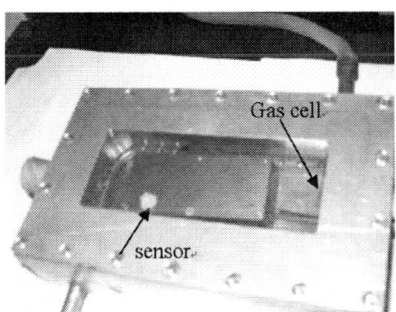

Fig.2: The picture of the well-packaged methane-sensor (inside the gas cell)

double-reflected light. As the concentration of the methane in the gas cell increases, the temperature around the catalyst rises and leads to a mis-match between the spectra of the sensing- and reference-FBG. (see the inset of Fig. 1). And consequently the light intensity received by the photodiode decreases. Since the reference-FBG is located near to the sensing-FBG (but

shielded from the catalyst), our system is sensitive to the methane concentration but insensitive to the variation of the environmental temperature.

Experiment results

Two identical FBGs (with reflection spectrum located around 1550.85 nm and bandwidth of 0.5 nm) are fabricated and packaged (with the methane catalyst) in a well-designed mechanical device (see Fig. 2). The concentration of the methane in the gas cell is adjustable by controlling the volume of the input CH4 with a high-precision flow-meter.

The relationship between the reflection wavelength of the sensing FBG and the methane concentration is measured by an OSA (ANDO AQ6317) and shown in Fig.3. The spectrum of the sensing-FBG shifts towards a longer wavelength as the

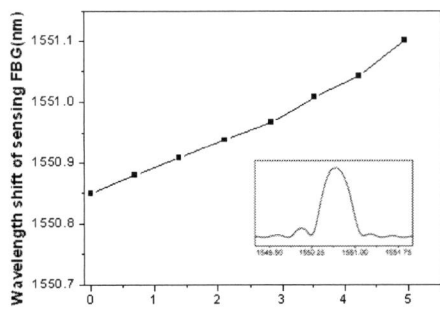

Fig.3: The reflection spectrum of the sensing FBG shifts towards the longer wavelength as the concentration of the methane in the gas cell increases. Inset shows the reflection spectrum of the sensing FBG

methane concentration in the gas cell increases and the sensitivity is about 0.051nm/1%. The resolution of measuring the methane concentration of this scheme is about 0.39%, which is limited by the wavelength-resolution (0.02nm) of the employed OSA.

We develop an edge-detection system as shown in Fig. 1. A broadband light passes through an isolator (avoid the reflective light to damage the light source) and a fiber coupler, and then is reflected by a sensing FBG close to a methane catalyst. The reflected light (with a narrow-band spectrum) will pass through another fiber coupler and meet a reference FBG (with a spectrum well-matched to that of the sensing FBG). Finally, we employ a photodiode to detect the intensity of the double-reflected light. As the concentration of the methane in the gas cell increases, the temperature around the catalyst rises and leads to a mismatch between the spectra of the sensing and reference FBG (see the inset of Fig. 1). And consequently the light intensity received by the photodiode decreases (see Fig.4). Since the reference FBG is located close to the sensing FBG (but shielded from the catalyst), our system is sensitive to the methane concentration but insensitive to the variation of the environmental temperature. The sensitivity of this system is 0.783V/1% and the resolution can be as high as 64ppm (which is 60 times higher than the resolution achieved by using an OSA). The stability of

our system to the environmental temperature variation is also tested in the experiment. The variation of the output voltage is less than 0.52 mV as the temperature in gas cell changes from 30℃ to 60℃.

Conclusions

A novel methane concentration measurement technology based on a pair of FBG is proposed. A methane catalyst is employed to transduce the concentration of methane into the temperature information, which can induce the shift of the spectrum of a sensing FBG. By employing another FBG as a reference, an edge-detection system is set up to improve the resolution (64ppm) of our sensing system. In the meanwhile, the crosstalk induced by the variation of the environmental temperature is also reduced by this technique. Compared with the sensing system employing complicated wavelength detecting instruments, our system is with low cost.

References

1 Okajima H., Kakuma S., Uchida K., Wakimoto Y., Noda K., "Measurement of methane gas concentration using an infrared LED", SICE-ICASE, 2006. International Joint Conference. p. 1652-1655, 2006.

2 K. Noda, M. Takahashi, R. Ohba and S. Kakuma, "Measurement of Methane Gas Concentration by Detecting Absorption at 1300 nm using a LaserDiode Wavelength Sweep Technique", Optical Engineering, Vol. 44, 2005.

3 S. J. Gentry, P. T. Walsh, "The influence of high methane concentrations on the stability of catalytic flammable-gas sensing elements", Sens. Actuators, A5, p.229-238, 1984.

4 W. W. Morey. Bragg-grating temperature and strain sensors[C] //Proc. OFS'89. Paris, France: [s.n.], 1989: 526.

5 T. Erdogan. Fiber grating spectra[J]. Journal of Lightwave Technology, 1997, 15(8): 1277-1294.

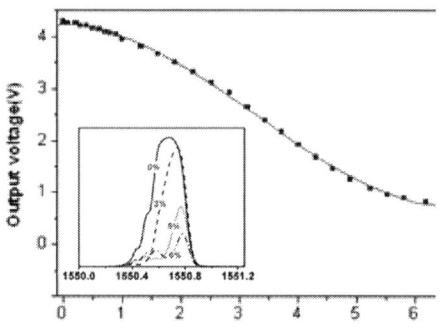

Fig.4: The output of the optical power meter decreases as the concentration of methane in the gas cell increases. Inset shows the evolution of the spectra (of the light received by the power meter) when concentration of methane is 0%, 3%, 5%, and 6%.

Emerging Fiber-optic Microendoscopy Technologies for High-resolution Biomedical Imaging

Xingde Li

Department of Bioengineering, University of Washington, Seattle, WA 98195, USA

Email: xingde@u.washington.edu

Abstract We present recent advances of micro-endoscopy technologies that are critical for clinical translation of high-resolution optical imaging modalities (such as OCT, SHG and two-photon fluorescence microscopy). Remaining challenges will also be discussed.

Introduction

High-resolution optical imaging technologies such as OCT, SHG and two-photon fluorescence (TPF) microscopy are new imaging modalities, capable of non-invasive assessment of biological tissue microanatomy *in vivo* and in real time. These imaging technologies can perform "optical biopsy" with a resolution approaching that of standard histopathology but without the need for tissue removal, paraffin embedding or staining. Except skin, imaging of other organs or systems, in particular internal organs including the gastrointestinal (GI) tract, respiratory tract, bile duct and cervix etc., requires a flexible miniature endoscope that can perform imaging beam delivery, scanning, and collection at the target site. Microendoscope has become a critical component for translating high-resolution optical imaging technologies to a clinically viable form. The challenges of developing microendoscopes include miniaturization of beam scanner, management of chromatic aberration over a broad spectrum (e.g. 400 -800 nm for SHG and TPF, and 650-950 nm for ultrahigh-resolution OCT), and small size of objective lenses with a tight focus and sufficient working distance.

OCT Microendoscope

Recently we have developed several innovative microendoscopes that overcome the above mentioned challenges and enable high-resolution imaging of internal organs with OCT, SHG or TPF. The first key component in a scanning endoscope is a miniature beam scanner. Figure 1A shows a schematic of a fiber-optic scanner which can perform 1D and 2D beam scanning [1]. High-speed beam scanning at 1-3 kHz is

achieved when a single-mode fiber-optic cantilever is driven at its mechanical resonance frequency using a tubular piezo actuator. The scanning fiber tip can be

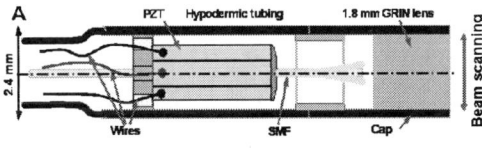

Figure 1. (A) Schematic of a fiber-optic scanner for lateral-priority beam scanning. (B) Photo of a super-achromatic micro lens.

imaged to a target by an imaging objective. A large scanning area (1-2 mm in diameter) can be easily achieved with a moderate piezo drive voltage (e.g. V_{p-p}~30-50V). The second key component of a microendoscope is a miniature, high-performance objective lens. Figure 1B shows a photo of a super-achromatic miniature compound objective lens we recently developed which has an outer diameter of 1.5 mm (or 2.0 mm when taking into account of the hypodermic housing). Using the two key components, a flexible, miniature and rapid scanning

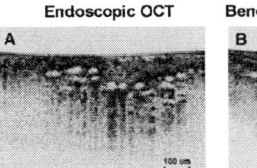

Figure 2. (A) High-speed endoscope OCT image of rat tail. (B) Ultrahigh-resolution OCT image of rat tail with a slow bench-top system.

978-0-9789217-3-6/07/$25.00

©2007 WEN GLOBAL SOLUTIONS

microendoscope has been engineered for high-speed, ultrahigh-resolution OCT imaging. Different from any other OCT microendoscopes, this rapid scanning endoscope permits a lateral-priority imaging sequence and thus real-time focus tracking. Figure 2A shows a representative real-time OCT image, of which the image quality is similar to the "standard" ultrahigh-resolution one acquired with a slow bench-top ultrahigh-resolution OCT system (Figure 2B).

Balloon Imaging Catheter
For some applications, a large working distance up to 8-9 mm is required. One example is assessment of the entire esophagus. Systematic surveillance of human esophagus for detection of surface and subsurface Barrett's epithelium/gland, and for detection of pre- and early neoplasia requires inflation of the esophagus for removing the natural esophageal folds. An endoscope/catheter would thus need to have a large working distance of 8-9 mm (i.e. the radius of an inflated human esophagus). Using the concept of compound micro objective lens, we have developed a double-lumen balloon OCT imaging catheter, which has a large working distance (9mm) while maintaining a small optics diameter (1.2 mm) to allow for endoscopic delivery of the balloon imaging catheter to the esophagus. The natural esophageal folds can be removed when the

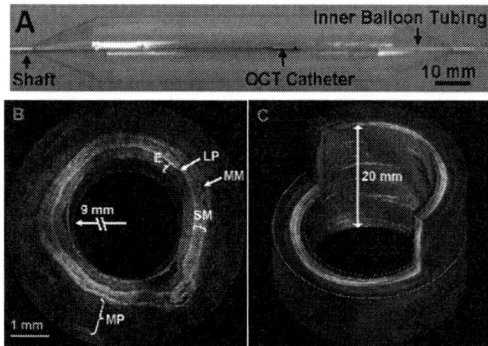

Figure 3. (A) Photo of a high-resolution OCT balloon imaging catheter. (B) 2D and (C) 3D OCT image of a pig esophagus, respectively, acquired with a balloon imaging catheter *in vivo* and in real-time. E: epithelium; LP: lamina propria; MM: muscularis mucosa; SM: submucosa; MP: muscularis propria.

balloon is inflated [2]. The imaging catheter resides within the inner lumen of the balloon and can be rotated for circumferential imaging. Figure 3A shows a photo of the high-resolution OCT balloon imaging catheter. Figures 3B and 3C show a representative real-time 2D and 3D images of a pig esophagus (*in vivo*), respectively. Normal layered structures can be clearly identified. High-speed 3D spiral OCT imaging was achieved by pulling back the imaging catheter while it was rotated. A swept-source OCT system was used for performing real-time balloon catheter imaging (12,000 A-scans/second).

SHG/TPF Microendoscope

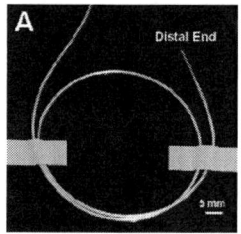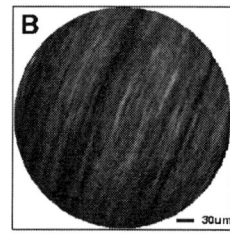

Figure 4. (A) Photo of a fiber-optic scanning microendoscope for SHG/TPF imaging. (B) Real-time SHG image of rat tail tendon acquired with the scanning fiber-optic microendoscope.

With further technology innovation, a rapid scanning microendoscope has been developed for SHG or TPF imaging [3, 4]. Instead of a single-mode fiber, a double-clad fiber is introduced for single-mode delivery of fs excitation laser pulses through the core of the double-clad fiber and multi-mode collection of the SHG or TPF signal through the inner clad (as well as the core). The use of a double-clad fiber improves the signal collection efficiency by a factor of more than 400. The material dispersion of the double-clad fiber can be conveniently managed by using a photonic crystal bandgap fiber. Figure 4A shows a photo of the flexible microendoscope with a scanner built-in at the distal end. A representative SHG image of rat tail tendon is shown in Figure 4B. Collagen bundles can be clearly identified.

Conclusions
In summary, innovative fiber-optics-based microendoscopes have been successfully developed. These microendoscopes

potentially enable clinical translation of the high-resolution optical imaging technologies for real-time imaging of biological tissues (in particular internal organs). It remains challenging to further reduce the size of microendoscopes, improve the beam scanning speed, and increase signal collection efficiency. Contributions to the reported work from group members and collaborators including Michael J. Cobb, Henry H. Fu, Dr. Kevin Hsu, Dr. Joo Ha Hwang, Dr. Yuxin Leng, Daniel J. MacDonald, Addie Warsen, Dr. Danling Wang and Jiefeng Xi and financial support from NIH, NSF, and Coulter Foundation are gratefully acknowledged.

References

1. X. M. Liu, M. J. Cobb, Y. Chen, M. B. Kimmey, and X. D. Li, "Rapid-scanning forward-imaging miniature endoscope for real-time optical coherence tomography," *Optics Letters* **29(15)**:1763-1765 (2004).

2. H. L. Fu, M. J. Cobb, Y. X. Leng, D. J. MacDonald, J. H. Hwang, and X. D. Li, "High-resolution OCT balloon catheter for systematic imaging of the esophagus," *Conference on Lasers and Electro-Optics (CLEO'07)*:Technical Digest (Paper CFL2) (2007).

3. M. T. Myaing, D. J. MacDonald, and X. D. Li, "Fiber-optic scanning two-photon fluorescence endoscope," *Optics Letters* **31(8)**:1076-1078 (2006).

4. X. L. Li, Y. X. Leng, D. J. MacDonald, D. Wang, M. J. Cobb, A. Warsen, and X. D. Li, "Flexible scanning micro-endoscope for two-photon fluorescence and SHG imaging," *Conference on Lasers and Electro-Optics (CLEO'07)*:Technical Digest (Paper CTu3) (2007).

Advances of Lighting Technologies

- From light bulbs to solid state light sources

Yung S. Liu

Center for Photonics Research &

Institute of Photonics Technologies

National Tsing Hua University, Hsinchu, 300, Taiwan

Tel: +886-3-516 2177, Fax: +886-3-575-1131, Email: liuys@mx.nthu.edu.tw

Abstract : Major breakthroughs that took place chronically in the light source development are: the invention of incandescence light bulbs, the invention of lasers, particularly semiconductor lasers, and the high-brightness blue/white LED. Two of these represent the major advances in the lighting technologies. It is further shown that what we called "nano-science and technology," today had played a critical role in commercialization of these great inventions.

Introduction:

In the history of light source development, the most significant technological breakthroughs that took place chronically are: the invention of incandescence light bulbs by Thomas Edison in the late 19th century, the invention of lasers, particularly semiconductor lasers in the mid-20th century, and the high-brightness blue and white GaN-based LED in the late 20th century. These breakthroughs make significant contribution to the advancement of modern sciences, as well as a profound impact upon how we live, work and communicate.

For example, the invention of incandescence light bulbs not only lit the world for the first time in the entire human history, but also created a totally new market for the electricity. Subsequently, it led to the development of modern electricity and the related infrastructure such as electricity generation, transmission .and conditioning, thus laid the foundation of a modern society that relies heavily on electricity today.

The invention of lasers, particularly semiconductor lasers was the key component used in the modern fiber optical communication (indirectly Internet), digital optoelectronics systems such as CD, DVD and laser printers. The invention of high-brightness white LED, still in the stage of development, has already made significant inroad to back light modules used in LCD displays, automobile illumination, and architecture lighting with a flurry of new applications coming out everyday.

In this talk, I would further point out that what we called "nano-science and technology" today have played a critical role in the subsequent commercialization of these major inventions.

Incandescence light

When Edison first invented the incandescence light bulbs, the filament was made of a thin long carbon which lasted only a few hours with an efficacy less than a few lumens per watt. Later, Irving Langmuir, a young chemist hired by Edison to work on the improvement of the life and efficiency of incandescence light bulbs, discovered that when an inert gas was introduced to the light bulb, the bulb life and efficiency were significantly improved.

In the course of his investigation, Langmuir realized the important role of the adsorbed layers played on the surface chemistry. He performed many studies of the modern surface chemistry and invented the apparatus for making mono-layer adsorbed films that later bears his name, Langmuir-Blodgett films. He went on to pioneer the study of the science of monolayer adsorbed films and invented the first gas-filled incandescence light bulb that is still used today.

Langmuir's pioneering work on surface chemistry of mono-layers becomes the foundation of modern surface sciences, today called, nano-sciences. He was awarded a Nobel Prize in chemistry in 1932 for his discoveries and investigation in surface chemistry. In fact, he was the first industrial scientist to receive a Nobel Prize.

In about the same time, the effort in seeking a new filament material for improving the efficiency and life of the incandescence light bulbs by another scientist, William Coolidge, led to the development of coiled tungsten filaments. In 1909, almost thirty years after the invention of light bulbs by Edison, the tungsten light bulb was marketed by GE as a commercial product with an efficacy reaching 10 lumen/Watt, about an order of magnitude improvement over that of the original Edison's light bulb. The incandescence light has since become a major light source used in the residential lighting.However, the efficacy of the incandescence lamp of about 16 lm/Watt has made little improvement since then in spite of much research effort devoted to it.

Fluorescence light

The fluorescence bulb first commercialized in 1938 by GE (U.S. Patent No. 2,259,040) represented another major advance in the light source technology. A fluorescence lamp is a discharged light source when turned on, the discharge current- induced electrons collide and excite the mercury atoms inside the tube. The UV emission from Hg atoms converts to visible light through fluorescence conversion of phosphors. The fluorescence light has an efficacy ranging from 60-100 lm/watt and is the work horse in commercial indoor lighting.

Semiconductor lasers

The invention of coherent light sources via stimulated emission processes is a truly remarkable scientific achievement.

Among all the coherent light sources developed to date, the semiconductor laser represents perhaps one of the most important coherent light sources ever developed.

When the first GaAs semiconductor laser was successfully operated at GE Research Labs by R.N. Hall and his team in the mid-twenty century, it was a rudimentary device operated at liquid nitrogen temperature with a relatively high threshold current density. This device was not considered a commercially viable product until the invention of hetero-junction structure which allows the carrier confinement to greatly reduce the threshold current.

In the decades that followed, the development of molecular beam epitaxial (MBE) process and molecular engineering which let precise material grow atomic layers by layers in the nano-meter scale. These critical processing techniques contributed significantly to the successful deployment of semiconductor lasers as commercial products.

GaAs-based semiconductor lasers are used in fiber communication, digital optoelectronics such as CD, DVD and laser printers. It has made profound changes to our life and to the way we communicate. Today, this type of semiconductor light emitting devices are able to produce hundred of watts output in infrared and are replacing discharged flash-lamps for pumping solid state lasers, and are used as pumping sources for fiber amplifiers; a critical component used in the optical communication.

GaN –based blue and white LED

For almost thirty years since GaAs LEDs were invented in the late 60's, the devices were used mainly as indicator lights in small display panels. In addition, lack of a high power LED in the blue and green colors has restricted LED from full color applications.

This was all changed in 1993, when the first high brightness blue GaN LED was commercialized by Nichia, and two years later, the green GaN LED. In the years that followed, Nichia further developed high brightness white LEDs by mixing of blue LED with yellow phosphors. Since then, the field was exploded and the use of LED for general lighting was no longer considered as an impossible task.

The application of LEDs has been rapidly expanded by the development of higher power LED devices employing various nano-technologies such as quantum-dots, quantum-rod, and photonic crystals. These sophisticated devices have been employed in the high power LED sources for LCD backlights, automobile illumination, and the RGB light sources used in the projection display. In the future, nano-science and technology will continue to play a critical role in realizing the ultimate goal of making a 150 Lumens/Watt device or beyond.

As Edison proclaimed when he invented the incandescence light bulb, "..Where this things going to stop, Lord only knows…"

Organic Light-Emitting Devices for Solid State Lighting

Jie Liu and Anil R Duggal

General Electric Global Research, 1 Research Circle, Niskayuna, NY 12309, U.S.A.
1: LIUJI@crd.ge.com; 2: DUGGAL@crd.ge.com

Abstract We present recent progress in developing efficient organic light-emitting devices and novel low-cost device fabrication for solid-state lighting applications.

Introduction

In the 1970's, Partridge built organic electroluminescent devices by depositing polyvinyl carbazole (PVK) based organic layers using a simple spin-coating process in a regular ambient atmosphere. [1] In the last two decades, tremendous progress has made in the development of organic electroluminescent materials and novel device designs. These fundamental developments in both the physics of charge transport and the chemistry of luminescent materials have led to the successful demonstration of highly efficient organic light emitting devices (OLEDs) and the commercialization of products that incorporate OLED technology. Most of this progress has been fueled both by a basic knowledge of electronic processes in organic materials and an applied interest in developing flat panel displays. If this knowledge can be directed to the continued improvement of OLED technology, the field has the potential to impact general lighting applications. In particular, a large area white-light emitting OLED could potentially provide a solid state diffuse light source that could compete with conventional lighting technologies in performance and cost. If OLED technology can be successfully transitioned to lighting applications the possible impact is enormous, and could result in substantial reductions in global energy use and carbon emissions.

In this presentation, we will first review the requirements necessary to create a lighting source using OLED technology, and discuss some of our previous efforts in building demonstration white light sources. In particular, we will focus on four performance attributes (illumination quality light, efficacy, device stability/lifetime and manufacturing cost) that are critical for the lighting applications. Moving beyond these initial prototypes to meet the stringent cost and performance requirements of the lighting market will require new approaches in not only OLED device architectures but also in materials and manufacturing techniques. We will illustrate these points using examples from some of our recent work in the development of electro-active polymer materials, advanced device architectures and the low-cost, high through-put manufacture of OLED devices.

Reference

1. R.H. Partridge, "Radiation Sources", US Patent 3 995 299, Nov. 30, 1976

978-0-9789217-3-6/07/$25.00
©2007 WEN GLOBAL SOLUTIONS

In-situ Fabrication of Highly-Fluorescent Nanohybrids Based on Carbon Nanotubes and Gold Nanoparticles

Renjia Zhou, Mang Wang, Hong-Zheng Chen*

Department of Polymer Science and Engineering, State Key Lab of Silicon Materials, Zhejiang University, Hangzhou 310027, P.R. China

Key Laboratory of Macromolecule Synthesis and Functionalization (Zhejiang University), Ministry of Education, Hangzhou 310027, P. R. China
(Tel: 0086-571-87952557; Fax: 0086-571-87953733; Email: hzchen@zju.edu.cn)

ABSTRACT Highly-fluorescent, water-soluble and ambient-stable nanohybrids based on multi-walled carbon nanotubes and gold nanoparticles (Au@MWNTs) with high-density and well-distributed Au nanoparticles were obtained assisted by organic optoelectronic active molecules as an interlinker and stabilizer via a facile in-situ fabrication method in aqueous solution, and were confirmed by TEM, SEM, EDX, UV-Vis measurements. The Au nanoparticle diameters of the hybrids can be controlled and decreased to a limited small range with high stability via this in-situ fabrication approach. The optical properties of the hybrids were studied, and it was found that the hybrid exhibited strong visible luminescence under the UV lamp irradiation, which might extend its potential application on light emitting sources, biological labeling etc.

978-0-9789217-3-6/07/$25.00
©2007 WEN GLOBAL SOLUTIONS

Planar Lightwave Circuits for FTTH and Photonic Networks

Katsunari Okamoto

Department of Electrical and Computer Engineering
University of California, Davis, California 95618, U.S.A.
E-mail : katsu@okamoto-lab.com

Abstract – Integrated-optic waveguide devices become more and more complicated to realize high functionality. Optical functional devices are important to solve electrical bottleneck issues. Ultra-compact and CMOS compatible silicon waveguides are important for the integration of an optical component and an electronic circuit aiming at higher level of functionalities.

This paper reviews the recent progress and future prospects of PLC devices and their applications to FTTH and photonic networks.

Optical splitters for FTTH

Single-mode 1xN splitters are essential optical components for PDS-PON (passive double star type passive optical network) systems. For greater levels of splitting such as 1x64 and 1x128, the excess loss and PDL of Y-branch splitters increase as more and more Y branches are cascaded. Funnel-type splitter configuration (Fig. 1) becomes advantageous for large port splitters [1]-[3]. In 1xN funnel splitter, an array of N output waveguides are positioned at the end of the slab region. The output waveguides are positioned so that they all point at the input waveguide. The width of the outer funnel waveguides is designed to be wider than that of the inner one to obtain a uniform coupling efficiency in all output waveguides. Experimental transmission spectra of 1x64 funnel splitter is shown in Fig. 2, where f = 6.1 mm, and W = 6.5~26 μm. Some of the radiated light leaks out from the outermost funnel waveguides. This causes excess loss in the funnel splitter (about 0.8 dB theoretical loss). Although the radiation pattern varies in accordance with the wavelength, an almost flat splitting spectrum is obtainable. The maximum insertion loss (IL) is 19.6 dB, IL uniformity is 1.22 dB, and the maximum polarization-dependent loss (PDL) is 0.17 dB, respectively.

Functional PLC devices for Photonic Networks

The aggregate transmission capacity of the current optical communications has increased dramatically owing to the WDM systems. Point-to-point throughput can exceed one Terabit/sec by employing, for example, 10 Gbit/sx128 channels or 40 Gbit/sx32 channels. However, since signal add/drop and cross-connects are carried out in electrical layer, signal processing at the node may become a bottleneck in the ultra-high bit rate systems. Normally, 20~30 % of the total signal is dropped at the node and rest of it is simply passed through. Therefore, 70~80 % of the signal is unnecessarily dropped into electrical layer.

If the through signal is processed in the optical layer, load to the electrical processing can be substantially alleviated. Then the node throughput can be matched with that of the transmission line. Reconfigurable optical add/drop multiplexer (ROADM) [4] for add/drop multiplexing and space division matrix switches [5] for cross-connect are quite important devices to realize the optical layer signal processing.

The lattice-form interleave filter, which is specially designed to separate adjacent WDM signals, is an attractive device because we can use it to double the channel count of existing WDM systems inexpensively and also relax the specifications of other optical devices. Typical lattice-form filter consists of four directional couplers and three delay lines [6]. The coupling ratios of the couplers are designed to be about 50%, 78%, 20%, and 3% to obtain a wide and flat-top passband. The path length differences of the three delay lines are L, 2L, and 2L, where L is about 2050 μm for 50-GHz channel spacing using waveguides with a refractive index difference Δ of 0.75%. Crosstalk of the fabricated lattice-form interleave filter strongly depends on the accuracy of coupling ratios of the directional couplers. It is known by the theoretical investigation that about 3% errors in the coupling ratios degrade the crosstalk down to -20 dB level. An effective way to eliminate the problem is to cascade lattice-form filters whose center wavelengths differ by half their free spectral range (FSR). Cascading filters improves the extinction ratio by doubling the extinction characteristics and also improves the chromatic dispersion characteristics by cancelling them each other.

A novel directional coupler configuration (called a stabilized coupler) was proposed to make possible the fabrication-tolerant coupler [6]. Fig. 3 shows a schematic of the stabilized coupler. The circuit has four individual directional couplers and three wavelength order delay parts. The delay values of the two ends are set at $\lambda g/4$ and $-\lambda g/4$, and the center value is set at between zero and $\lambda g/2$ depending on the desired coupling ratio. These values are so small that the two waveguides in these

978-0-9789217-3-6/07/$25.00
©2007 WEN GLOBAL SOLUTIONS

delay parts can be located close together. Consequently, these wavelength order delay parts are very stable as regards fabrication conditions. Then the total power coupling ratio of the circuit (η) with the amplitude coupling ratio (κ) of the individual coupler is obtained as

$$\eta = (8\kappa^2 - 24\kappa^4 + 32\kappa^6 - 16\kappa^8)\cos^2(\frac{\pi n_c}{\lambda}\Delta L)$$

When we set $\Delta L = \lambda g/6$, $\lambda g/4$, and $\lambda g/3$, coupling ratio η becomes to 0.75, 0.50, and 0.25, respectively. Coupling ratio η remains to be 0.49 – 0.50 for $\Delta L = \lambda g/4$ even when κ^2 varies from 0.3 to 0.7 (Fig. 4). The coupler also shows stable performance with respect to the spectral and polarization characteristics.

Extremely-High Index Contrast Waveguides

Silicon wire and ridge are very important EH-Δ waveguides. The major advantage of "Silicon Photonics" is the integration of an optical component and an electronic circuit, which can improve overall system performances. Bending radius of Si-wire waveguide could be 2~5 μm. Therefore a high level of photonic integration will be achieved with the goal of maximizing the level of optical functionality and optical performances.

On-chip optical buffers based on waveguide delay lines have significant implications for the development of optical interconnects in computer systems. Optical delay lines based on SOI sub-micrometer photonic wire waveguides that consist of up to 100 micro-ring resonators cascaded in either coupled-resonator or all-pass filter configurations [7] have been demonstrated [8]. Group delays exceeding 500 psec are obtained in a device with a footprint below 0.09 mm^2.

Novel Applications of PLCs

The use of millimeter-wave (MMW) signal at a frequency of >100 GHz is intensively investigated in the fields of wireless communications, radio astronomy, and MMW imaging. It is difficult for all-electronic technologies to generate MMW signal at >100 GHz because electronic device characteristics deteriorate as the operation frequency increases. Therefore, MMW signal generators using photonic technologies are attracting a great deal of interest because they can generate high-frequency signal more easily than one using all-electronic technologies. A frequency-tunable photonic MMW generator [9] consists of a single-mode laser, Mach-Zehnder optical modulator, a PLC on which an AWG and 3-dB optical combiner is integrated, and a high-speed photodiode. The two output channels of the AWG have a spacing of 120 GHz. Optical signal with two modes is converted into a MMW signal by a uni-traveling carrier (UTC) PD module [10]. The output power of the UTC-PD module is -10 dBm. A very sharp spectrum is obtained with the linewidth (FWHM) of less than 300 KHz (limited by resolution). Frequency of the MMW carrier signal can be varied from 90~125 GHz by varying optical modulator frequency. Demodulation experiment of 120-GHz MMW signal modulated at 10-Gbit/s was carried out by using the heterodyne method. Bit error rate of below 10^{-10} was obtained with a received power of -22 dBm. The photonic MMW generator that employs PLC can be applied for heterodyne detection systems, such as radio astronomy or MMW imaging, and the generated MMW signals show sufficient purity for the use of the 10-Gbit/s wireless links.

AWG that was originally developed for communication purpose can function as a compact spectroscopic sensor in which a broadband light source, sample injection cell, microfluidic pump, and detector array are integrated on a single substrate. Many of the sample materials in the environmental research and biomedical fields react to visible and near-infrared light. Aiming at the visible wavelength spectroscopic sensor, a "rainbow AWG" that operates from 400 nm to 700 nm wavelength region has been fabricated [11]. Core thickness and width are 2.9 μm and 3.0 μm with the index difference of 0.75%. Number of channels and the channel spacing are 8 and 37.5 nm, respectively. Path length difference is $\Delta L = 0.35$ μm, which makes the diffraction order of AWG to be m = 1. Chip size is 26 mm x 4 mm. AWG Extinction ratio larger than 20 dB is obtained in 600~700 nm wavelength region. Degradation of extinction ratio in 400~500 nm wavelength region is attributed to the weak output power of the incident light and the low sensitivity of the spectrum analyzer.

Optimization of Waveguide Structures

Waveguide optimization technique combined with accurate mode solvers will be important in designing novel compact structures with excellent transmission properties [12]-[14]. Wavefront matching (WFM) method is a novel optical circuit design method that provides us with a refractive index distribution that transfers a given input optical field to a desired output optical field. Refractive-index distribution $n(x, z)$ is piecewise modified by reducing the mismatch of wavefromts between the forward propagating wave and (hypothetically) backward propagating wave. The whole refractive index distribution is optimized by scanning the optimization point z from the input face to the output end. The coupling efficiency can be further optimized by repeated scanning. Waveguide crossing with small angle of less than 20 degree and a Y-branch splitter that offers both compactness and low loss have been realized by the WFM method.

Conclusions

Recent progress in PLCs has been reviewed.

Hybrid integration technologies with LiNbO3 [15], InP, MEMS, and polymer will further enable us to realize much more functional and high-speed devices.

References

[1] S.Day et al., *Electron. Lett.*, vol.28, p. 920, 1992.
[2] M.Zirngibl et al., *Electron. Lett.*, vol.28, p.1212, 1992.
[3] H.Takahashi et al., *PTL*, vol.5, p.58, 1993.
[4] K.Okamoto et al., *Electron. Lett.*, vol.31, p. 723, 1995.
[5] S.Sohma et al., *ECOC '06*, Tu4.4.3, Sept.24-28, 2006.
[6] M.Oguma et al., *JLT*, vol. 22, p. 895, 2004.
[7] J.B.Khurgin, *Opt. Lett.*, vol. 30, p. 513, 2005.
[8] F.Xia et. al., *Nature Photonics*, vol. 1, p. 65, 2007.
[9] A.Hirata et al.,*IEICE Electron.*,E88-C,pp.1458, 2005.
[10] H.Ito et al., *Electron. Lett.*, vol.38, pp. 1376, 2002.
[11] Y.Komai et al., *JJAP*, vol.45, pp.6742, 2006.
[12] T.Felici et al, *LEOS Annual Meeting* ThV5, 2002.
[13] T.Hashimoto et al., *Opt. Lett.*, vol. 30, pp.2620, 2005.
[14] Y.Sakamaki et al., *Electron. Lett.*, vol. pp. 217, 2007.
[15] Y.Miyamoto et al., *OFC '05*, OTuL2, 2005.

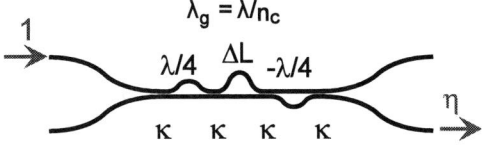

Fig. 3 Configuration of the stabilized coupler.

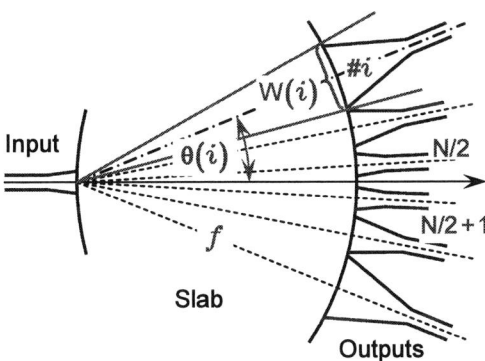

Fig. 1 Schematic configuration of 1xN funnel splitter.

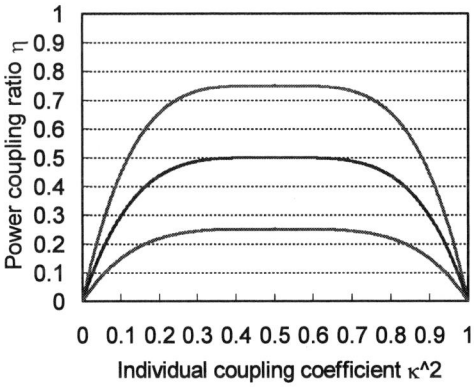

Fig. 4 Coupling ratios of the stabilized coupler.

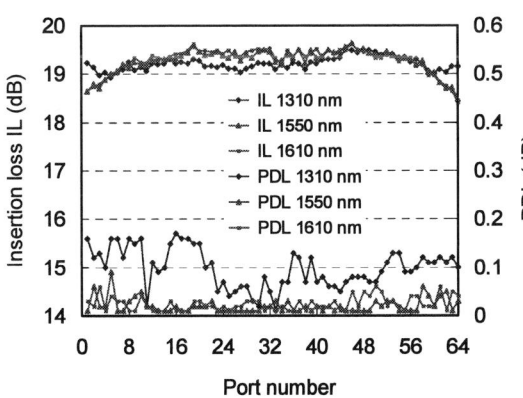

Fig. 2 Splitting loss spectra of 1x64 funnel splitter (Experiment).

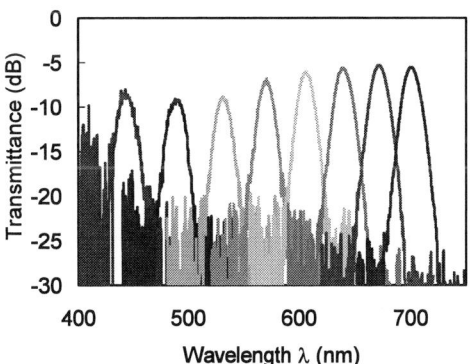

Fig. 5 Demultiplexing properties of the rainbow AWG.

305

Convergence of RoF and access systems employing dual-parallel modulator in the central station

Yikai Su (1), Qingjiang Chang (2)

1 : Shanghai Jiao Tong University, 800 DongChuan Rd, State Key Lab of Advanced Optical Communication Systems and Networks, Department of Electronic Engineering, Shanghai Jiao Tong University, Shanghai, 200240, China, yikaisu@sjtu.edu.cn
2 : Shanghai Jiao Tong University, changqj@sjtu.edu.cn

Abstract *We propose and demonstrate simultaneous delivery of wireless and baseband signals in a passive optical network architecture. The downstream baseband data and RF signals are generated based on a single-integrated dual-parallel modulator.*

Introduction

With the increasing bandwidth demand and possible frequency congestion in low frequency band, the emerging radio over fiber (RoF) transmission technique becomes a promising method in providing broadband wireless access services over wide areas. Meanwhile, future wideband services in access networks also require the fiber infrastructures to deliver wireline signals. Therefore, it is desirable to simultaneously transmit wireline and RoF signals based on the same network architecture in an integrated platform.

Among RoF systems in access areas, multi-band optical modulation technique is an attractive method, which simultaneously delivers baseband, micro-wave (MW) and millimeter-wave (MMW) signals. It exhibits flexible application potential in future multi-service access networks since wireline and multiple wireless services are seamlessly converged in a common network.

In addition, the convergence of video, voice and data into triple play service (TPS) in a single network is an effective solution for network service providers. Passive optical network (PON) technology is believed to be a promising scheme to provide TPS in an integrated platform with a cost-effective configuration.

In this paper, we perform two demonstrations based on converged RoF and access systems using a single-integrated dual-parallel Mach-Zehnder modulator (DPMZM) [1] in the central station (CS). The first one is a multi-band transmission system to deliver baseband data, 20-GHz, and 40-GHz RoF signals.

The second system is a PON which generates and sends video, voice and data from the central state. Both systems employ a single-integrated DPMZM, and upstream transmission is realized by carrier reuse.

Experimental setups and results

In previous demonstrations [2, 3], the need for high-frequency MMW signal generator and high-bandwidth modulator at the CS could result in complex transmitter architecture. Also, prior to our work [4], upstream data transmission was not demonstrated, while in practical access networks bidirectional data transmission is needed.

Here we show a full duplex RoF system to generate and transmit downstream wireline baseband data, wireless MW and MMW signals on a single wavelength. Upstream symmetric data transmission is obtained through re-modulating downstream differential phase-shift keying signal (DPSK) signal [5, 6], therefore, the base station (BS) configuration cost is greatly reduced since no additional light source and wavelength management are required at the BS. The scheme is scalable in frequency band if higher speed devices are available, thus optical MMW signal above 100 GHz can be envisioned using currently available electronic components.

The schematic diagram of the proposed RoF system is shown in Fig.1. The transmitter consists of a DPMZM [1] followed by a single-drive Mach-Zehnder modulator (SDMZM). The DPMZM consists of a pair of x-cut LiNbO3 MZMs (MZMA, MZMB) embedded in the two arms of a main MZM structure. The two sub-MZMs

978-0-9789217-3-6/07/$25.00

©2007 WEN GLOBAL SOLUTIONS

have the same structure and performance, and the main MZM combines the outputs of the two sub-MZMs. At the CS, a continuous wave (CW) laser is launched into the DPMZM. The MZMA is biased at its null point and driven by a 10-GHz RF signal loaded with data-1 of 1.25 Gb/s to generate a carrier suppressed optical sub-carrier multiplexed (SCM) signal , whose repetition rate is twice as the RF signal frequency. The MZMB is also biased at its null point and driven by another data-2 to produce a 1.25-Gb/s DPSK signal. The two optical signals are then added constructively by adjusting the bias of the main MZM, and they do not interfere with each other since the carrier of the optical SCM signal is suppressed. A following SDMZM biased at null point is driven by the same RF signal to shift the frequencies of the DPMZM outputs thus resulting in multi-band signals. The output of the MZMA is modulated with carrier suppression technique by the SDMZM to generate the optical baseband and frequency quadrupled MMW signals with data-1, therefore the wireline and wireless users can share the identical data service at the baseband and MMW frequency band, respectively, while the output of the MZMB goes through the same frequency shifting process to achieve frequency-doubled optical MW DPSK data.

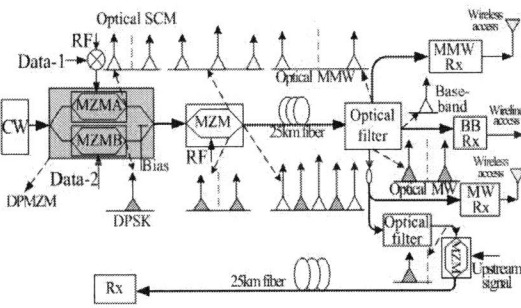

Fig.1. Schematic diagram of the proposed bidirectional RoF system

After the transmission, at the BS, an optical filter consisting of two fiber Bragg gratings (FBGs) with optical circulators are used to separate each band of the optical multi-band signals, which are detected respectively. The optical MW signal is split into two parts, one is detected by a MW receiver, and the other is filtered to obtain

its lower sideband, which is re-modulated by the upstream on-off-keying (OOK) signal. The upstream re-modulation signal is sent back to the CS and then detected by a low-speed receiver. Using this scheme, one can simultaneously deliver downlink multi-band signals and uplink data with a single wavelength in a bidirectional RoF system.

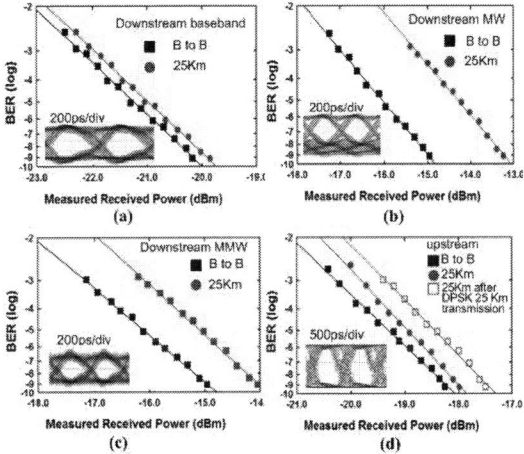

Fig.2. BER curves and electrical eye diagram. (a) Downstream baseband signal. (b) Downstream MW signal. (c) Downstream MMW signal. (d) Upstream re-modulation signal.

Fig.2 shows the measured bit error rate (BER) results. After transmission through a 25-km single mode fiber (SMF), for the downstream baseband data, the power penalty is about 0.2 dB as the chromatic dispersion effect is negligible at this rate; while for the MW and MMW signals, the power penalties are about 1.7 dB and 1.2 dB, respectively, which can be attributed to the chromatic dispersion of the transmission fiber in RF frequency band. The power penalty is less than 1 dB for the re-modulated upstream OOK signal. The electrical eye diagrams of the downstream multi-band signals and the upstream signal are provided as the insets in Fig.2, respectively.

Our second demonstration is a novel PON system to deliver video, voice and data using a single DPMZM [7]. In previous reports [8,9], only video, downstream data and upstream data were transmitted, simultaneous delivery of TPS including video, voice and data in the downstream were not demonstrated. To the best of our

307

knowledge, our scheme realizes the first centralized modulation of TPS signals with a single wavelength. Upstream data re-modulation based on downstream DPSK format is also achieved.

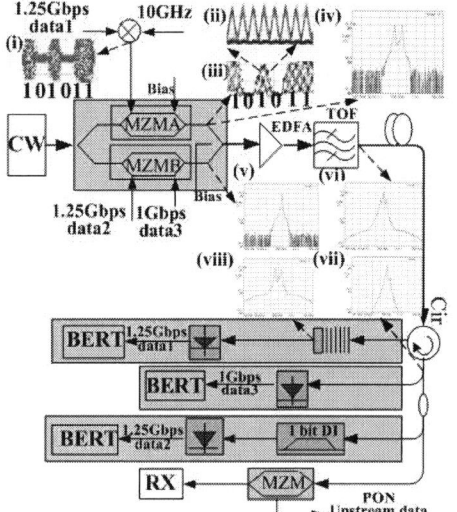

Fig.3. Schematic diagram of the proposed PON system.

The experimental setup of the proposed PON system is depicted in Fig.3. At the optical line terminal (OLT), the output of a CW laser is launched into the DPMZM. The MZMA is biased at its null point and driven by a 10-GHz RF signal loaded with a 1.25-Gb/s video signal to generate a carrier suppressed optical SCM signal. The MZMB is also biased at its null point and driven by 1.25-Gb/s data to obtain a DPSK signal. The 1-Gb/s voice signal is superimposed onto the DPSK signal to form ASK/DPSK format by modulating the bias point of the MZMB between the null and a small fraction of switching voltage [10]. The output of the MZMA and the MZMB are then added constructively by adjusting the bias of the main MZM. In this manner, the triple play signals are carried in the SCM, the ASK/DPSK formats, respectively.

After the transmission, an FBG with an optical circulator in an optical network unit (ONU) is used to reflect the ASK/DPSK signals while pass through the optical SCM signal. The passing optical SCM signal is directly detected by a 2.5-GHz receiver to retrieve the video signal. The reflected ASK signal is detected by an optical receiver to recover the voice signal, and the reflected DPSK signal is split into two parts, one is detected by a DPSK receiver to retrieve the data, and the other part is re-modulated by the upstream signal to the OLT.

Fig.4 shows the BER curves. For the downstream SCM signal and the ASK/DPSK signals, after transmission of 25 km, the power penalty is smaller than 0.8 dB. The electrical eye diagrams are shown in insets of Fig.4 (a), Fig.4(b) and inset (i) of Fig.4 (c), respectively, and the optical eye diagram for downstream DPSK after the interferometer is provided in Fig.4(c) as inset (ii). For the re-modulated upstream ASK signal, the power penalty for the DPSK signal is about 1.3 dB after 25-Km transmission at OLT, and the electrical eye diagram is shown in inset of Fig.4(d).

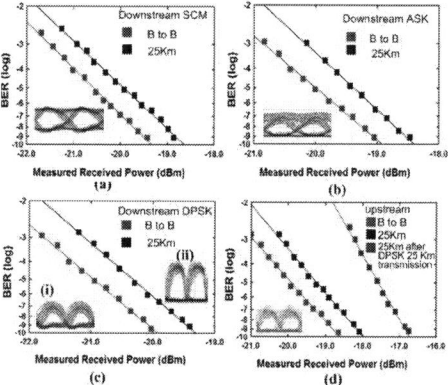

Fig.4. BER curves and eye diagrams for the downstream and upstream signals

Conclusions

We have proposed and demonstrated two RoF systems in a PON architecture. The first one generates baseband, MW, and MMW signals, and the second system delivers video, voice and data, all based on a DPMZM. The work is funded by the NSFC (60407008), and the 863 High-Tech program (2006AA01Z255).

References

1 K. Higuma et al *EL.*, vol. 37 (2001), 515
2 M. Bakaul et al *PTL*, vol. 18 (2006), 2311
3 K. Ikeda, *et al JLT.*, vol. 21(2003), 3194.
4 Q. Chang et al *ECOC* 2007 paper p100
5 Z. Jia, *et al PTL.*, vol. 19 (2007) 653.
6 W. Huang et al *PTL .*, vol. 15(2005) 1476
7 Q. Chang et al *ECOC* 2007 paper 4.4.7
8 J. Yu et al., *OFC* 2007, paper OWS4.
9 M. Khanal et al, PTL, Vol. 17(2005), 1992.
10 Yue Tian et al., ECOC 2006,Tu4.5.6.

160 Gbit/s/port Colored Optical Packet Switching System

Naoya Wada

National Institute of Information and Communications Technology (NICT),
4-2-1, Nukuikita-machi, Koganei, Tokyo 184-8795, Japan. E-mail:wada@nict.go.jp

Abstract A prototype 160Gbit/s/port colored optical packet switch with all-optical multiple label-processor and high-speed optical-switch is experimentally demonstrated. As related technologies, arrayed burst-mode Tx./Rx., 10GbE/optical-packet converter, and a novel transient response suppressed EDFA are also developed.

Introduction

In next-generation optical networks, high scalability and fine granularity will be essential, in addition to increased network capacity [1]. Although an approach using Internet protocol (IP) over generalized multiprotocol label switching (GMPLS) with WDM (IP/GMPLS/WDM) provides fine granularity [2], its slower electronic processing, such as memory access for header analysis at IP routers, will be a bottleneck in the network. To avoid such bottlenecks in commercial highend IP routers, electronic parallel processing technologies are often used. However, such large-scale parallel processing leads to serious power consumption problems.

Recently, despite the relative immaturity of optical technologies, many optical packet switching (OPS) systems have been developed to exploit the merits of OPS systems, such as high capacity, ultra-highspeed hopping, and fine physical granularity [3-10]. In 2005, based on many technologies [11-14], our group has proposed a prototype 160-Gbit/s/port optical packet switch with narrow-band optical code-label processing, optical switching, optical buffering, electrical scheduling, and optical MUX/DEMUX [15]. In order to extend granularity, transparency, and scalability of the OPS systems, we have introduced a colored OPS concept [16] and developed a novel interfaces (called as 10GbE-OP or OP-10GbE converters) between 80 Gbit/s OPS networks and 10 Gb Ethernet (10GbE) [17].

Here, a new experimental demonstration of field trial of 10GbE over 160 (16λ x 10) Gbit/s, fine granularity, colored-optical-packet switching network with >87km fiber transmission [18] is expressed in detail. In this demonstration, a new 320Gbit/s trough put OPS system demonstrator with 16x16 all-optical multiple label processors, high-speed optical switch, arrayed burst-mode transmitter/receivers (Tx./Rx.), and 10GbE-

OP/OP-10GbE converters, and transient-response-suppressed EDFAs is developed and used. In ingress edge nodes, we introduce the 80Gbit/s 10GbE-OP converters [17], packet pulse pattern generator (PPG) [13], and two 8-channel-arrayed burst-mode Tx. to encapsulate IP-packet over 10GbE-frame into an optical packets. In OPS node, optical packets are forwarded based on all-optical multiple code-label-processing performed by single decoding device [19, 20] and high-speed optical switches. In egress node, IP packet-loss-rate (PLR) and bit-error-rate (BER) are measured using arrayed burst-mode Rx. and packet error detector [13]. Error free (PLR $<10^{-6}$ and BER $<10^{-9}$) operation with >87km transmission by field installed fiber has been successfully achieved.

Key technologies

A proposed network architecture and layered structure are illustrated in Figs. 1(a) and 1(b). At the ingress edge node (EN) of the OPS network, an optical label is generated according to a look-up table and the IP address, and the IP packet is encapsulated into a wide-band optical packet. In OPS networks, we introduce large-capacity colored-optical-packets using dense WDM (DWDM) technology. A colored-optical-packet consists of 10 Gbit/s optical signals of 16 wavelengths with 50GHz channel spacing. Theoretically, the data rate of an optical packet is 160 Gbit/s and the packet length in time is 1/16 of the original 10 Gbit/s IP packet. Because of its fine granularity, we can fully use a bandwidth of OPS networks. Figure 2 shows the demonstration system (Fig. 2(a)) architecture of field trial of IP over 160 (16λ x 10) Gbit/s OPS and key technologies. The key technologies are addressed below.

All-optical multiple label processor (Fig.2(d)):
For multiple optical code-label processing, we first introduce 200 Gchip/s multiple optical

code (OC) encoder/decoder with an arrayed waveguide configuration. It can generate and recognize simultaneously sixteen different 16-chip optical phase shift keying codes. The processing rate is 13 Gpacket/s [19, 20].

Arrayed burst-mode Tx./Rx. (Fig.2(b) and 2(e)): The packet Tx. (Fig.2(b)) mainly consists of electro-absorption modulators (EAM) integrated distributed feedback (DFB) lasers. To be compatible with burst-mode packets, reverse bias voltage applied to EAM portions are optimized [17]. The Rx. (Fig.2(e)) mainly consists of specially tuned high-speed uni-traveling-carrier (UTC) photo-detector and a low jitter gated voltage controlled oscillator with digital ring phase lock loop (PLL) [14]. The gating circuit and the PLL enable very fast synchronization time (< 1 ns) of the output clock phase with the input data phase [17].

10GbE-OP/OP-10GbE converters (Fig.2(c)): Our developed 10GbE-OP converter divides an IP packet over 10GbE-frame to eight-segmented 10 Gbit/s payloads and generates an OC label selection signal from the IP address after header processing using look-up table. OP-IP converter recombines these eight-segmented 10 Gbit/s payloads into IP packets over 10GbE-frame again [17].

Transient-response-suppressed EDFA (Fig.2(f)): A transient response of EDFA distorts the waveform even in the case of short-term packet of ~400ns despite much longer lifetime of Er^{3+} about 10ms and it causes serious impairment depending on the link utilization. It is difficult to compensate transient response during such short time by electronic controlling. Alternatively, we developed a new class of EDFA which adopted EDF with enhanced active erbium

area [21] and successfully suppressed the transient response, as comparing commercially available one with our developed EDFAs in Fig.2 (g).

Ingress ENs #1 and #2 encapsulate asynchronous IP packet from client #1 and #2 into 160 Gbit/s optical packets using IP-OP converters and arrayed burst-mode Tx., respectively. Here, a half of payload is dummy data generated by packet PPG. In OPS node, transmitted 160 Gbit/s optical packets with different optical labels are switched into appropriate output ports by a multiple all-optical label processor. In each egress EN #3 and #4, 160 Gbit/s optical packets are reconverted into IP packets over 10GbE-frame using OP-IP converters and arrayed burst-mode Rx. We transmit optical packets through the field environment. The fiber lines are located between Keihanna and Nara with loop-back configuration as shown in Fig. 2(h).

Demonstration

IP packets over 10GbE-frames with different IP address were transmitted from network analyzers. The transmitted rate of 10GbE-frames are 5.3 Gbit/s and 3.2 Gbit/s, respectively. The frame length was 1500 byte. IN INs #1 and #2, 80 (8□ x 10) Gbps optical payloads were generated with 100 GHz channel spacing (odd wavelength channels) by 10GbE-OP converter and 8-arrayed burst-mode Tx.. In addition, we generated another 80 (8λ x 10) Gbps optical payloads with 100 GHz channel spacing between odd wavelength channels (even wavelength channels) by modulating 8-wavelength continuous wave lights with $LiNbO_3$ intensity modulator. These packets are launched into 32 km SMF, which is in the field, and 11.5 km RDF.

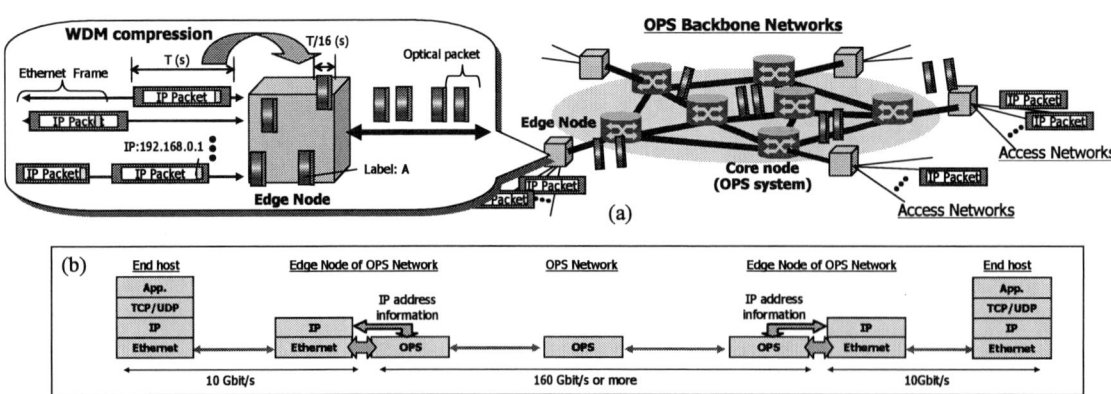

Figure 1. (a) 10GbE over OPS networks. (b) Layered structure in each node.

310

In the OPS system, we set the multiple label processor to recognize only a label *A*. Only optical packets with label *A* were switched into output port *A*. Through field lines, each optical packet is de-multiplexed into 16-wavelength channels and received at egress ENs #3 and #4. We measured the IP-PLR at even wavelength channels by a network analyzer and the BER of 8-optical payloads at odd wavelength channels. The error-free (PLR < 10^{-6} and BER < 10^{-9}) operation has been achieved in cases after transmission and that of back-to-back. The average power penalty of BER is about 4 dB [18].

Conclusions

We have proposed and experimentally shown a new demonstrator of 160 Gbit/s/port colored optical packet switched network with all-optical multiple label processor, arrayed burst mode Tx./Rx., 10GbE/optical packet converter, and a novel transient response suppressed EDFA. Error free (PLR < 10^{-6} and BER < 10^{-9}) operation with >87km transmission by field installed fiber has been successfully demonstrated.

Acknowledgements

The author would like to thank K. Ikezawa, A. Toyama, N. Itou, H. Shimizu of Yokogawa Electric, H. Iiduka and H. Fujinuma of NTT Electronic, E. Kong, P. Chan, R. Man of Amonics Ltd., G. Cincotti of Univ. Roma Tre, K. Kitayama of Osaka Univ., and H. Furukawa, H. Harai, Y. Awaji, T. Miyazaki, T. Makino, H. Sumimoto, Y. Tomiyama of NICT for their collaboration and valuable discussions.

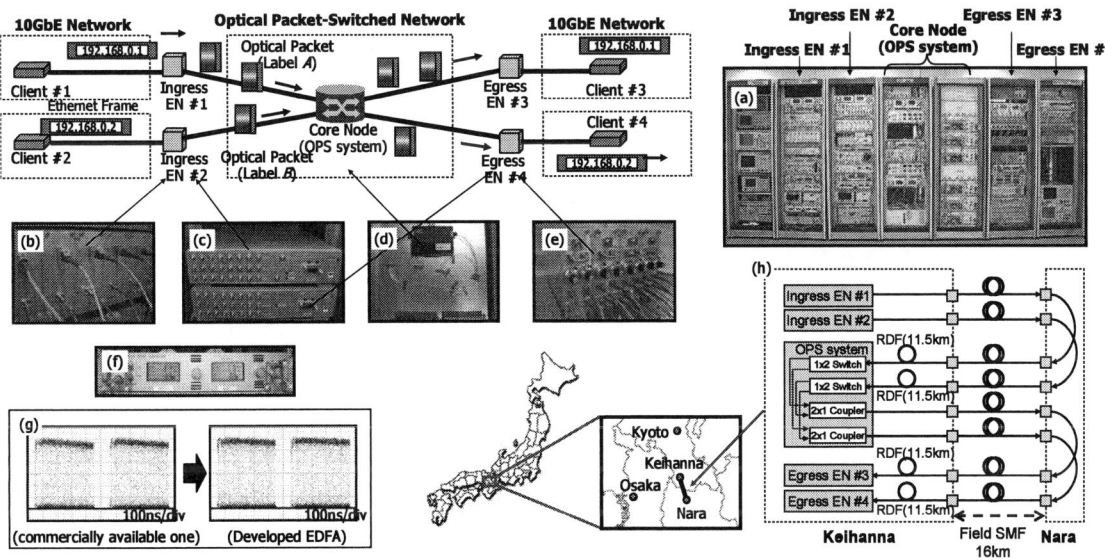

Figure 2. Demonstration system architecture, key technologies, and field trial configuration.

References

[1]   K. Fukuda, et al., ACM SIGCOMM Computer Commun. Review, vol. 35, no. 1, pp. 15-22, 2005.

[2]   K. Sato, et al., IEEE Commun. Magazine, vol. 40, no. 3, pp. 96-101, 2002.

[3]   D.J. Blumenthal, et al., IEEE Photon. Technol. Lett., vol. 11, no. 11, pp. 1497-1499, 1999.

[4]   K. Habara, et al., IEICE Trans. on Commun., vol. E83-B, pp. 2304–2311, Oct 2000.

[5]   S.B. Yoo, OFC2003, vol. 2, no. FS5, pp. 797-798, 2003.

[6]   D. Klonidis, et al., ECOC2004, no. PDP-Th4.5.5, 2004.

[7]   J. McGeehan, et al., OFC2003, vol. 2, no. FS6, pp. 798-801, 2003.

[8]   K. Kitayama, et al., IEEE Photon. Technol. Lett., vol. 11, no. 12, pp. 1689-1691, 1999.

[9]   N. Wada, et al., Proceedings of SPIE, vol. 4872, pp. 185-198, 2002.

[10] N. Wada, et al., OFC2003, vol. 2, no. FS7, pp. 801-802, 2003.

[11] H. Harai, et al., IEEE ICC 2002, pp. 2843-2847, May 2002.

[12] Y. Awaji, et al., ECOC2004, vol. 4, no. We3.5.1, pp. 430-431, 2004.

[13] N. Wada, et al., ECOC2004, no. PDP-Th4.5.4, 2004.

[14] N. Wada, et al., ECOC2004, no. Tu1.5.5, 2004.

[15] N. Wada, et al., ECOC2005, vol. 3, no. We1.4.1, Invited Paper, 2005.

[16] H. Harai, et al., OECC 2003, vol. 23, no. 0253-2239, pp. 703-704, November 2003.

[17] H. Furukawa, et al., OFC2007, OWC5, March 2007.

[18] H. Furukawa, et al., OFC2007, PDP4, March 2007.

[19] G. Cincotti, et al., Journal of Lightwave technology, vol. 24, no. 1, pp. 103-112, January 2006.

[20] N, Wada, et al., "Journal of Lightwave technology, vol. 24, no. 1, pp. 113-121, January 2006.

[21] Y. Awaji, et al., CLEO 2007, no. JTuA133, 2007.

Modified Duobinary Signals with Tunable Duty Cycle and its Application in a Label Switching Optical Network

Yufeng Shao (1), Shuangchun Wen(1), Lin Chen(1), Huiwen Xu(1), Jin He(1)
1 : Research Center of Laser Science and Engineering,
and School of Computer and Communication, Hunan University
Changsha 410082, China
Yufeng Shao:shaoyufeng2006@yahoo.com.cn Shuangchun Wen: scwen@vip.sina.com
Lin Chen: liliuchen@163.com Huiwen Xu: xhwth@163.com Jin He: hnu_jhe@hotmail.com

Abstract *A novel configuration to use modified duobinary signals with tunable duty cycle as optical labels in optical packet switching system is proposed. The performance of 10Gbit/s DPSK payload with 2.5Gbit/s MD-RZ label is analysed.*

Introduction

Optical packet switching (OPS) technology has been proposed worldwide as an effective way instead of electronic packet switching technology [1]. In an OPS network, the optical packets contain a payload and a label. Optical label switching (OLS) is an important aspect of OPS, and has been regarded as an efficient technique to route and forward packets transparently in the optical layer [2]. A few optical label switching technologies have been put forward. In different labelling techniques, orthogonal modulation has shown some advantages, such as its highly spectral efficiency [3]. One example is the combination of amplitude-shift-keying (ASK) label and differential phase-shift-keying (DPSK) payload [4]. However, modified duobinary signals with tunable duty cycle has not been used as optical label yet. In this letter, we propose and demonstrate a novel optical label switching scheme by the combination MD-RZ and DPSK for the first time. We use this modified duobinary label, because it has higher spectral efficiency than conventional binary signals. As we know that, in label switching optical networks, the label should be low-bit rate and easily detected. In our scheme, low-bit rate modified duobinary labels with tunable duty cycle can be directly detected by a regular receiver, so it is much practical and cost-effective.

Proposed Transmitter

Fig.1 shows the scheme setup for

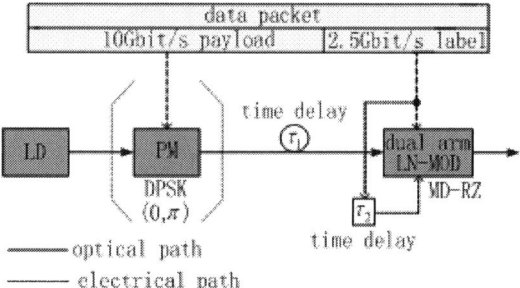

Fig.1 Scheme setup for Proposed Transmitter

generation of orthogonal labelling packet with MD-RZ label. The CW laser has a wavelength of 1550nm and an output power of 10dBm. The 10Gbit/s payload electrical signal is used to drive a phase modulator (PM) to generate 10Gbit/s DPSK payload optical signal. Subsequently, the DPSK payload is delayed (τ_1) and fed into the dual-arm LiNbO$_3$ modulator (LN-MOD). The 2.5Gbit/s label electrical signal is divided into two equal parts. One is delayed (τ_2) to drive the LN-MOD. The other is directly driven the LN-MOD from the other arm of the LN-MOD. The LN-MOD is biased such that there is the minimal output power when the data is turned off. In this way, the MD-RZ label is then impressed by the dual-arm LN-MOD. Depending on the time delay between the two arms, different duty cycle MD-RZ labels can be obtained. It should be noted that the alignment between the label and the payload is very crucial. Any timing misalignment between them may induce power penalty to data packet. Hence, the

978-0-9789217-3-6/07/$25.00

©2007 WEN GLOBAL SOLUTIONS

delayed time of the DPSK payload optical signal is equal to the delayed time between the two arms of the LN-MOD, that is $\tau_1 = \tau_2$.

In principle, any duty cycle MD-RZ labels can be generated by using this method; however, the duty cycle is limited by the rise–fall time of the data and the bandwidth of the LN-MOD.

Results and discussions

This MD-RZ label optical signal can be directly measured by a conventional binary intensity modulation direct detection (IM-DD) receiver. When the delayed time between the two arms of the LN-MOD is 100ps, 200ps or 300ps, the duty cycle of the MD-RZ label is 0.25, 0.5 or 0.75. Fig.2 shows some typical eye diagrams of different duty cycle MD-RZ label optical signals.

In Fig.1, we adjust the time delay between the two arms of the LN-MOD, different duty cycle MD-RZ labels can be obtained. Fig.3 shows the measured typical optical spectral diagrams of (a) the optical payload with 0.75 duty cycle label, (b) the optical payload with 0.5 duty cycle label, (c) t the optical payload with 0.25 duty cycle label. We can see that the smaller the time delay between the two-arm driven electrical signals, the smaller the duty cycle and the wider the spectrum. For bigger duty cycle, the peak power of optical packet is higher, and the spectral efficiency is higher. The MD-RZ label with bigger duty cycle can be utilized to increase the transmission capability. But the MD-RZ label with bigger duty cycle is more sensitive to nonlinear effects, because it has higher optical power. So, we must choose the MD-RZ label with appropriate duty cycle to meet the demand of a certain optical communication system.

Conclusions

Within the optical packet switching network, the label information is read, processed and updated, until the burst reaches the desired egress edge node. The label generation is expected to improve performance for the label switching. In this paper, we demonstrate superimposing of MD-RZ label with adjustable duty cycle over DPSK payload, which can be implemented rather easily.

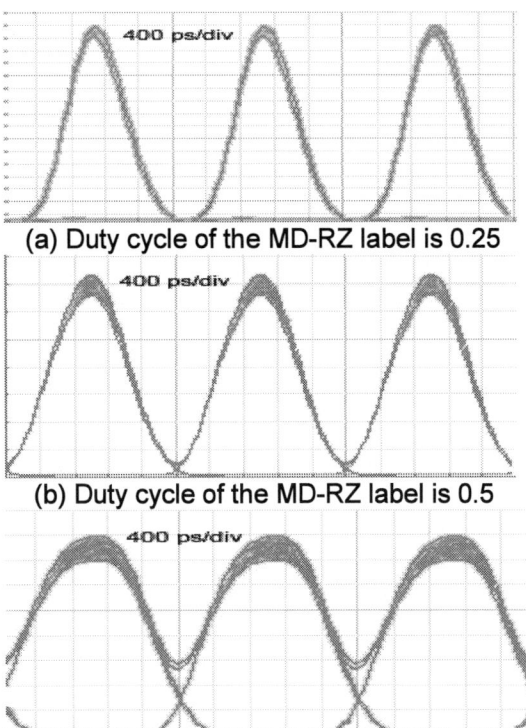

(a) Duty cycle of the MD-RZ label is 0.25

(b) Duty cycle of the MD-RZ label is 0.5

(c) Duty cycle of the MD-RZ label is 0.75

Fig.2 Measured typical eye diagrams of different duty cycle MD-RZ label optical signals

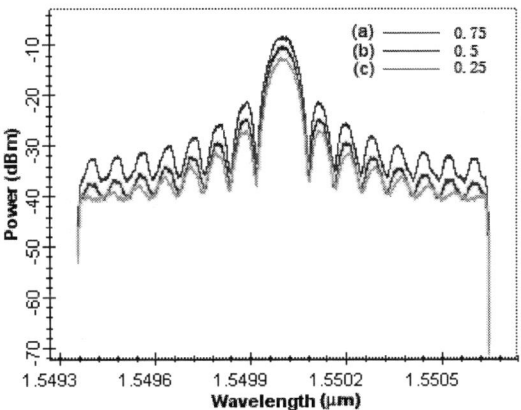

Fig.3 Measured optical spectral diagrams of the optical payload with different duty cycle labels

References

1 Gee-Kung Chang et al. J. Lightw.Technol., 23(2005), pp.3372-3387

2 D.J.Blumethal et al. J. Lightw.Technol., 18(2000), pp.2058-2075

3 N.Chi et al. Proc. ECOC(2002), Copenhagen, Denmark, Paper5.5.1

4 X.Liu et al. Proc. ECOC(2003), Rimini, Italy,Paper4.4

Dual Band Optical Receiver for Video Broadcasting Services over Fiber-To-The-Home Network

Young Cheol Kim(1), Young Ho Jang(2), and Hyun Deok Kim(3)

1 : School of Electrical Engineering and Computer Science, Kyungpook National University,
1370 Sankyuk-dong, Puk-gu, Daegu, 702-701 South Korea, email: yckim@ee.knu.ac.kr
2 : CoreTech Corporation, email: jyh@coretk.com
3 : Kyungpook National University, email: hyundkim@ee.knu.ac.kr

Abstract *We demonstrate a novel optical video receiver to provide dual band broadcasting services, CATV and SATV, by using a single photo-diode. The spectral flatness of the receiver was within ±1.5 dB over both VHF/UHF and L-bands with more than 36 dB band isolation.*

Introduction

The fiber-to-the-home (FTTH) network can easily provide multiple services such as voice, high-speed data and video streaming services [1]. Furthermore, it can provide an analog cable-television (CATV) and satellite -television (SATV) broadcasting services at the same time through an overlay scheme without intervening other services.

The frequency bands of the CATV and the SATV are usually VHF and UHF bands (50 ~ 870 MHz) and L- and S-bands (950 ~ 2500 MHz), respectively. The wide band optical video transmitter and receiver with flat frequency responses over VHF to low S-band are essential to realize the dual band broadcasting services. However, it is not easy to realize optical receiver for the dual band video broadcasting services since the wide impedance matching over CATV and SATV bands and a sufficient band isolation to prevent interferences between services are very difficult to realize at a reasonable cost.

Thus the dual band video broadcasting services are usually implemented base on a wavelength-division multiplexing technique [2]. Though the wide band analog optical receiver with a high linearity is available at this time it is not easy to provide a sufficient band isolation since the CATV signals are susceptible to the optical crosstalk [3].

In this paper, we propose and demonstrate a simple and low-cost optical video receiver to provide CATV and SATV broad-casting services over a fiber-to-the-home network, which provides a flat frequency and a sufficient band isolation between services.

Principle Description of Optical Receiver

The configuration of the proposed optical receiver is shown in Fig. 1. It is composed of a photo-diode (PD), two impedance matching components, Z_1 and Z_2, and two amplification and equalizing circuits, H_1 and H_2. Though the CATV and the SATV signals are inputted to the receiver at the same time, they are separated during the receiving process and provided through the output ports denoted as V_1 and V_2, respectively.

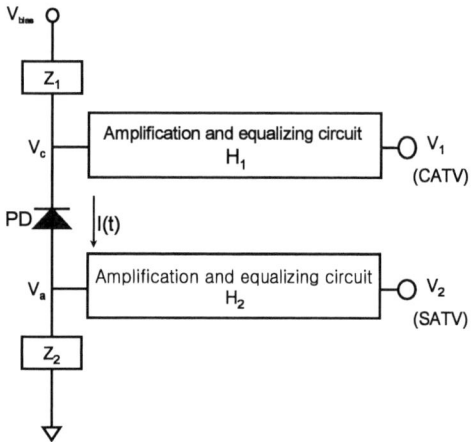

Fig. 1 The configuration of proposed optical receiver

The Z_1 and the H_1 are connected to the cathod of the PD while the Z_2 and the H_2 are connected to the anode of the PD. If a positive bias voltage (V_{bias}) is applied to the receiver, the PD generates a photo current, $I(t)$, which is proportional to the optical input to the PD and thus the inputs of the H_1 and the H_2 denoted as V_c and V_a are given as

978-0-9789217-3-6/07/$25.00

©2007 WEN GLOBAL SOLUTIONS

follows:

$$V_c = V_{bias} - Z_1 I(t)$$
$$V_a = Z_2 I(t)$$

Both Z_1 and Z_2 are composed of passive components such as resistors, capacitors and inductors and provide frequency-dependent spectral responses. The H_1 and the H_2 are composed of passive and active components such as amplifiers, diodes, attenuators and so on to perform signal amplifications, noise reductions, signal level adjustments and frequency response improvements.

The Z_1 and the Z_2 are used to convert the photo current of CATV and SATV bands to corresponding voltages, respectively. Namely, the Z_1 is a kind of low pass filter while the Z_2 is a high pass filter. Thus the unwanted signals are suppressed by the Z_1 and the Z_2, respectively. The pre-filtering scheme and the signal separation by using passive components (Z_1 and Z_2) provide several advantages. If the CATV and the SATV signals are simultaneously amplified in an amplifier as is used in the conventional receiver the beating noise due to the nonlinearity of the components may cause signal degradation. However, the pre-filtering of the signal in the proposed receiver reduces the interference between the signals of different bands. Furthermore, since each amplification and equalizing circuit only handles the CATV or the SATV signal the circuit designs become more flexible, which make it easy to satisfy the requirements on the frequency flatness, the band isolation and the impedance matching.

Simulation and Experimental Results
To design the proposed optical receiver, we used an RF circuit design tool, Agilent's ADS. The target frequency band of the CATV signal was 54~870 MHz and that of the SATV was 950~2500 MHz. We first designed the impedance matching components by using only passive components. And then we designed the amplification and equalizing circuits considering the parameters extracted from the actual components.

Fig. 2 shows the spectral response of the designed receiver through ADS simulation.

It was designed to provide RF amplification gains of 48 dB for CATV signal and 40.5 dB for SATV signal, respectively. The gain flatness over CATV band was within 1 dB and that of the SATV band about 2.7 dB.

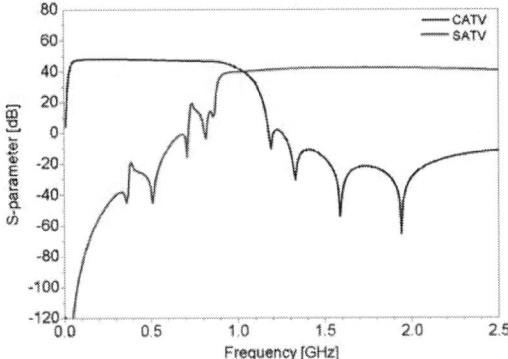

Fig. 2. Spectral response (simulation result)

To confirm the performance of the designed receiver, we implement the receiver as shown in Fig. 3.

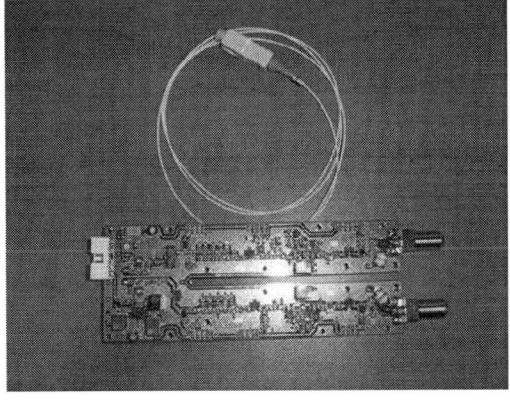

Fig. 3. Demonstrated optical receiver

We measured the characteristics of the implemented receiver by using a RF spectrum analyzer according to National Cable & Telecommunications requirements and Society of Cable Telecommunications Engineers requirements. Fig. 4 shows the frequency responses of the optical video receiver measured at the CATV output port (a) and the SATV output port (b). The measured frequency flatness of the receiver responses was about ±1.5 dB in both CATV and SATV bands. We also measured the band isolation between the CATV and the SATV bands. The measured band isolation was higher than 36 dB as shown in Fig. 5,

315

which satisfies the NCT and the SCTE requirements on the signal quality.

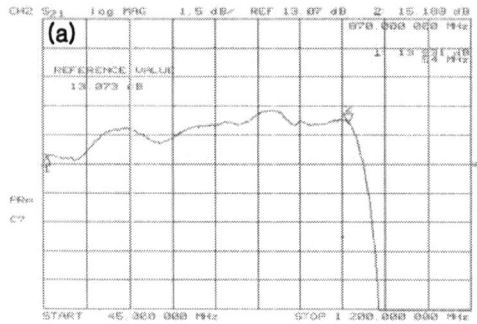

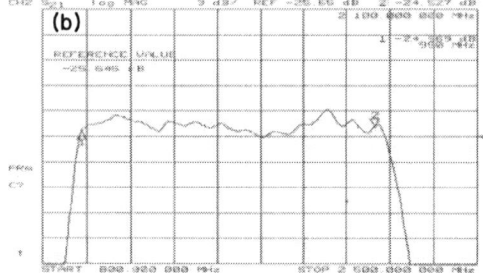

Fig. 4. Frequency response of implemented optical receiver measured at (a) the CATV and (b) the SATV ports, spectively

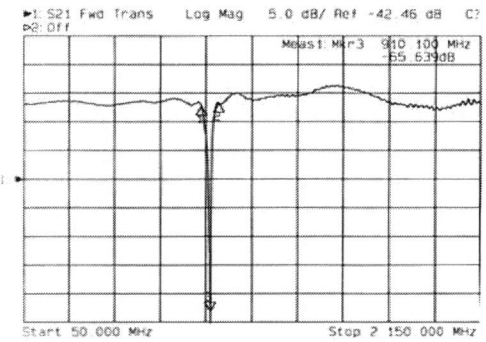

Fig. 5. The measured isolation between the CATV and the SATV bands

We also performed transmission experiment considering a FTTH network by using a 10-km single mode fiber between the receiver and the transmitter [4-5]. We used an Emcore's 3688 ULC subassembly as the transmitter and transmitted 77 channel NTSC signal in the CATV band and a SATV single in live service as shown in Fig. 6. The transmitter output power was about 10 dBm and the receiver input optical power was about -8 dBm, which offers an 18 dB system margin. We measured CNR, CSO and CTB of the CATV channels. All the

NTSC CATV channels satisfied the FCC requirements without intervening the live service of the SATV. The worst-case measurement results of the several channels are shown in Table 1.

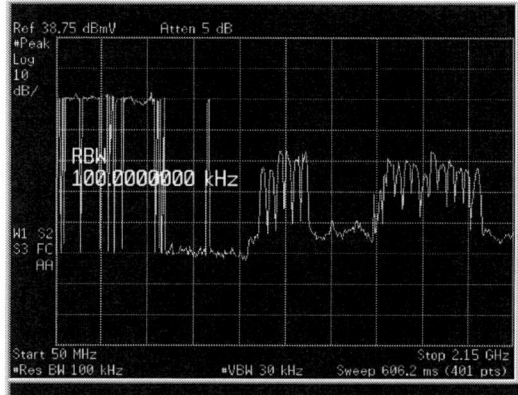

Fig. 6. The output spectrum of the receiver when the transmission length was 10 km.

Table 1. Measurement of RF distortion

Channel Number	Ch 14	Ch 47	Ch 77
Frequency (MHz)	121.25	325.25	541.25
CNR(dB)	44.1	45.6	46.3
CSO(dB)	61.9	62.8	58.2
CTB(dB)	64.8	65.3	67.2

Conclusion
The proposed optical receiver offers a cost-effective solution of the dual band video broadcasting service in the FTTH service. Though we used a single photo-diode in the receiver, it offered a superior performance thanks to the its novel configuration.

This work was supported by KOSEF under the NRL project and Korea Ministry of Education under the BK21 program.

References
1 T. E. Darcie et al., IEEE Trans. Microwave theory tech., vol. 38 (1990), pp. 524-533.
2 K. H. Han et. al., Journal of Optical Networking, vol. 1 (2002), pp. 338-343.
3 M. R. Phillips et. al., IEEE J. Lightwave technol. Lett., vol. 17 (1999), pp. 1782-1792.
4 M. R. Phillips et. al., IEEE Photon. Technol. Lett., vol. 4 (1992), pp. 790-792.
5 C. Y. Kuo, IEEE J. Lightwave Technol., vol. 10 (1992), pp. 235-243.

Novel Distributed All-optical Multicast WDM Fiber Network: Design and Implementation

Dan Lu (1), Xi Qin (2), Feng Zhang (3), Bo Lv, Ming Chen, Shui-sheng Jian

1: IEEE student member; Institute of Lightwave Technology, Beijing Jiaotong University, Beijing, 100044, China, danlucy@gmail.com

2: Institute of Lightwave Technology, biancaqin@263.net

3: Institute of Lightwave Technology, fengdieer_13@126.com

Abstract *A novel distributed all-optical multicast WDM fiber network based on a modified all-fiber multicast-capable OXC has been proposed and established. Configurable all-optical multicast including video and 10Gb/s data are experimentally demonstrated.*

Introduction

With the explosive growth of bandwidth-intensive applications such as high-definition television (HDTV), video-conference, there is increasing need to develop multicast in optical network. Multicast provides a means of a point-to-multipoint connection, in which a source can send a single data to several different destinations at one time, it's able to reduce the number of in-line and terminal facilities and improve the performance of network significantly compared to traditional unicast (point-to-point connection).

Recently, due to the bottleneck of electronic components, all-optical multicast has drawn more and more attentions for its transparent and high capacity[1]. Current efforts are mainly paid on the design of key components to support all-optical multicast, the data control scheme to reduce blocking rate, the multicast protection and survival etc.

In this paper, we review and compare the all-optical multicast-capable (AOMC) OXC proposed previously which is the key component to realize all-optical multicast at first, then a modified AOMC-OXC based on chirped fiber Bragg grating (CFBG) is presented in the section 3, which is simple, power-efficient, nonblocking and contrable. Furthermore, a novel distributed all-optical multicast WDM fiber network is established. Configurable Video and 10Gb/s data multicast are realized successfully. Last, the experiment result will be given and discussed.

Review of AOMC-OXC

AOMC-OXC is utilized to "copy" light and send them to designated multicast destinations. Until now, various AOMC-OXC have been reported.

The simplest AOMC-OXC is based on light splitter, such as DaC (Drop and Continue) element[2], SaD (Splitter and Delivery) switch [3]. They are nonblocking, but the SaD switch may not power efficient, as extra power loss introduced by light splitter leads to degradation of unicast signal. so MOSaD (Multicast-only Splitter and Delivery) switch was then proposed[4]. However, this architecture may cause blocking when there are several multicast requests but only one splitter. Hence, several multicast control algorithms are studied to reduce blocking rate.

In order to avoid split power loss completely, AOMC-OXC based on nonlinear effect (such as XPM, SPM) in SOA is proposed lately[5], it not only realizes all-optical multicast but also synchronously regenerates signal. In addition, dynamically controllable multicast technology using XGM in SOA is involves, which has ability to control the level of multicast (e.g. 1:2, 1:4 etc.), so optical signals no longer need to be wastefully dropped. Nevertheless, because of its complicate and low transition efficiency, it still needs further investigation.

In a word, the characteristic of designed AOMC-OXC is better to be simple and easy to implement; be able to distinguish between unicast and multicast signals in order to be power-efficient; and be nonblocking, and controllable.

Modified AOMC-OXC

As shown in Fig.1, DaC element is made up of a circulator, a fiber Bragg and two couplers. When multi-wavelengths (λ_1, λ_2) optical signals pass through, the appointed multicast wavelength (λ_2) is selected by fiber Bragg grating, then a coupler is employed to drop part of energy ($1-\alpha$) to local multicast receiver, the other part of energy (α) is re-coupled to keep on transmitting to next node. Hence, the unicast wavelength (λ_1) is free from power loss of coupler.

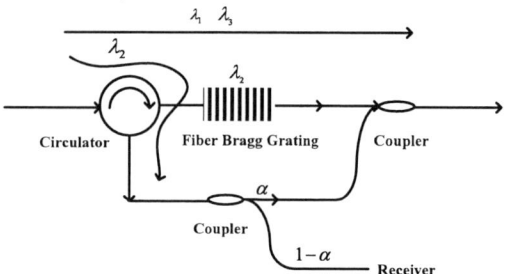

Fig.1 DaC element architecture

As a result, the all-fiber DaC element is simple without large numbers of switches as in SaD and MOSaD, and also power-efficient. In order to achieve nonblocking, a distributed wavelength routing network is designed based on DaC element. Different nodes have their own different multicast transmit wavelengths and corresponding receive wavelengths as shown in Table 1. For example, when node C is transmitter and node A is receiver, λ_2 is selected as unicast wavelength and λ_{12} is multicast wavelength. Each desired multicast wavelength in the node has one DaC element to carry out selection and split. Desired wavelengths are dropped, undesired ones just pass by, and therefore DaC elements distributed in all nodes of network together perform the function of all-optical cross connection (OXC). And all the nodes can multicast messages at the same time without blocking.

Besides, the performance of AOMC-OXC is optimized by using chirped fiber bragger grating (CFBG), the CFBG is totally designed and fabricated by ourselves, which has better quality with low delay ripple and steady temperature characteristic instead[6]. This modified AOMC-OXC can compensate dispersion in transmission and eliminate ASE caused by EDFA, so it's capable to improve the multicast

performance.

Table 1 Wavelength assignment regular

		Transmit wavelength			
		Node A	Node B	Node C	Node D
Receive wavelength	Node A		$\lambda_1,$ λ_8	$\lambda_2,$ λ_{12}	$\lambda_3,$ λ_{16}
	Node B	$\lambda_5,$ λ_4		$\lambda_6,$ λ_{12}	$\lambda_7,$ λ_{16}
	Node C	$\lambda_9,$ λ_4	$\lambda_{10},$ λ_8		$\lambda_{11},$ λ_{16}
	Node D	$\lambda_{13},$ λ_4	$\lambda_{14},$ λ_8	$\lambda_{15},$ λ_{12}	

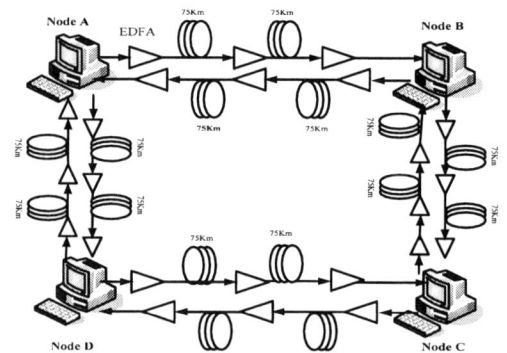

Fig.2 Topology of the network

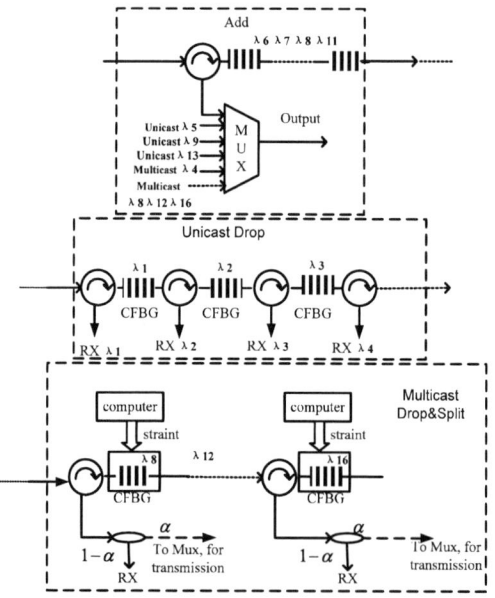

Fig.3 Nodal Architecture (Node A)

Meanwhile, strain is induced on CFBG controlled by computer lead to shift of the reflect wavelength because of the linear tuning of Bragg wavelength of CFBG. By

318

doing this, multicast signal don't pass the coupler in AOMC-OXC, so unnecessary power loss can be avoided.

Experimental Results

Fig.2 shows the topology of multicast network we constructed, which is ring topology and involves four nodes[7, 8], The distance of SMF between neighbours is 150km. Wavelength assignment regular is shown in Table 1. Of course, all wavelengths selected meet the demand of the ITU-T for WDM system. EDFA is used as amplifier, whose noise figure is less than 5 when used as a pre-amplifier and less than 6 in other cases. The delay ripple of CFBG employed is less than 15ps and the temperature coefficient is about $0.0002nm/\,^{\circ}C$, the reflectance spectrum ripple<1dBm, so cascaded grating can meet the practical demand. The Insert loss of 1×2 coupler with split rate 3:7 is about 3dB.

The nodal architecture is shown in Fig.3, Which contains three parts: Add, Unicast Drop, Multicast drop & split. Generally, signal transmission between nodes is clockwise, if any fiber is broken down, self-healing can be reacted quickly, and then signal transmission is anticlockwise[9]. From Fig.3, inside node A, the unicast wavelength λ_2 is dropped in the Unicast Drop part, while multicast signals (λ_{12}) is selected to pass through the modified AOMC-OXC in Multicast part. At the same time, if node A is about to send message, its transmit wavelengths, undesired wavelengths in node A and part of multicast wavelengths are all added in Add part to node B, finally go back to node C.

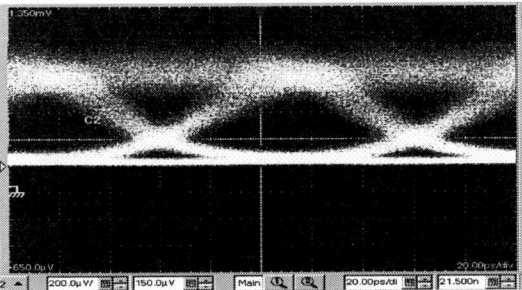

Fig.4 Eye diagram of received 10Gb/s multicast data in node D

Video multicast is realized successfully in this way, each node is able to receive vivid multicast video and configurable multicast is realized successfully by our method. To measure the performance of network, a 10Gb/s optical NRZ data modulated by $LiNbO_3$ modulator with $2^{23}-1$ pseudo-random bit sequence is performed. After transmission, the dispersion compensation in each channel is perfect; no error is detected in all channels. Fig, 5 shows the eye diagram of received signal in node A after 300km transmission when node C is the source and all the other nodes are the receivers.

In our experiment, the all-optical multicast network performs well and can be established easily. However, because the modified AOMC-OXC is still based on the light splitter, eye opening of multicast signal received in node A is inevitable lower than that received in node C.

Conclusions

A modified AOMC-OXC based on CFBG is proposed, which is simple, power-efficient, nonblocking and controllable. Based on it, configurable all-optical multicast including video and 10Gb/s data are experimentally demonstrated in the novel distributed wavelength routing WDM fiber network. The CFBG is made good use as multicast routing/control, multi-channel add/drop, dispersion compensation. The experimental result shows network performs well.

References

[1] A. Hamad, et al., Computer Networks, 50(2006), 3105-3164
[2] X. J. Zhang, et al., Journal of Lightwave Technology, 18(2000), 1917-1927
[3] V. Eramo, et al., Journal of Optical Networking, 5(2006), 82-96
[4] M. Ali, et al., Selected Areas in Communications, IEEE Journal on, 18(2000), 1852-1862
[5] G. Contestabile, et al., Photonics Technology Letters, IEEE, 18(2006), 181-183
[6] Y. Feng-Ping, et al., Chinese Physics, 16(2007),
[7] S. Jian, 2003' Forum of Optical Valley of China, 2003),
[8] Y. Chen, et al., APOC (2005),
[9] F. Zhang, et al., APOC (2006),

This page intentionally left blank.

Author Index

A

Adachi, Chihaya 187
Ai-zhen, Li... 58
Aiba, Takamichi 256
Akhmediev, Nail 152
Al-Mansoori, M. H. 390
Andalib, P. 222
Andersson-Engels, Stefan 234
Ansari, Lida 504
Anshi, Xu 599
Asatryan, A. 97
Ashihara, Satoshi 372
Awaji, Y. 210
Axelsson, Johan 234

B

Bae, Jeung-Hwan 131
Bae, Ji-Hong 140
Bae, Kwang-Soo 140
Barklund, Anna 113
Begum, F. 363
Bennion, Ian 85, 204
Bhatranand, Apichai 477
Borghsb, Gustaaf 249
Botten, L. 97
Bowers, John 1

C

Cada, Michael 56
Cai, Bo 196
Cai, Zhiping 387, 429
Cann, Robert 283
Cao, Lei 586
Cao, Wenbin 533
Chang, Guan-Liang 622
Chang, Li-Wen 280
Chang, Qingjiang 306
Chao, Peng 599
Che, Ming 459
Chen-Perdereau, X. 566
Chen, Antao 219
Chen, Baoquan 113
Chen, Biao 375
Chen, Chun-Ho 12
Chen, Daru 378, 402, 453, 542
Chen, Dingbo 277, 486
Chen, Fei 359
Chen, Hong-Zheng 302
Chen, Hongzheng 359
Chen, Hui 113

Chen, Jianping 104
Chen, Lili 82
Chen, Lin 312, 423
Chen, Ming 317
Chen, Q. 107
Chen, Shean-Jen 622
Chen, X.W. 79, 350
Chen, Xianfeng 67, 85, 204, 492, 495
Chen, Xue-Wen 444
Chen, Yan-feng 184
Chen, Yan-Ru 489
Chen, Yan 163
Chen, Yung-Sheng 347
Chen, Yuping 67, 492, 495
Chen1, Hong-Zheng 246
Chena, Hong-Zheng 249
Cheng, Alan 638
Cheng, Bing-Ying 456
Cheng, Yan 34, 551
Cheng, Yuanbing 486
Chenga, Yunan 249
Chiang, Chun-Ping 228
Chien, Liang-Chy 137
Chiou, Arthur 40
Chiu, Kuo-Chih 622
Chiu, Nan-Fu 237
Cho, Kwan Hyun 181
Choi, D-Y. 554
Choi, Ick Chang 450
Choi, Jae-Wan 353
Choi, Jeong-Hyeon 353
Choi, Kyung Cheol 181
Choi, Pyung-Suk 474
Choi, Yoonseuk 140
Chong-qing, Wu 625
Chou, C. 37
Chou, Chien 338
Choua, Ling-Chung 616
Choudhury, P. K. 390
Choy, Wallace C.H. 79, 240, 350
Chui, P.C. 79
Chung, Youngjoo 589
Cincotti , G. 98
Computer 303
Cui-luan, Wang 432
Cui, Kaiyu 586

D

Da-zheng, Wang 432
Dai, Daoxin 513, 563, 578
Dai, Jiajie 519
Damsgaard, Hans 18

Dang, Weirui 67, 492
Davies, Edward 85
Dennis, M. 162
Dinu, Raluca 113
Dong, X.Y. 341
Dong, Xiaoyi 204
Dong, Xinyong 134
Dong, Zuoren 524
Duggal, Anil R 301
Durouxab, L. 289
Durouxab, M. 289

E

Eggleton , B.J. 554
Enoki, Takatomo 64
Ersoy, Okan 507
Eun, Jae-Jeong 474

F

Fang, Hong 369, 381, 441
Fang, Qun 82
Fang, Yun 113
Fang, Zujie 524
Fanying, Meng 595
Faraji-Dana, R. 483
Fei, Yu 498, 545
Feng, Shuai 456
Feng, Tu 47
Feng, Yijun 163
Ferreira, J. 438
Finge, Norman 332
Fok, Mable P. 198, 638
Fong, H.H. 350
Fu, Hongyan 375, 378, 396, 453, 604
Fu, Shenggui 204
Fuerbach, A. 97
Fukuchi, Yutaka 70
Furukawa, H. 210

G

Gao, Junming 408
Gao, Lei 516
Gao, Shiming 378, 453
Gao, Yan 246
Gao, Ying 378, 453
García de Abajo, F.J. 162
Gity, Farzan 504
Gomes, N. J. 560
Goodall, Thomas 283
Grace, Warren 283
Granpayeh, N. 222, 530
Greatrex, Jeffrey 283
Grelu, Ph. 566
Grelu, Philippe 152

Guan, Bai-Ou 178
Guan, Zuguang 292
Guo, Bing 283
Guo, Tieying 369, 441

H

Hai, N. H. 363
Hai, Nguyen Hoang 384
Hainberger, Rainer 332
Han, Hongtao 557
Han, Sang-Kook 607
Han, Won-Taek 589
Hao, Jiaming 519
Hao, Yinlei 335
Hashida, H. 243
He, Jian-Jun 216, 329, 572
He, Jin 312
He, Jing 423
He, Ruei-Yu 622
He, S. 82, 378, 501
He, S.L. 79, 350
He, Sailing 329, 444, 604
He, Yonghong 580
Hine, Anna 85
Ho, H. P. 583
Ho, Ho-Pui 169
Hong, Fang 366
Hong, Xiaobin 320
Hossain, Md. Zahid 119, 619
Hsu, Chain-Shu 91
Hsu, Ya-Ting 12
Hu, Jing 329
Hu, Weisheng 414
Hu, Wen-Ping 246
Hu, Yonghua 480
Huang, Chaohong 387, 429
Huang, Chi-Feng 347
Huang, Chong-Jie 280
Huang, Diyun 113
Huang, Jia-Chi 246
Huang, Jun 462
Huang, Wei-Kai 12
Huang, Wei-Ping 274
Huang, Weihong 56
Huang, Weiping 320
Huang, Wencai 387, 429
Huang, Yi-Pai 12
Huang, Yidong 166, 471, 586
Huaqing, Wang 146
Huda, Md. Quamrul 119, 619
Huggard, P. G. 560
Hughes, Marcus 85
Huh, Jae-Hoon 116
Hui, Rongqing 253
Huo, Li 231

A-2

I

Ibrahim, Sahrul Hilmi 539
Ishii, Hiroyuki 213
Ives, Mark 283

J

James, J. 560
Jang, Nam-Young 474
Jang, Se-Jin 140
Jang, Young Ho 314
Janssen, A. P. 24
Jiabing, Du 595
Jian-liang, Zhang 625
Jian, Leng 143
Jian, Shui-sheng 317
Jian, Shuisheng 369, 381, 441
Jiang, Chun 575, 600
Jiang, D.S. 149
Jiang, Jun 128
Jiang, Li 15, 447
Jiang, Xiaoqing 335
Jiang, Xiaoyu 580
Jiang, Yurong 356, 465, 501
Jianqiang, Wang 595
Jie, Luo 47
Jin, Dan 113
Jin, L. 583
Jin, Xu 548
Jinwei, Li 548
Johansson, Ann 234
Johansson, Jonas 234
Jovanovic, N. 97
Jung, Jong-Wook 140

K

Kado, Yuichi 61
Kaijage, S. F. 363
Kaijage, Shubi 384
Kakitsuka, Takaaki 213
Kalogerakis , G. 201
Kamio, Yukiyoshi 157
Kan, D. 97
Kanai, N. 243
Kang, Jeung-Mo 353
Kao, Yee-Mou 405
Karasek, M. 420
Kawaguchi, K. 243
Kawakami, S. 610
Kawamura, Y. 243
Kawanishi, Tetsuya 110
Kazovsky, L. G. 201
Khijwania, Sunil K. 341
Kim, Byoung Yoon 601
Kim, Dong-Hyeon 607

Kim, Du-Hyun 353
Kim, Geun-Ho 353
Kim, Hyun Deok 314, 450
Kim, Jae-Hoon 140
Kim, Jae-Wook 353
Kim, Seung Hun 181
Kim, Sung-Ho 589
Kim, Sung-Jin 88
Kim, Tae Hoon 326
Kim, Young Cheol 314
Kimura, H. 243
Kinjo, T. 363
Kitamura, N. 193
Kitayama, K. 98
Kobayashi, M. 243
Kobayashi, Noriyo 64
Kokubun, Yasuo 323
Kong, S. K. 583
Kong, Siu Kai 231
Kosugi, Toshihiko 64
Koyama, Fumio 271
Kuan, C. H. 489
Kuan, C.-H. 616
Kuo, B. P. P. 201
Kuroda, Kazuo 372
Kuroki, Yasuhiko 116

L

Lamont, M.R.E. 554
Lang, Tingting 329, 513
Le, Yang 595
Lee , Ray-Kuang 405
Lee, Cheng-Kuang 228
Lee, Cheng-Ling 405
Lee, Chomsik 131
Lee, El-Hang 21
Lee, Hsiang-Chieh 228
Lee, Hyung Jae 326
Lee, J. 107
Lee, Jeong-Soo 353
Lee, Jiun-Haw 237
Lee, R.K. 583
Lee, Sin-Doo 88
Leung, Bernard 468
Li-jun, Zhang 625
Li, Bing 113
Li, C. 243
Li, Da-Hai 94
Li, Haifeng 34, 551
Li, Honglei 369, 381, 441
Li, Jinwei 356
Li, Lieming 533
Li, Mingyu 216
Li, Pengcheng 286
Li, Rong-jin 246

A-3

Li, Shangyuan 53
Li, Wei 580
Li, Xia 356, 465
Li, Xingde 295
Li, Xinwan 104
Li, Xun 274
Li, Y. 566
Li, Ying-Chang 338
Li, Yuquan 101
Li, Z. Y. 516
Li, Zhi-Yuan 456, 459, 536
Li1, Xin 82
Liang, C.J. 240
Liang, Song 277
Liao, Cheng-Yumr 12
Liao, Chi-Chang 31
Liao, Lin-Yao 12
Liao, Yung-Ming 91
Lin, Aoxiang 589
Lin, Chi-Hung 622
Lin, Chii-Wann 237
Lin, Chinlon 231, 583
Lin, Fang-Cheng 12
Lin, Rujian 393, 399
Lin, Xu 595
Lin, Yan Sen 411, 527
Lin, Yi-Hsin 31
Liu, Fang 166, 471
Liu, Fangfei 575, 639
Liu, Hongdu 468
Liu, Jie 301
Liu, Kou-Chen 237
Liu, L. 149
Liu, Rong-Juan 459
Liu, Tongyu 172
Liu, Weisheng 631
Liu, Xu 34, 551
Liu, Ya-Zhao 456
Liu, Ye 536
Liu, Yung S. 298
Liu, Zhihai 125
Liwen, Wang 366
Longfeng, Xiang 146
Lou, Shuqin 369, 381, 441
Lu, Chih-Feng 347
Lu, Chih-Wei 228
Lu, Dan 317
Lu, Feng 67, 495
Lu, Lin 101
Lu, Wenjie 67, 495
Lu, Yan-qing 184
Luo, Jingdong 219
Luo, Qingming 286
Luther-Davies, B. 554
Lv, Bo 317

M

Ma, Hongmei 15
Ma, Hui 580
Ma, Wen 283
Madden, S. 554
Maeda, Joji 70
Maeda, Yoshinobu 116
Mahdi, A. K. Zamzuri and A. 207
Main, Keith 557
Maki, Sayaka 116
Makino, R. 243
Maksymiuk, L. 259
Mao, Xiaoyu 586
Marhic, M. E. 201
Marshall, G. 97
Matsuo, Shinji 213
Matsushima, Toshinori 187
Matsushima, Yuichi 4
McDonald, Bill 265
Mei, Hui 498, 545
Meng, Qinghua 492
Miao, Xiangrui 453
Miller, Andrew 283
Miller, Eric 113
Ming, Nai-ben 184
Mino, Shinji 27
Miu, Lihong 67
Miyazaki, T. 98
Miyazaki, Tetsuya 157
Moghaddasi, M. Naser 504
Mohammad, Abu Bakar 539
Mohrdiek, Stefan 283
Mokhtari, A. 510
Moolayil, Merly 113
Morris, Jim 557
Moss, D.J. 554
Mou, Chengbo 204
Müllner, Paul 332
Murata, Hiroshi 64

N

Na, Yu-Jin 88
Nagatsuma , Tadao 61
Nakamura, Moriya 157
Nakanotani, Hajime 187
Nakayama, Hideki 116
Namihira, Y. 363, 384
Namihira, Yoshinori 268
Nango, Mamoru 634
Nasir, M. N. Mohd. 390
Neves-Petersenab, M. T. 289
Ngo, Nam Quoc 219
NianyZuo 268
Nien, Shou-Yu 237

Nilsson, Johan	360
Nozhat, N.	530

O

Ohashi, Masaharu	256, 262
Ohtera, Y.	610
Okamoto, Katsunari	303
Okamura, Yasuyuki	64
Oohashi, Hiromi	213
Ou, Haiyan	375, 604

P

Pan, Dandan	492
Park, Jaehee	131
Pelusi, M.D.	554
Peng, Hui	101
Peng, Jiangde	166, 471, 586
Petersenab, S. B.	289
Piao, Yin-Xing	607
Pliska, Tomas	283
Pojanasomboon, Pojamarn	507
Ponomarenko, Sergey A.	56
Pu, Tao	101, 417
Pyayt, Anan	219

Q

Qi, Xi	317
Qian, Jun	82
Qian, Song	498
Qin, Shan	402
Qinfeng, Zhou	344
Qinghao, Ye	595
Qingrong, Han	47
Qiu, Jianrong	613
Qiu, Min	160, 598
Qiu, Shaofeng	462
Qu, Ronghui	524
Quinlan, Shaun	283

R

R.Matai	47
Radil, J.	420
Rao, Y. J.	122
Rao, Yi	166, 471
Rapp, Ludovic	152
Rashid, H. A. Abdul	390
Razzak, S. M. A.	363
Ren, Jinjun	613
Rhee, Tae Hyung	326
Rigole, P-J	569
Roelens, M.A.F.	554
Roh, Sang-Kil	607
Ronda, Cees	637

Rongqiang, Cui	595
Rostami, G.	483
Rui, Geng	435

S

Sainidou, R.	162
Saito, T.	243
Samsuri, N. Md	207
Sato, T.	610
Sattmann, Ralph	50
Segawa, Toru	213
Shahabadi, M.	483
Shao, Li-Yang	134
Shao, Yufeng	312
Shen, Hao	468
Shen, P.	560
Shen, Wenjuan	283
Sheng, Zhen	329, 513
Shi, Min-Min	246
Shiao, Wen-Yu	347
Shibata, Nori	256
Shieh, Han-Ping D	12
Shimura, Tsutomu	372
Shinohara, Hiromichi	252
Shiquan, Yan	595
Shishegar, A. A.	510
Shu, Chester	198, 638
Shuqin, Lou	366
Siuzdak, J.	259
Skovsenab, E.	289
Somesfalean, Gabriel	628
Soto-Crespo, Jose M.	152
Stepniak, G.	259
Su-ping, Liu	432
Su, Yikai	306, 408, 575
Su, Yuan-Deng	622
Subrina, Samia	119, 619
Suen, Yick Keung	231
Sugden, Kate	85
Sun, Chi-Kuang	76
Sun, Weiqiang	414
Sun, Yu	486
Sun, Yubao	15, 447
Svanberg, Sune	234
Svensson, Tomas	234

T

Ta'eed, V.G.	554
Tam, H Y	341
Tam, Hwa-Yaw	134
Tam, HwaYaw	178
Tang, Jiansheng	477
Tang, Yongbo	521, 542
Teperik, T. V.	162

Terao, Y. 243
Tetsuro, Yabu 262
Thylén, Lars 7
Tiefeng, Xu 344
Tigang, Ning 435
Tolstikhin, Valery I. 190
Tong, L. 566
Tong, X.L. 149
Tong, Zhi 435
Towery, Chris 43
Tsai, C. C. 37
Tsai, Meng-Tsan 228
Tsai, Yao-Chou 237
Tsao, Cheng-Han 12

V

Vemagiri, Jeevan 113
Vienne, G. 566
Vojtech, J. 420
Von Der Weid, J. P. 438
Vydra, Jan 50

W

Wada, N. 98, 210
Wada, Naoya 309
Wai, P.K.A 444
Wan, Ruiyuan 166
Wang, Ai-Hong 94
Wang, Chuncan 435
Wang, Dan 533
Wang, Huan 277, 486
Wang, Hui-tian 184
Wang, Jun 216, 414
Wang, Liesong 486
Wang, Mang 246, 302, 359
Wang, Minghua 335
Wang, Qiong-Hua 94
Wang, Rong 417
Wang, Wei 486
Wang, Xin 468
Wang, Xiulin 387, 429
Wang, Xu 98
Wang, Yih-Ming 228
Wang, Yiping 104
Wang, Zhe-Chao 329
Wang, Zhechao 513, 563
Wang, Zhi 444
Wang, Ziqiang 533
Wanga, Mang 249
Wei, H. C. 37
Wei, Xue 143, 146, 548
Wei, Yizhen 542
Weiguo, Chen 366
Wen, H.Q. 149

Wen, Hong 423
Wen, Shuangchun 312, 423, 480
Wenzhong, Shen 595
Withford, M. 97
Won, Yong-Yuk 607
Wong, K. K. Y. 201
Wu, Botao 613
Wu, Hsieh-Ting 338
Wu, Hua-Lin 622
Wu, Jian 320
Wu, Rui 492
Wu, Shin-Tson 10, 31
Wu, Shu-Yuen 169
Wu, Weilei 101
Wua, Gang 249

X

Xi, Yanping 274
Xia, Yuxing 67, 492, 495
Xianfeng, Chen 372
Xiang, Liangzhong 225
Xianglong, Zeng 372
Xiao-ming, Feng 432
Xiao-yu, Ma 432
Xiao, Yun 396
Xie, Xuejuan 414
Xin-zhi, Sheng 625
Xin, Yan-Xia 94
Xing, Da 225
Xu, Chris 154
Xu, Hao 519
Xu, Huiwen 312
Xu, Huiying 387, 429
Xu, Ping 516
Xu, Xinyu 408
Xue, Jiangeng 592
Xue, Wei 356, 465, 501

Y

Yahiro, Masayuki 187
Yamashita*, Ikuo 262
Yang, C. C. 228, 347
Yang, Deng-Ke 30
Yang, Jhih-Ming 31
Yang, Jianyi 335
Yang, Jun 125
Yang, Li-Gong 246
Yang, Liu 329, 513, 578
Yang, Minwei 104
Yang, Shi 595
Yang, Shujun 477
Yang, Sidney S. 280
Yang, Su 175
Yao, Lei 369, 381, 441

Ye, Jiajun	393, 399
Ye, Qing	524
Ye, Yuqian	521
Yeh, Dong-Ming	347
Yeh, Yin	73
Yin, S.	107
Yokoyama, Daisuke	187
Yong-gang, Wang	432
Yong-gang, Zhang	58
Yong-sheng, Lan	432
You, Byung Gwon	326
Yu-quan, Li	175
Yu, Guomin	113
Yu, Yichuan	521
Yuan, Libo	125
Yuan, W.	583
Yue-jin, Zhao	498, 545
Yueyu, Xiao	426
Yun, Byoung-Ju	450
Yun, Tianliang	580
Yuqiong, Li	143
Yusoff, Z.	390

Zhou, Jun	219
Zhou, Kaiming	85
Zhou, Lei	519
Zhou, Renjia	302, 359
Zhou, Shifeng	613
Zhou(1), Kaiming	204
Zhu, Hongliang	277, 486
Zhu, Kun	396, 604
Zhu, Meiwei	393, 399
Zhu, Shi-ning	184
zhu, Weitao	462
Zhu, Ying-xun	417
Zhu, Ying	82
Zhu, Yong-yuan	184
Zhuo, Hui	480
Zou, N.	363

Z

Zeng, Nan	580
Zhan, Li	640
Zhang, Dao-Zhong	456, 536
Zhang, Fan	435
Zhang, Feng	317
Zhang, H.M.	240
Zhang, Junfeng	495
Zhang, Lin	85, 204
Zhang, Q.	107
Zhang, Tidd	43
Zhang, Wei	166, 471, 486, 586
Zhang, Yali	277
Zhang, Yang	178
Zhang, Yuan	501
Zhang, Yunxiao	486
zhang, Zhizhong	462
Zhao, Junming	163
Zhao, Ren-Liang	94
Zhao, Wu-Xiang	94
Zheludev, N. I.	162
Zheng, Benrui	387
Zheng, Lixin	113
Zheng, Wei	335
Zheng, Xiaoping	53
Zheng, Yuxin	166
Zhengbin, Li	599
Zhinong, Yu	143, 146, 548
Zhou, Bin	292
Zhou, Bingkun	53
Zhou, Haifeng	335
Zhou, Jianying	335

This page intentionally left blank.

Proceedings of the 2007 Asia Optical Fiber Communication and Optoelectronics Conference

Shanghai, China
17-19 October 2007

Pages 320-640

IEEE Catalog Number: CFP0739B-PRT
ISBN 10: 0-9789217-3-9
ISBN 13: 978-0-9789217-3-6

Copyright © 2007 by WEN GLOBAL SOLUTIONS
All Rights Reserved

IEEE Catalog Number: CFP0739B-PRT

ISBN 10: 0-9789217-3-9
ISBN 13: 978-0-9789217-3-6

Additional Copies of This Publication Are Available From:

IEEE Service Center
445 Hoes Lane
Piscataway, NJ 08854
Phone: (800) 701-4333
       (732) 981-1393
Fax: (732) 981-9667
E-mail: customer-service@ieee.org

AOE 2007 COMMITTEES

General Co-Chairs

Thomas L. Koch
Lehigh University
Bethlehem, PA, USA

Yi-Xin Chen
Shanghai Jiao Tong University
Shanghai, China

Program Chair
Sailing He
Zhejiang University
Hangzhou, China

SC1: Optical Fibers, Fiber Components and Subsystems

Lead Co-Chair: John Zyskind, *JDSU, , NJ, USA*

Co-Chair: Ping-kong Alexander Wai, *Hong Kong Polytechnic University, Kowloon, Hong Kong*

Committee Members: John Feng, *Avanex Corporation, Fremont, CA, USA*

Philippe Grelu, *Université de Bourgogne, Dijon, France*

Ruxiang Jin, *Lightelli Corporation, Shanghai, China*

Jie Luo, *Yangtz Optical Fiber and Cable Co., Ltd., Wuhan, China*

Sanjai Parthasarathi, *Avanex Corporation, Fremont, CA, USA*

Chester Shu, *Chinese University of Hong Kong, Shatin, Hong Kong*

Ping Shum, *National Technological University, , Singapore*

Yikai Su, *Shanghai Jiao Tong University, Shanghai, China*

Xu Wang, *Heriot-Watt University, Edinburgh, UK*

SC2: Optoelectronic Devices and Materials

Lead Co-Chair: Katsunari Okamoto, *University of California - Davis, Davis, CA, USA*

Co-Chairs: Jian-Jun He, *Zhejiang University, Hangzhou, China*

El-Hang Lee, *Inha University, Nam-ku, Korea*

Anders Olsson, *Finisar Corporation, Sunnyvale, CA, USA*

Committee Members: John E. Bowers, *UCSB, Santa Barbara, CA, USA*

Wei-Ping Huang, *McMaster University, Hamilton, ON, Canada*

Emil Koteles, *Lightip Technologies, Inc., Ottawa, ON, Canada*

Min Qiu, *KTH, Stockholm, Sweden*

Rang-Chen Yu, *Fiberxon, Inc., Santa Clara, CA, USA*

SC3: Optical Sensors and Biophotonics

Lead Co-Chair: Chinlon Lin, *Chinese University of Hong Kong, Shatin, Hong Kong*

Co-Chairs: Hui Ma, *Tsinghua University, Shenzhen, China*

Yun-Jiang Rao, *UESTC, Chengdu, China*

Hwa-Yaw Tam, *Hong Kong Polytechnic University, Kowloon, Hong Kong*

Committee Members: Chien Chou, *National Yang-Ming University, Taipei, Taiwan, R.O.C.*
Ho-pui Aaron Ho, *Chinese University of Hong Kong, Shatin, Hong Kong*
Hongdu Liu, *Peking University, Beijing, China*
Qingming Luo, *Huazhong University of Science and Technology, Wuhan, China*
Ammasi Periasamy, *University of Virginia, Charlottesville, VA, USA*
Steffen B. Petersen, *Aalborg University, Aalborg, Denmark*
Jianan Qu, *Hong Kong University of Science and Technology, Kowloon, Hong Kong*
Weihong Tan, *University of Florida, Gainesville, FL, USA*
Yin Yeh, *UC-Davis, Davis, CA, USA*
Libo Yuan, *Harbin Engineering University, Harbin, China*

SC4: Displays, Solid-State Lighting & Optoelectronics in Energy
Lead Co-Chair: Shin-Tson Wu, *University of Central Florida, Orlando, FL, USA*
Co-Chair: Wenzhong Shen, *Shanghai Jiao Tong University, Shangai, China*
Committee Members: Kyung Cheol Choi, *KAIST, Daejon, Korea*
Wallace C. H. Choy, *University of Hong Kong, , Hong Kong*
Rongqiang Cui, *Shanghai Jiao Tong University, Shangai, China*
Sin-Doo Lee, *Seoul National University, Seoul, Korea*
Yung S. Liu, *National Tsing Hua University, Hsinchu, Taiwan, R.O.C.*
Yong Qiu, *Tsinghua University, Beijing, China*
Franky So, *University of Florida, Gainesville, FL, USA*
Deng-Ke Yang, *Kent State University, Kent, OH, USA*

Industrial Forum on FTTH Technologies
Co-Chairs: Shoichi Hanatani, *Hitachi Communication Technologies, Ltd., Shinagawa, Japan*
Wei-Ping Huang, *McMaster University, Hamilton, ON, Canada*
Anders Olsson, *Finisar Corporation, Sunnyvale, CA, USA*

Joint Symposium on Enabling Technologies for Next Generation
Co-Chairs: Jian-Jun He, *Zhejiang University, Hangzhou, China*
John Zyskind, *JDSU, NJ, USA*

Slow-Light Workshop
Chair: Yikai Su, *Shanghai Jiao Tong University, Shanghai, China*

Table of Contents

Silicon Photonics And Lasers 1
    John Bowers

Recent Research Activities On Photonic Network Technologies...4
    Yuichi Matsushima

Prospects For Nanophotonics Circuits 7
    Lars Thylén

Advanced Liquid Crystal Displays 10
    Shin-Tson Wu

Advanced Technologies For High Quality LC Display 12
    Yi-Pai Huang, Fang-Cheng Lin, Cheng-Yumr Liao, Ya -Ting Hsu, Wei-Kai Huang,
    Cheng-Han Tsao, Lin-Yao Liao, Chun-Ho Chen, Han-Ping D Shieh

Response Times In Pi-Cell Liquid Crystal Displays..15
    Hongmei Ma, Li Jiang and Yubao Sun

Specialty Fibers As Key Components For Dispersion Management 18
    Hans Damsgaard

Micro/Nano-Scale Optical Circuits And Networks For Information And
Telecommunication Applications...21
    El-Hang Lee

The Transition From Discrete Optics To Optical Integration...24
    A. P. Janssen

Recent Progress On PLC Technologies For Large-Scale Integration...27
    Shinji Mino

Reflective Cholesteric Display: Principle And Progress 30
    Deng-Ke Yang

Polarizer-Free Liquid Crystal Displays...31
    Yi-Hsin Lin, Jhih-Ming Yang, Shin-Tson Wu, Chi-Chang Liao

The Color Temperature Adjusting Method For Multi-Primary Display Using Nonlinear
Programming Problems...34
    Yan Cheng, Xu Liu, Haifeng Li

Amplitude-Sensitive Interferometric Ellipsometer On TN-LCD Optical Parameters
Measurement 37
    H. C. Wei, C. C. Tsai, C. Chou

Biophotonics - A Tutorial Overview 40
    Arthur Chiou

Reduced Dispersion Fiber Extends Reach For Dispersion Tolerant Systems...........................43
    Chris Towery and Tidd Zhang

The Breakthrough Of Specialty Fiber Fabricated By PCVD Based Process 47
Han Qingrong, Tu Feng, Luo Jie and R.Matai

Online RIC Process For G.652.D Fiber Production 50
Ralph Sattmann and Jan Vydra

Optical Ultra-Wideband Pulse Generation Using Air-Guiding Photonic Bandgap Fiber And A Semiconductor Optical Amplifier 53
Shangyuan Li, Xiaoping Zheng and Bingkun Zhou

Polarization Changes Of Partially Coherent Pulses Propagating In Optical Fibers 56
Weihong Huang, Sergey A. Ponomarenko and Michael Cada

Mid-Infrared Optoelectronic Devices And Applications 58
Zhang Yong-gang and Li Ai-zhen

Microwave And Millimeter-Wave Photonic Devices For Communications And Measurement Applications 61
Tadao Nagatsuma and Yuichi Kado

Optically Controllable Millimeter-Wave Oscillator Using Inp-Based Hemts 64
Hiroshi Murata, Noriyo Kobayashi, Toshihiko Kosugi, Takatomo Enoki and Yasuyuki Okamura

Wavelength-Tunable Slow Light Of Fs Laser Pulse By Quadratic Nonlinear Cascading Process 67
Wenjie Lu, Yuping Chen, Lihong Miu, Weirui Dang, Feng Lu, Xianfeng Chen and Yuxing Xia

Characteristics Of All-Optical Ultra-Fast Retiming Switches Using Cascaded Second-Order Nonlinear Effect In Periodically Poled Lithium Niobate Waveguides 70
Yutaka Fukuchi and Joji Maeda

Overview Of Research Activities At The NSF Center For Biophotonics Science And Technology (CBST) 73
Yin Yeh

Least-Invasive Harmonic Generation Microscopy For Intravital Imaging 76
Chi-Kuang Sun

The Purcell Effect Of Silver Nanoshell On The Fluorescence Of Nanoparticles 79
Wallace C.H. Choy, X.W. Chen, S.L. He and P.C. Chui

Nanoparticle-Assisted DNA Nanosensor 82
Xin Li1, Jun Qian, Lili Chen, Ying Zhu, Qun Fang and S. He

DNA Hybridisation Biosensor Based On Dual-Peak Long-Period Grating 85
Xianfeng Chen, Kaiming Zhou, Marcus Hughes, Edward Davies, Lin Zhang, Anna Hine, Kate Sugden and Ian Bennion

Wettability Patterning Technology For Organic Displays 88
Yu-Jin Na, Sung-Jin Kim and Sin-Doo Lee

Synthesis Of High Birefringence Liquid Crystals For Display Application 91
Chain-Shu Hsu and Yung-Ming Liao

Stereo Viewing Zone In Autostereoscopic Display Based On Parallax Barrier...................94
Qiong-Hua Wang, Ren-Liang Zhao, Wu-Xiang Zhao, Da-Hai Li, Yan-Xia Xin and Ai-Hong Wang

Ultrafast Laser Direct-Writing Of Bragg -Glass Photonic Devices 97
G. Marshall, N. Jovanovic, D. Kan, A. Fuerbach, A. Asatryan, L. Botten and M. Withford

Advanced Modulation Techniques In OCDMA System 98
Xu Wang, N. Wada, T. Miyazaki, G. Cincotti and K. Kitayama

2.5Gbps 60km OCDMA Transmission Experiment Using EPS-SSFBG En/Decoder 101
Lin Lu, Weilei Wu, Hui Peng, Tao Pu and Yuquan Li

Experimental Study On The Spectral Behavior Of An Asymmetric Long Period Fiber Grating Via Erosion 104
Minwei Yang, Jianping Chen, Yiping Wang and Xinwan Li

A Review Of The Effects Of High Refractive Index Overlays On Tunable Long Period Fiber Gratings.......................107
J. Lee, Q. Chen, Q. Zhang, and S. Yin

High-Speed Versatile Modulator For Huge-Capacity Transmission.......................110
Tetsuya Kawanishi

Recent Advances In Commercial Electro-Optic Polymer Modulator 113
Bing Li, Raluca Dinu, Dan Jin, Diyun Huang, Baoquan Chen, Anna Barklund, Eric Miller, Merly Moolayil, Guomin Yu, Yun Fang, Lixin Zheng, Hui Chen and Jeevan Vemagiri

All-Optical Inverted Triode Based On Cross-Gain Modulation Using Inas Quantum Dot Semiconductor Optical Amplifiers.......................116
Yoshinobu Maeda, Sayaka Maki, Yasuhiko Kuroki, Hideki Nakayama and Jae-Hoon Huh

Modulation Properties Of Erbium Doped Silicon Laser Diode.......................119
Md. Zahid Hossain, Samia Subrina and Md. Quamrul Huda

In-Line Fiber-Optic Etalon Formed By Hollow-Core Photonic Crystal Fiber 122
Y. J. Rao

Two-Core Fiber Based In-Fiber Integrated Interferometers And Its Sensing Applications 125
Libo Yuan, Jun Yang and Zhihai Liu

A Nonimaging Optics Approach For Photoelectric Sensor Applications.......................128
Jun Jiang

Fiber-Optic Interferometric Temperature Sensor Using A Hollow Fiber 131
Jeung-Hwan Bae, Jaehee Park and Chomsik Lee

Transverse-Load Sensor Based On A Distributed Bragg Reflector Fiber Laser.......................134
Li-Yang Shao, Xinyong Dong and Hwa-Yaw Tam

Bistable Reflective Displays For Paper-Like Displays 137
Liang-Chy Chien

Fabrications Of Mechanically Stable Plastic Liquid Crystal Displays 140
Kwang-Soo Bae, Yoonseuk Choi, Se-Jin Jang, Ji-Hong Bae, Jong-Wook Jung and Jae-Hoon Kim

The Electrolytic Polishing Of Flexible Display Steel Substrate 143
Li Yuqiong, Yu Zhinong, Xue Wei and Leng Jian

The Bending Properties Of Flexible ITO Films 146
Yu Zhinong, Xiang Longfeng, Xue Wei and Wang Huaqing

Characterization Of Polymer Microtip Array Coated Gan Thin Film Using Femtosecond Pulsed Laser Deposition 149
X.L. Tong, H.Q. Wen, D.S. Jiang and L. Liu

Dissipative Solitons For Real World Optical Solitons 152
Philippe Grelu, Ludovic Rapp, Jose M. Soto-Crespo and Nail Akhmediev

Generation Of Energetic Wavelength Tunable Femtosecond Pulses In Higher-Order-Mode Fiber 154
Chris Xu

Phase Noise Tolerant & Real Time Multilevel Homodyne 157
Tetsuya Miyazaki, Moriya Nakamura and Yukiyoshi Kamio

Photonic Crystal And Plasmonic Devices For Photonic Integration 160
Min Qiu

Light Confinement At Interfaces And Talbot Effect Using Optical Surface Modes 162
F.J. García de Abajo, R. Sainidou, T. V. Teperik, M. Dennis and N. I. Zheludev

Optical Polarization Beam Splitting Through Anisotropic Metamaterial Slab Realized By Layered Metal-Dielectric System* 163
Junming Zhao, Yan Chen and Yijun Feng

Asymmetric Hybrid Three-Arm Coupler With Long Range Surface Plasmon Polariton And Dielectric Waveguides 166
Fang Liu, Ruiyuan Wan, Yi Rao, Yuxin Zheng, Yidong Huang, Wei Zhang and Jiangde Peng

Single-Beam Self-Referenced Phase-Sensitive Surface Plasmon Resonance Sensor With High Detection Resolution 169
Shu-Yuen Wu and Ho-Pui Ho

All Fiber Optic Coal Mine Safety Monitoring System 172
Tongyu Liu

Research On Optical Sensor For Pulsed Magnetic Field Measurement 175
Su Yang and Li Yu-quan

Distributed Bragg-Reflector Fiber-Laser Sensor For Lateral Force Measurement 178
Yang Zhang, Bai-Ou Guan and HwaYaw Tam

Application Of Microplasma Modes To A Highly Efficient Light Source For Displays ...181
Kyung Cheol Choi, Seung Hun Kim and Kwan Hyun Cho

Dielectric Superlattice And Its Potential Applications In Display Technology................................184
Yan-qing Lu, Shi-ning Zhu, Yong-yuan Zhu, Yan-feng Chen, Hui-tian Wang and Nai-ben Ming

Organic Light Emitting Devices From OLED To Organic Laser Diode................................187
Chihaya Adachi, Toshinori Matsushima, Hajime Nakanotani, Daisuke Yokoyama and Masayuki Yahiro

Integrated Photonics: Enabling Optical Component Technologies For Next Generation Access Networks................................190
Valery I. Tolstikhin

PLC Based Bi-Directional Optical Module For Access Fiber Networks................................193
N. Kitamura

The Low Cost Single Mode Laser Technology For Mass Deployment................................196
Bo Cai

Tunable Optical Delay Schemes Using All-Optical Processing In A Highly Nonlinear Bismuth Oxide Fiber................................198
Chester Shu and Mable P. Fok

Recent Advances In The Practical Fiber Optical Parametric Amplifiers................................201
K. K. Y. Wong, B. P. P. Kuo, M. E. Marhic, G. Kalogerakis and L. G. Kazovsky

Single Polarisation Fibre Ring Laser By Utilising Intracavity 45° Tilted Fibre Bragg Grating........204
Kaiming Zhou(1), Chengbo Mou, Xianfeng Chen, Lin Zhang, Ian Bennion, Shenggui Fu and Xiaoyi Dong

Brillouin/Erbium Fiber Laser With Pre-Amplified Brillouin Pump Using Ring-Cavity Configuration................................207
1N. Md Samsuri, A. K. Zamzuri and A. Mahdi

Impairment In Amplification Of Optical Packets Regarding The Gain Transient And Nonlinear Effect Depending On Peak Power Of NRZ Payload................................210
Y. Awaji, H. Furukawa and N. Wada

Inp-Based Photonic Integrated Devices................................213
Shinji Matsuo, Hiroyuki Ishii, Toru Segawa, Takaaki Kakitsuka and Hiromi Oohashi

Analysis Of Deep Etched Trench In Planar Optical Waveguide By FDTD Method................................216
Jun Wang, Mingyu Li and Jian-Jun He

Fabrication Of Polymer Integrated Optical Microring Resonator With Photobleaching Method................................219
Jun Zhou, Anan Pyayt, Jingdong Luo, Antao Chen and Nam Quoc Ngo

Image Resolution Analysis Of Different Super Lenses................................222
P. Andalib and N. Granpayeh

Photoacoustic And Thermoacoustic Imaging For Biomedical Applications................................225
Da Xing and Liangzhong Xiang

Optical Coherence Tomography For Oral Cancer Diagnosis................................228
Meng-Tsan Tsai, Hsiang-Chieh Lee, Chih-Wei Lu, Yih-Ming Wang, Cheng- Kuang Lee, Chun-Ping Chiang and C. C. Yang

Raman Signal Enhancement In A Liquid-Core Optical Fiber Based On Hollow-Core Photonic Crystal Fiber 231

Li Huo, Chinlon Lin, Yick Keung Suen and Siu Kai Kong

Time-Of-Flight Laser Spectroscopy In Biomedical Diagnostics 234

Stefan Andersson-Engels, Johan Axelsson, Ann Johansson, Jonas Johansson, Sune Svanberg and Tomas Svensson

Extraction Efficiency Enhancement Of An OLED Using Surface Plasmon Resonance 237

Shou-Yu Nien, Nan-Fu Chiu, Yao-Chou Tsai, Chii-Wann Lin, Kou-Chen Liu and Jiun-Haw Lee

Real-Time Voltage Controlled Color Tunable Oleds 240

Wallace C.H. Choy, C.J. Liang and H.M. Zhang

High-Performance Passive-Matrix OLED Display By Colour Conversion Method 243

Y. Terao, M. Kobayashi, N. Kanai, R. Makino, C. Li, Y. Kawamura, K. Kawaguchi, T. Saito, H. Hashida and H. Kimura

Improvement Of Electrical Characteristics Of Fluorinated Perylene Diimide Thin-Film Transistors By Gate Dielectric Surface Treatment 246

Li-Gong Yang, Jia-Chi Huang, Rong-jin Li, Min-Min Shi, Yan Gao, Mang Wang, Wen- Ping Hu and Hong-Zheng Chen1

Electrochemical Polyaniline/Polypyrrole Composite Film With Novel Nanostructure And High Biosensitivity 249

Yunan Chenga, Gang Wua, Gustaaf Borghsb, Mang Wanga and Hong-Zheng Chena

Overview Of Japanese FTTH Market And NTT's Strategies For Entering Full-Scale FTTH Era 252

Hiromichi Shinohara

On-Line Optical System Performance Monitoring Using Coherent Detection 253

Rongqing Hui

Waveguide Structure Evaluation Based On A Photon-Counting OTDR 256

Takamichi Aiba, Nori Shibata and Masaharu Ohashi

Measurements Of Multimode Fiber PON Bandwidth 259

L. Maksymiuk, G. Stepniak and J. Siuzdak

Novel Technique For Measuring Raman Gain Efficiency Distribution By Conventional OTDR 262

Masaharu Ohashi, Yabu Tetsuro and Ikuo Yamashita*

EPON Deployment Challenges - Now And In The Future 265

Bill McDonald

Fault Location For Fiber Links In PON By Means Of FSF Fiber Laser And Fbgs 268

NianyZuo and Yoshinori Namihira

VCSEL Photonics - Athermalization And Slowing Down - 271

Fumio Koyama

Threshold Analysis Of A Novel Dispersive Grating Distributed Feedback Laser Diode 274

Xun Li, Yanping Xi and Wei-Ping Huang

40 Ghz Self-Pulsation In Two-Section DFB Lasers With Varied Ridge Width 277
Dingbo Chen, Hongliang Zhu, Song Liang, Huan Wang and Yali Zhang

Emission Characteristics Of A Surface-Emitting Organic Photonic Crystal Laser 280
Sidney S. Yang, Li-Wen Chang and Chong-Jie Huang

Uncooled Submarine Pump Laser Module At 980 Nm 283
*Wenjuan Shen, Stefan Mohrdiek, Bing Guo, Tomas Pliska, Mark Ives, Shaun Quinlan,
Warren Grace, Andrew Miller, Thomas Goodall, Jeffrey Greatrex, Robert Cann and
Wen Ma*

**Investigating The Cortical Hemodynamics With High Spatiotemporal Resolution By Optical
Imaging Techniques** 286
Pengcheng Li and Qingming Luo

Photonics And Immobilisation Of Biomolecules 289
M. Durouxab, E. Skovsenab, M. T. Neves-Petersenab, L. Durouxab and S. B. Petersenab

Methane Concentration Monitoring System Based On A Pair Of Fbgs 292
Bin Zhou and Zuguang Guan

**Emerging Fiber-Optic Microendoscopy Technologies For High-Resolution
Biomedical Imaging** 295
Xingde Li

Advances Of Lighting Technologies - From Light Bulbs To Solid State Light Sources 298
Yung S. Liu

Organic Light-Emitting Devices For Solid State Lighting 301
Jie Liu and Anil R Duggal

**In-Situ Fabrication Of Highly-Fluorescent Nanohybrids Based On Carbon Nanotubes And
Gold Nanoparticles** 302
Renjia Zhou, Mang Wang and Hong-Zheng Chen

Planar Lightwave Circuits For FTTH And Photonic Networks 303
Katsunari Okamoto

**Convergence Of Rof And Access Systems Employing Dualparallel Modulator In The
Central Station** 306
Yikai Su and Qingjiang Chang

160 Gbit/S/Port Colored Optical Packet Switching System 309
Naoya Wada

**Modified Duobinary Signals With Tunable Duty Cycle And Its Application In A Label
Switching Optical Network** 312
Yufeng Shao, Shuangchun Wen, Lin Chen, Huiwen Xu and Jin He

**Dual Band Optical Receiver For Video Broadcasting Services Over
Fiber-To-The-Home Network** 314
Young Cheol Kim, Young Ho Jang and Hyun Deok Kim

**Novel Distributed All-Optical Multicast WDM Fiber Network: Design
And Implementation** 317
Dan Lu, Xi Qi, Feng Zhang, Bo Lv, Ming Chen and Shui-sheng Jian

Burst Mode Receiver Based On SOA 320
    Xiaobin Hong, Weiping Huang and Jian Wu

Microring Resonator Devices 323
    Yasuo Kokubun

Athermal AWG Multiplexer/Demultiplexer For E/C-Band WDM-PON Application 326
    Tae Hoon Kim, Byung Gwon You, Hyung Jae Lee and Tae Hyung Rhee

Experimental Demonstration Of Cross-Order Arrayed Waveguide Grating Triplexer 329
    Tingting Lang, Liu Yang, Jing Hu, Zhe-Chao Wang, Zhen Sheng, Jian-Jun He and
    Sailing He

Lateral Leakage In Symmetric SOI Rib-Type Slot Waveguides 332
    Rainer Hainberger, Paul Müllner and Norman Finge

Analysis On Curved Waveguide Grating (CWG) With Rowland Circle Construction 335
    Yinlei Hao, Jianyi Yang, Xiaoqing Jiang, Wei Zheng, Jianying Zhou, Haifeng Zhou and
    Minghua Wang

Paired Surface Plasmon Waves Biosensor 338
    Chien Chou, Hsieh-Ting Wu and Ying-Chang Li

**Temperature-Insensitive Pressure Sensor Using A Polarization-Maintaining Photonic
Crystal Fiber Based Sagnac Interferometer** 341
    H Y Tam, Sunil K. Khijwania and X.Y. Dong

Fiber Bragg Grating Interrogating System Employing An Arrayed Waveguide Grating 344
    Zhou Qinfeng and Xu Tiefeng

Phosphor-Free White-Light Light-Emitting Diodes Based On Ingan/Gan Quantum Wells 347
    Chi-Feng Huang, Chih-Feng Lu, Dong-Ming Yeh, Yung-Sheng Chen, Wen-Yu Shiao and
    C. C. Yang

**Comprehensive Investigation Of Light Emission Of Oleds: From Absolute Optical
Properties To The Purcell Effect** 350
    Wallace C.H. Choy, X.W. Chen, H.H. Fong and S.L. He

Mutual Thermal Effects Of Light-Emitting Diode With Wafer-Level Packages 353
    Jae-Wan Choi, Jeung-Mo Kang, Jae-Wook Kim, Jeong-Hyeon Choi, Du-Hyun Kim,
    Geun-Ho Kim and Jeong-Soo Lee

Enhance The Extraction Efficiency Of Zns:Mn TFEL By Photonic Crystals Structure 356
    Yurong Jiang, Jinwei Li, Xia Li and Wei Xue

A Facile Route To Synthesize Three-Dimensional Cds Nanocrystals 359
    Fei Chen, Renjia Zhou, Mang Wang and Hongzheng Chen

High Power Fiber Sources: More Than Kilowatts 360
    Johan Nilsson

**Dispersion Controlled In A Birefringent Modified Octagon Photonic Crystal Fiber For
Optical Communication Applications** 363
    S. F. Kaijage, Y. Namihira, N. H. Hai, F. Begum, S. M. A. Razzak, T. Kinjo and N. Zou

Full-Vector Effective Index Method For Modeling Endlessly Single-Mode And Large Mode Area Of Photonic Crystal Fiber 366
Wang Liwen, Lou Shuqin,Chen Weiguo and Fang Hong

High Negative Dispersion And Low Confinement Loss Photonic Crystal Fiber 369
Lei Yao, Shuqin Lou, Hong Fang, Tieying Guo, Honglei Li and Shuisheng Jian

Adiabatic Compression Of Quadratic Solitons And Frequency Shift By Using Cascading Nonlinearities 372
Zeng Xianglong, Satoshi Ashihara, Chen Xianfeng, Tsutomu Shimura and Kazuo Kuroda

New Microwave Up-Conversion Solution Using An Optical Phase Modulator In Radio-Over-Fiber Networks 375
Haiyan Ou, Hongyan Fu and Biao Chen

Photonic Frequency Down-Conversion For Millimeter-Wave-Band Radio-Over-Fiber Systems By Directly Modulating A Dual-Wavelength Fiber Laser 378
Shiming Gao, Ying Gao, Hongyan Fu, Daru Chen and S. He

Analysis Of Dispersion Properties In Highly Nonlinear Photonic Crystal Fibers 381
Honglei Li, Shuqin Lou, Hong Fang, Lei Yao and Shuisheng Jian

Micro-Structured Photonic Crystal Fibers With Large Mode Area And High Negative Dispersion 384
Nguyen Hoang Hai, Y. Namihira and Shubi Kaijage

C+L-Band Erbium-Doped Fiber ASE Source Using Dual-Forward Pumping Configuration 387
Wencai Huang, Chaohong Huang, Xiulin Wang, Benrui Zheng, Huiying Xu and Zhiping Cai

Enhancement Of Multi-Wavelength Brillouin-Erbium Fibre Laser Utilizing Fibre Bragg Grating Filter 390
M. N. Mohd Nasir, M. H. Al-Mansoori, H. A. Abdul Rashid, P. K. Choudhury and Z. Yusoff

A Novel Millimeter-Wave Generation Technique For Mm-ROF System Based On Harmonic Generation Principle 393
Meiwei Zhu, Jiajun Ye and Rujian Lin

Fiber Ring Based Microwave Photonic Filters Implemented In A Radio-Over-Fiber Link 396
Kun Zhu, Hongyan Fu and Yun Xiao

Study On Optical Digital Phase Modulation Applied To Millimeter-Wave Radio-Over-Fibre System 399
Meiwei Zhu, Rujian Lin and Jiajun Ye

A Simplified Model Of Multi-Wavelength Fibre Lasers Based On Hybrid Fibre Raman And Erbium Fibre Amplifications 402
Shan Qin and Daru Chen

Lagrange Multiplier Optimization Synthesis Of Long-Period Fiber Gratings 405
Cheng-Ling Lee, Ray-Kuang Lee and Yee-Mou Kao

Feasibility Study Of A Simple 100Gb/S Transmitter With Lowspeed Electronics And 0.8bit/S/Hz Spectral Efficiency 408
Junming Gao, Xinyu Xu and Yikai Su

Controlling Chaos In An Erbium-Doped Fiber Dual-Ring Laser Via Modulating Its Loss And Phase 411
Yan Sen Lin

A Shared Sub-Path Protection Strategy In Multi-Domain Optical Networks 414
Xuejuan Xie, Weiqiang Sun, Weisheng Hu and Jun Wang

Novel Multi-Channel Temporal Phase En/Decoder Used In OCDMA Over WDM PON 417
Ying-xun Zhu, Rong Wang and Tao Pu

Untraditional All-Optical Chromatic Dispersion Compensating Elements - Experimental Verification 420
J. Vojtech, M. Karasek and J. Radil

Simultaneously Realizing Optical Millimeter-Wave Generation And Photonic Frequency Down-Conversion Employing Optical Phase Modulator And Sidebands Separation Technique 423
Hong Wen, Lin Chen, Jing He and Shuangchun Wen

Analysis Of Photonic Band-Gaps Of A Novel PBGF Structure 426
Xiao Yueyu

Application Of Lambert W Function To Raman Fiber Laser 429
Chaohong Huang, Wencai Huang, Xiulin Wang, Huiying Xu and Zhiping Cai

A New Technique For Side Pumping Of Double-Clad Fiber Lasers 432
Wang Da-zheng, Feng Xiao-ming, Wang Yong-gang, Wang Cui-luan, Lan Yong-sheng, Liu Su-ping and Ma Xiao-yu

Design Of A Doubly Grooved Binary Metallic Diffraction Grating For Efficient Side-Pumping Of High-Power Fiber Lasers 435
Fan Zhang, Chuncan Wang, Zhi Tong, Geng Rui and Ning Tigang

Combined FEC/ SOP Scrambling With Delay Line PMD Mitigation Scheme 438
J. Ferreira and J. P. von der Weid

Tailoring Confinement Losses Of Photonic Crystal Fibers 441
Lei Yao, Shuqin Lou, Hong Fang, Tieying Guo, Honglei Li and Shuisheng Jian

Two-Stage Hermite-Gaussian Function Method With Perfectly Matched Layers For Analyzing Microstructured Optical Fibers 444
Xue-Wen Chen, Sailing He, Zhi Wang and P.K.A Wai

Liquid Crystal Optical Modulator Based On In-Plane Switching 447
Yubao Sun and Li Jiang

Analysis Of The Scalability Of The Video-Overlay System 450
Ick Chang Choi, Byoung-Ju Yun and Hyun Deok Kim

An All-Optical Frequency Down-Converter Based On Four-Wave-Mixing In A Highly Nonlinear Fiber For Radio-Over-Fiber Systems 453
Ying Gao, Shiming Gao, Hongyan Fu, Daru Chen and Xiangrui Miao

Shape Influence On The Two-Dimensional Photonic Crystal Devices 456
Ya-Zhao Liu, Shuai Feng, Zhi-Yuan Li, Bing-Ying Cheng and Dao-Zhong Zhang

Tunable Artificial Birefringence In Woodpile Photonic Crystals 459
Ming Che, Zhi-Yuan Li and Rong-Juan Liu

Wavelength Assignment Algorithm For Hybrid WDM/TDM Passive Optical Network 462
Shaofeng Qiu, Weitao zhu, Jun Huang and Zhizhong zhang

Effect Of Source Parameters On Beam Self-Collimation In 2D Photonic Crystal 465
Xia Li, Wei Xue and Yurong Jiang

Improvement Of Automatic Alignment Algorithm For Butterflylaser Module Packaging 468
Hao Shen, Xin Wang, Bernard Leung and Hongdu Liu

Low Loss Performances Of Long Range Surface Plasmon Polariton Waveguides With Buffer Layer Structures 471
Yi Rao, Fang Liu, Yidong Huang, Wei Zhang and Jiangde Peng

Measurement Of Small Aspheric Surface Using Interferometric System For Spherical Surface Test 474
Nam-Young Jang, Pyung-Suk Choi and Jae-Jeong Eun

Design And Realization Of Strip-Loaded Waveguide Electro- Optic Modulators In Barium Titanate 477
Jiansheng Tang, Shujun Yang and Apichai Bhatranand

Evolution Of Partially Coherent Solitons In Optical Lattices 480
Hui Zhuo, Shuangchun Wen and Yonghua Hu

A Proposal For Passive Optical Network Architecture (WDM-PON) Based On Array Of Ring Resonators 483
G. Rostami, R. Faraji-Dana and M. Shahabadi

An Improved Selective Area Growth Method In Fabrication Of Electroabsorption Modulated Laser 486
Huan Wang, Hongliang Zhu, Yuanbing Cheng, Dingbo Chen, Wei Zhang, Liesong Wang, Yunxiao Zhang, Yu Sun and Wei Wang

Experiments And Simulations Of Infrared Transmission By Transverse Electric Mode Through Au Gratings On Silicon With Various Au Widths 489
Yan-Ru Chen and C. H. Kuan

All-Optical Switch In Alkoxysilane Dye Doped Waveguides Based On M-Line Spectroscopy Technique 492
Weirui Dang, Yuping Chen, Rui Wu,Dandan Pan, Xianfeng Chen, Yuxing Xia and Qinghua Meng

Experimental Demonstration Of All Optical Wavelength Full Conversion Based On Quadratic Cascading Effect In Periodically Poled Mgo-Doped Lithium Niobate 495
Junfeng Zhang, Feng Lu, Yuping Chen, Wenjie Lu, Xianfeng Chen and Yuxing Xia

The Application Of The Wavelet Transform To The Continuous Wave Terahertz Imaging 498
Yu Fei, Hui Mei, Song Qian and Zhao Yue-jin

A Novel And Simple Power Splitter Utilizing Two-Branches Of Equal-Frequency Contours Of A Dielectric Periodic Structure 501
Yuan Zhang, Yurong Jiang, Wei Xue and S. He

Modelling And Numerical Analysis Of Carrier Transport Effects On The Wavelength Chirp Of SCH-QW Lasers..................504
Farzan Gity, M. Naser Moghaddasi and Lida Ansari

The Iterative Ranked Phased-Array Method 507
Pojamarn Pojanasomboon and Okan Ersoy

A Rigorous Vectorial Gaussian Beam Modeling Of Virtually-Imaged- Phased-Array 510
A. Mokhtari and A. A. Shishegar

Fabrication And Characterization Of Deeply-Etched Sio2 Waveguides..................513
Zhen Sheng, Liu Yang, Daoxin Dai, Tingting Lang and Zhechao Wang

Transmissive Properties And Faraday Rotation Of Tunable Photonic-Band-Gap System Containing Liquid Crystal..................516
Ping Xu, Lei Gao and Z. Y. Li

Resonance-Induced Transmissions Through Waveguides Below Cut-Off Frequencies: An Effective-Medium Model For Waveguide 519
Hao Xu, Jiaming Hao, Jiajie Dai and Lei Zhou

Modeling And Optimization For Segmented Transmission-Line Electroabsorption Modulators With Asymmetrical Electrodes..................521
Yongbo Tang, Yichuan Yu and Yuqian Ye

Study Of Optical Phased-Array Technology Based On PLZT Electro-Optic Ceramic 524
Qing Ye, Zuoren Dong, Ronghui Qu and Zujie Fang

Controlling Chaos In An Injection Multi-Quantum Well Laser Via Modulating The Injection Light 527
Yan Sen Lin

Analysis And Simulation Of A Channel Add-Drop Filter Composed Of Two Dimensional Photonic Crystal 530
N. Nozhat and N. Granpayeh

Numerical Research On Quality Factor Q Of 2D Photonic Crystal Microcavity With Modulation Of Localized States..................533
Ziqiang Wang, Lieming Li, Dan Wang and Wenbin Cao

Ultra-Fast All-Optical Switch And Its Nonlinear Dynamical Process 536
Ye Liu, Zhi-Yuan Li and Dao-Zhong Zhang

New PON Add/Drop Multiplexer To Support Next-Generation PON 539
Sahrul Hilmi Ibrahim and Abu Bakar Mohammad

A Novel Method To Measure Brillouin Frequency Shift For Brillouin-Based Sensing Application Incorporating A Dual- Wavelength Single-Longitudinal-Mode Fibre Laser..................542
Yizhen Wei, Yongbo Tang, and Daru Chen

Application Of Half-Cycle Phase-Stepping Algorithm In Eliminating Or Diminishing Errors Of Phase Measurement 545
Hui Mei, Yu Fei and Zhao Yue-jin

The Annealing Process Of R.F. Magnetron Sputtered Zno:Al Films 548
Yu Zhinong, Xu Jin, Xue Wei and Li Jinwei

Uniform Color Space For Color Storage..**551**
  Yan Cheng, Xu Liu and Haifeng Li

Chalcogenide Glass Photonic Chips For All-Optical Signal Processing.................................**554**
  V.G. Ta'eed, M.R.E. Lamont, M.D. Pelusi, M.A.F. Roelens, D.J. Moss, S. Madden,
  D-Y. Choi, B. Luther-Davies and B.J. Eggleton

Low Cost Integrated Optical Mux/Demux For LX4 Transceiver..**557**
  Hongtao Han, Jim Morris and Keith Main

**Tuneable Photonic Millimetre Wave Generation Using An Optical Phase Modulator And
DWDM Thin Film Filters**...**560**
  P. Shen, J. James, N. J. Gomes and P. G. Huggard

An Ultrasmall Polarization Rotator Based On Si Nanowire..**563**
  Zhechao Wang and Daoxin Dai

Bistable Device Based On The Kerr Effect In A Microfiber Resonator...................................**566**
  G. Vienne, Ph. Grelu, Y. Li, X. Chen-Perdereau and L. Tong

**Recent Progress In The Integration Of MGY-Based Tunable Lasers And
Mach-Zehnder Modulators**..**569**
  P-J Rigole

Wavelength And Space Switchable Semiconductor Laser..**572**
  Jian-Jun He

**Widely Tunable Slow-Light Delay Line Using Parametricamplification Assisted Silicon
Microring Resonator**..**575**
  Fangfei Liu, Chun Jiang and Yikai Su

Proposal Of A Thermally-Tunable Silicon-On-Insulator Microring Resonator Filter..........**578**
  Daoxin Dai and Liu Yang

**Rotating Linear Differential Polarization Imaging For Quantitative Characterization Of
Superficial Tissues**..**580**
  Xiaoyu Jiang, Wei Li, Tianliang Yun, Nan Zeng, Yonghong He and Hui Ma

**Applications Of Total Internal Reflection Fluorescence (TIRF) Microscopy In
Cellular Bio-Imaging**...**583**
  L. Jin, R.K. Lee, S. K. Kong, W. Yuan, H. P. Ho and Chinlon Lin

**Fluid Sensor Based On Transmission Dip Caused By Mini Stop-Bands In 2D Photonic
Crystal Waveguides**...**586**
  Lei Cao, Yidong Huang, Xiaoyu Mao, Kaiyu Cui, Wei Zhang and Jiangde Peng

**Fabrication And Photochromic Properties Of Ag/Ag+- Codoped Germano-Silicate
Glass Fiber**...**589**
  Aoxiang Lin, Sung-Ho Kim, Youngjoo Chung and Won-Taek Han

Organic Photovoltaics...**592**
  Jiangeng Xue

Development Of Solar Photovoltaic In China..**595**
  Cui Rongqiang, Wang Jianqiang, Ye Qinghao, Yan Shiquan, Shi Yang, Du Jiabing,
  Yang Le, Meng Fanying, Xu Lin and Shen Wenzhong

Control Of Slow Light In Coupled Resonator Optical Waveguide Structures With Highly Dispersive Media 598
  Min Qiu

Slow Light And Its Potential Applications 599
  Li Zhengbin, Peng Chao and Xu Anshi

Delay-Bandwidth Product Of A Novel Slow Light Waveguide 600
  Chun Jiang

All-Fiber Acousto-Optic Tunable Filters 601
  Byoung Yoon Kim

A Simple Implementation Of Tunable All-Optical Microwave Notch Filter With A Negative Tap Based On A Semiconductor Optical Amplifier 604
  Hongyan Fu, Haiyan Ou, Kun Zhu and Sailing He

Demonstration Of Optical Line Terminal For Full Colorless Bidirectional WDM-Passive Optical Networks Using Injection-Locked Fabry Perot Laser And Optical Carrier Suppression 607
  Yong-Yuk Won, Dong-Hyeon Kim, Sang-Kil Roh, Yin-Xing Piao and Sang-Kook Han

Patterned Photonic Crystals For Novel Applications 610
  Y. Ohtera, T. Sato and S. Kawakami

Novel Glasses And Glass-Ceramics For Broadband Optical Amplification 613
  Jianrong Qiu, Shifeng Zhou, Jinjun Ren and Botao Wu

Raman Enhancement Of TO-520cm-1 Mode Of Si By Off-Plane One-Dimensional Grating Etched On Si Substrate 616
  Ling-Chung Choua and C.-H. Kuan

Modeling Of Spontaneous Emission From Erbium Incorporated Silicon Nanocrystal 619
  Samia Subrina, Md. Zahid Hossain and Md. Quamrul Huda

Surface Plasmonic Microscopy For Live Cell Membrane Imaging 622
  Ruei-Yu He, Yuan-Deng Su, Kuo-Chih Chiu, Hua-Lin Wu, Chi-Hung Lin, Guan-Liang Chang and Shean-Jen Chen

Laser Ultrasound Detecting Experiment With Fiber Michelson Interferometer 625
  Zhang Jian-liang, Sheng Xin-zhi, Wu Chong-qing and Zhang Li-jun

Spectroscopic Applications To Environmental Monitoring And Nanobiophotonics 628
  Gabriel Somesfalean

Multiplexing Of Fiber Bragg Grating Pairs For Sensing Based On Optical Low Coherence Technology 631
  Weisheng Liu

Bio-Inspired Nanodevices For Artificial Solar Energy Conversion 634
  Mamoru Nango

Challenges In Luminescent Materials For Lighting And Medical Applications 637
  Cees Ronda

Spectrally Broadened Optical Pump Source Via Phase Modulation For Wideband SBS Slow Light 638
Chester Shu, Alan Cheng and Mable P. Fok

Slow Light In Silicon Nano-Waveguide 639
Fangfei Liu

Storage Capacity Of Slow-Light Based On Fiber Brillouin Amplifiers 640
Li Zhan

This page intentionally left blank.

Proceedings of the 2007 Asia Optical Fiber Communication and Optoelectronics Conference

Pages 320-640

Burst Mode Receiver Based on SOA

Xiaobin Hong (1), Weiping Huang (2), Jian Wu (3)

1: Key Laboratory of Optical Communication and Lightwave Technologies of MOE，Beijing University of Posts & Telecomm., Beijing, China 100876, xbhong@bupt.edu.cn

2: Department of ECE, McMaster University, Hamilton, ON, CA L8S 4K1, huang@ece.mcmaster.ca

3: Beijing University of Posts & Telecomm., Beijing, China 100876, jianwu@bupt.edu.cn

Abstract *In this paper, we propose a burst mode receiver model based on Semiconductor Optical Amplifier (SOA) and PIN photodiode, which operates as both a nonlinear amplifier when receiving a high power packet and a linear amplifier in small signal case. Experimental results show that a 21dB input dynamic range is reduced to 11dB power variation with proper SOA parameters and the signal quality is excellent in higher power operation. This is also a promise design in optical packet switching (OPS) and optical burst switching, where higher bit rate burst mode receiver is required. With this model, the high sensitivity demand to use Avalance Photodiode (APD) is also substituted by PIN.*

Introduction

Passive Optical Networks (PONs) is an effective solution for Fiber to the Home (FTTH). They could become the primary data path for existing and new multimedia services. In a PON system, the passive optical splitters connect an Optical Line Terminal (OLT) and decades of Optical Network Units (ONUs). In order to share the passive optical media, ONUs have to access OLT in Time Division Multiplexing Access (TDMA) mode. TDMA will introduce the optical burst operation mode to the upstream direction in PON [1], which is different from the traditional optical transmission. Burst mode receiver is also employed in Optical packet switching (OPS) [2] and optical burst switching (OBS) [3], which are both hot topic in optical network. Much higher bit rates are usually needed in OPS and OBS systems then PON systems.

In GPON, a sensitivity of −29dBm with 21dB dynamic range is required for the OLT burst mode receiver. APD is usually used to meet the high sensitivity demand and Auto Gain Control (AGC) circuit or auto-threshold level tracking circuit is adopted to adapt the amplitude difference between the different users [4,5,6]. Ota et al. described a receiver that has a 26dB dynamic range at bit rates up to 1.5 Gb/s [7]. A −30dBm sensitivity and 26dB dynamic range was achieved with PIN and AGC for a 1.25 Gbit/s PON system [6]. For higher bit rate, high sensitivity and dynamic range is limited by the thermal noise of PIN, charge and discharge effects of AGC or threshold recovery circuit. A faster AGC or threshold recovery circuit will lead to higher sensitivity penalty [8,9]. It is much more critical in high dynamic range case. So it is difficult to extend this type of receiver to high dynamic range, high sensitivity and high bandwidth at the same time.

SOA was widely studied as a linear preamplifier to improve the sensitivity of PIN. The high performance SOA up to 40Gbps was fabricated [6]. When talking about SOA as the preamplifier, people always try to improve its saturation power [7, 9] to operate it in linear mode. In this paper, the performances of SOA integrated with PIN and operating in nonlinear mode when a high power packet is received and in linear mode in small signal case are analyzed via simulations and experiments. We investigate the variations of the dynamic range and sensitivity of SOA-PIN receiver in the application of 2.5Gbps GPON. The SOA model in ref. [10] and noise model in ref. [11] are used in our simulations.

Results and Discussion

A SOA, Optical Filter and PIN integration is discussed in this paper as shown in Fig. 1. Here we focus on the nonlinear amplification of SOA to reduce the optical power variation in burst mode receiver. The input power varies from -29 dBm to -8 dBm and 2.5Gbps signal are assumed in reference point A in Fig. 1 as required in GPON. In our analysis, 30nm bandwidth of optical filter and 5dB insertion loss from reference point A to B are used in both simulation and experiment. Fig. 2 is the simulated power variation at reference point

978-0-9789217-3-6/07/$25.00

©2007 WEN GLOBAL SOLUTIONS

B. When a SOA with the saturation power of -10dBm and the small signal gain of 30dB is used, the Output power variation of SOA can be reduced to 7dB as shown in Fig. 2.

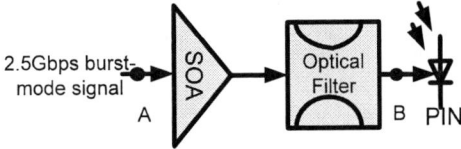

Fig. 1 Structure of burst-mode receiver

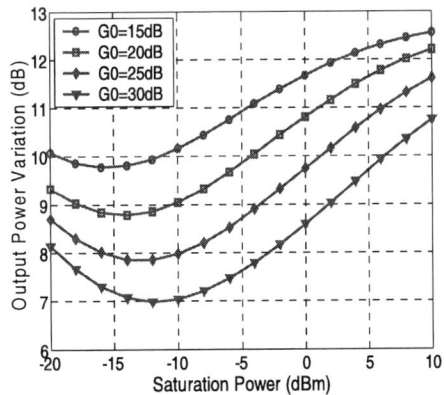

Fig. 2 Variation of signal power in different small signal gain and saturation power of SOA

Fig. 3(a) demonstrates the gain profile of SOA when insertion loss is considered, where small gain G0=25, saturation power=-8dBm are used in both simulation and experiment. Fig. 3(b) gives the change of the output signal power via the input signal power. When the burst signal is launched with a dynamic range of 21dB from -29dBm to -8dBm, the power variation in reference point B is about 11dB, which is less than the simulated result. This is due to the smaller gain in high power in our experiment compared to the simulated gain profile.

Since the SOA operates as a nonlinear amplifier in burst-mode receiver, we have more concerns on the signal quality. The sensitivity (small signal quality) of the SOA + PIN has been studied in [12]. Here we focus on the large signal. Fig. 4(a) shows the Q value and Extinction Ratio (ER) received by PIN. In this figure, ER in reference point A is 10dB and the noise

current in PIN corresponds to a sensitivity of -25dBm without SOA at 2.5Gbps. Note that although the ER decreases with higher input power, the Q value keeps above 6.36, which corresponds to a Bit Error Rate (BER) of 10^{-10}. The experimental ER is 2.5dB higher than the simulated in Fig. 4(a). Because the high power gain of SOA in our experiment is 2.5dB higher than that in simulation, which leads to a higher level of logic "1". As can be seen in Fig. 4(b), the experimental eye diagram in reference point B is excellent. The overshoot in the diagram is caused by the saturation of SOA.

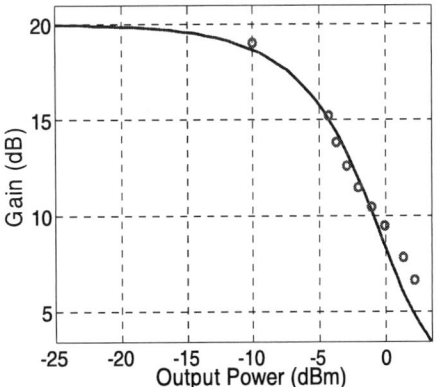

Fig. 3 (a) solid line: simulated gain of SOA, circle: experimental results

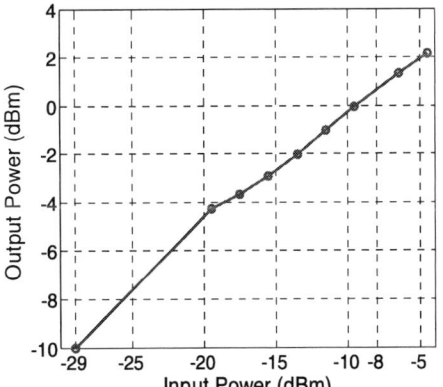

Fig. 3 (b) Experimental results: change of dynamic range before and after SOA

Conclusions

We analyzed a burst mode receiver model based on SOA and PIN photodiode. Simulated and experimental results show that the power variation can be reduced significantly. A signal with 21dB power variation can be reduced to less than 11dB with proper SOA parameters. It is much easier to design the burst-mode Integrate

Circuit (IC) with this model.

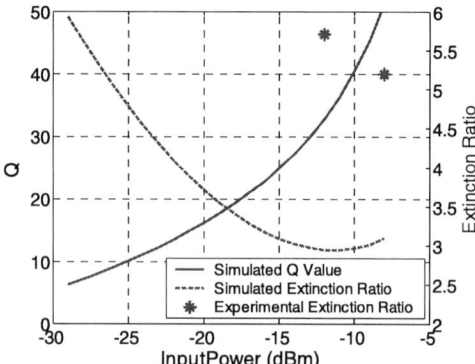

Fig. 4(a) Q value and ER received by PIN when the ER in reference point A is 10dB

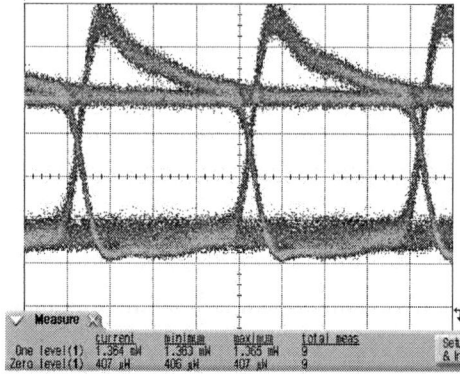

Fig. 4(b) Eye diagram in reference point B when the input power in reference point A is -8dBm.

References

1. ITU-T G.984.2 "Gigabit-capable passive optical networks (GPON): Physical media dependent (PMD) layer specification".
2. S. Yao, B. Mukherjee, and S. Dixit, "Advances in photonic packet switching: an overview," IEEE Commun. Mag., vol. 38, no. 2, pp. 84–94, Feb. 2000.
3. Efraim Rotem and Dan Sadot, Performance Analysis of AC-Coupled Burst-Mode Receiver for Fiber-Optic Burst-Switching Networks, Communications, IEEE Transactions on Volume 53, Issue 3, March 2005 Page(s):545 – 545.
4. Burst-mode receiver for 1.25Gb/s Ethernet PON with AGC and internally created reset signal, Quan Le; Sang-Gug Lee; Yong-Hun Oh; Ho-Yong Kang; Tae-

Hwan Yoo; Solid-State Circuits Conference, 2004. Digest of Technical Papers. ISSCC. 2004 IEEE International, 15-19 Feb. 2004 Page(s):474 - 540 Vol.1.
5. An 1.25Gbit/s -29dBm burst-mode optical receiver realized with 0.35um SiGe BiCMOS process using a PIN photodiode,Chen, Chun-, Tsai, Chia-Ming; Huang, Li-Ren, Proceedings of the 2004 IEEE Asia-Pacific Conference on Circuits and Systems, APCCAS 2004: SoC Design for Ubiquitous Information Technology, 2004, p 313-316.
6. A burst-mode optical receiver with high sensitivity using a PIN-PD for a 1.25 Gbit/s PON system, Nakamura, M.; Imai, V.; Umeda, V.; Endo, J.; Akatsu, Y.; Optical Fiber ommunication Conference, 2005. Technical Digest. OFC/NFOEC, Volume 5, March 6-11, 2005 Page(s):270 – 272.
7. Y. Ota, R. G. Swartz, V. D. Archer, III, S. K. Korotky, M. Banu, and A. E. Dunlop, "High-speed, burst-mode, packet-capable optical receiver and instantaneous clock recovery for optical bus operation," J. Lightwave Technol., vol. 12, pp. 325–331, Feb. 1994.
8. Chao Su, Lian-Kuan Chen, and Kwok-Wai Cheung, Theory of Burst-Mode Receiver and Its Applications in Optical Multiaccess Networks, JOURNAL OF LIGHTWAVE TECHNOLOGY, VOL. 15, NO. 4, APRIL 1997.
9. Ossieur, P.; Xing-Zhi Qiu; Bauwelinck, J.; Vandewege, Sensitivity penalty calculation for burst-mode receivers using avalanche photodiodes, J.; Lightwave Technology, Journal of Volume 21, Issue 11, Nov. 2003 Page(s):2565 – 2575.
10. G. P. Agrawal and N. A. Olsson, "Self-phase modulation and spectral broadening of optical pulses in semiconductor laser amplifiers," Quantum Electronics, IEEE Journal of, vol. 25, no. 11, pp. 2297-2306, 1989.
11. N. A. Olsson, "Lightwave systems with optical amplifiers," Lightwave Technology, Journal of, vol. 7, no. 7, pp. 1071-1082, 1989.

Microring Resonator Devices

Yasuo Kokubun

Yokohama National University, Graduate School of Eng., Tokiwadai, Hodogaya-ku,
Yokohama, Japan 240-8501, email ykokubun@ynu.ac.jp

Abstract *In this review, the fundamental characteristics of microring resonator filters using high index contrast (HIC) optical waveguides are introduced, and the recent progress in the hitless wavelength selective switch achieved by the author's group is mainly described.*

Introduction

High index contrast (HIC) optical waveguides, of which relative index difference exceeds 20%, have recently been attracting attention due to their compactness. We have proposed and demonstrated a vertically coupled microring resonator (VCMRR) as an Add/Drop filter using HIC waveguide[1-3], and recently realized a hitless wavelength channel selective switch[3-7] (hitless tunable Add/Drop filter) using Thermo-Optic (TO) effect of double and quadruple series coupled dielectric micro-ring resonator.

In this review, the principle and recent progress of microring resonator devices and wavelength selective switch (WSS) will be introduced.

Filter response synthesis

For the wavelength filters used in the DWDM system, a flat top pass band and sharp roll-off from pass band to stop band are required. Fortunately, the filter response of the ring resonator can be synthesized by the combination of coupled and cascaded topologies[1,2]. In addition, since the VCMRR is advantageous for dense integration due to its very compact element and crossgrid topology, it is easy to form the coupled and cascaded topologies

without deteriorating the compactness.

Fig.1(a) shows a vertically quadruple series coupled microring resonator with multilevel-crossing busline waveguides. This structure consists of two core layers; both layers involve the busline waveguide and the microring resonator and those waveguides are coupled to each other in a vertical direction. Therefore, the scattering will not occur at the crossing point. The measured filter response of the quadruple series coupled microring resonator is shown in Fig.1(b).

Buried Vacuum Cladding Structure

In the structure shown in Fig.1(a), there remains the problem that the radiation loss of the ring resonator increases due to the decrease of index contrast, because the ring core is covered by the cladding material. To solve this problem, we developed a novel fabrication process that enables the stacked dense integration of VC-MRR buried by SiO_2 cladding with a vacuum side cladding[2], as shown in Fig.2(a).

Using this technique, a single-ring resonator with a 5μm ring radius was fabricated and a clear drop port filter response was successfully observed. The numerical calculation shows that the radiation loss can be reduced to one-tenth of that of conventional buried structure.

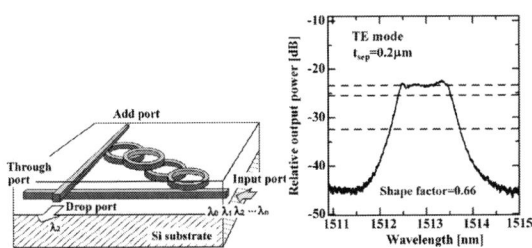

(a) Perspective view   (b) Measured filter response

Fig.1   Quadruple series coupled microring resonator filter with multilevel-crossing busline waveguide.

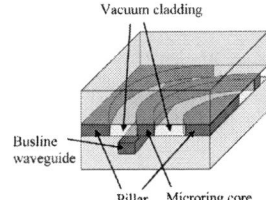 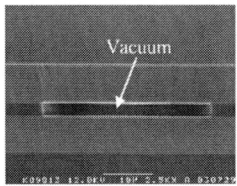

a) Perspective view   (b) SEM cross sectional view

Fig.2   Vertically coupled microring resonator with buried vacuum structure as side cladding.

978-0-9789217-3-6/07/$25.00

©2007 WEN GLOBAL SOLUTIONS

Center Wavelength Trimming by UV Irradiation

Since the center wavelength of 1x8 microring filter array was deviated by 3.7nm from the designed value, the accuracy of the center wavelength was not high enough for DWDM systems. To solve this problem, we developed a novel trimming technique by irradiating UV laser beam to the microring core made of SiN and SiON.

Utilizing this technique, we demonstrated wide-range (-12.1nm) trimming of the center wavelength of a VCMRR as shown in Fig.3(a). The long term stability of UV trimming was confirmed for 1,200hours as shown in Fig.3(b).

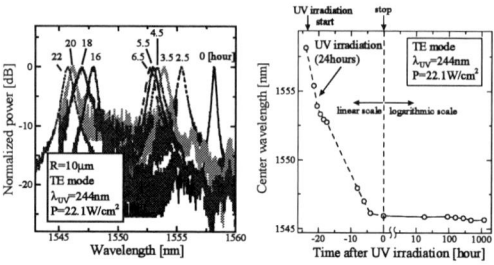

a) Drop port response   (b) Change of center wavelenfgth

Fig.3 UV trimming of resonant wavelength of VCMRR made of SiON.

Principle of hitless wavelength selective switch

Using the thermo-optic effect, we can control the resonant wavelength of microring resonator. We have demonstrated a wide range tuning of 9.4nm using a large thermo-optic effect of polymer material. However, this tunable filter has a problem of blocking other wavelength channels during the tuning as shown in Fig.4(a). In addition, the direction of wavelength shift is limited to shorter wavelength side, corresponding to the negative TO coefficient of polymer core.

To solve this problem, we proposed and demonstrated a hitless wavelength selective switch, of which the operation principle is shown in Fig.3(b). This device consists of a series coupled tunable microring resonators, as shown in Fig.5. When resonant wavelengths of resonators are made equal by controlling individual resonant wavelengths of a series coupled microring resonator, only the resonant wavelength channel can be transmitted to the drop port

(ON-state). On the other hand, when resonant wavelengths of resonators are different, no wavelength channel transmits to the drop port (OFF-state). Therefore, the resonant wavelength can be shifted to another wavelength channel without blocking other wavelength channels.

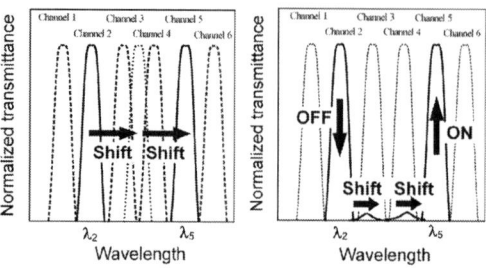

(a) Conventional tunable filter   (b) Hitless wavelength selective switch

Fig.4 Comparison of operational trajectory of resonant peaks.

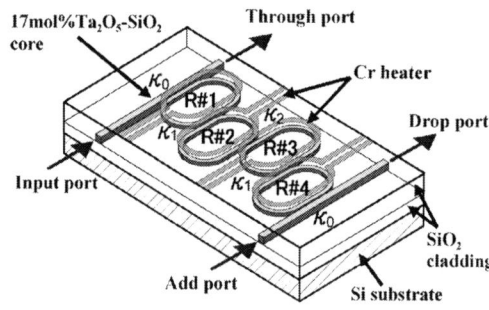

Fig.5 Structure of hitless wavelength selective switch using series coupled tunable microring resonators.

Demonstration of elements of WSS matrix

We are aiming at demonstrating N input, N output, and the M wavelength channel full matrix switch as shown in Fig.6. First, we fabricated a multi-wavelength channel selective switch by cascading three of the M WSS's using quadruple series-coupled microring resonators as shown in the inset of Fig.7.

By applying electric current to each Cr thin film heater above individual ring resonators separately, we measured multi-wavelength switching characteristics. When the electric current was changed, the change of measured through port responses are shown in Fig. 7. In this figure, measured through port responses

324

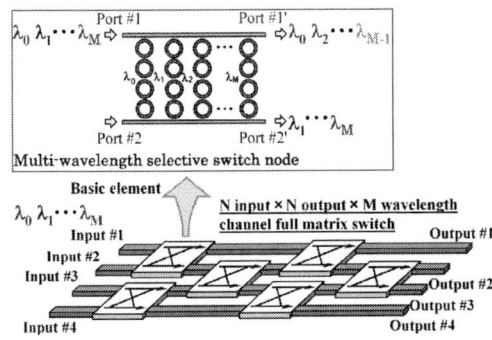

Fig.6 Full matrix optical switching circuit using hitless wavelength selective switch.

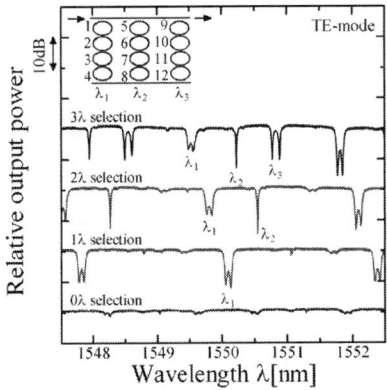

Fig.7 Multi-wavelength channel selective switching spectra from three wavelengths to zero wavelength of through port response.

are displayed by shifting four spectra vertically by 15dB to avoid the overlapping of curves. When each wavelength channel was selected (ON-state), the measured dip of through port response was greater than -10dB. This value was equal to the extinction ratio. In the measured zero wavelength selective switching characteristics, the loss of OFF-state in the through port response was about 1.35dB and this value was greatly improved compared with that of double series-coupled microring resonator (3dB-3.5dB).

On the other hand, in the drop port response, the extinction ratio and the switching crosstalk were 51.8dB and -23.9dB, respectively. These characteristics were also much better than those of double series-coupled microring resonators (27.2dB and -8.4dB)[6]. However, unexpected wavelength shift occurred for λ_1 ($\Delta\lambda_1$=0.58nm) and λ_2 ($\Delta\lambda_2$=0.30nm), resulting from the thermal interference. The

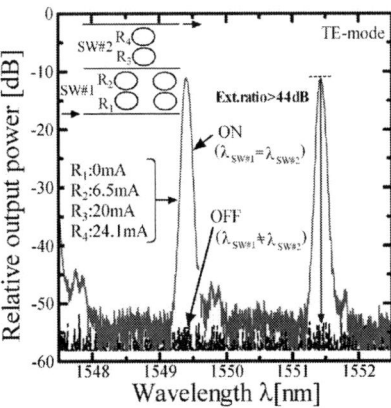

Fig.8 Three port switching spectrum from input port#1 to output port#3.

thermal interference will be reduced by optimizing the layout of electrode and by incorporating buried vacuum cladding structure[2] as the thermal isolation region.

Lastly, the measured the three port switching characteristics in the ON and OFF states of double series coupled WSS are shown in Fig.8. The extinction ratio was measured to be greater than 44dB. The shape factor was 0.3, and the FWHM bandwidth was 0.082nm, which was 20% narrower than that of single-stage response.

Conclusions

Since this hitless WSS has a feature of scalable integration, multi-wavelength and multi-port wavelength selective switch matrix with more than four wavelengths will be possible.

References

1 S. T. Chu, *et al*, Photon. Technol. Lett., vol.11, no.6, pp.691-693, 1999.

2 Y. Kokubun, (invited paper) IEICE Trans. on Electronics, vol.E88C, No.3, pp.349-362, 2005.

3 Y. Kokubun, (Invited paper), IEICE Trans. Electronics, vol.E90-C, no.5, pp.1037-1045, 2007.

4 S. Yamagata, *et al*, Electron. Lett., vol.41, No.10, pp.593-595, 2005.

5 Y. Goebuchi, *et al*, Photon. Technol. Lett., vol.18, no.3, pp.538-540, 2006.

6 Y. Goebuchi, *et al*, Photon. Technol. Lett., vol.19, no.9, pp.671-673, 2007.

7 Y. Goebuchi, *et al*, OECC/IOOC2007 Yokohama, 12E3-2, July 9-13, 2007.

Athermal AWG Multiplexer/Demultiplexer for E/C-Band WDM-PON Application

Tae Hoon Kim, Byung Gwon You, Hyung Jae Lee and Tae Hyung Rhee

POINTek, Inc., 506-1 Kumam, Seotan, Pyeongtaek, Gyeounggi Do, R. O. Korea, 451-852

Email: thrhee@pointek.co.kr

Abstract *Athermal AWG multiplexer/demultiplexer prepared for E/C-band WDM-PON application, by a novel and practical psudo-passive alignment packaging method, demonstrates <±30pm center wavelength shifts at ITU grids over the widest operating temperature range of -40°C~+80°C.*

Introduction

The AWG multiplexer/demultiplxer became a key device in the DWDM system [1]. However, thermal sensitivity of the conventional AWG limited wide spread application in the optical access system. Increasing demands of bandwidth and network flexibility have been seeking AWG with athermal characteristics and low cost. Thus, athermal AWG multiplexers are drawing considerable attractions from the Metro/ROADM systems. The growing worldwide trend in the convergence of communication and broadcasting has demanded broadband requirement to each home, resulting in renewed attention to a bi-directional athermal AWG for DWDM system based passive optical network (WDM-PON) [2].

To date, two basic concepts have been developed for athermal AWG: polymer-filled trench method and mechanical compensation method. In the former method, a specific part of AWC chip is trenched precisely and then these trenches are filled by a polymer with opposite thermal coefficients to the doped-silicate glass [3,4,5]. With the temperature-compenating polymer, the temperature dependence in silica waveguide AWG is cancelled out, thereby maintaining the center wavelength of the AWG. In the latter method, the center wavelength is maintained by the mechanical compensation adjustment of the input waveguide launching position to the slab waveguide [6,7]. This idea has long been applied in the AWG chip design to adjust the initial center wavelength easily by selecting a specific input waveguide position. The mechanical compensation method has been generally known to provide better athermal performance,

especially in the low operating temperatures, than the other method.

Experimental

For the athermal AWG development, we advanced the mechanical compensation method further in order to overcome the current technical difficulties of the athermalization processes. Our innovations include: 1) athermal operation within ±50pm (from ITU grids) center wavelength shift over the widest operating temperature range of -40°C~+80°C; 2) a novel method designed to minimize optical performance degradation of the AWG chip during the athermalization process; 3) use of conventional AWG chip for both Gausian and Flat-Top operation; 4) cost-effective pseudo-passive alignment packaging process; and 5) advanced assembly process leading the secured reliability.

The standard silica-on-silicon AWG chips were supplied from a leading PLC chip manufacturer. Each inspected AWG chip with 8° polished angles was pigtailed with the corresponding input and output fiber arrays on the automatic active alignment station. The pigtailed AWG chip was then diced at the interface between the input waveguide and the first slab waveguide to divide into two parts: the input waveguide sub-chip part and the AWG sub-chip part. The diced AWG sub-chip was attached on a top surface of a base plate, and the input waveguide sub-chip was placed on the same surface. Two thin plastic films, as spacers, were vertically inserted between the two sub-chips, and the input waveguide sub-chip was pushed to the AWG sub-chip as the thin films were being inserted. Then, the input waveguide sub-chip was moved

978-0-9789217-3-6/07/$25.00

©2007 WEN GLOBAL SOLUTIONS

laterally to adjust the center wavelength. With this scheme, the five optical axes among all six axes were aligned passively, and only one lateral axis was simply actively adjusted without using any alignment tool. Then, a metal rod part with a certain length was attached on both the base plate and the input waveguide sub-chip together to secure the center wavelength of the athermal AWG module. Finally, the athermal AWG assembly was placed in a package and the optical and thermal performances were measured. Fig. 1 shows the schematic drawing of the packaged athermal AWG.

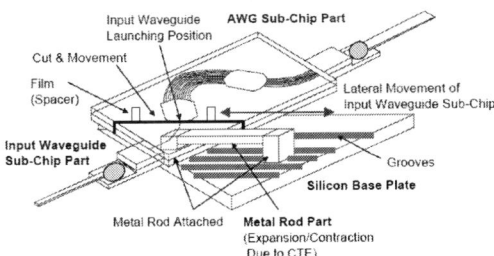

Fig. 1. Schematic Diagram of the Packaged Athermal AWG

Securing the reliability in terms of both center wavelength shift and insertion loss variation is one of the advantages resulted from the proposed packaging structure. The proposed assembly structure was reliable because: 1) the AWG sub-chip was attached on the base plate followed by being aligned with the input waveguide sub-chip on the surface of the base plate, and 2) one edge of the temperature compensation rod was attached on the top of the base plate while the other edge was attached at the top of the input waveguide sub-chip. As a result, two separate sub-chips were firmly aligned through the same base plate and the same temperature compensation metal rod. Therefore, the optical misalignment between two sub-chips at the launching position was minimized and repeatability of the launching position at the given temperatures was maintained because of the proposed assembly structure.

Results and Discussion
For the optical measurement, the following equipments were used: dBm Optics CAS 2004 spectrum analyzer with a built-in wave reference, Agilent 81640/81480 tunable

laser sources, and Agilent 8169 polarization controller. A 32-channel 100GHz Gaussian athermal AWG in Fig. 2 (Top) demonstrated <0.5pm/°C athermal performance at -30°C ~+70°C. It performed less than ±30pm center wavelength shifts of the temperature dependence from the ITU grid. Another 32-channel 100GHz Gaussian athermal AWG shown in Fig. 2 (Bottom) demonstrated less than ±30pm center wavelength shifts from the ITU grids over -40°C~+80°C operating temperature range. This is the first 100GHz AWG module to report the fully passive athermal operation over such widest operating temperature range, and it is able to cover the all applications including long-haul network, metropolitan network, WDM-PON and even extended temperature operations.

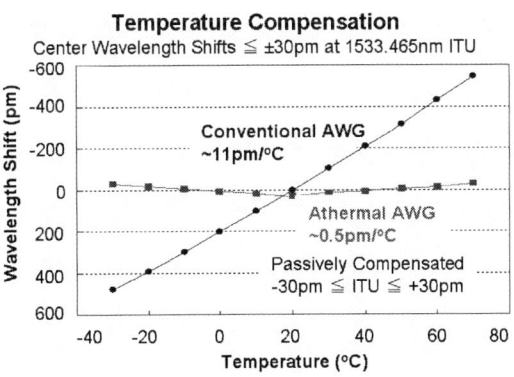

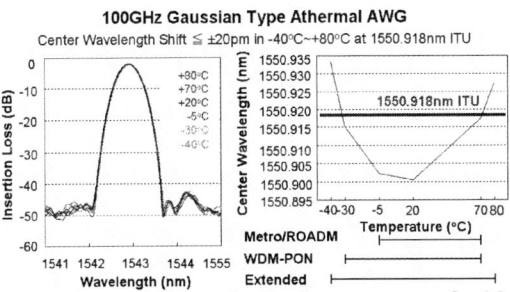

Fig. 2. Athermal Performance of 32-Channel 100GHz Gaussian Athermal AWG

A 16-channel 200GHz Semi-Flat AWG was prepared for E/C-band bi-directional WDM-PON application and was set for the operating temperature of -40°C~+80°C. The optical performances were excellent that 1dB and 3dB passbands in both C/E-bands were wider than 750pm and 1050pm, respectively. The worst-case insertion loss at the ±200pm clear passband was -4.4dB including SC/APC connectors in C-band and -5.1dB in E-band, respectively. Also,

327

the worst-case isolation characteristics of -32dB (adjacent), -35dB (nonadjacent), and -26dB (total) were measured. The worst-case center wavelength difference between measured value and ITU or specification in all 16 channels was -24pm in C-band, and -33pm in E-bands, respectively. The worst-case ripple was -0.35dB for both bands and PDL was -0.2dB. Center wavelength shifts in both E-and C-bands over -40°C~+80°C was maintained within ±50pm from the all ITU grids with the worst polarization case. Therefore, it is another first reported result that an E/C-band bi-directional WDM-PON 200GHz AWG multiplexer operates in the widest operating temperature range within ±50pm of center wavelength shift from ITU grids. Fig. 3 demonstrates the athermal operating performance of the 200GHz Semi-Flat athermal AWG multiplexer over wide -40°C~+80°C operating temperature range and the optical spectrum in both C/E-bands.

temperature range of -40°C~+80°C for the first time. The proposed packaging method turns out to be very suitable for building both Gaussian and Flat-Top athermal AWG modules with sufficient environmental reliability. In addition, this is the first report that a bi-directional WDM-PON 200GHz athermal AWG operates within ±50pm of center wavelength shift from all ITU grids over -40°C~+80°C temperature range in E/C-band. Therefore, the proposed method provides performance and reliability to any athermal AWG in order for meeting all DWDM related indoor and outdoor applications at wide -40°C~+80°C operating temperature range.

References

1 K. Okamoto, et al., Opt. Lett., 20 (1995), 43-45

2. S. J. Park, et al., Lightwave Technol., 22 (2004), 2582-2590

3. Y. Inoue, et al., Electron. Lett., 33 (1997), 1945-1946

4. A. Kaneko, et al., OFC-IOOC '99 Tecnical Digest, San Diego (1999), 204-206

5. K. Maru, et al., OFC '2000 Technical Digest, Baltimore, 2 (2000), 130-132

6. G. Heise, et al., Proc. ECOC '98, Madrid (1998), 319-320

7. T. Saito, et al., Furukawa Review, 24 (2003), 29-33

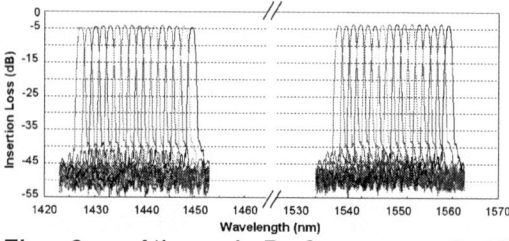

Fig. 3. Athermal Performance of 16-Channel 200GHz Semi-Flat Bi-directional Athermal AWG for C/E-Band WDM-PON

Conclusions

In this presentation, the athermal AWG multiplexers for outdoor applications, prepared by a novel and practical psudo-passive alignment packaging method, demonstrate <±30pm center wavelength shifts at ITU grids over the widest operating

Experimental Demonstration of Cross-Order Arrayed Waveguide Grating Triplexer

Tingting Lang, Liu Yang, Jing Hu, Zhe-Chao Wang, Zhen Sheng, Jian-Jun He, and Sailing He

State Key Laboratory of Modern Optical Instrumentation, Centre for Optical and Electromagnetic Research,
Zhejiang University, Hangzhou, PR China 310058
ttlang@coer.zju.edu.cn

Abstract *Experimental results on a cross-order arrayed waveguide grating triplexer for fiber-to-the-home applications are presented. The cross-order design allowed us to fabricate compact devices of a size of only 19mmx1.6mm. The measured spectra confirmed the operation principle of the device.*

Introduction

Triplexers are important components in fiber-to-the-home (FTTH) systems that employ an analog overlay channel for video broadcasting in addition to bidirectional digital transmissions [1,2]. According to ITU G.983 standards, a triplexer needs to multiplex or demultiplex three wavelength channels over a wide wavelength range, i.e. 1310nm for upstream digital signals, 1490nm for downstream digital signals and 1550nm for downstream analog signals.

Arrayed waveguide grating (AWG) multiplexers have been widely used in wavelength division multiplexing systems. They have many advantages such as high performance and high reliability. Unfortunately, for the triplexer application, the large spectral range requires the AWG to be designed and operated at a low diffraction order. Such a low-order AWG is difficult to realize, due to the small path length difference needed between adjacent arrayed waveguides. Recently, we proposed a cross-order design which can reduce the free spectral range (FSR) requirement, resulting in a smaller device size. In this paper, we report our first experimental results on the cross-order AWG triplexer.

Device design and simulation results

The cross-order AWG is designed so that Channel 1 at 1310nm works at a higher diffraction order than the other wavelength channels [3]. Suppose the diffraction order for Channel 2 at 1490nm and Channel 3 at 1550nm is m and that for Channel 1 at 1310nm is m', with $(m'-m)$ being a small integer. Because of the periodicity of the spectral response, the output waveguide for Channel 1 also has a response at the wavelength corresponding to the mth diffraction order, which is called the dummy channel. This wavelength is denoted by λ_1'. And the following equation is satisfied simultaneously for Channel 1 at λ_1 and the dummy channel at λ_1'

$$n_a(\lambda_1)\Delta L=m'\lambda_1 \qquad (1a)$$
$$n_a(\lambda_1')\Delta L=m\lambda_1' \qquad (1b)$$

where n_a and ΔL are, respectively, the effective index

and the path length difference of the arrayed waveguides. We choose the dummy channel to be substantially in the middle of Channels 2 and 3. In this case the FSR only needs to cover Channels 2 and 3. In order to have a sufficient margin in the FSR for reducing channel nonuniformity, we choose m' =14 and m =12. This leads λ_1' = 1524.2nm and ΔL = 12.6μm.

Fig. 1 is the layout of the designed AWG, and the inset is the expanded view of the output waveguides near the free propagation region (FPR) with the corresponding channel wavelengths indicated. The total size of the AWG is 19.2mm x 1.6mm.

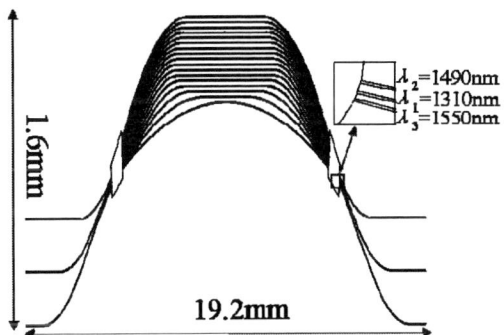

Fig. 1: Layout of the AWG with the vertical axis scaled up by a factor of 10.

The three-dimensional structure of the AWG is converted to a two-dimensional structure using effective index method (EIM). Two-dimensional beam propagation method (BPM) is then used to perform the numerical simulation. The spectral responses of the AWG for the three output waveguides are shown in Fig. 2. One can see that the central channel has a response peak at 1310nm as well as at the dummy channel wavelength of 1524.2nm.

The insertion loss, the peak wavelength and the 3-dB bandwidth of the three output waveguides simulated by BPM are shown in Table 1. The non-uniformity, as defined by the largest loss difference among the three operating wavelength channels, is 0.7dB.

978-0-9789217-3-6/07/$25.00

©2007 WEN GLOBAL SOLUTIONS   329

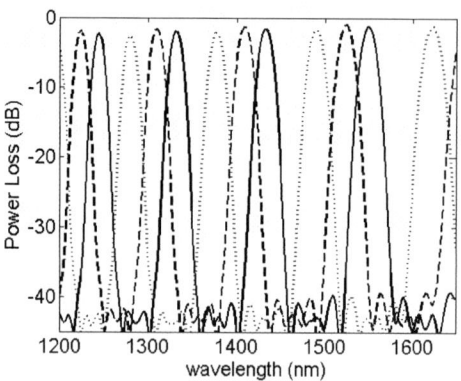

Fig. 2: Simulated spectral responses.

The bandwidth was increased by using a tapered input waveguide at the expense of an excess loss of about 0.6dB. The 3-dB bandwidths range from 12.4nm for 1310nm to 15.4nm for 1550nm. This variation is due to the fact that the width of the waveguide mode profile increases with the wavelength.

Table 1

Channel number	OW1	OW2	OW3
Wavelength (nm)	1490.0	1524.2	1550.0
Insertion loss (dB)	-1.69	-0.99	-1.21
3dB bandwidth(nm)	15.1	15.3	15.4
Wavelength (nm)	1281.5	1310.0	1332.5
Insertion loss(dB)	-2.60	-1.61	-1.97
3dB bandwidth(nm)	12.3	12.4	12.4

Fabrication

The device is fabricated in silica-on-silicon material systems with a 6μm x 6μm buried channel waveguide structure. First, a 12μm-thick undoped SiO_2 buffer layer is deposited on a 4-inch-diameter silicon wafer, followed by a 6μm-thick Ge-doped waveguide core layer. The layers are deposited by plasma enhanced chemical vapor deposition (STS Multiplex PECVD) at about 300°C using silane, germane and nitrous oxide as the processing gases. The deposition rate is approximately 170nm/min for the buffer and 185nm/min for the germanium (Ge) doped core layer. Using a prism coupler (Metricon model 2010), the refractive index of the buffer layer is measured to be 1.468 and 1.455 at 632.8nm and 1547nm wavelengths, respectively. The refractive index of the core layer is 0.87% higher than that of the undoped SiO_2. The dispersion coefficient of the effective index of the buried channel waveguide is derived to be -1.82×10^{-5}/nm.

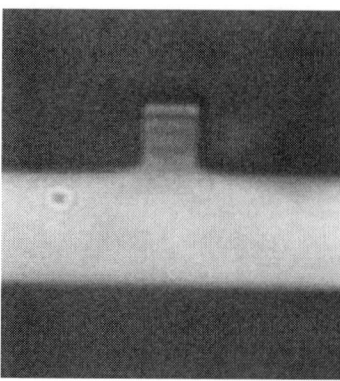

Fig. 3: Microscopic picture of the etched waveguide showing good sidewall verticality.

The AWG is etched by inductivity coupled plasma etching (STS Multiplex ICP) using 200nm Ti/Ni as the hard mask, which is patterned using the lift-off technique. The etch depth is about 6.5μm and the etch rate is 160nm/min. The cross-section view of the etched waveguide is shown in Fig. 3, which is observed using an optical microscope with 800x magnification. Good sidewall verticality has been achieved. After removing the remaining metal mask, a 9μm-thick undoped SiO_2 upper cladding layer is deposited by PECVD.

Measurement results and discussions

The insertion loss of a reference straight waveguides is measured to be about 3.0dB. The coupling loss between the waveguide and the single mode fiber is estimated to be 0.5dB/facet. Therefore, the propagation loss is about 1.0dB/cm. This is high because the upper cladding is undoped and no thermal annealing is used.

Fig. 4 shows the spectra responses of the triplexer, measured using an optical spectrum analyzer (OSA, Agilent 86142B) and two broadband light sources centred at 1280nm and 1550nm. The resolution and the sensitivity of the OSA used in the measurement are 0.06 nm and -80dBm, respectively. The loss of the reference straight waveguide has been subtracted. The responses below 1500nm in Fig. 4(a) and above 1340nm in Fig. 4(b) are distorted because the light sources are very weak in this wavelength region and the OSA has a limited dynamic range. The insertion loss, the central wavelength and the 3-dB bandwidth of the three output channels measured using the two broadband light sources are shown in Table 2. The relative wavelength positions of the three output waveguides (OWs) agree quite well with the simulation results.

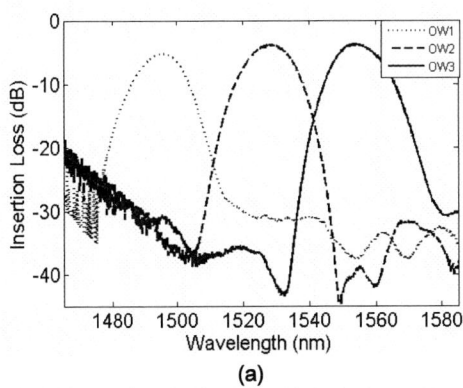

(a)

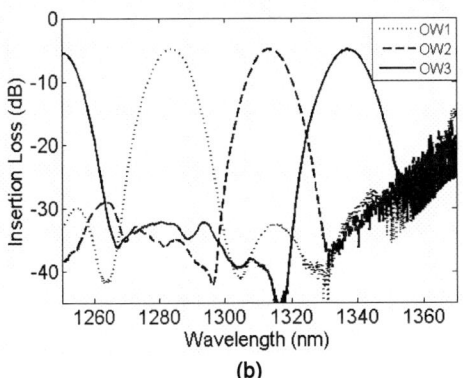

(b)

Fig. 4: Spectra responses of the triplexer measured using broadband light sources at 1550nm (a) and 1280nm (b).

Table 2

Channel number	OW1	OW2	OW3
Wavelength (nm)	1495.2	1528.2	1554.5
Insertion loss (dB)	-5.1	-3.7	-3.7
3dB bandwidth(nm)	14.0	14.1	14.5
Wavelength (nm)	1283.7	1313.4	1336.8
Insertion loss(dB)	-4.9	-4.9	-5.0
3dB bandwidth(nm)	10.4	10.5	11.2

The crosstalk is about -25B. The insertion loss of the triplexer is much larger than the propagation loss derived from the loss of the reference straight waveguide. In addtion to the fact that the input waveguide is tapered for passband flattening which introduces theoretical losses as given in Table I, there are several other possible reasons. First, there may be air gaps between the upper cladding and the core, especially near the interfaces between the arrayed waveguides and the slab waveguides. Another problem is that the channel waveguides seem to be narrower than the designed value due to photolithography inaccuracy, which results in the gaps between tapered arrayed waveguides at the grating circle to be larger than the designed value (2μm). This also causes a smaller 3dB bandwidth than the simulation results.

Conclusions

First experimental results on a cross-order arrayed waveguide grating triplexer are presented. The operating principle of the cross-order AWG design is demonstrated. The wavelength channel at 1310nm is successfully transposed to a dummy channel substantially in the middle of the channels at 1490nm and 1550nm. Central wavelengths can be adjusted easily with a new mask design. The device still needs to be improved in terms of loss and crosstalk.

Acknowledgement:

The authors would like to thank Dr. D.X.Dai, Dr. Liu Liu, Rui Hu, Ting Chen, and Y.X. Cui for their advice and help with the experiments. This work was partially supported by Natural Science Foundation of Zhejiang Province (under grant No. X106875).

References

1 X. Li et al IEEE Photon. Technol. Lett., 17(2005), page #1214-1216

2 C. L. Xu et al Optics Express, 14(2006), page #4675-4686

3 T. T. Lang et al IEEE Photon. Technol. Lett., 18(2006) page #232-234.

Lateral leakage in symmetric SOI rib-type slot waveguides

Rainer Hainberger, Paul Müllner, Norman Finger

Austrian Research Centers GmbH - ARC, Donau-City-Str. 1, 1220 Vienna, Austria

rainer.hainberger@arcs.ac.at

Abstract: We theoretically investigate the lateral leakage of rib-type slot waveguides caused by coupling between the TM-like slot mode and a TE slab mode using a finite element method eigenmode analysis and the variational mode-matching method.

Introduction

Slot waveguides have attracted much attention during the past few years [1]. In a slot waveguide structure, guided light can be strongly confined in a narrow low index gap between either two high index photonic wires (wire-type slot waveguide) or two high index rib waveguides (rib-type slot waveguide) [2] (see Fig. 1). The slot effect occurs only for the TM-like polarization, i.e. the polarization with the major electric field component lying perpendicular to the slot interface. The strong light confinement in the slot enables the realization of novel silicon photonic devices in which the characteristics of active optical low-index materials can be efficiently exploited.

In particular, the horizontal rib type slot waveguide is highly attractive because it facilitates electrical contacting of the upper and lower silicon layer. However, rib-type slot waveguides can suffer from leakage of light into the outer slot waveguide slab due to TM-TE mode coupling which renders them practically useless. In this study, we theoretically investigate this important issue of lateral leakage in rib-type SOI slot waveguides at the wavelength 1.55 μm.

TM-TE mode coupling

In conventional rib waveguides, the effect of lateral leakage of the TM-like rib waveguide modes occurs if their effective index is smaller than that of the TE slab mode

outside the rib [3,4]. In this case, the minor TE components of the TM-like mode are not totally reflected at the rib interface and leak into the slab. For TE-like rib modes, this effect does not occur at all, because their effective index is always larger than that of the TE- and TM-slab modes.

In rib-type slot waveguides the situation is slightly different. The major electric field component of the TM-like slot mode has symmetric orientation in vertical direction, whereas the orientation of the minor electric field component (TE) is antisymmetric (see Fig. 2). In a laterally infinite slot waveguide, later on called slot waveguide slab, there are two fundamental TE modes [5]: a symmetric mode with the electric field vectors in the upper and lower waveguide layer being parallel, and an antisymmetric mode with the electric field vectors in the upper and lower waveguide layer being anti-parallel.

Fig. 3a compares the effective indices of these modes. The effective index of the TM-like slot mode was calculated numerically using the variational mode-matching (VMM) method [6], whereas the effective indices of

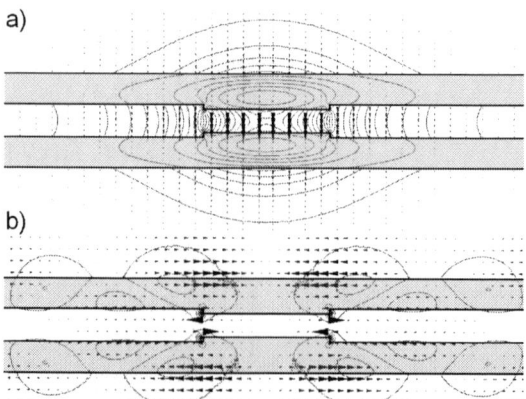

a)

b)

Fig. 2: Schematic optical power (contour) and electric field (arrow) distribution of the TM-like slot mode: a) overall, b) minor (TE) components.

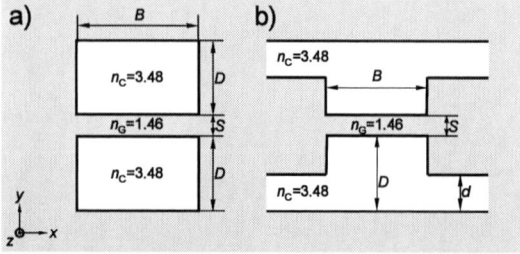

Fig. 1: a) Wire-type and b) rib-type slot waveguide.

the symmetric and antisymmetric TE slab modes were determined semi-analytically by solving the transcendent eigenmode equation of a five layer slot waveguide slab. For a relative slab thickness d/D larger than 0.6 the effective index of the antisymmetric TE slab mode is higher than that of the TM-like slot mode. Thus, the minor field components of the TM-like slot mode can couple to the antisymmetric TE and leakage occurs. Fig. 3b plots the corresponding leakage loss values obtained from the imaginary part of the effective index of the VMM simulations. An abrupt change of the leakage loss can be observed at the value of d/D at which the effective indices of the TM-like mode the TE slab mode become equal. Because of symmetry reasons for a perfectly symmetric rib-type slot waveguide, no coupling can occur to the symmetric TE slab mode, even though the effective index of that mode is always higher than the one of the TM-like slot mode.

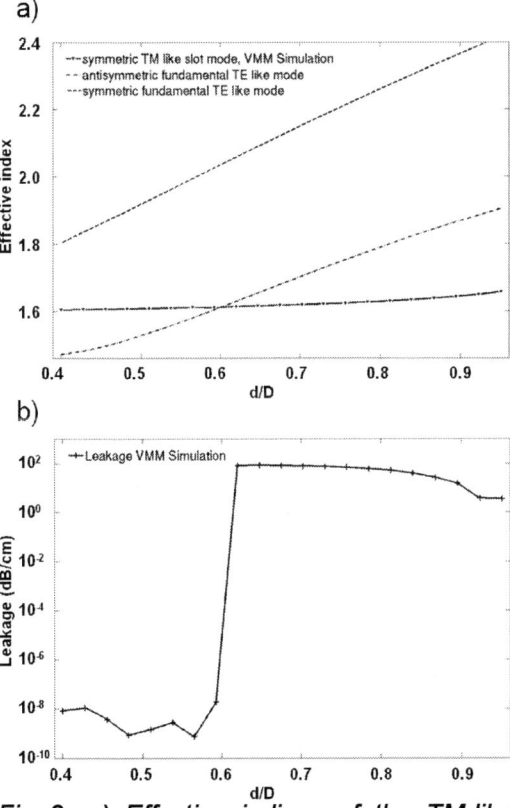

Fig. 3: a) Effective indices of the TM-like slot mode, and the symmetric/antisymmetric TE slab modes as a function of the relative slab thickness for D=110 nm, W=900 nm and S=150 nm at λ=1.55 µm; b) lateral leakage loss of the same structure.

Semi-analytic leakage criterion

The fact that the relationship between the effective indices of the TM-like slot mode and the antisymmetric TE slab mode solely determines whether leakage occurs allows the definition of the following leakage criterion for fully symmetric rib-type slot waveguides. Lateral leakage only occurs above a critical relative slab thickness d/D for which the effective index of the TM-like slot mode is equal to that of the antisymmetric TE-slab mode.

In order to avoid a time consuming fully vectorial 2D eigenmode analysis we approximate the effective index of the TM-like slot mode by solving the transcendent eigenmode equation of a slot waveguide slab for the inner and outer region of the rib-type slot waveguide structure and applying the well-known effective index method [7]. This approach is justified because of the high width-to-thickness ratio B/D of power optimized slot waveguides [8]. We compared this semi-analytical criterion with the numerical results of the VMM simulations in two ways: (i) directly with the leakage-values and (ii) by using the real part of the accurate effective index of the TM-like slot mode in the above stated leakage criterion. The results are shown in Fig. 4.

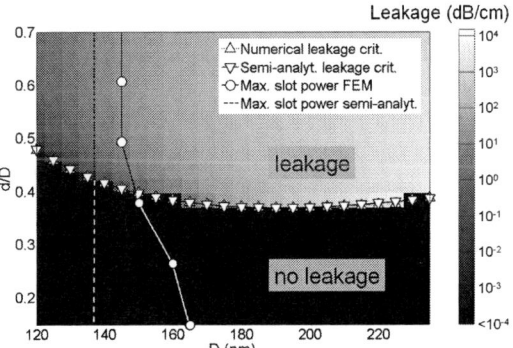

Fig. 4: Lateral leakage loss of a rib-type slot waveguide with W=400 nm and S=150 nm at λ=1.55 µm as a function of D and d/D; numerical and semi-analytical leakage conditions (upward and downward triangles); optimum d/D values determined by FEM simulations at which maximum light confinement in the slot region is achieved for a given D (circles); optimum D for maximum light confinement in the slot waveguide slab (dashed line).

The numerical criterion agrees well with the abrupt change in leakage loss values calculated by VMM simulations. Moreover, the semi-analytic criterion matches well with the numerical results even for the comparatively small waveguide width of W=400 nm.

Fig. 4 also shows the optimum relative slab thickenesses d/D determined by a fully vectorial 2D FEM eigenmode analysis at which maximum light confinement in the slot region is achieved for a given D. Although the thickness of a power optimized slot waveguide slab, which can be calculated by solving a single transcendent eigenmode equation, deviates from these values it represents a good approximation because the resulting power decrease in the slot region amounts to a few percent. Therefore, a fast and effortless design of power optimized leakproof rib-type slot waveguides can be performed in good approximation by using only semi-analytical methods.

Influence of geometry

The influence of the geometry parameters S and B on the leakage behaviour are summarized in Fig. 5. The strong dependence of the effective index of the TM-like slot mode on the slot thickness S causes a significant shift of the leakage criterion to smaller values of d/D for increasing S. The dependence on the waveguide width B is less pronounced

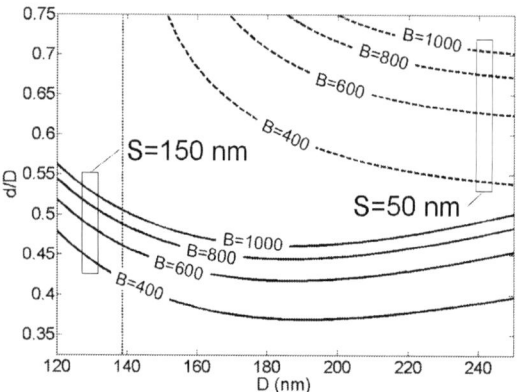

Fig. 5: Influence of slot thickness S and rib width B on the leakage behavior; the dotted line indicates the optimum D for maximum light confinement in slot waveguide slabs with S=50 nm / 150 nm at λ=1.55 μm.

because of smaller influence of the width on the effective index. With increasing width the effective index decreases and converges to that of the infinite slab system. It has to be mentioned that for certain widths resonance effects can strongly reduce the leakage loss [3,4].

Conclusions

We studied the lateral leakage mechanism in symmetric SOI rib-type slot waveguides, which hold the promise to enable a new category of active silicon photonic devices. We demonstrated that by solving the transcendent slot waveguide slab eigenmode equation and using the effective index method a leakage criterion for symmetric rib-type slot waveguides can be determined. We proofed the validity of this approach by VMM simulations. This semi-analytical leakage criterion facilitates the design of leakproof rib-type slot waveguides. In a strict sense, however, this criterion is only valid for a perfectly symmetric geometry and further investigations have to be performed on the influence of structural deviations caused by the fabrication process.

Acknowledgements

This research was supported through the grant PLATON Si-N (project no. 1103) funded by the Austrian NANO Initiative.

References

1 V. R. Almeida et al., Opt. Lett., 29 (2004), 1209

2 C. A. Barrios et al., Opt. Express, 13 (2005), 10092

3 A. A. Oliner et al., IEEE Trans. Microw. Theory, 29 (1981), 855

4 M. A. Webster et al., IEEE Photon. Technol. Lett., 19 (2007), 429

5 R. Hainberger et al., Proc. of the Symp. on Photon. Technol. for 7th Framework Program, Wroclaw (2006), 282

6 N. Finger et al., Conference on the Mathematics of Finite Elements and Applications MAFELAP 2006

7 V. Ramaswamy, Bell System Technical Journal, 53 (1974), 697

8 P. Müllner et al. IEEE Photon. Technol. Lett., 18 (2006), 2557

Analysis on Curved Waveguide Grating (CWG) with Rowland Circle Construction

Yinlei Hao (1), Jianyi Yang (2), Xiaoqing Jiang (3), Wei Zheng (4),
Jianying Zhou (5), Haifeng Zhou (6), Minghua Wang (7)

1 : Department of Information Science & Electronics Engineering,
Zhejiang University, Hangzhou, 310027, China
E-mail: haoyinlei@zju.edu.cn

2 : Zhejiang University, yangjy@zju.edu.cn
3 : Zhejiang University, iseejxq@zju.edu.cn
4 : Zhejiang University, zhengw@zju.edu.cn
5 : Zhejiang University, phd16@zju.edu.cn
6 : Zhejiang University, wuu@zju.edu.cn
7 : Zhejiang University, wangmh@zju.edu.cn

Abstract Configuration of Curved Waveguide Grating (CWG) with Rowland circle construction is presented, working mechanism of CWG is demonstrated, result show that CWG is a promising candidate wavelength demultiplexer in optical DWDM network.

Introduction

Optical Wavelength Division Multiplexing (WDM) components are indispensable in multi-wavelength optical network. Several mechanisms, together with corresponding device configurations, have been proposed for optical wavelength multiplexing/demultiplexing devices, namely optical interference filter, Fiber Bragg Grating (FBG)[1], Arrayed waveguide grating (AWG) [2,3], Etched Diffraction Grating (EDG)[4], as well as the planar waveguide optical spectrum analyzer proposed by Madsen[5]. Recently, we presented a novel dispersive and focusing device, Curved Waveguide Grating (CWG). A general analysis has been given, on wavelength demultiplexing mechanism of CWG, by means of Fourier optics approaches [6].

In this paper, Demonstration on dispersive mechanism is presented by a different approach, with channel waveguide dispersion being taken into consideration, for a specific CWG configuration based on Rowland circle construction.

Device description

Figure 1 shows the schematic layout of the CWG with Rowland circle construction. Light with various wavelength components is launched into an are-shaped single-mode channel waveguide, in which tilted grating lines are induced to tap the guided light out. The light taped out is captured by a slab waveguide adjacent to the arc-shaped

waveguide, and being guided to the output waveguides locate on circumference of the Rowland circle, focal line of the arc-shaped waveguide. The slab waveguide is offset from the channel waveguide by a reasonable distance, which avoids evanescent coupling between the arc-shaped waveguide and the slab waveguide, and minimizes the distance that the radiated light diffracts before being guided in the slab waveguide as well.

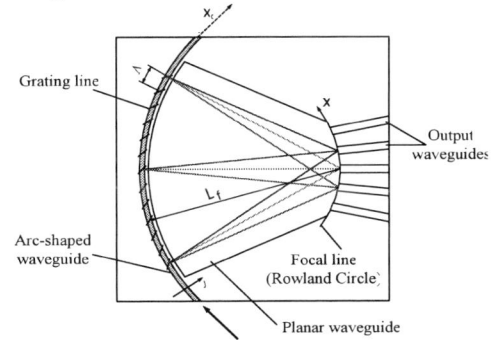

Fig. 1. Schematic of the CWG.

Theoretical Modeling

CWG is essentially a multiple beam interferometer, with multiple light beams generated by diffracted at grating lines on the arc-shaped waveguide. With diffraction theory employed, dispersive and focusing mechanism of CWG could be demonstrated.

A. Diffraction

According to the diffraction theory, and considering structure of the grating line in the arc-shaped waveguide, propagation

978-0-9789217-3-6/07/$25.00

©2007 WEN GLOBAL SOLUTIONS

direction, amplitude spatial profile, phase, as well as amplitude, of diffracted light beams from grating lines could be deduced.

A1. Propagation direction

Fig. 2 shows a local portion of the curved waveguide with grating. The longitudinal phase matching condition can be given as

$$\frac{2N_{eff}\pi}{\lambda} - \frac{2N_c\pi}{\lambda}\cos\theta = m\frac{2\pi}{\Lambda} . \qquad (1)$$

where N_{eff} is effective index of the guided mode, n_c is the refractive index of channel waveguide cladding layer, λ is wavelength in vacuum, Λ is the grating period, as shown in Fig. 2, m is known as grating order.

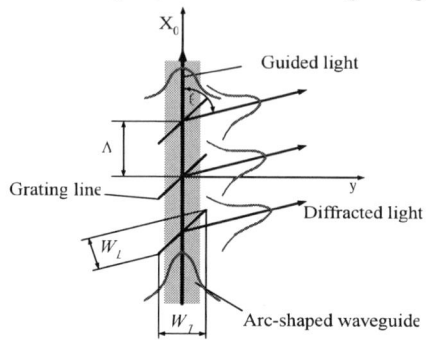

Fig. 2. Bragg diffraction for a local section of the CWG.

A2. Amplitude spatial profile

Consider the lightwave propagates in the curved waveguide with grating, represented by its slowly varying amplitude in the longitude direction, and spatial field profile in the transverse section as well.

$$E(x_0, y) = U_g(x_0)f_0(y)e^{-i\beta(x_0+N\Lambda)} . \qquad (2)$$

where y is the coordinate perpendicular to the channel waveguide in the wafer plane, as shown in Fig. 1 and Fig. 2, $f_0(y)$ is the power normalized spatial field profile, $U_g(x_0)$ is amplitude factor, β is propagation constant of the channel waveguide.

For the fundamental mode propagation in the channel waveguide, $f_0(y)$ can be approximated as a power normalized Gaussian function

$$f_0(y) = \left(\frac{2}{\pi\omega_0^2}\right)^{1/4} e^{-\left(\frac{y}{\omega_0}\right)^2} . \qquad (3)$$

where ω_0 is the mode field radius.

For a special case with a tilted angle of $\pi/4$, and grating lines with sufficient length, $g(x_0)$ can be reasonably approximated as a power normalized Gaussian function

$$g(x_0) = \left(\frac{2}{\pi\omega_g^2}\right)^{1/4} e^{-\left(\frac{x_0}{\omega_g}\right)^2} . \qquad (4)$$

where ω_g is the mode field radius of diffracted light.

A3. Phase

From Fig.2 we know that there exists phase delay between diffracted light from adjacent grating lines, the value being given as

$$\Delta\Phi = \beta\Lambda . \qquad (5)$$

A4. Amplitude

According to the coupled-mode theory for guided-wave optics, amplitude of the guided mode would decrease exponentially in the propagation direction along the channel waveguide, given a uniform coupling coefficient C. Amplitude factor of diffracted light of grating line at x_0 can be given as

$$U_d(x_0) = (2C\Lambda)^{1/2}U_g(-N\Lambda)e^{-C(x_0+N\Lambda)}$$

$$(x_0 = n\Lambda, \ -N \le n \le N, \ n \in \text{integer}). \qquad (6)$$

B. Focusing

From Fig. 1 it is seen that the diffraction angle θ resulting from a phase difference $\Delta\Phi$ between adjacent grating lines follow as

$$\beta\Lambda - \beta_s\Lambda\cos\theta = 2\pi m . \qquad (7)$$

in which β_s is the propagation constant of slab waveguide mode.

Diffracted light beam from grating line with same θ will be focused onto the same point on Rowland circle.

C. Dispersion and Free Spectral Range

The dispersion D of CWG is

$$D = \frac{dx_1}{dv} = -L_f \cdot \frac{d\theta}{dv} = \frac{1}{v_c} \cdot \frac{\tilde{N}_{eff}}{N_s} \cdot L_f . \qquad (8)$$

in which $v_c = c/\lambda_c$ is the central frequency, N_s is the mode index of the slab waveguide, and $\tilde{N}_{eff}$ is the group index of the waveguide mode.

From Eq. (7), it is seen that the response of the CWG is periodical. Spatial Free Spectral Range (SFSR), can be given as

$$\Delta x_{1,FSR} = \frac{N_{eff}L_f}{N_s m} . \qquad (9)$$

Discussions

CWG possess a channel waveguide with an input port and an output port, and several waveguides to collect and conduct diffracted light with different wavelength, a suitable layout for network performance monitoring in multi-wavelength optical communication network.

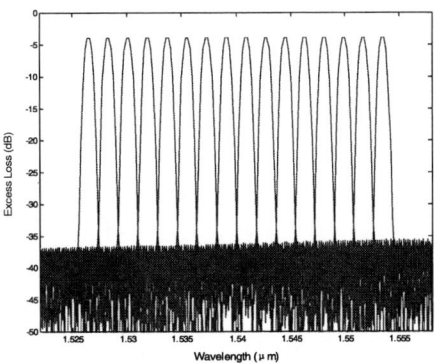

Fig. 4. Simulated spectral response of a CWG.

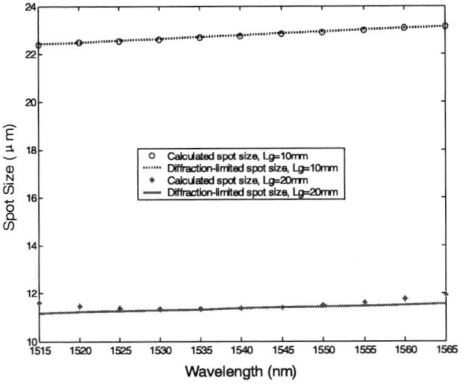

Fig. 3. Diffraction-limited and calculated focused spot size for Rowland circle configuration for grating length 10mm and 20mm.

CWG exhibits excellent imaging functionality, due to the induction of Rowland circle construction, which is one of most classical configuration, to accomplish focusing and imaging functionalities. By introducing this construction, focusing functionality can be expected to be diffraction-limited over the wavelength range utilized for optical communication, given proper configuration parameters. Fig. 3 gives a comparison between theoretical and numerical simulation result of focal point dimensions of CWG over the wavelength range 1515~1565nm, with theoretical dimension by means of Fourier approach. Main CWG parameters evolved are: focal length L_f=107mm, slab waveguide refractive index n_s=1.45, center wavelength λ_0=1540nm, grating length L_g=10mm, 20mm. As shown in Fig. 3, the spot sizes deviation from diffraction limited value over 1515~1565nm is sufficiently slight, a good reason for the CWG with Rowland circle construction to be regarded as diffraction-limited.

Potential application of CWG in DWDM optical network is further demonstrated by a numerical simulation on its spectral response, result being given in Fig. 4.

Conclusions

CWG with Rowland circle construction is presented, Diffraction, focusing as well as dispersion mechanism is demonstrated. Numerical simulation on CWG focusing capability and spectral response show that CWG has a potential application in DWDM optical network.

Acknowledgements

This work was supported in part by Science & Technology Program of Zhejiang Province under Grant 2007C21022, in part by the Natural Science Foundation of China under Grant 60436020 and Grant 60676028, and in part by the Major State Basic Research Development Program under Grant 2007CB307000.

References

1 Turan Erdagan et al IEEE Journal of Lightwave Technology, 15(1997)1276

2 Meint K. Smit et al IEEE J. Sel. Top Quantum. Electron., 2(1996)236

3 Jianyi Yang et al Optics Express 12(2004) 1084

4. Emilio Gini et al IEEE Journal of Lightwave Technol., 16(1998)625

5 C. K. Madsen et al IEEE J. Sel. Top Quantum. Electron., 4(1998)925.

6 Yinlei Hao et al Optics Express, 14(2006) 8630

7. P. Munoz et al IEEE J. Light. Technol., 20(2002)661

8 Amnon Yariv et al IEEE J. Quantum. Electron.,13(1977)233

Paired surface plasmon waves biosensor

Chien Chou (1,2), Hsieh-Ting Wu (1), Ying-Chang Li (2)
1: Institute of Radiological Sciences, National Yang-Ming University,
No. 155, Sec. 2, Li-Nong st.,Peitou, Taipei, Taiwan 112
e-mail: cchou@ym.edu.tw, d49120004@ym.edu.tw
2: Department of Optics and Photonics, National Central University,
No.300, Jhongda Rd., Jhongli City, Taoyuan County, Taiwan 320.
e-mail: s1226013@cc.ncu.edu.tw

ABSTRACT *To integrated paired surface plasma waves biosensor with heterodyne technique to construct a paired SPW biosensor in order to enhance the detection sensitivity is developed. It demonstrates $\Delta n_{eff} = 10^{-10}$ RIU of detection sensitivity between mouse IgG/rabbit anti-mouse IgG interaction in real time.*

Introduction

Among many techniques on the measurement of biomolecular interaction, surface plasmon resonance (SPR) biosensor is capable of real-time measurement with high detection sensitivity. In previous reports, the detection sensitivity is in the range from 10^{-6} to 10^{-8} RIU [1-3] via scanning resonance angle or phase detection. Some developments of the specific SPR biosensor which introduces optical heterodyne technique is able to detect the phase and amplitude of the heterodyne signal in temporal domain. The detection sensitivity has been improved by optical heterodyne technique successfully [4]. Nowadays, a biosensor with higher detection sensitivity becomes important in the fields of life science and drug discovery. This results in the development on a highly sensitive SPR biosensor.

In this study, we propose different method to integrate the feature of paired surface plasma waves (SPWs), which is excited by an evanescent wave with optical heterodyne technique to setup a novel optical heterodyne surface plasma waves biosensor (OHSPWB) [5] and a paired surface plasma waves biosensor (PSPWB) [6]. And then we compare these two biosensors each other by measuring the detection sensitivity in the interaction of low-concentrated mouse IgG and rabbit anti-mouse IgG..

Experimental setup and Results

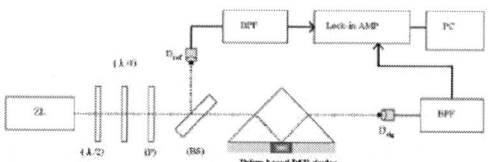

Fig. 1. Schematic of the optical heterodyne SPR biosensor: ZL, Zeeman He–Ne laser. Other abbreviations are defined in text.

Fig. 1 shows the schematic setup of OHSPWB, two orthogonal linearly polarized light waves (p wave and s wave) with slightly differential temporal frequency are generated by use of a Zeeman He-Ne laser, and then are converted into circularly polarized right (R) and left (L) light waves simultaneously. A beam splitter separates the light beam into two equal waves. One wave incidents into the SPR device and reflects the attenuated amplitudes $A_p^{\prime}$ and $A_s^{\prime}$ to the photo-detector, D_{sig}. Therefore the intensity of the light beam from D_{sig} becomes

$$I_{P_1+P_2}(\Delta\omega t) = \frac{1}{2}A_p^{\prime 2} + \frac{1}{2}A_s^{\prime 2} + A_s^{\prime}A_p^{\prime}\cos(\Delta\omega t) \quad (1)$$

where $A_p^{\prime}$ and $A_s^{\prime}$ are the amplitudes of p_1 and p_2 waves, $\Delta\omega = \omega_P - \omega_S$ is the beat frequency of the heterodyne signals. A polarizer (P) is used to select p-wave components from R and L waves simultaneously. Another part of incident beam into the photo-detector, D_{ref}, as the refeence. After both signals pass through band pass filter, $A_p A_s$ from D_{ref} and $A'_p A'_s$ from D_{sig}, are obtained simultaneously. The ratio of the amplitude $\chi = A'_p A'_s / A_p A_s$ is measured by using a lock-in amplifier. For more detailed explanations, Ref. 5 gives the complete discriptions on OHSPWB.

The experimental results from OHSPWB, Fig. 2 shows the time response of the variation of the amplitude unit (A.U.) This is the interaction of mouse IgG/rabbit anti-mouse IgG at the concentration of 100pg/ml of mouse IgG. It is in contrast to the method by Branfenberg et al [7] of experimentally demonstration the sensitivity at $\Delta n_{eff}=4.1\times10^{-6}$ RIU by testing 50 ng/ml mouse IgG. As results, the detection sensitivity in OHSPWB for measuring same biomolecule interaction at a lower concentration at 100pg/ml on mouse IgG, is improved significantly ($\Delta n_{eff}=8.2\times10^{-9}$ RIU).

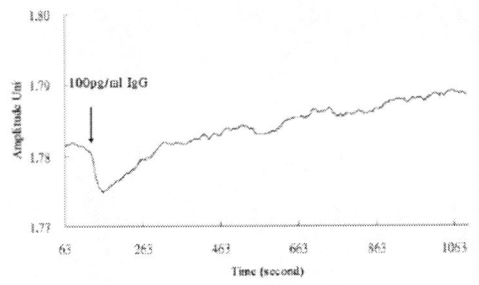

Fig. 2. Time response of 100pg/ml mouse IgG interacting with immobilized rabbit anti-mouse IgG on CM5 biochip in OHSPWB.

In order to enhance the detection sensitivity, we introduce another optical heterodyne system, which is paired SPW biosensor (PSPWB) as shown in Fig. 3. To compare it with OHSPWB, this novel system replaces the Zeeman He-Ne laser with a common-path polarized heterodyne interferometer, which consists of a single frequency stabilized linear polarized He-Ne laser in a Mach-Zehnder interferometer. The output beams from the heterodyne interferometer are two paired of highly spatial and temporal correlated linear polarized light waves, (p_1+s_1) with

frequency ω_1 and (p_2+s_2) with frequency ω_2. All these waves which pass through the SPR device and the PBS are separated into two paired linear polarized light waves, (P_1+P_2) and (S_1+S_2) and optical heterodyned. Both pairs of the polarized heterodyne signals have the same beat frequency $\Delta\omega = \omega_1 - \omega_2$. The ratio of the amplitude of the two polarized heterodyne signals, $\chi = (A'_{P_1} \cdot A'_{P_2} / A_{S_1} \cdot A_{S_2})$, is also measured by use of a lock-in amplifier.

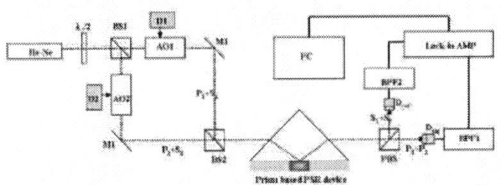

Fig. 3. Time response of 100pg/ml mouse IgG interacting with immobilized rabbit anti-mouse IgG on CM5 biochip

Thus, the setup of PSPWB has a common-path configuration and generates two pairs of orthogonal polarized linear waves. Both features are able to reduce the excess noise from laser intensity fluctuation and the background noise at the same time. This is a key point to retain the amplitude stability in PSPWB to ensure high detection sensitivity. The experiment on IgG/anti-mouse IgG interaction is also tested where the concentration of the mouse IgG is down to 10 pg/ml for comparison. Fig. 4 is the time response in this test. This result expresses that the PSPWB can monitor the interaction in ultra-low IgG concentration. By calculation, the experimental detection sensitivity of PSPWB for mouse IgG/anti-mouse IgG interaction becomes $\Delta n = 8.2\times10^{-10}$, which is improved 10^4 compared with the result by Brandenburg et al.[7]

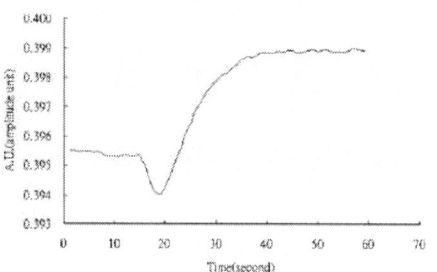

Fig. 4. Time response of 10pg/ml mouse IgG interacting with immobilized rabbit anti-mouse IgG on CM5 biochip in PSPWB

Conclusions

We have proposed two optical heterodyne SPR biosensors in order to enhance detection sensitivity of SPR biosensor. The common-path configuration in conjunction with the ratio of the amplitudes of polarized optical heterodyne signals provide paired SPW biosensor able to reduce laser noise fluctuation that results the sensitivity on effective refractive index change to $\Delta n_{eff} \approx 8.2 \times 10^{-10}$ successfully.

Reference

1. Homola J. et al Surface plasmon resonance sensors: review. Sens. Actuators B 54 (1999) 3
2. C.-M. Wu et al Sens. Actuators B 92 (2003) 133
3. Yu X. et al Sens. Actuators B 91 (2003) 285
4. S. Y. Wu et al Opt. Lett. 29 (2004) 2378
5. W.-C. Kuo et al Opt. Lett. 28 (2003) 1329
6. Chien Chou et al Opt. Express 14 (2006) 4307
7. Brandenburg A et al Appl. Opt. 39 (2000) 6396

Acknowledgment

This research was supported by The National Science Council, Taiwan.

Temperature-Insensitive Pressure Sensor Using a Polarization-Maintaining Photonic Crystal Fiber based Sagnac Interferometer

H Y Tam, Sunil K. Khijwania*, and X.Y. Dong

Photonic Research Center, Department of Electrical Engineering

The Hong Kong Polytechnic University, Hong Kong

*Department of Physics, Indian Institute of Technology, Guwahati, India

Abstract: A novel, highly sensitivity, fiber-optic pressure sensor with polarization-maintaining photonic crystal fiber (PM-PCF) based Sagnac interferometer (SI) is proposed and experimentally demonstrated. A wavelength-pressure sensitivity ~3.75 nm/MPa is achieved with a 79.6-cm PM-PCF as the sensing element. Owing to the birefringence insensitivity for the PM-PCF, the sensor is inherently temperature insensitive.

Keywords: Pressure sensor, Sagnac interferometer, polarization-maintaining photonic crystal fiber

Introduction:

Optical fiber Sagnac interferometers have been successfully employed for sensing aplications.[1,2] In particular, a polarization-maintaining fiber (PMF) based Sagnac interferometer (SI)[3-6] offers great advantages owing to their low insertion loss, polarization independence to input light, and an operation spanning across a broad spectral bandwidth to name a few. However, the performance of a PMF SI is seriously affected by environmental changes such as temperature fluctuations. This is due to a high thermal sensitivity of the birefringence of polarization-maintaining fibers (e.g. Panda and bow-tie PMFs).[3-4] A thermal sensitivity of ~0.99 nm/°C has been observed for PMF SI, which is about one to two orders of magnitude higher than that of a long period and fiber Bragg grating (FBG). Photonic crystal fibers (PCF), emerged recently with a wide range of inherent fabrication geometry including PM-PCF, have attracted lots of research activities due to a host of highly unusual and tailorable properties.[7] An extremely low temperature sensitivity (164 times less) of PM-PCF SI compared to the conventional PMF SI[8], birefringence tuning to a very high value over a very short length of PM-PCF[9] and the possibility of employing a bare and undoped fiber without any additional embedded structure (such as gratings etc.) have opened up lots of opportunity to realize robust, compact temperature insensitive sensors for various applications.

Pressure is one of the critical parameter for various engineering/industrial applications e.g. aerospace, petrochemical, structural monitoring etc. Most popular pressure sensors, employed presently, are based on Fabry-Perot interference or FBG. All such sensors are limited with low pressure sensitivity and a high temperature cross-sensitivity[10]. Recently, Bock *et al*[11] reported the first PCF based pressure sensor operating at 1618 nm. They spliced various segments of the PCF, which led to temperature self-compensation. Gahir *et al*[12] employed PM-PCF to realize a pressure sensor operating at 1550

nm. Relying on polarimetric format, both schemes were complicated in terms of realization and practical applications. In another application, Dong *et al*[13] employed PM-PCF SI to monitor strain. In this paper, a Sagnac interferometer pressure sensor has been proposed and demonstrated employing PM-PC, a 3-dB single-mode fiber coupler and a 1550 nm broadband light source. The short length of PM-PCF, employed in Sagnac loop served directly as pressure sensing element. Applied pressure is wavelength encoded from the wavelength shift of the transmission minimum. A simplified theoretical model for the sensor is presented to analyze the pressure induced wavelength shift of the transmission minimum. Sensitivity of ~3.75 nm/MPa for the pressure measurement is achieved with a 79.6-cm length of PM-PCF inserted in SI. Ultra low thermal coefficient for the birefringence of PM-PCF eliminated the need of extra and complex temperature cross-sensitivity compensation.

Principle:

A PM-PCF-SI is realized using a relatively short length of the PM-PCF, spliced between the throughput and the coupled arms of a 3-dB single-mode-fiber (SMF) coupler. Optical light launched into the Sagnac-loop via the input arm of the coupler, splits equally into the two counter-propagating waves. After traversing the loop, these counter-propagating waves recombine to interfere at the coupler because of the specific phase difference introduced across the two orthogonal modes by the PM-PCF. Calculating the transfer matrix for this structure using standard Jones matrix, the transmission ratio of the optical intensity injected to the Sagnac-loop can be written as:

$$T = [1 - \cos(\delta)]/2 , \qquad (1)$$

where δ is the total phase-retardance. Neglecting the temperature-induced retardance[11], δ can be simply given by:

$$\delta = \delta_0 + \delta_P . \qquad (2)$$

Here, δ_0 and δ_P are the phase retardance due to the intrinsic and pressure-induced birefringence over a length

---

978-0-9789217-3-6/07/$25.00

©2007 WEN GLOBAL SOLUTIONS

L of the PM-PCF with

$$\delta_0 = \frac{2\pi \cdot B \cdot L}{\lambda} \quad \text{and} \tag{3}$$

$$\delta_P = \frac{2\pi \cdot (K_P \Delta P) \cdot L}{\lambda}, \tag{4}$$

making T a periodic function of λ with a spacing between the adjacent transmission minima given by $S = \lambda^2/(BL)$. Here, B $(= n_s - n_f)$ is the birefringence of the PM-PCF, λ is the wavelength of the light, n_s and n_f are the effective refractive-indices along the slow and the fast axis of this fiber, ΔP is the amount of pressure variation, and K_P gives the birefringence-pressure sensitivity of the PM-PCF[12]:

$$K_P = \frac{\partial n_s}{\partial P} - \frac{\partial n_f}{\partial P} \tag{5}$$

The corresponding wavelength shift of the transmission minimum due to the pressure-induce retardance is given by

$$\Delta\lambda = S \cdot \delta_P / \pi \tag{6}$$

which can be simplified as

$$\Delta\lambda = \left(\frac{K_P \cdot \lambda}{B}\right) \cdot \Delta P \tag{7}$$

The above relation clearly depicts a linear relationship between the wavelength shift of the interference minima and the applied pressure variations.

Experimental Setup:

Experimental setup of the proposed PM-PCF based SI pressure sensor is shown in Fig. 1. A Sagnac loop was realized using a standard 3-dB single-mode fiber coupler and a 79.6-cm PM-PCF (PM-1550-01, Blaze-Photonics, beat length of < 4 mm at 1550 nm, polarization extinction ratio of >30 dB over 100 m). The intrinsic birefringence of the PM-PCF used in the experimental was ~8.15×10^{-4} at 1550 nm. The cross section of the PM-PCF is shown in

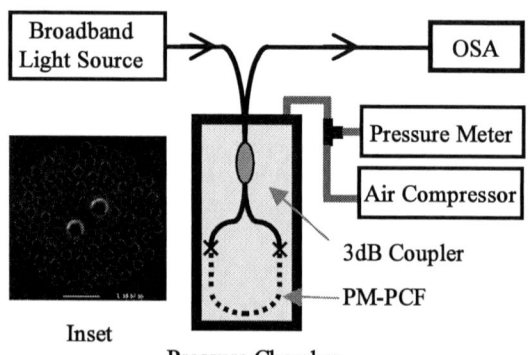

Inset

Pressure Chamber

Fig. 1: *Schematic diagram of the PM-PCF based Sagnac interferometer pressure sensor. (Inset: SEM of the cross section of the PM-PCF)*

the inset of Fig. 1. Mode-field-diameters of the two orthogonal polarizations for the PM-PCF were 3.6 and 3.1 μm respectively. The two ends of the fiber coupler were fusion spliced to the PM-PCF by repeated arc discharges technique.[13] Total loss of less than 4 dB was achieved for the two splicing points. A pressure chamber, capable to sustain high pressure was designed. The chamber had a provision for gas inlet (to increase/decrease pressure) and lead-in/lead-out fibers as shown in the schematics. The Sagnac-interferometer was fixed to the bottom of the pressure chamber. An air compressor with pressure-meter was used to adjust the air pressure in the chamber. The light was launched into the Sagnac-loop through the lead-in fiber of the 3-dB coupler by a broadband amplified-spontaneous-emission source, centered at 1550 nm and the transmitted output was observed via the lead-out fiber through an optical spectrum analyzer (Agilent 86140B).

Results and Discussion:

Figure 2 shows the transmission spectrum of PM-PCF based Sagnac interferometer within a wavelength range of

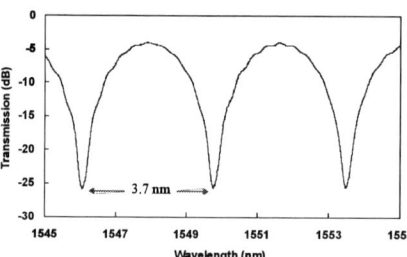

Fig. 2: *Transmission spectrum of the PM-PCF based Sagnac interferometer.*

10 nm with no applied pressure except the atmosphere (about 0.1 MPa). The wavelength spacing between the two adjacent transmission minima is observed to be ~3.7 nm. An extinction ratio of more than 20 dB is achieved for the transmission minima. Afterwards, pressure in the high-pressure chamber was slowly varied in a range of 0 – 0.4 MPa. It is important to mention that the applied pressure range was limited to the resources available in the lab and hence the sensor response is not limited to the pressure variations reported in this study. Figure 3 shows the transmission spectra around the transmission minimum at ~1549.62 nm when the sensor was subjected to the pressure variations. As can be observed from the figure, transmission minimum (originally at 1549.62 nm) shifts toward the longer wavelength as the pressure increases. A wavelength shift up to 1.12 nm is achieved for the transmission minimum by increasing the pressure to 0.3 MPa only. The wavelength corresponding to the transmission minimum is plotted against the applied

342

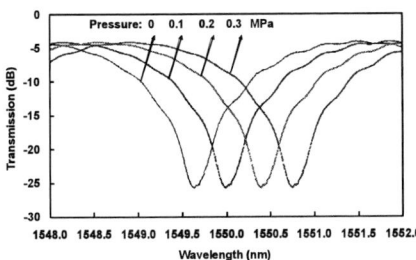

Fig. 3: *Measured transmission spectra under different pressures.*

pressure in Figure 4. A linear fitting to the experimental data reveals a wavelength-pressure sensitivity of ~3.75 nm/MPa. A high R^2 value of 0.999 indicates an excellent linearity in the sensor response as predicted by the theoretical analysis. The birefringence-pressure sensitivity calculated from (7) came out to be ~2.0×10^{-6} for the proposed sensor. The resolution of the pressure measurement, limited by the resolution (10 pm) of the optical spectrum analyzer (OSA), is ~2.7 kPa.

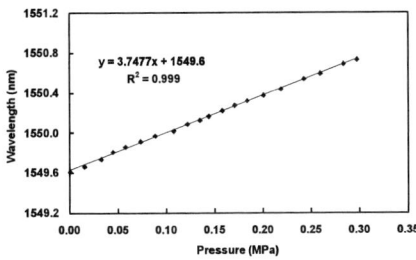

Fig. 4: *Wavelength shift of the transmission minimum at 1549.62 nm against the applied pressure.*

To reduce the effects of environmental interference (such as the one from vibration) and achieve a high degree of stability, Sagnac loop was properly and carefully fixed to the bottom on the chamber avoiding any relative vibration within the loop. Further, in order to apply the pressure uniformly and directly over the sensing element (PM-PCF) was laid down straight in the fixed loop avoiding any fiber-overloading on this element. Essentially, this increased the length to the entire sensing structure. Nevertheless, the wavelength-pressure sensitivity is independent of the length of the PM-PCF, as described in equation (7). Thus a very compact and stable sensor without compromising with the sensitivity can be achieved with a short length of PM-PCF. Thus, based on these merits and the inherent simplicity, proposed sensor would be of extreme importance for various applications requiring pressure monitoring.

Conclusion:

A novel all-optical pressure sensor employing Sagnac

interferometer is proposed and demonstrated with PM-PCF as the sensing element. Experimental results and simplified theoretical analysis of the pressure sensor are presented. The wavelength-pressure sensitivity of ~3.75 nm/MPa and a resolution of ~2.7 kPa is achieved. A linear dynamic range of the sensor is demonstrated with an excellent agreement from the theory. Compared to the reported pressure sensors based on conventional PMF-SI and FBG/LPG, the proposed sensor is inherently temperature insensitive and thus eliminates complex temperature compensation requirement. In addition, fiber as such being the sensing element makes the proposed sensor much simpler and easy to realize as compared to the rest where a particular and complex structure in needed to be embedded/inscripted. These inherent merits along with the merits of all-fiber-optical sensing make this sensor an ideal candidate for practical applications.

Acknowledgement:

The authors would like to thank L. Xiao and Dr. S. Y. Liu for fruitful discussions and timely help offered.

Reference:

1. V. Vali and R. W. Shorthill, Appl. Opt. **15**, 1099 (1976).
2. S. Knudsen and K. Blotekjaer, J. Lightwave Technol. **12**, 1696 (1994).
3. A. N. Starodumov, L. A. Zenteno, D. Monzon, and E. De La Rose, Appl. Phys. Lett. **70**, 19 (1997).
4. E. De La Rose, L. A. Zenteno, A. N. Starodumov, and D. Monzon, Opt. Lett. **22**, 481 (1997).
5. M. Campbell, G. Zheng, A. S. Holmes-Smith, and P. A. Wallace, Mes. Sci. Technol. **10**, 218 (1999).
6. Y. Liu, B. Liu, X. Feng, W. Zhang, G. Zhou, S. Yuan, G. Kai, and X. Dong, Appl. Opt. 44, 2382 (2005).
7. T. A. Birks, D. Mogilevtsev, J. C. Knight, and P. St. J. Russel, IEEE Photon. Technol. Lett., **11**, 674 (1999).
8. D.-H. Kim and J. U. Kang, Opt. Express **12**, 4490 (2004).
9. T. P. Hansen, J. Broeng, S. E. B. Libori, E. Knudosen, A. Bjarklev, J. R. Jensen, and H. Simonsen, IEEE Photon. Technol. Lett. **13**, 588 (2001).
10. G. Chen, L. Liu, H. Jia, J. Yu, L. Xu, and W. Wang, Opt. Commun. **228**, 99 (2003).
11. W. J Bock, J. Chen, T. Eftimov, and W. Urbanczyk, IEEE Trans. Intrum. Measur., **55**, 1119 (2006).
12. H. K. Gahir and D. Khanna, Appl. Opt. **46**, 1184 (2007).
13. X. Dong, H. Y. Tam and P. Shum, Appl. Phys. Lett. **90**, 151113 (2007).
14. L. Xiao, W. Jin, and M. S. Demokan, Opt. Lett., 32, 115 (2007).

Fiber Bragg grating interrogating system employing an arrayed waveguide grating

Zhou Qinfeng, Xu Tiefeng

Faculty of Information Science and Engineering, Ningbo University, Ningbo, 315211,
zhouqinfeng@yahoo.com

Abstract *Two types of interrogating schemes based on an waveguide grating for fiber Bragg grating sensors are reviewed. A method of extending the wavelength measuring range is proposed. And Experiment results match well with the theory.*

Introduction

Recently, fiber Bragg gating sensors (FBG) have demonstrated great advantage over electronic sensors for applications in smart structures, civil engineering, or harsh environments, because FBG's sensing information is encoded in an absolute parameter [1]. So, for a sensing system, the measurement of a small Bragg wavelength shift is most important. Many scientists and researchers dedicate themselves to this field. And many schemes for wavelength interrogation are reported such as interferometric detection [2], Mach-Zehnder interferometer [3], acousto-optic tunable filters [4], and Fabry-Perot interferometer [5], and a free-space CCD camera-based spectrometer [6]. However, none of them are enough satisfactory for response precision, accuracy, speed, sensor multiplexing and cost. In order to reduce cost and complexity of the system, other demodulation schemes is proposed, for instance, wavelength interrogator employing arrayed waveguide grating (AWG) [7-10], which potentially offer a low-cost, compact, and high-performance solution for the interrogation of FBG distributed sensors.

In this paper, we summarize the principles of the two types of interrogating schemes employing an AWG. The first scheme, the Bragg wavelength of the fiber Bragg grating sensors is interrogated by thermally scanning an AWG-based demultiplexer, which depends on the relationship of the AWG central wavelength with temperature. In the second scheme, AWG is used as an edge filter, by monitoring the output ratio of each pair of two-adjacent-channels of the AWG. Although they have a high precision and high speed, the measuring range is limited. Therefore, we propose a scheme of

monitoring three or more adjacent channels by adding a comparator and adjusting the photo detectors' output data in certain way. Both theoretical and experiment results show that the range of wavelength measurement is enlarged comparatively, with a high precision.

Theory and Experiment

A. Thermally scanning scheme [7]

According to the AWG operating principle, temperature has the same effect on all the AWG transmission channels. So we discuss only the temperature effects on the AWG center wavelength (λ_{ac}) below.

$$\Delta L n_a = m \lambda_{ac}, \quad (1)$$

where ΔL is the optical pathlength difference, n_a is the effective index of the waveguide. Since both ΔL and n_a are temperature dependent, from *Eq. (1)* we can determine the relationship of λ_{ac} with temperature:

$$\lambda_{ac}(T) = \lambda_{ac}(T_0) + K_{aT}\Delta T, \quad (2)$$

$$K_{aT} = \frac{1}{m}\frac{d(n_a\Delta L)}{dT}, \quad \Delta T = T - T_0$$

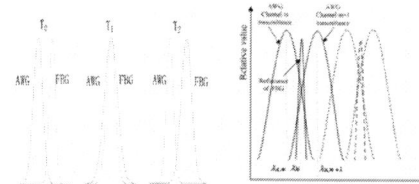

Fig. 1 Thermally   Fig. 2 Ratio-metric scanning interrogation   interrogation

As is shown in *Fig. 1.*, assuming that at temperature T_0, the transmission wavelength of AWG channel m ($\lambda_{a,m}$) is shorter than the FBG Bragg wavelength (λ_b), then the power of the detector m (PD_m) is low. If we increase the AWG temperature, the

AWG transmission wavelengths will move to long wavelengths, and at a certain temperature (suppose it is T_1) the transmission peak of AWG channel m will overlap with the reflection peak of FBG ($\lambda_{a,m} = \lambda_b$), in this case we will obtain a maximum optical power output ($PD_{m,\max}$) from the corresponding detector m. If we go on increasing the temperature, PD_m will decrease, because $\lambda_{a,m}$ becomes longer than λ_b , which leads to the two peaks don't overlap each other again. As a result, by monitoring the $PD_{m,\max}$, we can to obtain the λ_b via some mathematic conversion described above.

B. Ratio-metric scheme [8]
In this scheme, the shift of the Bragg wavelength can be precisely interrogated by the relative intensity reading of two-adjacent-channels of the AWG-based demultiplexer. Using the relative intensity technique, undesirable effects due to possible source fluctuation and micro-bend attenuations should be corrected (shown in *Fig. 2*).

The spectral reflectance of an apodized FBG has, to a good order approximation, a Gaussian profile [12].

$$R_{FBGi}(\lambda) = b_i \exp[-4\ln(2)\frac{(\lambda - \lambda_{bi})^2}{\Delta\lambda^2_{bi}}], \quad (3)$$

and the spectral transmittance of each channel of AWG can be approximated in similar way.

$$T_{AWGm}(\lambda) = a_m \exp[-4\ln(2)\frac{(\lambda - \lambda_{a,m})^2}{\Delta\lambda^2_{a,m}}], \quad (4)$$

so the output signal form the mth channel of the AWG is given by

$$I_m(\lambda_{bi}) = d_m \int_0^\infty S(\lambda)R_{FBGi}(\lambda)T_{AWGm}(\lambda)d\lambda + I_{noise}$$

$$= d_m S a_m b_i \Delta\lambda_{a,m}\Delta\lambda_{bi} \times \sqrt{\frac{\pi}{4\ln 2(\Delta\lambda^2_{a,m} + \Delta\lambda^2_{bi})}}$$

$$\times \exp[-4\ln 2\frac{(\lambda_{a,m} - \lambda_{bi})^2}{\Delta\lambda^2_{a,m} + \Delta\lambda^2_{bi}}] \quad (5)$$

where $S(\lambda)$ is an emission spectrum of the light source. For a commercial Broadband source, its light power within a very narrow range of wavelength can be consider as a constant of S.

Yasukazu Sano and Toshihiko Yoshino defined a wavelength interrogation function

$\rho(\lambda_{bi})$ by the logarithm of the output ratio between the adjacent two channel of an AWG as

$$\rho(\lambda_{bi}) = \log\frac{I_{m+1}(\lambda_{bi})}{I_m(\lambda_{bi})}, \quad (6)$$

Assuming that $|\Delta\lambda_{a,m+1} - \Delta\lambda_{a,m}| \ll \Delta\lambda_{bi}$, the interro-gation function comes

$$\rho(\lambda_{bi}) = A_m[\lambda_{bi} - \frac{\lambda_{a,m+1} + \lambda_{a,m}}{2}] + B_m, \quad (7)$$

$$A_m = \frac{8\ln 2(\lambda_{a,m+1} - \lambda_{a,m})}{[\frac{(\Delta\lambda_{a,m+1} + \Delta\lambda_{a,m})}{2}]^2 + \Delta\lambda^2_{bi}},$$

$$B_m = \log\frac{d_{m+1}a_{m+1}\Delta\lambda_{a,m+1}}{d_m a_m \Delta\lambda_{a,m}} + \frac{1}{2}\log\frac{\Delta\lambda^2_{a,m} + \Delta\lambda^2_{bi}}{\Delta\lambda^2_{a,m+1} + \Delta\lambda^2_{bi}}$$

which indicates that $\rho(\lambda_{bi})$ is in linear relationship with the FBG Bragg wavelength λ_{bi}, so that λ_{bi} can be readily determined from the measurement of $\rho(\lambda_{bi})$. This is the principle of wavelength interrogation of the present scheme.

C. Enlarging the measuring range
Obviously there is some drawback as the dynamic range of wavelength measurement is limited by the channel's transmission wavelength range (in the first scheme) or by the range between the central wavelengths of the adjacent two channels of AWG (in the second scheme). In order to extend the range, we can use three or more channels to determine one FBG by measuring the value of $\rho(\lambda_{bi})$, which can be easily achieved by taking a small operation on the placement of device shown in *Fig. 3*. adding comparison device between AWG channels, i.e. channel *m-1* and channel *m+1*.

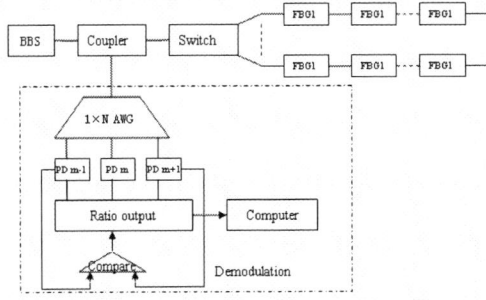

Fig. 3 Placement of device

In this case the interrogation function can be given by

$$\rho'(\lambda_{bi}) = \begin{cases} \log\dfrac{I_m(\lambda_{bi}) - I_M}{I_{m-1}(\lambda_{bi}) - I_M} & (I_{m-1} \geq I_{m+1}) \ (8) \\[3mm] \gamma \log\dfrac{I_{m+1}(\lambda_{bi}) - I_M}{I_m(\lambda_{bi}) - I_M} + P & (I_{m-1} < I_{m+1}) \end{cases}$$

$$r = \left|\frac{\lambda_{a,m} - \lambda_{a,m-1}}{\lambda_{a,m+1} - \lambda_{a,m}}\right| \frac{(\Delta\lambda_{a,m+1} + \Delta\lambda_{a,m})^2 + 4\Delta\lambda^2_{bi}}{(\Delta\lambda_{a,m-1} + \Delta\lambda_{a,m})^2 + 4\Delta\lambda^2_{bi}}$$

$$P = I_{m-1}(\lambda_{a,m}) + |I_m(\lambda_{a,m})|$$

$$\approx \frac{1+r}{4} A_{m-1}(\lambda_{a,m+1} - \lambda_{a,m-1})$$

where r is the modified factor of the slopes between A_{m-1} and A_m. If the ranges between the adjacent two channels of AWG have the same or nearly same space, and the FWHM of the channel $m-1$ equals to that of the channel $m+1$, r can be ignored. So we can educe that the relationship between λ_{bi} and $\rho'(\lambda_{bi})$, shown as Eq. (9)

$$\lambda_{bi} = \frac{\rho(\lambda_{bi}) - B_{m-1}}{A_{m-1}} + \frac{(\lambda_{a,m-1} + \lambda_{a,m})}{2} \ , \ (9)$$

To investigate the expected performance of such a system, the source is an ASE soure that has a peak power density of -12 dBm at 1556 nm with the -3 dB band of 40nm. The FBG Bragg wavelength shifted (Fig. 4) between the range of the 5th and 7th AWG channel's central wavelength (Ta. 1). The datum were plotted in Fig. 5 by black square as the measured datum and the linear fit beeline is also drawn. From which we can see that a very good agreement between the measured and the calculated curve (the central data shows high perfor-mance with the coefficient reaching 0.9996), although at the ends shows some mismatch, due to various noise existing the experiment. obviously the measuring range of 1.26 nm (nearly the total range between channel 5 and channel 7), is enlarged comparatively (with *Ref. 7*).

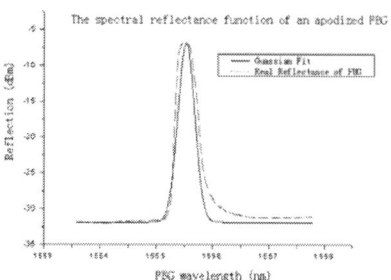

Fig. 4 the reflected spectra of FBG

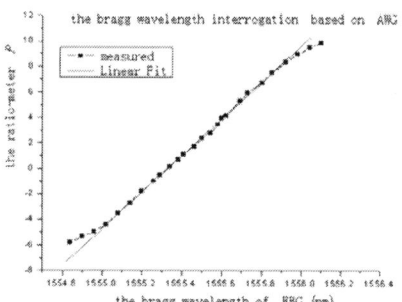

Fig. 5 large measuring range

Conclusions

The two inexpensive interrogating technique of fiber Bragg grating sensors based on an arrayed waveguide grating is discussed in this paper. As both of them have a limit of measuring range, we proposed a scheme based on the second scheme, and the high performance is observe. Furthermore, as is shown in *Fig. 3*, we can use the optical switch to realize STM/WTM sensor nets.

References

[1] S. Abad, et al. *Lightwave Technol*, vol. 21 (2003), pp. 127-131.

[2] A.D. Kersey, et al. *Electron. Lett.*, vol. 28 (1992), pp. 236-238.

[3] M. Song, et al. *Appl. Opt.*, vol. 39 (2000), pp. 1106-1111.

[4] M. G. Xu, et al. *Electron. Lett.*, vol. 29 (1993), pp. 822-823.

[5] A.D. Kersey, et al. *Opt. Lett.*, vol. 18 (1993), pp. 1370-1372.

[6] C. G. Askins, et al. *Proc. SPIE*, vol. 2444 (1995), pp. 257-266.

[7] G. Z. Xiao, et al. *Opt. Lett.*, vol. 29 (2004), pp. 2222-2224.

[8] Y. Sano, et al. *Lightwave Technol*, vol. 21 (2003), pp. 132-139.

[9] A. Fender, et al. *Applied Optics*, vol. 45 (2006), pp. 9041-9048.

[10] H. Su, et al. *Opt. Commun.*, vol. 275 (2007), pp. 196-200.

Table 1   The AWG parameters

No.	$\lambda_{a,m}$/nm	$\Delta\lambda_{a,m}$/nm	(a_m/dB)
1	1551.69	0.425	-4.2
2	1552.45	0.423	-4.4
3	1553.22	0.429	-4.2
4	1554.02	0.443	-4.0
5	1554.81	0.434	-4.1
6	1555.61	0.435	-4.0
7	1556.40	0.436	-4.1
8	1557.13	0.428	-4.2

Phosphor-free White-Light Light-emitting Diodes Based on InGaN/GaN Quantum Wells

Chi-Feng Huang, Chih-Feng Lu, Dong-Ming Yeh, Yung-Sheng Chen, Wen-Yu Shiao, and
C. C. Yang
Graduate Institute of Electro-Optical Engineering, National Taiwan University,
No. 1, Roosevelt Road, Section 4, Taipei, Taiwan, R.O.C.
e-mail: ccy@cc.ee.ntu.edu.tw

Abstract Two approaches for fabricating white-light light-emitting diodes are discussed, including the stacking of various InGaN/GaN quantum wells for mixing into white light and the use of II-VI nano-crystals for converting short-wavelength photons into long-wavelength light.

1. Introduction

The development of phosphor-free white-light light-emitting diodes (LEDs) has become an important issue because of the disadvantages of using phosphors for converting light colors, including lower efficiency, shorter lifetime, and patent restriction. Various device schemes have been proposed for such development. Since the band gap of InGaN can cover up to the near-infrared range, using InGaN/GaN quantum wells (QWs) of different parameters for emitting lights of various colors to mix into white light has been attempted. By controlling the QW width, different levels of piezoelectric field can lead to different emission wavelengths [1]. In this case, however, the strong quantum-confined Stark effect (QCSE) in a QW of a large well width will result in a significant blue shift when a plenty of carriers are injected to produce the screening effect. Although white-light LEDs of simultaneously emitting two or three colors by growing InGaN/GaN QWs of different indium contents have been reported [2,3], the internal quantum efficiencies of the long-wavelength components were quite low and the QCSE screening effect was still a major problem. In the first part of this paper, we introduce a prestrained MOCVD growth technique for enhancing indium incorporation in InGaN/GaN QWs such that they can effectively emit light in the yellow-red range. By mixing with the blue light from another QW, we can obtain high-quality white light.

For mixing colors to obtain white light of a high rendering index, currently the major difficulty is the low emission efficiency of

red-emitting InGaN compounds [4]. Although the high-efficiency red emission from an InGaN/GaN nano-column structure has been reported, the process procedures for fabricating such a white-light LED can be quite complicated [5]. Before efficient red-emitting InGaN compounds for easy integration with blue- and green-emitting structures can be available, a photon down-conversion material is still needed. In particular, a material for converting blue photons into red light is very useful. Although phosphors for converting UV photons into red light exist, that for efficiently converting blue photons into red light has not been reported yet. Recently, it has been proved that the use of CdZe/ZnS nano-crystals for such conversion is quite attractive [6,7]. Basically, such a crystal of a few nm in diameter functions as a quantum dot. It can efficiently absorb light in the range from UV through blue and re-emit red light. Its absorption and emission spectra can be easily tuned through controlling its size. In the second part of this paper, we first demonstrate the growth and fabrication of a blue/green two-wavelength LED by stacking four QWs of two different growth conditions. Then, we show white-light generation by coating CdSe/ZnS nano-crystals on such a two-wavelength LED for converting blue photons into red light.

2. Fabrication of White-light Light-emitting Diodes with Prestrained Growth

We grow a white-light InGaN/GaN QW LED epitaxial structure with its EL spectrum close to the ideal condition in Commission International de l'Eclairage chromaticity based on the presrained MOCVD growth

978-0-9789217-3-6/07/$25.00
©2007 WEN GLOBAL SOLUTIONS

technique. The prestrained growth leads to the efficient yellow emission from three InGaN/GaN QWs of increased indium incorporation. The color mixing for white light is implemented by adding a blue-emitting QW at the top of the yellow-emitting QWs. The blue shifts of the blue and yellow spectral peaks of the generated EL spectra are only 1.67 and 8 nm, respectively, when the injection current increases from 10 to 70 mA. Such small blue shifts imply that the piezoelectric fields in our QWs are significantly weaker than those previously reported. Fig. 1 shows the EL spectra of an orange LED at various injection current levels based on the prestrained growth. Fig. 2 shows the EL spectra at various injection current levels of the aforementioned white-light LED by mixing the blue and yellow lights. Fig. 3 shows the picture of the phosphor-free white-light LED.

3. White-light generation with CdSe/ZnS nano-crystals coated on an InGaN/GaN quantum-well blue/green two-wavelength light-emitting diode

First, we grew and processed a blue/green two-wavelength light-emitting diode (LED) based on the mixture of two kinds of quantum well (QW) in epitaxial growth. The X-ray diffraction and photoluminescence measurements indicated that the crystalline structure and the basic optical property of individual kinds of QW are not significantly changed in the mixed growth. The relative electro-luminescence (EL) intensity of the two colors depends on the injection current level, which controls the hole concentration distribution among the QWs. At low injection levels, the top green-emitting QW dominates in EL. As the injection current increases, the blue-emitting QWs beneath becomes dominating. We also coated CdSe/ZnS nano-crystals on the top of the two-wavelength LED for converting blue photons into red light. With the coating of such nano-crystals, the device emits blue, green, and red lights for white-light generation. Fig. 4 shows the EL spectra of the blue/green two-wavelength LED at various injection current levels. Here, one can see that at relatively higher injection currents, the blue intensity is higher than the green intensity. Fig. 5 shows the

photoluminescence excitation (PLE) and PL spectrum of the used CdSe/ZnS nano-crystals, representing the absorption and emission spectra, respectively. One can see that the nano-crystals can absorb blue photons and emit red light. Fig. 6 shows the output spectra of the LED after the nano-crystals are coated on its top surface. Here, the mixing of the three colors leads to white light.

4. Discussions and Conclusions

In summary, we have demonstrated an prestrained MOCVD growth technique to successfully enhance indium incorporation of InGaN/GaN QWs. Based on this technique, we have grown a white-light InGaN/GaN QW LED epitaxial structure with its EL spectrum close to the ideal condition in CIE chromaticity. The prestrained growth led to the efficient yellow emission from three InGaN/GaN QWs of increased indium incorporation. The color mixing for white light was implemented by adding a blue-emitting QW at the top of the yellow-emitting QWs. The blue shifts of the blue and yellow spectral peaks of the generated white-light EL spectra were only 1.67 and 8 nm, respectively, when the injection current increased from 10 to 70 mA. Such small blue shifts implied that the piezoelectric fields in our QWs were significantly weaker than those previously reported.

Also, we have grown and processed a blue/green two-wavelength LED based on the mixture of two kinds of QW in epitaxial growth. The XRD and PL measurements indicated that the crystalline structure and the basic optical property of individual kinds of QW were not significantly changed in the mixed growth. The relative EL intensity of the two colors depended on the injection current level, which controlled the hole concentration distribution among the QWs. At low injection levels, the top green-emitting QW dominated in EL. As the injection current is increased, the blue-emitting QWs beneath became dominating. We then coated CdSe/ZnS nano-crystals on the top of the two-wavelength LED for converting blue photons into red light. By coating such nano-crystals, the device emitted blue, green, and red lights for white-light generation.

Acknowledgements

This research was supported by National Science Council, The Republic of China, under the grant of NSC 95-2120-M-002-012 and NSC 95-2221-E-002-287, and by US Air Force Scientific Research Office under the contract AOARD-06-4052.

References

1. B. Damilano et al., Jpn. J. Appl. Phys. **40** (2001), L918.
2. M. Yamada et al., Jpn. J. Appl. Phys. **41** (2002), L246.
3. M. Yamada et al., IEICE Trans. Electronics **E88–C** (2005), 1860.
4. M. Yamada et al., Jpn. J. Appl. Phys. 41, L246-L248, 2002.
5. A. Kikuchi et al., Jpn. J. Appl. Phys. Vol. 43, No. 12A, pp. L1524-L1526, Nov. 2004.
6. M. Achermann et al., Nature 429, 642-646, June, 2004.
7. D. M. Yeh et al., IEEE Photon. Technol. Lett. 18, March 2006.

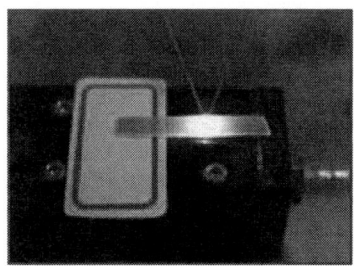

Fig. 3 Picture of the phosphor-free white-light LED.

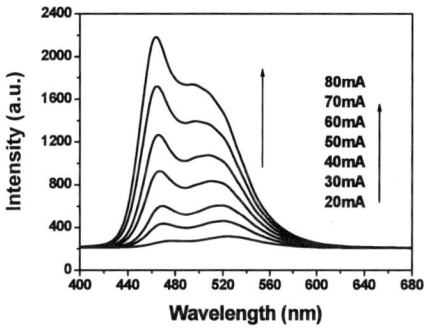

Fig. 4 Room-temperature EL spectra of the two-wavelength LED at various injection current levels.

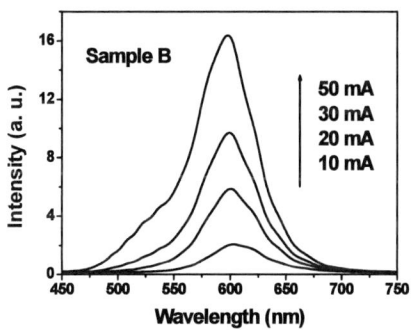

Fig. 1 EL spectra at various injection current levels of the orange LED.

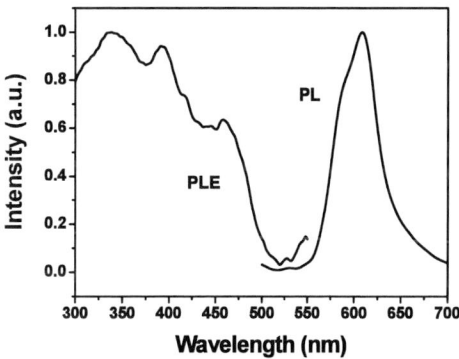

Fig. 5 Room-temperature PL and PLE spectra of the CdSe/ZnS nano-crystals.

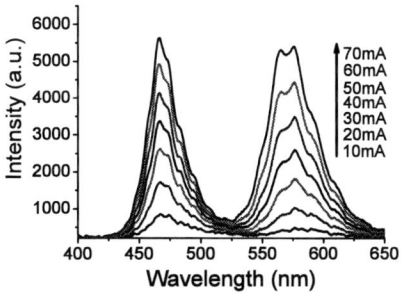

Fig. 2 EL spectra at various injection current levels of the white-light LED.

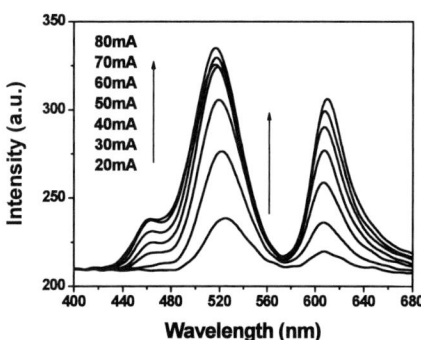

Fig. 6 Room-temperature EL spectra of the white-light device after nano-crystal coating at various injection current levels.

Comprehensive Investigation of Light Emission of OLEDs:
from Absolute Optical Properties to the Purcell effect

Wallace C.H. Choy (1*), X.W. Chen (1,2), H.H. Fong (3), S.L. He (2)

1 : Department of Electrical and Electronic Engineering, University of Hong Kong, Pokfulam Road, Hong Kong, China. *Corresponding author: chchoy@eee.hku.hk

2 : Centre for Optical and Electromagnetic Research, Zhejiang University; Joint Research Centre of Photonics of the Royal Institute of Technology (Sweden) and Zhejiang University, Zhijingang campus, Hangzhou 310058, China.

3. Department of Materials Science and Engineering, Cornell University, Ithaca, NY 14853, US

Abstract The dispersively absolute absorption coefficient and refractive index as well as the Purcell effect have been theoretically and experimentally studied for optimizing the light emission properties of organic light emtting devices.

Introduction

The performance of organic light-emitting diodes (OLEDs) depend strongly on the device structure. With the increasing complexity of the device configuration, efficient and rigorous modeling of light emission in OLEDs is desirable for optimizing device structures and getting a better understanding of the device physics. In classical approach of the OLEDs modeling, a number of schemes have been suggested. For instance, in the schemes described by Chance [1] and Kahen [2], the electric field and magnetic field over a surface surround the dipole are calculated to determine the total radiation power of the dipole. The scheme is rigorous but is inefficient. The half-space approach in [3] is simpler but cannot model OLEDs with two reflecting mirrors like top-emitting OLEDs. Besides, the effects of the change of radiative decay rate, which seems to be ignored in some recent studies, is important and should be considered in a rigorous model of the light emission in OLEDs. Meanwhile, the absolute optical properties of the organic materials are crucial for optimizing the layer sturcture of OLEDs. However, conclusive studies on the origins of the optical functions of the organic materials are limited.

In this paper, our targets are to determine and explain the absolute optical functions and develop an efficient and rigorous model of OLEDs with the consideration of the Purcell effect. The results will be experimentally and theoretically studied.

Experiment and Theory

In evaporating the organic thin film, the ITO and Si substrates are cleansed prior to loading into the evaporation chamber through scrubbing by detergent and soaking into de-ionized water for 10min in each step. The organic materials are purified by gradient sublimation prior to thin-film coating. The deposition rate is typically 1.0-2.0 Å/s and the substrate is Si. The evaporation chamber is operated at $\sim 10^{-6}$ Torr. Film thickness is monitored *in situ* using the quartz crystal monitor and *ex situ* by a stylus profilometer (Tencor α-step 500). The OLEDs

were biased a programmable Keithley source meter 2400.

The absolute optical functions are determined experimentally by using a VASE ellipsometer [a12]. The ellipsometer is operated with the antoretarder. The light source we used is a Xe lamp and the incident angle to the samples is selected as 65°, 70° and 75°. The diameter of the light spot is about 0.8 mm. The absolute optical properties were determined from the ellipsometry with a combined oscillator model of Lorentz, Guassian and Tauc-Lorentz oscillators. To investigate the origins of optical properties, the molecular orbital of Alq_3 and NPD are simulated using Gaussian 03 program [b5]. The molecular structure is optimized at the restricted Hartree–Fock (HF) level with the *6-31G(d)* basis set (*UHF/6-31G(d)*). The Kohn–Sham (KS) molecular orbital is calculated through Density functional theory (DFT) for the geometries obtained at the *UHF/6-31G(d)* level, employing the *B3LYP/6-31G(d)* functional. The vertical absorption gap is determined by finding the total energy difference between the excited state energy through the time-dependent density functional theory (TDDFT) and the DFT ground state energy with respect to the optimized geometry at the unrestricted HF level. In the model, the first 40 excited states have been calculated for the orgnanic materials. The HOMO and LUMO is considered to be Guassian shape while the width is evaluated from the time of flight mobility.

The light emission of multilayered OLED structures is rigorously modeled through classical electromagnetic approach with an emitting layer sandwiched between two stacks of films taking into account the Purcell effect [12] which will be strengthened in Fabry Perot structure [13] of OLEDs. The nonradiative losses due to the metal electrode and other materials used in the structure as well as the effects of thick glass substrate have been fully considered, which have been ignored by others [5, 14, 15]. The randomly oriented electric dipole with equal probability for all directions in space dipole

978-0-9789217-3-6/07/$25.00

©2007 WEN GLOBAL SOLUTIONS

locates in the recombination zone. The two stacks of films can be considered as two effective interfaces characterized by the total reflection and transmission coefficients. The total radiation power F, normalized by the radiation power of the dipole in an infinite medium, can be obtained [16]. Similarly, the normalized power U transmitted to the outmost region (air) can also be obtained. The radiative decay rate of excitons Γ_r^m is modified to be $F \cdot \Gamma_r^0$ by the Purcell effect and F is also commonly named as Purcell factor [18], where Γ_r^0 is the radiative decay rate in the infinite medium. With a typical assumption that the non-radiative decay rate Γ_{nr} is a constant, the internal quantum efficiency η_{int}^{cav} and exciton lifetime in the microcavity become

$$\eta_{int}^{cav} = \Gamma_r^m / (\Gamma_r^m + \Gamma_{nr}) = F / [(F-1)\eta_{int}^0 + 1] \cdot \eta_{int}^0 \quad (1)$$

where η_{int}^0 is internal quantum efficiency (QE) of the bulk emitting material. The photon out-coupling efficiency $\eta_{cp}(\lambda)$ is defined as $U(\lambda)/F(\lambda)$. Due to change of internal QE in the microcavity, the photon out-coupling efficiency should be modified as

$$\eta_{cp}^m(\lambda) = U(\lambda) / (1 + (F(\lambda)-1)\eta_{int}^0)] \quad (2)$$

where $\eta_{cp}^m(\lambda)$ is the modified photon out-coupling efficiency taking the Purcell effect into account. The emission characteristics of an OLED are obtained by averaging the dipole radiation over the recombination zone and wavelength of interest. Thus we define an integrated out-coupling efficiency η and angular intensity distribution $I(\alpha)$ as,

$$\eta = \int_{\lambda_1}^{\lambda_2} \eta_{cp}^m(\lambda) s_0(\lambda) d\lambda / \int_{\lambda_1}^{\lambda_2} s_0(\lambda) d\lambda \quad (3)$$

$$I(\alpha) = \int_{\lambda_1}^{\lambda_2} P(\alpha,\lambda) d\lambda / \int_{\lambda_1}^{\lambda_2} P_0(\lambda) d\lambda \quad (4)$$

where (λ_1, λ_2) is the considered wavelength range $P(\alpha,\lambda)$ and $P_o(\lambda)$ are the angular power density in outmost region (air) and the intrinsic emission spectrum of the emitting material respectively. The electroluminescence (EL) efficiency of the OLED (η_{ext}^0) equals $\eta_{int}^0 \cdot \eta$.

Results and Discussion

The experimental and theoretical absorption spectra of two robust materials of Alq3 and NPD are plotted together in Fig. 1. For Alq3, both the closely distributed excited states and the gradually changed oscillator strength make the overall absorption become single peaked at around 395nm as shown in Fig. 1(a). While the satellite peak at ~320nm has been described by the fast non-radiative relaxation from the ^1B_b to ^1L_a state [a1], our model can unveil the detail of the radiation transitions as the mainly transitions of the 10th, 11th and 12th excited states to ground state. The feature observed both in the experimental and theoretically results in Alq3 is not the same as that in NPD, where the first eight states

distribute into two small groups and make the absorption spectrum forms two peaks at ~320nm and ~370nm. Even though in experimental results, the two peaks have a strong broadening and merge, the shoulder which corresponds to the first overall absorption peak at ~370nm determined in theory can still be clearly resolved as show in Fig. 1(b). Consequently, the orbitals and the absorbance of Alq3 and NPD molecules determined from the density function approach can well describe the absorption features of the pristine organic films particularly the first absorption peak at long wavelength region which is important for the applications of organic materials such as the Förster energy transfer in guess-host OLEDs.

With the absolute optical properties, the layered OLED structures can be optimized by our device model. In order to intuitively illustrate the Purcell effect, we calculate the change of internal QE and exciton lifetime for all the four devices as shown in Table I and the schematic in Fig. 2. The results are listed in Table II. The results have been averaged over the wavelength of interest. We see that the increment of internal QE for fluorescent material in top emission OLED (TOLED) is considerably larger than in bottom emission OLED (BOLED). Device H_η has an increment of 84% in internal QE for fluorescent material. However, even in BOLED like device C, the Purcell effect cannot be neglected for analyzing and optimizing device structures. In fact, it has been reported that the internal QE is increased to 35%-45% for the conjugated polymers based emitters in a BOLED [3], which agrees to our calculation. The effect of internal QE increment was also reported in InGaN based emitters [d13]. As shown in Table II, the change of lifetime is also more prominent in TOLED than in BOLED. The results show that the conventional assumption of $\tau_{cav}/\tau_{con} \cong 1$ for the analysis of a TOLED [d7,d12] will overestimate the EL efficiency. Particularly for the phosphorescent TOLED with $\eta_{int}^0 = 1.0$, the overestimation is as large as 38% (see the lifetime of device A for $\eta_{int}^0 = 1.0$). But for fluorescent material, the over-estimation is smaller. For example, $\tau_{cav}/\tau_{con} = 0.816$ for device H_η and C. Another confirmation of our simulated result is the experiment reported in ref. [d7] where a nearly 20% reduction of the lifetime of the fluorescent emitter was observed. Consequently, apart from analyzing and optimizing TOLEDs for the maximum zero-degree luminance and EL efficiency in previously paragraphs, we also clearly show that the Purcell effect is important for accurate assessment and optimal design of not only TOLEDs but even weak cavity devices like conventional BOLEDs. Meanwhile, our simulated results are in good agreement to the experiments.

Conclusions

We have deterimined and described the absolute optical properties and presented an accurate analysis of TOLEDs taking the Purcell effect into account. The absolute otpical properties can be explained by the electrical transition of the molecular structures. The Purcell effect on the internal QE and the outcoupling QE as well as the photon lifetime has been comprehensively investigated and the model results show very good agreement to the experimental one for optimized the layered OLED strcutures.

References

[1] R. R. Chance, A. Prock, and R. Silbey, "Molecular fluorescence and energy transfer near interfaces," *Adv. in Chem. Phys.*, vol. 37, pp. 1–63, 1978.

[2] K. B. Kahen, "Rigorous optical modeling of multilayer organic light emitting diode devices," *Appl. Phy. Lett.*, vol. 78, pp. 1649–1651, 2001

[3] J.-S. Kim, P. K. H. Ho, N. C. Greenham, and R. H. Friend, "electroluminescence emission pattern of organic light-emitting diodes: implications for device efficiency calculations," *J. Appl. Phys.*, vol. 88, pp. 1073 -1081, 2000.

[4] John A. Woollam, Blaine Johs, Craig M. Herzinger, James Hilfiker, Ron Synowicki, and Corey L. Bungay, "Overview of Variable Angle Spectroscopic Ellipsometry (VASE), Part I: Basic Theory and Typical Applications", Critical Rev. of Optical Sci. and Technol., vol. CR72, pp. 3-28, 2002.

[5] http://www.gaussian.com

[6] E. M. Purcell, Spontaneous emission probabilities at radio frequencies," Phys. Rev. 69, 681 (1946).

[7] W. C. H. Choy and E. H. Li, "The applications of interdiffused quantum well in normally-on electro-absorptive Fabry-Perot reflection modulator," IEEE J. Quantum Electron., vol. 33, pp. 382-393, 1997.

[8] K. Neyts, P. D. Visschere, D. K. Fork and G. B. Anderson, "Semitransparent metal or distributed Bragg reflector for wide-viewing-angle organic light-emitting-diode microcavities," J. Opt. Soc. Amer. B, vol. 17, pp.114-119, 2000.

[9] O.H. Crawford, "Radiation from oscillating dipoles embedded in a layered system," J. Chem. Phys. 89, pp.6017-6027, 1989.

[10] V. Bulovic, V. B. Khalfin, G. Gu, P. E. Burrows, D. Z. Garbuzov, and S.R. Forrest, "Weak microcavity effects in organic light-emitting devices," Phys. Rev. B, vol.58, pp. 3730–3740, 1998.

[11] X. W. Chen, W. C. H. Choy and S. He,"Efficient and Rigorous Modeling of Light Emission in Planar Multilayer Organic Light-Emitting Diodes", IEEE J. Display Technol., vol.3, pp.110-117, 2007.

[12] D.Z. Garbuzov, V. Bulovic, P.E. Burrows, S.R. Forrest, "Photoluminescence efficiency and absorption of aluminum-tris-quinolate (Alq3) thin films", Chem. Phys. Lett., vol. 249, pp.433-437, 1996.

[13] K. Okamoto, I. Niki, A. Shvartser, Y. Narukawa, T. Mukai, and A. Scherer, Nature material, vol. 3, pp.601-603, 2004.

[14] C.L. Lin, H.W. Lin, and C.C. Wu, "Examining microcavity organic light-emitting devices having two metal mirrors", Appl. Phys. Lett. 87, pp.021101-1-3, 2005.

[15] R. H. Jordan, L. J. Rothberg, A. Dodabalapur, and R. E. Slusher, "Efficiency enhancement of microcavity organic light emitting diodes", Appl. Phys. Lett., vol. 69, vol. 1997-1999, 1996.

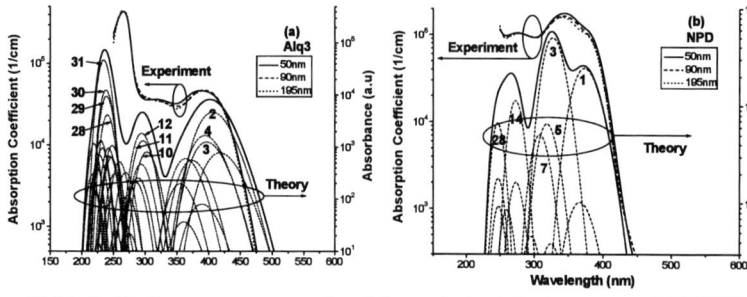

Fig. 1. The absorption coefficient determined from ellipsometry and the absorbance of Alq3 and NPD theoretically determined by orbital model. The number of the excited state corresponding to the major individual absorption spectra is given.

Table I. Design parameters and η of the optimized devices

	H_I_o	H_η	B	C	D
L(nm)	105	113	277	105	268
P(nm)	60	65	220	75	220
Q(nm)	68	68	60	/	/
η	0.44	0.52	0.41	0.28	0.27

Table II. Internal quantum efficiencies (IQE) and lifetime of excitons in devices H_η, B, C and D

	H_η		B		C		D	
	IQE	τ_{cav}/τ_0	IQE	τ_{cav}/τ_0	IQE	τ_{con}/τ_0	IQE	τ_{con}/τ_0
$\eta^0_{int}=0.25$	0.46	0.71	0.33	0.89	0.36	0.87	0.27	0.97
$\eta^0_{int}=1.0$	1.0	0.38	1.0	0.67	1.0	0.61	1.0	0.90

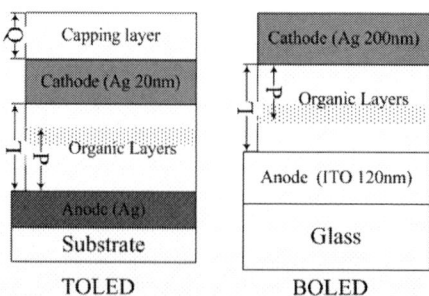

Fig. 2 Schematic diagrams of a TOLED and BOLED, where L, P and Q are the three design parameters.

Mutual Thermal Effects of Light-Emitting Diode with Wafer-Level Packages

Jae-Wan Choi, Jeung-Mo Kang, Jae-Wook Kim, Jeong-Hyeon Choi, Du-Hyun Kim,
Geun-Ho Kim, and Jeong-Soo Lee
LED R&D Lab., LG Electronics Institute of Technology, Seoul 137-724, Korea
jmokang@lge.com

Abstract *Wafer-level packaged LEDs are useful for the high power applications such as back light unit (BLU) and general solid state lighting due to the compactness and integrated fabrication process with Si-MEMS technology. In this paper, thermal characteristics of wafer-level packaged LEDs with four multi-chips are investigated including mutual thermal effects due to the adjacent chips using both serial and parallel measurement methods.*

Introduction
Light emitting diodes (LEDs) have superior characteristics to the conventional light bulbs in terms of a guaranteed long life, lower power consumption, higher brightness and less harmfulness. These merits of LEDs forced them to be used for various applications such as automotive lights, traffic signals, general lightings and back light units for a liquid crystal displays. However, the cost and the thermal design issues are key parameters in high power LED packages.

Wafer-level packages with Si-MEMS technology could be a solution for this problem [1]. Due to the well-developed Si-technology, a cost effective mass production will be possible with high yield. Unfortunately, the detailed thermal effects for the wafer-level packaged LEDs are not published up to now. In this paper, thermal characteristics of wafer-level packaged LEDs with multi-chips investigated including mutual thermal effects due to the adjacent chips using both serial and parallel measurement methods.

Fabrication of Waver-Level Packaged LED
Fig. 1 (a) illustrates the fabrication process of wafer-level packaged LEDs. Fig. 1(b) shows the schematic views of the wafer-level packaged LEDs (XiOB™-Silicon Optical Bench) including Zener diode to prevent high forward and reverse current. The produced XiOB with four blue chips was presented in Fig. 1(c).

The 3-dimensional shape of XiOB was produced by an anisotropy etching process. Then a reflector and side metals were deposited on the etched surface of XiOB to increase optical extraction efficiency by reflection and current driving, respectively. The

XiOB has merits in simple process, compactness and Zener-diode integration.

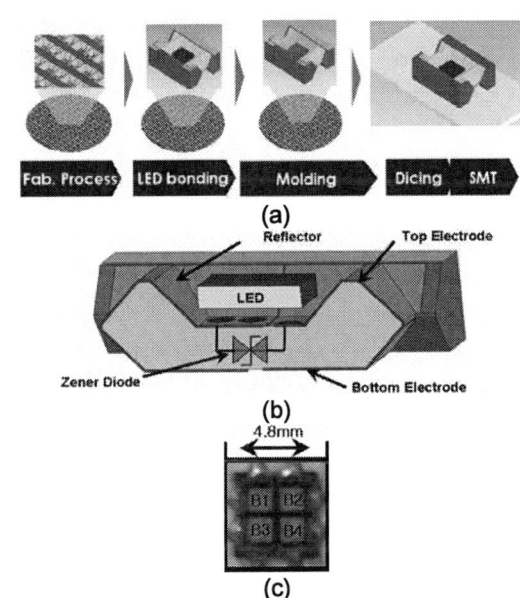

Fig. 1 (a) Fabrication process of wafer-level packaged LEDs
(b) Schematic views of the wafer-level packaged LEDs (XiOB™)
(c) The produced XiOB with four blue chips

Thermal Model of XiOB
The reliability of LEDs was related significantly with the junction temperature. And the thermal resistance was a key parameter for the proper thermal design in many applications. In JEDEC standards, for a single chip package, the relationship between thermal resistance and junction temperature was explained by following equation (1) [2].

$$R_{th,J-REF} = \frac{T_J - T_{REF}}{P} = \frac{\Delta T}{P} \qquad (1)$$

978-0-9789217-3-6/07/$25.00
©2007 WEN GLOBAL SOLUTIONS

where, $R_{th\ J-REF}$ is a thermal resistance between junction and reference point, T_J is a junction temperature, T_{REF} is a reference point temperature and P is the power dissipation from the package.

In case of multi-chip packages which consist of N identical chips, the R_{th} could be simplified as [3],

$$R_{th,Total\_pararllel} = \frac{T_J - T_{REF}}{P_{Total}} = \frac{R_{th,J-slug}}{N} + R_{th,slug-ambient} \quad (2)$$

where, $R_{th,\ Total-parallel}$ is the total junction-to-ambient thermal resistance of multi-chip packaged LED by parallel driving, $R_{th,\ J-slug}$ is the junction-to-slug thermal resistance of a LED chip, $R_{th,\ slug-ambient}$ is the slug-to-ambient thermal resistance and N is the number of chips, respectively.

Experimental Results and Discussions
A. Parallel Driving Thermal Measurements
The parallel thermal measurements are performed using a MicRed's thermal transient tester (T3Ster).

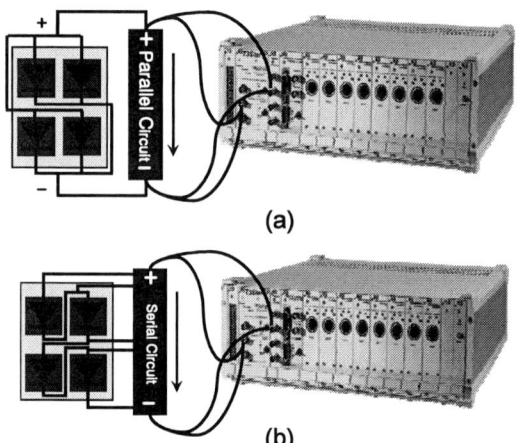

(a)

(b)

Fig.2 Experimental setup for XiOB
(a) Thermal measurements for parallel driving
(b) Thermal measurements for serial driving

Firstly, for the parallel thermal measurement, four blue chips on the XiOB were connected parallel as shown in Fig. 2 (a). We investigated following four cases to analyze the mutual thermal effects of multi-chips on a single XiOB package; i) single-chip driving (350mA), ii) two-chip driving (700mA), iii) three-chip driving (1,050mA) and iv) four-chip driving (1,400mA). In all cases, the operating current for each chip

was still remained by 350mA. Then, a calibration for each four case was performed with from 20°C to 60°C range using thermostat with a sensor current of 3mA. Obtained k-factors were -1.394 [mV/°C] for single-chip driving, -1.427 [mV/°C] for two-chip driving, -1.375 [mV/°C] for three-chip driving and -1.419 [mV/°C] for four-chip driving, respectively.

After calibration, each thermal transient curve of four cases was obtained with parallel circuit. The temperature of a thermostat was holed 25°C during thermal transient measurements. Fig. 3 shows the differential structure function curves of four parallel connection cases as mentioned above.

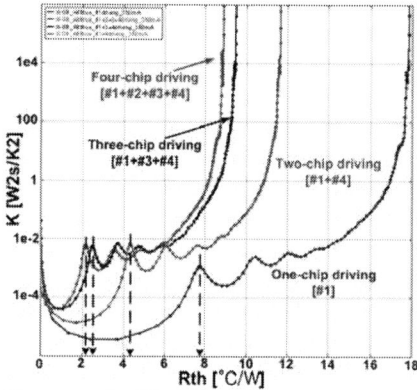

Fig.3 Measured differential structure functions of four cases by parallel driving thermal tests

Compared with thermal resistance of one-chip driving condition (7.76 [°C/W]), the thermal resistance of four-chip driving case (2.12 [°C/W]) were reduced greatly. The measured thermal resistances and junction temperatures of four cases were summarized in table 1. Measured thermal resistances coincided nearly with theoretical values using the equation (2). We think that the acceptable errors might come from the differences in four blue chips.

Table 1 Thermal resistance and junction temperature of parallel measurement methods

Parallel Method					
Configuration	K-factor [mV/°C]	Rth, J-slug [°C/W]	Rth, J-a [°C/W]	T_J [°C]	Current [mA]
1	−1.394	7.76	17.87	46.44	350
1+4	−1.427	4.32	11.66	53.28	700
1+3+4	−1.375	2.51	9.51	59.56	1050
1+2+3+4	−1.419	2.12	8.86	68.87	1400

B. Serial Driving Thermal Measurements
LED packages could be operated either in

parallel or serial driving according to the applications. In serial thermal measurements, individual k-factors and thermal transient curves of all four chips were obtained simultaneously. All four chips in a XiOB were connected serially as shown in Fig. 2 (b). Measured k-factors were -1.373 [mV/°C] for # blue 1, -1.429 [mV/°C] for # blue 2, -1.303 [mV/°C] for # blue 3 and -1.433 [mV/°C] for # blue 4, respectively. Fig. 4 shows the differential structure function curves of each chip which connected serially with driving current of 350mA. Thermal Characteristics of each four chip was nearly identical and has the range of thermal resistance from 33 [°C/W] to 35 [°C/W] except # blue 2 chip, which shown in redline in Fig. 4. Junction temperatures and thermal resistances of individual chips on XiOB using serial thermal measurements were listed in table 2.

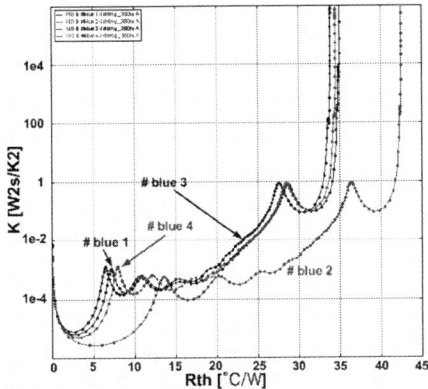

Fig.4 Measured differential structure functions of individual four chips using serial thermal tests

Table 2 Thermal resistance and junction temperature of each four chip on a XiOB using serial driving measurements

Serial Method			
# of Chip	K-factor [mV/°C]	Rth, J-a [°C/W]	T_J [°C]
# Blue 1	−1.373	34.84	66.57
# Blue 2	−1.429	42.46	76.82
# Blue 3	−1.303	33.79	66.1
# Blue 4	−1.433	34.42	67.79

In serial driving measurements, the mutual thermal effects due to other three adjacent chips were included. We measured both the amounts of self-heating due to the each powered-on chip itself and mutual thermal effects by other adjacent chips which were powered-off. Each individual chip's portion to

the junction temperature and thermal resistance could be calculated using matrix thermal measurement methods [4] and obtained values were listed in table 3. The diagonal terms in table 3, which mean the thermal effects due to the each powered chip, agree well to the thermal effects of one chip driving cases in table 1. Thus the serial thermal measurements methods of wafer-level packaged LEDs with multi-chip could be acceptable in thermal testing and design of LEDs.

Table 3 Thermal resistance and junction temperature due to the mutual thermal effects

Mutual Thermal Effect				
Rth	# Blue 1	# Blue 2	# Blue 3	# Blue 4
# Blue 1	19.05	6.12	4.87	5.68
# Blue 2	6.02	27.20	5.97	4.73
# Blue 3	4.70	6.03	17.71	5.99
# Blue 4	5.47	4.81	5.97	18.84
Rth_Total	35.24	44.17	34.51	35.24

delta T_J	# Blue 1	# Blue 2	# Blue 3	# Blue 4
# Blue 1	22.69	7.17	5.60	6.51
# Blue 2	7.46	33.14	7.35	5.86
# Blue 3	5.92	7.25	21.52	7.25
# Blue 4	7.05	5.87	7.44	23.39
Total Δ T_J	43.12	53.43	41.91	43.01
T_J	68.12	78.43	66.91	68.01

Conclusions

Thermal characteristics of wafer-level packaged LEDs with multi-chip are investigated and analyzed using both parallel and serial measurement methods. The serial measurement method which including mutual thermal effects was confirmed by matrix measurement analysis. These results are helpful to the thermal design and analysis for multi-chip LED packages.

References

[1] C. Tsou, et. al, "Silicon-Based Platform for Light-Emitting Diode," *IEEE Transaction on Advanced Packaging*, **Vol. 29**, 607-614 (2006).
[2] G. Farkas, et. al, "Electric and thermal transient effects in high power optical devices" *in Proc 20th IEEE SEMI-THERM Symposium, 168-176, (2004).*
[3] http://www.cree.com/products/pdf/XLampThermalManagement.pdf
[4] T. Treurniet, et. al, "Thermal Management in Color Variable Multi-Chip LED Modules" *in Proc 22th IEEE SEMI-THERM Symposium, 173-177, (2006).*

Enhance the extraction efficiency of ZnS:Mn TFEL by Photonic Crystals Structure

Yurong Jiang, Jinwei Li , Xia Li , Wei Xue

Department of Optical Engineering, School of Information Science and Technology, Beijing Institute
of Technology ,Beijing 100081, China

E-mail :yrkitty@bit.edu.cn

Abstract A photonic crystals(PC) structure is introduced in the Thin-film electroluminescence (TFEL) device to enhance the light extraction efficiency. FDTD method was used to simulate and get the optimized photonic crystals structure parameters.

Introduction

Thin-film electroluminescence (TFEL) device is a very promising flat panel display. But its light extraction efficiency is restrained by the high index materials in the device, especially in ZnS:Mn Thin-film electroluminescence device. In this paper, we introduce a photonic crystals (PC) structure between glass substrate and ITO film to enhance the light extraction efficiency of the TFEL, without causing any major change in their electrical properties.

Model

Our structure used for the simulation is illustrated in Fig. 1. In the structure, metal electrode is made of Al metal and assumed to have a complex refractive index (n=0.867+i6.49). Insulator is made of Y_2O_3 film and the refractive index is 1.92. Phosphor is made of ZnS:Mn film and the refractive index is 2.5. Anode is made of ITO film and the refractive index is 1.90. The light emitting excitions insides the TFEL are modeled as dipole pulse that has a center frequency of 570 nm and a full bandwidth at half-maximum of 50 nm. There are several potential positions where a PC slab can be inserted into a TFEL device. From the viewpoint of the structure and fabrication of device, it maybe the best choice that the PC slab is inserted into the interface between ITO and glass substrate. If the PC slab is inserted in other interface, it will cause undesirable effect about electrical properties.

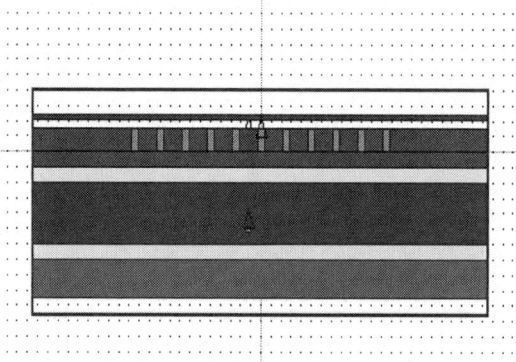

Fig. 1. Schematic diagram of stimulation model of TFEL device

Simulation and Results

FDTD method is used to stimulate the output extraction efficiency of device. During the stimulation, square and hexagonal lattice photonic crystals are inserted and calculated respectively. Firstly, the structure parameters of the common device and the device with PC slab are the same except the PC slab , which means that the thickness of film and calculation zone have no difference between the above two devices. Secondly a dipole source is positioned at the center of phosphor layer. Finally, the extraction power between common device and device with PC slab is compared. The output extraction efficiency of Zns:Mn TFEL device can be obtained.

We use Fullwave software of Rsoft Corporation to stimulate the extraction efficiency. The device size is too large for a direct FDTD calculation, so a $5 \times 5 \times 3$um volume is chosen as the working region and a PML boundary conditions is used.

978-0-9789217-3-6/07/$25.00

©2007 WEN GLOBAL SOLUTIONS

There are several parameters of Photonic crystals will affect the output efficiency, so a Orthogonal experiment design is the best choice for us to get the optimized structure parameters. Period, radius, height and refractive index of Photonic crystals are chosen as the four factor of the experiment and each factor has three levels, so that a L9 （3^4） Orthogonal experiment design can be achieved. Factors and levels are illustrated in Form Table 1 and L9 （3^4） Orthogonal experiment design is illustrated in Table2.

Table 1 Factors and Levels

Factor	Period (nm)	Diameter	Height (nm)	Refractive index
level1	400	0.3*P	200	1.5
level2	600	0.5*P	300	1.8
level3	800	0.7*P	400	2.1

Table 2 L9 （3^4） Orthogonal experiment design

	Period (nm)	Diameter	Height (nm)	Refractive index
1	400	0.3*P	200	1.5
2	400	0.5*P	300	1.8
3	400	0.7*P	400	2.1
4	600	0.3*P	300	2.1
5	600	0.5*P	400	1.5
6	600	0.7*P	200	1.8
7	800	0.3*P	400	1.8
8	800	0.5*P	200,	2.1
9	800	0.7*P	300	1.5

The calculation result is illustrated in form 3 and form 4. The output power of experiment 0 is the power of common device, while the output power of experiment 1-9 is the power of device with PC slab.

Table 3 Output power of common device and device with 2D square lattice Photonic crystal

Simulation	Output (Arbi-Unit)
0	2.762
1	2.521
2	2.626
3	2.347
4	2.349
5	5.111
6	3.237
7	2.692
8	2.814
9	4.025

Table 4 Output power of common device and device with 2D hexagonal lattice Photonic crystal

Simulation	Output (Arbi-Unit)
0	2.762
1	2.676
2	2.660
3	2.964
4	3.000
5	3.881
6	2.860
7	2.998
8	3.016
9	4.638

Conclusions

In summary, we have investigated the introduction of a 2D PC slab into the TFEL in order to enhance the output extraction efficiency and calculated the efficiency using FDTD method while output power of experiment while varying period, radius, height and refractive index of photonic crystals. In our calculation, 1.85 times of enhancement of light extraction efficiency has been predicted when a square PC slab with a period 800nm, a diameter of 300nm, a height of 400nm and refractive index of 1.5 is inserted into the TFEL device.

References

1 D-H. Kim et al, Appl .Phys.Lett., Vol.87(2005): 203508

2 M. Boroditsky et al, Appl. Phys. Lett.,

1999, Vol. 75(1999):1036-1038.

3 Yong-Jae Lee et al, Appl. Phys. Lett.,

Vol.82(2003): 3779-3881.

4 Young Rag Do et al, J. Appl. Phy, Vol.

96(2004): 7629-7636

5 Masayuki Fujita, et al, Proc. of SPIE,2005,

Vol. 5624(2005):142-151

6 Y. R. Do et al, Appl. Phys. Lett, Vol. 82(2003): 4172-4174.

A Facile Route to Synthesize Three-Dimensional CdS Nanocrystals

Fei Chen, Renjia Zhou, Mang Wang, Hongzheng Chen*

Department of Polymer Science and Engineering, State Key Lab of Silicon Materials,

Zhejiang University, Hangzhou 310027, P. R. China

Key Laboratory of Macromolecule Synthesis and Functionalization (Zhejiang

University), Ministry of Education, Hangzhou 310027, P. R. China

(Tel:+86-571-87952557; Fax: +86-571-87953733; Email: hzchen@zju.edu.cn)

Abstract: Synthesis of nanoscale inorganic crystals with controlled size, shape, and hierarchy has attracted intensive research interest, since they are potential building blocks for advanced materials and optoelectronic devices. As an important group II-IV semiconductor with extensive applications in optical and electronic nanodevices, complex three-dimensional nanocrystals CdS with well-defined structures have just received great attentions since they offer new opportunities as building blocks to fabricate or assemble into advanced materials for further generation of optoelectronic devices. we report for the first time to large-scale synthesize an interesting 3D structured CdS nanocrystals by a facile hydrothermal method. The photoluminescence results of the 3D CdS nanocrystal indicated they have the potential applications for the building blocks on optoelectronic devices.

High power fiber sources: more than kilowatts

Johan Nilsson

Optoelectronics Research Centre, University of Southampton
Southampton SO17 1BJ, England
jn@orc.soton.ac.uk

Abstract *Fiber lasers and amplifiers have reached several kilowatts of output power, and offer many attractions beyond raw power. This tutorial discusses basic scientific and practical issues of fiber sources, and advantages and limitations that derive from them.*

Introduction

The huge investment made by the tele-communications community in photonics over the last three decades brings a considerable opportunity to disrupt also other areas with new optical technology. The extraordinary level of optical control coupled with the extended reliability that is routine in telecoms is regarded as revolutionary in, e.g., lasers for industrial processing or the life sciences. Remarkably, the mW of tele-communications can be scaled with near-perfect beam quality to the kilowatt level using fiber amplifiers. These attributes and the ability to control high power through modulation of a small seed laser, or **M**aster **O**scillator, followed by a **P**ower **A**mplifier make these so-called MOPAs revolutionary light sources that can enable and dominate a range of application areas.

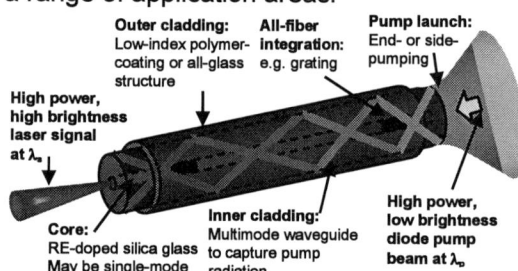

Figure 1. Schematic of cladding-pumped fiber laser.

This tutorial will cover the basics of lasers and amplifiers based on cladding-pumped silica fibers, as exemplified in Fig. 1. This includes the fundamental advantages and limitations of the fiber geometry and guided-wave operation, key properties of the fiber material, and important practical issues such as pump coupling. Thermal considerations, nonlinearities, and spectroscopy are examples of important issues, with different impact for different types of fiber sources made with different active dopants (Yb, Er, Tm, Nd) in different modes of operation.

Ytterbium-doped fiber lasers at 1.1 μm

Ytterbium-doped fiber lasers (YDFLs) exhibit a host of attractive features that make them superior for scaling to the highest powers. These features allowed them to be scaled beyond 1 kW and now up to 3 kW with (nearly) diffraction-limited output [1], [2], [3]. The key features behind this power scalability are the thermal management that the fiber geometry facilitates, the low quantum defect of Yb^{3+}, the high efficiency that can be achieved even at high Yb-concentrations, the high damage threshold and low loss of the silica host, and the excellent properties of suitable pump diodes in the 915 – 975 nm wavelength range.

The high Yb-concentration and the high pump absorption it enables bring many performance advantages. It allows the fibers to be shorter, which mitigates non-linear effects, and it also facilitates the use of fibers with the large inner claddings that are required for kilowatt-level operation. An important aspect of cladding-pumped fiber lasers is that they convert diode pump beams of relatively low brightness into a much brighter, even diffraction-limited, beam from the fiber laser. The high Yb-concentration is crucial also for this. Ytterbium-doped fibers pumped at the absorption peak at around 975 nm allow a pump beam of relatively low intensity launched into the inner cladding to be converted into a signal of roughly three orders of magnitude higher intensity, as averaged over the core. This is limited, roughly, to one order of magnitude less than the ratio of the pump absorption to the background loss. If we try to enhance the intensity conversion further, the efficiency will suffer as the background loss through the length of fiber increases. In addition, the pump-NA can be up towards one order of magnitude larger than the sig-

978-0-9789217-3-6/07/$25.00

©2007 WEN GLOBAL SOLUTIONS

nal-NA (say, 0.5 vs. 0.05). Thus, as the brightness is approximately proportional to the inverse of the square of the NA, an YDFL can improve the brightness by five orders of magnitude in total, with current fiber technology.

Er:Yb co-doped fiber lasers at 1.6 μm

Erbium:ytterbium co-doped fiber lasers (EYDFLs) are the preferred choice for high-power sources in the important 1.5 – 1.6 μm wavelength range. This range is attractive because of its "eye-safer" nature and for its compatibility with telecom components. Ytterbium sensitization increases the pump absorption, which is quite low in Yb-free Er-doped fibers. While power levels as high as 0.3 kW have been reached with EYDFLs [4], this is still an order of magnitude below that from YDFLs. The primary reason for this is the much more complex spectroscopy of EYDFLs. It involves energy transfer from Yb-ions to the lasing ions, and this constitutes a bottle-neck which hinders power-scaling. Figure 2 shows a measurement of the Yb fluorescence decay time of an EYDFL, which is closely related to the energy transfer time.

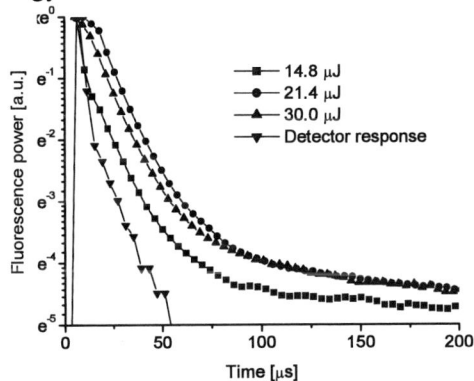

Figure 2. Yb fluorescence decay curves following excitation with 100 ns, 920 nm pulses of different energies. From [4].

It is difficult to reach transfer rates as high as those of Fig. 2, and further improvements would require very challenging developments of the Er:Yb gain medium. The thermal load can also be troublesome in high-power EYDFLs, but the fiber geometry helps in this respect, and if necessary additional fiber and device-level engineering can be used to further improve the thermal management.

Neodymium-doped fiber lasers at 0.9 μm

Neodymium-doped fiber lasers (NDFLs) typically emit at around 1060 nm, but because of factors such as a much lower permissible concentration and therefore pump absorption, they are out-performed by YDFLs at that wavelength. However, NDFLs can also emit at 0.9 μm, but this only works if the normally dominant 1060 nm emission is suppressed. In fibers, this is possible with a core that guides at 0.9 μm but not at 1060 nm. Such so-called waveguide filters can be realized with W-type fibers [5], [6] and hollow optical fibers [7]], both of which have a cut-off wavelength for the fundamental mode, as well as with photonic bandgap fibers, which only guide over a restricted wavelength range [8]. Figure 3 shows the transmission characteristics of a W-type fiber at different wavelengths with a clear loss difference between 920 nm and 1060 nm. Figure 4 illustrates the near complete suppression of the 1060 nm emission.

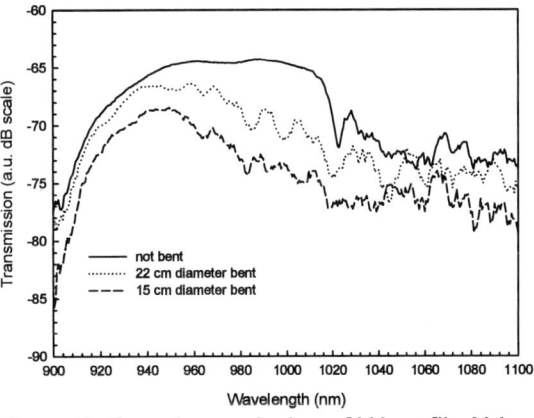

Figure 3. Core transmission of W-profile Nd-doped fiber at different bend radii. From [6].

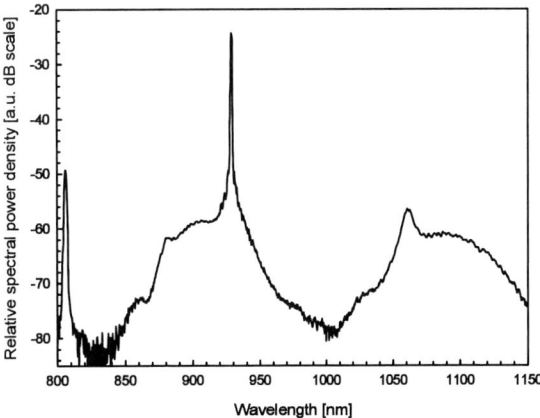

Figure 4. Spectrum of a 0.9 μm Nd-doped fiber laser. Resolution 1 nm. From [6].

This type of engineerability greatly adds to the attraction of fibers as laser gain media.

Single-frequency MOPAs

Fibers are attractive for high-power MOPAs because of their unique capability of high gain, high power, and high efficiency, all at the same time. Single frequency fiber MOPAs take advantage of this, and amplify a well-controlled but low-power seed source to high powers in a cascade of amplifiers providing 40 – 50 dB of gain. However, stimulated Brillouin scattering (SBS) limits the power that can be reached. SBS is a nonlinear effect, so it can be mitigated by a shorter effective length and a larger mode area. Broadening of the signal linewidth as well as the Brillouin gain bandwidth helps, too. In the absence of broadening, the Brillouin-limited output power is given by

$$\sqrt{\frac{I_p A_{dop} A_{eff} G_B^{max} \alpha_p^{core}}{g_B}}$$

where I_p is the pump intensity that can be launched into the fiber amplifier, A_{dop} is the doped area (typically the core area), A_{eff} the effective area of the mode, G_B^{max} the maximum allowable Brillouin gain, α_p^{core} the core pump absorption, and g_B is the unbroadened Brillouin gain coefficient. According to this, it is difficult to reach output powers of more than around 100 W in single-frequency Yb-doped fiber MOPAs. However, with thermal broadening of the Brillouin gain we have reached over 0.4 kW of output power with SBS-free linearly polarized operation. Other means of Brillouin suppression have also been demonstrated recently.

Many other types of sources, including pulsed ones, also benefit from the high level of control that fiber MOPAs provide.

Summary

These examples of devices illustrate some important issues for high-power fiber systems, as well as their potential. Other important issues include the pump launch and the pump brightness, and in particular the core size. It is hugely beneficial to use a large core. Many interesting concepts have been proposed, as will be discussed in the tutorial.

References

1 Y. Jeong, J. K. Sahu, D. N. Payne, and J. Nilsson, "Ytterbium-doped large-core fiber laser with 1.36 kW continuous-wave output power," Opt. Express **12**, 6088-6092 (2004)

2 V. Reichel, K. W. Moerl, S. Unger, S. Jetschke, H.-R. Mueller, J. Kirchof, T. Sandrock, A. Harschack, A. Liem, J. Limpert, H. Zellmer, and A. Tünnermann, "Fiber-laser power scaling beyond the 1-kilowatt level by Nd:Yb co-doping," Proc. SPIE **5777**, 404-407 (2005)

3 www.ipgphotonics.com (2007)

4 Y. Jeong, S. Yoo, C. A. Codemard, J. Nilsson, J. K. Sahu, D. N. Payne, R. Horley, R. Horley, P. W. Turner, L. M. B. Hickey, A. Harker, M. J. Lovelady, and A. N. Piper, "Erbium:ytterbium co-doped large-core fiber laser with 297 W continuous-wave output power", IEEE J. Sel. Top. Quantum Electron. **13**, 573-579 (2007)

5 T. J. Kane, G. Keaton, M. A. Arbore, D. R. Balsley, J. F. Black, J. L. Brooks, M. Byer, L. A. Eyres, M. Leonardo, J. J. Morehead, C. Rich, D. J. Richard, L. A. Smoliar, and Y. Zhou, "3-Watt blue source based on 914-nm Nd:YVO4 passively-Q-switched laser amplified in cladding-pumped Nd:fiber", presented at Advanced Solid-State Photonics, Santa Fe, NM, USA, February 2-5, 2004, paper MD7.

6 D. B S. Soh, S. W. Yoo, J. Nilsson, J. K. Sahu, K. Oh, S. Baek, Y. Jeong, C. A. Codemard, J. Kim, and V. N. Philippov, "Neodymium-doped cladding pumped aluminosilicate fiber laser tunable in the 0.9 μm wavelength range", IEEE J. Quantum Electron. **40**, 1275-1282 (2004)

7 J. Kim, P. Dupriez, D. B. S. Soh, J. Nilsson, and J. K. Sahu, "Core area scaling of Nd:Al-doped silica depressed clad hollow optical fiber and Q-switched laser operation at 0.9 μm", Opt. Lett. **31**, 2833-2835 (2006)

8 A. Wang A, A. K. George, and J. C. Knight, "Three-level neodymium fiber laser incorporating photonic bandgap fiber", Opt. Lett. **31**, 1388-1390 (2006)

9 Y. Jeong, J. Nilsson, J. K. Sahu, D. N. Payne, R. Horley, L. M. B. Hickey, and P. W. Turner, "Power scaling of single-frequency ytterbium-doped fiber master oscillator power amplifier sources up to 500 W", IEEE J. Sel. Top. Quantum Electron. **13**, 546-551 (2007)

Dispersion Controlled in a Birefringent Modified Octagon Photonic Crystal Fiber for Optical Communication Applications

S. F. Kaijage [1], Y. Namihira [2], N. H. Hai, F. Begum, S. M. A. Razzak, T. Kinjo, and N. Zou

Graduate school of Engineering and Science, University of the Ryukyus, 1 Senbaru, Nishihara
Okinawa, 903-0213 Japan

Email: 1) k068455@eve.u-ryukyu.ac.jp , 2) namihira@eee.u-ryukyu.ac.jp

Abstract *We report a dispersion controlling technique with a modified octagon photonic crystal fiber. The low flattened dispersion feature, as well as low leakage loss and high birefringence are the main advantages of the proposed PCF.*

Introduction

Photonic crystal fibers (PCFs) have in recent years attracted much scientific and technological interest. Broadly speaking, PCFs may be defined as optical fibers in which the core and/or cladding regions consist of microstructured rather than homogeneous materials. The most common type of PCF, which was first fabricated in 1996 [1], consists of a pure silica fiber with an array of air holes running along the longitudinal axis. A desirable property of PCFs is that, the additional design parameters of air-hole diameter d, air-hole rings and hole-to-hole spacing, the pitch Λ, offer much greater flexibility in the design of transmission properties to get the required applications. By manipulating circular air holes diameter and pitch, it is possible to control the PCF dispersion properties, for example to change the zero-dispersion wavelength or to engineer the dispersion curves to be ultra flattened [2].

Chromatic dispersion and losses in optical fibers limits the data carrying capacity in optical communication systems, causes optical pulse to spread resulting to inter-symbol interference, and becomes a critical problem as transmission rate exceeds 10 Gb/s. So far, several techniques to control dispersion and leakage properties have been reported [2, 3]. Various PCF designs have been proposed to control different transmission properties, such as hexagonal PCFs, square PCFs and octagonal PCFs [3, 4, 5]. These literatures demonstrate that periodic lattice arrangements for the air-hole are not absolutely necessary to achieve the guidance of the light based on the total internal reflection mechanism. Furthermore, octagonal PCFs (OPCFs) possess the advantages of wider wavelength range operating in single mode region, more circular-like field distribution, and significantly less confinement loss [5]. At the same time PCFs with polarization maintaining (PM) properties (with high birefringence characteristics) are also

required in many applications as they can eliminate the influence of polarization-mode dispersion or stabilize the operation of optical devices. However, current PM fibers, such as polarization maintaining and absorption reducing or bowtie fibers usually show modal birefringence of about 5×10^{-4}. In PCFs birefringence can be produced intentionally by using squeezed crystal lattice, where by the air hole along two orthogonal axes are different [6], and by introducing size change of few air holes[7].

In this paper, as a part of ongoing efforts to find simple PCFs designs, with a modified OPCF (M-OPCF) in a structure of omitting several air holes in the core of the conventional OPCF to form an elongated core, chromatic dispersion and birefringence characteristics can be controlled. All the transmission properties were calculated and analyzed using the finite difference method (FDM) with perfectly matched layer (PML) absorbing boundary conditions [8].

Design guidelines of the proposed PCF

Fig. 1 depicts a proposed index guiding modified OPCF with a background of the pure silica to control both dispersion and polarization properties. A, B and C are the missing air holes, Λ is the pitch, and d and d_1 are the air hole diameters for large and small air holes respectively as illustrated from Fig. 1. The air holes are arranged in an isosceles triangle lattice with a vertex angle of 45°, including the geometric parameters of the air-hole diameters d and d_1, the spacing between air holes on the same air-hole ring is Λ_1, which is related to spacing between the adjacent rings Λ (pitch) by the expression, $\Lambda_1 = \Lambda\sqrt{2-\sqrt{2}} \approx 0.765\Lambda$. The judicious choice of air-hole diameters and pitch facilitate low flattened dispersion, high birefringence and low leakage loss characteristics simultaneously.

978-0-9789217-3-6/07/$25.00

©2007 WEN GLOBAL SOLUTIONS

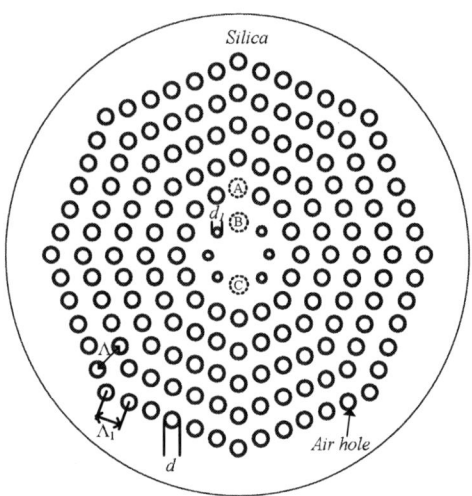

Fig. 1, Cross-section of the proposed OPCF. The background is a pure silica (n=1.45). A, B and C are the missing air-hole channels.

It is also worth noting that our proposed PCF does not have the doped core and hence lowers the possibilities of high optical losses raised by high doping concentration.

The chromatic dispersion is calculated by [4]:

$$D(\lambda) = -\frac{\lambda}{c} \frac{d^2 \, \mathrm{Re}[n_{\text{eff}}]}{d\lambda^2} \qquad (1)$$

in ps/(nm.km), whereby, λ is the operating wavelength, Re (n_{eff}) is the real part of the refractive index, n_{eff} and c is the velocity of light in a vacuum. The material dispersion given by Sellmeier's formula is directly included in the calculation.

Confinement loss is the light confinement ability within the core region. The resulting confinement loss L_c, is obtained from the imaginary part of n_{eff} as follows [4]:

$$L_c = 8.686 \, k_0 \, \mathrm{Im}[n_{\text{eff}}] \qquad (2)$$

in dB/m, where $k_0 = 2\pi/\lambda$ is the free space wave number. We evaluate phase modal birefringence B and polarization beat length L_B of the proposed PCF as in [7]:

$$L_B = \frac{2\pi}{|\beta_x - \beta_y|} = \frac{\lambda}{|n_x - n_y|} = \frac{\lambda}{B} \qquad (3)$$

Where β_x and β_y are propagation constants of two modes and n_x and n_y are the refractive index of these modes.

Results and discussion

Fig. 2 demonstrates the chromatic dispersion for the optimized proposed OPCF as a function of wavelength. The optimized set of design parameters have been obtained through numerical simulations to be Λ=1.65 µm, d=1.02

µm, d_1=0.5 µm. To quantify the performance of the proposed PCF we have computed the ultra low flattened chromatic dispersion from -0.45 to 0.09 ps/(nm.km) within 1.45 µm~1.7 µm wavelength range. At 1.55 µm wavelength, the chromatic dispersion of about 0.039 ps/(nm.km) and dispersion slope of about 2.25 x 10^{-3} ps/(nm^2.km) are obtained. Furthermore, it is revealed that it may be possible to shift the zero dispersion wavelengths to any wavelength in the telecommunication window while maintaining low flattened chromatic dispersion.

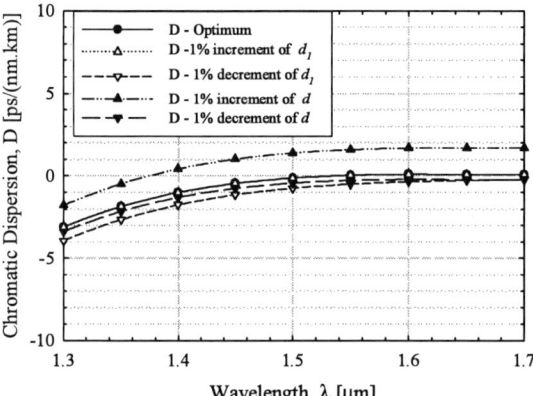

Fig. 2 Chromatic dispersion properties of the M-OPCF for the optimized design parameters, Λ=1.65 µm, d= 1.02 µm and d_1=0.5 µm, and for variation of design parameters d and d_1.

However, in standard fiber draw, 1% variations in fiber diameters may occur [3]. From this point of practical feasibility, we studied the influence of ±1% change in air-hole diameters d and d_1 while the pitch was fixed as depicted in Fig. 2. It is observed that there is an increment of dispersion figure (from 0.039 to 1.58 [ps/(nm.km)] at λ=1.55 µm) when the parameter d is increased by 1% from its optimum value. Also, a very small change was revealed when the diameters d and d_1 were reduced by 1% from the optimum design value. However, dispersion figure remains almost unchanged from its optimum dispersion value for the 1% increment of the parameter d_1. Therefore, it can be concluded that for ±1% variations in the parameters d and d_1 do not affect the chromatic dispersion drastically and are within the acceptable limit. The influence of air hole diameters gives an upper bound on the severity of the fabrication imperfection. In practical applications, only the low flattened dispersion behaviors may not be enough for justifying the usefulness of our proposed PCF. Low confinement loss and/or relatively small effective mode area are needed for some optical applications like non linear ones. Fig. 3 shows

364

confinement losses and effective area of the proposed OPCF for the optimized parameters against wavelength. Effective area A_{eff} of about 8.83 µm^2 at 1.55 µm wavelength is depicted for the optimized design parameters, which is small value compared to that of conventional optical fiber (about 85 µm^2). From these results, the proposed PCF is deemed suitable for non linear optical applications, such as super continuum generation etc. Low confinement loss of less than 10^{-5} dB/km at the entire telecommunication window is obtained as illustrated in Fig. 3. Fig. 4 shows the electric field distribution of the x-polarized mode at 1.55 µm wavelength. We can see the strong confinement of light in the core of the M-OPCF.

Birefringence emerges due to unidirectional extension of the core. Birefringence decreases with wavelength because the mode field diameter falls as the wavelength shortens and hence causes a reduction in a modal asymmetry as demonstrated in Fig. 5. At 1.55 µm, the birefringence value is about 8×10^{-4} which is larger than that of the conventional PM fibers. This means that the proposed PCF can be used as PM optical transmission fiber while exhibiting low-flattened dispersion as well as low leakage loss. Furthermore, the polarization beat length of M-OPCF is illustrated in Fig. 5.

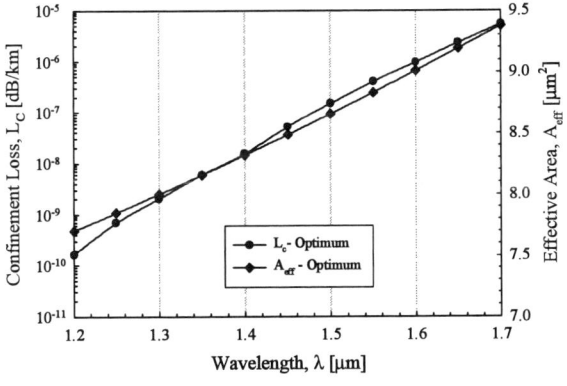

Fig. 3 Confinement loss and effective area against wavelength for the optimized design parameters, Λ=1.65 µm, d= 1.02 µm and d_1=0.5 µm.

At λ=1.55 µm, the beat length of about 1.98 mm is obtained for the optimized design parameters, far shorter than in conventional high birefringent fibers.

Conclusion

In this paper, dispersion and polarization properties of undoped modified octagon photonic crystal fiber have been investigated and FDM method applied to analyze them. A PCF with low flattened dispersion, low leakage loss and high birefringence has been proposed. The proposed modified OPCF can be potentially used where the ultra low flattened dispersion, low leakage loss, and polarization maintaining characteristics are simultaneously needed especially in WDM networks.

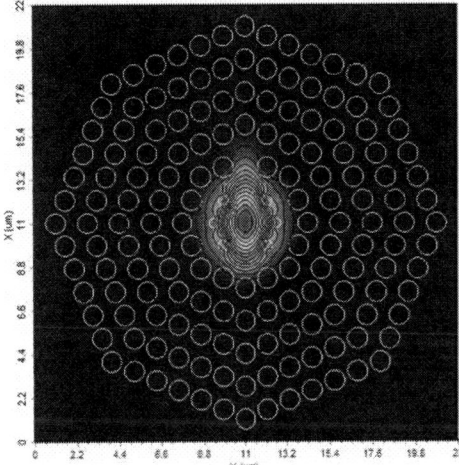

Fig. 4 Electric field distribution of the x-polarized fundamental mode at λ= 1.55 µm for the proposed OPCF.

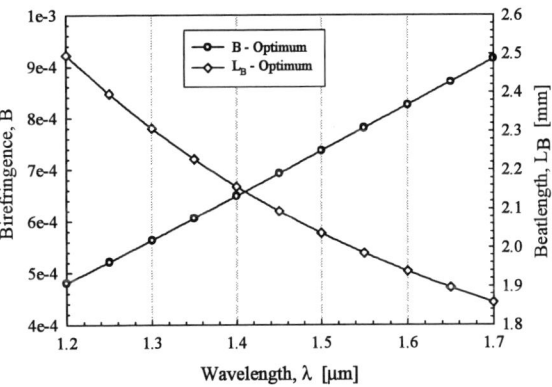

Fig. 5 Birefringence and polarization beat length characteristics of the proposed OPCF against wavelength for the optimized design parameters.

References

1 J. C. Knight et al,Opt. Lett, 21 (1996), 547.
2 S. M. A. Razzak, Y. Namihira, et al,Opt. Rev, 14(2007), 14.
3 W. H. Reeves et al, Opt. Express, 10(2002), 609
4 F. Begum, Y. Namihira et al, IEICE Trans. Electron., E90-C (2007), 607-612
5 J. Chiang et al.,Opt. Comm. 258 (2006),170-176
6 L. Zhan et al, Opt. Express, 11(2004),2371-2376
7 A. Ortigosa-Blanch et al, Opt. Lett, 25(2000), 1325
8 S. Guo et al, Opt. Express,12(2004),3341

Full-vector Effective Index Method for Modeling Endlessly Single-mode and Large Mode Area of Photonic Crystal Fiber

Wang Liwen (1), Lou Shuqin(1), Chen Weiguo(1), FANG Hong(1)

1 : Key Lab of All Optical Network & Advanced Telecommunication Network, Beijing Jiaotong University, Beijing, 100044, China and **email:**shqlou@bjtu.edu.cn

Abstract *Full-vector effective index method for modeling photonic crystal fiber is developed and the influence of structure parameters such as hole size, pitch and number removed in the center on endlessly single-mode and large mode area are discussed in the detail.*

Introduction

Photonic crystal fiber (PCF) is made from a single material such as silica [1]. A regular hexagonal array of air holes runs along the length of the fiber. Some holes in the center is removed, and the resulting solid region acts as a core, and light is guided in this core. These fibers exhibit many unusual properties, such as an endlessly single mode, highly tunable dispersion, and highly controllable effective mode areas for linear and nonlinear applications [2-7] and so on.

Large mode area single-mode fibers are essential to many application including high power delivery, fiber amplification, ultra-high-speed communication system and low non-linear data transmission. It is difficult for conventional single-mode fibers to maintain large mode area and single-mode operation simultaneously, but PCFs are the ideal candidates for designing the large mode area and single mode fibers, due to their endlessly single-mode properties [3]. In this paper, Full-vector effective index method (FVEIM) for modeling photonic crystal fiber is developed and the influence of structure parameters such as hole size, pitch and number removed in the center on endlessly single-mode and large mode area are discussed in the detail.

Methodology

Cross section of the triangular PCF is shown in Fig.1. In the cladding region, the air holes with the same diameter arranged in an array of the triangular lattice on the silica glass. The diameter of air hole and hole pitch is represented by d and Λ, respectively. In the center, one hole is removed and form the core of PCF.

Fig.2 shows the unit cell of single periodic structure. To apply the model analysis theory of fiber easily, the hexagonal unit cell is approximated by a circular one of radius b shown in Fig 2. The outer radius is obtained and $b= (3^{1/2}/2\pi)^{1/2}\Lambda$ with the equivalent area.

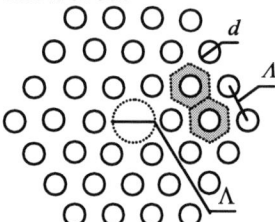

Fig.1 The cross section of PCF

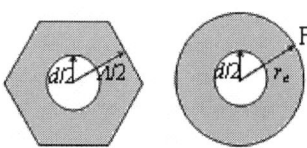

Fig. 2 The equivalent circular unit of a hexagonal one

In FVEIM, the effective cladding index is calculated using a fully vector approach. It can be express as below:

$$\left(\frac{P_l^{'}(\gamma a)}{\gamma a P_l(\gamma a)}+\frac{I_l^{'}(\kappa a)}{\kappa a I_l(\kappa a)}\right)\left(\frac{n_{silica}^2 P_l^{'}(\gamma a)}{\gamma a P_l(\gamma a)}+\frac{n_{air}^2 I_l^{'}(\kappa a)}{\kappa a I_l(\kappa a)}\right)$$

$$=l^2\left[\left(\frac{1}{\gamma a}\right)^2+\left(\frac{1}{\kappa a}\right)^2\right]^2\left(\frac{\beta}{k}\right)^2. \qquad (1)$$

where I, J, Y are Bessel functions, Defining

$J_l[\gamma(\omega)\rho]Y_l[\gamma(\omega)r_e]-Y_l[\gamma(\omega)\rho]J_l[\gamma(\omega)r_e]$ as $P_l[\gamma(\omega)\rho]$, $\kappa(\omega)$ and $\gamma(\omega)$ as

$$\kappa^2(\omega)=\beta^2(\omega)-n_{air}^2\omega^2/c^2, \qquad (2)$$

$$\gamma^2(\omega)=n_{silica}^2\omega^2/c^2-\beta^2(\omega), \qquad (3)$$

setting $l=1$ in the above equation, one can solve for effective index of cladding $n_{clading}=\beta_{FSM}/k_0$ of the fundamental space-

978-0-9789217-3-6/07/$25.00

©2007 WEN GLOBAL SOLUTIONS

filling mode. The effective index n_{clad} of the cladding region of PCF is defined by the index of the fundamental space filling mode (FSM) in the cladding.

After the effective index is obtained, PCF can be equivalent to the step fiber and the propagation constant β_e and fundamental mode field can also be solved through equation (1), and then the effective index of fundamental mode can be deduced by $n_{eff}=\beta_e/k_0$.

One of the advantages of PCF is the flexible choice of structure parameters such as the hole pitch Λ and the relative ratio of hole diameter to hole pitch d/Λ that the effective mode area A_{eff} can be tailored. The effective area of modal field is expressed as below :

$$A_{eff} = 2\pi \left[\int_0^\infty E^2(r)rdr \right]^2 / \int_0^\infty E^4(r)rdr \cdot (4)$$

Where, E(r) is the transverse electric field.

Mode cut-off
Similar to the conventional fibers, the effective index modes which the effective indexes is $n_{cladding}<n^j_{eff}<n_{silica}$ can propagate in the core of the fibers. To make the fiber operate at single mode, the high order mode must be cutoff.

In Fig.3, there are effective indexes of the fundamental mode and the four second order mode in the triangle-PCF, which show that not only the fundamental mode but also the second order modes including TE_{01}, TM_{01} and HE_{21} can be guided in the higher frequency. Effective index of the second order mode decrease with the normalized frequency decreasing. Cutoff point of the second order mode occur at $\Lambda/\lambda=2.3$, which is in accord with result of the literature [6].

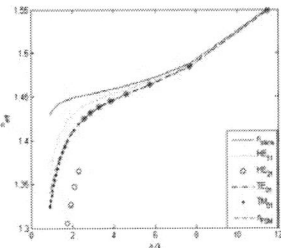

Fig.3 Effective Index of fundamental space-filling mode for the PCF at Λ=2.3um and d/Λ=0.6 .

Endlessly single mode properties

The endless single-mode properties mean that only the fundamental mode can propagate in a broad wavelength range in the fiber. The proprieties of the single mode in PCF can be analyzed by the cutoff frequency of the second order mode.

For the PCF whose core is formed by removing one hole in the center, the single mode properties are showed in Fig.4. Solid line represents the cutoff normalized frequency curve of the second order mode as the function of d/Λ. The dotted line d/Λ=0.41 is the asymptote of the cutoff normalized frequency curve of the second order mode. There are three regions in Fig.4. Region 1 is the region of endlessly single-mode which is located at the left side of the asymptote i.e. $d/\Lambda<0.41$ in Fig.4. When $d/\Lambda>0.41$, region 2 is single mode region of PCF and region 3 is multi-mode region.

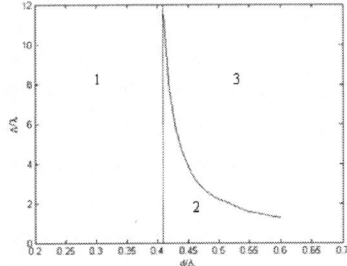

Fig.4. Cutoff frequency of the second order mode as the function of relative hole size for the triangular PCF with Λ=2.3um and one air hole was removed in the core region.

For the PCF whose core is formed by removing multi-holes in the center as shown in the Fig.5. The effective core radius is defined as $a=\Lambda(3^{1/2}\times N/(2\times\pi))^{1/2}$, where N is the number of air-holes removed in the center [2.3.7]. It is clear that the endless single mode region reduces with the removed holes number increase in the center of PCF. The fibers operate in the endlessly single-mode region with the relative hole size $d/\Lambda<0.23$ when three air-holes are removed and $d/\Lambda<0.07$ when seven air-holes are removed, i.e. the upper limit value of the d/Λ of the endlessly single-mode PCF is reduced when the number of the air-holes removed added.

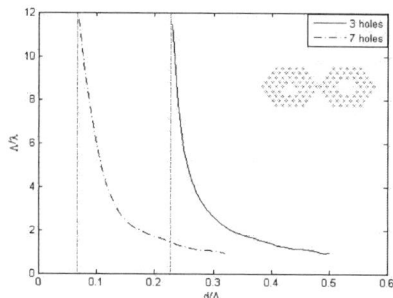

Fig.5 Cutoff frequency of the second order mode as the function of relative hole size for three hole in the center (solid line) and seven holes (dotted line) removed in the center of the triangular PCF with Λ=2.3um.

Effective area

Here, we only investigate the effective area at the telecommunication window around 1.55um while only single mode can be propagated in PCF. The variation of effective mode area as different pitches are investigated for the PCF with the one hole core, three holes core and seven holes core in the condition of single mode and the maximum hole size at the wavelength of 1.55um. Fig.6 shows the effective area as function of the pitch at the critical single mode structure. Numerical results demonstrate that effective mode area increases with an increase in the hole pitch, and that the larger the pitch and the more of the number of air holes removed at core are, the more obviously effective mode area increase.

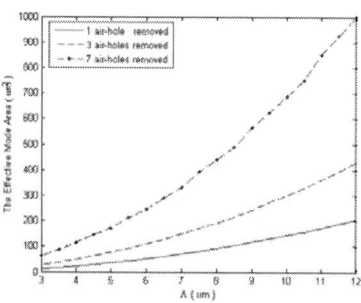

Fig. 6 Effective mode area of PCF with the one hole, three holes core and seven holes core, respectively. Here, the PCF even operate in the single-mode region at 1.55um.

The effective area of LEAF fabricated by Corning is 72um^2. However, according to the above analysis, the effective area can be up to 200um^2 easily, even to 600um^2 or 1000um^2. These results offer a possibility of design large mode area PCFs which have the number of potential application in the high-power delivery.

Conclusion

The influence of the structure parameters on the endlessly single-mode properties and the large mode area of PCF are investigated with the full-vector index model. According to the numerical result, the property of endlessly single-mode can be maintained when the relative hole size is below a certain upper limit. The upper limit is influenced by the number of the holes removed in the center. Based on the single-mode operation, increasing the pitch Λ or increasing the number of the air-holes removed in the core region could be increase the effective mode area to 200um^2, 600um^2, even 1000um^2 to meet the demand for the high power delivery. The conclusions are useful to guide the design of the large mode area fibers.

Acknowledge

This work is supported by China National Science Foundation and fthe Fund of Beijing Jiaotong University.

References

1. J.C. Knight, T.A. Birks, P.St.J. Russell, D.M. Atkins, Opt. Lett. 21 (1996) 1547.
2. T.A. Birks, D. Mogilevtsev, J.C. Knight, P.St.J. Russell, J. Broeng, P.J. Roberts, J.A. West, D.C. Allan, J.C. Fajardo, OFC98 FG4 (1998) 114
3. T.A. Birks, J.C. Knight, P.St.J. Russell, Endlessly single-mode photonic crystal fiber, Opt. Lett.22 (1997) 961
4. Li Shuguang, Liu Xiaodong, Hou Lantian. Acta Phys .Sin 53 2004 1973.
5. G.B. Ren, Z. Wang, SQ. Lou, SS. Jian, Acta optic sinica, 24 (2004) 2239
6. A. Bjarklev, J. Broeng, K. Dridi, S.E. Barkou, ECOC98(Madrid Spain),135-136
7. K. Tajima, J. Zhou, Y. Nakajima, et.al., J Lightwave Technology 22 (2004) 7.

High Negative Dispersion and Low Confinement Loss Photonic Crystal Fiber

Lei Yao(1), Shuqin Lou(1), Hong Fang(1), Tieying Guo(1), Honglei Li(1), Shuisheng Jian(1)

1 : Key Lab of All Optical Network & Advanced Telecommunication Network, Beijing Jiaotong University, Beijing, 100044, China and email: yangrou1369ok@sohu.com

Abstract *A novel design of high negative dispersion and low confinement loss photonic crystal fiber (PCF) was proposed. The PCF designed has a high dispersion of -1120 ps/nm/km and a very low confinement loss of $1.01*10^{-6}$ dB/km at 1.55µm.*

Introduction

Since the first working example of photonic crystal fibers (PCFs) was reported in 1996[1], scientists have paid great attention to developing various kind of PCFs with novel properties such as endlessly single mode[2], high birefringence[3], flexible dispersion[4], and so on. Recently, PCFs that have a high refractive index core and a defect ring in the cladding and similar dual-core fiber geometries (DCPCFs) are studied as dispersion compensating fibers[5,6]. Due to flexible structure of PCFs, higher negative dispersions than conventional dual core fibers' are achieved. The largest negative dispersion achieved by this design is -55000 ps/nm/km [7], but the refractive indices of the inner core are very high (1.5~2.8). It is unsuitable to apply in the actual fiber communication system. Another issue is that the air holes diameter is less than 1µm so that it is drawed difficultly.

Moreover, the defect ring using reduced holes should be avoided in the design of DCPCFs. Because the chromatic dispersion is sharply sensitive to the average refractive index of outer core and controlling the diameter of air holes is quite difficult in the actual fabrication[8].

Based on the summarization of the previous works in dual core PCF, we propose an improved dual core PCF showed in Fig. 1, which has a low Ge-doped core, a defect ring with pure silica and a outer ring of large holes. Due to defect ring in the third ring which increases distance between central core and ring core, so that higher negative dispersions are achieved than the defect ring in the second ring[9]. The hole pitch Λ is

7µm, the air-hole diameter d is 3.15µm, and the diameter of the Ge-doped core d_c is 10µm. The large hole d_L, which is 5.6µm, introduced to the most outer ring can efficiently reduce the confinement loss.

In this paper, a full vector finite element method (FVFEM) with transparent boundary conditions (TBC) is used[10] to model the PCF we proposed.

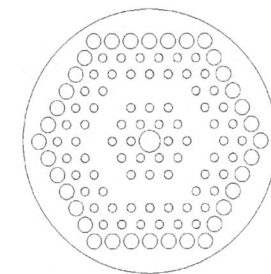

Fig.1. Cross section of the dual core PCF proposed

Methrodology

The main idea of FVFEM with TBC comes from reference [10]. Vector wave equation based on the magnetic field vector **H** can be derived from two curl equations of the Maxwell's equation. It is expressed as below:

$$\nabla\times\varepsilon_r\nabla\times\mathbf{H}=k_0^2\mathbf{H} \qquad (1)$$

Considering the transverse components of the magnetic field, Using the Galerkin procedure and discretizing the computational domain into triangular elements, the following discretized weak equation can be deduced:

978-0-9789217-3-6/07/$25.00

©2007 WEN GLOBAL SOLUTIONS

$$\sum_{BoundaryElement}\left\{-\int_{\Gamma_e}\frac{1}{n_{zz}^2}w_y(\partial_xH_y-\partial_yH_x)dy-\int_{\Gamma_e}\frac{1}{n_{zz}^2}w_x(\partial_xH_y-\partial_yH_x)dx\right.$$

$$\left.-\int_{\Gamma_e}\frac{1}{n_{yy}^2}w_x(\partial_xH_x+\partial_yH_y)dy+\int_{\Gamma_e}\frac{1}{n_{xx}^2}w_y(\partial_xH_x+\partial_yH_y)dx\right\}$$

$$+\sum_{InterfaceElement}\left\{-\int_{\Gamma_{int,e}}\frac{1}{n_{yy}^2}w_x\left(\partial_xH_x+\partial_yH_y\right)dy+\int_{\Gamma_{int,e}}\frac{1}{n_{xx}^2}w_y\left(\partial_xH_x+\partial_yH_y\right)dx\right\}$$

$$+\sum_{TriangularElement}\iint_{\Omega_e}\left\{\frac{1}{n_{zz}^2}\left(\partial_xw_y-\partial_yw_x\right)\left(\partial_xH_y-\partial_yH_x\right)\right.$$

$$+\left[\partial_x(\frac{1}{n_{yy}^2}w_x)+\partial_y(\frac{1}{n_{xx}^2}w_y)\right]\left(\partial_xH_x+\partial_yH_y\right)$$

$$\left.+k_0^2n_{eff}^2\left(\frac{1}{n_{yy}^2}w_xH_x+\frac{1}{n_{xx}^2}w_yH_y\right)-k_0^2\left(w_xH_x+w_yH_y\right)\right\}dxdy=0 \qquad (2)$$

The derivatives of the fields occurring in the boundary term in equation (2) will be handled through the BGT-like transparent boundary conditions (TBC) to reflect the properties of the fields in the exterior domain properly. Then the effective index can be obtained by calculating the equation above.

If the effective index is denoted by n_{eff}, the confinement loss α can be deduced from the imagine part of n_{eff} [11].

$$\alpha=\frac{20}{\ln(10)}\text{Im}\left(k_0n_{eff}\right)\quad dB/m \qquad (3)$$

Waveguide dispersion is expressed as below:

$$D\approx-\frac{\lambda}{c}\frac{d^2\text{Re}\left(n_{eff}\right)}{d\lambda^2}\quad ps/nm/km \qquad (4)$$

Dispersion

The structure to be studied in the following is shown in Fig.1, and Ge-doped concentration of the inner core is 1 mol% which is very low. The dispersion curve is shown in Fig.2. The highest dispersion reaches to -1120 ps/nm/km at 1.55μm.

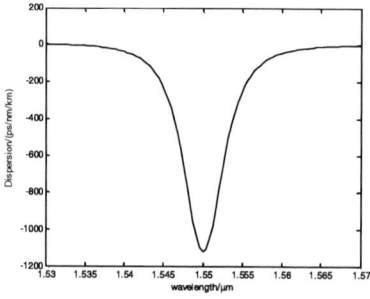

Fig.2 Chromatic dispersion of DCPCF proposed versus wavelength

Fig.3 below shows the change of dispersion characteristics along with Ge-doped concentration of the inner core. The corresponding refractive index of the doped inner core is 1.445356, 1.445378 and 1.4454 respectively.

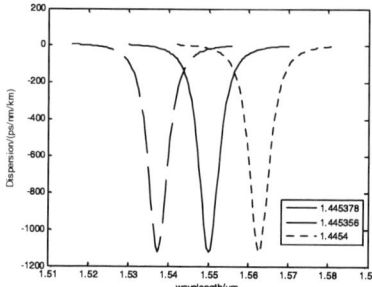

Fig.3 Change of diepersion coefficient along with Ge-droped concentration of the inner core

When the refractive index of the doped inner core increases from 1.445356 to 1.4454, the peak position is shifted from 1.537μm to 1.563μm. It is concluded that the high negative peak position increases with an increase in the refractive index of the doped inner core. This offer a possibility for designing high negative dispersion in various wavelength.

Confinement loss

It is known that the confinement loss of PCFs decreases while enlarging the air hole diameter d[12]. Here, we concentrate on investigating the confinement loss in the PCF without the large holes in the most outer ring and the structure we proposed.

Firstly, we analyze the dispersion of the PCF without the large holes in the most outer ring shown in Fig.4. Numerical result demonstrates that there is a slight variation in dispersion between the two kinds of structure. That means that introduction of the large holes into the most outer ring would slightly affect the dispersion.

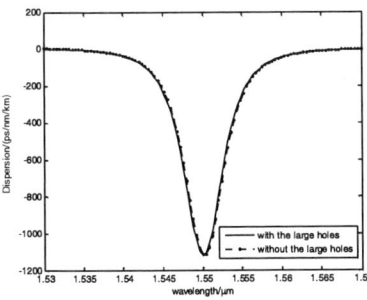

Fig.4 Comparison of the dispersion coefficient

Then, we discuss the effect of large holes in the most outer ring on confinement loss. Numerical result shown in Fig.5 exhibits that Confinement loss is 0.046 dB/km in the PCF without the large holes in the most outer ring and reaches to $1.01*10^{-6}$ dB/km in the structure we proposed . It is decreased by over four orders of magnitude through introducing the large holes in the most outer ring. A high negative dispersion of -1120 ps/nm/km and a low confinement loss of $1.01*10^{-6}$ dB/km at 1.55µm, can be obtained simultaneously in the structure in Fig. 1.

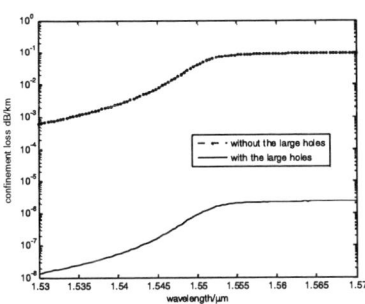

Fig.5 Comparison of the confinement loss

Conclusions

Based on *full vector finite element method*, we have proposed and analyzed a novel dual core PCF design with a high dispersion of -1120 ps/nm/km and a low confinement loss of $1.01*10^{-6}$ dB/km at 1.55µm. It was found from the analysis above that, when the refractive index of the doped inner core increases, the high negative peak position increases. Moreover, it was found that while the hole diameters of outer ring of PCFs are enlarged, the confinement loss is reduced with the dispersion characteristics keeping unchanged. These conclusions will offer a possibility of designing high negative dispersion PCFs with very low confinement loss.

Acknowledge

This work is supported by the national science foundation of China and the fund of Beijing Jiaotong University.

References

1 Knight J C et al. Opt. Lett., 21(1996), 1547
2 T.A.Birk et al. Opt. Lett., 22(1997), 961~963
3 A.Ortigosa-Blanch et al. Opt. Lett., 25(2000), 1325~1327
4 A.Bjarklev, et al. ECOC98,135~136.
5 Y.Ni et al. IEEE Photon. Technol. Lett., 16(2004), 1516 ~1518
6 F.Gerome, et al. Opt. Lett., 29(2004), 2725~2727
7 A.Huttunen, et al. Opt Exp., 13(2005), 627~635
8 Sigang Yang, et al. Opt Exp., 14(2006), 3015~3023
9 Takeshi Fujisawa, et al. Opt Exp., 14(2006), 893~900
10 H P Uranus et al. Opt Exp., 12(2004), 2795~2809
11 Olszewski J, et al. IEEE Tran. Opt. Net., 1(2004), 354
12 Masanori Koshiba, et al. IEEE Photon. Technol. Lett., 15(2003), 691 ~693

Adiabatic compression of quadratic solitons and frequency shift by using cascading nonlinearities

Zeng Xianglong [1], Satoshi Ashihara [2], Chen Xianfeng [3],
Tsutomu Shimura [4] and Kazuo Kuroda [4]

1: Communication and Information Engineering, Shanghai University, China
Email: zenglong@shu.edu.cn
2: Department of Applied Physics, Tokyo University of Agriculture and Technology, Japan
2: PRESTO, Japan Science and Technology Corporation, Japan
3: Physics department, Shanghai Jiao tong University, China
4: Institute of Industrial Science, the University of Tokyo, Japan

Abstract *We investigate that the equivalent Kerr self-phase-modulation (SPM) and cross-phase-modulation (XPM) effects introduced in cascaded second-order processes is dependent on the phase mismatch and group velocity mismatch. Adiabatic compression of quadratic solitons and frequency shift is reported.*

1. Introduction

Cascading of $\chi^{(2)}$ processes in quadratic media introduce a large nonlinear phase shift under phase-mismatched second harmonic generation (SHG), where the phases and the amplitudes of fundamental (FF) and second-harmonic (SH) waves are modulated through the reiterated up- and down conversions [1]. Equivalent third-order Kerr effect is self-induced for the fundamental pulse from cascaded nonlinearities, i.e., an intensity-dependent index change, which is of self-focusing or self-defocusing nonlinearities, depending on the sign of the phase mismatch. Because of the presence of group-velocity mismatch (GVM) between FF and SH pulses, cascaded nonlinearities become more attractive in the controllable magnitude and sign, especially exploited in ultrashort femtosecond region recently. Group-velocity control of the FF (SH) pulse is realized experimentally in the presence of suitable GVM, owing to mutual trap and drag between FF and SH pulse during the cascading interaction processes [2]. Strong GVM effects are particularly severe, since FF and SH pulses tend to walk off in the time domain after the effective interaction propagation length. The non-instantaneous feature of the cascading nonlinear phase shifts, caused by GV mismatch, has been highlighted as Raman-like nonlinearities [3]. It is demonstrated that the Raman-like nonlinearities, or self-steepening effect [4], work for frequency shift. The direction of frequency shift is dependence on the sign of

phase mismatch and such shift is enhanced by the local control of the cascaded nonlinearities. But these analyses are simply taken for granted the single generalized nonlinear Schrödinger equation for FF pulse, which is derived from the coupled equations governing the interaction of FF and SH field envelopes propagating in the medium with quadratic nonlinearity. SH is negligible for consideration. Here, we use the novel reduced model [5] to investigate the interaction behaviours between FF and SH pulses through cascaded nonlinearities. The nonlinear phase shift by the effective Kerr self-phase-modulation (SPM) and cross-phase-modulation (XPM) is dependent on the phase mismatch and group velocity mismatch. We also report numerical and experimental evidence for this property.

2. Theory background

Through the multiscale approach from Ref [5], coupled wave equations under the slowly varying envelope approximation in the medium with quadratic nonlinearity, can be generalized as a reduced model just including the leading-order (averaged) contributions of the electric field [5]

$$i\partial_z A_1 = \frac{1}{L_{D1}}\partial_{\tau\tau}^2 A_1 + \frac{1}{\Delta k}A_1\left[\left|A_1\right|^2 - \left|A_2\right|^2\right]$$

$$i\partial_z A_2 = \frac{1}{L_{D2}}\partial_{\tau\tau}^2 A_2 + i\frac{1}{L_{GVM}}\frac{\partial}{\partial\tau}A_2 - \frac{1}{\Delta k}\left|A_1\right|^2 A_2$$

(1)

where L_{GVM} and L_{Di} are the characteristic group-mismatch length and dispersion length. Here $L_{GVM} = \tau_0\delta$ and $\delta = 1/v_{g1} - 1/v_{g2}$ is temporal walk-off. $A_i(z, \tau)$ denotes the leading-order (averaged) contributions of the electric field, and the subscripts 1 and 2 correspond to FF and SH pulses, respectively. Time τ is measured in a frame of reference along with FF propagation. The wave-vector mismatch is determined by $\Delta k = k_2 - 2k_1$, k_1 and k_2 is FF and SH wave numbers.

The last nonlinear terms in the model come from the sequences of cascaded nonlinear processes, i.e. up- and back conversions in phase-mismatched SHG. This two-step process leads to an effective Kerr self-phase-modulation (SPM) and cross-phase-modulation (XPM) effects. The effective cubic nonlinearity is represented as $1/\Delta k$, which has controllable sign and magnitude. The sign of effective SPM and XPM terms reprsents the focusing or defocusing nonlinearity impressed on FF and SH pulses. In the presence of GVM, the nonlinear phase shift is influenced by the phase mismatch and GV mismatch, i.e. $\phi_i(\tau) \propto (1/\Delta k, \delta)$.

Since the effective SPM and XPM terms determined by FF and SH leading-order components have opposite signs for any given mismatch Δk, the overall focusing or defocusing nature of the process is not only dependent on the sign of phase mismatch but also on the competence between the effective SPM and XPM effects. The advantage of this model gives a simple sight into the interaction behaviour in the phase modulation between FF and SH pulses.

3. Adiabatic compression of quadratic solitons

Quadratic solitons in $\chi^{(2)}$ media are formed through the cascaded quadratic nonlinearities. GV matching between the interacting waves significantly enhances the cascaded nonlinearities and enables us to generate multi-colored temporal solitons. Soliton compression described here is to especially utilize the adiabatic process of group-velocity matching cascaded interaction. In the absence of GVM between FF and SH pulses, the quadratic solitons are compressed with soliton (-like) shape during propagation. The quality factor is not degraded. The detailed theoretic work and experimental propose was shown in Ref [6, 7].

In our experimental setup, the linearly-chirped aperiodic quasi-phase-matching grating is 50mm length. The fundamental pulse with a full-width at half-maximum (FWHM) duration of 95 fs (peak intensity 22 GW/cm^2) at the wavelength of 1570 nm, comes from a Spectra-Physics femtosecond optical parametric amplifier synchronously pumped by a mode-locked Ti:sapphire laser.

Fig.1 shows the dependence of pulse durations of FF and SH pulses on the effective wave-vector mismatch at the output position of the chirped QPM grating, $\Delta k_{\text{eff}} = k_2 - 2k_1 - 2\pi/\Lambda(L)$, $\Lambda(z)$ is the local domain reversal period.

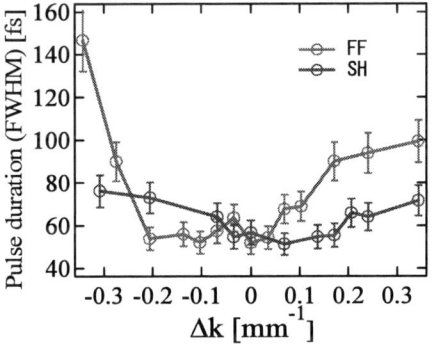

Fig.1 Dependence of pulse durations of the fundamental and the harmonic pulses on Δk_{eff}.

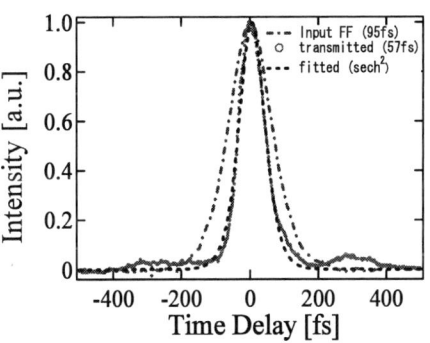

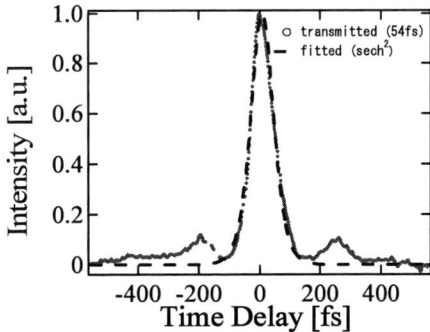

Fig.2 Measured intensity autocorrelation traces of (a) the input pulse, transmitted FF and (b) SH pulses and their sech2-shaped fitted functions at dk = 0.07mm^{-1}.

Figs.2 (a) and (b) show the measured intensity autocorrelation traces of input and output compressed FF and SH pulses in the chirped QPM grating. Here we see that the compressed temporal profiles show high-quality soliton-like shapes without obvious pedestals.

4. Spectral shift

In the presence of small GVM, FF and SH pulses mutually trap and drag each other and propagate with a velocity between the GV control of ultrafast pulses was demonstrated through $\chi^{(2)}$ cascaded interactions in QPM gratings. On the other hand, the nonlinear phase shift becomes asymmetry due to the effective XPM effects in FF and SH pulses, in the presence of GVM. This asymmetry nonlinear phase shift causes the spectra shift of each pulse. The simulations for spectra shift is shown in Fig. 3.

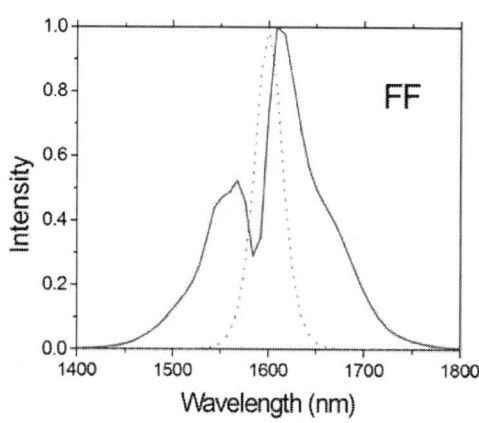

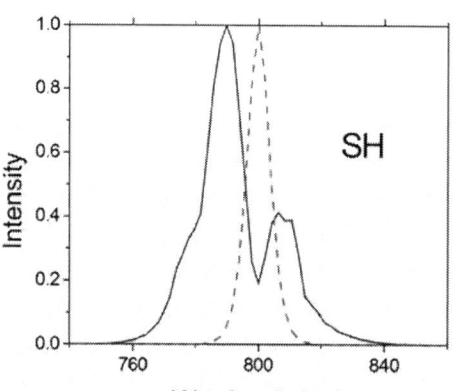

Fig.3 Calculated spectra shift of (a) FF amd (b) SH in the presence of group velocity mismatch

From the results, spectra of FF and SH shifts to opposite direction and the shift of FF is smaller than that of SH, because of the larger effective XPM term in SH, which is dominated by FF component. It is easy to confirm that the direction of spectra shift is decided not only by the phase mismatch, but also by the GV mismatch, which provide a flexible and efficient route to control frequency shift of femtosecond pulses.

5. Conclusions

Through the reduced coupled equations of leading-order components of FF and SH pulses, the effective Kerr self-phase-modulation (SPM) and cross-phase-modulation (XPM) effects have effects on the phase modulations both in FF and SH pulses, which can be controlled by the phase mismatch and group velocity mismatch. As a practical approach, adiabatic soliton compression of femtosecond pulses is very attractive because of their ability to efficiently generate two-color solitons with shorter pulse durations. Besides phase mismatch, GVM can be treated as another controllable parameter for spectra shift of femtosecond pulses

References

1. G. I. Stegeman, et al. J. Optical and Quant. Electron. 28, 1691-1740 (1996).
2. M. Marangoni, et al. Opt. Lett. 31, 534-536 (2006).
3. K. Beckwitt, et al. Opt. Lett. 29, 763-765 (2004).
4. F. Baronio, et al. Opt. Express 14, 4774-4779 (2006)
5. C. Conti, et al. J. Opt. Soc. Am. B 19, 852 (2002)
6. L. Torner, et al. Opt. Lett. 23, 903-905 (1998).
7. Zeng Xianglong et al. Opt. Express, Vol.14. 9358-9370 (2006);

New microwave up-conversion solution using an optical phase modulator in Radio-over-Fiber Networks

Haiyan Ou (1), Hongyan Fu (2), Biao Chen (3)

1 : Centre for Optical and Electromagnetic Research, Joint Research Centre of Optical Communications, Zhejiang University, Hangzhou 310058, China ; ouhaiyan@coer.zju.edu.cn 2 : fuhongyan@coer.zju.edu.cn , 3 : biaochen@coer.zju.edu.cn

Abstract *An approach for frequency up-conversion in Radio-over-Fiber system is demonstrated by simulation. The technique employs a phase modulator and an FBG. Error-free data transmission was demonstrated for downlink with frequency conversion from 3GHz to 53GHz.*

1. Introduction

Nowadays, there is growing interest in Radio-over-Fiber (RoF) technologies which are capable of providing the anticipated demand for broadband wireless access services over wide areas [1-2]. With regard to technique choices as radio frequency (RF) over fiber, intermediate frequency (IF) over fiber and baseband over fiber, IF over fiber, where lower frequency subcarriers are transmitted in the fiber and frequency conversion is needed in the Base Station (BS), is a compromise between the other two [3]. In contrast to RF over fiber, the effect of fiber chromatic dispersion on the distribution of IF signals is less severe; however the need for a millimeter-wave (mm-wave) local oscillator (LO) increases the complexity of the BS. For RoF systems employing IF over fiber technology, much work has considered the remote delivery of the LO in order to simplify the BS architecture. There are ways employing an expensive dual-electrode Mach–Zehnder Modulator (MZM) [4], where the LO is generated though double side band with optical carrier suppression (DSBCS) modulation schemes; other ways are based on single electrode MZM [5], which is biased at the minimum transmission point to realize the DSBCS modulation schemes; However, they all suffer from the direct current (dc) bias-drifting problem when an optical intensity modulator is used.

In this paper, we present an approach to generate remote LO mm-wave signals for frequency up-conversion by employing an optical phase modulator and an FBG. LO signals at a frequency equal to double of the electrical drive signal is generated at the BS without being severely affected by fiber

chromatic dispersion. Error-free data transmission was demonstrated for downlink with frequency conversion from 3GHz to 53GHz by simulation.

2. Principle of the Proposed system

The principle of the proposed approach is illustrated in Fig.1. A laser diode (LD) is directly modulated by a microwave signal at a frequency of f_{Rf} and then coupled to an optical phase modulator (PM). The PM is driven by an electrical signal at a frequency of f_{Lo}. The frequency of f_{Rf} and f_{Lo} are in IF band and mm-wave band respectively. After transmitting through a roll of dispersive fiber (SMF), the signal is injected into an FBG. The reflected signal is detected by a low-speed photodetector (PD) (below 5GHz) and the transmitted signal is detected by a high-speed PD, where the LO mm-wave signal at a frequency equal to double of the electrical drive signal is remotely generated.

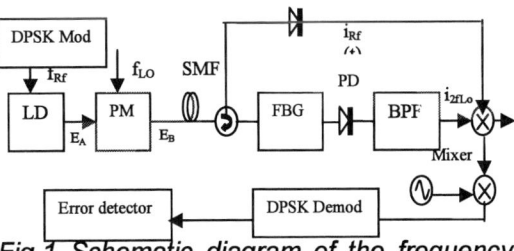

Fig.1 Schematic diagram of the frequency up-conversion method. (LD: laser Diode, PM: phase modulator; SMF: single mode fiber, FBG: Fiber Bragg Grating; PD: photodetector.)

The optical signal at the output of the LD is given by:

$$E_A(t) = \sqrt{1+m\cos(2\pi f_{Rf}t)}e^{j\beta_1\sin(2\pi f_{Rf}t)}e^{j2\pi f_0 t} \quad (1)$$

Where f_0 is the frequency of the optical carrier, f_{Rf} is the frequency of the electrical

978-0-9789217-3-6/07/$25.00

©2007 WEN GLOBAL SOLUTIONS

drive signal; m is the IM index and β_1 is the FM index.

According to the LD characteristic, in the linear part, it can be approximated as:

$$E_A(t) = (1 + \frac{m}{2}\cos(2\pi f_{Rf}t))e^{j\beta_1\sin(2\pi f_{Rf}t)}e^{j2\pi f_0 t} \quad (2)$$

After coupling into the PM, the optical signal at the output becomes:

$$E_B(t) = E_A(t)e^{j\beta_2\cos(2\pi f_{Lo}t)} \quad (3)$$

where β_2 is the modulation index of the PM and f_{Lo} is the frequency of the electrical signal used to drive the PM.

Considering (1) and (3), we can have

$$E_B(t) = (1 + \frac{m}{2}\cos(2\pi f_{Rf}t))e^{j2\pi f_0 t}$$
$$\times \sum_{n=-\infty}^{+\infty} J_n(\beta_1)\exp(jn\omega_{Rf}t) \quad (4)$$
$$\times \sum_{m=-\infty}^{+\infty} J_m(\beta_2)\exp(jm\omega_{Lo}t)$$

where J_n is the Bessel function of the first kind of order n.

The useful expressions of complex envelopes of the spectral lines are:

$$SL_{f0} = J_0(\beta_1)J_0(\beta_2)$$
$$SL_{f0\pm f_{Rf}} = \frac{m}{4}J_0(\beta_1)J_0(\beta_2)\pm J_1(\beta_1)J_0(\beta_2)$$
$$SL_{f0\pm f_{Lo}} = \pm J_0(\beta_1)J_1(\beta_2) \quad (5)$$
$$SL_{f0\pm(f_{Lo}-f_{Rf})} = \pm\frac{m}{4}J_0(\beta_1)J_1(\beta_2)-J_1(\beta_1)J_1(\beta_2)$$
$$SL_{f0\pm(f_{Lo}+f_{Rf})} = \pm\frac{m}{4}J_0(\beta_1)J_1(\beta_2)+J_1(\beta_1)J_1(\beta_2)$$

Fig.2 shows the spectrum lines in the optical field after the PM. The inset is the transmission spectrum of the FBG. The 3dB bandwidth of the FBG is about 0.2nm, which is about 25GHz. It can be seen that the optical components in the 3dB bandwidth (<12.5GHz) are mainly reflected by the FBG, while the others (>12.5GHz) are transmitted through it almost unaffected.

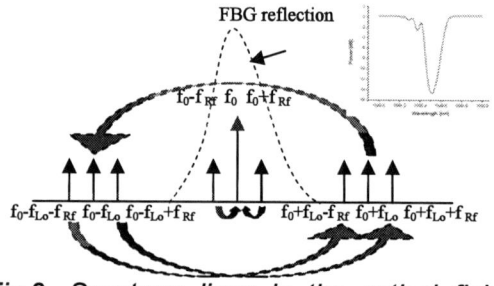

Fig.2. Spectrum lines in the optical field after the PM.

The transfer function of the dispersive fiber is given by [4]

$$H(f) = \exp[j\frac{\pi DLf^2\lambda_0^2}{c}] \quad (6)$$

where c is the velocity of the light in vacuum, D is the chromatic dispersion of the fiber (D=17ps/nm.km) and f is the modulation frequency.

When the signals are fed to PD, frequency components at f_{Rf} and f_{2Lo} are generated. The signals with frequency at f_{Rf} are generated by the dominant beating of SL_{f0} with $SL_{f0\pm f_{Rf}}$ as shown in Fig.2, which are reflected by the FBG. It can be expressed as follows:

$$i_{Rf} = \eta[\frac{m^2}{4}J_0^2(\beta_1)J_0^2(\beta_2)\cos(\frac{\pi DLf_{Rf}^2\lambda_0^2}{C})$$
$$+ J_0(\beta_1)J_1(\beta_1)J_0^2(\beta_2)\sin(\frac{\pi DLf_{Rf}^2\lambda_0^2}{C})] \quad (7)$$

where η is a constant decided by the responsivity of the PD.

As mentioned above, the signals at the frequency of f_{Rf} are in the IF band, where the effect of fiber chromatic dispersion is less severe than the signals at mm-wave band. Of course, these signals can be transported over longer fiber distance.

As the optical carrier reflected by the FBG, the remote LO mm-wave signal is generated with DSBCS modulation scheme which is tolerant to fiber chromatic dispersion. It can also be seen from Fig.2 that the signals at the frequency f_{2Lo} are resulted from the sum of the three dominant beating terms which transmitted through the FBG: (1) $SL_{f0+f_{Lo}}$ with $SL_{f0-f_{Lo}}$; (2) $SL_{f0+(f_{Lo}-f_{Rf})}$ with $SL_{f0-(f_{Lo}+f_{Rf})}$; (3) $SL_{f0+(f+f_{Rf})}$ with $SL_{f0-(f_{Lo}-f_{Rf})}$. Other beating terms caused by higher order intermodulation products are too small and can be neglected. The generated signals at f_{2Lo} can be expressed as:

$$i_{2f_{Lo}} = \eta[J_0^2(\beta_1)J_1^2(\beta_2)$$
$$+ \frac{m^2}{4}J_0^2(\beta_1)J_1^2(\beta_2)\cos^2(\frac{4\pi DLf_{Rf}f_{Lo}\lambda_0^2}{C}) \quad (8)$$
$$+ 4J_1^2(\beta_1)J_1^2(\beta_2)\sin^2(\frac{4\pi DLf_{Rf}f_{Lo}\lambda_0^2}{C})]$$

Fig.3 shows the power of the LO signal as a function of the modulation frequency after transmission through 25km SMF with double side band (DSB) modulation and DSBCS modulation respectively.

376

It can be seen from Fig.3 that with the DSBCS modulation scheme, the whole system is tolerant to fiber chromatic dispersion. LO signals at mm-wave band can be remotely generated without being severely affected by chromatic dispersion.

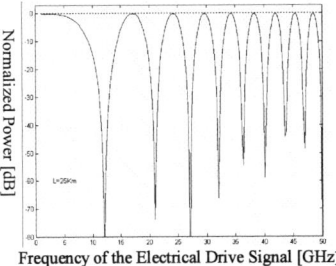

Fig.3 Simulated and predicted dispersion-induced RF power penalty for the received LO signal with 25km SMF (normalized to power at 0 GHz; dash line: DSB modulation; dotted line: DSBCS modulation.

3. Simulation setup and results

The simulation setup for the proposed frequency up-conversion solution is the same as Fig.1. The simulation results were obtained using Optiwave software. A direct modulated LD is used as the optical source, centered at 1449.5 nm. A microwave signal generator at 3GHz with 450Mb/s differential phase-shift keying (DPSK) data is used as a microwave source to drive the LD. The optical phase modulator is driven by a tunable microwave signal source. A 25-km SMF is used to distribute the signals from the PM to the FBG. The maximum reflection wavelength of the FBG is set to match the LD, with 3dB bandwidth equal to 20GHz. At the BS, the reflected and transmitted signals from the FBG are detected by two PD respectively. The generated mm-wave signals are observed by the electrical spectrum analyzer.

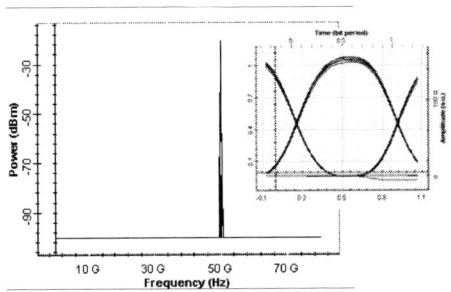

Fig.4 Electrical spectrum of the generated LO signals

Fig.4 shows the spectrum of the optical generated electrical signal with the electrical drive signal at 25GHz, measured at the output of the PD. It is easily seen that an mm-wave signal at twice the frequency of the drive signal at 50GHz is generated. When the electrical drive signal is tuned from 15 to 30GHz, the frequency-doubled mm-wave signal from 30~60GHz can be generated. Inset is the eye diagrams of the proposed system. The eye diagram is clear and wide open, demonstrating that an error-free up-conversion is realized. The simulation results agree well with the mathematical analysis above.

4. Conclusions

In summary, we have proposed a new method for frequency up-conversion in RoF networks. An optical phase modulator and an FBG are used together. The reflected and transmitted signals by the FBG are detected by two PDs respectively. The signals generated by the reflection part at f_{Rf} are in the IF band where chromatic dispersion are not severe; while the signals generated by the transmission part at f_{2Lo} are in the mm-wave band but with DSBCS modulation scheme, which is tolerant to the chromatic dispersion. The simulation results agree well with the mathematical results.

There are many advantages of the proposed approach. Besides simplicity and cost-effective, it can generates frequency-doubled LO mm-wave signals that can be continously tuned up to tens of gigerherts without being severely affected by fiber chromatic dispersion and due to the use of optical phase modulator, there's no dc bias drifting problem compared with approaches using intensity modulator. These properties give more flexibility to system configuration.

References

1. H. Ogawa etc IEEE Trans. Microwave Theory Tech., vol. 40, 1992, pp. 2285–2292
2. R. P. Braun etc Electron. Lett, vol. 32, no. 7, Mar. 28,1996, pp. 626–628.
3. Yannis Le Guennec etc JLT, VOL. 24, NO. 3, MARCH 2006
4. Christina Lim etc, JLT, VOL. 18, NO. 10, OCTOBER 2000, pp. 1355-1363.
5. Chun-Ting Lin etc, IEEE PTL, VOL. 18, NO. 23, December 1, 2006, pp. 2481-2483.

Photonic Frequency Down-Conversion for Millimeter-Wave-Band Radio-over-Fiber Systems by Directly Modulating a Dual-Wavelength Fiber Laser

Shiming Gao, Ying Gao, Hongyan Fu, Daru Chen, S. He

Centre for Optical and Electromagnetic Research, State Key Laboratory for Modern Optical Instrumentation, Zhejiang University, Hangzhou 310058, P. R. China

gaosm@zju.edu.cn

Abstract *A novel photonic frequency down-conversion scheme is proposed and demonstrated through directly modulating a dual-wavelength single-longitudinal-mode fiber laser. A 10.6 GHz microwave is down-converted to 1 GHz and a >7 dB signal-to-noise ratio is obtained.*

Introduction

Millimeter-wave-based radio-over-fiber (RoF) technique is considered as an effective solution to future broad-band wireless communication systems. In RoF systems, microwave signals are optically intensity-modulated and transmitted in fiber links between central office and base stations with low loss [1]. However, optically intensity-modulated millimeter-wave signals greatly suffer from fiber chromatic dispersion due to their high frequencies [2]. Since intermediate-frequency (IF) signals with low frequencies are hardly affected by fiber dispersion, a simple solution is to transmit IF signals in the fiber link between the central office and base stations, and perform frequency up- and down-conversion at the base stations.

Frequency up- and down-conversion has been demonstrated by using nonlinear photodetection techniques [3], cross-absorption modulation in electro-absorption modulators [4], photonic cascaded external modulation [5], four-wave mixing or cross-phase modulation in highly nonlinear fibers or semiconductor optical amplifier between optical carrier and local oscillation [6]-[8]. In these schemes performed in [5]-[8], either twice of the optical modulation or two light sources are required.

In this paper, we propose and demonstrate a novel, simple photonic frequency down-conversion scheme for millimeter-wave-based RoF systems by directly modulating a dual-wavelength single-longitudinal-mode

fiber laser. In this scheme, only a dual-wavelength light source is required and it is just modulated once.

Experimental setup

The principle of the proposed photonic frequency down-conversion is shown in Fig. 1. A dual-wavelength fiber laser provides two local oscillation (LO) lights v_{LO1} and v_{LO2} with a frequency difference of f_{LO}. When the dual-wavelength lights are modulated by a high-frequency RF signal f_{RF}, two sidebands with a frequency difference f_{RF} between the sideband and the LO light will be generated around each LO light. One can filter one LO light and the adjacent sideband from the other LO light, and then obtain a low-frequency IF signal, whose frequency is $f_{IF} = f_{RF} - f_{LO}$, through the heterodyne method since their phases are relatively fixed.

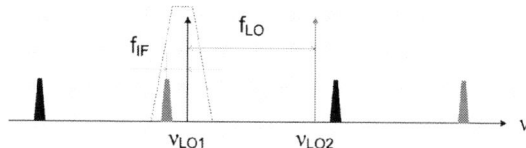

Fig. 1. Principle of photonic frequency down-conversion based on dual-wavelength laser external modulation.

Figure 2 shows the experimental setup for photonic frequency down-conversion. The structure of the presented dual-wavelength fiber laser is described in the block diagram. A pair of fibre Bragg gratings (FBG1 and FBG2) consist of a Fabry-Perot filter, which have two very narrow transmission peak in their reflective band. They form the laser cavity together with an optical circulator (OC1), a 2.24-m-long erbium doped fiber and FBG3. The reflective peak of FBG3 can be tuned to match that of the FBG pair. The erbium doped fiber serves as the gain medium, which is pumped by a 980 nm laser diode. This fiber laser can provide stable dual-wavelength signle-longitudinal-mode lasing with a frequency spacing of about 9.6 GHz at room temperature. It is intensity-modulated by the incident RF signal f_{RF} from an analog signal generator (Agilent E8257D) through an intensity modulator (IM). The required LO and sideband lights are filtered by using FBG4, and the converted low-frequency IF signal is obtained at the photodetector (PD) by optical heterodyne. The down-converted IF signal is observed through an electrical spectrum analyzer (ESA, Anritsu MS2687B).

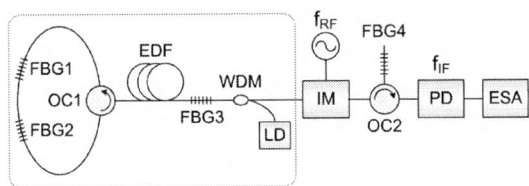

Fig. 2. Experimental setup of photonic frequency down-conversion.

Results and discussion

The spectrum of the dual-wavelength fiber laser is measured by using an optical spectrum analyzer (Agilent 86142B), as shown in Fig. 3(a). The output lasering is beaten at a photodetector and the measured frequency spectrum is shown in Fig. 3(b). It can be seen that the heterodyne spectrum only has a single line, which means that each of the two wavelengths created by the fiber laser only own a single frequency, i.e. single-longitudinal mode. By modulating such a dual-wavelength laser and beating, the obtained IF signal will also be single-frequency, which is very helpful to reduce unexpected noises.

When the dual-wavelength laser is externally modulated, two sidebands will be created around each wavelength. Multiple microwave frequency-components will emerge if the optical filter is not used. They

will affect the required IF signal as noises. When one of the local oscillation light and the adjacent sidebands are effectively filtered simultaneously, only the expected IF signal can be generated through optical heterodyne. The beating result is shown in Fig. 4. As shown in this figure, a single-frequency IF signal of 1 GHz is obtained. The other beating microwaves, including the microwaves of 9.6 GHz and 10.6 GHz, are suppressed to the noise level by optical filter. The down-converted IF signal is about 7 dB stronger than the noises nearby and the signal-to-noise ratio can be greater than 9 dB when the basis of the ESA is taken into account.

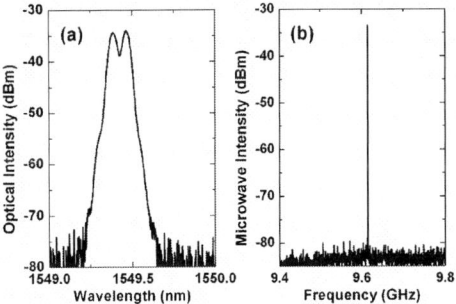

Fig. 3. (a) Output Spectra of the dual-wavelength fiber laser; (b) the heterodyne result between the two wavelengths of the dual-wavelength laser.

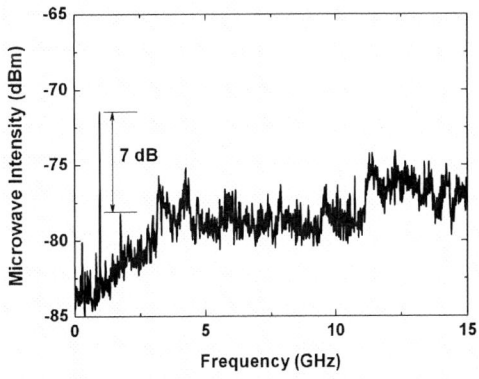

Fig. 4. Measured frequency down-conversion result of a 10.6 GHz microwave.

Conclusions

We have proposed and demonstrated a simple photonic frequency down-conversion scheme for millimetre-wave-based RoF systems. A 10.6 GHz RF signal is down-converted to a low-frequency IF signal with a single-frequency of 1 GHz through dual-wavelength laser external modulation and optical heterodyne. The high-frequency noises are effectively depressed and a signal-to-noise ratio of >7 dB is experimentally realized.

Acknowledgements
This work is supported by the Science and Technology Department of Zhejiang Province (grant No. R104154 and 2004C31095) and Science and Technology Bureau of Hangzhou municipal government (grant No. 20051321B14).

References
1 A. J. Seeds et al J. Lightwave Technol., 24 (2006), page 4628
2 G. J. Meslener IEEE J. Quantum Electron., 20 (1984), page 1208
3 M. Tsuchiya et al IEEE Trans. Microw. Theory Tech., 47 (1999), page 1342
4 C. S. Park et al IEEE Photon. Technol. Lett., 17 (2005), page 1950
5 T. Kuri et al IEEE Photon. Technol. Lett., 14 (2002), page 1163
6 J. Yu et al IEEE Photon. Technol. Lett., 17 (2005), page 1986
7 H.-J. Kim et al Opt. Express, 15 (2007), page 3384
8 J.-H. Seo et al IEEE Trans. Microw. Theory Tech., 54 (2006), page 959

Analysis of Dispersion Properties in Highly Nonlinear Photonic Crystal Fibers

Honglei Li(1), Shuqin Lou (1), Hong Fang(1), Lei Yao(1), Shuisheng Jian(1)

1. Key Lab of All Optical Network & Advanced Telecommunication Network of EMC, Beijing Jiaotong University, Beijing, 100044, China and email: lhl69822416@163.com

Abstract A fully vectorial effective index method and multiple-cladding method (FVEIM&MCM) is developed for modeling photonic crystal fibers. The large negative dispersion and nearly zero flattened dispersion in highly nonlinear photonic crystal fibers are analyzed numerically.

Introduction

Photonic crystal fibers (PCFs) have been widely modeled, studied, and fabricated due to their peculiar properties such as endlessly single-mode, high-nonlinearity, and overall controllable dispersion properties. The presence of air holes in the cladding makes them versatile for many dispersion-managed applications, e.g., flattened dispersion [1], large negative dispersion [2], and so on.

The properties of PCFs have been modeled by many methods, such as the effective index method (EIM), the localized basis function method, finite element method (FEM), finite difference method (FDM), and so on. A fully vectorial effective index method (FVEIM) [3, 4] for calculating properties of PCFs has been proposed and obtains higher accuracy than the conventional EIM does. We proposed the method of FVEIM&MCM for modeling the dispersion properties of PCFs with different hole-size in the cladding. FVEIM&MCM combines the advantages of those two algorithms, the accuracy obtained is almost the same as FEM or FDM does and computer time can be improved in a large magnitude.

In this paper, FVEIM&MCM is developed to model the dispersion of PCF. Through changing the different rings air-hole diameters and adjusting Ge-doped concentration in the core region, nearly zero flattened dispersion or large negative dispersion can be obtained. These properties can both be used to design large negative dispersion highly nonlinear PCFs for Raman amplifier and design flattened dispersion highly nonlinear PCFs for super-continuum generation and femtosecond pulse compression, and so on.

FVEIM&MC models

Firstly, the FVEIM is developed for calculating the effective cladding indices of different air-hole diameters in the different rings of PCFs. The equivalent circular unit cell of a hexagonal one in the periodic structure is shown in Fig.1 and a=d/2, $r_e = (\sqrt{3}/(2\pi))^{1/2}\Lambda$. Secondly, MCM is developed for obtaining effective index of fundamental mode in PCFs, then, dispersion properties can be derived from the effective index of fundamental mode.

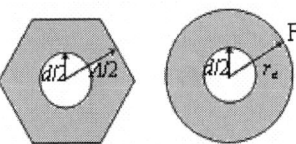

Fig. 1 The equivalent circular unit of a hexagonal one

In FVEIM, the effective cladding index is calculated using a fully vectorial approach. It can be express as below [3]:

$$\left(\frac{P_l'(\gamma a)}{\gamma a P_l(\gamma a)} + \frac{I_l'(\kappa a)}{\kappa a I_l(\kappa a)} \right) \left(\frac{n_{silica}^2 P_l'(\gamma a)}{\gamma a P_l(\gamma a)} + \frac{n_{air}^2 I_l'(\kappa a)}{\kappa a I_l(\kappa a)} \right)$$
$$= l^2 \left[\left(\frac{1}{\gamma a}\right)^2 + \left(\frac{1}{\kappa a}\right)^2 \right]^2 \left(\frac{\beta}{k}\right)^2. \quad (1)$$

Where I, J, Y are Bessel functions, Defining $J_l[\gamma(\omega)\rho]Y_l[\gamma(\omega)r_e] - Y_l[\gamma(\omega)\rho]J_l[\gamma(\omega)r_e]$ as $P_l[\gamma(\omega)\rho]$, $\kappa(\omega)$ and $\gamma(\omega)$ as

$$\kappa^2(\omega) = \beta^2(\omega) - n_{air}^2 \omega^2 / c^2, \quad (2)$$

$$\gamma^2(\omega) = n_{silica}^2 \omega^2 / c^2 - \beta^2(\omega), \quad (3)$$

Setting $l = 1$ in the above equation (1), one can solve for effective index $n_{clad}(\omega)$ of the fundamental space-filling mode which is also called the effective cladding index. When air-hole diameters change we can get different $n_{clad}(\omega)$.

In MCM, the schematic cross section of equivalent multi-cladding waveguide is shown in fig. 2; air-hole rings with different $n_{clad}(\omega)$

correspond to different layers in Fig. 2. Firstly, we use the vectorial wave equation to obtain the mode fields,

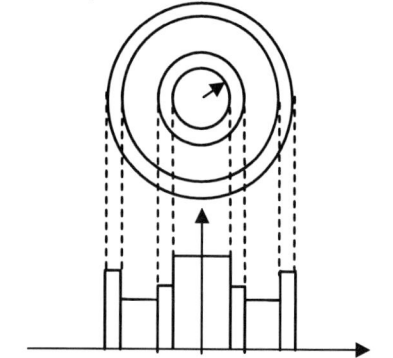

Fig. 2 Schematic cross section of equivalent multi-cladding waveguide

and then, apply the continuous boundary conditions of the field components to get the characteristic equation as below:

$$|SA \quad K| = 0, \qquad (4)$$

Where the SA and K are obtained as follows:

$$\begin{bmatrix} E_\varphi \\ H_\varphi \\ E_z \\ H_z \end{bmatrix}_{\rho=1} = S_N S_{N-1} \Lambda\, S_2 \begin{bmatrix} E_\varphi \\ H_\varphi \\ E_z \\ H_z \end{bmatrix}_{\rho=\rho 1} = SA \begin{bmatrix} a_1 \\ c_1 \end{bmatrix}, \qquad (5)$$

S_i is a 4×4 matrix which elements are J_l, K_l, Y_l etc. In the outmost layer, the field components as

$$\begin{bmatrix} E_\varphi \\ H_\varphi \\ E_z \\ H_z \end{bmatrix}_{\rho=1} = \begin{bmatrix} \frac{l\beta a}{W^2} K_l & -j\frac{\omega\mu_0}{W} a K_l' \\ -j\frac{\omega\varepsilon_2 a}{W} K_l' & -\frac{l\beta a}{W^2} K_l \\ K_l & 0 \\ 0 & K_l \end{bmatrix} \begin{bmatrix} b_a \\ d_a \end{bmatrix} \underline{def K(W)} \begin{bmatrix} b_a \\ d_a \end{bmatrix}, \qquad (6)$$

in the core,

$$\begin{bmatrix} E_\varphi \\ H_\varphi \\ E_z \\ H_z \end{bmatrix}_{\rho=1} = \begin{bmatrix} -\frac{l\beta r_e}{U^2} J_l & j\frac{\omega\mu_0}{U} r_e J_l' \\ j\frac{\omega\varepsilon_1 r_e}{U} J_l' & \frac{l\beta r_e}{U^2} J_l \\ J_l & 0 \\ 0 & J_l \end{bmatrix} \begin{bmatrix} a_0 \\ c_0 \end{bmatrix} \underline{def A(U)} \begin{bmatrix} a_0 \\ c_0 \end{bmatrix}, \qquad (7)$$

Setting $l=1$ in Eq. (4), we can get the propagation constant $\beta(\omega)$ of the fundamental mode. Effective mode index n_{eff} can be deduced from $n_{eff} = \beta(\omega)/k$. Dispersion of PCFs can be obtained as below:

$$D(\lambda) = D_m(\lambda) + D_w(\lambda) \qquad (8)$$

$$D_m(\lambda) = -\frac{\lambda}{c}\frac{d^2 n_m}{d\lambda^2} \qquad (9)$$

where, n_m can be computed by Sellmeier formula.

$$D_w(\lambda) = -\frac{\lambda}{c}\frac{d^2 n_{eff}}{d\lambda^2} \qquad (10)$$

where, λ is the operating wavelength and c is the velocity of light in the vacuum.

Numerical results

We set the hole-to-hole pitch Λ the same in one designed PCF, which can be fabricated easily in practical preform fabrication and fiber drawing processes. Firstly, only the air-hole diameter d1 in the first ring has been changed. As shown in Fig. 3, by increasing d_1/Λ to 0.5, 0.6, 0.7, 0.8 and 0.9, the dispersion decrease with the wavelengths increasing from 1.4μm to 1.6μm.

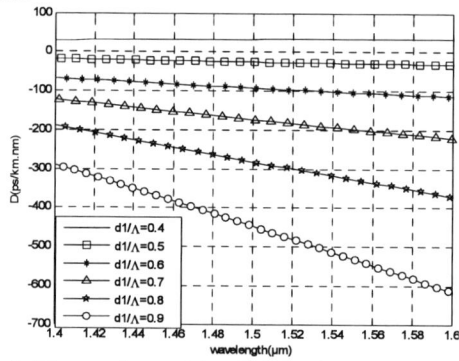

Fig. 3 Dispersion of the PCF with d/Λ=0.9 and Λ=1μm at different d_1/Λs

Bringing back d_1/Λ to 0.9, changing the air-hole diameter d2 in the second ring as reported in Fig.4, for d_2/Λ=0.9,0.8,0.7 and 0.6, the dispersion becomes more negative in the wavelength considered. Only with the air-hole diameter d_3 of the third ring changing, the dispersion of PCF is shown in Fig. 5. When wavelength is longer than 1.46μm and d3 is small, the dispersion value decreases much faster.

From the above phenomena, we got to know that larger air-hole diameters in the first ring and smaller ones in the second ring or third ring will sustain large negative dispersion properties.

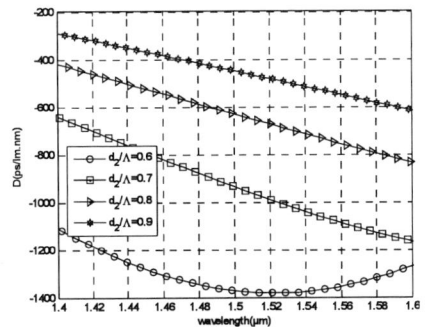

Fig. 4 Dispersion of the PCF with d/Λ=0.9 and Λ=1μm at different d_2/Λ

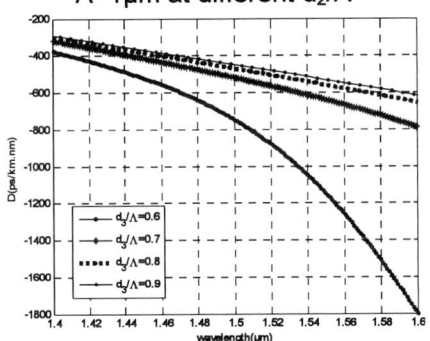

Fig. 5 Dispersion of the PCF with d/Λ=0.9 and Λ=1μm at different d_3/Λ

Alternatively, if smaller air-hole diameters in the first ring and larger ones in the other rings nearly zero flattened dispersion properties can be obtained. In order to verify this properties,

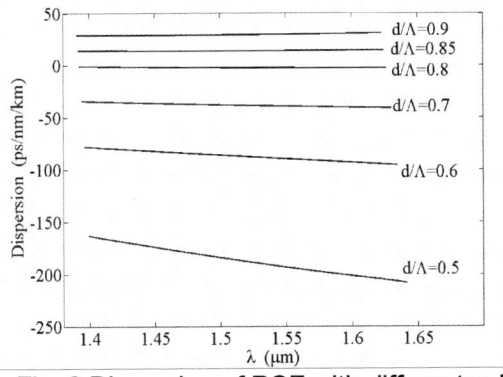

Fig. 6 Dispersion of PCF with different values of d/Λ. Here the effective core of PCF is undoped.

we set d_1=0.4μm, Λ=1μm and change the other diameters d from 0.5μm to 0.9μm as shown in Fig. 6. Numerical result demonstrates that d_1/Λ=0.4 and d/Λ=0.8 is better for flattened dispersion.

The Ge doped concentration in the core of PCF with d_1=0.4μm, Λ=1μm, d/Λ=0.8 is also effect the dispersion. As shown in Fig. 7, the dispersion will shift down with the Ge doped concentration increases and the shift is little more significant at longer wavelength than at lower wavelength. Therefore, we can do a slight adjustment in dispersion through controlling concentration of doped Ge in the core of PCF.

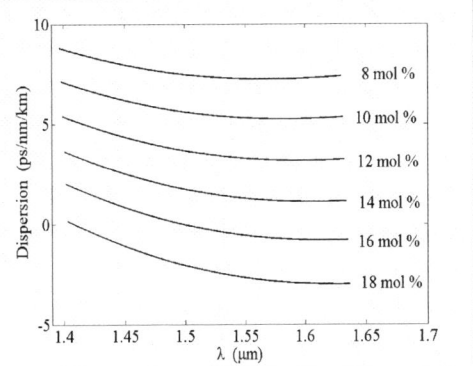

Fig. 7 Dispersion of PCF with different Ge concentration in the doped core.

Conclusions

In this paper, FVEIM&MCM is developed for modeling photonic crystal fibers. Based on the effect of different air-hole diameters in the first three rings and the Ge concentration in the doped core, large negative dispersion and nearly zero flattened dispersion PCFs can be designed for high nonlinear application.

Acknowledge

This work is supported by the national science foundation of China and the fund of Beijing Jiaotong University.

References

1 F. Poli et al. IEEE PHOTON TECHNOL LETT 16(2004),1065

2 Shailendra K. Varshney et al. IEEE PHOTON TECHNOL LETT 17(2005), 2026

3 Yan-feng Li et al. Opt. Comm. 238(2004), 29

4 XU Yong-Zhao et al CHIN.PHYS.LETT 23(2006), 2476

Micro-Structured Photonic Crystal Fibers with Large Mode Area and High Negative Dispersion

Nguyen Hoang Hai (1), Y. Namihira (2), Shubi Kaijage (3)

1: University of the Ryukyus, 1 Senbaru Nishihara Okinawa, Japan

Email: nhhaijp@ieee.org, namihira@eee.u-ryukyu.ac.jp

Abstract *We propose a novel PCF with Ge-rods at the center core. The authors designed a single mode PCF with large $A_{eff} > 200 \mu m^2$ over the wavelength above $1.2 \mu m$ and exhibits high negative dispersions (-186 to -158 [ps/(nm-km)]).*

Introduction

We have studied on a new LMA-PCF structure that has two cladding layers with 5% Ge rods in the inner core. This LMA-PCF can offer single-mode operation over an extended range of wavelength with very large effective area and high negative dispersion, which is essential for dispersion compensating in high bit rate transmission. In this paper the whole work aims at designing and numerical simulating the novel LMA and high negative dispersion PCFs. The basic idea for achieving all the above mentioned characteristics of the newly proposed LMA-PCF is to dope small areas as the one in the Fig. 1.

The advantage of the proposed LMA-PCF is that it should be much simpler to fabricate, since it has fewer elements and bigger silica/air interfaces in the final structure. To the best of our knowledge however, the task to design a such LMA-PCFs having all the above mentioned characteristics at the same time, particularly useful in high speed communication systems base on DWDM technology has not been reported in literature so far.

Structure Design of LMA-PCF

The schematic cross section of the PCFs structure is shown in the Fig. 1, the LMA-PCF has two cladding layers with different indices are shown in the right panel. The proposed LMA-PCF is composed of circular air-holes in the outer cladding arranged in a triangular array with lattice constant Λ_2 and diameter d_2. The host material is pure silica. The center core area of the fiber has been perturbed to have micro 5% Ge rods, where the unit cells are defined the same way as

in regular photonic crystal fiber structures. To enlarge mode area, 5%Ge rods are included in the center core area of the fiber with reduced hole diameter d_1 and lattice constant Λ_1.

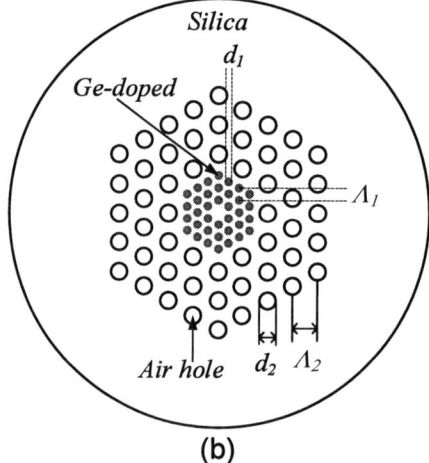

(b)

Fig 1. Schematic disign of proposed large mode area PCF with Ge Doped at center core

When the air-hole diameters are relatively small and the lattice constant is large, high positive dispersion characteristics can be achieved. The proposed LMA-PCF in Fig. 1 is an ideal candidate for our design purpose, since the PCF's core size is smaller than those reported in [1], [3]. Thus, by scaling up the physical size of the structure, the effective index contrast between two cladding can be made arbitrarily small, and the core size can be scaled up to the limit set by the tolerable micro- and macro-bending losses. In fact, due to the strong wavelength dependence of the effective cladding index, the fibers can be single-mode at all wavelengths.

978-0-9789217-3-6/07/$25.00

©2007 WEN GLOBAL SOLUTIONS

Results and Discussion

Fig 2(a) show the A_{eff} as a function of the wavelength of the proposed LMA-PCF for the optimum parameters Λ_1= 1.9µm, d_1=0.7µm, Λ_2=6.5µm, d_2=1.56µm were calculated. Also, A_{eff} with -1%, +2% and +4% variation of all parameters were obtained. Obviously, with a micro-adjustment of the design parameter d_1, d_2, Λ_1 and Λ_2 we can successfully enlarge the mode area. The A_{eff} of more than 200 µm² can be obtained over the whole wavelength range above 1.2 µm. The optimum effective area is about 218 µm² at 1.55 µm wavelength, a very large figure in comparison with those reported in [1], [3]. Fig 2(b) shows the impact of the design parameter d_2 of outer airhole rings to the effective mode area. In Fig. 2(c), the dispersion curve tend to increase at the longer wavelengths and dispersion properties of proposed PCFs are insensitive with -1%, +2% and +4% variation of all parameters (Λ_1=1.9µm, d_1=0.7µm, Λ_2=6.5µm, d_2=1.56µm). The simulation results show that the proposed LMA-PCF exhibits a large negative dispersion coefficient from −186 to −158 [ps/nm-km] in all wavelengths ranging from 1.2 µm to 1.8 µm.

It can be seen from Fig. 2(d) that although the field has been confined mostly in Ge doped core region, noticeable penetration are seen and which spreads to the second cladding, thus leading to the large effective area of the proposed PCF. There is no leakage outside of the first innermost air-hole ring because the contrast refractive index between the two cladding is very strong. From the simulation results, we can conclude that, by scale up and adjust the physical size of air-holes at the second cladding, larger effective area of the fiber can be achieved while maintaining single mode at all wavelength.

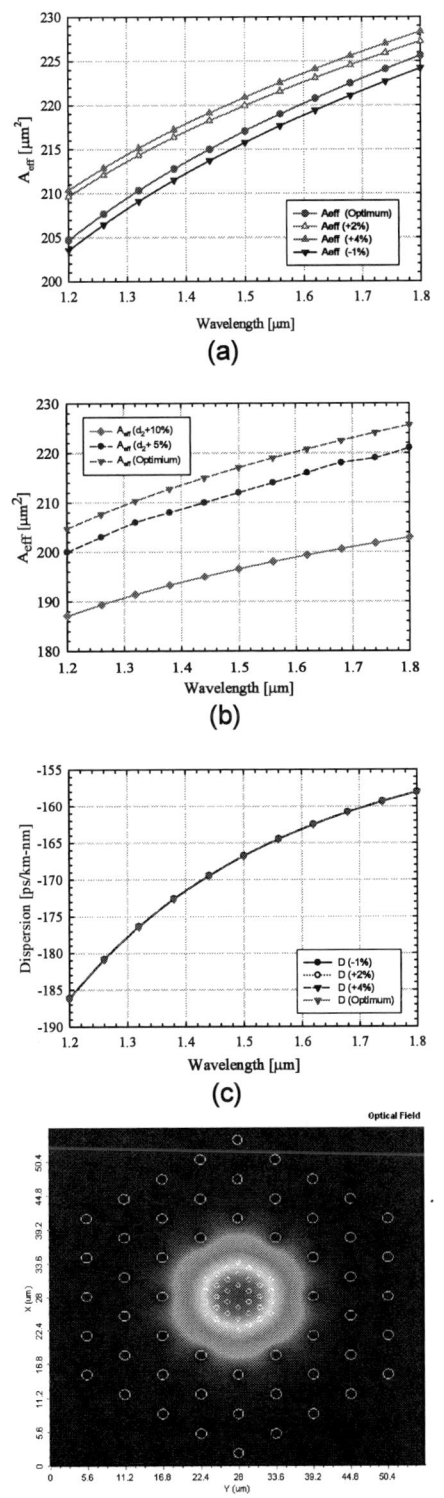

(a)

(b)

(c)

(d)

Fig. 2 (a), (b) Effective Mode Area, (c) Dispersion as a function of wavelength, (d) calculated intensity distribution of LMA-PCFs with Ge doping at wavelength of 1.5µm

Conclusions

Novel photonic crystal fibers with two different air-holes and 5% Ge rod-core claddings geometries have been investigated, in order to design single-mode large mode area fibers. We show that the proposed LMA-PCFs could archive an effective area of more than $200\mu m^2$ and with large negative dispersion coefficient from −186 to −158 [ps/nm-km] in all wavelengths ranging from 1.2 μm to 1.8 μm. We believe that the proposed LMA-PCF will be beneficial for dispersion compensating in future ultra broadband transmission application. The proposed LMA-PCF with flattened-dispersion is being investigated.

References

1. T. Matsui et al Journal of Lightwave Tech, Vol. 23 (2005), page 4178-4183
2. S. Guo et al Opt. Express, Vol. 12 (2004), page 3341-3352
3. K. S.Chiang et al OFC Proceeding (2002), page 620-621
4. W.J. W et al CLEO Proceeding(2001), page 319
5. N. Florous et al Optics Express, Vol. 14 (2006), page 901-913.

A C+L-band erbium-doped fiber ASE source using dual-forward pumping configuration

Wencai Huang[1], Chaohong Huang[1], Xiulin Wang[2,3], Benrui Zheng[1], Huiying Xu[1], Zhiping Cai[1]

1: Department of Electronic Engineering, Xiamen University, Xiamen 361005, P. R. China
2: Department of Physics, Xiamen University, Xiamen 361005, P. R. China
3: Department of Physics, Jimei University, Xiamen 361021, P. R. China
Email: huangwc@xmu.edu.cn

Abstract *We present a C+L-band ASE fiber source using double-pass dual-forward pumping configuration. The proposed ASE source offers a high pumping efficiency of 24.6%, output power of 52mW and a wide bandwidth of about 80nm.*

Introduction

Broadband amplified spontaneous emission (ASE) fiber sources have been not only intensively studied as the essential sources for passive optical networks, wavelength-division multiplexing (WDM) communication systems [1, 2] and test instruments for optical components [3], but also used as light sources for fiber-optic gyroscopes and spectral slicing WDM systems [4, 5]. In particular, ASE sources based on erbium-doped fiber (EDF) are considered to be the best candidates since they offer a broad spectrum, high output power, and high mean wavelength stability to meet the application requirements. Recently, several groups have reported their research and development of the ASE sources with a broad bandwidth covering both C-band and L-band for the immediate expansion of the fiber optic communication window from C-band to C+L-band. An ASE source with a bandwidth up to 80nm has been demonstrated using an un-pumped EDF acting as a secondary pump [6], two-stage of EDF ASE source scheme with a long period fiber grating [7], or one-stage configuration using hybrid pumping wavelength [8]. In ref.[9], an optical circulator is added between two stages to construct a double-pass and bi-directional pumping configuration with an output power up to 178mW and a bandwidth of 81nm. However, the output emission from these C+L-band ASE sources have a large spectral ripple, and an external spectral flattening filter is required. In this paper, we proposed a two-stage C+L-band ASE light source design to provide a spectral range covering both C-band and L-band with a low spectral ripple and high pumping efficiency.

Experimental setup

Figure 1 shows the proposed double-pass dual-forward pumping configuration for a low ripple and high power C+L-band ASE source. In the structure, the first stage is an L-band ASE seed source with the double-pass forward pumping scheme (DPF) which consists of a broadband fiber mirror (BBR) with a reflectance above 90%, a 980 nm pump laser diode (LD1), a piece of EDF (EDF1) and a wavelength division multiplexer (WDM). The fiber mirror is self-made using a 3dB broadband fiber coupler and used to reflect the backward ASE lights generated from the EDF1. The second stage, i.e. output stage consists of a piece of EDF (EDF2), which not only generates the C-band ASE source in the forward direction with forward pumping by a 980 nm pump laser diode (LD2) but also amplifies the weak L-band seed spectrum from the first stage. An optical isolator after EDF2 is used to prevent any feedback that might lead to lasing. The EDFs used here are both Nufern fibers. The EDF1 is EDFL-980-HP, with a peak absorption coefficient of 24.1 dB/m at 1530nm and 16.5 dB/m at 980nm, mode field diameter of 5.7μm, and a cutoff wavelength of 919 nm. The EDF2 is EDFC-980-HP, with a peak absorption coefficient of 6.3 dB/m at 1530nm and 3.9 dB/m at 980nm, mode field diameter of 5.8μm, and a cutoff wavelength of 964 nm.

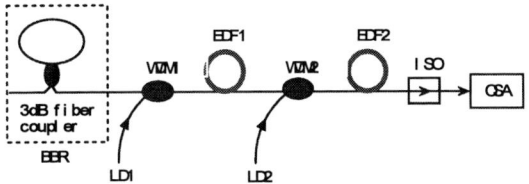

Fig.1 The proposed C+L-band ASE source.

Results and discussions

We have performed extensive simulations of the two-stage C+L-band configuration as shown in Fig.1 in order to gain insight into and optimize its output properties. Our simulation results show that the designed configuration is capable to generate a flat C+L-band ASE spectrum when the optimum fiber lengths and pump powers of the two stages have been used. By comparing the output spectra with different fiber lengths, we selected the fiber length of L1=18m, L2=14m for the two sections, respectively. The results show that a flat C+L-band ASE spectrum can be achieved with different ratio of pump powers for the two stages, P1/P2.

Then, the output characteristics of the ASE source were investigated experimentally. The output spectrum was measured using ADVANTEST Q8384 optical spectrum analyzer (OSA) with a resolution of 0.1 nm. Figure 2 shows the corresponding output spectra of the proposed ASE source for the cases of (a) P1 = 0mW and P2 = 70mW, (b) P1 = 26.8mW and P2 = 0mW, and (c) P1 = 26.8mW and P2 = 70mW. From the figure, we can see that the output ASE spectrum falls on C-band when only EDF2 was pumped by LD2 (case a). The ASE had a very weak spectrum within L-band when only EDF1 was pumped by LD1 (case b). When both DEF1 and EDF2 were pumped simultaneously, a flat ASE spectrum covering both C-band and L-band with nearly 80nm bandwidth was obtained (case c).

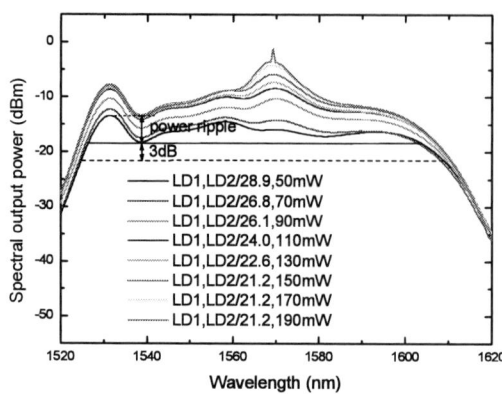

Fig.3 Optimum C+L-band ASE spectra for different LD1 and LD2 pumping powers.

The output spectra of C+L-band ASE source were measured for different pumping powers using the OSA. The pump power for EDF2, P2 was 50, 70, 90, 110, 130, 150, 170 and 190mW respectively. For each case, the optimum pump power for EDF1, P1 was measured to be 28.9, 26.8, 26.1, 24, 22.6, 21.2, 21.2 and 21.2mW, respectively. Figure 3 shows the measured spectra of the ASE for different values of P2 with the corresponding optimum value of P1. The experimental results are qualitatively consistent with those from simulations. Namely, a flat C+L-band ASE spectrum can be achieved with different pumping power levels. Both experimemental and simulation results show that when P1 is lower, the larger P2 is required to achieve the flat C+L-band ASE spectrum. The power ripple ΔPr in Fig.3 was measured from the valley power at 1537 nm to the peak power, and the spectral ripple is defined as a half of the output power variation. The bandwidth was measured for the output spectrum of ASE source, was defined as -3dB power level at 1537 nm as shown in Fig.3. Figure 4 shows the effects of pumping power on the measured C+L-band characteristics of bandwidth and spectral ripple for the optimum adjustment of LD1 power. All the cases of the pumping power of LD2 above 50mW incorporating all the pumping powers of LD1 can achieve a broadband C+L-band with -3dB bandwidth up to 80nm and spectral ripple less than 2.9dB. With the increase of LD2 pumping power, the bandwidth decreases firstly then becomes

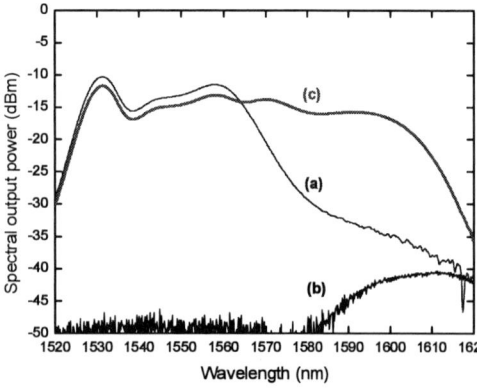

Fig.2 Spectra showing output of first and second stage and two-stage combined.

stable at about 80nm while the spectral ripple increases monotonously. In the best case, a spectral ripple of 2.4dB can be achieved for LD1 pumping power of 28.9mW and LD2 of 50mW. The results reported in this paper are better than those in ref. [7-9], where the spectral variations were of ~3dB, 4.5dB and 2.5dB over an 80nm bandwidth, respectively, without using external spectral filters. Figure 5 illustrates the dependence of the characteristics of pumping efficiency and output power against the pumping power of LD2 for the optimum adjustment of LD1 power. The highest pump efficiency of 24.6% was obtained for 21.2mW of LD1 and 190mW of LD2, which corresponding to an output power of 52mW.

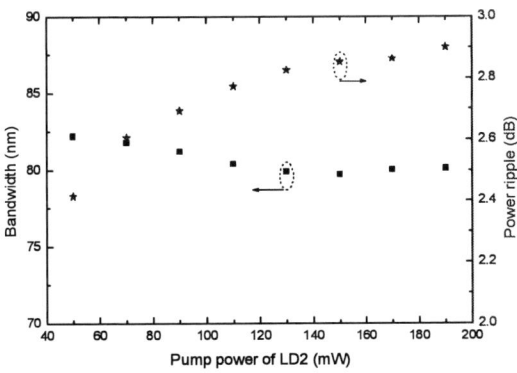

Fig.4 Bandwidth and power ripple against LD1 and LD2 pumping power.

Conclusions

We have proposed and experimentally demonstrated a low spectral ripple and high pump efficiency C+L band ASE source by using dual-forward pumping configuration. The best performances of spectral ripple and pumping efficiency can be obtained with proper pumping power allotment. The C+L-band ASE source with an output power of 52mW, pump efficiency of 24.6%, bandwidth of about 80 nm, and a power

ripple of 2.9 dB has been achieved.

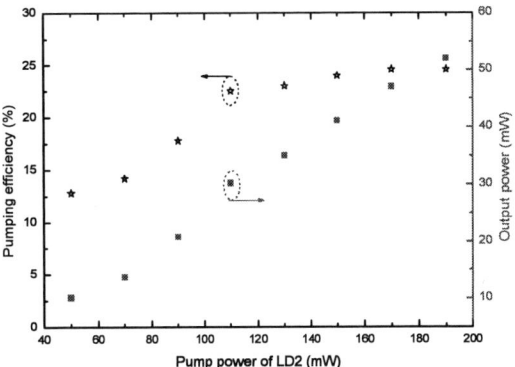

Fig.5 Pumping efficiency and output power against LD1 and LD2 pumping power.

Acknowledgments

This work was supported by the Key Scientific and Technical Innovation Project of Xiamen University under Grant No.K70007, the Natural Science Foundation of Fujian Provincial of China under Grant No.2006J0243, and the Program for New Century Excellent Talents in Fujian Province University.

References

1 J. S. Lee et al IEEE Photon. Technol. Lett., 5, (2002)1458.
2 Y. K. Chen et al Opt. Comms., 169, (1999) 245.
3 S. Yamashita et al IEEE J. Select. Topics Quantum Electron., 7, (2001) 41.
4 P. F. Wysocki et al J. Lightwave Technol.,12, (1994) 550.
5 M. Zirngibl et al IEEE Photon. Technol. Lett., 8, (1996) 721
6 J. H. Lee et al Opt. Lett., 24, (1999) 279.
7 R. P. Espindola et al Electron. Lett., 36, (2000)1263.
8 W. C. Huang et al Electron. Lett., 38, (2002) 2061.
9 H. Lin et al Opt. Exp., 12, (2004) 6135.

Enhancement of Multi-wavelength Brillouin-Erbium Fibre Laser Utilizing Fibre Bragg Grating Filter

M. N. Mohd Nasir, M. H. Al-Mansoori, H. A. Abdul Rashid, P. K. Choudhury, Z. Yusoff[*]

Institute of Photonics Research and Applications,
Faculty of Engineering, Multimedia University, 63100, Cyberjaya, Selangor, Malaysia.
*Email: zulfadzli.yusoff@mmu.edu.my

Abstract: *We demonstrate an efficient multi-wavelength Brillouin-erbium fibre laser incorporating an FBG filter. The self-lasing effect is reduced and subsequently the number of output channels increased. Furthermore, the stability of output channels is also improved.*

Introduction

Dense wavelength division multiplexing (DWDM) plays a great role in the future of optical communication networks. Multiple wavelength laser sources are an important application in DWDM systems and Brillouin-erbium fibre laser (BEFL) is capable in producing them [1], [2], [3]. The enabling technique of this type of laser is the combination of high gain from erbium doped fiber (EDF) and the narrow band nonlinear gain of stimulated Brillouin scattering (SBS) in optical fiber. Therefore, the total hybrid-gain performs inhomogeneous broadening property for multiwavelength oscillation. For Fabry-Perot cavity BEFLs, the self lasing cavity modes that results from high EDF pump power are the major foundation to the output instability. Thus, these unwanted modes compete with the cascaded Brillouin Stokes signals to be the dominant ones in the laser cavity. Therefore, Sufficient Brillouin pump (BP) power is required to suppress this effect so that the process of generating cascaded Brillouin Stokes signals (output channels) can take place [3].

Previous demonstrations have shown that improvements in the total number of Stokes signals were achieved when a Sagnac loop filter [4] or a band-pass filter [5] were added in standard ring BEFL design. The efficiency of generating the Stokes signals improved due to the suppression of amplified spontaneous emission (ASE) from erbium doped fibre (EDF). Hence, the Stokes signals get amplified more efficiently within the narrow bandwidth of the bandpass filter. However, the self-lasing effect also gets amplified as efficient, so a sufficiently large BP is still needed to suppress the self-lasing effect.

In this paper, we propose and experimentally demonstrate the utilization of fiber bragg grating (FBG) filter as one of the cavity mirrors to increase the number of Brillouin Stokes signals in a Fabry-Perot cavity of a BEFL. The main innovative step in our approach is the use of FBG to allow only a certain laser modes to oscillate by suppressing other modes. Thus, the stability and the number of output channels increases in the cavity.

Experimental Setup and Principles of Operation

Fig. 1 shows the experimental setup of a Fabry-Perot BEFL configuration. A 980 nm laser diode (LD) was used in the experiment as the primary pump light for the 12 m EDF. Wavelength selective coupler (WSC) was used to multiplex the pump and signal lights. The Brillouin gain media was provided by a 7 km long single mode fibre (SMF) and the BP was provided by an external cavity tunable laser source with 100 nm tuning range (1520–1620 nm) and 5 dBm maximum pump power. A 3-dB coupler (C) was used to couple the BP into the SMF.

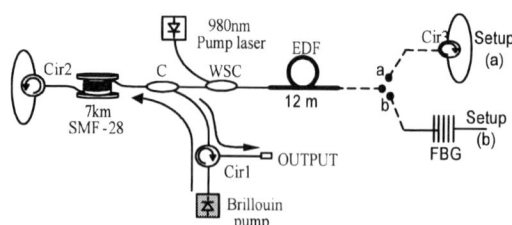

Fig. 1: Experimental configurations of BEFL with the Fabry-Perot cavity of the laser system produced by: two circulators in setup (a) and one circulator and FBG filter in setup (b).

In order to study the impact of utilized FBG filter to improve the BEFL performance, we compared the performance of the BEFL incorporating the FBG filter as one of the cavity mirrors, setup (a), with that of using circulators at both ends, setup (b). Compared to the circulators that reflect all wavelengths back into the cavity, the FBG filter only selectively reflects about 40 nm of wavelengths (1520–1560±2 nm) back into the cavity of the laser system. The wavelengths in the region where self-lasing occurs (>1560 nm) were not reflected by the FBG i.e. removed from the cavity.

At sufficiently high 980 pump power, BP light launched through the 3-dB coupler induced stimulated Brillouin scattering (SBS) in the SMF with the Brillouin Stokes signal downshifted 0.088 nm from the BP wavelength. The signal of the first order Brillouin Stokes signal would propagate in the opposite direction of the BP and travelled to the EDF for double-pass amplification in the linear cavity. When the total gain generated from the SBS and EDF are equal to the cavity loss, a laser oscillation is formed between the two cavity mirrors. The higher order Brillouin Stokes signals can be generated by a lower order Brillouin Stokes signals when it passed through the SMF twice in each round. The cascading of the Brillouin Stokes signals generation is continued until the total gain in the laser cavity is less than the cavity loss at the operating wavelength. A stable laser will be produced at the steady state condition which consist the BP and its Brillouin Stokes signals.

Results and Discussion

The Fabry-Perot BEFL cavity configuration would operate as bidirectional erbium-doped fibre laser (EDFL) without launching the BP and it operates at random oscillating modes (self-lasing cavity modes) at the EDF peak gain as depicted in Fig. 2. The 980 nm pump power is set to 50 mW. With low BP power, the self lasing cavity modes will not be suppressed and it results in sharing of operation between BEFL and EDFL cavity modes [3]. Therefore, instability of the BEFL would occur, especially at the peak gain, which is the most efficient region for BEFL operation. As shown in Fig. 2 (a) for BEFL with two circulators formed the Fabry-Perot cavity, the oscillation bandwidth of the self-lasing

cavity modes is about 2 nm in the region between 1562.5-1564.5 nm. On the other hand, incorporating the FBG filter in place of one of the circulators (Cir3) in the design as in setup (b) would make the self-lasing to be less efficient as some of its power is transmitted out of the cavity of the laser system. Therefore, the oscillation bandwidth of the self-lasing cavity modes is reduced to about 0.6 nm between 1560.8-1561.4 nm as illustrated on Fig.2 (b). In addition, the EDF peak gain is shifted to the region around 1561 nm. This means that less BP power is needed to suppress the self-lasing effect. At the same time this would let more energy of the EDF amplification to amplify the Brillouin Stokes signals rather than the self-lasing cavity modes.

For the same 980 nm pump, BP power and wavelength, we compared the performance of the BEFL using circulators at both ends, setup (a), with that of incorporating the FBG filter as one of the cavity mirrors, setup (b).

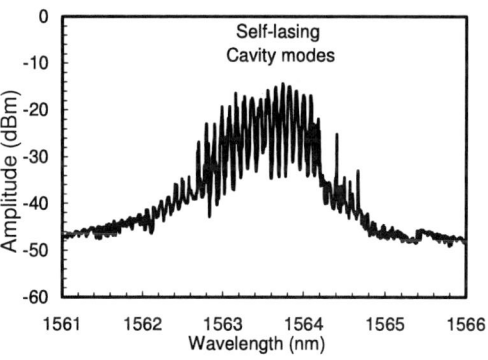

(a)

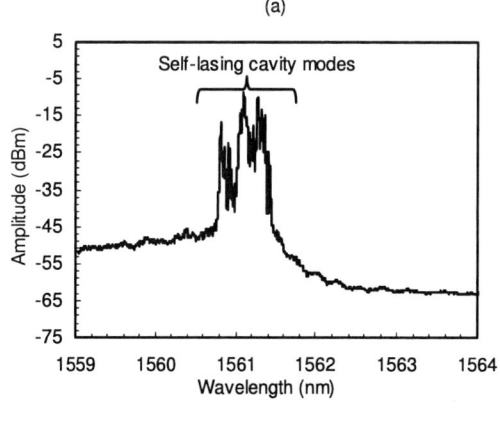

(b)

Fig. 2: Self-lasing cavity modes at 50 mW of 980 nm pump power for (a) two circulators (b) one circulator and FBG filter.

The output spectra obtained from this experiment are depicted in Fig. 3 with the optical spectrum analyzer's resolution bandwidth is set at 0.015 nm. The 980 nm pump and BP powers were set at 130 mW, and 5 dBm respectively. Fig. 3(a) shows the superimposed optical spectra of the BEFL for setup (a). The cascaded Brillouin Stokes lines are obtained when the BP wavelength is set at 1559 nm. The presence of self-lasing cavity modes with the Brillouin Stokes signals is clearly seen around 1562.5-1564.5 nm (EDFL peak gain). The self-lasing cavity modes compete with the Stokes signals to form laser in the cavity. The amount of energy is distributed between these two oscillations; self-lasing cavity modes and Brillouin Stokes signals. Therefore, the number of Brillouin Stokes signals generated was limited to 18 Stokes signals as can be seen in Fig.3 (a). On the other hand, these eccentrically oscillation cavity modes are completely suppressed

with the incorporation of the FBG filter as one of the cavity mirrors, as evidently shown in Fig. 3(b). In addition, a more efficient laser was produced with setup (b) which consists of 25 Brillouin Stokes signals with no self-lasing effect as shown in Fig. 3(b). Moreover, besides the improvement in terms of the total number of output Stokes, the reduction of the self lasing effect would also improve the stability of BEFL operation as there would be less sharing of operation with self-lasing cavity modes.

Conclusion

Multi-wavelength Brillouin-erbium fibre laser has been successfully demonstrated experimentally by using an FBG filter as one of the cavity mirrors. The bandwidth of the self lasing effect was reduced in the design, which in turn increase the total number output Stokes signals and also improves the stability of the BEFL system. Up to 26 maximum output channels including the BP power were obtained in the design with the FBG filter as compared to 19 channels using two circulators.

References

[1] G.J. Cowle et al, "Multiple Wavelength Generation With Brillouin/Erbium Fiber Lasers", IEEE Photon. Technol. Lett., Vol. 8 (1996), pp.1465 – 1467..

[2] D.S, Lim et al, "Generation of multiorder Stokes and anti-Stokes lines in a Brillouin erbium-fiber laser with a Sagnac loop mirror," Opt. Lett. 23, (1998), pp. 1671-1673.

[3] M. H. Al-Mansoori et al, "Multi-wavelength Brillouin-Erbium fibre laser in a linear cavity", Optics Communications, Vol. 242 (2004), pp.209 – 214.

[4] Y.J. Song et al, "Tunable Multiwavelength Brillouin-Erbium Fiber Laser With a Polarization-Maintaining Fiber Sagnac Loop Filter", IEEE Photon. Technol. Lett., Vol. 16 (2004), pp. 2015 – 2017.

[5] S. Saharudin et al, "Enhancements of Brillouin stokes powers in multiwavelength fibre laser utilizing band-pass filter", Microwave and Optical Technol. Lett., Vol. 40 (2004), pp. 408 – 410.

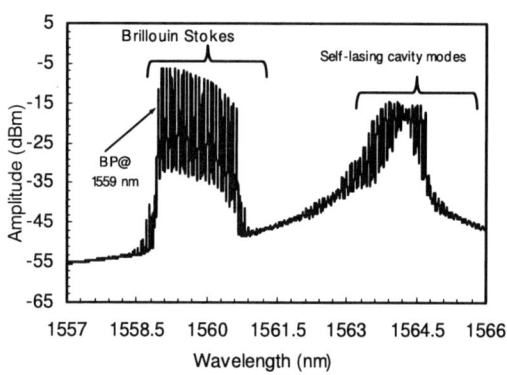

(a)

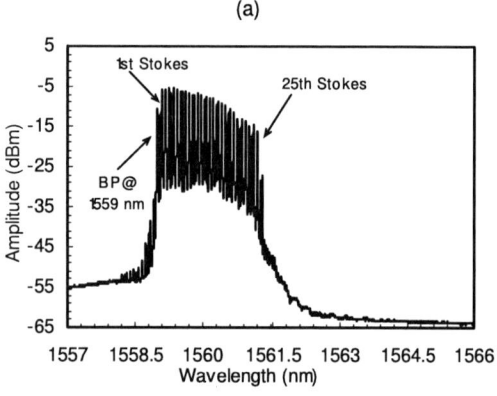

(b)

Fig. 3: BEFL output spectrum at 130 mW of 980 nm pump power, BP wavelength at 1559 nm with power of 5dBm for (a) two circulators (b) one circulator and FBG filter.

A Novel Millimeter-Wave Generation Technique for mm-ROF System Based on Harmonic Generation Principle

Meiwei Zhu (1), Jiajun Ye (1), Rujian Lin (1)

1 : School of Communication and Information engineering, Shanghai University,
149 Yanchang Road, Shanghai 200072, China. zmw1231@shu.edu.cn

Abstract *A novel millimeter-wave generation technique for mm-ROF system is proposed in this paper. The light-wave from single-mode laser is split into two optical carriers. One carrier is phase modulated by 5-GHz microwave signal and then combined with the other carrier. The mixed optical signal after transmission over 15-km-long fiber is photodetected to generate electrical signal consisting of harmonics of 5-GHZ microwave signal. The desired 40 GHz millimeter-wave signal can be obtained by bandpass filtering the harmonics. Data transmission on downlink system with the proposed generation method is demonstrated.*

Introduction

Millimeter-wave radio-over-fiber technology has been under much investigation over the last few years due to its high capacity, flexibility and cost-effectiveness, making it a potential candidate for future broadband pico-cell wireless access network. Radio signal is generated at central station (CS) and remotely delivered to several distributed base stations (BS) through optical fiber. Signal processing and equipment are mostly concentrated at central station while base stations are simplified with low complexity [1]. The millimeter-wave optical generation is obviously one of the key issues of mm-ROF technology. Many optical microwave generation techniques have been proposed, such as modulation-sideband-technique [2], dual-mode laser [3] and FM-IM conversion in dispersive fiber [4]. For different generation methods, the millimeter-wave displays different properties and is applicable to specific applications. However, the harmonic based generation like FM-IM conversion in dispersive fiber has advantages of low optical phase noise and generating high-frequency microwave by using low-frequency optoelectronic technology. A bidirectional mm-ROF system based on FM-IM conversion using periodic optical filter instead of dispersive fiber, which is called optical frequency multiplication (OFM), has been reported [5] and shows its high tolerance to fiber dispersion[6][7].

In this paper, we propose an even simpler

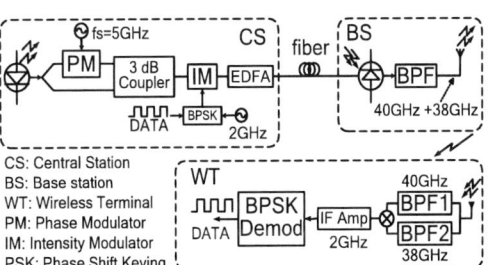

Fig.1.Downlink system with the proposed millimeter-wave generation technique.

and more cost-effective harmonic generation based method, where the periodic optical filter of OFM is omitted. As shown in Fig.1, the light-wave emitting from single-mode laser is split into two optical carriers, one of which is phase modulated while the other one is just combined with the phased modulated carrier. The two carriers after transmission over 15-km-long single-mode fiber are mixed at PD to produce harmonics containing the desired millimeter-wave signal. We describe the theoretical principle and how to optimize the system. Also, 100Mb/s baseband data transmission on downlink system is demonstrated.

Principle of Millimeter-Wave Generation and Data Modulation

As we all know, optical PM (or FM) signal has a constant envelope and will generate dc power when photodetected. But optical spectrum of PM signal already has a large number of sideband lines spaced by modulation tone around center carrier. If we shift the center frequency of PM spectrum to zero dc frequency by optical coherent

---

This work is supported by Natural Science Foundation of China (Grant 60377024).

978-0-9789217-3-6/07/$25.00

©2007 WEN GLOBAL SOLUTIONS

detection with an optical carrier at center frequency of PM signal, the sideband lines will be in electrical domain and the output of PD will not be DC power but a lot of harmonics of modulation tone. This is the basic conception of the proposed generation technique. Detailed analysis is as follow.

As depicted in Fig. 1, the lightwave from single-mode laser is split into two branches. One is phase modulated by 5-GHz sinusoidal microwave signal using external optical phase modulator (PM). After phase modulation, this optical carrier can be expressed in electrical field as

$$E_1(t) = E_{c1} \exp[\, j\omega_c t + j\beta \cos \omega_s t + j\Phi_{PN}(t)] \quad (1)$$

where ω_c is the angular frequency of optical carrier, β is the phase deviation; ω_s is the angular frequency of 5-GHz sinusoidal microwave signal; and $\Phi_{PN}(t)$ is the laser phase noise. However, the other optical carrier is not processed and just combined with the phase modulated carrier. Then, the mixed optical signal is given by

$$E_2(t) = E_{c1} \exp[j\omega_c t + j\beta \cos \omega_s t + j\Phi_{PN}(t)]$$
$$+ E_{c2} \exp[j\omega_c(t - \Delta\tau) + j\Phi_{PN}(t - \Delta\tau)] \quad (2)$$

where $\Delta\tau$ is the differential time delay between the two branches. To make sure the states of polarization of the two optical carriers are identical, the polarization controllers must be used before 3 dB coupler.

If the signal $E_2(t)$ is intensity modulated by data signal, the output of intensity modulator (IM) is

$$E_{out}(t) = \sqrt{1 + km(t)} E_2(t) \quad (3)$$

where $m(t)$ is 2 GHz PSK (or QAM) subcarrier with 100Mb/s data; k is modulation index. The output of photodetector is

$$i(t) \propto \frac{1}{2} E_{out}(t) E_{out}(t) *$$
$$= \frac{1}{2}[1 + km(t)]\{E_{c1}^2 + E_{c2}^2$$
$$+ 2E_{c1}E_{c2} \cos[\beta \cos \omega_s t + \omega_c \Delta\tau + \Delta\Phi_{PN}(t)]\} \quad (4)$$

with $\Delta\Phi_{PN}(t) = \Phi_{PN}(t) - \Phi_{PN}(t - \Delta\tau)$. Expanding (4) with Bessel functions of first kind, we can observe that the signal $i(t)$ contains the frequency components at every harmonic n of 5-GHz microwave signal as shown below

$$[1 + km(t)]\cos(\omega_c \Delta\tau + \Delta\Phi_{PN}(t)) \sum_{n=1}^{\infty} (-1)^n J_{2n}(\beta)\cos(2n\omega_s t)$$

and

$$[1 + km(t)]\sin(\omega_c \Delta\tau + \Delta\Phi_{PN}(t)) \sum_{n=1}^{\infty} (-1)^{n+1} J_{2n-1}(\beta)\cos[(2n-1)\omega_s t]$$

for the even and odd harmonics with relative amplitudes respectively. The 8th harmonic is the desired 40 GHz millimeter wave signal which has been AM modulated by $[1 + km(t)]$. Obviously we can obtain 38GHz (or 42GHz) PSK data signal and 40GHz sinusoidal signal. Besides the 38GHz radio data signal, the 40 GHz millimeter-wave signal as a pilot tone should also be radiated into air channel for wireless terminal (WT) reception and downconversion.

If the phase deviation β is properly set according to Bessel functions, the desired millimeter-wave signal can be maximized. From the analysis above, we note that the laser phase noise may be not completely correlated between the two optical carriers, which will lead to a linewidth broadening of the electrical mm-wave carrier resulting in peak power degradation [1] if the time delay $\Delta\tau$ is not zero. To ensure the phase noise be correlated as possible, the differential time delay $\Delta\tau$ should be zero or approach zero. By doing this, the laser phase noise $\Phi_{PN}(t)$ and the odd harmonics all disappear while the even ones are maximized.

If fiber dispersion is taken into consideration, FM-IM conversion will take place [4], which will contribute to the generation of higher order harmonics. Thus fiber dispersion does not seriously affect the system performance [6][7]. But further analysis is necessary for future work.

Verification by Simulation
The setup is shown in Fig1. In the central station (CS), the optical carrier at 1550nm with linewidth 10MHZ is split into two carriers. One carrier is phase modulated by 5-GHz sinusoidal signal with phase deviation β of 9.59 radian/v to maximize the desired 40GHz signal according to Bessel functions. The states of polarization of the two carriers are controlled to be identical and the time delay $\Delta\tau$ is set to be very close to zero. Firstly, the optical signal

outputting from 3dB coupler is directly transmitted over 15-km-long single-mode fiber and photodetected at base station (BS).

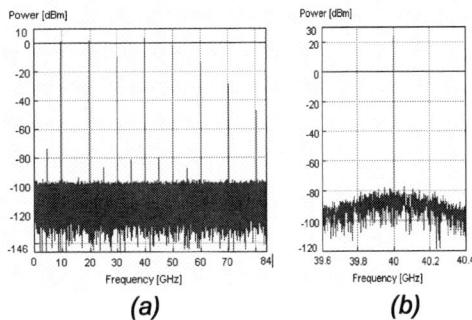

(a)   *(b)*

Fig. 2. Simulation results of millimeter-wave generation technique without modulation (a) RF spectrum of generated harmonics. (b) 40-GHz millimeter-wave signal with high purity after BPF and amplifier.

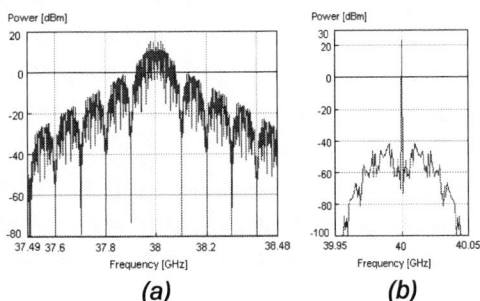

(a)   *(b)*

Fig. 3. (a) RF spectrum of 38 GHz BPSK signal (b) 40 GHz pilot tone.

Fig.2 shows the RF spectrum of harmonics generated at PD and 40 GHz millimeter-wave signal with high spectra purity without modulation. It produces only even harmonics of 5 GHz sinusoidal signal, among which the desired 8th harmonic is maximal while odd harmonics are all suppressed.

Then the output lightwave of 3dB coupler is intensity modulated by 2 GHz BPSK radio signal with 100Mb/s random binary sequence. Fig. 3 shows RF spectrum of 38 GHz modulated millimeter-wave signal and 40GHz pilot tone obtained at the output of bandpass filter (BPF). It can be seen from Fig.3 that the spectrum of 38-GHz signal is expanded with the first zero point bandwidth of 100MHz and the 40GHz pilot tone has high spectra purity. In WT, mixed with the

40GHz pilot tone, the 38 GHz millimeter-wave data signal is downconverted into 2 GHz intermediate frequency signal. Recovering the binary sequence by BPSK demodulation, we measure the bit-error-rate (BER) as a function of the received optical power at BS. The BER curve is shown in Fig.4.

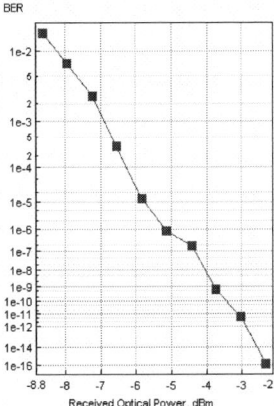

Fig. 4. BER vs Received optical power.

Conclusions

A novel millimeter-wave generation technique and modulation scheme have been proposed and verified by simulation successfully. The results agree well with the theory. The system mainly consists of only an optical phase modulator, an intensity modulator and a few passive optical components. It's very simple and cost-effective to implement millimeter-wave ROF system employing this generation technique. Optimum system performance can be achieved by proper setting. It shows that the scheme is a feasible solution for millimeter-wave radio-over-fiber system.

References

1　R. Hofstetter et al, Trans Microw Theory. Tech. 43(1995), 2263-2269.

2　J. O'Reilly et al, IEEE J. Lightwave Technol. 12 (1994), 369-375.

3　D. Wake et al, IEEE Trans Microw Theory. Tech. 43 (1995), 2270-2276.

4　N. G. Walker et al, Electron Lett. 28(1992), 2027-2028.

5　M. G. Larrode et al, IEEE photon Technol Lett. 18(2006), 241-243.

6　M. Xiu et al, SPIE 2005. v6022, 885-892.

7　M. G. Larrode et al, Electron Lett. 42(2006), 872-874.

Fiber Ring Based Microwave Photonic Filters Implemented in a Radio-over-Fiber Link

Kun Zhu*, Hongyan Fu, Yun Xiao

Center for Optical and Electromagnetic Research, Joint Research Center of Optical
Communications of Zhejiang University, Zhejiang University, Hangzhou 310058, China;
*zhukun@coer.zju.edu.cn, fuhongyan@coer.zju.edu.cn, xiaoyun@coer.zju.edu.cn

Abstract *A microwave photonic filter based on fiber ring implemented in a Radio-over-Fiber link with two carrier frequencies is demonstrated. System performance under the situation of different filter parameters and different frequency spacing is analyzed.*

Introduction

As the development of communication technology and people's increasing demand on communication quality, large-capability and high-speed mobile communication system has attracted much research attention. For instance, in the emerging broadband wireless access networks and standards spanning from universal mobile telecommunications systems (UMTS) to fixed access picocellular networks, including wireless local area network (WLAN), World Interoperability for Microwave Access, Inc. (WIMAX), local multipoint distribution service (LMDS), etc., there is a need to increase the capability by reducing the coverage area [1]. Radio-over-Fiber (RoF) system, where radio signals are processed at a central office, and distributed from the central location to inexpensive remote antenna units (RAUs) using fiber optic transmission, possessing characteristics of both mobile communication and fiber optic communication, obtains the above objective and is almost the best candidate for the next generation mobile communication system.

Microwave photonic filter, in which signals are processed in optical domain, can overcome "electronic bottleneck" in the traditional filters, and has the properties of low loss, high bandwidth, and immunity to electromagnetic interference (EMI). It can find applications in RoF systems based on both intensity modulator and phase modulator to improve system performance [2]-[3]. In this letter, a microwave photonic filter implemented in a Radio-over-Filer link with two carrier frequency is demonstrated.

This microwave photonic filter is simply based on fiber ring structure. In the experiment, we make several filters with different free spectral ranges (FSRs) and different filter rejection ratios by changing the length of fiber ring and coupling ratio of the fiber coupler. The frequency spacing between the two carrier frequencies should match the filter's FSR, and also the influence of their mismatch on the performance of the system is analyzed.

Principle

The experimental setup of the system is shown in Fig. 1. It is a simple Radio-over-Fiber link, except for the lack of antenna subsystem after the photo-detector. The light from a DFB laser is fed to an electro-optic modulator (EOM) after an isolator and a polarization controller. The intensity modulator is driven by two microwave signals at different frequencies via a power combiner. In our experiment, we choose the carrier microwave signals at frequency between 1 GHz and 3 GHz. Actually, it is called "IF-over-fiber" link in some literatures. After the two carriers are transmitted in a roll of fiber with the length of 25km, we use fiber ring based microwave photonic filter (formed by joining together one input port and one output port of a fiber coupler) to select one of the two carriers in one base station. By adjusting the parameters of the filter carefully, the amplitude of undesirable carrier can be suppressed effectively, and the desirable one can be filtered out. The output RF signal from the photodiode is observed through the spectrum analyzer.

978-0-9789217-3-6/07/$25.00

©2007 WEN GLOBAL SOLUTIONS

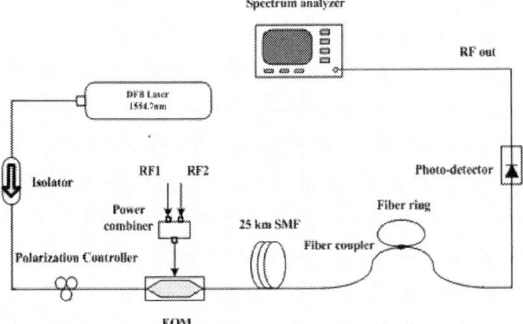

Fig. 1 Experimental setup of the system

The time delay provided by the filter ring per cavity recirculation is given by T. In theory, the number of samples is infinite. And only if the optical source has a coherence time much smaller than the basic filter delay, then there is a linear relationship between the input and output RF signals [4]. The FSR of the filter can be expressed as Eq. (1), which shows that FSR decreases with the inceasing length of the fiber ring.

$$FSR = \frac{1}{T} = \frac{c}{n\Delta L} \qquad (1)$$

Experimental results and Discussions

In the experiment, the DFB laser is at the wavelength of 1554.7 nm, and its output linewidth is less than 1 nm. The DC voltage bias of the EOM (SDL 10AP-MOD9140-F-F-0), whose working frequency can be as high as 10 GHz, is set at 3.7 V at its quasilinear point. And the bandwidth of photodiode used in the experiment is 10 GHz.

In the first experiment, the microwave photonic filter is based on the 20cm-length fiber ring, whose FSR is about 700 MHz. We set one carrier frequency at 1.096 GHz at the rejection band, and the other at 1.394 GHz at the peak of the passband.

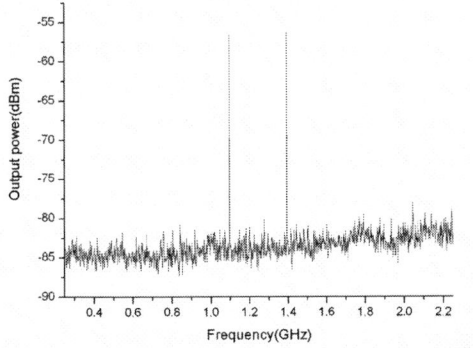

Fig. 2 Spectrum at output without filtering
The spectrum at the system output without

filtering is shown in Fig. 2. Both of the carrier signals have the same power amplitude. After we add a fiber ring made of a fiber coupler whose coupling ratio is 50:50, the spectrum is shown in Fig. 3. Obviously, the signal at the frequency of 1.096 GHz is suppressed by about 9 dB. The inset of Fig. 3 shows the frequency response of a microwave photonic filter with the FSR of 700 MHz and filter rejection ratio of 9 dB.

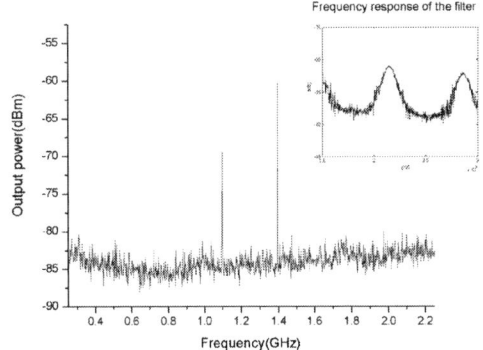

Fig. 3 Spectrum at output after a 700MHz-FSR microwave photonic filter with filter rejection ratio of 9 dB. Inset: the frequency response of the first microwave photonic filter

In the second experiment, the length of the fiber ring of the microwave photonic filter is changed to about 2 m to get a microwave photonic filter with FSR of 102 MHz, and the coupling ratio of the fiber coupler is 40:60 to obtain a lager rejection ratio compared to that of the filter based on 50:50 coupler [5]. The frequencies of two carriers are set at 1.5295 GHz at the peak of passband, and at 1.5805 GHz at the rejection band. Fig.4 shows the spectrum at the system output using this filter, and the inset of Fig. 4 shows the frequency response of the filter with the length of fiber ring of about 2m and the coupling ratio of the coupler of 40:60. One can see from the inset of Fig. 4 that the rejection ratio of the fiber is more than 15 dB. As can be seen From Fig. 4, the carrier at the frequecy of 1.5805 GHz is suppressed effectively by using the second filber ring based microwave photonic filter.

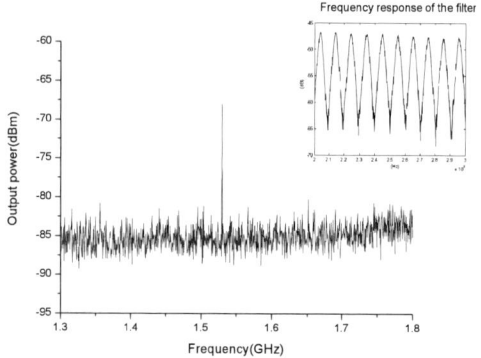

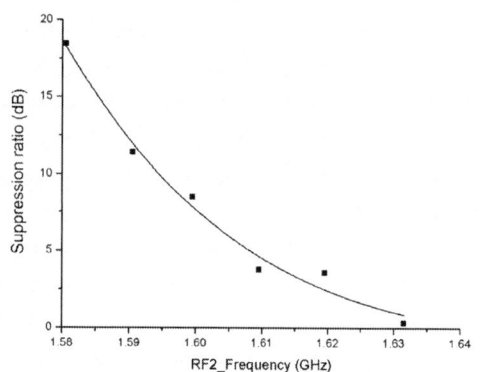

Fig. 4 Spectrum at output after a 102MHz-FSR microwave photonic filter with filter rejection ratio of more than 15 dB. Inset: the frequency response of the second microwave photonic filter

Fig. 5 Curve of suppression ratio under the situation of mismatch

Based on the second experiment, we further change frequency of one of the carriers to get the infulence of mismatch of the frequency spacing and the filter's FSR on system performance. For the two input carriers' frequencies, RF1 is fixed to 1.5295 GHz, RF2 is set to change from 1.5805 GHz to 1.6315 GHz discretely, and the amplitudes of both carriers are the same. Table 1 shows the output amplitudes of both carriers and the suppression ratio when the carrier RF2 is set at different frequency. One can see that the suppression ratio of RF2 in system decreases as its frequency is getting farther away from the notch point of frequency response of the 102MHz-FSR filter as shown in Fig. 5. Therefore, to get a better system performance, we should avoid mismatch of the frequency spacing and the implemented filter's FSR. The frequency of the carrier which is needed should be at the peak point of the passband, and others at the notch points in the frequency response of microwave photonic filter.

Table 1 The results of the output amplitudes and suppression ratio as the frequency of RF2 is changed

Conclusions
In this letter, we have demonstrated a fiber ring based microwave photonic filter implemented in a Radio-over-Fiber link with two carrier frequencies. Two microwave photonic filters with FSR of 700 MHz, filter rejection ratio of about 9 dB, and FSR of 102 MHz, filter rejection ratio of more than 15 dB, are implemented for channel selection application. The performance of channel selection in system is measured by employing both microwave photonic filters. The influence of mismatch of the frequency spacing of the carriers and the implemented filter's FSR on the supression ratio is also analyzed. The experimental results shows that, to get a better channel selection performance, the mismatch mentioned above should be avoided.

References
1 Jiunn-Shyen Wu et al IEEE Trans. Vehicul. Technol., 47(1998), page 84-94
2 Fei Zeng et al IEEE Photon. Technol. Lett., 17(2005), page 899-901
3 Jun Wang et al IEEE Photon. Technol. Lett., 17(2005), page 1737-1739
4 Jose Capmany et al J. Lightw. Technol., 24(2006), page 201-229
5 Hao Chi et al J. Optoelectronics Laser, 17(2006), page 17-19

RF2_frequency(GHz)	1.5805	1.5905	1.5995	1.6095	1.6195	1.6315
RF2_output(dBm)	-86.49	-80.3	-77.17	-73.35	-71.96	-68.66
RF1_output(dBm)	-68.03	-68.89	-68.67	-69.58	-68.38	-68.33
Suppression ratio(dB)	18.46	11.41	8.5	3.77	3.58	0.33

Study on Optical Digital Phase Modulation Applied to Millimeter-Wave Radio-over-Fibre System

Meiwei Zhu (1), Rujian Lin (1), Jiajun Ye (1)

1 : School of Communication and Information engineering, Shanghai University,
149 Yanchang Road, Shanghai 200072, China. zmw1231@shu.edu.cn

Abstract *This paper presents a novel millimeter-wave radio-over-fiber system based on harmonic generation technique employing optical PSK modulation. 60GHz BPSK millimeter-wave signal is generated at basestation by 6GHz sinusoidal signal scan at central station. And triangular periodic signal is used to implement multiphase PSK.*

Introduction

The millimeter-wave radio-over-fibre is a promising technology for future pico-cellular wireless broadband communication due to its high capacity, ultra-wide bandwidth and cost-effectiveness. A large number of mm-ROF schemes have been proposed based on different optical millimeter-wave generation techniques such as dual-mode laser [1], modulation-sideband-technique [2] and harmonic generation technique [3]. While the harmonic based generation technique has advantage of generating millimeter-wave by using low frequency optoelectronic technology. A bidirectional mm-ROF system employing optical frequency multiplication (OFM) technology which is based on harmonic generation has been reported [4]. But electrical modulator limits the maximum speed of baseband data signal in this kind of systems [5].

However, optical digital phase modulator can achieve very high baseband modulation speed (e.g.10Gb/s even more). In this paper, we apply high speed optical PSK modulation to the harmonic generation based mm-ROF system to improve baseband data rate. Fig. 1 shows the proposed mm-ROF system which consists of central station (CS), base station (BS) and optical fibre link. The optical PSK modulator is used in parallel with the phase modulator (PM). We first theoretically analyse how to generate millimeter-wave that is also PSK modulated at BS. It shows that sinusoidal wave scan can only generate BPSK millimeter-wave. By using triangular periodic signal at CS multiphase PSK modulation is available. Then the theory is verified by simulation.

This work is supported by Natural Science Foundation of China (Grant 60377024).

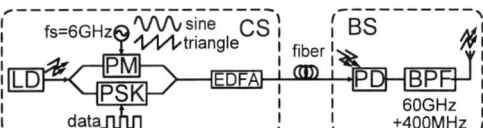

Fig.1. 60GHz ROF system with optical PSK modulation

Signal Analysis

As depicted in Fig.1, in the CS the lightwave emitting from laser diode is split into two optical carriers. The first one is phase modulated by 6 GHz periodic signal and the other one is digitally phase modulated by baseband signal. Then they are combined together and the mixed signal can be expressed in electrical field as

$$E_{out}(t) = E_c \exp[j2\pi f_c t + j\beta\phi(t) + j\varphi_{PN}(t)]$$
$$+ E_c \exp[j2\pi f_c t + j\Phi_N + j\varphi_{PN}t] \quad (1)$$

where f_c is the frequency of the optical carrier; β is the phase deviation; $\phi(t)$ is the 6 GHz periodic signal with unit amplitude; $\varphi_{PN}(t)$ is the laser phase noise; and Φ_N represents the N possible phases that convey the transmitted information.

After amplification and transmission over single mode optical fibre, the lightwave is photodetected at BS to generate electrical signal given by

$$i_d(t) \propto \frac{1}{2}E_{out}(t)E_{out}(t)*$$
$$= E_c^2\{1 + \cos[\beta\phi(t) - \Phi_N]\}$$
$$= E_c^2\{1 + \cos[\Phi_N]\cos[\beta\phi(t)]$$
$$+ \sin[\Phi_N]\sin[\beta\phi(t)]\} \quad (2)$$

where the laser phase noise disappears because of its inherently correlated noise. And if the 6 GHz periodic signal $\phi(t)$ is

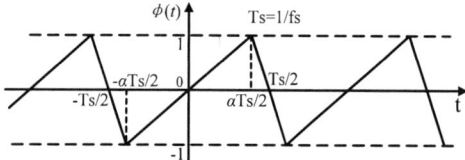

Fig.2. Triangular periodic signal.

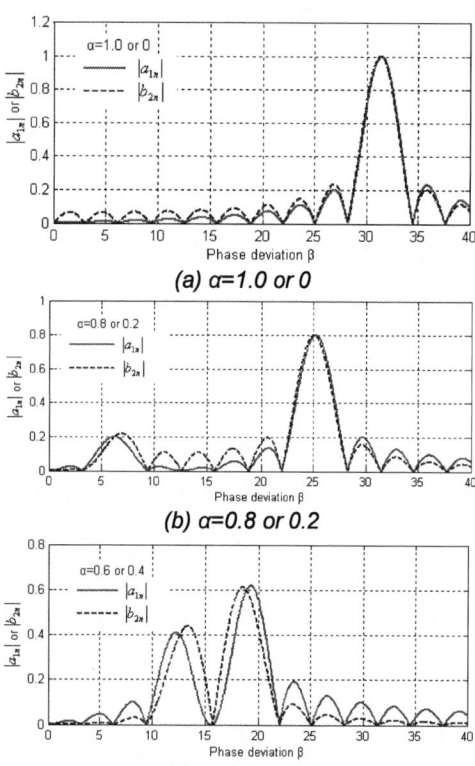

(a) α=1.0 or 0

(b) α=0.8 or 0.2

(c) α=0.6 or 0.4

Fig.3. Fourier coefficient a_{1n} and b_{2n} (n=10) as a function of β with α in six cases .

sinusoidal signal $\sin 2\pi f_s t$, expanding (2) with Bessel functions of first kind yields:

$$i_d(t) \propto E_c^2 \{1 + J_0(\beta)\cos\Phi_N$$

$$+ 2\cos\Phi_N \sum_{n=1}^{\infty} J_{2n}(\beta)\cos(2n \cdot 2\pi f_s t)$$

$$+ 2\sin\Phi_N \sum_{n=1}^{\infty} J_{2n-1}(\beta)\sin[(2n-1)\cdot 2\pi f_s t]\} \quad (3)$$

It contains even and odd harmonics of 6GHz periodic signal $\phi(t)$, which have been modulated by information phase Φ_N. The 10th harmonic is the desired 60GHz BPSK (N=2) millimetre-wave signal and can be launched into air channel for wireless terminal (WT) reception. But multiphase modulated millimetre-wave such as QPSK

is not available in this way because the mth harmonic component is one-dimensional.

To obtain two-dimensional harmonic which has in-phase and quadrature-phase carriers at same frequency, we use triangular periodic signal [3] instead of sinusoidal signal. As depicted in Fig. 2, the triangular wave can be written as

$$\phi(t)=\begin{cases} \dfrac{2}{(\alpha-1)T_s}(t+\dfrac{T_s}{2}-nT_s), & -\dfrac{T_s}{2}<t-nT_s<-\dfrac{\alpha T_s}{2} \\[2mm] \dfrac{2}{\alpha T_s}(t-nT_s), & -\dfrac{\alpha T_s}{2}<t-nT_s<\dfrac{\alpha T_s}{2} \\[2mm] \dfrac{2}{(\alpha-1)T_s}(t-\dfrac{T_s}{2}-nT_s), & \dfrac{\alpha T_s}{2}<t-nT_s<\dfrac{T_s}{2} \end{cases} \quad (4)$$

where α is a symmetric factor between 0 and 1. Expanding $\cos[\beta\phi(t)]$ and $\sin[\beta\phi(t)]$ terms in (2) with Fourier series gives:

$$\cos[\beta\phi(t)] = \sum_{n=0}^{\infty} a_{1n}\cos(n\cdot 2\pi f_s t) \quad (5)$$

$$\sin[\beta\phi(t)] = \sum_{n=1}^{\infty} b_{2n}\sin[n\cdot 2\pi f_s t]\} \quad (6)$$

$$a_{1n} = \frac{\beta}{\beta-n\pi(\alpha-1)}\cdot\frac{\sin(\beta-n\alpha\pi)}{\beta-n\alpha\pi}$$
$$+ \frac{\beta}{\beta+n\pi(\alpha-1)}\cdot\frac{\sin(\beta+n\alpha\pi)}{\beta+n\alpha\pi},$$

$$b_{2n} = \frac{\beta}{\beta-n\pi(\alpha-1)}\cdot\frac{\sin(\beta-n\alpha\pi)}{\beta-n\alpha\pi}$$
$$- \frac{\beta}{\beta+n\pi(\alpha-1)}\cdot\frac{\sin(\beta+n\alpha\pi)}{\beta+n\alpha\pi}\},$$

$$b_{1n} = 0, \ a_{2n} = 0.$$

where a_{1n} / b_{1n} and b_{2n} / a_{2n} are the calculated Fourier coefficient of $\cos[\beta\phi(t)]$ and $\sin[\beta\phi(t)]$ respectively. Inserting (5) and (6) into (2) and extracting the nth harmonic component in photocurrent, we obtain the desired millimetre-wave given by

$$F_n \propto a_{1n}\cos\Phi_N a\cos(n\cdot 2\pi f_s t)$$
$$+ b_{2n}\sin\Phi_N\sin(n\cdot 2\pi f_s t) \quad (7)$$

To obtain multiphase PSK modulation, $|a_{1n}|$ must be equal to $|b_{2n}|$ in F_n. Observing the analytical expression of a_{1n} and b_{2n}, we note that $|a_{1n}|$ equals $|b_{2n}|$ when $\beta+n\alpha\pi = m\pi, m=1,2,3\cdots$. For more intuitionistic explanation, $|a_{1n}|$ and $|b_{2n}|$ are shown in Fig. 3 as a function of phase deviation β with the factor α as parameter for the 10th harmonic (n=10) at 60GHz. It is seen that the peak amplitude increases as the factor α increases from 0.5 to 1.0 or decreases from 0.5 to 0 which means that the wave shape of $\phi(t)$ becomes more asymmetric. For α at 1.0(0), 0.9(0.1) or 0.8(0.2), $|a_{1n}|$ equals $|b_{2n}|$ at a wide range of phase deviation β and can easily reach the peak. While for α around 0.5 , $|a_{1n}|$ doesn't equals $|b_{2n}|$ well, especially when $\alpha = 0.5$, $|b_{2n}|$ becomes 0.

Simulation Results

The system setup is shown in Fig. 1. Firstly, we use 6 GHz sinusoidal signal to generate 60 GHz BPSK millimetre-wave signal. The phase deviation β is set to be 11.9 radian/v. Fig.4 shows the RF spectrum of the harmonics obtained at the output of PD, among which the 10th harmonic is maximal. Fig. 6(a) shows the IQ constellation diagram of the BPSK signal (200Mbit/s) recovered at 60 GHz after 15 km single-mode fibre.

Then we use 6 GHz triangular periodic signal to implement QPSK modulation. The symmetric factor α is chosen to be 1.0 and 0.8 for simulation and the phase deviation β is set to be 31.4 and 25.2 respectively to maximize the amplitude of 60GHz signal as designated by Fig. 3. The results in Fig. 5 show the harmonics obtained at the output of PD. The 10th harmonic is dominant over the others, especially in the case of $\alpha = 1.0$. Fig. 6(b) shows the IQ constellation diagram of the QPSK signal (400Mbit/s) recovered at 60GHz. The simulation results agree well with the theoretical analysis.

Conclusions

A novel mm-ROF system employing optical PSK modulation is proposed and verified by simulation successfully. BPSK millimeter-wave is generated by sinusoidal signal scan and QPSK modulation is implemented by using triangular wave. The IQ constellation diagrams of BPSK & QPSK data recovered at 60GHz are observed. In addition, differential channel coding is necessary in practical system. And fibre dispersion effect isn't taken into consideration here. Further analysis needs to be taken.

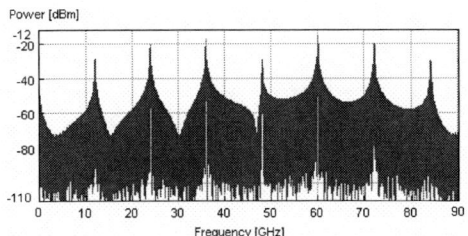

Fig.4. Harmonics generated by sinusoidal signal scan.

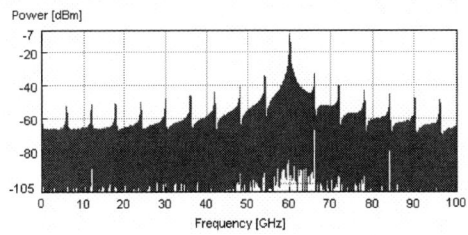

(a) α =1.0

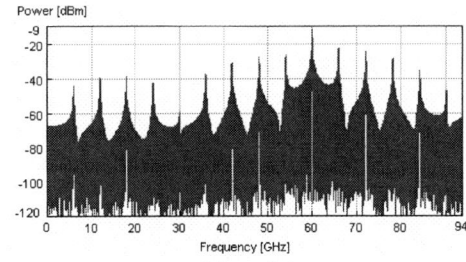

(b) α =0.8

Fig.5. Harmonics generated by triangular scan with symmetric factor α=1.0 and 0.8.

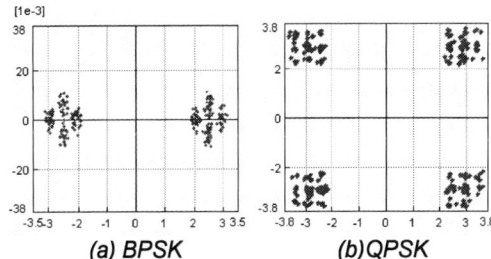

(a) BPSK   (b)QPSK

Fig.6. Constellation diagrams of 200Ms/s BPSK and QPSK data recovered at 60GHz.

References

1 D. Wake et al, IEEE Trans Microw Theory. Tech, 43 (1995), 2270-2276.
2 J. O'Reilly et al, IEEE J. Lightwave Technol. 12 (1994), 369-375.
3 M. Xiu et al, SPIE 2006. v6025, 294-300.
4 M. G. Larrode et al, IEEE photon Technol Lett. 18(2006), 241-243.
5 J. M. Fuster et al, Microw Opt Tech. Lett, 36(2003), 72-74.

A Simplified Model of Multi-wavelength Fibre Lasers based on Hybrid Fibre Raman and Erbium Fibre Amplifications

Shan Qin, Daru Chen

Centre for Optical and Electromagnetic Research, Joint Research Centre of Photonics of KTH (Sweden) and Zhejiang University,
Zhejiang University, Hangzhou 310058, China;
shan.qin@gmail.com

Abstract *A simplified model is established to explain the principle of stable multi-wavelength fibre laser based on hybrid fibre Raman and Erbium-doped fibre amplifications, the simulation results agree with the experimental results very well.*

Introduction

Multi-wavelength fibre lasers (MWFLs) are recently attracting much attention in wavelength-division-multiplexed (WDM) systems, fibre-optical sensors, optical component testing, and spectroscopy due to their various advantages such as low cost, compact structure, and compatibility to fibres [1-3]. Based on various gain mechanisms e.g. Erbium-doped fibre (EDF) amplification, semiconductor optical amplification [1, 2], and fibre Raman amplification [3-5], stable multi-wavelength lasing has been achieved. Especially, MWFLs using EDFs have been investigated widely due to their advantages such as the high power conversion efficiency and low threshold. However, since EDF has a strong homogenous line broadening and cross-saturation gain [6, 7] at room temperature, the multi-wavelength lasing is normally unstable[8]. To make a multi-wavelength EDF laser (MW-EDFL) stable at room temperature, various techniques have been utilized such as cooling the EDF to cryogenic temperature with liquid nitrogen, using some special fibres [9], utilizing a frequency-shifted feedback technique in a laser cavity [10], and employing a highly nonlinear fibre [8].

More recently, we have proposed and demonstrated a novel MWFL by introducing the Raman gain to suppress the strong gain competition in MW-EDFLs [11, 12], which possesses good performance (e.g. good stability, power equilibrium and wide output spectrum). However, the principle of stable multi-wavelength lasing of the fibre laser base on the hybrid gain has not been fully investigated. In this paper, we study further the stability principle and conditions for the stable MWFL based on the hybrid fibre Raman and EDF amplifications.

Theoretical Model

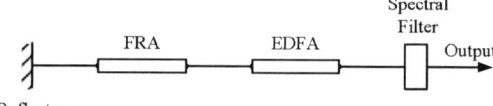

Fig.1 Schematic diagram of the MWFL based hybrid FRA and EDFA, FRA: Fibre Raman amplifier; EDFA: EDF amplifier.

Corresponding to the experimental setup of the MWFL in [11], Fig. 1 shows the simplified configuration, which is composed of four ideal devices i.e. a reflector, a fibre Raman amplifier (FRA), an EDF amplifier (EDFA) and a comb filter. According to our former experimental results, the cross-saturation gain effect is very strong in EDFAs at room temperature even for the wavelengths with the spacing over 20 nm, but it is hardly observed in FRAs even for the wavelengths with the spacing of only 0.1 nm [11]. The multiple wavelengths coupled intensity equations for an FRA can be expressed as

$$\frac{dI_i}{dt} = \alpha_{Ri} I_i - \beta_{Ri} I_i^2, \tag{1}$$

and for an EDFA, they have[6, 7]

$$\frac{dI_i}{dt} = \alpha_{Ei} I_i - \sum_{j=1}^{N} \kappa_{ij} I_i I_j, \tag{2}$$

where I_i (I_j) is the intensity of the ith (jth) wavelength, t is the time, α_{Ri} and α_{Ei} are the overall gains (including the trip loss of the cavity) of the ith wavelength for the FRA and the EDFA, respectively.

978-0-9789217-3-6/07/$25.00

©2007 WEN GLOBAL SOLUTIONS

$\kappa_{ij}(i=j)$ and $\kappa_{ij}(i \neq j)$ are the self-saturation gain coefficient of ith wavelength and the cross-saturation gain coefficient of ith and jth wavelengths for the EDFA, respectively. N is the number of the lasing wavelengths. For an ideal MWFL with both FRA and EDFA as gain elements (as shown in Fig.1), assuming that the system is in quasi-equilibrium state, the coupled intensity equations are

$$\frac{dI_i}{dt} = \alpha_i I_i - \beta_i I_i^2 - \sum_{j=1}^{N} \kappa_{ij} I_i I_j, \qquad (3)$$

where $\alpha_i = \alpha_{Ei} + \alpha_{Ri}$, and $\beta_i = \beta_{Ri}$. On account of the comb filter which is equivalent to the wavelength dependent loss (WDL), the overall trip gain α_i of the cavity can be expressed as αR_i, where α is the overall gain of both FRA and EDFA, and R_i is the reflectivity at the ith wavelength. Assuming that α, β_i and κ_{ij} are all independent on wavelengths in a small spectral range and they can be considered as constants (i.e., for an EDFA, the self-saturation gain coefficient κ_{ii} equals to the cross-saturation gain coefficient κ_{ij}), the equations are then

$$\frac{dI_i}{dt} = \alpha R_i I_i - \beta I_i^2 - \kappa I_i \sum_{j=1}^{N} I_j . \qquad (4)$$

If we select a typical comb filter [e.g., a fibre Mach-Zehnder Interferometer (MZI) filter] whose reflectivity accords with

$$R_i = \frac{1}{2}[1 - \cos(\frac{2\pi \nu_i n \Delta L}{c})], \qquad (5)$$

where ν_i is the lasing frequency, $n\Delta L$ is the optical path difference between MZI's two arms, and c is the light speed in free space, the numerical solution of Eqs.(4) can be obtained by selecting appropriate initial conditions [I_i(t=0)], parameters (α, β and κ) and the number of wavelengths (N).

Simulation results and analysis

Table 1 Parameters selection in our simulation for the strong and weak coupling situations

	R_i	α	β	κ	Situation	N
(1)	$R_i = \frac{1}{2}[1-\cos(\frac{1}{20}i)]$	12	4	2	$\beta > 0$	500
(2)	$R_i = \frac{1}{2}[1-\cos(\frac{1}{8}i)]$	4	0	1	$\beta = 0$	200
(3)		8	-2	4	$\beta < 0$	200

According to the two situations, i.e., weak

coupling ($\beta > 0$) and strong coupling ($\beta \leq 0$) [6, 7], we select three typical parameters (listed in Table 1) for modelling. The simulation results are shown in Fig.2, and the sub figures (B), (C) and (D) are relevant to the parameters (1), (2) and (3) listed in Table 1, respectively.

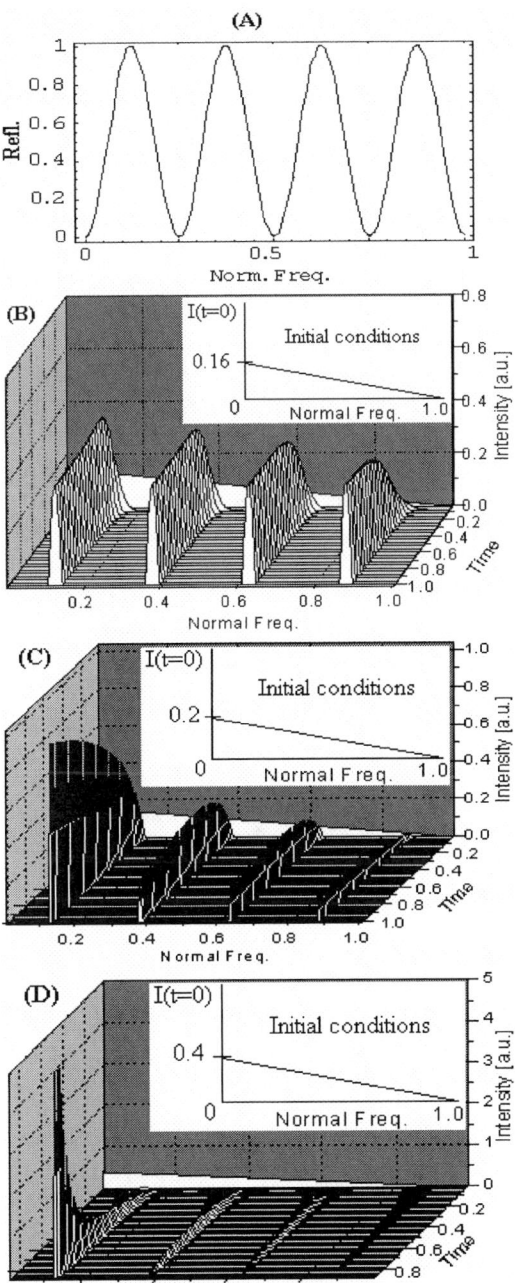

Fig.2 Simulation results for weak and strong coupling situations; (A) is the comb filter reflectivity, (B), (C) and (D) are weak ($\beta > 0$), strong ($\beta = 0$) and strong ($\beta < 0$) coupling situations, respectively.

Since the saturation effects generally resist the positive gain, β and κ can be regarded as not negative. Actually, it is the strong coupling situation ($\beta = 0$, i.e., there is no Raman cross-saturation gain) for a pure EDF laser, and it is just the weak coupling situation ($\beta > 0$) for a laser based on hybrid FRA and EDFA that we presented in [11].

From Fig.2 (B), one can see that the multi-wavelength lasing can finally be stable and uniform after some time even though the initial light intensities are not uniform, i.e., the MWFL based on hybrid FRA and EDFA is a self-stabilizing system, which can come back its stable state again after a perturbation. However, one can see from Fig.2 (C) and (D) that only one lasing can be obtained at the wavelength corresponding to the stronger initial intensity, that is to say, any perturbation will destroy the equilibrium and can not restore the original state automatically, so a MWFL based on pure EDF is generally unstable. The conclusions agree with the experimental results shown in Fig.3 (a) and (b), which are based on hybrid FRA and EDFA and pure EDF, respectively.

Conclusions

The homogeneous line broadening property of an EDFA results in very strong cross-saturation gain effect (i.e., strong coupling), which presents mode competition in an MW-EDFL and makes the lasing unstable. However, cross-saturation gain effect hardly exists in an FRA [11]. When an FRA is introduced into an MW-EDFL, the self-saturation gain effect exceeds the cross-saturation gain effect, and the laser is in weak coupling situation, so it can support stable multi-wavelength operation.

In summary, a simplified model has been established, and the simulation results can well explain the experimental results [11], which shows that an FRA is effective to make an MW-EDFL stable at room temperature. This model can help us to understand the stabilization principle of the MWFL based on hybrid FRA and EDFA.

References

1. Staring, A.A.M., et al., **IEEE Photon. Technol. Lett.**, **8**(1996), page 1139.
2. Zirngibl, M., et al., **IEEE Photon. Technol. Lett.**, **8**(1996), page 870.
3. Mermelstein, M.D., et al., **Electron. Lett.**, **38**(2002), page 636.
4. De Matos, C.J.S., et al., **Electron. Lett.**, **37**(2001), page 825.
5. Dong, X.Y., et al., **Optics Express**, **14**(2006), page 3288.
6. Lamb, W.E., **Physical Review**, **134**(1964), page A1429.
7. Siegman, A.E., *Lasers*, University Science Books, 1986.
8. Pan, S.L., et al., **Optics Express**, **14**(2006), page 1113.
9. Graydon, O., et al., **IEEE Photon.Technol. Lett.**, **8**(1996), page 63.
10. Bellemare, A., et al., **J. Lightwave Technol.**, **18**(2000), page 825.
11. Qin, S., et al., **Opt. Express**, **14**(2006), page 10522.
12. Chen, D., et al., **Optics Express**, **15**(2007), page 930.

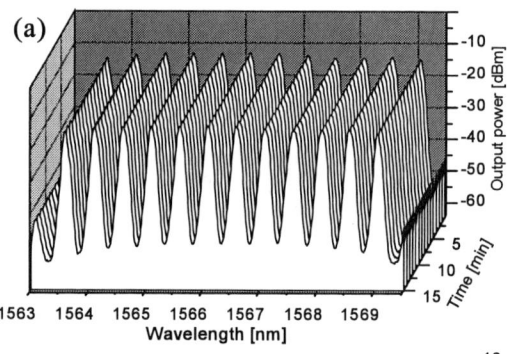

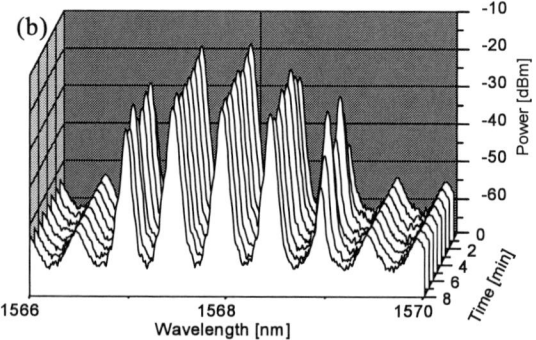

Fig.3 Experimental repeated scanning spectrum of the two type lasers based on (a) hybrid FRA and EDFA and (b) pure EDFA, respectively.

Lagrange Multiplier Optimization Synthesis of Long-Period Fiber Gratings

Cheng-Ling Lee[1,*], Ray-Kuang Lee[2] and Yee-Mou Kao[3]

[1,*] Department of Electro-Optical Engineering, National United University,No.1, Lien-Da,Kung-Ching Li, Miaoli 360, Taiwan, R.O.C. Email: cherry@nuu.edu.tw

[2] Institute of Photonics Technologies, National Tsing-Hua University, Hsinchu 300, Taiwan, R.O.C.

[3] Department of Physics, National Changhua University of Education, Changhua 500,Taiwan, R.O.C.

Abstract: *A Lagrange multiplier optimization method to the inverse design problems for complex LPGs was developed in the present study. For the first time, an EDFA gain flattening LPG is successfully synthesized by the proposed algorithm.*

Introduction

Long-period fiber gratings (LPGs) have found important applications in fiber communication and fiber sensing systems, in which they act as EDFA gain flattening filters, band-rejection filters, mode converters, and high sensitivity fiber sensors [1]. In the literature, the synthesis methods for LPGs can be classified into two categories: the inverse scattering methods [2]; and the optimization methods [3]. The inverse scattering methods directly calculate the required grating coupling coefficient profile from the LPG transmission spectra and the solution can be uniquely determined if an additional assumption about the filter properties (the under-coupled assumption) is used [2]. Although, theoretical LPGs with the required spectral transmission shaping can be inversely synthesized by using the DLP inverse scattering algorithm. In practice, there are still a number of disadvantages in designing such a high standard LPG filter using this method. These include: the required grating length is typically long, the spatial grating profiles (including amplitude and phase) are complicated, and the unknown phase transmission spectrum of the designed LPG. Therefore, several LPG synthesized optimization methods are known as to directly synthesize the LPG with some possible constraints in the

parameters of the gratings. The optimization methods just directly minimize the difference between the targeted and the synthesized spectra. The minimization can be performed by usual optimization algorithms or in particular the evolutionary algorithms (EA). Although the optimization method has some superiorities, however, the calculation time is always long. It is not very practical and efficient. Therefore, in this present paper, a new and simple approach to the solution of LPG (transmission-type gratings) synthesis problem is developed. The method is based on the Lagrange multiplier optimization (LMO) method, which can control various parameters of the designed devices for the practical application demands through a user-defined cost functional. We have demonstrated that the LMO optimization algorithm can be implemented for designing multi-channel FBGs (reflection-type gratings) for DWDM applications [4]. In this paper, we further extend the proposed algorithm in such a way that it can handle synthesis problems involving transmission-type grating filters, like LPGs. Besides, the convergence rate is fast and direct in the LMO algorithm when compare with evolutionary optimization approach. For the first time, we have demonstrated and analyzed, the transmission-type LPG: an EDFA gain flattening LPG (EDFAGF-LPG) filter with Gaussian apodization function

978-0-9789217-3-6/07/$25.00

©2007 WEN GLOBAL SOLUTIONS

initial guess coupling coefficient was successfully synthesized by using the proposed method. The designed results also prove that the presented algorithm is an effective method for designing not only reflection-type fiber grating but also transmission-type fiber grating devices.

LMO algorithm for synthesis of LPGs

The main idea of the LMO algorithm for synthesized LMO-LPGs is to find the complex spatially coupling coefficient ($\kappa(z)$) of the grating to obtain a spectrum ($T_{co}(\lambda)$) that close the given target spectrum ($T_d(\lambda)$). The proposed method is started with the coupled mode equations for LPGs. Therefore, the defined cost functional needs to be minimized, such as,

$$J = \frac{1}{2}\int_{-\infty}^{\infty}\left[T_{co}(\lambda)-T_d(\lambda)\right]^2 d\lambda + \frac{\beta}{2}\int_0^L \left[\kappa(z)\right]^2 dz$$

$$+ \int_0^L \int_{-\infty}^{\infty} \bar{\mu}^T \cdot \begin{bmatrix} \dfrac{dA^{co}}{dz} - i\delta A^{co} - i\kappa A^{cl} \\ \dfrac{dA^{co}}{dz} + i\delta A^{cl} - i\kappa^* A^{co} \end{bmatrix} dz d\lambda \quad (1)$$

Where $A^{co}(z)$ and $A^{cl}(z)$ represents the core and the cladding modes respectively, δ is the detuning parameter. The calculated spectrum of output core mode transmission for LPG is

$$T_{co}(\lambda) = \left|A^{co}(L)/A^{co}(0)\right|^2 \quad (2)$$

In Eq(1), the $T_d(\lambda)$ is the target transmission spectrum, L the total length of the grating, and β is a positive number acting as a weighting parameter for the constraint control. In the defined cost functional, Eq. (1), the spatially coupling coefficient $\kappa(z)$ was used to shape an output core mode transmission spectrum $T_{co}(\lambda)$ of a given target transmission spectrum $T_d(\lambda)$ and to minimize both the spectra difference and the norm of the coupling coefficient profiles simultaneously. In the presented LMO algorithm for synthesized LPGs, we introduced Lagrange multipliers μ^{co}, μ^{cl} for the core and cladding modes of LPGs, respectively. A variationed process was used, then; the resulting equations of motion for the Lagrange multipliers can be

found and are shown as follows Eq.(3) with the boundary conditions Eq.(4)

$$\frac{\partial \mu^{co}(z)}{\partial z} = i\delta \cdot \mu^{co}(z) + i\kappa(z)\mu^{cl}(z)$$

$$\frac{\partial \mu^{cl}(z)}{\partial z} = -i\delta \cdot \mu^{cl}(z) + i\kappa^*(z)\mu^{co}(z) \quad (3)$$

$$\mu^{co}(L) = -2 \cdot A^{co}(L) \cdot \Delta_t$$

$$\mu^{cl}(L) = 0 \quad (4)$$

where $\Delta_t = T_{co}(\lambda) - T_d(\lambda)$ is the discrepancy between the output and target transmission spectra. The cost functional J was variationed again with respect to real and imaginary parts of the coupling coefficient function κ_R and κ_I, respectively. Then, the parameters $\dfrac{\delta J}{\delta \kappa_R}$ and $\dfrac{\delta J}{\delta \kappa_I}$ can be obtained. Therefore, with an initial guess coupling coefficient function $\kappa(z) = \kappa_R(z) + i\kappa_I(z)$, the optimization of LMO algorithm for LPG was solved with the help of Eq. (1-4), and until the convergence was satisfied through the following iteration equations Eq(5-6):

$$\kappa_R'(z) = \kappa_R(z) - \alpha_R \frac{\delta J}{\delta \kappa_R};$$

$$\kappa_I'(z) = \kappa_I(z) - \alpha_I \frac{\delta J}{\delta \kappa_I} \quad (5)$$

where α_R and α_I are ad hoc constants for real and imaginary parts of the coupling coefficient, respectively. The new coupling coefficient can be expressed as follows

$$\kappa'(z) = \kappa_R'(z) + i\kappa_I'. \quad (6)$$

Design results and Discussions

In published papers, various design schemes of LPG devices have been proposed to equalize the gain spectrum of an EDFA. They are almost based on the phase shift LPGs, multiple different LPGs scraped together or especially arranged LPGs in order to match the spectral shaping for flattening EDFA gain spectrum. However, in this study, we can synthesized gain flattening LPGs directly just by using the coupled-mode equations.

In the following designed case, the well known EDFA gain flattening LPG (EDFA-GFLPG) was designed. We simply set the parameters $\alpha = \alpha_R = \alpha_I$, the parameter β is zero, and the grating length $L = 40mm$. Fig.1 shows the designed results for synthesizing LMO-EDFA-GFLPG filter, (a) the target and designed spectra of the designed LPG in dB scale, (b) Flattened gain-profiles of EDFA gain spectra by the designed LPG, (c) ripple deviations in the flattened wavelength region. From the Fig. 1 (a), the designed reflection spectrum meets well with the target spectrum with a little deviation within about 5nm wavelength range. One can also see the Fig. 1 (c), the studied spectrum can be flattened to less than ±1.45 dB variations within the entire C-band and the average ripple deviation in the flattened wavelength region is lower 0.5dB. Fig.2 displays the designed complex index modulation profiles of the designed GFLPG in Fig.1. Fig. 2(a) shows the Gaussian apodization of index modulation peak with $\Delta n_0 = 1.5 \times 10^{-5}$ for the initial guess and the α parameter is 5×10^3 (based on our experience). Fig. 2 (b) shows the finally designed profiles of index modulations, and Fig.2 (c) the amplitude and phase of the designed complex index modulation across the grating. From the figure, the designed coupling coefficient profile is very similar with the initial guess profile and which is not complicated with short grating length. The designed devices may have some advangtage for easily actual fabrication.

Conclusion

In this paper, for the first time, we have presented and investigated a simple and effective LPG synthesis method based on LMO optimization algorithm. An EDFA gain flattening LPG filter was successfully synthesized and analyzed by using the proposed method. Based on the simulation results, the evolutionary profiles of the final coupling coefficients are very similar with initial guesses. Therefore, an optimal solution might be obtained by adding a little perturbation on the partial profile of the index modulation that will be discussed and demonstrated in a future paper. From this first study, it appears that the proposed approach is an effective method which can converge quite well on designing not only

reflection-type FBG but also transmission-type LPG filters. Future research will include the extension of this approach to design more complex grating based devices.

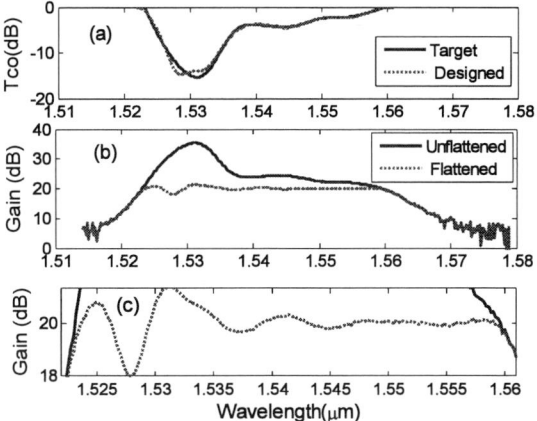

Fig.1 The synthesized core mode transmission spectra (dB scale) of LMO-EDFA GF-LPG filter

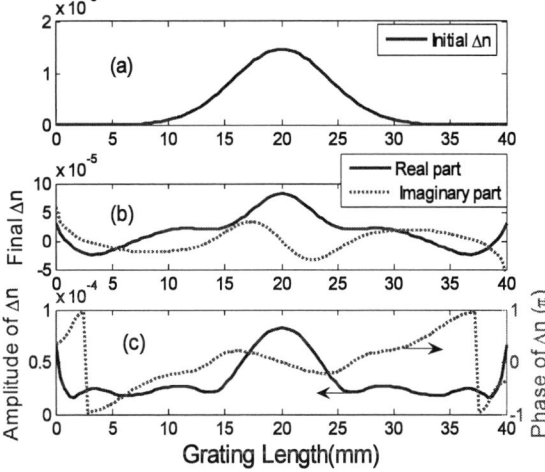

Fig. 2 The designed complex index modulation profiles of synthesized LMO-EDFA-GFLPG.

References

[1] A. M. Vengsarkar et. al, Opt. Lett., 21 (1996) 336-338.
[2] R. Feced et. al, J. Opt. Soc. Am. A 17 (2000) 1573-1582.
[3] C. L. Lee et. al, IEEE Photonics Tech. Lett., 14 (2002) 1557-1559.
[4] C.L. Lee et. al, Optics Express, 14 (2006) 11002-11011.

Feasibility Study of a Simple 100Gb/s Transmitter with Low-speed Electronics and 0.8bit/s/Hz Spectral Efficiency

Junming Gao, Xinyu Xu, Yikai Su

State Key Lab of Advanced Optical Communication Systems and Networks, Department of Electronic Engineering, Shanghai Jiao Tong University, Shanghai 200240, China,
Email: yikaisu@sjtu.edu.cn

Abstract *We propose a simple 100Gb/s transmitter using multi-wavelength generation scheme and DQPSK modulation format. Only devices of lower speed are needed to reduce the system cost.*

Introduction

Over the past years, data-based traffic has grown rapidly. This trend originates from the wide deployment and rapid growth of Ethernet. Currently 10 Gb/s Ethernet (GbE) has been employed in local area networks (LANs). With the emergence of IPTV, IP Video and other broad bandwidth service, higher capacity network will be desirable in the near future. Historically Ethernet has grown by a factor of 10 and it is wildly believed that this trend is going to occur towards the next generation Ethernet of 100GbE, which is also being considered to play an important role in the metropolitan area networks (MANs) and wide area network (WANs) [1].

Recently there have been several methods proposed for the 100GbE implementation which can be divided into two categories: one using wavelength division multiplexing (WDM) [2] and the other with optical time division multiplexing (OTDM) [3] or electrical time division multiplexing (ETDM) [4]. The first method needs a set of lasers and modulators at the transmitter side, which greatly increase the system cost. As for the OTDM method, ultra short optical pulse source, which is usually at a level of pico or sub-picosecond, and optical de-multiplexer, are indispensable. Currently the ETDM method is based on the immature ultrahigh speed electronic devices. In [5], a scheme based on the multi-wavelength generation was used, but devices of high bandwidth were still needed. In this paper we propose a simple 100Gb/s transmitter utilizing multi-wavelength generation and DQPSK modulation format. Only one stable laser is needed, and the electrical devices used here are of lower bandwidths of

25GHz and 10GHz.

Structure of the proposed transmitter

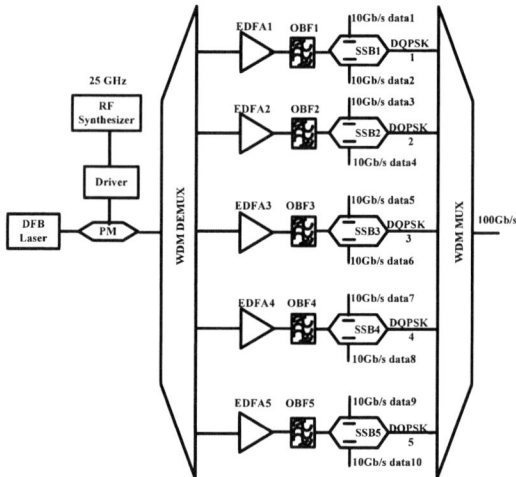

Fig 1 Structure of the 100Gb/s transmitter.

Figure 1 shows the schematic structure of the proposed transmitter. The 100Gb/s signal consists of five 20Gb/s DQPSK sub-channels. Only one highly-stable laser source is needed for the five sub-channels. This is implemented by phase modulating a continuous wave (CW) light by a 25GHz clock signal. With proper filtering and amplification, five stable, separated lightwaves are obtained with fixed channel spacing of 25GHz and similar powers. Each of the five channels is then injected into a single-side band (SSB) modulator, which modulated by the 20Gb/s DQPSK signal. After a WDM multiplexer (MUX), these five channels are combined together, resulting in a 100Gb/s signal. In this transmitter structure, only the bandwidths of phase modulator (PM) and the PM driver are 25GHz, all the other devices, including five SSBs are of 10GHz bandwidth.

978-0-9789217-3-6/07/$25.00

©2007 WEN GLOBAL SOLUTIONS

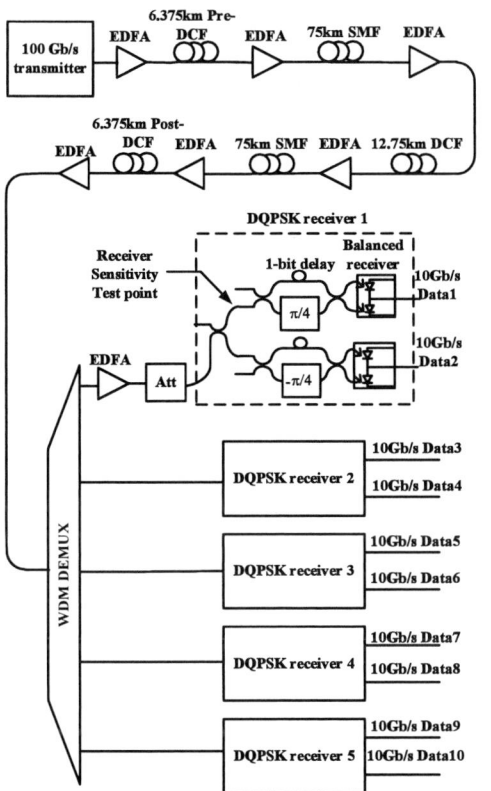

Fig 2 Simulation diagram of 100Gb/s signal generaton and transmission. Att: attenuator.

Results and discussion

Figure 2 shows the schematic diagram of the simulation for the 100Gb/s signal generation and transmission. The carrier wavelength from the DFB laser is 1553nm, with an output power of 10dBm.The modulation index of the PM is set to be 90 degree. A WDM demultiplexer (DEMUX) is used to separate the five sub-channels, including the zero-order, two first-order and two second-order mode lightwaves. In order to reduce the crosstalk between the neighbouring channels, the 3dB bandwidths of the WDM multiplexer and the demultiplexer are chosen to be 20GHz. The optical spectrum after the PM is shown in Fig.3 (a). With proper amplifications of the first and second-order mode lightwaves, we obtained five sub-channels with a comparative power around -5dBm and a fixed 25GHz channel spacing as shown in Fig.3 (b). The generated five channels are defined as C_{-2}, C_{-1}, C_0, C_1 and C_2, respectively. Each of the five channels is then separated and individually modulated by a SSB modulator with two branches of 10Gb/s PRBS signal with a word length of $2^7 - 1$ so that to carry a 20Gb/s DQPSK

signal. After combination of these five 20Gb/s QPSK signals, a 100Gb/s signal is generated whose spectrum is shown in Fig.3 (c).

The generated 100Gb/s signal is then amplified by an Erbium-doped fiber amplifier (EDFA) and sent into a 150km transmission line which is composed by two spans of 75km single mode fiber (SMF) with loss of 0.2dB/km and dispersion of 17ps/nm/km. In order to compensate the dispersion through transmission, dispersion compensation fiber (DCF) (0.5dB/km loss) of -100ps/nm/km is used. The total input powers into the DCFs and SMFs are 3dBm and 10dBm, respectively. The spectrum after transmission is shown in Fig.3 (d).

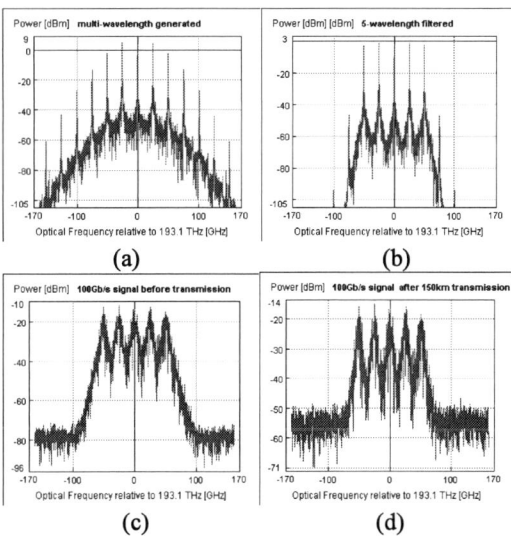

Fig 3 (a) Multi-channel generated by PM. (b) Filtered 5 channels. Optical spectra of 100Gb/s signal (c) before and (d) after transmission.

At the receiver side, a WDM DEMUX with the same characteristics as the one used in the transmitter is employed to separate the five sub-channels. The optical spectra of the filtered channels are shown in Fig. 4. From these diagrams one can see that the suppression ratio between the desired channel and its neighbouring channels is about 20dB. So the crosstalk between these channels can be sufficiently suppressed. After the filtering, each channel is sent to a DQPSK demodulator which consists of two balanced receivers for the detection of the in-phase and quadrature components. After demodulation, ten streams of 10Gb/s signal are recovered whose eye diagrams are provided in Fig.5, respectively. The received power defined at the input to the balanced receiver is set to be -20dBm. It can be seen that the eye diagrams of the ten branches of signal are widely opened, so error free operation can be expected. We also observe that there is amplitude fluctuation in the recovered data, which can be attributed to the crosstalk between these channels. As the dispersion compensation is mainly set for the original wavelength in the transmission span, inter-symbol interference is more evident in the channels of C_{-2}, C_2, C_{-1} and C_1, compared with C_0. This can be greatly alleviated by precise dispersion post-compensation at each of the receiver side.

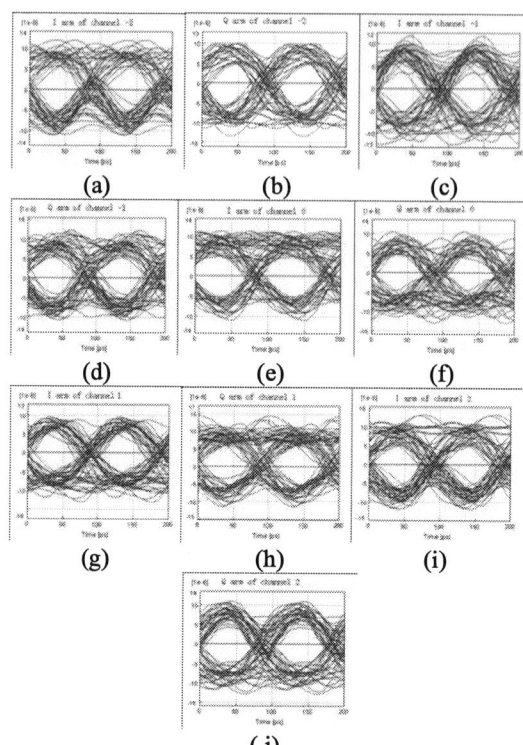

Fig 5 *Eye diagrams of the recovered electrical data at a received power of -20dBm. (a) I arm and (b) Q arm of channel C_{-2}, (c) I arm and (d) Q arm of channel C_{-1}, (e) I arm and (f) Q arm of channel C_0, (g) I arm and (h) Q arm of channel C_1, (i) I arm and (j) Q arm of channel C_2.*

Conclusion

We have proposed and simulated a simple transmitter with spectral efficiency of 0.8bit/s/Hz for 100Gb/s network, using five 20Gb/s DQPSK sub-channels generated from one DFB laser and one PM, with low speed optical and electrical devices. In our simulations, this novel transmitter shows good performance in a 150km transmission system.

References

1 M. Duelk et al, ECOC 2005,Tu3.1.2.
2 A. Kish et al, CLEO 2005, CMGG3.
3 B.Mikkelsen et al, EL 1999,Vol.35,No.21.
4 P.J.Winzer et al, ECOC 2005, Th4.1.1.47.
5 J.Yu et al, OFC 2007,JThA42.

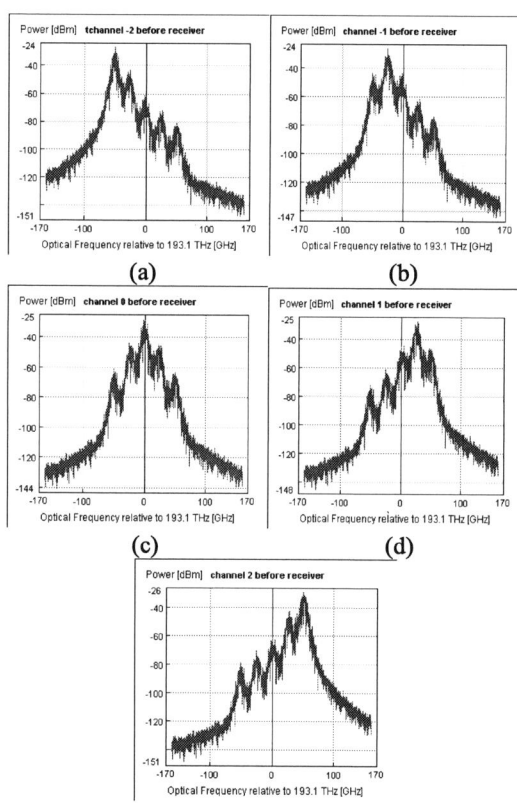

Fig 4 *Optical spectra of channel (a) C_{-2}, (b) C_{-1}, (c) C_0, (d) C_1, and (e) C_2 before detection.*

Controlling chaos in an erbium -doped fiber dual-ring laser via modulating its loss and phase

YAN Sen Lin

Physical Department, Nanjing Xiaozhuang College, Nanjing 210017, China, email:senlinyan@163.com

Abstract: *Chaos-control of an erbium-doped fiber dual-ring laser is studied. When optical attenuators and a phase modulator are used to modulate the loss and the phase shift of the laser, the chaotic laser can be suppressed in a stable state, a single-period, a dual-period and multi-periods, respectively.*

I. Introduction

Erbium-doped fiber laser has been the subject of intension owing to hisr potential in various fields [1-3]. People have great interested in fiber laser because of its slow oscillation relaxation times. The generation of chaos in a single-mode single-polarization erbium-doped fiber laser and single-mode erbium-doped fiber dual-ring laser were reported [4-6]. Up to now, controlling chaotic semiconductor lasers, flared lasers and other kind of lasers has widely been studied by the periodic perturbation and optical negative feedback methods [7,8]. However, controlling chaotic erbium-doped fibers has rarely been reported so far [9,10].

In this paper we present the phase and loss methods for controlling chaos in a single-mode erbium-doped fiber dual-ring laser. The methods can control efficiently chaos into stably states and all kinds of periodic states. The results are very significance for people to know the chaotic dynamics of the laser.

II. Physical Model

The dual-ring laser is shown in fig.1. Its fundamental system contains two rings of erbium-doped fiber lasers [4-6]. They are coupled via a coupler C_0. Both ring resonators are also coupled to a separate fiber through WDM couplers. The lasing fields in two rings produces phase change $\pi/2$ through the coupler C_0 from one ring to the other. The dynamics of two rings are described as following the equations [1,2,4-6]

$$\frac{d}{dt}E_a = -k_a(E_a + \eta_0 E_b) + g_a E_a D_a \quad (1)$$

$$\frac{d}{dt}E_b = -k_b(E_b - \eta_0 E_a) + g_b E_b D_b \quad (2)$$

$$\frac{d}{dt}D_a = -\left(1 + I_{pa} + |E_a|^2\right)D_a + I_{pa} - 1 \quad (3)$$

$$\frac{d}{dt}D_b = -\left(1 + I_{pb} + |E_b|^2\right)D_b + I_{pb} - 1 \quad (4)$$

Where E_a and E_b are the output fields and D_a and D_b are the population inversions in ring a and ring b, respectively. I_{pa} and I_{pb} are the pump intensities in ring a and ring b, respectively. The parameters k_a, k_b, g_a and g_b are the decay rates and the gain coefficients in ring a and ring b, respectively. η_0 is the coupling coefficient of the coupler C_0. We calculate the Lyapunov exponents to know the dynamical behavior of the system, the leading Lyapunov exponent of the system is 17.8, which indicates the system has chaotic characteristic. In numerical calculation, the normalized parameters are taken as $I_{pa}=I_{pb}=4$, $k_a=k_b=1000$, $\eta_0=0.2$, $g_a=4800$ and $g_b=10500$ [1,2,4-6]. A chaotic attractor

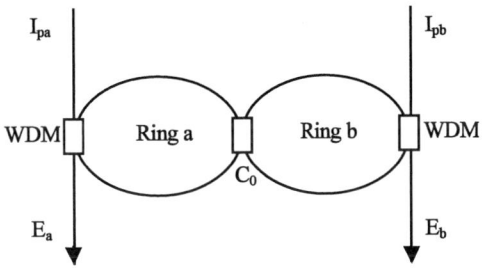

Fig.1. Erbium-doped fiber dual-ring laser

in the laser shows in fig.2

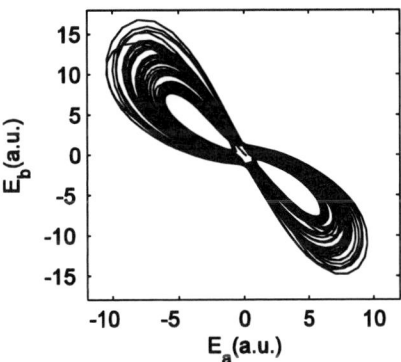

Fig.2 A chaotic attractor in the laser

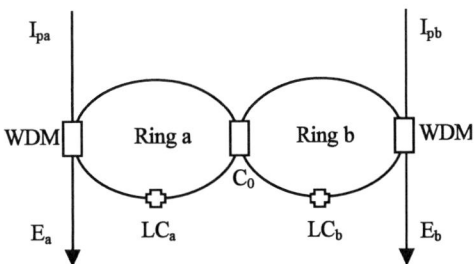

Fig.3 Scheme of control of chaos by controlling loss

Firstly, the method of loss to control chaos in the laser is introduced. The scheme is illustrated in fig.3,

978-0-9789217-3-6/07/$25.00

©2007 WEN GLOBAL SOLUTIONS   411

where LC_a in ring a and LC_b in ring b are optical attenuators, respectively. LC_a modulates lasing loss in ring a and LC_b modulates lasing loss in ring b such that chaotic laser can be controlled. When both LC_a and LC_b show in the system, ka in Eq.(1) must be rewritten as $M_a \times k_a$ and k_b in Eq.(2) must be rewritten

When LC_a don't show in the system, single LC_b can also control chaos. Let $M_b=0.72$, the chaotic laser can be conducted into a stable state after it takes 2ms in fig.7. For $M_b=0.92$, a dual-cycle is controlled to show in fig.8. We can find that both the dual-loss and the single-loss can control effectively chaos.

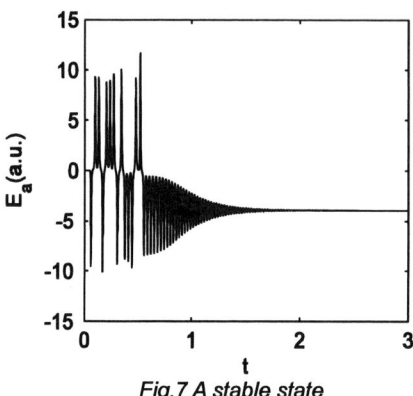

Fig.4 A dual-cycle

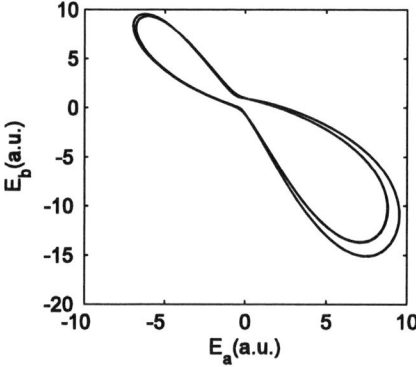

Fig.7 A stable state

as $M_b \times k_b$, where M_a and M_b are the control coefficients. For $M_a=0.9$ and $M_b=0.8$, a dual-cycle is illustrated in fig.4, which implies that chaotic laser has been stablized in a dual-periodic state. When $M_a=0.35$ and $M_b=0.65$, another dual-cycle is controlled to show in fig.5. When $M_a=0.4$ and $M_b=0.68$, a 5-cycle is controlled to show in fig.6, which implies that chaotic laser has been conducted in a multi-periodic state.

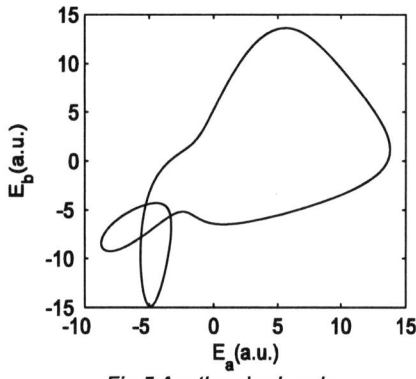

Fig.5 Another dual-cycle

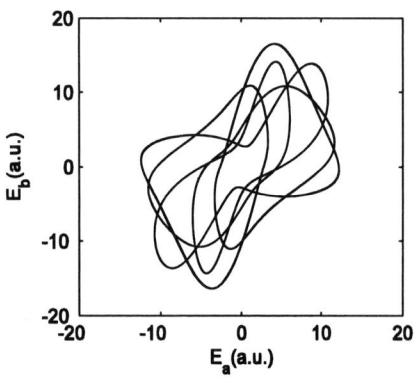

Fig.6 A 5-cycle

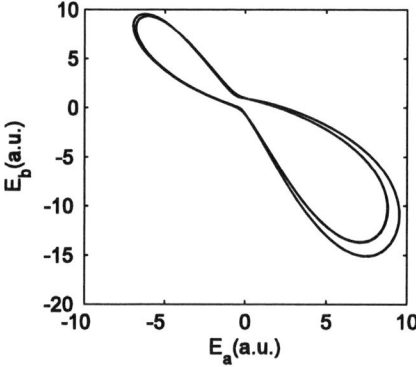

Fig.8 other dual-cycle state

Next, the method of phase to control chaos in the laser is introduced. This scheme is illustrated in fig.9, where PC_b in ring b is the phase modulator. PC_b can modulates lasing phase from ring b into ring a such that chaotic laser can also be controlled via the phase modulator. When PC_b shows in the system, E_b in Eq.(1)

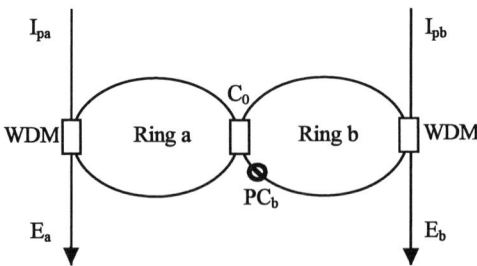

Fig.9 Scheme of control of chaos by modulating the phase

must be rewritten as $\eta_b E_b \exp(-jm\pi)$, where m_p is the phase control coefficient and let the loss coefficient $\eta_b=0.8$. Numerical results are as following.

For $m_p=1$, a stable state is controlled to to show in

412

fig.10, it takes 0.5ms, the chaotic laser is stablized. When m_p=0.1, the chaotic laser is conducted into a periodic state, where the laser oscillation frequency F_p is F_p=17.5kHz. We also find that the laser can also be controlled into some periodic states when m_p=0.1 to m_p=0.9. We liste as following: for m_p=0.2, F_p=23kHz, for m_p=0.3, F_p=22.8kHz, for m_p=0.4, F_p=22.5kHz, for m_p=0.5, F_p=20.5kHz, for m_p=0.6, F_p=17.5kHz, for m_p=0.7, F_p=14kHz, for m_p=0.8, F_p=9.8kHz, for m_p=0.9, F_p=5kHz.

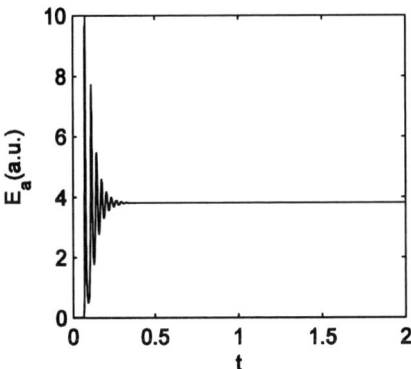

Fig.10 Chaotic laser is conducted into a stable state.

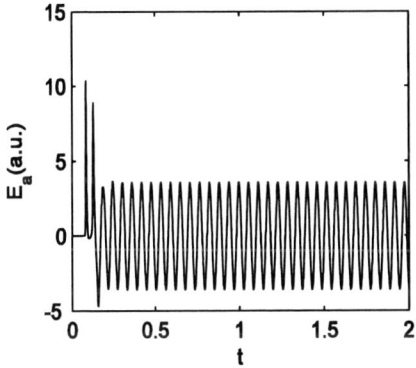

Fig.11 Chaotic laser is conducted into a periodic state.

Optical attenuators and a phase modulator are used to modulate the loss and phase shift of the laser,

VI. Conclusion

The optical attenuators and phase modulator can be performed to realize successfully control of chaotic laser. Chaos can be stabilized in a stable state and a variety of multi-periodic cycles. It is also important that The two methods are helpful for study of the fiber laser.

References

1 L. G. Luo and Chu J.Opt.Soc.Am.B, vol.15(1998), pp.2524-2530
2 R. Wang, K. Shen IEEE J.QE, vol.37(2001), pp.960-963
3 F. Zhang and P. L. Chu J. Lightwave Technology, vol21(2003),pp.3334-3343
4 E. Lacot, F. Stoeckel, and M. Chenevier Phys.Rev.A, vol.49(1994), pp.3997-4008
5 J. Daniel, J. M. Costa, P. LeBoudec, G. Stephan, and F.Sanchez J.Opt.Soc.Am.B, vol.15(1998), pp.1291-1294
6 L. Luo, T. J. Tee, and P. L. Chu J.Opt.Soc.Am.B, vol.15(1998), pp.972-978
7 Y.Liu, N. Kikuchi and J. Ohtsubo Phys.Rev.E, vol.51(1995), 2697-2700
8 G. Levy and A. A. Hardy IEEE J.QE, vol.34(1998), pp.1-6
9 L.G. Luo and P. L. Chu Opt.Lett., vol.22(1997), pp.2274-2276
10 S.L.Yan, et al CHIN. J.of Lasers, vol.35(2005), pp.642-746

This work was supported by The Education Department of Jiangsu Province (No. 06KJD140111).

A shared sub-path protection strategy in multi-domain optical networks

Xuejuan Xie, Weiqiang Sun, Weisheng Hu, Jun Wang

State Key Lab of Advanced Optical Communication System and Networks, Shanghai Jiao Tong University, 800 Dongchuan Road, Shanghai 200240, P.R.China

E-mail: sunwq@sjtu.edu.cn

Abstract *In this paper, a shared sub-path protection mechanism for multi-domain optical networks is proposed. It divides the computed working path into several working sub-paths which span two domains. For each working sub-path, two intra-domain and an inter-domain backup sub-paths are computed to protect intra-domain and inter-domain failures. Simulation results show it has shorter failure recovery time and is resource efficient.*

Introduction

In order to conform to the requirements of the expanding network, the next generation optical network will be divided into many routing domains. For reasons of scalability and security, only aggregated information is exposed across domains. As a result, no node has the global information of the multi-domain network, thus the issue of improving the end-to-end reliability is more difficult than that of single domain networks.

A few valuable works has been done to address this issue. In [4], a two-step shared path protection strategy is proposed. It can get a close to optimal working/backup path pair and has a satisfying resource utilization ratio. But the backup path routed across the entire network may lead to long switching time [1]. Failures are notified to all the nodes on the working path. This is not acceptable if the domains belong to competitive operators. In [5], a p-cycle based protection scheme is proposed. It offers high availability of protection and acceptable resource usage without the knowledge of the working paths. But it can't handle node failures. A no-sharing segment protection mechanism was introduced in [6]; it was then improved in [7]. In both cases, the dividing of the working path and the routing model are not addressed.

None of these works focuses on failure hiding and switching time saving, which are also important. Because failures within domains run by competitive operators should be hidden for commercial reasons. Shorter failure recovery time makes failures more transparent to users. So in this paper, we propose a shared sub-path protection strategy. Its main idea is to orderly divide the computed working path into several sub-paths; and then compute two intra-domain backup sub-paths and an inter-domain backup sub-path for each working sub-path. It can save resources, can protect link and node failures, can

hide failure and can switch fast.

Notations and strategy description

For clarity, we first introduce some notations in the proposed approach. A multi-domain network can be represented by a graph $D=(N,L)$ composed of M connected domains $D_i=(N_i,L_i)$, $i=1,2,...,M$. When the topology aggregation is applied to D_i, $i=1, 2..., M$, a virtual topology $G_i = (N_i^{BORDER}, E_i^{VIRTUAL})$ is obtained. *Fig. 2* shows the aggregated network topology of the network in *Fig. 1*.

A method of underestimating the working and backup cost of links in G was proposed in [4]. The working cost of a virtual link e in G was underestimated by α_e:

$$\alpha_e = \begin{cases} \|e\|d & if \quad d \leq \gamma_e^{res}, e \in E^{VIRTUA} \\ d & if \quad d \leq \gamma_e^{res}, e \in L^{INTER} \\ \infty & otherwise \end{cases} \quad (1)$$

Where d is the requested bandwidth, $\|e\|$ is the minimal number of hops of the physical path between the two end nodes of e. γ_e^{res} is the maximum bandwidth that can be routed over an instance of e.

The backup cost of a link e' in G to protect a working path π is underestimated to the maximum cost for the virtual link e' to protect paths passing through an inter-domain link l in the working path.

$$\beta_{e'}^{\pi} = \max_{e \in \pi} \beta_{e'}^{e} \approx \max_{e \in \pi, l \in L^{Inter}} \beta_{e'}^{l} \quad (2)$$

In formula (2), L^{Inter} is the set of the inter-domain links in the working path. $\beta_{e'}^{l}$ is then underestimated by

$$\beta_{e'}^{l} \approx \| e' \| \min\{\max\{0, B_{max}^{l} + d - \overline{B}\}, d\} \quad (3)$$

Where B_{max}^{l} is the maximum backup bandwidth reserved on a link to protect working path passing through link l. $\overline{B}$ is the maximum backup bandwidth reserved on a link in the domain to

978-0-9789217-3-6/07/$25.00

©2007 WEN GLOBAL SOLUTIONS

which e' belongs. When e' is an inter-domain link, $\overline{B}$ in formula (3) is replaced by $B_{l'}$ which is the maximum backup bandwidth reserved on inter-domain link l'.

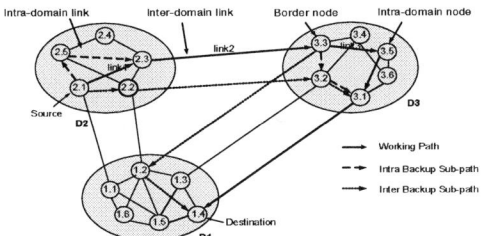

Fig.1 Multi-domain network

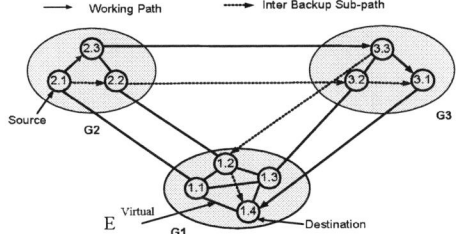

Fig.2 Virtual network of Fig.1

To get the physical working/backup path pair, the working cost of a physical link l is defined as formula (4) in [4].

$$a_l = \begin{cases} d & if \quad d < c_l^{res} \\ \infty & otherwise \end{cases} \quad (4)$$

Where c_l^{res} is the residual bandwidth on link l. The backup cost of link l' is formulated as

$$b_{l'}^P = \min\{\max\{ \quad 0, B_{\max}^P + d - B_{l'}\}, d\} \quad (5)$$

$B_{\max}^P$ is the maximum backup bandwidth reserved on a network link to protect working paths passing through links on the working path. The following is the elaboration of our strategy.

1. Working path computing

After receiving a request, a loose working path is routed on the virtual network G using formula (1). It may transverses many domains. We call the nodes through which the working path enters domain D_i a Domain Ingress Node (DIN_i) and the nodes through which the working path exits domain D_i a Domain Egress Node (DEN_i).Then a physical path is computed between two border nodes on the loose working path using formula (4) as the link working cost.

Take the network in *Fig. 1* as an example, the loose working path 2.1-2.3-3.3-3.1-1.4 is first computed in *Fig. 2*. Then the physical paths 2.1-2.3, 3.3-3.5-3.1 is obtained in *Fig.1* between border nodes 2.1, 2.3 and 3.3, 3.1. The *DINs* along the working path are 2.1, 3.3 and 1.4; the *DENs* are 2.3, 3.1, and 1.4.

2. Working path dividing

Divide the working path computed in step.1 into several sub-paths, each sub-path begins at DIN_i and ends at DEN_{i+1}, i= (1, 2......M) and is represented as $p_{i,i+1}$. Path on $p_{i,i+1}$ in domain D_i (D_{i+1}) is represented as p_i (p_{i+1}) For example, the working path in *Fig.2* can be divided into two sub-paths: $p_{2,3}$ 2.1-2.3-3.3-3.5-3.1 and $p_{3,1}$ 3.3-3.5-3.1-1.4. 2.1-2.3 is p_2 and 3.3-3.5-3.1 is p_3.

3. Intra-domain backup sub-path obtaining

In domain $D_i(D_{i+1})$, an intra-domain backup sub-path p_i' (p_{i+1}') is computed from DIN_i (DIN_{i+1}) to DEN_i (DEN_{i+1}) protecting failures on p_i (p_{i+1}). p_i' (p_{i+1}') is obtained by using formula (5) with $B_{\max}^p$ replaced by the maximum backup bandwidth reserved on a link in D_i (D_{i+1}) to protect paths passing through links on p_i' (p_{i+1}'). In Fig.1, the intra-domain backup sub-paths of sub-path 2.1-2.3-3.3-3.5-3.1 are p_2' 2.1-2.5-2.3 in domain D_2 and p_3' 3.3-3.2-3.1 in domain D_3.

4. Inter-domain backup sub-path obtaining

In D_i and D_{i+1} an inter-domain backup sub-path $p_{i,i+1}'$ is computed for working sub-path $p_{i,i+1}$ from DIN_i to DEN_{i+1} protecting failure on border nodes DIN_{i+1}, DEN_i and the inter-domain link between them. The worst scenario of inter-domain failure is that border nodes DEN_i, DIN_{i+1} fail at the same time, this will affect three links in the working path, the inter-domain link between DIN_{i+1} and DEN_i, the intra-domain link on border node DEN_i on the working path, the intra-domain link on border node DIN_{i+1} on the working path. $link_1$, $link_2$ and $link_3$ in *Fig.1* are three affected links on working path 2.1-2.3-3.3-3.5-3.1-1.4 because of failures on border nodes 2.3 and 3.3. So in our strategy, set L^{Inter} in formula (2) is replaced by set $L^{Affected}$. Then by using the modified formula (2) we get the rough inter-domain backup sub-path of $p_{i,i+1}'$. By replacing p in $B_{\max}^P$ with $L^{Affected}$, physical backup sub-path $p_{i,i+1}'$ can then be obtained. Take *Fig.1* as an example, the inter-domain backup sub-path of working sub-path 2.1-2.3-3.3-3.5-3.1 is $p_{2,3}'$ 2.1-2.2-3.2-3.1. If any of the four steps fails, the request is blocked.

Simulation results and Analysis

We perform our algorithm on the network showed in *Fig. 1*. We assume that all links are unidirectional; information aggregation is carried out once a request arrives; all nodes in the network are equipped with wavelength converters. The arrival distribution is poison distribution, the bandwidth a request requires is uniformly distributed among OC-{1,3,4,8}, the bandwidth of the inter-domain links is 100 times as much as OC-1, which is ten times as much as that of the intra-domain links.

We evaluate the performance of the proposed mechanism (SSPPMD) with the shared path protection strategy in single domain (SPPSD) and multi-domain (SPPMD) networks in three metrics: blocking probability, backup resources utilization ratio and the failure recovery time. The backup resources utilization ratio is defined as the mean backup bandwidth allocated for protecting a unit working bandwidth (shorted as MPB/WB). The failure recovery time is indicated by the mean number of backup path's hops divided by the working path's length. In the following paragraphs, we'll compare these metrics of SSPPMD, SPPSD and SPPMD.

Fig. 3 shows MPB/WB vs. load. SPPMD consumes more backup bandwidth than SPPSD, which is caused by the underestimation in SPPMD; SSPPMD consumes the most because there are overlaps in the working sub-paths in order to make sure border nodes' failures are protected. The heavier the load, the more the backup bandwidth can be shared, so MPB/WB decreases. *Fig.3* also shows that when the load is heavy (>100), SSPPMD and SPPMD consume almost same MPB/WB.

Fig. 4 shows blocking probability vs. load. As load increases, blocking probability increases. Mechanism which consumes more bandwidth gets a higher blocking probability.

Fig.5 shows that the failure recovery time vs. load. SSPPMD can switch faster than SPPSD and SPPMD because it has limitations in finding the backup sub-paths. When the working path passes through more domains, it becomes clearer.

Conclusions

In this paper, we proposed a shared sub-path protection mechanism for multi-domain optical networks. It can protect node and link failures. Path division eliminates the need of disseminating failure information across domain boundary. It can reduce failure recovery time since each inter-domain backup sub-path only spans two domains. Simulation results reveal

that when the load is heavy, it has similar resource efficiency with existing SPP approaches for multi-domain networks.

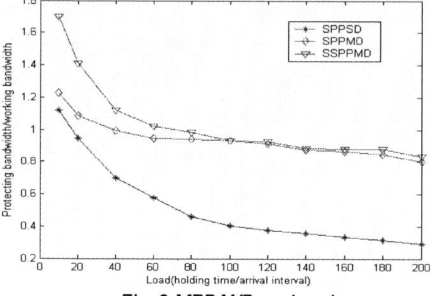
Fig. 3 MPB/WB vs. Load

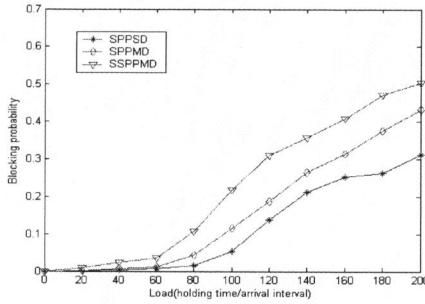
Fig.4 Blocking probability vs. Load

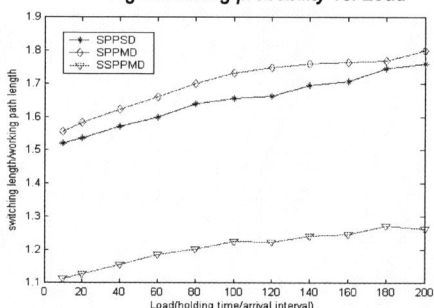
Fig.5 Failure recovery time vs. load

References

1 T. Miyamura, T. Kurimoto, International Conference on Broadband Networking, (2004).

2 J. Cao, L. Guo, International Conference on Communications, Circuits and Systems Proceedings, (2006).

3 X. Su, C-F. Su, IEEE International Conference on Communications (2001).

4 D. L. Truong, B. Thiongane, Journal of Optical Networking (2006), pp. 58–74.

5 Arnold Farkas, Janos Szigeti, 5th International Workshop on Design of Reliable Communication Networks (2005).

6 C. Ou, H. Zang, Technical Digest, Optical Fiber Communications Conference (Mar. 2002).

7 A. A. Akyamac, S. Sengupta, presented at the National Fiber Optic Engineers Conference, (2002).

8 G. Bernstein, V. Sharma, and L. Ong, Journal of Optical Networking, (2002).

Novel Multi-Channel temporal phase En/Decoder used in OCDMA over WDM PON

Ying-xun ZHU, Rong WANG, Tao PU

PLA Univ. of Sci. & Tech., Institute of Communication Engineering,

No.2 Biaoying, Yudao Street

Nanjing, 210007 China

E-mail: ying_xun319@163.com

Abstract: *A novel SSFBG based multi-channel OCDMA en-decoder is proposed. By sinc sampling the FBG, uniformity and good performance of each channel are obtained. Simulation shows that it is very suitable to an OCDMA over WDM PON.*

Introduction

Optical en-decoder is a key technology in Optical Code Division Multiple Access (OCDMA) system. Among the numerous realizable en-decoding schemes, the SSFBG based temporal phase en-decoder plays a more important role for its high performance, large capacity and low cost[1, 2], and therefore is very suitable to OCDMA based WDM PON. However, in a WDM over OCDMA PON, each user is assigned a pair of wavelength-code channel and en/decoders for different wavelength channels should be manufactured with different phase masks, which raise the cost and complexity of system. A solution is to use a multi-channel en/decoder to accommodate different wavelength channels simultaneously, so that a large amount of single channel en/decoders can be removed.

In this paper, we proposed a novel SSFBG based multi-channel OCDMA temporal phase en-decoder. During design process, we can adjust the center wavelength and bandwidth according to the system requirement and the same performance is guaranteed for each channel. A scheme using the proposed en-decoder for WDM over OCDMA PON is also discussed.

Principle of multi-channel en-decoder

A uniform SBG can be regarded as several superimposed ghost gratings and each ghost grating corresponds to a reflection peak of the spectrum of the SBG. Based on Fourier theory, the sampling function $S(z)$ can be expended to Fourier series:

$$S(z) = \sum_m F_m \exp(j2m\pi z / P) \quad (1)$$

F_m is the Fourier coefficient and P is the sampling period. Then the reflection index of the mth ghost can be expressed as:

$$\Delta n_m(z) = A(z)F_m \exp(j2\pi z / \Lambda + j2m\pi z / P) \quad (2)$$

$A(z)$ is the apodization of the seed grating and Λ is grating period. According to equivalent phase shift theory[3], when the sampling period of the SBG changes ΔP at some point along z, the phase of the index modulation of the mth ghost grating changes to:

$$\Delta n_m(z) = A(z)F_m \exp(j2\pi z / \Lambda + j2m\pi z / P + j\theta) \quad (3)$$

Where $\theta = 2m\pi\Delta P/P$. When $\Delta P = P/2m$, θ equals to π. Thus an equivalent π shift is induced to the mth ghost grating. An EPS based en-decoder is thus constructed by changing the sampling period of different point along z.

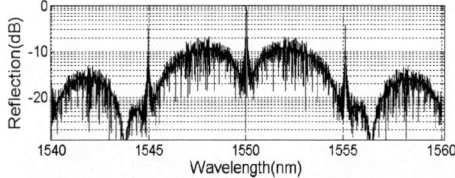

Fig. 1 Reflection of an equivalent phase shift encoder with cycle duty 0.4

978-0-9789217-3-6/07/$25.00

©2007 WEN GLOBAL SOLUTIONS 417

Fig. 1 is a typical reflection of a rectangular sampling EPS-BPSK en-decoder. It can be seen that the spectrum is composed of several peaks and encoding channels. A problem shown in Fig.1 is that two dips occur at 1543nm and 1552nm in $\pm 3^{rd}$ channel, and therefore the $\pm 3^{rd}$ channel can not used as en-decoding channel. The dips are caused by the Fourier coefficient F_m of the sampling function. When the sampling function is a rectangle, the absolute value of F_m becomes

$$|F_m| = \frac{\gamma \sin(m\pi\gamma)}{m\pi\gamma} = \gamma Sinc(m\gamma) \qquad (4)$$

Where γ is the duty cycle of the sampling function. Hence, the two dips are just the zeros of $sinc$ function. The dips can be diminished by changing the sampling function. When the sampling function has a $sinc$ profile, uniform channels can be achieved. Then we can set the apodization as[4]:

$$A(z) = \sin(N\frac{L}{b}\pi(\frac{z}{L}+\frac{b}{2L}) \bigg/ N\sin(\frac{L}{b}\pi(\frac{z}{L}+\frac{b}{2L}))) $$
$$(N=1,2...) \qquad (5)$$

Where, L is length of the sampling grating, b is the sampling length and P/b is an even. Then we can rewrite the sampling function as:

$$A_s(z) = \sum_{i=0}^{M} A(z)S_i(z-iP_i) = \sum_{i=0}^{M} A_{si}(z-iP_i) \qquad (6)$$

Here, M is the length of code, P_i is the period of ith chip, $P_i = P-(\varphi_i-\varphi_{i-1})P/2\pi$, φ_i is the phase of the ith chip. In the weak SSFBG limit, the reflection of SSFBG can be obtained simply by the Fourier transform of the spatial refractive index modulation profile. Moreover, the spatial position z and the time t comply with the following relationship[5]: $t=2zn_{eff}/c$ (c is velocity of light, n_{eff} is the effective refractive index of the media), so $A_s(z)$ can be regarded as a temporal function:

$$A_s(t) = \sum_{i=0}^{M} A_{si}(t-iT_i) \qquad (7)$$

Here, $T_i = 2P_i n_{eff}/c$. Therefore, the reflection of the grating can be expressed as:

$$A_s(f) = Fourier[A_s(t)] = A_{si}(f)\sum_{i=0}^{M} e^{-j2\pi f i T_i} \qquad (8)$$

From (8), we can see that the reflection profile only has relations with $A_{si}(f)$. $A_{si}(t)$ has a $sinc$ profile, then $A_{si}(f)$ has a rectangular frequency spectra profile. From (5) to (8), we know the bandwidth of $A_{si}(f)$ is $Nc/2bn_{eff}$. As the channel interval of the SBG is $\lambda_B^2/2n_{eff}P$, then the $Sinc$ sampling grating has NP/b channels.

Simulation and analysis

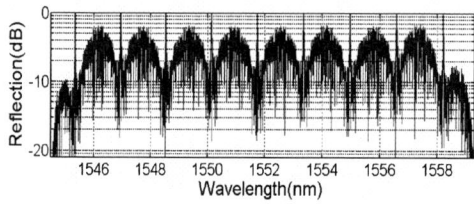

Fig. 2 Reflection of an 8-channel en-decoder

Fig.2 shows the reflection spectra of an 8-channel en-decoder. The narrow reflection peak is the non-encoding channel constructed by even level ghost gratings. The spectra between peaks are the encoding channels constructed by odd level ghost gratings. In order to be compatible with the standard of WDM channel interval of ITU-T, we set the Bragg wavelength of the en/decoder 1551.72nm (corresponds to 193.2THz) and channel interval 1.6nm (corresponds to 200GHz). The other parameters are: $P=1.0281\times10^{-3}$m，refractive index modulation $\Delta n=3\times10^{-4}$, effective refractive index $n_{eff}=1.4475$, and the address code is a 127 Gold sequence. The center wavelength of each channel can be calculated by:

$$\Lambda_m = \frac{1}{1/\Lambda+m/P} \qquad (9)$$

The calculated center wavelength of each channel is listed in table 1. In order to validate the performance of each encoding channel, a Gaussian signal is injected into the encoder as an optical pulse source. The FHWM and the power of the Gaussian signal are 10ps and 1mw. In order to perfectly match the encoding channel, we let the center wavelength of the injecting signal equal to the values listed in table 1. The calculated peak-wing-ratio (P/W)

418

of the auto-correlation output is listed in table2. From the results in table2, we can conclude that similar performance of the en/decoding channels is achieved. The P/Ws of each channel is not less than 17.6dB, and the fluctuation of P/Ws is not less than 0.213dB.

Table 1 Center wavelength of each encoding channel in Figure 1

channel	+1	+3	+5	+7
CW(nm)	1550.92	1549.31	1547.71	1546.11
channel	-1	-3	-5	-7
CW(nm)	1552.52	1554.13	1555.75	1557.36

Table 2 Performance of each encoding channel

channel	+1	+3	+5	+7
PA	0.1354	0.1424	0.1380	0.1602
P/W(dB)	17.701	17.795	18.026	17.695
channel	-1	-3	-5	-7
PA	0.1322	0.1366	0.1286	0.1377
P/W(dB)	17.674	17.784	17.936	17.945

Application of the en-decoder

The multi-channel en-decoder mentioned above can be applied in OCDMA over WDM PON. Fig. 3 shows the structure of a PON based optical access network. In the OLT end, $m \times n$ users are divided into m groups, users with different wavelength in one group share the same address code, while in the receiving end, only one decoder is needed for recognition of n users. Hence, a total of $2m(n-1)$ en-decoders can be saved, and therefore the cost and complexity is dramatically reduced.

Conclusion

According to equivalent phase shift theory, a novel SSFBG based multi-channel temporal phase en-decoder is proposed. Uniform

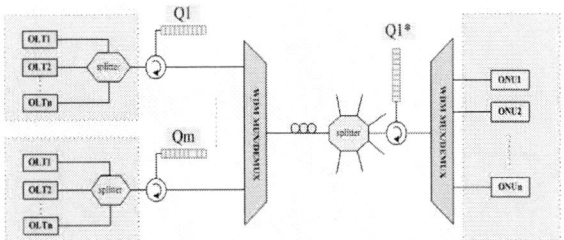

Fig. 3 The structure of a PON based optical access network

performance of each channel of the en-decoder is obtained by *sinc* sampling the FBG. An 8-channel en-decoder is analyzed to validate the performance and results show that the P/Ws are not less than 17.6dB and the fluctuation of P/Ws is not less than 0.213dB. A new scheme of PON access network using the proposed multi-channel en-decoder is also demonstrated. Compared to traditional one, a total of $2m(n-1)$ en-decoders are saved. Therefore, the cost and complexity of the network are markedly reduced.

Acknowledgments

This work was supported by the National Natural Science Foundation of China under GRANT NSFC-60502003 and 60132020, and NNSF of Jiangsu Province under GRANT BK2007501.

References

[1] Taro Hamanaka et al., J. Lightwave Technol., 2006, **24**(1):95-102.
[2] Yitang Dai et al., Optics Lett., 2006, 31(11):1618-1620
[3] Yitang Dai et al., 2004, IEEE Photon. Technol. Lett, 16(10): 2284-2286
[4] Morten Ibsen, IEEE Photon. Technol. Lett., 1998, 10(6): 842-844.
[5] P. Chiong Teh. et al., J. Lightwave technol., 2001,19(9): 1352-1365

Untraditional All-Optical Chromatic Dispersion Compensating Elements - Experimental Verification

J. Vojtech (1), M. Karasek (1, 2), J. Radil (1)

1 : CESNET a.l.e., Zikova 4, 160 00 Prague 6, Czech Republic, josef.vojtech@cesnet.cz
2 : Institute of Photonics and Electronics, Academy of Sciences of the Czech Republic,
karasek@ufe.cz

Abstract

This contribution deals with experimental investigation of usability of untraditional all-optical chromatic dispersion elements. Results are compared with traditional dispersion compensating fibres; emphasis was also given to tuneability and broadband characteristics of elements.

Introduction

Standard single mode fibres (SSMF) according ITU-T specification G.652 represent the majority of already installed fibres. SSMFs were original designed for operation in O band and thus their wavelength of zero chromatic dispersion (CD) is about 1310 nm. Their low loss in C band and the availability of reliable and relatively cheap erbium doped fibre amplifiers (EDFAs) makes the 1550 nm window attractive for multi channel high speed transmission. Unfortunately within this window SSMF shows relatively large CD, about +16.8 ps/(nm*km), severely limiting transmission distance unless compensated. In the past fibre vendors tried to overcome CD issue by introducing non-zero dispersion shifted fibres (NZDF). Nevertheless these fibres represent a minority of installed fibre base and their usage under high channels counts (induces high powers in fibre) is considered as controversial. At transmission rates 10 Gb/s and higher the effect of CD can be mitigated by various means. These range from electrical to all-optical ones. As electrical processing is performed per wavelength and therefore unsuitable for wavelength division multiplex (WDM) systems, we will focus all optical methods.

Very typical approach represents application of dispersion compensating fibres (DCF) [1], showing negative CD typically of between -90 and -100 ps/(nm*km). Unfortunately they also have higher insertion loss (IL) compared with SSMF. To compensate 80km of SSMF it is necessary about 13km of DCF having IL about 13 dB. Although DCF is a broadband element, dispersion slope of SSMF and DCF are not exactly balanced so in WDM systems only one channel can be compensated exactly.

Next quite often used element is a chirped fibre Bragg grating (FBG) [2]. At first channelized FBGs becomes available, compensating CD only in 100 GHz or 50 GHz spaced channels – typically of ITU-T grid. Nevertheless broadband FBG are now also available. Main advantages of FBG CD compensating modules are low IL (about 3 dB for device compensating 80km of SSMF) and possibility of designing the DC modules to exactly match dispersion slope of compensated fibre. The disadvantage is that fixed FGB module must be tailored to compensate certain distance.

Tuneability of CD compensation becomes crucial with increasing transmission bit rate because the tolerance of receivers to accumulated CD decreases with square of transmission rate. For example existing 40 Gb/s NRZ receivers can tolerate residual CD of about 100 ps/nm, only. Furthermore in reconfigurable networks accumulated CD can change due to rerouting of optical path. There are several possibilities how to achieve tuneable CD compensation, including: differential thermal tuning of nonlinearly chirped FBG [3, 4], thermal tuning of free space or FBG coupled-cavities Gires-Tournois etalons (GTE's) [5, 6] and virtually-imaged phase-array (PA's) [7, 8].

978-0-9789217-3-6/07/$25.00

©2007 WEN GLOBAL SOLUTIONS

Experimental setup

In the area of research and educational networking REN (typically based on leased dark fibres) it is sometimes uneconomical or even impossible to deploy inline amplification and CD compensation.

Our experimental setup takes into account this fact - using so called nothing in line (NIL) approach, where transmission equipment is placed in terminal nodes, only [9,10]. In our experiment signals from eight 10 GE DWDM XFP transceivers were combined in a multiplexer (MUX), amplified in a high-power C-band EDFA and launched into the test fibre link. The test link consisted of 225 km of SSMF on spools with granularity 50 km and 25 km. The chromatic dispersion of the link was about +3780 ps/nm. At the receiver side signals were first amplified in a low noise preamplifier EDFA. Accumulated CD was compensated and signals were finally de-multiplexed. Bit-error rate tests were performed by Packet Blazer 10GigE FTB-5810G module.

Experimental results

In the reference experiment CD was compensated by DCF modules. It was necessary to use additional EDFAs after DCFs to overcome their high IL of up to 18 dB. Transmission tolerance to total input power is shown in Fig 1 with the amount of compensated CD as parameter.

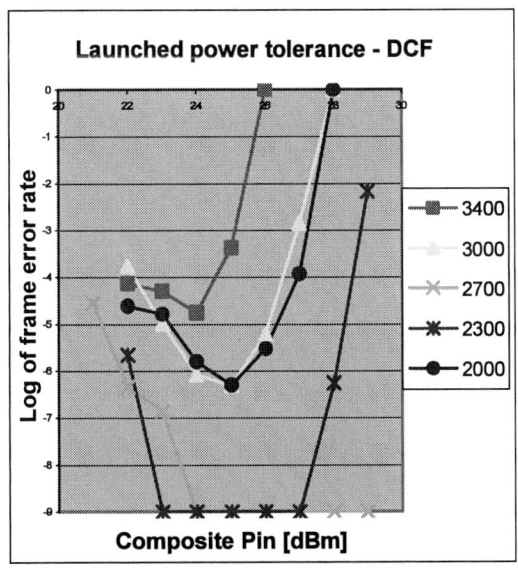

Fig 1: Transmission tolerance to launched input power – DCF

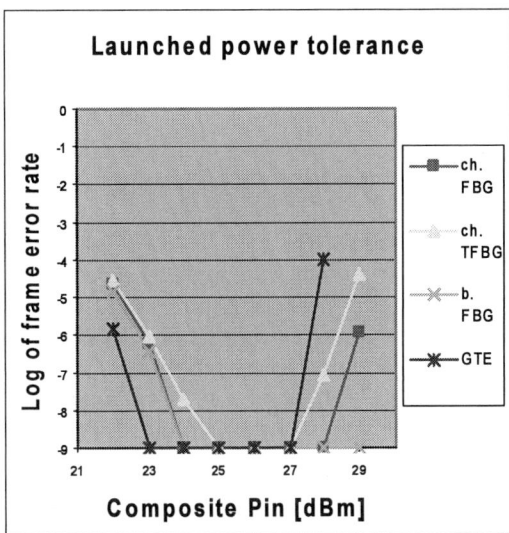

Fig 2: Transmission tolerance to launched input power – untraditional elements

In following experiments CD was compensated by one of following all optical methods: channelized FBGs, channelized tuneable FBGs (TFBG), broadband FBGs and tuneable GTEs. Fig 2 demonstrates the tolerance to input power for different compensating methods, when the value of compensated CD was fixed at 3400 ps/nm.

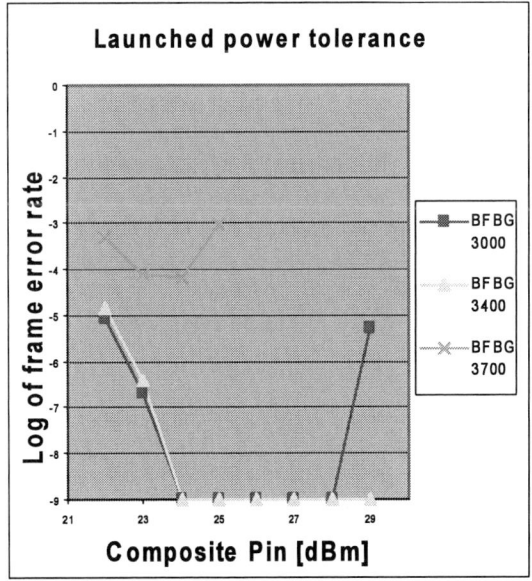

Fig 3: Transmission tolerance to launched input power – broadband FBGs

Next, tolerance to input power for different values of compensated CD using tuneable elements (channelized TFBGs, GTEs) and broadband FBGs (different modules were available) was examined. For example in Fig 3 power tolerance for broadband FBGs is shown. For tuneable compensators launched power was fixed to +25 dBm and operational range over compensated CD was experimentally investigated too.

Conclusions

We experimentally investigated different CD compensation techniques in NIL setup. Under test were traditional DCFs, fixed channelized and broadband FBGs, tuneable FBGs and tuneable GTEs. Tests were performed by transmitting 8x10 GE channels over 225 km of SSMF, signals were amplified at transmitter and receiver sides by standard C-band EDFAs only. Experiments confirmed that unconventional elements allow implementation of simpler setups due to lower IL compared with DCFs. Furthermore we can summarize that for error-free transmission the GTEs allow launching of lower signal powers in contrast to broadband FBGs, which can tolerate higher launch powers. When comparing tuneable devices we can state that FBGs allow error free operation over broader ranges of input power and compensated CD, comaped with GTEs. However, GTEs offer tuneable CD compensation with very low IL at reasonable prices. Experimental setups based on partial CD pre-compensation proved that small CD ($\approx$-340 ps/nm) pre-compensation can dramatically increase transmission CD tolerance range (2 times), we believe due to suppression of self-phase modulation in the link. On other hand higher CD pre-compensation prevented error-free transmission completely. In the future we would like to experimentally verify applicability of virtually-imaged phase-array CD compensators, both fixed and tuneable.

Acknowledgement

This research has been supported by the Ministry of Education, Youth and Sport of the Czech Republic under research plan no. MSM6383917201 *"Optical National Research Network and It's New Applications"*.

References

[1] L. Grüner-Nilesen et al Optical Fibre Technology, 6, (2000), 164-180

[2] W.H. Loh et al IEEE Phot. Technol. Lett., 8, (1996), 944-946

[3] Y.W. Song et al IEEE phot. Technol. Lett., 14, (2003), 1193-1195

[4] R. Lachance et al Proc. Optical Fiber Communication (OFC 2003), Paper TuD3, (2003), 164-162

[5] B.J. Vakoc et al IEEE Phot. Technmol. Lett., 17, (2005), 1043-1045

[6] S. Doucet et al IEEE Phot. Technol. Lett., 16, 2004, 2529-2531

[7] M. Shirasaki et al Proc. Optical Fiber Communication (OFC 2001), Paper TuS1, (2001)

[8] G.H. Lee et al Proc. Optical Fiber Communication (OFC 2006), Paper OThE5, (2006).

[9] J. Radil et al Proc. TERENA Networking Conference, Experimental transmission of 10 GE over G.652 without in-line amplifiers, (2004)

[10] M. Karasek et al Proc. Conference on Optical Network Design and Modelling, (2005), 55-58

Simultaneously Realizing Optical Millimeter-wave Generation and Photonic Frequency Down-conversion Employing Optical Phase Modulator and Sidebands Separation Technique

Hong Wen (1), Lin Chen (2), Jing He (3), Shuangchun Wen * (4)
Research Center Laser Science and Engineering, and School of Computer and
Communication, Hunan University, Changsha 410082, China
1: niubihwen@yahoo.com.cn 2: liliuchen12@vip.126.com
3: hnu_jhe@hotmail.com 4: scwen@vip.sina.com

Abstract *A novel scheme for simultaneously realizing optical millimeter-wave generation and photonic frequency down-conversion by employing optical phase modulator and sidebands separation technique is proposed and verified by simulation.*

Introduction

Radio-over-fibre (RoF) has become very attractive in present and future wireless access networks since it can provide a truly broadband access to end user units while warranting the mobility. All-optical generation of the millimeter-wave (mm-wave) signal is one of the key techniques in RoF system, and many approaches to implement it have been reported recently. Among them, optical mm-wave generation employing phase modulator is the simplest and the most accurate one and shows great potential application for producing high-frequency mm-wave signals [1].

Moreover, the design of simplification of fibre uplink transmission should be considered for accomplishing a full-duplex RoF system. Up to now, three schemes have been reported for optical uplink connection, which include the RF-over-fibre, IF-over-fibre, and baseband-over-fibre, respectively. Baseband-over-fibre requires the using of high-speed demodulator and synchronization at each base station. RF-over-fibre is most simple configure, but the effect of serious fibre chromatic dispersion limits its transport distance very short. Low-frequency intermediate frequency (IF) electronic signal transmission over fibre can solve this problem [2]. However, expensive mm-wave mixers and phase-locked local oscillators (LOs) are needed for electrical frequency down-conversion from mm-wave frequency to IF. To eliminate this problem, it

is preferable to generate a low IF directly from the optical signal, avoiding the use of high frequency RF signal processing. In previous studies, optical single sideband (SSB) transmitter or SSB filtering has been proposed in the central station (CS) to generate optical SSB signal, in which the complexity and cost of systems is increased [3]. Furthermore, additional LO signal have to be used for realize bidirectional transmission.

In this paper, we will propose a novel scheme for simultaneously realizing optical mm-wave generation and photonic frequency down-conversion by using optical phase modulator and sidebands separation technique. In CS, optical double sidebands (DSB) signals are created by employing an optical phase modulator. One of the optical sidebands is filter out to carry downstream data signal. In base stations (BSs), the sideband signal carried with downstream data beats with part of optical carrier, and thereby generate mm-wave signal, while another sideband along with the other part of optical carrier is injected into the second optical phase modulator to implement photonic frequency down-conversion for uplink transmission.

Principle

Fig.1 shows the principle of photonic frequency down-conversion employing optical phase modulator.

---

* E-mail: scwen@vip.sina.com, Fax: 86-731-8823474.

978-0-9789217-3-6/07/$25.00

©2007 WEN GLOBAL SOLUTIONS

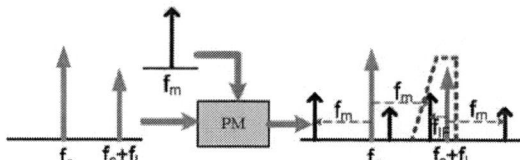

Fig.1 Principle of photonic frequency downconversion

Suppose that an electric field has an optical carrier at f_c which is injected into a phase modulator (referred to PM-1), which is driven by a LO signal $V_L \cos \omega_L t$, where V_L is the modulation voltage, $\omega_L = 2\pi f_L$ is the LO modulation angular frequency. The waveform at the output of the optical phase modulator can be given by

$$E_{out}(t) \cong E_o \{ J_0(\xi)\cos(\omega_c t) - J_1(\xi)\sin(\omega_c + \omega_L)t$$
$$- J_1(\xi)\sin(\omega_c - \omega_L)t + ... \} \qquad (1)$$

here E_o and $\omega_c = 2\pi f_c$ are the amplitude and angular frequency of the optical carrier; and $J_n(\xi)$ is a Bessel function of the first kind of order n with argument of ξ, $\xi = \pi V_L / V_\pi$ is related to the phase-modulation depth, where V_π is the half-wave voltage of the optical phase modulator. By adjusting the value of ξ all other high order optical sideband can be omitted in real operation. The first term in (1) is the optical carrier, and the second and the third terms are the upper sideband (USB) and lower sideband (LSB) respectively. In our proposed system, we only use LSB to carry downstream data and then generate an mm-wave by beating part of optical carrier, while the USB along with another part of optical carrier is used to implement remote photonic down-conversion for uplink

connection. The generated mm-wave in downlink can be express by

$$I(t) \propto J_0(\xi)J_1(\xi)\sin(\omega_L t) \qquad (2)$$

Now we consider another PM (referred to PM-2) which is used for optical frequency down-conversion, driven by a RF signal $a(t)V_m \cos(\omega_m t)$, where a(t) is the carried upstream data signal, V_m is the modulation voltage, and $\omega_m = 2\pi f_m$ is the RF modulation angular. When the USB along with optical carrier emitted from the first PM (PM-1) is injected into the PM-2, the output is then approximated by

$$E_{out}(t) \propto$$
$$J_0(\beta)\{ J_0(\xi)\cos \omega_c t - J_1(\xi)\sin(\omega_c + \omega_L)t \}$$
$$- J_1(\beta)\{ J_1(\xi)[\cos(\omega_c + \omega_L - \omega_m)t + \cos(\omega_c + \omega_L + \omega_m)t]$$
$$- J_0(\xi)[\sin(\omega_c - \omega_m)t + \sin(\omega_c + \omega_m)t]\}$$

(3)

where β presents the modulation index of PM-2. Note that the terms in the first brace represent the input optical carrier and its USB signal, and the terms in the second brace represent the frequency down-conversion and up-conversion of optical carrier and its USB signal. The optical spectra of the electric fields with (3) are shown in Fig.1. It is observed that the frequency up-converted optical carrier is about a few of gigahertz far from the input USB signal, the same as the down-converted USB signal far from the input carrier. The frequency up-converted optical carrier is filtered out by a band-pass filter, and then be sent to photondetection. Finally, the electrical signal with an IF is obtained.

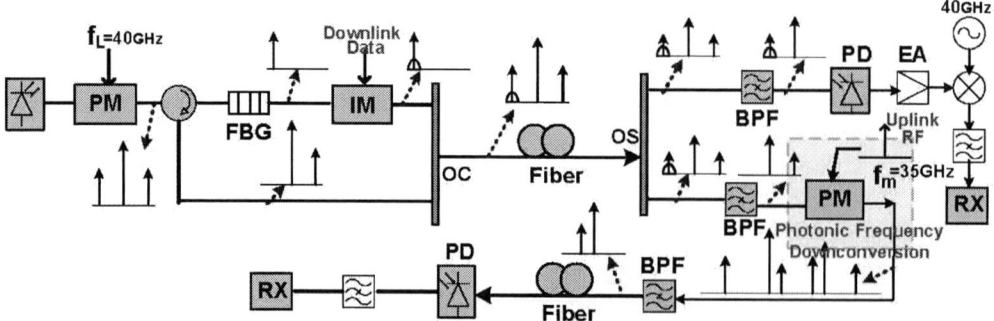

Fig.2. Verification setup for simulation

424

System Description and Simulation Result

The proposed system configuration is shown in Fig.2. Optical DSB signal is created by using an optical phase modulator (PM-1) driven a 40GHz sine wave. An optical circuit along with a FBG is used to separate the USB signal from other parts. The separated USB signal is modulating with 1.25Gbit/s downstream NRZ data, and then recombined with optical carrier and LSB before transmit over 40km signal mode fiber (SMF). In the BS, we use an optical splitter to divide the recombined signals into two identical parts. One divided part is detected by a PIN photo-detector (PD) and converted into 40GHz mm-wave electrical signal after it is cancelled the USB signal by an optical bass-pass filter. The spectra of generated mm-wave electrical signal is shown in Fig.4. For the lower path, another optical band-pass filter is used to suppress the LSB. The remained optical carrier and USB signal are injected into the second PM (PM-2) for modulating 35GHz upstream RF signal, as well as the photonic down-conversion is implemented. The optical specra of output of PM-2 is shown in Fig.3. As we can observer that the frequency space between down-converted USB signal and input carrier is reduced to 5GHz. We use another optical pass-band filter to filter out the USB and the frequency up-shifted carrier. The filtered uplink optical signal was distributed back to central station over 40km SMF, where a PD is used converted it to electronic signal. Finally, the electrical signal with a 5GHz IF is obtained. The received eye diagrams for both up-link and down-link over 40km SMF transmission are shown in Fig.5(a) and Fig.5(b) respectively.

Conclusion

A novel scheme for simultaneously realizing optical mm-wave generation and photonic frequency down-conversion employing optical modulator and sidebands separation technique has been proposed and verified by simulation. By using the scheme, we can simplify the configuration of base station, and moreover, the additional LO signal or RF signal is not needed since we make full use of the optical sidebands of DSB signal generated in CS, i.e. one of the sidebands

was used to generate mm-wave signal, while the other sideband was used to implement photonic frequency down-conversion for up-link connection.

References

1 J. Yu, et al IEEE Photon. Technol. Lett., 19 (2007), 140-142

2 A. Kaszubowska, et al IEEE Photon. Technol. Lett., 18 (2006), 562-564

3 T. Kuri, et al IEEE Trans. Microwave Theory Technol., 47 (1999), 1332-1337

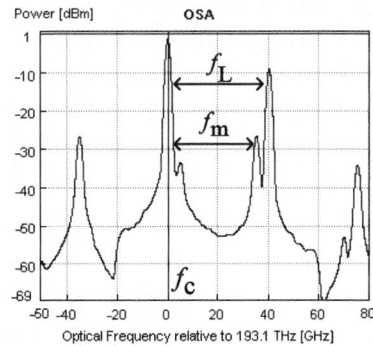

Fig.3. Optical spectra of output of PM-2

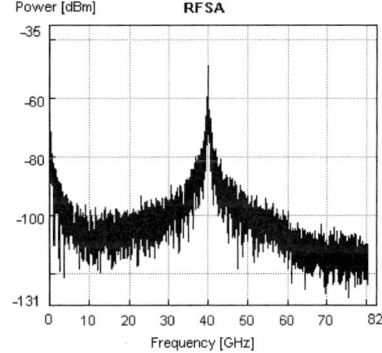

Fig.4. Spectra of generated mm-wave signal after 40km SMF transmission

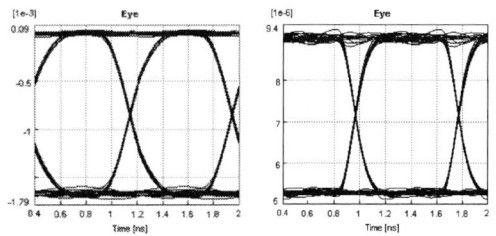

Fig.5. Eye diagram after down-conversion at the receiver (a) uplink (b) downlink

Analysis of Photonic Band-gaps of a Novel PBGF Structure

Xiao Yueyu

Institute of Fiber Research, ShangHai University, ShangHai, P. R. China.

yyx@staff.shu.edu.cn

Abstract A novel PBGF structure of hexagonal symmetry is presented. Numerical investigation shows that the optimized air-guiding PBG is almost as large as that of triangular structure, while the range of normalized frequency allowing aie-guiding is much wider.

Introduction

In recent years, photonic band-gap fibers (PBGFs) have attracted much attention for their unconventional guiding mechanism and promising applications [1~2]. In the design of PBGFs, not only the core structure but also the cladding structure is very important. A novel photonic band-gap structure is presented. Differed from all existed structures of hexagonal symmetry [3~5], elliptical holes are added on the edges of the hexagonal. Numerical investigation shows that the optimized air-guiding PBG is almost as large as that of triangular structure, while the range of normalized frequency is much wider.

In the following, we first introduce the theoretical basis and give the out-of-plane photonic band structure of the fiber. Secondly, the dependence of the PBGs on structure parameters is analyzed. Finally, optimized structure parameters are given.

Theoretical Basis

In order to model the cladding band structure of the fiber, a standard plane-wave method is employed [6]. The Maxwell's equation of the magnetic field is:

$$\nabla \times \left[\frac{1}{\varepsilon(r)} \nabla \times \vec{H}(r) \right] = \left(\frac{\omega}{c} \right)^2 \vec{H}(r) \qquad (1)$$

According to the Bloch's theorem, the magnetic field $\vec{H}(r)$ and the dielectric constant $\varepsilon(r)$ can be expanded in plane waves, and the following matrix equations are
obtained:

$$\sum_{\vec{G}'} |\vec{k} + \vec{G}| |\vec{k} + \vec{G}'| \kappa_{\vec{G} - \vec{G}'} M \begin{bmatrix} h_{1,\vec{k}+\vec{G}'} \\ h_{2,\vec{k}+\vec{G}'} \end{bmatrix} = \left(\frac{\omega}{c} \right)^2 \begin{bmatrix} h_{1,\vec{k}+\vec{G}} \\ h_{2,\vec{k}+\vec{G}} \end{bmatrix}$$
(2)

where, $\vec{k}$ is the wave vector, $\vec{G}$ is the reciprocal-space vector, and $\kappa_{\vec{G}-\vec{G}'}$ is the Fourier transform of $1/\varepsilon(r)$.

$$M = \begin{bmatrix} e_2 \cdot e_2' & -e_2 \cdot e_1' \\ -e_1 \cdot e_2' & e_1 \cdot e_1' \end{bmatrix},$$

with e_1, e_2 are unit vectors perpendicular to the vector $\vec{k} + \vec{G}$.

The structure presented is shown in Fig.1. Differed from other hexagonal structure, we add six elliptical hole on the edges. R_e is the length of an ellipse axis and g is the

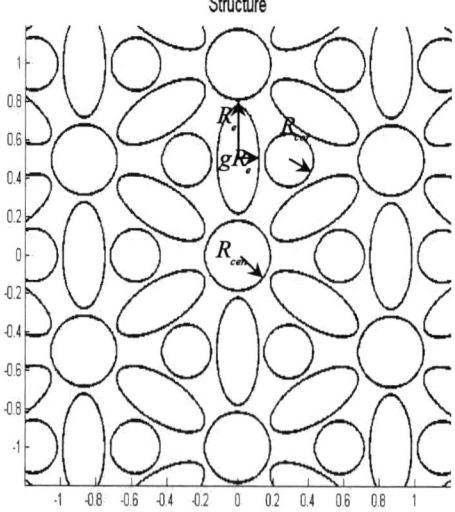

Fig. 1 Refractive index distribution of the structure

ratio of the two axes of the ellipse. R_{cen} is the radius of the central hole, and R_{cor} is the radius of the hole in the corner. Hence, $\kappa_{m,p}$ can be given:

$$\kappa_{m,p} = \frac{2}{A}\left(\frac{1}{\varepsilon_a} - \frac{1}{\varepsilon_b}\right)\left\{\kappa_{ellip} + \kappa_{cen} + \kappa_{cor}\right\}_{m,p} + \frac{1}{\varepsilon_b}\delta_{m,p}$$

(3)
Where,

$$\kappa_{ellip} = 2\pi R_e^2 g \begin{bmatrix} (-1)^{m+p} J_{m,p}(R_e, \frac{2\pi}{3}) + \\ (-1)^p J_{m,p}(R_e, \frac{\pi}{3}) \\ +(-1)^m J_{m,p}(R_e, 0) \end{bmatrix},$$

$$\kappa_{cen} = 2\pi R_{cen}^2 J_{m,p}(R_{cen}, 0),$$

$$\kappa_{cor} = 4\pi R_{cor}^2 \cos(\frac{2\pi}{3}(m+p)) J_{m,p}(R_{cor}, 0)$$

In the above formulas:

$$J_{m,p}(R, \varphi) = \frac{J_1(R\sqrt{A^2(\varphi) + B^2(\varphi)})}{R\sqrt{A^2(\varphi) + B^2(\varphi)}},$$

Where J_1 is the first-order Bessel function, and

$$\begin{cases} A(\varphi) = \frac{2\pi m}{a}\cos\varphi + \frac{2\pi(-m+2p)}{\sqrt{3}a}\sin\varphi \\ B(\varphi) = -\frac{2\pi m}{a}g\sin\varphi + \frac{2\pi(-m+2p)}{\sqrt{3}a}g\sin\varphi \end{cases}$$

By solving Eq. (2), the photonic band structure of the fiber can be obtained.

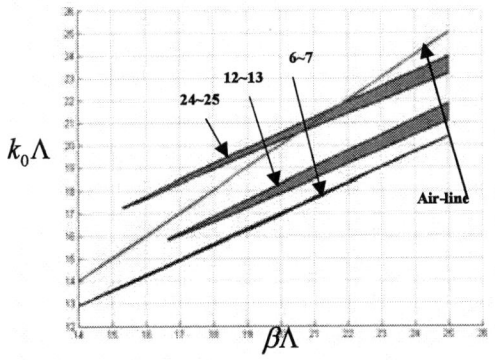

Fig. 2 Variation of PBGs with normalized propagation constant.

Parameters Dependence

In addition to the period Λ, there are four geometrical parameters in this structure, R_{cen}, R_{cor}, R_e and g. Fig. 2 shows the PBGs for the structure with $R_{cen} = 0.185\Lambda$, $R_{cor} = 0.10\Lambda$, $R_e = 0.25\Lambda$ and $g = 0.65$. From Fig.2, we found that in high normalized frequency, the lower band-gaps, such as gap 6~7 and gap 12~13 can't cover the air-line. According to [7], air-line across the band-gap is an indispensable condition for air-core PBG guidance, we tried to maximize the band-gap 24~25 above the air-line. It's well-known that structures exhibiting larger band-gaps allow a stronger confinement of the localized mode within the core region. As a stronger confinement results in a more robust and bend insensitive fiber, large band-gaps are of significant importance for the realization of ultra-low loss PBGFs.The criteria of optimization is the normalized band-gap $\Delta k / k_{mid}$ (Δk is gap size and k_{mid} is the center k value in the gap).

Firstly, we varied R_e and kept the other

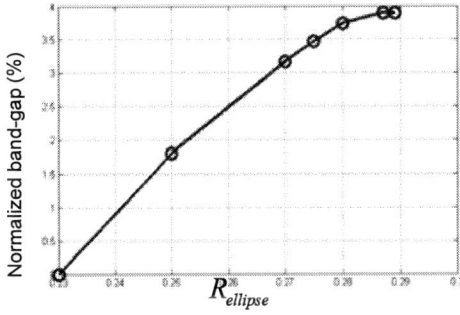

Fig. 3 Variation of normalized band-gap with $R_{ellipse}$

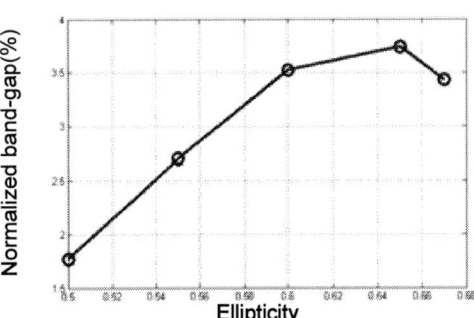

Fig. 4 Variation of normalized band-gap with ellipticity

three parameters unchanged. The results are shown in Fig. 3. It can be seen that air-guiding PBGs increase with R_e.

However, there is a limit for the normalized band-gap. The reason for the limit is that the band-gap is moving parallel with the air-line.

Secondly, the ellipticity g is varied, while the other three parameters are unchanged. The results are shown in Fig. 4. We could find that there is a best ellpticity for the largest normalized band-gap.

Finally, the radius of corner and the ellipticity are changed simultaneously, and the results are shown in Fig. 5.

From the above analysis, the best

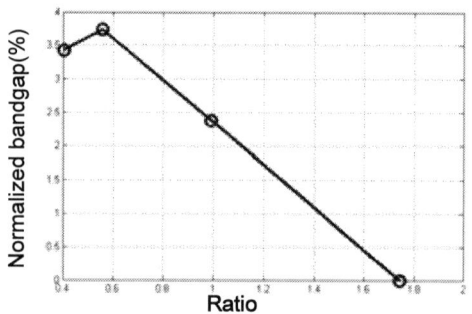

Fig. 5 Variation of normalized band-gap with the ratio of R_{cor} and gR_e

parameters of the structure we got is $R_{cen} = 0.185\Lambda$, $R_{cor} = 0.10\Lambda$, $R_{ellip} = 0.287\Lambda$, $g = 0.65$, with normalized band-gap equals to 3.9% at the normalized propagation constant $\beta\Lambda = 24.44$. The range of normalized frequency allowing for air-guiding is about 8.4. The maximum normalized band-gap of the triangular structure is 3.5%, the range of normalized frequency allowing for air-guiding is about 3.2 [8].

Conclusions

A novel photonic band-gap structure is presented. With the elliptical holes added on the edges of the hexagonal, it changes the PBGs property. A large air-guiding PBG lying in a very wide normalized frequency is obtained. With more detailed research, it is promising to offer some useful properties for the design of PBGFs.

References

1 Konorov S. O, *et al.* Appl. Opt., 43(2004): 2251-2256.

2 Luan F, *et al.* Opt. Express, 12 (2004): 835-840.

3 T. A. Birks, *et al.* Electronics Letters. 26(1995): 1941-1943.

4 M. Yang, *et al.,* Photonics Technology Letters, 17(2005):64~66.

5 M. Y. Chen, *et al.,* Photonics Technology Letters, 16(2004): 819-821.

6 K. M. Ho, *et al.,* Phys. Rev. Lett., 65(1990): 3152.

7 J. Broeng, *et al.,* Optics Letters, 25(2000):96-98.

8 H. K. Kim, *et al.,* Quantum Electronics, 40(2004): 551-556.

Application of Lambert W function to Raman fiber laser

Chaohong Huang[1], Wencai Huang[1], Xiulin Wang[2], Huiying Xu[1], Zhiping Cai[1]

[1]Department of Electronic Engineering, Xiamen University, Xiamen 361005, China

[2]Department of Physics, Jimei University, Xiamen 361021, China

E-mail: hchmail@xmu.edu.cn

Abstract *With the assistance of Lambert W function, we derive an explicit expression for output power of Raman fiber laser without using the depleted-pump approximation. Moreover, the optimal fiber length and FBG reflectivity are obtained analytically.*

Introduction

Recently significant progresses in Raman fiber lasers(RFLs) have been made due to the use of ultralow-loss P_2O_5-doped silica fibers(PDF), high power Yb-doped dual-cladding fiber lasers and high-reflectivity fiber Bragg gratings(FBG)[1-2]. It was found difficult to seek for the exact analytical solution for the RFLs as a result of the nonlinear power coupling between pump and the Stokes wave inside cavity. To optimize the RFLs, some numerical[3,4] or approximate analytical[5-8] methods have been developed. For odd-order RFLs, one couldn't obtain explicit solution for output power unless considering the depleted-pump approximation which is not always valid to a really RFL. With the assistance of Lambert W function, we obtain an explicit analytical solution for first-odd RFL without using the depleted-pump approximation. Lambert W function, the inverse function of $f(x) = xe^x$ [9], has been applied in many fields[10,11] such as physics, semiconductor, graph theory etc. The proposed explicit solution for the RFLs provides us a clear physical understanding to the optimal design of the lasers.

Theoretic analysis

Fig.1 shows the schematic diagram of the Raman fiber laser. FBG1 and FBG2 form F-P cavity for Stokes wave. FBG1 is highly-reflective and FBG2 is partially-reflective at the Stokes wavelength.

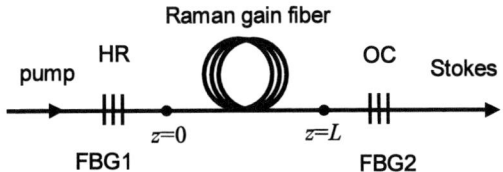

Fig. 1. Configuration of Raman fiber laser

The forward- and backward-propagating pump and Stokes powers in Raman gain fiber obey the well-known differential equations

$$\frac{1}{P_0}\frac{dP_0}{dz} = -\alpha_0 - g\frac{\lambda_1}{\lambda_0}(P_1^+ + P_1^-) \quad (1a)$$

$$\frac{1}{P_1^\pm}\frac{dP_1^\pm}{dz} = \mp\alpha_1 \pm gP_0 \quad (1b)$$

where the subscripts represent pump(0) and Stokes wave(1). The superscripts denote forward(+) and backward(-) propagating beams. λ_i is the wavelengths of pump(i=0) and Stokes radiation(i=1) and α_i is the loss coefficient of Raman fiber at wavelength λ_i. The parameter g refers to the Raman gain coefficient in unit $W^{-1}m^{-1}$.

At $z = 0$ and $z = L$, Eq.(1a)-(1b) meet the boundary conditions

$$\begin{cases} P_0(0) = P_{in}' \\ P_1^+(0) = R_1^0 P_1^-(0), P_1^-(L) = R_1^L P_1^+(L) \end{cases} \quad (2)$$

where $P_{in}' = 10^{-0.1\delta_F - 0.1\delta_s}P_{in}$, $R_1^0 = 10^{-0.2\delta_s}R_1$, $R_1^L = 10^{-0.2\delta_s}R_2$. R_1 and R_2 are the reflectivity of FBG1 and FBG2, respectively. δ_s and δ_F (in dB) are the splicing and insert losses of FBGs, respectively.

Eq.(1a)-(1b) can be rewritten as follows by introducing new variables $u_0 = \ln\left(P_0 / P_{in}'\right)$ and $u_1 = \ln(P_1^+ / P_1^-)/2$

$$\frac{du_0}{dz} = -\alpha_0 - 2g\frac{\lambda_1}{\lambda_0}\sqrt{c_1}\cosh(u_1) \quad (3a)$$

$$\frac{du_1}{dz} = -\alpha_1 + gP_{in}'\exp(u_0) \quad (3b)$$

where $\sqrt{c_1} = \sqrt{P_1^+P_1^-}$ is an invariable on the coordinates z[12]. One can refer to $\sqrt{c_1}$ as

978-0-9789217-3-6/07/$25.00

©2007 WEN GLOBAL SOLUTIONS

the geometric mean power for the Stokes wave and u_i as the gain factors for the pump and Stokes wave. Thus, Eq.(3a)-(3b) represent the evolvement of the gain factors along the fiber with undetermined constant $\sqrt{c_1}$. The boundary conditions for the new variables u_i are

$$\begin{cases} u_0(0)=0 \\ u_1(0)=\frac{1}{2}\ln R_1^0, u_1(L)=-\frac{1}{2}\ln R_1^L \end{cases} \quad (4)$$

The steady-state conditions for laser oscillation can be obtained by integrating (3a)-(3b) from z=0 to z=L

$$2g\sqrt{c_1}\lambda_1 L l_1^{eff} = \lambda_0(\ln P_{in}' - \ln P_r - \alpha_0 L) \quad (5a)$$

$$gP_{in}' L l_0^{eff} = \delta_1 \quad (5b)$$

where $\quad l_0^{eff}=\frac{1}{L}\int_0^L e^{u_0}dz \quad$ and

$l_1^{eff}=\frac{1}{L}\int_0^L \cosh(u_1)dz$ are referred to as the normalized effective fiber length for the pump and Stokes wave, respectively. $\delta_1 = \alpha_1 L - \ln(R_1^0 R_1^L)/2$ is the single-pass loss factor for the Stokes wave. P_r is the residual pump power at $z=L$.

From Eq.(3a)-(3b), the following relation can be deduced by eliminating z and integrating

$$\sqrt{c_1}=\frac{\lambda_0}{\lambda_1}\frac{g(P_{in}'-P_r)-\alpha_1 \ln \frac{P_{in}'}{P_r} - \alpha_0[u_1(L)-u_1(0)]}{2g[\sinh(u_1(L))-\sinh(u_1(0))]} \quad (6)$$

The equation states the fact that the number of input pump photons is equal to the total number of output pump and Stokes photons from the RFL plus photon number dissipated in cavity.

The threshold pump power can be derived from (5a) and (6) while considering $\sqrt{c_1}=0$.

$$P_{th}'=\frac{\alpha_0 \delta_1}{g(1-e^{-\alpha_0 L})} \quad (7)$$

Although pump beam attenuate faster while $P_{in}' > P_{th}'$, we can still assume pump beam propagate linearly along the fiber with a larger attenuate constant(A_0) than α_0. The approximation, which has proved to be valid to the second-order RFL [8], leads to

$$l_0^{eff}=\left(1-e^{-A_0 L}\right)/\left(A_0 L\right) \quad (8)$$

Substituting (8) into (5b), we can obtain the A_0 as a function of input power

$$A_0=\left[\gamma+W_0\left(-\gamma e^{-\gamma}\right)\right]/L \quad (9)$$

where $\gamma = P_{in}'/[\delta_1/(gL)]$. W_0 is the principal branch of Lambert W function. The residual pump power at z=L is

$$P_r = P_{in}' e^{-A_0 L} \quad (10)$$

According (10), one can determine whether pump power is depleted or not by comparing P_{in}' and $\delta_1/(gL)$. If $P_{in}' \gg \delta_1/(gL)$, then $P_r \ll P_{in}'$, i. e., pump power is depleted.

Substituting (6) and (10) into 5(a), we can express the l_1^{eff} using the boundary conditions

$$l_1^{eff}=\frac{\sinh(u_1(L))-\sinh(u_1(0))}{u_1(L)-u_1(0)} \quad (11)$$

Thus $\sqrt{c_1}$ and the output of Stoke wave can be explicitly expressed as follow from (5a)

$$\sqrt{c_1}=\frac{\lambda_0}{\lambda_1}\frac{1}{2l_1^{eff}}\frac{1}{g}\left(A_0-\alpha_0\right) \quad (12)$$

$$P_1^{out}=2\eta_{out}\sqrt{c_1}\sinh(u_1(L)) \quad (13)$$

where $\eta_{out}=10^{-0.1\delta_F-0.1\delta_s}$. In pump-depleted approximation, $\sqrt{c_1}$ and P_1^{out} is dependent linearly on input pump power and the Slope efficiency is

$$\eta_{slope}=\eta_{out}\frac{\lambda_0}{\lambda_1}\frac{1}{l_1^{eff}}\frac{1}{\delta_1}\sinh(u_1(L)) \quad (14)$$

The typical parameters of P-doped fiber fabricated by Fiber Optic Research Center of Russia are selected for calculation: g=1.28×10⁻³W⁻¹m⁻¹, λ_0=1.06μm, λ_1=1.24μm, α_0=1.8dB/km, α_1=1.16dB/km. The splicing loss is 0.02dB and insert loss of FBG is 0.1dB. Any spectral broadening for pump and Stokes waves are not considered in the classic model depicted by Eq.(3a)-(3b). However, the RFL always suffer from the spectral broadening effect. While considering this effect, the reflectivity of FBGs should refers to its effective reflectivity[13]. For convenience, the effective reflectivity of FBG1 is assumed to be 95%.

Figure 2 shows the output pump and Stokes power as a function of input pump power while L=500m, R₁=30%. The results of numerical simulation are also plotted as a comparison. The discrepancy between analytical results and numerical simulation is less than 1.5% up to P_{in}=10W. When P_{in}>5W, the slope efficiency equals to a constant 65%, which is in excellent agreement with the value calculated from Eq.(14).

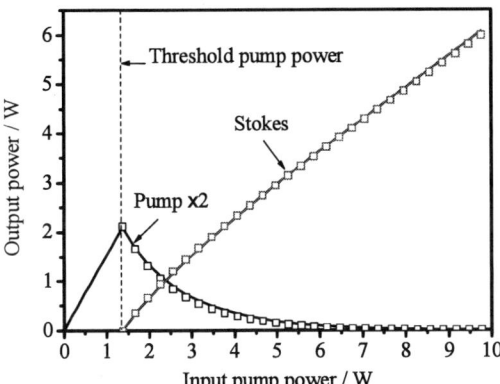

Fig. 2. *The output power of the Stokes wave and residual pump power versus input pump power (L=500m, R_2=30%, squares: numerical; lines: analytic).*

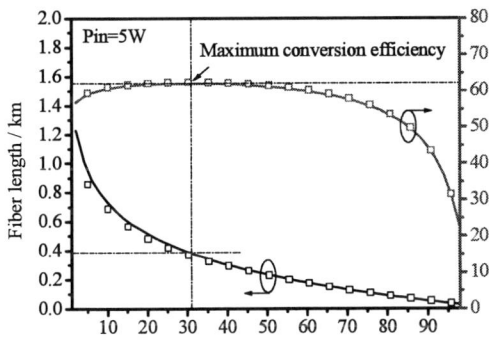

Fig. 3. *The optimal fiber length and conversion efficiency versus R_2 (P_{in}=5W, squares: numerical; lines: analytical).*

Let $\partial P_1^{out}/\partial L = 0$, we can readily deduce the optimal fiber length as a function of pump power and reflectivity of output FBG

$$L = \frac{1}{2}\ln\frac{1}{R_1^0 R_1^L}\ln\frac{P_{in}'g}{\alpha_1}\left/\left(P_{in}'g - \alpha_1 - \alpha_1\ln\frac{P_{in}'g}{\alpha_1}\right)\right. \quad (15)$$

Using (15) and (13), we plot the optimal fiber length and conversion efficiency (P_1^{out}/P_{in}) as the functions of R_2 when P_{in}=5W in fig.3. The numerical optimal results are also plotted as a comparison with analytical results in this figure. The

results agree well with the numerical optimal results. From this figure, we can find that the conversion efficiency is maximized (about 62.2%) when L=385m and R_2=31.5%.

Conclusion

With the assistance of Lambert W function, we propose an explicit analytic solution for the Raman fiber laser without using the depleted-pump approximation. The proposed solution shows good agreement with numerical simulation. According the solution, the depleted-pump approximation is valid only while $P_{in}' \gg \delta_1/(gL)$. The optimal design of the RFL is discussed using the proposed solution. The optimal fiber length and reflectivity of output FBG are obtained analytically.

Acknowledgements

The work is partially supported by the Innovation Fund of Xiamen University (XDKJCX20041003).

References

1 N. S. Kim et al Opt.Comm.,176(2000) 219

2 E. M. Dianov et al IEEE Journal of Selected Topics in Quantum Electronics, 6(2000)1022

3 M. Rini et al IEEE J.of Quant.Electro. , 36(2000)1117

4 S. Cierullies et al Opt.Comm., 217(2003) 233

5 I. A. Bufetov et al Quant.Electro., 30(2000) 873.

6 S. A. Babin et al Opt.Commun., 226(2003)329

7 B. Burgoyne et al J. Opt. Soc. Am. B, 22(2005)764

8 C.H. Huang et al Opti. Fiber. Technol., 13(2007)22

9 E. M. Wright, Bull. Amer. Math. Soc., 65(1959)89.

10 Banwell et al Electro. Lett., 36(2000) 291

11 Valluri et al Canad. J. Phys., 78(2000)823

12 J. AuYeung et al J.Opt. Soc. Am. B, 69(1979)803

13 J. C. Bouteiller, IEEE Photonic. Tech. Lett., 15(2003)1698

A New Technique for Side Pumping of Double-Clad Fiber Lasers

WANG Da-zheng(1), FENG Xiao-ming(1), WANG Yong-gang(1),
WANG Cui-luan(1), LAN Yong-sheng(1), LIU Su-ping(1), MA Xiao-yu(1)

1: Group of High Power Lasers, Institute of Semiconductors,
Chinese Academy of Sciences, Beijing 100083, China
Address: Institute of Semiconductors, Chinese Academy of Sciences,
P.O.Box 912, Beijing 100083
E-mail: dzhwang04@163.com

Abstract *We report a new technique, called SAP, for side pumping of double-clad, rare-earth-doped fiber lasers using fiber-coupled pump sources. The highest coupling efficiency can even exceed 92% in theory with this structure.*

Introduction

Rare-earth-doped fiber lasers are unique optical sources that offer a variety of advantages, particularly for practical applications[1]. In a double-clad (DC) fiber, the core is surrounded by a large, multimode (MM) inner cladding into which the pump light is launched, allowing the use of MM pump sources[2,3]. The advent of DC fiber has enabled rare-earth-doped fiber lasers to be scaled to high average powers[4] and pulse energies[5,6]. Although several methods have been developed to couple pump light into the inner cladding, pumping of DC fiber remains a significant issue for many important applications.

In this paper, we report a new technique for side pumping of double-clad, rare-earth-doped fiber lasers using fiber-coupled pump sources. In this technique, called "side-adhesive pumping" (SAP), the MM fiber is fixed to the DC fiber paralleled by index-matching adhesive. In order to keep light beam confined within the two optic fibers, the two parts are wrapped up together by another index-matching adhesive, gilded on the outside of the whole, or just bared in the air. The pump beam is launched by both reflections from the end of the MM fiber that is angle-polished and gilded and coupling from the side of the inner cladding. Through optimizing the polished-angle, the length contacted and choosing suitable index-matching adhesive, the highest coupling efficiency can even exceed 92% in theory with this structure. Additional advantages include no destruction of the inner-cladding, high coupling efficiency, little loss of power in coupling the pump beam, relatively low sensitivity to misalignment, simplicity (low parts count), compact and rugged packaging, scalability to high power, low cost and easy industrialization. Because of the large coupling area and low coupling power density at the junction of the two kinds of fibers, the technique is uniquely well solve the problem of power limitation due to the heat resistance of index-matching adhesive and suited to the direct coupling of the output of a high power LDs into DC fibers through index-matching adhesive.

Description of the technique

SAP has a simple structure. Fig. 1 (a) shows a schematic diagram of one embodiment of the method, and Fig. 1 (b) shows a three-dimensional view of the components.

We use the core of plastic-clad silica multimode fiber which is shucked off the cladding and jacket by means of chemical corrosion. The diameter of the core is 125um, and is quite suitable to joint the end with the pigtailed LD which is also 125um of the cladding diameter. Using the same method of chemical corrosion, we shuck off the outer cladding and jacket of the DC fiber and then clean the surface of the inner cladding by absolute alcohol and nitrogen blowing. We use precision finishing processing a small angle at the end of the MM fiber (In this way, the beam remains will be reflected by the angle-polished gilt face). Then we use appropriate index-matching adhesive to fix the MM fiber to the DC fiber side-by-side by index-matching adhesive.

The length of the joint should be high lighted, because it is one of the most critical factors that impact the efficiency of coupling. There are a few point of length at which the efficiency of coupling can reach maximum points. These length-points should be found out by theoretical analyses, analogue and laboratory experiments. The index-matching adhesive should be mentioned, since unsuitable index of the adhesive will badly reduce the efficiency of coupling or even stop the coupling.

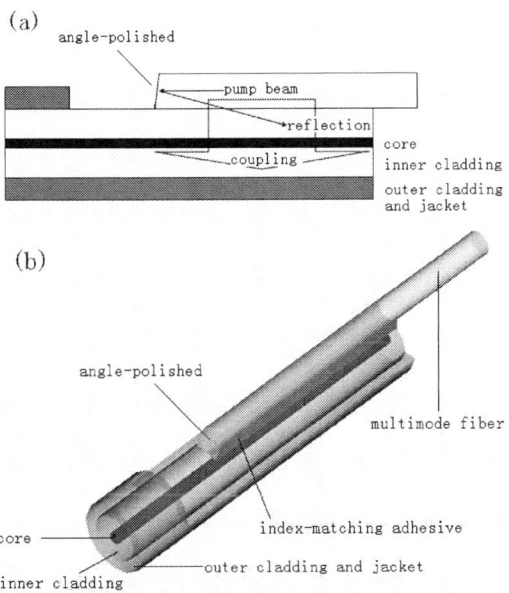

Fig.1 Schematic diagrams of this technique. (a) Cross-sectional side view showing the multimode fiber side-adhesive to the double-cladding fiber, the angle-polished end, the pump beam and the beam paths-- reflection and coupling. (b) Three-dimensional view of the SAP components.

In order to keep light beam confined within the two optic fibers, the two parts are wrapped up together by another index-matching adhesive, gilded on the outside of the whole, or just bared in the air (we just do the primary research by using this structure bared in the air as the first step of

the experiment). The pump beam is launched by both reflections from the end of the multimode fiber that is angle-polished and gilded, and coupling from the side of the inner cladding.

Through computing and optimizing the length contacted and choosing suitable index-matching adhesives, we can find the maximum of coupling efficiency.

Computing experiment and analysis
We establish a two-dimensional model, in which the 400um×1m rectangle stands for the inner cladding of DC fiber and 125um× 1m rectangle stands for the MM fiber. We compute the change of coupling efficiency with length contacted and find out the law.

As Fig.2 (1) shows to us, with a 1 meter contact length, there are three maximum points at the length around 0.2m, 0.7m and 0.9m respectively, especially at the length around 0.2m. The maximum coupling efficiency at 0.20m is up to 97.6%.

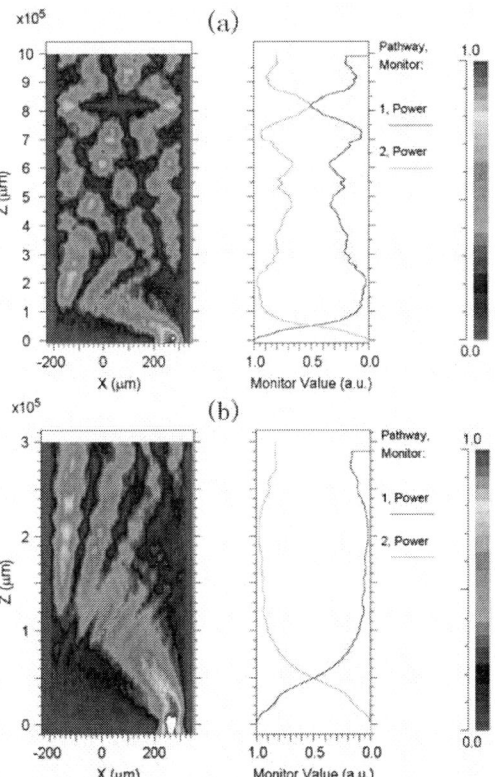

Fig.2 Graphs of two-dimensional computing model. (a) 1 meter length contacted model. (b) 30 cm length contacted model. The curves in blue stand for the normalized

power in MM fiber, and the curves in green stand for the normalized power in DC fiber. Pictures on the left side show the two-dimensional distributing of power in MM and DC fiber. The colourful bars on the right stand for the value of power.

In three-dimensional computing model, we focus on the length contacted below 30cm. From Fig.3, we can find out that at the length around 15cm, the coupling efficiency reach the maximum value, which is up to 92.5%.

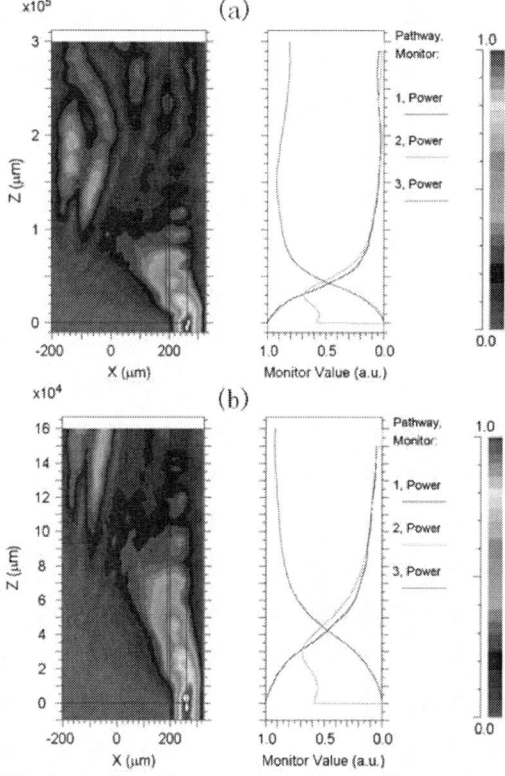

Fig.3 Graphs of three-dimensional computing model. (a) 30cm length contacted model. (b) 16cm length contacted model. The curves in blue stand for the normalized power in MM fiber. The curves in green stand for the normalized power in

index-matching adhesive. And the curves in red stand for the normalized power in DC fiber. Pictures on the left side show the three-dimensional distributing of power in MM, DC fiber and adhesive. The colourful bars on the right stand for the value of power.

Conclusions

We invent a new method, called SAP, for side pumping of double-clad, rare-earth-doped fiber lasers using fiber-coupled pump sources. Through computing just on coupling, the highest coupling efficiency exceed 92% in theory with this structure. If cooperated with reflection from the angle-polished face and gilt surface, this structure can provide higher pumping efficiency.

References

1 M. J. F. Digonnet, Rare Earth Doped Fiber Lasers and Amplifiers. New York: Marcel Dekker (1993).

2 E. Snitzer, H. Po, F. Hakimi, R. Tumminelli, and B. C. McCollum, Double-clad, offset core Nd fiber laser. Conf. Optical Fiber Sensors (1988), p. PD5.

3 L. Zenteno, High-power double-clad fiber lasers, IEEE J. Lightwave Technol., vol. 11 (1993), pp.1435-1446.

4 V. Dominic, S. MacCormack, R. Waarts, S. Sanders, S. Bicknese, R. Dohle, E. Wolak, P. S. Yeh, and E. Zucker, 110W fiber laser, Electron. Lett., vol. 35 (1999), pp. 1158-1160.

5 A. Galvanauskas, Mode-scalable fiber-based chirped pulse amplification systems, IEEE J. Select. Topics Quantum Electron., vol. 7 (2001), pp. 504-517.

6 F. Di Teodoro, J. P. Koplow, S. W. Moore, and D. A. V. Kliner, Diffraction-limited, 300-kW peak-power pulses from a coiled multimode fiber amplifier, Opt. Lett., vol. 27 (2002), pp. 518-520.

Design of a Doubly Grooved Binary Metallic Diffraction Grating For Efficient Side-Pumping of High-Power Fiber Lasers

Fan Zhang[1], Chuncan Wang, Zhi Tong, Geng Rui, Ning Tigang

1. Institute of Lightwave Technology, Beijing Jiaotong University, Beijing 100044, China

Abstract: A metallic binary grating with two grooves per period is applied to side-pump high-power fiber lasers for the first time. By combining a gradient algorithm and micro-genetic algorithm, the grating is optimized and the maximum coupling efficiencies of 94.77% and 77.27% for TM and TE polarization waves respectively are demonstrated.

Introduction

The advent of DCF has enabled rare-earth-doped fiber lasers and amplifiers to be scaled to high average powers and pulse energies. Single DCF laser have already achieved hundred watts power and even kilowatt power through end pumping [1,2]. However, the end pumping scheme has several disadvantages. When using the end face pumping technique, the output power is limited due to pumping through maximal two end faces and both of the fiber ends are not free. Sophisticated and expensive beam-shaping techniques are necessary to couple high-power laser diode bars and laser diode stacks into a DCF. Besides, a single injection point on the fiber end facet can lead to thermal loading of the fiber and damage the coating. Therefore various methods have been developed for launching high power pump lights into the DCF through its side, such as the V-groove side-coupling technique [3], the embedded-mirror method[4], the fiber angle-polished method[5], the fusing taper fiber method[6] and multi-fiber arrangements technique[7]. However, these techniques are quite delicate and complicated and imply the risk of damaging the core. R. Herda et al. [8] introduced a highly efficient side-coupling into multimode DCF by using a binary reflection grating which is not written into the fiber surface, but placed behind the fiber without a modification of the pump core itself. However, This method has a main drawback that the structure of this kind of gold diffraction grating is polarization-dependent and a high diffraction efficiency can be achieved only for TM polarization waves. Therefore, a doubly grooved binary metallic diffraction grating is introduced for side pumping high power fiber lasers.

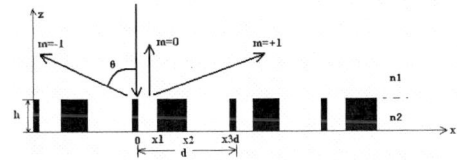

Figure.1 A metallic binary grating with two grooves per period, defined by different line-widths x1, x2, x3 and the same depth h.

Figure.2 The experiment setup of side-pumping high-power fiber lasers by using doubly grooved binary metallic diffraction gratings

Grating Design and Analysis

The grating illustrated in Fig.1 consists of the periodically modulated rectangular grooves with same thickness h and two grooves per period d in the range of the pump wavelength[9], defined by the transition points x1, x2 and x3. The gold is chosen as the material of metal grating and its complex refractive index is 0.09-6.12i [8] when the pump wavelength λ_p is 976nm. And the period of the grating, d is chosen between λ/n_1 and $2\lambda/n_1$ so that only orders m=-1, m=0 and m=+1 can propagate.

The basic idea of this side-pump scheme is shown in Fig.2; the doubly-grooved metal (gold) reflection grating is fabricated and placed behind the inner-clad of Yb^{3+}-doped DCF without a modification of the pump core itself. The incident pump lights of

---
[1] zwenjunf@163.com   phone:86-01-5168-3834

978-0-9789217-3-6/07/$25.00
©2007 WEN GLOBAL SOLUTIONS

a high-power diode bar collimated by a cylindrical lens are focused through the DCF on the metal reflection grating. Between the fiber and the grating is an index matching substance. Pump lights will be guided in the DCF if its angle θ with respect to the interface coating-core is smaller than the critical angle of total internal reflection θ_{max}. Therefore the diffraction angles $\varphi=\pi/2-\theta$ in the ±1 orders have to be larger than $\pi/2-\theta_{max}$, to be guided in the fiber. Thus a doubly-grooved metal grating with a diffraction angle φ with high efficiencies in the m=±1 orders has to be designed. Because the grating period is in the range of the wavelength a rigorous electromagnetic theory [10] which considers the vectorial character of the light has to be used.

Optimized Results

The maximum coupling efficiencies η_{max} and the corresponding optimal grating parameters (x1, x2, and x3, h) are presented in Table.1.

Table 1 Grating profiles (transition points x1, x2 and x3) and diffraction characteristics (diffraction efficiencies η_{-1}, η_0, η_{+1}, $\eta_{coupling}$) of metallic reflection gratings with two grooves per period optimized for TE- and TM-polarized are presented .

Parameter	TE	TM
x1	0.2471	0.2286
x2	0.3053	0.419
x3	0.6244	0.6287
h	0.5825	0.1159
η_{-1}	0.3860	0.4881
η_0	0.1225	0.0044
η_{+1}	0.3867	0.4596
$\eta_{coupling}$	0.7727	0.9477

Table.1 shows that the maximum coupling efficiencies of pump lights with TE- and TM-polarized are 94.77% and 77.27% respectively by using a doubly-grooved metal reflection grating. The optimized results are better than that of the binary metal grating for side-pumping DCF in [8].

Fabrication Tolerance Analysis

We assume that the grating is fabricated by electron-beam (EB) lithography and the minimum feature size of grating $\Delta x/\lambda=0.05$ is close to the EB spot diameter w [9]. The effect of fabricated errors γ of   the three critical design parameters h/λ, x1/λ and x2/λ are calculated respectively when the narrowest feature x3=d-Δx has been fabricated correctly.

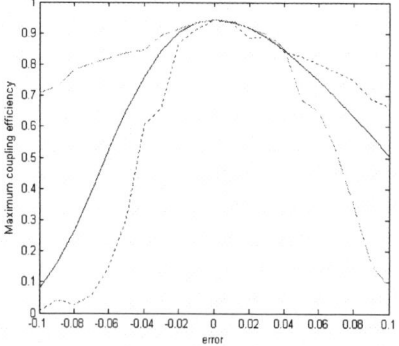

Figure.3 Effect of fabrication error γ for maximum coupling efficiency of h/λ (solid curve),x1/λ (dashed curve) and x2/λ (dotted curve) respectively.

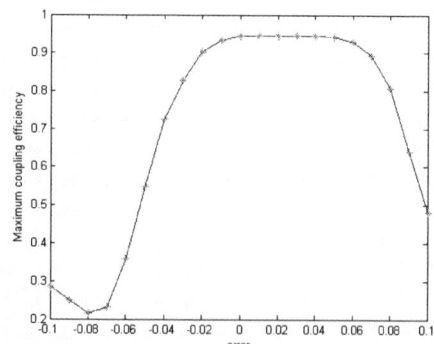

Figure.4 Effect of fabrication error γ for maximum coupling efficiency of x1/λ (solid curve) and x2/λ (star curve) when the pulse width w is correct.

It is clear from Fig.4, that the fabricated error tolerances are strict and almost same for parameters x1/λand x2/λ, and a little loose for h/λ. This could be expected by physical reasoning because the positions of the transition points are directly linked to the breaking of the inversion symmetry of the design [9]. Also, we assume that the spot diameter w of the electron-beam can be controlled accurately and let w=x2-x1. Therefore, the effects of x1 and x2 positional errors on the maximal coupling efficiency η are calculated when the w is fixed in Fig.4. Fig.4 shows that the variation ranges of the maximal coupling efficiency η are almost same when the coordinate positions of x1 and x2 have ±0.1λ error respectively under the condition that the spot diameter w of the electron-beam is controlled accurately. And when -0.02<x1/λ,x2/λ<0.07, the maximal coupling efficiency can be retained above 90%. According to Fig.3 and 4, it is found that the depth of the grooves is the most important parameter when the spot diameter of the beam pulse is assured to be a constant during the process of the fabrication and the maximal coupling efficiency more than 90% can be reached as long as the fabrication error of the groove depth h is limited in the range of ±0.02λ.

Power Handling

The power handling of the grating coupler is decided by the thermo-physical properties of the grating material. For the doubly-grooved metal grating, it is fabricated and placed behind the inner-clad of Yb^{3+}-doped DCF by an optical adhesive, whose refractive index matches to that of the inner clad of DCF. Therefore, heat-resistant-property of the optical adhesive decides the power handling of the grating coupler. Fluorinated epoxy resin adhesives have showed excellent refractive index (controllable from 1.45 to 1.59) matching to the inner-clad of DCF and further characteristics include high adhesion strength and strong thermal resistance due to the effect of fluorinated substituents [11]. The laser diode bars with output power of 10-20w can be applied when the grating coupler is placed behind the inner-clad of DCF by fluorinated epoxy resin adhesive.

Conclusion

A novel side-pump scheme by using a doubly grooved metal binary reflection grating is presented. This scheme has a good conservation of brightness of pump lights because they don't need any intervening optical element. Besides, they have no obstruction at both of fiber end, which good for further applications of fiber lasers. And the optimized results show that the maximum side-coupling efficiencies of pump lights with TE- and TM-polarized are 94.77% and 77.27% respectively by using a doubly-grooved metal reflection grating.

References

[1] Y. Jeong, J.K. Sahu, S. Baek, C. Alegria, D.B. Soh, C. Codemand, J. Nilsson, Opt. Commun. 234 (2004) 315.

[2] Y. Jeong, J.K. Sahu, D.N. Payne, J. Nilsson, Opt. Express. 12 (2004) 6088.

[3] D.J. Rippin, L. Goldberg, Electron. Lett. 31 (1995) 2204.

[4] J.P. Koplow, S.W. Moore, A.V. Kliner, Electron. Lett. 39 (2003) 529.

[5] J. Xu, J. Lu, G. Kuma, J. Lu, K. Ueda, Opt. Commun. 220 (2003) 389.

[6] G.P. Valentin,G. Samartsev, US. Patent. (1999) 5,999,673.

[7] A.B. Grudinin, P.A. Borisovich, T.D. Neil, N.P. William, Z.L. Johan, I.M. Nickolaos, D. Morten, K. Michael, US Patent. (2000) 6,826,335.

[8] R.Herda, A.Liem, B.Schnabel, A.Drauschke, H.-J Fuchs, E.-B Kley, H. Zellmer, A. Tuennermann,

Electron. Lett. 39 (2003) 276.

[9] J. Saarinen, E. Noponen, J. Turunen, Appl. Opt. 33 (1995) 2401.

[10] J. Turune, F. Wyrowski Diffraction theory of microrelief gratings in: Taylor& Francis (Ed.), London, 1997.

[11] N. Kouzaburou, M. Tohru, Rev. Eletr. Commun. Lab. 35 (1987) 253.

Combined FEC/ SOP Scrambling with Delay Line PMD Mitigation Scheme

J. Ferreira (1), J. P. von der Weid (2)

1 : Inmetro – National Institute of Metrology, Standardization and Industrial Quality – Av. Nossa Sra. Das GRaças, 50 – Xerém, Duque de Caxias, Brazil - jferreira@inmetro.gov.br

2 : Center for Telecommunications Studies – Pontifical Catholic University of Rio de Janeiro Marquês de S. Vicente 225 – Rio de Janeiro 22453-900, Brazil - vdweid@cetuc.puc-rio.br

Abstract *Based on simulations of the time evollution of PMD-related variables a PMD mitigation technique combining FEC and polarization scrambling with DGD control is presented. An improvement of 40% for the robustness to PMD was obtained*

Introduction

Polarization mode dispersion (PMD) is a dispersive effect in optical fibers and devices that occurs when a lightwave signal splits into two delayed signals according to their polarization state. A number of different known techniques have been proposed and developed to overcome the residual dispersion due to PMD. Addition of a polarization controlled delay line [1], and forward error correction (FEC) combined with input SOP scrambling [2] are popular examples of PMD mitigation methods. Among these, FEC has proven to be a very robust and simple technique which can be used in multichannel transmissions as well.

PMD simulations normally make use of the frequency-time equivalency to predict the statistical behavior of the polarization mode delay (DGD) and related quantities in single mode fibers. Although very efficient, these methods provide only the overall statistical behavior, correlations, probability distributions (pdf) etc, but no information or simulation on the actual evolution of quantities such as polarization states (SOP), differential group delay (DGD) or principal states of polarization (PSP). Measurements of PMD and related quantities in installed fibers have been object of increasing interest [3], [4]. Simulations of the time evolution of DGD, signal distortion, polarization states and any PMD-affected variable help in understanding or predicting polarization effects in optical fibers and are very useful in the development of polarization control or PMD mitigation algorithms. A dynamic model based on random temperature

fluctuations was previously proposed and tested by comparison with experimental measurements [5]. Different dynamic models were also proposed for this purpose, either considering random variable coupling angles between fiber sections [6] or an Ornstein-Uhlenbeck random walk process [7]. Here we use our temperature based simulation tool on the temporal evolution of PMD variables to propose and test a PMD mitigation scheme for high bit-rate transmissions, combining the well known fast polarization scrambling and forward error correction technique [8] with the 1st order polarization delay line mitigation scheme [9].

Simulations

To simulate a system that evolves in temporal steps, the fiber is described as a random concatenation of N fiber segments each one with fixed length h_n and birefringence axes. Each segment has a given initial birefringence $b_n(0)$ and is coupled to the following one by a fixed rotation matrix whose angle α_n is random from segment to segment. Temperature is the main driver for the evolution of PMD variables. Birefringence acts on these variables always through the product length x birefringence so that length variations were neglected and all the effect of the temperature was included in birefringence variations. We assumed the temperature evolving randomly along the fiber length, increasing or decreasing with time, such as what usually happens in a real installed fiber. Using the Jones matrix formalism, the Jones matrix $T(\omega, t)$ describing the fiber at time t and optical frequency ω can be

978-0-9789217-3-6/07/$25.00

©2007 WEN GLOBAL SOLUTIONS

calculated by N successive products of two matrices [5]:

$$T(\omega,t) = \prod_{n=1}^{N} B_n(\omega,t) R_n(\alpha)$$

$$= \prod_{n=1}^{N} \begin{bmatrix} e^{j[b_n(t)\omega h_n]/2} & 0 \\ 0 & e^{-j[b_n(t)\omega h_n]/2} \end{bmatrix} \begin{bmatrix} \cos\alpha_n & \sin\alpha_n \\ -\sin\alpha_n & \cos\alpha_n \end{bmatrix} \quad (1)$$

where $B_n(\omega,t)$ represents the birefringence matrix of the nth segment with length h_n. $R_n(\alpha_n)$ is the matrix of a rotator that represents the random coupling at each splice or mode coupling point.

The temporal evolution of the fiber was simulated by stepwise linear variations of the birefringence $b_n(t)$, the step δ_n for each segment being chosen at random within a limited birefringence range. This emulates different thermal variations for each section of the fiber [5].

$$b_n(t) = b_n(0) + \delta_n \cdot t \quad (2)$$

After a certain time a new random choice for the steps δ_n is done, always with positive or negative values, so that the overall variation of the birefringence is bounded and has zero mean for an infinite sequence. In order to keep the section birefringence within reasonable values, the slope of the variation is inverted each time the birefringence differs 5% from the initial value. This roughly corresponds to the effect of ambient temperature variations within ~ -10 to 40 degrees Celsius [10]. Calculating the evolution of the Jones Matrix describing the fiber birefringence is obtained for each set $\{b_n(t)\}$, and a continuous evolution of the fiber DGD and PSP can thus be calculated, as well as the evolution of the output polarization state for any desired fiber PMD or signal wavelength. Once the PSP are determined, projected the output SOP in the PSP basis to calculate the received optic al signal level including the delay between the two PSP. Because different field situations may lead to very different time variations, the time scale of the model is adjusted to real situations by means of the autocorrelation function of any measured PMD-related variable, such as the polarization state or DGD.

PMD mitigation

Figure 1 displays the proposed PMD mitigation scheme. As in the FEC/scrambling method, Polarization scramblers are placed in the transmitter preventing the system from remaining at a bad situation for a long time [8]. At the receiver, a polarization controller is placed in series between the fiber link and a polarization delay line (DL). The delay line was chosen to be equal to the mean fiber PMD. After the delay line, part of the optical power is launched into a detector, giving a signal corresponding to the detected RF power for a feedback loop which drives the polarization controller in such a way that the mean RF signal is set to a maximum.

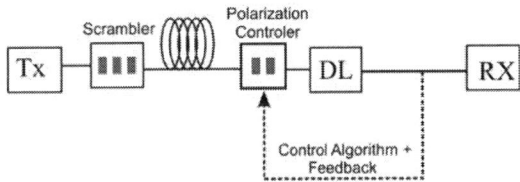

Figure 1 – PMD Mitigation scheme

When scrambling the SOP in a single channel transmission, the combination of the delay line and the link PMD in such a way as to maximize the received signal will give rise to a minimum total DGD as a unique solution because the input SOP is being scrambled. Hence PMD mitigation will be achieved by 1st order DGD compensation and FEC will account for remaining effects of residual 1st order, as well as 2nd or higher order PMD. In a multichannel transmission, the addition of the delay line will be adjusted to maximize the total RF power and minimize its peak-to-peak variation due to the scrambling frequency.

Results

Figure 2 displays a typical time evolution of the received RF power in a scrambled 10 Gb/s NRZ transmission as a function of time, with (our scheme) and without the DGD control scheme proposed. The RF power fluctuates very fast, as expected from a scrambling input SOP facing a time evolving PMD vector. The black covered area represents the RF power excursions between maxima and minima. It is clear that the received RF power drops are much deeper in the uncompensated mode. The input SOP scrambling minimizes the time during which the distortion is intense, thus avoiding burst errors that would limit the capacity of the FEC algorithm to regenerate the bit stream, so that only deep distortions are to be considered. Indeed, there are situations where the fiber DGD is small and the delay line DGD degrades somewhat the signal but the FEC algorithm keeps the system running. However, in deep distortion situations the addition of the delay line will always improve system performance.

The limits of the SOP scrambling method are given by the trade-off between the minimum frequency needed to avoid burst errors and the maximum acceptable frequency jitter induced by the scrambling frequency [11]. Power drops are strongly reduced so that burst errors are better controlled and the scheme is more robust to high PMD values. Jitter is also smaller in the combined FEC/SOP scrambling + DGD control scheme.

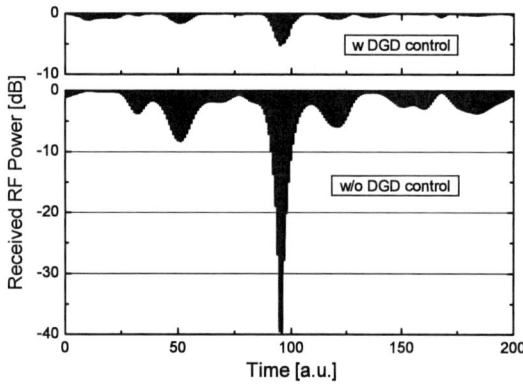

Figure 2 – Detected RF power with and without DGD control

Figure 3 presents the relative required OSNR (optical signal to noise ratio) at BER = 10^{-15} with only FEC/SOP scrambling and with added DGD control. Results from our simulation are compared to calculations performed with a diferent statistical model with only FEC/SOP scrambling [8]. The margin at low PMD is ~6.5 dB, given essentially by the FEC algorithm, regardless the SOP scrambling. As the mean DGD increases, the margin decreases but the added DGD control clearly improves the robustness of the scheme. At 2-dB penalty the PMD tolerance is increased from 0.24 T to 0.34 T, an improvement of 40%.

Conclusions

Based on time domain simulations of PMD-related variables in an optical fiber we propose a PMD mitigation scheme that combines two different techniques to improve the performance of PMD mitigation. Our results show an important increase of PMD tolerance when DGD control is added to the FEC/SOP-scrambling mitigation technique.

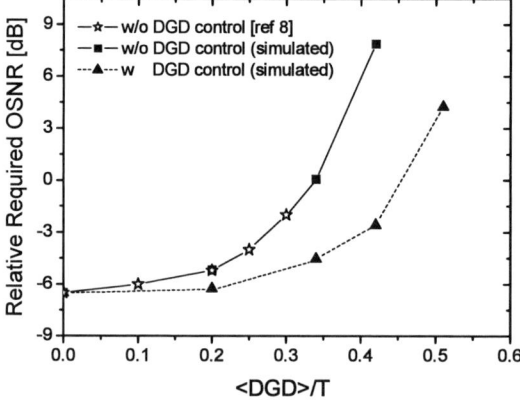

Figure 3 – Relative required OSNR at BER=10^{-15} vs PMD measured with only FEC/SOP-scrambling (stars) [8] and simulated (squares). Triangles represent the result for the combined FEC/SOP scrambling with DGD control

References

1 F. Heismann, et al, US Patent number: 5930414

2 B. Wedding and C. N. Haslach, Proc. OFC'2001, Vol 3, paper. WAA1-1, (2001)

3 M. Karlsson et al, J. Lightwave Tecnnol., Vol.18 (2000), page 941.

4 M. Brodsky et al, IEEE Photonics Technol. Lett., Vol. 16 (2004) page 209.

5 J. F. Macêdo and J. P. von der Weid, IEEE/LEOS Workshop of Fibres and Optical Passive Components, (2005), page 176

6 C. Xie et al, Proc. OFC'2006, paper OFL5, (2006)

7 C. Antonelli et al, Proc. OFC'2006, paper OWJ4, (2006)

8 X. Liu at al, Proc. OFC'2004, Vol 1 (2004), page 23

9 H. Sunnerud et al, J. Ligthwave Technology Vol 20 (2002), page 368.

10 M. J. Marrone at al, Journal of Lightwave Technology, LT-2(2) (1984), page 155.

11 Z. Li et al, Proceedings OFC´2004, Vol. 1 (2004), page 936.

Tailoring confinement losses of photonic crystal fibers

Lei Yao(1), Shuqin Lou(1), Hong Fang(1), Tieying Guo(1), Honglei Li(1), Shuisheng Jian(1)

1 : Key Lab of All Optical Network & Advanced Telecommunication Network, Beijing Jiaotong University, Beijing, 100044, China and email: yangrou1369ok@sohu.com

Abstract *Confinement losses in photonic crystal fibers (PCFs) with different structure are discussed by a full vectorial finite element method. Based on numerical results, confinement losses in high birefringence PCFs and high negative dispersion PCFs are tailored.*

Introduction

Since the first working example of photonic crystal fibers (PCFs) was reported in 1996[1], scientists have paid great attention to developing various kind of PCFs with novel properties such as endlessly single mode[2], high birefringence[3], flexible dispersion[4], and so on. Usually pure silica is adopted as a background material of core region and the outer cladding region in PCFs, so the guided modes are inherently leaky. To lessen confinement losses is important to design high performance PCFs. Through arranging different hole size in different ring, low confinement loss PCFs can be achieved[5][6]. In this paper, a full vectorial finite element method (FVFEM)[7] is used to analyze different micro-structured PCFs. Through arranging air hole array, hole pitchs Λ, the relative ratio of hole diameters to hole pitch d/Λ, confinement losses of different micro-structured PCFs are discussed. Based on numerical results, the confinement losses of two kinds of PCFs, which are high negative dispersion PCF and high birefringence PCF, are tailored, respectively.

1. Methrodology

The main idea of FVFEM with TBC comes from reference [7]. Vector wave equation based on the magnetic field vector **H** can be derived from two curl equations of the Maxwell's equation. It is expressed as below:

$$\nabla \times \varepsilon_r \nabla \times \mathbf{H} = k_0^2 \mathbf{H} \qquad (1)$$

Considering the transverse components of the magnetic field, Using the Galerkin procedure and discretizing the computational domain into triangular elements, the following discretized weak equation can be deduced:

$$
\begin{aligned}
\sum_{\substack{Boundary\,Element}} & \left\{ -\int_{\Gamma_s} \frac{1}{n_{zz}^2} w_y (\partial_x H_y - \partial_y H_x) \mathrm{d}y - \int_{\Gamma_s} \frac{1}{n_{zz}^2} w_x (\partial_x H_y - \partial_y H_x) \mathrm{d}x \right. \\
& \left. -\int_{\Gamma_s} \frac{1}{n_{yy}^2} w_x (\partial_x H_x + \partial_y H_y) \mathrm{d}y + \int_{\Gamma_s} \frac{1}{n_{xx}^2} w_y (\partial_x H_x + \partial_y H_y) \mathrm{d}x \right\} \\
+ \sum_{\substack{Interface\,Element}} & \left\{ -\int_{\Gamma_{int,s}} \frac{1}{n_{yy}^2} w_x (\partial_x H_x + \partial_y H_y) \mathrm{d}y + \int_{\Gamma_{int,s}} \frac{1}{n_{xx}^2} w_y (\partial_x H_x + \partial_y H_y) \mathrm{d}x \right\} \\
+ \sum_{\substack{Triangular\,Element}} & \iint_{\Omega_s} \left\{ \frac{1}{n_{zz}^2} (\partial_x w_y - \partial_y w_x)(\partial_x H_y - \partial_y H_x) \right. \\
& + \left[\partial_x (\frac{1}{n_{yy}^2} w_x) + \partial_y (\frac{1}{n_{xx}^2} w_y) \right] (\partial_x H_x + \partial_y H_y) \\
& \left. + k_0^2 n_{eff}^2 \left(\frac{1}{n_{yy}^2} w_x H_x + \frac{1}{n_{xx}^2} w_y H_y \right) - k_0^2 (w_x H_x + w_y H_y) \right\} \mathrm{d}x\mathrm{d}y = 0 \qquad (2)
\end{aligned}
$$

The derivatives of the fields occurring in the boundary term in equation (2) will be handled through the BGT-like transparent boundary conditions (TBC) to reflect the properties of the fields in the exterior domain properly. Then the effective index can be obtained by calculating the equation above.

If the effective index is denoted by n_{eff}, the confinement loss α can be deduced from the imagine part of n_{eff} [8].

$$\alpha = \frac{20}{\ln(10)} \mathrm{Im}(k_0 n_{eff}) \quad dB/m \qquad (3)$$

Waveguide dispersion is expressed as below:

$$D \approx -\frac{\lambda}{c} \frac{d^2 \mathrm{Re}(n_{eff})}{d\lambda^2} \quad ps/nm/km \qquad (4)$$

For high birefringence PCF, the modal birefringence can be calculated by equation (5).

$$B = \mathrm{Re}(n_{eff}^x - n_{eff}^y) \qquad (5)$$

Where n_{eff}^x and n_{eff}^y represent the effective indices for the polarization modes of the same order.

978-0-9789217-3-6/07/$25.00

©2007 WEN GLOBAL SOLUTIONS

2. Numerical results

2.1 Relationship between confinement losses and the structure of PCFs

In this part, the index of silica is assumed to be 1.45, and the wavelength λ is 1.55 μm while calculating. The PCF studied as Fig.1.

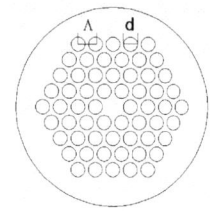

Fig.1 Cross section of triangular PCF

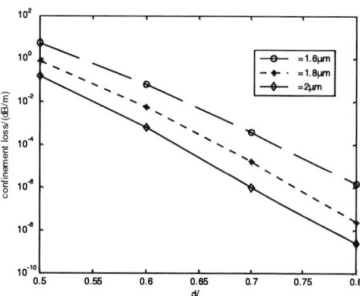

Fig.2 Relationship between confinement losses and d/Λ, with Λ are 1.6μm, 1.8μm and 2μm respectively.

To analyze the relationship between confinement losses and the arrangement of air holes in detail, each of the four rings air holes will be changed respectively with other rings fixed.

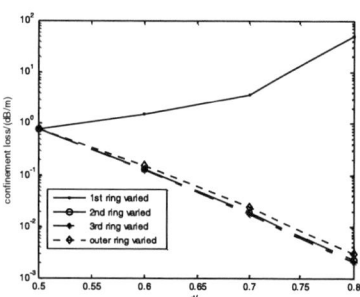

Fig.3 The varieties of confinement losses along with the each of the four rings air holes changed respectively

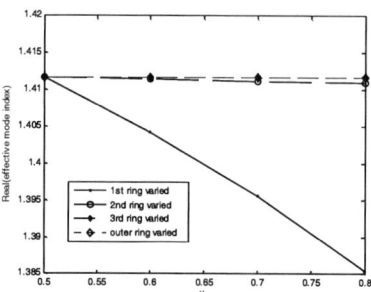

Fig.4 The varieties of the real part of the effective mode indices along with the four rings of air holes changed respectively

From the Fig.3 and Fig.4 above, it is seen that confinement losses declined sharply and the real part of the effective index are nearly unvaried through enlarging the most outer ring of air holes. And this helps us to maintain other anticipant characters of PCFs while reducing the confinement losses.

2.2 Tailoring high negative dispersion PCFs

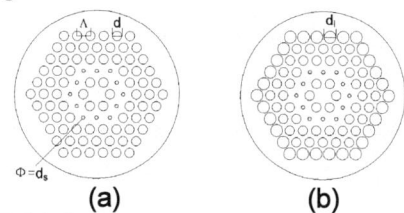

(a)   (b)

Fig.5 (a) is a schematic diagram of a high negative dispersion PCF with the same structure as reference [9] (d/Λ=0.7, ds/Λ=0.261 with Λ=2μm); (b) changes the diameter of the outer ring of air hole into dl/Λ=0.9

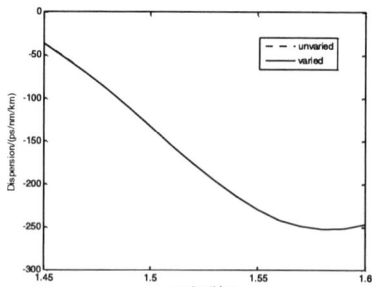

Fig.6 Comparison of the dispersion coefficient of PCFs in Fig.5

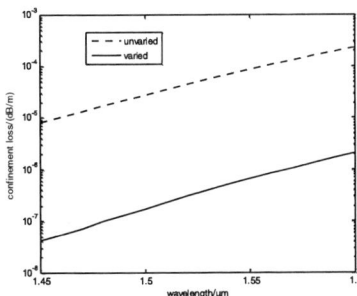

Fig.7 Comparison of the confinement loss of PCFs in Fig.5

The varieties from Fig.5 (a) to (b) lead to confinement losses declined by two orders of magnitude and dispersion coefficient unvaried.

2.3 Tailoring high birefringence PCFs

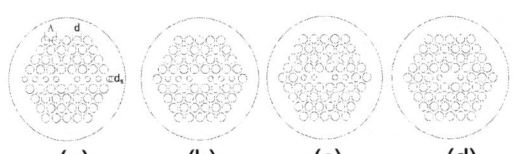

(a)　　(b)　　(c)　　(d)

Fig.7 (a) a kind of high birefringence PCF with a row of small holes located in the middle of it along x-axis. (b) to (d) enlarge the smaller holes from outside to inside gradually.

In Fig.7, The hole pitch Λ is 2μm, the air-hole diameter d is 1.6μm, the smaller air-hole diameter d_s is 1μm

	HE11x	HE11y
a	1.4043614-1.04E-12i	1.4019641-3.81E-14i
b	1.4043652-6.28E-14i	1.4019646-5.19E-15i
c	1.4043625-2.44E-15i	1.4019647-5.34E-16i
d	1.4043509-3.12E-16i	1.4019586-7.45E-17i

From the table above, the values of the birefringence of four kinds of PCFs are almostly the same, which is 2.4×10^{-3}. However, from (a) to (d), the confinement losses are reduced obviously.

Conclusions

It can be concluded that while the hole diameters of the most outer ring of PCFs

are enlarged, the confinement losses are reduced. As the same time, the other characters such as dispersion and birefringence are keeping unchanged. This conclusion will give us some good references for design some other novel PCFs with anticipant characters and very low confinement loss.

Acknowledge

This work is supported by the national science foundation of China and the fund of Beijing Jiaotong University.

References

1 Knight.J.C et al. Opt. Lett., 21(1996), 1547
2 T.A.Birk et al. Opt. Lett., 22(1997), 961~963
3 A.Ortigosa-Blanch et al. Opt. Lett., 25(2000), 1325~1327
4 A.Bjarklev, et al. ECOC98, 135~136.
5 Suzuki.K et al. Opt. Exp., 9(2001), 676
6 Yi Ni et al. IEEE. Photon. Technol. Lett., 16(2004), 1516
7 H.P.Uranus et al. Opt. Exp., 12(2004), 2795~2809
8 Olszewski.J et al. IEEE. Tran. Opt. Net., 1(2004), 354
9 Varshney.S.K et al. IEEE. Photon. Technol. Lett., 17(2005), 2062

Two-Stage Hermite-Gaussian Function Method with Perfectly Matched Layers for Analyzing Microstructured Optical Fibers

Xue-Wen Chen (1), Sailing He (1), Zhi Wang (2) and P.K.A Wai (3)

1 Centre for Optical and Electromagnetic Research, Zhejiang University, Zhijingang Campus, Hangzhou 310058, China. E-mails: xwchen@coer.zju.edu.cn sailing@zju.edu.cn
2 School of science, Beijing Jiaotong University, Beijing, China. E-mail: zhiwang@bjtu.edu.cn
3 Photonics Research Centre and Department of Electronic and Information Engineering, The Hong Kong Polytechnic University, Hong Kong, China. E-mail: enwai@polyu.edu.hk

Abstract *A Two-stage Hermite-Gaussian function expansion method incorporated with anisotropic perfectly matched layers (PMLs) is developed to efficiently calculate the complex propagation constant of a microstructured optical fiber (MOF).*

Introduction

Microstructured optical fibers (MOFs) have been the subject of intensive research due to their impressive properties, including endless single mode, high birefringence and high nonlinearity [1]. Since the guiding mechanism of an MOF is very different from that of a conventional fiber, developing an efficient and accurate mode solver is of considerable interest for many applications. Existing mode solvers, such as the finite difference method (FDM) [2], finite element method (FEM) [3], and multipole method [4] are useful numerical tools for designing MOFs for different kinds of applications. However, these methods have their own drawbacks. Both the FEM and the FDM are computationally intensive since one need to solve the eigenvalue problem of a general complex sparse matrix of large scale. Only a few eigenvalues (closest to a trial eigenvalue) and eigenvectors are calculated to reduce the computation time. The multipole method [4] now can only treat MOFs with circular or elliptical inclusions and becomes very inefficient when the total number of inclusions is large.

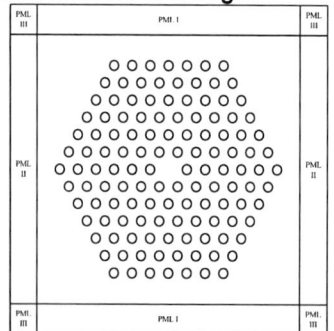

Fig. 1 An MOF supercell with PMLs

The Hermite-Gaussian (H-G) function expansion method [5] is an efficient method for analyzing the modal profile of an MOF with arbitrary transverse geometry. In the H-G method, a few expansion terms suffice and thus the dimension of the eigen system is much smaller than that in the FEM or FDM. However, the standard H-G method can only be used to calculate the real part of the complex propagation constant. In this paper, we propose an extended H-G method which can calculate the complex propagation constant and retain all the advantages of the standard H-G method. The extended H-G method includes the anisotropic perfectly matched layers (PMLs) to calculate the confinement loss of an MOF of arbitrary structure. A two-stage implementation is proposed to improve considerably the convergence and accuracy of the H-G method.

Theoretical Formulation

To calculate the confinement loss, one can enclose the open waveguide structure with an artificial PML [6] which is treated as an anisotropic medium and can absorb any outgoing wave without reflection. As shown in Fig. 1, we construct a supercell containing an MOF structure and the anisotropic PMLs. The full-vectorial wave equation in terms of the magnetic field $\vec{h}$ in the supercell can be expressed as:

$$\nabla \times \left(\frac{1}{\varepsilon_r \vec{\varepsilon}} \nabla \times \vec{h} \right) - k_0^2 \mu_r \vec{\mu} \vec{h} = 0 , \qquad (1)$$

Where k_0 is the free space wavenumber and ε_r, μ_r are the scalar relative permittivity and permeability. The permittivity and

978-0-9789217-3-6/07/$25.00

©2007 WEN GLOBAL SOLUTIONS

permeability tensors are given by

$$\ddot{\varepsilon}=\ddot{\mu}=\begin{bmatrix} s_y/s_x & 0 & 0 \\ 0 & s_x/s_y & 0 \\ 0 & 0 & s_x s_y \end{bmatrix}=\frac{1}{\varepsilon_r}\begin{bmatrix} \varepsilon_x & 0 & 0 \\ 0 & \varepsilon_y & 0 \\ 0 & 0 & \varepsilon_z \end{bmatrix}=\frac{1}{\mu_r}\begin{bmatrix} \mu_x & 0 & 0 \\ 0 & \mu_y & 0 \\ 0 & 0 & \mu_z \end{bmatrix} \quad (2)$$

The values of s_x and s_y are summarized in Table I with the parameter s defined as

$$s = 1 + j\sigma/\omega\varepsilon_0 \quad (3)$$

where ω is the angular frequency and σ is the conductivity which is set to be a constant. By assuming an $\exp(-j\beta z)$ dependence for the field and eliminating h_z (in terms of h_x and h_y), one could obtain the following wave equation for h_x

$$\left(\frac{\varepsilon_y}{\varepsilon_z}\frac{\partial^2}{\partial y^2}+\frac{\varepsilon_x}{\varepsilon_z}\frac{\partial^2}{\partial x^2}+\frac{\varepsilon_y}{\varepsilon_z}\frac{\partial \ln \varepsilon_z}{\partial y}\frac{\partial}{\partial y}+k_0^2\varepsilon_x\mu_y\right)h_x-\left(\frac{\varepsilon_y}{\varepsilon_z}\frac{\partial \ln \varepsilon_z}{\partial y}\frac{\partial}{\partial x}\right)h_y=\beta^2 h_x \quad (4)$$

The wave equation for h_y can be obtained similarly. By introducing $f_x=\varepsilon_x/\varepsilon_z-1$ and $f_y=\varepsilon_y/\varepsilon_z-1$, Eq. (4) can be re-arranged as,

$$\left(\frac{\partial^2}{\partial y^2}+\frac{\partial^2}{\partial x^2}+\frac{\partial \ln \varepsilon_z}{\partial y}\frac{\partial}{\partial y}+k_0^2\varepsilon_x\mu_y\right)h_x-\left(\frac{\partial \ln \varepsilon_z}{\partial y}\frac{\partial}{\partial x}\right)h_y+\left(f_y\frac{\partial^2}{\partial y^2}+f_x\frac{\partial^2}{\partial x^2}\right)h_x=\beta^2 h_x \quad (5)$$

The values of f_x and f_y in different regions of a supercell are shown in Table I. One can see that the third term on the left side of Eq. (5) is nonzero only in the PML region and is independent of the dielectric profile of the MOF. This term is essential to determine the confinement loss.

Table 1 Values of s_x, s_y, f_x and f_y at different regions of the supercell

	PML I	PML II	PML III	Other area
s_x	1	s	s	1
s_y	s	1	s	1
f_x	0	$1/s^2-1$	$1/s^2-1$	0
f_y	$1/s^2-1$	$1/s^2-1$	$1/s^2-1$	0

Two-Stage Implementation
To solve Eq. (5) and its counterpart for h_y, we employ the H-G function method described in [5]. The magnetic field is expanded in terms of H-G functions:

$$h_x(x,y)=\sum_{a,b=0}^{N-1}\varepsilon_{ab}^x\psi_a(x)\psi_b(y) \quad (6)$$

$$h_y(x,y)=\sum_{a,b=0}^{N-1}\varepsilon_{ab}^y\psi_a(x)\psi_b(y) \quad (7)$$

where the H-G function is defined by

$$\psi_k(s)=2^{-k/2}\pi^{-1/4}(k!W)^{-1/2}\exp(-s^2/2W^2)H_k(s/W) \quad (8)$$

Here W is the intrinsic width of the H-G function and $H_k(x)$ is the Hermite polynomial of the k-th order. The dielectric

function is expanded in terms of Fourier series. Eq. (5) and its counterpart for h_y can be transformed into algebra equations:

$$\begin{bmatrix} I_{abcd}^{(1)}+k^2 I_{abcd}^{(2)}+I_{abcd}^{(3x)}+I_{abcd}^{PML} & I_{abcd}^{(4x)} \\ I_{abcd}^{(4y)} & I_{abcd}^{(1)}+k^2 I_{abcd}^{(2)}+I_{abcd}^{(3y)}+I_{abcd}^{PML} \end{bmatrix}\begin{bmatrix} \varepsilon^x \\ \varepsilon^y \end{bmatrix}=\beta^2\begin{bmatrix} \varepsilon^x \\ \varepsilon^y \end{bmatrix} \quad (9)$$

where I_{abcd}^{PML} can be evaluated by

$$I_{abcd}^{PML}=\iint_\infty \psi_a(x)\psi_b(y)\left[f_y\left(\frac{\partial^2\psi_d(y)}{\partial y^2}\psi_c(x)\right)+f_x\left(\frac{\partial^2\psi_c(x)}{\partial x^2}\psi_d(y)\right)\right]dxdy \quad (10)$$

$$=\left(\frac{1}{s^2}-1\right)\left\{\iint_{I+III}\psi_a(x)\psi_b(y)\frac{\partial^2\psi_d(y)}{\partial y^2}\psi_c(x)dxdy+\iint_{II+III}\psi_a(x)\psi_b(y)\frac{\partial^2\psi_c(x)}{\partial x^2}\psi_d(y)dxdy\right\}$$

Equation(10) can be calculated analytically [7]. The other matrix elements in Eq. (9), also can be obtained analytically [5]. Equation (9) is a standard matrix eigenvalue problem and can be solved using existing algorithms. However, the result of this single-stage implementation is usually unsatisfactory. We observe that the intrinsic width W is an important parameter and significantly influences the accuracy of the H-G function expansion method. Therefore, an optimal width W is desirable to improve the convergence and accuracy. We propose a two-stage implementation. In the first stage, we choose a large value W_0 for the intrinsic widths (W_x and W_y) of the H-G functions and calculate the field profile of the mode interested. Then the calculated dominant component (h_x or h_y) of the modal field is used to determine the optimal values (through a simple scanning search) for intrinsic widths W_x and W_y by maximizing the overlap integration between the calculated dominant component of the modal field (determined with W_0) and the first few order expansion functions with intrinsic widths W_x and W_y. We find the optimal widths are not sensitive to W_0. Thus only a little (or very rough) prior knowledge of the value W_0 is needed. In the second stage, the H-G functions with the optimal widths (W_x and W_y) are used for the expansion, and the complex propagation constant is calculated.

Numerical Results and Discussion
In this section, we apply the proposed two-stage H-G method to analyze the modal properties of some MOFs. To validate our method, we analyze the same MOF considered in Ref. [4], i.e. a six-hole MOF with hole radius of 1.13 µm and a hole spacing of 4.0 µm.

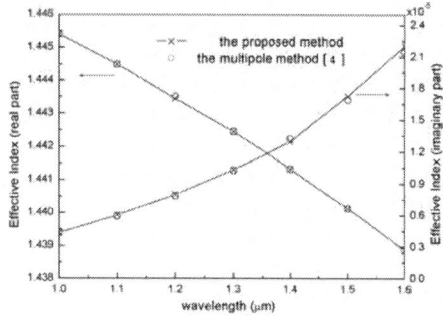

Fig. 2 Comparison of the effective indices calculated with the proposed method and the multipole method

Fig. 2 shows the effective indices obtained by the multipole method (indicated with the circles) and the present method (indicated with the crosses; 20 expansion terms are used). One sees clearly a good agreement between the two methods.

To illustrate the efficiency of the present two-stage H-G method, we study a more practical MOF with six air-hole rings as shown in Fig.1 at the wavelength of 1.3μm. The hole spacing and radius are 2.3μm and 0.5μm, respectively. The optimal values of the intrinsic widths of the H-G functions for the present MOF are found to be W_x =1.1μm and W_y =1.08μm, respectively. Fig. 3 show the calculated field profiles for a y-polarized fundamental mode. The CPU time for analyzing the present MOF (20 expansion terms are used) is 135 seconds on a PC of Pentium IV (3.0 GHz). The FEM with PMLs [3] is also employed to calculate n_{eff} for the same MOF. Although only a quarter of the MOF is considered in the calculation, a large number of elements are required to get a result with similar accuracy. Due to the large dimension of the eigen system in the FEM, only a few eigenvalues (closest to a trial eigenvalue) and eigen-vectors are calculated for the concern of computation time. Thus prior knowledge on the trial eigenvalue, which may not be available for a new MOF, is required to reduce the computation time. The CPU time for calculating 10 eigenvalues (closest to the trial eigenvalue) and eigen-vectors (one of which corresponds to the fundamental mode) at a similar accuracy is 383seconds It is 135 seconds for the present two-stage H-G method without the need of any prior

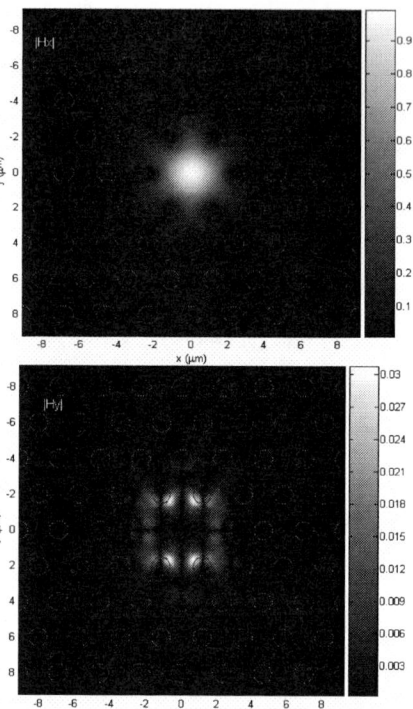

Fig. 3 Field profile of a y-polarized fundamental mode of the MOF in Fig.1.

knowledge. In addition, all the eigenvalues are computed in a single run. For the MOF with six air-hole rings, the CPU time for the multipole method is tens of minutes on the same PC.

Conclusions

We have proposed a two-stage H-G function method incorporated with anisotropic PMLs to efficiently calculate the complex propagation constant of an MOF. The two-stage implementation improves considerably the convergence and accuracy of the H-G method.

References

1. J.C. Knight, et al, Opt. Lett., 21(1996), 1547
2. S.Guo,et al,Opt.Express12(2004), 3341
3. K. Saitoh, et al, IEEE J Quantum Electron. 38(2002), 927
4. T.P. White, et al, Opt.Lett. 26(2001), 1660
5. W. Zhi, et al, Opt.Express 11(2003), 980
6. S. D. Gedney, IEEE Trans. Antennas Propagat.46 (1996), 1630.
7. I.S. Gradshtein, et al, "Tables of integrals, series and products," (New York: Academic, 1994).

Liquid Crystal Optical Modulator Based on In-Plane Switching

Yubao Sun, Li Jiang

Department of Applied Physics, Hebei University of Technology, Tianjin, 300401, P. R. China,
Email: hmtj450@vip.sina.com

Abstract The in-plane switching liquid crystal optical modulator is simulated. The critical frequency of the liquid crystal rotation depends on the strength of electric field, the rotational viscosity and the dielectric anisotropy of liquid crystal.

Introduction

In recent years, there is a growing concern over polarization mode dispersion in fiber optic cables, polarization mode dispersion induce optical birefringence and result in two different polarization group velocities for the signal along two orthogonal states of polarization[1]. This causes signals to become increasing distorted and unreadable as they travel along the fiber, to repair this signal degradation, endless polarization controllers are developed which are fast enough to combat bottlenecks that appear where signal manipulation is performed to convert optical signals to electronic signals. These polarization controllers are able to transform any input polarization state into any other polarization state. A nematic liquid crystal active waveplate device acts as an electrical polarization controlling waveplate, which switches at suitable speeds and does not require resetting. The device comprises two glass plates with four electrodes are sandwiched a thin layer of nematic liquid crystal material, as shown in figure 1. The surfaces are treated to induce homeotropic alignment of the liquid crystal material [2].

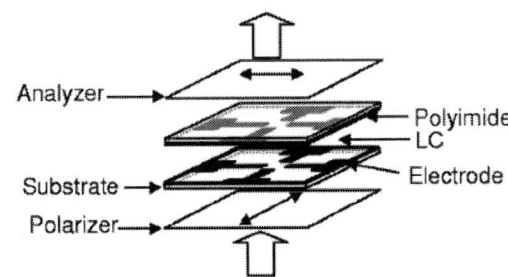

Fig.1 The geometry of the cell. Corresponding electrodes on the top and the bottom substrates are electrically connected together.

Principle and theory

By applying a voltage to these electrodes, it is possible to induce in-plane switching at the centre of the cell, providing that the nematic liquid crystal material has a positive dielectric anisotropy. This increases the birefringence of the cell, and so the polarization of light passing through the cell is changed. Therefore, by choosing suitable voltages on the electrodes, the polarization of light passing through the device can be controlled. As the surfaces are treated to have homeotropic anchoring, the director configuration can rotate freely as the electric field is rotated, providing an endless polarization controller device.

As the centre of the cell is typically 25μm (which is more large than cell thickness, about 5 micron) away from the electrodes producing the in-plane switching. Therefore, we can use the one-dimensional model because the light only transits the centre of the cell. The director is defined as:

$$n = \left(\cos\theta\cos\phi, \cos\theta\sin\phi, \sin\theta \right), \quad (1)$$

Where θ and ϕ are the commonly used director angles which are dependent on the position through the cell, z and time, t.

The assumption can be made that the electric field which has no component in the z direction. Therefore, the electric field can be written as $E = \left(E_x, E_y, 0 \right)$. To operate the device, the electric field is rotated around by varying the magnitude of the voltage on each of the electrodes and the central "thread" is controlled by the overall component of these voltages. The rotation of the electric field can be seen as the rotation of the cell. So the electric field can be further simplified to $E = \left(E_x, 0, 0 \right)$. To

978-0-9789217-3-6/07/$25.00

©2007 WEN GLOBAL SOLUTIONS

model mathematically a rotating cell, $\Delta\phi_E(t)$ must be added to $\phi(z,t)$, where $\Delta\phi_E(t)$ is the angle of rotation of the cell as time progresses.

As neglecting flow and assuming that $\phi(z,t)$ is uniform throughout the z-axis and so $\phi(z,t)=\phi(t)$. Therefore, the governing equations are [3]

$$\gamma\frac{\partial\theta}{\partial t}=\left(K_1\cos^2\theta+K_3\sin^2\theta\right)\frac{\partial^2\theta}{\partial z^2}$$

$$+\left(K_3-K_1\right)\sin\theta\cos\theta\times\left(\frac{\partial\theta}{\partial z}\right)^2,\quad(2)$$

$$-\varepsilon_0\Delta\varepsilon E_x^2\cos^2\phi\sin\theta\cos\theta$$

And

$$\frac{\partial\phi}{\partial t}=-\frac{\varepsilon_0\Delta\varepsilon E_x^2}{\gamma}\sin\phi\cos\phi\quad(3)$$

Here, γ is the rotational viscosity of the liquid crystal, K_1, K_2 and K_3 are the splay, twist and bend elastic constants, respectively, ε_0 is the permittivity of vacuum, $\Delta\varepsilon$ is the dielectric anisotropy of the liquid crystal.

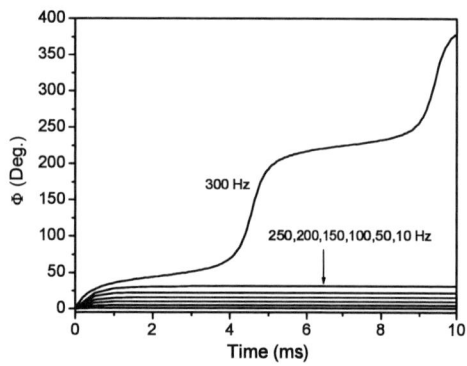

Fig. 2 The angular difference between the electric field and the director field, Φ, change with the frequency of electric field.

When the cell driven by a fixed rotational electric field frequency, the overall twist angle of the electric field, ϕ_E, at any on time was different from the overall twist angle of the director field, ϕ_d, which was seen to lag behind the electric field. The angular difference between the two profiles, $\Phi=\phi_E-\phi_d$, remained constant as long as

the frequency of electric field rotation also remained constant. Increasing the rotation frequency of the electric field, the angle Φ increasing as the director field fell further behind the electric field. However, once Φ had reached approximately $\pi/4$, the electric field broke away from the director profile. This was because the torque acting on the director profile by the electric field as it rotates increases as Φ increases, but after $\Phi=\pi/4$, the torque decreases as seen in equation 3. In figure 2, the angle Φ is plotted from different rotational frequencies. Here, the parameters we used are: cell thickness is $5\mu m$, $\Delta\varepsilon=10$, $\gamma=0.1N/(m^2s)$, and $E_x=2V/\mu m$. The angle Φ increases from 0 but quickly reaches an equilibrium at low frequency, but continues to increases but no longer reaches an equilibrium as the frequency is large enough. There must be a critical frequency. We analyse the critical frequency in the following.

As mentioned above, the rate of change of Φ is

$$\frac{\partial\Phi}{\partial t}=\frac{\partial\phi_E}{\partial t}-\frac{\partial\phi_d}{\partial t}.\quad(4)$$

When the field is being rotated and the director profile is still smoothly following the electric field, it can be assumed that $\partial\phi/\partial z=0$ and $\partial\phi/\partial t=0$, equation 3 can be rewritten as

$$\frac{\partial\phi_d}{\partial t}=-\frac{\varepsilon_0\Delta\varepsilon E_x^2}{\gamma}\sin\phi_d\cos\phi_d,$$

And equation 4 can be rewritten as

$$\frac{\partial\Phi}{\partial t}=\frac{\partial\phi_E}{\partial t}+\frac{\varepsilon_0\Delta\varepsilon E_x^2}{\gamma}\sin\phi_d\cos\phi_d.\quad(5)$$

For the low frequency, $\partial\Phi/\partial t\to0$, and so

$$\frac{\partial\phi_E}{\partial t}=-\frac{\varepsilon_0\Delta\varepsilon E_x^2}{\gamma}\sin\phi_d\cos\phi_d$$

The frequency of electric field rotation can be written as $F=(1/2\pi)\partial\phi_E/\partial t$ and note that $\phi_d=\phi_E-\Phi$, so the frequency can be written as

$$F=-\frac{\varepsilon_0\Delta\varepsilon E_x^2}{2\pi\gamma}\sin\left(\phi_E-\Phi\right)\cos\left(\phi_E-\Phi\right)$$

As the assumption above, the system with a rotating cell rather than a rotating field, ϕ_E

must be subtracted from the above equation, giving

$$F = \frac{\varepsilon_0 \Delta \varepsilon E_x^2}{2\pi\gamma} \sin\Phi\cos\Phi . \qquad (6)$$

For the stable rotational frequency, the max of the angle Φ is $\pi/4$, so the critical frequency is obtained

$$F = \frac{\varepsilon_0 \Delta \varepsilon E_x^2}{4\pi\gamma} . \qquad (7)$$

From above equation, it can be obtained the needed parameters for a suitable frequency. For example, the cell is operated at F=1kHz, the needed parameters can be calculated and shown in figure 3.

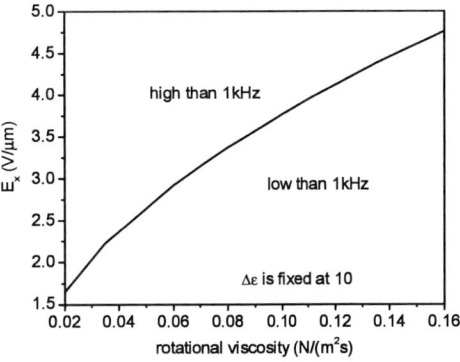

Figure 3 The combinations of dielectric anisotropy, rotational viscosity and electric field for the director profile maintain a smooth rotation under an electric field roation at 1kHz.

Conclusions

In this paper, we analyse the frequency of the director and electric field in an active waveplate device using a one-dimensional model. We explain the limited range of electric field rotation frequencies available in a real device. By analysing the governing equations, the maximum rotation frequency of the electric field was linked to the magnitude of the electric field applied to the cell, the rotational viscosity and the dielectric anisotropy of the liquid crystal material for the director field still maintained a smooth rotation. To achieve high rotational frequencies, a large applied electric field, a large dielectric anisotropy and a low rotational viscosity is considered.

This research was supported by Key Subject Construction Project of Hebei Province University and Natural Science Foundation of Hebei Province (No. A2006000675), P. R. China.

References

1 M. le Gall et al Opt. Commun. 176 (2000), 113-119.

2 B. R. Acharya et al Appl. Phys. Lett., 81 (2002), 5243-5245.

3 A. J. Davidson et al J. Phys. D: Appl. Phys. 38 (2005), 1470-1477.

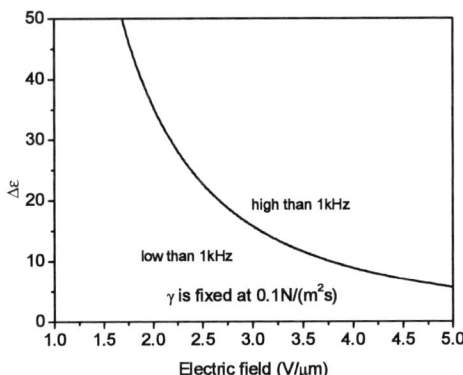

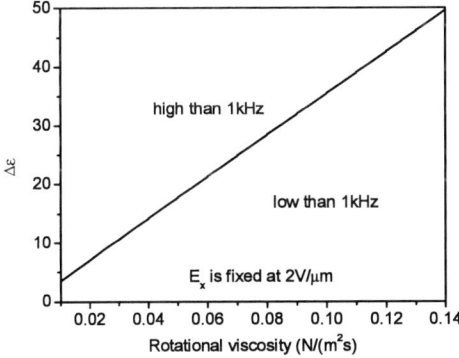

Analysis of the scalability of the video-overlay system

Ick Chang Choi (1), Byoung-Ju Yun(2), and Hyun Deok Kim (3)

1 : School of Electrical Engineering and Computer Science of Kyungpook National University,
1370, Sankyuk-dong, Puk-gu, Daegu, 702-201, South Korea, Email: choic@ee.knu.ac.kr
2 : Kyungpook National University, Email: bjisyun@ee.knu.ac.kr
3 : Kyungpook National University, Email: hyundkim@ee.knu.ac.kr

Abstract *We have derived an analytic expression for the scalability of the video-overlay system in the FTTH network and shows that the scalability strongly depends on both the optical amplifier noise and the configuration of the distribution network.*

Introduction

Fiber-To-The-Home (FTTH) is an exciting technology for telecom operators because of its capability to deliver the voice, data and video "triple play". The passive optical network (PON) is one of the promising architectures to deliver these services [1]. Video broadcasting is one of the important services of the PONs, and an overlaying subcarrier-modulated video signal with a different wavelength (i.e. a video overlay PON) is a most commonly used way to achieve service functionality [2].

When the PON architecture is deployed with the 1550 nm wavelength for video overlay, one of the issues about the operator may encounter is optical amplifier noise. Since the video-overlay system distributes the amplified output of an optical transmitter the use of optical amplifiers and splitters is essential. Therefore, optical amplifier noise may degrade signal quality and thus limit the expandability of the whole system.

In this paper, we consider various signal degradation factors in the video-overlay system which transmits an analog CATV signals. Using the equivalent input noise (EIN) parameter, we drive an analytic expression for deciding the quality of a received signal in the video-overlay system and analyze the scalability of the FTTH network.

Derivation of system noise figure

In general, several factors can degrade the quality of the optical signal in an optical analog transmission system. For example, relative intensity noise (RIN) of an optical transmitter, thermal noise and shot noise of an optical receiver are the representative degradation factors in an optical CATV

transmission system. However, the optical amplifier used for signal amplification may cause signal degradation since the optical amplfier noise can generate a beating noise at the receiver. Fig. 1 partially shows an example of video-ovlay distribution network with a *4MN* subscriber with L km from the central office.

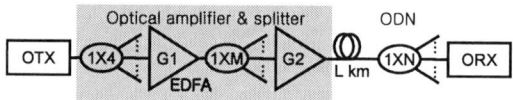

Fig. 1 Example of the vidoe-overlay network

To model the limiting factor due to the various noises including otpcal amplifier noise, we describe noise fator as the EIN. Then the total EIN of the video-overlay system is given as follows,

$$EIN_{Total} = EIN_{RIN} + EIN_{Thermal} + EIN_{Shot}$$
$$+ EIN_{Sig-ASE} + EIN_{ASE-ASE} \qquad (1)$$
$$+ EIN_{ASE-Shot}$$

Where EIN_{RIN} is the EIN due to RIN of a laser, and $EIN_{Thermal}$ and EIN_{Shot} are the EINs due to the thermal noise and the shot noise, respectivley. Each noise EIN is given by

$$EIN_{RIN} = \frac{R_{in}}{\eta_{Tx}^2} \cdot RIN \cdot M_{dc}^2 \left(\frac{I_{RF}}{m} \right)^2 \qquad (2)$$

$$EIN_{Thermal} = \frac{4kT}{\eta_{Tx}^2 \cdot \eta_{Rx}^2} \cdot \frac{1}{G^2} \cdot \frac{R_{in}}{R_s} \qquad (3)$$

$$EIN_{Shot} = 2q \cdot \frac{P_{Tx} \cdot R_{in}}{\eta_{Tx}^2 \cdot R} \cdot \frac{1}{G} \qquad (4)$$

where R_{in} and η_{Tx} are the resistance and the efficiency of the optical transmitter, respectively. η_{Rx} is the efficiency of the receiver. M_{dc} is the dc modulation gain given

978-0-9789217-3-6/07/$25.00

©2007 WEN GLOBAL SOLUTIONS

by the slope efficiency of the L-I curve and m is the optical modulation (OMI). I_{RF} is the input current of the transmitter and P_{Tx} is the optical power at transmitter. k is the Boltzmann's constant, T is the absolute temperature and q is the electron charge quantity. R_s is the output resistance, R is the responsivity of the photo detector and G is the total gain of the link.

The last three terms of the EIN in the equation (1) describes the effects of the optical amplfier noise. The optical amplfier noise is generated from the amplified spontaneous emission (ASE) and the power of ASE in each optical amplifier with a gain of G is given by [3]

$$P_{ASE} = 2n_{sp}(G-1)hvB_o \qquad (5)$$

where n_{sp} is the population inversion factor, h is Plank constant, v is optical frequency and B_o is optical bandwidth. Since the video overlay system generally utilize concatenated optical amplifier we assume the amplifier chain as shown in Fig. 2 to model the overall optical noise inputted to the receiver.

Fig. 2 Optical amplfier chain model

Then the total link gain with N amplifiers is given by

$$G = \frac{G_1 \times G_2 \times \cdots \times G_n}{L_1 \times L_2 \times \cdots \times L_n \times L_{n+1}} \qquad (6)$$

From equation (5) and (6) the accumulated ASE noise ($P_{Rx\text{-}ASE}$) inputted to the receiver is given by

$$P_{Rx-ASE} = 2n_{sp} \cdot hv \cdot B_o \cdot G$$
$$\times \left\{ \sum_{i=1}^{n} \left((G_i - 1) \prod_{k=1}^{i} \frac{L_k}{G_k} \right) \right\} \qquad (7)$$

where G_k and L_k is the gain and loss of the kth amplifier and the link, respectively. The accumulated ASE noise generates beating noises in the optical receiver and they are given by

$$\hat{\sigma}_{Sig-ASE} = 8n_{sp} \cdot hv \cdot R^2 \cdot G \cdot P_s$$
$$\times \left\{ \sum_{i=1}^{n} \left((G_i - 1) \prod_{k=1}^{i} \frac{L_k}{G_k} \right) \right\} \cdot B_e \qquad (8)$$

$$\hat{\sigma}_{ASE-ASE} = 8n_{sp}^2 \cdot (hv)^2 \cdot R^2 \cdot G^2$$
$$\times \left\{ \sum_{i=1}^{n} \left((G_i - 1) \prod_{k=1}^{i} \frac{L_k}{G_k} \right) \right\} \qquad (9)$$
$$\times (2B_o - B_e) \cdot B_e$$

$$\hat{\sigma}_{ASE-Shot} = 4q \cdot n_{sp} \cdot hv \cdot B_o \cdot G \cdot R$$
$$\times \left\{ \sum_{i=1}^{n} \left((G_i - 1) \prod_{k=1}^{i} \frac{L_k}{G_k} \right) \right\}^2 \cdot BW \qquad (10)$$

where P_s is the receiver input optical power and BW is the bandwidth of the ASE noise. From equations (8) ~ (10), we can calculate the EINs due to the ASE-induced beating noises.

$$EIN_{Sig-ASE} = 8n_{sp} \cdot \frac{1}{\eta_{Tx}^2} \cdot R_{in} \cdot \frac{1}{G} \cdot P_s$$
$$\times \left\{ \sum_{i=1}^{n} \left((G_i - 1) \prod_{k=1}^{i} \frac{L_k}{G_k} \right) \right\} \qquad (11)$$

$$EIN_{ASE-ASE} = 8n_{sp}^2 \cdot (hv)^2 \cdot \frac{1}{\eta_{Tx}^2} \cdot R_{in}$$
$$\times \left\{ \sum_{i=1}^{n} \left((G_i - 1) \prod_{k=1}^{i} \frac{L_k}{G_k} \right) \right\}^2 \qquad (12)$$
$$\times (2B_o - 1)$$

$$EIN_{ASE-Shot} = 4q \cdot n_{sp} \cdot hv \cdot B_o \cdot \frac{R_{in}}{\eta_{Tx}^2} \cdot \frac{1}{R} \cdot \frac{1}{G}$$
$$\times \left\{ \sum_{i=1}^{n} \left((G_i - 1) \prod_{k=1}^{i} \frac{L_k}{G_k} \right) \right\} \qquad (13)$$

Since the total EIN is the sum of each EIN we can calculate it from the equations (2)~(4) and (11)~(13). The overall system noise figure (NF) can be derived from the EIN and given by [4]

$$NF = 10\log\left(EIN_{mW/Hz} / K_{T_0} \right) \qquad (14)$$

where K_{T0} is the noise which is generated by load at the temperature of 290 K. From equation (14) we can calculate the CNR degradation due to the various noise factors of video-overlay system and also analyze the effects of the factors on the scalability of the system since the CNR degradation should not a certain value if the system requirement is given.

Scalability analysis

Generally, the CNR of the input signal at the optical transmitter is about 51 dB since it is connected to the trunk system and the CNR at the receiver output should be higher than 43 dB to satisfy FCC requirements. To analyze the scalability of the video-overlay system we assume conventional system parameters. The gains of the amplifier used at the transmitter side as shown in Fig. 1 were assumed to 17 dB and 20 dB, respectively and the loss coefficient of the optical fiber was assumed to 0.275 dB/km.

Fig. 3 shows the total EIN and the receiver output CNR as a function of the fiber length when M is 8 and N is 32. As the fiber length increases, the EIN increases and the CNR decreases. Thus the fiber length should be limited to a certain value to satisfy the required CNR.

As we increases the number of the splitter port (M) at the transmitter side we can accommodate more subscriber but the effect of the amplifier noise increases. When the number of the distribution port of the passive node (N) is 32, the CNR decreases as M is increases as shown in Fig. 4. Thus M should be less than 8 to guarantee 20 km coverage, which means the number of subscribers can not exceed 1024(=4x8x32). On the other hands, the geographic coverage should be less than 3 km if we accommodate more than 2048 subscribers (i.e. M is 16).

If we assume M is fixed to 16, the coverage can not exceed 14 km to accommodate 1024 subscribers (i.e. N is 16) as shown in Fig. 5.

Conclusions

The influence of optical amplifier noise in the video-overlay system has analyzed considering other degradation factors. The analytic expression given by in terms of the noise figure and EIN can be used to analyze the scalability of the video-overlay system and optimize the distribution network as well.

Acknowledgement

This work was partially supported by KOSEF under the NRL project and Ministry of Education under BK21 Project.

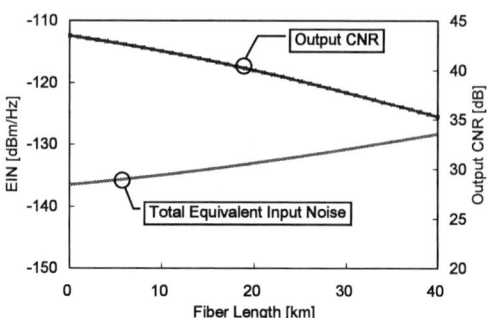

Fig. 3 The total EIN and the CNR as a function of the fiber length

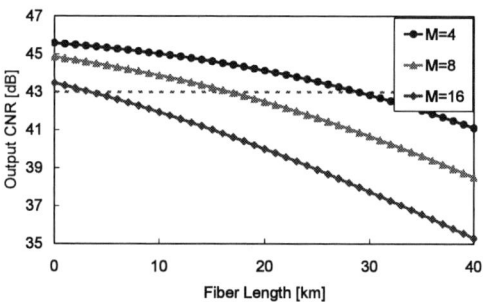

Figure. 4 Receiver output CNR for the number of the transmitter splitter ports

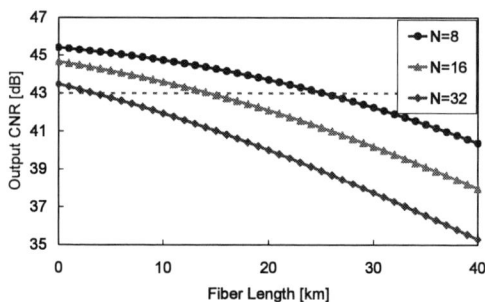

Fig. 5 Receiver output CNR for the number of the distribution port of the passive node

References

1 Fred Coppinger et al, OFC2006, OThK7 (2006)

2 Akira Agata et al, ECOC 2005 Proceedings, Vol. 2 (2005), page 145-146

3 N. A. Olsson, Journal of Lightwave Technology, Vol. 7, (1989), pp. 1071-1082

4 EMCORE, Application Note, (2003)

An All-Optical Frequency Down-Converter Based on Four-Wave-Mixing in a Highly Nonlinear Fiber for Radio-over-Fiber Systems

Ying Gao, Shiming Gao, Hongyan Fu, Daru Chen, Xiangrui Miao

Centre for Optical and Electromagnetic Research, State Key Laboratory for Modern Optical Instrumentation, Zhejiang University, Hangzhou 310058, P. R. China

gaoying@coer.zju.edu.cn, gaosm@zju.edu.cn

Abstract *An optical frequency down-converter based on four-wave mixing in highly nonlinear fibers is proposed and experimentally realized for radio-over-fiber systems. A down-converted 1.5 GHz intermediate-frequency signal is obtained from a received 11.1 GHz millimeter-wave signal.*

Introduction

Modern communication systems require broad band, high speed, good mobility, low transmission loss, and effective cost. Radio-over-fiber (RoF) systems satisfy all these requirements and have been actively investigated recently [1], [2]. In RoF systems, radio signals are transmitted through single-mode fibers (SMFs) between the central and base stations, which have low transmission loss. But large dispersion will be introduced to destroy the signal properties when optically intensity-modulated millimeter (mm)-wave signals are transmitted in SMFs. We can use dispersion compensation techniques to solve this problem, but this method is difficult and complex design process has to be demanded [3]. One can use frequency conversion techniques to convert the mm-wave signals to intermediate frequency (IF) before transmission in remote base stations, since IF signals is hardly influenced by fiber dispersion [4], [5].

Optical frequency conversion techniques can be used to avoid the requirement of the expensive mm-wave band oscillators and mixers, which make the architecture of base station simple and cost effective. Up to now, optical frequency down-converter has already been realized by using the nonlinearity of an electroabsorption modulator in which the modulation format is limited [4]. In this paper we proposed a novel optical frequency down-converter utilizing four-wave-mixing (FWM) effect in a highly nonlinear fiber (HNLF). This method will potentially provide a large bandwidth and a transparent modulation format.

Principle

The schematic structure of the optical frequency down-conversion based on FWM is shown in Fig. 1.

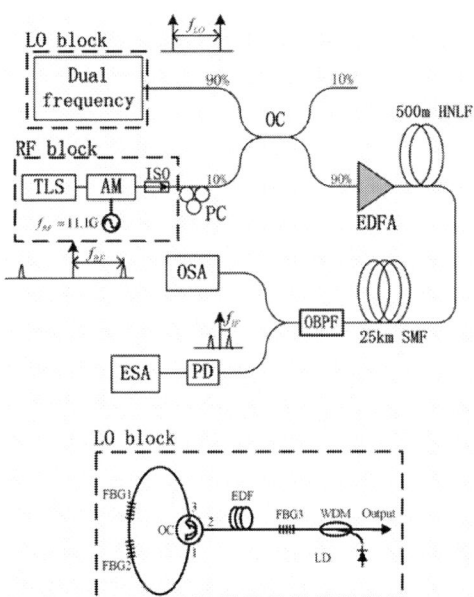

Fig. 1. Schematic configuration of frequency down-conversion. OC, optical coupler; OSA, optical spectrum analyzer; EDF, erbium-doped fiber; EDFA, erbium-doped fiber amplifier; TLS, tunable laser source; AM, amplitude modulator; OBPF, optical band pass filter; ESA, electrical spectrum analyzer.

978-0-9789217-3-6/07/$25.00

©2007 WEN GLOBAL SOLUTIONS

The received external mm-wave, radio frequency (RF) signal (f_{RF}), is modulated on the optical carrier centerd at v_{RF} by an amplitude modulator at base station. The optical heterodyne local oscillator (LO) signal (f_{LO}) centerd at v_{LO} is generated at the central station and transmitted to the base station through SMF. When one injects the LO optical signal at v_{LO} along with the modulated optical carrier signal at v_{RF} into the HNLF after high power erbium-doped fiber amplifier (EDFA), the FWM effect will occur between the LO optical signal (v_{LO1} and v_{LO2}) and the RF-modulated optical carrier (v_{RF}). As shown in Fig. 2, three different optical waves (v_{IF1}, v_{IF2} and v_{IF3}) will be created through degenerate or non-degenerate FWMs. Their frequencies can be expressed as $v_{IF1} = 2v_{LO1} - v_{RF}$, $v_{IF2} = v_{LO1} + v_{LO2} - v_{RF}$, and $v_{IF3} = 2v_{LO2} - v_{RF}$. At the same time, the modulated sideband will also be created through the FWM effects. After Filtering through these generated optical signals and detecting them with a photodetector (PD), the frequency down-converted signal at $f_{IF} = f_{RF} - f_{LO}$ will be generated as a beating result between the converted optical carrier and the adjacent modulated sideband.

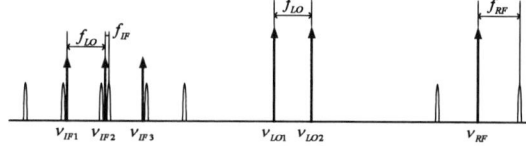

Fig. 2. Principle of the optical frequency down-converter.

Experiment and results

In Fig. 1, the LO optical signal is provided by a dual-wavelength laser, which consists of a fiber Bragg grating (FBG)-based Fabry-Perot filter (FBG1 and FBG2) as one reflector, FBG3 as the other reflector, and a 2.24-m EDF as the gain medium. The output lasering includes two single-frequency wavelengths, whose frequency difference, i.e. f_{LO}, is about 9.6 GHz. The optical carrier from a tunable laser at 1544.94 nm is modulated by an 11.1 GHz mm-wave signal through an intensity modulator. The LO and the carrier are amplified to around 1 W using a high-power EDFA. The FWM process occurs in a 500-

m HNLF, whose zero-dispersion-wavelength and nonlinear coefficient are 1553.35 nm and 11.5 $W^{-1}km^{-1}$, respectively, and three optical signals centerd at 1553.93 nm are generated. Fig. 3 shows the optical spectra of the carrier, LO, and converted optical signal, respectively. The wavelength span, resolution and sensitivity of the OSA (Agilent 86142B) are set to be 1 nm, 0.06 nm and -65 dBm, respectively. Here, the sideband generated by amplitude modulation can be created simultaneously. It is noticeable that more than three optical wavelengths are generated in Fig. 3(c), because the input power is large enough to support high-order FWMs.

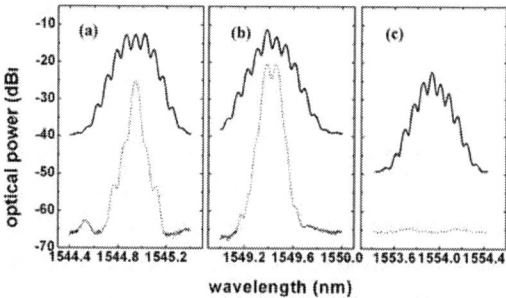

Fig. 3. Optical spectrum before (dotted line) and after (solid line) the FWM process. The spectra (a)-(c) are for the optical carrier, the LO, and the generated optical signals, respectively.

After Filtering through the created optical signals and sidebands through FWM by using an optical band-pass filter, the beating electrical signal will be produced on PD. The electrical spectrum is shown in Fig. 4. The frequency span and resolution bandwidth of the ESA (Advantest R3261A) are set to be 15 GHz and 3 MHz, respectively. Besides the required IF signal of 1.5 GHz, other mm-waves at 9.6GHz and 11.1G Hz are also created. The peak at 9.6 GHz comes from the heterodyne between the converted LO and the heterodyne between their converted sidebands. Similarly, the peak at 11.1 GHz comes from the heterodyne between the converted LO and its sideband. They can be eliminated by using a low-response-speed PD and the cost will be cut down. Another way is to filter through a single converted optical carrier and the adjacent sideband directly, but the ultra-narrow bandwidth of the filter may be a

454

challenge. The detailed measurement without or with a long distance transmission (a 25-km SMF, as shown in Fig. 1) is shown in the inset (a) or (b) with a span and resolution bandwidth of 400 MHz and 100 kHz. Without SMF transmission, the signal-to-noise ratio of the down-converted IF signal is larger than 41 dB. With transmission in a 25-km SMF, the signal-to-noise ratio is larger than 31 dB, which means our proposed frequency down-conversion scheme is effective even though the signal transmission between base and central station in real RoF systems.

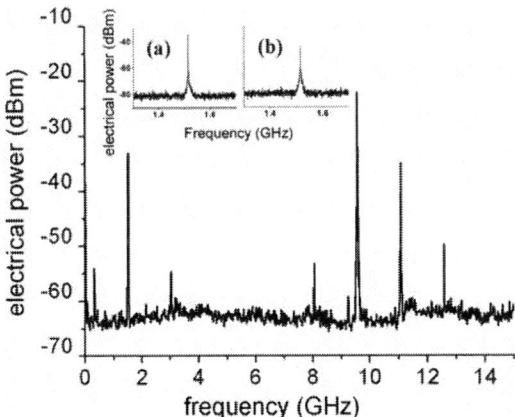

Fig. 4. Electrical spectrum of the heterodyne between the generated optical signals through FWM. The insets (a) and (b) show the down-converted IF signal without and with a 25-km SMF transmission.

Conclusions

An effective all-optical frequency down-converter based on FWM in HNLF has been proposed and experimentally demonstrated. The signal-to-noise ratio of the down-converted IF signal is larger than 41 dB. This frequency down-conversion scheme is effective in real RoF systems. the frequency down-converter based on the FWM effect in HNLF have many advantages, such as fast response, large bandwidth, and transparent modulation format, it is very competitive in the future RoF applications.

Acknowledgements

This work is supported by the Science and Technology Department of Zhejiang Province (grant No. R104154 and 2004C31095) and Science and Technology Bureau of Hangzhou municipal government (grant No. 20051321B14).

References

1 J. Yao, G. Maury et al, IEEE Photon. Technol. Lett., 17 (2005), page 2427.
2 Z. Jia et al, IEEE Photon. Technol. Lett., 17 (2005), page 2724.
3 F. Ramos et al, IEEE Photon. Technol. Lett., 11 (1999), page 1171.
4 J.-H. Seo et al, IEEE Photon. Technol. Lett., 17 (2005), 1073.
5 J.-H. Seo et al, IEEE Trans. Microw Theory Tech., 54 (2006), page 959.

Shape Influence on the Two-Dimensional Photonic Crystal Devices

Ya-Zhao Liu(1), Shuai Feng(2), Zhi-Yuan Li(3), Bing-Ying Cheng (4), Dao-Zhong Zhang(5),
1:Laboratory of Optical Physics, Beijing National Laboratory for Condensed Matter Physics,
Institute of Physics, Chinese Academy of Sciences, Beijing 100080, China,
liuyazhao@aphy.iphy.ac.cn 2: fs@aphy.iphy.ac.cn 3: lizy@aphy.iphy.ac.cn 4:
bycheng@aphy.iphy.ac.cn5: dzzhang@aphy.iphy.ac.cn

Abstract *The influence of the irregular shape of atom on the optical characteristics of photonic crystal devices was demonstrated. A multi-channel filter based on this theroy was designed to demonstrate the shape tuning capability.*

Introduction

Small size, high precision and ease in fabrication [1,2] these advantages made two-dimensional photonic crystal (PhC) devices, espacially the multi-channel filters attractive in recent years. It has been well known that structure is the kernel of filter designing. In this article, the shape of the cavities is systematically investigated, and we prove that the characters of the ellipse air holes , including the ellipticity, the orientation, and the area, can have a great influence on the localized cavity modes. Moreover,we will demonstrate the application of shape influence in design a multi-channel filter based on 2D photonic crystal slab.

Optical characteristics of the filter with shaped air holes

Our 2D PhC slab is made of a triangular lattice of air holes on a thin slab. A major channel is created by removing a single line of air holes along the $\Gamma - K$ direction (W1 waveguide), a cavity is formed by leaving three air holes unetched along another $\Gamma - K$ direction [3]. Usually the air holes in the PhC filter have circular shape, as is the case in Fig. 1(a). However, in principle they can take other shapes, such as an ellipse. Figure 1(b) shows an enlarged picture of an elliptical air hole filter in the region around the cavity. One of its axes is oriented anticlockwise by an angle θ to the x axis, namely, the input light pro-pagation direction. The sizes of the axes parallel and per-pendicular to this orientation are a and b, respectively. The ellipticity of the air holes can be well described by a parameter as $p=a/b$.

To show the influence of the air hole shape on the filtering functionality, we tried several kinds of shape. In the first one that we consider, the elliptical air holes have the same area as the regular circular holes of $0.56a$ (a is the lattice constant set to 430nm) in diameter, but their ellipticity changes from $p=0.8$ to $p=1.3$. The orien-tation of the ellipse keeps parallel to the x axis, namely, $\theta=0°$. Figure 2 illustrates the simulation result of the transmission spectrum obtained from 2D FDTD calculations. The maximum shift of the resonant peak exceeds 15 nm, ranging from 1581 to 1596 nm.

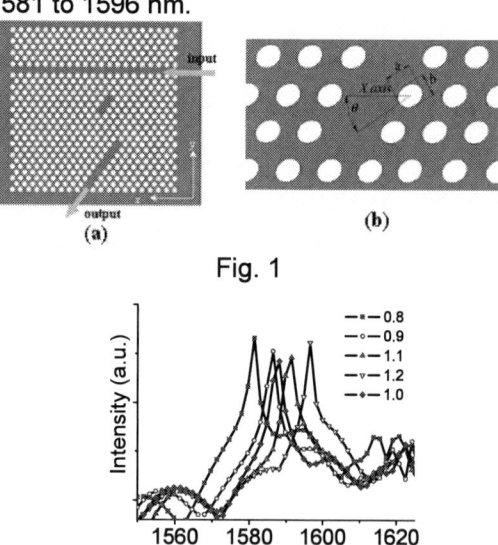

Fig. 1

Fig. 2

In the second kind of filter structures that we consider, the orientation of the elliptical air holes is rotated clockwise with a step of 15° relative to the filter structures discussed in Fig. 1. The resulting transmission spectra are displayed in Fig. 3(a)-(e). Each picture corresponds to a particular ellipticity of air holes, which ranges from 0.8 to 1.3 for panel (a) to (e), respectively. In panel Fig. 3(a), the ellipticity is 0.8 and the peaks have a narrow range of shift within 4 nm when the air hole orientation changes. The filter with an ellipticity of 0.9 has an even narrower range of shift of peak within 1.7 nm. The reason is simple. When the ellipticity becomes smaller, the air hole gets closer to a regular circle, and is subject to weaker influence from the orientation of the air hole. The spectra displayed in Figs. 3(c)-(e) also confirm this expectation. When the ellipticity increases from 1.1 to 1.2, and to 1.3, the range of the peak shift grows from 4 nm to 7 nm, and to 10nm.

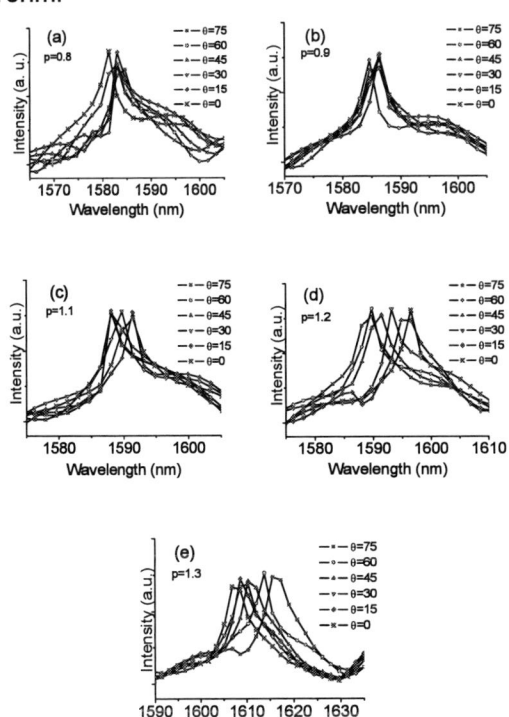

Fig. 3

Another enlarged structures whose area of air holes are 1.3 multiple of the previous one has also been tested. Figure 4 and Fig. 5 show the resulting transmission spctra of it. The distributions of the resonant peaks are similar to the previous structure, despite the shift toward the shorter wavelength. The resonant peaks shift in a range of 23 nm as the ellipticity changes from 0.8 to 1.3, the range of peaks caused by rotation increase from 3nm to 8nm as the angle changed. All the simulations and discussions above clearly indicate that the shape of the "atoms" (here the air holes) comprising the cavity is an important element to affect the optical properties of a PhC filter in addition to the

usual "atom" position tuning.

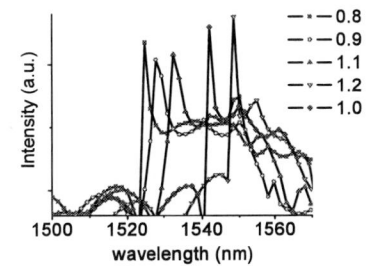

Fig 4

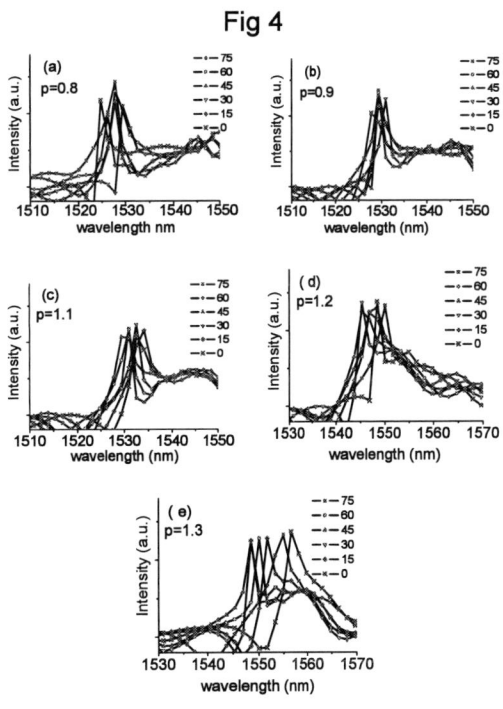

Fig. 5

Experiment and simulation

A four-channel PhC filter was fabricated by focused ion beam lithographic technique on the SOI wafer. The thickness of the silicon slab is 260nm. Both of the silicon oxide layers beneath and above the silicon slab were removed to form an air-bridge structure. The input and output channels connect the input and output ridge waveguides [4] to connect the outside devices. The SEM picture of the four-channel filter is displayed in Fig. 6. Four cavities are located on the two sides of the central linear W1 waveguide. The whole filter is made from a hetero-geneous triangular lattice PhC structure where the first part (connecting channel 2 and channel 4) has a lattice constant of 430nm, while the second part (connecting channel 1 and channel 3) has a lattice constant of 420nm [5]. This two-segment geometric configuration can make sure that the optical wave can pass through the filter device without being disturbed by the photonic band gap. The air holes in the four channels have different axes, but no rotation is exerted so that $\theta=0°$ for all the four channels. The cavities' parameters are described in Table 1.

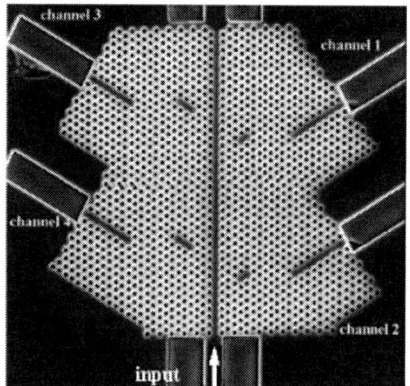

Fig. 6

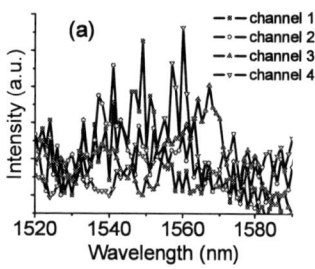

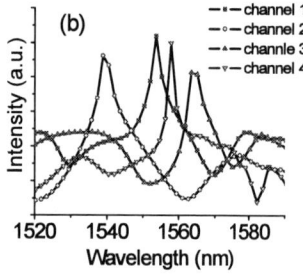

Fig. 7

Light came from a continuous wave tunable semi-conductor laser with the wavelength ranging from 1500 to 1640 nm launched into the ridge waveguide. The experimental result of the transmission spectra for the four channels is displayed in Fig. 7(a). Although significant noise exists, a resonant peak can be clearly found for each channel. The peaks are located at 1549, 1541, 1567 and 1560 nm for channel 1, 2, 3, and 4, respectively. Numerical simulations for the transmission spectra were also

performed by means of the 2D FDTD technique. The results are plotted in Fig. 7(b). A resonant peak is found for each channel, and they are located at 1553, 1539, 1563, and 1558 nm. The small discrepancy of the resonant peak positions (with a maximum deviation of 4 nm) can be attributed to the inevitable inaccuracy in the process of fabrication.

Table 1

	Number of missing air holes in cavities	Long axis a (nm)	Short axis b (nm)	Theoretical resonant peak (nm)	Measured resonant peak (nm)	Deviation (nm)
Channel 1	2	240	200	1553	1549	4
Channel 2	2	260	240	1539	1541	2
Channel 3	3	240	220	1563	1567	4
Channel 4	3	280	240	1558	1560	2

Conclusions

theoretically and experiment indicated that in addition to the usual size influence, the shape of the air holes has a great influence on the optical characteristics of devices built in 2D PhC slab. The holes with shaped ellipticity and orientation angle have the ability to control the filtering functionality. changing the ellipticity of the elliptical air holes can induce a rough tuning of the resonant wavelength, while changing the orientation of the elliptical air holes brings a fine tuning of the resonant wavelength (within 1.7 nm) of the filters. Based on this principle, a four-channel filter were built via focused ion beam lithographic technique and their optical properties were measured. The results confirm the air-hole shape influence on the functionality of the PhC filter devices.

References

1 E. Yablonovitch, Phys. Rev. Lett. **58**, 2059 (1987).

2 S. G. Johnson, S. Fan, P. R. Villeneuve, J. D. Joannopoulos, and L. A. Kolodziejski, Phys. Rev. B **60**, 5751 (1999).

3 J.D. Joannopoulos, Photonic crystals (Princeton University Press, 1995) .

4 A. Yariv, P. Yeh, Optical Waves in Crystals (New York:John-Wiley, 1984).

5 B.S. Song, T. Asano, Y. Akahane, Y. Tanaka, and S. Noda, Appl. Phys. Lett. **85**,4591 (2004).

Tunable Artificial Birefringence in woodpile photonic crystals

Ming Che (1), Zhi-Yuan Li (2), Rong-Juan Liu (3)

1: Beijing National Laboratory for Condensed Matter Physics, Institute of Physics, Chinese Academy of Sciences, Beijing 100080, China Email:cheming@aphy.iphy.ac.cn

2: Email:lizy@aphy.iphy.ac.cn 3: Email:liurj@aphy.iphy.ac.cn

Abstract A large relative anisotropy can be realized in three-dimensional woodpile photonic crystals in the long-wavelength limit. We find that the anisotropy can be tuned. Transition from positive anisotropy to negative anisotropy can be widely achieved.

Introduction

Nature offers us a large category of anisotropic crystals, either negative or positive [1-3]. However, it also imposes a great obstacle and limitation to tunable and accessible anisotropy. For instance, the anisotropy is usually not as large as desired, and it is difficult for a crystal to transit from positive to negative anisotropy.

Recently, there appear some efforts to mimic natural optical birefringence by means of artificial microstructures, for instance, photonic crystals (PCs), which have periodic modulation of dielectric function [4,5].There are also some efforts to study the homogenization and effective medium properties of PCs in the long-wavelength limit[6-11]. However, most are theoretical studies on two-dimensional PCs perhaps because of the ease to achieve accuracy and analyticity for these simple structures.

In this paper we will show that three-dimensional (3D) woodpile PC can offer much greater freedom to design optical anisotropy and can behave as an artificial uniaxial crystal of positive or negative anisotropy, or even an isotropic crystal by simply tuning the geometry of its building block. Very large anisotropy can be readily achieved.

Results and analysis

The configuration of the 3D woodpile PC is schematically depicted in Fig. 1. The crystal is formed by stacking rectangular dielectric rods layer by layer consecutively along the (001) direction (the z axis) in an air background. Rods in each layer are arrayed into a 1D lamellar grating with a pitch of a. Rods in one layer are perpendicular to those in the next layer,

while rods in one layer are shifted by $a/2$ with respect to those in the next two layers. Each rod has a width w, a thickness h, and a dielectric constant ε.

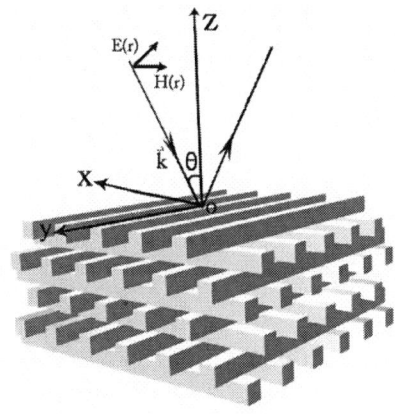

Fig. 1. Schematic configurations of a 3D woodpile photonic crystal composed of rectangular dielectric rods in air.

The 3D woodpile PC should behave as a uniaxial effective medium in the long-wavelength with the optic axis coincident with the (001) stacking direction. Light propagating inside the crystal will witness the ordinary and extraordinary index of refraction n_o and n_e when the electric field (E-field) is polarized perpendicular and parallel to the optic axis respectively. The effective refractive index of this crystal is calculated from the reflectivity spectrum of an electromagnetic plane-wave incident on the (001) surface of a semi-infinite crystal in the long-wavelength by the plane-wave-based transfer-matrix method, Fresnel's formula and Snell's formula[12]. A plane-wave light propagating in the *xoz* plane and polarized at the TM mode (with the E-field lying within the *xoz* plane) is used to probe

978-0-9789217-3-6/07/$25.00

©2007 WEN GLOBAL SOLUTIONS

the reflectivity of the crystal, as is depicted in Fig. 1. The incident angle with respect to the optic axis is θ.

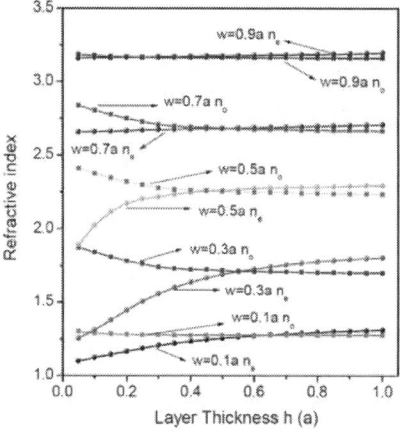

Fig. 2. (Color online) Calculated results of n_o (solid square dot) and n_e (solid circle dot) as a function of h for different values of w.

Figure 2 displays the calculation results of n_o (square dot) and n_e (circle dot) as a function of the thickness of the dielectric rod h, which is taken from $0.05a$ to $1.00a$ for different values of the width of the dielectric rod w ranging from $0.10a$ to $0.90a$. Several major features can be seen from the figure. First, n_o and n_e both increase when w, or equivalently the filling fraction f ($f=w/a$) of the high dielectric rod increases. The reason is simple: More high-dielectric media are involved. Second, $n_o(n_e)$ decreases (increases) as h becomes larger. n_e has a much more significant dispersion than n_o. Third, n_o and n_e is more sensitive to the variation of h at small values of h than at larger values of h. Fourth, the artificial optical anisotropy can change from positive with $n_e > n_o$ to negative with $n_e < n_o$ for every value of w by simply tuning the value of h, with the existence of a transition point where $n_e = n_o$. n_e is almost equal to n_o at $h=0.60a$ for $w=0.10a$, at $h=0.55a$ for $w=0.30a$, at $h=0.45a$ for $w=0.50a$, at $h=0.45a$ for $w=0.70a$, and at $h=0.20a$ for $w=0.90a$. At these parameters, the structurally anisotropic 3D woodpile PC behaves like an optically isotropic medium. This is an interesting while somewhat puzzling phenomenon that is quite beyond the conventional wisdom and one's naive imagination.

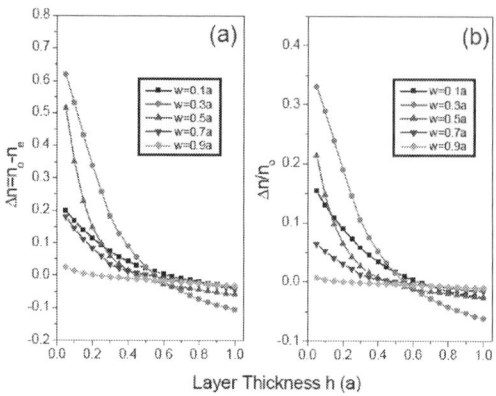

Fig. 3. (Color online) Calculation results for (a) $\Delta n = \left(n_o - n_e\right)$ and (b) $\Delta n/n_0$ as a function of h for different values of w.

To have a clarified picture of the extent of the artificial anisotropy, we display in Fig. 3(a) $\Delta n = \left(n_o - n_e\right)$ as a function of h for different values of w. The corresponding picture of the relative anisotropy $\Delta n/n_0$ is illustrated in Fig. 3(b). A significant point is that both negative and positive anisotropy of the 3D woodpile PC is most significant for a modest filling fraction of the dielectric rods as $f=0.30$. In addition, a small f more benefits the generation of a strong anisotropy than a large f. In fact, at a very large value as $f=0.90$, the anisotropy is negligibly small irrespective of the variation of h. At $w=0.30a$ and $h<0.55a$, $\Delta n > 0$, the 3D woodpile PC exhibits negative anisotropy. The magnitude of anisotropy $|\Delta n|\left(|n_e - n_o|\right)$ grows monotonically with the decrease of h. At $w=0.30a$ and $h=0.05a$, $|\Delta n|$ can reach a very significant value of 0.619, and the corresponding relative anisotropy $\Delta n/n_0$ can be as high as 33%. At $w=0.30a$ and $h>0.55a$, $\Delta n<0$, the 3D woodpile PC exhibits positive anisotropy. $|\Delta n|$ becomes greater with the increase of h, however, $|\Delta n|$ has a much slower variation compared with the small h situation. At $h=a$, $|\Delta n|$ is about 0.1, and $\Delta n/n_0$ is about 6%. Positive anisotropy is far weaker than negative anisotropy for the small h case. It seems that the 3D woodpile structure can facilitate creation of negative anisotropy rather than positive anisotropy. In nature, positive anisotropy seems to be more common than negative anisotropy. For instance, a very significant anisotropy is found for tellurium with n_e=6.2 and n_o=4.8. One can also find from Fig. 3 that an appropriate value of layer thickness h (about 0.5a) can result in small anisotropy or even isotropy. All these features clearly indicate that an optimum design of the nanostructure is of vital importance for artificial optical birefringence.

Conclusions

In summary, we have systematically investigated the optical birefringence in woodpile PC structures with a large variety of geometric parameters by determining the ordinary and extraordinary index of refraction from calculation of the reflection coefficient in the long-wavelength limit with the use of the plane-wave-based transfer-matrix method, Fresnel's formula and Snell's formula. We find that the anisotropy can be widely tuned by simply changing the width and thickness of the dielectric rod. Transition from positive anisotropy to negative anisotropy can be achieved, and at certain parameters, a structurally anisotropic medium can behave like an optically isotropic medium. Our study opens a window to use artificial nanostructures to create an arbitrary optical anisotropy that is not possible in natural crystals. The rapid development of nanoscience and nanotechnology will allow successful synthesis of these nanostructures.

Reference

1 C. Huygens, Traite de la Lumiµere (Van der Aa, Leiden, 1690).

2 W. Nicol, Edinburgh New Phil. J. **6**, (1829) 83.

3 M. Born and E. Wolf, Principles of Optics (Cambridge University Press, Cambridge, England, 1999).

4 J. D. Joannopoulos, R. D. Meade, and J. Winn, Photonic Crystals (Princeton University Press, Princeton, New Jersey, 1995).

5 E. Yablonovitch, Phys. Rev. Lett. **58**, (1987) 2059.

6 S. Datta, C. T. Chan, K. M. Ho, and C. M. Soukoulis, Phys. Rev. B **48**, (1993) 14936.

7 M. C. Netti, A. Harris, J. J. Baumberg, D. M. Whittaker, M. B. D. Charlton, M. E. Zoorob, and G. J. Parker, Phys. Rev. Lett. **86**, (2001) 1526.

8 F. Genereux, S. W. Leonard, H. M. van Driel, A. Birner, and U. Gosele, Phys. Rev. B **63**, (2001) 161101 (R).

9 A. A. Krokhin, P. Halevi, and J. Arriaga, Phys. Rev. B **65**, (2002) 115208.

10 A. A. Krokhin and E. Reyes, Phys. Rev. Lett. **93**, (2004) 023904.

11 R. Reyes, A. A. Krokhin, and J. Roberts, Phys. Rev. B **72**, (2005) 155118.

12 Z. Y. Li and K. M. Ho, Phys. Rev. B **68**, (2003) 155101.

Wavelength assignment algorithm for hybrid WDM/TDM passive optical network

Shaofeng Qiu, Weitao zhu, Jun Huang, zhizhong zhang

Chongqing Univ. of Posts and Telecommun., Chongqing, PRC; qsf@sjtu.org

Abstract *The guideline of wavelength assignment for hybrid WDM/TDM passive optical network is studied and a novel wavelength assignment algorithm is proposed. The numerical and simulations results demonstrate that this algorithm can improve the system's performance.*

Introduction

Whit the development of emerging applications, such as plain old telephone service (POTS), high definition TV, video conferencing and interactive game, the demand of bandwidth for subscribers will keep increasing1-4. Nevertheless, the infrastructure of current access networks suffers from limited bandwidth, which obstructs the networks from delivering integrated services to end-users. Among several types of broadband access technologies, wavelength division multiplexing passive optical network (WDM PON) is the most favorable for the requited bandwidth in the near future, but the application of this system is limited by the high cost of DWDM apparatus and maintenance.

In this paper, the authors propose a novel hybrid WDM/TDM optical access network architecture called wavelength shared WDM PON. The architecture is based on a two-stage distribution tree connecting the optical line terminal (OLT) and optical network units (ONU), which enable ONUs share upstream channel to accommodate more users. Therefor, a scheduling algorithm must be used to enable the leaves to share the available wavelengths dynamically based on changing traffic demands1-4.

Wavelength shared WDM PON architecture

The architecture of wavelength shared WDM PON is shown as Fig. 1. Optical distribute network (ODN) consists of two-stage optical splitter devices, thus ONUs can be divided into w groups (w is the number of downstream data wavelength). At downstream transmission ONUs in a group are assigned a fixed wavelength and TDM technology is used. At upstream link hybrid WDMA/TDMA is used, thus wavelength can be shared by all ONUs. Wavelength shared WDM PON include 1) enabling a set of ONUs to use one downlink wavelength in downstream transmission by hybrid WDM/TDM technology; 2) uplink wavelength shared by all ONUs in upstream transmission using dynamic wavelength-timeslot allocated scheme. 3) the numbers of ONU can be much more than that of available wavelength; 4) minimizing the system cost by adopting CWDM technology.

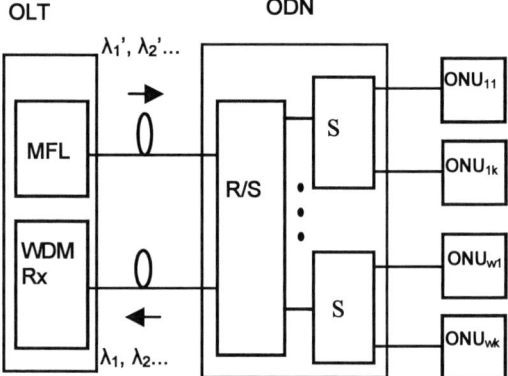

MFL: Multi-frequency laser, R: router, S: splitter
Fig. 1 Structure of wavelength shared WDM-PON

Guidelines of wavelength assignment for WDM PON

The simple system's performance such as average packet delay, throughput, fairness, is clearly superior to the complex system is they have the same capacity. Multi-wavelength system can be considered as complex system, and single wavelength system with the same capacity as the Multi-wavelength system, which is referred as corresponding single wavelength system (CSWS), is a simple system. However, line rate must be higher in corresponding single wavelength system, which leads to the cost of transceiver increasing and will suffer from

978-0-9789217-3-6/07/$25.00

©2007 WEN GLOBAL SOLUTIONS

electronic bottleneck. Multi-wavelength system, instead of CSWS becomes more feasible. From above, Guidelines of wavelength assignment can be denoted as wavelength assignment algorithm is a mode of organized channel, which aim is to eliminate the decline of performance caused by multi-channel. Thus we can evaluate wavelength assignment algorithm through compared with corresponding single wavelength system.

Wavelength assignment algorithm

The algorithm called earliest available time wavelength assignment based frame (EATWA), attempts to schedule the transmission of a packet at the earliest possible time, by selecting a data channel which is available the soonest and transmitting collisionless. In order to achieve the algorithm, the scheduler must maintain two global information: channel available time (CAT), which records the free time of each channel and transmitter available time (TAT), which records the ideal time of each ONU's transmitter. The algorithm works as follows:
(1) Initializing global information at the beginning of a frame.
(2) Selecting the channel with the minimal CAT and transmitting collisionless at the same time when assign wavelength for a given ONU according to the global information.
(3) Updating the global information and repeating (2) until finish of the frame.

Performance analysis

We assume that packets arrive to each of the N ONUs according to Bernoulli distribution of rate λ, thus the number of packets (R) arriving to the system submits to binomial distribution:

$$P\{R = r\} = \binom{Nl}{r} \lambda^r (1-\lambda)^{Nl-r} \qquad (1)$$

where l is the length of slot.
We assume that a frame contains D slots, where d slots used for control meassage, radom variable X and Y denot the packet number in queue at the begining of a frame and after a control slot arriving. The equilibrium probability mass functions of X and Y can be denoted as:

$$P\{X = x\} = \begin{cases} \sum_{k=0}^{w(D-d)} P\{Y = k\} & x = 0 \\ P\{Y = x + w(D-d)\} & \\ \quad x = 1, 2, \cdots s - w(D-d) \\ 0 & x > s - w(D-d) \end{cases} \qquad (2)$$

$$P\{Y = y\} = \begin{cases} \sum_{k=0}^{y} P\{X = k\} P\{R = y-k\} & y = 0,1,\cdots,s-1 \\ \sum_{k=0}^{s-(D-d)/W} P\{X=k\} \sum_{r=s-k}^{\infty} P\{R=r\} & y = s \\ 0 & y > s \end{cases} \qquad (3)$$

Where w is the wavelength number and s is the length of buffer.
The equilibrium probability of the system can be solved form equation (1)-(3). The average length of queue is x Σ P{X} and the packet delay decreases with this value decreasing.

Simulation and results

Given D=53, d=3, w=8, N=64, s=300, defining load of system as N λ D/(D-d), iterate to max {P{X(k)-P{X(k-1)}} < 10 −7, we obtain the average length of queue VS. load of system shown as figure 2. For contrast, we calculate the average length of queue for corresponding single wavelength system and static wavelength assignment (SWA). In SWA, the wavelength is preassigned to each leaf. From fig. 1., it is clear that the length of queue in EATWA is much shorter than that of in SWA, close to that of csws.

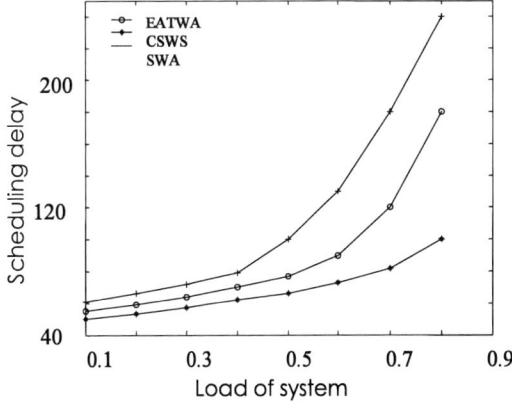

Fig. 2 Average length of queue VS. load of system

In order to validate theory analyses,

Extensive simulations are conducted and the results shown as fig. 3. From Fig.3, it can been seen that the scheduling delay of EATWA is near to that of CWSW, much less than that of SWA, and is not sensitive to load of system.

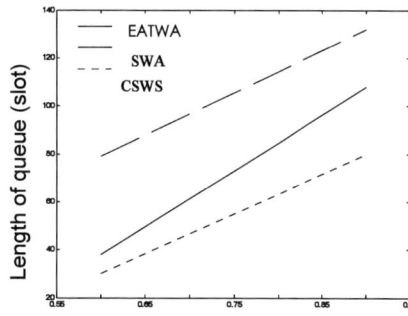

Fig. 3 Scheduling delay VS. load of system

Conclusions

In this paper, we study the guideline of wavelength assignment for hybrid WDM/TDM passive optical network and present a novel wavelength assignment algorithm: earliest available time wavelength assignment based frame. Both theory analysis and simulation demonstrate that this new algorithm can improve the system's performance.

References

1. F. T. An, K. S. Kim and D. Gutierrez, SUCCESS: a next-generation hybrid WDM/TDM optical access network architecture. J. Lightwave Technol., 2004, 22(11): 2557-2569.
2. S. J. Park, C. H. Lee and K. T. Jeong, Fiber-to-the-Home services based on wavelength division multiplexing passive optical network. J. Lightwave Technol., 2004, 22(11): 2582-2591.
3. Y. L. Hsueh, M. S. Rogge and W. T. Shaw, Success-DWA: a highly scalable and cost effective optical access network. IEEE optical Communication, 2004, No. 8, pp524-530.
4. R.D.Feldmen . An evaluation of architectures incorporating wavelength division multiplexing for broad-band fiber access. J. Lightwave Technol., 1998, 16(9):1546-1559.
5. Kataumi Iwatsuki et al J. Lightwave Technol., 22(2004), page 2623.
6. Anthony C. Kam et al IEEE J. Select. Areas Commum., 18 (2000), page 2029.
7. Eytan Modiano et al J. Lightwave Technol., 18(2000), page 461.
8. Yu-Li Hsueh et al IEEE optical communications, (2004), page 524.

Effect of Source Parameters on Beam Self-collimation in 2D Photonic Crystal

Xia LI (1), Wei XUE (1), Yurong JIANG (1)

1 : Department of Photo-electronic Engineering, School of Information Science and Technology, Beijing Institute of Technology, Beijing, 100081, China, email: xiafeng717@bit.edu.cn

Abstract *The influence of source parameters on self-collimated Gauss beam in two-dimensional photonic crystal is analyzed by Finite-difference time-domain method. The transmission fluctuation with source shift, the stable beam-expansion suppression with increased source width are observed and explained.*

Introduction

Recently, dispersion-based self-collimation phenomenon in photonic crystal (PC) has attracted great attention due to its application in light guiding and routing [1-5]. The factors affected self-collimated-beam, like incident beam angle, structure disorder, dielectric constant contrast, etc [1, 6-7], are studied in many literatures. However, there are little attentions paid to effect of source. What influence on propagated beam will be brought as source change? The three parameters of source: horizontal shift dx, perpendicular shift dz, and source width dw, has been chosen in this paper to analyze the effect on self-collimated beam in 2D PC.

2D self-collimation system

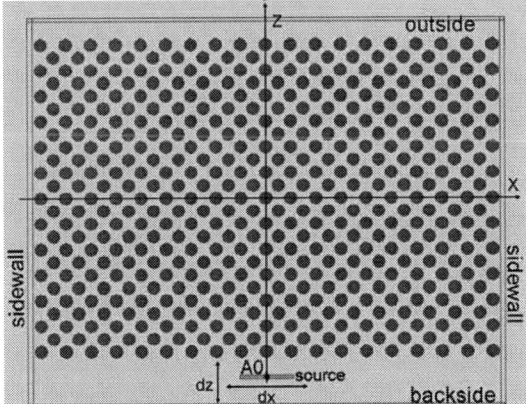

Fig.1 Schematic view of 2D PC system in XZ plane.The Gauss beam of source normally incident to PC. Four monitors marked as outside, backside and sidewall, are located to detect the power passed and describe transmission, reflection and side-leak (sum of two sidewall monitor value), respectively.

Here the 2D system consisted square lattice of air hole in dielectric background

(refractive index is 3.4) in XZ region of $20\sqrt{2}a \times 16\sqrt{2}a$ is considered. See Fig.1, air hole radius is $0.35a$, where a is lattice constant. The source center is located at A0 point $(0, -7\sqrt{2}a)$ and its shift from $A0$ in X and Z direction is described as dx and dz, respectively. The source is Gauss type with dw, the lateral width along X of the optical input field, to be launched. TM mode (the magnetic field is perpendicular to XZ plane) is studied in discussion.

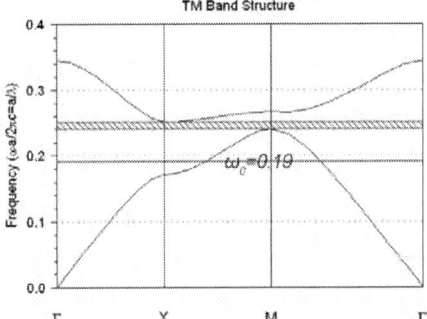

Fig.2 TM band diagram of the 2D square lattice PC in the reduced Brilliouin zone

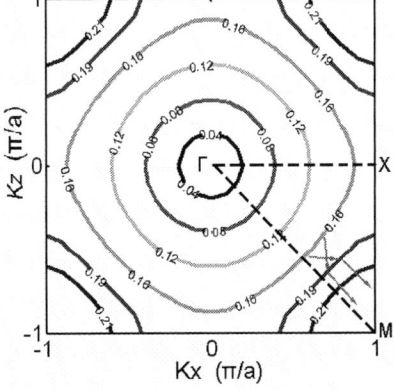

Fig.3 The EFCs at $\omega = 0.02 - 0.21(2\pi c / a)$ (normalized frequency, from inner to outer) in the K-space of the band0 for TM mode.

With given set, the dispersion and equifrequency contours (EFCs) of this 2D PC can be calculated by plane wave expansion method, and the results are shown in fig.2 and fig.3. We take the normalized frequency $\omega_0 = 0.19$ to discuss. It is clear that the light with $\omega_0 = 0.19$ is not fall in PBG and can propagate. Due to the wave vector curve for $\omega_0 = 0.19$ is completely flat and normal to $\Gamma - M$ direction, the incident Gauss beam will exhibit obvious self-collimation phenomenon (see fig.4).

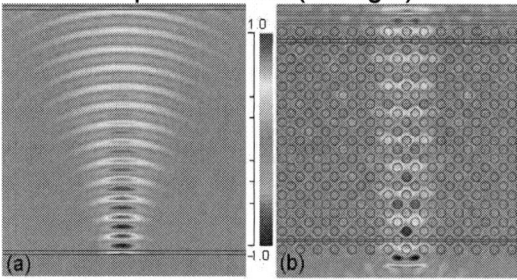

(a) (b)

Fig.4 The propagating magnetic field map for Gauss beam with normalized frequency of 0.19 (a) in silicon, (b) in 2D PC

Effect of source parameters on self-collimation

Firstly, the position change of source is discussed. Due to the periodicity of system, dx is ranged in $[-\sqrt{2}a/2, \sqrt{2}a/2]$.

Fig.5 illustrated that the transmission increases symetrically as source far away from A0 along with X axis.

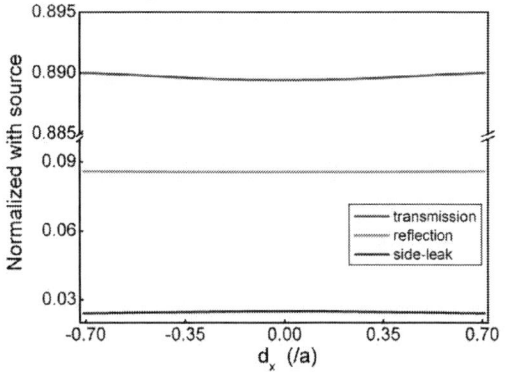

Fig.5 Normalized spectra at outside VS source shift in X direction. $dw = 3a, dz = 0$.

When source vertically shift in Z direction, the transmission wavely decreases like shown in fig.6. The reflection from PC boudary monotonously slight decreases in both two cases but the side-leak reversely varies with transmission. That is to say, the

source position change have more influence on the transmission and side-leak but little on reflection for the collimated beam.

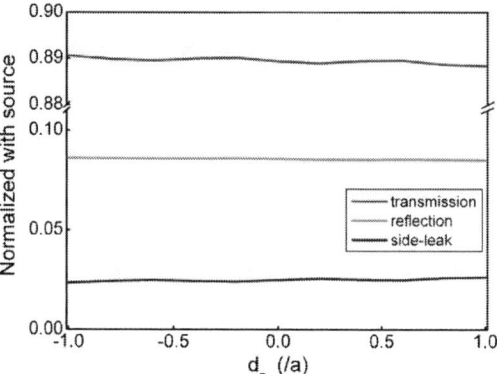

Fig.6 Normalized spectra at outside VS source shift in z direction, dz range $[-a,a]$, $dw = 3a, dx = 0$

Secondly, fix the source at A0 and change the source width. From fig.7, it is found that the transmission increases with dw increase and the reflection decreases simultaneously. Especially, when the source width is very narrow, the light is most reflected in the incident plane of PC and obviously leaked from the sidewall.

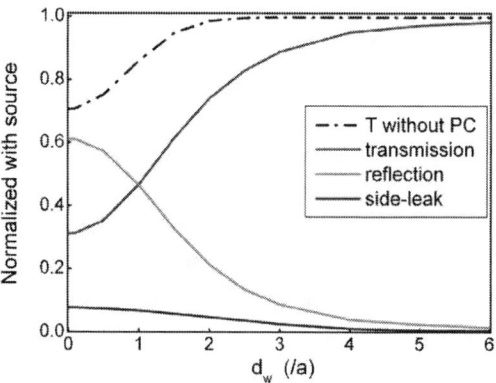

Fig.7 Normalized spectra at outside VS source width, Black dash-dot is the transmission without PC. dw from $0.01a$ to $6a$, $dx = 0, dz = 0$.

We can explain the above difference in K-space. Because the normalized source frequency is fixed to 0.19 and the wave vector components satisfy $k_x^2 + k_y^2 = k_0^2$, the propagating wave can represented with a circle in the K-space (see fig.8). Limited by EFCs shape of $\omega_0 = 0.19$, only parts of directions (filled with green in fig.8) light can be collimated and transmit out PC. Furthermore, the amplitude distribution of

466

this part of directions can be obtained by Fourier transform (illustrated in fig.9, simplified in one dimension due to the dependent of k_x and k_z). Obviously, the amplitude of wave vector components included in $[-k_0, k_0]$ for narrow source width is less than for wider, it means that the carried energy is low and results transmission of collimated beam fall down. If the source width is fixed, of cause the transmission keeps stable.

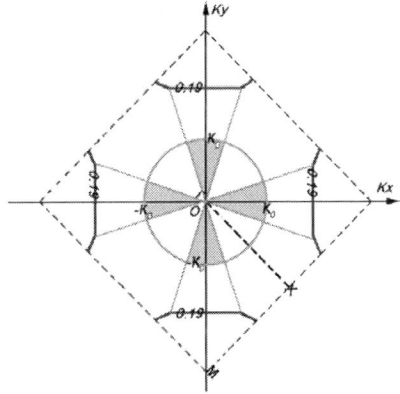

Fig.8 Wave vector circle and EFCs in K-space for $\omega_0 = 0.19$. $\Gamma - X - M$ indicates the irreducible Brillouin zone and k_x can be considered an direction factor in this zone. eg. $k_x = 0$,normally incident to the flat QFCs.

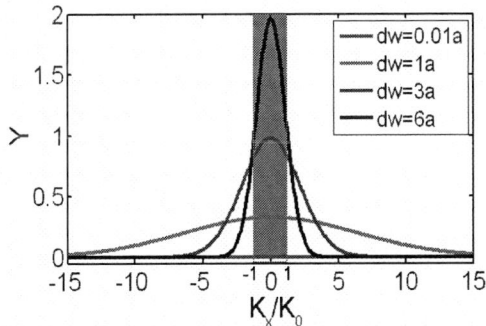

fig.9 Space Fourier transform of Gauss source in one dimension. The gray rectangle shows the region $[-k_0, k_0]$.

Finally, we investigate the influence of source width on self-collimated beam. As before-mentioned in fig.4, the Gauss beam with $\omega_0 = 0.19$ expanded propagates in uniform dielectric but the beam expansion is suppressed in 2D PC. The FWHM of source and beam in outside transmitting with PC or without PC is shown in fig.10. Obviously, the Gauss beam is excellently collimated in

2D PC, especially when the source width is narrow. Furthermore, the beam collimation is not weakening with propagation distance increasing. However, the width of collimated beam is limit close to $2.55a$ and the beam expansion can not be suppressed little more.

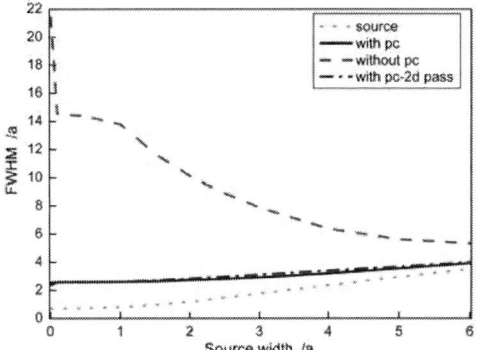

Fig.10 The FWHM contrast of source, out-beam in silicon, in 2d PC and in 2D PC with double propagating distance

Conclusions

To summarize, we have analyzed the effect of source parameters on self-collimated beam in 2D PC by employing FDTD method. The transmission efficiency has little change with different source position, but is strongly impacted by source width. We also give well explain to the difference in K-space. The FWHM comparison results demonstrate the perfect self-collimation to Gauss beam in 2D PC. But there is an utmost value of FWHM for collimated beam, $2.55a$, in our set system. Besides, there are other phenomena relative to self-collimated beam in 2D PC, like out-beam sidelobes and reflection loss at incident boundary, which are not mentioned in this paper but will be studied in next work.

References

1 H. Kosaka, et al, Appl. Phy. Lett. 74(1999), 1212-1214
2 J. Witzens, et al, IEEE J. Quantum Electron. 8(2002), 1246-1257.
3 K.C. Kwan, et al, Appl. Phy. Lett. 82(2003), 4414.
4 S.Y.Shi, et al,Opt. Lett. 29(2004), 617-619.
5 Binglin Miiao, et al, IEEE PHOTONICS TECH. LET. 17(2005), 61-63.
6 Haibo Chen, et al, Physica E 35(2006), 64–68
7 T. Yamashita, et al, IEEE J. Communications, 23(2005), 1341-1347

Improvement of Automatic Alignment Algorithm for Butterfly Laser Module Packaging

Hao Shen (1), Xin Wang (1), Bernard Leung (2), Hongdu Liu (2)

1 : Harbin Institute of Technology Shenzhen Graduate School, Shenzhen, 518055, China,
yzshenhao@163.com

2 : Photonic Manufacturing Service Ltd. Shenzhen, 518038, China,

Abstract *This paper introduces two kinds of approach to improve the efficiency and stability of butterfly laser module packaging. One is the 3-D simplex optical fiber alignment algorithm; the other is named the crash preventing method.*

Introduction

Butterfly laser modules are now widely used in fiber communication and sensing. High coupling efficiency and high stability are the main challenges of butterfly packaging. As we known, Laser Welding 4200 (Newport Corp.) is the most popular alignment and welding machine for butterfly packaging. But this machine still remains something unsolved:

1. The alignment algorithm has a narrow searching field and somehow mistakes a local peak as a maximum optical power point.
2. Short of crash protection. When coupling the Z-axis, the fiber tip is very likely to crash the semiconductor laser diode.

To improve the above-mentioned issues to some degree, we employ the simplex algorithm to improve the coupling efficiency and use the image processing method to prevent the crash. The following sections describe the two kinds of automatic improvement respectively.

Simplex Alignment Algorithm

Because the hill-climbing method is a "one dimensional-at-a-time" gradient searching method, it can only deal with the alignment on one direction at one time, so it is usually time-consuming and often the real peak cannot be detected because of local peak trapping. Another disadvantage of the hill-climbing method is that its searching filed is very narrow, before hill-climbing there should be a coarse alignment first. So a novel approach based on the application of the Nelder–Mead Simplex optimization method was presented.

Although this method has been widely applied in various fields as a promising optimization tool, its first application in photonics packaging automation has been addressed by R. Zhang and F. G. Shi [1]. The fiber-optic alignment process corresponds to the optimization of the 3 DOF coordinates. The essential point of the approach is illustrated using a simple example, for the case of the optimization of two parameters, i.e., X and Y coordinates. In this case, three vertices, corresponding to three different X and Y coordinates, are initially selected to form a triangle, the optical power are measured at these three positions, it is easy to determine the highest, lowest, and middle value of optical power. The next step is to go through the iteration. The purpose of all iterations is to discard the lowest-power vertex, and meanwhile to locate a better vertex by reflecting or contracting the triangle. The iteration ends when the stopping criterion is reached, e.g., some threshold of the optical power is detected, or the standard derivation of the last several measurements is less than a certain value [2].

To test the simplex alignment algorithm, we align two collimators as the experiment results. Before doing the alignment, the light power distribution was scanned by the coupling stage. Seen from Figure 1, a 60 microns offset in X and Y axes (represented by the red and yellow curve) will cause a 50 percent decline in light power, while a 1000 microns offset in Z axis (represented by the green curve) will only cause a 80 percent decline. So X and Y are very sensitive while Z is not.

978-0-9789217-3-6/07/$25.00

©2007 WEN GLOBAL SOLUTIONS   468

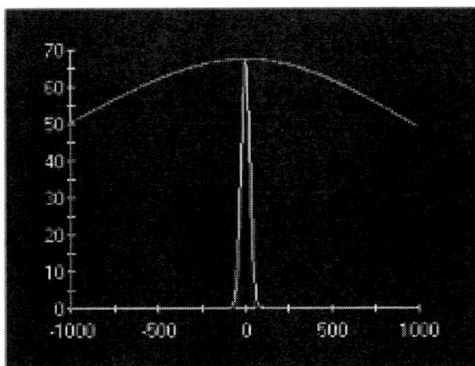

Figure 1. Light power distribution is plotted for 3 axes respectively

In this test, simplex was applied in the alignment of X and Z axes.

We chose 4 different starting points to test the simplex alignment algorithm, and alignment process was repeated 3 times for each starting point. All the testing data was recorded in Table 1. From these data we can conclude that:

1. Simplex has a good repeatability. Every time the peak power was convergent to 65~67 µw, the X coordinate of end points were convergent to 9866 µm. Since the Z axis is not sensitive, its convergent range is a little bigger (seen from Figure 2).

2. Simplex has a huge search range. Even the first light power is 10 nw, simplex still can find the summit.

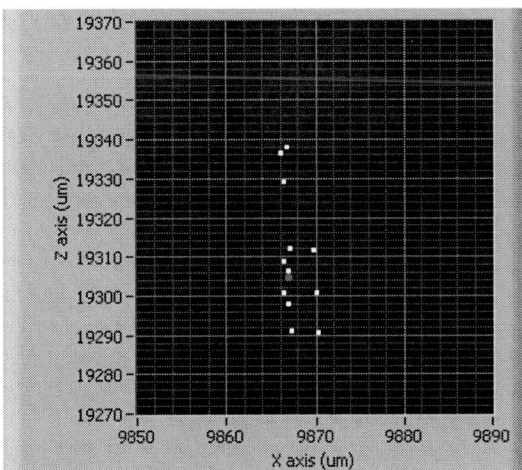

Figure 2. End points of every searching, red point represents the real peak point.

Also, there rose a new problem: crash. Simplex is a nonrestraint optimization

algorithm. If there is no limitation in the motion of Z-axis, the transmitting collimator will crash the receiving one.

So we decide to appeal for the help of image processing method.

Table 1. Testing data

First Light Power (µw)	Peak Power (µw)	Coupling Time (s)
0.60	65.65	46
	65.59	58
	65.65	54
1.67	65.53	52
	65.53	59
	65.47	55
1.99	65.41	53
	65.35	58
	65.35	63
0.0105	65.16	69
	65.04	68
	65.04	62

Crash Prevention

The semiconductor laser diode is very sensitive to the surface damage. Any slight scratch will damage the emitting surface (cavity mirror) and cause the emitting optical power decreasing sharply.

Laser Welding 4200 provides a 3-D hill-climbing algorithm. Before alignments, technicians should set the feeding steps. Large step may cause a crash while little step is time-consuming. In case of sample building, someone prefer to first employ hill-climbing to couple the lensed fiber and the laser chip along X-axis and Y-axis; then drive the fiber close to the laser diode step by step manually. In order to prevent crashing the chip, technicians need to use their experience to feed the fiber with a smaller step, generally 0.5 micron one step, until the detected optical power is decreasing. This is time-consuming and also depends on technicians' skills.

This is a semiautomatic procedure and the most unfortunate thing is when the fiber tip is very close to the laser diode, the transmitting laser will be a disturbing light source for the technicians to estimate the distance between the fiber and the chip.

So we employ the image processing method to detect the gap distance between

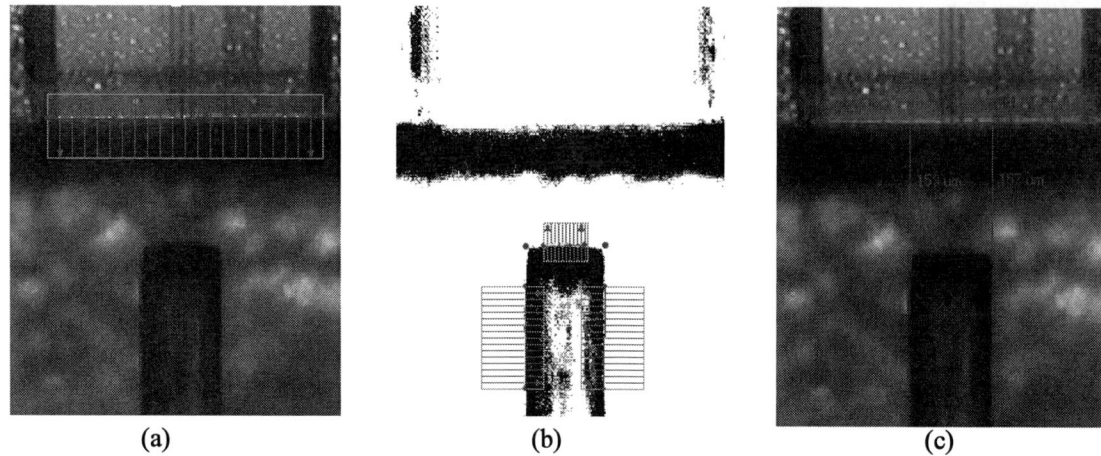

| (a) | (b) | (c) |

Figure 3. Image processing sequence, (a) represents step 2, detecting laser chip edge; (b) represents step 3 and 4, enhancing image and detecting corner; (c) represents step 5, calculating the real distance.

the fiber and the laser diode chip, calculate the moving limitation of the fiber, when the fiber's position will move over the limitation, algorithm will automatically stop the fiber's moving.

The image processing method sequences are divided into 5 steps:

Step 1: Calibrate the image. Get the relation between the image coordinate and the real world coordinate. In our research the scale between them is 1.6 micron/pixel.

Step 2: Search the edge of the laser chip. Detect a mass of points on the edge, fit these points into a line.

Step 3: Enhance the image contrast. Since the contrast around the fiber tip is very low, we first do an image minus operation and then a multiply operation, so the fiber tip can be segmented easily from the background.

Step 4: Fit the edge line of the fiber tip and calculate their intersections as the corners of the fiber tip.

Step 5: Calculate the distances between the laser chip and the two corners. Choose the smaller one as the gap distance. The limited position equals to the current position minus the gap distance.
Figure 3 describes the main image processing sequence.

With the help of image processing method, not only the crash can be avoided, but also the coupling time is reduced remarkably.

Conclusions

These two improvements in automatic alignment are both applied on the Newport 4200 laser welding machine. Newport has provided a software system named PCS for users to operate and also some base functions for engineers to do the improvements [3]. We used some of these basic functions and also add some new codes to realize the whole improvement.

Compared to the semiautomatic Hill-climbing alignment algorithm, new simplex method remarkably improves the coupling speed. Based our experience, technicians need spend 180s or even more to finish the coupling process before while now 80s is enough.

Crash prevention method has been verified to be effective for avoiding crashing the laser diode chip. Therefore it improves the yield-rate remarkably.

Reference

1 R Zhang and F. G. Shi, "A Novel Algorithm for Fiber-optic Alignment Automation," 53rd Electronic Components and Technology Conference. Piscataway, NJ, IEEE. 2003, pp. 256-260
2 J. A. Nelder and R. Mead, "A Simplex Method for Function Minimization," Computer J., vol. 7, pp. 308-313, 1965.
3 LaserWeld PCS™ Process Control System User's Manual. Newport Corp.

Low Loss Performances of Long Range Surface Plasmon Polariton Waveguides with Buffer Layer Structures

Yi Rao, Fang Liu, Yidong Huang, Wei Zhang, and Jiangde Peng

State Key Lab. of Integrated Optoelectronics, Department of Electronic Engineering,
Tsinghua University, Beijing 100084, China
Email: yidonghuang@tsinghua.edu.cn

Abstract *Long range surface plasmon strip waveguide with low refractive index dielectric buffer layers is proposed in this paper. Simulation results show that ultra low propagation loss can be obtained.*

Introduction

Surface plasmon polariton (SPP) is a kind of transverse-magnetic surface electromagnetic excitation that propagates in a wavelike fashion along the interface between metal and dielectric medium.[1,2] Long range surface plasmon polariton (LRSPP) mode supported by a thin metal waveguide shows high potential to be applied in the photonic integrated circuit because it not only has a very low propagation loss[3,4], but also can carry optical signals and electrical signals simultaneously.[5]

We have developed LRSPP waveguides[6] and researched the combination between the LRSPP and traditional dielectric devices.[7,8] In this paper, we report a new LRSPP waveguide structure by sandwiching the metal strip between two low refractive index buffer layers. Comparing with the simulation results reported in Ref. 10, where metal layer embedded in buffer dielectric layers was discussed in 1-dimension,[9,10] the buffered stripe LRSPP waveguides proposed here have further low propagation loss, and can confine the electromagnetic field in two dimensions. This makes them more practical for integrated optical devices.

Structure and Simulation Method

Figure 1 shows the proposed structure. It is consisted of a metal (Au) strip with thickness t and two buffer layers with thickness d_b and refractive index n_b, here n_b is less than that of the surrounding dielectrics n_s. The width of the sandwich strip is w.

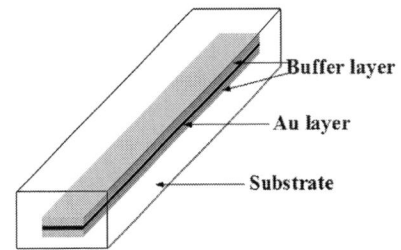

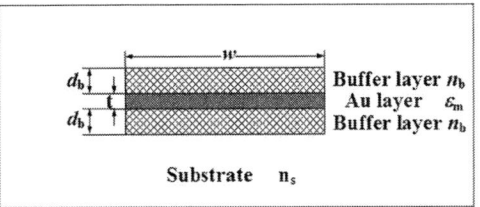

Fig.1 structure of low refractive index buffered LRSPP stripe waveguide with finite width.

The finite elements mehod (FEM), which is very convenient to solve the surface plasmon polariton modes, was adopted here. The propagation constant of LRSPP mode is $\beta = \beta_R + \beta_I$, thus the mode index is β/k_0, where k_0 is the free space propagation constant, $k_0 = 2\pi/\lambda_0$, and wavelength is fixed at 1550nm. All the results shown following is based on $n_b < n_s$ for realizing ultra-low propagation loss (the reason will be discussed in the next section). The substrate is assumed as BCB, a kind of widely used optical polymer, whose refractive index is 1.535. Dielectric constant of Au used in our simulation is $\varepsilon_m = -132 + 12.65i$.[11]

Results and Discussions

The LRSPP mode field distributions are shown in Fig.2 with different d_b when $t = 20$nm and $w = 8\mu$m. The refractive index of

978-0-9789217-3-6/07/$25.00

©2007 WEN GLOBAL SOLUTIONS

the buffer layers is set as n_b=1.5. It is known that, the fewer overlap area between Au strip and mode field, the lower propagation loss of LRSPP is.

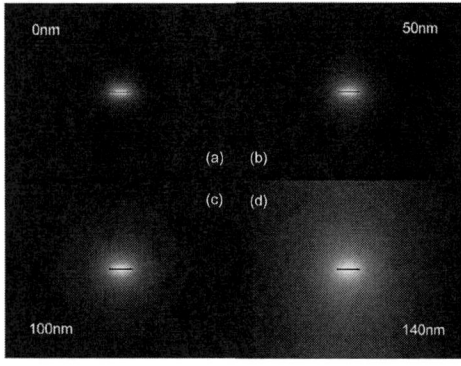

Fig. 2 LRSPP mode field distribution with buffer layer thickness d_b equals (a) 0nm, (b) 50 nm, (c) 100 nm, and (d) 140 nm. Here t = 20nm, w = 8μm, n_b = 1.5, and n_s = 1.535.

As shown in Fig.2, the mode distribution enlarges when thickness of buffer layer increases. It means that the mechanism of reducing the propagation loss by adding buffer layer is the same as reducing Au strip thickness. Thus there should exist a cut-off point of the mode when the thickness of buffer layer increases. At this point, LRSPP mode tends to cut off and propagation loss approaches to zero, which means the propagation distance can be rather large.

The relationship between propagation constant β / k_0 and d_b is shown in Fig. 3. The thickness of Au strip t = 20nm, 30nm, 40nm and the width w is fixed at 8μm. The refractive index of the buffer layer n_b = 1.5, 1.528 and 1.531, respectively. It can be seen that, all the modes is cut-off when n_b equals 1.5, while the 40nm thick Au strip is not cut-off when n_b is equals 1.528 and 1.531, 30nm thick Au strip not cut-off when n_b is equals 1.531.

All the β_R / k_0 are asymptotic to the specific value β_R^∞ / k_0 when d_b goes infinite. If $\beta_R^\infty / k_0 < n_s$, β_R / k_0 is lower than n_s and the LRSPP mode is cut-off when the buffer layer is thick enough. On the other hand, if β_R^∞ / k_0 is larger than substrate refractive index n_s, the LRSPP mode is not cut-off.[10] For instance, when n_b = 1.531 and t = 40nm, β_R^∞ / k_0 is

1.538, which is larger than n_s = 1.535, so this mode is not cut-off no matter how thick the d_b is.

When the curve goes to cut-off point, β_R / k_0 is asymptotic to the substrate refractive index n_s (Fig.3 (a)), and β / k_0 goes to zero at the cut-off point (Fig.3 (b)). When n_b = 1.5, t = 20nm, w = 8μm and d_b = 150nm, β / k_0 is as low as 1.57e-6, which means the propagation distance can be about 160mm. These results are much lower compared with the loss reported in Ref. 11.

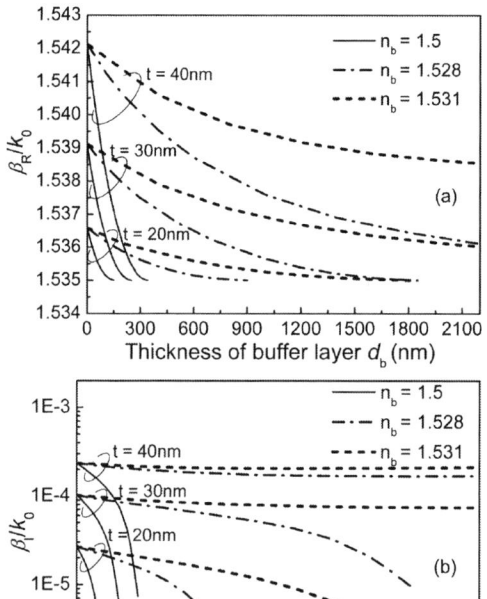

Fig.3 Propagation constant of LRSPP mode supported by buffered strip waveguide varies versus thickness of buffer layer. Here n_b equals 1.5, 1.528 and 1.531, respectively. (a) real part, (b) imaginary part.

According to Fig.3 (a), when the thickness of Au strip t enlarges or the refractive index of buffer layer n_b increases, the LRASPP mode goes hard to be cut-off, because the restriction on mode field distribution is stronger. If $n_b > n_s$, there is no cut-off point no matter how thin the t is. This is why n_b must smaller than n_s.

The propagation loss changes with t and d_b. Figure 4 shows the calculation results. It can be seen that, when there is no buffer layer, the propagation loss is the highest compared with other curves. With the same thick Au strip, propagation loss is lower in thicker buffer layer until the LRSPP mode goes to be cut-off. The propagation loss of 25nm thick Au strip with buffer layer is equivalent to 15nm thick Au strip without buffer layer. Thus the propagation loss can be reduced without decreasing the thickness of Au strip. This result is very useful in fabrication process because it is hard to get Au film thinner than 15nm.[12]

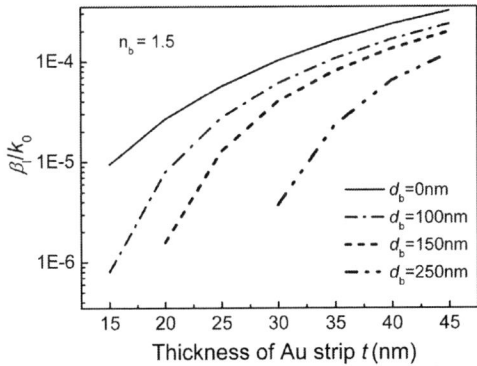

Fig.4 Propagation loss (β/k_0) varies versus thickness of Au strip t and buffer layer d_b. Here $n_b = 1.5$

Conclusions

A new structure of low refractive index dielectric materials buffered LRSPP strip waveguide is reported in this paper. Mode characteristics of propagation and cut-off are analyzed. The simulation results show that the propagation loss can be reduced efficiently and long propagation distance as long as 160mm can be estimated. This kind of waveguide provides a new method to realize ultra-low propagation loss LRSPP waveguide instead of challenging the fabrication of ultra-thin metal film.

Acknowledgement

This work was supported by the National Basic Research Program of China (973 Program) under Contract No. 2007CB307004. The authors would like to thank D. Ohnishi, H. Takasu, and A. Kamisawa of ROHM Corporation for their valuable discussions and helpful comments.

References

1 H. Raether *Surface Plasmons* (Springer, Berlin, 1988), p. 4–13.

2 A. V. Zayats, et al, Phys. Rep. 408 (2005), 131

3 P. Berini, Phys. Rev. B 61 (2000), 10484

4 T. Nikolajsen, et al, Appl. Phys. Lett. 82 (2003), 668

5 T. Nikolajsen, et al, Appl. Phys. Lett. 85 (2004), 5833

6 Y. Rao, et al, Chin. Phys. Lett. 24 (2007), 1626

7 F. Liu, et al, Appl. Phys. Lett. 90 (2007), 141101

8 F. Liu, et al, Appl. Phys. Lett. 90 (2007), 241120

9 J. P. Guo, et al, Opt. Express 14 (2006), 12409

10 R. Adato, et al, Opt. Express 15 (2007), 5008

11 E. D. Palik *Handbook of Optical Constants of Solids* (Academic, Orlando, 1985), p. 286–297.

12 L. Holland *Vacuum Deposition of Thin Films* (Chapman and Hall, 1996).

Measurement of Small Aspheric Surface Using Interferometric System for Spherical Surface Test

Nam-Young Jang (1), Pyung-Suk Choi (2), Jae-Jeong Eun (2)

1: School of Mechatronics Engineering, Changwon National University, Korea,
optofiber@changwon.ac.kr

2: Department of Electronics Engineering, Changwon National University, Korea,
choips@sarim.changwon.ac.kr, jjeun@changwon.ac.kr

Abstract *Using the aspheric analysis algorithm, a Fizeau-type phase shifting interferometer (FPSI)-based spherical surface testing system is developed for analyzing small aspheric surface. Measurement result of an aspheric surface is compared to the conventional aspheric testing system.*

Introduction

In recent years, a demand of optical systems has increased rapidly because of the advance of opto-electronics and optical information instruments. Especially, the more compact and lighter optical systems have been very popular due to the increased demand and marketing and expansion of mobile phone cameras, digital cameras etc.. Therefore, with reduction of the number of lenses in the system and without loss of the optical performance, aspheric lenses come into the spotlight.

Also, mass production of aspheric lenses and its cost-down were allowed by precision molds and injection molding technology. By the way, deformation of aspheric shapes which can degrade their performance is induced by changes of susceptible injection condition or abrasion of molds. This problem requires a technology of testing aspheric surface in a quantitative and in simpler and easier than conventional counterpart, e.g., the computer generated hologram (CGH) [1-2] and null lens test [3]. Typically, the CGH or null lens test has limits for testing different aspheric lenses because each aspheric lens requires its own unique null lens or CGH which are expensive.

The purpose of this paper is to present the measurement results of a small aspheric lens by a simple and cost effective interferometric system using the aspheric analysis algorithm [4].

FPSI-based Spherical Surface Testing System

For measuring the small aspheric lens, the FPSI-based spherical surface testing system employing the aspheric analysis algorithm was designed and developed. Fig. 1 shows a schematic diagram of the FPSI system, and this system requires the computer-controlled system that is incorporated an optical processing part and an image signal processing part, to automate the measurements and reduce the time consumption. The optical processing part was implemented by applying the phase shifting interferometric method to Fizeau interferometer. In the Fizeau interferometer, a surface error $s(x,y)$ tested in reflection is followed [5].

$$s(x,y) = \frac{\lambda}{4\pi}\phi(x,y) \qquad (1)$$

Where (x,y) is a spatial coordinates and λ is a wavelength of the light source. To analyze the acquired interferogram from the optical processing part, the 5-step algorithm [5] with 5 interferograms was applied. In this algorithm, a linear phase shift of $\pi/2$ is assumed between each step, so that the phase wavefront or phase difference $\phi(x,y)$ is given by

$$\phi(x,y) = \tan^{-1}\left[\frac{2(I_2 - I_4)}{2I_3 - I_5 - I_1}\right] \qquad (2)$$

Where I_1, I_2, I_3, I_4 and I_5 is the five measured intensity interferograms and (x,y) is the two-dimensional coordinates at each point in the interferogram. For the phase shifting interferometric method, piezo-electric transducer (PZT) was used to

978-0-9789217-3-6/07/$25.00

©2007 WEN GLOBAL SOLUTIONS

realize time varying characterization for phase shift and Itoh's method [6] which is the one-dimensional phase unwrapping method was also applied for recovering of the successive phase wavefront.

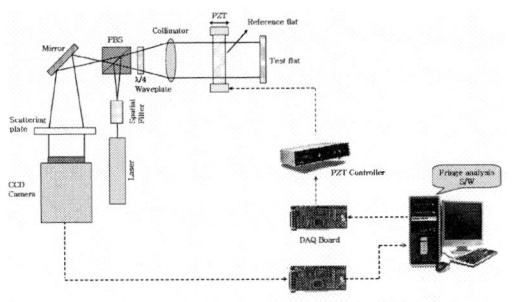

Fig.1. Schematic diagram of the FPSI-based spherical surface testing system

The Aspheric Analysis Algorithm

Measurement of an aspheric surface using the spherical surface testing system causes a systematic error, as shown in Fig. 2, including the shape error of the aspheric surface due to the difference between the reference spherical surface and the testing aspheric surface.

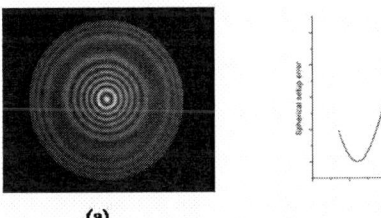

Fig.2. General interferogram of an aspheric surface (a) and systematic error of an aspheric surface (b) in the spherical surface testing system

The aspheric analysis algorithm applied to the FPSI-based spherical surface testing system is a method that separates an aspheric surface error from the systematic error. In this algorithm, the process of aspheric surface analysis is following in steps. First, if an aspheric surface is regarded as a poor spherical surface, it can be aligned in the spherical test setup with minimized the number of fringes. Then, the systematic error is measured. Second, using the aspheric equation (3) and Fig. 3, a theoretical systematic error can be calculated by obtaining the distance D

between the focal point of the reference spherical surface and the origin of the aspheric surface depending on the position of the aspheric surface to be tested. The optimal test position D is a distance to be minimizing the difference between aspheric surface and reference spherical surface. Finally, the shape error of aspheric surface can be determined by subtracting the measured systematic error from the theoretical error calculated in above step. Generally, aspheric surface with rotational symmetry are described by the aspheric equation which is represented as a high-order polynomial [7].

$$z(Y) = \frac{CY^2}{1+\sqrt{1-(A_2+1)C^2Y^2}} + \sum_{n=2}^{5} A_{2n}Y^{2n} \quad (3)$$

Where $z(Y)$ is sagitta value of the surface, i.e., the distance at height Y ($Y^2 = x^2 + y^2$) to the x, y–plane along the z-axis, $C = 1/R_O$ is the curvature of the surface and R_O is the radius of reference spherical surface. Also, A_{2n} is the aspheric coefficient and A_2 is the conic coefficient.

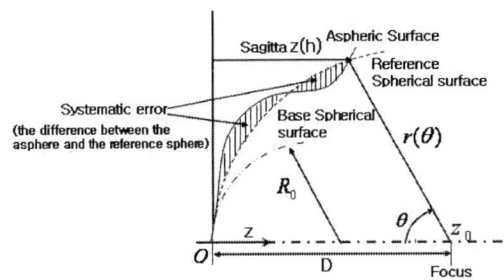

Fig.3. Relation between the aspheric surface and reference spherical surface in a spherical test setup

Measurement Results

In this section, the spherical and aspheric surface measurements were demonstrated in the present system with a small spherical and aspheric lens. First, the spherical surface was measured and the result is shown in Fig. 4. Also, the result was compared to those of Mark IV produced by Zygo. Fig. 4 shows the reconstructed 3-D spherical surface, wrapped wavefront and 5 interferograms for the small spherical lens. PV and RMS of the reconstructed surface

475

are 0.724 wave and 0.1074 wave respectively. In Mark IV, PV and RMS are 0.707 wave and 0.114 wave respectively. For both systems, the measured error of RMS was less than $\lambda/100$ and the measurement results for spherical surface are summarized in table 1.

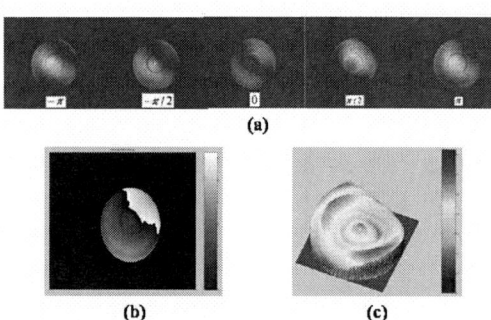

Fig.4. Reconstructed spherical surface using the FPSI-based spherical surface testing system: (a) Interferograms, (b) wrapped wavefront and (c) Reconstructed 3-D spherical surface for the small spherical lens

Table 1. Comparision of the measurement results for spherical surface

Interferometer	PV	RMS	ΔPV	ΔRMS	Repeativity
Zygo Mark IV	0.707	0.114	•	•	•
FPSI	0.724361	0.10741	0.0174	0.0066	0.005

Next, the measurment result of the aspheric surface shows in Fig. 5. In Fig. 5(a), the dotted curve represents the measurement data, the full line is the theoretical systematic error and the dashed line is the shape error of aspheric surface remaining after subtracting the theoretical error from the measurment data. Also, the optimal test position D is calculated at -3.009mm. PV and RMS value of the reconstructed aspheric surface are 1.897 wave and 1.107 wave respectively.

Conclusions
In this paper, the FPSI-based spherical surface testing system applying the aspheric analysis algorithm was developed. Also, the spherical and aspheric surfaces were measured. The result of the measurement of spherical surface was compared to conventional system.
Also, for measurement of aspheric surface,

the result of the measurement of aspheric surface was presented and will be compared to that of conventional aspheric measurement system.

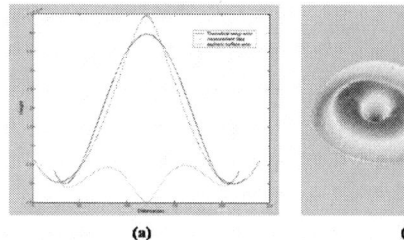

Fig.5. Reconstructed aspheric surface using the FPSI-based spherical surface tesing system : (a) Elimination of the systematic error and (b) Reconstructed 3-D aspheric surface for the small aspheric lens

References
1 A. J. MacGovern et al, Appl. Opt., 10(1971), 619
2 J. C. Wyant et al, Appl. Opt., 11(1972), 2833
3 R. T. Holleran, Appl. Opt., 2(1963), 1336
4 T. Blumel et al, Proc. SPIE, 5965(2005), 351
5 D. Malacara, *Optical Shop Testing*, John Wiley & Sons, Inc., New York (1992)
6 K. Itoh, Appl. Opt., 21(1982), 2470
7 M. Laikin, *Lens Design*, Marcel Dekker, Inc., New York (1995)

Design and Realization of Strip-loaded Waveguide Electro-Optic Modulators in Barium Titanate

Jiansheng Tang (1), Shujun Yang (2), Apichai Bhatranand (3)

1 : Department of Mathematics and Physics,
Hunan First Normal College, Changsha, Hunan 410002, P.R. China
E-mail: yztjs@163.com

2 : Applied Materials, Inc.
974 E. Arques Ave, Sunnynale,CA 94085, U.S.A.

3 : Department of Electronics and Telecommunication Engineering
King Mongkut's University of Technology Thonburi, Bangkok 10140, Thailand

Abstract *The design and realization of Si_3N_4 strip-loaded waveguide electro-optic modulators in barium titanium ($BaTiO_3$) are demonstrated. Low half-wave voltage length product of 2.3 V-cm and low loss of 0.5 dB/cm are achieved at 1.55 μm.*

Introduction

The increasing use of wavelength division multiplexing in high-speed and long-haul fiber communication systems necessitates the use of low driving-voltage and high-speed electro-optic modulators. Significant advances have been made in the design and performance of lithium niobate electro-optic modulators. In recent years, Barium Titanate ($BaTiO_3$) material has attracted increased interests for integrated optics and communication systems due to its even larger linear and non-linear electro-optic coefficients, compared to that of lithium niobate. The large electro-optic coefficients of barium titanate makes it an important material for low driving-voltage electro-optic modulators [1]. It is preferred that electro-optic waveguide modulators have low propagation loss and low half-wave voltage length product. Ridge waveguide electro-optic modulators in $BaTiO_3$ thin film, fabricated by wet-etching or dry-etching, were reported, but their waveguides had relatively large propagation loss due to a large surface and sidewall roughness during etching [2,3]. Strip-loaded thin film $BaTiO_3$ electro-optic waveguide modulators were reported also, however, due to multi-domain structure in the $BaTiO_3$ thin film, its largest electro-optic coefficients are not realized. In this paper, we report low loss silicon nitride strip-loaded strain-induced waveguide electro-optic modulators in $BaTiO_3$ crystal. The strain-induced waveguide are produced through increasing refractive index of the crystal from a thin film deposited on the surface without changing the crystal's composition [4]. When a Si_3N_4 film is deposited on the $BaTiO_3$ substrate at an elevated temperature, the $BaTiO_3$ substrate tends to expand much more than the silicon nitride film. The thermal expansion coefficient of Si_3N_4 is 3.3 x 10^{-6} and that of $BaTiO_3$ is 6.2 x 10^{-6} (c-axis crystal). After cooling down to room temperature, the silicon nitride film on the top of the $BaTiO_3$ substrate is compressed by the contraction the substrate. In this way, a compressive strain is induced in the film. After delineating the waveguide pattern through etching, the refractive index undergoes a change via the photo-elastic effect due to static shear strain. The strain-induced waveguide shows low propagation loss of 0.5 dB/cm at 1550nm. Strip-loaded strain-induced waveguide electro-optic modulator in $BaTiO_3$ is demonstrated for the first time and has a half-wave voltage length product of 2.3 V-cm.

Structure and Fabrication

The structure of the Si_3N_4 strip-loaded strain-induced $BaTiO_3$ waveguide is shown schematically in Fig. 1. The barium titanate substrates used here are commercially available one-side-polished, (100)-oriented single crystal. A 2 μm Si_3N_4 layer was deposited on the $BaTiO_3$ crystal using plasma enhanced chemical vapor deposition (PECVD) at an elevated temperature of 300 °C. A 8μm wide waveguide was delineated defined using photolithography. The strain waveguides were formed by etching the Si_3N_4 layer using a CF_4-based chemistry. The strain is resulted from a thermal expansion mismatch between the crystal and the Si_3N_4 film, after cooling down from 300 °C (PECVD) to room temperature. The refractive index undergoes a change via photo-elastic effect due to a static shear strain between Si_3N_4 and $BaTiO_3$. The increased refractive index under Si_3N_4 film forms a channel waveguide. A lithographic lift-off process with subsequent deposition of a metal layer of 200 nm NiCr by E-beam evaporation was used to obtain the electrodes. The electrodes gap is 15 μm and length is 4.5 mm. The $BaTiO_3$ substrate end-faces were polished for light coupling.

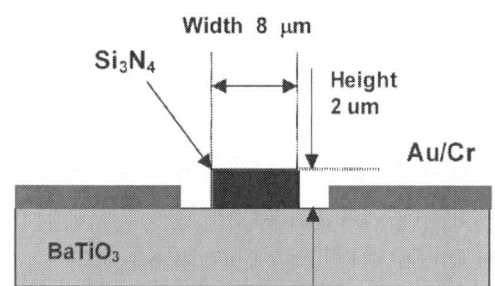

Fig. 1 Schematic structure of strain-induced waveguide modulator.

Fig. 2 shows simulated mode profile based on the above fabrication parameters. It is clearly shown that single mode property is achieved for this structure.

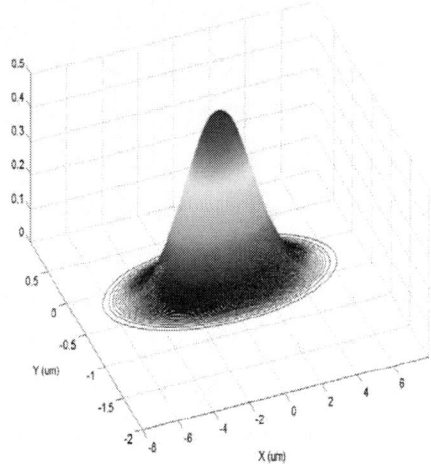

Fig. 2 Waveguide mode profile for the strain-induced waveguide.

Fig. 3 shows the simulated electric field distribution along the waveguide. Uniform electric field can be obtained along the waveguide center.

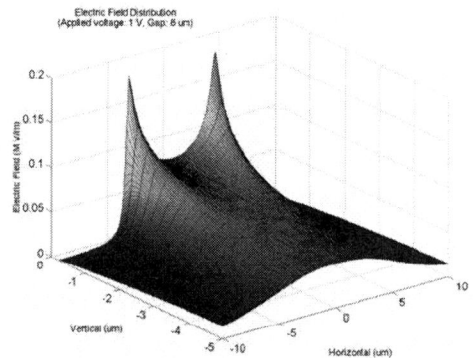

Fig. 3 Electric field distribution along waveguide.

Experimental Results

Single mode waveguide characteristics were carried out at a wavelength of 1553 nm by butt coupling from a pigtailed distributed feedback (DFB) laser diode. Propagation losses of the Si_3N_4 strip-loaded strain-induced waveguides were measured by a standard cutback techniques. The waveguide couping endface remained unaltered for the whole measurement process. TE and TM polarizations were chosen by a fiber polarization controller. The waveguide propagation losses are 0.7 dB/cm and 0.5 dB/cm for TM and TE mode, respectively. The polarization dependent loss (PDL) is 0.2 dB/cm. Only single mode characteristics were observed for both TE and TM polarizations.

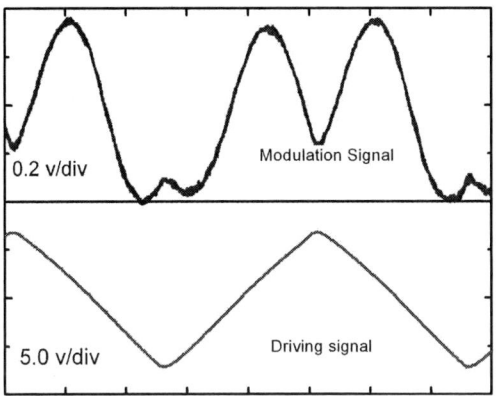

Fig. 4 Electro-optic response of the strain-induced electro-optic waveguide modulator. Top trace is the modulation output signal from the electro-optic modulator. Bottom trace is applied driving signal on the electrodes of the modulator.

Electro-optic modulation tests to measure half-wave voltage V_π were carried out by coupling light from 45 degree to the <100> crystal direction and detecting the output signal through a crossed linear analyzer. A 1 kHz modulation frequency was directly applied on the electrodes. electrodes. The applied electric signal and the output signal acquired by LabVIEW from the detector are shown in Fig. 4. V_π for an EO modulator of 4.5 mm length is 5.2 V. The V_π-L product is, therefore, 2.3 V-cm.

Conclusions

In summary, low loss strip-loaded strain-induced waveguide electro-optic modulators in $BaTiO_3$ crystal are demonstrated for the first time. Only single mode characteristics were observed for TE and TM modes. The propagation losses for TM and TE modes are 0.7 dB/cm and 0.5 dB/cm, respectively. The PDL is 0.2 dB/cm. Low half-wave voltage length product of 2.3 V-cm is achieved. The half-wave voltage length product can be further decreased through an optimizing waveguide structure.

References

1. M. Zgonik et al, *Phys. Rev. B,* **50** (1994), pp. 5941-5945.
2. D. M. Gill et al, *Appl. Phys. Lett.,* **71**, (1997), pp. 1783-1785.
3. A. Petraru et al, *Appl. Phys. Lett.* **81** (2002), pp.1375-1377.
4. J. M. Marx et al, *Appl. Phys. Lett.,* **67**(1995), pp. 1381-1383.

Evolution of partially coherent solitons in optical lattices

Hui Zhuo (1), Shuangchun Wen (2), Yonghua Hu (3)

Research Center of Laser Science and Engineering and School of Computer and Communication, Hunan University, Changsha 410082, China

1: *zhuohuitxh@gmail.com*,

2: Corresponding author: *scwen@vip.sina.com*,

3: *hyhyt@126.com*

Abstract *we investigate the propagation of a partially coherent beam in optical lattice. The condition for formation of partially coherent solitons is obtained using mutually coherent function, and the influence of partial coherence on the evolution of light beam is identified.*

Introduction

Periodic nonlinear lattices occur in a large variety of systems, recently, those periodic photonic microstructured media and photonic crystals attracted a lot of interest due to they are able to capture and hold optical radiation and propagation [1,2] and are essential for applications such as all-optical signal routing and switching. It is a microstructured modulation providing for a powerful tool for the study of nonlinear dynamics. When the propagation of light is in the nonlinear media with optical lattices along the transverse direction, there are two key parameters, namely, the beam width and transverse modulation period [3]. The above papers are studied the propagation of full coherence beam, however, all beams are partial coherence and actually, it is not possibility to full coherence or full incoherence. So, for the aim of all-optical controlled, among recent advances in the field of optical solitons, demonstration of the formation of partially coherent spatial solitons [4,5] has attracted particularly strong attention as it opens the possibility of using light sources with degraded or poor coherence in soliton-based all-optical signal processing. Typically, spatial solitons are created by self-trapping coherent optical beam, i.e., beams whose phase at any two points is fully correlated. Such beams differ significantly from those generated by an incoherent light source in which these is no correlation between light emitted from two different points. This results in some level of randomness (or partial correlation) in the phase across the beam. The weaker the phase correlation, the stronger the incoherence. As a result, a partially coherent beam spreads faster than its coherent counterpart of the same width. However, the optical lattices microstructured media have the properties of nonlinearity and can suppress the diffraction effect, and can still keep the balance between diffraction and nonlinearity. Simultaneity, it has turned out that self-focusing and soliton formation are still possible provided the nonlinear medium is inertial and responds much slower than the time scale characterizing the random phase variation. In such cases the medium will respond to the time averaged intensity, which, being a smooth function of the spatial variables, will induce a smooth waveguidelike structure trapping the beam.

There are several approaches as far as the theoretical description of propagation of partially coherent beams in a slow nonlinear medium is concerned: the coherent density approach [6], a multimode decomposition of the field [7], however, the most natural way of treating the propagation of a partially coherent beam is to use the so-called mutual coherence function, which gives a measure of the correlation between the amplitude of the field at two different points in the beam. In this paper we show that the evolution equation for the coherence function in a logarithmically nonlinear and optical lattices medium has an exact analytical stationary solution for partially coherent beam.

Analysis of the evolution of partially coherent beams

Let us consider the propagation of a partially coherent beam in a slow nonlinear bulk medium which responds to the time averaged intensity and along the z axis with periodic modulation of the linear refractive index in the x direction. We start with a paraxial wave equation describing propagation of a one-dimensional quasimonochromatic partially beam with the amplitude $\psi(x,z)$:

$$i\frac{\partial \psi}{\partial z} + \frac{1}{2k}\frac{\partial^2 \psi}{\partial x^2} + \delta n(I)\psi + hp(x)\psi = 0 \qquad (1)$$

where $k = n_0\omega/c = 2\pi n_0/\lambda$, where n_0 is the linear part of the refractive index, λ is the wavelength of the input light beam, h is the refractive index modulation depth, $p(x) = \cos(2\pi x/T)$ describes the refractive-index profile, T embodies the modulation period.

We study the propagation of partially coherent beam using the mutual coherence function $\Gamma(x_1,x_2)$, defined as

$$\Gamma(x_1,x_2,z) = \langle \psi(x_1,z)\psi^*(x_2,z)\rangle \qquad (2)$$

where brackets denote temporal or ensemble averaging. We assume the nonlinear term δn is

$$\delta n(x,I) = n_2 \ln\left[\int R(x-\xi)I(\xi)d\xi\right] \qquad (3)$$

We consider only response functions $R(x)$ that are real (i.e., no nonlinear loss or gain). We further assume that the response function can be written the following relation:

$$R(x) = 1/(\pi\sigma^2)\exp(-x^2/\sigma^2). \qquad (4)$$

where $\int R(x)dx = 1$, σ determines the degree of nonlocality. It is straightforward to show that the mutual coherence function satisfies the differential equation:

$$i\frac{\partial \Gamma}{\partial z} + \frac{1}{2k}\frac{\partial^2}{\partial p\partial q}\Gamma + \left[\delta n(x_1,I) - \delta n(x_2,I)\right]\Gamma$$
$$+ h\left[\cos\left(\frac{2\pi x_1}{T}\right) - \cos\left(\frac{2\pi x_2}{T}\right)\right]\Gamma = 0 \qquad (5)$$

where $p = (x_1+x_2)/2$, $q = x_1-x_2$ In order to find solutions to this equation, we assume that the incident beam $\psi(x,0)$ possesses Gaussian statisices which implies that the coherence function initially has form

$$\Gamma(p,q,z=0) = \exp\left(-\frac{p^2}{\rho_0^2} - \frac{q^2}{\theta_0^2}\right), \qquad (6)$$

Where ρ_0 is the initial beam width, θ_0 is the effective coherence length and there is a relation between θ_0 and the initial coherent length r_e:

$$\frac{1}{\theta_0^2} = \frac{1}{r_e^2} + \frac{1}{4\rho_0^2} \qquad (7)$$

Due to the Gaussian response function formation of the nonlinearity, the envelope function $\psi(x,z)$ will maintain the Gaussian statistics during propagation, and thus the coherence function will keep the form of equation (6). We can therefore look for solutions to Eq. (5) using the Gaussian function

$$\Gamma(p,q,z) = A(z)\exp\left(-\frac{p^2}{\rho^2(z)} - \frac{q^2}{\theta^2(z)} + ipq\mu(z)\right), \qquad (8)$$

where $A(z)$ and $\mu(z)$ reqresent the amplitude and phase variation of the coherence function, and $\rho(z)$ and $\sigma(z)$ its diameter and coherence radius, respectively. The initial conditions are $A(0) = 1$, $\rho(0) = \rho_0$, $\sigma(0) = \sigma_0$, and $\mu(0) = 0$. Inserting these expressions and Eq. (3), (8) into Eq. (5), we can obtain:

$$i\frac{\partial \Gamma}{\partial z} + \frac{1}{2k}\frac{\partial^2 \Gamma}{\partial p\partial q} - \frac{2n_2 Apq\Gamma}{\rho^2+\sigma^2} + 2h\left(\frac{\pi}{T}\right)^2(x_2^2-x_1^2)\Gamma = 0. \qquad (9)$$

Note that we have approximated to second-order in the last term of Eq. (5). We obtain a set of ordinary differential equations for the parameters of the coherence function using Eq. (8):

$$\frac{\partial A}{\partial z} = -\frac{2A\mu}{k} \qquad (10)$$

$$\frac{\partial \rho}{\partial z} = \frac{\rho\mu}{k} \qquad (11)$$

$$\frac{\partial \theta}{\partial z} = \frac{\theta\mu}{k} \qquad (12)$$

$$\frac{\partial \mu}{\partial z} = \frac{4}{k^2\rho^2\theta^2} - \frac{\mu^2}{k} - \frac{2n_2}{\rho^2+\sigma^2} - \frac{4\pi^2 h}{T^2} \qquad (13)$$

Combining Eq. (10)- (13), we obtain the evolution equation

$$\frac{d^2\rho}{dz^2} = \frac{4\rho_0^2}{k^2\rho^3\theta_0^2} - \frac{2n_2\rho}{k(\rho^2+\sigma^2)} - \frac{4\pi^2 h\rho}{kT^2}, \qquad (14)$$

describing the dynamics of the width $\rho(z)$ of a partially coherent beam (with Gaussian statistics) in nonlocal nonlinear medium. Clearly, the dynamics is determined by a competition free spreading and nonlinearity and optical lattices. We finally integrate Eq. (14) once. This yields an equation for the variation of $\rho(z)$, which is analogous to that of a particle moving in a potential well and it is convenient to introduce the normalization $y(z) = \rho(z)/\rho_0$. This yields

$$\frac{1}{2}\left(\frac{dy}{dz}\right)^2 + \Pi(y) = 0. \qquad (15)$$

The potential field $\Pi(y)$ is given by

$$\Pi(y) = \frac{2}{k^2\theta_0^2\rho_0^2 y^2} + \frac{n_2}{k\rho_0^2}\ln(\rho_0^2 y^2 + \sigma^2) + \frac{2\pi^2 hy^2}{kT^2} + c', \qquad 16)$$

where $c' = -2/(k^2\theta_0^2\rho_0^2) - n_2\ln(\rho_0^2+\sigma^2)/(k\rho_0^2) - 2\pi^2 h/(kT^2)$.

Thus, the solution of partially coherent

solitons problem is reduced to solving Eq. (16) for the beam width variation, since once $\rho(z)$ is know, the others can be determined by Eqs. (10)-(13).

The potential well description

We assume that the "particle starts from rest," i.e., that the beam at $z=0$ has $\rho(0)=\rho_0$ and $(d\rho/dz)_{z=0}=0$. For the soliton solution, we can obtain a special case of $\Pi'(1)=0$ from Eq. (18). In this case, the potential well has degenerated into a single point with y and a particle released at this point stays there (see Fig. 1). Under this condition, the diffraction of light beam is balanced with the nonlinear and optical lattice and partial coherence effects. This should correspond to the case of a bright soliton solution, in which the beam is trapped to form a stable spatial soliton. From Eq. (18), the soliton condition is

$$n_2 = 2\left(\rho_0^2 + \sigma^2\right)\left[\frac{1}{k\rho_0^2}\left(\frac{1}{r_e^2} + \frac{1}{4\rho_0^2}\right) - \frac{\pi^2 h}{T^2}\right]. \tag{19}$$

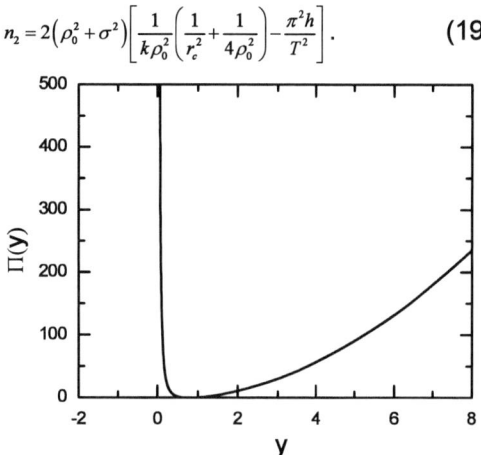

Fig.1 Qualitative plot of potential function $\Pi(y)$ under the condition of Eq. (19). Clearly, for $\Pi'(1)=0$, a soliton can be formed.

Physically, this condition means that soliton existence requires the nonlinearity-induced focusing and optical lattices effect to balance for beam spreading due to diffraction and partial coherence. Firstly, the degree of nonlocality influences the formation of soliton: the larger the σ becomes, the stronger the soliton requires the nonlinearity. So, we should reduce the degree of nonlocality to satisfy the soliton condition (see Fig. 2). Secondly, the role of partial coherence make the beam spreading, in the Eq. (19), we note that for

full coherent beam, i.e., for $r_e \to \infty$, we recover the known solution for coherent solitons in nonlinear media (see Fig. 2). Thirdly, optical lattices compensate the influence of partial coherence. From Eq. (19), we can suppress the partial coherence by increasing the refractive index modulation depth h and reducing the modulation period T.

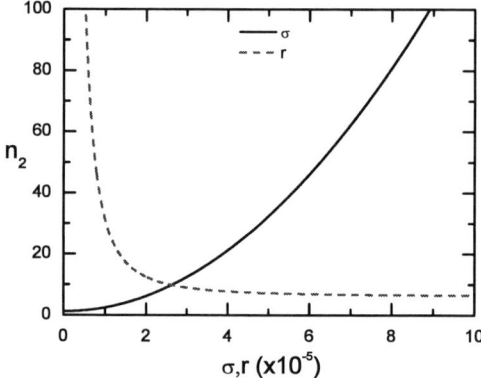

Fig.2 Variation of nonlinear index n_2 with the degree of nonlocality σ and the initial coherent length r_e.

Conclusion

We have presented an analysis of the propagation of partially coherent beam in nonlocality medium. The conditions of soliton formation are obtained by using approach of the evolution of the mutual coherence function. It can be fund that the degree of nonlocality and the initial coherent length influence the formation of soliton. Especially, optical lattices can compensate the influence of partial coherence.

References

1. D. N. Christodoulides, et al., Opt. Lett. 13 (1988)794.
2. A. Trombettoni, et al., Phys. Rev. Lett. 86 (2001) 2353.
3. Y. V. Kartashov , et al., J. Opt. Soc. Am. B 22 (2005)1356.
4. M. I. Carvalho, et al., Phys. Rev. E 59 (1999), 1193.
5. A. W. Snyder et al.,a Phys. Rev. Lett. 80 (1998),1422.
6. D. N. Christodoulides, et al., Opt. Lett. 22 (1997), 1080.
7. D. N. Christodoulides, et al., Phys. Rev. Lett. 80 (1998), 2310.

A Proposal for Passive Optical Network Architecture (WDM-PON) Based on Array of Ring Resonators

G. Rostami, R. Faraji-Dana and M. Shahabadi

School of Electrical and Computer Engineering, University of Tehran, Tehran, Iran

E-mail: gh.rostami@ece.ut.ac.ir

Abstract- *In this paper, we propose a novel and manageable passive optical network (PON) based on array of ring resonators. The proposed structure includes a compact chip to operate as multiplexer, demultiplexer, add-drop multiplexer and remote node.*

Introduction

The passive optical network (PON) is an optical fiber based network architecture, which can provide much higher bandwidth in the access network compared to traditional copper-based networks. A PON is a point-to-multipoint optical network, where an optical line terminal (OLT) at the central office (CO) is connected to many optical network units (ONUs) at remote nodes (RN) through one or multiple 1:N optical splitters. The network between the OLT and the ONU is passive; it does not require any power supply [1]. Wavelength-Division-Multiplexed (WDM) passive optical networks (PONs) offer many advantages including large capacity, easy management, network security, and upgradeability. The WDM-PON will become a revolutionary and scalable broadband access technology that will provide high bandwidth to end users [2]. In wavelength division multiplexing (WDM) systems with large number of channels, there is basic demand to insert a large number of channels in a single optical fiber. For this purpose, several devices have been developed, and the arrayed-waveguide grating (AWG) has been indicated as a proper choice. This is due to suitable characteristics of this device. However, there are limitations in the use of *AWG* devices in dense WDM (DWDM) systems. One of these limitations is related to value of the *free spectral range* (FSR), which reduces the number of channels multiplexed or demultiplexed. Another limitation is related to small isolation between channels. Taking this into account, it is desirable to have a method of determining the *AWG* physical

dimensions that lead to the channel spacing and *full width half maximum* (FWHM) required for all the channels of the DWDM system [1], [3]. The AWG's central wavelength shift of 0.01 nm/°C makes it difficult to be used in the RN of a WDM-PON, since the RN is located in the harsh temperature environment such as from −40°C to +85°C. This dependency originates from the index of refraction and physical size. There is another common scheme for multiplexing/demultiplexing, called thin-film or multilayer interference filters. By positioning cascaded filters in the optical path, wavelengths can be demultiplexed and vice versa. The filter is designed to transmit a standard wavelength while reflecting others [1]. So, in this paper, we propose a novel structure for WDM-PON using array of ring resonators (ARR). In WDM-PON based on ARR each subscriber is assigned to a separate WDM channel and these channels are routed by array of ring resonator. Also, passive wavelength routing device located in the remote nodes. In recent WDM PON, the demultiplexing function is implemented by microring resonator (MRR's). The proposed device is used as remote node in WDM-PON. It can become the basic building block of WDM PON's. The MRR permits flexible high-speed point-to-point switching and point-to-multipoint broadcast connections by virtue of the periodicity of the MRR's passband. The monolithic capability within optical integrated circuits, different wavelengths for sending and receiving, bidirectional capability for simultaneous operation, wavelength division multiplexing

978-0-9789217-3-6/07/$25.00

©2007 WEN GLOBAL SOLUTIONS

operation, large number of channels, inherent filtering operation, upgradeable, extendable and amplification properties can be mentioned [4]. So, the introduced idea can be applied for realization of large distance access network. The organization of this paper is as follows.

In section II, basic principle for network structure is presented. Simulation results are presented and discussed in section III. Finally the paper ends with a short conclusion.

Network Structure- In this section for realization of passive optical network and large distance access network two strategies are used. These strategies are named as single channel and band division based sending and receiving. In first part, we describe very narrow band filter design within DWDM standard using ring resonators (Fig. (2-a)). In the second part, first we divide whole band into some pre-specified bands and then into each band we extract individual channels (Fig. (2-b)).

a) Single Channel case: In this method network structure is illustrated in Fig. (3).

b) Band dividing method: The topology is presented in Fig. (4).

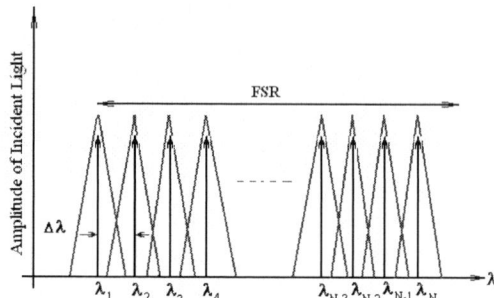

Fig. (2-a) Transfer function and channel dividing method for DWDM Systems using an Array of Ring-Resonator.

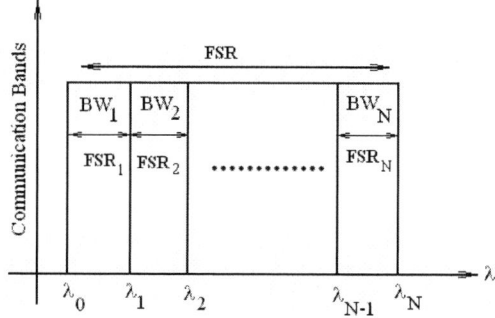

Fig. (2-b): Transfer function and Band dividing method for DWDM Systems using an Array of Ring-Resonator.

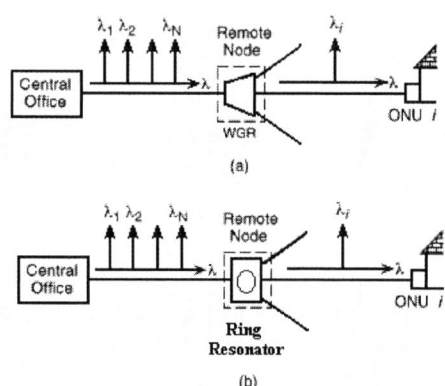

Fig.(1). PON Architectures: (a) WDM-PON based on AWG (b) WDM-PON based on ARR.

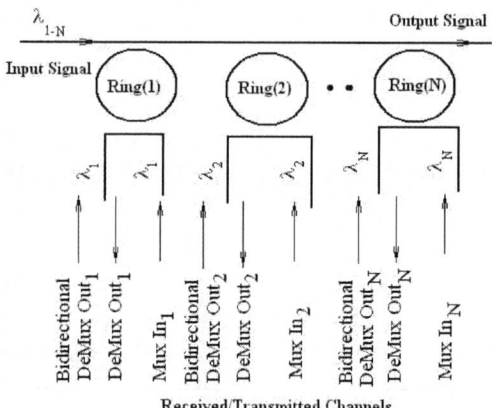

Fig. (3): Single Channel sending/receiving method and routing node architecture.

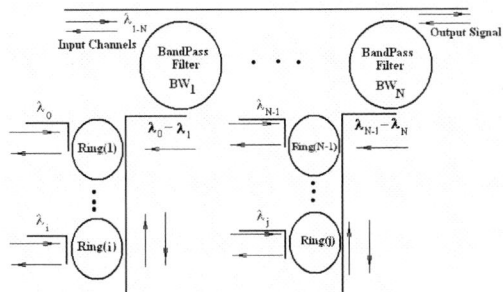

Fig. (4): Array of Ring Resonators for Realization of band dividing method and routing node architecture.

Results and Conclusion - In this section the proposed idea for realization of the integrated passive optical network block is simulated and results are discussed. First the proposed idea based on single ring resonator operation as receiving and transmitting ports at single wavelength is simulated and illustrated in Fig. (5). In the figure the index of refraction is changed for realization of DWDM. The index of refraction resolution used in this paper is 0.001. In the presented simulation, we consider band including 100 nm duration. For each channel 1 nm spacing is considered. For realization of band extraction, we consider 3 cascaded ring resonators. For example in the following figure 20 nm band are realized using 3 cascaded ring resonators (Fig. (6)). It should be mentioned that the band duration and central wavelength can be managed using the parameters of ring resonators.

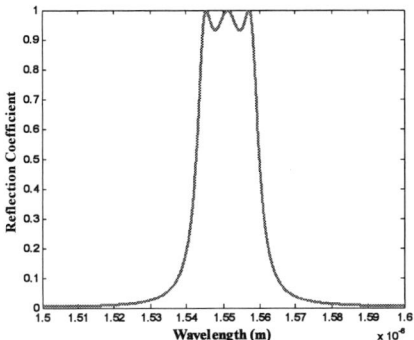

Fig. (6): Reflection Coefficient Vs. Wavelength for Band Selection.

Typically simulation within 20 nm band, DWDM channels including one nanometer distance is illustrated in Fig. (7). so, the proposed idea in Fig. (2) can be implemented based on ring resonators.

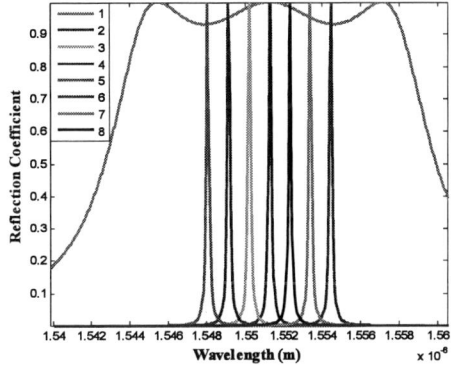

Fig. (7): Reflection Coefficient Vs. Wavelength for the proposed building block for realization of passive optical network in the case of Band Selection.

References

1 Amitabha Banerjee et al, Journal of optical networking, Vol. 4, No. 11(November 2005), page 737-758.

2 Guido Maier et al, Journal of Lightwave technology, Vol. 18, No. 2 (February 2000), page 125-143.

3 Herbert Venghaus, Wavelength Filters in Fiber Optics (Springer 2006).

[4]. A. Rostami et al, Journal of Lightwave Technology, Vol. 23, No. 1 (January 2005), page 446-460.

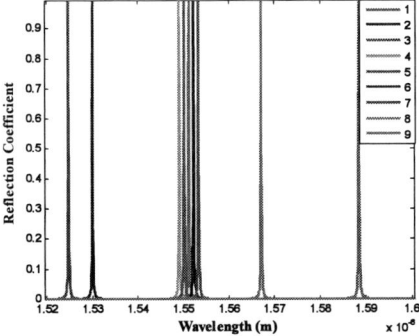

Fig. (5): Reflection Coefficient Vs. Wavelength for channel selection.

An Improved Selective Area Growth Method in Fabrication of Electroabsorption Modulated Laser

Huan Wang(1), Hongliang Zhu(2), Yuanbing Cheng(3), Dingbo Chen(4),
Wei Zhang(5), Liesong Wang(6), Yunxiao Zhang(7), Yu Sun(8) and Wei Wang(9).
State Key Laboratory on Integrated Optoelectronics
and Key Laboratory of Semiconductors Materials, CAS
Institute of Semiconductors, CAS, P. O. Box 912, Beijing 100083, P. R. China
Email: whuan21@semi.ac.cn

Abstract *An improved selective area growth (SAG) method is proposed to better the fabrication and performance of the Electroabsorption modulated laser. The typical threshold current of the EML is 18mA, and the output power is 5.6mW at EAM facet.*

Introduction

Electroabsorption modulated laser (EML), is a promising light source in the wavelength-division-multiplexing(WDM) optical transmission system owing to its potential for high-speed modulation, high extinction ratio at low driving voltage, low chirp as well as small size. Various fabrication techniques like e.g., Butt-joint method[1,2], selective area growth(SAG) method[3] and stacked active layer technology[4,5] have been proposed and demonstrated for the monolithic integration of EAM and DFB laser diodes.

SAG is one of the simplest methods to integrate the DFB laser and EA modulator on the same wafer. But in our experiences of fabricating the EML by SAG, several drawbacks are found out. DFB grating is difficult to fabricate stably on the upper Separate Confining Heterostructure(SCH) because of the height difference between the LD and EAM sections. The far-field property is worse and the light absorption loss in the separate section is high.

In this paper, some improvements in fabrication of SAG method are proposed and demonstrated. And in order to realize our novel design easily, we just fabricate the EML with low frequency for our experiment. The elementary results have demonstrated our new design.

The novel fabrication of an EML by SAG involves three metal-organic vapor phase epitaxy(MOVPE) growth steps, which needs one more time than general SAG method. InP buffer layer and low SCH layer are grown directly on a (100) n-InP substrate by MOVPE. Then a 150nm thick SiO_2 dielectric film is deposited by PECVD on the low SCH layer and the parallel SiO_2 mask stripes are defined along the [110] direction, shown in Fig.1(1). The width of the opening growth region (W_o) between the mask stripes and the width of mask stripes (W_s) is both set to be 15µm. The goal of depositing SiO_2 mask after the growth of those two layers instead of depositing it directly on the InP wafer is to reduce the height difference between the LD and EAM sections after the growth of Multiple Quantum Qell(MQW). The reduced height difference makes it easy and stable to form the Bragg grating on the LD section, and hope to improve the far-field property of the diode.

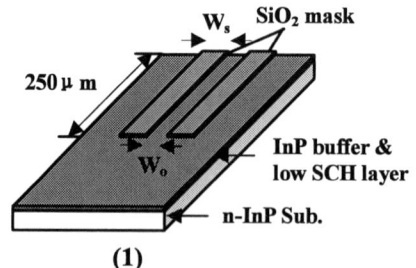

(1)

Device design and fabrication

978-0-9789217-3-6/07/$25.00

©2007 WEN GLOBAL SOLUTIONS

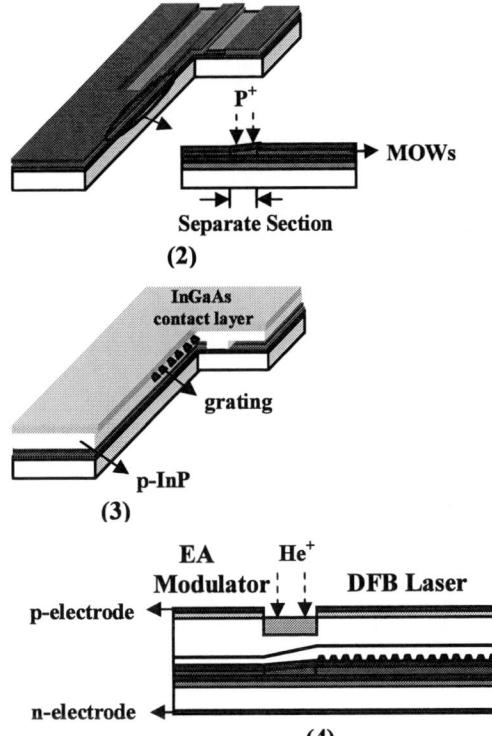

Fig.1 Novel fabrication process of the EML by SAG method: (1)SAG preparation after the growth of InP buffer and low SCH layer; (2)selective area MOVPE and P+ implantation; (3)growth of p-InP and p+-InGaAs contact layers; (4)schematic cross section of final device.

An eight-pair InGaAsP MQW active layer structure is grown on the SiO_2-masked low SCH layer, as indicated in Fig.1(2). The MQW layer structure of devices consists of 6nm tensile strained wells, which are separated by 5.5nm compressively barriers (λ_g=1.2µm). The crystal quality and MQW bandgaps of the grown layers are characterized by microscope photoluminescence (PL). The PL peak wavelength at the centre of the SAG opening growth region (λ_{PL-LD}) is 1.532µm, while at the flat unmasked region (λ_{PL-EAM}) is 1.492µm. P^+ ions are then implanted into the 50µm wide separate section. The bandgap of this section is enlarged by the diffusion of the impurity ions in annealing step after implantation, which can reduce the absorption loss in this section. The PL wavelength of the separation section ($\lambda_{PL-Sep.}$) is about 140nm blue shift from

the LD section. The stable Bragg grating is easy to form on the LD section, shown in Fig.2. Then the common p-InP confining layer and p^+-InGaAs contact layer are grown over the entire structure, indicated in Fig.1(3).

The schematic view of the device structure is shown in Fig.1(4). The lengths of the laser and modulator are 300µm and 150µm, respectively. A 2µm wide ridge is etched to the up SCH layer. Over 50kΩ electrical isolation between LD and EAM is realized by etching off the InGaAs contact layer and deep He^+ ion implantation into the separation section.

Fig.2 Photograph of Bragg grating by Scan Electronic Microscope (SEM)

Experimental results

The blue shift of the wavelength in LD section and separation sections after annealing is shown in Fig.3. The wavelength shift in LD section is 8nm, and in separation section is 140nm. The wavelength shift leads to a low absorption loss in the separation section, which is only about 0.2mW from the LD facet to the EAM facet in the diode.

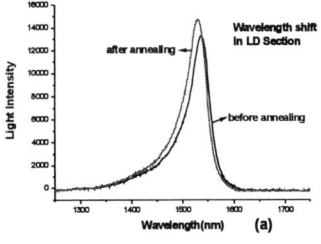

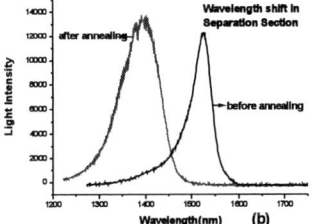

Fig.3 The blue shift of the wavelength in LD section (a) and separation section (b)

The typical characteristics of the output power at the EAM facet and LD forward voltage versus the current for the EML is shown in Fig.4. The typical threshold current of the EML is 18mA and the output power at the modulator facet is 5.6mW at a laser operation current of 100mA and a modulator bias voltage of 0V. The lasing spectrum of the integrated device is shown in the inset of Fig.4, and the Side Mode Suppression Ratio(SMSR) is over 33dB.

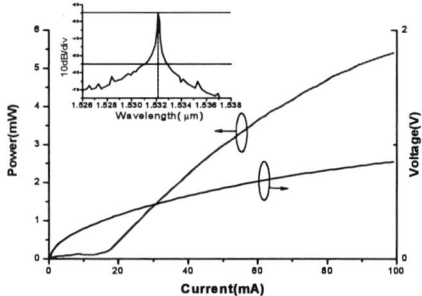

Fig.4 Output power from the modulator facet and LD forward voltage versus laser drive current. The inset illustrates the lasing spectrum of the integrated device.

Furthermore, we investigate the far-field property of the integrated device, shown in Fig.5. The Full Width at Half Maximum (FWHM) angle of the horizontal direction is 37.86° and the vertical direction is 43.59°. It is not good enough, but better than the devices we fabricated before.

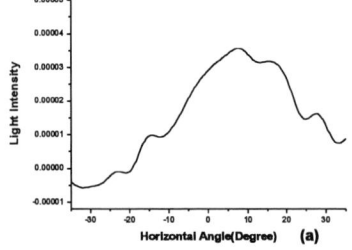

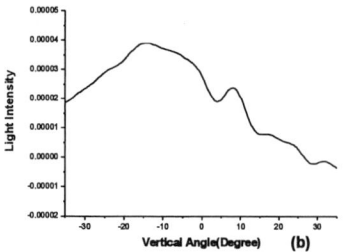

Fig.5 The far-field property of the integrated device, horizontal in (a) and vertical in (b)

Injured Active layers, which lead to high threshold current and low output power, are found by SEM in the LD, EAM and separate sections. This is the result of the P^+ ions implantation at the room temperature and the partial failure of the resist mask on the LD and EA sections. P^+ ions implantation at a higher temperature, 200ºC for example, and applying the Au deposited layer as the resist mask to protect the LD and EAM sections while implantation may reduce the injury. The loss and threshold current of EML will be reduced if AR and HR coating are applied to the facets of the EAM and DFB laser respectively.

Conclusions

We have proposed and demonstrated an improved SAG method to better the fabrication and performance of the EML. The typical threshold current of the EML is 18mA, and the output power is 5.6mW at a laser gain current of 100mA and a modulator bias voltage of 0V.

References

1 K. T. Hiroaki Takeuchi et al IEEE Journal of Selected Topics in Quantum Electronics, 3(1997), 336-347
2 H. S. Kawanishi et al in optical Fiber Communications Conference, (2003), 270-271
3 Zhao Qian et al Chinese Physics Letter, 22(2005), 2016-2019,.
4 Bernhard Stegmueller et al IEEE Photonics Technology Letters,14(2002), 1647-1649
5 Saravanan B.K et al in Indium Phosphide and Related Materials, (2004), 236-238.

Experiments and Simulations of Infrared Transmission by Transverse Electric Mode through Au Gratings on Silicon with various Au widths

Yan-Ru Chen(1), and C. H. Kuan(2)

(1)(2) Affliation: Graduate Institute of Electronics Engineering and Department of Electrical Engineering, National Taiwan University, Taipei, Taiwan, Republic of China

(1)(2) Full address: Department of Electrical Engineering, National Taiwan University, No. 1, sec. 4, Roosevelt Rd., Taipei, Taiwan, R.O.C.

(1) email:d93943029@ntu.edu.tw (2) email:kuan@cc.ee.ntu.edu.tw

Abstract *We have measured infrared (2.5~25µm) transmission by the transverse electric (TE) mode through Au gratings on silicon substrate with various Au widths. Simulation is used by surface impedance boundary condition (SIBC) method.Simulation results agree well with experiments and offer an evanescent-wave model.*

Introduction

Transmission through one-dimensional metallic gratings has been researched for decades because of their optical characteristics and potential applications in various fields[1], including beam splitting polarizers and photodetectors. In this paper, we have investigated infrared transmittance (2.5~25µm) by transverse electric (TE) mode through various Au gratings on silicon substrate.

Fabrication and measurement

At first, electron beam lithography system (ELS-7500EX), which was provided by ELIONIX, was used to pattern various gratings on silicon substrate. The ZEP520A was used as resist. 20nm Au film was evaporated on the sample after developing. Then ZDMAC was used to lift off the unpatterned parts. The fabrication procedures are shown in Figure 1. In our experiment, the period of Au gratings is always 4µm. The thickness of Au is 20nm, which is larger than the skin depth of Au (~10 nm). The total area of the gratings is 3.6mm×3.6mm. Figure 2 shows the scanning electron microscopy (SEM) picture of Au gratings on silicon substrate. The FTIR system (IFS 66v/S), which was provided by Bruker, was finally used to investigate the TE-mode infrared transmittance of the sample, as shown in the schematic set-up of Figure 3. It is noticed that bare silicon substare is as our reference, so the grating transmission was divided by that of bare silicon.

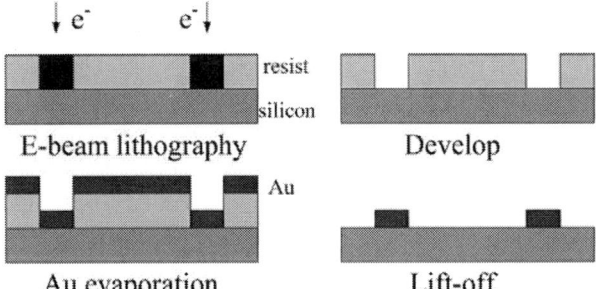

Figure 1: Fabrication process of Au gratings on silicon substrate

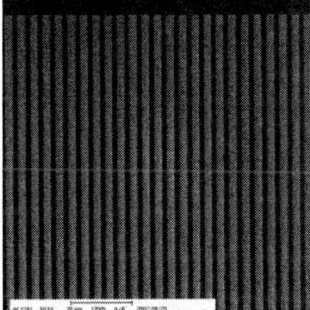

Figure 2: SEM picture of Au gratings on silicon substrate

Results and simulation

In Figure 4 shows the experimental transmission of Au gratings with various Au widths, divided by that of bare silicon. The x-axis represents the wavelength (2.5µm~25µm), and the y-axis represents transmission. The percentage of the air-slot width over the period, from 25% to 91% is the parameter for the various Au width.

978-0-9789217-3-6/07/$25.00

©2007 WEN GLOBAL SOLUTIONS   489

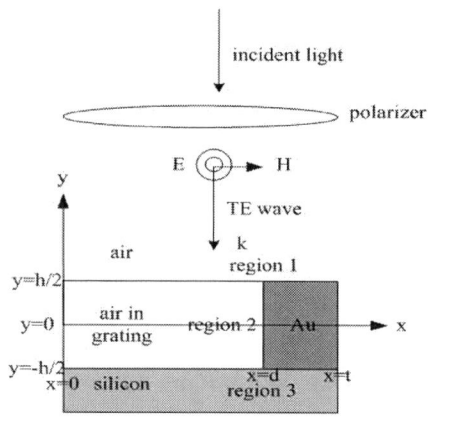

Figure 3: Illustration of infrared transmission through Au gratings and definition of coordinate and parameters in silulation

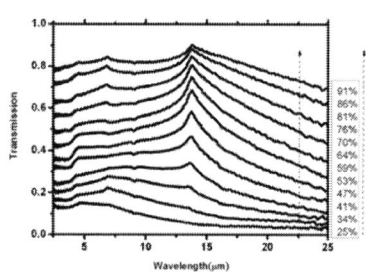

Figure 4: Experiment data of transmission through Au gratings

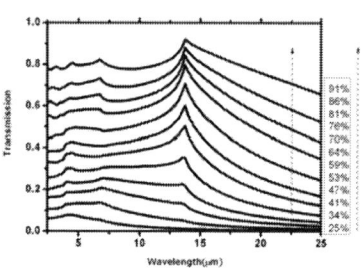

Figure 5: Simulation of transmission through Au gratings

The surface impedance boundary condition (SIBC) method[2] is used to simulate the TE-mode infrared transmittance through Au-gratings. Figure 3 defines the coordinate and parameters in calculation. Boundary condition $E_{\parallel}=Z\hat{n}\times H_{\parallel}$ is applied to relating tangential components of electric and magnetic fields at Au/dielectric interface, where Z is. $1/n_{metal}$, with n_{metal} being the refraction index of the metal. The tangetial electric fields in various parts are expressed

as a linear combination of orthogonal modes as follows:

tangetial electric fields in region 1:

$$\exp(-ik_{y,0,1}(y-h/2))+$$

$$\sum_{n=-\infty}^{\infty} R_n \exp\{i[2n\pi x/p+k_{y,n,1}(y-h/2)]\}$$

tangetial electric fields in region 2:

$$\sum_{m=1}^{\infty} X_m(x)Y_m(y)$$

$$=\sum_{m=1}^{\infty}\{[d_m\sin(k_{x,m,2}x)+\cos(k_{x,m,2}x)]\}$$

$$\{[a_m\exp(i(k_{y,m,2})y)+b_m\exp(-i(k_{y,m,2})y)]\}$$

tangetial electric fields in region 3:

$$\sum_{n=-\infty}^{\infty} T_n \exp\{i[2n\pi x/p-k_{y,n,3}(y+h/2)]\}$$

where

p is the period of gratings

$$k_{y,n,1}=\sqrt{\varepsilon_1 k_0^2-(2n\pi/p)^2}$$

$$k_{y,n,3}=\sqrt{\varepsilon_3 k_0^2-(2n\pi/p)^2}$$

$$k_{x,m,2}^2+k_{y,m,2}^2=k_0^2\varepsilon_2$$

where $k_0=2\pi/\lambda$ with λ being wavelength of incident light and ε_1, ε_2, and ε_3 are respectively the dielectric constant of region 1, 2, and 3. In our simulation, region 1 and 2 are air, and region 3 is silicon.

Applying SIBC to the left-hand and right-hand side of Au/air interface results:

$$d_m=(k_0/iZ)/k_{x,m,2} \text{ and}$$

$$\tan(dk_{x,m,2})=2k_{x,m,2}(k_0/iZ)/[k_{x,m,2}^2-(k_0/iZ)^2]$$

where $k_{x,m,2}$ can be found by the Newton method

Equating the electric and magnetic fields and applying the SIBC conditions at $y=h/2$ and $y=-h/2$ yields the following four equations.

$$1+\sum_{n=-\infty}^{\infty} R_n \exp(i2n\pi x/p)$$

$$=\sum_{m=1}^{\infty} X_m(x)(\varphi_m a_m+\varphi_m^{-1}b_m), 0\leq x\leq d$$

490

$$-ik_{y,0,1} + \sum_{n=-\infty}^{\infty} ik_{y,n,1} R_n \exp(i(2n\pi x/p))$$

$$= \begin{bmatrix} \sum_{m=1}^{\infty} X_m(x) k_{y,m,2}(\varphi_m a_m - \varphi_m^{-1} b_m), 0 \le x \le d \\ (k_0/iZ)\{1 + \sum_{n=-\infty}^{\infty} R_n \exp[i(2n\pi x/p)]\}, d \le x \le p \end{bmatrix}$$

$$\sum_{n=-\infty}^{\infty} T_n \exp(i(2n\pi x/p))$$

$$= \sum_{m=1}^{\infty} X_m(x)(\varphi_m^{-1} a_m + \varphi_m b_m), 0 \le x \le d$$

$$\sum_{n=-\infty}^{\infty} -ik_{y,n,3} T_n \exp(i(2n\pi x/p))$$

$$= \begin{bmatrix} \sum_{m=1}^{\infty} iX_m(x) k_{y,m,2}(\varphi_m^{-1} a_m - \varphi_m b_m), 0 \le x \le d \\ -(k_0/iZ)\sum_{n=-\infty}^{\infty} T_n \exp[i(2n\pi x/p)], d \le x \le p \end{bmatrix}$$

where

$$\varphi_m = \exp(ik_{y,m,2} h/2)$$

Then multiplying the first and the third equation by $X_m(x)$, and integrating over the region $0 \le x \le d$, and multiplying the second and fourth equation by $\exp(i2q\pi x)/p$ and integrating over the region $0 \le x \le p$ yields a series of equations that are used to determine the unknown coefficients R_n, T_n, a_n, and b_n.

$$\sum_{n=-\infty}^{\infty} G_{nm} R_n - \sum_{n=1}^{\infty} N_{nm} \varphi_n a_n - \sum_{n=1}^{\infty} N_{nm} \varphi_n^{-1} b_n = -\delta_{0m} G_{nm}$$

$$\sum_{n=-\infty}^{\infty} [ik_{y,n,1} - (k_0/iZ)J_{nq}]R_n$$

$$-\sum_{m=1}^{\infty} iK_{mq} k_{y,m,1} \varphi_m a_m + \sum_{m=1}^{\infty} iK_{mq} k_{y,m,1} \varphi_m^{-1} b_m$$

$$= \delta_{0q}[ik_{y,0,1} + (k_0/iZ)J_{nq}]$$

$$-\sum_{n=1}^{\infty} N_{nm} \varphi_n^{-1} a_n - \sum_{n=1}^{\infty} N_{nm} \varphi_n b_n + \sum_{n=-\infty}^{\infty} G_{nm} T_n = 0$$

$$-\sum_{m=1}^{\infty} iK_{mq} k_{y,m,2} \varphi_m^{-1} a_m + \sum_{m=1}^{\infty} iK_{mq} k_{y,m,2} \varphi_m b_m +$$

$$\sum_{n=-\infty}^{\infty} [-ik_{y,n,3} + (k_0/iZ)J_{nq}]T_n = 0$$

where

$$G_{nm} = \int_0^d X_m(x) \exp(i2n\pi x/p) dx$$

$$N_{nm} = \delta_{nm}\{[(k_0/iZk_{x,n,2})^2 + 1]d/2 + k_0/(iZk_{x,n,2}^2)\}$$

$$K_{mq} = (1/p) \int_0^d X_m(x) \exp(i2q\pi x/p) dx$$

$$J_{nq} = (1/p) \int_d^p \exp[i2(q+n)\pi x/p] dx$$

Finally $|T_0|^2$ is divided by air/silicon theoretical transmission. The results are shown in Figure 5. From Figure 4 and 5, it is observed that experiment and simulation results are well matched. The results obtained above are checked using another method that assumes that the Au/dielectric interface is perfectly conducting. It is found that $|T_n|$ will approach to zero as mode number is very large. So SIBC method is suitable for calculating infrared transmission through Au gratings.

Conclusions

To sum up, Au gratings on silicon substrate are fabricate by EBLS. We measure the infrared (2.5μm~25μm) transmission through Au gratings on silicon substrate by FTIR. We also use the surface impedance boundary condition (SIBC) method to simulate the experiment results. Simulation results agree well with experiments and offer an evanescent-wave model.

Acknowledge

We are pleased to acknowledge the NTU Center for Information and Electronics Technologies support for our use of electron beam lithography system from which our work extended.

References

[1] F. J. Garcia-Vidal and L. Martin-Moreno, " Transmission and focusing of light in one-dimensional periodically nanostrucutred metals," Phys. Rev. B **66**, 155412(1) - 155412(10) (2002)

[2] David Crouse and Pavan Keshavareddy, "Polarization independent enhanced optical transmission in one-dimensional gratings and device applications," OPTICS EXPRESS, Vol. 15, No. 4, 1415

All-optical switch in alkoxysilane dye doped waveguides based on m-line spectroscopy technique

Weirui Dang （1）, Yuping Chen （1）, Rui Wu （1）, Dandan Pan （1）, Xianfeng Chen （1）, Yuxing Xia （1）, Qinghua Meng （2）

1:Institute of Optics & Photonics, Department of Physics, Shanghai Jiao Tong University, 800 Dongchuan Rd., Shanghai 200240, China
2:School of Chemistry and Chemical Engineering, Shanghai Jiao Tong University, 800 Dongchuan Rd., Shanghai 200240, China
Email: dwown@sjtu.edu.cn

Abstract *We use the m-line spectroscopy technique to characterize the guiding modes in alkoxysilane dye (ASD) doped polymethyl methacrylate film waveguide. All-optical switching phenomenon with response time of ps was observed with improved prism-waveguide coupling.*

Introduction
Over the past two decades, nonlinear optical materials have been attracted great interest due to their potential applications in the fields of all-optical signal-processing devices [1]. Several chemical methodologies such as inorganic crystal, poled polymers and inorganic-organic hybrid materials have been developed. In the latter, the nonlinear optical chromophores are incorporated into the inorganic matrix by one or more covalent bonds. These hybrid materials capitalize on the unique properties offered by the two components to generate novel materials with desired remarkable nonlinear optical characteristics.

In this Letter, based on m-line spectroscopy technique, and one improved prism-waveguide coupling configuration, we observed nonlinear optical-Kerr switching phenomenon within 50 ps in *alkoxysilane dye (ASD) doped polymethyl methacrylate film waveguide*

ASD materials and experimental setup
The molecular structure of ASD is shown in Figure 1. Optical properties are determined by the chromophoric molecules. These azobenzene type chromophores which exhibited high transparency in the blue range are synthesized and further react with 3-isocyanatopropyl-triethoxysilane (ICTES) to give alkoxysilane dyes via a urethane reaction. Followed by a sol-gel process of the alkoxysilane dyes, the inorganic-organic hybrid nonlinear optical films are

successfully prepared. It exhibits excellent thermal stability of dipole alignment at elevated temperature, with starting decomposing temperature above 290 Centigrade. And the hybrid films also exhibit large optical nonlinearity (d_{33} is 14.3 pm/V) and full transparency in the visible range.

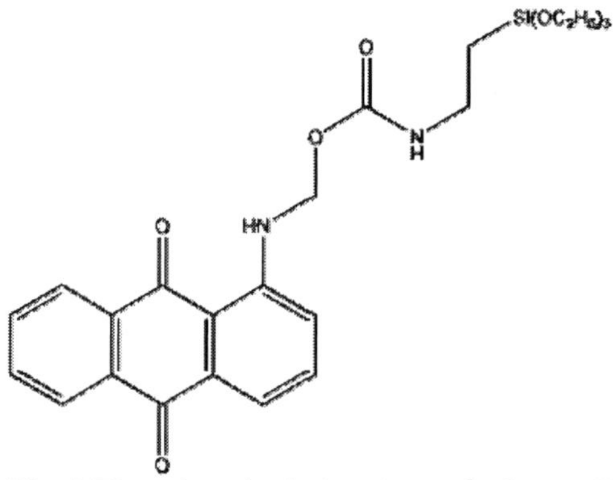

Fig.1 The chemical structure of alkoxysilane dye (ASD)

In the experiment we used the m-line spectroscopy technique to characterize the guided modes in the waveguide. From the angular shift of the m-line caused by the strong pump light is switched on, the optical Kerr effect could be observed. The experimental setup of the improved two-wavelength prism coupling method is shown in Figure 2 [2, 3]. The prism, made of ZF7, is symmetric trapezoid shape with base angles of 60° and its refractive index n_p is 1.798 at 650 nm. The linearly polarized monochromatic probe light of 650 nm wavelength from a laser diode, with transverse magnetic (TM) polarization is incident into the waveguide from the hypotenuse of the prism. The pump light of the mode-locked Nd:YAG laser (λ = 1064nm ; 50-ps pulse width; maximum repeat frequency, 10 Hz; maximum output power, 100mW) is incident upon the upper base of the prism perpendicularly. The merit of this method is to guarantee that the pump light is always coupled into the loading force point with the motorized rotary assembly rotate. The coupling pressure of the waveguide system was adjusted by a micrometer fixture holding the prism and waveguide. The waveguide and the coupling prism are rotated on a high-precision motorized rotary table on which they are mounted under computer control. Guided intensity is measured on the opposite hypotenuse side of the waveguide by a Si photodetector as a function of the incident angle. A narrow seam is used to minimize the noise before the Si photodectector. Because the change of the refractive index of the film caused by nonlinear optical Kerr effects is too small to detect, we use a 500MHz oscillograph to observe the pulse signal [2].

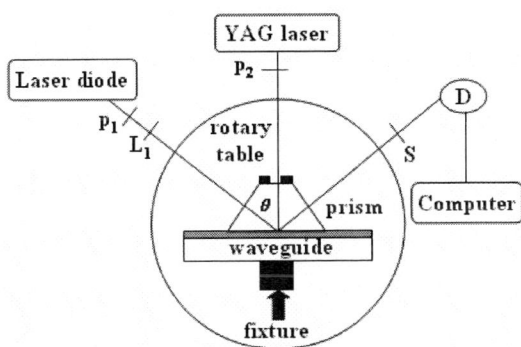

Fig.2 Experimental setup of the improved two-wavelength prism coupling method to observe the optical Kerr effect: P1, P2, polarizer; L1, lens; D's, detectors; S, narrow seam, θ is the incident angle after the probe light is incident into the prism.1 The chemical structure of alkoxysilane dye (ASD).

Initially, at the angle in the midst of the mount-up resonance dip, the pump light of the Nd:YAG laser, which is incident upon the upper base of the prism perpendicularly, is then switched on. As the pump light is switched on or off, an intense pump-pulse beam induces nonlinear optical Kerr effects

that originate from the intensity dependence of the refractive index, so that the changes of refractive index of the ASD film waveguide cause the change of probe beam intensity. The repeatability of the measured coupling angle has been checked and found to be less than ± 0.01°.

Results

The experimental resonance dip spectrum of the improved two-wavelength prism coupling method is displayed in Figure 3.

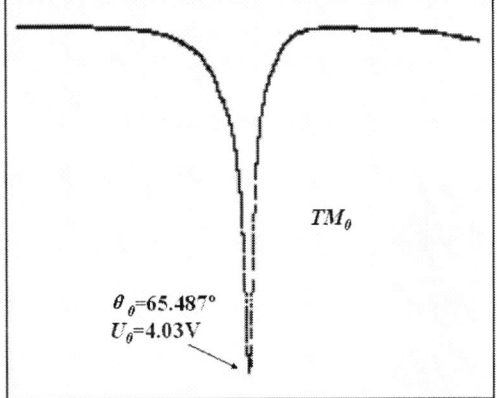

Fig. 3 The experimental resonance dip spectrum of the improved two-wavelength prism coupling method, showing the TM_0 guiding mode.

Since ASD was dissolved in tetrahydrofuran (THF), the solution deposited onto a substrate is not thick enough by the spin coating method, so only one guiding mode was observed in the resonance dip spectrum. The resonance dip is the TM0 guiding mode in the waveguide, where the evanescent field is phase-matched to a guided wave in the hybrid guiding layer, and the energy is coupled from the incident light into the guided wave. As we all know, the resonance dip is characterized by two main parameters which are associated with the performance of the modulator closely. And they are the position of the minimum and the value of the minimum of the dip. As shown in Figure 3, the value of the minimum of the dip is 4.03 V at 65.487°.

Figure 4 is the signal pulse detected from the probe beam as the pump light is switched on. We can obtain the change of probe beam level ΔL is about 90 mV, that is 2.04 mW in intensity. Because the response time of Si photodetector is of ns order magnitude, and laser pulse duration is 50 ps, which is faster than detector's response time, then ps of switching time of all-optical switch is expected in this paper.

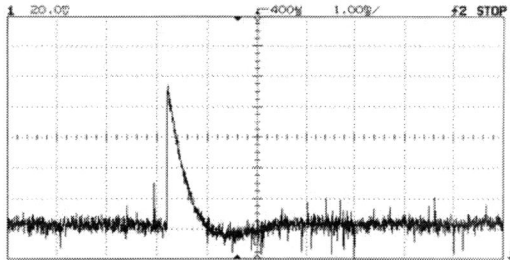

Fig. 4 The signal pulse detected from the probe beam as the pump light is switched on.

Conclusions

In summary, we have observed a nonlinear optical Kerr switching phenomenon within 50 ps in alkoxysilane dye (ASD) doped waveguides by improved prism-waveguide coupling based on the m-line spectroscopy technique. The result makes alkoxysilane dye (ASD) films as potential candidates for future application in all optical switching.

Acknowledge

This research is supported by the National Natural Science Foundation of China (60407006 and 60477016).

References

1 Jun Zhou, Zhuangqi Cao, Yingli Chen, Optics Letters, Vol. 22, No. 19, 1-2 (1997)

2 Rui Wu, Yuping Chen, Dandan Pan, Xianfeng Chen, Proc. of SPIE Vol. 5646, 5-7 (2005)

3 Dandan Pan, Yuping Chen, Rui Wu, Xianfeng Chen, Qinghua Meng, Zhicheng Sun, All-optical light modulation in anthraquinone dye doped waveguides based on an improved prism coupling method, Synthetic Metals , Vol.157, Issues 4-5, 186-189(2007)

Experimental demonstration of all optical wavelength full conversion based on quadratic cascading effect in periodically poled MgO-doped lithium niobate

Junfeng Zhang, Feng Lu, Yuping Chen, Wenjie Lu, Xianfeng Chen and Yuxing Xia

Department of Physics, The State Key Laboratory on Fiber Optics Local Area Communication Networks and Advanced Optical Communication Systems, Shanghai Jiao Tong University, 800 Dongchuan Rd., Shanghai 200240, China
Email: jfzhang@sjtu.edu.cn

Abstract: *A broad pump wavelength band up to 27 nm has been obtained in a 20-mm long periodically poled MgO:LiNbO$_3$. Arbitrary wavelength conversion of signal waves at communication band has been successfully demonstrated through nonlinear quadratic cascading process.*

1. Introduction

All optical wavelength conversion is essential for wavelength division multiplexed (WDM) optical networks. Within the various approaches, wavelength converter based on quasi-phase-matching (QPM) engineered nonlinear materials is more attractive, for this kind of the wavelength converter has the advantage of ultra-high speed and transparency to the signal format and the bit rates owing to its pure optical nonlinear processes [1-4].

In this paper, we demonstrated that arbitrary wavelength conversion can be realized in periodically poled MgO:LiNbO$_3$ (MgO:PPLN) through cascaded effect of second harmonic generation (SHG) and difference frequency generation (DFG). It is due to the broadband SHG process which has broad pump wavelength band up to 27 nm.

2. Experiments

The MgO:PPLN sample for the wavelength converter with the length of 20-mm is fabricated by the electrical poling method. The poling period is 20.4 μm. Because we use $o+o\rightarrow e$ scheme [5] in this experiment, a polarization controller is used to control the polarization of input waves. By carefully tuning the pump wavelength and temperature, we obtained the SHG power versus pump wavelength, as showed in Fig.1.

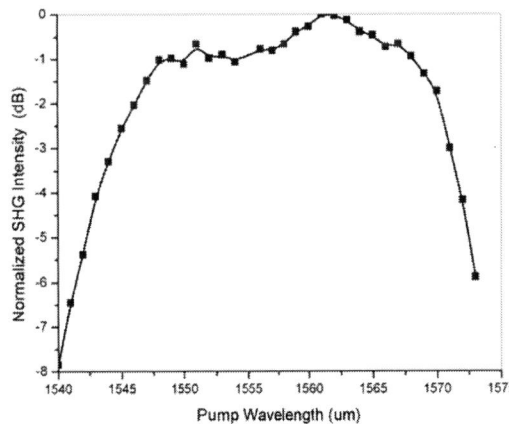

Fig.1. The second harmonic generation of pump light, the wavelength band is 27nm.

We found that a 3 dB pump wavelength broadband up to 27 nm is realized. The peak

SHG intensity is located at the wavelength of $\lambda_p = 1562nm$ when the temperature of the sample is 38 ℃.

Fig.2. shows the experiment setup. Two tunable lasers are acted as the signal and the pump sources, respectively. The sources

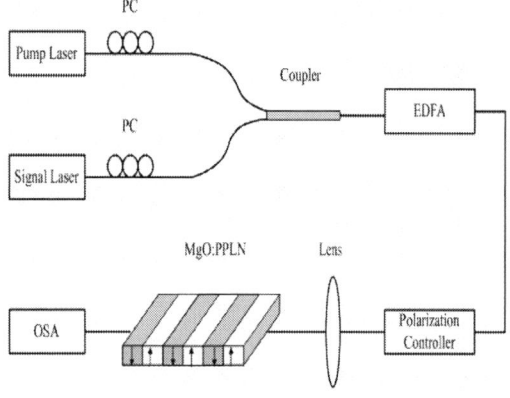

Fig.2. Experimental setup

are amplified by an erbium-doped fiber amplifier (EDFA) and then combined to be injected into the wavelength converter through a coupler. The temperature of MgO: PPLN sample is controlled by an oven. The converted wavelength is measured by an optical spectrum analyzer (OSA). Firstly, the pump laser ω_p is converted to SHG light $2\omega_p$. Then, the signal laser is converted to conversion wave $\omega_c = 2\omega_p - \omega_s$ by DFG.

By simultaneously coupling the pump and the signal laser into MgO: PPLN, the converted light is generated, which is presented in Fig. 3. The wavelengths of signal pump and converted are 1546.92nm, 1562.23nm and 1577.86nm, respectively.

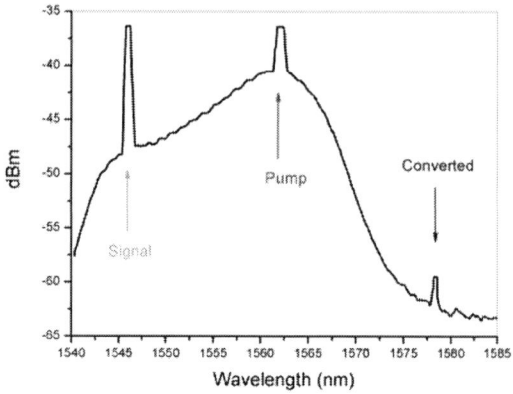

Fig.3. Measured optical wavelength conversion in MgO:PPLN

Due to the broadband of pump wavelength, multiple channels can be selected for pump light. Consequently, the signal light can be freely switched to different channels of converted light by differently selected pump light. Fig.4. shows that 3 signal waves (selected according to ITU grid) are arbitrarily switched to 3 converted waves by 9 pump waves.

The frequencies of pump waves are 191.7, 191.75, 191.80, 191.90, 191.95, 192.00, 192.1, 192.15, 192.2 THz, corresponding to the wavelengths of 1563.86, 1563.45, 1563.05, 1562.23, 1561.83, 1561.42, 1560.61, 1560.20, 1559.79nm. It is noted that a pump wavelength band of 4.07nm is employed in the experiment.

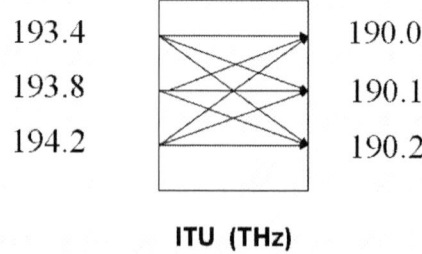

ITU (THz)

Fig.4. Demonstration of 3×3 arbitrary wavelength conversion

3. Conclusion

In summary, we have demonstrated experimentally that the arbitrary wavelength conversion can be obtained through nonlinear quadratic cascading process in MgO: PPLN. A broadband wavelength of pump light up to 27nm has been yielded in the 20-mm long sample. A band of 4.07nm for 9 pump channels has been employed to realize the 3×3 arbitrary wavelength conversion. This kind of MgO:PPLN converter provides a component, alternative approach for all-optical communication networks.

Acknowledge

This research is supported by the National Natural Science Foundation of China (60407006 and 60477016).

References

1. Masaki Asobe et al.,'"Multiple quasi phase-matched LiNbO3 wavelength converter with a continuously phase-modulated domain structure",Opt. Lett. 28, 558-560, (2003).

2. Brener, et.al.. " Efficient wideband wavelength conversion using cascade second-order nonlinearities in LiNbO3 waveguides", OFC' 2000, FB6.

3. Y. L. Lee, et al., 'Channel Selective Wavelength Conversion and Tuning in periodic poled Ti:PPLN Channel Waveguides,' Opt.Express12, 2649,(2004).

4. C.-Q. Xu, et al., cascaded wavelength conversions on sum-frequency generation and difference-frequency generation,"Opt.Lett 29, 292,(2004).

5. N.E.Yu,et al.,"Broadband quasi-phase-matched second-harmonic generation in MgO-doped periodically poled LiNbO$_3$ at the communications band," Opt.Lett. 27, 1046-1048, (2002).

The Application of the Wavelet Transform to the Continuous Wave Terahertz imaging

YU Fei (1), HUI Mei (2), SONG Qian (3), ZHAO Yue-jin(4)

1 : Department of Photo-electric Engineering, School of Information Science and Technology, Beijing Institute of Technology Beijing 100081 yfvcf2005@bit.edu.cn
2 : Department of Photo-electric Engineering, School of Information Science and Technology, Beijing Institute of Technology huim@bit.edu.cn, 3 : Department of Photo-electric Engineering, School of Information Science and Technology, Beijing Institute of Technology songqian@bit.edu.cn, 4 : Department of Photo-electric Engineering, School of Information Science and Technology, Beijing Institute of Technology yjzhao@bit.edu.cn

Abstract *The wavelet transform is used for restoration of the continuous wave THz images which have the appearance of sinusoidal waves covering the images with standing wave patterns for the THz imaging system.*

Introduction

A persistent problem in CW (Continuous Wave) THz imaging, especially in reflection modes, is the presence of standing wave patterns within the images. The patterns result from reflected inside of the imaging system interfering with the images coming from the sample. They have the appearance of sinusoidal waves covering the image, as shown in Fig. 1. The pattern interferes with contrast enhancement techniques, and can make the analysis of small areas difficult.

A method dealing with this problem is Frequency Domain Filtering proposed by Nick Karpowicz[1]. Using Fourier Transform the image can be analyzed in frequency space. Near the center of frequency space, there are the low frequency components. A vector drawn from the center to each pole gives the direction of the interference fringes. Eliminating the frequency components corresponding to these poles will remove the fringes. But in this method it can't find each of the frequency components corresponding to the interference fringes. When the quality of the image is bad, or the object is big, the result of the image restoration in this method isn't ideal.

In this paper, a method of the Wavelet Transform is mentioned. In an image, the frequency of the interference fringes and the part with objects' information are different. So after the wavelet transform, it can be easy to find which area corresponds to the interference fringes and which area corresponds to the part of object's information. Then, the image can be divided into two parts, one contains the whole fringes and the other contains the whole objects. Using image fusion method based on wavelet, the result without interference fringes can be gained. For the wavelet can exactly describe the frequency components of each area, the whole interference fringes can be found and eliminated.

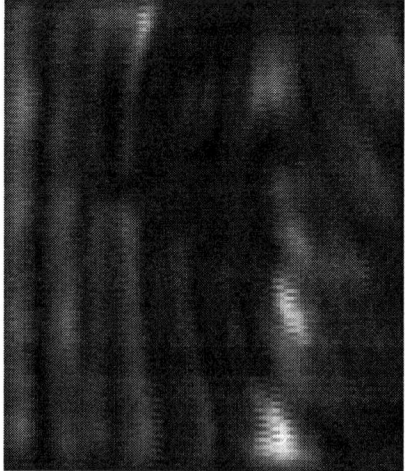

Fig. 1:
CW THz image with interference fringes

1. Analysis of eliminating interference fringes in frequency domain

The method which is proposed by Nick

978-0-9789217-3-6/07/$25.00

©2007 WEN GLOBAL SOLUTIONS

Karpowicz is through frequency-domain analysis of the acquired images. If frequencies whose amplitudes approach 1 are picked out, the poles responding to the interference fringes will be eliminated without touching the rest of the data. The original image is shown as Fig. 2(a), and the image after filtering is shown as Fig. 2(b).

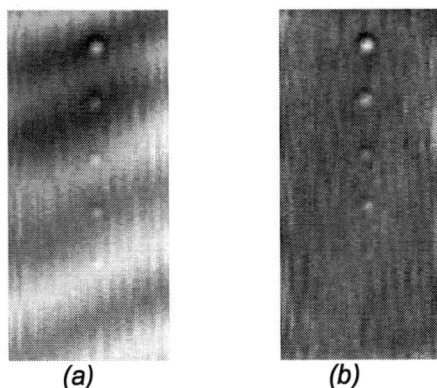

(a)   (b)

Fig. 2:
(a) Unfiltered Image (b) filtered Image

From Fig. 3 it can be seen that although each pole has been eliminated, the low-frequency components corresponding to the interference fringes can not be eliminated absolutely. And if we use low-frequency filter, the components corresponding to the object's information and the image's background will be eliminated too. So, like Fig 4 when the size of the object is not less than one interference fringe, the result of eliminateing interference fringes in frequency domain is not ideal.

From Fig. 3 it can be seen that although each pole has been eliminated, the low-frequency components corresponding to the interference fringes can not be eliminated absolutely. And if we use low-frequency filter, the components corresponding to the objects' information and the image's background will be eliminated too. So, like Fig. 4 when the size of the object is not less than one interference fringe, the result of eliminateing interference fringes in frequency domain is not ideal.

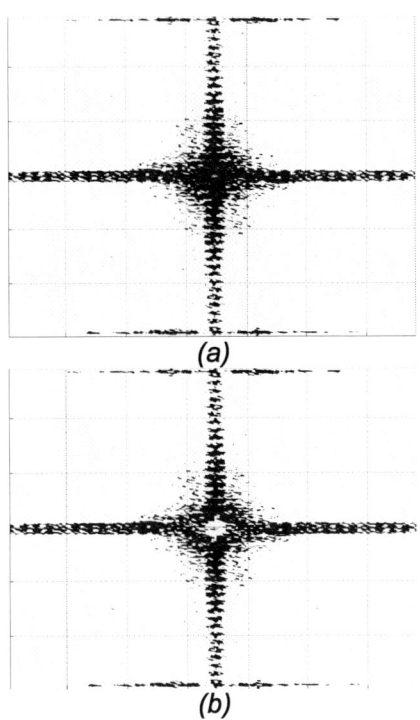

(a)

(b)

Fig. 3:
(a) Unfiltered Image in the frequency domain (b) filtered Image in the frequency domain

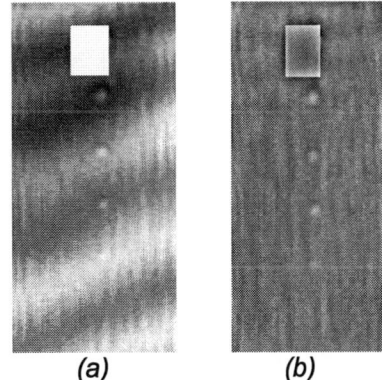

(a)   (b)

Fig. 4:
(a) Unfiltered Image (b) filtered Image

2. Wavelet Transform in CW THz image processing

We choose Wavelet Transform to analyze the THz images and eliminate interference fringes. In this method, when the size of object is larger than one interference fringe, the character of the object can also be reserved. With Wavelet Transform, the scope of CW THz imaging's application can be extended and clearer images can be acquired.

When digital images are to be viewed or processed at multiple resolutions, the Wavelet Transform is the mathematical tool of choice. In addition to being an efficient and highly intuitive framework for the representation, the Wavelet Transform provides powerful insight into an image's spatial and frequency characteristics. The Fourier Transform, on the other hand, reveals only an image's frequency attributes.

We cut a deltoid hole at an iron plate, and make the hole as the object which we want to detect. The image detected by CW THz is shown as Fig. 1. It can be seen that the interference fringes cover the image.

After using 1-D Wavelet Transform to process each row of the original image, we can acquire the interference fringes and the components responded to the object. Then using image fusion basing on Wavelet Transform we can gain the image shown as Fig. 5, which doesn't contain the interference fringes.

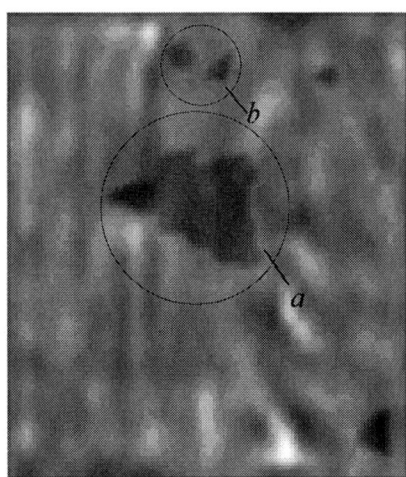

Fig. 5:
image after image fusion based on Wavelet Transform

Conclusions
Using Wavelet Transform can differentiate between the interference fringes and the information of the object. After attenuating the components corresponding to the interference fringes and enhancing the components corresponding to the object, the character of object can be seen clearer. With image fusion based on Wavelet Transform, the CW THz images can be restored from interference fringes caused by standing wave patterns in the THz imaging system.

Part *a* of the Fig. 5 is the character of object, and part *b* of the Fig. 5 is the misjudgment in the 1-D Wavelet Transform. In the actual imaging system, the interference fringes are not standard sinusoidal waves, so there may be misjudgment in the course of differentiating the object from interference fringes.

In the farther research, we can use row-column or column-row passes of the 1-D Wavelet Transform and the dilation and erosion of gray-scale morphology to reduce errors and eliminate the misjudgment area.

References
1 Nick Karpowicz, "Removal of Standing Wave Patters from CW THz Images Through Spatial Frequency Filtering", project report of standing wave filtering , 15th May 2006, pp 1-8

2 David M. Sheen, Douglas L. McMakin, and Thomas E. Hall, "Three-Dimensional Millimeter-Wave Imaging for Concealed Weapon Detection," *Proc IEEE, Transactions On Microwave Theory And Techniques*, vol. 49, No. 9, SEPTEMBER 2001, pp 1581-1592

3 I.S. Gregory, W.R. Tribe, B.E. Cole, C. Baker, M.J. Evans, I.V. Bradley, E.H. Linfield, A.G. Davies and M. Missous, "Phase sensitive continuous-wave THz imaging using diode lasers," *Proc IEEE, Electronics Letters*, Vol. 40 No. 2, 22nd January 2004, pp. 1716-1717

4 T. Kleine-Ostmann, P. Knobloch, M. Koch, S. Hoffmann et al, "Continuous-wave THz imaging," *Proc IEEE, Electronics Letters*, Vol. 37 No. 24, 22nd November 2001, pp. 1460-1461

A novel and simple power splitter utilizing two-branches of equal-frequency contours of a dielectric periodic structure

Yuan Zhang, Yurong Jiang, Wei Xue and S. He

Department of Photo-electronics Engineering, School of Information Science and Technology, Beijing Institute of Technology, Beijing, 100081, China.
Email: zhydxx@bit.edu.cn

Abstract *With a special design, the equal-frequency contours of a special one-dimensional dielectric periodic structure can have two branches indicating two different propagating directions. Based on this property a novel power splitter is designed.*

Introduction

Periodic structures have been an active research subject, and attracted great interest. Representative works are 1D, 2D or 3D photonic crystals (PCs) [1]. These structures can have very specific spatial dispersion, and beam propagating in these structures can have abnormal properties [2-4]. One kind of potential applications based on these structures is a beam splitter [5, 6]. So far, main works of these splitting phenomena are based to 2D dielectric PCs [5], or metallic PCs (MPCs) [6]. However, there are some disadvantages for these splitters: for a 2D PC splitter, the structure is complex; for a MPCs splitter, the metal layers should be very thin in order to decrease the metal absorption loss at visible/infrared range [6], which is not easy to control in fabrication.

In this paper, we will investigate the spatial dispersion characteristics of a 1D dielectric periodic structure, and explain how this can be used to realize a novel and simple power splitter without metal loss.

The structure and its special EFCs

The structure we designed is shown in Fig.1. The dielectric layer (refractive index n=3.4) and air layer (n$_0$=1) are arrayed alternately, and the thickness of the dielectric layer is *w=0.25a*, where *a* is the period. We considered that this structure have infinite period along X axis and finite length'd' in Z direction. The incident plane is XZ plane, and the beam incident into the structure with angle θ.

For this 1D dielectric period structure, we plot its band diagram and corresponding equal-frequency contours (EFCs), as shown

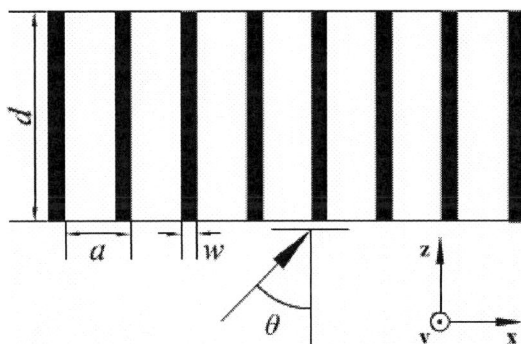

Fig.1. Dielectric layer (black strips, n=3.4) and air layer (white strips, n$_0$=1) are arrayed periodically with a finite length'd' along Z axis. The period is a, and the thickness of dielectric layer is w=0.25a.

in Fig.2. Fig.2 (a) shows the first two bands for TE waves (electric filed parallels Y axis) in first Brillouin zone, and those white lines and curves on band surfaces represent the horizontal section at normalized frequency $\omega a/2 \pi c=0.6$, which frequency we will study in rest of this paper. For more details, we plot the EFCs of the first band and the second band in Fig.2 (b) and (c), respectively. For $\omega a/2 \pi c=0.6$, the EFC of the first band is two flat lines, but two curves for the second band. By choosing the parameters carefully, we prevent the third band to disturb our plan. Next we will explain how do we use these tow bands combined to realize a power splitter.

Mechanism and simulation results

We pick up the contours of $\omega a/2 \pi c=0.6$ from Fig.2 (b) and (c), and put them in one figure as shown in Fig.3 (notice that we just plot half part of these EFCs for symmetry.)

978-0-9789217-3-6/07/$25.00

©2007 WEN GLOBAL SOLUTIONS

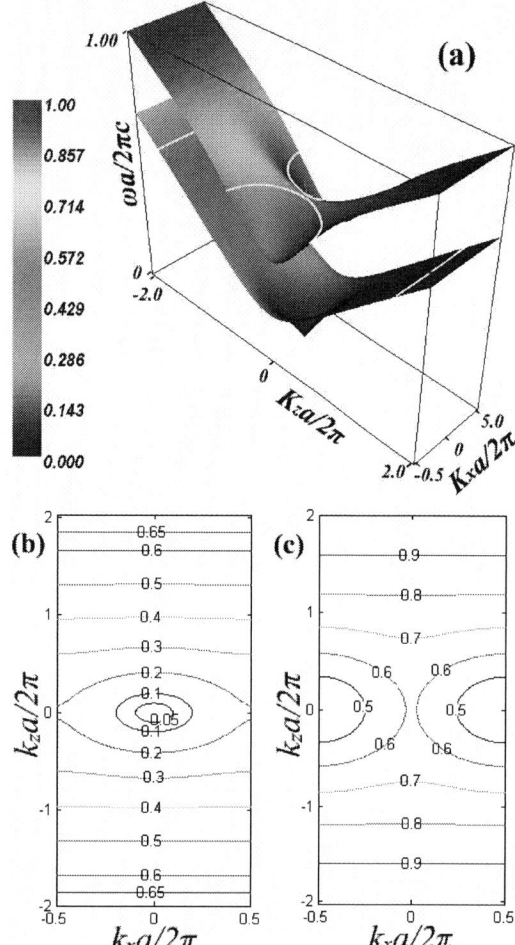

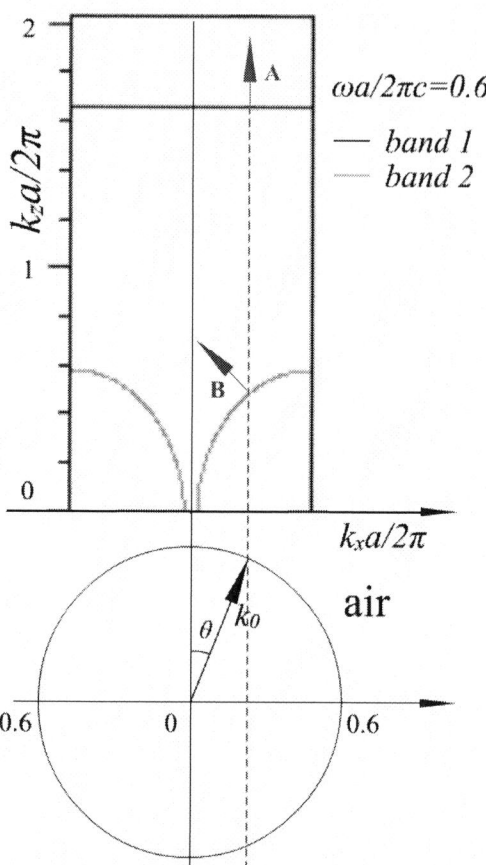

Fig.3. Wave vector diagram (for TE waves) for our 1D dielectric periodic structure. The dash line represents the case of θ =20 °.

Fig.2. Band diagram for the first two bands for TE waves of the structure in Fig.1. (a), and the corresponding EFCs for the first band (b), and the second band (c).The white lines and curves in (a) represent horizontal section at normalized frequency $\omega a/2 \pi c$=0.6.

To explain the Mechanism, we also put the wave vector in air (k_0) at the bottom, θ representing the incident angle. Considering TE waves incident from air into the structure with an incident angle θ, for the conservation of the X component of wave vector at the interface, we could find the two corresponding crosspoints on the two branches of EFCs. For the reason that group velocity is always along the normal direction of the EFCs, this will determine the propagating direction of the refractive wave. As shown in Fig.3, there will be differences between the two refractive directions corresponding to the first band (arrow A) and the second band (arrow B). That means

if a Gaussian beam incident into the structure with incident angle θ, there will be two different refractive directions in the structure: one is vertical to the interface and the other will have a abnormal refraction angle namely negative refraction (the incident wave and the refractive wave are at the same side of the normal). When the two beams propagate out of the structure from the top interface, they will be split to two parallel beams, and Fig.4 is a schematic plot for this situation. Fig.5 shows the simulation result of a Gaussian beam (with incident angle θ =20°) propagating through this structure utilizing FDTD method. We can see one refractive beam goes vertical to the interface, and one negative refracted beam appears in the structure clearly. But at the top interface, the reflection is so strong that one of the output beams is weakened badly. To reducing the reflection,

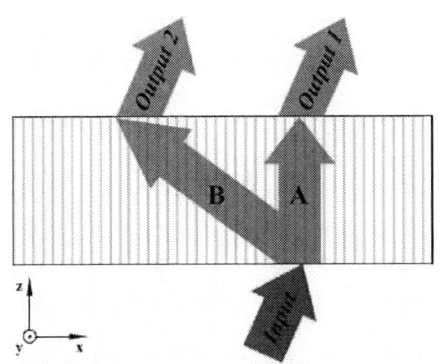

Fig.4. Schematic plot for a Gaussian beam transmitting through the structure.

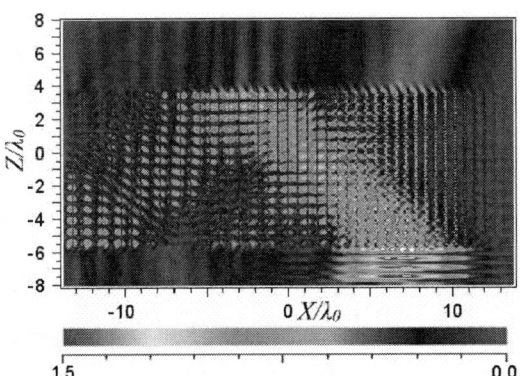

Fig.5. Electric field (Ey) distribution for a TE Gaussian beam incident into the structure with an incident angle $\theta =20^{\circ}$.

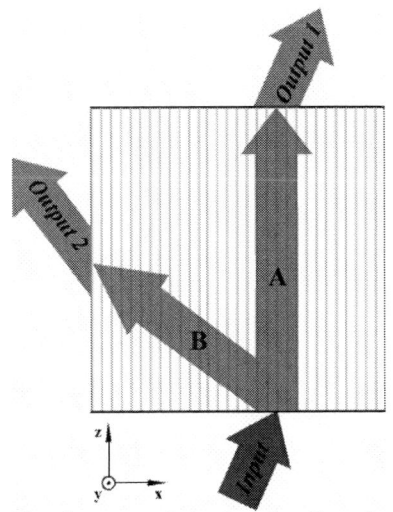

Fig.6. The improved design for the structure We make the beam B exit out of the structure from the left side.

we improve the structure geometrically by removing some layers of the left of the structure and increasing the length 'd', and this will make the beam 'B' exits out of the

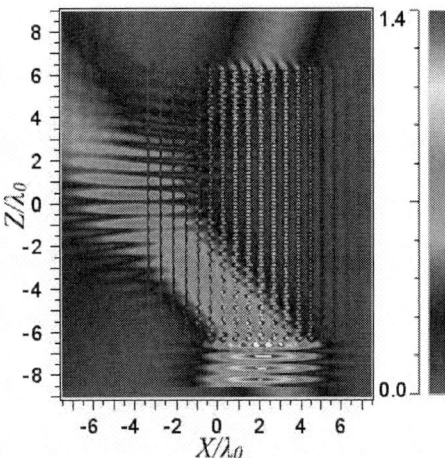

Fig.7. Electric field (Ey) distribution for a Gaussian beam incident into the improved structure with an incident angle $\theta =20^{\circ}$

structure from the left side, shown in Fig.6. Fig.7 shows the simulation result of this improved design, and we can see clearly that the incident beam is split to two beams when propagating through the structure.

Conclusions

To summarize, we study the special EFCs of TE waves for the one-dimension dielectric periodic structure we designed, and the propagating of a Gaussian beam through this structure. Based on these special characteristic, we design a novel power splitter, which is simple in structure and avoid the metal absorption loss.

Acknowledgements

This work is supported partially by the National Basic Research Program (973) of China (2004CB719801).

References

1 Sajeev John. Phys. Rev. lett. 58(1987). 2486.
2 S. Foteinopoulou. et al. Phys. Rev. B. 72(2005). 165112.
3 M. Notomi. Phys. Rev. B. 62(2000). 10696.
4 Xiebin Fan. Phys. Rev. lett. 97(2006). 073901.
5 Xiaofang Yu. et al. Appl. Phys. lett. 83(2003). 3251.
6 Linfang Shen. et al. Phys. Rev. lett. 90(2007). 251909.

Modelling and Numerical Analysis of Carrier Transport Effects on the Wavelength Chirp of SCH-QW Lasers

Farzan Gity (1), M. Naser Moghaddasi (2), Lida Ansari (3)

1 : dept. of Electrical Engineering, Islamic Azad University - Science and Research Branch, Tehran, Iran P.O. Box: 14185-748, Farzan.Gity@gmail.com

2 : Member of Central Commission for Scientific, Literacy & Art Societies, Islamic Azad University - Science and Research Branch, Tehran, Iran, moghadasi@iaucss.org

3 : dept. of Electrical Engineering, Amirkabir University of Technology (Tehran Polytechnic), Tehran, Iran, LidaAnsari@yahoo.com

Abstract *We drive a numerical model, based on physical principles, for a separate confinement heterostructure-quantum well (SCH-QW) laser. Effects of carrier transport on the transient responses and chirp characteristic is analyzed by means of finite-difference method.*

Introduction

Frequency deviation or chirp of a laser is a parasitic phenomenon that leads to a dynamic shift of the emission wavelength. The seriousness of the chirping induced performance degradation increases with the transmission bit rate and can ultimately limit the performance of the lightwave communication systems [1]. Thus, to model the reliability of a directly modulated laser and to achieve higher bit rates, especially in recent years that the transmission rates have reached more than 40 Gbit/s [2], it is important to estimate its chirp characteristics which relates to the carrier-dependence of the laser's linewidth enhancement factor. In the analysis and design of optical and microwave circuits using directly modulated laser diodes, it is often necessary to determine the dynamic response of the laser [3].

Theory of the model

In this paper we have considered a SCH-QW laser structure as illustrated in Fig. 1. The output characteristics of SCH-QW lasers mainly depend on the carrier transport across the SCH region and within QW. Variations of the carrier density in the SCH region have a significant impact on frequency chirping of high speed lasers under direct modulation. In QW lasers which generally have fairly small optical confinement factors, carrier density variations and the resultant changes in refractive index become important.

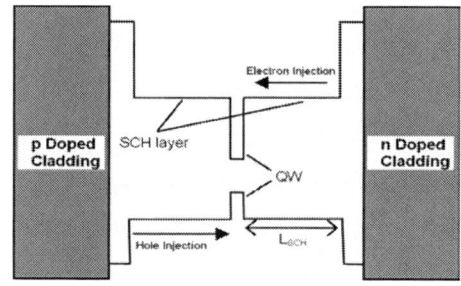

Fig. 1. Schematic representation of a single quantum-well laser with a separate confinement heterostructure modeled in this paper.

Therefore, this paper calculates self-consistently the time-dependent variations of the optical output phase through solving the phase equation coupled with the rate equations of photon, SCH and QW carrier densities, including carrier transport effects using an accurately calculated gain function, taking advantage of the Fermi's Golden Rule, and the carrier-dependent linewidth enhancement factor for QW lasers. We have incorporated different parameters that have remarkable influence on the high frequency direct modulated behavior of a SCH-QW laser.

The rate equations for a SCH-QW laser considering chirp characteristic are as follows [4]:

$$\frac{dN_b}{dt} = \frac{\eta_i I(t)}{q V_{SCH}} - R_b(N_b) - \frac{N_b}{\tau_{cap}} + \frac{V_{well}}{V_{SCH}}\frac{N}{\tau_{esc}} \quad (1)$$

$$\frac{dN}{dt} = \frac{V_{SCH}}{V_{well}}\frac{N_b}{\tau_{cap}} - \frac{N}{\tau_{esc}} - R_w(N) - \Gamma V_{gr} G(E,N)S \quad (2)$$

$$\frac{dS}{dt} = -\frac{S}{\tau_p} + R_{w\beta}(N) + \Gamma V_{gr} G(E,N)S \quad (3)$$

978-0-9789217-3-6/07/$25.00

©2007 WEN GLOBAL SOLUTIONS

$$\frac{d\phi}{dt} = \frac{1}{2}\alpha_H(N)\left[\Gamma V_{gr}G(E,N) - \frac{1}{\tau_p}\right]S \quad (4)$$

where N_b is the carrier density in the SCH layer, N is the QW carrier density, S is the photon density, ϕ is the optical output phase. $G(E,N)$ is the gain and $\alpha_H(N)$ is the exact carrier-dependant linewidth enhancement factor.

If we assume the steady state frequency as $f_0 = \omega_0/2\pi$, frequency shift occurs during the modulation as follows [1]:

$$\Delta f = \frac{1}{2\pi} \times \frac{d\phi}{dt} \quad (5)$$

In (1) - (3), $R_b(N_b)$, $R_w(N)$ and $R_{w\beta}(N)$ are indicating the effects of nonradiative and Auger recombinations, carrier and photon lifetimes and spontaneous emission as [4]:

$$R_b(N_b) = A_b N_b + B_b N_b^2 + C_b N_b^3 \quad (6)$$

$$R_w(N) = AN + BN^2 + CN^3 \quad (7)$$

$$R_{w\beta}(N) = \beta_A AN + \beta_B BN^2 + \beta_C CN^3 \quad (8)$$

Other parameters of our model and their descriptions are listed in Table 1.

For numerical treatment, (1) to (8) are written in dimensionless form by the introduction of a new independent variable $\tau = t / \tau_0$. The dependent variables are $\sigma = S / S_0$, $\xi = N / N_0$, $\xi_b = N_b / N_0$ and $\theta = f / f_0$. To let the variables be order of one, $S_0 = 10^{14}$ cm^{-3}, $N_0 = 10^{18}$ cm^{-3} and $f_0 = 10^9$ Hz are chosen [5].

In order to apply a precise carrier dependence, exact gain and linewidth enhancement factor calculations must be used. We use Fermi's golden rule in our model as [5], [6]:

$$g(E,N) = \frac{g_0}{E}|M|^2 \rho_{red}$$
$$\times (f_c(E,N) + f_v(E,N) - 1) \quad (9)$$
$$g_0 = q^2\hbar\pi/\varepsilon_0 m_0^2 cn$$

where $|M|$ is the optical matrix element, ρ_{red} is the reduced density of state, f_c and f_v are the conduction and valence bands Fermi-Dirac distribution functions, respectively, and E is energy.

Spectral broadening lowers the peak gain and shifts the emission wavelength to a shorter one. It is included through a Lorentzian-shaped broadening function [6]:

$$G(E,N) = \int_0^\infty g(W,N)\frac{(1/\pi)(\hbar/\tau_{in})}{(\hbar/\tau_{in})^2 + (W-E)^2}dW \quad (10)$$

For the exact calculations of the linewidth enhancement factor, the following relation for $\alpha_H(N)$ is used in our model [7]:

$$\alpha_H(N) \approx -\frac{\{\frac{1}{2}\log[\frac{(E_{g1}-E_{c0})^2+\Gamma_c^2}{(E_g-E)^2+(\hbar/\tau_{in})^2}] + \frac{(E_{c0}-E)}{\Gamma_c}[\frac{\pi}{2}-\tan^{-1}\frac{(E_{g1}-E_{c0})}{\Gamma_c}]\}}{\frac{\pi}{2}-\tan^{-1}\frac{(E_{g1}-E)}{\hbar/\tau_{in}}} \quad (11)$$

where $\Gamma_c = 2kTm_c/m_r$ and $E_{c0} = E_{g1}+(m_c/m_r)E_{fc}$. In the above relations E_{g1} is the energy separation between the first sub-bands in the conduction and valence bands, E_{fc} is the conduction band Fermi energy, m_c and m_r are the electron effective mass and the reduced mass [8], respectively.

Table 1: Laser parameters

Symbol	Description
η_i	Current injection efficiency
τ_{esc} / τ_{cap}	QW escape / capture lifetime
τ_p	Photon life time
τ_{in}	Interaband relaxation time
V_{SCH} / V_{well}	Volume of SCH layer / QW region
Γ	Optical confinement factor
V_{gr}	Medium group velocity
A / A_b	QW / SCH unimolecular recombination rate coefficient
B / B_b	QW / SCH radiative recombination rate coefficient
C / C_b	QW / SCH Auger recombination rate coefficient
$\beta_A / \beta_B / \beta_C$	Unimolecular / Radiative / Auger recombination coupling term

Simulation results and discussion

In this paper we have considered three different bias currents with the same pulse current. The transient responses for three different bias points are presented in Figs. 3 to 5. Transport of carriers along the confinement region, the carrier capture into, and the carrier escape out of QW region are the limiting processes affecting the transient response of SCH-QW lasers.

Figs. 6 and 7 illustrate the output power response and the chirp characteristic of the laser to a modulation pulse with a frequency of 2.5 Gb/s. As can be seen in Fig. 7, the value of frequency deviation increases with the bias current. These effects are related to the carrier transport time while passing through the SCH region. Our results are in good agreement with the experimental and simulation results [9].

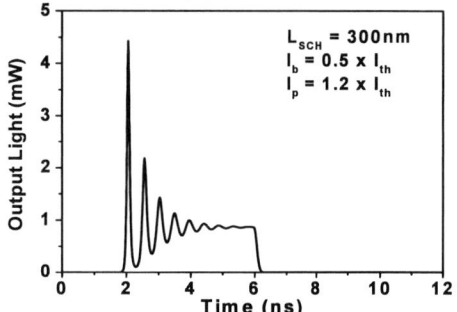

Fig. 2. Transient response of a SCH-QW laser ($I_b=0.5I_{th}$ and $I_p=1.2I_{th}$).

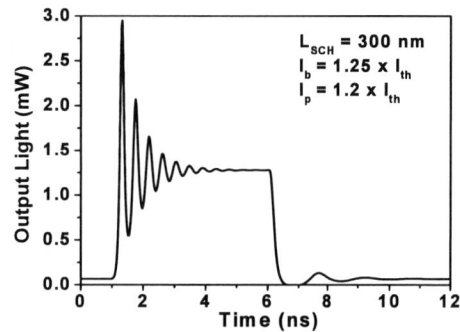

Fig. 3. Transient response of a SCH-QW laser ($I_b=1.25I_{th}$ and $I_p=1.2I_{th}$).

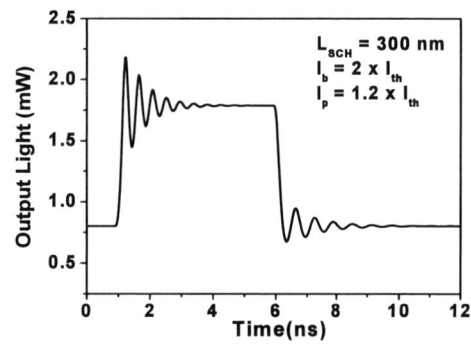

Fig. 4. Transient response of a SCH-QW laser ($I_b=2I_{th}$ and $I_p=1.2I_{th}$).

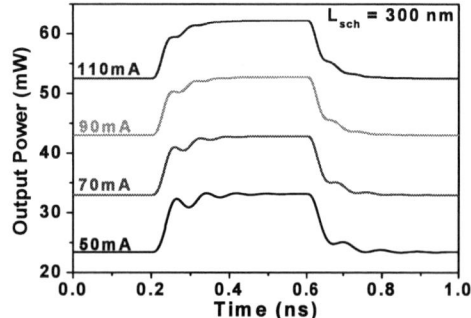

Fig. 5. Output power of a SCH-QW laser.

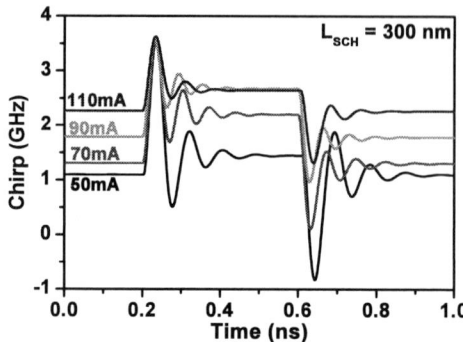

Fig. 6. Chirp feature of a SCH-QW laser.

Conclusions

This paper has proposed a precise numerical model to simulate the behavior of the SCH-QW laser. In this model which is based on the interaction between electrical and optical properties of the device, we have investigated the physical effects of carrier transport through a self-consistent solution of the phase shift equation coupled with the carrier and photon density equations, by means of finite difference method, utilizing Fermi's Golden Rule for the gain of QW laser and an exact carrier-dependent linewidth enhancement factor. This model clearly shows the significant impact of carrier transport on the functional parameters of the SCH-QW laser such as transient response and wavelength chirp of the laser.

References

1 D. McDonald et al, IEEE J. Quantum Electron., 31 (1995), 1927-1936

2 L. Billia et al, IEEE Photon. Technol. Lett., 17 (2005), 49-51

3 M. R. Salehi et al, IEEE J. Quantum Electron., 38 (2002), 1510-14

4 L. Ansari et al, 4th Int. Conf. Optical Communications and Networks (ICOCN), (2005), 332-5

5 F. Gity et al, Solid-State Electron., in press.

6 P. S. Zory, Quantum Well Lasers, San Diego: Academic Press (1993)

7 L. D. Westbrook et al, IEE Proc., 135 (1988), 223-5

8 S. L. Chuang, Physics of Optoelectronic Devices, New York: Wiley (1995)

9 K. Czotscher et al, IEEE J. Select. Topics Quantum Electron., 5 (1999), 606-612

The Iterative Ranked Phased-Array Method

Pojamarn Pojanasomboon (1), Okan Ersoy (2)

1 : Broadband Telecommunication Lab, Assumption University, Bangkok, THAILAND
2 : School of Electrical and Computer Engineering, Purdue University, West Lafayette, Indiana, USA

Abstract *Based on the Ranked Phased-Array Method (RPAM), the proposed iterative technique can improve the performances of the designed device by normalizing the output intensities with equivalent signal-to-noise ratio characteristics.*

Introduction

The Ranked Phased-Array Method (RPAM) is an algorithm to design an optical device that can function as a multiplexer / demultiplexer in a dense wavelength division multiplexing (DWDM). The RPAM principles can be described as follow [1]:

Let an optical field at (x, z) be represented by $U(x, z)$.

The subscript "*i*" and "*o*" specify the coordinates on the *input* and *output* planes as illustrated in Figure 1, respectively.

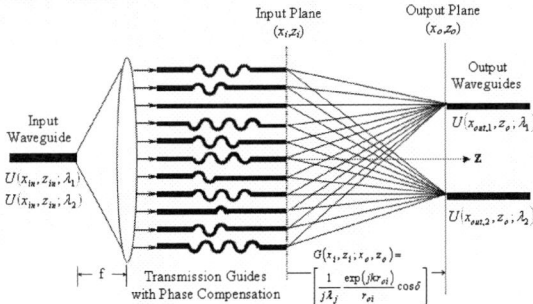

Fig.1. Phase compensation designed by the Ranked Phased-Array Method (RPAM)

With input N_i points of a specific wavelength λ_j, the output field can be expressed with *Huygens-Fresnel approximation* [2] for free-space propagation as

$$U(x_o, z_o) = \sum_{i=1}^{N_i} U(x_i, z_i) \left[\frac{1}{j\lambda_j} \frac{\exp(jkr_{oi})}{r_{oi}} \cos\delta \right] \Delta x_i \quad (1)$$

Where $k = \frac{2\pi n}{\lambda_j}$ is the propagation constant in a medium with refraction index n, and δ is the angle between the vector $r_{oi} = \sqrt{(x_o - x_i)^2 + (z_o - z_i)^2}$ and the z-axis. Because $(z_o - z_i) \gg (x_o - x_i)$, the resulting $r_{oi} \cong (z_o - z_i)$ and $\delta \cong 1$ are considered constants for all coordinates $(x_i, z_i; x_o, z_o)$.

For an optical system with *uniform-amplitude* input field $U(x_i, z_i) = \mathbf{U}\exp(\varphi_i)$, Eq. (1) becomes

$$U(x_o, z_o) = \left(\frac{\mathbf{U}\Delta x_i}{j\lambda_j r_{oi\alpha}} \right) \sum_{i=1}^{N_i} \exp(\varphi_i)\exp(jkr_{oi}) \quad (2)$$

According to principle of superposition, the amplitude of the output field at (x_o, z_o) will be maximized if

$$kr_{oi}(x_i, z_i; x_o, z_o) + \varphi_i(x_i, z_i) = 2\pi M + \varphi_o(x_o, z_o) \quad (3)$$

where M is an integer. $\varphi_i(x_i, z_i)$ is the initial phase at input coordinate (x_i, z_i), and, $\varphi_o(x_o, z_o)$ is a specific output phase at the output coordinate (x_o, z_o).

To accomplish the intensity focusing at an arbitrary point (x_o, z_o), the RPAM will modify $\varphi_i(x_i, z_i)$ by adjusting the input waveguide length $\hat{\delta}$ at the specific coordinate (x_i, z_i).

Although Eq. (2) shows that different λ_j's demand different sets of initial phases, there can only be *one common set of adjusted waveguide lengths to work for all different wavelengths* in one single structure. To achieve the optimal $\hat{\delta}_i$, the relationship between the input phase $\varphi_i(x_i, z_i; \lambda_j)$ and the corresponding adjusted waveguide length $\delta_i(\lambda_j)$ is established as

$$\varphi_i(x_i, z_i; \lambda_j) = \frac{2\pi n_g}{\lambda_j} \delta_i(\lambda_j) \quad (4)$$

where n_g is the refractive index of input waveguides.

Then, RPAM *ranks and picks* only some of the input positions x_i that can generate *least phase deviations* from $\overline{\varphi}_i(x_i, z_i)$ as shown in the flowchart in Figure 2:

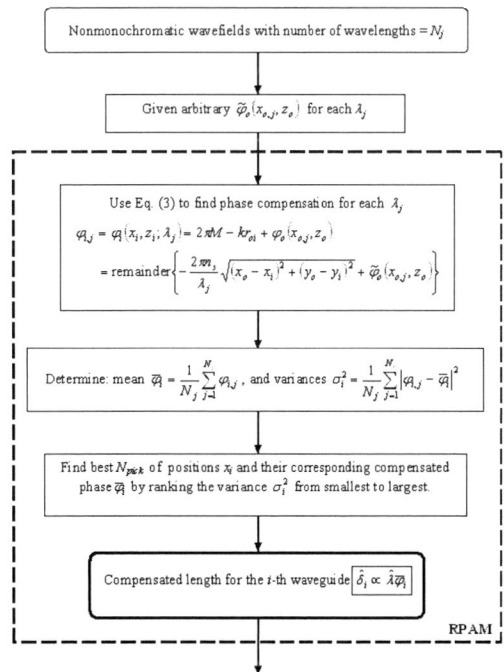

Fig.2. Flowchart of the Ranked Phased-Array Method

The Iterative Ranked Phased-Array Method

At the beginning of the conventional RPAM, the output phases are randomly selected or assigned as zeros. In the iterative-RPAM, the output phases are adjusted at the end of each iteration so that the RPAM gives better correspondence between the estimated output phases $\widetilde{\varphi}_o\left(x_{o,j}, z_o\right)$ and the actual output phases $\varphi_o\left(x_o, z_o\right)$ that are obtained after the application of the RPAM.

The flowchart of the iterative-RPAM shown in Figure 3 and can be described as follow:

i) Using the conventional RPAM to obtain the *optimal waveguide lengths* $\hat{\delta}_i = \dfrac{\hat{\lambda}}{2\pi n_g}\overline{\varphi}_i\left(x_i, z_i\right)$ for those N_{pick} input positions that yields the least phase variations, where the *mean input phase* and its *input phase variance* are defined as

$$\overline{\varphi}_i = \frac{1}{N}\sum_{j=1}^{N_j}\varphi_{i,j} \quad \text{and} \quad \sigma_i^2 = \frac{1}{N}\sum_{j=1}^{N_j}\left|\varphi_{i,j} - \overline{\varphi}_i\right|^2 \quad (5)$$

ii) Simulate the output waveforms by launching Gaussian beams as input fields for each operating wavelength through the design.

iii) Check whether the variations of peak intensities are within *the specified tolerances*. If so, terminate. If not, use the computed output phases $\varphi_o\left(x_o, z_o\right)$ for the current iteration as the assigned output phases $\widetilde{\varphi}_o\left(x_{o,j}, z_o\right)$ for the next iteration.

iv) Repeat the process until the criteria of normalized intensities are met.

Simulation Results

All computer simulations are considered for the case of implementing *256 input waveguides* out of possible 600 locations on the input plane with waveguide width of $4\ \mu m$ and the minimum separation of adjacent waveguides $\Delta x_i = 1\ \mu m$.

Optical signals are launched through the input waveguides, whose refractive index (n_g) at the average wavelength equals 1.5, and, the substrate with the refractive index (n_s) of 1.45.

The operating wavelengths used in the simulations are within the optical-fiber low-loss bandwidths, which are approximately 120 *nm* around 1550 *nm*, and are to be demultiplexed in arbitrary order according to the output locations on the plane (z_0 = 18 *mm*).

The simulation results are summarized in Table 1 with the signal-to-noise ratio (SNR) defined as

$$\mathrm{SNR} = \frac{\text{Power of the desired output}}{\sum \text{Power from the other wavelengths}} \quad (6)$$

Table 1 Comparison of the simulation results by (a) the conventional RPAM, and (b) the iterative-RPAM

Wavelength @ Output Location x_o	Intensity		SNR	
	(a)	(b)	(a)	(b)
1525.0 *nm* @ -900 μm	1.1587	1.3728	1.7097	1.4480
1600.0 *nm* @ -300 μm	1.6445	1.4636	2.1006	2.0356
1570.0 *nm* @ -100 μm	1.4153	1.3882	2.2153	2.3179
1550.0 *nm* @ +400 μm	1.6182	1.5610	2.9664	2.1548
Mean Value	**1.459**	**1.446**	**2.248**	**1.989**
Standard Deviation	**0.225**	**0.086**	**0.526**	**0.379**

The simulation results shown in Figure 4 verify that the iterative-RPAM yields equalized output intensities, which exhibit much less standard deviation of 0.086 with similar mean value of 1.446 to 1.459 of the conventional RPAM. Additionally, the performances in terms of SNR are not much affected as 2.248 and 1.989 for the conventional RPAM and iterative-RPAM, respectively.

Figure 5 illustrates the designed optical components with the waveguide lengths $\hat{\delta}_i$ in corresponding to the selected input positions x_i.

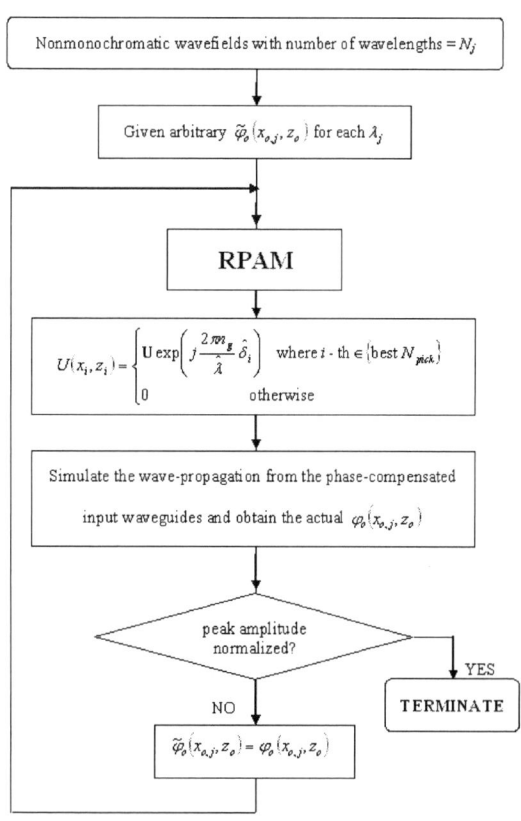

Fig.3. Flowchart of the Iterative-RPAM

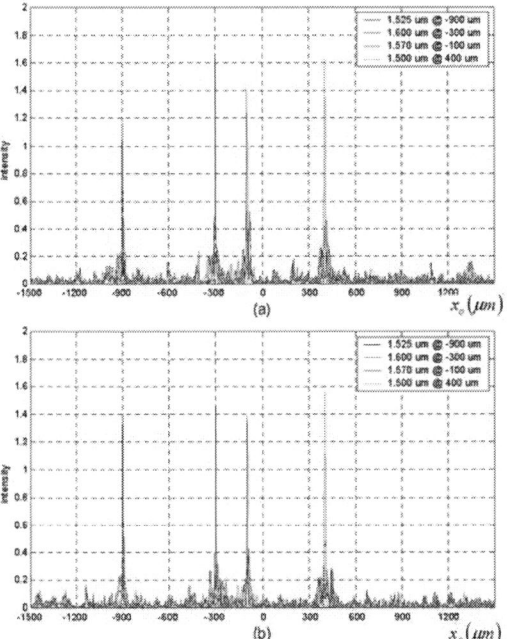

Fig.4. The output intensities vs. the output positions after demultiplexing 4 operating wavelengths by (a) the conventional RPAM, and (b) the iterative-RPAM.

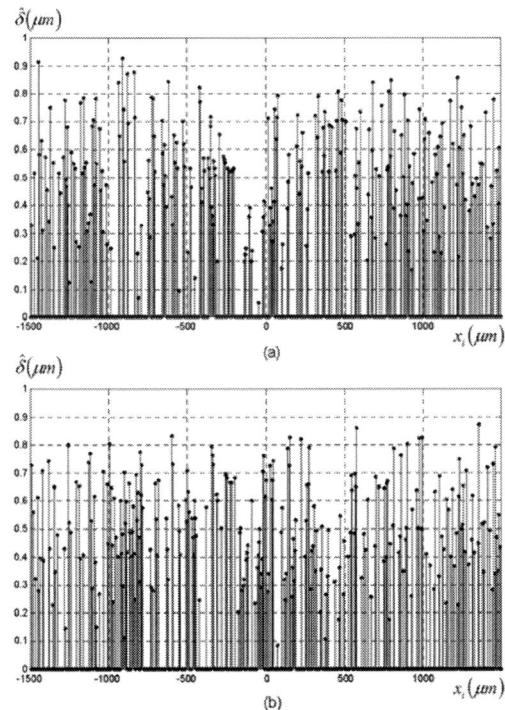

Fig. 5. The *resulting optimal waveguide lengths* $\hat{\delta}$ vs. selected input positions for demultiplexing 4 operating wavelengths by (a) the conventional RPAM, and (b) the iterative-RPAM.

Conclusions

The RPAM has been found to be an effective method to design optical components that can separate multiple wavelengths from each other *with arbitrary arrangement of the operating wavelengths for any specific output locations*. Moreover, the iterative-RPAM can enhance the performances by normalizing the output levels without much trade-off for the SNR performances.

References

1 Pojamarn Pojanasomboon, et al, *Technical Digest of 5th International Conference on Optics-photonics Design & Fabrication*, December 2006, Japan, page 109-110.
2 Joseph W Goodman, *Introduction to Fourier Optics*, McGraw-Hill, 1998, page 48-54.

A Rigorous Vectorial Gaussian Beam Modeling of Virtually-Imaged-Phased-Array

A. Mokhtari, A. A. Shishegar

Department of Electrical Engineering, Sharif University of Technology, Tehran, Iran
amokhtari@ee.sharif.edu, shishegar@sharif.edu

Abstract *The Virtually-Imaged-Phased-Array (VIPA) is a modified version of etalon used as a hyperfine spectral disperser. We have developed an analytical Gaussian beam tracing formulation which can model the VIPA output pattern and the effects of polarization changes.*

Introduction

Spectral dispersers play a key role in optical signal processing by spatially separating the optical frequencies. They have a wide variety of applications such as wavelength demultiplexing, dispersion compensation, femtosecond pulse shaping. Well-known spectral dispersers like prism and diffraction grating do not provide sufficient resolution required for thin channel spacing of dense wavelength multiplexing [1, 2, 3]. To overcome this problem, new spectral dispersers like Arrayed Waveguide Grating (AWG) and VIPA have been proposed. Shiarasaki [1] demonstrated that the VIPA-based multiplexer /demultiplexer has better performance, simple and compact design, and polarization insensitivity. He also introduced the VIPA with graded reflectivity to improve the performance later [2]. Novel applications of VIPA in optical pulse shaping such as OCDMA encoder/decoder [3], photonic-microwave arbitrary waveform generation [4], and programmable optical burst manipulation [5] have also been reported in the literature.

The VIPA principal system is depicted in fig1. The VIPA operation can be explained based on the Fabry-Perót etalon. It consists of two high reflective coated plates. The input (entry) side has a reflectivity factor close to 1. A window remains uncoated or coated with anti-reflection (AR) material to allow light beam entrance. There are two types of VIPA depending on the dielectric material that the etalon cavity is filled with. If the VIPA medium between two plates is air, it is called air-filled VIPA; otherwise it is called solid VIPA.

The laser source beam is focused on the output plate of the etalon. The collimated beam enters the etalon through the anti reflection (AR) coated window and reflects back and forth. During each reflection from output side, a portion of the light power emits out of the VIPA. The reflection continues until all injected power leaks out of the VIPA. The VIPA operation can be viewed best as the interference of an infinite number of waves from virtual sources (VS) of progressively smaller amplitudes and equal phase differences [6].

The multiple diverging beams from virtual sources interfere with each other and form a collimated beam. The phase differences between the diverging beams are highly sensitive to the wavelength variations. As wavelength changes, the collimated beam's output angle varies and results in hyperfine angular dispersion.

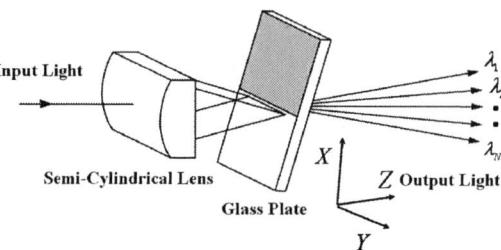

Fig1: VIPA as a spectral disperser [2]

There are a few analytical treatments in the literature suffering from restricting assumptions. Long Yang [7] modeled VIPA with 1-D Gaussian beam as a periodic filter and calculated the maxima of pass bands as a function of diffraction angle. Vega et al [8] proposed a grating equation based on a plane wave theory for relatively large incident angles. Xiao et al [9] took into account the paraxial waves and generalized the Vega approach to the small incident angles.

Vectorial Gaussian beam derivation

Consider the problem of three homogeneous lossless media n_1, n_2 and n_3 separated by two parallel planes as illustrated in fig2. A laser output beam considered as an elliptic vectorial Gaussian beam (1) passes distance Z_0 to intercept the output plane. This interception results in two new beams: reflected (2) and transmitted (3) and the same process will go on. Incident elliptic Gaussian beam in the primed (beam-based) coordinate (x'_1, y'_1, z'_1) with origin at P_1 on the source is [10]: $k_1 = \omega \sqrt{\varepsilon_1 \mu_1}$

$$\underline{E}^i(x'_1, y'_1, z'_1) = \begin{pmatrix} E_{0x'_1} \\ E_{0y'_1} \end{pmatrix} \frac{1}{\sqrt{W_{x'_1}(z'_1)W_{y'_1}(z'_1)}} \quad (1)$$

$$\times \exp\{-jk_1z'_1 + j\eta_1(z'_1) - \frac{jk_1}{2}(\frac{x'^2_1}{q_{x'_1}(z'_1)} + \frac{y'^2_1}{q_{y'_1}(z'_1)})\}$$

Where $W_i(z'_1)$, $E_{0,i}$ and $q_i(z'_1)$ are the Gaussian beam radius, electric field's phasor and q at z'_1 on the related polarization respectively $i = (x'_1, y'_1)$.

978-0-9789217-3-6/07/$25.00

©2007 WEN GLOBAL SOLUTIONS

$\eta_1(z'_1)$ is the Gaussian beam phase correction factor. The primed coordinate system changes to the incident point local coordinate system (x_1, y_1, z_1) at P_0 by a rotation α_1 (denoted by $Rot(\alpha_1)$) and a shifting z_0. The reflected beam in its beam-based coordinate system is: $k_2 = \omega\sqrt{\varepsilon_2\mu_2}$

$$\underline{E}^r(x'_2, y'_2, z'_2) = \begin{pmatrix} E_{0x'_2} \\ E_{0y'_2} \end{pmatrix} \frac{1}{\sqrt{W_{x'_2}(z'_1)W_{y'_2}(z'_1)}} \quad (2)$$

$$\times \exp\{-jk_1 z'_2 - j\eta_2(z'_2) - \frac{jk_1}{2}(\frac{x'^2_2}{q_{x'_2}(z'_2)} + \frac{y'^2_2}{q_{y'_2}(z'_2)})\}$$

We follow a rigorous vectorial Gaussian beam tracing method used in [11]. There are a few unknown parameters in the reflected and transmitted beams that should be extracted from the incident known parameters through satisfying boundary conditions at the incident point. Reflected beam in its local coordinate system (x_2, y_2, z_2) is [11]:

$$\underline{E}^r(x_2, y_2, z_2) = Rot(\alpha_2)\begin{pmatrix} E_{0x'_2} \\ E_{0y'_2} \end{pmatrix} \frac{1}{\sqrt{W_{x_2}(z'_1)W_{y_2}(z'_1)}}$$

$$\times \exp\{-jk_1 z_2 - j\eta_2(z_2) - \frac{jk_1}{2}\begin{pmatrix} x_2 \\ y_2 \end{pmatrix}^t \underline{Q}^r \begin{pmatrix} x_2 \\ y_2 \end{pmatrix}\} \quad (3)$$

$$\underline{Q}^r = Rot(\alpha_2)\begin{pmatrix} \dfrac{1}{q_{x_2}(z_2)} & 0 \\ 0 & \dfrac{1}{q_{y_2}(z_2)} \end{pmatrix} Rot(-\alpha_2)$$

The similar expression is valid for the transmitted beam (3) by changing $r \to t, 1 \to 2, 2 \to 3$.

Reflected beam calculations
After a lengthy mathematical calculation, the matrix of reflected beam becomes:

$$\underline{Q}^r = \begin{pmatrix} A & B \\ B & C \end{pmatrix} \quad (4)$$

$$A = \frac{\cos^2(\alpha_1)}{q_{x_1}(Z_0)} + \frac{\sin^2(\alpha_1)}{q_{y_1}(Z_0)}$$

$$B = \frac{\sin(2\alpha_1)}{2}(\frac{-1}{q_{x_1}(Z_0)} + \frac{1}{q_{y_1}(Z_0)})$$

$$C = \frac{\sin^2(\alpha_1)}{q_{x_1}(Z_0)} + \frac{\cos^2(\alpha_1)}{q_{y_1}(Z_0)}$$

The next step is to find the eigenvalues and eigenvectors of this matrix to derive the q-parameters and rotation angle of the reflected beam:

$$\begin{cases} q_{x_2}(0) = q_{x_1}(Z_0) \\ q_{y_2}(0) = q_{y_1}(Z_0) \end{cases}, \quad \alpha_2 = -\alpha_1 \quad (5)$$

Transmitted beam calculations

Following the same approach we used for reflected beam, the transmitted beam matrix $\underline{Q}^t$ becomes:

$$\underline{Q}^t = \frac{k_1}{k_2} \times \begin{pmatrix} b^2 A & bB \\ bB & C \end{pmatrix}, \quad b = \frac{\cos(\theta_i)}{\cos(\theta_o)} \quad (6)$$

$q_{x_3}(0)$ and $q_{y_3}(0)$ are complicated functions of both polarizations' q-parameters of incident beam i.e. $q_{x_1}(Z_0)$ and $q_{y_1}(Z_0)$. Other unknowns can be extracted from the Gaussian beam's q-parameters.

Following the same approach, the q-parameters for the N*th* virtual source are derived. Then all unknown electric field phasors are extractable in terms of geometrical and incident Gaussian beam parameters for the N*th* virtual source:

$$\begin{pmatrix} E_{0x'_{3N}} \\ E_{0y'_{3N}} \end{pmatrix} = Rot(-\alpha_3)\begin{pmatrix} \dfrac{k_1}{k_2}T_m & 0 \\ 0 & T_e \end{pmatrix} Rot(\alpha_1) \quad (7)$$

$$\times (Rot(-\alpha_2)\begin{pmatrix} R_m & 0 \\ 0 & R_e \end{pmatrix} Rot(\alpha_1))^{2(N-1)}\begin{pmatrix} E_{0x'_1} \\ E_{0y'_1} \end{pmatrix}$$

$$\times \exp\{-j(k_1 Z_0 + 2(N-1)k_1 l\}$$

$$\times Amplitude\ and\ Phase\ correction\ factors$$

Where (R_e, T_e) and (R_m, T_m) are the reflection and transmission amplitude coefficients for transverse electric field in TE and TM polarizations respectively.

Simulation method
Electric field's phasors at the VIPA output side are calculated Using equation (7) in its local coordinate system, and then we compute each virtual source's field on arbitrary vertical plane concerning the propagation effects on Gaussian beam parameters. Total field is obtained by shifting the local coordinate systems related to the main coordinate system and summation over all the resulted fields.
This Process yields an analytical closed form formula that is used to compute output VIPA pattern for the arbitrary scheme parameters.
This formulation is simulated for both uniform and graded Solid VIPA using reference [2] scheme based on the derived analytical relations.

Numerical results
We have used $\lambda = 1550\,nm, n_1 = n_3 = 1, n_2 = 1.5$

$\theta_i = 4.3°, W_{x'_1}(0) = W_{y'_1}(0) = 5.7\,\mu m, Z_0 = 50\,\mu m,$

VIPA Thickness $= 100\,\mu m, \alpha_1 = 0° \varphi = 87.5°$
The number of virtual sources is 100. Suppose $R_e = R_m = \sqrt{.95} = 0.9747$ for uniform VIPA and

$R_e = R_m = \sqrt{1 - (0.5X)^2}$ for graded VIPA where X is between 0 and 2mm. Fig3 shows the comparison between the results in [2] and current work. As it is depicted, fig3 (a) and (b) are the output beams with uniform and graded reflectivity respectively. Two works have a great similarity. We consider a 3-D model and took into account more details so that the model differences are irresistible. The method overview is as follows: 1) The Source beam waists are finite in both polarizations that are modeled as a generalized elliptic vectorial Gaussian beam.2) 3D analysis in the transverse plane 3) Precise modeling of the coating films by reflection and transmission coefficients. 4) Compatibility to treat both types of VIPA

Conclusion

We proposed a rigorous vectorial Gaussian beam formulation to derive a 3-D general model for the VIPA. The advantages and potentials of this method are also investigated.

Acknowledgement: This work is supported by the Iran Telecommunication Research Center (**ITRC**). Their support is appreciated.

References

1. M. Shirasaki, *Opt. Lett.*, vol. 21 (1996), pp.366–368
2. M. Shirasaki et al. *IEEE Photon. Technol. Lett.*, vol. 11 (1999), pp. 1443–1445
3. S. Etemad, et al. *OFC 2004*, FG5, 2004.
4. S. Xiao et al, *IEEE Photon. Technol. Lett.*, vol. 16 (2004), no. 8
5. G. Lee et al., *IEEE JLT*, vol. 23 (2005), no. 11
6. E. A. Bahaa Saleh, C. T. Malvin, *"Fundamentals of Photonics"*, pp 70-72, 1991 John Wiley & Sons
7. Long Yang, *OFC 2002,* 320-321
8. A. Vega et al., *Appl. Opt.*, vol. 42 (2003), no. 20,
9. S. Xiao et al., *IEEE JQE*, vol. 40 (2004), no. 4
10. A. Rohani et al, *Optics Communications*, vol. 232 (2004), pp 1-10
11. M. Akhavan-Bahabadi et al, AP-S International Symposium (*Digest*), vol. 3 (2002), pp 332-335

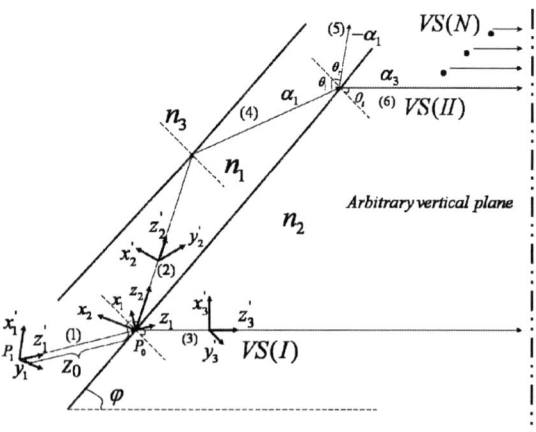

Fig2: The VIPA equivalent problem

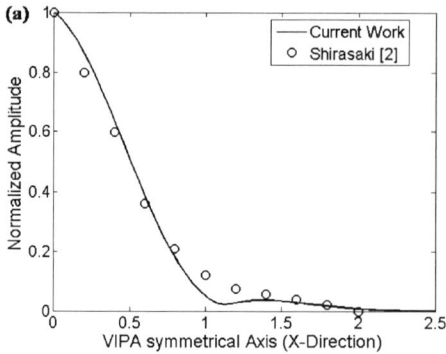

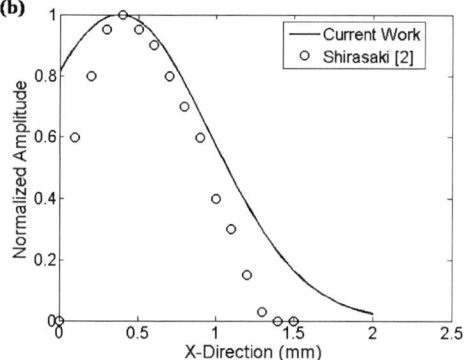

Fig3: Simulated output 2-D pattern of VIPA on the symmetrical axis of VIPA
(a) uniform with Circular Gaussian Beam (GB)
(b) Graded with Circular GB

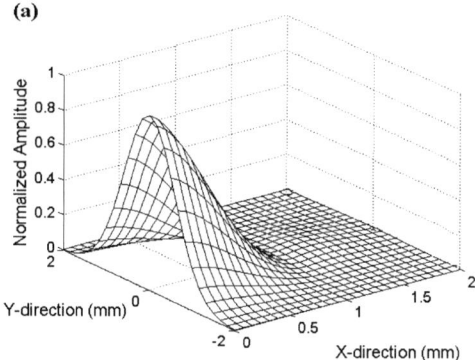

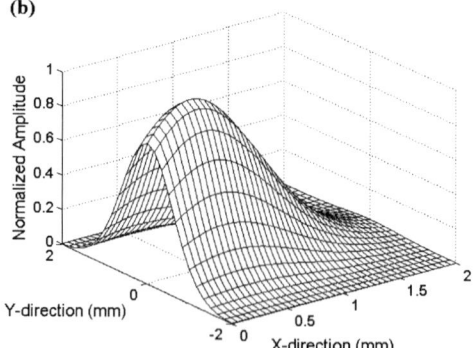

Fig4: Simulated output 3-D pattern of VIPA
(a) uniform with Circular Gaussian Beam (GB)
(b) Graded with Circular GB

Fabrication and characterization of deeply-etched SiO$_2$ waveguides

Zhen Sheng, Liu Yang, Daoxin Dai, Tingting Lang, and Zhechao Wang

Centre for Optical and Electromagnetic Research,
State Key Laboratory for Modern Optical Instrumentation, Zhejiang University,
Hangzhou 310058, China (E-mail: dxdai@zju.edu.cn).

Abstract *Deeply-etched SiO$_2$ waveguide is fabricated and characterized. The fabrication process is easier than the conventional buried SiO$_2$ waveguide. The fabricated waveguide exhibits good performances, such as low propagation loss, broad band.*

Introduction

In the past several decades, photonic integrated circuits (PICs) have been developed drastically because of their excellent performances and compact sizes. Among various kinds of waveguides with different materials or structures, the SiO$_2$-on-Si buried waveguides [1] is one of the most popular choices because of their excellent performances such as low cost, mature fabrication process, small propagation loss, and good matching to a single-mode fiber (SMF). However, the bending radius of such a silica buried waveguide is very large because of its small index contrast and the corresponding devices are usually very large in size. In order to achieve ultra-small photonic integrated circuits, the silicon nanowire waveguide [2] developed in recent years has become very popular because of its large index contrast. However, there are several challenges for Si nanowires and devices, e.g., the serious sidewall scattering loss, the large coupling loss with SMFs, the expensive and difficult fabrication. Recently the deeply-etched silica waveguide has been demonstrated for sharp bends (about 10μm) [3]. Such a waveguide is easy to fabricate and has an efficient coupling with SMF.

Fig. 1 shows the cross section of this novel waveguide. The three silica layers (cladding, core and buffer) are etched through to achieve a small bending loss as well as a small leakage loss to the Si substrate. Furthermore, by etching to the substrate, the stress induced by the deposited silica layer is almost eliminated. This deeply-etched silica waveguide can be used in a wide wavelength range (from visible to infrared) due to the large

transparent widow of silica. Apart from applications in optical communication and interconnects, the deeply-etched silica waveguide can be conveniently used for the application of sensing since its core is in direct touch with the sample and therefore expects a good sensitivity.

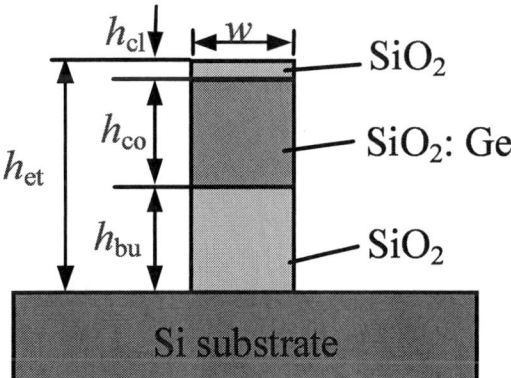

Fig. 1 Cross section of the deeply-etched SiO$_2$ waveguide.

Fabrication

The fabrication process is shown in Fig. 2. First, we use the STS PECVD (Plasma Enhanced Chemical Vapour Deposition) machine to deposit the layers of buffer (6μm), core (6μm) and cladding (1μm) in sequence without interruption by any other process [Fig. 2 (a)]. The relative refractive index contrast Δn is 0.75%. Then UV lithography is carried out to form the photoresist patterns [Fig. 2 (b)]. A lift-off process is then performed to form a metal mask [Fig. 2 (c)]. In our experiment, a 300nm Ti/Ni metal layer is used. Then an ICP (Inductively Coupled Plasma) deep etching follows. It takes about 90 minutes to have a more than 13μm etching depth. After removing the residual metal mask by acid, the process is completed and the final

978-0-9789217-3-6/07/$25.00

©2007 WEN GLOBAL SOLUTIONS   513

waveguide structure is shown in Fig. 2 (d).

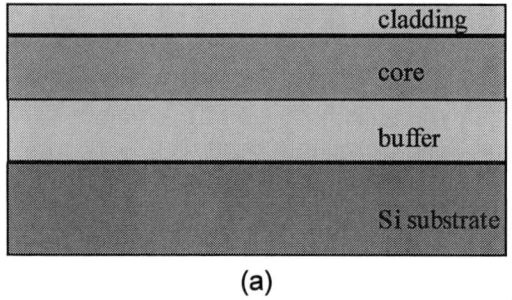

(a)

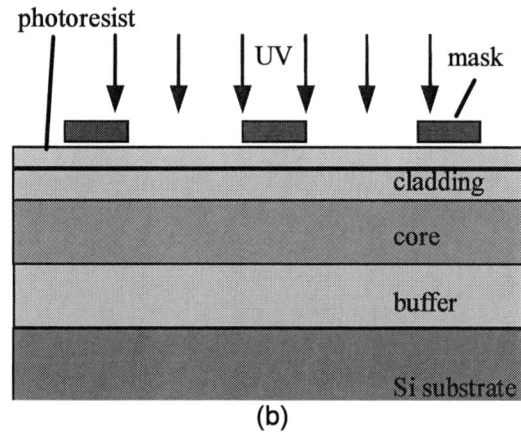

(b)

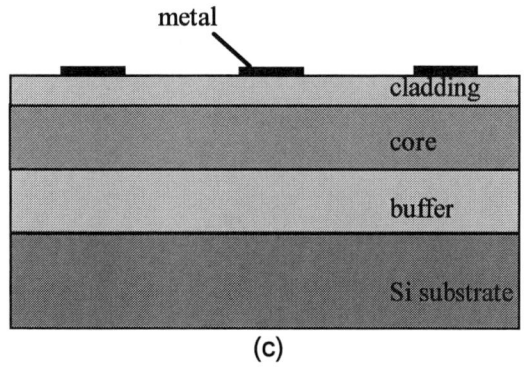

(c)

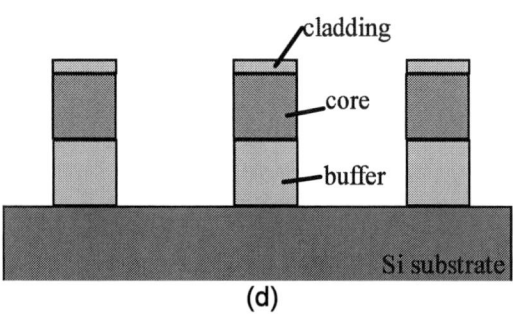

(d)

Fig. 2 The fabrication process of the deeply-etched SiO₂ waveguide.

In comparison with the conventional silica buried waveguide, the present deeply-etched SiO₂ waveguide has an easier fabrication process since the conventional one needs a process of deposition–etching–redeposition.

Fig. 3 shows the cross section of the fabricated waveguide. One can see that the three SiO₂ layers are etched through and Si substrate is also etched a little. The present sidewall angle is about 84°. Further improvement of process is required to ensure a more vertical sidewall. Here we use a previous mask which only includes the patterns with a linewidth of 6μm. In this case, the width of the fabricated waveguide is a little more than 6μm due to the fabrication error and thus it is actually multi-mode. Fabrication of deeply-etched SiO₂ waveguides with a smaller width for a single-mode operation is undertaken.

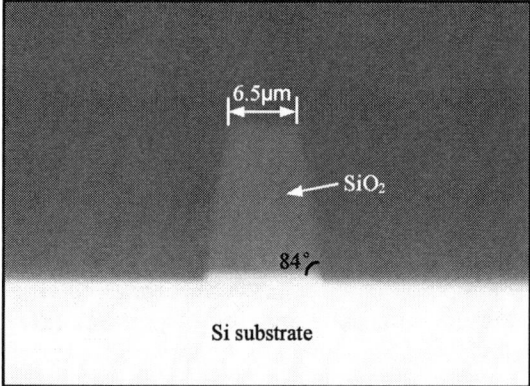

Fig. 3 Microscopical photograph of the cross section of the present deeply-etched SiO₂ waveguide.

Measurements and discussions

First we measured the propagation characteristics of the straight waveguide with different wavelengths. The transmission spectrum of a 6.5cm-long straight waveguide is shown in Fig. 4. The measured loss includes the propagation loss of the waveguide and the coupling loss at two interfaces between waveguide and fibers. One can see that there are small ripples on the transmission curve. Apart from the experimental uncertainty (e.g. unstable light source), the multi-mode effect is one of the reasons for such ripples since the present waveguide is multi-mode (as what we mentioned above). A semi-Victoria finite difference (FD) mode solver [4] was used to evaluate the present waveguide. The obtained modes are summarized in Table 1.

514

Table 1 Modes obtained by an FD simulation. TE_{mn} (TM_{mn}) means the mode field has m peaks in the lateral direction and n peaks in the vertical direction.

Mode	Effective index	Leakage loss (dB/cm)
TE_{11}	1.4626	0.0019
TE_{21}	1.4460	0.0020
TE_{12}	1.4549	0.13
TE_{13}	1.4487	2.3
TM_{11}	1.4631	0.028
TM_{21}	1.4483	0.029
TM_{12}	1.4553	2.7
TM_{13}	1.4496	40

Due to multi-mode interference in the straight waveguide, the field distributions at the output end of the waveguide will be wavelength-sensitive, which consequently introduces a wavelength-dependent coupling efficiency with the output single-mode fiber. At 1550nm, the coupling loss between the waveguide and the fiber is about 1.5dB per facet on average. So the propagation loss is about 0.5dB/cm. One can also see that the loss is small over a large wavelength range (from 1210nm to 1600nm). The propagation loss is mainly caused by the scattering at the sidewall and a little substrate leakage. This can be improved further by improving the uniformity of the SiO_2 films and by reducing the roughness of the sidewalls.

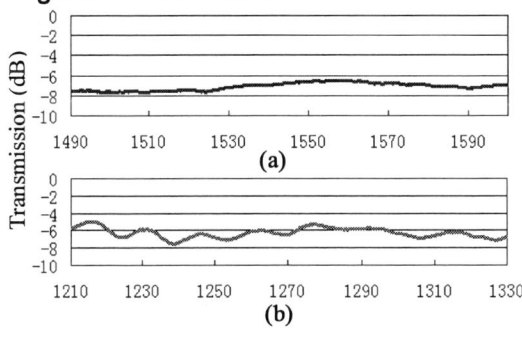

Wavelength (nm)
Fig. 4 The transmission spectrum of a 6.5cm-long straight waveguide with (a) 1490-1600nm and (b) 1210-1330nm.

Fig. 5 shows the spectral responses of an arrayed waveguide grating (AWG) (de)multiplexer [5] based on the present deeply-etched waveguides. Due to the birefringence and the multimode effects in

the optical waveguides, the spectral responses have several small peaks (as shown in Fig. 5). A narrow optical waveguide will be used to avoid the multimode effects in our future work.

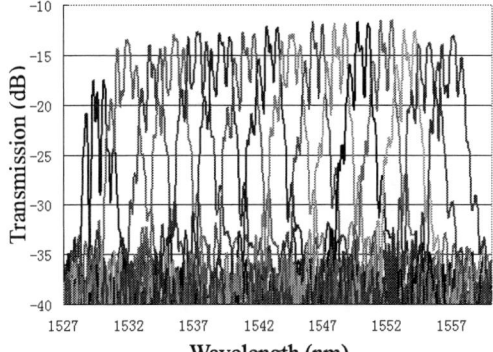

Fig. 5 The measured spectral responses of an AWG (de)multiplexer based on the present deeply-etched waveguides.

Conclusions
Deeply-etched SiO_2 waveguide has been fabricated and measured. Because of the easy fabrication and the ability for sharp bending, this kind of waveguide has the potential to realize compact low-loss PICs. Further work will be done to achieve a low-loss single-mode deeply-etched SiO_2 waveguide and devices.

Acknowledgement
The authors would like to thank Mr. Xinsong Hu for performing lithography.
This project was supported by research grants (No. 20061343, and 2006R10011) of the provincial government of Zhejiang Province of China, the National Science Foundation of China (Nos. 60607012 and 60688401).

References
1 T. Miya, et al, IEEE J. Sel. Top. Quantum Electron., vol. 6 (2000), pp. 38-45
2 T. Tsuchizawa, et al IEEE J. Sel. Topics Quantum Electron., vol. 11 (2005), pp. 232-240
3 D. Dai, et al, J. Lightw. Technol., vol. 24 (2006), pp. 5019-5024
4 C. L. Xu, et al, IEE Proc. –Optoelectron., vol. 141 (1994), pp. 281-286
5 M. K. Smit, et al, IEEE J. Sel. Top. Quantum Electron., vol. 2 (1996), pp. 236-250

Transmissive properties and Faraday rotation of tunable photonic-band-gap system containing liquid crystal

Ping Xu[a,b], Lei Gao [a,b] and Z. Y. Li[a,b*]

[a] *Department of Physics, Suzhou University, Suzhou, 215006, China* [†]
[b] *Key Laboratory of thin films in Jiangsu Province, Suzhou, 215006, China*

The optical and magneto-optical properties of the defect mode in a periodic structure have been demonstrated using a liquid crystal as a defect layer. The results show that the defect mode in the photonic band gap is sensitive to the thickness scale and a significant enhancement of the Faraday rotation is observed.

I. INTRODUCTION

Photonic crystals (PCs) having periodic structure have attracted many attentions during the past decades due to their unique properties in controlling the propagation of electromagnetic waves. The PCs are characterized by electromagnetic stop bands or photonic band gaps (PBGs), within which the propagation of electromagnetic waves is forbidden[1,2]. In particular, the study of a defect mode is one of the most attractive subjects since PCs with controlled defects exhibit highly localized defect modes and are useful for construction of filters with a narrow transmission band, resonant cavities, mirrors, and waveguides.

Kubo et al. reported the recording of the photonic band gap liquid crystal (LC) materials[3]. The difference in the light scattering intensity between LC phase and isotropic phase leads to change in the reflected light intensity. Recently, several works on tunable photonic band gap materials containing liquid crystals (LCs) have been reported. They mainly focused on the theoretical and experimental investigation of tunable optical characteristics by introducing LCs. LCs have anisotropic optical, electrical, and magnetic properties that are sensitive to external fields as well as to other external stimuli.

More recently, optical and magneto-optical effects in magnetophotonic crystals(MPCs) composed of yttrium iron garnet YIG($Y_3Fe_5O_{12}$) have been theoretically investigated by several groups, taking into account the magnetization-induced off-diagonal components of the permittivity tensor[5,6]. A remarkable property of YIG is its bigyrotropy, which is characterized by the fact that in the transparency regime its permittivity tensor and its permeability tensor have comparable contribution to the magneto-optical(MO) response. Inclusion of magnetic materials in PCs leads to peculiarities in the propagation of electromagnetic waves.

To achieve angles of magneto-optical Faraday rotation as high as possible in samples and high optical transmittance has been a major goal in the development of transparent magneto-optical materials. We have not found

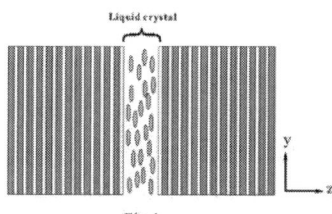

FIG. 1: One-dimension photonic crystal with liquid crystal defect layer.

any reports on optical and MO effects of PCs containing LC defect. The purpose of the present work is to investigate the effects of the coexistence of LC defect on light propagation in metal-dielectric photonic crystals and magnetophotonic crystals.

II. MODEL AND THEORY

Let us consider a one-dimensional periodic structure composed of a defect LC layer. In Fig. 1, the layer structure is symmetric with respect to its center layer (LC layer). The incident electromagnetic wave propagates along the z-direction. The long molecular axis of the LC is aligned parallel to the y-direction. The fundamental equations of light are given by Maxwell's equations,

$$\nabla \times \vec{E}(\vec{r}, t) = i\omega\mu_0\tilde{\mu}\vec{H}(\vec{r}, t)$$
$$\nabla \times \vec{H}(\vec{r}, t) = i\omega\varepsilon_0\tilde{\varepsilon}\vec{E}(\vec{r}, t)$$

where $\tilde{\epsilon}$ and $\tilde{\mu}$ are the dielectric permittivity tensor and magnetic permeability tensor.

For the calculation of the transmission characteristics, we used the $4*4$ transfer matrix method[7]. In our multi-layer structure, components of the permittivity and the permeability tensor are not taken to 1 as usual. The role of permeability becomes important. Therefore, it is necessary for us to use the generalized expressions of the $4*4$ transfer matrix method. The state vector of light is defined by $\phi = (e_x, e_y, h_x, h_y)^t$, which are components of the electric field vector $e(= \epsilon_0 E)$ and the magnetic field vector $h(= \mu_0 H/c)$ of the light. ϵ_0, μ_0 and c are the permittivity, permeability and the light vacuum velocity. The subscript t denotes the transpose operator.

---

[*]Corresponding author.
[†]E-mail: zyli@suda.edu.cn

978-0-9789217-3-6/07/$25.00

©2007 WEN GLOBAL SOLUTIONS

Using front and back travelling light with right and left circular polarization, the state of the incident light is described by

$$\Phi(z) = A \begin{bmatrix} 1 \\ -i \\ i\sqrt{\frac{\varepsilon_1 + \varepsilon_2}{\mu_1 + \mu_2}} \\ \sqrt{\frac{\varepsilon_1 + \varepsilon_2}{\mu_1 + \mu_2}} \end{bmatrix} e^{ik_1 z} + B \begin{bmatrix} 1 \\ -i \\ -i\sqrt{\frac{\varepsilon_1 + \varepsilon_2}{\mu_1 + \mu_2}} \\ -\sqrt{\frac{\varepsilon_1 + \varepsilon_2}{\mu_1 + \mu_2}} \end{bmatrix} e^{-ik_1 z}$$

$$+C \begin{bmatrix} 1 \\ i \\ -i\sqrt{\frac{\varepsilon_1 - \varepsilon_2}{\mu_1 - \mu_2}} \\ \sqrt{\frac{\varepsilon_1 - \varepsilon_2}{\mu_1 - \mu_2}} \end{bmatrix} e^{ik_2 z} + D \begin{bmatrix} 1 \\ i \\ i\sqrt{\frac{\varepsilon_1 - \varepsilon_2}{\mu_1 - \mu_2}} \\ -\sqrt{\frac{\varepsilon_1 - \varepsilon_2}{\mu_1 - \mu_2}} \end{bmatrix} e^{-ik_2 z}$$

where $k_1 = \frac{\omega}{c}\sqrt{(\varepsilon_1 + \varepsilon_2)(\mu_1 + \mu_2)}$ and $k_2 = \frac{\omega}{c}\sqrt{(\varepsilon_1 - \varepsilon_2)(\mu_1 - \mu_2)}$ are the wave numbers of right and left circular polarization. And A, B, C and D are the coupling coefficients. If the single layer whose permittivity and permeability are tensor occupies the spatial region of $Z_0 < Z < Z_0 + d$, where d is the thickness of the layer, the state vector can be given as

$$\Phi(z_0 + d) = T\Phi(z_0)$$

where T is a $4 * 4$ state matrix. In the exterior space $z < z_0$, the state vector is the sum of the ingoing TM wave and the reflected TM and TE wave. While in the exterior space $z > z_0 + d$, the state vector is the sum of the outgoing TM and TE wave. For a given incident wave, the reflectance and transmittance of the electromagnetic wave can be directly obtained by the coupling coefficients.

III. CALCULATION AND DISCUSSION

A. Metal-dielectric photonic crystal(MDPC) with a LC defect

It has been shown experimentally that the width of the gap in metal-dielectric PBG crystal can far exceed the width of the gap in dielectric/dielectric PCs because of the large reflectivity at a metal-dielectric interface. In the GHz frequency range, the skin depth of metals is on the order of microns. We might expect that a single 40nm layer of Ag should be transparent to microwave radiation. The lowest energy pass band is shut off due to the dispersion properties of metals. We have theoretically calculated the transmission spectra of 1D metal-dielectric PC Ag/Ta_2O_5 with a LC defect.

Fig.2(a) shows the theoretical transmittance for a four-pair multilayer without LC defect, which contains a total of 104nm of Ag. The transmission resonance at 360nm is near the plasma frequency for Ag. A 15dB deep stop band ranging from 450nm to 750nm is observed. Prominent pass bands are evident at 425nm and 745nm. The width of the stop band is 50% of the center frequency,

which is wider than that of the dielectric-dielectric PBGs. The lowest energy pass band shuts off at 870nm due to the large dispersion in the extinction coefficient of Ag. The reflectance is larger than 99% for a wide range of frequency in the stop band, which means that the metallic absorption in the stop band is very small.

Fig.2(b) shows the calculated transmission spectrum of a 1D MDPC with a LC defect of 100nm thickness. In order to simplify the geometry of the system, the LC optical axis orientation is set normal to the incidence plane. In this way, for p- and s- polarized waves, the LC layer exhibits the ordinary and extraordinary indices, respectively. The refractive index anisotropy δn is 0.209 at room temperature. The upper layer with an index higher than the LC extraordinary index and the bottom layer one lower than the LC ordinary index. The pass bands appear at 425nm, 635nm and 825nm. The change of the bandwidth of the MDPC with LC defect is caused by a redistribution of the displacement field within the photonic crystal when a third component is inserted into the photonic crystal. Accordingly, the transmittance through a MDPC with LC defect is smaller than that through a MDPC without LC defect. As indicated in Fig.2(c), thickness of LC defect is 200nm. The defect mode peak of LC shifts toward shorter wavelength with the thickness change. The peak shift originates from the increase in the optical length of the LC defect layer. With the increase of the LC layer thickness there appears more defect mode peaks in PBG shown in Fig2.(d). The transmittance of defect mode peaks in the PBG are smaller. This result indicates that the photons of the defect mode wavelengths are strongly confirmed in the defect layer.

B. Magnetophotonic crystal(MPC) with a LC defect

Let us consider a periodic structure composed of YIG layers of thickness d_1 alternating with nonmagnetic GGG($Gd_3Ga_5O_{12}$) layers of thickness d_2. For the introduction of the LC defect layer, an LC is sandwiched between two substrates with multilayers of alternating YIG and GGG. The sum of the thickness of an YIG layers is 600nm.

In Fig.3(a) the transmittance for the MPC without LC defect in the case of perpendicular incidence is plotted against the wavelength. The corresponding figures for MPC with LC defect of different thickness are quite similar. The change of optical length leads to absorption changing. For sharp resonances, the decrease in transmittance is severe. In Fig 3(b) and 3(c) we present the transmittance for LC defect $d = 80nm$ and $d = 60nm$, respectively. With a change in thickness of the defect layer, transmittance is small. Furthermore, the results of magneto-optical FR are presented in Fig.4. FR spectra are normalized to the sum of MPC layer thicknesses. Nonmagnetic garnet crystal GGG has no contribution to the FR effect. Compared with FR in MPC without

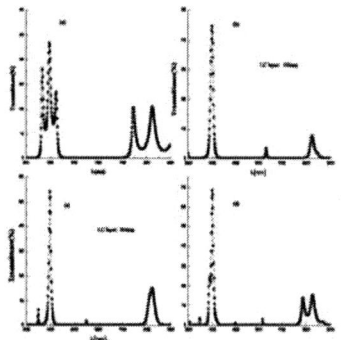

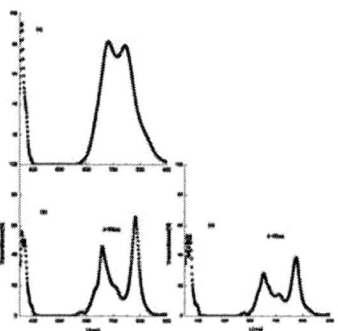

FIG. 2: (a) Calculated transmission spectra of 1D MDPC without LC defect at normal incidence, 3.5 periods of $Ag(26nm)/Ta_2O_5(140nm)$. (b) Calculated transmission spectra of 1D MDPC with LC defect of 100nm thickness. (c) The thickness of LC defect layer is 200nm. (d) The thickness of LC defect layer is 500nm.

FIG. 3: (a) Theoretical transmission spectra of 1D MPC without LC defect at normal incidence, 3.5 periods of YIG(150nm)/GGG(175nm). (b) Theoretical transmission spectra of 1D MPC with LC defect of 80nm thickness. (c) The thickness of LC defect layer is 60nm.

LC defect as shown in Fig.4(a), we get an prominent enhancement of magneto-optical effect in Fig.4(b). Faraday rotation angle increases from -1.27 to -87.43 deg/um at λ=640nm for MPC without LC defect and with LC defect, respectively. While at λ=745nm FR angle increases from -0.71 to -89.06 deg/um. It clearly turns out that MPC with LC defect exhibits higher FR than that without LC layer. Enhanced FR in MPC with LC defect is

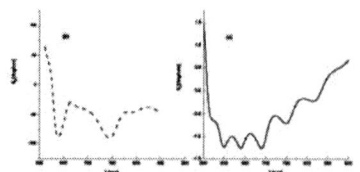

FIG. 4: (a) Spectra of Faraday rotation for 1D MPC without LC defect. (b) Theoretical magneto-optical properties of 1D MPC with LC defect.

attributed to the anisotropic optical, electrical and magnetic properties of LCs. The study of MPCs containing LC film deserves attention because it means that there is an enhancement of the MO effect.

IV. SUMMARY

We have considered transmissive properties and Faraday rotation of MDPC and MPC containing LC defect using a generalized $4*4$ transfer matrix method. We compared MDPC and MPC system with and without LC defect in order to identify the effect of LC on the transmittance and magneto-optical properties of PCs. We demonstrated the defect mode in a MDPC PBG by using a liquid crystal defect layer. With a change in thickness of LC defect layer, the defect mode peak shifts to shorter wavelengths due to the increase in the optical length, and also appears several defect mode peaks. On the other hand, LCs have a large optical anisotropy. Based on such highly anisotropy, it has turned out the enhancement of magneto-optical effect. We consider our investigation as a proof of feasibility and expect that PCs possessing high magneto-optical Faraday rotation as well as high optical transmittance will soon been applied to optoelectronics and MO devices.

V. ACKNOWLEDGMENT

This work was supported by the National Natural Science Foundation of China under grant No. 10547128 and the National Science Foundation of Jiangsu Education Committee of China under grant No. 05KJB140112, and partly No. 10474069.

---

[1] E. Yablonovitch, Phys. Rev. Lett. **58**, 2059 (1987).

[2] S. John, Phys. Rev. Lett. **58**, 2486 (1987).

[3] S. Kubo, Z. Z. Gu, K. Takahashi, Y. Ohko, O. Sato and A. Fujishima, J. Am. Chem. Soc. **124**, 10950 (2002)

[4] D. Kang, J. E. Maclennam, N. A. Clark, A. A. Zakhidov and R. H. Baughman, Phys. Rev. Lett. **86**, 4052 (2001).

[5] M. Inoue and T. Fujii, J. Appl. Phys. **85**, 5768 (1999).

[6] S. Kahl and A. M. Grishin, Appl. Phys. Lett. **84**, 1438 (2004).

[7] A. Figotin and I. Vitebskiy, Phys. Rev. E **63**, 066609 (2001).

Resonance-induced transmissions through waveguides below cut-off frequencies: An effective-medium model for waveguide

Hao Xu*, Jiaming Hao, Jiajie Dai, and Lei Zhou**

Surface Physics Laboratory (State key laboratory) and Physics Department, Fudan University, Shanghai 200433, P. R. China

Abstract

It is well known that electromagnetic (EM) waves can not transmit through a metallic waveguide at frequencies well below its cutoff frequency, since only evanescent waves are allowed inside the waveguides under such conditions. However, through inserting local resonance structures of either electric or magnetic type into the waveguide, we find extraordinary transmissions of EM waves with different polarizations through the waveguide at frequencies well below the waveguide's cut-off value [1] [2]. We have identified two different mechanisms for such unusual transparencies, in which the medium at transparency can be characterized either by doubly positive ε_{eff} and μ_{eff} [see Fig. 1(left)] or by doubly negative ε_{eff} and μ_{eff} [see Fig. 1(right)] [see also Ref. 2 for such a mechanism]. We also demonstrated that while a hollow metallic waveguide can be viewed as an effective medium with $\varepsilon_{\text{wg}} = 1 - f_p^2 / f^2, \mu_{\text{wg}} = 1$ for the transverse-electric (TE) polarized waves, it should be viewed as a different effective medium with $\varepsilon_{\text{wg}} = 1, \mu_{\text{wg}} = 1 - f_p^2 / f^2$ for the transverse-magnetic (TM) polarized waves, where f_p is the cut-off frequency of the waveguide. We have designed realistic electric/magnetic resonance structures, and have performed accurate finite-different-time-domain (FDTD) simulations on such realistic structures to successfully demonstrate all theoretical predictions based on the model calculations.

978-0-9789217-3-6/07/$25.00

©2007 WEN GLOBAL SOLUTIONS

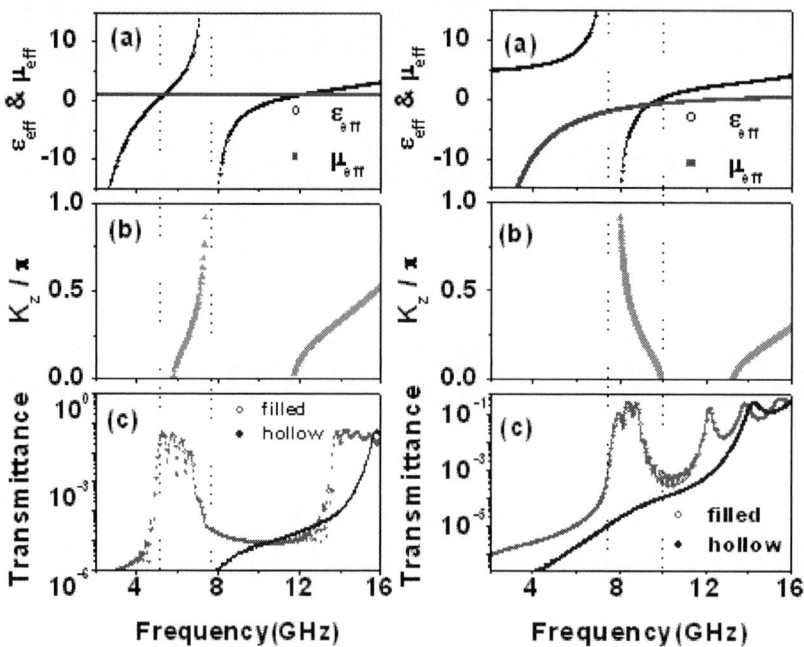

Figure 1. For TE- (left panel) and TM-polarized wave (right panel), (a) calculated ε_{eff} and μ_{eff} of the waveguide filled with a periodic array (For TE mode lattice constant = 2.9mm; For TM mode lattice constant = 2.5mm) of metamaterial layers (thickness = 1mm) with permittivity and permeability given by

$$\varepsilon_{xx} = \varepsilon_{yy} = 1 + \frac{300}{7.66^2 - f^2} + \frac{1900}{21.65^2 - f^2}, \varepsilon_{zz} = 1 \text{ and } \mu = 1 \text{ ;}$$

(b) calculated Bloch wave-vector for the above mentioned waveguide; (c) calculated EM wave transmissions through the hollow waveguide (solid circles) and the above-mentioned filled waveguide (open circles), based the FDTD simulations on realistic structures.

* **Presenting author; ** Corresponding author, Email: phzhou@fudan.edu.cn**

[1] H. Xu, J. M. Hao, J. J. Dai, and L. Zhou, unpublished.

[2] R. Marques, J. Martel, F. Mesa and F.Medina Phys. Rev. Lett, 89, 183901(2002)

Modeling and Optimization for Segmented Transmission-Line Electroabsorption Modulators with Asymmetrical Electrodes

Yongbo Tang (1), Yichuan Yu (2), and Yuqian Ye (3)

Center for Optical and Electromagnetic Research, Joint Research Centre of Photonics of the Royal Institute of Technology (Sweden) and Zhejiang University, Zhejiang University, Hangzhou 310058, China.

(1) tyb@coer.zju.edu.cn (2) yuyich@coer.zju.edu.cn (3) yeyq@coer.zju.edu.cn

Abstract *Normalized RF link gain is derived for a segmented transmission-line electroabsorption modulator. Genetic algorithm is used to optimize the electrode structures which are asymmetrical and non-periodic. Performance of the optimized EAM design is analyzed.*

Introduction

High speed electroabsorption modulators (EAMs) have been widely used in high-speed fiber-optic links due to their high extinction ratio, low driving voltage, and easy integration with semiconductor laser. In order to achieve high bandwidth and high modulation efficiency, the travelling wave (TW) configuration is proposed to overcome the RC time-constant bandwidth limitation inherent to lumped design. However, TW-EAMs usually have very low characteristic impedance, close to 25Ω or even lower. Such devices are not compatible with the standard 50Ω RF connections. Impedance mismatch will result in the increase of the driving voltage because of multiple electrical reflections.

Different methods have been presented to achieve more effective impedance match. A simple way is to utilize a low impedance terminator, though it will lead much high low-frequency RF link gain sacrifice [1]. A bandwidth of 43GHz was achieved for a 450μm-long device integrated with a 13Ω termination resistor [2]. More complex structures such as the segmented electrode design were also introduced. The bandwidth extrapolated from the measured response curve for a segmented TW-EAM with 165μm total active length has surpassed 80GHz [3]. However, the symmetrical and periodic electrode design adopted in this design may not be the best choice as the RF signal is travelling from one side to the other and the voltage distribution along the electrode is not symmetrical, so is the

optical wave. Hence, the potential of the bandwidth enhancement in segmented transmission-line structure could be explored further.

In this paper, an effective modeling approach based on conventional TML theory is developed to analyze the performance of the segmented transmission-line EAMs, which could take into account the transmission-line segments with arbitrary lengths. By using genetic algorithm (GA), the optimized segmented transmission-line electrodes have been obtained. The optimized electrode structures are asymmetrical and non-periodic.

DEVICE MODELING

The electrode of the segmented EAMs, with a structure alternating between passive microstrip and modulation segment [3], could be treated as a cascaded connection of several two-port networks, which could be easily modeled by the ABCD matrices. For the uniform lossy transmission line with a physical length l, the ABCD matrix could be expressed as

$$\begin{bmatrix} A & B \\ C & D \end{bmatrix} = \begin{bmatrix} \cosh(\gamma l) & Z_c \sinh(\gamma l) \\ Z_c^{-1} \sinh(\gamma l) & \cosh(\gamma l) \end{bmatrix} \quad (1)$$

where Z_c and γ are the characteristic impedance and propagation constant, respectively.

978-0-9789217-3-6/07/$25.00

©2007 WEN GLOBAL SOLUTIONS

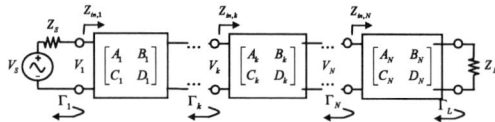

Fig. 1. *Cascaded network representation for the electrode structure of segmented EAM.*

In Fig. 1., the segmented electrode is represented as a cascaded network. Each electrode segment can be described as a block. The index denotes the block's order number marked from the microwave signal generator. The voltage, reflection coefficient and input impedance looking toward the load at the kth discontinuities are denoted as V_k, Γ_k and $Z_{in,k}$, respectively. $Z_{in,k}$ and Γ_k could be readily obtained by using ABCD matrices.

The voltage along the electrode is the superposition of forward and backward traveling voltage waves. For the first segment, whose length is l_1, the voltage $V(z)$ along the segment could be calculated [4].

$$V(z) = V_1^+ \times (e^{-\gamma(z-l_1)} + \Gamma_2 e^{\gamma(z-l_1)}) \quad (2)$$

$$V(0) = V_1 = \frac{Z_{in,1}}{Z_{in,1} + Z_S} V_S \quad (3)$$

where V_1^+ is the forward voltage at the first discontinuities after microwave signal source. As a result, $V(z)$ could be obtained as

$$V(z) = V_1 \times \frac{(e^{-\gamma z} + \Gamma_2 e^{\gamma(z-2l_1)})}{1 + \Gamma_2 e^{-2\gamma l_1}} \quad (4)$$

Then, the voltage V_2, which represents the voltage at the end of the first segment, could be calculated. So the voltage along the next segment could be readily obtained following the similar way, so is that along the rest.

Assuming the optical small signal frequency response is linearly to the voltage integral over all modulation segments, taking velocity mismatch and the RC-penalty included through the active layer capacitance C_{im} and the semiconductor

material resistance R_{sm} into consideration [1], the optical modulation response could be derived as

$$I_{ac}(\omega) = \frac{1}{1 + j\omega R_{sm} C_{im}} \sum_{i=1}^{(N-1)/2} \frac{V_{2i} e^{j\varphi_i}}{1 + \Gamma_{2i+1} e^{(-2\gamma_m l_{2i})}}$$
$$\times \left\{ \frac{e^{\{(j\beta_O - \gamma_m)l_{2i}\}} - 1}{(j\beta_O - \gamma_m)} + \Gamma_{2i+1} e^{-2\gamma_m l_{2i}} \frac{e^{\{(j\beta_O + \gamma_m)l_{2i}\}} - 1}{(j\beta_O + \gamma_m)} \right\} \quad (5)$$

where, ω is the frequency, $\beta_O = \omega / v_O$, which is associated with the optical wave packet group velocity v_O. And φ_i is the initial phase of optical wave pocket at the start of the ith modulation segment and related to the optical path length.

If the optical modulation response is normalized by the ideal case with lossless microwave propagation, perfect impedance match and velocity match, normalized RF link gain for the general segmented case could be expressed as [5]

$$LG_{norm} = \left| I_{ac}(\omega) / (\sum l_m V_S / 2) \right|^2 \quad (6)$$

OPTIMIZATION DESIGN

In the above section, the RF link gain has been derived for the general segmented transmission-line EAMs. The expression is complex due to many variables. It is difficult to optimize the structures with traditional optimization methods. Here we resort to the genetic algorithm, which is a robust global search method and could deal with highly nonlinear and high-dimensional problems.

A fitness function for GA in order for the maximum 3dB bandwidth of the EAMs could be derived from (6). The lengths of the segments including modulation, passive and uncovering optical segments were discretized with binary-coded form and they formed the GA's input parameter. Taking into account the fabrication factors, some actual considerations were applied to the algorithm. First, 1μm was the discrete intervals for the modulation segment while 5μm was discrete intervals for the passive or optical segment. Second, it is inevitable to introduce discontinuities between the modulation and passive segments, which were modeled as a 10μm additional passive section [3]. Third, the uncovering optical

522

segment must be shorter than the actual length of corresponding passive electrode segment to keep the electrode's continuity. Fourth, the first and last passive segments were both added 5μm length for the pads' effect.

The characteristic impedances and propagation constants of the segmented sections were calculated with the equivalent-circuit parameters presented by Lewen [3], which are reliable since they were extracted from straight test structures. A small modification was applied so that the DC resistance R_{dc} was added to the conduction resistances for the modulation section, as well as the passive TML.

The performances of the 165μm-length EAMs with five different structures were analyzed. Three types are segmented structures optimized by GA. The device having three modulator segments (l_{m1}, l_{m2}, l_{m3}), four passive TML segments (l_{p1}, l_{p2}, l_{p3}, l_{p4}) and two uncovering optical segments (l_{o1}, l_{o2}), is denoted as SEAM3m. Similarly, the case with two and one modulation sections are labeled as SEAM2m and SEAM1m, respectively. The traditional EAM and the semented device with the structures presented in [3] were also simulated for comparison. The parameters of the segmented devices were summrized in Table 1. The results show that the optimized structures are asymmetrical and non-periodic. For all the simulation cases, the impedances of microwave signal source and load were fixed as 50Ω and the optical group index is 3.5 [1].

Table 1 (unit: μm)

	l_{p1}	l_{m1}	l_{o1}	l_{p2}	l_{m2}	l_{o2}	l_{p3}	l_{m3}	l_{p4}
SEAM3m	15	51	215	315	69	260	505	45	480
SEAM2m	15	74	220	255	91	--	440	--	--
SEAM1m	45	165	--	440	--	--	--	--	--
LEWEN[3]	124	55	138	248	55	138	248	55	124

The E/O response and microwave reflection for EAMs with different structures were shown in Fig. 2. The bandwidths reach 96GHz for SEAM3m and SEAM2m, while in both two cases the reflection coefficients are below -10dB. The SEAM2 is preferred since it has a shorter optical waveguide that

will lead to a less optical loss.

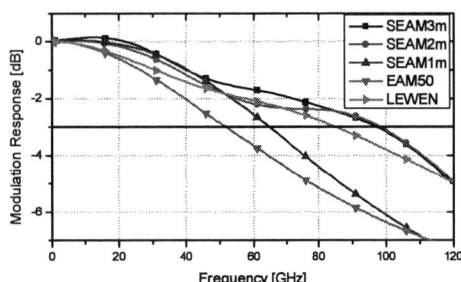

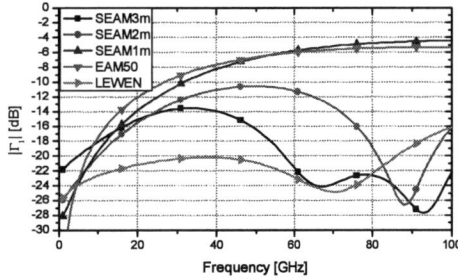

Fig. 2. Calculated modulation response and reflection coefficient magnitudes at the input connection.

Conclusions

We have investigated the simulation method of the segmented Transmission-Line EAMs and deeply explored the potential of the bandwidth enhancement in this design via genetic algorithm. The simulated results show that a significant improvement of EAMs' bandwidth has been achieved by optimizing the segmented EAMs with unsymmetrical and non-periodic electrode design.

References

1 G. L. Li et al, *IEEE Trans. MTT*, 47 (1999), pp. 1177–1783.
2 S. Irmscher et al, *IEEE PTL*, 14 (2002), pp. 923–925.
3 R. Lewen et al, *J. Lightwave. Technol.*, 22 (2004), pp. 172–179.
4 R. E. Collin, *Foundations for Microwave Engineering*, 2nd edition, Singapore: McGraw-Hill, 1992, ch. 8.
5 G. L. Li et al, *J. Lightwave. Technol.*, 22 (2004), pp. 1789-1796.

Study of optical phased-array technology based on PLZT electro-optic ceramic

Qing Ye, Zuoren Dong, Ronghui Qu, Zujie Fang

Shanghai Institute of Optics and Fine Mechanics, CAS,

P.O. Box 800-211, Shanghai 201800, P.R. China

Abstract: Based on the optical characteristics of PLZT electro-optic ceramic, a novel optical phased-array beam deflector with the transverse electro-optic effect is proposed. The deflection characteristic and mechanism of the deflector is analyzed theoretically and experimentally.

1. Introduction

Spatial optical beam scanning has a good application foreground in optical communication, optical memory and optical interconnections. Optical phased-array technology can provide a versatile means for fast, solid-state and random access beam scanning. At present, the studies of optical phased-array technology based on electro-optical effect focus on mainly liquid crystal[1-2] and electro-optical material technology[3-8]. For the electro-optical material technology, AlGaAs material is selected as a phase modulator in the earlier studies, however, PLZT (Lead Lanthanum Zirconate Titanate)[9] electro-optical ceramic has made a large progress to instead of AlGaAs for its larger electro-optical coefficient, wide transmission spectrum, small transmission loss and low cost. The material has been used widely in the area of optical memory, electro-optical switch [10] and optical phased-array technology [7-8].

In this paper, a novel optical phased-array technology with up-down electrode structure's transverse electro-optic effect is proposed based on the optical characteristics of PLZT electro-optic ceramic. Theoretically, the electro-optic deflection characteristic and mechanism of the deflector is analyzed; experimentally, a

systematic scheme is also designed to verify the theoretical results.

2. Basic principle

Figure 1 shows the proposed optical phased-array technology basic principle scheme, and it is composed of multi-independent phase-modulation unit. An incident plane interference beam is input the front surface of phase modulator, and the output beam can become a continuous linear phase distribution by the action of each phase modulation unit for different applied voltages, then the far-field diffraction pattern will form deflection.

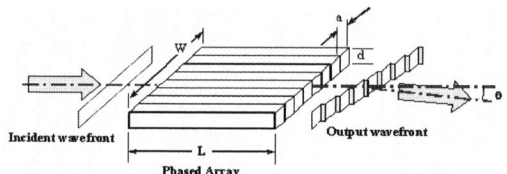

Fig.1 The basic scheme of optical phased-array beam deflector

When a plane interference beam is input the phase modulation array, the intensity of far-field interference pattern can be described as [7]

$$I(\theta) = C \cdot EF(\theta) \cdot AF(\theta) \qquad (1)$$

where C is constant of proportionality and the element factor and the array factor are defined as

$$EF(\theta) = \left[\left(\frac{w}{a} \right) \cdot \sin c \left(\frac{w\theta}{\lambda} \right) \right]^2 \qquad (2a)$$

978-0-9789217-3-6/07/$25.00

©2007 WEN GLOBAL SOLUTIONS

$$AF(\theta) = \left(\frac{\sin\left[N\pi\left(\frac{a}{\lambda}\right) \cdot (\theta - \theta_f) \right]}{N \cdot \sin\left[\pi\left(\frac{a}{\lambda}\right) \cdot (\theta - \theta_f) \right]} \right)^2$$

(2b)

$$\theta_f = \left(\frac{\Delta\varphi}{2\pi}\right) \cdot \left(\frac{\lambda}{a}\right)$$

(2c)

$\Delta\varphi$ is the phase difference between adjacent phase modulation unit, and Table 1 gives other parameters illumination. For a quadratic electro-optical material, i.e. the refractive index change is proportional to the electrical field square $|E|^2$, the phase difference $\Delta\varphi$ can be obtained as

$$\Delta\varphi = \frac{1}{\lambda d^2} \pi L n_0^3 R_{12}(V_{i+1}^2 - V_i^2)$$

(3)

In order to obtain the output phase face linear distribution, the application voltages for each phase modulation unit must satisfy the following condition

$$V_i \propto \sqrt{i} = \sqrt{i} \cdot V_1 , V_N = V$$

(4)

Table 1: Parameter illumination

参数	注释	相关参数取值
$EF(\theta)$	调相阵列的单元器件因子	
$AF(\theta)$	调相阵列的阵列因子	
λ	入射光的波长	$633\,nm$
a	调相单元周期	$350\,\mu m$
w	调相单元遮光的宽度	$150\,\mu m$
d	调相阵列的厚度	$800\,\mu m$
L	调相阵列的长度	$1\,cm$
N,i	调相单元的个数和第几个	8
W	调相阵列的宽度	
θ,θ_f	角变量和光束偏转的角度	
$\Delta\varphi$	相邻调相单元的相位差	
n_0	调相阵列的材料的折射率	2.5
k	入射光的波矢量($=2\pi/\lambda$)	
R_{12}	调相阵列的材料的电光系数	$-2.42\times10^{-17}\,m^2/V^3$
Δn_i	电光效应引起的第 i 调相单元折射率改变	
V_i	施加在第 i 调相单元上的电压	
V	接入的外加电压	

Figure 2 show the corresponding numerical simulation results for above theoretical analysis and parameters. From this figure to see, the interference stripe will deflect with the application voltage increasing. When the voltage is 400V, the deflection angle of

center stripe is 1.2mrad. Figure 3 shows the deflection angle varies with different application voltage, and the curve satisfies a quadratic relation.

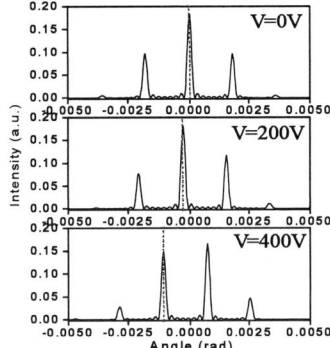

Fig.2 The far field interference pattern with the different application voltage.

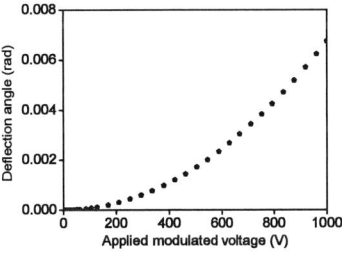

Fig.3 The deflection angle varies with the different application voltages.

3. Experiment

Based on the above theoretical analysis, an experiment setup is shown in Fig.4. The optical beam from a 633-nm He-Ne laser is transmitted to optical attenuator, polarization controller, collimating lens, cylindric lens, diaphragm and amplitude mask, and is input a PLZT electro-optical ceramic (9/65/35) phased-array modulator. The far-field interference pattern (7.2m) is received by a screen. Figure 5 shows the photograph of PLZT modulated-phase array structure.

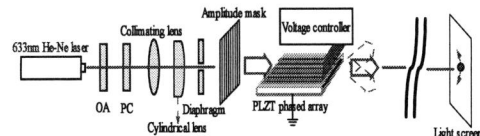

Fig.4 The experimental setup.

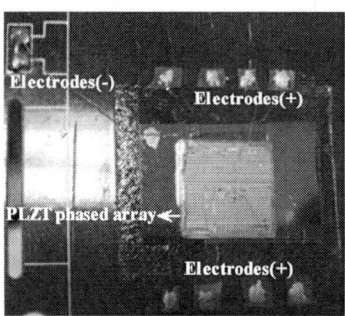

Fig.5 The photograph of PLZT modulated-phase array structure

In experiment, the eight phase modulation units is selected, and each unit has a 350 μm period and 150 μm passing-light aperture (i.e. electrode area). In order to keep the beams transmitting in passing-light aperture, an amplitude mask is applied, which possesses the same period with phase-array modulator. Figure 6 shows the phase of different modulated-phase unit variation with the different applied voltage, and Fig.7 shows the corresponding beam deflection. It is very easily to see from these figures that the far-field interference pattern obtains an obvious deflection as the application voltage increasing. When the voltage are 283V, 385V, 470V and 560V, the center stripe deflection angle are 0.59mrad, 1.14mrad, 1.57mrad and 2.35mrad, respectively. The experimental result is accord with the above theoretical analysis.

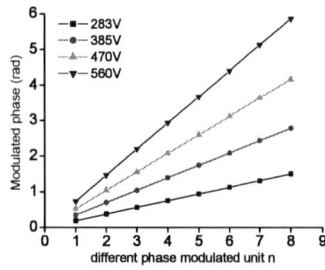

Fig.8 The phase of different modulated-phase unit varies with the different application voltage.

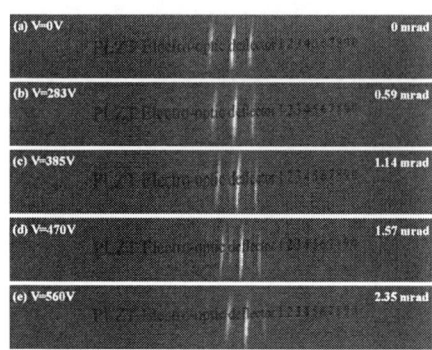

Fig.9 The beam deflection of modulated-phase array varies with the different applied voltage.

4. Conclusion

A novel optical phased-array technology with up-down electrode structure's transverse electro-optic effect has beem proposed based on the optical characteristics of PLZT electro-optic ceramic. The deflection characteristic and mechanism of the deflector is analyzed theoretically and experimentally in detail.

The authors want to thank Prof. Aili Ding for affording the high-quality PLZT electro-optical ceramic.

References

[1] Paul.F.Mcmanamon, et al, Proceedings of the IEEE , 84(1996), 268.

[2] G. D. Love, et al Opt. Lett., 19(1994), 1170.

[3] R. A. Meyer, Appl. Opt.,11(1972), 613.

[4] F.Vasey, et al, Appl. Opt., 32(1993), 3220.

[5] Shi Shunxiang，et al., Acta Optica Sinica，22(2002),1318. (In chinese)

[6] Liang Huawei, et al, Acta Photonica Sinica, 35 (2006), 1654. (In chinese)

[7] J.A.Thomas，et al, Appl. Opt., 37(1998): 6196.

[8] Qiwang Song, et al, Appl. Opt., 35(1996), 3155.

[9] D.Goldring, et al, Appl. Opt., 42(2003): 6536.

[10] Feng Liu, et al, J. Opt. Soc. Am. B, 23(2006), 709.

Controlling chaos in an injection multi-quantum well laser via modulating the injection light

YAN Sen Lin

Physical Department, Nanjing Xiaozhuang College, Nanjing 210017, China, email:senlinyan@163.com

Abstract: *Chaos-control of an injection multi-quantum-well (MQW) laser is studied. A frequency modulator is used to modulate the frequency of an injection light, the chaotic laser can be suppressed in a stable state, a single-periodic state, a dual-periodic and multi-periodic states, respectively.*

I. Introduction

Injection semiconductor laser systems with a master laser and a slave laser have been widely investigated [1-3] because of their promising applications in coherent detection, spectral narrowing, modulation bandwidth widening, and reduction of partition noise as well as frequency conversion. On the other hand, the dynamical behavior of a slave laser can be complicated and exhibits chaotic light fluctuation when the intensity of the optical injection signal is strong oscillating. Since the OGY method was presented [4], chaos-control techniques have been successfully demonstrated to convert a chaotic motion to a periodic regular motion [5,6]. Chaotic oscillation is stabilized to one of the unstable periodic orbits embedded in a chaotic attractor by the application of a small perturbation or occasional proportional feedback [9-11]. It has been recently shown, both experimentally and numerically, that chaotic laser can be controlled or suppressed by the current modulation or the optical negative feedback method [7,8].

In this paper we will present novel chaos-control methods of an injection graded-index separate confinement heterostructure (GRIN-SCH) multi-quantum-well (MQW) laser by adjusting or modulating the injection light.

II. Chaotic MQW laser

A model block of an injection laser is illustrates in Fig.1. A chaotic laser system consists of a slave (S) MQW laser diode (L) and a master (M) MQW laser diode (L). E_m is the injection light while E is the output light at wavelength 1.55μm. ML output light injecting into SL, SL output light becomes of chaotic state [3-5]. We can use the lang-kobayashi equations to describe the dynamics of injection SL [3-5]

$$\frac{dE}{dt} = \frac{1}{2}(G - \gamma_p)E + \frac{k}{\tau_L}E_m\cos(\phi) \qquad (1a)$$

$$\frac{d\phi}{dt} = \frac{1}{2}\beta_c(G - \gamma_p) + \frac{k}{\tau_L}\frac{E_m}{E}\sin(-\phi) - \Delta\omega_m \qquad (1b)$$

$$\frac{dN_B}{dt} = \eta_i\frac{I}{q} - \gamma_{BQ}N_B + \gamma_{QB}N \qquad (1c)$$

$$\frac{dN}{dt} = \gamma_{BQ}N_B - (\gamma_e + \gamma_{QB})N - GV_pE^2 \qquad (1d)$$

where E and Φ are the amplitude and phase of the optical field, N and N_B are carriers in the barrier region and in the active region, respectively. E_m is the amplitude of the injection light and let E_m=0.2529E_s.

The nonlinear gain is $G = [(\Gamma g_0 v_g)/(1 + E^2/E_s^2)] \times$ $\lg\{(N + N_s)/(N_{th} + N_s)\}$, where Γ=V/V_p is the mode confinement coefficient, V is the volume of laser cavity, V_p is the mode volume of laser, g_0 is the gain constant, v_g is the group velocity of photon in laser cavity, E and E_s are normalized in such a way that E=$(P/V_p)^{1/2}$ is the amplitude of the optical field with photon number P and E_s=$(P_s/V_p)^{1/2}$ is the amplitude of the optical field at saturation with saturation photon number P_s, N_s=n_sV is the parameter of logarithmic gain with its density n_s, N_{th}=$n_{th}V$ is the carrier number at transparency with its density n_{th}. $\gamma_p = v_g(\alpha_m + \alpha_{int})$ is the total photon loss with the group velocity v_g, α_m is the cavity loss, α_{int} is the internal loss. $\tau_L = 2n_gL/c$ is the round-trip time in the cavity with its length L, c is the light velocity in vacuum, n_g=c/v_g is the group refractive index. η_i is the internal quantum efficiency. I is the drive current. q is the unit charge. β_c is the optical linewidth enhancement factor. γ_{BQ} is the loss of carriers from

Table 1

Symbol	Value	Unit
L	1200	μ m
V	50.4	μ m^3
Γ	0.045	
n_g	3.6	
n_s	0.1×10^{18}	
α_m	11.5	cm^{-1}
α_{int}	20	cm^{-1}
n_{th}	2.1×10^{18}	cm^{-3}
A_{nr}	2.5×10^8	s^{-1}
B_r	1.0×10^{-10}	cm^3/s
C_A	5.0×10^{-29}	cm^6/s
P_s	2.2×10^7	
g_0	2700	cm^{-1}
β_c	3	
I	50	mA
n_i	0.8	
γ_{BQ}	2.5×10^{10}	s^{-1}
γ_{QB}	5×10^9	s^{-1}
k	0.1559	
$\Delta\omega_m$	4π×10^9	rad/s

the SCH region to the quantum wells and γ_{QB} is the loss of carriers escaping from the active region to the SCH layer. $\gamma_e = A_{nr} + B_{nr}(N/V) + C_A(N/V)^2$ is the

---

978-0-9789217-3-6/07/$25.00

©2007 WEN GLOBAL SOLUTIONS

total carrier loss in the active layer, A_{nr} is the nonradiative recombination rate, B_{nr} is the radiative recombination coefficient, C_A is the Auger recombination coefficient. k is the optical injection factor. $\Delta\omega_m$ can be regarded as the frequency detuning between the external injection light and the output light. A chaotic attractor in the laser is shown in fig.2.

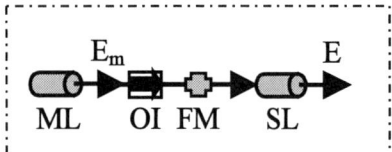

Fig.1. Control-chaos model block where OI is the optical isolator, and FM is the frequency modulator.

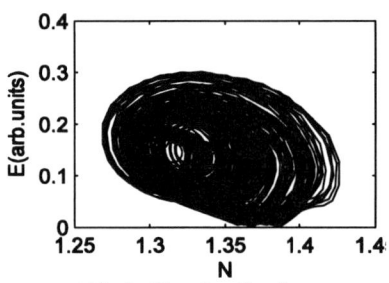

Fig.2. Chaotic attractor

III. Frequency shift control and its results

Based on Fig.1, a frequency modulator (FM) is used to be placed between ML and SL to shift the frequency of the injection light from ML to SL. So the frequency detuning should be adjusted or governed by FM while SL dynamical behavior should be affected. In this case, we can shift the frequency of the injection light to realize chaos-control of SL by using FM. A chaos-control term is introduced in the right expression of (1b) as following

$$\frac{d\phi}{dt} = \frac{1}{2}\beta_c(G - \gamma_p) + \frac{k}{\tau_L}\frac{E_m}{E}\sin(-\phi) - \Delta\omega_m - \Theta \quad (2)$$

where Θ is the chaos-control term determined by FM.

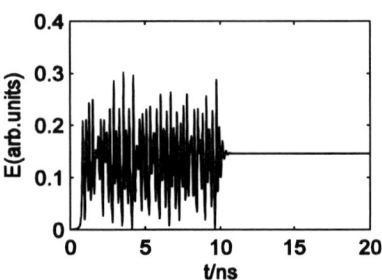

Fig.3. Chaos is suppressed to a stable state after 10ns

Fig.3 shows that chaos is suppressed into a stable state by performing FM at 10ns for Θ=-6.774GHz, which implies that chaos has been effectively controlled via FM adjusting the frequency of

the injection light. When Θ is taken to be -3.871 GHz, it is found to take about 8ns before the chaotic laser

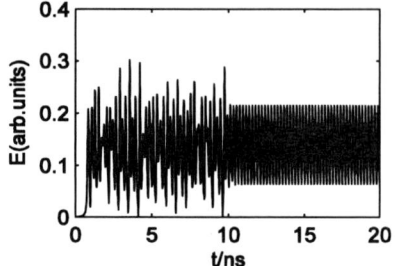

Fig.4. Chaos is stabilized to a single-periodic state after 10ns.

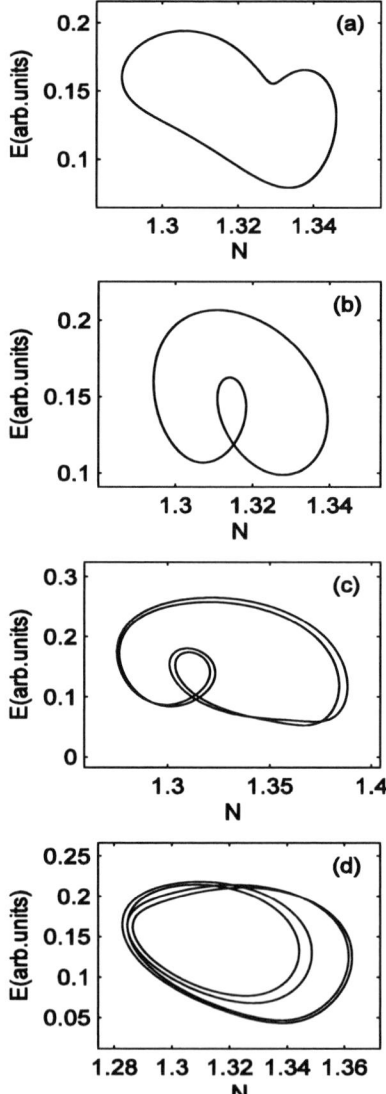

Fig.5. Mulit-periodic states. (a) a dual-periodic cycle, (b) another dual-periodic cycle, (c) a period-4 cycle, and (d) another period-4 cycle.

can be stabilized in another stable state. We also find

that the chaotic laser can be stabilized other stable states for Θ being taken from -3.483GHz to -7.161GHz while it indicates that the laser can exhibit continual stable light output.

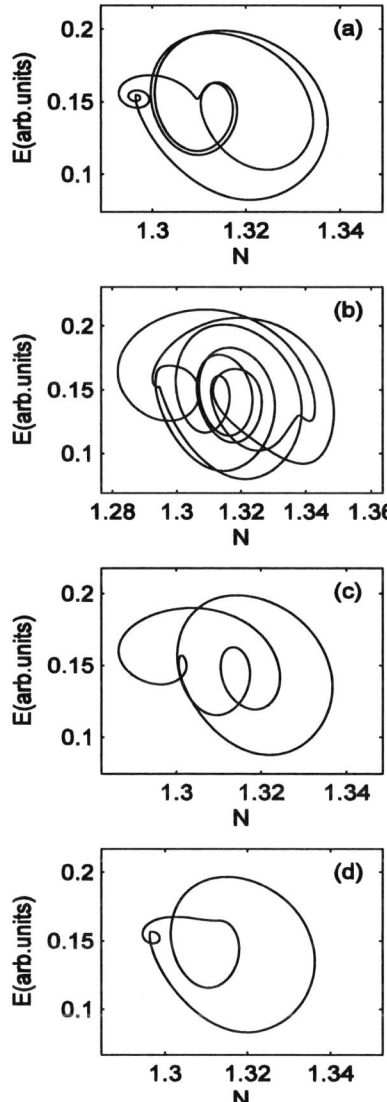

Fig.6. High cycles of periodic states. (a) a period-7 cycle,(b) a period-8 cycle, (c) a period-4 cycle, and (d) a tri-period cycle .

When Θ is adjusted to be 1.602GHz, chaos can be stabilized in a single-periodic state with an oscillation frequency of 5GHz shown in fig.4. Let Θ=7.742GHz, the laser output can be brought into another single-periodic state with an oscillation frequency of 10GHz. And Θ=-13.548 GHz, the laser output shows other single-periodic state with an oscillation frequency of 12.8GHz. Further, It is found that the chaotic laser can be effectively stabilized in single-periodic states for Θ from -100GHz to –13GHz while it indicates that the laser can emit periodic regular optical pulses.

If the frequency of the injection light can be modulated continually via FM, chaos-control is also realized. The control function is introduced in such a way that we let Θ be rewritten as $2\pi f_0 A_f \times square(2\pi f_f t)$ where f_0 is the frequency of the laser output, A_f is the modulation depth and *square* is the square-wave function with a modulation frequency of f_f. In this case, many high cycles of periodic states are simulated to show in Fig.6. Fig.6 (a) shows a period-7 state for A_f=0.03 and f_f=1GHz, Fig.6 (b) shows a period-8 state for A_f =0.03 f_f=1.5GHz, Fig.6 (c) illustrates a period-4 state for A_f =0.03 f_f=1.8GHz, and Fig.6 (d) exhibits a tri-periodic state for A_f =0.03 f_f=2GHz. We find that these multi-periodic states don't exist in the attractor in Fig.2 and they may newly appear in the chaos-control.

VI. Conclusion

Adjustment and modulation of the frequency of the injection light can be performed to realize successfully control of chaotic laser. Chaos can be stabilized in a stable state and a variety of multi-periodic cycles. It is also important that the schemes will be experimentally easily realized. The method is helpful for study of chaos-control and injection mode-locked lasers.

References

1 L.Li IEEE J.Quantum Electron., vol.30(1994), pp.1701-1707
2 V.Annovazzi-Lodi, et al IEEE J.Quantum Electron., vol.34(1998), pp.2350-2357
3.Troger, et al IEEE J.Quantum Electron., vol.35(1999), pp.32-38
4 E. Ott, et al Phys.Rev. Lett.,vol.64(1990), pp.1196-1199
5 Y.Liu et al Opt.Lett., vol.19(1994), pp. 448-450
6 A.Uchida et al Phys.Rev.A, vol.58(1998), pp. 7249-7253
7 C.G.Lim, et al IEEE J.Quantum Electron., vol.37(2001), pp.699-706
8 Y.Liu, et al IEEE J. Quantum Electron., vol.39(2003), pp.269-278.

This work was supported by The Education Department of Jiangsu Province (No. 06KJD140111).

Analysis and Simulation of a Channel Add-Drop Filter Composed of Two Dimensional Photonic Crystal

N. Nozhat and N. Granpayeh

Faculty of Electrical Engineering, K. N. Toosi University of Technology, Tehran, Iran.
Emails: nozhat@ee.kntu.ac.ir; granpayeh@kntu.ac.ir

Abstract

In this paper, the properties of the resonant modes in the two dimensional (2-D) photonic crystal (PC) cavities and channel add-drop filters have been presented. Both square and triangular PC lattices have been analyzed. We have used an effective numerical method based on the finite-difference time-domain (FDTD) scheme for computing resonant modes. We have clarified coupling between the single point cavity and a line defect waveguide as a channel add-drop filter.

Introduction

Recently, photonic crystals, either two or three dimensional periodic structures, have attracted many researchers' attentions. These structures have a photonic band gap (PBG), the range of frequency that the light cannot be propagated in them. Because of their attractive properties, some optical communication devices incorporating photonic crystals, such as thresholdless laser diodes (LDs), Mach-Zehnder interferometers, endlessly single mode fibers, low-loss and sharp bend waveguides, and channel add-drop filters have been proposed and fabricated [1]-[3].

If a single point defect is created in the structure, some resonant modes appear in the band gap, which performs as a cavity with high quality (Q) factor. By introducing a point defect near a line defect waveguide in a 2-D PC the trapping and emission of photons from a cavity can be occurred, which the structure can be used as a channel add-drop filter, an important component in the optical wavelength division multiplexing (WDM) systems and networks [2].

The fabrication of 3-D photonic crystals is difficult. Therefore, it is more convenient to derive a complete photonic band gap with 1-2 dimensional PCs.

In this paper, we have studied the resonant frequency and field distribution of the cavity modes. Also, a channel add-drop filter has been proposed and analyzed.

Numerical Analysis

Many numerical methods can b applied for analysis of photonic crystals, including plane-wave expansion (PWE) method, exact Green's function method, transfer matrix method, and the finite-difference time-domain (FDTD) method. In our analysis we have used a 2-D FDTD method for demonstration of evolution of the electromagnetic fields in the photonic crystals, because comparing to the PWE method, that the computational time growth is in the order of N^3 (where N is the number of plane waves), this method is in the order of N (where N is the number of discretization points), therefore, the computation time and memory requirements are reduced [4], [5].

We have used Berenger's perfectly matched layer (PML) as absorbing boundary condition.

Numerical Results and Discussion

First, we have analyzed the properties of a microcavity in a 2-D photonic crystal. Consider a 2-D PC with a square lattice of dielectric rods in air. The radius of the rods is r=0.2a, where a is the lattice constant, and the relative permittivity of the dielectric rods is $\varepsilon_r = 11.56$. In our FDTD simulation, the unit cell includes 2500 (50×50) grid points, and the total number of time steps is 32768. We have used a 7×11 rod photonic crystal, similar to what is shown in Fig. 1, but without defect. The source is a Gaussian pulse with central frequency of 0.35 (c/a) and a width of 0.2 (c/a), as demonstrated in Fig. 2. The selected photonic crystal structure has a band gap for TM modes, but not for TE modes. As shown in Fig. 2, the band gap associated with this structure varies in the range of $f = 0.29(c/a)$ to $f = 0.42(c/a)$ for TMz polarization.

978-0-9789217-3-6/07/$25.00

©2007 WEN GLOBAL SOLUTIONS

For the same structure, but with triangular lattice, the band gap for TM modes is much wider, because of its better symmetry and smoother Brillouin zone. It varies in the range of $f = 0.27(c/a)$ to $f = 0.44(c/a)$.

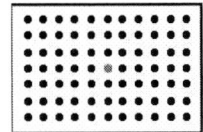

Fig.1 Schematic of a photonic crystal consisting of dielectric rods and a central defect, as a cavity.

A defect can be presented in the structure, either by changing the refractive index or the radius of a rod.

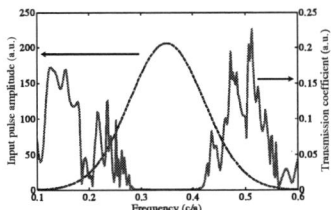

Fig. 2 Gaussian pulse as an incident wave to the PC (dashed line) and transmission coefficient calculated at the output of the structure (solid line).

Figure 3 illustrates the transmission spectrum of a PC with a single defect, by removing the central rod. There is a defect mode in the normalized frequency of $f = 0.38(c/a)$. Since the defect involves removing dielectric material in the crystal, the mode appears near the higher edge of the gap, because the effective refractive index of the cavity has been reduced [6].

The same as microwave cavities, in photonic crystal cavities, based on the shape and resonant frequency of the cavity, some modes are generated. These modes are named according to the number of maximum field points in the cavity. Figure 4 demonstrates the distribution of the electromagnetic fields in the cavity. In this case we have used a single mode sinusoidal excitation source.

Depending on the location of the feeding point, some particular modes are excited in the cavity, because of the symmetry [7].

When the radius of the cavity is $r = 0.1a$, the electric field becomes maximum in one point of the cavity and called monopole. By increasing the defect radius, some other modes appear.

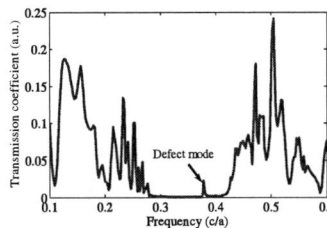

Fig. 3 Transmission spectrum of the photonic crystal with defect, as shown in Fig. 2, the defect mode is created at frequency of 0.38(c/a).

For $r = 0.3a$ two degenerate modes, named dipole, appear in the cavity. For $r = 0.6a$ the number of resonant modes increases, which are called monopole, quadrupole and hexapole, as shown in Fig. 4.

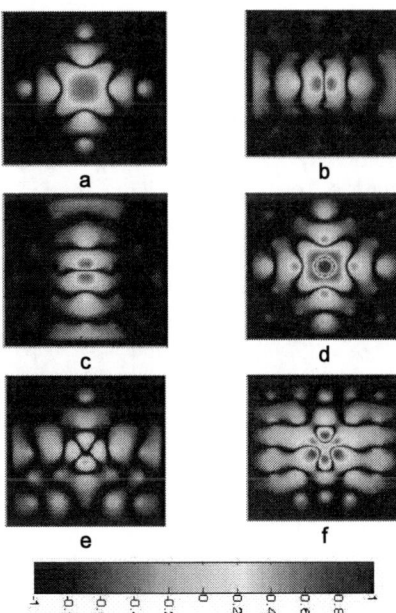

Fig. 4 TMz electric-field distribution of the defect in a square lattice of dielectric rods located in air. (a) Monopole, r=0.1a, f=0.33(c/a) (b) and (c) Doubly degenerated modes, r=0.3a, f=0.33(c/a), (d) Second-order monopole, r=0.6a, f=0.343(c/a), (e) Quadrupole, r=0.6a, f=0.319(c/a), and (f) Hexapole, r=0.6a, f=0.3916(c/a).

When the crystal includes air holes in dielectric material such as GaAs with $\varepsilon_r = 12.96$, the band gap exists for TE modes, but not for TM polarization.

The triangular lattice of air holes in GaAs has been analyzed. The radius of air holes is $r = 0.3a$ and the refractive index of the dielectric is $n = 3.6$. The band gap associated with this structure is from $f = 0.2(c/a)$ to $f = 0.272(c/a)$ for TE modes.

Channel Add-Drop Filter

Figure 5 shows a photonic crystal consisting of cylindrical dielectric rods in air with two point defects near a line defect waveguide, acting as a channel add-drop filter.

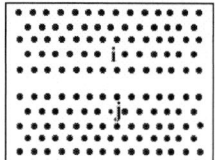

Fig. 5 Arrangement of the channel add-drop filter.

In order to show the function of a channel drop filter, we have launched a sinusoidal wave to the waveguide. Two cavities of the add-drop filter of Fig. 5 having the radii $r = 0$ and $r = 0.1a$, respectively. When the input lightwave having the normalized frequencies of $f = 0.38(c/a)$ and $f = 0.33(c/a)$, the wave is dropped from the waveguide by cavities i and j, respectively, and emitted from them, as illustrated in Fig. 6.

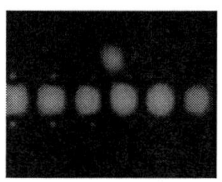 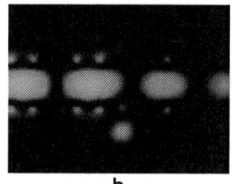

a   b

Fig. 6 The drop filter performance from two point cavities simulation besides waveguide, shown in Fig. 7. When the cavity rods have radii r=0 and r=0.1a, the frequencies of 0.38 (c/a) and 0.33 (c/a) are emitted from the defects i and j, respectively.

The transmission spectrum of the channel add-drop filter for drop operation has been illustrated in Fig. 7. The launched input wave is a Gaussian pulse, and for example, as shown in Fig. 7, when the frequency corresponding to the resonant frequency of the cavity i drop from it. Then we have investigated the function of an add operation in 2D PC, by launching two sinusoidal waves with frequencies 0.38(c/a) and 0.33(c/a) into two distinctive cavities i and j. Intensity of the

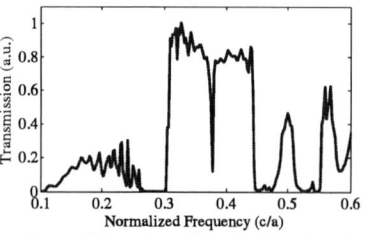

Fig. 7 Spectrum of the drop filter for the structure of Fig. 5.

lightwave at the output of the waveguide has been shown in Fig. 8. Two lightwaves are coupled from two defects to the waveguide.

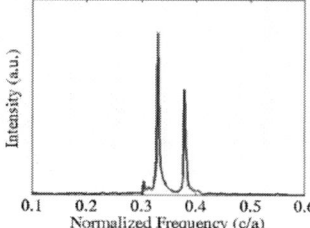

Fig. 8 Spectrum of add filter at the end of the waveguide of Fig. 5..

It has been shown that the emission wavelength from the cavity in a channel add-drop filter depends on the defect radius. Also, variations of the refractive index of the cavities affect the resonant, and therefore the add-drop frequency of the filter.

Conclusion

In this paper, we have presented the existence of a large PBG in a square lattice PC. This PBG becomes wider in a triangular lattice, because of better symmetry of the structure. We have analyzed and simulated the properties of the photonic crystal channel add-drop filter consisting of a line defect and a point defect in the PC structure. We have illustrated the channel add-drop performance, and have shown that the add-drop functionality depends on the radius and refractive index of the defect.

Acknowledgment

The authors would like to thank Iran Telecommunication Research Center (ITRC) for their help and financial support of the project.

References

1 M. Imada, S. Noda, A. Chutinan, M. Mochizuki, and T. Tanaka, IEEE J. Lightwave Technol., Vol. 20 (2002), page 873-878.

2 B. S. Song, T. Asano, Y. Akahane, Y. Tanaka, and S. Noda, IEEE J. Lightwave Technol. Vol. 13 (2005), page 1449-1455.

3 H. Ren, C. Jiang, W. Hu, M. Gao, and J. Wang Opt. Commun. Vol. 266 (2006), page 342-348.

4 M. Qiu and S. He, Phys. Rev. B, Vol. 61 (2000), page 12871-12876.

5 M. Qui and S. He, Phys. Lett. A, Vol. 278 (2001), page 348-354.

6 P. R. Villeneuve, S. Fan, and J. D. Joannopoulos, Phys. Rev. B, Vol. 54 (1996), page 7837-7842.

7 J. M. Lopez-Alonso, J. M. Rico-Garcia, and J. Alda Opt. Exp. Vol. 12 (2004), page 2176-2186.

Numerical research on quality factor Q of 2D photonic crystal microcavity with modulation of localized states

Ziqiang Wang[1], Lieming Li[2], Dan Wang[1], Wenbin CAO[1*]

[1]Department of Inorganic Nonmetallic Materials, School of Materials Science and Engineering, University of Science and Technology Beijing, Beijing 100083, China

[2]Department of Physics, Tsinghua University, Beijing 100084, China

Abstract *A theoretical model of a microcavity composed of a photonic crystal with five-point defects is proposed. The effect of the number of scatters on the quality factor Q and transmission T have been investigated. Modulation of localized states was introduced to try to increase the Q and T of the microcavity.*

Introduction

2D photonic crystal (PC) microcavity with high quality factor (Q) has the properties of effective field confinement, resulting in very high Purcell effect, which makes them suitable candidates, in particular, for cavity quantum electrodynamic [1] [2]. Mode-matching is one of the most commonly used methods to simulate the Q by changing the position or dimension of the scatters located at both ends of the defect in the cavity [3].

In this paper, the Q and transmission T of a proposed microcavity have been theoretically studied and finally high T and large Q have been achieved simultaneously by introducing the modulation of localized states [4].

PC microcavity design and calculation method

The 2D PC microcavity investigated in this paper consists of Al_2O_3 rods in a square lattice with a rod radius of r=0.3a, and the rod dielectric constant ε=11.4, the width *w* of the optical source is about 5a, and the distance from the light source to the arrays l=5a(a, the lattice constant). Five rods were removed from the perfect structure to create a microcavity. Figure 1 is the schematic plot

of the proposed 2D PC microcavity with five-point defects.

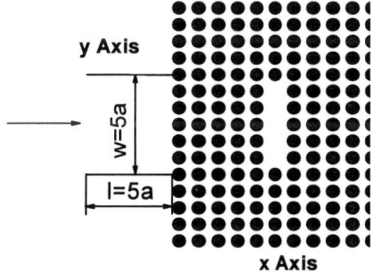

Fig. 1. Schematic plot of the 2D PC microcavity. The arrow indicates the incident direction of the slit light

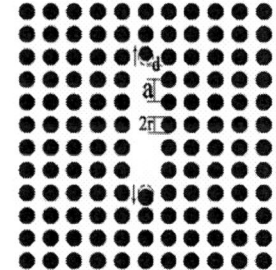

Fig. 2. Schematic plot of the two-dimensional PC microcavity with shifted rods located at both ends of the defect

The number of rods located along the x-axis is marked as Lx, and that located along y direction is marked as Ly.

The T and Q of microcavities with different number of arrays were calculated. The transmittance spectra of the PC microcavities is calculated by using the

---

* Corresponding author, E-mail: wbcao@mater.ustb.edu.cn, Tel/fax:86-10-62332457

978-0-9789217-3-6/07/$25.00

©2007 WEN GLOBAL SOLUTIONS

multiple-scattering computation method published in reference[6]. The modulation of localized states was realized by slightly shifting the two cylinders located at both ends of the defects as shown in Fig .2. The position offset d of the two rods was varied from 0.1a to 0.20a step by step during the simulation.

Results and discussions

The factor Q was increased from 155 to 163 when Ly was increased from 9 to 15 while fixing the Lx=7. With the increasing of Ly the transmission T has been increased from 71.2% to 83%. If Ly was fixed at 15, the calculated Q would be increased from 163.1 to 12937.6 when Lx increased from 7 to 13, and the T was increased from 71% to 95%. From the results we can see that the leakage along the x- and y- direction was restrained due to the strong Bragg reflection.

Figure 3 is transmitance spectra of photonic crystal microcavity with different number of rods located along y-axis while Lx was fixed as 7. And when the Ly was fixed as 15, the calculated transmitance spectra of PC microcavities with different number of rods distributed along x-axis were shown in Fig. 4.

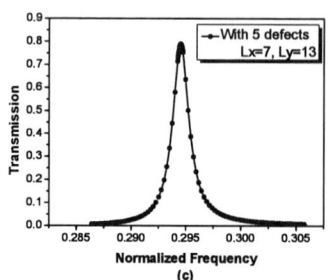

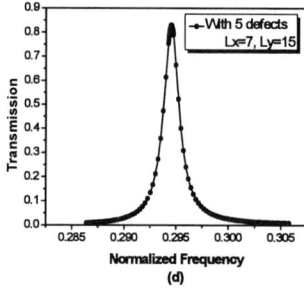

Fig. 3. Transmittance spectra of PC microcavity with different number of rods in y axis. (a) For 9 rods array in y axis. (b) For 11 rods (c) For 13 rods (d) For 15 rods

Although high T and large Q have been realized successfully, the mode volume of the microcavity was enlarged with more and more scatters added in the structure. That is not advisable.

So, modulation of localized states was introduced to try to achieve high T and large Q simultaneously to avoid increasing the mode volume of the microcavity at the same time.

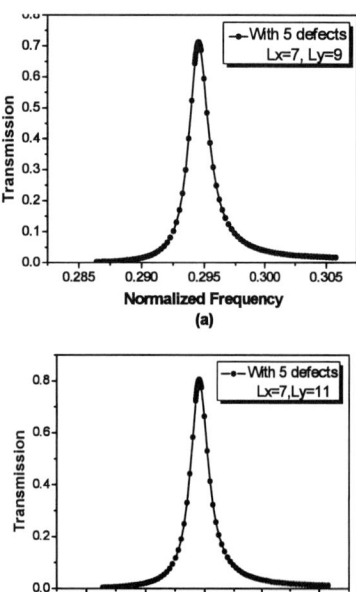

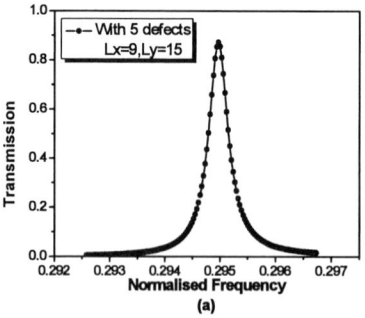

534

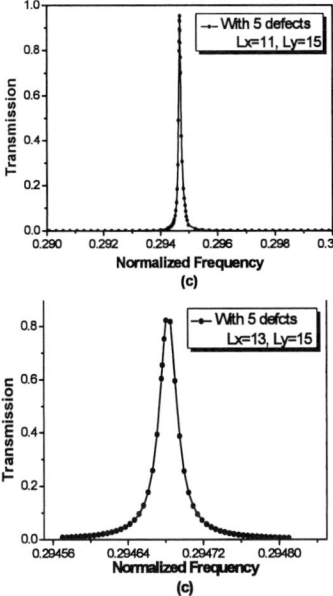

Fig. 4. Transmitance spectra of photonic crystal microcavity with different number of rods in x axis. (a) For 9 rods array in x axis. (b) For 11 rods array. (c) For 13 rods array

As indicated in Fig. 4(c), the Q is about 3096.2 and T is 95.34% when the Lx and Ly was fixed as 11, 15 respectively. If the two rods located at both ends of the defect was shifted 0.1a (d=0.1a) outerward, the Q and T would be increased to 3108.87 and 92.7%. With further increase of the d to 0.15a, the largest Q with the value of 3146.15 is achieved but the T is dropped to 93.6%. After that, Q is decreased to 3110.2 and T is 94.4% when the offset d=0.2a.

Once the rods position was shifted, the in-plane Bragg condition was destroyed, and Bragg reflector for the phase-matching condition has been broken. At this case, local optical field will penetrate into mirrors outside the microcavity, the limitation of the light propagation in perpendicular direction can be enhanced effectively and the Q factor can be improved obviously.

Conclusions

A theoretical model of a microcavity composed of a photonic crystal with five-point defects is proposed. The effect of the number of scatters on the quality factor Q has been investigated. 2D square lattice photonic crystal microcavity with modulation of localized states was proposed to increase the factor Q and T simultaneously by shifting the rods located at both ends of the defects. The largest Q factor and transmission achieved in this mode is 3146.15, 93.6% respectively when Lx=11, Ly=15, and d=0.15a. Results show that factor Q can be increased without increasing the volume of microcavity.

References

[1] G. Manzacca, D. Paciotti, A. Marchese, M.S. Moreolo, G. Cincotti, 2D Photonic Crystal Cavity-Based WDM Multiplexer, Photonics and Nanostructures Fundamentals and Applications (2007), doi:10.1016/j.photonics.2007.03.003.

[2] Evelin Weidner, Sylvain Combrié, Nguyen-Vi-Quynh Tran, Alfredo De Rossi, Julien Nagle, and Simone Cassette, Achievement of ultrahigh quality factors in GaAs photonic crystal membrane nanocavity, APPLIED PHYSICS LETTERS 89, 221104 (2006).

[3] Simon Frédérick, Dan Dalacu, Jean Lapointe, Philip J. Poole, Geof C. Aers, Robin L. Williamsa, Experimental demonstration of high quality factor, x-dipole modes in InAs/InP quantum dot photonic crystal microcavity membranes, APPLIED PHYSICS LETTERS 89, 091115 (2006).

[4] A.S. Jugessur, P. Pottier, R.M. De La Rue, Microcavity filters based on hexagonal lattice 2-D photonic crystal structures embedded in ridge waveguides, Photonics and Nanostructures–Fundamentals and Applications 3 (2005) 25–29.

[5] Ping Jiang, Chengyuan Ding, Xiaoyong Hu, Qihuang Gong, Tunable double-channel filter based on two-dimensional ferroelectric photonic crystals, Physics Letters A 363 (2007) 332–336.

[6] Lieming Li, ZhaoQing Zhang, Multiple-scattering approach to finite-sized photonic band-gap materials, PHYSICAL REVIEW B 58 (1998) 9587.

Ultra-Fast All-Optical Switch and its Nonlinear Dynamical Process

Ye Liu(1), Zhi-Yuan Li(2), Dao-Zhong Zhang(3)

1: Beijing National Laboratory for Condensed Matter Physics, Institute of Physics, Chinese Academy of Sciences, Beijing 100080, China. liuye@aphy.iphy.ac.cn
2:lizy@aphy.iphy.ac.cn,3: zhangdz@aphy.iphy.ac.cn

Abstract *Nonlinear optical interaction in a nonlinear Kerr photonic crystal is considered theoretically. We find field enhancement at some wavelengths where the transmission ratio exceeds one. Then we explain some interesting phenomena on optical switching observed in our previous experiments.*

Introduction

Nowadays, nonlinear photonic crystals (PCs) have received growing studies both theoretically [1,2] and experimentally[3,4,5] with their attractive properties. As a key device of ultra-fast information processing, all-optical switch[4,5] is a good application example of nonlinear photonic crystals. In order to fabricate more efficient switch devices, we should fully understand the intrinsic mechanism of nonlinear photonic crystals. However, the studies of all-optical switch mainly lie in experiments, and the few theoretical studies in literatures are mainly based on the "stable" model which cannot represent the practical process. Now we propose a numerical method to simulate the dynamic process of nonlinear photonic crystal interacting with the pump and probe pulses. With this method, we obtain some interesting phenomena.

Results and analysis

At first, we only consider the interaction of pump pulse and nonlinear photonic crystal. The nonlinear photonic crystal sample is fabricated with the material of polystyrene, which possesses a large third-order nonlinear susceptibility and femto-seconds response time. With the strong pump pulse propagating through the PC, the refractive index of nonlinear photonic crystal changes dynamically which may lead to the dynamical gap shift[4,5]. We calculate the transmission spectrum of pump pulse, and find two main effects. One is that the transmission spectrum is widened obviously. The other is the field enhancement effect at the gap edge, where the transmission field will significantly exceed the incident field. These results can be found in figure 1. Comparing with the third-order nonlinear effect of bulk nonlinear medium, we think the widened transmission spectrum is still mainly caused by the self-phase modulation effect[7] (nonlinear frequency shifts) in a third-order nonlinear susceptibility medium. However, the widening effect is more pronounced in a nonlinear PC than in a bulk medium. In addition, we think the gap edge field enhancement effect comes from the enhanced nonlinearity of the photonic crystal band gap due to low group velocity of light and enhanced

978-0-9789217-3-6/07/$25.00
©2007 WEN GLOBAL SOLUTIONS

field intensity. Figure 1 shows the case of a very strong Kerr nonlinearity, which may change the refractive index by 5%. These two main results will be observed when nonlinear interaction is weaker as well.

From figure 1, we can also discover that not only the spectrum width is widened, but also the gap width. And a "defect mode" appears at the original linear gap edge. This is a very interesting phenomenon which has not been reported before.

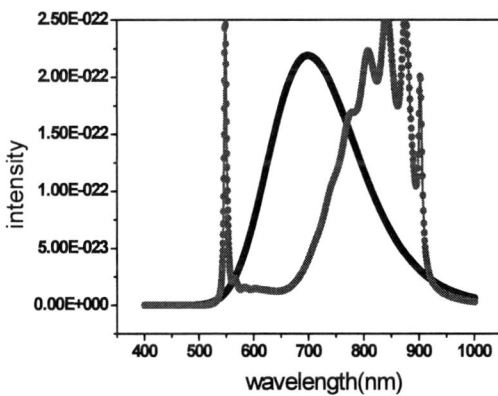

Figure1. Calculated transmission spectrum of a nonlinear photonic crystal with strong Kerr nonlinearity. The black and red lines are the incident and transmission spectrum respectively.

Then, we take the probe pulse into account in the above model. So we consider the interaction between the nonlinear photonic crystal and pump pulse and the interrelationship of pump and probe pulse. This is the pump-probe method, which has been adopted in our experiment. Because the nonlinearity of polystyrene is weak in our experiment which can only lead to a change of refractive index as no more than 1%, we calculate with the condition of weak nonlinearity.

As a switch, one important thing is to obtain the time response or the time delay curve at which the horizontal coordinate is time delay between pump and probe pulse and the vertical coordinate is the transmissivity. We calculate the time response with the introduced delay between pump and probe pulse. The results are shown in figure 2. There are several characters we can find in this figure. The first is that the maximum of the curve is not the exact "zero delay" between pump and probe pulse, but a few femto-seconds delay of pump pulse. The second is the rising period is equal to the incident pulse width, but the dropping period is a little longer than the incident pulse width. The third is that there is a small peak at the dropping period which we have observed in our previous experiments. All these phenomena can be explained by considering the transmission process of a femto-second pulse through the PC. A series of pulses can be found due to multiple scattering effect within the PC, and there exists a "phase cancellation" effect between these pulses.

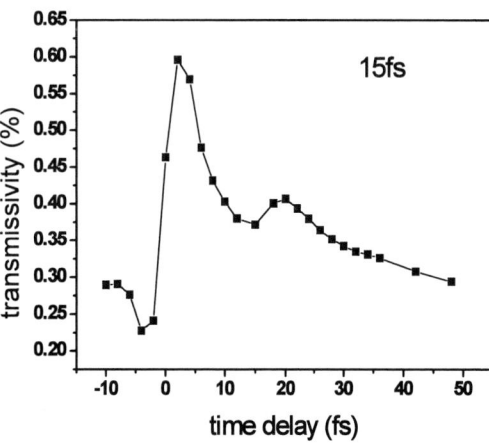

Figure 2. Time response of PC optical switch.

More important, we have found that the transmissivity change of the PC switch not only depends on the gap shift effect, but also the weak field enhancement effect at the gap edge (figure 3). The gap enhancement effect can also be explained by the "phase cancellation picture".

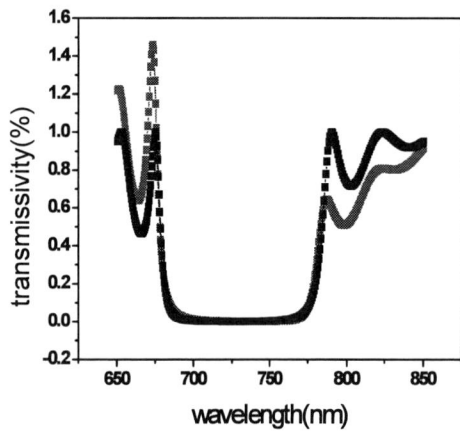

Figure 3. Comparison of the transmissivity under "mean refractive index" (black line) method and the "dynamical nonlinear process" method (red line).

Conclusions
We have studied the dynamical process of few femto-seconds PC switch in detail.

We find the mechanism of transmissivity change is the combination of gap shift and gap edge enhancement effect. With weak nonlinearity, the gap edge enhancement effect is not very obvious, but it is enough for the application of our switch. With stronger nonlinearity, more detail will appear such as the obvious widened transmission spectrum, the stronger gap edge enhancement effect, and the widened gap effect.

References
1. Alejandro Rodriguez, J.D.Joannopoulos et al Optics Express vol.15(2007), 7303-7318
2. Sheng Lan et al Physical Review B 71(2005), 125122
3. Ma, G. H. et al Physica B **305**(2001), 147-154 .
4. Liu, Y. H., Hu, X. Y., Zhang et al Appl. Phys. Lett. **86**(2000), 151102
5. Xiaoyong Hu et al Appl. Physics Letters 83(2003) 2518
6. Wonjoo Suh and Shanhui Fan Optics Letters 28(2003), 1763
7. Shen Y. R. The Principles of Nonlinear Optics. John Wiley & Sons, Inc., 1984,4

New PON Add/Drop Multiplexer to support Next-Generation PON

Sahrul Hilmi Ibrahim (1,2), Dr. Abu Bakar Mohammad (2)
1. Access Network Technology Program, TM Research & Development Sdn Bhd, Selangor
2. Department of Optic & Telemetic, Faculty of Electrical Engineering, Universiti Teknology Malaysia, 81310 UTM Skudai, Johor, Malaysia
Email: hilmi@tmrnd.com.my, bakar@utm.my

Abstract *We propose a novel WDM-PON architecture which employs multiple new optical add/drop multiplexer which is made possible configuration for a next-generation PON. The architecture provides WDM traffic in both directions and its fully operating at 2.5 Gb/s for triple play services.*

Introduction

To keep up with the explosive demand for Internet access, the commercial introduction of GE-PON (Gigabit Ethernet Passive Optical Network) systems with a transmission rate of 1.25 Gbps has been proceeding as a means of building cost-effective high-speed optical subscriber networks. To enhance the speed and capacity, focus is being placed on the research and development of next-generation PON systems such as WDM (Wavelength Division Multiplexing)-PON systems. For next-generation PON systems, targets include the capturing of FTTH subscribers such as high-end users, business users and multifamily-housing users and the provision of bandwidth-guaranteed services and high-definition video services among others. To accommodate such users and provide such services, WDM-PON systems using WDM technology that allow each user to occupy the bandwidth of a single wave using TDM (Time Division Multiplexing) technology, which increases the transmission rate from the current 1 Gbps/s to 10 Gbps/s, hold promise [1,2].

Most of the WDM-PON architectures presented in [3,4], only allow wavelength channel separation at the arrayed waveguide grating (AWG) router, which then goes onto its individual output ports or creating multiple WDM-based light tree PONs on the same physical fibre. To the place where the population density is very high, it is difficult to re-configure the PON network layout once the fiber been installed in the field as it will cause high operating cost. If need arises to provide another PON in between OLT and the AWG router, laying new fiber and installing new WDM-PON access network is the first thing the network operator would like to avoid.

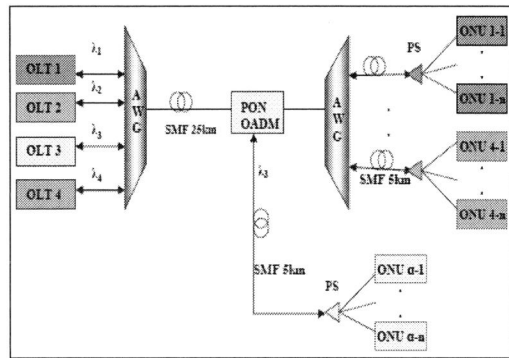

Figure 1: Proposed a novel FTTH-PON architecture

This paper proposes a novel Dynamic WDM-PON architecture by adding a new PON OADM which specifically achieve cost effective, highly scalable multiple PON configuration that allows smooth migration path and bandwidth management. It can be used to serve users in order of hundreds/thousands, and able to cover 30 km in radius which cannot be served from one split point. We also like to report a new PON OADM concept, which consist of passive optical components and a new version of current OADM normally used in Metro ring and ring based PON topologies [5].

978-0-9789217-3-6/07/$25.00

©2007 WEN GLOBAL SOLUTIONS

Experimental Results

The proposed FTTH PON is shown in Figure 1. It is a symmetric multistage architecture based upon the coarse WDM (CWDM) FTTH-PON since the technology offer a network configuration using devices with lower costs compared to dense WDM (DWDM). All OLTs and ONUs consists of transmitter and receiver based on CWDM wavelength range. The wavelengths used are within the S-band and C-band. The experiment was conducted only for 4 PON. Each PON tree was connected to1x32 splitter and evaluated using 2^7-1 PRBS sequence at 2.5Gbps and was used to externally modulate a fixed laser source at 0 dBm.

The novelty of the proposed architecture lies in the usage of the new design PON OADM. The Optical signals at different wavelength from separate transmitter or OLT are combined in a multiplexer and travel through the same fiber from the multiplexer to an OADM which then route one wavelength to create new PON tree or individual output port. The advantage of this new PON OADM is, a wavelength transmitted back to (add) into the PON OADM will be routed to the original OLT as shown in figure 2.

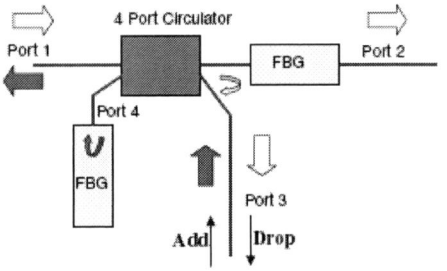

Figure 2: Configuration of new PON OADM

Depending on the type of Fiber Bragg Grating FBG) used, multiple wavelength can be routed to an OADM to have more wavelength per PON. By having more than one light tree in the same physical fibre link, service providers have the flexibility to introduce new services (e.g. faster bit rate or new content) with minimal or no interruption to the legacy systems that has already been deployed. The infinite number of wavelengths in a single physical PON allows infinite scalability in the future. On the other hand,

initially the upstream traffic from users will be aggregated by the ONUs and transmitted upstream towards the OLT via a TDMA technique. Initial upstream traffic will be transmitted via the time domain multiple access (TDMA) technique but as the amount of upstream traffic increases, an extra WDM TDMA link can be created using another wavelength that falls on the free spectral range range.

The ability to create a new light tree on the existing network allows the possibility to introduce new services while maintaining the legacy network users (i.e. low speed home users vs. high speed business users). Use of this device makes it possible to freely add or drop signals with arbitrary wavelengths over multiplexed optical signals by assigning a wavelength to each destination. Moreover, it is possible to simplify the component configuration of optical amplifiers through reduction of the optical attenuation for the express channels (optical channels neither add nor drop at nodes) in OADMs, thereby decreasing the total cost of networks.

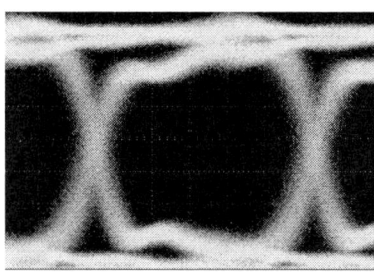

Figure 3: Error-free eye-diagram of received signal at -25dBm after the PON OADM

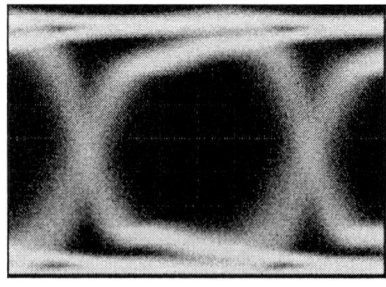

Figure 4: Error-free eye-diagram of received signal at -27dBm after the AWG

Following the first AWG, all signals are combined and transmitted over 25 km of standard single-mode fiber before applied to the AWG router (dedicated for PON1, 2 and 4). Only wavelength for PON 3 had been drop from main fiber to create new PON along 5 km of standard single-mode fiber. Based on the test results, the introduction of the PON OADM only increases the optical loss by an average of 1dB per model, compared to 5dB for the AWG. Error-free transmission is still achievable at the receiver. Eye diagram of the received signal at -25 dBm for the end-to-end optical link with a PON OADM with a 1x32 splitter is shown below in Fig. 3 whereas for the AWG with a 1x32 splitter is shown in Figure 4.

Discussion and Conclusion

This paper started by describing the needs and advantages of WDM-PON to the next generation access network. The trend is anticipated to change due to the high demand for large bandwidth from high density population area such as business, commercial, education and entertainment areas. But, to have a promising access network layout is not an easy matter especially when it comes to cost effectiveness in the design architecture. Since one split point PON only can cover horizontally, to provide vertical area will require another fiber installation which will double the cost. A novel WDM FTTH–PON architecture has been introduced in this paper. The newly proposed architecture uses additional PON OADM which has the capability to virtually limitlessly scale the network with minimal or no alteration to the existing network. Service providers have the flexibility to introduce new services while maintaining the current legacy services since several logical light-tree shares the same physical fibre. The additional PON OADM in the network only imposes an extra 1dB power penalty but still maintaining an error free 2.5 Gbps transmission along a 30 km passive optical link with a 1x32 splitter installed.

The scalability of the proposed architecture is its main advantage over the traditionally proposed WDM-PON. Multiple logical wavelength light trees can be created on the same physical fibre link by using the new design PON OADM. Nevertheless the addition of a PON OADM into the optical link introduced an additional loss of 1dB but it is still able to deliver an error-free transmission over a 30 km link with 32 optical splitting (approximately 27dB optical loss). With careful selection of launch wavelengths at the OLT, the scalability of the proposed FTTH-PON is virtually unlimited both in terms of wavelength and bit rate. The proposed architecture also has the potential of offering a wide variety of services on the same physical network.

Figure 2 shows the configuration of the novel PON OADM using fiber Bragg grating (FBG). It consists of one 4-port circulator and two FBG. Similar to normal OADM module, when the wavelength-multiplexed signals are input to port 1 of the circulator it flows through port 2 to a FBG that reflect only certain wavelength. The wavelength reflected from the FBG is returned to port 3 and output (or dropped) from port 3. The wavelength added to port 3 goes to port 4 which is then reflected from FBG and returned as output from port 1. The sequence is then: (Port 1 → Port 2 → Port 3 → Port 4 → Port 1). By using a low-loss optical circulator which has insertion loss less than 0.5 dB, it can significantly improve OADM performance. Since the development of the new PON OADM is still on going and more evaluation is still being carried out.

References

[1] Chang Hee Lee et al., 'Fiber-to-the-Home Using a PON Infrastructure', Journal of Lightwave Technology, Vol. 24, No. 12, December 2006.

[2] J.-I. Kani, M. Teshima, K. Akimoto, N. Takachio, H. Suzuki, and K. Iwatsuki, "A WDM-Based Optical Access Network For Wide-Area Gigabit Access Services," IEEE Optical Communications, vol. 41, pp.S43-S48, 2003

[3] Amitabha Banerjee et al., 'Wavelength-division-multipled passive optical network (WDM-PON) technologies for broadband access: a review' Journal of Optical Networking, November 2005, Vol. 4, No. 11

[4] Sahrul Hilmi, Bernard H. Lee, Kaharudin Dimyati, "Scalable FTTH-PON Architecture for Unlimited User and Flexible Services", Proceedings of SPIE, Volume 6022, Shanghai, China, December 2005

[5] David Gutierrez et al., "SUCCESS-HPON: A Next-Generation Optical Access Architecture for Smooth Migration from TDM-PON to WDM-PON" Proceedings of ECOC 2006, September 2006

A Novel Method to Measure Brillouin Frequency Shift for Brillouin-based Sensing Application Incorporating a Dual-Wavelength Single-Longitudinal-Mode Fibre Laser

Yizhen Wei, Yongbo Tang, and Daru Chen

Centre for Optical and Electromagnetic Research, State Key Laboratory for Modern Optical Instrumentation, Zhejiang University, Hangzhou 310058, China

weiyizhen@coer.zju.edu.cn

Abstract A method to measure Brillouin frequency shift incorporating a dual-wavelength single-longitudinal-mode fibre laser for sensing application is proposed. Brillouin frequency shift at 10.64 GHz has been shifted and measured at low frequency of 1.03 GHz.

Introduction

In recent years, Brillouin-based distributed fibre sensor (BDFS) [1] has attracted considerable attention due to their important role in applications such as the monitoring and real-time surveillance of long distance vessels, fire detection and so on. Since both the frequency shift and the power of the Brillouin scattering signal are dependent on the strain and temperature on the fibre, it is possible to measure the temperature and strain simultaneously by measuring the Brillouin scattering signal. Several technologies have been developed to realize BDFS, including Brillouin optical fibre frequency domain analysis [2], Brillouin optical fibre time domain analysis [3], [4] and Brillouin optical fibre time domain reflectometry (BOTDR) [5-7]. Comparing with other schemes, BOTDR can be used in applications which require single-end measurement with a simple configuration and relatively larger range.

The measurement of the Brillouin frequency shift (the frequency spacing between Brillouin signal and Rayleigh backscatter) is extremely important for the above-mentioned sensing applications. High-performance optical filters are necessary in the measurement since the Brillouin frequency shift is often very small (~11 GHz at 1550 nm in single-mode fibre (SMF)).in optical domain. BOTDR using fibre Bragg grating notch filter has been reported [5]. This type of filter achieves the recovery of the Brillouin signal by suppressing the Rayleigh scattering light. Coherent detection offers greater dynamic range than direct detection, by mixing the backscatter and a local reference light [6]. Microwave heterodyne detection has been employed to create a spontaneous Brillouin distributed temperature sensor [7]. However, a high speed electro-optic modulator (EOM) is always required to introduce a frequency shift between the reference light and sensing pulse for coherent detection.

In this letter, a novel approach to measure Brillouin frequency shift using a dual-wavelength single-longitudinal-mode (SLM) fibre laser is proposed. The dual-wavelength SLM fibre laser with a spacing of ~0.08nm is used as a probe light. By beating Brillouin scattering light of the shorter wavelength and Rayleigh backscatter of the longer one at a photo-detector (PD), Brillouin frequency shift at 10.64 GHz is shifted to 1.03 GHz which contains the information carried by Brillouin scattering light. It is much easier to deal with this low-frequency signal in post-signal-processing system, and higher accuracy can be expected. This scheme suggests a much simpler configuration to measure the Brillouin frequency shift, in which EOM is not required. Using this method, microwave signals lower to several hundreds MHz could also be obtained by tuning the parameters of the dual-wavelength SLM fibre laser to produce a dual-wavelength lasing with a frequency spacing near the Brillouin frequency shift.

Experimental Arrangement

The schematic configuration of the dual-wavelength SLM fibre laser is shown in Fig.1 (a). An FBG pair which is composed of two uniform FBGs (FBG1 and FBG2) is used as a Fabry-Perot filter. The FBG pair companied with an optical circulator, and a

978-0-9789217-3-6/07/$25.00

©2007 WEN GLOBAL SOLUTIONS

third FBG (FBG3) form a linear cavity. The gain medium is a section of erbium doped fibre with a length of 2.24 m. Pump light from a laser diode of 980 nm wavelength and 140 mW peak power is injected into the linear cavity by a WDM. This fibre laser can provide stable dual-wavelength SLM lasing with a wavelength spacing of ~0.08 nm at room temperature.

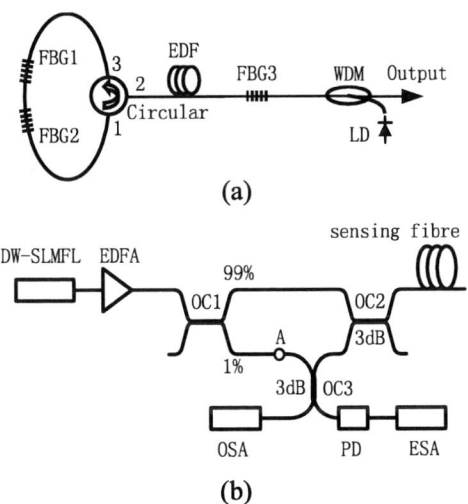

(a)

(b)

Fig. 1. Schematic configuration of the dual-wavelength SLM fibre laser (a) and experimental setup (b). FBG: fibre Bragg grating. WDM: wavelength division multiplexer. LD: Laser diode. DW-SLMFL: dual-wavelength SLM fibre laser. OC: optical coupler. EDFA: erbium-doped fibre amplifier. OSA: optical spectrum analyzer. PD: photo detector. ESA: electrical spectrum analyzer.

Fig. 1 (b) shows the experimental setup. The dual-wavelength SLM fibre laser described above is used as the light source, whose output is amplified by an EDFA. An optical coupler (OC) divides the output of the laser into two parts: One part with 99% power, used as the probe light, is injected into the sensing fibre through OC2, from which the backscatter light could be abstracted, while the 1% part is used as the local reference light. The backscatter comprising both Rayleigh scattering light and Brillouin scattering signal is received by a photo-detector.
Note that both the two SLM wavelengths in probe light produce Brillouin and Rayleigh scattering. So without reference light, there

are still four different frequency components received by PD and three microwave signals with different frequency will be generated. The signal with lowest frequency，which contains the information of Brillouin scattering，is suitable for the sensing application. These microwave signals are measured by an electrical spectrum analyzer. And the output optical spectrum could be observed via an optical spectrum analyzer.

Results
A 10-km long SMF whose Brillouin frequency shift is ~10.6 GHz is used as the sensing fibre. When the point A is disconnected, only backscatter is received by PD. The electrical spectrum of the microwave signal is experimentally measured. As shown in Fig. 2 (a), three microwave signals are generated and their frequencies are 1.03 GHz, 9.61 GHz and 10.64 GHz, respectively. The 9.61 GHz signal is contributed by both the mixing of two Rayleigh scatter and two Brillouin scatter; 10.64 GHz signal is generated by Rayleigh scatter and the corresponding Brillouin scatter; and the 1.03 GHz microwave signal is produced by mixing the Brillouin scatter excited by the shorter wavelength and Rayleigh scatter by the longer one. The 1.03 GHz signal enlarged in Fig. 2 (b) contains the information from Brillouin scattering, which present the temperature and strain applied to the sensing fibre. We also consider the condition of mixing the backscatter and the reference light by connecting point A. Fig. 3 shows the spectrums which are similar to Fig. 2. Compared with the case in Fig. 2, the intensity of 1.03 GHz signal is 16 dBm stronger, and the noise is smaller. Dispersion compensation fibre is also employed as sensing fibre, and 307 MHz microwave signal is achieved.

543

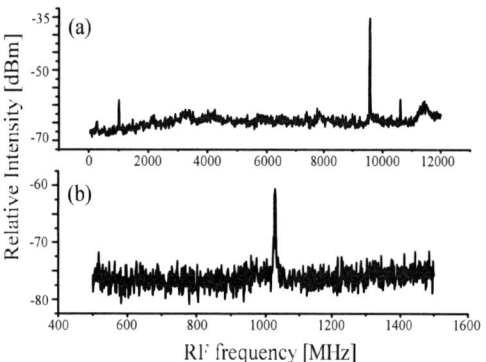

Fig. 2. Electrical spectrum of the microwave signals when point A is disconnected. (a) Electrical spectrum with a frequency range of 0-12 GHz; (b) Enlarged electrical spectrum of the 1.03 GHz signal.

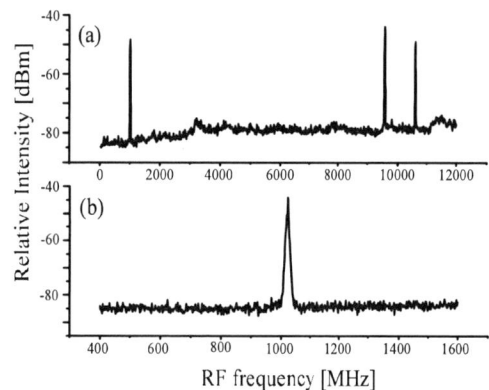

Fig. 3. Electrical spectrum of the microwave signals when point A is disconnected. (a) Electrical spectrum with a frequency range of 0-12 GHz; (b) Enlarged electrical spectrum of the 1.03 GHz signal.

Conclusions

In summary, a novel method to measure Brillouin frequency shift for Brillouin-based sensing application incorporating a dual-wavelength SLM fibre laser is proposed and demonstrated. Brillouin frequency shift at 10.64 GHz has been shifted and measured by detecting the 1.03 GHz microwave signal. It is believed that this frequency can be much lower when the dual-wavelength SLM fibre laser is carefully designed to generate a dual-wavelength lasing with a frequency spacing further closer to the Brillouin frequency shift.

References

1 Tsuneo Horiguchi, Kaoru Shimizu, Toshio Kurashima, and Mitsuhiro Tateda, J. Lightwave Technol. **13** (1995), 1296.

2 Dieter Garus, Torsten Gogolla, Katerina Krebber, and Frank Schiep, J. Lightwave Technol. **15** (1997), 654.

3 Kazuo Hotate, and Masato Tanaka, IEEE Photon. Technol. Lett. **14** (2002), 179.

4 Anthony W. Brown, Bruce G. Colpitts and Kellie Brown, IEEE Photon. Technol. Lett. **17** (2005), 1501.

5 P. C. Wait and A. H. Hartog, IEEE Photon. Technol. Lett. **13** (2001), 508.

6 Mohamed N Alahbabi, Hicholas P Lawrence, Yuh T Cho, and Trevor P Newson, Measurement Science and Technology. **15** (2004), 1539

7 Sally M. Maughan, Huai H. Kee, and Trevor P. Newson, IEEE Photon. Technol. Lett. **13** (2001), 511

Application of half-cycle phase-stepping algorithm in eliminating or diminishing errors of phase measurement

HUI Mei (1), YU Fei (2), ZHAO Yue-jin(3)

1 : 1 : Department of Photo-electric Engineering, School of Information Science and Technology, Beijing Institute of Technology Beijing 100081 huim@bit.edu.cn
2 : Department of Photo-electric Engineering, School of Information Science and Technology, Beijing Institute of Technology yfvcf2005@bit.edu.cn, 3 : Department of Photo-electric Engineering, School of Information Science and Technology, Beijing Institute of Technology yjzhao@bit.edu.cn

Abstract *A half-cycle phase-stepping algorithm based on the character of trigonometric function is proposed for diminished the phase delay error and azimuth error of 1/4 plate and analyzer corner error in phase measurement.*

Introduction

In the course of actual measurement, the phase delay error and azimuth error of 1/4 plate and analyzer corner error will influence on accuracy of phase measurement. A new method based on the half-cycle phase-stepping algorithm is proposed for eliminating or diminishing these error in phase measurement.

Original theory of the half-cycle phase-stepping algorithm

We rewrite phase measurement error due to 1/4 plate phase delay error r、 azimuth angle error $\Delta\alpha$, polarimeter angle error $\Delta\theta_i$ as follows:

$$\Delta\varphi_r = \frac{1}{2}\sin 2\varphi(\cos r - 1) \qquad (1)$$

$$\Delta\varphi_\alpha = \Delta\alpha^2 \sin 2\varphi - 2\Delta\alpha \qquad (2)$$

$$\Delta\varphi_\theta = \frac{1}{4}\sum_{i=1}^{4}\Delta\theta_i(1-\cos 2\varphi) \qquad (3)$$

It is clearly made out from this three equations that $\Delta\varphi_r, \Delta\varphi_\alpha and \Delta\varphi_\theta$ are all related to measured phase φ, which is a cycle error changing according to trigonometric function. Besides of constant term, the error is changing with twice of measured phase φ in the sine and cosine relationship. So, the error has the character of trigonometric function which is changing the positive and negative symbols after a half of a cycle. We make several groups of sampling for the light intensity of interference field. In each group we acquire

N frame images. The original value of reference phase in each group is increasing in π/2 with each time. Then, the value of $\Delta\varphi$ implied at the expressions of phase φ in each adjacent group has different numerical symbols. During the half-cycle phase-stepping algorithm belonging to the eliminating system error methods, after we make the number of groups even and figure out their average value, we can diminish these recurrent error effectually.

With more than four frame interference pattern, applying traditional four frame fast phase-stepping algorithm, front four frame interference pattern as one cycle to calculate the phase. Adding phase $\pi/2$ for next cycle, calculate another phase. For 2M+3 frame interference pattern，make overlap calculation for 2M times，and approach 2M φ_k value。Then, the requiring phase $\varphi(x,y)$ is obtained through count-averaging. Then M group $\Delta\varphi$ can be used to counteract the offset of positive and negative error of M group $\Delta\varphi$. Following the increase of *M,* the effect of eliminating or diminishing errors will be better. Now the expression of $\Delta\varphi_r$, $\Delta\varphi_\alpha$,and $\Delta\varphi_\theta$ is:

$$\Delta\varphi_r = \frac{1}{2M}\frac{1}{2}\sin 2\varphi(\cos r - 1) \qquad (4)$$

$$\Delta\varphi_\alpha = \frac{1}{2M}\Delta\alpha^2 \sin 2\varphi - 2\Delta\alpha \qquad (5)$$

978-0-9789217-3-6/07/$25.00

©2007 WEN GLOBAL SOLUTIONS

$$\Delta \varphi_\theta = \frac{1}{2M} \frac{1}{4} \sum_{i=1}^{4} \Delta \theta_i (1 - \cos 2\varphi) \qquad (6)$$

Experimental result of this algorithm

To validate the availability of this method, Ra 0.09 sine-form roughness plate is measured in small region with sampling windows. 2M+3 frames are sampled at sampling interval $\pi/2$, assuming $M=2$, 7 frames are sampled, as Fig.1 Fig. 2(a) is the topography of Ra 0.09 roughness plate applying traditional four frame phase-stepping algorithm, Fig. 2(b) is the topography of Ra 0.09 roughness plate applying half-cycle ($M = 2$) four frame phase-stepping algorithm. The distinguish between them is unconspicuous because of axis scale in topography figures.

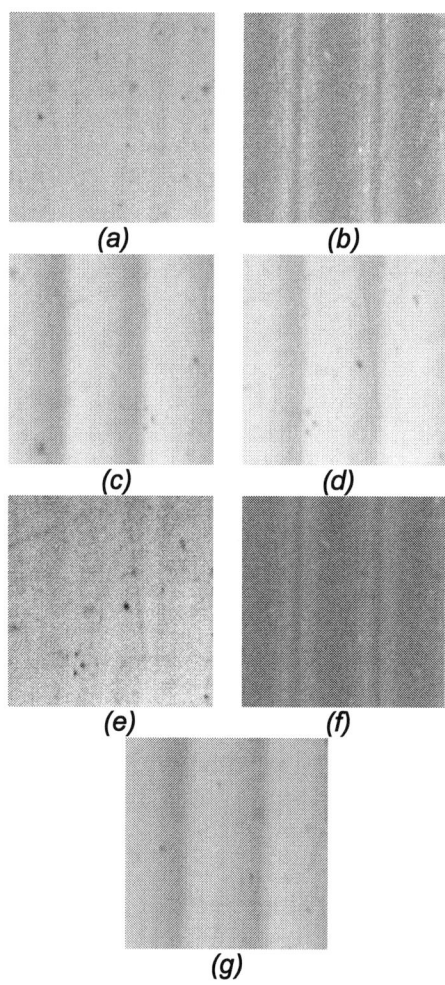

(a)   *(b)*

(c)   *(d)*

(e)   *(f)*

(g)

Fig.1:
The phase-stepping interference image of

Ra 0.09 roughness plate
(a)   phase-stepping 0 ;   (b) phase-stepping $\frac{\pi}{2}$; (c) phase-stepping π ; (d) phase-stepping $\frac{3\pi}{2}$; (e) phase-stepping 2 π ; (f) phase-stepping $\frac{5\pi}{2}$; (g) phase-stepping 3 π ;

Fig.3(a)(c)is profiles of Fig.2(a)at y=50 and y=100；Fig.3(b)(d)is profiles of Fig.2(b)at y=50 and y=100.

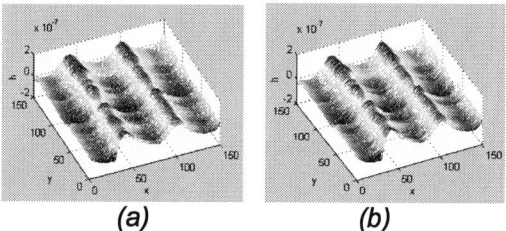

(a)   *(b)*

Fig. 2:
Topography of Ra 0.09 roughness plate
(a)traditional   four   frame   phase-stepping algorithm
(a)half-cycle   (M = 2)four   frame   phase-stepping algorithm

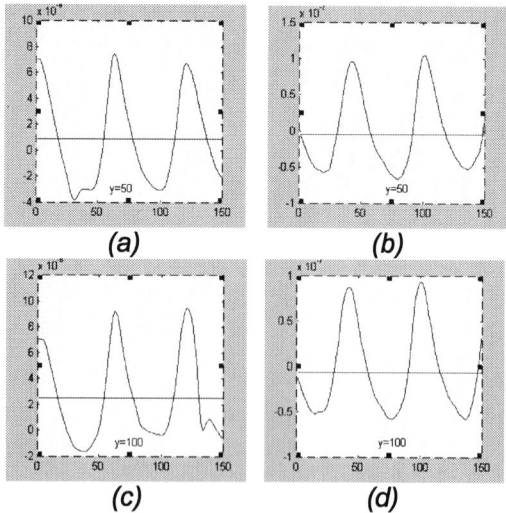

(a)   *(b)*

(c)   *(d)*

Fig.3:
Profile of Ra 0.09 roughness plate
(a)section line of Fig.2(a) at y = 50 ;
(b)section line of Fig.2(b) at y =50;
(c)section line of Fig.2(a) at y = 100 ;
(d)section line of Fig.2(b) at y =100

Conclusions

With magnifying y-axis in profiles of measured object, the distinguish between traditional and half-cycle four frame phase-stepping algorithm is obviously made out from the two-two corresponding four figures. After applying half-cycle four frame phase-stepping algorithm, the non- sinusoid caused by errors is improved in roughness plate measurement.

References

1 Qian Kemao. Comparison of some phase shifting algorithms with a phase step of π/2 [J]. Advanced Photonic Sensors and Applications II, 2001, 310-313

2 Shouhong Tang. Self-calibrating five-frame algorithm for phase shifting interferometry [J]. SPIE Vol. 2860:91-98

3 Carles Pizarro, Josep Arasa, Ferran Laguarta, Nu´ ria Toma` s, and Agustı´ Pinto. "Design of an interferometric system for the measurement of phasing errors in segmented mirrors" APPLIED OPTICS Vol. 41, No. 22, 1 August 2002:4562-4570

4 C. L. Koliopoulos. Simultaneous phase shift interferometer. Proc SPIE, 1513, 1991: 119-127

The annealing process of R.F. magnetron sputtered ZnO:Al films

YU Zhinong[1]*, XU Jin[2], XUE Wei[3], LI Jinwei[4]

1. Department of optical Engineering, School of Information Engineering, Beijing Institute of Technology, Beijing 100081, P.R.China，*Corresponding author, Tel.: 010-68913259-11; fax: 010-68912550. E-mail address: znyu@bit.edu.cn
2. Beijing Institute of Technology, xujin214@gmail.com
3. Beijing Institute of Technology, xuewei@bit.edu.cn
4. Beijing Institute of Technology, lijingwei@bit.edu.cn

Abstract The ZnO:Al films were annealed at different atmosphere. The films annealed in vacuum showed perfect properties, compared with in air and Ar atmosphere, and the low resistivity of the order of $10^{-3}\,\Omega\cdot cm$ was obtained in the condition of vacuum annealing at 220℃.

1. Introduction

The transparent conducting oxides such as Sn doped In_2O_3 (ITO), SnO_2 (NESA) and ZnO have been widely studied[1]. By considering the better doping effect, ZnO films are usually doping with $Si^{[2]}$, $Ga^{[3]}$, $Al^{[4]}$, and so on. The Al-doped ZnO (AZO) transparent conducting film has high transmittance in the visible region and low resistance. It has advanced applications such as displays, solar cells, optoelectronic devices, electrochromic devices, etc. Non-toxic, low cost, with high thermal/chemical stability and resource availability, doped ZnO thin films are intensively studied to replace ITO thin films in the above mentioned applications[5].

Annealing is an important way to modify the properties of films. The properties of films generally were changed at high temperature, long time and in different atmosphere[6]. It was shown that wide-band-gap semiconductors such as SnO_2 and ZnO tend to be extremely sensitive to the ambient atmosphere. Their electrical properties, which are governed by the intrinsic electronic defects formation, can be considerably altered by the exposure to a specific gas[7]. This effect, which depends on the surface-to-volume ratio of the material, is enhanced when the metal oxide is deposited in the form of a thin layer. In this paper, we report the properties of R.F. magnetron sputtered ZnO:Al films.

2. Experimental

The ZnO:Al films were deposited on glass substrates in a r.f. magnetron sputtering system with a base pressure of 2×10^{-3}Pa. The substrates were polished with special powder and cleaned by the alcohol, and then placed inside the chamber. A sintered ceramic target with a mixture of ZnO (purity 99.99%) and Al_2O_3 (purity 99.99%) was used. The percentage of Al_2O_3 in the ZnO target was 2 wt%.

Annealing of the films was carried out in a vacuum, in air and Ar at a specified

978-0-9789217-3-6/07/$25.00

©2007 WEN GLOBAL SOLUTIONS

temperature and for a specified period. The resistivity of the film was measured by a four-probe method. The transmittance spectrum and the crystal structure, were examined by spectral photometer, X-ray diffraction apparatus, respectively.

3. Results and discussion

Table 1 shows the electrical properties (resistivity) after annealing in the atmosphere, Ar and vacuum. The annealing time is 1 hour and the annealing temperature is 220℃. The result in the atmosphere shows the resistance increases sharply, the main reason is that the oxygen reduces the oxygen vacancies and Al

donors in the film. So the oxygen can make the resistivity increase. The result in the Ar shows the resistance hasn't changed the resistance much. It can be concluded crystallization and structure haven't changed much in this temperature. The result in the vacuum shows the resistance is the best data. It is known that the conductivity of AZO film is related to carrier density, Hall mobility, oxygen vacancies, and so on. During annealing in vacuum, there is oxygen desorption from the grain boundaries. This desorption produces a great deal of oxygen vacancies and leads to a decrease of the resistivity.

Table1. The result (resistivity) of the annealing experiments.

	The as-grown film	Annealed in air	Annealed in Ar	Annealed in vacuum
Resistivity $(10^{-3}\Omega\cdot cm)$	5.97	5078	5.19	3.36

The change in transmittance properties, compared with that in electrical properties, is small. The transmittance within the visible and the near infrared region is always higher than 80% (including the glass substrate), which reveals the superior optical properties

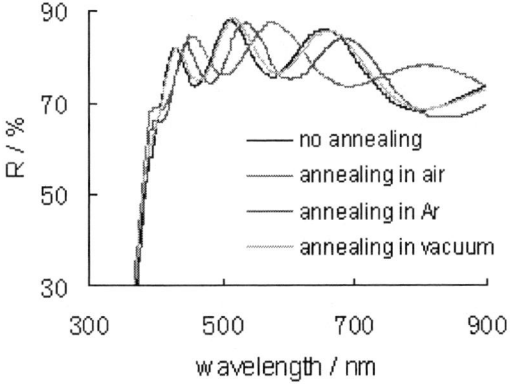

Fig.1 the transmittance of AZO films annealed in air, Ar, and vacuum, compared with the as-grown film

exhibited by the ZnO:Al thin films produced in this work, as shown in Fig.1. However, the tiny change of transmittance among the films annealed in different atmosphere and the as-grown film indicates the structure departure of the films after annealing, compared to that of the as-grown films. Especially, the change of transmittance curve of the annealed film in air maybe implies the oxidization of Al donor or the input of oxygen in the structure of film, which leads to an increase of sheet resistivity after annealing in air.

Fig.2 shows the XRD spectra of the annealed film in vacuum. The intensity exceeds 2.5×10^{5}cps, and one of the as-grown film is about 2×10^{5}cps (not shown in the Fig.2), It reveals that the film crystallinity has been improved by annealing in vacuum. Only the (002) peak, which has been observed at $2\theta \approx 34.26°$, was very close to the preferred orientation of the standard ZnO crystal. This implies that aluminum replaces zinc in the hexagonal lattice or aluminum segregates to the non-crystalline region in grain boundary and the obtained film has a preferred orientation with the c-axis perpendicular to the substrate.

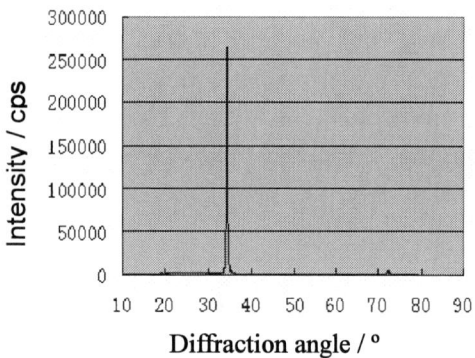

Fig.2 the XRD spectra of the annealed film in vacuum

4. Conclusions

R.F. magnetron sputtered ZnO:Al films were annealed at different atmosphere. The annealed films in vacuum showed perfect properties, compared with in air and Ar atmosphere, After vacuum annealing at 220℃, the film resistivity changes to 3.36 $\times 10^{-3}\Omega.cm$, and the average transmittance rate in the visible wave band is higher than 82%

References

[1] K.H. Kim, K.C. Park, D.Y. Ma, J. Appl. Phys. 81 (12) (1997) 7764.

[2] T. Minami, H. Sato, H. Nanto, S. Takata, Jpn. J. Appl. Phys. 1125(1986) L776.

[3] B.H. Choi, H.B. Im, Thin Solid Films 193–194 (1990) 712.

[4] G.A. Hirata, J. Mckittrik, T. Cheeks, J.M. Siqueiros, J.A. Diaz, O.Contreras, et al, Thin Solid Films 288 (1996) 29.

[5] S. Fujihara, C. Sasaki, T. Kimura, Appl. Surf. Sci.180 (2001) 341.

[6] D.H. Zhang, D.E. Brodie, Thin Solid Films 238 (1994) 95.

[7] G.N. Advani, A.B. Jordan, J. Electron. Mater. 9 (1980) 29.

Uniform Color Space for Color Storage

Yan CHENG, Xu LIU, Haifeng LI

1 : Yan CHENG: Doctor candidate, chinacheng404@etang.com
2 : Xu LIU (correspond author): Professor, liuxu@zju.edu.cn
3 : Haifeng LI: Professor, lihaifeng@zju.edu.cn

Abstract *The RGB color space has been widely used, but has limitations. A new presentation method is put forward. It uses floating-point to characterizing colors in L*a*b* color space. Due to high precision of floating-point and uniform perception of L*a*b* color space, it can present better color perception.*

Introduction

Digital display technology has been developed and widely used in many fields. But the method to characterize the color data has become a limitation of the usage and development of the image store, display and transfer. Currently we represent color using three values R, G and B. Each of them uses 1 byte to represent the color value, which is 8 bits, to store data, which ranges between 0 and 255.

For a color value, for example red RGB: 255/0/0, displays in different devices will present difference in appearing. More precisely, if we measure the tristimulus values of red color RGB: 255/0/0 represented on CRT and LCD, we will get the different tristimulus values. That means using RGB to describe the color is device-dependent [1] [3].

Another disadvantage of this method is the problem of precision. As mentioned before, each value occupies 8 bits. Not only RGB, but every color space representations, such as YPbPr, YUV have same digital length, uses 8 bits. This is usually not enough for the application in the high quality display system.

So it is necessary to design a new way for color data digitalization characterizing. A new format of color data image will put forward in this paper. It uses floating-point, which has enough power to deal with the precision. And it uses the color values in L*a*b* color space [2]. The L*a*b* color space has the advantage of the uniform

perception. So that the same color perception for the same image between different devices becomes possible.

New Color Format

The format of digital data to present the color image is composed of two segments, information segment and data segment. Information segment is a summary of the image file. Data segment is the actual image data. Information segment contains all the information of the file shown in the Table 1.

Table 1. The structure of information segment

Name	Size (byte)	Description
Mark	3	Indicate this is a "Uniform Color Space Store" file. The value is 0x756373. 0x75 is the ASCII value of "u". 0x63 is the ASCII value of "c". 0x73 is the ASCII value of "s".
Height	2	Indicate the height of the image in pixel.
Width	2	Indicate the width of the image in pixel.
Size	5	Indicate the size of the file in byte.

Data segment contains the actual color data. Data segment is organized in "pixel unit". Each pixel unit represents one pixel. It takes 16 bytes, 12 bytes for describing L*a*b* color space values, which is 4 bytes for each dimension, and 4 bytes for describing brightness degree.

The reason using 4 bytes for each L*a*b* dimension and the transparent degree is that we need 4 bytes to present a floating-point data in single format according to IEEE Standard 754 for Binary Floating-Point Arithmetic. And floating-point has many advantages when dealing with L*a*b* color space values. It also adapts or matches to almost every platform. "pixel unit" is organized in Table 2.

Table 2 The structure of pixel unit

transparent degree				b*			
15	14	13	12	11	10	9	8
a*				L*			
7	6	5	4	3	2	1	0

The right side is low order, and the left side is high order. There are totally 16 bytes in one pixel unit. Each of L*, a*, b* and transparent degree occupies 4 bytes and its content conforms to IEEE Standard 754. Here is an example.

	L*			
byte order	3	2	1	0
hex	0x3F	0x80	0x00	0x00
bin	0011 1111	1000 000	0000 0000	0000 0000

Let L*=1.0. Then it will be 0x3F800000 in hex mode as showing in the second line. The third line is the binary form.

"transparent" is used for alpha bending. The range is between 0 and 1.0. If we want

From	To	Accuracy
0.000000	0.000008	0.0000000000000909494701772928238
0.000008	0.000015	0.000000000001818989403545856476
0.000015	0.000031	0.00000000003637978807091712952
0.000031	0.000061	0.0000000000727595761418342590 3
0.000061	0.000122	0.0000000001455191522836668518 07
0.000122	0.000244	0.0000000002910383304567337036 13
0.000244	0.000488	0.0000000005820766091346740722 7
0.000488	0.000977	0.000000001164153218269348144 53
0.000977	0.001953	0.000000002328306436538696289 06
0.001953	0.003906	0.000000004656612873077392578 12
0.003906	0.007812	0.000000009313225746154785156 25
0.007812	0.015625	0.00000001862645149230957031 25
0.015625	0.031250	0.0000000372529029846191406 25
0.031250	0.062500	0.0000000745058059692382812 5
0.062500	0.125000	0.000000014901161193847656 25
0.125000	0.250000	0.0000000298023223876953125
0.250000	0.500000	0.0000000596046447753906 25
0.500000	1.000000	0.0000001192092895507812 5
1.000000	2.000000	0.0000002384185791015625
2.000000	4.000000	0.000000476837158203125
4.000000	16.000000	0.00000190734863281 25
16.000000	32.000000	0.0000038146972656 25
32.000000	64.000000	0.00000762939453125
64.000000	128.000000	0.0000152587890625
128.000000	256.000000	0.000030517578125

a brightness effect, we can set this one less than 1.0, then the image will be blended with the background.

This new format uses L*a*b* color space to measure the value of color; it is device-independent, which means the observer can obtain the same feeling from the different devices. L*a*b* is a uniform color space according to the Munsell color system.

We can find that even though the CIE 1960 chromaticity diagram is a uniform one, but it has no luminance value. If the application refers luminance "Y", it will show a non-linear relation (approach to an exponential relation) with Munsell luminance value "Vy", as showed the red line in Figure 1.

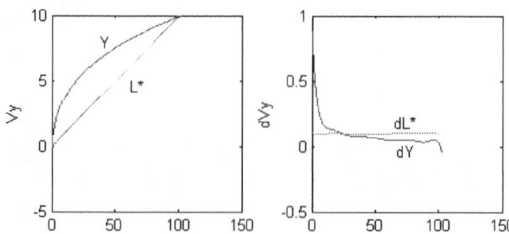

Figure 1 The relation of Vy-Y. Figure 2 The Derivative of Vy vs. Y

The derivative result is showed in Figure 2.

If we have an error at Y=1, we will have a big error perception. And we change the luminance from 0 to 100 linearly, we won't get the linear perception.

In stead, the L*a*b* is a totally uniform in both chromaticity and luminance, as showed in Figure 1 with green line. L* has a linear relationship with Munsell luminance value "Vy". Linear has the advantage of minimize the error. Figure 2 green line is the dVy-dL* relationship. The line is more or less a horizontal line, which means we won't get a very huge error on a particular spot [4] [5].

Another advantage is high precision. As mentioned before, RGB uses integer, which only has 256 levels, to format data. Floating-point is more accurate. And for smaller color value, the interval grade will be more accurate.

Conclusion

A new color format for the digital color image system is proposed in this paper. It will overcome the drawbacks of the conventional RGB color system, and will be possible to offer the display devices-independent color image, and have much more high color precision, especially in the dark color state.

References

1. Huang Qingmei. Analysis on color characterization and calibration of LCD. OPTICAL TECHNIQUE
2. Lv Xinguang. Study on the Uniformity of CIE LAB Chromaticity Space. JOURNAL OF ZHENGZHOU UNIVERSITY
3. Liu Mingliang. Determination of the Color CRT Display Gamuts. Journal of Wuhan Technical University of Surveying and Mapping
4. Liu Weiqi. The Relation Between Munsell Color System and Visual Perception Brightness. OPTICS AND PRECISION ENGINEERING
5. Principles of Color by Riy S. Berns, published by John Wiley & Sons. Inc.

Chalcogenide glass photonic chips for all-optical signal processing

V.G. Ta'eed (1), M.R.E. Lamont (1), M.D. Pelusi (1), M.A.F. Roelens (1), D.J. Moss (1),
S. Madden (2), D-Y. Choi (2), B. Luther-Davies (2) and B.J. Eggleton (1)
CUDOS, Centre for Ultrahigh-bandwidth Devices for Optics Systems (CUDOS)
1 : School of Physics, University of Sydney, NSW 2006, Australia
2 : Laser Physics Centre, Australian National University, Canberra ACT 0200, Australia
e-mail: vahid@physics.usyd.edu.au

Abstract *The large, ultra-fast Kerr nonlinearity and low nonlinear loss of chalcogenide glasses are attractive for chip based all-optical signal processing. We highlight successes including optical time demultiplexing at 160 Gb/s in As_2S_3 planar waveguides.*

Introduction

All-optical signal processing (e.g. signal regeneration, wavelength conversion and switching) exploiting the nonlinear behaviour of optical devices potentially resolves the communications bottleneck caused by opto-electronic devices and photo-detectors. Chalcogenide glasses [1] appear to resolve the issues associated with other potential platforms as they provide strong third-order nonlinearities over two magnitudes greater than silica, as well as offering an ultra-fast response unhampered by free-carrier effects commonly associated with semiconductor optical amplifier based devices. Chalcogenide glasses also exhibit low to moderate nonlinear absorption and strong photosensitivity for inscribing waveguide Bragg gratings.

In this paper we examine cross-phase modulation (XPM) based wavelength conversion and optical demultiplexing using four-wave mixing (FWM) in As_2S_3 chalcogenide glass planar waveguides. This latter experiment demonstrates virtually penalty free demultiplexing of a 160 Gb/s signal. Examination of both devices, in terms of power requirements and optical bandwidth limitations, points to simple steps for achieving further performance improvements.

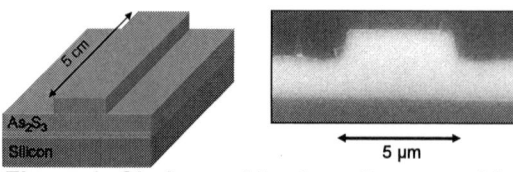

Figure 1. Chalcogenide glass rib waveguide

Planar waveguides

Figure 1 shows the transverse profile of the As_2S_3 rib waveguide. The waveguides were fabricated by ultra-fast pulsed laser deposition of an As_2S_3 layer on to a thermally oxidized Silicon wafer [2]. The nonlinear refractive index of As_2S_3 is 3.05×10^{-14} cm^2/W; over two orders of magnitude greater than silica. Moreover, this composition exhibits negligible two photon absorption at telecom wavelengths [3]. Depending on the design dimensions, the waveguide nonlinear coefficients are in excess of 2000 W^{-1}km^{-1}.

Wavelength conversion

XPM is an effective technique for achieving wavelength conversion of a data signal [4]. Pump pulses carrying data are directed though the nonlinear medium along with a CW probe beam near the desired output wavelength. The pump pulses induce a transient chirp on the probe via XPM through the Kerr nonlinearity. This broadens the probe spectra generating sidebands, and when a single sideband is selected using an optical filter, the output wave at the converted wavelength is modulated in time similarly to the pump pulse.

Figure 2 shows the experimental setup. The signal (pump) was generated from a 10 GHz active mode-locked fiber laser (MLFL) producing 2.1 ps pulses. An external electro-optic MZ modulator encoded a pseudorandom pattern onto the pulse train before being combined through a 50/50 coupler with a CW probe generated by an amplified laser diode.

978-0-9789217-3-6/07/$25.00

©2007 WEN GLOBAL SOLUTIONS

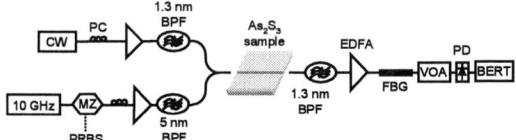

Figure 1. Experimental setup for the wavelength conversion.

Figure 3 shows the XPM spectral broadening of the probe at the output of the waveguide. A significant portion of the CW probe remains due to the low signal duty cycle. A 1.3 nm band pass filter and fiber Bragg grating notch filter selected the longer wavelength XPM sideband and removed residual CW probe. The XPM spectral broadening is independent of the pump-probe wavelength offset as the short device length ensures the dispersive pulse walk-off between the pump and probe is negligible, allowing for wide wavelength conversion. The bit-error rates (BER) of the converted pulses were measured at all three probe wavelengths and corresponding Q-factors calculated using the margin measurements technique [5]. The converted signal showed a Q-factor penalty of only 2.3 dB for wavelength conversion compared to back-to-back.

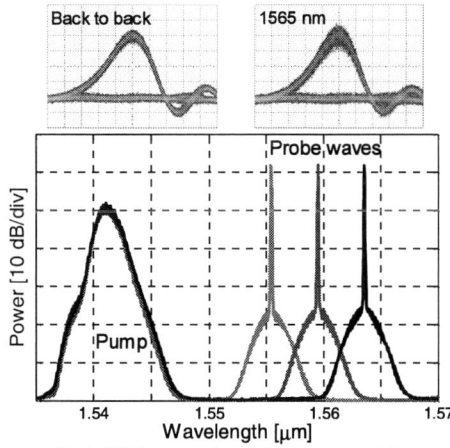

Figure 3. XPM spectra at waveguide output and electrical eye (30 GHz bandwidth) for back-to-back and 1565 nm converted signal.

Optical time demultiplexing
While chalcogenide glasses exhibit strong normal dispersion at telecommunication wavelengths [3], coherent phase-matching can be achieved provided the nonlinear interaction length is shorter than the coherence length as determined by the pump-probe wavelength offset [6]. Using this scheme a high bit-rate data signal can be optically demultiplexed within a chalcogenide waveguide. The signal (the probe) is combined with a pulsed pump operating at a sub-harmonic repetition of the signal bit-rate and degenerate FWM produces an idler wave containing a tributary of the initial signal.

Figure 4 shows the experimental setup used to achieve demulitplexing of a 160 Gb/s OTDM signal. The 160 Gb/s signal was generated from a 40 GHz repetition rate active MLFL. An external electro-optic MZ modulator encoded data on the pulses at 40 Gb/s, before being optically multiplexed to 160 Gb/s. The pulse width at the waveguide input was 1.9 ps. The pump source was a 10 GHz MLFL emitting pulses at 1550 nm wavelength and was synchronized to the 160 Gb/s signal centred at 1560 nm. Signal and pump were amplified then combined with a coupler and launched into the waveguide. The pump pulses at coupler output were 1.5 ps wide with average power set to 150 mW.

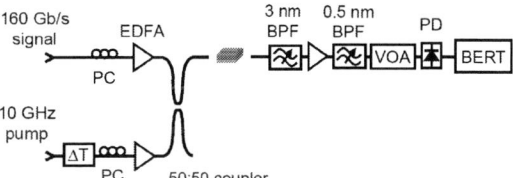

Figure 2. Experimental setup for the demultiplexing

The measured optical spectra at the waveguide output is shown in Fig. 5. For the operating conditions described, FWM between pump and signal pulses generates a 10 Gb/s idler signal at ~1540 nm, which is extracted using optical BPF of bandwidth 3 nm and 0.55 nm. The FWM idler power and is maximized by optimizing both the inter-pulse delay and polarization states of pump and signal pulses. Despite the large dispersion value compared to standard optical fiber, the high nonlinearity enabled a device length shorter than the calculated coherence length ensures efficient FWM-DEMUX of the 160 Gb/s signal. The measured frequency conversion efficiency of the 10 Gb/s channel from 1560 to ~1540 nm was calculated to be 12 % from integration of the optical spectra. Bit-error rate (BER) measurement in Fig. 6 reveal that the power penalty is <1 dB at a BER of 10^{-9} for all 16 channels of the 160 Gb/s signal compared to the back-to-back measurement.

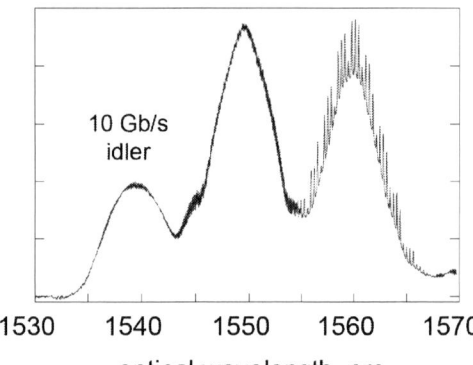

Figure 5. FWM spectra at waveguide output

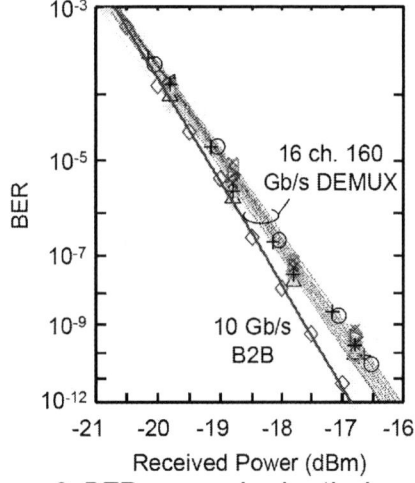

Figure 6. BER vs. received optical power.

Conclusions

We have shown that all-optical signal processing of high data rate signals is achievable in chalcogenide glass based photonic chips. As both experiments rely on the large nonlinear coefficients of the As_2S_3 rib waveguides, it is important to note that greater nonlinear coefficients can be achieved through tailoring the material nonlinear index and reducing the effective waveguide area. This second step would also result in significant waveguide dispersion which could offset the material dispersion. This would allow devices with longer waveguides (for even further power reduction) and increased operating bandwidth. The virtually instantaneous nonlinear process does not limit the operating data rate to 160 Gb/s and all-optical processing of even high data date signals is feasible.

References

[1]   V. Ta'eed et al., Optics Express, 15 (2007) 9205-9221.

[2]   Y. L. Ruan et al., Optics Express, 12 (2004) 5140-5145.

[3]   M. Asobe, Optical Fiber Technology, 3 (1997) 142-148.

[4]   B. E. Olsson et al., IEEE Photonics Technology Letters, 12 (2000) 846-848.

[5]   N. S. Bergano et al., IEEE Photonics Technology Letters, 5 (1993) 304-306.

[6]   G. P. Agrawal, Nonlinear Fiber Optics, 3rd ed. San Diego: Academic Press, 2001.

Low Cost Integrated Optical Mux/Demux for LX4 Transceiver

Hongtao Han, Jim Morris, Keith Main

Tessera North America, 9815 David Taylor Drive, Charlotte, NC 28262,

hhan@tessera.com; jmorris@tessera.com; kmain@tessera.com

Abstract: This presentation reviews the development of an integrated, optical Mux/Demux based on wafer scale micro-optics and optical wavelength filters. This small form factor design enabled the LX4 transceiver to satisfy X2-MSA for 10G Ethernet applications.

Introduction

Fiber optics has been very successfully adopted for many applications because of its competitive advantages, which includes bandwidth and immunity from electromagnetic interference. For many applications, such as data communication, cost is the main driver. Cost reduction may be achieved through reductions in component cost, assembly cost, testing cost, system cost, deployment cost, etc. As a low-cost solution for system vendors, the LX4 optical transceiver incorporates 4-channel long wavelength (1310 nm) CWDM technology, with each channel transmitting 3.125Gbit/s for 10 Gbit/s Ethernet applications. LX4 allows legacy fiber to transmit data at distances up to 300 m. It also works for both SMF and MMF [1,2].

The optical Mux and Demux can be made by a fused fiber technique, but the form factor is too big, especially for an X2 package. In this paper, we present an integrated Mux and Demux characterized by a small form factor and low manufacturing cost.

Design

Both the optical Mux and Demux are applicable for 4-channel CWDM, with central wavelengths of 1275.7 nm, 1300.2 nm, 1324.7 nm, and 1349.2 nm. The pitch between each channel of active devices is 0.5 mm. Figure 1 is an illustration of both a Mux and a Demux design (the red die represents laser die for Mux and photo diode for Demux). Figure 2 is a photo image of a Demux assembly. Both the Mux and Demux are comprised of collimating optics (silicon refractive), thin film wavelength band-pass filters, and relay/focus optics (reflective and diffractive). To reduce the manufacturing cost, both the collimating

optics and the relay/focus optics are fabricated on a wafer scale using photolithography, deposition, and etching techniques. Additionally, wafer scale metrology is implemented to characterize the micro optics to ensure known good die for assembly. Regarding the wavelength filters, we have two designs: one is based on discrete filters; the other is based on wafer scale integrated filters where different wavelength filters are fabricated on the same wafer using photolithography and deposition techniques. The latter design is appropriate for the optical Mux, because the Mux design is very sensitive to the filter tilt around both the x-axis and y-axis.

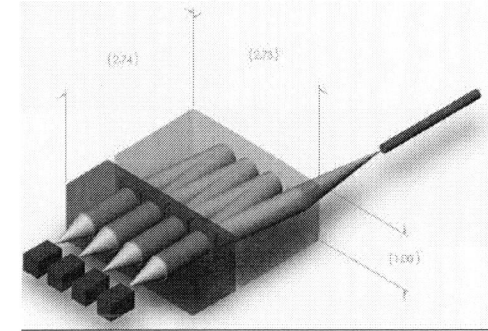

Figure 1. Illustration of Mux/Demux design

Figure 2. Photo image of 4-channel Optical Demux for CWDM

978-0-9789217-3-6/07/$25.00

©2007 WEN GLOBAL SOLUTIONS

One of the advantages of this design is the small size of the Mux and Demux. According to the current design, the Mux size is 3.07 mm (W) x 3.14 mm (L) x 1.45 (H), and the Demux size is 3.00 mm (W) x 2.40 mm (L) x 1.35 mm (H). Figure 3 illustrates the small form factor of a Transmit Optical Sub-Assembly (TOSA) and Receive Optical Sub-Assembly (ROSA) using the integrated Mux and Demux designs. The Mux/Demux easily fits into a TOSA/ROSA size of 12 mm (width) x 27 mm (length including fiber receptacle) x 5 mm (height). This enables the manufacture of a TOSA/ROSA which can fit into an X2 package. With this technology, a further reduced TOSA/ROSA can be fabricated which will fit into an XFP form factor.

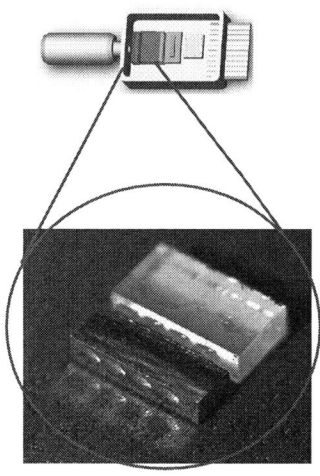

Figure 3. Small form factor of TOSA/ROSA based on integrated Mux and Demux designs

Assembly
First, both silicon refractives and glass diffractives are fabricated on a wafer scale using photolithography and etching techniques. After the completion of fabrication, micro optics are characterized using metrology on wafer scale. A wafer mapping is generated to identify known good die. Then, the micro optics are singulated using conventional dicing techniques.

For the discrete filter design, there are four wavelength filters. Assembly for the discrete filter design involves a total of six components. First, wavelength filters are bonded onto a glass optic block using a pick & place tool, then a silicon optic block is bonded onto the wavelength filters (as shown in Figure 2) using the pick & place tool.

For the integrated filter design, all wavelength filters are fabricated on the wafer scale. Then, the filter die are singulated using conventional dicing techniques. The assembly of the integrated filter design only involves three components, which is similar to, but quicker than, the assembly of the discrete filter design. After the Mux and Demux are assembled, all are functionally tested to ensure known good dies for the TOSA and ROSA assemblies.

Functional Test & Performance
To ensure the performance of products, both optical Mux and Demux are tested functionally for insertion loss. The test setup is similar to the assembly process for a TOSA/ROSA. For Mux testing,, the laser array is passively aligned to the Mux; for Demux testing, the detector array is passively aligned to the Demux. The output fiber (a single mode fiber for Mux) and input fiber (a 62.5 µm multimode fiber for Demux) are actively aligned to the optical Mux and Demux, respectively. In order to reduce the manufacturing cost, these active alignments are 2-axis alignment (x-axis and y-axis only). The angular alignment and z-distance between the fiber and Mux/Demux in each setup are fixed. This greatly reduces the assembly time compared to a conventional 6-axis active alignment method. For optical Demux functional testing, we also use a mode scrambling device (offset launch patch cord) to excite higher order of modes to simulate real applications. The product specification of insertion loss for the Demux is < 2.5 dB. Because of design tolerances, the typical insertion loss is about 2.1 dB. The insertion loss specification for the Mux is < 7.0 dB.

Conclusions
In summary, we designed and fabricated a 4-channel integrated optical Mux and Demux for an LX4 transceiver for 10 Gb/s Ethernet applications. These Mux and

Demux designs are based on wafer scale micro optics (refractive, diffractive and reflective) and thin film filter technologies. These designs feature a small form factor for making a small size and low cost TOSA/ROSA. To further reduce the assembly cost, these designs allow TOSA/ROSA manufacturers to passively align the laser array and Mux for a TOSA or the detector array and Demux for a ROSA. The active alignment of the fiber to the Mux and the fiber to the Demux only involve two axes (x-axis and y-axis). The typical insertion loss for a Mux is < 7.0 dB, and insertion loss for a Demux is <2.5 dB.

To reduce manufacturing costs and improve the product performance of the optical Mux design, we also developed an integrated filter technique where processing is completed on a wafer scale using photolithography and filter patterning techniques.

References

1 S Maruo, et al Hitachi Cable Review, No. 25, (2006), page 1-4

2 Deyu Zhou, LX4 ready to drive 10-Gigabit Ethernet market on Lightwave Direct, December 23, (2004), available at: http://lw.pennnet.com/Articles/Article_Displa y.cfm?Section=OnlineArticles&SubSection= Display&PUBLICATION_ID=13&ARTICLE_ ID=218178&pc=ENL

Tuneable photonic millimetre wave generation using an optical phase modulator and DWDM thin film filters

P. Shen, J. James, N. J. Gomes (1) and P. G. Huggard (2)

1: Department of Electronics, University of Kent, Canterbury, CT2 7NT, UK

2: Space Science and Technology Department, Rutherford Appleton Laboratory, Chilton, Didcot OX11 0QX, UK

Abstract *we report on the generation of continuously tuneable W-band millimetre-wave signals with low phase noise based on sideband filtering of a phase modulated optical signal by using a novel cost optical notch filter*

Introduction

Demand for high data rate wireless communications has led to the development of millimetre-wave (MMW) over fibre technology[1]. In such a system, a high quality MMW signal is often required for providing a local oscillator. One method to generate such a MMW signal is the sideband filtering technique. This technique has been used in directly modulated and externally modulated systems [2-6], and for an optical frequency comb generator (OFCG) [7].

In general, direct modulation schemes are limited by the less efficient high-frequency response of the lasers. The OFCG, on the other hand, provides a large bandwidth, but, the system tends to be complex and in some cases limited in tunability. External modulation is more promising; it provides the advantage of good frequency coverage up to 160 GHz, as well as offering a simple system. Both intensity and phase modulators can be used in such a system. The advantage of using a phase modulator is that it is free from bias point drift, therefore providing a stable output.

In this paper[1] we report W-band MMW signal generation based on sideband filtering of an externally phase modulated optical signal. It is a low cost MMW generation scheme as inexpensive thin film optical filters are used for sideband selection. By using a novel configuration of the optical filters, there is no need for the polarization alignment of the selected optical sidebands, and the light propagates in a common optical path.

This work is supported by the UK EPSRC under the grant COMCORD

Principle of operation

A phase modulated signal can be expressed as $E_c e^{-j\omega_r t} e^{-j\beta \sin(\omega_m t)}$ where E_c and ω_r are the amplitude and the angular frequency of the optical carrier, ω_e is the angular frequency of the RF drive signal. The modulation index β determines the power of the sidebands. By adjusting the RF drive signal of the phase modulator, the modulation index can be set such that the optical power in the desired sidebands is dominant. By increasing the modulation index the higher order sidebands can be made more significant.

In contrast to intensity modulation, the amplitude of an angularly modulated signal (FM or PM) always remains constant. The spectral components are so related in amplitude and phase that they cancel each other when directly detected by a photodiode. As a result, if detected by a square law photodiode, no beat signal can be detected. By altering the amplitude or phase relation between the optical components, one can generate a detectable signal. A simple method of achieving this is to use optical filters to filter out two spectral components that are separated by the desired MMW frequency. When these signals are detected by a photodiode, a clean MMW signal is produced, due to the high coherence between the optical sidebands.

There are many possible configurations for the selection of the optical sidebands [2-6]. When the selected sidebands propagate through different optical paths it is important to maintain the length match between differential paths, as small path changes can be converted into large phase

978-0-9789217-3-6/07/$25.00

©2007 WEN GLOBAL SOLUTIONS

deviations in the generated MMW signal. Another factor that determines the quality of the MMW generation is the match between the polarizations of the selected sidebands. To achieve good efficiency, the state of polarization dispersion between the two selected sideband must be small. A polarization controller is often employed to match the polarizations.

Experimental setup

The proposed system (Fig. 1) consists of an external cavity laser to generate an optical carrier which is modulated by a high speed optical phase modulator. An optical notch filter is employed to select the upper and lower 2nd sidebands of the modulated lightwave. The notch filter is made by cascading a 100 GHz band-pass filter and 50 GHz band-stop filter to simultaneously select the two second order sidebands. Both filters are standard DWDM thin film filters (with 200GHz and 100 GHz channel spacing) in ITU wavelength grid C29. The selected sideband is amplified by an Erbium Doped Fiber Amplifier (EDFA) and then feed into a WR-10 waveguide based photomixer [8].

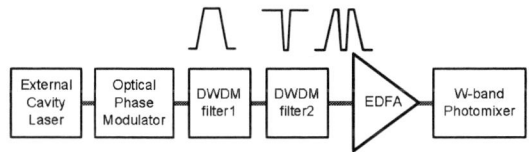

Fig.1: Experimental setup

For the experiment, a tuneable microwave source from 18.75GHz to 27.55GHz is applied to the phase modulator, resulting in fourth harmonic generation covering the whole WR-10 band (75 GHz to 110 GHz). The laser wavelength is adjusted so that it is at the centre of both filters. This ensures equal amplitude for the upper lower 2nd sidebands that are selected by the filters.

In such a configuration, the selected sidebands propagate through the same optical fibre and optical components with no significant polarization mode dispersion (PMD). This common path configuration eliminates the phase drift caused by differential path length changes, and results in a stable MMW phase. Furthermore, the polarization matching of the two sidebands

is maintained as they propagate in the fibre, so there is no need to control their relative polarizations.

Experimental results

Fig. 2. shows the transmission profile of the cascaded DWDM thin film filters. In the profile, the two transmission peaks are separated by around 1 nm and each has a passband of around 0.25 nm.

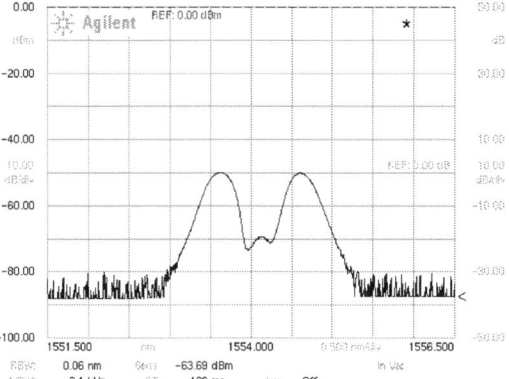

Fig. 2: Optical response of cascaded band-pass and a band-stop DWDM thin-film filters

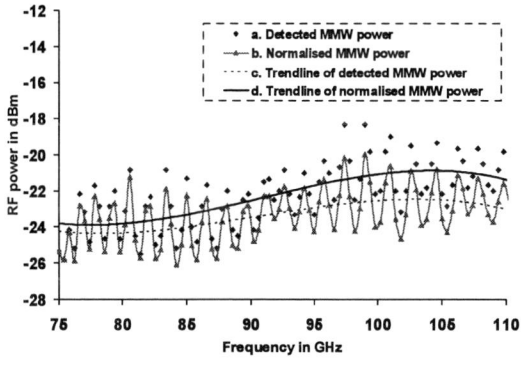

Fig. 3: Generated MMW power vs. frequency

Fig. 3 shows the variation of the power of the generated MMW signal as a function of frequency from 75GHz to 110GHz. Line a. and c. show the detected MMW power and its trendline, respectively. It is observed that across the frequency band, the photocurrent fluctuates a bit. Line b. and d. show the MMW power after the normalising to the corresponding photocurrent and its trendline. The ripples with a period of around 1 GHz in the detected MMW signal power actually come from ripples in the

561

response of the particular photomixer and are not due to the heterodyning process of the filtered sidebands of the phase modulated signal. The trendline d. has a peak to peak deviation of 3 dB across the W-band. It can be concluded that the technique is suitable for generating a MMW LO signal covering the W-band from 75 to 110 GHz.

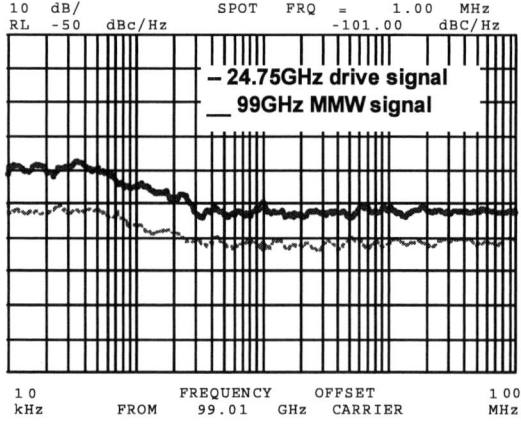

.a.

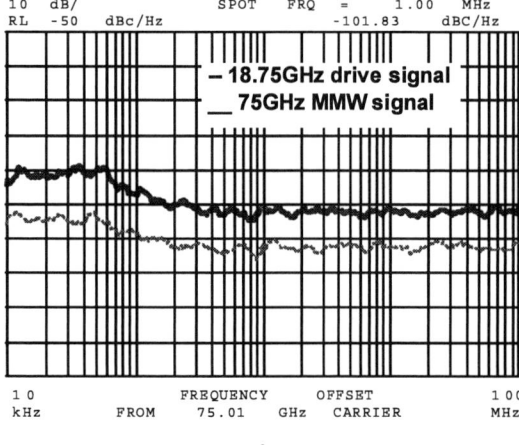

b

Fig. 4 Measured phase noise of the MMW signals and of the corresponding drive signal at 75GHz (a) and 99GHz(b).

Fig. 4. a and 4.b depict the detected phase noise of a 75GHz and 99GHz signals respectively. It can be concluded that the phase noise is around -95dBc/Hz at 100kHz offset from the carrier, whereas the phase noise floor is at around 101dBc/Hz (limited by the sensitivity of the measurement system). The inserted dashed lines are the phase noise of the corresponding drive

signal when it is attenuated to a similar power level to the MMW signals. The detected phase noise level is around -107dBc/Hz at 100 kHz offset. In both plots, it is clear that the signal phase noise has degraded by around 12 dB compared to the drive signal, which follows the theoretical prediction of the increase in the phase noise by 20LogN (dB) for the Nth harmonic. This comparison shows that the phase noise of the generated MMW signal is not affected by the increase in the frequency and there is no additional phase noise caused by the generation system.

Conclusion
A low cost MMW photonic LO generator based on sideband filtering of an externally phase modulated signal is presented. Utilizing a novel notch filter consists of two thin film filters cascaded together, the system benefits from having all the optical spectral components passing through a common optical path, therefore avoiding the large phase deviations due to the differential path length changes. This configuration also has the advantage of eliminating the polarization controller in matching the polarization of the two selected sidebands, therefore present a much simpler and stable system. It has been demonstrated successfully the generation of a continuously tuneable W-band MMW signal, with a phase noise of -95dBc/Hz at 100 kHz offset from the carrier.

References
1 J.J. O'Reilly et al, Phil. Trans. R. Soc. Lond, (2000), 2297-2307
2 P. Shen et al, Int. Topical Mtg on Microwave Photonics, 2003, 189-192
3 Guohua Qi et al, J. Lightwave Technol., Vol 23, (2005), 2687-2695
4 J. Menders et al, Proc. SPIE, Vol.4112, (2000), 91-100
5 G.Qi et al, Proc. SPIE, 5579, (2004), 673-679
6 H. Suzuki et al, Electronics Lett, 41, (2005), 355-356
7 P. Shen et al, Int. Topical Mtg on Microwave Photonics (2002), pp. 101 –104,
8 P.G. Huggard et al, IEEE Photon. Technol. Lett.,14, (2002), 197-199,

An Ultrasmall Polarization Rotator Based on Si Nanowire

Zhechao Wang, Daoxin Dai

Centre for Optical and Electromagnetic Research, Zhejiang University;
Joint Research Centre of Photonics of the Royal Institute of Technology (Sweden) and
Zhejiang University, Zijingang campus, Hangzhou 310058, PR China.
Email: zcwang@coer.zju.edu.cn

Abstract *An ultrasmall polarization rotator based on asymmetrical Si nanowires is presented. Almost 100% polarization rotation is achieved with a very small beat length (~10μm). And a broad 3-dB bandwidth (~120nm) is obtained.*

Introduction

As the demand of higher bit rates in the next generation of optical communication systems increases, it is becoming more and more important to develop advanced polarization-manipulating devices [1]. Among these devices, a polarization rotator is one of the most essential components. Various methods have been developed to realize polarization rotation, such as using electro-optical or thermal strain effects [1]. However, a passive polarization rotator is preferred due to the relatively easy fabrication and the cost-effectiveness.

Several types of passive rotator have been presented, e.g., rotators with periodic asymmetrical loaded sections, rotators with cascaded bend sections and rotators with slante sidewalls. However, The total length of these rotators is still about several hundred microns. For the sake of a high integration density, micro-scale polarization rotators are desirable. In this paper, we propose an ultrasmall polarization rotator based on Si nanowires with an asymmetrical cross section. A vertical sidewall is used, which is different from those given in ref. [2] and can be fabricated by using the process of dry etching. Due to the strong confinement and the asymmetry of the specific Si nanowire, the propagation constant difference between these two modes is very large, which helps to reduce the beat length L_π greatly (according to the expression of the beat length $L_\pi = \pi / (\beta_0 - \beta_1)$, where β_0 and β_1 are the propagation constants of the fundamental and the first order modes, respectively).

Analysis and Design

Fig. 1 shows the cross section of the asymmetrical Si nanowire as well as the schematic configuration of the present polarization rotator (see the inset). By etching a corner of the Si nanowire, the asymmetry of the structure is introduced to realize the polarization rotation. H and W denote the total height and the total width of the silicon Si nanowire, respectively. $H =$ 500nm and W=500nm are fixed in the following analysis. H_e and W_e denote the height and the width of the etched corner, respectively. The wavelength λ_0=1550nm is considered in the rest of this paper. The refractive indices of SiO_2 insulator layer and Si guiding layer are n_1=1.46 and n_2=3.45, respectively.

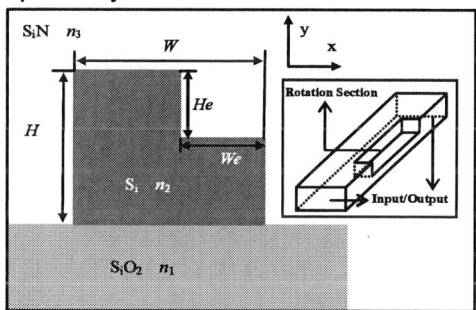

Fig. 1 Cross section of the asymmetrical Si nanowire for the polarization rotation section. Inset: the schematic configuration of the present polarization rotator.

In the following analysis, the finite-element-method (FEM) [3] is used, since with triangular elements, it is more adaptive and gives a more accurate calculation for the field distributions at the discontinuities (such as the corners).

Here, we consider the case of the polarization conversion from TE-polarization to TM-polarization (the two foundation modes of the input Si nanowire) , and define a feature i.e., the polarization conversion efficiency (PCE). It is given as PCE=P_{TM} / $(P_{TM}+P_{TE})$ ×100%, where, P_{TM}, P_{TE} are the

978-0-9789217-3-6/07/$25.00

©2007 WEN GLOBAL SOLUTIONS   563

powers of the two output polarized modes, respectively. PCE indicates the percentage of power transferred to the orthogonally polarized mode from the polarized input field. From ref. [2], the equation of PCE is showed as follows:

$$PCE = \sin^2 2\theta \sin^2(\frac{\pi L}{2L_\pi}) \times 100\%, \quad (1)$$

where θ is the polarization rotation angle, L_π is the beat-length, and L is the actual length of the polarization rotator section. Obviously, for a single section rotator, the polarization rotation angle $\theta=45°$ is required to realize 100% polarization rotation. There are several ways to define the optical axis rotation angle θ [4,5]. Here we choose the following definition [4]:

$$\tan(\theta) = R = \frac{\iint_\Omega n^2(x,y) \cdot H_x^2(x,y)dxdy}{\iint_\Omega n^2(x,y) \cdot H_y^2(x,y)dxdy} \quad (2)$$

where R is the polarization rotation parameter, $n(x, y)$ is the refractive index distribution, $H_x(x, y)$ and $H_y(x, y)$ are the transverse and horizontal components of an eigenmode, respectively. Since the two lowest order modes are orthogonal, it is easy to obtain $\theta=\tan^{-1}(R_1) = \tan^{-1}(1/R_2)$ (R_1 and R_2 are the rotation parameters of the two lowest order modes). In the rest of this paper, only R_1 is considered. As analyzed above, for a single section polarization rotator, the geometrical parameters (such as the etching width W_e and etching height H_e) should be optimized to make $R=1$.

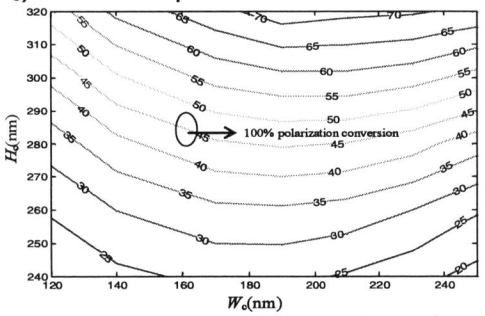

Fig. 2 Optical axis rotation angle θ as a function of the etching width W_e and etching height H_e. The cladding layer index $n_3=2.36$

Fig.2 shows the optical axis rotation angle θ as a function of the etching width W_e and etching height H_e. SiN is chosen as the material for the cladding layer, whose refractive index could be controlled in a wide range [6], and we choose $n_3=2.36$. On the 45° contour line in fig.2, the optimal

parameters of the asymmetrical Si nanowire are chosen as follows: W_e=170nm, H_e=280nm. FEM analysis shows that the PCE>99.95%, and the two fundamental modes of the rotation section have almost the same amplitude distribution of transverse components H_x, H_y, which indicates $R\approx1$ (θ=45°). If an x-polarized field is used as the input field, both modes will be excited. These two excited modes will interfere and there is a beat length $L_\pi=\pi/(\beta_0-\beta_1)$. After a propagation length of L_π, the propagation mode will be y-polarized. For the present case, the effective refractive indices of the two lowest order modes are N_{eff0}=2.884 and N_{eff1}=2.849, and the corresponding beat length L_π=22.1μm. This beat length is much smaller than the size of the conventional single section polarization rotator [2] and thus it is useful for realizing an ultra-high integration density.

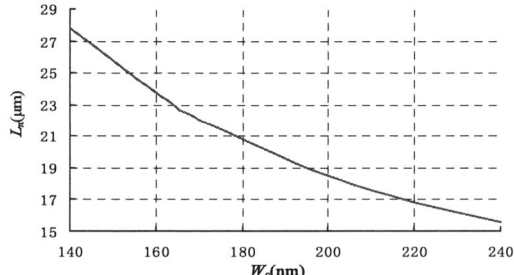

Fig. 3 Beat-length L_π as a function of etching width W_e

Fig. 3 shows the beat length as a function of the etching width W_e. It shows that the beat length decreases and becomes almost stable as the etching width W_e increases. When W_e>220nm, the beat length L_π stabilizes to about 15μm. However, at the junction between the input/output section and the rotation section, there is a coupling loss due to the mode mismatching, and it increases as the etching width increases. When W_e>220nm, the loss becomes larger than 0.3dB. Thus, there is a tradeoff between the beat length and the coupling loss. Furthermore, a connection taper could be used to reduce the coupling loss.

We note that the beat length L_π can be further reduced by choosing a smaller cladding refractive index n_3 since the difference $\beta_0-\beta_1$ is larger. For example, we use a SiO$_2$ cladding (e.g. n_3=1.46) and choose the other parameters as follows: W_e=240nm, H_e=240nm. The FEM analysis shows that the optical axis $\theta \approx 45°$, and $L_\pi \approx 7.2 \mu m$. By using 3D FDTD method (Δx=15nm, Δy=15nm, and Δz=25nm), we simulate the propagation of E_x and E_y fields in the designed polarization rotator as shown in fig. 4. From fig. 4(a), one sees that the x-polarized input field (e.g. E_x) fades over the propagation distance while in fig.4 (b), the y-polarized field gathers head. At the conversion distance $L = L_\pi$, the x-polarized input field disappears and almost all of the power is transferred to the orthogonal polarized mode.

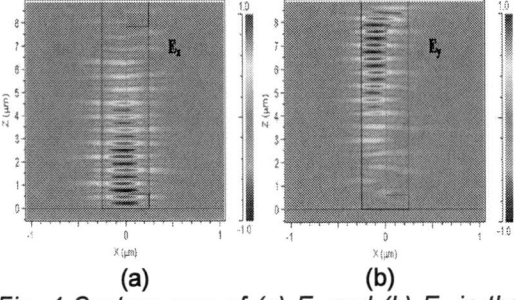

(a)   (b)

Fig. 4 Contour map of (a) E_x and (b) E_y in the polarization rotation section when W_e=240nm, H_e=240nm, n_3=1.46.

In order to achieve a 100% polarization rotation, single mode operation is required to avoid any undesirable multimode interference. However, in order to realize single mode operation, the geometry of the Si nanowire is required to be so compact that the fabrication tolerance is very small, which makes it hard to realize. The parameters we choosed in fig.2 is not strictly single mode operation. Vectorial integral between the fundamental mode of the input Si nanowire and the fundamental modes of the asymmetrical Si nanowire shows that more than 95% of the input power is transferred into the fundamental modes of the asymmetrical Si nanowire. And by using an FDTD analysis, most of the high order modes leak out along the propagation direction. At the end of the rotation section, the extinction ratio between the fundamental and high order modes is high. Thus, the degradation of the performance is small. About 0.4dB insertion loss is introduced because of the non-single mode operation, however, it is worthwhile to expand the fabrication tolerance.

It is well known that the fabrication tolerance of Si nanowire is usually a tough issue. We determine the fabrication tolerance of the present device with a

condition of PCE>90%. Simulation shows that the fabrication tolerance of the etching width is about ±30nm. And the performance is even more sensitive to the variation of the etching height. Consequently one has to control the core width and the etching depth very precisely.

By calculating the values of PCE as a function of the wavelength, we explore the wavelength dependence of the present polarization rotator. The 3dB bandwidth is about 120nm. One has PCE >97%, when the operating wavelength is in the range of 1490nm-1600nm,. Thus the wavelength-insensitive polarization rotator we present is suitable for the wavelength-division-multiplexing (DWDM) systems.

Conclusions

In this paper, we have presented an ultrasmall polarization rotator based on asymmetrical Si nanowires. Numerical simulation has shown that the PCE>99.95% is achieved by optimizing the etching width W_e and etching height H_e. Because of the strong confinement and asymmetrical geometry of the Si nanowire, the beat length of the present polarization rotator is as small as about 22.1μm, which is beneficial to realize ultra-high integration density. We also explore the wavelength dependence of the present polarization rotator, and obtain a broad 3-dB bandwidth (~120nm).

Acknowledgement

The work is supported by the National Science Foundation of China (Nos. 60688401 and 60607012) and research grants of Zhejiang Province (Nos. 20061343 and 2006R10011).

References

[1] R. Alferness, et al., IEEE J. Quantum Electron. QE-17, 946–959 (1981).

[2] H. Deng, et al., IEEE J. Lightwave Technol., vol. 23, no. 1, Jan 2005.

[3] N. Somasiri, et al., J. Lightw. Technol., vol. 20, no. 4, pp. 751–757, Apr. 2002.

[4] V. P. Tzolo, et al., Opt. Commun., vol. 127, pp. 7–13, Jun. 1996.

[5] B. M. A. Rahman, et al., J. Lightw. Technol., vol. 19, no. 4, pp. 512–519, Apr. 2001.

[6] T. Fujisawa, et al., IEEE Photon. Technol. Lett., vol. 18, no. 11, Jun 1, 2006

Bistable Device based on the Kerr Effect in a Microfiber Resonator

G. Vienne (1), Ph. Grelu (2), Y. Li (1), X. Chen-Perdereau (2), L. Tong (1)

1 : Department of Optical Engineering, Zhejiang University, 310027 Hangzhou, China
guillaumevienne@gmail.com

2 : ICB, UMR 5209 CNRS, Université de Bourgogne, 21078 Dijon, France
philippe.grelu@u-bourgogne.fr

Abstract *We propose a bistable device based on the Kerr effect in a microfiber resonator. Our simulations show that low switching powers (in the order of a few tens of mW) are expected with tellurite microfibers.*

.

Introduction

Bistable optical devices are among the key elements needed for the development of optical data processing. Bistability requires an adequate conjunction of feedback and nonlinearity. A prerequisite for short switching times is a fast nonlinear response. The Kerr effect is particularly suitable due to its femtosecond response time. Recirculation in a resonator is favorable to low switching powers but it may limit the response time. Another important consideration is the competing thermal effects. Minimization of thermal nonlinearity relies on optimizing the resonator geometry in order to efficiently dissipate the heat produced by light absorbed in the resonator, as well as on using processes faster than the thermal response time. A glass microsphere providing a Q factor as high as 10^9 can only provide a response time in the order of a microsecond. In addition, to demonstrate Kerr-based optical bistability in microspheres, the use of a superfluid helium cryogenic environment was necessary [1]. The moderate Q-factor of microfiber resonators, typically in the order on 10^4 [2], provides a subnanosecond response time, and we will show that this fast response time does not always come at the expense of high switching powers because of the high confinement and the large nonlinear refractive index possible in glass microfibers. Moreover, the fiber geometry offers superior heat dissipation. Finally, as microfiber resonators can be compatible with the use of nanosecond pulses, each nanosecond pulse can be processed individually, which is crucial for optical data processing. Nanosecond pulses are also more immune to Brillouin scattering, which limits the efficiency of long fiber resonators. Indeed, previous bistability experiments in meter-long resonators used short pulses to enhance nonlinear effects while reducing Brillouin scattering [3,4].

Optical bistability in a microfiber resonator

The geometry of the microfiber resonator and the related notations are presented in Fig. 1. Kerr-based optical bistability in a knot resonator has been studied theoretically in Refs. 5 and 6. However, these studies did not consider resonance optimization which takes place for critical optical coupling, nor did they consider the role of critical detuning.

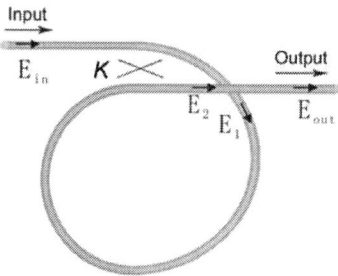

Figure 1. Skematic of a microfiber resonator indicating the electric field notation. K is the cross-coupling coefficient.

The manifestation of bistability requires a negative detuning with respect to the low power resonance wavelength. This detuning should be, in absolute value, larger than the critical detuning. For low input powers, the cavity is non resonant. However, in a low-loss knot resonator, since there is no

978-0-9789217-3-6/07/$25.00

©2007 WEN GLOBAL SOLUTIONS

reflected wave, the output intensity remains close to the input intensity. When the input power increases significantly, the nonlinear phase shift pushes the cavity towards resonance. It results in an enhanced intracavity field, which increases the net amount of power losses, so that the output power tends to drop abruptly, but stabilizes at a value which corresponds to a positively detuned cavity. This process is shown by the red curve of Fig. 2. Decreasing the input power allows reaching the resonance, for which the output power vanishes, if the critical coupling condition is met. Further reducing the input power, the reduction of nonlinear phase shift pushes quickly the cavity out of resonance, so that the transmission increases abruptly since cavity losses are no longer enhanced. This corresponds to following the green (backward) curve of Fig. 2. Between the switching powers P_{in}- and P_{in}+, the microresonator can act as a bistable device.

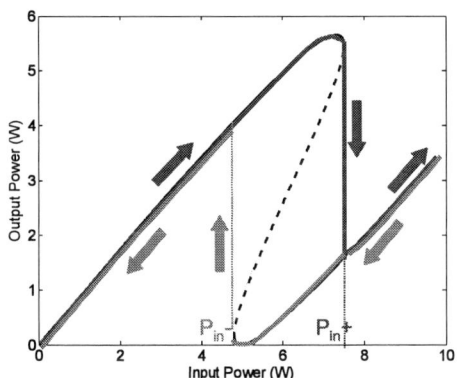

Fig. 2 Illustration of the bistable behavior. The instable branch is in dotted line. The switching powers are denoted as P_{in}- and P_{in}+. The calculation is performed for a silica microfiber resonator, at 1.6 times critical detuning. The fiber and the resonator are 700 nm and 3 mm in diameter, respectively.

Switching Power Requirements

To give a starting point, Figure 2 was computed for a 3-mm long resonator made of a silica fiber 700 nm in diameter.

We see that in that case the switching powers are several watts, which may be practical. The design of a microresonator based on a highly nonlinear material maybe necessary. Among the oxide glasses,

tellurite glasses offer third-order nonlinear coefficients one order of magnitude above that of silica. The larger linear index of refraction of tellurite glasses (2.0 as opposed to 1.45 for silica) also has the beneficial effect of increasing the field confinement. Computations of bistability curves for a tellurite microfiber knot resonator are shown in Fig. 3. In order to show the importance of the critical detuning, several curves are plotted, corresponding to various cavity detunings, namely 0.8, 1.0, 1.2, 1.4, 1.6, 2.0 times the critical cavity detuning. In practice one should only slightly exceed the critical detuning, in which case the switching power should remain below 100 mW.

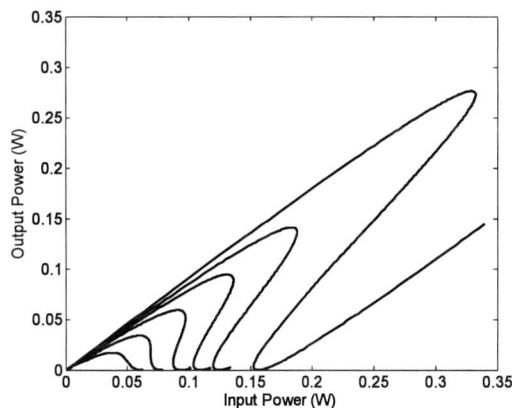

Fig. 3 Curves of input/output power relationships, for 0.8, 1.0, 1.2, 1.4, 1.6, 2.0 times the critical cavity detuning. Only the interesting parts of the five first curves are shown in order to simplify the graph appearance.

Conclusions

We have proposed a new configuration for the realization of an optical bistable device. The microfiber resonator geometry, with its high surface to volume ratio and short cavity length, is particularly suitable to limit both the thermally induced phase shift and the competition from non-linear effects other than the Kerr effect, such as the Brillouin effect. Switching power requirements are estimated to be in the order of several Watts for silica, whereas they could be maintained below 100 mW for tellurite fibers. The switching powers could be further reduced with the use of non-oxide glasses, which can provide both higher linear and nonlinear refractive indices, and therefore enhanced nonlinearity.

References

1 F. Treussart et al., Eur. Phys. J. D **1** (1998) 235.
2 X. S. Jiang et al., Appl. Phys. Lett. **88** (2006) 223501.
3 H. Nakatsuka *et al.*, Phys. Rev. Lett. **50** (1983) 109.
4 S. Coen *et al.*, Opt. Lett. **24**, (1999) 80.
5 B. Crosignani *et al.*, Opt. Comm. **59** (1986) 309
6 F. Sanchez *et al.*, Opt. Commun. **142** (1997) 211.

Recent progress in the integration of MGY-based tunable lasers and Mach-Zehnder modulators

P-J Rigole' *Syntune AB, Torshamnsgatan 30A, S-164 40 Kista, Sweden*

1. Introduction

Tunable transmitters that are based on a monolithically integrated laser and modulator offer a compact, efficient, and inexpensive alternative to tunable lasers that are based on the assembly of external cavities or discrete elements. Fully integrated solutions are attractive because of their mechanical robustness, and also because they provide a reduced part count, reduced operational power dissipation, and reductions in packaging costs (principally as a result of fewer active optical alignment steps) when compared with discrete assemblies. Moreover the compactness provided by an integrated chip is critical for deployment within the smallest form factor transmitters, namely XFP transmitters.

We report recent progress on the link transmission performance of a zero chirp Mach Zehnder modulator monolithically integrated with a widely tunable modulated grating Y branch laser and SOA, operating at 10 Gb/s, which was first described in Ref. [1]. The module-packaged device reported here features a modulated output power of +4.5 dBm over the C-band [2], which, to our knowledge, is a record for a monolithic tunable laser transmitter. The device is hermetically packaged with a wavelength locker within a 12.7 x 15.24 mm module. With a fixed receiver threshold and fixed voltage modulation depth, the extinction ratio exceeds 12 dB and the receiver penalty for dispersion in the range of ±800 ps/nm is below 1.5 dB on all channels. Simulations of transmission performance and corresponding transmission measurements show good agreement.

2. Device Description and Determination of Operational Settings

The MG-Y laser is compatible with rapid tuning for agile networks through the use of injection-current-based tuning elements, in contrast with much slower thermal tuning based approaches. The MG-Y laser consists of a gain section, a phase tuning section, and a multi-mode interferometer (MMI) connected to two reflectors having modulated gratings. A wide tuning range is obtained by utilizing the additive Vernier effect through the relative tuning of the two reflectors, each with a comb-like reflectivity spectrum. Static characteristics of MG-Y lasers were presented in Ref. [2]. This device has been integrated monolithically with an SOA and a MZ modulator [1][2].

The splitters and combiners in the MZ modulator are constructed using the same type of MMI as is used in the MG-Y laser. The MZ active region consists of a core of sixteen InGaAsP quantum wells and barriers that are designed to achieve efficient phase modulation using the quantum-confined Stark effect. The device has been hermetically packaged with an isolator and an internal, etalon-based wavelength locker. The hermetic case has an area of only 12.7 x 15.24 mm, which is, to our knowledge, the most compact implementation of full-band tunable transmitter module achieved to date. A plot of the MZ facet output power vs. the voltage delta between the arms for several wavelengths over the tuning range is given in Figure 1.

The MZ DC bias settings over the C-band are obtained from simple fits to set-points that are identified for a small subset of channels. The tunable laser set-points are separately obtained from a fully automated scan. After the MZ and laser set-points have been independently obtained, there are no subsequent adjustments required to ensure that all channels meet SMSR and dispersion penalty specifications. The ability to generate the laser and MZ set-points independently indicates that there is no significant optical, thermal, or electrical cross talk within the chip.

978-0-9789217-3-6/07/$25.00

©2007 WEN GLOBAL SOLUTIONS

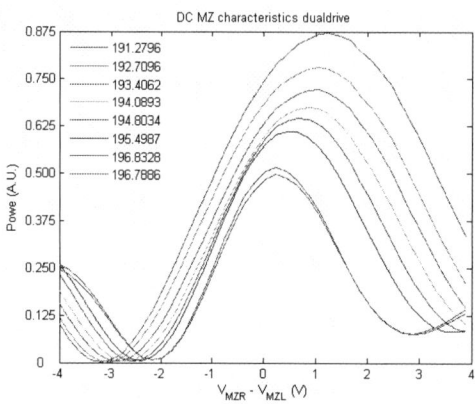

Figure 1: MZ DC transmission curves for optical frequencies in the C-band.

3. Link tests

A 9.95328 Gb/s PRBS pattern (pattern length 2^{31}-1) was applied to the modulator via an OKI KGL4126HA differential driver having a peak-to-peak voltage swing of 2.9 V on each arm. With a bias of 152 mA on the SOA and 98 mA on the laser gain section, the average modulated fiber-coupled power was >4.5 dBm over the tuning range. The optical signal was transmitted through a dispersive fiber span (with either negative, zero, or positive dispersion) on its way either to a sampling oscilloscope or to a receiver via a variable attenuator and a power meter. Back-to-back and transmitted eye diagrams were recorded with a Tektronix CSA8000 sampling oscilloscope after dark level calibration. For all wavelengths, the ER of the filtered back to back eyes exceeded 12 dB.

BER curves were obtained with an NEL 10 Gb/s optical receiver having a sensitivity at BER = 10^{-12} of approximately -16.6 +/- 0.4 dBm, depending on the back-to-back ER, operating wavelength, and optical crosspoint, which was held between 45% and 55%. The electrical decision threshold reference voltage was set to -55 mV and was held constant for all wavelengths and all dispersion cases. The link test was conducted at 0 km, over 45 km of dispersive fiber, or through one or two negative dispersion modules of -550 ps/nm each. The dispersion of the 45 km span varies with wavelength between 709 and 832 ps/nm, while for the -1100 ps/nm module combination, the wavelength dependence results in dispersion in the range of -1056 to -1181 ps/nm. The BER curves that were obtained are displayed in Figure 2..

The simulated and measured receiver penalty vs. dispersion for different wavelengths is displayed in Figure 3. The MATLAB-based simulation program is described in Refs [1-2]. Within the dispersion range ±800 ps/nm, the receiver penalty (at BER = 10^{-12}) vs. back to back transmission is below 1.5 dB for all wavelengths. The simulated penalty vs. dispersion curve is qualitatively similar to the corresponding measured curve. The penalty-dispersion curves for different wavelengths differ in the sense that the short wavelength curve has a negative penalty for -548 ps/nm. This indicates that the modulator introduces a small amount of positive chirp at that operation point, which enhances transmission over negatively dispersive fiber. It is possible to adjust the modulator settings to achieve less chirp and, hence, a more symmetric penalty characteristic.

4. Conclusion

We have demonstrated a MZ modulator monolithically integrated with a widely tunable MG-Y laser and SOA that delivers over +4.5 dBm of average fiber-coupled modulated power, a dynamic ER exceeding 12 dB, and less than 1.5 dB dispersion penalty at 10 Gb/s for link dispersion in the range of -800 to +800 ps/nm, under the constraint of operation with a constant RF amplitude and fixed receiver settings. Transmission performance for zero-chirp operation is in good agreement with a MATLAB-based link model. The integrated chip enables inexpensive packaging within a hermetic module that features record compactness (12.7 x 15.24 mm). The module performance and footprint are highly suitable for the deployment of flexible wavelength provisioning within dispersion-managed long-haul transmission systems.

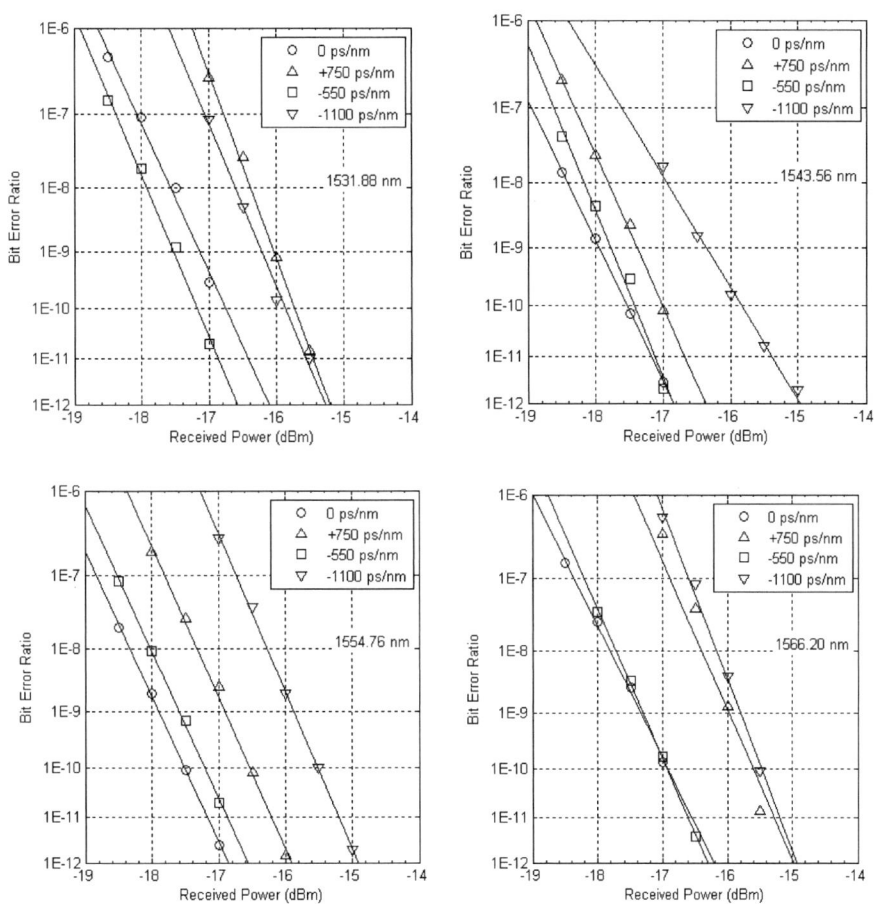

Figure 2: BER vs. received power at 1531.88 nm, 1543.56 nm, 1554.76 nm and 1566.20 nm, for 0 ps/nm and for links having a nominal dispersion of -1100 ps/nm -550 ps/nm, and +750 ps/nm.

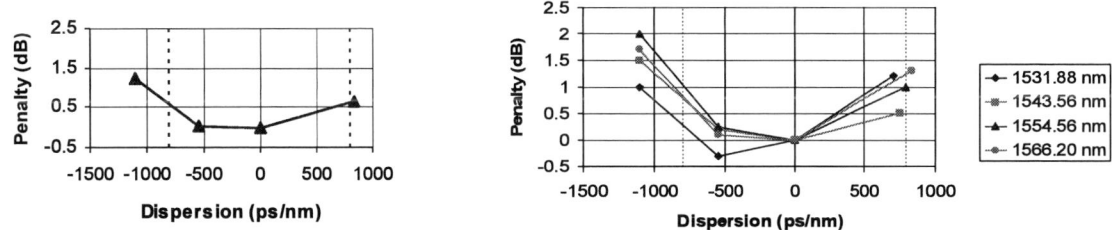

Figure 3: Penalty vs. dispersion for simulated 1554 nm (left) and measured four different wavelengths (right). Dashed lines indicate ±800 ps/nm.

References

[1] Adams, D.M.; Isaksson, M.; Wesström, J.-O.; Eriksson, U.; Hammerfeldt, S.; Stoltz, B.; Granestrand, P.; Goobar, E.; Lewén, R.; Rigole, P.-J.; Bardyszewski, W., "Transmission Performance of a Monolithically Integrated Y Branch Tunable Laser with a Zero Chirp Mach Zehnder Modulator", Elec. Letters, Vol. 43, No. 9, April 2007, pp 522-524

[2] Isaksson, M.; Adams, D.M.; Wesström, J.-O.; Eriksson, U.; Hammerfeldt, S.; Stoltz, B.; Granestrand, P.; Goobar, E.; Lewén, R.; Rigole, P.-J.; Bardyszewski, W., "Zero-Chirp Transmission Performance of a Mach Zehnder Modulator Monolithically Integrated with a Modulated Grating Y-Branch Tunable Laser", NOC 2007 Proceedings, pp. 539-544.

[3] Wesstrom, J.-O.; Sarlet, G.; Hammerfeldt, S.; Lundqvist, L.; Szabo, P.; Rigole, P.-J., "State-of-the-art performance of widely tunable modulated grating Y-branch lasers", OFC 2004 Digest, Vol. 1, Paper TuE2

Wavelength and Space Switchable Semiconductor Laser

Jian-Jun He

Centre for Integrated Optoelectronics,
State Key Laboratory of Modern Optical Instrumentation,
Zhejiang University, Hangzhou, China 310027
jjhe@zju.edu.cn

Abstract *A novel monolithic integrated wavelength and space switchable laser is presented. It consists of two V-coupled cavities with optimized coupling coefficient and phase for high single-mode selectivity. The waveguide branch structures inside and outside the cavities allow the wavelength and space switching functionalities to be realized simultaneously.*

Introduction

As one of the key enabling technologies for next generation intelligent optical networks, low-cost and reliable wavelength switchable semiconductor lasers are of great interest. For example, the combination of wavelength switchable lasers with wavelength routers can provide large format-independent space switches and reconfigurable optical add/drop functions.

A conventional tunable semiconductor laser consists of a multi-electrode distributed Bragg reflector (DBR) laser with a phase shift region [1][2]. When the reflectivity peak wavelength of the DBR grating is tuned by injecting a current, the phase shift region must be adjusted simultaneously in order to prevent the laser from mode-hopping. The tuning range of such a laser is limited to about 10nm due to the limitation of commonly achievable refractive index change in semiconductor materials.

More sophisticated tunable lasers with wider tuning ranges have been developed in the forms of sampled grating distributed Bragg reflector (SGDBR) laser [3][4], superstructure grating (SSG) DBR laser [5], grating-assisted codirectional coupler (GACC) laser [6] and modulated grating Y-branch (MG-Y) laser [7]. In addition to the fabrication complexity involving non-uniform gratings and multiple epitaxial growths, they usually use at least four electrodes for wavelength tuning and power control. Complex electronic circuits with multi-dimensional current control algorithms and look-up tables are required. Such complexities reduce the fabrication yield and operational reliability, and increase the cost. These problems have been the major roadblocks to the wide deployment of tunable lasers in commercial systems.

A widely tunable or wavelength switchable laser can also be realized by using two serially coupled cavities of slightly different lengths. The tuning range can be greatly increased by using the Vernier effect. The coupled-cavity laser can be fabricated either by etching a groove inside a cleaved Fabry-Perot laser [8], or by using a cleaved-coupled-cavity structure [9]. However, the mode selectivity of such a coupled-cavity laser is very poor, which results in very limited use in practice.

Coupled-cavity lasers have also been investigated in the form of a Y-laser [10]. The Y-laser has the advantage of being monolithic without the fabrication requirement for gratings or deep etched trenches. However, the mode selectivity of the Y-laser is similar to serially coupled cavities, which is insufficient for stable single-mode operation with high side mode suppression ratio.

In this paper, we present a novel wavelength switchable laser structure using V-coupled cavities. It allows an optimized coupling coefficient and phase between two cavities to be realized to achieve significantly improved single-mode selectivity, while allowing the lasing wavelength to be switched over a wide range using the Vernier effect. It also allows the laser output to be switched between two output ports.

Device structure and operation principle

Fig. 1 shows the schematic of the monolithically integrated wavelength and space switchable laser based on two V-coupled cavities. The two cavities comprises waveguides of slightly different optical lengths with deeply etched trenches on both ends as partially reflecting mirrors.

978-0-9789217-3-6/07/$25.00

©2007 WEN GLOBAL SOLUTIONS

The two waveguides are disposed side-by-side on the substrate to form V-shaped waveguide branches with a predetermined cross-coupling at the closed end and no cross-coupling at the open end. The length of one of the cavities (i.e. the fixed gain cavity) is chosen so that its resonance frequency interval Δf matches the spacing of the operating frequency grid (e.g. 100GHz spacing as defined by ITU). The second cavity (i.e. the channel selector cavity) has a slightly different optical length and thus slightly different resonance frequency interval Δf '. It has two sets of electrodes deposited on the waveguide, one for injecting a fixed current to produce an optical gain and the other for applying a variable current (or voltage) to change the effective index in order to switch the laser wavelength. The waveguide segment of variable index is disposed away from the coupling region so that its refractive index change does not affect the coupling coefficient between the two cavities. The light emitted from the two waveguide branches at the open end is coupled to an output port by a 2x2 MMI coupler (or a directional coupler). An electrode is deposited on the space switching section outside the cavity to adjust the phase and to switch the laser from one output port to another. On the other side of the laser, a tapered combiner can be used for power monitoring.

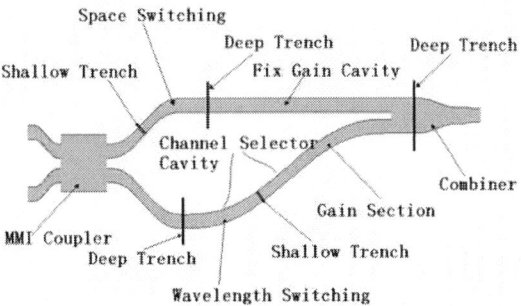

Fig. 1: Schematic of monolithically integrated wavelength and space switchable laser

The laser frequency is determined by the resonant peak of the fixed gain cavity that coincides with a resonant peak of the channel selector cavity. A shift of $|\Delta f - \Delta f'|$ in the resonant peaks of the channel selector cavity results in a jump of a channel interval Δf in the laser frequency. Therefore, the laser wavelength can be digitally switched using a single electrode control. The change of the laser frequency is amplified by a factor of $\Delta f / |\Delta f - \Delta f'|$ with respect to the resonant frequency change of the chanel selectror cavity. This increased tuning range is the advantage of the Vernier effect similar to the case of a SG-DBR laser.

Consider an example in which Δf =100GHz, and Δf ' =90GHz. The range of the laser frequency variation is increased by a factor of 10 with respect to what can be achieved by the index variation directly. For this numerical example, assuming the effective group refractive index of the waveguide is 3.215, the lengths of the fixed gain cavity and the channel selector cavity are L=466.24µm and L'=518.31µm, respectively. The cavity lengths are therefore comparable to conventional DFB or FP lasers.

Optimization of mode selectivity

The mode selectivity, which can be characterized by the threshold difference between the lowest threshold mode and the second lowest threshold mode, is a critical parameter in the design of the laser. The mode selectivity of the V-coupled cavity laser (VCCL) can be optimized by appropriately designing the coupling structure to achieve a certain coupling coefficient and phase. A typical coupling structure is a MMI coupler folded with respect to the deep trench which acts as a mirror. The optimal phase difference between the cross-coupling and self-coupling is a multiple of π. Fig. 2 shows the threshold difference as a function of the cross-coupling coefficient when the phase difference is optimum. We can see that a cross-coupling coefficient around 0.1 gives the largest threshold difference.

By increasing the length difference between the fixed gain cavity and the channel selector cavity, the threshold difference between the lowest threshold mode and the second lowest threshold mode can be increased. For example, if the length of the channel selector cavity is increased to L' = 570.38µm (Δf ' = 81.7GHz) in the above example, the maximum achievable threshold difference is increased from 7.4 cm^{-1} to 14.5 cm^{-1}, and the optimum cross-coupling coefficient at which the maximum threshold difference occurs is increased from 0.1 to 0.26.

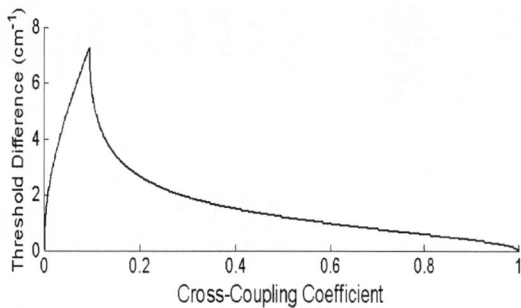

Fig. 2: Threshold difference as a function of the cross-coupling coefficient for V-coupled cavity laser

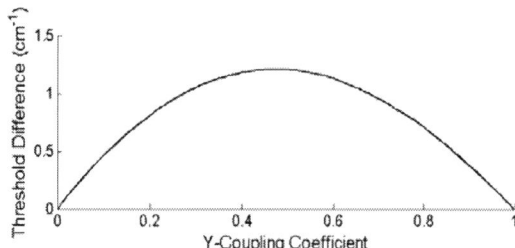

Fig. 3: Threshold difference as a function of the Y-coupling coefficient for Y-laser

It is also found that the maximum threshold difference is achieved when the simple round trip gains in the two cavities are equal. The maximum achievable threshold difference decreases as the pumping condition deviates away from this optimal condition. Under this condition, the output powers emitted from the two waveguide braches are also equal, resulting in a high extinction ratio for the space switch outside the cavity. Therefore, it is preferable that gain variations be avoided when the refractive index of the channel selector cavity is changed to switch the laser wavelength. This can be realized by using a substantially passive wavelength switching section with little gain or loss, using etch-and-regrowth method or post-growth quantum well intermixing technique [11]. This also allows flexible output power control independent of the wavelength switching.

An important advantage of the V-coupled cavity laser compared to the previously investigated Y-coupled cavity laser [10] is that the achievable threshold difference is much higher for the same cavity lengths. Fig. 3 shows the threshold difference as a function of the coupling coefficient from the common waveguide segment to one of the waveguide branches of the Y-laser. The largest threshold difference is only 1.2 cm^{-1} and it occurs when the Y-branch is an equal power splitter. This compares to a maximum achievable threshold difference of 7.4 cm^{-1} for the V-coupled-cavity laser of the same cavity lengths. Therefore, the V-coupled cavity laser can produce a much larger side-mode threshold difference and consequently a much higher side-mode suppression ratio.

Conclusions

A novel monolithic integrated wavelength and space switchable semiconductor laser is presented. The active region consists of two V-coupled cavities with optimized coupling coefficient and phase. It is shown that a much better single-mode selectivity can be achieved compared to Y-coupled cavity laser. The waveguide branch structures inside and outside the optical cavities allow the wavelength and space switching functionalities to be realized simultaneously. The digital wavelength switching only requires a single electrode control. The added functionality for switching the laser output port is useful for network protection, for example. Alternatively, the extra-cavity space switch section can also be designed and used as a variable optical attenuator or high speed Mach-Zehnder modulator.

References

1 S. Murata et al, Electron. Lett., 23 (1987), 403.
2 T. L. Koch et al, Appl. Phys. Lett., 53 (1988), 1036.
3 V. Jarayman et al, Appl. Phys. Lett, 60 (1992), 2321.
4 L. A. Coldren, IEEE JSTQE 6 (2000), 988.
5 Y. Tohmori et al, IEEE J. Quantum Electron., 29 (1993), 1817.
6 R. C. Alferness et al, Appl. Phys. Lett., 60 (1992), 3209.
7 J.-O. Wesström et al, OFC'2004, paper TuE2.
8 L. A. Coldren et al, Appl. Phys. Lett. 38 (1981), 315.
9 W. T. Tsang, Semiconductors and Semimetals, 22 (1985), 257.
10 O. Hildebrand et al, J. of Lightwave Tech., 11 (1993), 2066.
11 S. Charbonneau et al, IEEE JSTQE 4 (1998), 772.

Widely Tunable Slow-light Delay Line Using Parametric-amplification Assisted Silicon Microring Resonator

Fangfei Liu (1), Chun Jiang (1), Yikai Su (1)

1: State Key Lab of Advanced Optical Communication Systems and Networks, Department of Electronic Engineering, Shanghai Jiao Tong University, 800 Dongchuan Rd, Shanghai 200240, China, yikaisu@sjtu.edu.cn

Abstract *We propose a widely tunable slow-light delay element based on silicon microring resonator assisted by parametric amplification. This scheme provides flexible adjustment of the delay time and bandwidth to adapt to different data rates.*

Introduction

On-chip slow-light delay line based on silicon-on-insulator (SOI) platform is an important candidate technology for future integrated telecommunication and computer systems due to its small footprint, and compatibility with the IC manufacturing. Several SOI based slow light mechanisms have been investigated experimentally and numerically, including microring resonator [1][2], photonic band gap structures [3] and Stimulated Raman Scattering (SRS) induced slow light [4][5]. Microring resonator based on-chip slow-light devices show advantages in relatively simple structure and large delay compared to other slow-light mechanisms. Another merit of microring resonator is its enhanced nonlinear effect due to the increased intra-resonator intensity and effective path length [6].in InP and GaAs / GaAlAs based microring resonator [7].

Telecommunication applications require that the slow-light elements have large delay-tuning range with variable bandwidth adapted to different data rates while achieving maximum delay. Most of the previous methods realized tunable delay by tuning the resonance spectrum using electro-optic or thermo-optic methods, which increase the manufacturing complexity. Recently, a method to achieve tunable delay based on SRS in microring resonator was proposed [5], however, the operating wavelength is not within telecommunication window.

Here we propose an SOI wire waveguide based microring resonator assisted by the parametric amplification. We find that even a weak round-trip parametric gain is sufficient to compensate the round-trip loss so that the transmission spectrum can be continuously tuned from a deep notch to unit gain, thus the effective bandwidth and the delay can be adjusted according to the data rate of the signal. In addition, we quantify the system performance of the return-to-zero (RZ) signals passing through the slow-light element using a new figure of merit (FOM).

Theory and model

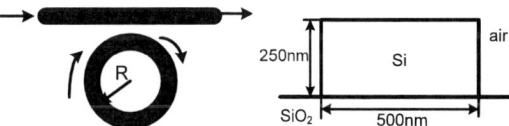

Fig. 1 single side coupled microring resonator and its vertical structure

In our study, we consider a single side coupled SOI-wire based microring resonator with a width of 500 nm and a height of 250 nm (Fig. 1). This waveguide supports a TE mode and the effective mode area is estimated to be 0.12μm^2. The effective index is about 2.54 at 1550 nm and the corresponding group index is 4.024. We assume that free carrier absorption effect is negligible by employing certain techniques such as applying an external electric field or introducing non-radiative centers [8]. A signal and a pump with optical frequencies ω_s and ω_p at the two resonant frequencies of the resonator are launched into the bus waveguide, another wave denoted as idler would be generated at the frequency $\omega_i = 2\omega_p - \omega_s$, which is also at the resonant frequency. The transfer function of a ring resonator can be written as [6]:

$$T(\omega) = \frac{t - a\exp(i\phi)}{1 - t\exp(i\phi)} \qquad (1)$$

where $t = \sqrt{1-r^2}$ is the transmission coefficient; a contains the per-turn loss, parametric gain and two photon absorption (TPA) induced loss, Φ contains both the linear and the amplification-induced nonlinear phase shifts. The typical TPA coefficient and nonlinear index of refraction are about 7×10^{-14} cm^2/W and 0.45 cm/GW. As the pump intensity is much larger than the signal intensity, there is only SPM, TPA effect and linear loss for the pump while the signal experiences additional parametric amplification and XPM induced by the pump.

Design of the widely tunable slow-light element

To design such a tunable slow-light element, some important characteristics of the resonator should be taken into account, including the delay tuning range, the largest achievable bandwidth and the pump power required to tune the gain spectrum from a deep notch to unit gain. These characteristics are determined by the combination of the resonator parameters such as the coupling coefficient r, the circumference length L, and the linear loss of the resonator. As the largest achievable bandwidth is mainly determined by the coupling coefficient r and the circumference length L, we first fix the coupling coefficient r to be 0.35 and the circumference length L to be 130 um in order to obtain a relatively large bandwidth. Therefore the signal and the pump wavelengths are located at two resonances of 1530.7 nm and 1576.8 nm, respectively. We plot the pump power coupled in the ring versus the input pump power by varying the linear loss (Fig. 2). We find that resonator with smaller linear loss has larger intracavity pump power at the same input pump power, which facilitates the generation of parametric gain however limits the maximum delay. We choose the linear loss to be 6/cm in the following simulations. Both the resonant wavelengths of the pump and the signal are red shifted due to the SPM effect and the XPM effect, respectively, which are shown in Fig. 3(a). Fig. 3(b) depicts the typical signal spectrum

when the intracavity power is 6 W, 11 W and 18.68 W, respectively.

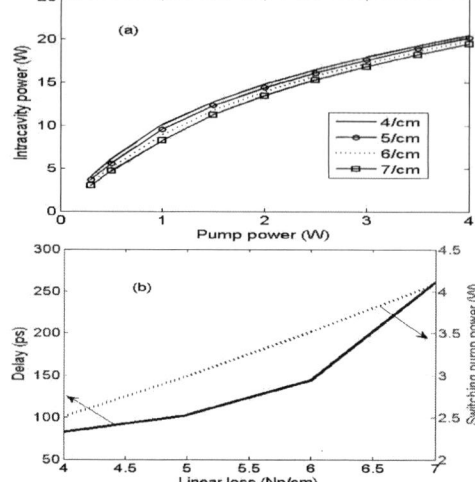

Fig. 2 (a) Intracavity power vs. input pump power when the linear loss is 4/cm, 5/cm, 6/cm and 7/cm. (b) Maximum delay and switching pump power vs. linear loss.

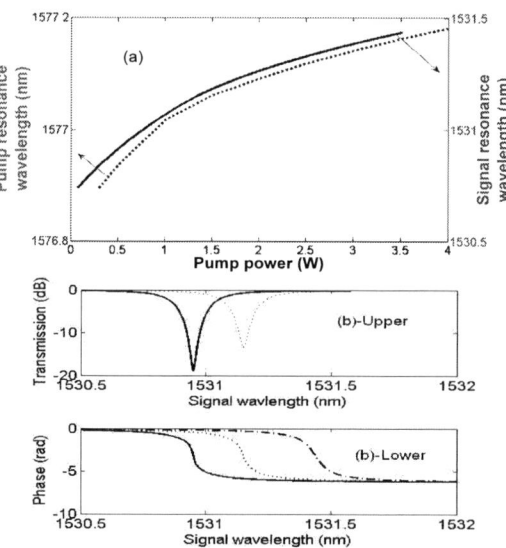

Fig. 3 (a) pump and signal resonance wavelength vs. the input pump power. (b) Signal spectrum when the intracavity power is 6W (solid), 11W (dashed) and 18.68W (dot-dashed).

One can obtain the delay of the slow-light element by deriving the phase-frequency curve. The 3-dB bandwidth of the transmission spectrum is much wider than the delay bandwidth. Therefore, we use the 3-dB bandwidth of the delay-frequency curve to find out the optimum data rate so that the output signal does not suffer much distortion.

576

We plot the maximum delay, the bandwidth and their product versus the input pump power in Fig. 4. We find an increase in the delay when the pump power is low, due to the fact that TPA exceeds parametric gain when the pump power is low and vice versa when the pump power is high. Fig. 4 shows that the the largest delay reaches a maximum of ~150ps when the pump power is ~0.6 W. Further increasing the pump power leads to a saturation of the delay to ~50 ps. The trends of the bandwidth and the delay-bandwidth product variation are opposite to that of the maximum delay. The delay can be tuned from 53 ps with a bandwidth of ~15 GHz to about 150 ps with a bandwidth of ~2 GHz.

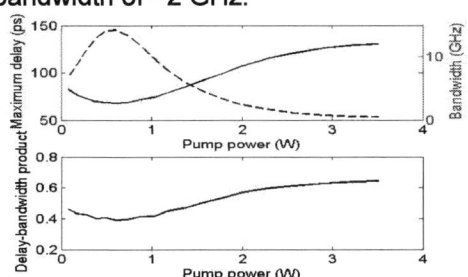

Fig. 4-Upper: Maximum delay and bandwidth vs. the pump power; 4-Lower: delay-bandwidth product vs. pump power.

System performance of the slow-light element

As the parametric-amplification assisted microring resonator is a dispersive medium, the output signal is distorted, including broadening caused by the group velocity dispersion, and asymmetric oscillation caused by the third order dispersion. Here we use a FOM defined as the average difference between the normalized output RZ pulse and the input RZ pulse to quantify the distortion induced by the slow-light element. Fig. 5 plots this FOM and the delay time for different data rates when the intracavity power is 6 W, 11 W and 18.68 W, respectively. The RZ pulses are 33% duty-cycle Gaussian shape with a length of 2^7-1 pseudo random bit sequence (PRBS) as used in [9]. From these curves, one can choose the suitable operating condition (pump power) to satisfy different delay and

signal-quality requirements. For comparison, we also show the relation between the largest accommodated data rate and the delay as well as the corresponding FOM by detuning the resonance spectrum. The advantage of the parametric-amplification assisted tunable slow-light resonator over the detuning method is evident in terms of tuning range and delay time with low distortion especially at high data rate.

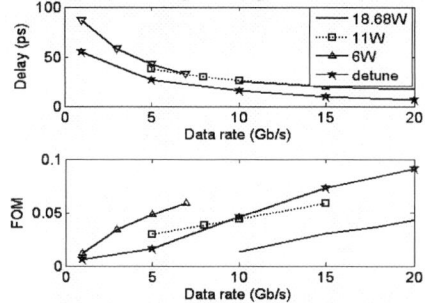

Fig. 5 Delay (upper) and distortion FOM (lower) vs. data rate, and comparison with the method of shifting the signal wavelength.

Conclusions

We have investigated parametric amplification in microring resonator and proposed an optically tunable slow-light element based on this mechanism. Simulations show that large delay ranging from ~18 ps to ~90 ps for data rates from 20 Gb/s to 1 Gb/s with low signal distortion can be achieved.

References

1 F. Xia et al Nature Photonics,1(2007), 65
2 J. E. Heebner et al J. Mod. Opt., 49(2002), 2629
3 H. Gersen et al Phys. Rev. Lett., 94(2005), 073903
4 Y. Okawachi et al Opt. Express, 14(2006), 2317
5 S. Blair et al Opt. Express, 14(2006), 1064
6 Y. Chen et al J. Opt. Soc. Am. B., 20(2003), 2125
7 S. Mikroulis et al PTL, 17(2005), 1878
8 Q. Lin et al. Opt. Express, 14(2006), 4787
9 F. Liu et al. OFC 2007, paper OWB5

Proposal of a thermally-tunable silicon-on-insulator microring resonator filter

Daoxin Dai, Liu Yang

Centre for Optical and Electromagnetic Research, State Key Laboratory for Modern Optical Instrumentation,
Zhejiang University, China

dxdai@zju.edu.cn

Abstract *An ultrasmall microring resonator (MRR) based on silicon-on-insulator nanowires is presented. The metal circuit along the microring is used as a submicron heater. For the fabrication of the present thermally-tunable MRR, only a single lithography process is needed.*

Introduction

Micro-ring resonators (MRR) have received much attention recently because of their compactness, and flexiblility. There are various MRR-based functional components, e.g., add-drop filters [1], optical modulators [2], optical switches [3]. It is becoming more and more important to develop optical components with wide-range tunability [4]. Using the well-known thermo-optical (TO) effect is an effective method to realize tunable components [5, 6]. In this paper We propose an improved design for a thermally-tunable MRR filter based on Silicon-on-insulator (SOI) ridge nanowires, which have a large thermooptical coefficient and the ability of ultra-sharp bending [7-12]. In the present structure, a submicron heater with the same width as the Si nanowire waveguide is included. For the fabrication of the present Si nanowire MRR, one only needs a standard fabrication process with only single lithography process

Structure and fabrication process

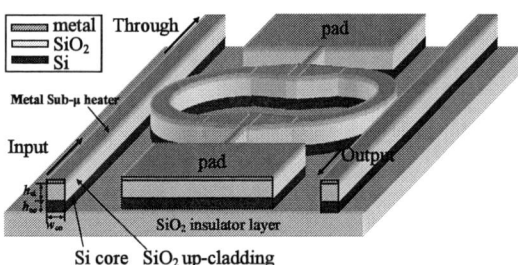

Fig. 1. The 3D view of schematic configuration of the present tunable MRR filter;

Fig. 1 shows the 3D view of the schematic configuration of the present ultrasmall MRR filter with four ports. Two 100μm×100μm pads are placed at the two sides of the microring as the contact points for the probes connecting to the electrical power and each pad is connected to the microring circuit with a T-junction. This excess loss due to the T-junction can be minimized by expanding the singlemode Si nanowire waveguide at the T-junctions with two inversed cosine tapers [13, 14]. The thickness of the SiO$_2$ up-cladding should also be thick enough to minimize the absorption due to the metal heater.

The fabrication procedure is described as follows: (1) a SiO$_2$ up-cladding layer and a metal thin film are deposited on a commercial SOI wafer in sequence; (2) an E-beam (or deep-UV) lithography is implemented to form the photoresistor pattern, which is then transformed to the metal layer by using a process of wet-etching; (3) Use an ICP etching process and then a ridge nanowire waveguide is formed; (4) remove the remained photoresistor and the metal layer left on the top of the microring is used as the submicron heater. For our designed structure, one only needs one photomask in the fabrication for the lithography process (instead of two photomasks needed for conventionally designed TO devices).

Results and discussions

First we consider a quasi-singlemode Si nanowire with a core of 500nm×300nm. The refractive indices for the core and cladding layers are 3.455 and 1.46 at the wavelength of 1.55μm. Here we use a 100nm thick gold heater (with a complex refractive index of 0.55+10i). For the present Si nanowire with a metal heater, a SiO$_2$ up-cladding with a thickness of 450nm is used to reduce the absorption of the metal heater.

For the MRR filter, we choose the parameters as follows. A small bending radius of R=4μm is used and the gap between the microring and the straight waveguide is w_g=100nm to have a good coupling. By optimizing the two reversed cosine tapers at the T-junction, the theoretical excess loss is reduced to about 0.08dB per T-junction.

Our thermal simulation based on thermal-conduction equation shows that the effective index increases by about 0.0435 under the condition of P=10mW. The corresponding wavelength shift $\Delta\lambda$ is about 26nm estimated by the formula of $\Delta\lambda=(\Delta n_{eff}/n_{eff})$.

Fig. 2 shows the calculated spectral response of the TE polarization from the through and drop ports at different heating powers (the TM polarization has similar results). This spectral response is obtained by using the analytical formula given in ref. [15]. When the heating power increases from 0mW to 10mW, the resonator wavelength is tuned from 1557.0nm to 1584.4nm. One sees that a largely tunable MRR filter with low power consumption is obtained. The corresponding Q-factor ($Q=\lambda/\Delta\lambda$) is about 1000. One can achieve a much higher Q-factor by reducing the coupling between the input/drop waveguides and the microring. For example, when using a weak coupling by increasing the gap w_g to 250nm, one obtains a

978-0-9789217-3-6/07/$25.00

©2007 WEN GLOBAL SOLUTIONS

smaller 3dB bandwidth of 0.148nm and a higher Q-factor of more than 10000.

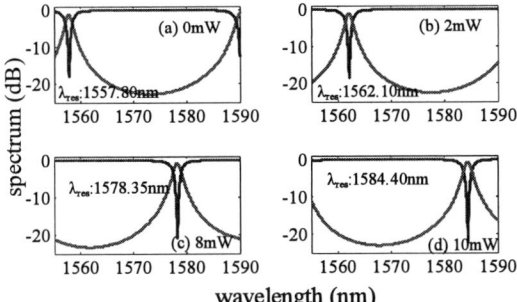

Fig. 2. Spectral responses of the TE polarization from the through and drop ports for different hearting powers. (a) P=0mW; (b) P=2mW; (c) P=8mW; (d) P=10mW;

Conclusions

In this paper, an improved design for an ultrasmall thermally-tunable MRR filter based on Si ridge nanowires has been proposed. The submicron metal circuit along the microring is used as the submicron heater. With the present design, one can use a standard fabrication process with only a one-time lithography process (unlike the double lithography processes required in the fabrication of conventional TO components). The calculated result has shown that the resonator wavelength has a wide tuning range of about 27nm when the heating power varies from 0 to 10mW.

Acknowledgement

This project was supported by research grants (No. 20061343, and 2006R10011) of the provincial government of Zhejiang Province of China, the National Science Foundation of China (Nos. 60607012 and 60688401).

References

1. Y. Kokubun, et al, "Fabrication technologies for vertically coupled microring resonator with multilevel crossing busline and ultracompact-ring radius," IEEE J. Sel. Topics Quantum Electron., 11(2005): 4.
2. Q. F. Xu, et al, "Micrometre-scale silicon electro-optic modulator," Nature, 435 (2005): 325.
3. Y. Goebuchi, et al, "Fast and stable wavelength-selective switch using double-series coupled dielectric microring resonator," IEEE Photonics Technology Letters, 18 (2006): 538.
4. S. Yamagata, et al, "Wide-range tunable microring resonator filter by thermo-optic effect in polymer waveguide," Japanese Journal of Applied Physics, 43(2004): 5766.
5. T. Tsuchizawa, et al, "Low-loss Si wire waveguides and their application to thermooptic switches", Japanese Journal of Applied Physics Part 1, 45 (2006): 6658.
6. G. Sekiguchi, et al, "Polarization-independent tuning of widely tunable vertically coupled microring resonator using thermo-optic effect," Japanese Journal of Applied Physics, 44(2005): 1792.
7. T. Tsuchizawa, et al, "Microphotonics devices based on silicon microfabrication technology," IEEE J. Select. Top. Quantum Electron., 11(2005), 232.
8. W. Bogaerts, et al, "Nanophotonic waveguides in silicon-on-insulator fabricated with CMOS technology," J. Lightwave Technol. 23(2005), 401.
9. D. Dai, et al, "Characteristic analysis of nanosilicon rectangular waveguides for planar light-wave circuits of high integration," Applied Optics, 45(2006): 4941.
10. K. Sasaki, et al, "Arrayed waveguide grating of 70×60 μm^2 size based on Si photonic wire waveguides," Electron. Lett. 41 (2005), 801.
11. D. Dai, et al, "Design and fabrication of an ultra-small overlapped AWG demultiplexer based on α-Si nanowire waveguides," Electron. Lett. 42(2006), 400.
12. B. E. Little, et al. "Ultra-compact Si–SiO$_2$ microring resonator optical channel dropping filters," IEEE Photonics Technology Letters, 10(1998): 549.
13. L. Martinez, et al, "High confinement suspended micro-ring resonators in silicon-on-insulator," Optics Express, 14(2006): 6259.
14. T. Fukazawa, et al, "Low loss intersection of Si photonic wire waveguides," Jpn. J. Appl. Phys. 1, 43(2004), 646.
15. A. Yariv, "Universal relations for coupling of optical power between microresonators and dielectric waveguides," Electronics Letters, 36(2000): 321.

Rotating linear differential polarization imaging for quantitative characterization of superficial tissues

Xiaoyu Jiang, Wei Li, Tianliang Yun, Nan Zeng, Yonghong He, Hui Ma

Laboratory of Optical Imaging and Sensing
Graduate School at Shenzhen, Tsinghua University, Shenzhen 518055

Abstract Differential polarization images corresponding to different incident and exit polarizations are recorded and fitted to an analytical expression. The fitted parameters correlated quantitatively to the structural and optical properties of the superficial tissues.

Introduction

There has been increasing interests in the studies of polarization imaging in turbid media[1-3]. It is an effective technique for rejecting multiple scattered photons from the deep layers and retrieving structural and optical information on superfacial tissues. However, polarization images tend to be dependent on the angles of incident and exit polarizations and on the orientation of the sample. Also, in many applications, the connections between observables and the structural or optical properties of the sample are not apparent. Experimental data usually contains entangled information and are difficult to interpret quantitatively. Using rotating linear differential polarization (LDP) imaging technique, we are able to characterize quantitatively the degree of disorder and orientation of the fibrous tissues.

Experiment and data processing

The experimental setup is typical for reflective polarization imaging. Collimated 650nm LED output passes though a linear polarizer before illuminates the sample at $30°$ to the normal. Backscattered photons pass though a linear analyzer and recorded by a 12bit CCD. Both the polarizer and the analyzer can rotate around their optical axis to vary the incident and exit polarization angles, θ_i and θ_s. The sample can also rotate around its normal. During each measurement, a series of LDP images are taken and then fitted at each pixel to an analytical expression:

$$LDP(\theta_i, \theta_s) = \sqrt{A\cos(4\theta_s - \varphi_1) + B}\cos(2\theta_i - \varphi_2(\theta_s)) + C\cos(2\theta_s - \varphi_3)$$

The fitting results in a set of parameters, A, B, C, φ_1, and φ_3. Each of them forms a new image which can be used to characterize the structural or optical properties of biological samples.

Physics origin of the fitted parameters

We investigated the relation between sample orientation and the fitted parameters. The sample is a piece of porcine skeletal muscle of 2cm thickness with clearly visible fibrous structure running parallel to one another on the surface. As we rotate the sample, we take a series of LDP images corresponding to each orientation angle and obtain the set of fitting parameters at each pixel. We then choose arbitrarily a fixed point at the images and calculate the average of A, B, C^2, A/B and $\varphi_3/2$ over a 20 pixel circle around the point. It is immediately apparent (Fig.2) that A/B is independent on and $\varphi_3/2$ is proportional to the orientation angle of fibrous sample. The results suggest that both parameters originated from the structural features of the sample.

978-0-9789217-3-6/07/$25.00
©2007 WEN GLOBAL SOLUTIONS

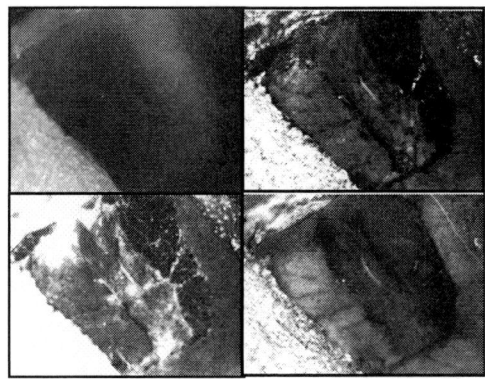

Fig 1. Normal image (top left), images of A (top right), B (bottom left) and C (bottom right) of tissues

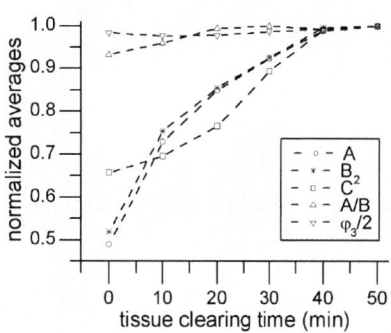

Fig 3. Effects of tissue clearing.

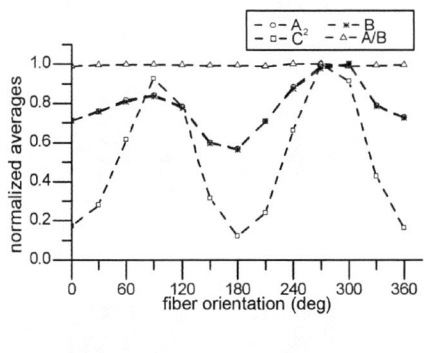

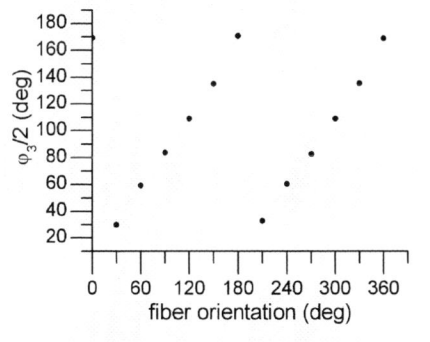

Fig 2. Effects of sample rotation.

We carried on further tests using tissue clearing technique[4]. A drop of glycerol is added on the surface of the sample. As glycerol penetrates deeper, it reduces the scattering coefficient by dehydration or refractive index marching, but causes little changes in the macro structure. The fact that both A/B and $\varphi_3/2$ are independent on tissue clearing time (Fig.3) also indicates

that they describe intrinsic structural properties of the sample.

To investigate the connection between the macro structure of the sample and the fitted parameters, we take average A/B and $\varphi_3/2$ for different biological tissues. The results show that A/B is correlated to the disorder of the fibrous structure. For tissues with clear fibrous structure, such as bovine porcine and chicken skeleton muscles and tandems, A/B are usually higher than 0.9. For structureless tissues, such as fat and liver, A/Bs are less than 0.2. The results also show that $\varphi_3/2$ equals approximately to the orientation angle for fibrous tissues. For samples without apparent fibrous structure, there is an uncertainty in the fitted value of $\varphi_3/2$.

More discussions on the parameters

More solid evidences for the physics origin of A/B and $\varphi_3/2$ are provided by Monte Carlo simulations. Fibrous tissues can be approximated by an idealized phantom which consists a mixture of solid spheres and infinitively long cylinders suspended in liquid. Orientation of the cylinders can fluctuates around a fixed direction. Both the ratio of scattering coefficients of the two types of scatterers and the spread of fluctuation can vary independently to change the anisotropy of the phantom. A Monte Carlo program has been used to simulate the propagation of polarized photons in the phantom and generate images corresponding to the actual experimental conditions. Simulated image series are then processed to obtain the fitted parameter set. The results show that when the fraction of cylindrical scatterers in the mixture increase or their orientation angle fluctuate stronger, both mark the increase of anisotropy, A/B increases monotonically. The simulation also show that $\varphi_3/2$ does represent the most probable orientation of the cylindrical scatterers.

Conclusions

Rotating linear differential polarization reflectometry can be used to retrieve a set of parameters which are correlated to the structural and optical properties of tissues. We proved by both experiments and Monte Carlo simulations that A/B measures the structural anisotropy and $\varphi_3/2$ measures the orientation of the fibrous tissue. The technique is capable of quantitative characterization of biological tissues and may become a powerful tool in diagnosis and pathology.

References

1. S. G. Demos, R. R. Alfano, Appl. Opt., 36, 150 (1997).
2. S. L. Jacques, J. R. Roman, K. Lee, J. Biomed. Opt., 7, 329 (2002).
3. S. P. Morgan, I. M. Stockford, Opt. Lett. 28, 114 (2003).
4. V.V. Tuchin, J. Phys. D: 38, 2497 (2005).

Applications of Total Internal Reflection Fluorescence (TIRF) Microscopy in Cellular Bio-imaging

L. Jin, R.K. Lee*, S. K. Kong*, W. Yuan, H. P. Ho and Chinlon Lin
Center for Advanced Research in Photonics and Department of Electronic Engineering
* Department of Biochemistry
The Chinese University of Hong Kong, Shatin, Hong Kong SAR, China
ljin@ee.cuhk.edu.hk

Abstract Evanescent field based imaging and sensor technologies have been widely used for a long time, but only recently has TIRF been introduced for bio-imaging. This paper will discuss the reasons and give a brief summary of the recent development of TIRF which is also based on evanescent wave excitation, and which is well suited for ultra thin optical sectioning (hundreds of nanometres) at cell-substrate regions with an unusually high signal-to-noise ratio and image quality.

Introduction

The concepts underlying Total internal Reflection Fluorescence (TIRF) are not new, since evanescent field based imaging [1] and sensor [2] technologies have been developed for decades, and have many important applications in non-life-science research fields. TIRF Microscopy which is also utilizing evanescent wave, provides a means to selectively excite fluorophores in an aqueous or cellular environment very close to substrate interface without exciting fluorescence from regions farther from the interface. So it's very suitable for studying biochemical kinetics and single biomolecule dynamics at surfaces in a direct manner that was not previously possible.

It has long been recognized that TIRF Microscopy could potentially become a powerful tool in answering a number of biological questions, and although utilized for over 20 years, the technique has not received a considerable amount of attention until recently. One possible reason is that most of the TIRF systems have been constructed with prism to evoke the evanescent field. This configuration requires unusually high complexity and precision to adjust the prism, and may bring on geometric constraints in manipulating living specimen. Investigators who wanted to utilize the technique were required to build their own systems. This difficulty combined with the necessity of setting up and maintaining an open laser on an optical bench, meant that

earlier users of TIRF with prism were more often physicists than biologists.

Not until high NA objectives have been used in TIRF configuration, and the emergence of commercial available turnkey system, have TIRF Microscopy been widely adapted by biologists, and many interesting research work been carried out, such as sectioning visualization of cell membrane, visualization as well as spectroscopy of single molecule fluorescence very close to or near a surface, micromorphological structures and dynamics on living cells, measurements of the kinetic rates of binding of extra-cellular and intracellular proteins to cell surface receptors and artificial membranes[1,3-5].

Physical basis of TIRF Microscopy

The physical phenomenon of total internal reflection relies upon the incident angle of light beam striking on an interface of two media. Total internal reflection is only possible in situations in which the propagating light encounters a boundary to a medium of lower refractive index. Its refractive behavior is governed by Snell's Law:

$$n(1) \cdot \sin\theta(1) = n(2) \cdot \sin\theta(2)$$

where $n(1)$ is the higher refractive index and $n(2)$ is the lower refractive index. The angle of the incident beam, with respect to the normal to the interface, is represented by $\theta(1)$, while the refracted beam angle within the lower-index medium is given by $\theta(2)$. When light strikes the interface of the two

978-0-9789217-3-6/07/$25.00
©2007 WEN GLOBAL SOLUTIONS

materials at a sufficiently high angle, termed the critical angle ($\theta(c)$), its refraction direction becomes parallel to the interface (90 degrees relative to the normal), and at larger angles it is reflected entirely back into the first medium.

Although light no longer passes into the second medium when it is incident at angles greater than the critical angle, the reflected light will generate a highly restricted EM (electromagnetic) field that is adjacent to the interface in the lower-index medium. This electromagnetic field is the 'evanescent field', and is capable of exciting fluorescent molecules that might be present near the surface. The range over which excitation is possible is limited by the exponential decay of the evanescent wave energy in the z direction (perpendicular to the interface). The penetration depth (d) is dependent upon the wavelength of the incident illumination ($\lambda(i)$), and is typically hundreds of nanometres, according to the equation:

$$d = \lambda(i)/4\pi \cdot (n(1)^2\sin^2\theta(1) - n(2)^2)^{-1/2}$$

If we replace the second medium with stained living cell, then only in the very thin optical section, could the fluorophores be excited by the electromagnetic energy, resulting in images with very low background fluorescence, with virtually no out-of-focus fluorescence, and minimal exposure of cells to light at any other planes in the sample.

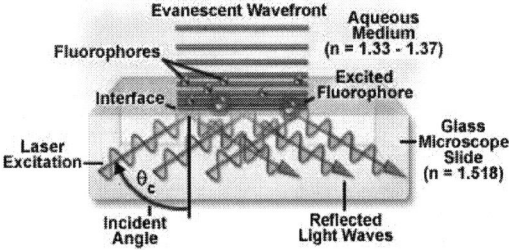

Fig 1: Schematic illustration of TIRF

Basic instrument configuration of TIRF Microscopy

There are two kinds of basic configuration of TIRF instrument: the prism approach and the objective-lens approach. In the prism-based technique, a laser beam is introduced into the microscope coverslip by means of a prism attached to its surface, and the beam

incidence angle is adjusted to the critical angle to evoke evanescent field. Though there are many variations of the prism configuration, most of them restrict access to the specimen, especially making it difficult to perform living samples with incubation media. In the prism approach, laser beam and collection lens are on the opposite sides of the prism, making it difficult to align and adjust the angle of laser beam. Few biological research groups can afford the efforts of maintaining such a complicated and precision experiment setup.

In the objective lens approach, incident light beam and the collection lens use the same microscope objective. The critical angle is achieved by the use of objective of high numeric aperture (ideally 1.45 or higher). Once the laser beam can be adjusted by the setup shown in Fig.2, further fine alignment of incident angle will be easily achieved to reduce the evanescent field penetration depth in a smooth and reproducible manner. The manipulation of living samples under the objective will also be much easier than that in the prism approach.

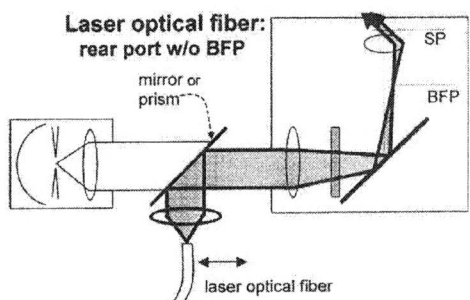

Fig.2 TIRF configuration on objective lens approach with laser optical fiber. SP refers to sample plane, and BFP refers to the objective's back focal plane [3].

TIRF Microscopy's applications in bio-imaging

Since evanescent field decays exponentially with distance from the interface, TIRF can be used qualitatively to observe the position, extent, composition, and motion of cell/substrate contact regions, especially TIRF can be used to study cellular cross-membrane signal transduction, and membrane molecular dynamics. Although

TIRF cannot view deeply into thick cells, it can display with high contrast the fluorescence-marked submembrane filament structures as shown in Fig.3.

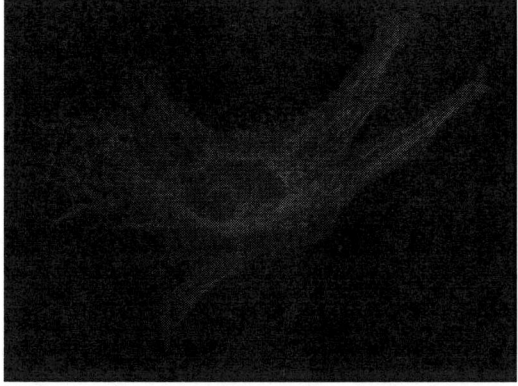

Fig.3 TIRF image of cell filament stained with FITC. Different colors represent different parts (depths) of the cell sample.

TIRF is an ideal tool for investigation of both the mechanisms and dynamics of many of the proteins involved in cell development, Due to the high signal-to-noise ratio property afforded by the evanescent wave excitation, the visualization of molecule fluorescence with sufficient temporal resolution for dynamic studies is possible with TIRF. Sequential time lapse frames may provide insight into abnormal cell growth that occurs in diseases such as cancer, as shown in Fig.4. The TIRF technique is compatible with a wide range of illumination modes, including bright-field, dark-field, phase-contrast, and differential-interference-contrast (DIC), as well as conventional epi-fluorescence as seen in Fig.4c.

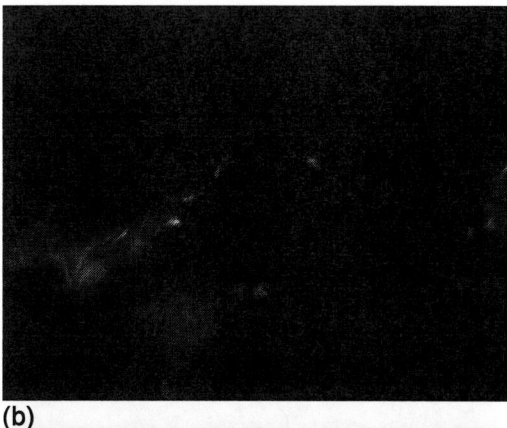

(b)

(c)

Fig.4 (a) and (b) are time lapse fames showing cancer cell development and cell-cell connection as a function of time. The cancer cells were stained with JC-1. (c) is the DIC (differential interference contrast) image at the same time.

Conclusions

TIRF Microscopy based on objective lens approach is well suited for biomedical research work. Further development of TIRF can be combined with other complementary techniques, such as atomic force microscopy (AFM) and FRAP, FRET, to provide more specific detailed information of the various living samples at the molecular level.

References
1 D. Axelrod et al J Cell Biol, Vol.89 (1981), p. 141.
2 H.P. Ho, et al Applied Optics, Vol.45, No.23 (2006), p. 5819.
3 Daniel Axelrod, et al, Traffic, 2 (2001), p. 764
4 Robin M. Shaw, et al, Cell, Vol.128 (2007), p. 547.
5 Jonne Helennius, et al, Nature, Vol.441 (2006),p.115.

(a)

Fluid Sensor Based on Transmission Dip Caused by Mini Stop-Bands in 2D Photonic Crystal Waveguides

Lei Cao, Yidong Huang, Xiaoyu Mao, Kaiyu Cui, Wei Zhang, Jiangde Peng

State Key Lab. of Integrated Optoelectronics, Department of Electronic Engineering

Tsinghua University, Beijing, 100084, P.R. China

Email: yidonghuang@tsinghua.edu.cn

Abstract *In this paper, we propose a fluid sensor based on transmission dip caused by mini stop-band in 2D photonic crystal waveguides. Simulation results shows that it has large detective range and relative high sensitivity.*

1. Introduction

Various fluid sensors have been proposed based on photonic crystal (PC) structures[1-3]. The porous structure in PC permits the analyte to go into the high optical field directly, thus high spectroscopic sensitivity to ultrasamll quantity of analyte can be obtained. However, a number of problems and difficulties still remain in their design, fabrication, and measurement. In this paper, we propose a kind of fluid sensors based on the transmission dip caused by mini stop-band (MSB) in 2D PC waveguides (PCWGs). Simulation results show that the central frequency of the transmission dip is sensitive to the refractive index of the substance filled in the holes of PC structure. Using this method, relative high detective precision of 73nm/RIV and large detective range can be estimated.

2. Simulation and Discussions

MSB, appearing as a mini gap in the frequency region of defect modes and a cute dip in their transmission spectrum, was observed and investigated in 2D PCWGs[4-6]. It originates from the mode-coupling of propagation modes with the following three factors: these two modes propagate in anti-direction; the wave vectors of them differ by an integer number of reciprocal lattice vectors; they have the same symmetry[5].

Figure 1 shows simulation model of the PCWGs, with triangular lattice of air holes in GaAs-based dielectric slab sandwiched by AlGaAs. The refractive index of background and substance filled in the holes is n_2 and n_1, respectively. The lattice constant is a and the radius of the air holes is r. Because

the air-filling factor of PCs[7] is f =1.155$\pi(r/a)^2$, r/a is taken as a key structure factor influences MSBs. The waveguides are along the ΓK direction and symmetric with respect to the mirror plane in the center of them. The width of PCWG, denoted as d_m in Fig. 1, can be expressed as d_m=0.866(m+1)a, and the waveguide is denoted as W_m. Here, m can be fraction with continuous d_m by adjusting the location of the two rows holes.

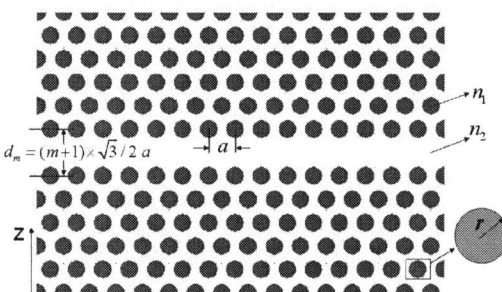

Fig.1 Simulation model of PCWGs.

Here, we use the supercell method[7] to calculate the dispersion relations of W_3 PCWG with n_1=1, n_2=3.32, and r/a=0.32. Figure 2 (a) shows the results. It can be found that there is a photonic band gap for TE polarization covering the normalized frequency range of u=a/λ=0.22-0.3, between the air band and dielectric band (dark areas). Within the band gap, there are several guided modes (defect modes) and MSBs. The fundamental mode (FM) signed by black line couples with one of the high-order modes (HOM) and generates the corresponding MSB. Zooming in this MSB is shown in Fig.2 (b). In order to discuss the calculation results conveniently, we define

978-0-9789217-3-6/07/$25.00

©2007 WEN GLOBAL SOLUTIONS   586

the central frequency of MSB as $U_0=(U_++U_-)/2=(\omega_++\omega_-)a/4\pi c$, where $U_0=\omega_0 a/2\pi c=a/\lambda_0$ is normalized frequency, and ω_0, ω_+ as well as ω_- are central, upper, and lower frequency of MSB, respectively. Applying the finite-difference time-domain (FDTD) method[8], we compute the transmission spectrum of this waveguide, and the simulation results are shown in Fig.2 (c). It should be noticed that the frequency where dip located corresponds to the central frequency U_0 in Fig. 2 (a).

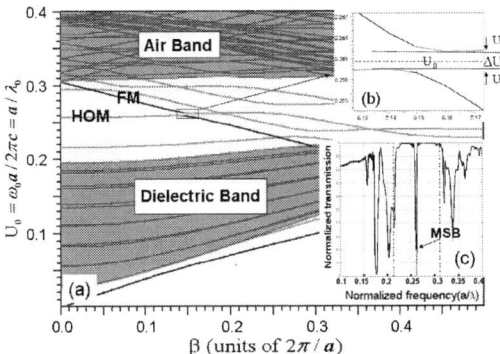

Fig. 2 Dispersion relations of W_3 waveguide. MSB and transmission spectrum are shown in the inserts.

The sharp dip caused by small width of MSB is sensitive to the refractive index of substance filled in the holes of PC (n_1). Figure 3 shows the calculation results, where $n_2=3.32$ and $r/a=0.36$. It is can be seen that, the transmission dip experiences red-shift as n_1 increases. This is because that the effective refractive index of 2D PCs becomes larger with n_1 increasing, thus the bands of PCWGs and the MSBs, as well as their central frequency move downwards to lower frequency. This phenomenon can be used to develop a fluid sensor. Compared with other types of sensors based on PC structure[1-3], proposed fluid sensor here has follow advantages. Firstly, in the proposed sensor, analyte only fills in the holes or distributes around the waveguide, it scarcely influences the light transmitting in the waveguide. No matter how large the absorption of the detected medium is, light propagates in the waveguide can still keep low loss. Secondly, the width of MSB can be controlled by adjusting the parameters of the PCWGs. The sensitivity and detective range can be designed freely. Thirdly,

compared with the sensor based on PC micro-cavity, the sensor proposed here is easy to couple the light in and out.

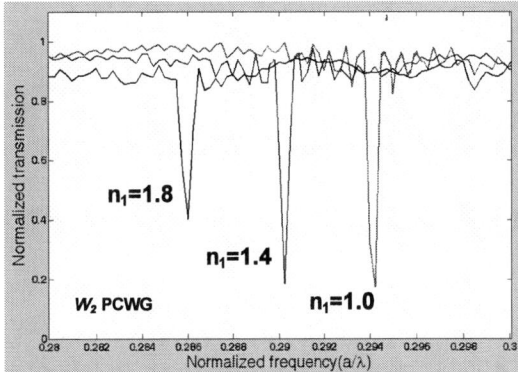

Fig.3. Cute dip shifts with the varieties of n_1 for W_2 waveguide.

The calculation results of the sensitivity for the central frequency to n_1 are shown in Fig.4. The simulation is based on W_2 waveguide with fixed $n_2=3.32$. It is can be seen that, the central frequency of MSBs decreases nearly linearly with the increase of n_1. Taking one line in Fig.4 as an example, for the line of $r/a=0.36$, the linear fittings $U_0=-0.0114\times n_1+0.259$ could be derived. We can estimate the detective precision of sensors working by detecting wavelength shift. For instance, for PCWGs in Fig.4 illuminating just above, setting working wavelength $\lambda=1550$nm, lattice constant $a=376$nm, the detective precision could be up to 73nm/RIV.

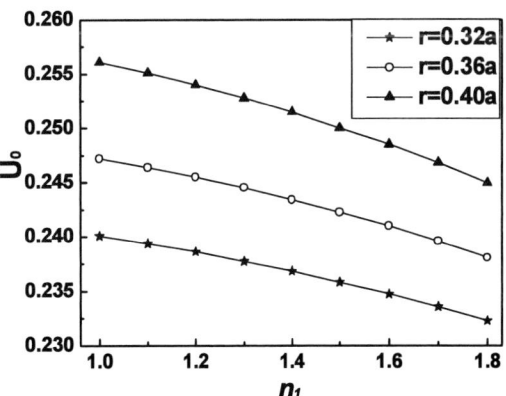

Fig. 4 Calculation results of the sensitivity of the central frequency to n_1 with different air-filling factor

Also considering the middle line in Fig.4, for PC structure with $n_2=3.32$ and $r/a=0.36$, the photonic band gap is u=a/λ=0.23-0.35. If the refractive index of analyte varies from 1.0 to 2.0, the central frequency of MSB shifts from 0.259 to 0.2476 and still lies in the photonic band gap, which means that sensors based on this structure have large detective range more than 1.0. It should be noticed the detective range is related to the slope of line between central frequency of MSB and n_1. Therefore, those three almost parallel lines in Fig.(4) mean sensors based on them will have the similar large detective range.

In addition, as all structure parameters of PCWG influence the dispersion relation of waveguide as well as the property of MSB, by adjusting the structure parameters of waveguide, we can design appropriate work frequency, optimize the performance, and tune the range of detecting spectrum for this type of fluid sensor. These would be investigated in further work.

3. Conclusions

Based on MSB phenomenon in 2D PCWGs, we propose a kind of novel fluid sensor with advantages of compact structure and possibilities for integration. Simulation results show that, proposed sensor has large detective range and relative high sensitivity.

Acknowledgment

This work is supported by the National Natural Science Foundation of China (NSFC-60537010), and the National Basic Research Program of China (973 Program) under Contract No. 2006CB302804.

References

1 Victor S.-Y. Lin et al Science, **278** (1997), 31

2 M Loncar et al, IEICE TRANS. ELECTRON., **E87-C** (2004), 3

3 Tushar Prasada et al Optical Materials, **29** (2006) 56 – 59

4 S. Olivier et al Phys. Rev. B., **63** (2001), 113311

5 S. Olivier et al Opt. Express, **11** (2003), 1490

6 E. Viasnoff-Schwoob et al Appl. Phys. Lett., **86** (2005), 101107

7 H. Benisty, J. Appl. Phys., **79** (1996), 7483

8 A. Lavrinenko et al Opt. Express, **12** (2004), 234

Fabrication and Photochromic Properties of Ag/Ag+-Codoped Germano-Silicate Glass Fiber

Aoxiang Lin, Sung-Ho Kim, Youngjoo Chung, and Won-Taek Han*

Department of Information and Communications, Gwangju Institute of Science and Technology (GIST), 1 Oryong-dong, Buk-gu, Gwangju, 500-712, Republic of Korea
*E-mail: wthan@gist.ac.kr

Abstract *Ag/Ag+-codoped germano-silicate fiber was developed and the photocatalytic reduction from Ag+ ions to Ag nanoparticles inside the fiber core by UV light was found.*

Introduction

Ultraviolet (UV) light, although invisible to the human eye, is the component of sunlight that has the most effect on skin. To meet the need of removing excessive UV light, UV protection window, automobile glass, sunglasses and films were developed [1,2]. On the other hand, to make good use of UV light or to detect the intensity of UV light, accurate and low-cost UV dectors become necessary. Compared with glass or microchip type of UV detector [3], a fiber-type UV detector has obvious advatages, such as being relatively inexpensive, easy to install, resistant to moisture damage while being relatively nonflammable.

In this study, we attempted to develop a fiber-type UV detector by using the modified chemical vapor depostion (MCVD) technique and the solution doping process. Time- and power-dependent photochromic properties including coloration and bleaching of the fiber doped with Ag/Ag+ under UV light illumination were investigated.

Experiments

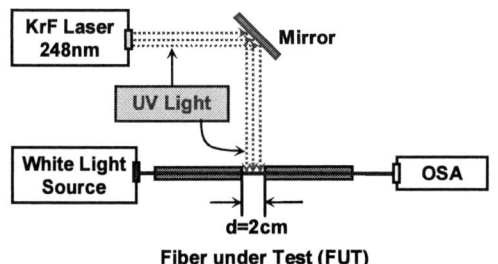

Fig.1. Schematic setup for the photochromism measurement of the Ag/Ag+-codoped fiber.

The fiber doped with Ag/Ag+ was fabricated in house using the MCVD and solution doping processes. After the deposition of the germano-silicate core layers in the silica glass tube by using the MCVD process, the porous deposition layers were soaked with the doping solution of AgNO$_3$ (1mol%, in distilled water) for two hours. Then the tube was sintered and sealed, and finally the preform was drawn into fibers with 125μm in diameter using the draw tower at 1900℃.

To examine photochromic properties of the fiber, the stripped portion of 2-cm of the fiber was illuminated with a 248nm KrF laser at room temperature. Figure 1 illustrates the experiment setup in which the 248nm KrF UV light laser beam was reflected by a mirror and then illuminated onto the stripped portion. In order to obtain accurate transmission signal, high sensitive optical spectrum analyzer (OSA) was used.

Results and discussion

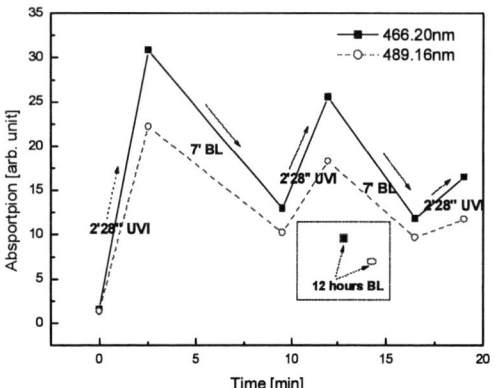

Fig.2. *Time dependence of the absorption of the Ag/Ag+-codoped fiber at 489.16nm and 466.20nm after the UV illumination (UVI) and self-bleaching (BL) processes*

Figure 2 shows the time dependence of the absorption at 489.16nm and 466.20nm of the Ag/Ag$^+$ codoped fiber during the UV illumination (UVI) and self-bleaching (BL) processes. This result indicates the stability and repeatibilty of this UVI-BL characterized photochromic processes happened inside the Ag/Ag$^+$-codoped fiber.

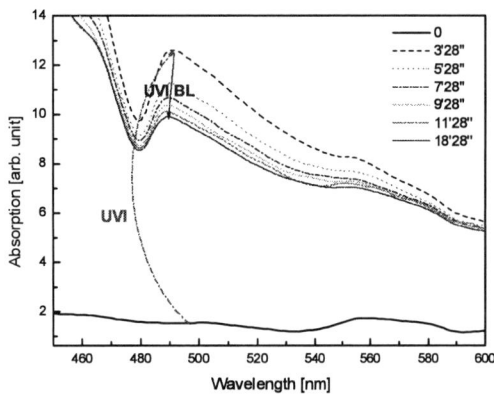

Fig.3. *Bleaching (UVI BL) process of the Ag/Ag$^+$- codoped fiber with the increase of the UV illumination.*

Under the appropriate UV light illumination (E=4.98eV, $\lambda = 248$nm, less than 2min), the electrons from the trapping centers such as the adsorbed oxygens O_1^- (E$_g$=2.0eV) around Ag$^+$ ions [4] and inevitable impurity of H$_2$O molecules [5] inside the fiber core combine with Ag$^+$ ions to form Ag NPs which are responsible for the surface plasmon resonance (SPR) absorption peak appeared at 489.16nm. Removing the UV light or applying the excessive UV light illumination, the subsequent BL process happened, including the slef-bleaching process (BL) in Fig.2 and UVI-induced BL (UVI BL) in Fig.3. During these BL processes, these electrons in the conduction band of Ag NPs were captured by the trapping centers again and therefore a gradual depletion of the particle-plasmon band was found. The recovery process can be accelerated by excessive UV irradiation or just after 12 hours self-bleaching process, and the fiber can be recovered to the original state of the fiber in the end as shown in the insert of Fig.2. These recovery phenomena are opposite to the bleaching process found in TiO$_2$ photocatalyzed

AgNO$_3$ adsorbed film [6]. With the increasing time of UVI, Fig. 3 shows the change of the absorption spectra of the fiber, describing the UV-induced BL process (UVI BL), in which the SPR absorption peak of Ag NPs at 489.16nm are found decreased rather than increased with the increasing time of UVI as we expected as usual before this experiment.

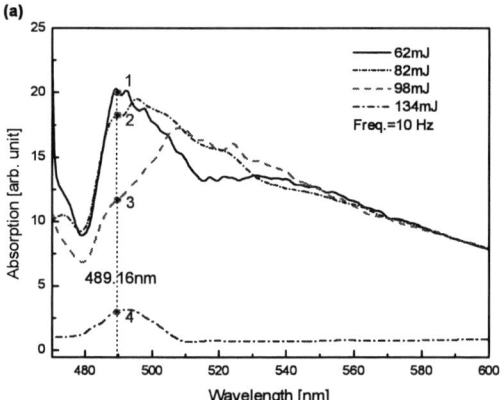

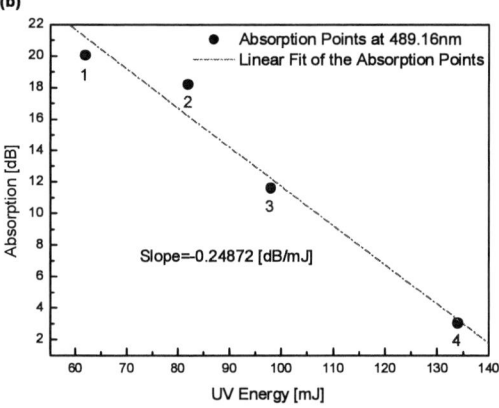

Fig.4. *(a) Asborption change with the increase of UV light energy; (b) Linear dependence of the absortpion intensity at the SPR peak of 489.16nm.*

The reactions inside the fiber core during the UVI and BL processes can be written in Eq. (1)-(3).

$$2H_2O \underset{BL}{\overset{UV}{\rightleftarrows}} O_2 + 4H^+ + 4e^- \qquad (1)$$

$$O_1^- \underset{BL}{\overset{UV}{\rightleftarrows}} O_1 + e^- \qquad (2)$$

$$Ag^+ + e^- \underset{BL}{\overset{UV}{\rightleftarrows}} Ag \qquad (3)$$

Figure 4 (a) shows the absorption spectra with the increase of the UV illumination energy and the absorption intensity at 489.16nm was plotted in Fig. 4 (b). The absorption intensity was found to linearly decrease with the UV energy, showing the stable and applicable photochromic property of the Ag/Ag$^+$-codoped fiber. However, the mechanism of this minus linear dependence on UV light energy is not clearly known yet.

Conclusions

Ag/Ag$^+$-codoped fiber was developed by MCVD and solution doping processes. The photochromic properties, such as the absorption change and the bleaching process with the UV illumination, were observed and the absorption peak at 489.16nm was found to decrease with the UV energy.

Acknowledgements

This research was partially supported by Korea Science and Engineering Foundation (KOSEF) through grant No.R01-2004-000-10846-0, by the National Core Research Center (NCRC) for Hybrid Materials Solution of Pusan National University, and by BK-21 Information Technology Project, Ministry of Education and Human Resources Development, Republic of Korea.

References

1. B.L.D. DSc et al J. Am. Acad. Dermatol. 43 (2000) 1024.
2. M. Guerin et al Trends Biotechnol. 21 (2003) 210.
3. M.M. Abdel-Aziz et al Appl. Surf. Sci. 252 (2006) 8716.
4. E.P. O'Reilly et al Phys. Rev. B 27 (1983) 3780.
5. B. Ohtani et al J. Phys. Chem. 91(1987) 3550.
6. J. Okumu et al J. Appl. Phys. 97 (2005) 09403.

Organic Photovoltaics

Jiangeng Xue

Department of Materials Science and Engineering, University of Florida
PO Box 116400, Gainesville, FL 32611, USA
Tel: 1-352-846-3775; FAX: 1-352-846-3355; E-mail: jxue@mse.ufl.edu

Abstract *Organic photovoltaic cells have the potential to offer low cost solar energy conversion, and are considered as a promising candidate to replace fossil fuels in the future. This tutorial will describe the fundamental operation principles of these devices and review the major advances in this field.*

Introduction

Fossil fuels, which include oil, coal, and natural gas, have been the major energy source powering the modern society for decades. However, the finite supply of fossil fuel sources and the adverse environmental effect caused by burning fossil fuels, coupled with the rising world demand for energy, have made finding sufficient supplies of clean energy urgent and of utmost importance in the next half century. One of the most promising, yet vastly under-utilized, alternative energy source is solar energy, which is clean, renewable, safe, ubiquitous, and abundant. Using photovoltaic (PV) cells or solar cells to directly convert sunlight to electricity is one of the major methods to capture and convert solar energy; however, such-generated solar electricity only provided a very small portion of the world's electricity, less than 0.1% in 2001, despite its growth of 35-40% per annum recently. The economics has been the limiting factor, as solar electricity is about five to ten times more expensive than electricity generated from fossil fuels (especially coal).

Compared with the conventional, i.e. inorganic, semiconductors, organic materials generally have significantly lower material costs [1]. They can also be easily made into thin films through inexpensive room-temperature processes. They are compatible with flexible substrates that can be light weight and inexpensive, which makes them suitable for high throughput, low cost roll-to-roll processing and conducive to reduce the module installation costs. Moreover, many properties of organic materials can be tailored to suit particular applications through chemical synthesis. Hence, organic semiconductors are particularly suited for large area, low cost, light weight, and flexible device applications, such as photovoltaics. Organic photovoltaic (OPV) cells have been heralded as a third-generation PV technology that is expected to play a very important role in supply sufficient amount of clean energy in the foreseeable future [2]. In this tutorial, the fundamental operation principles of OPV cells will be described and some major advances in this field will be reviewed.

Donor-Acceptor Heterojunctions

Unlike the covalent bonding among atoms in inorganic semiconductors, organic semiconductors are held together by the much weaker van der Waals type intermolecular interactions, resulting in highly localized charges. Absorption of an incident photon thus leads to a bound electron-hole pair, or an *exciton*, on one single molecule or on neighboring molecules. Hence, almost all efficient OPV cells are based on the donor-acceptor (DA) heterojunction (HJ) structure, first demonstrated by Tang in 1986 [3]. As shown in Fig. 1, with appropriate energy level alignment at the DA interface, the exciton binding energy can be overcome and efficient dissociation of excitons can occur through a rapid charge transfer process, leading to holes in the highest occupied molecular orbital (HOMO) of the donor material (which has a smaller ionization potential) and electrons in the lowest unoccupied molecular orbital (LUMO) of the acceptor (which has a larger electron affinity). Driven by the built-in

978-0-9789217-3-6/07/$25.00

©2007 WEN GLOBAL SOLUTIONS

electric field or the concentration gradients, these holes (electrons) are subsequently transported through the donor (acceptor) molecules towards the anode (cathode) where they are collected, generating a photocurrent or photovoltage. Using copper phthalocyanine (CuPc) as the donor and a perylene derivative as the acceptor, a power conversion efficiency (PCE) of approximately 1% was achieved [3].

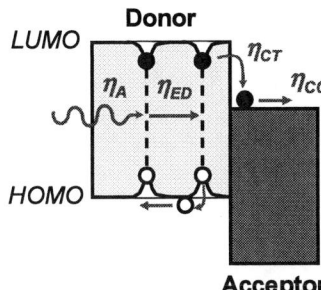

Figure 1. Schematic energy level diagram illustrating the photovoltaic processes in an organic donor-acceptor heterojunction.

Exciton Diffusion Bottleneck

One key process limiting the performance of OPV cells is the short exciton diffusion lengths (L_D) in organic semiconductors, typically ~10 nm or less. This is much shorter than the necessary thickness needed to absorb a large portion of the incident photons (~100 nm). Hence one way to improve the overall efficiency of an OPV cell is to use materials with long L_D. By replacing the acceptor material used in the Tang cell [3] with C_{60}, whose L_D was found to be approximately 40 nm, the PCE was improved from 1% to 3.6% [4]. The very long L_D in C_{60} was attributed to the much longer lifetime of triplet exciton states existing in this material.

Series Resistances of OPV Cells

The charge carrier mobility and conductivity of organic semiconductors are generally much lower than those in inorganic semiconductors. Hence it has been long believed that OPV cells will have high series resistances, which is detrimental to PV device performance. It has been shown, however, that in certain cases the organic layers may have negligible contribution to the series resistance of an OPV cell. For example, in CuPc/C_{60} cells, due to the small layer thicknesses (20-40 nm) and the relatively high carrier mobility in these materials (10^{-4}-10^{-2} cm^2/V·s), the device series resistance can be below 0.1 Ω·cm^2, and is mostly dominated by the limited conductance of the transparent conducting oxide anode [5]. In such devices, the PCE increases logarithmically with the illumination intensity due to the increase in the open-circuit voltage (V_{OC}), and reaches a maximum of 4.2% at 4-12 suns (1 sun = 100 mW/cm^2, AM1.5) [5].

Exciton Blocking Layer

In OPV cells with a single DA HJ, the cathode (typically a reflecting metal such as Al or Ag) is usually thermally evaporated onto the organic active region. This could have several adverse effects. The energetic metal atoms can damage the organic molecules, and may diffuse into the organic layers, leading to defect states in both cases. The metal cathode can also quench nearby photogenerated excitons. Moreover, the optical field of the incident light vanishes at the metal surface, leading to low absorption in the active region near the metal surface. A transparent, wide-gap exciton blocking layer (EBL) was thus proposed to be inserted between the organic active region and the metal cathode, leading to a double-heterojunction structure [6]. This EBL can act as a damage absorber upon deposition of the cathode, confine the excitons in the active region, and provide an optical spacer to enhance the optical intensity in the active region.

Different DA Heterojunction Structures

As described earlier, in OPV cells with a bilayer DA HJ (also known as "planar" HJ), the exciton diffusion problem leads to inefficient absorption of the incident photons. One way to circumvent this problem is to blend or mix the donor and acceptor molecules (or polymers) to form the so-called "bulk" or "mixed" HJ structures [7], so that the DA interface is spatially distributed in the three dimensional space. If the phase separation between the two species is not very significant, the DA interface will be within reach of every exciton generation site. Using this approach, PCE's ranging from 3-5% have been achieved by blending polymers such as MDMO-PPV and P3HT with a soluble fullerene derivative (PCBM) [8].

The intermixing of materials in the bulk HJ also has significant consequences on charge transport. The randomly formed DA mixtures may present islands, bottlenecks, and cul-de-sacs, so that charges may encounter difficulties in finding a pathway to the respective electrodes, leading to reduced charge collection efficiency. A hybrid planar-mixed HJ (PM-HJ) was thus proposed to combine the high exciton diffusion efficiency in a mixed HJ and the good charge transport properties of a planar HJ [9]. Using the $CuPc/C_{60}$ material system, a PCE of 5% was achieved.

Tandem Devices

Multi-junction, or tandem cells have been used to produce the most efficient PV cells demonstrated so far [10]. Using multiple subcells to cover different portions of the solar spectrum is more advantageous than using a single cell by reducing thermalization losses (or heat generation) during the exciton or charge generation process. Furthermore, tandem organic solar cells [11, 12] utilize one important technological advantage of organic materials, i.e. the growth of high quality organic films does not require epitaxy or lattice matching, which makes it really flexible to incorporate different materials.

Solar Spectrum Matching

Many of the existing OPV materials can efficiently absorb ultraviolet and a significant portion of the visible photons, but has very limited absorption in the infrared (IR) region. For example, P3HT has an absorption edge of around 600 nm; CuPc strongly absorbs from 550 nm to 750 nm. There has been work focusing on developing low gap molecular and polymeric materials for use in OPV cells. For example, OPV cells based on SnPc utilize near IR photons up to 1 μm [13]. Using a low gap polymer PCPDTBT which has absorption up to 850 nm, a PCE of 3.2% was achieved [14].

The absorption of organic materials generally exhibit bands with a couple hundred nm width. Thus a combination of materials will be needed to cover the broad solar spectrum. The tandem cell geometry provides an excellent platform to do so. Efficiencies around 6% have been achieved in small molecules or polymers tandem cells [12, 15], in which one subcell absorb shorter wavelength light whereas the other was tuned to absorb longer wavelength light.

Conclusion and Outlook

The efficiency of OPV cells has been steadily improved over the last decade by introducing new device architectures and employing new materials. State-of-the-art OPV cells have PCE's around 6%. Further improvement in efficiency is needed to make this technology a viable candidate for commercialization, which will rely on a better understanding on the fundamental operation principles of these devices and the development of new materials with desired properties. Besides the efficiency, other issues such as lifetime/reliability and manufacturability will need to be addressed.

References

1 S. R. Forrest, *Nature*, 428 (2004), 911.
2 S. E. Shaheen et al., *MRS Bull.*, 30 (2005), 10.
3 C. W. Tang, *Appl. Phys. Lett.*, 48 (1986), 183.
4 P. Peumans et al., *Appl. Phys. Lett.*, 79 (2001), 126.
5 J. Xue et al., *Appl. Phys. Lett.*, 84 (2004), 3013.
6 P. Peumans et al., *Appl. Phys. Lett.*, 76 (2000), 2650.
7 J. J. M. Halls et al., *Nature*, 376 (1995), 498; G. Yu et al., *Science*, 270 (1995), 1789; P. Peumans et al., *Nature*, 425 (2003), 158; S. Uchida et al., *Appl. Phys. Lett.*, 84 (2004), 4218.
8 C. J. Brabec et al., *Adv. Funct. Mater.*, 11 (2001), 15; W. L. Ma et al., *Adv. Funct. Mater.*, 15 (2005), 1617.
9 J. Xue et al., *Adv. Mater.*, 17 (2005), 66.
10 R. R. King et al., *Appl. Phys. Lett.*, 90 (2007), 183516.
11 A. Yakimov et al., *Appl. Phys. Lett.*, 80 (2002), 1667; P. Peumans et al., *J. Appl. Phys.*, 93 (2003), 3693.
12 J. Xue et al., *Appl. Phys. Lett.*, 85 (2004), 5757.
13 B. P. Rand et al., *Appl. Phys. Lett.*, 87 (2005), 200508.
14 D. Muhlbacher et al., *Adv. Mater.*, 18 (2006), 2884.
15 J. Y. Kim et al., *Science*, 317 (2007), 222.

Development of Solar Photovoltaic in China

Cui Rongqiang, Wang Jianqiang, Ye Qinghao, Yan Shiquan, Shi Yang, Du Jiabing, Yang Le, Meng Fanying, Xu Lin, Shen Wenzhong

Solar Energy Institute of Shanghai Jiao Tong University, 800 Dongchuan Road, Shanghai, China 200240

Abstract: With the rapid development of global solar photovoltaic (PV) industries, China's PV is growing at a surprising high-speed. China's renewable energy law was promulgated on Jan. 1, 2006. The law provides the legal endorsement to the development of all types of renewable energy, including solar photovoltaic (PV) power. The present development of solar photovoltaic in China will be briefly introduced in this paper.

1 Introduction

In China's power network, nearly 24% electricity is generated from waterpower while the rest 76% is still generated from firepower. China is rich in sunshine resources. It is absolutely possible to use solar energy as an alternative to the firepower. The proactive development of solar photovoltaic application has already set a solid foundation in further utilizing the great potential of solar energy in wider areas in China.

2 An Ideal Model of Chinese PV Growth

SE (Solar Energy) + ES (Energy in Storage) = WED (Whole Energy Demand)

It's an ideal energy industry development model in China. We believe that it could be absolutely feasible in the context of China's rapid, continuous technological and economic development.

3 Five-Stages of PV Application in China

Stage 1: Off-grid PV System designed to supply power to remote regions. The main advantage is such system does not need power transmission equipments. The ideal capacity can be 100 W to 100 kW and the ideal module price can be US$4-6 per Wp.

Stage 2: On-grid roof mounted PV systems, which is called as Building Integrated PV System (BIPV). Such system can be installed on building's roof and walls to save cost and land usage. The ideal capacity can be 1 kW to 10 MW and the ideal module price can be US$4-6 per Wp.

Stage 3: PV Power Station in desert, which could take advantage of rich sunshine in desert areas. The ideal capacity can be 1 MW to 1 GW and the ideal module price can be US$3-4 per Wp.

Stage 4: PV Power Station in coastal areas. China's coastal urban areas cities are in great demand for electricity. PV stations there can meet such demand and save power transmission cost. The ideal capacity can be 10 MW to 10 GW and module price can be US$2-3 per Wp.

Stage 5: PV Power roof over streets. PV roof set up over the streets to supply electric power to electricity powered cars or trains. With such technology, we can replace traditional gas engines with solar powered electro-motors and enter into an era of solar power. The ideal capacity can be 10 GW to 1,000 GW and the ideal module price can be less than US$1 per Wp.

4 Photovoltaic Technology Research &Development

More than 30 Universities or Institutions take PV research programs.

Laboratory Efficiency and Size of Photovoltaic Cell mainly developed in China

978-0-9789217-3-6/07/$25.00

©2007 WEN GLOBAL SOLUTIONS

Cell Type	Highest Efficiency (%)	Maximum Size (cm^2)	Research Organization
Mono-Si	20.4	2×2	Tianjin Inst. of Electrical Source
	18	12.5×12.5	Suntech
Poly-Si	16.5	12.5×12.5	Suntech
GaAs	29.25	1×1	Tianjin Inst. of Electrical Source
CIGS	14.3	0.87	Nankai Univ.
CdTe	13.38	0.502	Sichuan Univ.
DSC	7.4	10.2	Inst. of Plasma Physics
a-Si	9.2	20×20	Nankai Univ.
μ-Si/a-Si	11.8	10×10	Nankai Univ.
HIT	17.27	1.2	College of Physics Sciences
nc-Si	8.87	1.0	Shanghai Jiao Tong Univ.
Thin film pc-Si	15.1	1.05	Beijing Solar Energy Institute

5 PV Manufacturer

5.1 C-Si Solar Cell Output and Capacity in China

	Company	Output in 2005 (MW)	2006 (MW)	
			Output	Capacity
1	Suntech Power	82	160	300
2	Solarfun	1	26	60
3	Ningbo Solar Cell Factory	25	43	100
4	CEEG Nanjing PV-Tech	5	47	190
5	Trina			50
6	Baoding Yingli	3	38	60
7	Yunnan Tianda		25	50
8	Shanghai Solar Energy S&T		7	35
9	Shanghai Topsolar Green Energy	7	20	45
10	CSI			25
11	Jing Ao Solar		30	75
12	Yangzhou Tianbao		10	25
13	Shenzhen Topray		10	20
	Others	15	34	235
	Total	139	450	1270

By the end of 2006, China's manufacturing capacity and output of C-Si solar cell is 1,270 MW and 450 MW respectively, The PV module capacity and output in 2006 is 2,000 MW and about 800 MW respectively. In 2007, there will be about 2000 MWp PV cells and more than 1600 MWp PV modules manufactured in China.

5.2 Expansion of a-Si Solar Cell Capacity

There're five thin film solar cell manufacturers in China:Shenzhen Topray, Tianjin Jinneng, Shenzhen Trony, Shenzhen Sumoncle, Harbin-Chronar

The total capacity and output in 2006 is 30.5 MW and 17.5 MW. In Year 2007 Suntech invests to build up a 400 MWp amorphous and microcrystalline silicon manufacture in Shanghai.

5.3 Development of PV Material Production

(1) they are three high purity poly-silicon manufacturers in China and they are as follow:

Emei Semiconductor，Sichuan Xin-Guang，Luoyang Silicon High-Tech

The total output of high purity poly-silicon is 301 ton in 2006. In 2007, there are more then 10 projects of high purity silicon have already been initiated. By 2010, we estimate the capacity and the output of high purity silicon will be 20,000-50,000 and 10,000-25,000 ton per year respectively.

(2) China has around 1,200 crystal silicon furnaces by 2006, mainly supplying wafer to Cell manufactures in mainland China and Taiwan, and newly 800 crystal silicon furnaces will be set up in China this year. The Main CZ Silicon manufactures are as follow:

Ningjin Jinglong, Shunda, Shanghai Jingyong, Jiangsu Huariyuan, Comtec LTD. Jinzhou Xinri Silicon, Songgong, Zhejiang Yuhui, Trina, Shanghai JiuJing.

(3)China has four poly-silicon ingot and wafer manufacturers:

LDK, Baoding Yingli, SIPV, Jinggong-Preiss-Daimler

The total capacity and output in 2006 is about 338 MW and150 MW. We estimate that the capacity of LDK and Yingli will be around 1,600 MW by 2009 and 600 MW by 2010 respectively.

5.4 Localized PV manufacturing equipments

(1)More than 10 equipment manufacturers in China can assemble and produce crystal silicon furnace, polycrystalline silicon furnace and cutting machine to this year.

(2)More than 20 manufacturers in China such following equipments: wafer rinse machines, diffusion furnace, etching machine, PECVD deposition equipments, drying furnace, belt furnace, and semiautomatic screen printer and so on. PV testers are mainly made by Shanghai New-Bridge Solar energy equipment, SJTU and others.

5.5 PV production materials and peripheral equipments

Most PV production materials and peripherals, such as frame, junction box, EVA, glass and inverter are now can be made in China, then it can greatly decreases the PV products cost and makes it more competitive in international PV market.

6 Application of PV in China

By 2006, China has installed about 85 MW PV systems, about 15 MW installed in 2006.

(1)In the government and World Bank/GEF sponsored project, PV and PV-wind hybrid systems are installed to supply electricity to remote regions. Around 5 MW systems were installed in 2006.

(2)Off-grid PV systems are installed in relay stations, railway stations, weather monitoring stations and navigation lighthouses. Around 5 MW systems were installed in 2006.

(3)Many On-grid roof mounted demo PV systems have been installed for Beijing Olympics 2008 and Shanghai 2010 World Expo. Total installed capacity was about 1MW in 2006. Solarfun Co. has wined the 1MW on-grid PV station project in Shanghai Chongming Island, and the PV electric price of this project is under negotiation.

(4)About 1800MWp PV application target by 2020 provided from NDRC and detailed information is on consideration.

7 Conclusion

Although China PV industry has made great progress, it still falls behind its counterparts Japan, US, Germany and European countries. In order to expand Chinese PV market, we still truly need to acquire more support policies from the government and learn more experiences from PV developed countries, and then Chinese people can really benefit from this green, clear and safe PV energy.

Control of slow light in coupled resonator optical waveguide structures with highly dispersive media

Min Qiu

Department of Microelectronics and Applied Physics, Royal Institute of Technology (KTH)
Electrum 229, 164 40 Kista, Sweden
Email: min@kth.se

Abstract *Dispersive material background may enable dynamic tunability of slow light of the coupled resonator optical waveguides. It may also results in an enhanced transmission spectrum with a time-dependent tuning of the dispersive medium.*

Many interesting optical phenomena, including the ultraslow light propagation and electromagnetically induced transparency (EIT), have been observed in highly dispersive media [1, 2]. Additionally, the introduction of a highly dispersive media into an optical cavity results in dramatic narrowing of spectral features as well as enhancing cavity lifetimes [3].

Figure 1 shows a schematic of such a coupled-resonator optical waveguide (CROW) formed by an infinite straight line of defect cavities in a two-dimensional (2D) photonic crystal (PhC). This structure can also provide slow light propagation through the wave coupling between high-Q resonators [4]. Yet, for realizing active components in an optical circuit, a dynamic control over the group velocity is desirable. The problem then becomes how one can manipulate the coupling between resonators in a seemingly fixed system. Here we show that a tunable, highly dispersive process, such as EIT, will facilitate control over the propagation velocity in the CROW structure by modifying the coupling factor between cavities.

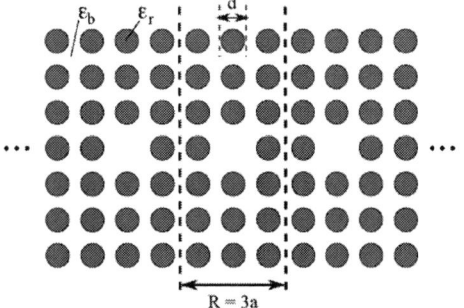

Fig. 1 Schematic of a CROW structure formed in a 2D PhC square lattice of high index rods where single rods were removed

to create the resonant cavities.

We have successfully modeled a PC CROW structure embedded in a three-level EIT medium with the FDTD method [5], which was employed to incorporate the frequency dependence of ε of the highly dispersive EIT background into the time-stepping algorithm. We found that by adjusting the control field strength, dynamic control the shape of the CROW band was enabled, whereas the shape of the CROW band is typically fixed since it is controlled by the coupling strength of the cavities, eg. distance between the defects.

We find that the coupling strength between nearest neighbor cavities in the CROW decreases with increasing steepness of the background dispersion, which is continuously tunable as it is directly related to the control field Rabi frequency. The weaker coupling decreases the speed of pulse propagation through the waveguide. In addition, due to the dispersive nature of the EIT background, the CROW band shape is tuned around a fixed k-point. Thus, the EIT background enables dynamic tunability of the CROW band shape and the group velocity in the structure at a fixed operating point in momentum space.

References

1 K.-J. Boller et al, Phys. Rev. Lett., 66 (1991), 2593.
2 L. V. Hau et al, Nature, 397 (1999), 594.
3 M. Soljacic et al, Phys. Rev. E, 71 (2005) 026602.
4 A. Yariv et al, Opt. Lett. 24 (1999), 711.
5 C. W. Neff et al, New J. Physics, 9 (2007) 48.
6 C. W. Neff et al, Opt. Express, 15

Slow light and its potential applications

Li Zhengbin, Peng Chao, Xu Anshi

Lizhengbin@pku.edu.cn

State Key Laboratory of Advanced Optical Communication Systems and Networks

Peking University, Beijing, China

Slow light has achieved great progress in waveguide and fibers in room temperature, which could be accelerating the transition of this technique to applications. This talk will review devising methods for tailoring the dispersion of optical materials, such as electromagnetically induced transparency, photonic crystals, and nano-optic resonators over the last decade. And the potential applications of slow light for optical buffering, data synchronization, optical memories, and optical signal processing will be cataloged.

Then the interaction between slow light and Sagnac effect is discussed. The concept of the Sagnac effect in a slow-light medium and resonator structure with high group dispersion is investigated. It is found that a slow-light medium can be utilized for relative motion sensing, and a slow-light resonator structure is suitable to detect absolute rotation for navigation purposes. It is noted that the high group dispersion leads to a huge enhancement of the rotation sensor's sensitivity in a resonating structure, and an approach to evaluate and design resonator devices with slow-light property is proposed. Moreover, a folded loop-lattice-based structure is numerically simulated to verify the concept.

References

1. C. Peng, Z. Li, and A. Xu, " Rotation sensing based on a slow-light resonating structure with high group dispersion," Appl. Opt. **46**, 4125-4131 (2007)

2. Chao Peng, Zhengbin Li, and Anshi Xu "Optical gyroscope based on a coupled resonator with the all optical waveguide " www.opticsexpress.org/viewmedia.cfm?id=131606&seq=0

3. R.W. Boyd and D.J. Gauthier, " 'Slow' and 'Fast' Light," in *Progress in Optics, Vol. 43*, E. Wolf, Ed. (Elsevier, Amsterdam, 2002), Ch. 6, pp. 497-530.

4. R.W. Boyd, D.J. Gauthier, and A.L. Gaeta, "Applications of slow light in telecommunications," Optics and Photonics News **7**, 18-23 (2006).

Delay-bandwidth product of a novel slow light waveguide

Chun Jiang

State Key Laboratory of Advanced Optical communication Systems and Networks, Shanghai Jiao Tong University, Room 5-209, SEIEE Buildings, 800 Dong Chuan Road, Shanghai 200240, China,
cjiang@sjtu.edu.cn

Abstract *A optimally-designed novel slow light waveguide is present, the calculated dispersion curve, delay and bandwidth of it reveal that the waveguide has a ultra-slow group velocity and negative dispersion mode and giant delay-bandwidth product.*

Slow light devices have recently been attracting interest of researcher in opt-electronics field. Their potential applications include optical buffers by reducing the length of light path effectively, and integrated photonic circuit with effective reduction of the scale and operational power of elect-optical device due to pulse compression in slow light waveguide [1]. Enhancement of light amplification based on stimulated emission of active ions and stimulated Raman scattering were theoretically investigated in photonic crystal waveguides with abnormal group velocity [2,3]. In this summary, based on side-wall grating waveguide [4], we present the numerical results of delay and bandwidth of an optimally-designed slow light wave-guides composed of an uniform waveguide and a longitudinal grating waveguide. The results show that the waveguide has ultra-slow group velocity mode with zero or negative dispersion, leading to larger delay-bandwidth product.

The schematic structure and its field are shown in Fig.1. The longitudinal grating has a lattice constant of 0.55 a ,and the thickness and length of the single side-wall grating is 0.46a, 1.05a, respectively, the width of the central strip waveguide is 0.5a, where a is arbitrary unit, and the refractive index n of the all material is 3.5

Fig.1 the schematic structures and field distribution of the designed waveguide.

The dispersion relation of the waveguide is calculated by plane wave method [5], and the temporal spectra and electrical field pattern are calculated using finite difference time domain (FDTD) method with perfectly matched layer as absorbing boundary condition [6]. The excitation source is Gaussian pulse with given width, and its amplitude is set as 1.0 (normalized unit).

The calculated dispersion curve shows that the first-order derivative (group velocity) and second-order derivative (group velocity dispersion) of angular frequency to wave vector may be very ultra slow, and zero or negative at non-zero wave-vector, respectively. The results may predict that the waveguide has large delay-bandwidth product, and will be validated using numerical experiments with FDTD method. A ultra-short pulse with optimal frequency and bandwidth is sent from source at the input end of waveguide and transmits to the other end, its delay in the waveguide and bandwidth at the output end are calculated. The numerical results indicate that the pulse duration become shorter or invariant when passing through the slow light waveguide, and the delay depends on specific group velocity. The ultra-slow or zero group velocity mode with zero or negative dispersion may result in the giant delay-bandwidth product of the waveguide.

References

1. Marin Soljacic, et al., J. Opt. Soc. Am. B **19** (9) (2002), 2052-2059.
2. Kazuaki Sakoda, et al, Optics Express, 4(5) (1999), 167-176.
3. McMillan J F, et al. Opt Lett, 31(2006), 1235-1237.
4 M. L. Povinelli, et al, Optics Express, **13** (18) (2005), 7145-7159.
5. S.G.Johnson, Optics Express, 8 (3) (2001), 173-190.
6. Ardavan Farjadpour, et al, Optics Letters, **31** (20) (2006), 2972–2974.

978-0-9789217-3-6/07/$25.00

©2007 WEN GLOBAL SOLUTIONS

All-Fiber Acousto-Optic Tunable Filters

Byoung Yoon Kim

Novera Optics & KAIST, 373-1 Guseong-dong, Yuseong-gu, Daejeon, Korea

yoon.kim@noveraoptics.com

Abstract *Several types of all-fiber tunable filters are described where acoustic wave and optical wave are guided along the same optical fiber.*

Summary

In this presentation, several types of all-fiber tunable filters are described where acoustic wave and optical wave are guided along the same optical fiber. The acoustic waves utilized are fundamental modes of flexural [1] and torsional waves [2] that are generated by coaxial acoustic horn transducers. Bandpass [3] and notch filter types [4, 5] are demonstrated utilizing core mode blockers [6]. The acoustic wave propagating along the fiber produces periodic optical coupling between two spatial (with flexural wave) [4, 5] or polarization (with torsional wave) modes [7] guided in the fiber core and cladding of standard telecom fibers, high birefringence (HB) fibers [7] and photonic crystal fibers [8, 9]. The tunable filters demonstrated high coupling efficiency, fast tuning speed, wide tuning wavelength range, and very low insertion loss [4, 5]. Many unique optical characteristics that are not achievable by other means will be discussed in some detail. The tunable filters are expected to find wide applications in optical fiber communications and sensors.

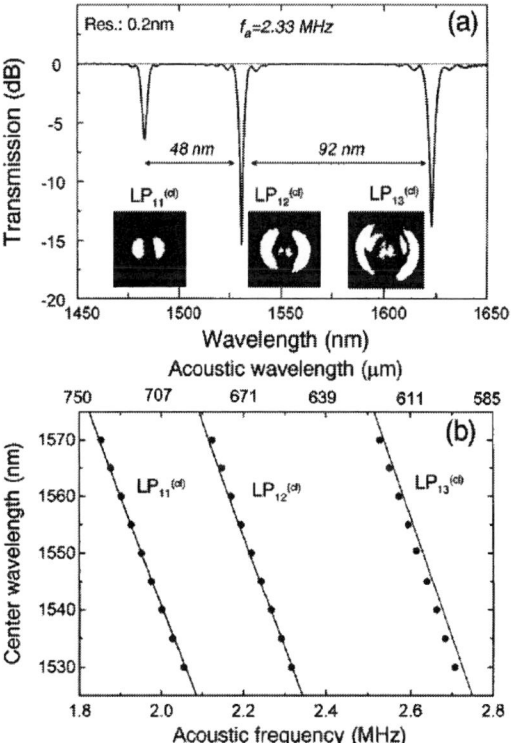

Fig. 2 (a) Transmission spectrum showing coupling to three different cladding modes and their far-field radiation patterns (b) Center wavelengths of notches as a function of acoustic frequency for couplings to three different cladding modes.

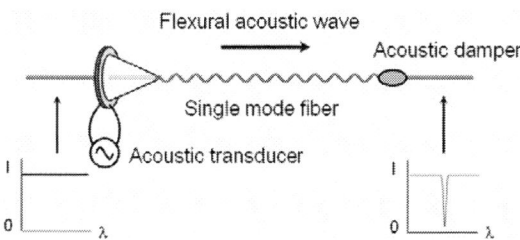

Fig. 1 Schematic of a novel all-fiber acousto-optic tunable filter (Notch type).

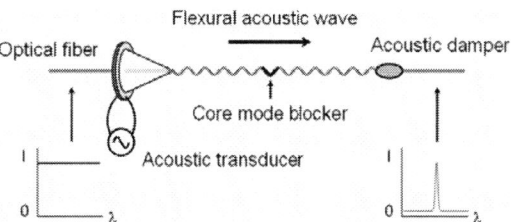

Fig. 3 Schematic of an all-fiber acousto-optic tunable bandpass filter using flexural acoustic wave.

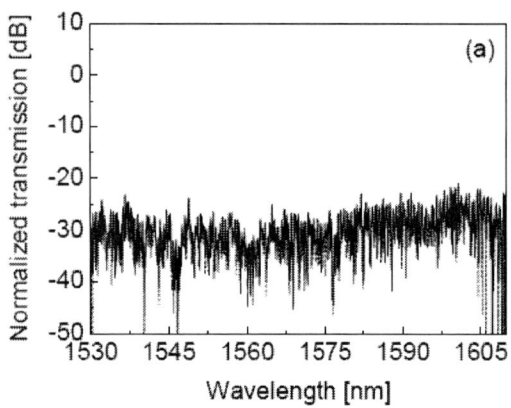

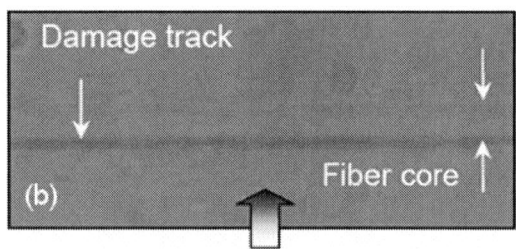

Direction of UV illumination

Fig. 4 (a) The spectral characteristic of the fabricated core mode blocker, (b) Microscopic image of the damage track at the core-cladding boundary of the UV exposed optical fiber.

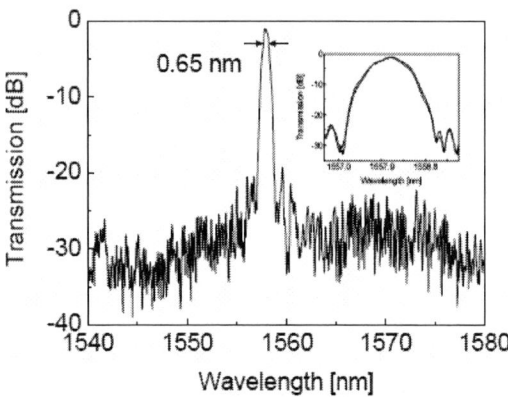

Fig. 5 Measured transmission spectrum of the acousto-optic tunable bandpass filter. The Inset shows optical transmission spectra for two orthogonal input polarizations.

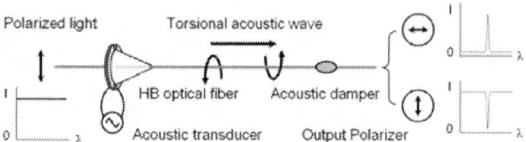

Fig. 6 Schematic of an all-fiber acousto-optic tunable polarization filter (AOTPF) using torsional acoustic wave.

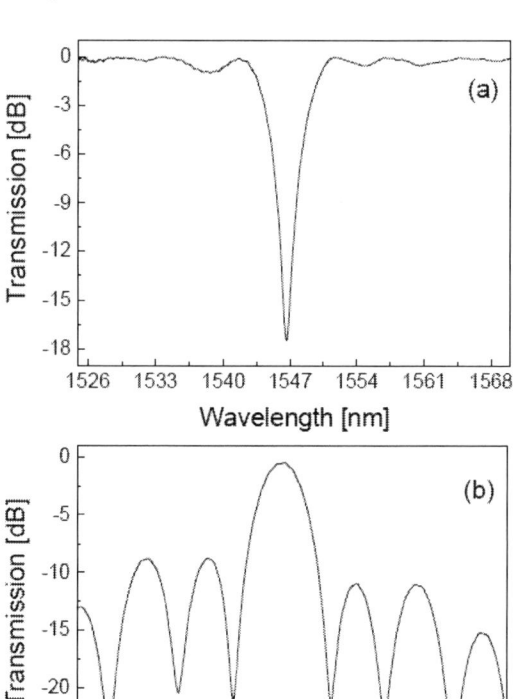

Fig. 7 Measured transmission spectra of the all-fiber AOTPF operating as (a) the notch type and as (b) the bandpass type.

References

1 H. E. Engan et al, J. Lightwave Technol., 6 (1988), 428.

2 H. E. Engan, J. Opt. Soc. Am. A, 13 (1996), 112.

3 K. J. Lee et al, Opt. Express, 15 (2007), 2987.

4 H. S. Kim et al, Opt. Lett., 22 (1997), 1476.

5 S. H. Yun et al, IEEE Photon. Technol. Lett., 16 (2004), 147.

6 K. J. Lee et al, in Proceedings of IEEE CLEO Pacific Rim, (2005) 1082.

7 K. J. Lee et al, Opt. Express, in process (2007).

8 Magnus W. Haakestad et al, J. Lightwave Technol., 24 (2006), 838.
9 K. S. Hong et al, Opt. Express, in process (2007).

A Simple Implementation of Tunable All-Optical Microwave Notch Filter with a Negative Tap Based on a Semiconductor Optical Amplifier

Hongyan Fu, Haiyan Ou, Kun Zhu and Sailing He

Center for Optical and Electromagnetic Research, State Key Laboratory of Modern Optical Instrumentation, Joint Research Center of Optical Communications of Zhejiang University, Zhejiang University, Hangzhou 310058, China; E-mail: fuhongyan@coer.zju.edu.cn

Abstract *A simple implementation of tunable all-optical microwave notch filter with a negative tap based on a semiconductor optical amplifier is demonstrated. The present filter shows good tunability, and is easy to implement and cost-effective.*

Introduction

In recent years, microwave photonic filters have attracted much research interest, due to the inherent advantages of photonics, such as high bandwidth, low loss and immunity to electromagnetic interference (EMI). Moreover, microwave photonic filters can overcome as the electronic bottleneck problem and other sources of degradation of traditional microwave filters [1-3]. Microwave photonic filters have a good potential in the application of promising Radio-over-Fiber systems [4]. Many approaches have been proposed to realize microwave photonic filters, most of which are under incoherent operation, which exhibits some limitations, e.g., the shape of the frequency response is quite restricted since in general it is hard to realize any negative tap (coefficient) in this way; and there is always a filter resonance at the baseband.

Many approaches to implementing negative coefficients have been proposed. The first one is an optoelectronic approach based on differential detection [5] (which is in fact not an all-optical configuration). Microwave photonic filter with negative coefficients based on a specially designed 2×1 integrated Mach–Zehnder modulator [6] was proposed. However, it needs a special modulator which increases the cost. Some implementations were proposed by using the cross gain modulation (XGM) effect in a semiconductor optical amplifier (SOA) [7, 8], which takes the advantage of SOA that is frequently used in the optical link of a RoF system. In these configurations, optical filters and accurate optical delay line are essential, and the coherence time of the laser sources must be taken into consideration. These in turn make the filters complicated and not cost-effective. In this paper, we present a tunable all-optical microwave notch filter with a negative tap based on an SOA and a dispersive medium. The negative tap of the notch filter is realized through the XGM effect in the SOA. Tunability is achieved by tuning the wavelength of one laser source. This filter configuration does not need any optical filter or accurate optical delay control, and does not need to consider the coherent time of the laser sources. The present configuration is very easy to implement and cost-effective.

Operation principle and experimental setup

The frequency response for a multi-wavelength and dispersive medium-based microwave photonic filter can be expressed as

$$H(\Omega) \propto \cos(\frac{\pi \lambda_0^2 D}{c} \frac{\Omega^2}{2}) \left| \sum_{m=1}^{N} A_m \exp[-j\pi(m-1)\Omega D \Delta \lambda] \right| \quad (1)$$

where N is the total number of wavelengths, A_m is the coefficient of the mth wavelength, $\Delta \lambda$ is the wavelength spacing, and D, c and λ_0 are the dispersion (ps/nm) of the dispersive medium, the light velocity in the medium and the central wavelength, respectively. The free spectral range (FSR) of the filter can be expressed as

$$FSR = \frac{1}{D\Delta \lambda} \quad (2)$$

978-0-9789217-3-6/07/$25.00

©2007 WEN GLOBAL SOLUTIONS

Obviously, *FSR* decreases as the wavelength spacing increases. A microwave photonic notch filter can be realized when two carrier wavelengths are used. There is no baseband resonance (at zero frequency) for the notch filter with a negative tap, while there is baseband resonance for the filter with two positive taps.

The experimental setup for our tunable all-optical microwave notch filter with a negative coefficient is shown in Fig. 1. The light from a DFB laser at λ_1 is modulated with amplified RF signal from port 1 of the network analyzer (NA) via the electro-optic modulator (EOM), and then combined with the un-modulated light from a tunable laser at λ_2 through an optical coupler. They are then launched into an SOA after being amplified in the EDFA. XGM effect will occur in the SOA which makes the inverted copy of signal on the un-modulated light from the tunable laser. Then both the non-inverted (at wavelength λ_1) and inverted (at wavelength λ_2) lights go into a dispersive medium, which in our experiment is a roll of single mode fiber (SMF). Due to the different time delays in the dispersive medium at different wavelengths, two taps with one positive and the other negative are produced, and the light with the non-inverted and inverted signals represent the positive and negative taps, respectively. The output RF signal from the photodiode is measured by the network analyzer.

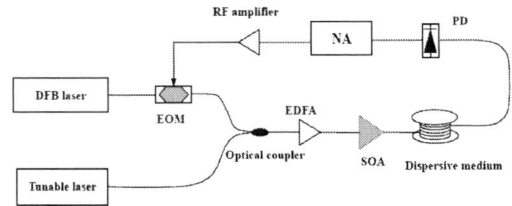

Fig. 1 Experimental setup of the tunable all-optical microwave notch filter with a negative tap

Experimental results and discussion
In our experiment, the DFB laser is at the wavelength of 1554.7 nm and the DC current bias of the SOA (SOA-L-OEC-1550 from CIP) is set at 140mA. A roll of SMF

with length of 25 km is used as the dispersive medium in the experiment. The photodiode (PD) used in the experiment is SIR5-FC from Thorlabs Inc., and the network analyzer used to measure the frequency response of the notch filter is HP R3765C. The optical spectrum analyzer (OSA, Agilent 86142B) is used in the experiment to monitor the spectra of the two lasers.

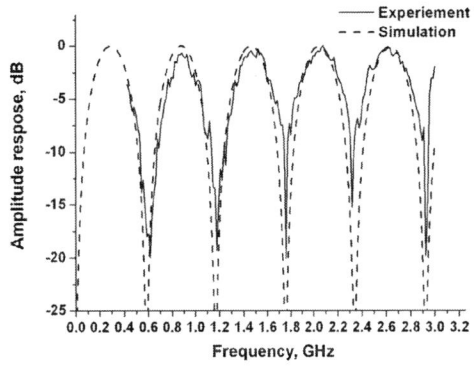

(a)

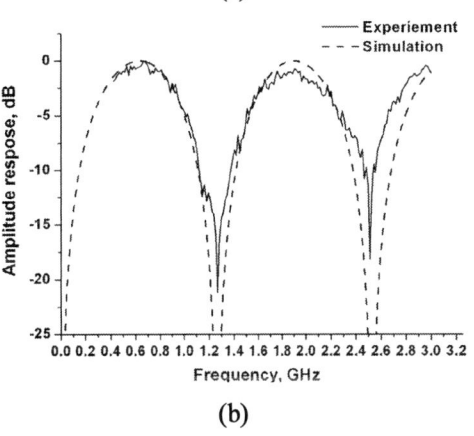

(b)

Fig. 2 Measured (solid line) and theoretical (dashed line) frequency responses of our all-optical microwave notch filter with a negative tap when the FSRs are 0.61 GHz (a) and 1.27 GHz (b), respectively

The measured (solid line) and theoretical (dashed line) frequency responses of our all-optical notch filter with a negative tap are shown in Fig. 2. Due to the cutoff frequency of the microwave amplifier, the response at low frequency can not be measured. However, one can still observe that there are no baseband resonances from the tendency of the measured frequency response, which agree very well with the theoretical calculation. From the experimental results, one can see that the all-optical notch filter with a negative tap implemented in our experiment exhibits the extinction ratio of about 20 dB. To show the tunability, the wavelength of the tunable laser is adjusted to get different free spectrum ranges (FSR), and then different notch positions. In Fig. 2(a) and (b), the wavelength spacing are 3.85 nm and 1.85 nm, respectively (corresponding to the FSR of the notch filter of 0.61 GHz and 1.27 GHz, respectively). One can see from Fig. 2 that the extinction ratio of the measured frequency response is not as large as the theoretical results. That is because in the theoretical calculation, the mismatch between the two lasers, including the amplitude and the shape mismatch, and also the undesirable noise of devices are not taken into consideration.

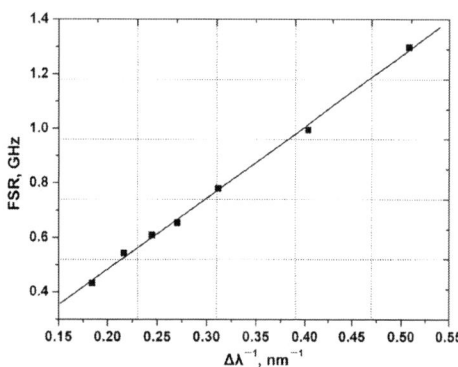

Fig. 3 Experimental result of relationship between the FSR of the all-optical notch filter and the reciprocal of the wavelength spacing of the two lasers

Fig. 3 shows the experimental result of the relationship between the FSR of our all-optical notch filter and the reciprocal of the wavelength spacing of the two lasers as we change the central wavelength of the tunable laser in the experiment. One can see that the experimental result agrees very well with the relationship given by Eq. (2) and shows an excellent tuning linearity.

Conclusions

We have demonstrated a simple implementation of tunable all-optical microwave notch filter with a negative tap based on XGM in an SOA and dispersive medium. An un-modulated light together with the light at a different wavelength modulated with RF signal are launched into an SOA, where the XGM effect occurs to make an inverted copy of the RF signal at the un-modulated wavelength, which represents a negative tap. A fiber roll with the length of 25 km is used as the dispersive medium in the experiment. The frequency responses of the notch filter shows that no baseband resonance appears, and the measured frequency response agrees well with the theoretical results. By tuning the wavelength of one laser source, FSR of the notch filter is changed. Tuning of the FSR of the notch filter in the experiment shows good agreement with the expectation and good tuning linearity. The presented tunable all-optical microwave notch filter with a negative tap is easy to implement and cost-effective, and shows potential application in the Radio-over-Fiber systems.

References

[1] Jose Capmany et al, IEEE Journal of Lightwave Technology, 24(2006), page.201-229

[2] David B., et al, IEEE Photonics Technology Letters, 11(1999), page.874-876

[3] B. Ortega et al, IEE Electronics Letters, 41(2005), page.1133-1135

[4] Jun Wang et al, IEEE Photonics Technology Letters, 17(2005), page.1737-1739

[5] S. Sales et al, Electron. Letters, 31(1995), page. 1095–1096

[6] Jose Capmany et al, Optics Express, 13(2005), page. 1412-1417

[7] F. Coppinger et al, IEEE Transactions on Microwave and Techniques, 45(1997), page.1473-1477

[8] J. Mora et al, Electronics Letters, 41 (2005), page. 53-54

Demonstration of Optical Line Terminal for Full Colorless Bidirectional WDM-Passive Optical Networks using Injection-Locked Fabry Perot Laser and Optical Carrier Suppression

Yong-Yuk Won(1), Dong-Hyeon Kim(2), Sang-Kil Roh(3), Yin-Xing Piao(4), and Sang-Kook Han(5)

Department of Electrical and Electronic Engineering, Yonsei University, Seoul 120-749, Korea

(1)bluejerry@yonsei.ac.kr, (2)kimdhy2k@yonsei.ac.kr, (3)wizard1052@yonsei.ac.kr, (4)espark@yonsei.ac.kr, (5)skhan@yonsei.ac.kr

Abstract *A new architecture of optical line terminal for fully colorless bidirectional wavelength division multiplexed passive optical networks is proposed and experimentally demonstrated. It is implemented by injection-locked Fabry-Perot laser diode and optical carrier suppression. Error free transmissions (bit error rate of 10^{-9}) of both 1.25-Gb/s downstream signal and upstream signal are executed at three representative wavelengths (1530.364 nm, 1540.14 nm, and 1556.713 nm). Accordingly, both optical line terminal and optical network unit are colorlessly operated within 25 nm range.*

Introduction

Wavelength division multiplexed-passive optical network (WDM-PON) has been a very promising architecture for near future broadband access networks. This system can handle a large bandwidth and can provide a good data security[1]-[7]. There is also no need for complicated protocols.

However, this architecture can be relatively expensive because of high-cost optical network unit (ONU) due to location-specific optical source and temperature control circuit. The implementation of optical source-free ONU can effectively reduce the cost of the whole system. Various schemes have been proposed for the centralized control of ONU. For example, reflective semiconductor optical amplifier (RSOA), injection –locked Fabry-Perot (FP) laser, and dual optical sources were presented[2]-[5]. These methods have the weak points such as the power penalty of upstream signal due to downstream one[2]-[4], additional optical sources[4]-[5], and complex transceiver architecture[5]-[7]. Also, these schemes should have all inventories of wavelength-specific optical modules available in multiple channels at optical line terminal (OLT) in terms of the maintenance and repair of system. Accordingly, it is important to realize colorless-OLT as well as colorless-ONU.

In this paper, a new OLT architecture for both colorless-bidirectional WDM-PON is proposed to reduce the cost of the system, to effectively manage, and to implement centralized systems. Light sources for both downstream and upstream are simultaneously generated by the use of single optical source at OLT. Specifically, downstream one is produced due to optical carrier suppression and injection locking of directly modulated FP laser and upstream one is also generated by a split single optical source. Therefore, OLT can be operated regardless of the wavelength of optical source by simply controlling the wavelength of FP laser due to

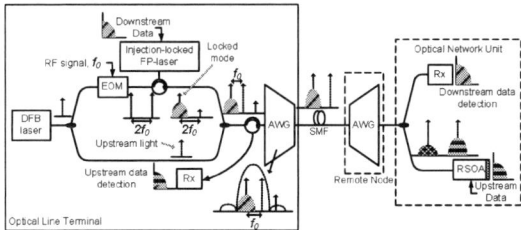

Fig. 1. Proposed bidirectional colorless WDM-PON architecture.

the adjustment of operating temperature and using colorless optical and electrical modules with the same specifications. A colorless-ONU is also implemented using an RSOA because it modulates and amplifies irrespective of the wavelength of input optical signal within its optical gain region.

We experimentally demonstrate that the proposed scheme can be implemented as a bidirectional WDM-PON architecture with colorless-OLT and ONU through 1.25-Gb/s data transmission

Bidirectional WDM-PON architecture with colorless OLT and ONU

The bidirectional colorless WDM-PON architecture with the proposed colorless OLT scheme is shown in Fig. 1. The light from DFB-laser is split into two parts. The light of upper arm is converted into optical carrier suppressed (OCS) one with two wavelengths using electro-optic modulator (EOM) such as a LiNbO$_3$ Mach-Zehnder modulator (MZM). An OCS light is injected into directly modulated FP-laser by downstream data and then one of two wavelengths is locked at specific FP-mode. They are consisted of locked mode with downstream data and unlocked mode. Locked mode functions as a downstream optical signal. Two modes and the continuous wave (CW) light of lower arm used as an upstream optical signal are combined and then moved to an arrayed waveguide grating (AWG). Unlock mode is automatically filtered out due to the limitation of the channel bandwidth of AWG. Locked mode and upstream light are transmitted to an ONU through single mode fiber (SMF). The downstream data of locked mode is detected at a receiver of ONU. The CW upstream light is injected into an RSOA and modulated by upstream data. It is retransmitted back to an OLT and detected at a receiver. A colorless-OLT can be implemented because a locked FP-mode can be easily generated regardless of the wavelength of DFB-laser by controlling the temperature of FP-laser when optical sources of each DFB-laser from multiple channels exist within the range of several FP-modes.

Experimental setup

The experimental setup for proposed scheme is shown in Fig. 2. Insets of each point were measured using optical spectrum analyzer with resolution bandwidth of 0.065 nm and video bandwidth of 1 kHz. A tunable light source (TLS) was used to check if colorless-OLT can be implemented or not. Its launched optical power was 8 dBm. CW light from TLS was divided into two parts by 3-dB coupler. One was modulated by 30-GHz RF signal at MZM. An OCS light was generated with optical carrier suppression ratio of ~10 dB as shown in inset (b) of Fig. 2. The optical beat interference between remained suppressed optical carrier and the other CW light from TLS (CW upstream optical signal) was eliminated by the use of polarization controller (PC) 2 because noise due to this effect adversely affected both downstream and upstream data. An OCS source was injected into FP-laser with 1.1-nm spaced FP-modes from 1530 nm to 1560 nm. Its intensity was adjusted for optimal injection ratio by the use of erbium doped fiber amplifier and optical attenuator. The output power of FP-laser was 5 dBm above threshold level (Threshold current, I_{th} = 8 mA). It was directly modulated by 1.25-Gb/s baseband data with 2^{23}-1 pseudo-random binary sequences (PRBS) and 2 V_{p-p} swing depth. Optical spectra of injection locked FP-laser at wavelength of 1540. 14 nm was measured as shown in inset (a) of Fig. 2. They were measured repeatedly at two wavelengths (1530.364 nm, 1556.713 nm). We obtained very similar results compared to that of 1540.14 nm. The specific FP-mode was locked by the only first sideband, not the second sideband (Unlocked mode). The side mode suppression ratio between locked mode and other FP-modes was about 30 dB. The locking range of directly modulated FP-laser was 50 GHz at the state of error free transmission showing the bit error rate (BER) of 10^{-9}. The light from injection locked FP-laser and CW optical carrier of lower arm were combined by 3-dB coupler. The combined lights were fed into an AWG with an insertion loss of 5 dB and 3-dB bandwidth of 0.3 nm. Unlocked mode of them was suppressed because it exists out of channel bandwidth of AWG

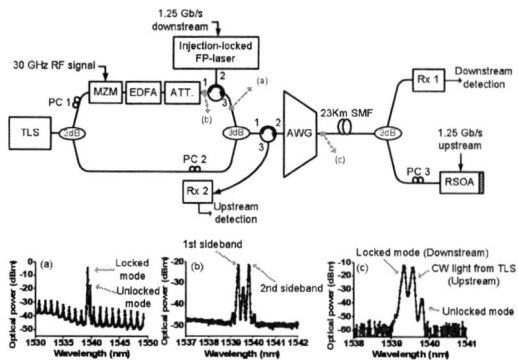

Fig. 2. Experimental setup of proposed scheme and optical spectra at each point. TLS: Tunable light source, MZM: Mach-Zehnder modulator, EDFA: Erbium doped fiber amplifier, ATT.: Optical attenuator. PC: Polarization controller, Rx: 1.25 GHz optical receiver.

(~37.5 GHz) as shown in inset (c) of Fig. 2. They were transmitted through 23-km SMF and then detected at 1.25-GHz optical receiver (Rx 1). They were also injected into an RSOA and modulated by 1.25-Gb/s baseband data with 2^{23}-1 PRBS and 2 V_{p-p} swing depth. The total propagation loss of combined lights was -15 dB, including an AWG, 3-dB coupler, circulator, and an SMF of 23 km. Polarization controller (PC 3) was used to maximize the optical gain of the input signal injected into an RSOA. The polarization dependent gain (PDG) of RSOA used in our experimental setup was about 3 dB at gain saturation region, when the input optical power was -12 dBm. The output power from an RSOA was 0 dBm. The modulated lights at an RSOA were retransmitted and then detected to the 1.25-GHz receiver at OLT (Rx 2).

Based on this experimental setup, the BER curves of both downstream and upstream data at three representative wavelengths (1530.364 nm, 1540.14 nm, and 1556.713 nm) were measured repeatedly to verify the realization of colorless-OLT.

Results and discussion

To verify that proposed scheme can be operated as a colorless-OLT and ONU, the BER curves of both 1.25-Gb/s downstream and upstream data were measured as the wavelength of light from TLS changed as shown in Fig. 3. Three representative wavelengths were selected according to the range of FP-mode (1530.364 nm, 1540.14 nm, and 1556.713 nm). Error free transmission (BER of 10^{-9}) of both downstream and upstream was observed at three channels. The power margins of both downstream and upstream were 12dB and 6dB respectively at BER of 10^{-9}. The power penalty of ~1.5 dB per each channel was also observed at downstream case. This can be mainly attributed to the well known Rayleigh backscattering noise. In case of upstream, we observed that the power penalty of ~2 dB was generated at each wavelength. This is also because the same noise as a transmission of downstream adversely affected 1.25-Gb/s upstream data. As shown in Fig. 3, we observed that the range of wavelength which both OLT and ONU are capable of operating was over 25 nm, from 1530 nm to 1556 nm. This means that they can be implemented as colorless OLT and ONU by the use of the proposed WDM-PON architecture.

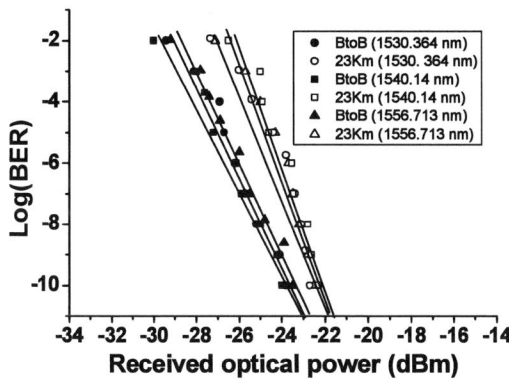

(a)

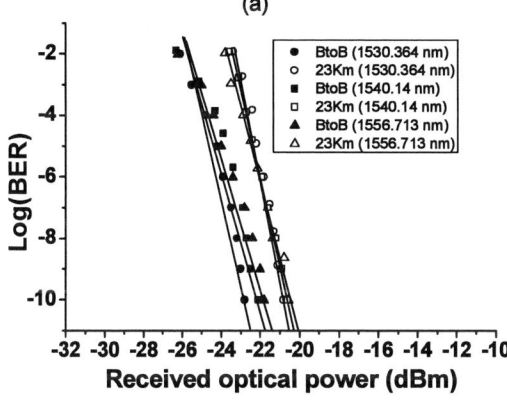

(b)

Fig. 3. Measured BER curves of 1.25-Gb/s downstream and upstream data after 23 Km transmission against the variation of wavelength of optical source from TLS. (a): Downstream BER, (b): Upstream BER.

Conclusion

A new OLT architecture for fully colorless WDM-PON was proposed in this paper. It was colorlessly implemented by the use of injection locking of FP-laser and optical carrier suppression. An RSOA was employed as colorless-ONU. Error free transmission (BER of 10-9) of both downstream and upstream was accomplished at three representative wavelengths (1530.364 nm, 1540.14 nm, and 1556.713 nm). The power penalties of both 1.25 Gb/s downstream and upstream data were 1.5 dB and 2 dB respectively at BER of 10-9. These were mainly generated by Rayleigh backscattering noise. Based on these

experimental results, we can say that the colorless operation of OLT and ONU was realized at the range of 25 nm.

References

1 N.J. Frigo et al., "A wavelength-division multiplexed passive optical network with cost-shared components," IEEE Photon. Technol. Lett., vol. 6 (1994), pp. 1365-1367.

2 P. Healey et al., "Spectral slicing WDM-PON using wavelength-seeded refletive SOAs," Electron. Lett., vol. 37 (2001), pp. 1181-1182.

3 W. Lee et al., "Bidirectional WDM-PON Based on Gain-Saturated Reflective Semiconductor Optical Amplifiers," IEEE Photon. Technol. Lett., vol. 17 (2005), pp. 2460-2462.

4 L. Y. Chan et al., "Upstream traffic transmitter using injection-locked Fabry-Perot laser diode as modulator for WDM access networks," Electron. Lett., vol. 38 (2002), pp. 43-45.

5 W. Hung et al., "An optical network unit for WDM access networks with downstream DPSK, and upstream remodulated OOK data using injection-locked FP laser," IEEE Photon. Technol. Lett., vol. 15 (2003), pp. 1476-1478.

6 O. Akanbi et al., "A new scheme for bidirectional WDM-PON using upstream and downstream channels generated by optical carrier suppression and separation technique," IEEE Photon. Technol. Lett., vol. 18 (2006), pp. 340-342.

7 J. Yu et al., "Demonstration of a Novel WDM Passive Optical Network Architecture With Source-Free Optical Network Units," IEEE Photon. Technol. Lett., vol. 19 (2007), pp. 571-573.

Patterned Photonic Crystals for Novel Applications

Y. Ohtera (1), T. Sato (2), S. Kawakami (3)

1 : TUBERO, Tohoku Univ., ohtera@tubero.tohoku.ac.jp
2 : Photonic Lattice, Inc., sato@photonic-lattice.com
3 : Photonic Lattice, Inc., SFAIS, kawakami@photonic-lattice.com

Abstract *Using patterned photonic crystals fabricated by Autocloning, novel functional elements, devices, and systems are realized. Applications of axisymmetric polarization, photoelasticity imaging, and spectroscopic imaging are introduced.*

1. Introduction

We have developed a fabrication method of multilayer-type two- and three-dimensional photonic crystals(PhCs) on a basis of rf sputtering process (autocloning method[1]). The horizontal lattice geometry of the resulting periodic structure can be designed with a large degree of freedom, which enables various applied optical elements for polarization and spectroscopic imaging.

2. Elements for generating axisymmetric polarization

Recently, beams with a spatially inhomogeneous polarization have attracted a great deal of attention as a new optical technology. Axisymmetricly polarized beams such as the radial and azimuthal polarization shown in Fig. 1, are particularly promising for optical applications such as laser cutting, high-resolution microscopy, and optical trapping. In this section, a simple method of generating such polarizations is demonstrated by using patterned photonic crystal polarizers.

Figure 2 shows an illustration of, and a surface SEM of a photonic crystal with a concentric ring pattern. The multilayer structure is composed of Nb_2O_5/SiO_2 with an in-plane pitch of 500 nm. Layers are deposited to a thickness appropriate to function as a mirror the reflectance of which depends on whether the polarization is radial or azimuthal. The reflectances of the radial polarization (local TE wave) and azimuthal polarization (local TM wave) are 90% and 10% at 1064nm, respectively. This element is deployed as an output coupler of a Nd:YAG laser resonator. Figure 3 shows the intensity distribution of the output laser beam. The intensity distribution of the beam after passing through a linear polarizer shows that the beam is radial polarized.

Similarly, an azimuthally polarized beam has been generated using a concentrically patterned photonic crystal with reflectance values of 90% and 10% for azimuthal and radial polarizations respectively.

Fig. 1 Illustration of radial polarization (left) and azimuthal polarization (right).

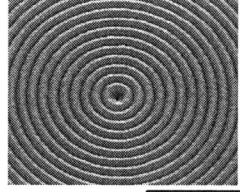

$3\mu m$

Fig. 2 SEM cross-section and surface view of photonic crystal with concentric patterned.

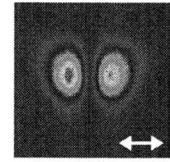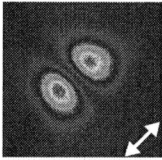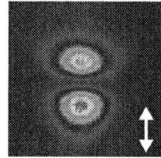

Fig. 3 Beam profiles after passing through a linear polarizer. Arrows show the orientation of the analyzer.

3. Imaging of 2D distribution of retardation and photoelasticity

We have developed a polarization imaging camera consisting of an arrayed photonic crystal polarizer and a CCD sensor [2]. Commonly polarization imaging requires the sequential acquisition of several images of different polarization states. This may be achieved by rotating a linear polarizer in front of the sensor, or by similar means. Our camera does not require a rotating polarizer because it is able to acquire intensities corresponding to four polarization states simultaneously, from which the major axis of any polarization can be determined. The advantages of our camera are real-time acquisition, and a simple configuration suited to applications such as metrology for in-line processes.

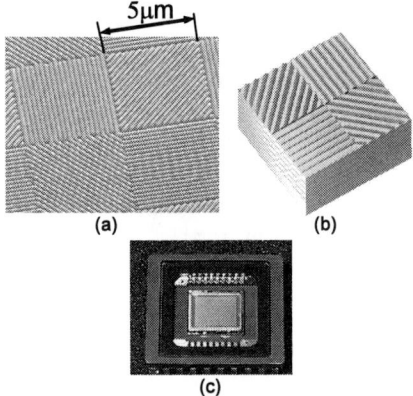

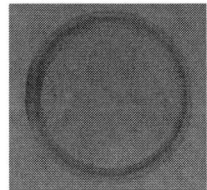 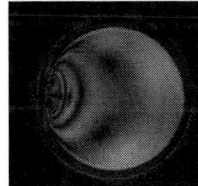

Fig. 5 (a) Illustration of the arrayed photonic crystal polarizer, (b) surface SEM of the polarizer, and (c) image of the arrayed polarizer integrated with a CCD.

Fig. 6 Birefringence imaging of a lens,(a) intensity image, and (b) birefringence image.

Figure 5 is a surface SEM of the polarizer array with a pitch of 4.6 microns, which corresponds to the pixel size of the CCD. The polarizer array consists of an array of polarizers with transmission axes of four different orientations; 0, 45, 90 and 135 degrees, which is repeated periodically. The output of a pixel is expressed generally as I = M + A cos(2 (α-θ)), where M, A, α, and θ are the mean intensity, amplitude of the orientation dependence, the direction of the major axis, and the orientation variable respectively. The polarization parameters M, A, and α, are computed and displayed at 9 frames per second.

We have recently developed a system to image the 2D distribution of retardation or photoelesticity, based on the polarization imaging camera. A circularly polarized light source is employed so that small amounts of retardation or photoelasticity with an arbitrary optical axis can be measured in a single shot. Figure 6 indicates the measured results for a plastic lens. It shows photoelasticity inhomogeneity originating in the injection molding process. The system has applications in in-line inspection of

injection molded optical products such as plastic lenses.

4. Application to edge filters

Autocloning-type photonic crystals are also useful as wavelength filters[3,4]. Transmission spectrum of a PhC is slightly shifted from that of ordinary flat multilayer due to the in-plane structural modulation of the refractive index. An example of the transmission characteristics for unmodulated (flat) and modulated mutilayers are plotted in Fig. 7. Autocloned PhC represents deeper photonic bandgaps(PBGs). The edge wavelength of the PBGs depends on both in-plane and vertical lattice constants of the periodic structure. For typical Nb_2O_5/SiO_2 materials, the sensitivity of the edge wavelength(λ_e) of the second PBG to the in-plane lattice constant(p) is about $d\lambda_e/dp \sim 1.3$. Fig. 8 shows multi-channel edge filters for $\lambda_e \sim 800nm$, in which twelve square edge filter regions with almost evenly spaced cut-off wavelengths are arranged on one substrate.

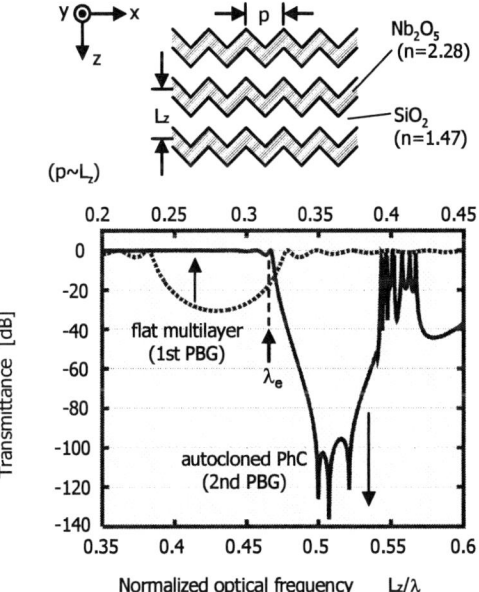

Fig. 7 Calculated transmission spectra of flat(dotted) and autocloned(solid) multilayers. The number of the layers is 20 for both cases.

5. Mosaic wavelength filters for spectroscopic imaging sensors

By minimizing the individual wavelength filter regions in Fig. 8(a) and arrange them in such a manner that a set of small unit filters is repeated in the horizontal direction, a mosaic filter for the imaging spectroscopy ca be created. The size of each miniature filters should be about 5μmx5μm to be directly integrated with image sensors as CCD/CMOS. The picture of the PhC mosaic wavelength filters is shown in Fig. 9.

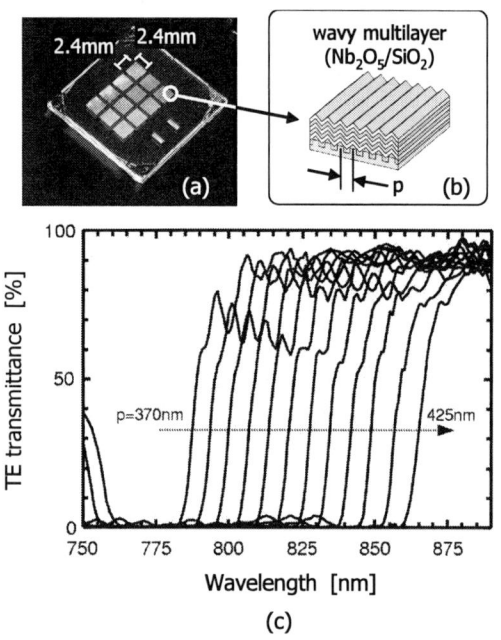

Fig.8 *Photonic crystal multi-channel wavelength filters.(a) picture of the filters, (b) schematic view of each element, (c) measured transmission spectra.*

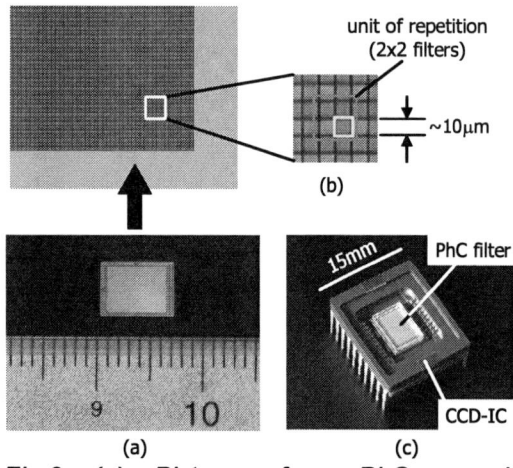

Fig.9 *(a) Picture of a PhC mosaic wavelength filter. (b) magnified view. (c) CCD image sensor integrated with the filter.*

This filter element is designed to photograph four-channel spectroscopic images at λ~800nm at a same time and is applicable to, for example, the biomedical imaging of a human body. Fig. 10(a) and (b)

show the average reflectivity map of a human skin(hand) and the calculated distribution of the hemoglobin concentration existing in the microvessels. The merits of using such autocloning-type PhCs as wavelength filters for spectro-scopic sensors are the capability of the parallel picturing of the multiple wavelength images, low spectral cross-talk, and the compactness of the sensor structure. Such PhC-type spectroscopic sensors are also useful in various industrial fields.

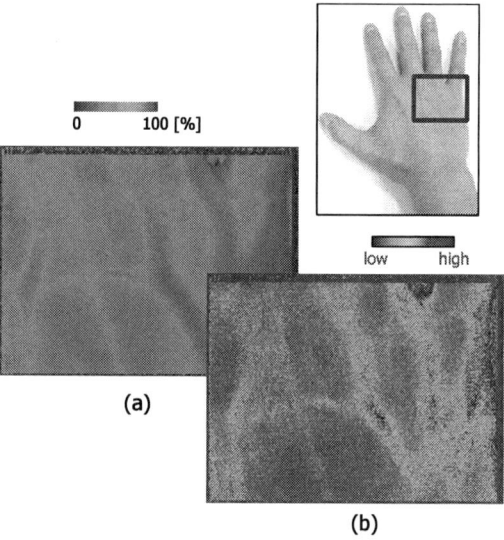

Fig.10 *Example spectroscopic images of a human body(skin) captured by the PhC filter-mounted image sensor. (a) reflectivity at λ~800nm, (b) map of the hemoglobin concentration index(estimated by using the reflectivity of the four wavelength bands).*

Conclusions

Several applied optical elements created by the autocloning-type PhCs are reviewed. By making use of the flexibility of the design of the horizontal periodic structure, a number of functions that cannot be achieved by the conventional flat multilayers are realized.

References

1. S. Kawakami, Electron. Lett., <u>33</u>(14), 1260-1261(1997).
2. S. Kawakami, IEICE trans. Electronics, <u>E90-C</u>(5), 1046-1048(2007).
3. Y. Ohtera et al., J. Lightwave Technol., <u>25</u>(2), 499-503(2007).
4. Y. Ohtera et al., Jpn. J. Appl. Phys., Part .1, <u>46</u>(4A), 1511-1515(2007).

Novel glasses and glass-ceramics for broadband optical amplification

Jianrong Qiu[1,*], Shifeng Zhou[1], Jinjun Ren[2] and Botao Wu[2]

1 : State Key Laboratory of Silicon Materials, Zhejiang University, Hangzhou 310027, China

2 : Graduate School of Chinese Academy of Sciences, Shanghai Institute of Optics and Fine Mechanics, Shanghai 201800, China

*E-mail address:qjr@zju.edu.cn

Abstract *Optical fiber amplifier plays an important role in the optical communication system. Recently, considerable effort has been devoted to develop optical fiber amplifiers which can be used to produce optical gains at different communication bands. In this paper, we review recent research progress on the development of novel materials for optical amplification. We will introduce Bi-doped glasses and transparent Ni^{2+}-doped glass-ceramics, which are very promising for the realization of ultrabroad band optical amplification.*

Introduction

As an optical device used to compensate for the signal loss in the transmission fiber, the fiber amplifier takes an important role in the development of modern optical communication networks. There are many types of fiber amplifiers for providing gain for wavelength division multiplexed (WDM) systems, for example, erbium-doped silica/tellurite-based fiber amplifiers, thulium-doped fluoride fiber amplifier and praseodymium-doped fiber amplifiers.[1-4] One of the problems unsolved today in communication technology is the limited bandwidth of rare-earth (RE) ion-doped amplifiers since the luminescent bandwidth of the 4f-4f optical transition is narrow by nature. Although fiber Raman amplifiers (FRAs) can realize broadband amplification, they require multiwavelength pumping schemes and have complex structures. If broadband amplification of the whole telecommunication window (1.26-1.65 μm) is realized in a single fiber amplifier excited by a single wavelength pumping source, a drastic evolution could be expected to occur in the communication technology. In this paper, we introduce recent research progress on the development of novel materials for ultrabroad band optical amplification. [5-10]

1. Bi-doped glasses

Recent research has revealed that Bi-doped glasses are promising as broadband fiber amplifier medium since they show broadband emission, optical amplification, and laser operation in the second telecommunication window. However, there was no investigation about optical amplification properties in the whole 1.3-1.6 μm region owing to the barrier of acquiring ultrabroadband emission in the Bi-doped glasses and there were only a few investigations related to the materials.

Recently, we observed ultrabroadband and efficient optical amplification covering the whole 1.3-1.6 μm region telecommunication windows in Bi-doped germanium silicate glasses. Additionally, relative flat emission and gain are observed in the germanium silicate glasses.

The germanium silicate glass samples with compositions of $96.5GeO_2-3Al_2O_3-0.5Bi_2O_3$ (GAB) and $79.5GeO_2-17SiO_2-3Al_2O_3-0.5Bi_2O_3$ (GSAB) (in mol%) exhibit broad emission in the 1000-1700 nm wavelength region. The emission peak position of GSAB glass show some redshift compared to GAB glass. The full width of half maximum of the luminescence of GAB and GSAB is 332 nm and 390 nm, respectively.

It is attractive that broadband emission from Bi-doped GAB and GSAB glasses covers the O, E, S, C and L bands (1260-1625 nm) pumped by commercially available 980 nm LD. Actually, as an efficient amplifier materials for WDM operation, the broadband and flat luminescence characteristics are two premises for the demand of compensation and maintenance of the gain at each channel, so that the gain excursion over the whole spectral width can be minimized. The

978-0-9789217-3-6/07/$25.00

©2007 WEN GLOBAL SOLUTIONS

single-pass optical amplification was measured in Bi-doped GAB and GSAB glasses for the purpose of investigating their potential application in broadband fiber amplifiers.

The traditional two-wave mixing configuration was adopted for internal gain measurement. We measured the wavelength-dependent internal gain from 1272 to 1348 nm and the gain at 1560 nm was also measured. The excitation power was 1.12 W. We observed that the measured spectral dependence of the optical gain resembles the fluorescence spectrum. The highest gain at 1272 nm of GAB and GSAB samples reached to 3.65 dB and 6.73 dB, respectively. The values are larger than the reported results in Bi-doped silica glass. [5] The efficient gain from 1272 to 1348 nm demonstrates that the GAB and GSAB glasses can be used as a broadband amplifier especially at the second telecommunication window since there is no appropriate amplifier except Pr-doped fluoride fiber amplifier in this window as yet. Furthermore, it is significant that optical amplification at 1560 nm was also observed. The single-pass gain at 1560 nm of GAB and GSAB samples were 0.95 and 2.3 dB, respectively. It is significant since there is no report of realizing optical amplification at O and C bands simultaneously excited by a single pumping source in Bi-doped materials. According to the broadband emission and wavelength-dependent gain properties of our samples, the acquired gain at O and C bands was just limited by our available seed source and broader gain covering the whole O, E, S, C and L bands is expected. Such a broadband amplification characteristic is seldom observed in other kinds of gain materials. We also measured the internal gain at 1300 and 1560 nm as a function of pumping power. The results show that the gain increases linearly with excitation power up to 1.12 W for both samples and it is expected that the optical gain can be improved by increasing the pumping power.

2. Ni-doped glass-ceramics

A glass sample with composition 13.3MgO-28.9Al_2O_3-10.2TiO_2-46.7SiO_2-0.9Li_2O (wt.%) (MATS) doped with 0.1 wt.% NiO was prepared by conventional melt-quenching method. We measured infrared fluorescence spectra of the MATS glass and glass ceramic sample (GCs) excited by 980 nm LD. A broad emission band centered at 1320 nm covering the 1100-1700 nm range was observed in the MATS GCs. The full width at half maximum (FWHM) is more than 300 nm. Such an emission is characterized as the typical emission of the $^3T_2(F) \rightarrow {}^3A_2(F)$ transition of octahedral Ni^{2+} ions. The MATS glass doesn't show any emission in the near-infrared region and it can be concluded that Ni^{2+}-doped crystals which were embedded in the glass matrixes contribute the infrared luminescence. The emission decay curve of MATS GCs was measured and the mean decay lifetime was calculated as 122 μs. Non-exponential character of the decay curves of MATS GCs might be ascribed to the variation of environment surrounding the Ni^{2+} ion and the complicated energy transfer process in GCs materials. The detailed mechanism is under investigation. It is interesting to point that the emission of Ni^{2+}-doped MATS shows some red-shift compared with Ni^{2+}-doped lithium-galium – silicate (LGS) GCs and Ga_2O_3 GCs which show broad emission at 1300 nm and 1200 nm, respectively. The infrared luminescence is determined by the energy interval between the $^3T_2(F)$ and $^3A_2(F)$ levels and the smaller energy interval may result in lower energy emission. The varies of Dq/B parameter shift the positions of the $^3T_2(F)$ level and the lower crystal field strength of Ni^{2+} in MATS GCs will result in the smaller energy interval and longer wavelength emission. MATS GCs with a broadband infrared luminescence at 1320 nm has a potential application as a broadband optical amplifier.

The change in optical signals at telecommunication bands was also measured when the seed beam passes through the bulk MATS GCs with or without 980 nm excitation. Neither spontaneous emission nor leakage of the excitation beam was detected. The observed enhancement of signal might be ascribed to the optical amplification. But, it is not easy for Ni^{2+}-doped glasses and crystals to realize optical amplification and lasing at room temperature due to the strong excited absorption of Ni^{2+}. Further investigations will be carried out to check whether the enhanced signal is due to the amplified spontaneous emission. The relationship between the change in optical signal and the launched pump power was discussed. The change in signal increases with excitation power up to 1.12 W. The spectral dependence of the changes in signal resembles the fluorescence spectrum and the bandwidth is expected to be more than 77 nm.

The broad luminescence peaked around 1300 nm pumped by common commercial 980 nm LD for Ni^{2+}-doped MATS GCs has great significance since it lies at the second telecommunication window. It is known that Pr-doped fluoride fiber is used as amplifier in this window at present. However, the fluoride fiber is difficult to connect to silica fiber and the gain width of Pr-doped fluoride fiber is small. Furthermore, the broadband emission covering almost the whole telecommunication window has potential application in broadband fiber amplifier.

Conclusion

In conclusion, ultrabroadband and efficient gain was realized in Bi-doped germanium silicate glass excited with a commercially available 980 nm LD. Broadband optical amplification covering the O, E, S, C and L bands and bridging the second and third telecommunication windows has great significance in modern telecommunication networks. In addition, broadband infrared luminescence was observed in Ni-doped glass-ceramics. The mean decay lifetime was calculated as 122 µs. These materials are promising for the realization of broadband optical amplification.

References

[1] M. Yamada, H. Ono, and Y. Ohishi, Electron. Lett. **34**, 1490 (1998).

[2] L. Huang, A. Jha, S. Shen, and X. Liu, Opt. Express **12**, 2429 (2004).

[3] T. Kasamatsky, Y. Yano, and H. Seller, Opt. Lett. **24**, 1684 (1999).

[4] Y. Miyajima, T. Sugawa, and Y. Fukasaku, Electron. Lett. **27**, 1706 (1991).

[5] Y. Fujimoto and M. Nakatsuka, Appl. Phys. Lett. **82**, 3325 (2003).

[6] M. Peng, J. Qiu, D. Chen, X. Meng, I. Yang, X. Jiang, and C. Zhu, Opt. Lett. **29**, 1998 (2004).

[7] X. Meng, J. Qiu, M. Peng, D. Chen, Q. Zhao, X. Jiang, and C. Zhu, Opt. Express, **13**, 1635 (2005).

[8] T. Suzuki, G. S. Murugan, and Y. Ohishi, Appl. Phys. Lett. **86**, 131903 (2005).

[9] S. Zhou, H. Dong, G. Feng, B. Wu, H. Zeng, and J. Qiu, Opt. Express **15**, 5477 (2007).

[10] S. Zhou, G. Feng, B. Wu, N. Jiang, S. Xu, and J. Qiu, J. Phys. Chem. C **111**, 7335 (2007).

Raman Enhancement of TO-520cm^{-1} Mode of Si by Off-Plane One-Dimensional Grating Etched on Si Substrate

Ling-Chung Chou[a,b,1] and C.-H. Kuan[a,b,2,*]

Graduate Institute of Electronics Engineering and Department of Electrical Engineering, National Taiwan University, Taipei, Taiwan, Republic of China[a]

Department of Electrical Engineering, National Taiwan University, No. 1, sec. 4, Roosevelt Rd., Taipei, Taiwan, R.O.C.[b]

d94943031@ntu.edu.tw[1], Kuan@cc.ee.ntu.edu.tw[2]

Abstract *We have observed the relationship between the intensity of Raman scattering of TO-520cm^{-1} mode of Si and variations of the geometrical structure of Si gratings as well as the polarizations of incident waves.*

Introduction

There have been many efforts done about the behavior of light in gratings, or, in another words, photonic crystals, theoretically (1-5). Some of them have even predicted the enhanced phenomena of Raman as well as the transmission of light by evaluations. However, there were not too many people whom we known who to discuss Raman enhancement by Si lateral crystals (6) experimentally,. That is, there are not plenty enough of systematic research in experiment on this region. We have observed Raman scatterings from Si lateral gratings fabricated on substrate surfaces by changing the polarizations of incident waves as well as the geometrical structure of gratings systematically. And the authors think that it would be helpful for us to comprehend the behavior of lights in the grating and to apply them to other regions such as optical detection, etc. That target we measured is the TO-520cm^{-1} mode of Si. Up to this article, we only present the relationships between the intensity of Raman scattering and different parameters in our experiment. Other dependences such as the full width of half maxima (FWHM), etc., will be studied later.

Experiment

We fabricated the grating by the E-beam wrier (ELS-7500EX, ELIONIX Inc., Japan) with ZrO/W thermal emission (Schottky emission) electron gun and the inductively coupled plasma reactive ion etching system, ICP-RIE (EIS-700, ELIONIX Inc., Japan) on (1 0 0) Si wafer directly. Fig. 1 shows the SEM picture of one of our gratings with

period 400nm, line width 180nm and etching depth 250nm. The spectrometer, T64000 (diffraction grating: 1800 lines/mm), made by JOBIN YVON-SPEX and the detector, CCD-3000, made by JOBIN YVON consist the main optical equipments

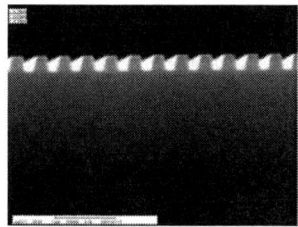

Fig. 1: The SEM picture of the grating with period 400nm, line width 180nm and etching depth 250nm.

in our experiment. Furthermore, we use a micro-optical system, BX-41, OLYMPUS and a 100×, NA=0.9 object to achieve 1μm incident beam size. An Nd-YAG 532nm laser is used to be the light source and the

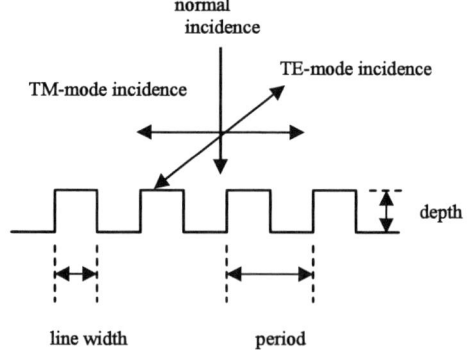

Fig. 2: the terminologies including line, period and depth used to describe the geometry of the grating in this article

incident direction is normal to the substrate. The incident power is 3mW and the CCD integration time is 1 sec. We say that the incidence with polarization parallel to the direction the grating as the TM-mode and with polarization vertical to it as the TE-

---

978-0-9789217-3-6/07/$25.00

©2007 WEN GLOBAL SOLUTIONS

mode. Then the scattering excited by the TM-mode incidence is called as the TM mode, which excited by the TE-mode incidence is called as the TE mode, in this

line width (nm) \ period (nm)	80	100	120	150	200
160	•				
200	•	•	•	•	
250		•	•	•	
300				•	•
400	•	•	•	•	•
500	•	•	•	•	•
600	•	•	•	•	•
800	•	•	•	•	•
1000	•	•	•	•	•
1200		•			
1500				•	
2000					•

Table 1: A list of all grating structures in this experiment.

article; respectively. We change the geometry of the grating systematically: line width is from 80nm to 200nm and period is designed from 160nm to 2000nm. Table 1 lists all the structures in this paper. We do not have modified any data in all discussions during the next section.

Fig. 3 represents the Raman spectrums of TO-520cm^{-1} mode of Si from polished plane Si substrate (curve a) and a grating etched

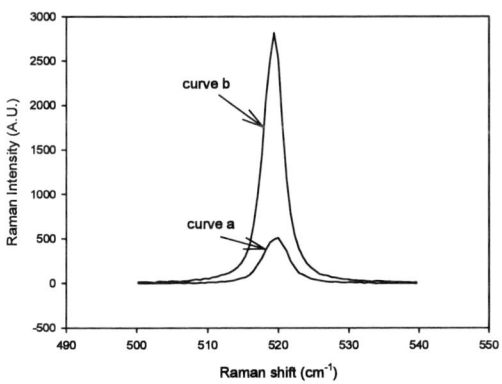

Fig. 3: Raman spectrums of Si's TO-520cm-1 mode from plane Si substrate (a) and from that with grating etched on it (b); where the grating has 100nm line width, 500nm period and 250nm etching depth.

on the same substrate as the former one (curve b), in which the geometric parameters of that grating is: period=500nm, line width=100nm and etching depth=250nm. Obviously, one can see that plane substrate is less than that from the

other one. Hence, we refer the Raman scattering from the polished plane Si substrate outside that mesh of grating etched as our reference. And the Raman intensities discussed through this entire article are all normalized.

Results and Discussions

Fig. 4 shows the Raman responses excited by the TM-mode incidence; where the horizontal axis labels the grating period in unit of nanometer (period=80nm (O), 100nm (Δ), 120nm (▽), 150nm (+) and 200nm (□)), and the vertical axis states the normalized intensity of Raman scattering. According to these data, one could find that there seems to exist some period for specified line width such as the Raman scattering has the extreme, obviously. The magnitude of period corresponding to the location of this extreme differs very small as period changing although it seems that the

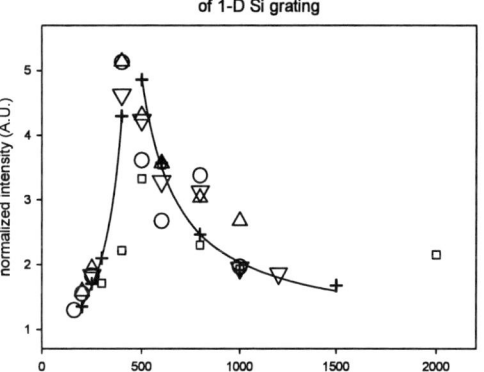

Fig. 4: Raman responses excited by the TM-mode incidence; where the horizontal axis labels the grating period in unit of nanometer (period=80nm (O), 100nm (Δ), 120nm (▽), 150nm (+) and 200nm (□)), and the vertical axis states the normalized intensity of Raman scattering.

scattering of the grating with less period is stronger than that of the grating with larger period. On the other hand, our data show that the positions of this extreme location in period concentrate during a very small region in period, although the position appears to lie at more left while the period of grating be smaller. Hence, as the period of grating increasing indefinitely, the Raman intensity becomes smaller. And we predict that it will approaches to 1 since the structure of our sample will tends to a plane substrate. Oppositely, while the period decreases away from that with maximum scattering, one can find the scattering intensity decays much more rapidly than the other side.

The Raman responses excited by the TE-mode incidence are shown in Fig. 5; where the horizontal axis labels the grating period in unit of nanometer (period=80nm (O), 100nm (Δ), 120nm (∇), 150nm (+) and 200nm (□)), and the vertical axis states the normalized intensity of Raman scattering. Similarly, there exist an extreme on some period region for specific line width. However, it is apparently to find that the intensity of Raman scattering excited by TE-mode incidence is much weaker than that induced by the TM-mode incidence. It is not alike to those cases of TM mode, we

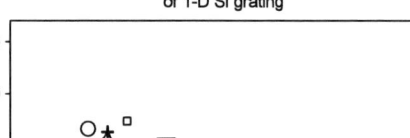
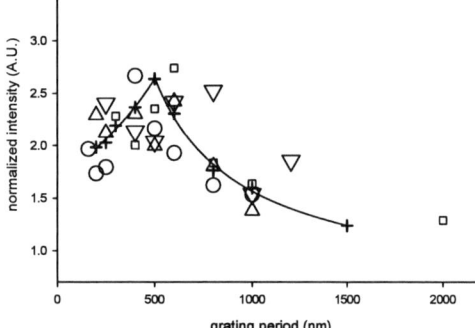

Fig. 5: Raman responses excited by the TE-mode incidence; where the horizontal axis labels the grating period in unit of nanometer (period=80nm (O), 100nm (Δ), 120nm (∇), 150nm (+) and 200nm (□)), and the vertical axis states the normalized intensity of Raman scattering.

observe that the intensity of Raman scattering decays more rapidly on the right side of the region which the maxima occurs than it does on the other side. At the same time, one predicts the scattering intensity will approaches to 1 as the period of grating approaching to infinity.

Through Fig. 4 and 5, we may be interesting to the variation of the ratio of TM mode and TE mode as the grating period varies for a specific line width. Fig. 6 shows that the dependence of the ratio, TM/TE, upon the grating period; in which the horizontal axis represents the grating period in unit of nanometer (period=80nm (O), 100nm (Δ), 120nm (∇), 150nm (+) and 200nm (□)), and the vertical axis states the value of TM over

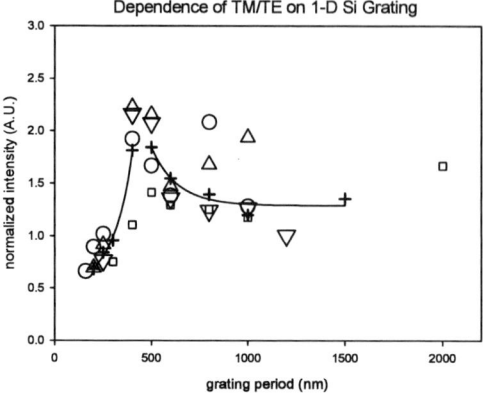

Fig. 6: The dependence of the ratio, TM/TE, upon the grating period; in which the horizontal axis represents the grating period in unit of nanometer (period=80nm (O), 100nm (Δ), 120nm (∇), 150nm (+) and 200nm (□)), and the vertical axis states the value of TM over TE.

TE. The pattern of the variation of that intensity ratio, TM/TE, can consist with Fig. 4 and 5. It is worthy to notice the magnitude of TM mode is even weaker than that of TE mode.

Conclusions

The intensity of Raman scattering is quietly dependent on that geometrical structure of grating. In general, the scattering induced by TM-mode incidence is much stronger than that excited by TE-mode incidence. However, this phenomena is opposite while the period is very small for a specific line width; namely, TM mode will be weaker than TE mode as the slit width is smaller enough. Furthermore, it seems to exit an optima geometrical arrangement of a grating such that the intensity of Raman scattering will have a maxima.

Reference

1 Se-Heon Kim et al., Physical Review B 73 (2006), 235117

2 A. E. Serebryannikov et al., Physical Review E 74 (2006), 066607

3 Estsban Moreno et al., Physical Review B 69 (2004), 121402(R)

4 Jeremy Witzens et al., Physical Review E 69 (2004), 046609

5 Ovidiu Toader et al., Physical Review E 70 (2004), 046605

6 Vladimir V. Poborchii et al., Physics E 7 (2000), 545-549

Modeling of Spontaneous Emission from Erbium Incorporated Silicon Nanocrystal

Samia Subrina (1), Md. Zahid Hossain (2), Md. Quamrul Huda (3)

Department of Electrical and Electronic Engineering, Bangladesh University of Engineering and Technology (BUET), Dhaka 1000, Bangladesh

Email: 1: samiasubrina@eee.buet.ac.bd

2: zahidhossain@eee.buet.ac.bd, 3: mqhuda@eee.buet.ac.bd

Abstract *Mathematical model has been developed for erbium (Er) luminescence in silicon nanocrystal (nc-Si) embedded in SiO_2. Performances of Er luminescence under different conditions have been analyzed for both steady state and time varying case.*

Introduction

The indirect energy bandgap nature of silicon greatly limits its potential in optoelectronic arena. Many alternatives are recently investigated and one viable approach is the doping of silicon with erbium, because Er^{3+} ions incorporated into silicon produce stable sharp luminescence from the intra-4f shell transition at around 0.81eV, which corresponds to the absorption minimum in silica based fiber optic communication. However, luminescence from erbium in bulk silicon suffers strong temperature quenching even at room temperature and as a result, practical operation of on-chip emission in silicon has not been achieved yet [1-3].

Recently, erbium incorporated nanometer sized silicon has been recognized as a quite efficient method to overcome this problem with increased erbium luminescence at room temperature. The band structure of silicon is strongly modified in the nanocrystal environment, providing silicon nanocrystals with functions that do not occur in bulk Silicon. The quantum size effect becomes prominent for nanocrystals smaller than about 10nm in diameter. The band structure shows "direct like" behavior due to quantum confinement effect and this provides the recombination process more efficient [4]. Moreover, non-radiative processes like Auger with free carriers, energy back transfer, etc. are strongly suppressed in erbium incorporated silicon nanocrystal [5-8]. In this paper, we present a model describing the spontaneous emission from erbium in silicon nanocrystals and also show the effect of different parameters on erbium luminescence.

Erbium Excitation

Excitation of erbium in silicon nanocrystals is believed to originate through electron hole mediated process. The excited erbium atom then returns to the ground state through radiative or non-radiative transitions. In this paper, we include all possible routes of erbium excitation and de-excitation and develop the model for luminescence.

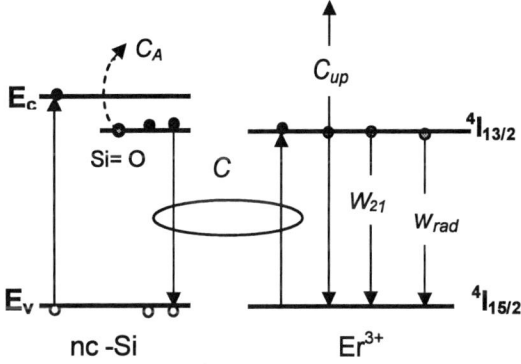

Fig 1 Energy level scheme for the system of interacting silicon nanocrystal and erbium ions in our proposed model.

Mathematical Modeling

Let N_{er} be the number of active erbium atoms per unit volume incorporated in silicon nanocrystal and among them, $n_{er}{}^*$ erbium atoms are in the excited state at steady state. Hence $N_{er} - n_{er}{}^*$ be the number of active erbium atoms per unit volume available for excitation. Let n_{si} be

978-0-9789217-3-6/07/$25.00

©2007 WEN GLOBAL SOLUTIONS

the carrier density in nanocrystal silicon at steady state. At steady state, the rate of excitation of erbium atoms is given by

$$P = Cn_{si}(N_{er} - n_{er}^*) \qquad (1)$$

where C is the coefficient of coupling between the recombination mechanism in nanocrystals and the corresponding erbium atoms. Once excited, the Er atom can de-excite from $^4I_{13/2}$ to the ground state ($^4I_{15/2}$) either radiatively or non-radiatively. The rate of de-excitation is given by

$$D = (w_{21} + \frac{1}{\tau_{rad}})n_{er}^* + 2C_{up}(n_{er}^*)^2 + C_A n_{si} n_{er}^* \qquad (2)$$

Here τ_{rad} is the radiative decay lifetime of erbium, C_A is the coefficient of Auger process, C_{up} is the coefficient of cooperative up conversion process and w_{21} is the rates of additional non-radiative decay processes.

At steady optical excitation, the rate of erbium excitation and de-excitation must be equal. Equating (1) and (2), we have

$$n_{er}^* = \frac{-(Cn_{si} + C_A n_{si} + w_{21} + \frac{1}{\tau_{rad}})}{2C_{up}}$$

$$+ \frac{\sqrt{(Cn_{si} + C_A n_{si} + w_{21} + \frac{1}{\tau_{rad}})^2 + 8C_{up}Cn_{si}N_{er}}}{2C_{up}}$$

The luminescence from erbium doped silicon nanocrystal is given by

$$I \propto \frac{n_{er}^*}{\tau_{rad}}$$

During switching transition of the optical source, both excitation and de-excitation rates are time dependent and the rate of change of excited erbium is given by

$$\frac{dn_{er}^*(t)}{dt} = C n_{si} (N_{er} - n_{er}^*(t)) - [(w_{21} + w_{rad})n_{er}^*(t) + C_A n_{si} n_{er}^*(t) + 2C_{up}(n_{er}^*(t))^2]$$

This differential equation is solved using forth order Runge-Kutta method.

Results and Discussion

From Fig 2 it is clear that there is no significant impact of Auger process on erbium luminescence incorporated in silicon nanocrystal as Auger process provides a

very small percentage (less than 1%) of the total non-radiative decay processes. This is due to the quantized nature of both the excitonic and the erbium related levels. Moreover it is considered that the optically active erbium ions are located in SiO_2 or at the Si/SiO_2 interface, so Auger effect is suppressed due to the presence of oxide barrier layer between erbium and silicon nanocrystal.

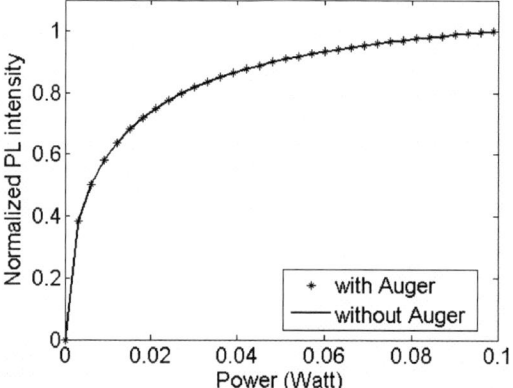

Fig 2 Erbium luminescence under variable pump power to observe the effect of Auger process.

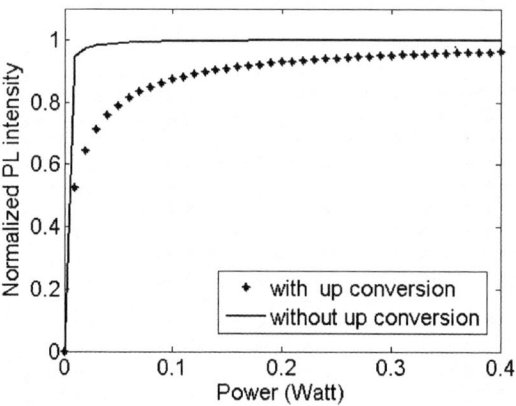

Fig 3 Erbium luminescence under variable pump power to observe the effect of cooperative up conversion process.

At high concentration, interaction between Er ions is an important gain limiting effect. This happens as the radiative efficiency decreases due to the increased non-radiative decay processes. In Fig 3 our observations also resemble these phenomena.

Fig 4. shows that rise time of erbium luminescence varies with excitation power. This can be explained by the fact that at higher excitation level, increased carrier density increases Er excitation rate. This high carrier density also increases non-radiative decay processes but under constant excitation, radiative decay process dominates over non-radiative processes. Hence less time is required to establish a certain luminescence.

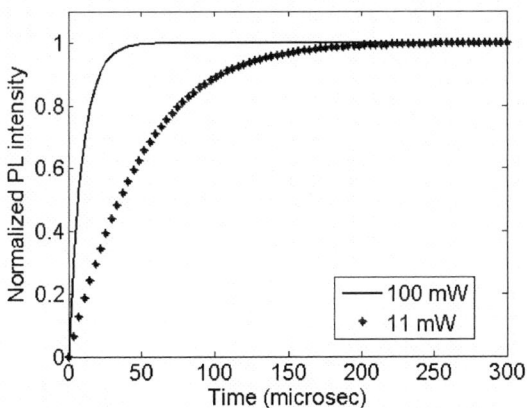

Fig 4 PL intensity profile at two different excitation levels. Both the curves have been normalized to peak values.

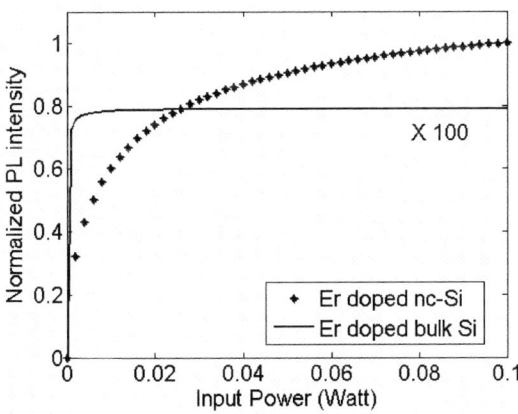

Fig 5 Photoluminescence (PL) Intensity of erbium vs. pump power profile. [luminescence from erbium doped bulk silicon system is multiplied by 100].

In silicon nanocrystal, the effective absorption cross section of the intra-4f shell transition of Er^{3+} is enhanced by 2-4 orders of magnitude because of the large absorption cross section of silicon nanocrystal and because of the efficient energy transfer from nc-Si to Er^{3+}. These effects highly enhance the luminescence from erbium in silicon nanocrystal. Moreover, the non-radiative de-excitation processes typically limiting erbium luminescence in bulk silicon, namely, Auger process and energy back transfer, are strongly reduced in this case further improving the luminescence yield. From Fig 5 it is clear that the erbium luminescence from silicon nanocrystal is much higher than that from bulk silicon. Similar phenomena are also experimentally shown by Franzo et al. [5]

Conclusion

An analytical model has been developed for spontaneous emission from erbium incorporated in silicon nanocrystal. We have shown that the impact of Auger process is negligible in erbium luminescence and since nanometer sized silicon provides wider bandgap which causes reduced energy back transfer process, LED operation from erbium doped silicon nanocrystal is feasible with increased luminescence at room temperature with switching time of the order of micro second.

References

1 Xie et al *J. Appl. Phys,* Volume 70 (1991), page # 3223.

2 Franzo et al " Appl. Phys. Lett., Volume 64 (1994), page# 2235.

3 Gregorkiewicz et al Phys. Rev. B, volume 61 (2000), page # 5369.

4 Fujii, et al *J. Appl. Phys.*, vol. 95 (2004), page # 272.

5 Franzo et al *Appl. Phys. Lett.,* vol. 76 (2000), page # 2167.

6 Samia et al *J. of Chemical Physics,* volume 120 (2004), page # 8716.

7 Kik et al *J. Appl. Phys.,* Volume 88 (2000), page # 1992

8 Pacifici et al Phys. Rev. B, volume 67 (2003), page# 245301.

9 Samia Subrina M.Sc., Dept. of EEE, BUET, Thesis 2006.

Surface Plasmonic Microscopy for Live Cell Membrane Imaging

Ruei-Yu He[1], Yuan-Deng Su[2], Kuo-Chih Chiu[2], Hua-Lin Wu[3], Chi-Hung Lin[4],
Guan-Liang Chang[1], and Shean-Jen Chen[2,*]

[1] Institute of Biomedical Engineering, National Cheng Kung University, Tainan 701, Taiwan
[2] Department of Engineering Science, National Cheng Kung University, Tainan 701, Taiwan
[3] Department of Biochemistry, National Cheng Kung University, Tainan 701, Taiwan
[4] Institute of Microbiology and Immunology, National Yang-Ming University, Taipei 112, Taiwan
* sheanjen@mail.ncku.edu.tw

Abstract *This study presents a surface plasmon-enhanced total internal reflection fluorescence microscopy (TIRFM) and a surface plasmon polariton (SPP) phase microscopy techniques to image live cell membranes. In the surface plasmon-enhanced TIRFM, the developed microscopy technique is successfully applied to the real-time observation of the thrombomodulin proteins of live cell membranes.*

Introduction

Previously reported surface plasmon-enhanced fluorescence measurement techniques were designed only for single-spot sensing applications[1]. However, this study proposes a surface plasmon-enhanced fluorescence measurement technique capable of imaging the molecular interactions between a cell membrane and a biosurface in real-time. Thrombomodulin (TM), an integral membrane glycoprotein, is a thrombin receptor identified originally on the endothelium which acts as a natural anticoagulant[2]. Using membrane protein TM fused to green fluorescent protein (GFP) to study the prevention of cancer formation is a crucial research topic in the cell biology field[3]. Therefore, this study first employs the proposed surface plasmon-enhanced TIRFM technique to dynamically image the interaction of membrane protein TM with a collagen biosurface. Also, we develop a wide-field surface plasmon polariton (SPP) phase microscope system with subwavelength grating structure for real-time dynamic measurement of biomolecular interaction. The SPP phase imaging system with the advantages of the common-path phase-shift interferometry (PSI) technique can overcome the phase drifts due to environmental change, mechanical vibration, and light source fluctuation to provide long-term stability and high resolution[4]. Therefore, we can achieve a surface phase imaging system with high sensitivity, long-term stability, and high lateral spatial resolution according to the advantages of

the SPP sensing, common-path PSI technique, and normal-incidence in-plane subwavelength grating coupling, respectively.

Materials and methods

A. Optical system setup

Figure 1(a) shows the optical configuration of the proposed prism-coupled surface plasmon-enhanced TIRFM, in which the attauated-total-reflection method is used to excite the SPs in order to enhance the local EM field and to increase the SNR of the fluorescence. In this configuration, a thin silver film via a sputtering deposition process, a chemical self-assembly monolayer (SAM), and a biomolecular layer are sequentially deposited on an SF-11 slide in accordance with an optimal design which is known to enhance the intensity of the fluorescence and therefore to increase the attainable frame rate. A schematic of the SPP phase microscope system is illustrated in Fig. 2. The He-Ne laser beam passes through a linear polarizer which can adjust the optical axis to control the weighting of the reference and signal beams. The beam then passes through a nematic liquid crystal which is a birefringence material with two difference optical refractive index. By using that, a five-step PSI technique can facilitate high-resolution inspection of the spatial distribution of the two-dimensional phase variation. The light then passes through a spatial filter and a lens focusing. The focusing beam then passes the beam

978-0-9789217-3-6/07/$25.00

©2007 WEN GLOBAL SOLUTIONS

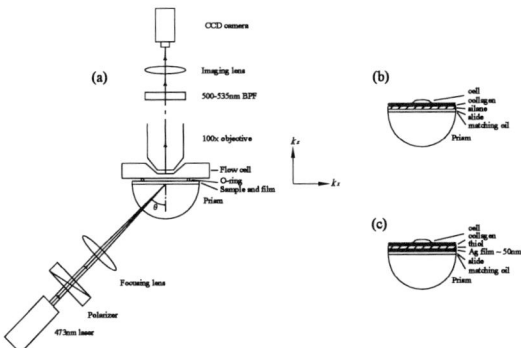

Fig. 1. (a) Schematic illustration of experimental configuration employed for live cell imaging using conventional and surface plasmon-enhanced TIRFM techniques. (b) Conventional TIRFM chip: cell is cultured on collagen-coated slide modified with silane. (c) Enhanced SPR chip, cell is cultured on collagen-coated silver thin film modified with thiol.

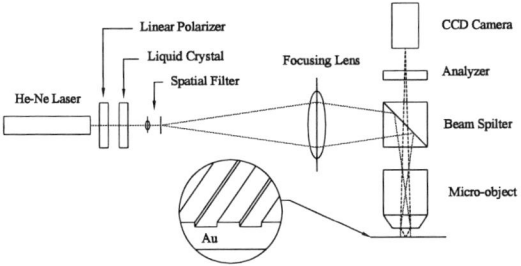

Fig. 2. Schematic diagram of the SPP phase imaging configuration.

splitter and the reflection beam focusing on the back focal plane of the micro-objective. The output laser light from the objective forms a collimate beam to normal incident the subwavelength sensor. The refraction light then passes through an analyzer to make the signal beam and reference beam interference and the interference fringes are grabbed by the CCD camera.

B. Cell culture protocol

TM has an apparent molecular weight of 78 kDa and consists of 557 amino acid residues arranged in five distinct domains. In this study, cultured human melanoma cells were stably transfected with DNA plasmids encoded with the target protein TM fused to GFP. The human melanoma-GFP-tagged TM cells were maintained in a Dulbecco's Modified Eagle Medium (DMEM) supplemented with 0.292 g/liter L-glutamine, 2% sodium bicarbonate, 1% sodium pyruvate, 1% penicillin, and 10% fetal bovine serum. To maintain the live cells

stably expressing TM-GFP fusion proteins, the cells were cultured in DMEM supplemented with 300 µg/ml antibiotic G418. The cell is cultured on a collagen-coated silver thin film modified with a thiol SAM. The metal film was immersed in 1 mM 2-aminoethanethiol hydrochloride solution to form a dense SAM on its surface. Covalent activation was then performed by immersing the chip in a solution containing EDC[N-(3-dimethylaminopropyl)- N'-ethylcarbodimide hydrochloride, 2 mM] and NHS(N-hydroxysuccinimide, 5 mM) for 6 hours to immobilize the protein collagen.

Experimental results and discussions

Live cell membrane images for cell on, attachment, and migration biosurfcae (Fig. 3) are grabbed by utilizing a surface plasmon-enhanced TIRFM[5]. The experimental results and the simulation results demonstrate that the live cell membrane images obtained in the proposed surface plasmon-enhanced TIRFM technique are approximately one order of magnitude brighter than those provided by conventional TIRFM. In order to analyze molecular interactions on the cell membrane with biomolecules having no additional fluorescence labeling, the developed SPP phase microscope is used to study the cell adhesion and migration on the biosurfaces. (See Fig. 4.)

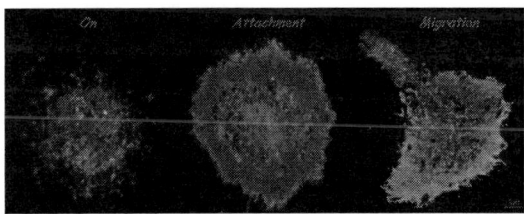

Fig. 3. Enhanced living cell membrane imaging by plasmon-controlled total-internal-reflection fluorescent microscope.

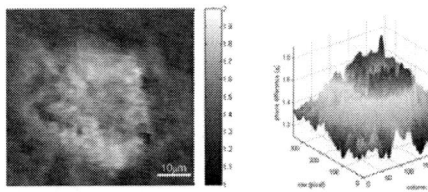

Fig. 4. Left: Reconstruction phase image in 350×350 pixel region covering 56×56 µm² physical area; Right: Phase spatial distribution recovered from cell adhesion.

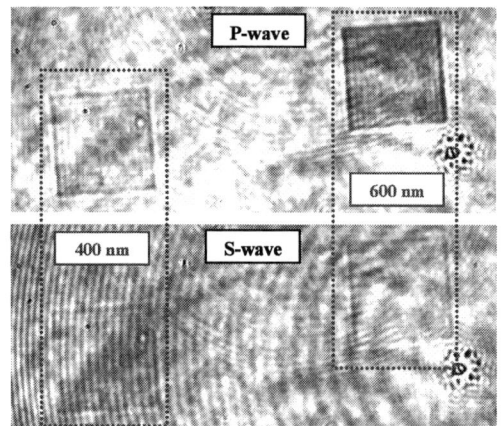

Fig. 5. Top: P-wave excitation in air; Bottom: S-wave excitation in air.

The long-term stability of the SPP phase microscope is found to be better than $10^{-3}\pi$ root mean square. Applying an interpolation analysis method to the measurement data with a phase resolution of $10^{-3}\pi$ under two different gas media, it is found that the detection limit of the biosensing system for changes in the medium's refractive index approaches 10^{-7} RIU. Figure 5 demonstrate that the reflectivity images of the P-wave (top) and S-wave (bottom) when a SPP chip with subwavelength grating of 400 nm (left) and 600 nm (right) period is measured under medium air when the wavelength of the incident light is 632.8 nm. Based on the above simulation, the SPR condition is near to the period of 600 nm for P-wave. Therefore, the reflectivity image of the P-wave for the period of 600 nm is darker than that for the period of 400 nm.

Conclusions
The developed surface plasmon-enhanced TIRFM technique has been successfully employed to observe the interactions of melanoma-GFP-tagged TM cells with a collagen biosurface. Also, the SPP phase microscope with a subwavelength grating structure is designed for high-resolution in-plane image measurement.

Acknowledgments
Funding the National Science Council (NSC 95-2221-E-006-322-MY3) of Taiwan is gratefully acknowledged.

References
1 K. Aslan et al., Curr. Opin. Biotechnol. 16 (2005) 55.
2 Y. Tezuka et. al., Cancer Res. 55 (1995) 196.
3 Y. Zhang et al., J. Clin. Invest. 101 (1998) 1301.
4 Y.-T. Su et al., Opt. Lett. 30 (2005) 1488.
5 L.-Y. He et al., Opt. Exp. 14 (2006) 9307.

Laser Ultrasound Detecting experiment with

Fiber Michelson Interferometer

ZHANG Jian-liang (1), SHENG Xin-zhi (2), WU Chong-qing (3),
ZHANG Li-jun (4),

1,2,3,4 : Key lab of Luminescence and Optical Information Technology, EM; Institute
of Optical Information, Beijing Jiaotong University, Beijing 100044, PR China.
(05121802@bjtu.edu.cn)

Abstract *A laser ultrasound stably detecting has been demonstrated by a new type fiber optic Michelson interferometer with 3×3 direction coupler. The experimental result suggests that it's able to sense an ultrasound vibration frequency from 10 KHz to 1.5MHz.*

Introduction

The basic suggestion that optical fiber technology could be useful in sensors dates back well over 30 years. From that time, fiber optical sensor is investigated widely as its attraction combined a non-contact measurement with immunity to electromagnetic interference, factors which still continue to feature in the benefits list of fiber sensor technology [1]. As one of the fiber optical sensor, the fiber Michelson interferometer becomes more and more attractive, and it is applied diffusely in the fields of industry and scientific research, such as dynamic parameters

measurement, fluid control, perimeter security, medicinal examination etc [2-3]. In this paper we describe investigation results of fiber laser ultrasound vibration measurement based on fiber Michelson interferometer with 3×3 direction coupler. Compare with conventional optical fiber interferometer with 2×2 coupler, the interferometer is able to work steadily.

Experimental set

The experimental schematic is indicated in Fig.1. We use the transfer matrix to analyze our experiment. It is assumed that the three waveguides in the coupling region are identical, and the coupling ratio of the 3×3 couple is 1:1:1 and the electric field E of a laser beam is injected into 1st arm of the 3 ×3 coupler. So the electric fields output to the detector D1 and D2 are $E_{5'}$ and $E_{6'}$:

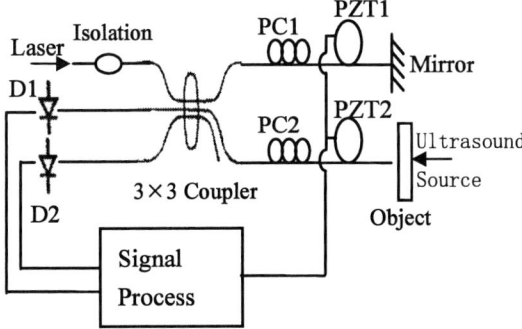

$$\begin{pmatrix} E_{5'} \\ E_{6'} \end{pmatrix} = \frac{E}{3} e^{2j\beta_1 L_1} \begin{pmatrix} e^{j\frac{2}{3}\pi} + e^{j\frac{4}{3}\pi + 2j\gamma} \\ e^{j\frac{2}{3}\pi} + e^{j\frac{2}{3}\pi + 2j\gamma} \end{pmatrix} \quad (1)$$

Where β_1 and β_2 is respectively the propagation factors in the signal and reference arms of this interferometer, L_1 and L_2 respectively length of the signal

Fig.1. Schematic of the Fiber Michelson Interferometer with 3×3 Couple for detection.

and reference arms. And the phase difference between the signal and reference arms of this interferometer

$$\gamma = \beta_2 L_2 - \beta_1 L_1 = \Delta\varphi + \Delta f \qquad (2)$$

Where $\Delta\varphi$ presents the slow change part and Δf the fast part of the phase difference between the signal and reference arms. So the optical intensity output to D1 and D2 are:

$$I_{5'} = E_{5'}^{*} E_{5'} = \frac{2}{9} P_{in} \left[1 + \cos 2\left(\frac{\pi}{3} + \gamma\right) \right], \quad (3)$$

$$I_{6'} = E_{6'}^{*} E_{6'} = \frac{2}{9} P_{in} \left[1 + \cos 2\gamma \right] \quad . \ (4)$$

The difference is

$$\Delta I = I_{6'} - I_{5'} = \frac{2\sqrt{3}}{9} P_{in} \sin\gamma . \quad (5)$$

The Michelson Interferometer experimental set with 3×3 direction coupler is schematically shown in Fig 2. A DBF laser operating at 1551nm is used as the light source of the interferometer. Two polarization controllers (PC1 and PC2), but no accurate adjustment is required, are utilized only to avoid null visibility of the fringes. A section of the reference fiber arm of the interferometer is wound around a PZT cylinder, and the reference beam is reflected by a mirror which is assembled with the fiber. Another arm, the signal arm, is around the PZT2 and irradiate to the object which sticks on a laser ultrasound base. The reflected beam from the object and the reference beam interference in the 3×3 couple and the signal will be detected by D1 and D2.

The phase difference signal is processed and $\Delta\varphi$ signal is send to control PZT1 and PZT2 in order to adjust the difference between the signal and reference arms, in such way that $\Delta\varphi \to 0$ and the interferometer counteract the infection from the environment. The adjusting signals to the two PZTs are contrary, in this way the PZT steadies the interferometer quickly. When the $\Delta\varphi \to 0$, the difference optical intensity to the detector D1 and D2 is

$$\Delta I = \frac{2\sqrt{3}}{9} P_{in} \sin\Delta f \qquad (6)$$

Therefore, the interferometer is working in the linear area.

Experimental results

1. Continuous Signal Prompting

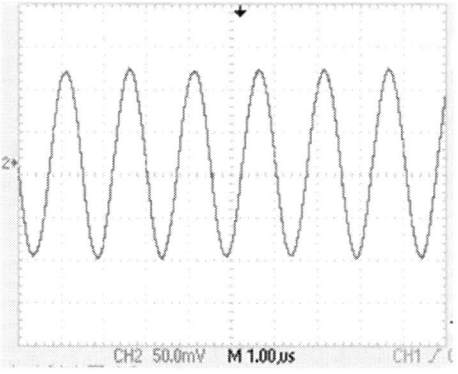

Fig.2 The amplitude of the output signal is obviously, about 230mV. The output signal is the same frequency with the input signal, so this fiber Michelson interferometer can detect the ultrasound vibration successfully.

To verify the experimental set validity, a continuous signal is used to prompt a continuous sinusoidal ultrasound wave. As shown in Fig.2, the ultrasound wave is reproduced perfectly so that the interferometer is working in the linear area. When the frequency prompting ultrasound is changed, the signal amplitude of ultrasound changed too. The experimental result is shown in the

Fig.3 for the cases in same prompting power and the ultrasound frequency changed from 10 KHz to 1.6 MHz with a pace 10 KHz, and a resolution <20KHz is got in ultrasound frequency range.

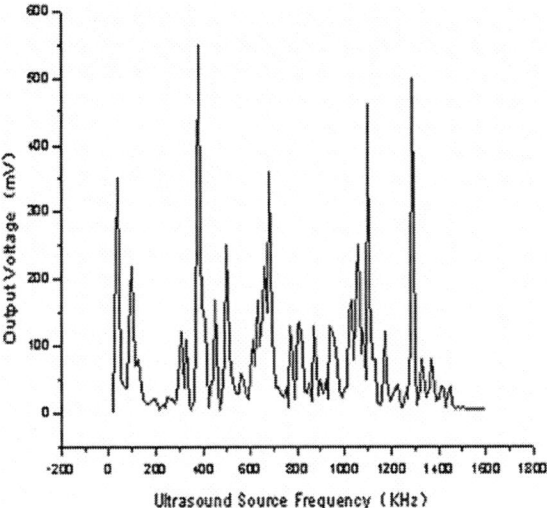

Fig.3 When the frequency of the ultrasound source from 10 KHz to 1.6 MHz, the amplitude of the output signal changes, and the pace is 10 KHz.

2. Pulsing Signal Promoting

Detecting ultrasound pulse is where the shoe pinches for a laser ultrasound study. Fig.4 shows the experimental results of the Michelson Interferometer experimental set with 3×3 direction

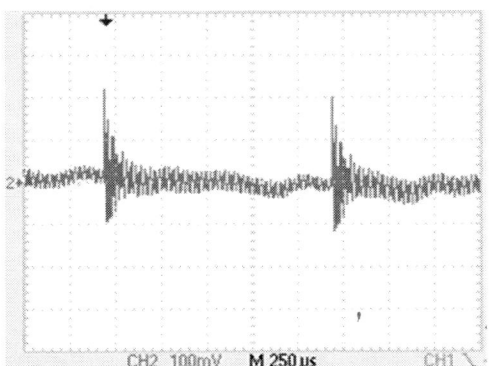

Fig.4. The amplitude of the output of the signal is about 300mV, and the frequency is the same. with the ultrasound signal, it is 0.8 KHz.

coupler in the case of ultrasound prompting with a pulse signal. It is a real-time signal. The amplitude of the output of the signal is about 300mV. From this case, the output signal reflect the laser ultrasound vibration of the object commendably.

Conclusions

The feasibility of stable detecting laser ultrasound by a fiber optic Michelson interferometer with 3×3 direction coupler has been demonstrated. The result suggests that the experimental set works steadily, and has a high resolution, about 20 KHz. It is able to sense an ultrasound wave frequency from 10 KHz to 1.5MHz, and can reveal the vibration of the object successfully.

References

[1] Brian Culshaw, "Optical fiber sensor technologies: opportunities and—perhaps—pitfalls", *Journal of Lightwave Technology,* vol.22, No.1, 2004, Page(s):39-50.

[2] Wojtek J. Bock, Magdalena S. Nawrocka, Waclaw Urbanczyk, "Dynamic Calibration of the Fiber-optic Pressure Sensor Based on Side-hole Fiber", *Sensors, 2003. Proceedings of IEEE*, vol.1, 22-24 Oct. 2003,.

[3] Szustakowski,M. Ciurapinski, W. Palka, N. Zyczkowski, M. , "Recent development of fiber optic sensors for perimeter security" , Security Technology, 2001 IEEE 35th International Carnahan Conference on 16-19 Oct. 2001 Page(s):142-148.

Spectroscopic Applications to Environmental Monitoring and Nanobiophotonics

Gabriel Somesfalean

Joint Research Center of Photonics of KTH and Zhejiang University, 310058 Hangzhou, P.R. China
Department of Physics, Lund Institute of Technology, P.O. Box 118, SE-221 00 Lund, Sweden
Phone/Fax: +86-571-88206513, Electronic mail: gabriels@fysik.lth.se

Atmospheric gas sensing can help us understand the complex processes governing global warming and the impact of pollutant emissions related to human activity. Spectroscopic techniques, especially based on robust and affordable diode lasers, have a great potential to become an industrially well-established technology for environmental monitoring.

The accessible spectral range of tunable diode lasers can be extended by use of non-linear frequency mixing [1-2]. High-resolution ultraviolet (UV) spectroscopy of mercury isotopes around 254 nm was performed on low-pressure cells as well as at atmospheric pressure (see Fig. 1). UV radiation around 300 nm for sulfur dioxide monitoring was produced using a sum-frequency generation scheme employing a blue and a near infrared (NIR) diode laser (see Fig. 2).

The detection sensitivity can be improved by several orders of magnitude by employing frequency modulation techniques. Traffic-generated emission of nitrogen dioxide was monitored *in situ* by using long path absorption at a wavelength around 635 nm [3].

A novel temporal gas-correlation scheme was developed to overcome the intrinsic multimode and mode-jump behaviour of diode lasers [4-5]. The gas concentration is determined by wavelength tuning the diode laser across an absorption band of the gas and by simultaneous temporal correlation of the detected signal with the signal from a known reference gas. No knowledge of the exact spectrum is needed. The method was applied to oxygen diffusion and acetylene gas monitoring.

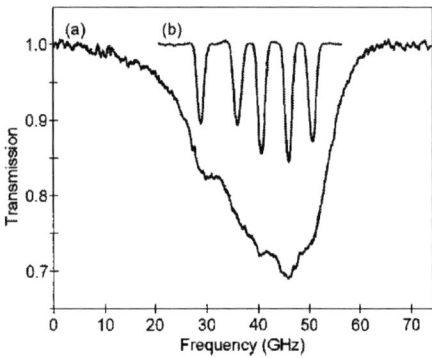

Fig. 1 (a) Mercury absorption signal at atmospheric pressure in a 4 mm cell and (b) isotopically resolved reference lines from a low pressure cell.

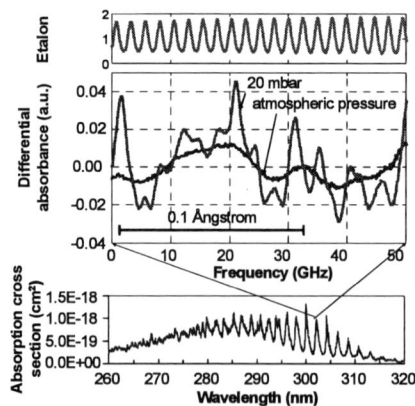

Fig. 2 Sulfur dioxide spectra recorded around 300 nm at low pressure (20 mbar) and at ambient pressure. For comparison, a spectral overview of the broad UV absorption of SO_2 is presented.

978-0-9789217-3-6/07/$25.00
©2007 WEN GLOBAL SOLUTIONS
628

Diode-laser-based spectroscopic techniques have applications also in the field of biophotonics. Fluorescent labeling is widely used for, e.g., single virus tracing, DNA sequencing, and frequency resonance energy transfer. Multicolor labels are needed for simultaneous imaging and tracking of multiple molecular targets.

Upconversion (UC) fluorescent labels show high detection sensitivity by converting NIR absorbed radiation into visible and stable fluorescence. The excitation can be performed with a compact and cost effective continuous wave (cw) NIR diode laser.

Pure green (525/545 nm) and red (655 nm) emissions were obtained in Er^{3+}-ion-doped ZrO_2 nanocrystals (see Fig. 3) which arise from the transitions $^2H_{11/2}/^4S_{3/2} \rightarrow ^4I_{15/2}$ and $^4F_{9/2} \rightarrow ^4I_{15/2}$ of the Er^{3+} ions, respectively [6-7]. The spectrally designed fluorescence is very different from the usual UC luminescence where several bands are emitted simultaneously. A general route to enhance the UC radiation was demonstrated [8], facilitating further application of the nanocrystals.

UV lasers have a wide range of applications including environmental monitoring, biophotonics, optical data storage, etc. Generation of UC UV fluorescence by NIR lasers is of great interest from several aspects: (a) fundamentally, more detailed information about the energy transfer processes and interactions in solids can be obtained; (b) practically, it can lead to the discovery of new UV laser materials. Since high power and cost-effective commercial diode lasers in the NIR range are readily available, the costs of the UV laser instrumentation can be greatly reduced.

Room temperature UV emissions of Tm^{3+} ions at 298 ($^1I_6 \rightarrow ^3H_6$), 364 ($^1D_2 \rightarrow ^3H_6$), and 391 ($^1I_6 \rightarrow ^3H_5$) nm were obtained by intra-4f electron transitions of Tm^{3+} ions in Y_2O_3:Yb^{3+}-Tm^{3+} via cw diode laser excitation of 980 nm (see Fig. 4). This corresponds to five- and six-photon UC processes – never reported previously in rare-earth-doped materials [9]. High multi-photon processes are also expected for other trivalent rare-earth ions.

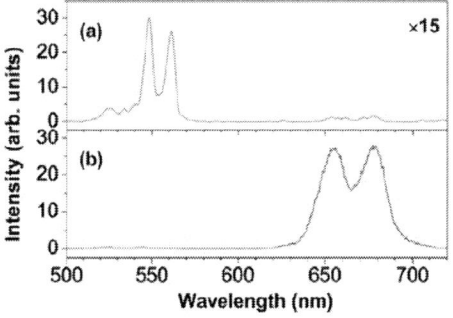

Fig. 3 Tailored fluorescence spectra of ZrO_2 nanocrystals (a) doped with Er^{3+} 1mol%, and (b) doped with Er^{3+} 1mol% and Yb^{3+} 10mol%.

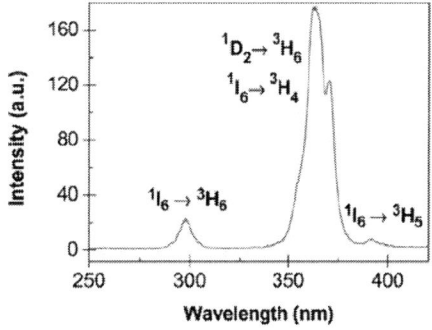

Fig. 4 UV UC fluorescent spectrum obtained in Y_2O_3 powders doped with 0.2 mol % Tm^{3+} and 3 mol % Yb^{3+}.

References

[1] J. Alnis, U. Gustafsson, G. Somesfalean, and S. Svanberg, "Sum-frequency generation with a blue diode laser for mercury spectroscopy at 254 nm", Appl. Phys. Lett. **76**, 1234-1236 (2000).

[2] G. Somesfalean, Z.G. Zhang, M. Sjöholm, and S. Svanberg, "All-diode-laser ultraviolet absorption spectroscopy for sulfur dioxide detection", Appl. Phys. B **80**, 1021-1025 (2005).

[3] G. Somesfalean, J. Alnis, U. Gustafsson, H. Edner, and S. Svanberg, "Long-path monitoring of NO_2 with a 635 nm diode laser using frequency modulation spectroscopy", Appl. Opt. **44**, 5148-5151 (2005).

[4] G. Somesfalean, M. Sjöholm, L. Persson, H. Gao, T. Svensson, and S. Svanberg, "Temporal correlation scheme for spectroscopic gas analysis using multimode diode lasers", Appl. Phys. Lett. **86**, 184102 (2005).

[5] X.T. Lou, G. Somesfalean, F. Xu, Y.G. Zhang, and Z.G. Zhang, "Gas sensing by fast tunable multimode diode laser absorption spectroscopy", to be published (2007).

[6] G.Y. Chen, Y.G. Zhang, G. Somesfalean, Z.G. Zhang, Q. Sun, and F.P. Wang, "Two-color upconversion in rare-earth-ion-doped ZrO_2 nanocrystals", Appl. Phys. Lett. **89**, 163105 (2006).

[7] G.Y. Chen, G. Somesfalean, Y. Liu, Z.G. Zhang, Q. Sun, and F.P. Wang, "Upconversion mechanism for two-color emission in rare-earth-ion-doped ZrO_2 nanocrystals", Phys. Rev. B **75**, 195204 (2007).

[8] G.Y. Chen, H.C. Liu, G. Somesfalean, Z.G. Zhang, Q. Sun, F.P. Wang, "A general route to enhance upconversion radiation in rare-earth-ion-doped oxide nanocrystals", to be published (2007).

[9] G.Y. Chen, G. Somesfalean, Z.G. Zhang, Q. Sun, and F.P. Wang, "Ultraviolet upconversion fluorescence in rare-earth-ion doped Y_2O_3 induced by infrared diode laser excitation", Opt. Lett. **32**, 87-89 (2007).

Multiplexing of Fiber Bragg Grating Pairs for Sensing Based on Optical Low Coherence Technology

Weisheng Liu

Centre for Optical and Electromagnetic Research, Zhejiang University, East Buliding No.5, Zijingang campus, Zhejiang University, Hangzhou 310058, China.

liuweisheng@coer.zju.edu.cn

Abstract *A distributed sensing system based on fiber Bragg grating pairs is presented. Multiplexing is achieved by using optical low coherence technology. A measurement of strain is demonstrated to verify good performance of the system.*

Introduction

Fiber Bragg gratings (FBGs) which are integrated into the light guiding core of the fiber are excellent optical sensing elements. They have attracted a lot of research interests due to their good properties, such as electrically passive operation, electromagnetic immunity, high sensitivity, compact size, and especially large multiplexing capability.

In the developed multiplexing systems for FBG sensors, wavelength division multiplexing (WDM) technique and time division multiplexing (TDM) technique are of the most popular [1]-[3]. However, both these two techniques require a high-cost light source. In a WDM system, a broadband light source is required while a narrow-pulsed light source is necessary for a TDM system. We have successfully interrogated the multiplexed sensors based on FBG pairs with low reflectivity using optical low-coherence reflectometry (OLCR) technology recently [4]. However, in that system, since all FBG pairs are arranged in line, there is cross talk between different FBG pairs which limits the sensitivity and the signal-to-noise ratio of the system.

In this work, we propose a new coherence multiplexing scheme for sensing FBG pairs. In this system, all FBG pairs are parallel arranged which avoids cross talk between FBG pairs. A measurement of strain is demonstrated to show good performance of our system.

Principles and analysis

The system setup is as shown in fig. 1. The interrogation scheme in this system is actually an OLCR although it is quite different from a conventional one. In the presenting system, the OLCR is actually a scannable Michelson interferometer (MI) with one arm incorporated a tunable optical delay line (TODL). The device under test (DUT) is not in any arm of the MI as the conventional OLCR but actually related to several F-P interferometers formed by the FBG pairs which induce specific additional optical path differences (OPDs) to the input light entering the MI. By scanning the TODL, the OPDs induced by the F-P interferometers will be compensated respectively and then interference occurs. The signals corresponding to different FBG pairs (F-P interferometers) can be obtained and well distinguished. The interference intensity corresponding to the *s-th* FBG pair received at the detector is [4]

$$I(\omega) \propto \frac{1}{n} \langle \gamma_{ref-s}^*(\omega) E_0^*(\omega) \cdot \gamma_{sens-s}(\omega) \tau^2_{ref-s}(\omega)$$

$$E_0(\omega) \exp(-j(\Delta\varphi - \Delta\phi_s)) \rangle$$

$$(1)$$

Where n is the multiplexed number of FBG pairs, $E_0(\omega)$ is the field of the input light-source, $\gamma_{ref-s}(\omega)$ and $\tau_{ref-s}(\omega)$ are the reflectivity and the transitivity of the reference FBG of the *s-th* FBG pair, $\gamma_{sens-s}(\omega)$ is the reflectivity of the sensing FBG of the s-th FBG pair. $\Delta\varphi$ and $\Delta\phi_s$ are the phase differences induced by the unbalanced MI and the F-P interferometer formed by the s-th FBG pair respectively, and they can be expressed as

$$\Delta\varphi = 2n_{eff}z \qquad (2a)$$

$$\Delta\phi_s = 2n_{eff}l_s \qquad (2b)$$

When the induced OPD is completely compensated by the scanning OLCR, that is $z = l_s$, we have $\Delta\varphi - \Delta\phi = 0$, and the

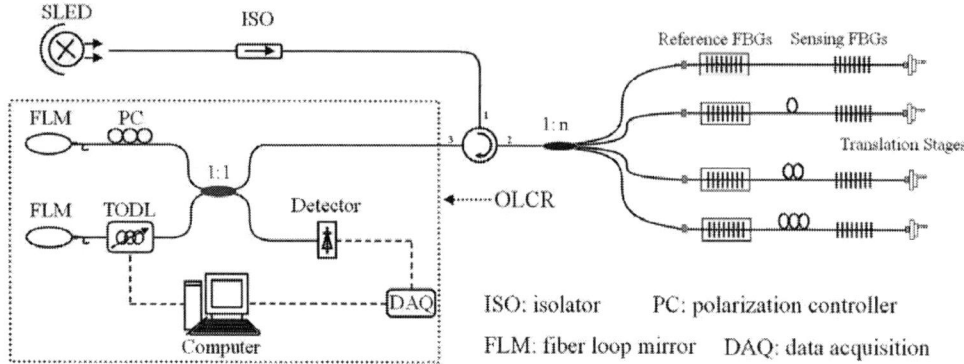

Fig. 1: Schematic diagram of the present coherence multiplexing system for FBGP sensors. The insets show the sensing mechanism of the presented system.

maximal interference intensity for all frequency of the *s-th* FBG pair can be achieved as

$$I_{total-s} = \int_{-\infty}^{+\infty} I(\omega)d\omega$$

$$\propto \frac{1}{n}\left\langle \int_{-\infty}^{+\infty} \gamma_{ref-s}(\omega)\gamma_{sens-s}(\omega)\tau^2_{ref-s}(\omega)\cdot\left|E_0(\omega)\right|^2 d\omega\right\rangle$$

$$(3)$$

We can see from Eq. (3) that the interference intensity is roughly proportional to the overlap integral of γ_{ref-s} and γ_{sens-s}.

When the spectra of the reference FBG and the sensing FBG of the *s-th* FBG pair are well matched, we can obtain the highest interference intensity. By keeping the spectrum of the reference FBG stable, the shift (induced by some measurand around this sensing FBG) of the central reflection wavelength of the sensing FBG will reduce the corresponding interference intensity. In other words, the intensity decreases rapidly as the mismatch of the spectra of the FBGs forming the pair increases. The sensing information of the measurand can thus be obtained.

Experiment results

We fabricate four identical FBG pairs to form the sensing system. As we can see from Eq. (3), for a perfect FBG pair, the reflectivity of the reference and sensing FBGs should be 38.2% and 100% respectively. However, we fabricated identical FBGs to form the FBG pairs for convenient.

The FBGs were fabricated by using a KrF excimer laser (TuiLaser Ltd., Germany) with

a phase-mask grating-writing technique. The phase mask we employed is with a pitch of 1070.6 nm. The fiber used in the experiments is a commercial boron/Ge co-doped photosensitive fiber (Fibercore Ltd., U.K.) with an effective index n_{eff} = *1.448* for the guided fundamental mode at 1550nm. All the FBGs we fabricated are identical (with a reflective wavelength of 1549.706 nm and a reflectivity of about 43%). The grating length is 2mm (controlled by an optical slot). The measured reflection spectra of a single FBG and a FBG pair with grating-distance of 28mm are shown in fig.2.

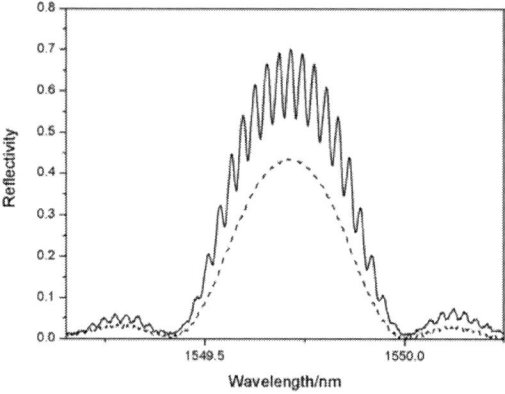

Fig. 2: Reflection spectra of a single FBG (dashed line) and an FBG pair with grating-distance of 28mm (oscillating solid line)

The system setup is as shown in fig. 1. Four pairs of FBGs form four F-P interferometers. The first FBG of each FBG pair works as a reference (shielded to be free of strain) and the second works as a sensing FBG. The grating-distances of FBG pairs are 28mm, 34mm, 40mm and 46mm respectively.
The system is illuminated by a super light emitted diode (SLED) with a central

wavelength around 1550nm and a bandwidth of about 80nm. An optical isolator is employed to eliminate unwanted reflection. The light reflected from the sensing array enters into an OLCR, which is formed by a 2×2 fiber coupler, two fiber-loop mirrors (used here as totally reflective mirrors), a computer-controlled motorized TODL (General Photonics Corp.), a detector, a data acquisition (DAQ) and an computer. In the other arm of the MI, a polarization controller (PC) is used to keep the lights propagating in the both arms are with the same polarization state. Finally the optical signal will be translated into the electrical signal by a detector and then acquired by the DAQ which is installed in the computer.

When there is no strain applied on any of the sensing FBGs, the interference intensities according to all FBG pairs are equal and this value is set as one unit, as shown in fig. 3.

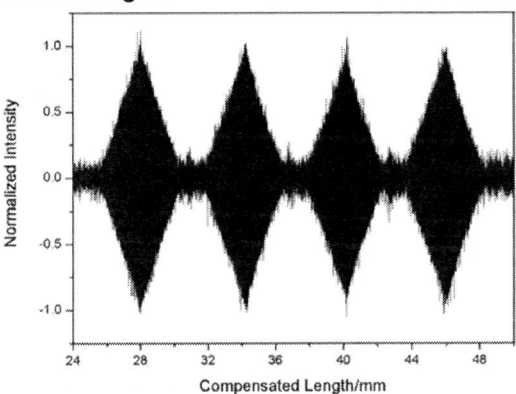

Fig. 3: Measured interferograms of the distributed sensing system when no strain is applied.

Then we increase the strain added onto the third FBG pairs (with a grating-distance of 40mm) gradually, the interference intensity corresponding to it decreases rapidly (see fig. 4) whereas the intensities corresponding to the other FBG pairs keep unchanged as expected. When the applied strain increases from 0 to 350με, the interference intensity corresponding to the third FBG pair decreases by 63.7%, and a decrease of 0.182%/με for average is achieved.

Conclusions
A new scheme of Coherent multiplexing of distributed FBG pairs for sensing is presented. An experiment of stain

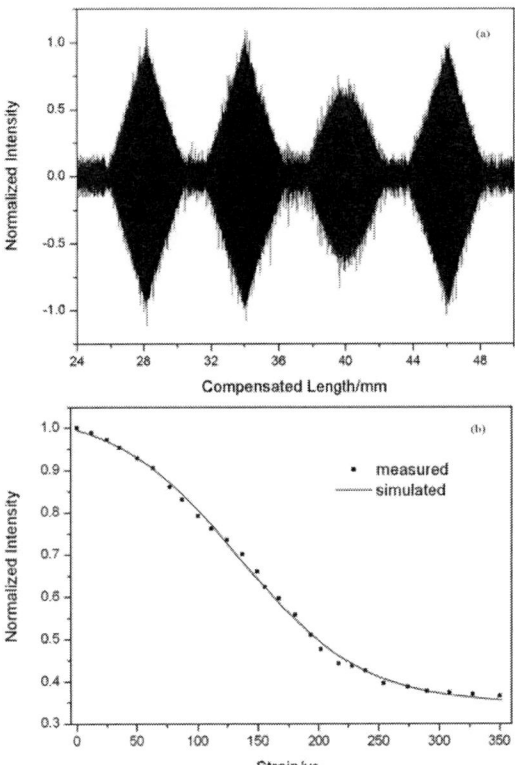

Fig. 4: Interference intensity decrease when strain is applied: (a) Measured interferograms of the system when the strain applied on the third FBG pair is 137με; (b) The interference intensity corresponding to the third FBG pair decreases with the strain applied on this FBG pair increases.

measuring is demonstrated which shows good performance of the scheme. Compared with other multiplexing sensing system for FBGs such as WDM and TDM, the cost of our system can be cut down greatly since the interrogation device is simply based on a mechanical scan. And compared with our former work, cross talk between FBG pairs is avoid since all FBG pairs are parallel connected in this scheme.

References
1 A. D. Kersey, M. A. Davis, H. J. Patrick, M. LeBlanc, and K. P. Koo, J. Lightwave Technol. vol. 15 (1997), pp. 1442-1463.
2 Jackson DA, Ribeiro ABL, Reekie L, et al., Opt. Lett., vol. l8 (1993), pp. 1192-1194.
3 Y. Yu, L. Lui, H. Tam, and W. Chung, IEEE Photon. Tech. Lett., vol. 13 (2001), pp. 702–704.
4 Z.G. Guan, D.R. Chen and S. He, J. Lightwave Technol. 2007 (in press).

Bio-inspired nanodevices for artificial solar energy conversion

Mamoru Nango

Department of Applied Chemistry, Nagoya Institute of Technology,
Gokiso-cho, showa-ku, Nagoya 466-8555, Japan   E-mail: nango@nitech.ac.jp

Abstract

Photocurrent response of photosynthetic light harvesting core complex (LH1-RC) and its model complex on electrodes were performed to attempt the construction of an artificial photosynthetic antenna complex toward developing useful nanodevices of solar energy conversion.

1. Introduction

At the early stages of purple bacterial light-harvesting complexes, light-harvesting complexes, called LH1 and LH2, absorb solar energy and transfer it to the reaction center (RC), whereupon the absorbed energy is efficiently converted into electrochemical energy (Figure 1) [1]. These reactions take place within a 'core complex' consisting of a RC located inside the LH1 complex, where pigment complexes play important roles on these reactions. We are interested in the rapid and efficient energy transfer between pigments in these complexes, photosynsthetic units (PSUs) [2-4]. These pigment complexes have been aiming to construct an artificial solar energy device based on a natural solar energy conversion system such as the core complex. Recently, the X-ray crystal structure of the LH1-RC core complex has been reported and revealed that it is oval rather than circular as shown in Figure 1[1].

Integration of photosynthetic proteins or protein-mimics with solar energy devices for tasks of light-harvesting and charge separation will expand current solar energy device technology with novel and inexpensive bio-components. Our goal is to use modified photosynthetic light-harvesting complex as a light harvester of the well-established cell to convert light energy in the ultraviolet and visible region into that in the near infrared region for the development of energy conversion materials. The advantage of the light-harvesting complex is its high efficiency of light-energy conversion throughout the near UV to near IR region and much higher durability than ordinary isolated pigments supported by its inherent photo-protective function. Thus, the results of the above grounds can be directly applied to the development of solar cells using modified photosynthetic light-harvesting materials [2-7]

We have recently reported that LH1-RC core complexes isolated from *Rs. rubrum* can be assembled a cationically-modified transparent indium tin oxide (ITO) electrode, which exhibits photoinduced current generation [6]. Our current understanding of energy transfer and charge separation reactions in the LH2 and LH1-RC complexes has enabled the first step to be taken towards generating artificial systems from them that convert light energy into usable electrical current. Previous attempts to produce an artificial, energy-converting electrode system used either the LH1 complexes. Until now, there have only been a few attempts to immobilize intact core complexes, consisting of both the LH1 complex and the RC components together, onto an electrode [6,7]. We have recently developed a procedure to create a self-assembled monolayer (SAM) of reconstituted LH1 complexes on a transparent indium tin oxide (ITO) electrode modified with 3-aminopropyltriethoxysilane (APS-ITO) between the electrode surface and the anionic LH1 polypeptides at pH 8.0 [5,6]. The NIR absorption spectrum showed that the LH1 complex was stable

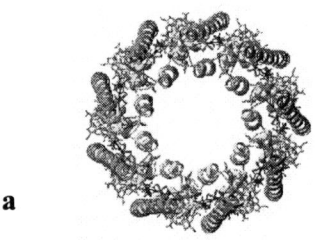

a

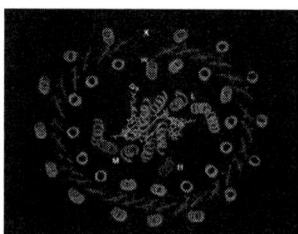

b

Figure 1. Structure of light-harvesting anntena complex (a) LH2 complex of *Rps. acidophilla* 10050LH2 and (b) LH1-RC complex of *R. palustris*.

When immobilized onto these electrodes. This study was extended using native LH1-RC complexes [6]. LH1-RC complexes isolated from *Rb. sphaeroides* were successfully assembled on APS-ITO. Efficient energy transfer and photocurrent responses could be observed upon illumination at 880 nm.

Further, we assembled PSUs, LH1-RC core complex, LH1, RC, and admixed complex of LH1 and RC on a modified Au electrode to investigate assembling manner and to develop photocurrent generation system for these assemblies to attempt the construction of an artificial antenna complex toward developing useful nanodevices.

2. Self-Assembled Monolayer of Light-harvesting Core Complexes of Photosynthetic Bacteria on an Amino-Terminated ITO Electrode

Figure 2 shows the NIR absorption spectra of the isolated *Rb. sphaeroides* core complexes in 20 mM Tris HCl buffer pH 8.0 OG micelle (dotted line) and assembled onto an APS-ITO electrode (solid line), respectively. These spectra show that these core complexes have the absorption maximum at 880 nm with two smaller peaks at 800 nm and 760 nm. The former peak is attributable to the overlap of bacteriochlorophyll *a* (BChl*a*) in the LH1 complex (880 nm) and the reaction center Bchl*a* dimer 'special pair' (870 nm) and the latter two peaks to the BChl*a* called 'accessory' (800 nm) and bacteriopheophytin (760 nm) in the RC, respectively. The NIR absorption spectra of these core complex on the electrode indicate that these complexes are stable when assembled onto an APS-ITO. In the previous study it was apparent that when the RC of *Rb. sphaeroides* was assembled, by itself, on the electrode it was relatively labile. Whereas in present study the complete core complex, when assembled onto the electrode, provided to be quite stable. The enhanced stability of the RC surrounded by the LH1 complex probably results from supportive interactions between the two complexes. Interestingly, when illuminating at 880 nm the fluorescence emission of BChl*a* molecules in the LH complex of *Rb. sphaeroides* on the APS-ITO was strongly quenched, due to the presence of the RC of *Rb. Sphaeroides*. This indicates that an efficient energy transfer from BChl*a* in the LH1 complex to the RC in the core complex is still occurring on the electrode (data not shown). FT-IR spectra of the LH complex and the LH1-RC core complexes assembled on the APS-ITO show the absorptions at 1650 cm^{-1} and 1550 cm^{-1}. These bands can be assigned to the amide I and amide II bands, respectively. These results indicate that the LH polypeptides are in the same *a* helical configurations on the ITO electrode as in OG micelles.

The time course of the photocurrent generated

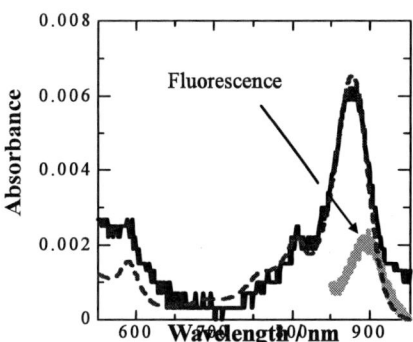

Figure 2. NIR absorption spectra of isolated *Rb. sphaeroides* in 20 mM Tris HCl buffer pH 8.0 OG micelle (dotted line) and assembled onto an APS-ITO electrode (solid line)

from the core complex or the RC of *Rb. sphaeroides* assembled onto an APS-ITO showed that an enhanced photocurrent was observed for the core complex when the electrode was illuminated with a pulse of light at 880 nm. In contrast no photocurrent was observed for either LH complexes or the RC. Under our experimental conditions a cathodic photocurrent was observed, implying that one-way electron transfer from pigments in the core complex (special pair of BChl*a*, SP in RC) to methyl viologen was occurring as shown in Figure 3.

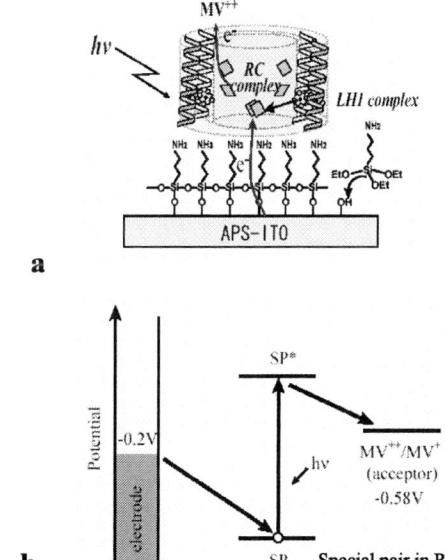

Figure 3. (a) Schematic drawing of LH1-RC core complexes on an APS-ITO electrode , which shows the electron flow from the complex to methyl viologen. (b) Energy diagram for cathodic photocurrent generation by the LH1-RC core complex.

Thus, the enhanced photocurrent observed at 880 nm in the assembled core complex can be ascribed to energy transfer from the LH1 to the RC and then electron transfer from the RC to the electrode as shown in Figure 3. This data indicates that the core complex was well organized on the ITO and the photocurrents were driven by light that was initially absorbed by the LH components.

3. Self-assembled Monolayer of Light-harvesting Core Complexes from Photosynthetic Bacteria on a Gold Electrode Modified with Alkanethiols

LH1-RC complexes isolated from *Rb. sphaeroides* were self-assembled on a gold electrode modified with self-assembled monolayers (SAMs) of alkanethiols, NH_2-$(CH_2)_n$-SH; n = 2, 6, 8, 11, HOOC-$(CH_2)_7$-SH, and CH_3-$(CH_2)_7$-SH, respectively to attempt the construction of an artificial antenna core complex towards developing useful nanodevices as shown in Figure 4. The adsorption on a gold electrode modified with SAMs of NH_2-$(CH_2)_n$-SH, n = 2, 6, 8, 11 depended on the methylene chain length, where the adsorption increased with increasing the methylene chain length. The clear presence of the

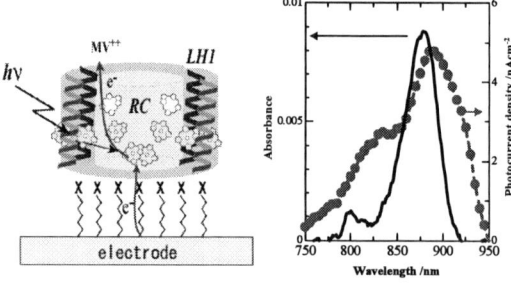

Figure 4. Schematic drawing of LH1-RC core complex on a gold electrode, and the absorption and action spectra of *Rb. sphaeroides* LH1-RC assembled onto a gold electrode modified with NH_2-$(CH_2)_6$-SH.

well known LH and RC peaks NIR spectra of the LH1-RC complexes indicates that these complexes were only fully stable on the SAM gold electrodes modified with the amino group. In the case of modification with the carboxyl group the complexes were partially stable while in the presence of the terminal methyl group the complexes were extensively denatured. An efficient photocurrent response of these complexes on the SAMs of NH_2-$(CH_2)_n$-SH; n =2, 6, 8, 11 was observed upon illumination at 880 nm when n = 6 as shown in Figure 4. This corresponds to a distance between the amino terminal group in NH_2-$(CH_2)_6$-SH and the gold surface of 1.0 nm.
In conclusion, the SAM method is clearly successful in allowing assembly of functional core complexes

on the electrode. This has been confirmed by NIR absorption spectroscopy, demonstrating that the photocurrent response, which is derived from electron transfer between the RC and the electrode, is enhanced by illumination at 880 nm. These results provide useful methodology to better understand the suprastrucure of LH1-RC complex as well as to gain knowledge about building an artificial fabrication of LH1-RC complex on solid substrates toward useful nanodevices. Various combinations of these complexes are being tested for their usefulness in constructing nanodevices for artificial solar energy conversion.

ACKNOWLEDGMENT

The present work was partially supported by AOARD.

REFERENCES

[1] A.W. Roszak, T.D.Howard, J.Southhall , A. Gardiner, C.J. Law, N.W.Issacs, R.J. Cogdell, *Science* 302 (2003) 1969.

[2] M. Nango, *Chlorophylls and Bacteriochlorophylls, Biochemistry, Biophysics and Biological Functions of Chlorophylls*, Chapter **28** Editors: B. Grimm, R. Porra, W. Rüdiger, H. Scheer, Springer, 2006.

[3] T. Dewa, T. Yamada, M. Ogawa, K. Yoshida, Y. Nakao, M. Kondo, K. Iida, K. Yamashita, T. Tanaka, M. Nango, *Biochemistry*, **44**, (2005), 5129-5139.

[4] K. Iida, J. Inagaki, K. Shinohara, Y. Suemori, M. Ogawa, T. Dewa, and M. Nango, *Langmuir*, **21**, (2005), 3069-3075.

[5] Ogawa M, Shinohara K, Nakamura Y, Suemori Y, Nagata M, Iida K, Gardiner AT, Cogdell RJ and Nango M, *Chem Lett*, **33**, (2004) 772-773.

[6] Y.Suemori, M.Nagata, Y. Nakamura, K.Nakagawa, A.Okuda, J. Inagaki, K.Shinohara, M.Ogawa, K.Iida, T.Dewa, K.Yamashita, A. Gardiner, R.J.Cogdell, M.Nango, *Photosynthesis Res*. **90**, (2006),17-21.

[7] Das R, Kiley PJ, Segal M, Norville J, Yu AA, Wang L, Trammell SA, Reddick LE, Kumar R, Stellacci F, Lebedev N, Schnur J, Bruce BD, Zhang S and Baldo M, *Nano Lett* **4**, (2004) 1079-1083.

Challenges in luminescent materials for lighting and medical applications

Cees Ronda, Philips Research Laboratories, Weisshaussstrasse 2, D-52066 Aachen, Germany

Utrecht University, Ornstein Laboratory, P.O. Box 80.000, 3508 TA Utrecht
The Netherlands

Centre for Optical and Electromagnetic Research, Zijingang Campus, Zhejiang University, Hangzhou 310058, China

Abstract In this paper, we will outline how new requirements have revitalised research on luminescent materials. Using selected examples from applications in lighting and the medical field, we will show how materials science on luminescent materials results in new and improved phosphors.

Introduction

Research on luminescent materials has seen a strong revival. Energy conservation considerations have resulted in the need for stable fluorescent lamp phosphors. Blue/near UV LEDs have induced research on new emitting compositions and form factors. Finally medical applications require new phosphors, also with absorption and emission features outside the visible spectral range.

Title section

Compact fluorescent lamps can easily replace incandescent lamps. In this way a considerable reduction in energy consumption for lighting can be realised. In compact fluorescent lamps, however, the luminescent materials interact strongly with the Hg-plasma, which is used to excite the phosphors. This can lead to sizeable Hg-consumption and in this way to a loss in light output. We will show which factors lead to Hg-consumption and how Hg consumption can be reduced.

The advent of blue/near UV LEDs induced a search for completely new luminescent compositions, which are much more covalent than compositions used in fluorescent lamps, to be able to excite them with blue/near UV light. In addition, the materials used are subjected to much higher excitation densities and higher temperatures. In this lecture, we will address these issues in some detail and give examples of how materials science is used to meet the requirements LED phosphors have to fulfil.

Finally, we will address the emerging field of luminescent materials for medical applications.

Conclusions

New problems require new solutions. We have identified new problems and in the lecture we will present new solutions. In doing so, we will also show that material science on luminescent materials is a challenging and rewarding research area.

978-0-9789217-3-6/07/$25.00
©2007 WEN GLOBAL SOLUTIONS

Spectrally Broadened Optical Pump Source via Phase Modulation for Wideband SBS Slow Light

Chester Shu, Alan Cheng, and Mable P. Fok

Department of Electronic Engineering and Center for Advanced Research in Photonics,
The Chinese University of Hong Kong, Shatin, N.T., Hong Kong.
E-mail: ctshu@ee.cuhk.edu.hk

Abstract We demonstrated slow light of 1 GHz and 10 GHz pulses using stimulated Brillouin scattering in a single mode fiber. The cw pump sources are spectrally broadened by phase modulation at different bit rates.

Recently, much attention has been focused in developing slow light techniques [1] for applications in optical memories, optical signal processing, and optical buffer for communication networks. Stimulated Brillouin scattering (SBS) [2,3] slow light is particularly attractive for communications since it is a fiber-optic approach and is ready to operate at any arbitrary wavelength. A limitation of SBS is its relatively narrow gain bandwidth in the order of tens of MHz. To solve the problem, one can apply external modulation to broaden the linewidth of the pump in delaying short pulses [4]. While direct intensity modulation of the pump has resulted in 12-GHz bandwidth for SBS slow light [5], phase modulation of the pump has only been demonstrated to delay signal up to 1 Gb/s [6].

In this paper, we demonstrated slow light of 1 GHz and 10 GHz pulse trains using SBS with a phase-modulated pump. We also investigate the effect of the bit rate of modulation. The phase-modulated pump provides a constant power and thus no synchronization is needed between the pump and the signal pulses. Hence, the approach offers a practical means in amplifying and delaying high-bit-rate pulses and true data.

In our experiment, a 1551.318-nm laser diode is modulated by an electro-optic intensity modulator at 1 and 10 GHz to produce the input signal pulses. Slow light takes place in an 8-km SMF-33 single mode fiber via SBS. The pump source is a tunable CW laser at 1551.230 nm such that the Stokes gain peak is aligned with the signal wavelength. The pump is phase-modulated by a 1-Gb/s bit pattern of "1011" or a 10-Gb/s pseudorandom bit sequence (PRBS) depending on the input signal pulse width. In the experiment, we first generate 386-ps signal pulses at 1 GHz. The pulses are delayed using two separate sets of modulated pumps. Using a phase-modulated pump at 10 Gb/s driven with a PRBS (231-1), a 30-ps delay is obtained when the average pump power is 27 dBm. The delay can be increased to 100 ps simply by changing the pump modulation to a 1-Gb/s pattern "1011". The result indicates that excessive broadening of the pump linewidth is unfavorable for low-bit-rate communication as it reduces the SBS

gain coefficient. An optimum linewidth of the pump is crucial in achieving a maximum amount of delay. To demonstrate the delay of shorter pulses for high-bit-rate communication, we increase the modulation frequency to 10 GHz to generate 26.6-ps signal pulses. With a 10-Gb/s PRBS phase-modulated pump of 27-dBm average power, a delay of 10.6 ps is obtained. The result is shown in Fig. 1. The optical pulse maintains its original shape; however, a slight pulse broadening is observed. The measured output width is 31.8 ps, corresponding to a broadening factor of 1.2.

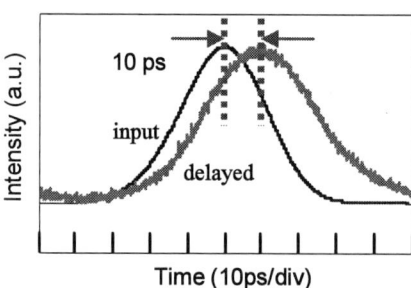

Figure 1: Temporal profiles of the 10-GHz input pulse and the delayed pulse using a 10-Gb/s PRBS pump modulation. A 10-ps delay is obtained.

References

[1] R. W. Boyd and D. J. Gauthier, Progress in Optics 43, Elsevier, Amsterdam, Chap. 6, pp. 497 – 530, 2002.

[2] K. Y. Song, M. G. Herraez, and L. Thevenaz, Opt. Express 13, 82-88, January 2005.

[3] Y. Okawachi, M. S. Bigelow, J. E. Sharping, Z. Zhu, A. Schweinsberg, D. J. Gauthier, R. W. Boyd, and A. L. Gaeta, Phys. Rev. Lett. 94, 15, 153902, April 2005.

[4] M. G. Herraez, K. Y. Song, and L. Thevenaz, Opt. Express 14, pp. 1395- 1400, February 2006.

[5] Z. Zhu, A. M. C. Dawes, and D. J. Gauthier, OFC 2006, PDP1, March 2006.

[6] L. Yi, L. Zhan, Y. Su, W. Hu, L. Leng, Y. Song, H. Shen and Y. Xia, ECOC 2006, We3.P.30, September 2006.

Slow light in Silicon Nano-waveguide

Fangfei Liu

State Key Lab of Advanced Optical Communication Systems and Networks, Department of Electronic Engineering, Shanghai Jiao Tong University, 800 Dongchuan Rd, Shanghai 200240, China, lfflys@sjtu.edu.cn

Abstract *We propose a widely tunable slow-light delay element based on nano-scale silicon microring resonator assisted by parametric amplification. This scheme provides flexible adjustment of the delay time and bandwidth to adapt to different data rates.*

Introduction

Slow lights can be achieved based on linear and nonlinear optical signal processing techniques [1]-[5]. Nonlinear effects in resonators including self phase modulation (SPM) and Raman effect have attracted much interest recently due to the enhanced intra-resonator intensity and the effective path length [5][6]. Here, for the first time to the best of our knowledge, we propose a microring resonator assisted by parametric amplification in nano-silicon waveguide to realize widely tunable slow-light delay and variable bandwidth for different data rates. We find that a weak parametric gain per round trip in the ring is sufficient to compensate the loss so that the transmission spectrum of the resonator can be continuously changed from the deep notch to a unit gain, thus the effective bandwidth and the delay of the slow-light resonator can be adjusted according to the data rate of the signal. Furthermore, we use a figure of merit (FOM) to quantify the signal quality of return-to-zero (RZ) signal under different data rates and delay requirements.

Parametric amplification in microring resonator

In this study, we consider a single coupled nano-Silicon microring resonator with a circumference of 60 µm (see the inset in Fig. 1). The nano-SOI waveguide used in the simulation is designed to be 250-nm height and 500-nm width, with an effective refractive index of 2.68 and a mode area of 0.12 µm^2. These lead to dispersion parameters of β_2 = -0.4 ps^2/m and β_4 = 5.2×10^{-7} ps^4/m at a pump wavelength λ_p in the vicinity of the resonance wavelength at 1576.47 nm. The signal wavelength is λ_s = 1531.42 nm, three times of the free spectral range (FSR) away from the pump wavelength. We assume that free carrier absorption effect is negligible by employing certain techniques such as applying an external electric field or introducing non-radiative centers [7]. The nonlinear refractive index n_2 and the two photon absorption (TPA) coefficient β_T are 7×10^{-14} cm^2/W and 0.45 cm/GW [7]. In order to achieve large tuning range in delay, the resonator is operated near the critical coupling point with a coupling coefficient r = 0.22 and an attenuation of 3.5 cm^{-1}. The linear transfer function of a ring resonator can be written as [6]:

$$T(\omega) = \frac{t - a\exp(i\phi)}{1 - t\exp(i\phi)} \quad (1)$$

where $t = \sqrt{1-r^2}$ is the transmission coefficient; a contains the per-turn loss, parametric gain and TPA induced loss, Φ contains both the linear and the amplification-induced nonlinear phase shifts.

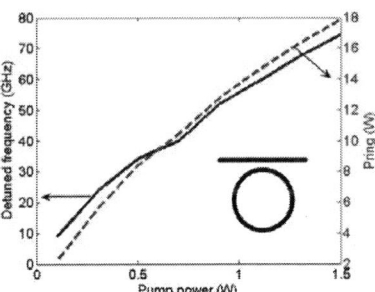

Fig. 1: Detuned frequency of the pump from the resonance (solid curve) and peak power coupled in the resonator, i.e. Pring (dashed curve) vs. λ_p. Inset: single coupled microring resonator.

As the nonlinear phase shift of the pump induced by SPM would red-shift the pump off the resonance, the pump should be detuned from the resonance so that the generation of parametric gain is maximized. Fig. 1 shows that both the detuned frequency and the power coupled in the resonator increase with the pump power. When the peak pump power is 0.1 W, the parametric gain is negligible and there exist deep notch in the transmission and sharp transition in the phase. We then utilize parametric amplification in microring resonator to achieve an optically tunable slow-light element. Simulations show that large delay ranging from ~18 ps to ~150 ps for data rates from 20 Gb/s to 1 Gb/s with low signal distortion [8] can be achieved.

References

1 F. Xia et al. Nature Photonics,1(2007), 65

2 J. E. Heebner et al. J. Mod. Opt., 49(2002), 2629

3 H. Gersen et al. Phys. Rev. Lett., 94(2005), 073903

4 Y. Okawachi et al. Opt. Express, 14(2006), 2317

5 S. Blair et al. Opt. Express, 14(2006), 1064

6 Y. Chen et al. J. Opt. Soc. Am. B., 20(2003), 2125

7 Q. Lin et al. Opt. Express, 14(2006), 4787

8 F. Liu et al. OFC 2007, paper OWB5

978-0-9789217-3-6/07/$25.00

©2007 WEN GLOBAL SOLUTIONS

Storage capacity of slow-light based on fiber Brillouin amplifiers

Li Zhan

State Key Lab of Advanced Optical Communication Systems and Networks, Department of
Physics, Shanghai Jiao Tong University, 800 Dongchuan Rd, Shanghai 200240, China,
lizhan@sjtu.edu.cn

Abstract *We study the the storage capacity of slow-light tunable optical buffers based on fiber Brillouin amplifiers (FBA) through theoretical analysis. Gain saturation and pulse broadening are two key factors which limit the buffer capacity. It is shown that the maximum buffer capacity varies with data bit rate.*

Tunable optical buffer is a key component in future all-optical routers, but it is still a bottleneck of all-optical communication network. In the past years, controlling the group velocity of light had been report to be a possible appoach to realize the tunbable buffer. SBS slow-light seems to be the most promising one for its large fractional delay, room-temperature operation, low threshold, possibility working at any wavelength, and compatibility with existing optical communication systems. To date, SBS slow-light experiment extend the delay to several pulse widths by employing cascaded scheme and increase the bandwidth to 12GHz by using a modulated pump, even exceed 25GHz by employing a double pump method.

However, no public reports up to date have evidently shown the delay exceeded one bit, although it exceeded one or two pulse widths in rare experiments by employing short pulses or cascaded scheme. The physics origin of this limitation has not been clarified. The pulse distortion is inevitable in all SBS slow-light shemes, although it can be reduced. Most of the slow-light experiments and theoretical analyses concentrated on single pulse case, in which pulse distortion is not a serious problem. For the real signals, the distortion has a fatal affection on the buffer capability and

system performance. However, a detailed theoretical model of SBS slow-light for bit streams has not been proposed yet and the problem of the real storage capacity of the TOB based on FBA remains unsolved.

We study the storage capacity of slow-light tunable optical buffers based on fiber Brillouin amplifiers through theoretical analysis, in which the pulse broadening has been considered. The storage capacity is discussed for two modulation formats: return-to-zero (RZ) and non-return-to-zero (NRZ). Gain saturation and pulse broadening are two key factors that limit the buffer capacity. This is maybe one possible reason why no delay over one bit was reported. It is shown that the maximum buffer capacity varies with data bit rate. We also investigate the optimum data bit rate to achieve the highest storage capacity.

In spite of small buffer capacity, the tunable SBS slow-light buffer can play important roles in some circumstances such as accurate data synchronization and data bit equalization. In addition, we believe similar method can be used to analyze the resonance induced slow-light buffers including stimulated Raman scattering, optical parametric amplification, electromagnetically induced transparency etc.

978-0-9789217-3-6/07/$25.00

©2007 WEN GLOBAL SOLUTIONS

Author Index

A

Adachi, Chihaya 187
Ai-zhen, Li 58
Aiba, Takamichi 256
Akhmediev, Nail 152
Al-Mansoori, M. H. 390
Andalib, P. 222
Andersson-Engels, Stefan 234
Ansari, Lida 504
Anshi, Xu 599
Asatryan, A. 97
Ashihara, Satoshi 372
Awaji, Y. 210
Axelsson, Johan 234

B

Bae, Jeung-Hwan 131
Bae, Ji-Hong 140
Bae, Kwang-Soo 140
Barklund, Anna 113
Begum, F. 363
Bennion, Ian 85, 204
Bhatranand, Apichai 477
Borghsb, Gustaaf 249
Botten, L. 97
Bowers, John 1

C

Cada, Michael 56
Cai, Bo 196
Cai, Zhiping 387, 429
Cann, Robert 283
Cao, Lei 586
Cao, Wenbin 533
Chang, Guan-Liang 622
Chang, Li-Wen 280
Chang, Qingjiang 306
Chao, Peng 599
Che, Ming 459
Chen-Perdereau, X. 566
Chen, Antao 219
Chen, Baoquan 113
Chen, Biao 375
Chen, Chun-Ho 12
Chen, Daru 378, 402, 453, 542
Chen, Dingbo 277, 486
Chen, Fei 359
Chen, Hong-Zheng 302
Chen, Hongzheng 359
Chen, Hui 113

Chen, Jianping 104
Chen, Lili 82
Chen, Lin 312, 423
Chen, Ming 317
Chen, Q. 107
Chen, Shean-Jen 622
Chen, X.W. 79, 350
Chen, Xianfeng 67, 85, 204, 492, 495
Chen, Xue-Wen 444
Chen, Yan-feng 184
Chen, Yan-Ru 489
Chen, Yan 163
Chen, Yung-Sheng 347
Chen, Yuping 67, 492, 495
Chen1, Hong-Zheng 246
Chena, Hong-Zheng 249
Cheng, Alan 638
Cheng, Bing-Ying 456
Cheng, Yan 34, 551
Cheng, Yuanbing 486
Chenga, Yunan 249
Chiang, Chun-Ping 228
Chien, Liang-Chy 137
Chiou, Arthur 40
Chiu, Kuo-Chih 622
Chiu, Nan-Fu 237
Cho, Kwan Hyun 181
Choi, D-Y. 554
Choi, Ick Chang 450
Choi, Jae-Wan 353
Choi, Jeong-Hyeon 353
Choi, Kyung Cheol 181
Choi, Pyung-Suk 474
Choi, Yoonseuk 140
Chong-qing, Wu 625
Chou, C. 37
Chou, Chien 338
Choua, Ling-Chung 616
Choudhury, P. K. 390
Choy, Wallace C.H. 79, 240, 350
Chui, P.C. 79
Chung, Youngjoo 589
Cincotti , G. 98
Computer 303
Cui-luan, Wang 432
Cui, Kaiyu 586

D

Da-zheng, Wang 432
Dai, Daoxin 513, 563, 578
Dai, Jiajie 519
Damsgaard, Hans 18

Dang, Weirui 67, 492
Davies, Edward 85
Dennis, M. 162
Dinu, Raluca 113
Dong, X.Y. 341
Dong, Xiaoyi 204
Dong, Xinyong 134
Dong, Zuoren 524
Duggal, Anil R 301
Durouxab, L. 289
Durouxab, M. 289

E

Eggleton , B.J. 554
Enoki, Takatomo 64
Ersoy, Okan 507
Eun, Jae-Jeong 474

F

Fang, Hong 369, 381, 441
Fang, Qun 82
Fang, Yun 113
Fang, Zujie 524
Fanying, Meng 595
Faraji-Dana, R. 483
Fei, Yu 498, 545
Feng, Shuai 456
Feng, Tu 47
Feng, Yijun 163
Ferreira, J. 438
Finge, Norman 332
Fok, Mable P. 198, 638
Fong, H.H. 350
Fu, Hongyan 375, 378, 396, 453, 604
Fu, Shenggui 204
Fuerbach, A. 97
Fukuchi, Yutaka 70
Furukawa, H. 210

G

Gao, Junming 408
Gao, Lei 516
Gao, Shiming 378, 453
Gao, Yan 246
Gao, Ying 378, 453
García de Abajo, F.J. 162
Gity, Farzan 504
Gomes, N. J. 560
Goodall, Thomas 283
Grace, Warren 283
Granpayeh, N. 222, 530
Greatrex, Jeffrey 283
Grelu, Ph. 566
Grelu, Philippe 152

Guan, Bai-Ou 178
Guan, Zuguang 292
Guo, Bing 283
Guo, Tieying 369, 441

H

Hai, N. H. 363
Hai, Nguyen Hoang 384
Hainberger, Rainer 332
Han, Hongtao 557
Han, Sang-Kook 607
Han, Won-Taek 589
Hao, Jiaming 519
Hao, Yinlei 335
Hashida, H. 243
He, Jian-Jun 216, 329, 572
He, Jin 312
He, Jing 423
He, Ruei-Yu 622
He, S. 82, 378, 501
He, S.L. 79, 350
He, Sailing 329, 444, 604
He, Yonghong 580
Hine, Anna 85
Ho, H. P. 583
Ho, Ho-Pui 169
Hong, Fang 366
Hong, Xiaobin 320
Hossain, Md. Zahid 119, 619
Hsu, Chain-Shu 91
Hsu, Ya-Ting 12
Hu, Jing 329
Hu, Weisheng 414
Hu, Wen-Ping 246
Hu, Yonghua 480
Huang, Chaohong 387, 429
Huang, Chi-Feng 347
Huang, Chong-Jie 280
Huang, Diyun 113
Huang, Jia-Chi 246
Huang, Jun 462
Huang, Wei-Kai 12
Huang, Wei-Ping 274
Huang, Weihong 56
Huang, Weiping 320
Huang, Wencai 387, 429
Huang, Yi-Pai 12
Huang, Yidong 166, 471, 586
Huaqing, Wang 146
Huda, Md. Quamrul 119, 619
Huggard, P. G. 560
Hughes, Marcus 85
Huh, Jae-Hoon 116
Hui, Rongqing 253
Huo, Li 231

I

Ibrahim, Sahrul Hilmi.................................539
Ishii, Hiroyuki 213
Ives, Mark..283

J

James, J...560
Jang, Nam-Young.......................................474
Jang, Se-Jin...140
Jang, Young Ho...314
Janssen, A. P..24
Jiabing, Du..595
Jian-liang, Zhang......................................625
Jian, Leng...143
Jian, Shui-sheng.......................................317
Jian, Shuisheng...........................369, 381, 441
Jiang, Chun.......................................575, 600
Jiang, D.S...149
Jiang, Jun...128
Jiang, Li...15, 447
Jiang, Xiaoqing..335
Jiang, Xiaoyu...580
Jiang, Yurong...........................356, 465, 501
Jianqiang, Wang..595
Jie, Luo...47
Jin, Dan...113
Jin, L..583
Jin, Xu...548
Jinwei, Li...548
Johansson, Ann...234
Johansson, Jonas.......................................234
Jovanovic, N..97
Jung, Jong-Wook.......................................140

K

Kado, Yuichi..61
Kaijage, S. F..363
Kaijage, Shubi..384
Kakitsuka, Takaaki.....................................213
Kalogerakis , G...201
Kamio, Yukiyoshi.......................................157
Kan, D..97
Kanai, N...243
Kang, Jeung-Mo...353
Kao, Yee-Mou...405
Karasek, M..420
Kawaguchi, K..243
Kawakami, S...610
Kawamura, Y...243
Kawanishi, Tetsuya.....................................110
Kazovsky, L. G..201
Khijwania, Sunil K.....................................341
Kim, Byoung Yoon.......................................601
Kim, Dong-Hyeon..607

Kim, Du-Hyun...353
Kim, Geun-Ho...353
Kim, Hyun Deok..................................314, 450
Kim, Jae-Hoon..140
Kim, Jae-Wook..353
Kim, Seung Hun...181
Kim, Sung-Ho...589
Kim, Sung-Jin..88
Kim, Tae Hoon..326
Kim, Young Cheol.......................................314
Kimura, H...243
Kinjo, T...363
Kitamura, N...193
Kitayama, K...98
Kobayashi, M..243
Kobayashi, Noriyo..64
Kokubun, Yasuo...323
Kong, S. K..583
Kong, Siu Kai..231
Kosugi, Toshihiko..64
Koyama, Fumio..271
Kuan, C. H..489
Kuan, C.-H..616
Kuo, B. P. P..201
Kuroda, Kazuo..372
Kuroki, Yasuhiko.......................................116

L

Lamont, M.R.E...554
Lang, Tingting....................................329, 513
Le, Yang..595
Lee , Ray-Kuang..405
Lee, Cheng-Kuang.......................................228
Lee, Cheng-Ling..405
Lee, Chomsik...131
Lee, El-Hang..21
Lee, Hsiang-Chieh......................................228
Lee, Hyung Jae...326
Lee, J...107
Lee, Jeong-Soo...353
Lee, Jiun-Haw..237
Lee, R.K..583
Lee, Sin-Doo...88
Leung, Bernard...468
Li-jun, Zhang...625
Li, Bing...113
Li, C..243
Li, Da-Hai..94
Li, Haifeng..34, 551
Li, Honglei.......................369, 381, 441
Li, Jinwei...356
Li, Lieming...533
Li, Mingyu..216
Li, Pengcheng..286
Li, Rong-jin..246

Li, Shangyuan 53
Li, Wei 580
Li, Xia 356, 465
Li, Xingde 295
Li, Xinwan 104
Li, Xun 274
Li, Y. 566
Li, Ying-Chang 338
Li, Yuquan 101
Li, Z. Y. 516
Li, Zhi-Yuan 456, 459, 536
Li1, Xin 82
Liang, C.J. 240
Liang, Song 277
Liao, Cheng-Yumr 12
Liao, Chi-Chang 31
Liao, Lin-Yao 12
Liao, Yung-Ming 91
Lin, Aoxiang 589
Lin, Chi-Hung 622
Lin, Chii-Wann 237
Lin, Chinlon 231, 583
Lin, Fang-Cheng 12
Lin, Rujian 393, 399
Lin, Xu 595
Lin, Yan Sen 411, 527
Lin, Yi-Hsin 31
Liu, Fang 166, 471
Liu, Fangfei 575, 639
Liu, Hongdu 468
Liu, Jie 301
Liu, Kou-Chen 237
Liu, L. 149
Liu, Rong-Juan 459
Liu, Tongyu 172
Liu, Weisheng 631
Liu, Xu 34, 551
Liu, Ya-Zhao 456
Liu, Ye 536
Liu, Yung S. 298
Liu, Zhihai 125
Liwen, Wang 366
Longfeng, Xiang 146
Lou, Shuqin 369, 381, 441
Lu, Chih-Feng 347
Lu, Chih-Wei 228
Lu, Dan 317
Lu, Feng 67, 495
Lu, Lin 101
Lu, Wenjie 67, 495
Lu, Yan-qing 184
Luo, Jingdong 219
Luo, Qingming 286
Luther-Davies, B. 554
Lv, Bo 317

M

Ma, Hongmei 15
Ma, Hui 580
Ma, Wen 283
Madden, S. 554
Maeda, Joji 70
Maeda, Yoshinobu 116
Mahdi, A. K. Zamzuri and A. 207
Main, Keith 557
Maki, Sayaka 116
Makino, R. 243
Maksymiuk, L. 259
Mao, Xiaoyu 586
Marhic, M. E. 201
Marshall, G. 97
Matsuo, Shinji 213
Matsushima, Toshinori 187
Matsushima, Yuichi 4
McDonald, Bill 265
Mei, Hui 498, 545
Meng, Qinghua 492
Miao, Xiangrui 453
Miller, Andrew 283
Miller, Eric 113
Ming, Nai-ben 184
Mino, Shinji 27
Miu, Lihong 67
Miyazaki, T. 98
Miyazaki, Tetsuya 157
Moghaddasi, M. Naser 504
Mohammad, Abu Bakar 539
Mohrdiek, Stefan 283
Mokhtari, A. 510
Moolayil, Merly 113
Morris, Jim 557
Moss, D.J. 554
Mou, Chengbo 204
Müllner, Paul 332
Murata, Hiroshi 64

N

Na, Yu-Jin 88
Nagatsuma , Tadao 61
Nakamura, Moriya 157
Nakanotani, Hajime 187
Nakayama, Hideki 116
Namihira, Y. 363, 384
Namihira, Yoshinori 268
Nango, Mamoru 634
Nasir, M. N. Mohd 390
Neves-Petersenab, M. T. 289
Ngo, Nam Quoc 219
NianyZuo 268
Nien, Shou-Yu 237

Nilsson, Johan 360
Nozhat, N. 530

O

Ohashi, Masaharu 256, 262
Ohtera, Y. 610
Okamoto, Katsunari 303
Okamura, Yasuyuki 64
Oohashi, Hiromi 213
Ou, Haiyan 375, 604

P

Pan, Dandan 492
Park, Jaehee 131
Pelusi, M.D. 554
Peng, Hui 101
Peng, Jiangde 166, 471, 586
Petersenab, S. B. 289
Piao, Yin-Xing 607
Pliska, Tomas 283
Pojanasomboon, Pojamarn 507
Ponomarenko, Sergey A. 56
Pu, Tao 101, 417
Pyayt, Anan 219

Q

Qi, Xi 317
Qian, Jun 82
Qian, Song 498
Qin, Shan 402
Qinfeng, Zhou 344
Qinghao, Ye 595
Qingrong, Han 47
Qiu, Jianrong 613
Qiu, Min 160, 598
Qiu, Shaofeng 462
Qu, Ronghui 524
Quinlan, Shaun 283

R

R.Matai 47
Radil, J. 420
Rao, Y. J. 122
Rao, Yi 166, 471
Rapp, Ludovic 152
Rashid, H. A. Abdul 390
Razzak, S. M. A. 363
Ren, Jinjun 613
Rhee, Tae Hyung 326
Rigole, P-J 569
Roelens, M.A.F. 554
Roh, Sang-Kil 607
Ronda, Cees 637

Rongqiang, Cui 595
Rostami, G. 483
Rui, Geng 435

S

Sainidou, R. 162
Saito, T. 243
Samsuri, N. Md. 207
Sato, T. 610
Sattmann, Ralph 50
Segawa, Toru 213
Shahabadi, M. 483
Shao, Li-Yang 134
Shao, Yufeng 312
Shen, Hao 468
Shen, P. 560
Shen, Wenjuan 283
Sheng, Zhen 329, 513
Shi, Min-Min 246
Shiao, Wen-Yu 347
Shibata, Nori 256
Shieh, Han-Ping D 12
Shimura, Tsutomu 372
Shinohara, Hiromichi 252
Shiquan, Yan 595
Shishegar, A. A. 510
Shu, Chester 198, 638
Shuqin, Lou 366
Siuzdak, J. 259
Skovsenab, E. 289
Somesfalean, Gabriel 628
Soto-Crespo, Jose M. 152
Stepniak, G. 259
Su-ping, Liu 432
Su, Yikai 306, 408, 575
Su, Yuan-Deng 622
Subrina, Samia 119, 619
Suen, Yick Keung 231
Sugden, Kate 85
Sun, Chi-Kuang 76
Sun, Weiqiang 414
Sun, Yu 486
Sun, Yubao 15, 447
Svanberg, Sune 234
Svensson, Tomas 234

T

Ta'eed, V.G. 554
Tam, H Y 341
Tam, Hwa-Yaw 134
Tam, HwaYaw 178
Tang, Jiansheng 477
Tang, Yongbo 521, 542
Teperik, T. V. 162

A-5

Terao, Y. 243
Tetsuro, Yabu 262
Thylén, Lars 7
Tiefeng, Xu 344
Tigang, Ning 435
Tolstikhin, Valery I. 190
Tong, L. 566
Tong, X.L. 149
Tong, Zhi 435
Towery, Chris 43
Tsai, C. C. 37
Tsai, Meng-Tsan 228
Tsai, Yao-Chou 237
Tsao, Cheng-Han 12

V

Vemagiri, Jeevan 113
Vienne, G. 566
Vojtech, J. 420
Von Der Weid, J. P. 438
Vydra, Jan 50

W

Wada, N. 98, 210
Wada, Naoya 309
Wai, P.K.A 444
Wan, Ruiyuan 166
Wang, Ai-Hong 94
Wang, Chuncan 435
Wang, Dan 533
Wang, Huan 277, 486
Wang, Hui-tian 184
Wang, Jun 216, 414
Wang, Liesong 486
Wang, Mang 246, 302, 359
Wang, Minghua 335
Wang, Qiong-Hua 94
Wang, Rong 417
Wang, Wei 486
Wang, Xin 468
Wang, Xiulin 387, 429
Wang, Xu 98
Wang, Yih-Ming 228
Wang, Yiping 104
Wang, Zhe-Chao 329
Wang, Zhechao 513, 563
Wang, Zhi 444
Wang, Ziqiang 533
Wanga, Mang 249
Wei, H. C. 37
Wei, Xue 143, 146, 548
Wei, Yizhen 542
Weiguo, Chen 366
Wen, H.Q. 149

Wen, Hong 423
Wen, Shuangchun 312, 423, 480
Wenzhong, Shen 595
Withford, M. 97
Won, Yong-Yuk 607
Wong, K. K. Y. 201
Wu, Botao 613
Wu, Hsieh-Ting 338
Wu, Hua-Lin 622
Wu, Jian 320
Wu, Rui 492
Wu, Shin-Tson 10, 31
Wu, Shu-Yuen 169
Wu, Weilei 101
Wua, Gang 249

X

Xi, Yanping 274
Xia, Yuxing 67, 492, 495
Xianfeng, Chen 372
Xiang, Liangzhong 225
Xianglong, Zeng 372
Xiao-ming, Feng 432
Xiao-yu, Ma 432
Xiao, Yun 396
Xie, Xuejuan 414
Xin-zhi, Sheng 625
Xin, Yan-Xia 94
Xing, Da 225
Xu, Chris 154
Xu, Hao 519
Xu, Huiwen 312
Xu, Huiying 387, 429
Xu, Ping 516
Xu, Xinyu 408
Xue, Jiangeng 592
Xue, Wei 356, 465, 501

Y

Yahiro, Masayuki 187
Yamashita*, Ikuo 262
Yang, C. C. 228, 347
Yang, Deng-Ke 30
Yang, Jhih-Ming 31
Yang, Jianyi 335
Yang, Jun 125
Yang, Li-Gong 246
Yang, Liu 329, 513, 578
Yang, Minwei 104
Yang, Shi 595
Yang, Shujun 477
Yang, Sidney S. 280
Yang, Su 175
Yao, Lei 369, 381, 441

Ye, Jiajun 393, 399
Ye, Qing 524
Ye, Yuqian 521
Yeh, Dong-Ming 347
Yeh, Yin 73
Yin, S. 107
Yokoyama, Daisuke 187
Yong-gang, Wang 432
Yong-gang, Zhang 58
Yong-sheng, Lan 432
You, Byung Gwon 326
Yu-quan, Li 175
Yu, Guomin 113
Yu, Yichuan 521
Yuan, Libo 125
Yuan, W. 583
Yue-jin, Zhao 498, 545
Yueyu, Xiao 426
Yun, Byoung-Ju 450
Yun, Tianliang 580
Yuqiong, Li 143
Yusoff, Z. 390

Z

Zeng, Nan 580
Zhan, Li 640
Zhang, Dao-Zhong 456, 536
Zhang, Fan 435
Zhang, Feng 317
Zhang, H.M. 240
Zhang, Junfeng 495
Zhang, Lin 85, 204
Zhang, Q. 107
Zhang, Tidd 43
Zhang, Wei 166, 471, 486, 586
Zhang, Yali 277
Zhang, Yang 178
Zhang, Yuan 501
Zhang, Yunxiao 486
zhang, Zhizhong 462
Zhao, Junming 163
Zhao, Ren-Liang 94
Zhao, Wu-Xiang 94
Zheludev, N. I. 162
Zheng, Benrui 387
Zheng, Lixin 113
Zheng, Wei 335
Zheng, Xiaoping 53
Zheng, Yuxin 166
Zhengbin, Li 599
Zhinong, Yu 143, 146, 548
Zhou, Bin 292
Zhou, Bingkun 53
Zhou, Haifeng 335
Zhou, Jianying 335

Zhou, Jun 219
Zhou, Kaiming 85
Zhou, Lei 519
Zhou, Renjia 302, 359
Zhou, Shifeng 613
Zhou(1), Kaiming 204
Zhu, Hongliang 277, 486
Zhu, Kun 396, 604
Zhu, Meiwei 393, 399
Zhu, Shi-ning 184
zhu, Weitao 462
Zhu, Ying-xun 417
Zhu, Ying 82
Zhu, Yong-yuan 184
Zhuo, Hui 480
Zou, N. 363

This page intentionally left blank.